Studies in Computational Intelligence

Volume 760

Series editor

Janusz Kacprzyk, Polish Academy of Sciences, Warsaw, Poland
e-mail: kacprzyk@ibspan.waw.pl

About this Series

The series "Studies in Computational Intelligence" (SCI) publishes new developments and advances in the various areas of computational intelligence—quickly and with a high quality. The intent is to cover the theory, applications, and design methods of computational intelligence, as embedded in the fields of engineering, computer science, physics and life sciences, as well as the methodologies behind them. The series contains monographs, lecture notes and edited volumes in computational intelligence spanning the areas of neural networks, connectionist systems, genetic algorithms, evolutionary computation, artificial intelligence, cellular automata, self-organizing systems, soft computing, fuzzy systems, and hybrid intelligent systems. Of particular value to both the contributors and the readership are the short publication timeframe and the world-wide distribution, which enable both wide and rapid dissemination of research output.

More information about this series at http://www.springer.com/series/7092

Preface

Econometrics is a branch of economics that uses mathematical (especially statistical) methods to analyze economic systems, to forecast economic and financial dynamics, and to develop strategies for achieving desirable economic performance.

An extremely important part of economics is finances: A financial crisis can bring the whole economy to a standstill and, vice versa, a smart financial policy can drastically boost economic development. It is therefore crucial to be able to apply mathematical techniques of econometrics to financial problems. Such applications are a growing field, with many interesting results—and with an even larger number of challenges and open problems.

This book contains both related theoretical developments and practical applications of econometric techniques to finance-related problems. The main objective of econometric analysis is to predict the effect of different financial strategies on the economics. To be able to make successful predictions, we need to understand the causal structure of economic and financial phenomena, develop quantitative models of these phenomena, and test these models—by making sure that they provide correct predictions of observed phenomena. In solving all these problems, additional challenges emerge from the need to take into account the data-rich character of the current information environment. The resulted issues of testing, prediction, and cause are handled in several chapters of this book.

In many situations, it is possible to design adequate models by using existing mathematical techniques—usually techniques from mathematical statistics. However, often, models constructed by using the traditional techniques do not allow accurate predictions. In such situations, new techniques are needed. A similar situation happened in physics in the early twentieth century, when the traditional statistical techniques turned out to be not very adequate for describing microscale phenomena. To adequately describe these phenomena, physicists came up with techniques of quantum mechanics. Recently, it has been shown that ideas motivated by quantum physics can also help in the description of economic phenomena; several related chapters are also included in this book.

While physics-motivated ideas can be very helpful, these ideas can rarely be directly applied to economic phenomena, because our objectives in physics and economics applications are usually very different: While in physics applications, we aim for revolutionary changes—such as transistors, space exploration—in economics, we usually want to avoid drastic changes and oscillations, we want to achieve a solid robust sustainable growth. We want to reach a dynamic state of economics in which external influences should not lead to drastic changes. In mathematics, a state that does not change under a certain operation is known as a fixed point. Thus, the study of fixed points is an important part of econometrics. Several related chapters form a special section of this book.

This book also contains applications of both traditional and novel econometric techniques to real-life economic problems, with a special emphasis on financial and finance-related problems.

We hope that this volume will help practitioners to learn how to apply various state-of-the-art econometric techniques to finance-related problems, and help researchers to further improve the existing econometric techniques and to come up with new techniques for financial econometrics.

We want to thank all the authors for their contributions and all anonymous referees for their thorough analysis and helpful comments.

The publication of this volume is partly supported by the Banking University of Ho Chi Minh City, Vietnam. Our thanks go to the leadership and staff of the Banking University, for providing crucial support. Our special thanks go to Prof. Hung T. Nguyen for his valuable advice and constant support.

We would also like to thank Prof. Janusz Kacprzyk (Series Editor) and Dr. Thomas Ditzinger (Senior Editor, Engineering/Applied Sciences) for their support and cooperation in this publication.

January 2018

Ly H. Anh
Le Si Dong
Vladik Kreinovich
Nguyen Ngoc Thach

Contents

Fixed-Point Theory

Applications

General Theory

Testing, Prediction, and Cause in Econometric Models

William M. Briggs$^{(\boxtimes)}$

340 E 64 St. Apt. 9A, New York, NY 10065, USA
matt@wmbriggs.com

Abstract. Classical statistical approaches used widely in econometrics centering around parameter estimation, hypothesis testing, and p-values should be abandoned. In their place, predictive modeling should be used. A predictive model answer the question all users of statistics have: if I change x, or leave it out of my model, what does this do to the uncertainty in y? Classical methods never answer that question directly. The reason why this is so, and why the predictive approach does, is shown.

Keywords: Cause · Decision · Hypothesis testing · Models
P-values · Priors · Prediction · Model verification

1 Introduction

This paper will use language about probability unfamiliar to most, who may be used to thinking of probability solely in mathematical terms, as functions of measure spaces, say. There is, of course, nothing wrong with the mathematical theory of probability; its lemmas and theorems remain as strong as ever. The difficulty comes in applying these constructs to the real world, and even to other-world, events. There, the mathematics does not so much break down as become misapplied. We begin to expect that nature behaves along the same strict, rigorous quantitative lines as our theories, but this is not so; and it is especially not so in complex systems. And there is nothing more complex than human interactions, the subject matter of econometrics.

A century and more earlier, scholars of probability such as Laplace [15] and Keynes [14] took a broader view. Probability was the specification of the uncertainty of propositions given assumptions. The modern mathematical view of probability fits into this as a subset, but its insistence on quantification limits it.

For instance, in logic, everybody knows the proposition S = "George wears a hat" is true *given* it is accepted that "All Martians wear hats and George is a Martian." We know S is a local, i.e. conditional, truth, just as we know there are no Martians and thus no hat-wearing Martians. The status of the proposition given the assumptions is true. Given *different* assumptions, the propositions can take different statuses.

For example, given "My brother always wears a hat and George is my brother" then S is still true. But given "A lot of Martians wear hats and George

© Springer International Publishing AG 2018
L. H. Anh et al. (eds.), *Econometrics for Financial Applications*, Studies in Computational Intelligence 760, https://doi.org/10.1007/978-3-319-73150-6_1

is a Martian" then S is only likely. There is no number for this "likely", because there is in the premises no indication of how to map "a lot" to any number. Of course, premises can be *added* (as subjectivists do), but this is in a sense cheating. Just as in logic, with probability, we must always deduce the qualities of S with only the given information.

As should be clear, there is no truth or falseness or probability of the unadorned, unconditional proposition S. "George wears a hat" has no epistemic status without conditions, without presumptions or observations or assumptions. This is the same in math, where the symbol "x" has no value without presuppositions (say, $y = x$ and $y = 7$, etc.). This is a long-winded way of saying, as in math and logic, *all* probability is conditional: no probability is unconditional.

And that is a round-about way of saying what Bruno de Finetti said so many years ago, and in all caps, "PROBABILITY DOES NOT EXIST", [17].

The reason for the emphasis on these matters is that they have profound consequences for the creation, interpretation, and use of probability models, the lifeblood of econometrics.

2   Probability Models

Everybody knows that correlation is not causation. Just as everybody knows that hypothesis testing proves correlations are causes—or something like causes, causes in all but name. Everybody is a little mixed up.

Consider the simple "time series"

$$y_t = y_{t-1} + \epsilon_t. \tag{1}$$

This model says that the value of y_t will be equal to the value of y_{t-1}, the time point before, with the addition of ϵ. Specifics about ϵ are not known. The best that can be said is the value ϵ takes can be ascribed by some parameterized probability distribution. This probability distribution is almost always *ad hoc*, and its parameters are also unknown.

The interpretation of the model is this. Given we are at time t, the uncertainty we have in values of y_t are specified by a certain parameterized probability distribution which is a function of y_{t-1} and the probability distribution of ϵ. There is no notion whatsoever that y_{t-1} is the efficient *cause* of the value of y_t; nor is any ϵ the efficient *cause* of y_t. It should be clear that if we knew the efficient cause of y_t, the model would not be probabilistic but a simple statement, "y_t will equal a be*cause* it will be caused to be a."

It is helpful to have an example in mind, so think of stock prices. The value of y_{t-1} does not cause y_t, even in those cases when $y_{t-1} = y_t$. Many, many things cause the price of the stock: the rules of the exchange, the mind sets of innumerable people who buy or sell the stock directly or indirectly, the state of their finances and finances of others. There is no one cause, but a myriad of them, with most or (usually) all unknown. Those that might be known are the rules of the exchange, which might limit prices in certain ways. Whether these (limited) causes are put in the eventual probability model is an open question; typically they are not.

Model (1) is purely correlational. It only says we know this-and-such about the uncertainty in y_t given some other assumptions. The model itself is an assumption, and a major one. Let that pass for the moment and concentrate on the second set of assumptions about ϵ.

In shorthand, people will say something akin to "ϵ is normal", when what they mean is the uncertainty in this additive value is characterized by a normal distribution, specified with some central and spread parameter.

But it is here that reification—the substitution of fiction for reality—enters the first time. It is almost impossible to resist the temptation to say that our models describe the genuine underlying reality. Some will say "ϵ *is* normal" as if this mysterious parameter has physical existence and possess characteristics unique to a "normal." Yet nothing in the world "is" normal. Nothing in the world is any probability. Probability is purely a matter of epistemology; it is a measure, not necessarily quantitative, of uncertainty of a proposition given certain assumptions. These assumptions must always be there: as above, this is another way of saying there is no such thing as unconditional probability: all probability is conditional. This is proved and expanded upon in [4].

What the model itself says is that the *uncertainty* in y_t can be characterized by some distribution, such as a normal, with central parameter a function of y_{t-1} and the central parameter of ϵ, and a spread parameter a function of t and the spread parameter of ϵ.

This is the epistemological, and thus correct, interpretation. There is no notion of cause in the model, and none is needed. The model make may reasonable, but approximate, predictions, but the model itself offers no explanation *why* anything happens. The predictions, incidentally, are always approximations because the distribution for ϵ lives on the continuum, which forces the probability of any value of y_t to be infinitesimal, and (depending on the distribution) to have values tailing off to infinity. Real-world values of y_t will, of course, take specific values with non-vanishing probability.

About this specific model (1), there is great concern. It is said to be a "random walk", and the series itself is said to contain a "unit root" and is non-stationary, which means that the uncertainty in y_t depends (in some way) on t. These facts do not make the series un-predictable, nor is the epistemological interpretation invalid. The concern arises because of the approximation of ϵ, which gives non-zero probabilities to values which are very large, and therefore impossible in practice, and which causes the uncertainty in y_t to become great. There are various "fixes" for this situation (e.g. differencing), none of which involve abandoning the impossibilities of continuum-based models. Another is to propose a more complicated model, such as this:

$$y_t = \beta y_{t-1} + \epsilon_t, \tag{2}$$

where a $\beta \neq 1$, and usually $|\beta| < 1$, indicates an auto-regressive (the term is borrowed from regression, which is explored next). The question becomes whether $\beta = 1$; if so, the model reverts to (1); if not, the uncertainty in y_t is considered tame enough for use.

There is no information in the *ad hoc* model on the value of β, and therefore its value must be forever unknown. This is key. To discover something about β, the model itself must be augmented by premises which allow inference of β. It should be clear by now that β does not exist in any real or ontic sense; no parameter exists, because probability does not exist. This parameter is only part of the mathematical apparatus, and therefore a specific value is not *per se* needed. This is examined later in the predictive approach for modeling. For now, the belief that β has a value, in real life, leads to the idea of hypothesis testing, which is discussed in full below. There is also the sense, in simulations, that a β with a fixed value can "exist". This is so, in a certain sense, but not in the sense hoped for, which is also explained below.

First, beside time series, regression is the most-used probability model in econometrics. Models for regression look like this:

$$y = \gamma_0 + \gamma_1 x_1 + \cdots + \gamma_p x_p + \epsilon, \tag{3}$$

where the y is the quantifiable observable of interest, the x also quantifiable observables, the *gamma* parameters relating the x to the uncertainty in y, and not *in* y itself, and ϵ is as above.

As with time series, most know the actual interpretation of (3), which is why is so odd that in practice this interpretation is dropped in preference for a glaring error. What the model says, of course, is that the uncertainty in y is conditional on the values of x, γ, and the parameters of ϵ. The model is silent as the tomb on the cause or causes of y. Yet people cannot help themselves but to interpret the γs as sort of causes. This error is found almost every time hypothesis testing is used.

3   The Predictive Approach

Before we discuss hypothesis testing, we need to decide just models are, and what to do about them. In this section is the proposed replacement for hypothesis testing, and indeed for all parameter-centric methods of classical statistics, both Bayesian and frequentist.

Suppose there are two rival models for y_t:

$$y_t = \beta_1 y_{t-1} + \beta_2 y_{t-2} + \epsilon_t, \tag{4}$$

$$y_t = \beta_1 y_{t-1} + \epsilon_t. \tag{5}$$

Both will give *ceteris paribus* different probabilities for y_t. Which of these models is correct?

Both. Unless one has specific evidence or assumptions that say one or the other is false or necessary, then because all probability is conditional on the information assumed, both of these models are correct *given our assumptions*.

These assumptions must, of course, exist. If they did not, then these model forms would not exist. They have to be suggested by some evidence, however loose or tacit. It is these assumptions that are the true model; the mathematical form is deduced via these assumptions.

One popular assumption has this form: M = "People before me used auto-regressive models on data like this, and I too will try a lag-1 and lag-2 model." M is loose and not well justified, but it (and ones like it) is surely the most common of all assumptions. From M, we *deduce* (4) and (5).

We know from above that there is no notion of cause in M, or therefore none in (4) and (5). These models, like all probability models, are correlational. The probabilities deduced via (4) and (5), given M, are therefore conditionally correct and true (assuming no errors in calculation). Now suppose that we do these calculations and arrive at:

$$\Pr(y_t \in s | M, \text{lag1}) = a$$
$$\Pr(y_t \in s | M, \text{lag2}) = a + \delta$$

where s is some set of values of y of interest and $a, \delta > 0$. Since $\delta > 0$ (or $\delta < 0$, with the obvious bounds respected), the two sub-models (both are children of the parent M), are different. There is, at this point anyway, no sense in which one model is superior to another. The two probabilities are, it is emphasized, correct, because each is produced given different assumption. There is nothing to choose between them save δ.

To get the formulas to work, we have to supply at least y_{t-1} and y_{t-2}. Any assumed or observed values will do. Properly, then, the correct way to write the probabilities is this:

$$\Pr(y_t \in s | y_{t-1}, M, \text{lag1}) = a$$
$$\Pr(y_t \in s | y_{t-1}, y_{t-2}, M, \text{lag2}) = a + \delta$$

where the lagged values of y either assumed or observed. This can represent data in the usual sense of observations, but it need not. No data is needed to make these models work; only assumptions.

Knowing a and δ implies we know the values of all parameters, β_1 etc. Supposing this is true (this assumption is loosened below) if one wants to decide between models, *how* do we decide between models? There is only one way. It is if the value of δ, for a given s of importance, is sufficiently large that we would change a decision we would make about the value of y_t. If the probability y_t is in s is a we would act one way, and if it equalled $a + \delta$ we would act another way, then we must select the lag-2 model, there being *no other information about which version to prefer in the assumption M*. Remember: probability, like math and logic, is only calculated on the information assumed, and none other.

Now since men different decisions, given s, a δ that is important to one man might be ignorable or trivial to another. Therefore, *based on the same evidence*, one man might opt for the lag-1 model, and the second for the lag-2. M says nothing about which version is preferred, therefore the only distinguishing characteristic are the different predictions.

Of course, since s can vary, and therefore a and δ will differ, the same man could at times prefer the lag-1 version and at other times the lag-2 version. And so on for all model users. Keep in mind that, at this point, the *only* information we have is that assumed in M. There are, as yet, no past observations to which we can compare the model.

One does not have to choose between model versions. The evidence in M—the sole evidence—is that either lag-1 or lag-2 better represents the uncertainty in y_t. This is not to say that either lag-1 or lag-2 is true: they are both conditionally or locally true given M. For a model to be true in the absolute sense, it must represent the efficient causes of y_t. Only (correct, of course) causal models are true. Correlational models are not false, however. To be false means we have proof, in the strict rigorous mathematical sense that one or more of the model premises is false. We obviously do not have that information in M, which is a bald assumption.

Recognizing this, we can say that, given only M, either lag-1 or lag-2 better describes the uncertainty in y_t, and that only one of these is best, then via the statistical syllogism (a means of deducing probabilities) there is a probability of 0.5 the lag-1 model is best, and a corresponding probability of 0.5 lag-2 is best. Then via the obvious calculation the probability, given M etc., $y_t \in s = a + 0.5\delta$.

The idea that either lag-1 or lag-2 is best, or that only one can be chosen, is not in M itself. Both are external assumptions. Which is right? Again, both. M is already *ad hoc*, and so are both of these assumptions. There is nothing to judge between them, save what we bring to the problem. The probability for $y_t \in s$ is not unique because the assumptions are not unique; and there is certainly nothing unique in M. There are various *ad hoc* criteria, such as AIC, that people sometimes use to picking models, and this is fine, but these criteria are nothing but (*ad hoc*) assumptions added to M: *any* assumptions may be added to M!

To recapitulate: a set of assumptions are given, and given these assumptions the probabilities $y_t \in s$ are calculated. This s is chosen because it is important to some decision maker. Not all s are interesting. For instance, for a stock price, $s < -1e6$ would give, for almost any time series model in current use, nearly identical (in any useful sense) probabilities; therefore, there would be nothing to distinguish between these models. And it really is true that there is no practical difference for s like this! The models *are* practically the same. The s chosen, then, must be important for decision makers. Different s lead to different model preferences. Probability is not decision.

As it might not be obvious, this is the predictive approach. Below, the same method is expanded for those times when we make assumptions about the parameters other than asserting their values, as was done here ("priors" and estimates). We bring in the observations and clarify their role. The idea of model verification is also introduced, which allows us to *a posteriori* assess model usefulness.

These concepts do not change the "guts" of the predictive approach. If we have a set of assumptions, observations, presuppositions or whatever, that lead to only one model form, then we are done. We make the predictions and act on them. This is, after all, what civil engineers and physicists do. But if the assumptions allow for model choice, as in this current example, then we either have to assume we want to combine the assumptions, leading to a melded prediction, weighted according to the probabilities of each model version deduced from M, or we pick "the best" model, where best is in reference to a living decision we make. Different decisions will lead to different best models.

There is no contradiction. Different assumptions lead to different conclusions in logic and math, just as they lead to different probabilities. Since probability is not decision, and that decision is prior to or above probability, it should be no surprise that different decisions lead to differing model choice.

Since choice and decision are so varying, it would be impossible to specify in advance a value of δ that should be universally adopted, as people have done with the magic number with p-values. That arbitrary choice has led to a world of grief, as all know, and p-values are in any case the wrong thing to use.

What is most beautiful about the predictive approach is the concentration on observables. We do not speak about obscure test statistics, or on unobservable non-causal parameters, but about probabilities of real things, things we can measure.

Customer walks in the door and asks, "If I change x in my model, what is the probability that y does this?" The predictive modeler says, "It's p." And we're done. Everybody is happy.

The classical statistician fixated on parameters says, "Well, the confidence interval on the third γ in your model associated with that x widens a tad". The hypothesis tester says, "Your p-value is wee: officially wee." Either way, the customer leaves scratching his head. His question has not been answered. He asked for a probability so that he could make a rational decision and he was given persiflage. Worse, he goes away more certain than he should be.

The example above focused on the time series model, but it works equally well on the regression; and indeed works on any probability model. As a quick example, compare the two models

$$y = \gamma_0 + \gamma_1 x_1 + \cdots + \gamma_{p-1} x_{p-1} + \epsilon,$$

and

$$y = \gamma_0 + \gamma_1 x_1 + \cdots + \gamma_p x_p + \epsilon,$$

where the comparison is between a "reduced" and "full" model (the order of linear regressors of course do not matter). Again, the probabilities of some relevant set s can be calculated:

$$\Pr(y \in s | M, \text{reduced}) = a$$
$$\Pr(y \in s | M, \text{full}) = a + \delta$$

and the comparison goes on as before. Whether to include x_p in the model depends on the decision one makes, and on whether one wants "the best" model or a weighted average of models, both of which as before are additional *ad hoc* assumptions.

One need not progress through M x by x, or lag by lag, including those parameters associated with moving averages, or "garch" parameters. One can examine the probabilities in "batches", examining various collections of parameters depending on what decisions are important. Decision is king in the predictive approach.

For instance, it may be a matter of controversy whether sex is important; rather, there is public concern over how sex relates to some y, say income. Models with and without sex are used to compute the probability y is in various s. These probabilities can be plotted along the x-axis, the probabilities along the y-axis, or, the human eye being flawed in perceiving differences, the differences between the probabilities themselves may be plotted. Experience shows that these differences vary by s and are not constant; but that depends on the actual observations. Then, no single decision must be made, and all can view the importance of conditioning on sex.

The predictive approach is no panacea. Using it does not save one from the usual mistakes, such as bad or misapplied data, Simpson's paradox, and all that. But, as we shall see when it comes to hypothesis testing and parameter estimation, it is a much fairer approach, and one which gives the best sense of uncertainty given the assumptions. Usually this uncertainty is much wider than in the other classical approaches, which is also a recommendation in its favor, the other methods producing over-certainty of pandemic proportion, as lamented by such authors as [16, 20].

4   Model Parameters

Last section, it was assumed the model parameters were known. Here, as is more usual, it is assumed they are unknown.

We learned above that the data do not possess the characteristics implied by the model. M is not ontic; its parameters are not ontic. There is no "true" probability value unconditional of the assumptions M. There is therefore only that which we can calculate given M.

The wrinkle, or rather seeming wrinkle, is the values of the parameters, β_1 and so on. If these are known, i.e. deducible from M, then there is no difficulty finding a and δ in the above examples. If not, then we are stuck: a and δ cannot be discovered. In these cases, M *must* be augmented in some way, usually depending on whether one follows frequentist or Bayesian theory.

Before that, it is well to highlight that parameters are not needed. To whit: in this box are m_0 black balls and m_1 white. Knowing only this, and nothing more, the probability that a white is selected (where we do *not* have information on the selection mechanism) is deduced as $m_1/(m_0 + m_1)$. Similar deductions can be made assuming only the possibility of black and white; predictions can be made how many of each will be seen, conditioning on whether no observations have thus been made, and so on.

All is discrete and finite here, as all actual observations and measurements in real life are. Taken to the limit, these simple, even bare, assumptions lead directly to the binomial model; and its predictive sense, the beta-binomial. No parameters ever appear, except in the limit, where, naturally, interesting mathematical things can be said. This kind of thing can be done for any set of assumptions and measurements, though it almost never is.

The usual recourse, because of a strong sense of custom, is to embrace models parameterized on the continuum. In order to guess values of these parameters, or at least incorporate into predictions, additional assumptions to M must be made.

The Bayesian way to do this is to assume a probability distribution for the parameters, in the exact same way an assumption was made for ϵ. These distributions are called, as we all know, "priors." For a reason that will become clear, there is great angst conveyed in doing this. It is discovered that prior A leads to one result, but that prior B leads to another. This is seen a bug, when, as we now know, it is a feature of probability models. Probability is calculated on given assumptions: change the assumptions, change the probability.

An unfortunate terminology has arisen in the search for priors. People want to find "non-informative" or "ignorance" priors. This is impossible: it is not unlikely: it is impossible. Any assumption added to M is informative if it is not irrelevant. If irrelevant information about the parameters is added to M, then this information gives no insight into the probability the parameters take any values, and therefore it is useless information. This is a general statement, incidentally. Information added to M that does not change the probability of y is irrelevant by definition. Whether any piece of information is relevant depends on what is already in M, of course.

The terminology is also flawed because the greatest source of information about y has already been given, and that is M itself, which is almost always *ad hoc*. Add to that the *ad hoc* nature of ϵ, we're already two levels deep in (if I may) *ad hociness*. Why quail about an additional layer for priors?

The answer to this question is that people mistakenly believe the data "have" a probability, a true probability that can be discovered, if only the search is carried on with sufficient assiduity. This is false: probability doesn't exist. We might discover the *cause* of y, as is ever the goal in science, but once we have it, we no longer have need of probability. This is why quantum mechanics are "stuck" with probability: we know that we *cannot* discern the cause of individual events. (In QM, M is usually deduced given simpler assumptions).

Very well, we have posited a prior for our parameters. This can be used, as above, to make initial predictions of y, or the uncertainty in the values of the parameters themselves may be discovered once observations have been taken. This produces the "posteriors", around which great interest centers. It shouldn't.

Knowing that it's likely a parameters lies in some range is of little use, though that it what current practice focuses on. Instead, what should happen is that the uncertainty in these parameters be "integrated out", so that predictions about the observable can be made. We want this:

$$\Pr(y \in s | x, \mathrm{M}, \mathrm{D}), \tag{6}$$

where x are assumed, or new, values for all those x we have in the model (the x may be lagged values of y, etc.), M is understood to be augmented with the assumptions about the prior, and D the past observations of y and x. This schema is for all probability models, time series, regression, or whatever. Equation (6) is called the "predictive-posterior" which, if we do not know cause, is *the* goal of modeling.

As above, if we are positing two different priors for the parameters, say a maximum entropy (maxent) or a Jeffreys, then we can calculate:

$$\Pr(y \in s | x, \mathrm{M}, \mathrm{D}, \text{maxent}) = a$$

and

$$\Pr(y \in s | x, \mathrm{M}, \mathrm{D}, \text{Jefferys}) = a + \delta.$$

We are in the same situation as above. Given s, is δ interesting? Would we make different decisions based on its size? Do we want the "best" single model or an average? And so on.

Contrast this with the frequentist approach. Frequentist theory is deeply flawed, as best argued by the philosopher Hájek, [11, 12], in which he presents 30 different counter-arguments. Most of these counter-arguments are unknown to econometricians, and they are worth investigating. For our purposes, we will mention only one, a fatal flaw. Frequentist theory assumes probability exists, and that no probability can be known with certainty until after an infinity of observations are taken. We now see that this is false. Probability is not unique; it is always conditional and never unconditional (the goal of all frequentist analysis), and probability can be deduced given any set of assumptions, though this deduction may not give a quantitative values.

But, ignoring that, or at least brushing aside these objections, what happens in a frequentist analysis? Either a hypothesis test, which is discussed next, or an estimate of the parameter, in the presence of data, is made. Frequentist theory does not allow predictions to be made pre-data, as it were.

The estimate of the parameter is made via routine calculations, and then perhaps a confidence interval around that parameter is given. There are similarities, and even exact equivalences in some cases, with Bayesian parameter posterior distributions. The meaning of the confidence interval is nearly always wrong. All we can say, following frequentist theory, is that the "true" value of parameter lies in the given interval or it does not. A probabilistic interpretation of the interval is forbidden; indeed, the interval itself only says things about what would happen were the "experiment" which gave rise to the calculated interval were repeated an infinite number of times.

Since that it too confusing, everybody lapses into a semi-Bayesian interpretation. If the interval is narrow, in some relative sense, in practice this is taken to convey large probability the parameter really is in that interval. The converse is also believed: wide intervals are taken to mean there is only a small probability the parameter lies in the interval.

Whether this is true or false (and it is false, given the theory's actual definition of confidence intervals), we have stopped short of our goal. When we give an estimate of an unobservable parameter, we have not done much. We have not learned how changes in any observable x changes the uncertainty in the observable y.

We just have an interval. Worse, the confidence people have in the value of the parameter is taken to be the confidence we have in the relation between the x and y, when this is not at all the case. We can be certain, absolutely certain, in the value of a parameter (we might have supposed it, or that it was given to us as an assumption, or even via deduction as in QM), but that does not translate into uncertainty in y. To discover that uncertainty, the predictive approach is necessary.

This will become clearer when hypothesis testing is discussed.

5   Hypothesis Testing

Let's now contrast the predictive, decision-based approach with hypothesis testing. All are familiar with the ritual, but it's as well to spell it out once more. Hypothesis testing, whether it be done in accordance with Neyman-Person rules or via Fisherian p-values, always involves a fallacy. It is always a confusion between probability and decision. Here is the proof of that contention.

A non-unique *ad hoc* function of the observations, i.e. a test statistic, is calculated. Then the probability of larger (in absolute value) values of this statistic are calculated given M and D and assuming one or more parameters of the expanded model are set equal to some value, usually 0. The reduced model with fixed parameters is called "the null." It is, as all know, usually, but of course not always, a straw man, a model that is not believed, but which the model author cannot bear to dismiss out of hand.

There are, of course, an infinite number of models for any y that can be considered. The number of possible x that could say something about y is unlimited. In M, almost all of them are dismissed at the beginning. According to frequentist theory, these dismissals must have taken place, because an unconditional probability for all "events", i.e. all possible x, must exist. This is no small point and it is everywhere unappreciated. If there really is an unconditional probability of y, then every possibility for y that exists must be examined or otherwise accounted for. What this means is that, at the start, an infinite number of hypothesis tests must have been done. What's left for us are only a handful of xs. This is obviously absurd.

In the predictive approach, which assumes all probability is conditional on just the assumptions we have made and none other, there is no paradox. Change the assumptions, change the probability. Whether to include an x depends on whether decisions would be different.

At any rate, in hypothesis test, a p-value for some x is at hand. It will be wee (less than the magic number) or not wee. If the p-value is not wee, the null has "failed to be rejected", i.e. the x is tossed from the model *regardless* of what changes this x makes to the predictions of $y \in s$, whether these predictions are important in any decision or not. If the p-value is wee, the x is kept, again regardless what this x does to the probability and decisions. P-values thus make one-size-fits-all decisions.

There are other problems. The next three paragraphs are a paraphrase of what I have written elsewhere [5].

Fisher [7] said: "Belief in null hypothesis as an accurate representation of the population sampled is confronted by a logical disjunction: *Either* the null is false, *or* the p-value has attained by chance an exceptionally low value." Something like this is repeated in every elementary textbook.

Yet Fisher's "logical disjunction" is evidently not one, since his either-or describes different propositions, i.e. the null and p-values. A real disjunction can however be found. Re-writing Fisher gives: *Either* the null is false and we see a small p-value, *or* the null is true and we see a small p-value. Or just: *Either* the null is true or it is false and we see a small p-value. Since "*Either* the null is true or it is false" is a tautology, and is therefore necessarily true no matter what, and because prefixing any argument with a tautology does not change that argument's logical status, we are left with, "We see a small p-value." The p-value thus casts no light on the truth or falsity of the null. Everybody know this, but this is the formal proof of it.

Frequentist theory claims, assuming the truth of the null, we can equally likely see any p-value whatsoever, i.e. the p-value under the null is uniformly distributed. To emphasize: assuming the truth of the null, we deduce we can see any p-value between 0 and 1. And since we always do see any value, all p-values are logically evidence *for* the null and not against it. Yet practice insists small p-value are evidence the null is (likely) false. That is because people argue: For most small p-values I have seen in the past, I believe the null has been false; I now see a new small p-value, therefore the null hypothesis in this new problem is likely false. That argument works, but it has no place in frequentist theory (which anyway has innumerable other difficulties). It is the Bayesian-like interpretation.

The decisions made using p-values are thus an "act of will", as Neyman criticized [18], not realizing his own method of not-rejecting and rejecting nulls had the same flaw.

There isn't the space here to detail all the arguments against p-values. That they are logically un-equipped to do the job asked of them is sufficient evidence to abandon their use. Nevertheless, a few more words on their weaknesses are in order.

Given a fixed M, p-values are not unique, because of the possibility of different statistics and different methods of calculating the distribution of those statistics assuming the null's truth. That means different decisions can be made with the same model. It is also too easy to pass the p-value's test. If an author has not discovered a wee p-value, it means he has not tried hard enough.

P-values encourage ritualized, even magical, thinking, [8]. If a p-value is wee, it is as if for many a spell has been successful. We must remember that the vast number of users of statistical methods are not cautious academics, but people wanting quick answers; p-values are too ready to supply these.

A colleague of mine (Roy Spencer), once did an informal "analysis" which showed (something like) the number of annual UFO reports correlated with the global average temperature anomaly. This analysis passed all the classical statistical tests he could think of. Wee p-values etc. Conclusion? UFOs cause global warming.

There was nothing wrong with his analysis. Spencer followed all the classical rules. It must then be the case that we are forced to say, at the least, that UFOs are "linked to" global warming, which is what the more cautious users of p-values say. But what does "linked to" mean? Tautologically, it means a wee p-value was seen. But everybody takes it as proving cause. How? Perhaps UFOs did not directly cause global temperatures to rise, but something the UFOs did caused something else to cause a rise in temperatures.

Wee p-values are far too seductive, and the temptation to say that one has shown a cause, or something vaguely like a cause (i.e. "linked to"), is scarcely resisted. Probability models simply cannot show cause, not ever (this is proved in [4]). The predictive approach is not immune to this temptation. A δ of sufficient size is bound to lead some to conclude that cause has been discovered. But since evidence in the predictive model for or against a "null" is guaranteed to be weaker, since knowledge in the observables is always weaker than knowledge in the parameters, there will fewer misascriptions of cause.

The problem of p-values is again a topic of lively debate. A collection of authors, [1], made a recent splash by saying p-values should be kept but that the magic number should be reduced tenfold to 0.005. This fixes none of the aforementioned problems. It will still be a logical fallacy to conclude—to decide, that is—the null is true or false based on a p-value. P-values conflate decision and probability. One universal number is picked, such that *all* nulls are considered true (rather, not proven false, because of Fisher's insistence on using Popper's curious doctrine of falsifiability) if the p-value is greater than this number, and *all* nulls are false if the p-value is less than this number.

In the predictive approach, the null is not decided to be true or false *per se*. All available evidence is used and whether to accept the evidence about the null either way depends on the decision that will be made. If there are no immediate decisions to be made, if for instance the results of the probability model are for the general public where a range of decisions are possible, the (conditional) probability of $y \in s$ is given with and without the null.

Why is this is important? It is common knowledge that weer p-values can be had by increasing the sample size. This is one reason why the size of p-value does not correlate (in the plain-English sense) with the value of δ. As the sample size increases, changes to δ become less likely, whereas the p-value can shrink indefinitely. Wee p-values therefore do *not* imply large values of δ, just as large p-values do *not* imply small values of δ. P-values are misleading, while the predictive approach gives the direct, desired probability based on the agreed upon assumptions (M, D, and so on).

6   Example

Annual income (in thousands) was measured for some 1,000 persons (from \$0K to \$563K), along with Sex, Age, Marriage status, Hours Worked, Race, USA Citizen status, presence of Health Insurance, and whether English was spoken, [6]. No other details are present, such as when the data was taken. Only those 18 years-old and older were analyzed, leaving $n = 787$. A simple linear regression with $y =$ "Income" on the remaining measurements was calculated. I make no claims for the value of this model except that its form is common and that analyses like this comprise the bulk of actual statistical practice. Reminder: there is no true model to be sought here, because a true model would identify the exact causes of each of the incomes of the 787 individuals. The *ad hoc* model here is not weak on that front, because all probability models would share this same weakness.

The estimated coefficient for male Sex was $10.4K, with a 95% confidence interval of ($2.1K, $18.7K); the p-value was 0.015. Almost everywhere it would be announced that this was "statistically significant", which is only another way of saying the p-value was wee. But it would be taken by many that it has been proved "Men make more than women". Indeed, it would be a strong man who could resist announcing "Women make less than men." This is, of course, not so. In this data, some women made more than some men (and we do not know why either way). But, having made the initial error, it would be irresistible to think that some women made less than some men be*cause* they were women.

If women made less than men because they were women, then, at least in some instances, it must be that the hiring manager, or whomever, looked at the woman and said, at least tacitly, "If you were a man, I would pay you ten thousand four hundred dollars more, but because you are woman, I will not." Of course, that is precisely the claim that will be made; that the wee p-value proves this "bias."

Let's not become distracted by the politics, except to note that this data says only *what* happened, not *why* anything happened. This is proved, if it is not already clear, by noting that the estimated coefficient for Health Insurance was $20.4K, with a p-value of 0.0006. If p-values prove cause, then it *must* be, as it was for women vs. men, that awarding somebody Health Insurance would *cause* a hefty increase in salary!

Contrast this with the predictive approach, illustrated in part in Fig. 1. These are probabilistic predictions: the marks in the figure represent the difference between men and women in actual probabilities for each value, a window of $1K centered on each mark. These are deduced from the predictive posterior distribution of the model, which used a "flat" prior, which for the special case of normals makes the results of the classical and predictive analyses the same, see [2].

The probabilities were calculated assuming values for each of the measurements, as is necessary in the predictive approach. M said each of these measurements were important, after all, so each must always be included in every prediction. If we do not care about the influence of, say, English language on the uncertainty of Income, then we should not have put it in the model. It is there, so we have to specify a value. The values were chosen were the observed medians, i.e. Age = 49, Married, Hours Worked = 40, USA Citizen, Health Insurance present, English spoken. Race and sex were separately specified for the plot with the other measurements held fixed.

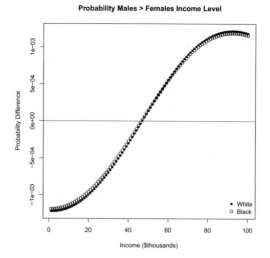

Fig. 1. The difference in probability that a man with the given characteristics will earn more than a woman with the same characteristics, at each of several income windows (each $1k), and for whites (dots) and blacks (open circles).

The first thing to note (not shown on the figure), is that, given M and values of the measurements, the probability of Income less than $0 for men, white or black, is about 13%, and for women about 19% (again, about the same for white and black).

This is a glaring error—and an exceedingly common one which I have elsewhere called *probability leakage*, [3]. This error is never noticed because in the classical approach the concern is on the parameters and p-values, and not the implied predictions. This model really does give substantial probability to values which we know to be impossible. The problem begins in not including these impossibilities in M. This can, of course, be done, and careful modelers will do so, but carefulness is not a property of most analyses; worse, the possibility of this error is rarely taught. And, again, it cannot be seen using parameter-centric and hypothesis testing methods.

Passing by the leakage, the next thing to note is that, for whites or blacks, the probability difference for sex is near 0 for Incomes around $50K—for Hours Worked 40, etc. The probability (for the given measurements) white men make between $40–$60K is 0.183, and for women 0.180. For blacks, the probabilities are 0.181 and 0.178. These are differences of only 0.003. Are we do bold as to say we can predict probabilities to the nearest thousandth place? Well, obviously we are, since these types of models are ubiquitous.

The window $40–$60K is chosen because I thought it important; others, as emphasized above, might find other windows of greater relevancy. In the data, for those who worked 40 h a week, and regardless of the other measurements, the observed median Income was $36K, with a mean of $41K, and an innerquartile range of $19K–$55K. So the window of $40–$60K represents roughly typical data.

Do men and women make different incomes? Yes. But we knew that before we started, by glancing at the data. The probability white men not in the sample with the given characteristics will make the most common salaries is 18.3%, and for white women 18.0%. Is this difference important? Statistics cannot tell us. Only the decision somebody would make on this can. There is also a 0.005 probability white men will make greater than $100K, and for white women 0.004. Is that difference important? We do not know; not from the data, at any rate.

It should be clear that the pleasing certainty that came with the classical parameter estimate and p-value has been greatly tempered in the predictive approach. This will always be the case; it is mathematically guaranteed. Certainty in the parameters does not translate into certainty in the predictions.

P-values and values of δ do not track, as promised above. For white women who are married and have the other fixed measurements, the probability of Incomes in $40–$60K is 0.180, while for unmarried women it is 0.176. This difference is *greater* than it was for Sex. The p-value for Marriage status was not however wee; it was 0.28. Yet marriage, for women, was of greater predictive value than Sex. The same is true for English spoken. There, married white women (etc.) have a probability of 0.182 of Incomes in $40–$60K, while for non-English speakers it is 0.180. The p-value for English was 0.32. The difference is even greater for US Citizens: married white women who speak English have a probability of 0.182 for Incomes in $40–$60K, while for non-Citizens it is 0.176. The p-value for Citizen status was 0.37. P-values do not translate to direct evidence.

There is, as is obvious, much more than can be done with this data, and that is in part the problem. The predictive approach is more work than the classical approach, sometimes much more. One cannot just glance at p-values and be done with it. One must take the model one assumed quite seriously, and then examine in detail the predictions made from that model. The decisions one wants to make must be made clear and be at the forefront of any analysis. The benefit, or payoff for this extra labor, is that the end result is much fairer, simpler, and usable than the classical approach.

One last word about model verification; see [4,9,10] for details on proper scores, calibration, and all that. The model makes Income probability predictions for those who have the specified measurements. That means we can directly check the predictions against new observations. We could also verify the model on the past observations, but since these past observations built the model, so to speak, we would be running in circles. It is always possible to, as all know, find a model which predicts past observations to any degree of accuracy one likes. The true test is performance on data *never before seen or used in any way*. This means it is difficult to trust any researcher who says he has fit his model on a portion of his data and verified it on the rest. The temptation to reuse "the rest" to improve the model on the initial portion is far too great: nobody can resist it.

That there is a difference between the sexes and so forth in prediction probabilities does *not* mean that these differences will manifest in new data. The only, the sole, the lone way to know is to apply the model to brand-new data. Nothing else suffices. This is, after all, the way civil engineers verify their models: if the bridge does not fall, their model is verified. In this way, all statistical analysis, as they are usually presented in scientific journals, must be seen a preliminary, as highly speculative, even. That these initial results are taken too seriously accounts for the so-called replication crisis, see e.g. [19]. The predictive approach allows models to be published as actual, verifiable predictions. The predictions are there and can be verified without recourse to any sophisticated apparatus. It is a very open approach.

Fitting models with the classical approach is far too easy. Anybody can do it; and anybody does.

References

1. Benjamin, D.J., Berger, J., Johannesson, M., Nosek, B.A., Wagenmakers, E.-J., Berk, R., Bollen, K., et al.: Redefine Statistical Significance. PsyArXiv, psyarxiv.com/mky9j (2017)
2. Bernardo, J.M., Smith, A.F.M.: Bayesian Theory. Wiley, New York (2000)
3. Briggs, W.M.: On probability leakage. arXiv, arXiv.org/abs/1201.3611 (2013)
4. Briggs, W.M.: Uncertainty: The Soul of Modeling. Probability & Statistics. Springer, Cham (2016)
5. Briggs, W.M.: The substitute for p-values. JASA **112**(519), 897–898 (2017)
6. St Clair, K.: Textbook Data for Math 215. http://people.carleton.edu/~kstclair/Math215.html. Accessed 27 Aug 2017
7. Fisher, R.A.: Statistical Methods for Research Workers, 14th edn. Oliver and Boyd, Edinburgh (1970)
8. Gigerenzer, G.: Mindless statistics. J. Socio Econ. **33**, 587–606 (2004)
9. Gneiting, T., Raftery, A.E., Balabdaoui, F.: Probabilistic forecasts, calibration and sharpness. J. R. Stat. Soc. Ser. B Stat. Methodol. **69**, 243–268 (2007)
10. Gneiting, T., Raftery, A.E.: Strictly proper scoring rules, prediction, and estimation. JASA **102**, 359–378 (2007)
11. Hájek, A.: Mises Redux–Redux: fifteen arguments against finite frequentism. Erkenntnis **45**, 209–227 (1997)
12. Hájek, A.: Fifteen arguments against hypothetical frequentism. Erkenntnis **70**, 211–235 (2009)
13. Jaynes, E.T.: Probability Theory: The Logic of Science. Cambridge University Press, Cambridge (2003)
14. Keynes, J.M.: A Treatise on Probability. Dover Phoenix Editions, New York (2004)
15. Laplace, P.S.: A Philosophical Essay on Probabilities. Dover, New York (1996)
16. Lash, T.L.: The harm done to reproducibility by the culture of null hypothesis significance testing. Am. J. Epidemiol. **186**(5), 1–9 (2017)
17. Nau, R.F.: De Finetti was right: probability does not exist. Theor. Decis. **51**, 89–124 (2001)
18. Neyman, J.: Outline of a theory of statistical estimation based on the classical theory of probability. Philos. Trans. R. Soc. Lond. A **236**, 333–380 (1937)
19. Peng, J.: The reproducibility crisis in science: a statistical counterattack. Significance **12**, 30–32 (2015)
20. Ziliak, S.T., McCloskey, D.N.: The Cult of Statistical Significance. University of Michigan Press, Ann Arbor (2008)

Information Criteria for Statistical Modeling in Data-Rich Era

Genshiro Kitagawa[⊠]

Meiji Institute for Advanced Study of Mathematical Sciences,
Meiji University, Tokyo 164-8525, Japan
kitagawa@ism.ac.jp

Abstract. Due to the dramatic development of measuring instruments in recent years, a huge amount of large-scale data has been acquired in all research areas. Along with this, research method has changed, and data-driven methods are becoming important as the fourth scientific methodology. In the data-driven approach, the model is built according to the theory, knowledge, data, and further the purpose of the analysis. Once a model is built, useful information can be extracted from the data through the fitted model. In this data-driven method, it is crucial to use a good model and thud the evaluation of the model is essential in the success of the data-driven approach. This paper outlines the model evaluation criteria such as AIC, GIC, EIC, and so on, focusing on information criteria for evaluating prediction accuracy based on statistical models. Since L_1 regularization is important in recent data analysis, the evaluation of the regularized model is also outlined.

1 Introduction

Due to recent development of information and communication technologies, human society is changing very rapidly. Actually, by the development of sensor devices, huge amount of data are now accumulating in various fields of scientific research, such as in life science, marketing, finance, environmental science, seismology, meteorology, astronomy and high-energy physics, etc.

Various changes occurred in this background. Firstly, the objects of scientific research were expanded (Fig. 1). Until the 19th century, the main target of the research was the static physical world. However, by the impact of Darwinism, evolutionary and changing world such as the life, economy becomes important objects in the 20th century. Further in this 21st century, owing to the development of ICT, we are facing to the so-called cyber-physical world. Secondly, objective of the research changed from the "quest for the truth" to the "prediction, simulation, knowledge creation or decision making." Thirdly, model itself was changed from physical model derived from the first principle to the modeling to achieve the objective of the research.

In parallel to the academic area, big data also appear in various aspects of our society. Actually it emerged from internet communications, sensor, drone, transaction, multi-media and various logs. And the emergence of the big data

© Springer International Publishing AG 2018
L. H. Anh et al. (eds.), *Econometrics for Financial Applications*, Studies in Computational Intelligence 760, https://doi.org/10.1007/978-3-319-73150-6_2

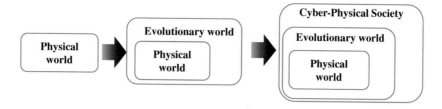

Fig. 1. Expansion of the objects of scientific research

is quickly changing our society. As examples, we can consider personalized medicine, marketing, recommendation system, data-driven industry and smartification of social infrastructure, and more recently, brilliant achievements of artificial intelligence in games, image analysis, automatic driving and so on.

In the book entitled "post-capitalist society," Drucker (1993) wrote

Every few hundred years in Western history there occurs a sharp transformation. We cross what I called a "divide." Within a few short decades, society rearranges itself, its worldview, its basic values, its social and political structure, its arts; its key institutions. Fifty years later, there is a new world. And the people born then cannot even imagine the world in which their grandparents lived and into which their own parents were born. We are currently living through just such a transformation.

In the past history, the science has changed the society by expanding its fields of applications and many area that used to be treated by the intuition and experience of experts at one time became the objects of scientific approach. As such examples, we may imagine the astrology, navigation, alchemy, production process, management, marketing, finance, risk management. Further, in recent years, service and policy making, even the scientific discovery became the object of scientific research.

One typical transition is the emergence of data-driven society. In the book entitled "Super Crunchers," Ayres (2007) asserts that the "big data analysis" surpasses the "experience and intuition" of experts in many area of decision making, and showed many examples such as the evaluation of wine quality, recruiting baseball players, airline customer service, individual pricing of premium and online sales and so on.

This shows that cyber intelligence comes close to a human being in the intellectual labor and it reminds us of the historic moment of the match between horsecar and steam locomotive held at Baltimore & Ohio Railroad in 1830, when the machine has caught up with an animal's physical labor. We may say that a data-centric society will appear before long and also that all research will become data science.

From the viewpoint of the inductive inference, in the 20th century, the main objective used be the exact reasoning based on well designed small number of experimental data. Now, by the advent of the big data, an important problem is the knowledge discovery or information extraction based on big data.

However, although the big data may contain enormous knowledge and value, it is usually difficult to extract them by the current methods and technologies because it is mostly unstructured, has low value density, large scale, sparse and further it is heterogeneous in terms of precision, form, observation frequency.

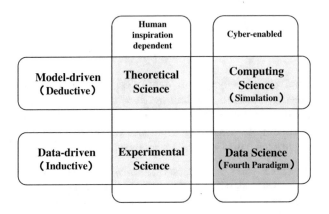

Fig. 2. Fourth Science: Data Science

To fully utilize the information contained in the big data, it is necessary to develop the fourth scientific methodology (Fig. 2). Until the 20th century, science was driven by two scientific methodologies, namely, the experimental science and the theoretical science. However, in the latter half of the 20th century, the computing science was developed for understanding or prediction of complex nonlinear systems. Now by the advent of big data, it is necessary to develop the fourth scientific methodology, namely the data science.

The basic technologies for the data science are big data processing, visualization and data analysis (Manyika et al. 2011). Big data processing is the techniques to handle scattered big data and consists of various information processing technologies such as distributed processing, parallel computation, etc. Visualization is the technologies to grasp high-dimensional data and computing results such as dimension reduction, feature extraction, pattern recognition, image processing. Data analysis is the method for obtaining deep knowledge from big data and is related to statistical modeling, Bayes inference, machine leaning, data mining, web information analysis, natural language processing and optimization.

In the data-driven approach, the model is built according to the theory, knowledge, data, and further the purpose of the analysis. Once a model is built, useful information can be extracted from the data through the fitted model. In this data-driven method, it is crucial to use a good model. Therefore, the problem of developing good model evaluation criteria is a very important.

This paper is organized as follows. In Sect. 2, we will consider the role of statistical modeling and viewpoint of predictive ability. Section 3 outlines the information criteria AIC, TIC and AIC_c which are obtained as the approximately unbiased estimates of the expected log-likelihood of the model whose parameters are estimated by the maximum likelihood method. Section 4 outlines the GIC for the evaluation of any types of estimators defined by statistical functional, such as M-estimator and Bayes model. In Sect. 5, the bootstrap information criterion EIC is outlined which can be applied to wide class of models and situation. In Sect. 6, evaluation criteria for the models obtained by regularization methods are considered. Finally, Sect. 7 summarized the paper.

2   Statistical Modeling and Predictive Model Evaluation

In statistical modeling, model is built by properly combining the information from the theory, empirical knowledge and data and even the objective of the problem (Fig. 3). In general context, it can be formulated by using Bayes model. Once the model is obtained, we can extract useful information from data, do prediction and simulation, and decision making based on the model. So the knowledge is provided through the model and the knowledge improves the model. And thus it constitutes the spiral of knowledge development.

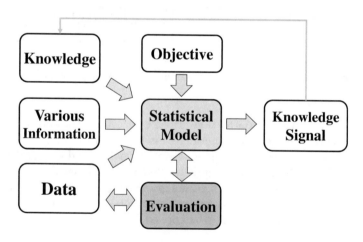

Fig. 3. Statistical modeling.

In statistical modeling, it is not necessarily assumed that the model is true or a close replica of the truth and we rather use it as a tool to extract useful information from data. Therefore, it is important to build a model by properly combining the information from the data and the prior information and knowledge on the subject and objective of the problem (Fig. 3).

In this situation, it is obvious that if we use a good model, then we can get good results but if we use a poor model, we will not be able to get meaningful results. Therefore, the use of good model is essential in statistical modeling and statistical knowledge extraction, and the evaluation of the estimated model is one of the most important problems in the data-driven approach. To achieve this, development of criteria for evaluating the goodness of statistical model is indispensable.

In developing a model evaluation criterion, Akaike advocated the predictive point of view. In the conventional statistical procedure, the objective of model fitting and parameter estimation is to obtain a good model that can reasonably reproduce the true model as precise as possible (Fig. 4). In contrast to this, in the predictive point of view, the estimated model is evaluated by the prediction ability. Akaike (1973, 1974) measured this ability by the Kullback-Leibler information between the predictive distribution and future data distribution. The AIC is obtained as an estimate of (the essential part of) the Kullback-Leibler information.

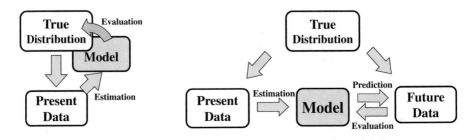

Fig. 4. Conventional statistical modeling (left) and predictive modeling (right).

Akaike's (1973, 1974) information criterion provides a useful tool for evaluating models estimated by the method of maximum likelihood and a number of successful applications of AIC in statistical modeling and data analysis have been reported (Bozdogan 1994; Kitagawa and Gersch 1996; Akaike and Kitagawa 1998). By extending Akaike's basic idea, several attempts have been made to relax the assumptions imposed in the derivation of AIC and obtained information theoretic criteria which may be applied to the various types of statistical models.

In recent years advances in the performance of computers enables us to construct models for analyzing data with complex structure, and consequently more flexible criteria are required for model evaluation and selection problems. The purpose of the present paper is to overview information criteria which yield more refined results than previously proposed criteria and may be applied to a variety of statistical models. The use of the bootstrap in model evaluation problems is also investigated from theoretical and practical points of view.

3   Information Criteria for ML Models

3.1   Estimation of Kullback-Leibler Information

Assume that the observations are generated from an unknown "true" distribution function $G(x)$ and the model is characterized by a density function $f(x)$. In the derivation of AIC (Akaike 1973, 1974; Konishi and Kitagawa 2008), the expected log-likelihood $E_Y \log f(Y) = \int \log f(y) dG(y)$ is used as the basic criterion to evaluate the closeness of a model to the true model, which is equivalent to the Kullback-Leibler information (1951). Here E_Y denotes the expectation with respect to the true distribution $G(y)$.

In actual statistical problems, the true distribution $G(x)$ is unknown and only a sample $\boldsymbol{X} = \{X_1, \ldots, X_n\}$ drawn from $G(x)$ is given. We then use the log-likelihood $n^{-1}\ell = \int \log f(x) d\hat{G}_n(x) = n^{-1} \sum_{i=1}^{n} \log f(X_i)$ as a natural estimator of the expected log-likelihood. Here $\hat{G}_n(x)$ is the empirical distribution function, having mass $1/n$ on each observation.

For a parametric model $f(x|\theta)$ with a parameter $\theta = (\theta_1, \ldots, \theta_m)^T$, it naturally leads to the maximum likelihood estimator, $\hat{\theta} = \hat{\theta}(\boldsymbol{X})$, which is the maximizor of the log-likelihood function

$$\ell(\theta) = \sum_{i=1}^{n} \log f(X_i|\theta) \equiv \log f(\boldsymbol{X}|\theta). \tag{1}$$

Interestingly, although the log-likelihood is a good estimate of the expected log-likelihood, $E_Y \log f(Y|\theta)$, the maximum log-liklihood $\log f(X|\hat{\theta})$ is NOT an unbiased estimate of $E_Y \log f(Y|\hat{\theta})$. Namely, $(n^{-1}$ times of) the maximum log-likelihood, $n^{-1}\ell(\hat{\theta}(\boldsymbol{X}))$, has a positive bias as an estimator of the expected log-likelihood, $E_Y \log f(Y|\hat{\theta}(\boldsymbol{X}))$, and it cannot be directly used for model selection.

This bias occurs because the same data set $\boldsymbol{X}$ was used twice for the estimation of the parameter and the expected log-likelihood. By correcting the bias

$$b(G) = nE_X \left\{ \frac{1}{n} \log f(\boldsymbol{X}|\hat{\theta}(\boldsymbol{X})) - E_Y \log f(Y|\hat{\theta}(\boldsymbol{X})) \right\}, \tag{2}$$

an unbiased estimator of the expected log-likelihood is obtained by $n^{-1}\{\log f(\boldsymbol{X}|\hat{\theta}(\boldsymbol{X})) - b(G)\}$. Therefore, considering the definition of AIC, generic information criteria is defined by

$$- 2 \log f(\boldsymbol{X}|\hat{\theta}(\boldsymbol{X})) + 2\hat{b}(G), \tag{3}$$

where $\hat{b}(G)$ is a properly defined approximation to $b(G)$.

3.2   Information Criteria: AIC, TIC and AIC$_c$

In a general setting, it is difficult to obtain the bias $b(G)$ in a closed form. Under some setting, Akaike evaluated an asymptotic bias as $b(G) = m$, and advocated the information criterion

$$\text{AIC} = -2 \log f(\boldsymbol{X}|\hat{\theta}_{ML}) + 2m, \tag{4}$$

where m is the number of estimated parameters (Akaike 1973, 1974; Konishi and Kitagawa 2008). Numerous successful application of the statistical modeling based on AIC have been reported (Bozdogan 1994; Kitagawa and Gersch 1996; Akaike and Kitagawa 1998).

Using the properties of the maximum likelihood estimators $\hat{\theta}_{ML}$, for incorrectly specified models (Huber 1976), the asymptotic bias can be evaluated as (Takeuchi 1976)

$$b_T(G) = \mathrm{tr}\{I(G)J(G)^{-1}\}, \tag{5}$$

where $I(G)$ and $J(G)$ are respectively the Fisher information matrix and the expected Hessian defined by

$$I(G) = E_Y \left[\frac{\partial \log f(Y|\theta)}{\partial \theta} \frac{\partial \log f(Y|\theta)}{\partial \theta^T} \right],$$

$$J(G) = -E_Y \left[\frac{\partial^2 \log f(Y|\theta)}{\partial \theta \partial \theta^T} \right]. \tag{6}$$

By correcting the asymptotic bias of the log likelihood, TIC is defined by Takeuchi (1976)

$$\mathrm{TIC} = -2 \log f(\boldsymbol{X}|\hat{\theta}_{ML}) + 2\mathrm{tr}\{\hat{J}(G)^{-1}\hat{I}(G)\}, \tag{7}$$

where $\hat{J}(G)$ and $\hat{I}(G)$ are consistent estimates of $J(G)$ and $I(G)$, respectively.

If the model contains the true distribution such that $g(x) = f(x|\theta)$ for some θ, it holds that $I(G) = J(G)$, and the asymptotic bias becomes $b_A(G) = m$, where m is the dimension of the parameter vector θ. Thus we obtain the Akaike information criterion, AIC (Fig. 5).

Further, for some specific models, it is possible to evaluate the bias directly and obtain a more precise bias correction term without resorting to asymptotic

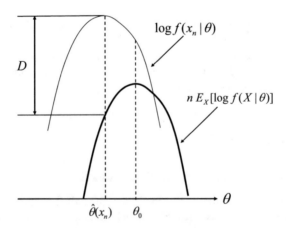

Fig. 5. Bias of the maximum log-likelihood as an estimator of the expected log-likelihood. (Konishi and Kitagawa 2008)

theory. As the simplest example, consider the normal distribution model, $y_n \sim N(\mu, \sigma^2)$. Then the log-likelihood of the model based on the data, $\{y_1, \ldots, y_n\}$, is given by

$$\ell(\mu, \sigma^2) = -\frac{n}{2}\log(2\pi\sigma^2) - \frac{1}{2\sigma^2}\sum_{\alpha=1}^{n}(y_\alpha - \mu)^2.$$

By substituting the maximum likelihood estimators $\hat{a}_j$ and $\hat{\sigma}^2$ into this expression, we obtain the maximum log-likelihood $\ell(\hat{a}_j, \hat{\sigma}^2) = -\frac{n}{2}\log(2\pi\hat{\sigma}^2) - \frac{n}{2}$. If the data set is obtained from the same normal distribution $N(\mu, \sigma^2)$, then the expected log-likelihood is given by $E_G\left[\log f(Z|\hat{\mu}, \hat{\sigma}^2)\right] = -\frac{1}{2}\log(2\pi\hat{\sigma}^2) - \frac{1}{2\hat{\sigma}^2}\left\{\sigma^2 + (\mu - \hat{\mu})^2\right\}$, where $G(z)$ is the distribution function of the normal distribution $N(\mu, \sigma^2)$. Therefore, the difference between the two quantity is $\ell(\hat{\mu}, \hat{\sigma}^2) - nE_G\left[\log f(Z|\hat{\mu}, \hat{\sigma}^2)\right] = \frac{n}{2\hat{\sigma}^2}\left\{\sigma^2 + (\mu - \hat{\mu})^2\right\} - \frac{n}{2}$. By taking the expectation with respect to the joint distribution of n observations distributed as the normal distribution $N(\mu, \sigma^2)$, and using $E_G\left[\frac{\sigma^2}{\hat{\sigma}^2(\boldsymbol{x}_n)}\right] = \frac{n}{n-3}$, $E_G\left[\{\mu - \hat{\mu}(\boldsymbol{x}_n)\}^2\right] = \frac{\sigma^2}{n}$, we obtain the bias correction term for the finite sample as

$$b_{cA}(G) = \frac{n}{2}\frac{n}{(n-3)\sigma^2}\left(\sigma^2 + \frac{\sigma^2}{n}\right) - \frac{n}{2} = \frac{2n}{n-3}. \tag{8}$$

Here, we used the fact that for a χ^2 random variable with degrees of freedom r, χ_r^2, we have $E[1/\chi_r^2] = 1/(r-2)$. Therefore, an information criterion (corrected AIC) for the normal distribution model is given by

$$\mathrm{AIC}_c = -2\ell(\hat{\mu}, \hat{\sigma}^2) + \frac{4n}{n-3}. \tag{9}$$

Similarly, for a linear regression model $y_n = \sum_{j=1}^{m} a_j x_{nj} + \varepsilon_n$, $\varepsilon \sim N(0, \sigma^2)$, where y_n and $x_{nj}, j = 1, \ldots, m$ are the objective variable and the regressors, respectively, the bias is evaluated as

$$b_{cA}(G) = \frac{(m+1)n}{n-m-2}. \tag{10}$$

4 Information Criteria for Wider Class of Models

4.1 Generalized Information Criterion GIC

This method of bias correction for the log-likelihood can be extended to a more general estimator defined by a statistical functional such as $\hat{\theta} = \boldsymbol{T}(\hat{G}_n)$, where $\boldsymbol{T}(\cdot) = (T_1(\cdot), \ldots, T_m(\cdot))^T$ is a functional on the space of distribution functions. For such a general estimator, the asymptotic bias is given by Konishi and Kitagawa (1996, 2008)

$$b_1(G) = \mathrm{tr}\left\{\int T^{(1)}(y; G)\frac{\partial \log f(y|\boldsymbol{T}(G))}{\partial \theta^T}dG(y)\right\}, \tag{11}$$

where $\boldsymbol{T}^{(1)}(\boldsymbol{Y};G) = (T_1^{(1)}(\boldsymbol{Y};G),\dots,T_m^{(1)}(\boldsymbol{Y};G))^T$ and $T_i^{(1)}(\boldsymbol{Y};G)$ is the influence function defined by

$$T_i^{(1)}(\boldsymbol{X};G) = \lim_{\varepsilon \to \infty} \{T_i((1-\varepsilon)G + \varepsilon\delta_\alpha) - T_i(G)\}/\varepsilon \tag{12}$$

with δ_α being a point mass at X_α. By subtracting the asymptotic bias estimate from the log-likelihood, we have (Fig. 6)

$$\text{GIC} = -2\log f(\boldsymbol{X}|\theta) + 2b_1(\hat{G}). \tag{13}$$

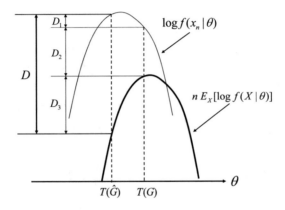

Fig. 6. Bias correction by GIC. (Konishi and Kitagawa 2008)

Example: GIC for the normal distribution model. Consider a simple normal distribution model with unknown mean μ and the variance σ^2

$$f(y|\mu,\sigma^2) = \frac{1}{\sqrt{2\pi\sigma^2}} \exp\left\{-\frac{(y-\mu)^2}{2\sigma^2}\right\}. \tag{14}$$

The maximum likelihood estimators of μ and σ^2 are given by statistical functionals,

$$T_\mu(G) = \int x\,dG(x), \quad T_{\sigma^2}(G) = \int (x - T_\mu(G))^2\,dG(x), \tag{15}$$

respectively. For these estimators, the derivatives of the functionals are given by

$$T_\mu^{(1)}(x;G) = x - \mu, \quad T_{\sigma^2}^{(1)}(x;G) = (x-\mu)^2 - \sigma^2. \tag{16}$$

Using these results, the bias correction term (11) is explicitly obtained by

$$b_1(G) = \frac{1}{2}\left(1 + \frac{\mu_4}{\sigma^4}\right), \tag{17}$$

where μ_4 denotes the fourth central moments of the true distribution G. In particular, when the true distributions are standard normal distribution ($\mu_4 = 3$) and Laplace distribution ($\mu_4 = 6$), they are given by $b_1(G) = 2$ and 3.5, respectively.

4.2   Maximum Likelihood Method: Relationship Among AIC, TIC and GIC

Assume that the maximum likelihood method is used for the estimation of a model $f(x|\boldsymbol{\theta})$ based on the observed data from $G(x)$. The maximum likelihood estimator, $\hat{\boldsymbol{\theta}}_{ML}$, is defined as a solution of the equation

$$\sum_{\alpha=1}^{n} \frac{\partial \log f(x_\alpha|\boldsymbol{\theta})}{\partial \boldsymbol{\theta}} = \mathbf{0}, \tag{18}$$

which can be expressed as $\hat{\boldsymbol{\theta}}_{ML} = \boldsymbol{T}_{ML}(\hat{G})$ using the p-dimensional functional $\boldsymbol{T}_{ML}(G)$ implicitly defined by

$$\int \frac{\partial \log f(x|\boldsymbol{\theta})}{\partial \boldsymbol{\theta}}\bigg|_{\boldsymbol{\theta}=\boldsymbol{T}_{ML}(G)} dG(x) = \mathbf{0}. \tag{19}$$

The influence function for the maximum likelihood estimator can be obtain as follows: By replacing the distribution function G in (19) with $(1-\varepsilon)G + \varepsilon\delta_x$, we have

$$\int \frac{\partial \log f(y|\boldsymbol{T}_{ML}((1-\varepsilon)G+\varepsilon\delta_x))}{\partial \boldsymbol{\theta}} d\{(1-\varepsilon)G(y)+\varepsilon\delta_x(y)\} = \mathbf{0}. \tag{20}$$

Differentiating both sides with respect to ε and setting $\varepsilon = 0$ yield

$$\int \frac{\partial \log f(y|\boldsymbol{T}_{ML}(G))}{\partial \boldsymbol{\theta}} d\{\delta_x(y)-G(y)\} \tag{21}$$

$$+ \int \frac{\partial^2 \log f(y|\boldsymbol{T}_{ML}(G))}{\partial \boldsymbol{\theta} \partial \boldsymbol{\theta}^T} dG(y) \cdot \frac{\partial}{\partial \varepsilon}\left\{\boldsymbol{T}_{ML}((1-\varepsilon)G+\varepsilon\delta_x)\right\}\bigg|_{\varepsilon=0} = \mathbf{0},$$

Noting that

$$\int \frac{\partial \log f(y|\boldsymbol{T}_{ML}(G))}{\partial \boldsymbol{\theta}} d\delta_x(y) = \frac{\partial \log f(x|\boldsymbol{T}_{ML}(G))}{\partial \boldsymbol{\theta}} \tag{22}$$

and using (19), from (21), we obtain the influence function for the maximum likelihood estimator $\hat{\boldsymbol{\theta}}_{ML} = \boldsymbol{T}_{ML}(\hat{G})$

$$\boldsymbol{T}_{ML}^{(1)}(x;G) \equiv \frac{\partial}{\partial \varepsilon}\left\{\boldsymbol{T}_{ML}((1-\varepsilon)G+\varepsilon\delta_x)\right\}\bigg|_{\varepsilon=0} = J(G)^{-1}\frac{\partial \log f(x|\boldsymbol{\theta})}{\partial \boldsymbol{\theta}}\bigg|_{\boldsymbol{\theta}=\boldsymbol{T}_{ML}(G)}, \tag{23}$$

where $J(G)$ is a $p \times p$ matrix given by

$$J(G) = -\int \frac{\partial^2 \log f(x|\boldsymbol{\theta})}{\partial \boldsymbol{\theta} \partial \boldsymbol{\theta}^T}\bigg|_{\boldsymbol{\theta}=\boldsymbol{T}_{ML}(G)} dG(x). \tag{24}$$

By replacing the influence function $\boldsymbol{T}^{(1)}(x;G)$ in (11) with (23), we obtain the asymptotic bias of the log-likelihood for the estimated model $f(x|\hat{\boldsymbol{\theta}}_{ML})$

$$b_{ML}(G) = \text{tr}\left\{J(G)^{-1}\int \frac{\partial \log f(x|\boldsymbol{\theta})}{\partial \boldsymbol{\theta}}\frac{\partial \log f(x|\boldsymbol{\theta})}{\partial \boldsymbol{\theta}^T}\bigg|_{\boldsymbol{\theta}=\boldsymbol{T}_{ML}(G)} dG(x)\right\},$$

$$= \text{tr}\left\{J(G)^{-1}I(G)\right\} \tag{25}$$

where the $p \times p$ matrix $I(G)$ is given by

$$I(G) = \int \frac{\partial \log f(x|\boldsymbol{\theta})}{\partial \boldsymbol{\theta}} \frac{\partial \log f(x|\boldsymbol{\theta})}{\partial \boldsymbol{\theta}^T}\bigg|_{\boldsymbol{\theta}=\boldsymbol{T}_{ML}(G)} dG(x). \tag{26}$$

Therefore, for the model $f(x|\hat{\boldsymbol{\theta}}_{ML})$ estimated by the maximum likelihood method, GIC in (13) is reduced to

$$\text{TIC} = -2\sum_{\alpha=1}^{n} \log f(x_\alpha|\hat{\boldsymbol{\theta}}_{ML}) + 2\text{tr}\left\{J(\hat{G})^{-1}I(\hat{G})\right\}. \tag{27}$$

4.3   GIC for the Models Estimated by M-estimators

In this subsection we derive an information criterion for evaluating a statistical model estimated by M-estimators, using the generalized information criterion GIC in (13).

Suppose that $f(x|\hat{\boldsymbol{\theta}}_M)$ is the model of the true distribution $G(x)$, where $\hat{\boldsymbol{\theta}}_M$ is a p-dimensional M-estimator defined as the solution of the system of implicit equations

$$\sum_{\alpha=1}^{n} \boldsymbol{\psi}(x_\alpha, \hat{\boldsymbol{\theta}}_M) = \mathbf{0}. \tag{28}$$

Here, $\boldsymbol{\psi} = (\psi_1, \psi_2, \cdots, \psi_p)^T$ and $\psi_i(x, \boldsymbol{\theta})$ is a real-valued function defined on the product space of the sample and parameter spaces. The M-estimator $\hat{\boldsymbol{\theta}}_M$ is given by $\hat{\boldsymbol{\theta}}_M = \boldsymbol{T}_M(\hat{G})$ for the p-dimensional functional $\boldsymbol{T}_M(G)$ defined as the solution of the implicit equations

$$\int \boldsymbol{\psi}(y, \boldsymbol{T}_M(G))dG(y) = \mathbf{0}. \tag{29}$$

Then the influence function for the M-estimator is obtained by the same method as for the MLE as

$$\boldsymbol{T}_M^{(1)}(x;G) \equiv \frac{\partial}{\partial \varepsilon}\left\{\boldsymbol{T}_M((1-\varepsilon)G + \varepsilon\delta_x)\right\}_{\varepsilon=0} = R(\boldsymbol{\psi}, G)^{-1}\boldsymbol{\psi}(x, \boldsymbol{T}_M(G)), \tag{30}$$

where $R(\boldsymbol{\psi}, G)$ is a $p \times p$ matrix whose (i, j)-components is given by

$$R(\boldsymbol{\psi}, G)(i, j) = -\int \frac{\partial \psi_j(x, \boldsymbol{\theta})}{\partial \theta_i}\bigg|_{\boldsymbol{\theta}=\boldsymbol{T}_M(G)} dG(x), \qquad (i, j = 1, \cdots, p). \tag{31}$$

Substituting this influence function $\boldsymbol{T}_M^{(1)}(x;G)$ into (11), we have the asymptotic bias of the log-likelihood of the model $f(x|\hat{\boldsymbol{\theta}}_M)$ in estimating the expected log-likelihood in the form

$$b_M(G) = \text{tr}\left\{R(\boldsymbol{\psi}, G)^{-1}\int \boldsymbol{\psi}(x, \boldsymbol{T}_M(G))\frac{\partial \log f(x|\boldsymbol{\theta})}{\partial \boldsymbol{\theta}^T}\bigg|_{\boldsymbol{\theta}=\boldsymbol{T}_M(G)} dG(x)\right\}$$

$$= \text{tr}\left\{R(\boldsymbol{\psi}, G)^{-1}Q(\boldsymbol{\psi}, G)\right\}, \tag{32}$$

where $Q(\psi, G)$ is a $p \times p$ matrix defined by

$$Q(\psi, G) = \int \psi(x, \boldsymbol{T}_M(G)) \frac{\partial \log f(x|\boldsymbol{\theta})}{\partial \boldsymbol{\theta}^T}\bigg|_{\boldsymbol{\theta}=\boldsymbol{T}_M(G)} dG(x). \tag{33}$$

Then, GIC for evaluating the statistical model $f(x|\hat{\boldsymbol{\theta}}_M)$ with the M-estimator $\hat{\boldsymbol{\theta}}_M$ is given by

$$\mathrm{GIC_M} = -2 \sum_{\alpha=1}^{n} \log f(x_\alpha|\hat{\boldsymbol{\theta}}_M) + 2\mathrm{tr}\left\{ R(\psi, \hat{G})^{-1} Q(\psi, \hat{G}) \right\}, \tag{34}$$

where $R(\psi, \hat{G})$ and $Q(\psi, \hat{G})$ are $p \times p$ matrices given by

$$R(\psi, \hat{G}) = -\frac{1}{n} \sum_{\alpha=1}^{n} \frac{\partial \psi(x_\alpha, \boldsymbol{\theta})^T}{\partial \boldsymbol{\theta}}\bigg|_{\boldsymbol{\theta}=\hat{\boldsymbol{\theta}}},$$

$$Q(\psi, \hat{G}) = \frac{1}{n} \sum_{\alpha=1}^{n} \psi(x_\alpha, \hat{\boldsymbol{\theta}}) \frac{\partial \log f(x_\alpha|\boldsymbol{\theta})}{\partial \boldsymbol{\theta}^T}\bigg|_{\boldsymbol{\theta}=\hat{\boldsymbol{\theta}}}. \tag{35}$$

4.4   GIC for Bayes Models

The basic predictive distribution model based on Bayesian approach is defined by the parametric model $\{f(x|\theta); \theta \in \Theta\}$ and the prior distribution $\pi(\theta)$ of the parameter θ as follows

$$h(z|\boldsymbol{X}_n) = \int f(z|\theta)\pi(\theta|\boldsymbol{X}_n)d\theta, \tag{36}$$

where $\boldsymbol{\pi}(\boldsymbol{\theta}|\boldsymbol{X}_n)$ is the posterior distribution of the θ based on the sample $\boldsymbol{X}_n$ and the prior distribution $\pi(\theta)$ and is given by

$$\pi(\theta|\boldsymbol{X}_n) = \prod_{\alpha=1}^{n} f(X_\alpha|\theta)\pi(\theta) \bigg/ \int \prod_{\alpha=1}^{n} f(X_\alpha|\theta)\pi(\theta)d\theta. \tag{37}$$

By substituting the posterior distribution (37), the predictive distribution is obtained by

$$h(z|\boldsymbol{X}_n) = \int \exp\left[n\left\{ q(\theta|\boldsymbol{X}_n) + \frac{1}{n} \log f(z|\theta) \right\} \right] d\theta \bigg/ \int \exp\left\{ nq(\theta|\boldsymbol{X}_n) \right\} d\theta. \tag{38}$$

Here, $\boldsymbol{q}(\boldsymbol{\theta}|\boldsymbol{X}_n)$ is given by

$$q(\theta|\boldsymbol{X}_n) = \frac{1}{n} \sum_{\alpha=1}^{n} \log f(X_\alpha|\theta) + \frac{1}{n} \log \pi(\theta). \tag{39}$$

For this density function, by obtaining the asymptotic expansion with respect to the sample size n based on the Laplace approximation of integrals (Tierney and Kadane 1986; Davison 1986), it becomes possible to apply information criterion GIC.

Assume that $\hat{\theta}_q$ and $\hat{\theta}_q(z)$ are the modes of $q(\theta|\boldsymbol{X}_n)$ and $q(\theta|\boldsymbol{X}_n) + n^{-1}\log f(z|\theta)$, respectively.

In the Laplace's method of integrals, the integrand is Taylor expansion around the mode, and obtain an approximation formula. For example, by applying the Laplace's approximation to the denominator of Eq. (38), we obtain

$$\int \exp\left\{nq(\theta|\boldsymbol{X}_n)\right\} d\theta = \frac{(2\pi)^{p/2}}{n^{p/2}\left|J_q(\hat{\theta}_q)\right|^{1/2}} \exp\left\{nq(\hat{\theta}_q|\boldsymbol{X}_n)\right\}\left\{1+O_p(n^{-1})\right\}. \quad (40)$$

Here, $J_q(\hat{\theta}_q) = -\partial^2\{q(\hat{\theta}_q|\boldsymbol{X}_n)\}/\partial\theta\partial\theta^T$. Similarly, by the Laplace approximation of the integrals in the numerator of (38), we obtain the approximation of the predictive distribution

$$h(z|\boldsymbol{X}_n) = (|J_q(\hat{\theta}_q)|/|J_{q(z)}(\hat{\theta}_q(z))|)^{1/2} \exp\left[n\left\{q(\hat{\theta}_q(z)|\boldsymbol{X}_n) - q(\hat{\theta}_q|\boldsymbol{X}_n) + \frac{1}{n}\log f(z|\hat{\theta}_q(z))\right\}\right]$$
$$\times \{1 + O_p(n^{-2})\},$$

where $J_{q(z)}(\hat{\theta}_q(z)) = -\partial^2\{q(\hat{\theta}_q(z)|\boldsymbol{X}_n) + n^{-1}\log f(z|\hat{\theta}_q(z))\}/\partial\theta\partial\theta^T$. From this Laplace approximation, we obtain the following asymptotic expansion of the Bayesian predictive distribution model

$$h(z|\boldsymbol{X}_n) = f(z|\hat{\theta})\left\{1 + \frac{1}{n}a(z|\hat{\theta}) + O_p(n^{-2})\right\}. \quad (41)$$

The estimator of the model $\hat{\theta}$ depends on whether the prior distribution $\pi(\theta)$ depends on the sample size n or not. Here, we consider the following two cases for the prior distribution (i) $\log\pi(\theta) = O(1)$, and (ii) $\log\pi(\theta) = O(n)$. In the case of (i), $\hat{\theta}$ becomes the maximum likelihood estimator $\hat{\theta}_{ML}$. On the other hand, for the case (ii), it becomes the mode of the posterior distribution $\hat{\theta}_B$. The statistical functionals corresponding to these estimators are respectively given as the solutions to

$$\int \frac{\partial}{\partial\theta}\log f(x|\boldsymbol{T}_{ML}(G))dG(x) = \boldsymbol{0}, \quad \int \frac{\partial}{\partial\theta}[\log\{f(x|\boldsymbol{T}_B(G))\pi(\boldsymbol{T}_B(G))\}]dG(x) = \boldsymbol{0}.$$

Therefore, in (34), by putting $\psi(\boldsymbol{x},\hat{\theta}) = \partial\log f(x|\hat{\theta}_{ML})\partial\theta$ and $\psi(\boldsymbol{x},\hat{\theta}) = \partial\left\{\log f(x|\boldsymbol{T}_B(\hat{G})) + \log\pi(\boldsymbol{T}_B(\hat{G}))\right\}\partial\theta$, we obtain the following evaluation criterion for the Bayes predictive distribution model $h(z|\boldsymbol{X}_n)$

$$\text{GIC}_B = -2\sum_{\alpha=1}^n \log h(X_\alpha|\boldsymbol{X}_n) + 2\text{tr}\left\{J(\psi,\hat{G})^{-1}I(\psi,\hat{G})\right\}. \quad (42)$$

4.5 Higher Order Bias Correction

The information criteria proposed previously are based on large-sample theory to obtain approximately unbiased estimators for the expected log-likelihood or equivalently the Kullback-Leibler information number.

We consider the statistical model $f(y|\hat{\theta})$, where $\hat{\theta}$ is defined by $\hat{\theta} = \boldsymbol{T}(\hat{G}_n)$ with $\boldsymbol{T}(\cdot)$ being a suitably defined m-dimensional functional. Hence by taking the expectation of $E_Y \log f(Y|\hat{\theta}(\boldsymbol{X}))$ over the sampling distribution G of $\boldsymbol{X}$, we have an expectation of the form

$$E_X E_Y \log f(Y|\hat{\theta}(\boldsymbol{X})) = \int g(y) \log f(y|\boldsymbol{T}(G)) dy + \frac{1}{n} a_1(G) + \frac{1}{n^2} a_2(G) + O(n^{-3}). \quad (43)$$

Information criteria based on the asymptotic bias-corrected log-likelihood is second order correct for $E_Y \log f(Y|\hat{\theta})$ in the sense that the expectations of $n^{-1} \left\{ \log f(\boldsymbol{X}|\hat{\theta}) - b_1(\hat{G}) \right\}$ and $E_Y \log f(Y|\hat{\theta})$ are in agreement up to and including the term of order n^{-1}, while the expectations of $n^{-1} \log f(\boldsymbol{X}|\hat{\theta})$ and $E_Y \log f(Y|\hat{\theta})$ differ in term of order n^{-1}.

We consider the bias of $\log f(\boldsymbol{X}|\hat{\theta}) - b_1(\hat{\theta})$, as the estimator of the expected log-likelihood, defined by

$$E_X \left[\log f(\boldsymbol{X}|\hat{\theta}) - b_1(\hat{G}) - nE_Y \log f(Y|\hat{\theta}) \right]$$
$$= E_X \left[\log f(\boldsymbol{X}|\hat{\theta}) - nE_Y \log f(Y|\hat{\theta}) \right] - E_X \left[b_1(\hat{G}) \right]. \quad (44)$$

The first term in the right-hand side of the above equation can be expanded as

$$b(G) = E_X \left[\log f(\boldsymbol{X}|\hat{\theta}) - nE_Y \log f(Y|\hat{\theta}) \right] = b_1(G) + \frac{1}{n} b_2(G) + O(n^{-2}), (45)$$

where $b_1(G)$ is the first order bias correction term given in (11) and $b_2(G)$ is the second order bias correction term.

The expectation of the asymptotic bias estimate $b_1(\hat{G})$ is given by

$$E_X \left[b_1(\hat{G}) \right] = b_1(G) + \frac{1}{n} \Delta b_1(G) + O(n^{-2}). \quad (46)$$

Hence noting that the bias of $\log f(\boldsymbol{X}|\hat{\theta}) - b_1(\hat{G})$ is

$$E_X \left[\log f(\boldsymbol{X}|\hat{G}) - b_1(\hat{G}) - nE_Y \log f(Y|\hat{\theta}) \right] = \frac{1}{n} \{ b_2(G) - \Delta b_1(G) \} + O(n^{-2}), (47)$$

we have the second order bias corrected information criterion in the form

$$\mathrm{GIC}_2(\hat{G}_n) = -2\ell(\hat{G}) + 2 \left\{ b_1(\hat{G}_2) + \frac{1}{n} \left(b_2(\hat{G}_n) - \Delta b_1(\hat{G}_n) \right) \right\}. \quad (48)$$

It might be noted that GIC_2 is third-order correct for the expected log-likelihood. However, analytic expression of $b_2(G)$ and $\Delta b_1(G)$ are very complicated (Kitagawa and Konishi 2010).

Example: Second order bias correction for the normal distribution model. For normal distribution model, these correction terms are explicitly given by

$$b_2(G) - \Delta b_1(G) = \frac{1}{2} \left(\frac{\mu_4}{\sigma^4} + \frac{\mu_6}{\sigma^6} \right), \tag{49}$$

$$b_1(G) - \frac{1}{n} \Delta b_1(G) + \frac{1}{n} b_2(G) = \frac{1}{2} \left(1 + \frac{\mu_4}{\sigma^4} \right) + \frac{1}{2n} \left(\frac{\mu_4}{\sigma^4} + \frac{\mu_6}{\sigma^6} \right), \tag{50}$$

where μ_j is the j-th cumulant of the true distribution.

These show that the estimated bias correction term $b_1(\hat{G}_n)$ is biased as an estimator of $b_1(G)$, and the difference may not be negligible for small n. One of the merit of AIC is that the bias correction term does not depend on G and thus $\Delta b_A(\hat{G}_n) = 0$.

5   Bootstrap Information Criterion EIC

The bootstrap method provides an alternative method for the evaluation of the bias of the log-likelihood (Cavanaugh and Shumway 1997; Ishiguro et al. 1997; Konishi and Kitagawa 1996; Shibata 1997). The advantage of the method is that the calculation does not require the exact form of bias correction term. In the bootstrapping, the true distribution function $G(x)$ is replaced by the empirical distribution function $\hat{G}_n(x)$ defined from the observations. Therefore, in the bias term in (2), the samples $\boldsymbol{X}$ and Y from $G(x)$ are replaced by $\boldsymbol{X}^*$ and Y^* from bootstrap sample $\hat{G}_n(x)$, and the expectation $E_Y \log f(Y|\cdot)$ by $E_{Y^*} \log f(Y^*|\cdot)$. Here E_{Y^*} denotes the expectation with respect to the empirical distribution function $\hat{G}_n(y)$. The bootstrap estimate of the bias $b_B(\hat{G}_n)$ is obtained by (Fig. 7).

$$b_B(\hat{G}_n) = n E_{\boldsymbol{X}^*} \left\{ \frac{1}{n} \log f(\boldsymbol{X}^*|\tilde{\theta}(\boldsymbol{X}^*)) - E_{Y^*} \log f(Y^*|\tilde{\theta}(\boldsymbol{X}^*)) \right\}, \tag{51}$$

where $\tilde{\theta}(\cdot)$ is an arbitrarily defined estimator of θ. In the simple i.i.d. case, we have

$$E_{Y^*} \log f(Y^*|\tilde{\theta}(\boldsymbol{X}^*)) = \int \log f(y^*|\tilde{\theta}(\boldsymbol{X}^*)) d\hat{G}_n(y^*) = \frac{1}{n} \log f(\boldsymbol{X}|\tilde{\theta}(\boldsymbol{X}^*)), \tag{52}$$

and the bootstrap estimate of the bias becomes simply

$$b_B(\hat{G}_n) = E_{\boldsymbol{X}^*} \left\{ \log f(\boldsymbol{X}^*|\tilde{\theta}(\boldsymbol{X}^*)) - \log f(\boldsymbol{X}|\tilde{\theta}(\boldsymbol{X}^*)) \right\}. \tag{53}$$

In actual computation, the bootstrap bias correction term $b_B(\hat{G}_n)$ is estimated by

$$b_B^*(\hat{G}_n) = \frac{1}{M} \sum_{i=1}^{M} \left\{ \log f(\boldsymbol{X}_{(i)}^*|\tilde{\theta}(\boldsymbol{X}_{(i)}^*)) - \log f(\boldsymbol{X}|\tilde{\theta}(\boldsymbol{X}_{(i)}^*)) \right\}, \tag{54}$$

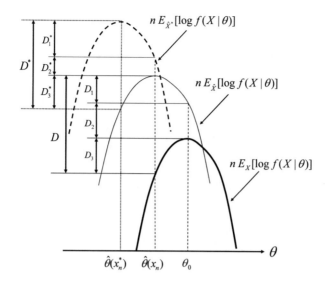

Fig. 7. Bias correction by EIC. $nE_X[\log f(X|\theta)$, $\log f(x_n|\theta)$ and $\log f(x_n^*|\theta *)$ are the expected log-likelihood, log-likelihood and the bootstrap log-likelihood, respectively. The expectation of D is the bias and that of D^* is the bootstrap bias. The expectation of D_2 is known to be 0 (Konishi and Kitagawa 2008).

where M is the number of bootstrap replication, $\boldsymbol{X}^*_{(1)}, \cdots, \boldsymbol{X}^*_{(M)}$ are M independent bootstrap resamples of size n from $\hat{G}_n(\boldsymbol{X})$. The bootstrap information criterion EIC then is defined by Ishiguro et al. (1997)

$$\text{EIC} = -2\log f(\boldsymbol{X}|\tilde{\theta}(\boldsymbol{X})) + 2b_B^*(\hat{G}_n). \tag{55}$$

This method of bootstrap bias correction can be easily extended to a predictive distribution of a Bayesian model defined by $p(y|\boldsymbol{X}) = \int p(y|\theta)\pi(\theta|\boldsymbol{X})d\theta$ where $\pi(\theta|\boldsymbol{X})$ is the posterior distribution of θ given data $\boldsymbol{X}$ (Konishi and Kitagawa 2008).

5.1 Decomposition of the Bias Term and the Reduction of the Variance in Bootstrapping

A practically important problem with the bootstrap method for the model selection is the reduction of the variance of the bias estimate. If the variance in the bootstrap simulation is large, a large M in (54) is necessary to obtain precise bootstrap estimate $b_B^*(\hat{G}_n)$ requiring long computing time especially when the model is very complicated. The variance of the bootstrap estimate of the bias defined in (54) can be reduced by the decomposition of the bias term $D(\boldsymbol{X};G)$ into three terms as follows (Fig. 7, Konishi and Kitagawa 1996, 2008; Ishiguro et al. 1997):

$$D(\boldsymbol{X};G) = D_1(\boldsymbol{X};G) + D_2(\boldsymbol{X};G) + D_3(\boldsymbol{X};G) \tag{56}$$

where

$$D_1(\boldsymbol{X};G) = \sum_{i=1}^{n} \log f(X_i|\boldsymbol{T}(\hat{G}_n)) - \sum_{i=1}^{n} \log f(X_i|\boldsymbol{T}(G))$$

$$D_2(\boldsymbol{X};G) = \sum_{i=1}^{n} \log f(X_i|\boldsymbol{T}(G)) - n\int \log f(y|\boldsymbol{T}(G))dG(y) \qquad (57)$$

$$D_3(\boldsymbol{X};G) = n\int \log f(y|\boldsymbol{T}(G))dG(y) - n\int \log f(y|\boldsymbol{T}(\hat{G}_n))dG(y).$$

Note that if $\hat{\theta}$ is the MLE, then $\boldsymbol{T}(G)$ and $\boldsymbol{T}(\hat{G}_n)$ are the maximizer of $\int \log f(y|\boldsymbol{T}(G))dG(y)$ and $\sum_{i=1}^{n} \log f(X_i|\boldsymbol{T}(\hat{G}_n))$, respectively.

For a general estimator defined by a statistical functional $\hat{\theta} = \boldsymbol{T}(\hat{G}_n)$, each term can be evaluated. See Kitagawa and Konishi (2010) for details.

Further, it can be seen that $Var\{D\} = O(n)$ and $Var\{D_1 + D_3\} = O(1)$. Therefore by estimating the bias by

$$b^*(\hat{G}_n) = E_{X^*}[D_1 + D_3], \qquad (58)$$

a significant reduction of the variance can be achieved for any estimators defined by statistical functional especially for large n.

5.2   Second Order Bootstrap Bias Correction

The bias of the log-likelihood shown in (2) can be expressed as

$$\frac{1}{n}b(G) = \frac{1}{n}b_1(G) + \frac{1}{n^2}b_2(G) + \frac{1}{n^3}b_3(G) + \cdots , \qquad (59)$$

where $b_j(G)$ is the jth order bias correction term. Therefore, the expected value of the bootstrap estimate of the bias term is given by

$$E_X[b_B(\hat{G}_n)] = E_X\left[b_1(\hat{G}_n) + \frac{1}{n}b_2(\hat{G}_n)\right] + o(n^{-1})$$

$$= b_1(G) + \frac{1}{n}\Delta b_1(G) + \frac{1}{n}b_2(G) + o(n^{-1}), \qquad (60)$$

where $\Delta b_1(G)$ is the bias of the first order bias correction term $b_1(G)$. This means that if $\Delta b_1(G) = 0$, the bootstrap estimate automatically yields the second order correction, namely it is the third order correct for the expected log-likelihood.

It is interesting to note that, in contrast to the above, the expected value of (11) in the GIC and (5) in TIC for the MLE are given by

$$E_X[b_1(\hat{G}_n)] = b_1(G) + \frac{1}{n}\Delta b_1(G) + o(n^{-1}). \qquad (61)$$

In actual situations for which unbiasedness $\Delta b_1(G)$ is not assumed, we can estimate the second order correction term by bootstrapping. If an analytic expression for $b_1(G)$ is available, it is given by

$$\frac{1}{n}b_2^*(\hat{G}_n) = E_{X^*}\left[\log f(\boldsymbol{X}^*|\boldsymbol{T}(\hat{G}_n)) - b_1(\hat{G}_n) - nE_{Y^*}\log f(Y^*|\boldsymbol{T}(\hat{G}_n))\right]. \quad (62)$$

On the other hand, if an analytic expression is difficult to compute, then we can obtain the second order correction by double bootstrapping (Kitagawa and Konishi 2010),

$$\frac{1}{n}b_2^{**}(\hat{G}_n) = E_{X^*}\left[\log f(\boldsymbol{X}^*|\boldsymbol{T}(\hat{G}_n)) - b_B^*(\hat{G}_n) - nE_{Y^*}\log f(Y^*|\boldsymbol{T}(\hat{G}_n))\right], \quad (63)$$

where $b_B^*(G)$ is the bootstrap estimate of the first order correction term given by (19).

6   Regularization, L_1 Sparse Modeling and Bridge Regression

In recent years, the regularization method is used for the modeling of big data in many fields. In this section, we first consider application of GIC for the penalized log-likelihood method or the L_2 regularization problem. We then consider the generalization of the Bayesian information criterion BIC for the application to L_1 regularization and the bridge regression which involves a more general L_p regularization.

6.1   GIC for Penalized Log-Likelihood Method

The method based on maximizing the penalized log-likelihood function was originally introduced by Good and Gaskins (1971) in the context of density estimation. The Bayesian justification of the method and application to Bayesian modeling have been investigated by many authors such as Wahba (1978), Akaike (1980), Kitagawa and Gersch (1984), Silverman (1985) and Shibata (1989).

Here, we consider a penalized log-likelihood of the form

$$\ell_\lambda(\theta) = \sum_{\alpha=1}^{n} \log f(x_\alpha|\gamma,\sigma) - \frac{n}{2}\lambda\gamma'K\gamma, \qquad (64)$$

where $\theta = (\gamma,\sigma)$ and K is a non-negative definite matrix. If we put $K = I_p$, $k \times k$ identity matrix, we obtained the simple L_2 regularization term.

Given the data $x_1, \ldots, x_n$, the maximum penalized log-likelihood estimates $\hat{\theta}$ is obtained as the solution to the implicit function

$$\sum_{\alpha=1}^{n} \psi(X_\alpha, \hat{\theta}) = \boldsymbol{0}, \qquad (65)$$

where $\psi = (\psi_1, \cdots, \psi_p)^T$. Note that the penalized maximum likelihood estimator $\hat{\theta}_{PL}$ is obtained by putting

$$\psi(X_\alpha, \hat{\theta}) = \frac{\partial}{\partial\theta}\left\{\log f(X_\alpha|\hat{\theta}_{PL}) - \frac{\lambda}{2}\hat{\gamma}^T K\hat{\gamma}\right\}. \qquad (66)$$

In the framework of the generalized information criterion GIC, the information criterion for the model $f(z|\hat{\theta})$ with the estimator $\hat{\theta}$ obtained as the solution of the (65) is given by

$$\text{GIC}_M = -2 \sum_{\alpha=1}^{n} \log f(X_\alpha|\hat{\theta}) + 2\text{tr}\left\{J(\psi, \hat{G})^{-1}I(\psi, \hat{G})\right\}, \tag{67}$$

where

$$J(\psi, \hat{G}) = -\frac{1}{n}\sum_{\alpha=1}^{n}\frac{\partial\psi(X_\alpha, \hat{\theta})^T}{\partial\theta}, \quad I(\psi, \hat{G}) = \frac{1}{n}\sum_{\alpha=1}^{n}\psi(X_\alpha, \hat{\theta})\frac{\partial\log f(X_\alpha|\hat{\theta})}{\partial\theta^T}. \tag{68}$$

6.2   Generalized BIC for Regularization Method

The BIC (Bayesian Information Criterion) proposed by Schwarz (1978)

$$\text{BIC} = -2\log f(\boldsymbol{x}_n|\hat{\boldsymbol{\theta}}) + k\log n \tag{69}$$

$$\approx -2\log p(\boldsymbol{x}_n) = -2\log\left\{\int f(\boldsymbol{x}_n|\boldsymbol{\theta})\pi(\boldsymbol{\theta})d\boldsymbol{\theta}\right\}$$

is a model evaluation criterion based on the posterior probability of a model. Here, $\hat{\boldsymbol{\theta}}_i$ is the maximum likelihood estimator of the k-dimensional parameter vector $\boldsymbol{\theta}$ of the model $f(x|\boldsymbol{\theta})$. Consequently, from the r models that are estimated using the maximum likelihood method, the model that minimizes the value of BIC can be selected as the optimal model.

Konishi et al. (2004) developed generalized Bayesian information criterion, GBIC, for the evaluation of the models obtained by the maximum penalized likelihood method. In this subsection, a simplified version of GBIC is shown briefly. Let $f(x|\hat{\boldsymbol{\theta}}_P)$ be a statistical model estimated by the regularization method for the parametric model $f(x|\boldsymbol{\theta})$, and $\hat{\boldsymbol{\theta}}_P$ is obtained by maximizing the penalized log-likelihood function

$$\ell_\lambda(\boldsymbol{\theta}) = \log f(\boldsymbol{x}_n|\boldsymbol{\theta}) - \frac{n\lambda}{2}\boldsymbol{\theta}^T K\boldsymbol{\theta}, \tag{70}$$

where K is a $p \times p$ matrix. The penalized log-likelihood function can be rewritten as

$$\ell_\lambda(\boldsymbol{\theta}) = \log\left\{f(\boldsymbol{x}_n|\boldsymbol{\theta})\exp\left(-\frac{n\lambda}{2}\boldsymbol{\theta}^T K\boldsymbol{\theta}\right)\right\}. \tag{71}$$

Then, $\exp(-n\lambda/2\boldsymbol{\theta}^T K\boldsymbol{\theta})$ in the above equation can be thought of as a prior distribution in which the smoothing parameter λ is a hyper-parameter,

$$\pi(\boldsymbol{\theta}|\lambda) = \frac{(n\lambda)^{p/2}|K|^{1/2}}{(2\pi)^{p/2}}\exp\left(-\frac{n\lambda}{2}\boldsymbol{\theta}^T K\boldsymbol{\theta}\right). \tag{72}$$

Given the data distribution $f(\boldsymbol{x}_n|\boldsymbol{\theta})$, and the prior distribution $\pi(\boldsymbol{\theta}|\lambda)$ with hyper-parameter λ, the marginal likelihood of the model can be rewritten as

$$p(\boldsymbol{x}_n|\lambda) = \int f(\boldsymbol{x}_n|\boldsymbol{\theta})\pi(\boldsymbol{\theta}|\lambda)d\boldsymbol{\theta}$$

$$= \int \exp\left\{nq(\boldsymbol{\theta}|\lambda)\right\} d\boldsymbol{\theta}, \tag{73}$$

where

$$q(\boldsymbol{\theta}|\lambda) = \frac{1}{n}\log\left\{f(\boldsymbol{x}_n|\boldsymbol{\theta})\pi(\boldsymbol{\theta}|\lambda)\right\} = \frac{1}{n}\left\{\log f(\boldsymbol{x}_n|\boldsymbol{\theta}) + \log\pi(\boldsymbol{\theta}|\lambda)\right\} \tag{74}$$

$$= \frac{1}{n}\left\{\log f(\boldsymbol{x}_n|\boldsymbol{\theta}) - \frac{n\lambda}{2}\boldsymbol{\theta}^T K\boldsymbol{\theta}\right\} - \frac{1}{2n}\left\{p\log(2\pi) - p\log(n\lambda) - \log|K|\right\}.$$

We note here that the mode, $\hat{\boldsymbol{\theta}}_P$, of $q(\boldsymbol{\theta}|\lambda)$ coincides with a solution obtained by maximizing the penalized log-likelihood function (70). By approximating it using Laplace's method for integrals, we have

$$\int \exp\{nq(\boldsymbol{\theta})\}d\boldsymbol{\theta} \approx \frac{(2\pi)^{p/2}}{n^{p/2}|J_\lambda(\hat{\boldsymbol{\theta}}_P)|^{1/2}} \exp\left\{nq(\hat{\boldsymbol{\theta}}_P)\right\}. \tag{75}$$

where

$$J_\lambda(\hat{\boldsymbol{\theta}}_P) = -\frac{1}{n}\frac{\partial^2 q(\boldsymbol{\theta}|\lambda)}{\partial\boldsymbol{\theta}\partial\boldsymbol{\theta}^T}\bigg|_{\hat{\boldsymbol{\theta}}_P} = -\frac{1}{n}\frac{\partial^2 \log f(\boldsymbol{x}_n|\boldsymbol{\theta})}{\partial\boldsymbol{\theta}\partial\boldsymbol{\theta}^T}\bigg|_{\hat{\boldsymbol{\theta}}_P} + \lambda K. \tag{76}$$

Taking the logarithm of this expression and multiplying it by -2, we obtain the generalized Bayesian information criterion GBIC (Konishi et al. 2004; Konishi and Kitagwa 2008),

$$\text{GBIC} = -2\log f(\boldsymbol{x}_n|\hat{\boldsymbol{\theta}}_P) + n\lambda\hat{\boldsymbol{\theta}}_P^T K\hat{\boldsymbol{\theta}}_P + \log|J_\lambda(\hat{\boldsymbol{\theta}}_P)| - p\log\lambda - \log|K|. \tag{77}$$

In the modeling by regularization method, the selection of the smoothing parameter λ is crucial and we select the λ that minimizes the GBIC as the optimal smoothing parameter.

By interpreting the regularization method based on the above argument from a Bayesian point of view, it can be understood that the regularized estimator agrees with the estimate that is obtained through the maximization (mode) of the following posterior probability depending on the value of the smoothing parameter;

$$\pi(\boldsymbol{\theta}|\boldsymbol{x}_n;\lambda) = \frac{f(\boldsymbol{x}_n|\boldsymbol{\theta})\pi(\boldsymbol{\theta}|\lambda)}{\displaystyle\int f(\boldsymbol{x}_n|\boldsymbol{\theta})\pi(\boldsymbol{\theta}|\lambda)d\boldsymbol{\theta}}, \tag{78}$$

where $\pi(\boldsymbol{\theta}|\lambda)$ is the density function resulting from (72) as a prior probability of the p-dimensional parameter $\boldsymbol{\theta}$ for the model $f(\boldsymbol{x}_n|\boldsymbol{\theta})$. For the Bayesian justification of the maximum penalized likelihood approach, we refer to Silverman (1985) and Wahba (1990).

Example: Regularization for the regression models. Suppose that n observations $\{(\boldsymbol{x}_\alpha, y_\alpha); \quad \alpha = 1, 2, \cdots, n\}$ are observed in terms of a p-dimensional explanatory variable $\boldsymbol{x}$ and a response variable Y, and consider a simple regression model

$$y_\alpha = \sum_{j=1}^{p} \beta_j x_{\alpha j} + \varepsilon_\alpha, \quad \varepsilon_\alpha \sim N(0, \sigma^2), \tag{79}$$

where $\beta^T = (\beta_1, \ldots, \beta_m)$, $\boldsymbol{\theta} = (\beta^T, \sigma^2)^T$ and $(y_\alpha, x_{\alpha 1}, \ldots, x_{\alpha, p})$, $\alpha = 1, \ldots, n$. If we estimate the parameter vector $\boldsymbol{\theta}$ by maximizing the penalized log-likelihood function (70), the estimators for $\boldsymbol{\beta}$ and σ^2 are respectively given by

$$\hat{\boldsymbol{\beta}} = (X^T X + n\lambda\hat{\sigma}^2 K)^{-1} X^T \boldsymbol{y}, \quad \hat{\sigma}^2 = \frac{1}{n}(\boldsymbol{y} - X\hat{\boldsymbol{\beta}})^T (\boldsymbol{y} - X\hat{\boldsymbol{\beta}}), \tag{80}$$

where X is an $n \times m$ matrix given by $X = (\boldsymbol{x}_1, \boldsymbol{x}_2, \cdots, \boldsymbol{x}_n)^T$ and $x_\alpha = (x_{\alpha 1}, \ldots, x_{\alpha p})$.

By applying GBIC in (77), the evaluation criterion for the regularized regression model $f(y_\alpha | \boldsymbol{x}_\alpha; \hat{\boldsymbol{\theta}}_P)$ estimated by the regularization method is given by

$$\text{GBIC} = n \log \hat{\sigma}^2 + n\lambda \hat{\boldsymbol{\beta}}^T K \hat{\boldsymbol{\beta}} + n + n \log(2\pi) + \log |J_\lambda(\hat{\boldsymbol{\theta}}_P)| - \log |K| - m \log \lambda, \tag{81}$$

where $J_\lambda(\hat{\boldsymbol{\theta}}_P)$ is the $(m+1) \times (m+1)$ matrix

$$J_\lambda(\hat{\boldsymbol{\theta}}_P) = \frac{1}{n\hat{\sigma}^2} \begin{bmatrix} X^T X + n\lambda\hat{\sigma}^2 K & \dfrac{1}{\hat{\sigma}^2} X^T \boldsymbol{e} \\ \dfrac{1}{\hat{\sigma}^2} \boldsymbol{e}' X & \dfrac{n}{2\hat{\sigma}^2}, \end{bmatrix} \tag{82}$$

with the n-dimensional residual vector $\boldsymbol{e} = \left(y_1 - \hat{\boldsymbol{\beta}}^T \boldsymbol{x}_1, y_2 - \hat{\boldsymbol{\beta}}^T \boldsymbol{x}_2, \cdots, y_n - \hat{\boldsymbol{\beta}}^T \boldsymbol{x}_n \right)^T$.

6.3 L_1 Regularization and Bridge Regression

In recent years, with the advent of big data, modeling based on the L_1 regularization method has been widely used in many fields of science and technologies. The feature of the L_1 regularization method is that parameter estimation and variable selection can be performed at the same time and it is important as a method of extracting essential information from high dimensional data.

In this subsection, we will consider the evaluation of the bridge regression model. The bridge regression model (Frank and Frieman 1993; Fu 1998) has an L_p regularization term

$$\ell_{\lambda, p}(\beta, \sigma^2) = \ell(\beta, \sigma^2) - \frac{n\lambda}{2} \sum_{j=1}^{p} |\beta_j|^p, \tag{83}$$

and it becomes the ridge regression for $p = 2$ and Lasso for $p = 1$. For $0 < p \leq 1$, bridge regression method can perform the selection of variable and parameter estimation simultaneously. Therefore, the bridge regression can be considered as an estimation method that encompasses many estimation methods.

Kawano (2014) presents the GBIC for the bridge regression model

$$\text{GBIC} = n \log \hat{\sigma}^2 + n\lambda \sum_{j \in A} |\hat{\beta}_j|^p + n + n \log(2\pi) + \log|J_\lambda| - 2|A| \log p$$

$$+ 2|A|(1 + \frac{1}{p}) \log 2 - \frac{2|A|}{p} \log(n\lambda) + 2|A| \log \Gamma \left(\frac{1}{p}\right), \qquad (84)$$

where $A = \{j; \hat{\beta}_j \neq 0\}$, and J is the $(|A| + 1) \times (|A| + 1)$ matrix given by

$$J_\lambda(\hat{\boldsymbol{\theta}}_P) = \frac{1}{n\hat{\sigma}^2} \begin{bmatrix} X^T X + n\lambda\hat{\sigma}^2 p(p-1)K & \frac{1}{\hat{\sigma}^2} X^T e \\ \frac{1}{\hat{\sigma}^2} e' X & \frac{n}{2\hat{\sigma}^2}, \end{bmatrix}. \qquad (85)$$

For $p < 1$, the influence function cannot differentiate, so GIC can not be directly applied. Matsui and Konishi (2011) use the SCAD penalty function to derive GIC and BIC. In addition, Umezu et al. (2015) derived AIC for the bridge regularization for $1 \geq p < 1$.

7   Summary

Due to the dramatic development of measuring instruments in recent years, a huge amount of large-scale data has been acquired in all research areas. Along with this, research method has changed, and data-driven methods are becoming important as the fourth scientific methodology. In the data-driven approach, the model is built according to the theory, knowledge, data, and further the purpose of the analysis. Once a model is built, useful information can be extracted from the data through the fitted model. In this data-driven method, it is crucial to use a good model. Therefore, the problem of developing good model evaluation criteria is a very important.

This paper outlined the model evaluation criteria such as AIC, GIC, EIC. Which are obtained by bias-correction of the log-likelihood of an estimated model. In particular, GIC can be applied to wide class of estimation procedures such as M-estimators, Bayes models and penalized likelihood methods. Bootstrap based information criterion EIC can be applied to various situation for which analytic methods are difficult to apply. Since L_1 regularization is important in recent data analysis, the evaluation of regularization model is also outlined.

References

Akaike, H.: Information theory and an extension of the maximum likelihood principle. In: Petrov, B.N., Csaki, F. (eds.) 2nd International Symposium in Information Theory, pp. 267–281 (1973). (Reproduced in Breakthroughs in Statistics, vol. I, Foundations and Basic Theory, S. Kots and N.L. Johnson, eds., pp. 610–624. Springer-Verlag, New York, 1992)

Akaike, H.: A new look at the statistical model identification. IEEE Trans. Autom. Control AC **19**, 716–723 (1974)

Akaike, H.: Likelihood and the Bayes procedure. In: Bernardo, N.J., DeGroot, M.H., Lindley, D.V., Smith, A.F.M. (eds.) Bayesian Statistics, Valencia, Spain, pp. 141–166. University Press (1980)

Akaike, H., Kitagawa, G. (eds.): The Practice of Time Series Analysis. Springer-Verlag, New York (1998)

Ayres, I.: Super Crunchers: Why Thinking-By-Numbers is the New Way To Be Smart. Bantam Books, New York (2007)

Bozdogan, H.: Proceeding of the First US/Japan Conference on the Frontiers of Statistical Modeling: An Informational Approach. Kluwer Academic Publishers, Netherlands (1994)

Cavanaugh, J.E., Shumway, R.H.: A bootstrap variant of AIC for state-space model selection. Statistica Sinica **7**, 469–473 (1997)

Davison, A.C.: Approximate predictive likelihood. Biometrika **73**, 323–332 (1986)

Drucker, P.F.: Post-capitalist Society. Routledge, London (1993)

Frank, L.E., Friedman, J.H.: A statistical view of some chemometrics regression tools. Technometrics **35**(2), 109–135 (1993)

Fu, W.J.: Penalized regressions: the bridge versus the Lasso. J. Comput. Graph. Stat. **7**(3), 397–416 (1998)

Good, I.J., Gaskins, R.A.: Nonparametric roughness penalties for probability densities. Biometrika **58**, 255–277 (1971)

Huber, P.J.: The behavior of maximum likelihood estimates under nonstandard conditions. In: Proceedings of the Fifth Berkley Symposium on Statistics, pp. 221–233 (1976)

Ishiguro, M., Sakamoto, Y., Kitagawa, G.: Bootstrapping log likelihood and EIC, an extension of AIC. Ann. Inst. Stat. Math. **49**(3), 411–434 (1997)

Kawano, S.: Selection of tuning parameters in bridge regression models via Bayesian information criterion. Stat. Pap. **55**(4), 1207–1223 (2014)

Kitagawa, G., Gersch, W.: A smoothness priors-state space modeling of time series with trend and seasonality. J. Am. Stat. Assoc. **79**(386), 378–389 (1984)

Kitagawa, G., Gersch, W.: Smoothness Priors Analysis of Time Series. Lecture Notes in Statistics, vol. 116. Springer-Verlag, Heidelberg (1996)

Kitagawa, G., Konishi, S.: Bias and variance reduction techniques for bootstrap information criteria. Ann. Inst. Stat. Math. **62**(1), 209–234 (2010)

Konishi, S., Ando, T., Imoto, S.: Bayesian information criteria and smoothing parameter selection in radial basis function networks. Biometrika **91**(1), 27–43 (2004)

Konishi, S., Kitagawa, G.: Generalized information criteria in model selection. Biometrika **83**(4), 875–890 (1996)

Konishi, S., Kitagawa, G.: Information Criteria and Statistical Modeling. Springer Series in Statistics. Springer, New York (2008)

Kullback, S., Leibler, R.A.: On information and sufficiency. Ann. Stat. **22**(22), 79–86 (1951)

Manyika, J., Chui, M., Bughin, J., Brown, B., Dobbs, R., Roxburgh, C., Byers, A.H.: Big Data: The Next Frontier for Innovation, Competition, and Productivity. McKinsey Global Institute, Washington (2011)

Matsui, H., Konishi, S.: Variable selection for functional regression models via the L1 regularization. Comput. Stat. Data Anal. **55**(12), 3304–3310 (2011)

Schwarz, G.: Estimating the dimension of a model. Ann. Stat. **6**, 461–464 (1978)

Shibata, R.: Statistical aspects of model selection. In: Willems, J.C. (ed.) From Data to Model, pp. 215–240. Springer-Verlag, New York (1989)

Shibata, R.: Bootstrap estimate of Kullback-Leibler information for model selection. Statistica Sinica **7**, 375–394 (1997)

Silverman, B.W.: Some aspects of the spline smoothing approach to nonparametric regression curve fitting (with discussion). J. Roy. Statist. Soc. Ser. B **47**, 1–52 (1985). Akaike's criterion

Takeuchi, K.: Distributions of information statistics and criteria for adequacy of models. Math. Sci. **153**, 12–18 (1976). (in Japanese)

Tierney, L., Kadane, J.B.: Accurate approximations for posterior moments and marginal densities. J. Am. Stat. Assoc. **81**, 82–86 (1986)

Umezu, Y., Shimizu, Y., Masuda, H., Ninomiya, Y.: AIC for non-concave penalized likelihood method. arXiv preprint arXiv:1509.01688 (2015)

Wahba, G.: Improper priors, spline smoothing and the problem of guarding against model errors in regression. J. Roy. Statist. Soc. Ser. B **40**, 364–372 (1978)

Wahba, G.: Spline Models for Observational Data. Philadelphia (1990)

An Invitation to Quantum Econometrics

Hung T. Nguyen[1,2(✉)] and Le Si Dong[3]

[1] New Mexico State University, Las Cruces, NM, USA
hunguyen@nmsu.edu
[2] Chiang Mai University, Chiang Mai, Thailand
[3] Banking University,
Ho-Chi-Minh City, Vietnam
lesidong@gmail.com

Abstract. We elaborate on the possibility to considering quantum probability calculus to improve statistical methods in economics in general, and in quantitative finance, in particular. A tutorial on the analogy between quantum mechanics and models in econometrics, using Kolmogorov probability theory, is given. Several research issues are mentioned.

Keywords: Density matrix · Hilbert spaces
Kolmogorov probability calculus · Observables · Quantum finance
Quantum mechanics · Quantum probability
Schrodinger wave equation · Self adjoint operators · Spectral measures

1 Introduction

This invitation aims mainly at calling your attention to an emerging effort to possibly improve the way we do econometrics. In fact, it has started at the dawn of econometrics by the man who created it (as a synthesis of mathematics, economic theory and statistics), Jan Tinbergen (obtaining the first Nobel Memorial Prize in Economics in 1969). He was a physicist turned economist. He proposed the gravity model of international trade by a formula similar to Newton's law of gravity in which mass is replaced by GDP. This connection with physics, or more precisely with mechanics, seems natural as both mechanics and econometrics, especially finance, are concerned about models and predictions of (uncertain) dynamical systems. Earlier, to "capture" (explain) the observed fluctuations of stock returns, Louis Bachelier in his Ph.D. thesis (1900) proposed a continuous time model based on the Brownian motion which later forms the foundations for financial mathematics (through works of Black, Scholes and Merton, 1973, where diffusion models are based on Brownian motion). But Brownian motion, as explained by Albert Einstein, in 1905, is a motion of minuscule pollen particles suspended in water (which can be seen to wiggle and wander when examined under a strong microscope), i.e., in the realm of quantum mechanics! (laws of motion of extremely "small" objects). Thus, a shift from Newtonian mechanics

© Springer International Publishing AG 2018
L. H. Anh et al. (eds.), *Econometrics for Financial Applications*, Studies in Computational Intelligence 760, https://doi.org/10.1007/978-3-319-73150-6_3

to quantum mechanics seems obvious in this context? More specifically, a shift from Kolmogorov probability to quantum probability seems desirable?

Remark. In fact, Brownian motion is modeled probabilistically as "limits" of random walks within Kolmogorov's probability theory. The same situation happened with *Statistical Mechanics* (see e.g., Sethna [12]. Surrounding the famous Black-Scholes option pricing formula are stuff such as PDE, Ito stochastic calculus, martingale method, and various extended models related to volatility. If quantum probability is to replace Kolmogorov probability, then we should turn to *Quantum Stochastic Calculus* (see, e.g., [10]).

It is well known that Kolmogorov probability theory is not appropriate to use in quantum mechanics (as exemplified by the two-slit experiment), especially the failure of the additivity property of probability measures. In fact, a radically different formalism for probability has been developed to calculating probabilities in quantum mechanics, with great successes (i.e., confirmed by experiments).

This is essentially *a lesson learned from physics*. It is not just importing stuff from physics to economics (in particular) but looking as physics as an evolutive science with great successes (as testified by what we got from engineering in our daily life!). When we are uncertain (epistemic or random) about some phenomenon, e.g. in "classical" mechanics (Newtonian and Einstein's relativity theories) or quantum mechanics, we propose *models*, based, of course, on "evidence" from observations, measurements, and "imagination". This is common in physics and *statistics* (used in, say, econometrics). Since physics has an advantage in natural science over social sciences (such as economics) as we can perform experiments to predict phenomena by our models and see if the predictions match the observations, the evolution of physics (from one model to another) proceeded peacefully, as opposed to statistical debates on modern methodologies! Let's give a striking example:

For the purpose of "improving" statistical methods (which are used in various applied fields), at least three things surfaced recently:

(i) The questionable use of P-values in hypothesis testing,
(ii) The seemingly realistic prediction methodology based on calibration vs estimation (especially when big data are available),
(iii) The possible used of quantum probability calculus in applied statistics.

Let's "compare" reactions of statisticians (to the above 3 proposed "innovative" things) with three models in quantum physics: It was discovered that a hydrogen atom consists of a single proton at the center, and a single electron orbiting around the proton. The problem is: How the electron moves around the proton? Since we cannot "see" the electron movement, we must propose models (then verifying if such models reflect "reality", or compatible with "observations"/some possible measurements).

(1) First model (Ernest Rutherford): Just like the earth rotating around the sun, the electron could just follow the "solar system". The "reality" is this. The solar system is stable (that's why we are still alive today! the earth does not collapse by falling to the sun, despite the existence of gravitational force between the earth and the sun), and so is the electron-proton system.

But, unlike the solar system, subatomic particles have electric charges (of opposite signs), and as such, the Rutherford model is unstable: the electron spiraling into the proton in the center, hence this model does not correspond to reality.

(2) Bohr's model: Thus the first model has to be replaced. To explain the stability of hydrogen atoms, Bohr proposed the following model. The electron rotates around the proton, not in a continuous fashion, but in "discrete" levels, i.e., there are countable numbers of orbits that the electron can travel, between which it can jump, so that the atom does not collapse. However, this model is only good for the hydrogen atom, and not for other particles.

(3) Schrodinger's model: Not only to modeling dynamics of all particles, but also to explain Bohr's model for hydrogen atom, Schrodinger proposed that the electron is in many places at once, in an "electron cloud" whose shape is given by a wave function (in Schrodinger's fundamental equation).

This evolving understanding of hydrogen atom dynamics is a "peaceful" and productive phenomenon! New proposed models were received with open mind. As Box has said, "all models are wrong, but some are useful", an open-minded attitude is helpful in sciences. Tradition should not be an obstacle to scientific progress.

Now, with respect to the main theme of this paper, namely, the proposal to see if quantum probability (a generalization of Kolmogorov probability calculus), viewing as a "new model for probability calculus" (not the meaning of probability per se) could be used in social sciences (e.g. economics), of course, when appropriate, the situation is this. Again, by "tradition" (like the issues (i), (ii) listed above), it's a slow motion, as usual! Perhaps, only a handful of statisticians is aware of the proposal, let alone taking a closer look at it.

Let's quote a recent opinion of some prominent statisticians on this proposal, namely Andrew Gelman and Michael Betancourt [5]:

"Does quantum uncertainty have a place in everyday applied statistics?"

(a) Open mind: "We are sympathetic to the proposal of modeling joint probabilities using a framework more general than standard model by relaxing the law of conditional probability".

"The generalized probability theory suggested by quantum physics might very well be relevant in the social sciences".

Remark. An obvious research issue arises right here: Beyond copulas? Interference vs correlation.

(b) A closer look at a new proposal: "Some of our own applied work involves political science and policy, often with analysis of data from opinion polls, where there are clear issues of the measurement affecting the outcome".

Remark. "Measurement affecting the outcome" is the main real phenomenon in quantum physics, as expressed by Heisenberg's uncertainty principle (responsible to the lack of a phase space in quantum mechanics). The point is this. It's all about data (observations): the data dictate the methods to use for analyzing them, and not the other way around.

(c) Some possible gains: "Just as psychologists have found subadditivity and superadditivity of probability in many contexts, we see the potential gain of thinking about violating of the conditional probability law".

Remark. To apply quantum probability calculus to social science problems, one needs to have clear evidence of the failure of classical probability theory. Remember "The ultimate challenge in statistics is to solve applied problems".

Finally, and this is important (!), to say loud and clear: If quantum probability calculus seems to be useful for applied statistics, it does not mean that we have to "ignore" standard probability theory, i.e., replace the latter by the former. This is important for two reasons:

(i) "Traditional" researchers should not be worry about abandoning what they used to work with until now! since Kolmogorov probability theory could remain appropriate for many situations,

(ii) Quantum probability calculus may be only suitable for some situations, but not all.

This is completely similar to the situation in mechanics: The discovery of quantum mechanics did not ignore Newton's mechanics: Newtonian mechanics is still valid in macrophysics.

So, assuming that we have an open mind, so that we love to understand the new proposal before making our own judgement of whether it could be used in, say, financial econometrics. Thus, tradition aside, let's find out why in quantum physics the calculus of probabilities is different than classical Kolmogorov's one.

Again, mathematical finance was founded on the Black-Scholes option pricing PDE which was based upon the modeling of financial returns as diffusion processes in the context of probability theory. In this modeling approach, the return distributions are classical probability distributions. The basic question of "econophysicists" is this:

"Should we model return distributions with distributions which reflect the data in a much closer way?"

Clearly, predictions would be improved if the models are better! In fact, research reported in the literature showed that this quantum approach can be of potential benefit.

By the very nature of Brownian motion, should we study finance *in the context of quantum mechanics, instead?* with the hope that "quantum probability distributions" will supply a reasonable answer to the above question. The attempt to put the Black-Scholes pricing formula in the quantum context was discussed by [11] who rationalized the use of quantum principles in option pricing context:

"A natural explanation of extreme irregularities in the evolution of prices in financial markets is provided by quantum effects".

See also [7].

At a technical level, the difference between classical modeling approach (i.e., based on Kolmororov's probability theory) and the "quantum approach" can be explained as follows.

(i) Kolmogorov probability formalism includes both objective and subjective probabilities which are *man-made* uncertainty (i.e. imposed uncertainty by men), e.g., in Von Neumann's game theory for economics/mixed strategies; and Savage expected utility theory, whereas uncertainty in quantum physics is due to the nature itself,

(ii) While we still have the same interpretation of the concept of probability ("chance"), the calculus of these types of probabilities is different. For example, quantum probability measures are not additive (due to quantum interference of waves of particles). This is clearly affecting our attempt on making financial predictions!

In any case, the quantum approach to finance in particular, and to econometrics in general, is an ongoing research direction. For an introduction to Quantum Finance, see [1].

In this introductory lecture to quantum econometrics, we will only focus on the main ingredient, namely the concept of quantum probability (and the context giving rise to it) which plays a crucial role in uncertainty analysis of quantum mechanics and possibly in social sciences. While *Feynman path integral* is useful for solving the initial value problem for the Schrodinger equation, it will not be discussed in this introductory lecture. Curious readers could read Keller and McLaughlin [6].

As such, in Sect. 2, a bit of quantum mechanics is given. Section 3 presents the uncertainty analysis in quantum context. Section 4 presents a mathematical formulation for quantum probability together with a comparison with Kolmogorov probability theory. Section 5 concludes the paper by discussing econometrics issues. Along the way, research issues will be mentioned.

2   A Bit of Quantum Mechanics

Unlike statistical mechanics, quantum mechanics reveals the randomness believed to be caused by nature itself. As we are going to examine whether economic fluctuations can be modeled by quantum uncertainty, we need to take a quick look at quantum mechanics. For a good and enjoyable reading on quantum mechanics, consult Feynman [3,4].

The big picture of quantum mechanics is this. A particle with mass m, and potential energy $V(x_o)$ at a position $x_o \in \mathbb{R}^3$, at time $t = 0$, will move to a position x at a later time $t > 0$. But unlike Newtonian mechanics (where moving objects obey a law of motion and their time evolutions are deterministic trajectories, with a state being a point in $\mathbb{R}^6$/position and velocity), the motion of a particle is not deterministic, so that at most we can only look for the probability that it could be in a small neighborhood of x, at time t. Thus, the problem is: How to obtain such a probability? According to quantum mechanics, the relevant probability density $f_t(x)$ is of the form $|\psi(x,t)|^2$ where the (complex) "probability amplitude" $\psi(x, t)$ satisfies the *Schrodinger equation* (playing the role of Newton's law of motion in macrophysics)

$$ih\frac{\partial \psi(x, t)}{\partial t} = -\frac{h^2}{2m}\Delta_x\psi(x, t) + V(x)\psi(x, t)$$

where h is the Planck's constant, $i = \sqrt{-1}$, and Δ_x is the Laplacian $\Delta_x\psi = \frac{\partial^2 \psi}{\partial x_1^2} + \frac{\partial^2 \psi}{\partial x_2^2} + \frac{\partial^2 \psi}{\partial x_3^2}$, $x = (x_1, x_2, x_3) \in \mathbb{R}^3$.

Solutions of the Schrodinger equation are "wave-like", and hence are called wave functions of the particle (the equation itself is called the wave equation). Of course, solving this PDE equation, in each specific situation, is crucial. Richard Feynman [2] introduced the concept of path integral to solve it.

For a solution of the form $\psi(x, t) = \varphi(x)e^{it\theta}$, $|\psi(x, t)|^2 = |\varphi(x)|^2$ with $\varphi \in L^2(\mathbb{R}^3, \mathscr{B}(\mathbb{R}^3), dx)$, in fact $||\varphi|| = 1$. Now, since the particle can take any path from $(x_o, 0)$ to (x, t), its "state" has to be described probabilistically. Roughly speaking, each φ (viewed as a "vector" in the complex, infinitely dimensional Hilbert space $L^2(\mathbb{R}^3, \mathscr{B}(\mathbb{R}^3), dx)$) represents a state of the moving particle. Now $L^2(\mathbb{R}^3, \mathscr{B}(\mathbb{R}^3), dx)$ is separable so that it has a countable orthonormal basis, φ_n, say, and hence

$$\varphi = \sum_n <\varphi, \varphi_n> \varphi_n = \sum_n c_n\varphi_n = <\varphi_n|\varphi|\varphi_n>$$

where $< ., . >$ denotes the inner product in $L^2(\mathbb{R}^3, \mathscr{B}(\mathbb{R}^3), dx)$, and the last notation on the right is written in popular Dirac's notation, noting that $||\varphi||^2 = 1 = \sum_n |c_n|^2$, and

$$\sum_n |\varphi_n><\varphi_n| = I \text{ (identity operator on } L^2(\mathbb{R}^3, \mathscr{B}(\mathbb{R}^3), dx))$$

where $|\varphi><\psi|$ is the operator: $f \in L^2(\mathbb{R}^3, \mathscr{B}(\mathbb{R}^3), dx) \to <\varphi, f><\psi| \in L^2(\mathbb{R}^3, \mathscr{B}(\mathbb{R}^3), dx)$.

From the solution $\varphi(x)$ of Schrodinger equation, the operator

$$\rho = \sum_n c_n|\varphi_n><\varphi_n|$$

is positive definite with unit trace $(tr(\rho) = \sum_n <\varphi_n|\rho|\varphi_n> = 1)$.

Thus it plays the role of the classical probability density function. By separability of $L^2(\mathbb{R}^3, \mathscr{B}(\mathbb{R}^3), dx)$, we are simply in a natural extension of finitely dimensional euclidean space setting, and as such, the operator ρ is called a *density matrix* which represents the "state" of a quantum system.

This "concrete setting" brings out a general setting (which generalizes Kolmogorow probability theory), namely, a complex, infinitely dimensional, separable, Hilbert space $H = L^2(\mathbb{R}^3, \mathscr{B}(\mathbb{R}^3), dx)$, and a density matrix ρ which is a (linear) positive definite operator on H (i.e., $<f, \rho f> \geq 0$ for any $f \in H$, implying that it is self adjoint), and of unit trace. A *quantum probability space* is simply a pair (H, ρ).

Remark. At a given time t, it is the entire function $x \to \psi(x,t)$ which describes the state of the quantum system, and not just one point! The wave function $\psi(x,t)$ has a probabilistic interpretation: its amplitude gives the probability distribution for the position, a physical quantity of the system, namely, $|\psi(x,t)|^2$.

Now, observe that for $\varphi(p)$ arbitrary, where $p = mv$ is the particle momentum, a solution of Schrodinger's equation is

$$\psi(x,t) = \int_{\mathbb{R}^3} \varphi(p) e^{-\frac{i}{\hbar}(Et - <p,x>)} dp/(2\pi h)^{\frac{3}{2}}$$

(where $E = \frac{\|p\|^2}{2m}$), i.e., ψ is the Fourier transform of the function $\varphi(p) e^{-\frac{i}{\hbar}(Et)}$, and hence, by Parseval-Plancherel,

$$\int_{\mathbb{R}^3} |\psi(x,t)|^2 dx = \int_{\mathbb{R}^3} |\varphi(p)|^2 dp$$

Thus, it suffices to choose $\varphi(.)$ such that $\int_{\mathbb{R}^3} |\varphi(p)|^2 dp = 1$ (to have all wave functions in $L^2(\mathbb{R}^3, \mathscr{B}(\mathbb{R}^3), dx)$, as well as $\int_{\mathbb{R}^3} |\psi(x,t)|^2 dx = 1$. In particular, for stationary solutions of Schrodinger' equation $\psi(x)e^{-iEt/h}$, describing the same stationary state. Here, note that $\|\psi\| = 1$.

Three things come up:

(i) With addition of waves and square integrability, the state space in quantum mechanics is a complex, infinitely dimensional Hilbert space,

(ii) Unlike Newtonian mechanics, the dynamics of particles are random in nature (in the sense that, under the same "state" (initial conditions), results are different), thus we cannot talk about "the trajectory" of a moving particle,

(iii) We need to be able to find the probability distribution of possible "trajectories". A plausible suggestion is $|\psi(x,t)|^2$ for probability density of the position. But then, while the meaning of probability remains the usual one (e.g. as a frequency interpretation), its calculus based on this formalism is different than Kolmogorov's probability calculus, e.g., additivity property breaks down (in, say, interference of waves).

Remark. Note that when dealing with uncertainty (ordinary or quantum), it is necessary to evoke its underlying logic, for purpose of "reasoning" (inference which is based on logic, and not on mathematical theorems, like the way to carry out statistical hypothesis testing problems using p-values!). It turns out that quantum logic is non Boolean, but seems to have a pleasant connection with the so-called Conditional Event Algebra. See the recent paper by Nguyen [9].

In summary, quantum mechanics concerns motions of particles. Particles moves like waves with a random behavior. The law of quantum mechanics is given by the Schrodinger's equation whose solution is the wave function describing the motion of a particle States of quantum systems are determined by quantum probabilities. Quantum mechanics does not predict a single definite outcome (observed), it predicts a number of different possible outcomes and tells us how likely each of these is (somewhat similar to coarse data in classical statistics). Interference occurs with particles by duality wave/particle.

3   Measuring Physical Quantities

Physical quantities are numerical values associated to a quantum system, such as position, momentum, velocity, and functions of these, such as energy.

In classical mechanics, the result on the measurement of a physical quantity is just a number at each instant of time. In quantum mechanics, at a given time, repeated measurements under the same state of the system give different values of a physical quantity A: There should exist a probability distribution on its possible values, and we could use its expected (mean) value.

For some simple quantities, it is not hard to figure out their probability distributions, such as position x and momentum p (use Fourier transform to find the probability distribution of p) from which we can carry out computations for expected values of functions of then, such as potential energy $V(x)$, kinetic energy (function of p alone). But how about, say, the mechanical energy $V(x) + \frac{p^2}{2m}$, which is a function of both position x and momentum p? Well, its expected value is not a problem, as you can take $E(V(x) + \frac{p^2}{2m}) = EV(x) + E(\frac{p^2}{2m})$, but how to get its distribution when we need it? Also, if the quantity of interest is not of the form of a sum where the knowledge of $E(x), E(p)$ is not sufficient to compute its expectation?

If you think about classical probability, then you would say this. We know the marginal distributions of the random variables x, p. To find the distribution of $V(x) + \frac{p^2}{2m}$, we need the joint distribution of (x, p). How? *Copulas* could help? But are we in the context of classical probability!?

We need a general way to come up with necessary probability distributions for all physical quantities, from the knowledge of the wave function $\psi(x, t)$ in the Schrodinger's equation. It is right here that we need mathematics for physics!

For a spacial quantity like position X (of the particle), or $V(X)$ (potential energy), we know its probability distribution $x \to |\psi(x, t)|^2$, so that its expected valued is given by

$$EV(X) = \int_{\mathbb{R}^3} V(x)|\psi(x, t)|^2 dx = \int_{\mathbb{R}^3} \psi^*(x, t)V(x)\psi(x, t)dx$$

If we group the term $V(x)\psi(x, t)$, it looks like we apply the "operator" V to the function $\psi(., t) \in L^2(\mathbb{R}^3)$, to produce another function of $L^2(\mathbb{R}^3)$. That operator is precisely the multiplication $A_V(.) : L^2(\mathbb{R}^3) \to L^2(\mathbb{R}^3) : \psi \to V\psi$. It is a bounded, linear map from a (complex) Hilbert space H to itself, which we call, for simplicity, an operator on H.

We observe also that $EV(X)$ is a real value (!) since

$$EV(X) = \int_{\mathbb{R}^3} V(x)|\psi(x, t)|^2 dx$$

with $V(.)$ being real-valued. Now,

$$\int_{\mathbb{R}^3} \psi^*(x,t)V(x)\psi(x,t)dx = <\psi, A_V\psi>$$

is the inner product on $H = L^2(\mathbb{R}^3)$. We see that, for any $\psi, \varphi \in H$, $<\psi, A_V\varphi> = <A_V\psi, \varphi>$, since V is real-valued, meaning that the operator $A_V(.) : \psi \to V\psi$ is *self adjoint*.

For the position $X = (X_1, X_2, X_3)$, we compute the vector mean $EX = (EX_1, EX_2, EX_3)$, where we can derive, for example, EX_1 directly by the observables of $Q = X_1$ as $A_{X_1} : \psi \to x_1\psi$ (multiplication by x_1), $A_{x_1}(\psi)(x,t) = x_1\psi(x,t)$.

Remark. The inner product in the (complex) Hilbert space $H = L^2(\mathbb{R}^3)$ (complex-valued functions on $\mathbb{R}^3$, squared integrable wrt to Lebesgue measure dx on $\mathscr{B}(\mathbb{R}^3)$) is defined as

$$<\psi, \varphi> = \int_{\mathbb{R}^3} \psi^*(x,t)\varphi(x,t)dx$$

where $\psi^*(x,t)$ is the complex conjugate of $\psi(x,t)$. The *adjoint operator* of the (bounded) operator A_V is the unique operator, denoted as A_V^*, such that $<A_V^*(f), g> = <f, A_V(g)>$, for all $f, g \in H$ (its existence is guaranteed by Riesz theorem in functional analysis). It can be check that $A_V^* = A_{V^*}$, so that if $V = V^*$ (i.e., V is real-valued), then $A_V^* = A_V$, meaning that A_V is self adjoint. Self adjoint operators are also called *Hermitian* (complex symmetry) operators, just like for complex matrices. The property of self adjoint for operators is important since eigenvalues of such operators are real values, and as we will see later, which correspond to possible values of the physical quantities under investigation, which are real valued.

As another example, let's proceed directly to find the probability distribution of the momentum $p = mv$ of a particle, at time t, in the state $\psi(x,t)$, $x \in \mathbb{R}^3$, and from it,.compute, for example, expected values of functions of momentum, such as $Q = \frac{\|p\|^2}{2m}$.

The Fourier transform of $\psi(x,t)$ is

$$\varphi(p,t) = (2\pi h)^{-\frac{3}{2}} \int_{\mathbb{R}^3} \psi(x,t) e^{-\frac{i}{h}<p,x>} dx$$

so that, by Parseval-Plancherel, $|\varphi(p,t)|^2$ is the probability density for p, so that

$$E(\frac{\|p\|^2}{2m}) = \int_{\mathbb{R}^3} \frac{\|p\|^2}{2m} |\varphi(p,t)|^2 dp$$

But we can obtain this expectation via an appropriate operator A_p as follows. Since

$$\psi(x,t) = (2\pi h)^{-\frac{3}{2}} \int_{\mathbb{R}^3} \varphi(p,t) e^{\frac{i}{h}<p,x>} dp$$

with $x = (x_1, x_2, x_3)$, we have

$$\frac{h}{i}\frac{\partial}{\partial x_1}\psi(x,t) = (2\pi h)^{-\frac{3}{2}}\int_{\mathbb{R}^3} p_1\varphi(p,t)e^{\frac{i}{h}<p,x>}dp$$

i.e., $\frac{h}{i}\frac{\partial}{\partial x_1}\psi(x,t)$ is the Fourier transform of $p_1\varphi(p,t)$, and since ψ is the Fourier transform of φ, Parveval-Plancherel implies

$$E(p_1) = \int_{\mathbb{R}^3}\varphi^*(p,t)p_1\varphi(p,t)dp = \int_{\mathbb{R}^3}\psi^*(p,t)[\frac{h}{i}\frac{\partial}{\partial x_1}](\psi(x,t)dx$$

we see that the operator $A_p = \frac{h}{i}\frac{\partial}{\partial x_1}(.)$ on H extracts information from the wave function ψ to provide a direct way to compute the expected value of the component p_1 of the momentum vector $p = (p_1, p_2, p_3)$ (note $p = mv$, with $v = (v_1, v_2, v_3)$) on one axis of $\mathbb{R}^3$. For the vector p (three components), the operator $A_p = \frac{h}{i}\nabla$, where $\nabla = (\frac{\partial}{\partial p_1}, \frac{\partial}{\partial p_2}, \frac{\partial}{\partial p_3})$.

As for $Q = \frac{\|p\|^2}{2m}$, we have

$$EQ = \int_{\mathbb{R}^3}\psi^*(x,t)[(\frac{-h^2}{2m})\Delta](\psi(x,t)dx$$

where Δ is the Laplacian. The corresponding operator is $A_Q = (\frac{-h^2}{2m})\Delta$.

Examples, as the above, suggest that, for each physical quantity of interest Q (associated to the state ψ of a particle) we could look for a self adjoint operator A_Q on H so that

$$EQ = <\psi, A_Q\psi>$$

A such operator extracts information from the state (wave function) ψ for computations on Q. This operator A_Q is referred to as the *observable* for Q.

Remark. If we just want to compute the expectation of the random variable Q, without knowledge of its probability distribution, we look for the operator A_Q. On the surface, it looks like we only need a weaker information than the complete information provided by the probability distribution of Q. This is somewhat similar to a situation in statistics, where getting the probability distribution of a *random set* S, say on $\mathbb{R}^3$ is difficult, but a weaker and easier information about S can be obtained, namely it coverage function $\pi_S(x) = P(S \ni x)$, $x \in \mathbb{R}^3$, from which the expected value of the measure $\mu(S)$ can be computed, as $E\mu(S) = \int_{\mathbb{R}^3}\pi_S(x)d\mu(x)$, where μ is the Lebesgue measure on $\mathscr{B}(\mathbb{R}^3)$. See e.g., Nguyen [8].

But how to find A_Q for Q in general? Well, a "principle" used in quantum measurement is this. Just like in classical mechanics, all physical quantities associated to a dynamical systems are functions of the system state, i.e., position and momentum (x, p), i.e., $Q(x, p)$, such as $Q(x, p) = \frac{\|p\|^2}{2m} + V(x)$. Thus, the observable corresponding to $Q(x, p)$ should be $Q(A_x, A_p)$, where A_x, A_p are observables corresponding to x and p which we already know in the above analysis. For example, if the observable of Q is A_Q, then the observable of Q^2 is A_Q^2.

An interesting example. What is the observable A_E corresponding to the energy $E = \frac{\|p\|^2}{2m} + V$?

We have $A_V = V$ $(Q_V(f) = Vf$, i.e., multiplication by the function V: $(A_V(f)(x) = V(x)f(x))$.

$$A_p = \frac{h}{i}\nabla = \frac{h}{i}\begin{pmatrix} \frac{\partial}{\partial x_1} \\ \frac{\partial}{\partial x_2} \\ \frac{\partial}{\partial x_3} \end{pmatrix}$$

so that

$$A_p^2(f) = (A_p \circ A_p)(f) = A_p(A_p(f)) = A_p \begin{bmatrix} \frac{h}{i}\frac{\partial f}{\partial x_1} \\ \frac{\partial f}{\partial x_2} \\ \frac{\partial f}{\partial x_3} \end{bmatrix}$$

$$= \left(\frac{h}{i}\right)^2 \begin{bmatrix} \frac{\partial f^2}{\partial x_1^2} \\ \frac{\partial f^2}{\partial x_2^2} \\ \frac{\partial f^2}{\partial x_3^2} \end{bmatrix} = -h^2 \begin{bmatrix} \frac{\partial f^2}{\partial x_1^2} \\ \frac{\partial f^2}{\partial x_2^2} \\ \frac{\partial f^2}{\partial x_3^2} \end{bmatrix}$$

Thus, the observable of $\frac{\|p\|^2}{2m}$ is $\frac{-h^2}{2m}\Delta$, and that of $E = \frac{\|p\|^2}{2m} + V$ is $A_E = \frac{-h^2}{2m}\Delta + V$, which is an operator on $H = L^2(\mathbb{R}^3)$.

By historic reason, this observable of the energy (of the quantum system) is called the *Hamiltonian* of the system (in honor of Hamilton, 1805–1865) and denoted as

$$\mathcal{H} = \frac{-h^2}{2m}\Delta + V$$

Remark. Since

$$E(V) = \int_{\mathbb{R}^3} \psi^*(x,t)V(x)\psi(x,t)dx = \int_{\mathbb{R}^3} \psi^*(x,t)(M_V\psi)(x,t)dx$$

it follows that $A_V = V$.

The Laplacian operator is

$$\Delta f(x) = \frac{\partial f^2}{\partial x_1^2} + \frac{\partial f^2}{\partial x_2^2} + \frac{\partial f^2}{\partial x_3^2}$$

where $x = (x_1, x_2, x_3) \in \mathbb{R}^3$.

Now, if we look back at Schrodinger's equation

$$ih\frac{\partial}{\partial t}\psi(x,t) = -\frac{h^2}{2m}\Delta\psi(x,t) + V(x)\psi(x,t)$$

with (stationary) solutions of the form $\psi(x,t) = \varphi(x)e^{-i\omega t}$, then it becomes

$$(-i^2)h\omega\varphi(x)e^{-i\omega t} = -\frac{h^2}{2m}\Delta\varphi(x)e^{-i\omega t} + V(x)\varphi(x)e^{-i\omega t}$$

or

$$h\omega\varphi(x) = -\frac{h^2}{2m}\Delta\varphi(x) + V(x)\varphi(x)$$

With $E = h\omega$, this is

$$-\frac{h^2}{2m}\Delta\varphi(x) + V(x)\psi(x) = E\varphi(x)$$

or simple, in terms of the Hamiltonian,

$$\mathcal{H}\varphi = E\varphi$$

Putting back the term $e^{-i\omega t}$, the Schrodinger's equation is written as

$$\mathcal{H}\psi = E\psi$$

i.e., the state ψ (solution of Schrodinger's equation) is precisely the *eigenfunction* of the Hamiltonian $\mathcal{H}$ of the system, with corresponding *eigenvalue* E. In other words, the wave function of a quantum system (as described by Schrodinger's equation) is an eigenfunction of the observable of the system energy.

In fact, the Schrodinger equation is

$$ih\frac{\partial}{\partial t}\psi(x,t) = \mathcal{H}\psi(x,t)$$

with $\mathcal{H}$ as an operator on a complex Hilbert space H in a general formalism, where the wave function is an element of H: The Schrodinger's equation is an "equation" in this "Operators on Complex Hilbert spaces" formalism. This equation tells us clearly: It is precisely the observable of the energy that determines the time evolution of states of a quantum system. On the other hand, being an element in a separable Hilbert space, a wave function ψ can be decomposed as a linear superposition of stationary states, corresponding to the fact that energy is quantified (i.e., having discrete levels of energy, corresponding to stationary states). Specifically, the states (wave functions in the Schrodinger's equation) of the form $\psi(x,t) = \varphi(x)e^{-i\omega t}$ are *stationary states* since $|\psi(x,t)| = |\varphi(x)|$, independent of t, so that the probability density $|\varphi(x)|^2$ (of finding the particle in a neighborhood of x) does not depend on time, resulting in letting anything in the system unchanged (not evoluting in time). That is the meaning of stationarity of a dynamical system (the system does not move). To have motion, the wave function has to be a linear superposition of stationary states in interference (as waves). And this can be formulated "nicely" in Hilbert space theory! Indeed, let φ_n be eigenfunctions of the Hamiltonian, then (elements of a separable Hilbert space have representations with respect to some orthonormal basis) $\psi(x,t) = \sum_n c_n\varphi_n(x)e^{-iE_nt/h}$, where $E_n = h\omega_n$ (energy level). Note that, as seen above, for stationary states $\varphi_n(x)e^{-iE_nt/h}$, we have $\mathcal{H}\varphi_n = E_n\varphi$, i.e., φ_n is an eigenfunction of $\mathcal{H}$. Finally, note that, from the knowledge of quantum physics where energy is quantified, the search for (discrete) energy levels $E_n = h\omega_n$ corresponds well to this formalism.

We can say that *Hilbert spaces and linear operators* on them form the *language of quantum mechanics*.

Thus, before continuing, let's put down an abstract definition: *An observable is a bounded, linear, and self adjoint operator on a Hilbert space.*

We have seen that multiplication operator $M_f : g \in H = L^2(\mathbb{R}^3) \to M_f(g) = fg$ is self adjoint when f is real-valued. In particular, for $f = 1_B$, $B \in \mathscr{B}(\mathbb{R}^3)$, M_{1_B} is a (orthogonal) *projection* on H, i.e., satisfying $M_{1_B} = (M_{1_B})^2$ (idempotent) $= (M_{1_B})^*$, which is a special self adjoint operator. This will motive the space $\mathscr{P}(H)$ of all projections on H as the set of "events".

Each observable A is supposed to represent an underlying physical quantity. So, given a self adjoint operator A on H, what is the value that we are interested in, in a given state ψ? Well, it is $< \psi, A\psi >$ (e.g., $\int_{\mathbb{R}^3} \psi^*(x, t)A(\psi)(x, t)dx$), with, by abuse of language, is denoted as $< A >_\psi$. Note that $< \psi, A\psi > \in \mathbb{R}$, for any $\psi \in H$, since A is self adjoint, which is "consistent" with the fact that physical quantities are real-valued.

Remark. If we view the observable A as a random variable, and the state ψ as a probability measure on its "sampling space" H, in the classical setting of probability theory, then $< A >_\psi$ plays the role of expectation of A wrt the probability measure ψ. But here is the fundamental difference with classical probability theory: as operators, the "quantum random variables" do not necessarily commute, so that we are facing a *noncommutative probability theory*. This is compatible with the "matrix" viewpoint of quantum mechanics, suggested by Heisenberg, namely that numerical measurements in quantum mechanics should be matrices which form a noncommutative algebra.

4   Distributions of Observables

Let's look back at the finitely dimensional case. This is in fact the origin of the so-called *spectral theory* (of operators).

For simplicity, and for concreteness, consider the euclidean space $\mathbb{R}^n$. This is a vector space over the scalar field $\mathbb{R}$. Moreover, it has an binary form which is called an inner product: $< x, y > = \sum_{j=1}^n x_j y_j$, where $x_j's$ are the coordinates of $x \in \mathbb{R}^n$ with respect to an orthonormal (canonical) basis of $\mathbb{R}^2$. When we consider infinitely dimensional spaces with similar properties, we will call them Hilbert spaces. Thus euclidean spaces $\mathbb{R}^n$ are finitely dimensional Hilbert spaces.

A (real) $n \times n$ matrix $A = [a_{jk}]$ is a linear transformation (we will call it an operator) on $\mathbb{R}^n$ (i.e., $A : \mathbb{R}^n \to \mathbb{R}^n$). If the matrix A is symmetric, i.e., $A = A^t$ (the transpose of A, i.e., $a_{jk} = a_{kj}$), then by changing coordinate systems (principal axes theorem in analytical geometry), we represent A in a "nice" form, namely diagonal, where the nonzero diagonal entries are roots of the characteristic polynomial $\det(A - \lambda I) = 0$, called the eigenvalues of A. If we let $\sigma(A)$ be the set of all eigenvalues of A, called the *spectrum* of A, then A is written as $A = \sum_{\lambda \in \sigma(A)} \lambda P_\lambda$, where P_λ is the (orthogonal) projections on $\mathbb{R}^n$ onto the eigensubspace $S(\lambda) = \{x \in \mathbb{R}^n : Ax = \lambda x\}$, i.e., the set of eigenvectors associated with the eigenvalue λ. This is referred to as the *spectral decomposition* of the matrix (operator) A.

Remark. The term "spectrum" (or spectral) is used possibly in relation of spectra of atoms in physics. Spectral theory was named after D. Hilbert (1910). But of course, "Hilbert space" was not named by Hilbert!

When we need to consider matrices with complex entries, e.g., linear operators on $\mathbb{C}^n$, symmetry is extended to Hermitian (or self adjoint) property, i.e., $A = A^*$ (transpose of complex conjugate matrix). Even in this case, the remarkable fact is that eigenvalues of *self adjoint operators* are real: $\sigma(A) \subseteq \mathbb{R}$. In particular, when $\sigma(A) \subseteq \mathbb{R}^+$, A is said to be a *positive operator,* which is equivalent to $< x, Ax >$ is positive for all $x \in \mathbb{C}^n$.

Look at the spectral decomposition $A = \sum_{\lambda \in \sigma(A)} \lambda P_\lambda$ of the symmetric matrix A. Since $\sigma(A) \subseteq \mathbb{R}$, we can define a a map $\xi_A : \mathscr{B}(\mathbb{R}) \in \mathscr{P}(\mathbb{R}^n)$ (space of projections on $\mathbb{R}^n$) by $\xi_A(B) = \sum_{\lambda \in B} P_\lambda$. Then, $\xi_A(\mathbb{R}) = \sum_{\lambda \in \mathbb{R}} P_\lambda = I$ (when $tr(A) = 1$), and for any pairwise disjoint $B_j \in \mathscr{B}(\mathbb{R})$, $\xi_A(\cup_j B_j) = \sum_j \xi_A(B_j)$. The set function $\xi_A(.)$ looks like a probability measure, but with $\mathscr{P}(\mathbb{R}^n)-$ valued, instead of $[0, 1]$. Such a set function is called a *spectral measure,* and $\xi_A(.)$ is the (discrete) spectral measure of the matrix A. In fancy notation (but useful when considering infinitely dimensional setting), we write $A = \int_{\sigma(A)} \lambda d\xi_A(\lambda)$.

We see that the study of (random) physical quantities Q on a quantum system, in a state ψ, is via its observable A_Q. Observables (in quantum context) play the role of random variables in Kolmogorov's probability theory.

For simplicity, let's elaborate on this in the finitely dimension case where, for "concreteness", observables are taken as $n \times n$ matrices with complex entries. These are linear, Hermitian operators on $\mathbb{C}^n$.

Recall that a matrix $A = [a_{jk}]$, as an operator on $\mathbb{R}^n$, gives rise to a quadratic form $< Ax, x > = \sum a_{jk} x_j x_k$. If A is symmetric, i.e., $a_{jk} = a_{kj}$, then, using an orthogonal transformation (leaving invariant Euclidean metric on $\mathbb{R}^n$), in analytic geometry, it can be rewritten in a normal form $< Ax, x > = \sum \lambda_j x_j^2$. Sylvester, in 1852, showed that the $\lambda_j's$ are roots of the characteristic polynomial $\det(\lambda I - A)$, i.e., eigenvalues of the matrix (operator) A. This form reduction corresponds to a diagonalization process on the matrix A: for some orthogonal matrix B, the matrix $D = B^{-1}AB$ is in diagonal form. The diagonal entries of D are eigenvalues of A. The set of eigenvalues of A is called the *spectrum* of A, and is denoted as $\sigma(A)$. Thus, there exists an orthonormal basis of $\mathbb{R}^n$, $\{e_1, e_2, ..., e_n\}$, with respect to it, A is diagonal with diagonal entries being eigenvalues of A. In other words, $A = \sum_{\lambda \in \sigma(A)} \lambda P_\lambda$, where P_λ is the (orthogonal) projection onto the eigensubspaces $S_\lambda = \{x : Ax = \lambda x\}$. This is referred to as the *spectral decomposition* of the symmetric matrix A.

Remark. In quantum mechanics, certain physical quantities cannot be measured simultaneously. This fact is interpreted as their observables (e.g. Hermitian matrices) do not commute (since the algebra of matrices is noncommutative). The set of possible values of a quantity Q is the spectrum of A_Q. Thus, the spectrum of the Hamiltonian of energy (energy levels, recalling that energy is quantified) of an atom is precisely the *spectrum of the atom.*

With this spectral decomposition of an observable (i.e., a self adjoint operator) in the finite case, let's point out right away that observables play the role of random variables in Kolmogorov's setting. First, observe that a projection operator is an "event" in quantum setting: for example P_λ is the event that the underlying physical quantity, represented by A, takes the value λ.

In Kolmogorov's setting, given a measurable space $(\Omega, \mathscr{A})$, an event is an element of the $\sigma-$ algebra $\mathscr{A}$ of subsets of Ω. We identify $B \in \mathscr{A}$ with its indicator function $1_B : \Omega \to [0,1]$ which, in turn, is identified with the multiplication operator on the Hilbert space $L^2(\Omega, \mathscr{A}, P)$; $f \to 1_B f : (1_B f)(\omega) = 1_B(\omega)f(\omega)$ (so that if B happens, i.e., $\omega \in B$, then $(1_B f)(\omega) = f(\omega)$, otherwise, it's 0). This operator on $L^2(\Omega, \mathscr{A}, P)$ is an orthogonal projection, and hence self adjoint. In other words, in quantum setting, projections correspond to events. Note also that, two quantum events (projections) p, q are compatible when pq is also an event (a projection): in this case, $pq = (pq)^* = q^* p^* = qp$, i.e., p and q commute. The counterpart of $\mathscr{A}$ is the set $\mathscr{P}(\mathbb{R}^n)$ of all projections on $\mathbb{R}^n$.

Now the spectral decomposition $A = \sum_{\lambda \in \sigma(A)} \lambda P_\lambda$ is similar to a "simple random variable" in classical probability. A simple random variable X is of the form $X(\omega) = \sum x_j 1_{B_j}(\omega)$, where $B_j = \{\omega : X(\omega) = x_j\}$, so that when the event B_j occurs, $X = x_j$. The probability density of X is $P(X = x_j) = P(B_j) = P[X^{-1}(\{x_j\})]$.

What is the probability density of the observable A? i.e., $P(P_\lambda)$ in quantum formalism? The counterpart of the probability measure P on $(\Omega, \mathscr{A})$, is the state ψ of the Schrodinger equation. Let ρ be a positive operator on $\mathbb{R}^n$ with unit trace (i.e., $tr(\rho) = \sum_{j=1}^{n} <e_j, \rho e_j> = 1$). Note that a positive operator is necessarily self adjoint. The triple $(\mathbb{R}^n, \mathscr{P}(\mathbb{R}^n), \rho)$ is called a (finite dimentional) *quantum probability space*, the "state" ρ is called a "density matrix".

For $B \in \mathscr{P}(\mathbb{R}^n)$, we have $tr(\rho B) = \sum_{j=1}^{n} <u_j, \rho u_j>$, where the $u'_j s$ is an orthogonal basis for the range of the projection B, so that $tr(\rho B) \in [0,1]$, and for $B_1, ..., B_k$, pairwise orthogonal (for $j \neq m$, $B_j B_m = 0$), so that $B_1 + ... + B_k$ is the event that at least one of the $B'_j s$ occurs, and $tr(B_1 + ... + B_k) = \sum_{j=1}^{k} tr(\rho B_j)$. Thus the map $tr(\rho \cdot (.)) : \mathscr{P}(\mathbb{R}^n) \to [0,1]$ acts like a probability distribution, with $tr(\rho B)$ being the probability of the event B under the state ρ.

For $A = \sum_{\lambda \in \sigma(A)} \lambda P_\lambda$, $\Pr(A$ takes the values $\lambda) = tr(\rho P_\lambda)$. Thus, the observable A on $\mathbb{R}^n$ is a discrete (finite) random variable with a probability mass function.

In summary, let H be a (finite dimensional) complex Hilbert space. representing states of a quantum system. Let $\mathscr{P}(H)$ denote the set of all projections on H (playing the role of events), and ρ a positive operator of H with unit trace. The Triple $(H, \mathscr{P}(H), \rho)$ is a quantum probability space.

In such a quantum probability space, under the state ρ, an observable A, with spectral decomposition $A = \sum_{\lambda \in \sigma(A)} \lambda P_\lambda$, has a probability distribution given by $\Pr(P_\lambda) = tr(\rho P_\lambda)$. The converse to this construction from a given ρ is Gleason's theorem, which says that any probability distribution $\mu : \mathscr{P}(H) \to [0,1]$ is of this form, i.e., has a density ρ.

Remark. There are several different definitions of quantum probability space in the literature, depending of levels of generality, e.g., in terms of C^*−algebra. Here we consider a low level in terms of Hilbert spaces.

Let A_Q the observable of the quantity Q. What are the possible values of Q? In fact, what are the values of Q that we can actually measure? And what is the probability distribution of A_Q?

To answer this, observe that if the state $\psi \in H$ is an eigenfunction of A_Q, i.e., there is some scalar a (corresponding eigenvalue), here real since A_Q is self adjoint, such that $A_Q(\psi) = a\psi$, then

$$EQ = \int_{\mathbb{R}^3} \psi^*(x,t)(A_Q\psi)(x,t)dx = \int_{\mathbb{R}^3} \psi^*(x,t)a\psi(x,t)dx$$
$$= a \int_{\mathbb{R}^3} |\psi(x,t)|^2 dx = a$$

$$E(Q^2) = \int_{\mathbb{R}^3} \psi^*(x,t)(A_Q)^2\psi)(x,t)dx = \int_{\mathbb{R}^3} \psi^*(x,t)(A_Q)(a\psi(x,t))dx$$
$$= a^2 \int_{\mathbb{R}^3} |\psi(x,t)|^2 dx = a^2$$

so that $Var(Q) = EQ^2 - (EQ)^2 = 0$. i.e., for sure, Q will take the value a (no uncertainty involved). Thus, measurements of a quantity Q are precisely the eigenvalues of its observables, i.e., the *spectrum* of the observable representing it, denoting as $\sigma(A_Q)$. Note that, since every A_Q is self adjoint, $\sigma(A_Q) \subseteq \mathbb{R}$, consistent with the fact that measured values of physical quantities have to be real (not complex numbers!).

Here, again, is a theory extending what we known from matrix theory.

With the interests in transforming symmetric quadratic forms ($(Ax, x) = \sum_{j,k} \alpha_{jk} x_j x_k$, the matrix A is symmetric) to normal form ($\sum_j \beta_j x_j^2$) via an orthogonal transformation $T : \mathbb{R}^n \to \mathbb{R}^n$ ($||Tx|| = ||x||$, for any $x \in \mathbb{R}^n$, norm invariant), back to the times of Descartes (1637), it was known that any symmetric matrix A ($Tx = Ax$) is orthogonally equivalent to a diagonal matrix D. i.e., $D = B^{-1}AB$, for some orthogonal matrix B ($||Bx|| = ||x||$). Note that B is orthogonal iff its columns form an orthonormal basis for $\mathbb{R}^n$. The diagonal entries of D are the *eigenvalues* of A, i.e., roots of the polynomial equation $\det(A - \lambda I) = 0$. The set $\sigma(A)$ of eigenvalues of a matrix A is called the *spectrum* of the operator (matrix) A. It x is a non zero vector such that $Ax = \lambda x$, then x is called an *eigenvector* (with associated eigenvalue λ). Thus, a symmetric matrix A can be written as $\sum_j \lambda_j u_j$.

As we will see, when we generalize matrix A on $\mathbb{R}^n$ is to an "nice" bounded operator (e.g. compact) on a "nice" Hilbert space H (separable), we will have countable eigenvalues and vectors, the latter form an orthonormal basis for H, so that each $h \in H$ can be written as an infinite series, and hence any operator on H can be represented as an "infinite matrix".

Let's start with matrices to bring out things we wish to generalize. For $n \times n$ matrices with complex entries (i.e., operators on $\mathbb{C}^n$, a finitely dimensional Hilbert space), it known from matrix algebra that, a self adjoint matrix A has real eigenvalues $\lambda_j, j = 1, 2, ..., n$ (i.e., $A - \lambda_j I$ are not invertible). The set $\sigma(A)$ of eigenvalues is called the spectrum of A. The eigenspaces associated with eigenvalues λ_j (i.e., $S(\lambda_j) = (x \in \mathbb{C}^n : Ax = \lambda_j x\})$ are orthogonal (for $\lambda_j \neq \lambda_k, S(\lambda_j) \perp S(\lambda_k)$). Moreover, $A = \sum_{\lambda \in \sigma(A)} \lambda P_\lambda$, where P_λ is the projection onto $S(\lambda)$. This is referred to as the *spectral decomposition* of A.

Remark. If A is an operator on an Hilbert space H, then its spectrum $\sigma(A) \subseteq \mathbb{C}$ is, by definition, the set complements of its "resolvent" $\{\lambda \in \mathbb{C} : (A - \lambda I)^{-1}$ exists$\}$. In general, the spectrum could be uncountable. The spectral decomposition of a self adjoint operator will be defined as an "integral wrt a spectral measure on $\mathscr{B}(\mathbb{C})$". In Quantum mechanics, quantities which are measured are matrices (more generally, operators) rather than real numbers. Also, "observables" may be functions of other observables, such as $f(A)$ where $f : \mathbb{R} \to \mathbb{R}$. As such, we need to make sense of $f(A)$ as an operator: this is the problem of *functional Calculus*. If $Au = \lambda u$, then we could set $f(A)u = f(\lambda)u$, so that clearly there is a connection between spectral theory and functional calculus. Both are related to quantum mechanics. When $A = \sum_j \lambda_j P_j$, we set $f(A) = \sum_j f(\lambda_j) P_j$. For Hilbert space H, an observable (i.e., a self adjoint operator A on H), has its spectral measure ξ_A on $\mathscr{B}(\mathbb{C})$, such that $A = \int_{\sigma(A)} \lambda d\xi_A(\lambda)$, we set $f(A) = \int_{\sigma(A)} f(\lambda) d\xi_A(\lambda)$.

The spectral decomposition of a self adjoint operator A in the finitely dimensional case is obtained from a "resolution of identity" $\{P_\lambda; \lambda \in \sigma(A)\}$. The map $\lambda \in \sigma(A) \subseteq \mathbb{C} \to P_\lambda \in \mathscr{P}(H)$, space of projections on H, acts like a finite probability density where probability values are projections! Note that, like $[0, 1]$, $\mathscr{P}(H)$ is not a Boolean lattice. For a "random variable X" taking values in $\sigma(A)$, formally, $\Pr(X = \lambda) = P_\lambda$. When H is of infinite dimensions, this "density" should be replaced by a measure on $\mathscr{B}(\mathbb{C})$. Thus, a *spectral measure* is defined as $\xi(.) : \mathscr{B}(\mathbb{C}) \to \mathscr{P}(H)$ having analogous properties of a numerical measure, namely, $\xi(\mathbb{C}) = I$, $\xi(\cup_n B_n) = \sum_n \xi(B_n)$ for any sequence of pairwise disjoint $B_n \in \mathscr{B}(\mathbb{C})$, where the infinite sum is taken in the sense of convergence wrt to the norm. This is a *projection-valued probability measure*.

The upshot is that any (bounded) self adjoint operator A on a Hilbert space H admits a unique spectral measure ξ_A such that it has the spectral decomposition (in the infinite case) as $A = \int_{\sigma(A)} \lambda d\xi_A(\lambda)$ (*von Neumann's spectral theorem*) which is the extension of $A = \sum_{\lambda \in \sigma(A)} \lambda P_\lambda$, in the finite case. The *spectral integral* is defined as a Lebesgue-Stieltjes integral, here, of the function $f(\lambda) = \lambda$, wrt ξ_A.

Now, $\mathscr{P}(H)$ is the set of events, i.e., special $\{0,1\}-$ valued random variables, general random variables (observables) are represented by (bounded) self adjoint operators on H. The spectral measure of a self adjoint operator thus plays the role of the probability law of a random variable in the quantum context, its existence and uniqueness are guaranteed by von Neumann's spectral theorem.

Remark. Why the spectral integral representing A is over its spectrum $\sigma(A)$? First, note that $\sigma(A)$ needs not be discrete, as it is $\{\lambda \in \mathbb{C}; A - \lambda I$ is not invertible$\}$.

The "support" of the spectral measure ξ_A is $\Lambda(\xi_A) = \mathbb{C}\backslash \cup_k B_k$ where the union is over all open set B_k in $\mathbb{C}$ for which $\xi_A(B_k) = 0$. The measure ξ_A is said to be compact if, by definition, its support $\Lambda(\xi_A)$ is compact in $\mathbb{C}$. It turns out that for compact spectral measures, $\sigma(A) = \Lambda(\xi_A)$. That answers our question.

We close this technical discussions with the concept of distribution of observables.

Let A be an observable, i.e., a self adjoint operator on a Hilbert space H. Let ρ be a density matrix, i.e., a positive operator on H with unit trace. Let ξ be the spectral measure of A. Let $\mu : \mathscr{B}(\mathbb{R}) \rightarrow [0, 1]$ be $\mu(B) = tr(\rho\xi(B))$. Then $\mu(.)$ is a probability measure, and it is called the "law" or probability distribution of the observable A, under the state ρ.

The interpretation is this. The distribution of A in the quantum framework is the same as the usual probability distribution of a random variable on $(\mathbb{R}, \mathscr{B}(\mathbb{R}), \mu)$.

Kolmogorov's theory of probability is a special case of quantum probability: a commutative theory within an arbitrary (commutative or not) theory: Each random variable $X : (\Omega, \mathscr{A}, P) \rightarrow \mathbb{R}$ is identified with the multiplication by it, acting on $L^2(\Omega, \mathscr{A}, P)$, i.e., a special self adjoint operator, where multiplication operators commute; whereas in quantum uncertainty analysis, observables are arbitrary self adjoint operators which might not commute. Among other things, noncommutativity of observables (meaning that they cannot be observed simultaneously) is characteristic for quantum modeling in applications, such as finance.

5   Quantum Modeling and Probability Calculus for Econometrics

Like the attempt of econophysicists to use statistical physics to model and analyze financial time series, an obvious rationale for using quantum mechanic formalisim is in the force driving their fluctuations. Specifically, the Hamiltonian of a quantum dynamical system (the observable total energy) controls the time evolution of the system. The Black-Scholes' equation in option pricing can be converted into a quantum system with a given Hamiltonian (see, e.g., [1]). It is about modeling, say, financial time series as quantum dynamical systems to gain new insights into the behavior of financial markets, for predictions among other purposes.

The heart of statistical analysis of time series data is models. Usually, having in mind just one theory of probability, namely Kolmogorov (in fact, one calculus of probabilities), all models are based on it. In particular, joint or conditional distributions, and correlations among variables are derived, including the use of *copulas*. It is perhaps time to ask "Are we using the right calculus of probabilities so far in financial data analysis?". Note that, Kolmogorov probability theory has no problem at all in games of chance! "All models are wrong, but some are useful" (G. Box) has a neat interpretation in quantum mechanics! Schrodinger equation is just our best guess of how nature behaves (as verified by experiments). But how to find the "useful models"? Of course, that is the main task of statisticians using all their statistical tools, such as model fitting on data, cross validation methods, etc.

Now with the knowledge of quantum mechanics which not only provides us with a sense of dynamics (what causes the financial data to fluctuate?), but also a way to conduct uncertainty analysis based on a new calculus of probabilities (nonadditive and noncommutative), we could reexamine, where appropriate, the ways to do econometrics so far. For example, in analyzing the factors which affect the fluctuations of a financial time series, we could discover a Hamiltonian driving these fluctuations and then examine whether we are in a quantum context. When it seems to be the case, we have found a "useful" model! A quantum model for a financial data set. And the familiar follow up tasks involve the use of quantum probability calculus.

References

1. Baaquie, B.E.: Quantum Finance. Cambridge University Press, Cambridge (2007)
2. Feynman, R.: Space-time approach to non-relativistic quantum mechanics. Rev. Mod. Phys. **20**, 367–387 (1948)
3. Feynman, R.: The concept of probability in quantum mechanics. In: Berkeley Symposium on Mathematical Statistics and Probability, pp. 533–541 (1951)
4. Feynman, R.: The Feynman Lectures on Physics, Volume III: Quantum Mechanics. Basic Books, New York (1965)
5. Gelman, A., Bethancourt, M.: Does quantum uncertainty have a place in everyday applied statistics? Behav. Brain Sci. **36**(3), 285 (2013)
6. Keller, J.B., MacLaughlin, D.W.: The Feynman integral. Am. Math. Mon. **82**(5), 451–465 (1975)
7. Khrennikov, A.Y.: Classical and quantum mechanics on information spaces with applications to cognitive, psychological, social and anomalous phenomena. Found. Phys. **29**, 1065–1098 (1999)
8. Nguyen, H.T.: An Introduction to Random Sets. Chapman and Hall/CRC Press, Boca Raton (2006)
9. Nguyen, H.T.: Conditional Events Algebras: The State-of-the-Art. Springer's Kreinovich Festschrift (2017, to appear)
10. Parthasarathy, K.R.: An Introduction to Quantum Stochastic Calculus. Birkhauser, Basel (1992)
11. Segal, W., Segal, I.E.: The Black-Sholes pricing formula in the quantum context. Proc. Natl. Acad. Sci. U.S.A. **95**, 4072–4075 (1998)
12. Sethna, J.P.: Statistical Mechanics: Entropy, Order Parameters, and Complexity. Clarendon Press, Oxford (2016)

GL^+ and GL^- Regressions

Charles Andoh[1(✉)], Lord Mensah[1], and Francis Atsu[2]

[1] Department of Finance, University of Ghana Business School, Legon, Ghana
{candoh,lmensah}@ug.edu.gh
[2] Department of Accounting and Finance, GIMPA School of Business, Accra, Ghana
fatsu@gimpa.edu.gh

Abstract. Regression analysis for which the dependent variable is binary has typically been modelled by the Logit and the Probit models. We propose two new regression models GL^+ and GL^- regressions based on the function of [5,6] and the function of [4] for binary dependent variables. These models allow for possible asymmetries in the underlying mechanisms governing the binary output variable and make allowance for the independent variables to determine its shape. Our simulation results of the univariate regression indicate that the expected average mean square error is smallest for the GL^+ model than the Logit or the Probit models. On the other hand, the expected correlation between the outcome and the predicted probabilities is greatest for the GL^- model than the Logit and Probit models. Therefore, the GL^+ having higher predictive power over the Logit and Probit, should be more useful to researchers, economists and scientists that rely on the Logit and Probit models for their work.

1 Introduction

The outcomes of most real-life occurrences are usually discrete/qualitative in nature rather than being continuous/quantitative [12,16]. For instance, a customer either defaults or honours a contractual obligation, a presidential candidate wins or loses an election, a football team either wins or loses a game, it will rain on Christmas day or it does not, among others. In these cases, it is scientifically prudent to employ models whose dependent variables are discrete. Against this background, the Logit and Probit models have been employed extensively in various fields [1,10–12,15,18,21–24, among others]. Specifically, [10] assessed reasons why 980 respondents living in Xian, China did not wish to buy a car. They used the probit model to identify households who did not have the intention to buy a car based on their socio-demographic profile and living conditions and then used the multinomial Logit model to estimate the probability of a particular reason for not buying a car mentioned. The results indicated that, socio-demographic variables, type and size of the house and non-availability of parking space have significant effects on car purchase decisions. However, the main reasons for not buying a car are related to costs considerations, parking difficulties and congestion. [15] also examined the relationship between participating in the Chicago Child-Parent Centre (CPC) Preschool Program and higher

© Springer International Publishing AG 2018
L. H. Anh et al. (eds.), *Econometrics for Financial Applications*, Studies in Computational Intelligence 760, https://doi.org/10.1007/978-3-319-73150-6_4

educational attainment (high school completion, highest grade completed, and college attendance) at age 22. This was done by using the probit regression to examine the group differences in educational attainment controlling for child and family characteristics, including gender, race/ethnicity, and family risk status. Results found showed a significant positive association between CPC preschool participation and higher education attainment. [17] adopted the discrete choice models to predict the winners of entertainment awards. Their analysis reveals how some of the earlier results might be considered truly surprising, thus nominees with low probability of winning who have overcome nominees who were strongly favored. [19] studied residential mobility, quality of neighborhood and life course events using limited dependent variable models and find that not all life course events that are associated with moves leads to neighborhood quality adjustment.

From the above studies, the shape and the skewness of the data set were implicitly imposed and these could lead to suboptimal decisions. Because real life data set rarely emanate from elliptic distribution (i.e. class of all symmetric distributions with support on the entire real line [13], the GL^+ and the GL^- models that give allowance for possible asymmetries and allows the independent variables determine its shape should be favored. [4] used these distributions in combination with artificial neural networks to overcome the estimation difficulty in stochastic variance models for discrete financial time series. In addition, when studying the distribution among competing distributions (namely, normal and the heavier-tailed distributions, GG^+, GG^-, Student-t, Normal Inverse Gaussian, GL^+ and GL^-, these distributions were employed in the estimation of the value-at-risk for some German stocks [3]. His findings were that among these competing distributions, the GL^+ and GL^- gave best estimate of the value-at-risk for a properly chosen skewness parameter.

Further, the Probit and Logit models assume that the processes that govern the binary (multinomial) outcome follow the elliptical distribution. These imply that the above models tend to impose a given symmetry and shape irrespective of the actual data structure. However, many social phenomena rarely follow the elliptical distribution (see for example [9]). Therefore, the imposition of the assumption of Logit and Probit models being symmetry may lead to biased parameter estimates. For instance, in corporate finance, firm fundamentals during distressed market periods are typically skewed; as such, imposing a different data structure may likely produce misleading results. In this regard, the paper makes some contributions to the literature of discrete outcome models. First, we propose and derive the maximum likelihood of two novel models, GL^+ and GL^- regressions, which are based on the GL^+ distribution of [5,6], and GL^- distribution of [4]. Second, we empirically show that the GL^+ and GL^- models are likely to produce more accurate estimates as compared to those of logit and probit models. This stems from the fact that the GL^+ and GL^- regression models are more general and do not impose any structure but extract the shape and symmetry from the data set. The Logit model is a special case of GL^+ and GL^- if the symmetry and shape parameters in these functions are set to 1 and $e = 2.7183$, respectively. Specifically, we explore the forging argument by comparing the performance of our models with the benchmark models using their goodness of fit measures such as average mean square error (AMSE, hereafter), expected mean square error, the correlation between the outcome and the predicted probabilities and the expected correlations. Following [2,7,20], we compute the above measures for GL^+ and GL^-, Logit and Probit using simulated data from a random sample emanating from a mixture of ten different distributions.

For the empirical analysis, we proceed as follows. First, we simulate a data set of size 50, 100, 150, 200, 250, 300, 350, 400, 450, 500, 550, 600, 650 and 750 each of length 1000 from a random sample from a mixture of ten distributions. Further, we estimate the parameters of the four models for each of these data sizes. Second, we compute the AMSE for each path, and the expected AMSE (E(AMSE)) for each of data size for each model. Finally, we evaluate the prediction power of the four models using the loss function (E(AMSE)) for each path and over all the data sizes. Next, we compute the expected correlation between the outcome and the predicted probabilities for each data size for each model. We performed similar exercise by computing the Akaike Information Criteria (AIC), Bayesian Information Criteria (BIC) for each path and the expected AIC and the expected BIC for each data size and each model. In all cases, we evaluate the predictive power of the four models using expected AIC, expected BIC for each path and over all the data sizes.

The results show that, on the average, GL^+ produce smaller $E(AMSE)$ as compared to those of the Logit and Probit models. On the hand, GL^- produced the greatest expected correlation between the outcome and the predicted probabilities than the Logit and the Probit. The probit produced the smallest expected BIC compared with the competing models. This imply that when studying any phenomenon whose outcome is binary (multinomial), it is more appropriate to use the GL^+ since this model do not impose any structure on the data set as compared with their counterparts: the Logit and Probit models.

Thus the objectives of the study are to determine whether asymmetries and the shape of the independent variables do matter in the modeling of regression analysis for which the dependent is binary.

The rest of the study is organized as follows. Section 2 is the methodology where the main results of the paper are presented. This section discusses the odds ratio version of the GL^+ and GL^- models, the log-likelihood functions and the marginal effects of these models. Section 3 presents the simulation results whereas Sect. 4 concludes the study.

2   Methodology

This section presents generic binary outcome maximum likelihood, the GL^+ and GL^- regressions and their corresponding maximum likelihood functions. We proceed as follows. First, we present the generic maximum likelihoods for models whose outcome are binary. Second, we derive the GL^+ and GL^- models using the GL^+ and GL^- functions and construct their respective maximum likelihoods. Finally, we deduce the marginal effects of our models.

2.1   Generic Maximum Likelihood Function for Binary Output Variables

Define the conditional probability of a binary output variable Y as follows: $P(Y = 1|X = x_1, X = x_2, ..., X = x_k)$ as a function of $X = (x_1, x_2, ..., x_k)$ incorporating possible asymmetries in the data of interest X. Assume that

$$P(Y = 1|X = x_1, X = x_2, ..., X = x_k) = F(X, \Theta) \tag{1}$$

for some function F parameterized by Θ. Let the observations $x_i's$ be independent of each other. Then the conditional likelihood function, L is given by

$$L(\Theta|X) = \prod_{i=1}^{n} P(Y = 1|X = x_1, X = x_2, ..., X = x_k)$$

$$= \prod_{i=1}^{n} F(x_i, \Theta)^{y_i}(1 - F(x_i, \Theta))^{1-y_i} \tag{2}$$

where the observed y_i is either 0 or 1.

In a sequence of Bernoulli trials $y_1, y_2, ..., y_n$ with a constant probability of success p, the likelihood function is defined by

$$L(p|Y) = \prod_{i=1}^{n} p^{y_i}(1 - p)^{1-y_i} \tag{3}$$

where $Y = y_1, y_2, ..., y_n$. The log-likelihood function is

$$l(p|Y) = log L(p|Y) = \sum_{i=1}^{n} [y_i log p + (1 - y_i) log(1 - p)] \tag{4}$$

which is maximized when $\hat{p} = \frac{\sum_{i=1}^{n} y_i}{n} = \bar{Y}$.

2.2   Our Models: GL^+ and GL^-

We derive the GL^+ and GL^- regressions by presenting the following definitions.

Definition 1: A random variable X follows the GL^+ distribution with parameters μ, ν^2, a and b (for short $GL^+(\mu, \nu^2, a, b)$) if its density function is given by

$$f(x) = b\frac{loga}{\nu}\frac{a^{-(\frac{x-\mu}{\nu})}}{\left[1 + a^{-(\frac{x-\mu}{\nu})}\right]^{b+1}} \tag{5}$$

where $-\infty < x < +\infty$, $\mu \in (-\infty, +\infty)$, $\nu^2, a \in R^+$, and $b > 0$. The distribution is negatively skewed if $b \in (0, 1)$, positively skewed if $b \in (1, +\infty)$, and the distribution is symmetric when $b = 1$. The logistic distribution is just a special case of this distribution with $b = 1$ and $a = e$.

Definition 2: A random variable X follows the GL^- distribution with parameters μ, ν^2, a and b (for short $GL^-(\mu, \nu^2, a, b)$) if its density function is given by

$$f(x) = b \frac{\log a}{\nu} \frac{a^{(\frac{x-\mu}{\nu})}}{\left[1 + a^{(\frac{x-\mu}{\nu})}\right]^{b+1}} \tag{6}$$

where $-\infty < x < +\infty$, $\mu \in (-\infty, +\infty)$, $\nu^2, a \in R^+$, and $b > 0$. The distribution is negatively skewed if $b \in (0, 1)$, positively skewed if $b \in (1, +\infty)$, and the distribution is symmetric when $b = 1$. The logistic distribution is just a special case of this distribution with $b = 1$ and $a = e$.

The GL^+ is the Type I Generalized logistic distribution of [5,6]. On the other hand, GL^- is the distribution of [4].

The cumulative distribution functions of GL^+ and GL^- distributions are respectively

$$F_+(x) = \left[1 + a^{-(\frac{x-\mu}{\nu})}\right]^{-b} \tag{7}$$

and

$$F_-(x) = \left[1 + a^{(\frac{x-\mu}{\nu})}\right]^{-b} \tag{8}$$

When Y is defined by X and if we let Y follow the GL^+ distribution, we can write

$$\hat{p}_i = F(\hat{z}_i) = \frac{1}{\left(1 + \hat{a}^{-\hat{z}_i^{\hat{b}}}\right)} \tag{9}$$

where $i = 1, ..., n$, $\hat{z}_i = \hat{\beta}_0 + \hat{\beta}_1 x_1 + \hat{\beta}_2 x_2 + ... + \hat{\beta}_k x_k$, $\hat{\epsilon}_i = z_i - \hat{z}_i$ and k the number of independent variables. From Eq. (9), we see that

$$\hat{z}_i = \frac{1}{\log \hat{a}} \log \left(\frac{\hat{p}_i^{\frac{1}{b}}}{1 - \hat{p}_i^{\frac{1}{b}}}\right) \tag{10}$$

Call the quantity $\frac{1}{\log \hat{a}} \log \left(\frac{\hat{p}_i^{\frac{1}{b}}}{1 - \hat{p}_i^{\frac{1}{b}}}\right)$, the GL^+ of p with skewness parameter b (compare with [14, p. 26] for the case when $b = 1$). It reports the ratio of the probability of success to the probability of failure for an included regressor. If this quantity is less than one, then there is less likelihood of the independent variable causing a success outcome, all other factors being held at a constant. If it is greater than one, then there is greater likelihood of the independent variable causing a success outcome when all the other variables are held constant. Hence the GL^+ regression model is

$$\hat{z}_i = \frac{1}{\log \hat{a}} \log \left(\frac{\hat{p}_i^{\frac{1}{b}}}{1 - \hat{p}_i^{\frac{1}{b}}}\right) = \hat{\beta}_0 + \hat{\beta}_1 x_1 + \hat{\beta}_2 x_2 + ... + \hat{\beta}_k x_k \tag{11}$$

Solving for $\hat{p}$, we have

$$\hat{p}_i = F(\hat{z}_i; \hat{\Theta}) = [1 + \hat{a}^{-(\hat{\beta}_0 + \hat{\beta}_1 x_1 + \hat{\beta}_2 x_2 + ... + \hat{\beta}_k x_k)}]^{-\hat{b}} \tag{12}$$

On the other hand, if Y is defined by X and if we let Y follow the GL^- distribution, we can write

$$\hat{p}_i = F(\hat{z}_i) = 1 - \frac{1}{1 + \hat{a}^{\hat{z}_i)\hat{b}}} \tag{13}$$

where $i = 1, ..., n$, $\hat{z}_i = \hat{\beta}_0 + \hat{\beta}_1 x_1 + \hat{\beta}_2 x_2 + ... + \hat{\beta}_k x_k$, $\hat{\epsilon}_i = z_i - \hat{z}_i$ and k the number of independent variables. Hence from Eq. (13), we can write

$$\hat{z}_i = \frac{1}{log\,\hat{a}} log\left(\frac{1}{1 - \hat{p}_i^{\frac{1}{b}}} - 1\right) = \hat{\beta}_0 + \hat{\beta}_1 x_1 + \hat{\beta}_2 x_2 + ... + \hat{\beta}_k x_k \tag{14}$$

Call the quantity $\frac{1}{log\,\hat{a}} log\left(\frac{1}{1 - \hat{p}_i^{\frac{1}{b}}} - 1\right)$, the GL^- of p with skewness parameter b.

Hence the GL^- regression model is

$$\hat{z}_i = \frac{1}{log\,\hat{a}} log\left(\frac{1}{(1 - \hat{p})_i^{\frac{1}{b}}} - 1\right) = \hat{\beta}_0 + \hat{\beta}_1 x_1 + \hat{\beta}_2 x_2 + ... + \hat{\beta}_k x_k \tag{15}$$

where $\hat{\epsilon}_i = z_i - \hat{z}_i$ and k is number of independent variables. From (15), we can write

$$log\left(\frac{1}{(1 - \hat{p}_i)^{\frac{1}{b}}} - 1\right) = log\,\hat{a}^{\hat{\beta}_0 + \hat{\beta}_1 x_1 + \hat{\beta}_2 x_2 + ... + \hat{\beta}_k x_k} \tag{16}$$

Solving for $\hat{p}$, we get

$$\hat{p}_i = F(\hat{z}_i; \hat{\Theta}) = 1 - [1 + \hat{a}^{(\hat{\beta}_0 + \hat{\beta}_1 x_1 + \hat{\beta}_2 x_2 + ... + \hat{\beta}_k x_k)}]^{-\hat{b}} \tag{17}$$

2.3   Log-Likelihood Function for GL^+ and GL^- Regressions

We first consider the GL^+ distribution.[1] Given an independent observational data $X = (x_1, x_2, ..., x_k)$ and an observed binary output variable y_i, the conditional log-likelihood function, $l(\Theta|X)$, from Eq. (2) is

$$l(\Theta|X) = \sum_{i=1}^{n} [y_i log F(x_i; \Theta) + (1 - y_i) log(1 - F(x_i; \Theta))] \tag{18}$$

where $\Theta = (a, b, \beta_0, \beta_1, ..., \beta_k)$ is the parameter space. The term $l(\Theta|X)$ can be written as

$$l(\Theta|X) = \sum_{i=1}^{n} \left[log(1 - F(x_i; \Theta)) + y_i log\left(\frac{F(x_i; \Theta)}{1 - F(x_i; \Theta)}\right) \right] \tag{19}$$

---

[1] For parsimony, the log-likelihood functions and marginal effects of the logit and probit models are not presented here. Interested readers can refer to [20], and [7] for better treatment on the likelihoods and the marginal effects of these benchmark models.

It should be noted from Eq. (13) that $F(x_i; \Theta) = [1 + \hat{a}^{-(\beta_0 + \beta_1 x_1 + \beta_2 x_2 + ... + \beta_k x_k)}]^{-b}$ and the parameters of interest have to be solved numerically by solving $\max_{s.t.\theta \in \Theta} l(\Theta|X)$. Equivalently, we can solve $-\min_{s.t.\theta \in \Theta} l(\Theta|X)$ to obtain the parameter of interest. For the case of the GL^- distribution, the conditional log-likelihood function is

$$l(\Theta|X) = \sum_{i=1}^{n} \left[log(1 - F(x_i; \Theta)) + y_i log\left(\frac{F(x_i; \Theta)}{1 - F(x_i; \Theta)}\right) \right] \qquad (20)$$

where $F(x_i; \Theta) = 1 - [1 + \hat{a}^{(\beta_0 + \beta_1 x_1 + \beta_2 x_2 + ... + \beta_k x_k)}]^{-b}$ from Eq. (18) and $\Theta = (a, b, \beta_0, \beta_1, ..., \beta_k)$. The parameters can be obtained by solving the optimization problem $-\min_{s.t.\theta \in \Theta} l(\Theta|X)$.

2.4 Marginal Effects for the GL^+ and GL^- Distributions

In the GL^+ model, we hypothesize that the probability of the occurrence of an event is determined by the function

$$p_i = F(z_i) = \frac{1}{(1 + a^{-z_i})^b} \qquad (21)$$

The marginal effect of z is given by

$$\frac{dp}{dz} = \frac{F(z)}{dz} = f(z) = bloga\frac{a^{-z}}{(1 + a^{-z})^{b+1}} \qquad (22)$$

The marginal effect of the $x_i's$ is given by

$$\frac{\partial p}{\partial x_i} = \frac{\partial p}{\partial z}\frac{\partial z}{\partial x_i} = f(x)\beta_i = \beta_i bloga\frac{a^{-z}}{(1 + a^{-z})^{b+1}} \qquad (23)$$

Compare [7, pp. 300–301].

In the GL^- model, we hypothesize that the probability of the occurrence of the event is determined by the function

$$p_i = F(z_i) = \frac{1}{(1 + a^{z_i})^b} \qquad (24)$$

The marginal effect of z is given by

$$\frac{dp}{dz} = \frac{F(z)}{dz} = f(z) = bloga\frac{a^z}{(1 + a^z)^{b+1}} \qquad (25)$$

and the corresponding marginal effect of the $x_i's$ is given by

$$\frac{\partial p}{\partial x_i} = f(z)\beta_i = \beta_i bloga\frac{a^{-z}}{(1 + a^{-z})^{b+1}} \qquad (26)$$

$\frac{\partial p}{\partial x_i}$ in Eqs. (24) and (27), explain the positive parameter as 'a unit increase in the related variable would lead to an absolute value β probability increase in the dependent variable'. The negative β is the opposite of the positive, in that it reports the reduction in the probability of the occurrence of the dependent variable with the absolute variable of the β. The marginal effect measures the expected instantaneous change of the dependent variable as a function of a change of an independent variable while all other independent variables are held constant.

3   Simulation Results

The simulation part of the study answers how well do GL^+ and GL^- models fit any data as compared with the logit and their probit models. We use only one independent variable emanating from a random sample of a mixture of ten distributions of size n.

Remark: It should be noted that the random sample could have been chosen from any reasonable number of mixture of distributions, not necessarily ten. We wanted to ensure that the independent variable X is not biased toward any particular distribution.

For our binary output variable Y, we generate random variables from the Bernoulli distribution of size n. Next, we generate M paths as follows: $[(X_n^{(1)}, Y_n^{(1)}), (X_n^{(2)}, Y_n^{(2)}), ..., (X_n^{(M)}, Y_n^{(M)})]$, $m = 1, ..., M$. For each path, we estimate the parameters $\Theta = (a, b, \beta_0, \beta_1)$ and the predicted probabilities $\hat{p}_i (i = 1, ..., n)$ for GL^+, GL^-, logit models and also for the probit model. We compute the loss function, average mean square error of path m ($AMSE_m$), for each model as

$$AMSE_m = \frac{1}{n} \sum_{i=1}^{n} (Y_i - \hat{p}_i)^2 \qquad (27)$$

Compare with [7], and [3, p. 35]. For M paths, we obtain $AMSE_1$, $AMSE_2, ..., AMSE_M$ for each model and derive the expected $AMSE$ ($E(AMSE)$) approximated by

$$E(AMSE) = \frac{1}{M} \sum_{i=1}^{n} AMSE_i \qquad (28)$$

as $M \to \infty$. Compare with [3, p. 38]. The model that produces the smallest $E(AMSE)$ should be preferred.

Next, in the spirit of [2,20], we compute the sample correlation coefficient between Y_i and $\hat{p}_i$, $r_{Y_i \hat{p}_i}$ for each path given by

$$r_{Y_i \hat{p}_i} = \frac{\sum_{i=1}^{n} (Y_i - \bar{Y})(\hat{p}_i - \bar{p})}{\sum_{i=1}^{n} (Y_i - \bar{Y}) \sum_{i=1}^{n} (\hat{p}_i - \bar{p})} \qquad (29)$$

for each model. The $r_{Y_i\hat{p}_i}$ is a consistent estimator of the correlation between Y_i and $\hat{p}_i$, $Corr(Y_i, \hat{p}_i)$. Then the expected $r_{Y_i\hat{p}_i}$, $E(r_{Y_i\hat{p}_i})$, can be approximated by

$$E(r_{Y_i\hat{p}_i}) \approx \frac{1}{M} \sum_{i=1}^{M} (r_{Y_i\hat{p}_i})_i \tag{30}$$

The model that gives the greatest $E(r_{Y_i\hat{p}_i})$ is superior in terms of prediction as compared to the other competing models (see [2]) on the measures for assessing the goodness of fit of Logit models.

Table 1. The expected mean square errors $(E(AMSE))$ for 1000 paths for various sample sizes

Sample size	$E(AMSE)$			
	GL^+ model	GL^+ model	Logit model	Probit model
50	0.2388	0.4843	0.2394	0.3043
100	0.2450	0.4664	0.2453	0.3001
150	0.2466	0.4620	0.2468	0.2939
200	0.2473	0.4562	0.2474	0.2835
250	0.2479	0.4456	0.2480	0.2480
300	0.2482	0.4452	0.2483	0.2659
350	0.2485	0.4461	0.2486	0.2577
400	0.2487	0.4411	0.2488	0.2617
450	0.2489	0.4377	0.2489	0.2539
500	0.2490	0.4306	0.2491	0.2491
550	0.2490	0.4235	0.2490	0.2540
600	0.2491	0.4276	0.2492	0.2688
650	0.2492	0.4226	0.2492	0.2585
700	0.2493	0.4219	0.2493	0.2653
750	0.2493	0.4181	0.2493	0.2493

Notes: Some initial values chosen for the optimization routine resulted in objective function values undefined at the initial value. These are caused by outliers in the data generated. One way to handle this problem is to repeatedly change the initial values for that data set. Alternatively, the outlier can be removed and there would be no problem with the estimating of parameters of interest. For our data set, when we generate about 1000 paths, we get about 10% of the paths generated having extreme values. These paths we removed and the optimization routine proceeded without problem. [8, pp. 43–44] raised a similar problem in the estimation of conditional volatility.

For the simulations exercise, we choose 1000 paths ($M = 1000$) for the sample sizes: $n = 50, 100, 150, 200, 250, 300, 350, 400, 450, 500, 550, 600, 650,$ and 700.

The expected $AMSE$ and expected correlations estimates are presented in Tables 1 and 2, respectively. Figure 1 depicts the $E(AMSE)$ against the sample size for different combinations of the four models, whereas Fig. 2 shows the enlarged plot of the lower left plot of Fig. 1 (i.e. $EAMSE$ for GL^+ (solid line) and logit (dashed line)). Figure 3 shows the $E(r_{Y_i \hat{p}_i})$ for different combinations of all four models.

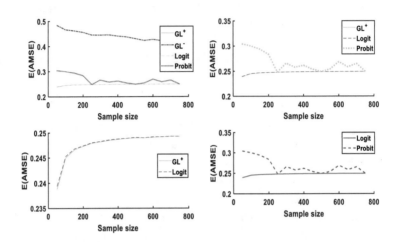

Fig. 1. $E(AMSE)$ for different combinations of all four models. Upper left: $E(AMSE)$ for GL^+ (solid line), GL^- (dashed line), Logit (dash-dot line) and Probit (dotted line). Upper right: $E(AMSE)$ for GL^+ (solid line), Logit (dashed line), and Probit (dotted line). Lower left: $E(AMSE)$ for GL^+ (solid line), Logit (dashed line). Lower right: $E(AMSE)$ for Logit (solid line), and Probit (dashed line).

It is clear from Table 1 that with data of size 50, 100, 150, 200, 250, 300, 350, 400, 500 and 600, the smallest $E(AMSE)$ was obtained with the GL^+ model. Observe that when the data size is 450, 550, 650, 700 and 750, the logit and the GL^+ models gave the same value. There is no case for which the $E(AMSE)$ of the Logit or was smaller than the GL^+. This can be seen in Fig. 2 where the Logit (dashed line) dominates the GL^+ (solid line). The GL^- model was worse among the four models comparing their $E(AMSE)$.

On the other hand, surprisingly, $E(r_{Y_i \hat{p}_i})$ is greatest for the GL^- model for data size 150, 200, 200, 300, 350, 400, 500, 550, 600, 650 and 700. This outcome suggests that, although the GL^- produce worse in-sample estimates, it is more likely to produce superior out-sample estimates as compared to the Logit and Probit models. For data size 750, the two models and are indistinguishable in terms of the $E(r_{Y_i \hat{p}_i})$. The Probit and Logit gave the same value of $E(r_{Y_i \hat{p}_i})$ for all data size except 150, 300, 350 and 450 for which the Probit was higher. For consistency in the estimation of the $E(AMSE)$ and $E(r_{Y_i \hat{p}_i})$, the GL^+ model is to be preferred among the four competing models.

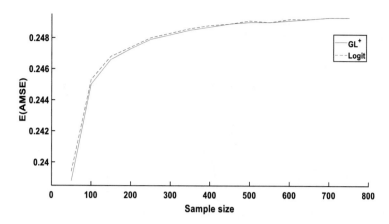

Fig. 2. $E(AMSE)$ for GL^+ (solid line) and Logit (dashed line) enlarged.

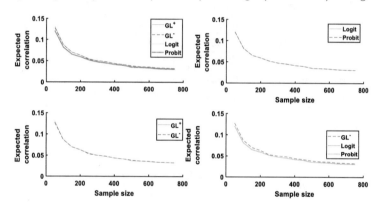

Fig. 3. $E(r_{Y_{\hat{p}}})$ for different combinations of all four models. Upper left: $E(r_{Y_{\hat{p}}})$ for GL^+ (dotted line), GL^- (dashed line), Logit (solid line), Probit (dotted line). Upper right: $E(r_{Y_{\hat{p}}})$ for Logit (dashed line), Probit (dashed line). Lower left: $E(r_{Y_{\hat{p}}})$ for GL^+ (solid line) and GL^- (dashed line). Lower right: $E(r_{Y_{\hat{p}}})$ for GL^+ (dotted line), Logit (solid line), Probit (dashed line)

Finally, we verify how in the information criteria, namely Akaike Information Criteria (AIC) and the Bayesian Information Criteria (BIC) perform under each of these models. For each path m, we compute the AIC as follows:

$$AIC_m(K) = log\left(\frac{SSR(K)}{n}\right) + \frac{2}{n}K \tag{31}$$

where K is the number of coefficients in the regression model including the constant term that has to be estimated and $SSR(K)$ is the sum of squares residuals of the regression model with K coefficients. See [20, p. 554], The BIC for path m is estimated in a similar fashion by

Table 2. The expected correlation between the predicted probabilities and the binary outcome variable for paths for various sample sizes

Sample size	$E(AMSE)$			
	GL^+ model	GL^+ model	Logit model	Probit model
50	0.1307	0.1273	0.1203	0.1203
100	0.0880	0.0872	0.0814	0.0814
150	0.0697	0.0699	0.0650	0.0651
200	0.0628	0.0630	0.0591	0.0591
250	0.0549	0.0533	0.0518	0.0518
300	0.0499	0.0500	0.0471	0.0472
350	0.0474	0.0480	0.0447	0.0448
400	0.0436	0.0438	0.0413	0.0413
450	0.0416	0.0415	0.0393	0.0394
500	0.0368	0.0382	0.0350	0.0350
550	0.0362	0.0364	0.0344	0.0344
600	0.0351	0.0358	0.0334	0.0334
650	0.0333	0.0337	0.0316	0.0316
700	0.0327	0.0331	0.0310	0.0310
750	0.0323	0.0323	0.0306	0.0306

Notes: Refer to Table 1 for notes.

$$BIC_m(K) = log\left(\frac{SSR(K)}{n}\right) + \frac{log(n)}{n}K \tag{32}$$

The expected AIC can be estimated by and

$$E(AIC) \approx \frac{1}{M}\sum_{i=1}^{M} AIC_i \tag{33}$$

Similarly, the expected BIC can also be estimated by

$$E(BIC) \approx \frac{1}{M}\sum_{i=1}^{M} BIC_i \tag{34}$$

The results are given in Tables 3 and 4. It can be seen from the Table 3 that GL^+ model is superior for the AIC for all the competing models. AIC is unable to distinguish any difference between the Logit and the Probit models. However, BIC elect the Probit as the superior model among the competing models with GL^+ in the second place.

Table 3. The expected AIC ($E(AIC)$) for 1000 paths for various sample sizes

Sample size	$E(AIC)$			
	GL^+ model	GL^+ model	Logit model	Probit model
50	−1.3532	−0.6635	−1.3506	−1.3506
100	−1.3667	−0.7419	−1.3655	−1.3655
150	−1.3734	−0.7661	−1.3727	−1.3727
200	−1.3773	−0.7854	−1.3768	−1.3768
250	−1.3788	−0.8194	−1.3784	−1.3784
300	−1.3801	−0.8197	−1.3798	−1.3798
350	−1.3810	−0.8176	−1.3807	−1.3807
400	−1.3815	−0.8330	−1.3813	−1.3813
450	−1.3819	−0.8425	−1.3817	−1.3817
500	−1.3822	−0.8653	−1.3821	−1.3821
550	−1.3830	−0.8795	−1.3828	−1.3828
600	−1.3831	−0.8700	−1.3830	−1.3830
650	−1.3833	−0.8840	−1.3832	−1.3832
700	−1.3835	−0.8865	−1.3834	−1.3834
750	−1.3838	−0.8977	−1.3836	−1.3836

Table 4. The expected BIC ($E(BIC)$) for 1000 paths for various sample sizes

Sample size	$E(BIC)$			
	GL^+ model	GL^+ model	Logit model	Probit model
50	−1.2767	−0.5870	−1.2741	−1.6170
100	−1.3146	−0.6898	−1.3134	−1.6082
150	−1.3332	−0.7259	−1.3325	−1.5870
200	−1.3443	−0.7524	−1.3438	−1.5400
250	−1.3506	−0.7912	−1.3502	−1.4946
300	−1.3554	−0.7950	−1.3551	−1.4513
350	−1.3589	−0.7955	−1.3586	−1.4090
400	−1.3616	−0.8130	−1.3613	−1.4321
450	−1.3636	−0.8242	−1.3634	−1.3907
500	−1.3654	−0.8484	−1.3652	−1.7830
550	−1.3673	−0.8638	−1.3672	−1.3945
600	−1.3685	−0.8554	−1.3683	−1.4764
650	−1.3696	−0.8702	−1.3694	−1.4201
700	−1.3705	−0.8735	−1.3704	−1.4581
750	−1.3714	−0.8854	−1.3713	−2.0596

4   Summary, Conclusion and Recommendations

The paper proposes two novel regressions, GL^+ and GL^- models, using the GL^+ function by [5,6], and GL^- function by [4]. The logit and probit models are used as benchmark models, where the GL^+ and GL^- models are compared using the following goodness of fit measures: average mean square errors, expected average mean square errors, the correlation between the predicted probabilities and the outcome variable, the expected correlations, the Akaike Information Criteria, the expected Akaike Information Criteria, the Bayesian Information Criteria and the expected Bayesian Information Criteria.

In the estimation exercise, we simulate 1000 paths with lengths 50, 100, 150, 200, 250, 300, 350, 400, 450, 500, 550, 600, 650, 700 and 750 for the binary outcome variable from a Bernoulli distribution and the independent variable from a mixture of ten distributions. For each of the four models, the parameters are estimated for all the 1000 paths for each of the above sample sizes. The average mean square error and the correlation measures are computed for each path. Further, the expected average mean square error and the expected correlation metrics are approximated by averaging the average mean square error and the correlation between the outcome variable and the predicted probabilities over the 1000 paths for each model. Similarly the expected Akaike Information Criteria and the expected Bayesian Information Criteria are approximated by averaging the Akaike Information Criteria and Bayesian Information Criteria over the 1000 paths for each model. The results are as follows. The expected average mean square error is smallest for the GL^+ model and the expected correlation between the outcome and the predicted probabilities is greatest for the GL^- model. Akaike Information Criteria is smallest for the GL^+ distribution than the all four models and place second among the four models with the Bayesian Information Criteria. Hence, allowing the data to determine its shape and skewness cannot be ignored in the data generating process of binary output variable. Thus asymmetries in the independent variable do matter in the modelling of regression models for which the dependent variable is binary. As the data size increases from 750 and beyond, the probit, logit and GL^+ are indistinguishable in its expected average mean square estimation.

The GL^+ and GL^- models are motivated relative to Probit and Logit regression, but there are variety of alternative link functions including methods for estimating the link function non-parametrically. It is unclear how these models competes with these alternative link functions and will be an interesting area to do further research.

References

1. Allison, P.D., Christakis, N.A.: Logit models for sets of ranked items. In: Sociological Methodology, pp. 199–228 (1994)
2. Amemiya, T.: Qualitative response models: a survey. J. Econ. Lit. **19**(4), 1483–1536 (1981)
3. Andoh, C.: Garch family models under varying innovations (2010)

4. Andoh, C.: Stochastic variance models in discrete time with feed forward neural networks. Neural Comput. **21**(7), 1990–2008 (2009)
5. Balakrishnan, N.: Handbook of the Logistic Distribution. Marcel Dekker, New York (1992)
6. Balakrishnan, N., Leung, M.Y.: Order statistics from the type I generalized logistic distribution. Commun. Stat. Simul. Comput. **17**(1), 25–50 (1988)
7. Dougherty, C.: Introduction to Econometrics. Oxford University Press, Oxford (2007)
8. Danielsson, J.: Financial Risk Forecasting: The Theory and Practice of Forecasting Market Risk with Implementation in R and MATLAB. Wiley, Hoboken (2011)
9. Embrechts, P., McNeil, A., Straumann, D.: Correlation and dependence in risk management: properties and pitfalls. In: Risk Management: Value at Risk and Beyond, pp. 176–223 (2002)
10. Gao, Y., Rasouli, S., Timmermans, H., Wang, Y.: Reasons for not buying a car: a probit-selection multinomial logit choice model. Procedia Environ. Sci. **22**, 414–422 (2014)
11. Horioka, C.Y.: Tenure choice and housing demand in Japan. J. Urban Econ. **24**(3), 289–309 (1988)
12. Hosmer Jr., D.W., Lemeshow, S., Sturdivant, R.X.: Applied Logistic Regression, vol. 398. Wiley, Hoboken (2013)
13. Klugman, S.A., Panjer, H.H., Willmot, G.E.: Loss Models: From Data to Decisions, vol. 715. Wiley, New York (2012)
14. Lehmann, E.L., Casella, G.: Theory of Point Estimation. Springer, New York (2006)
15. Ou, S., Reynolds, A.J.: Early childhood intervention and educational attainment: age 22 findings from the Chicago longitudinal study. J. Educ. Stud. Placed Risk **11**(2), 175–198 (2006)
16. Pampel, F.C.: Logistic Regression: A Primer, vol. 132. Sage Publications, Thousand Oaks (2000)
17. Pardoe, I., Simonton, D.K.: Applying discrete choice models to predict academy award winners. J. R. Stat. Soc. Ser. A (Stat. Soc.) **171**(2), 375–394 (2008)
18. Pregibon, D.: Logistic regression diagnostics. Ann. Stat. **9**(4), 705–724 (1981)
19. Rabe, B., Taylor, M.: Residential mobility, quality of neighbourhood and life course events. J. R. Stat. Soc. Ser. A (Stat. Soc.) **173**(3), 531–555 (2010)
20. Stock, J.H., Watson, M.W.: Introduction to Econometrics. Addison-Wiley, Boston (2007)
21. Swaminathan, H., Rogers, H.J.: Detecting differential item functioning using logistic regression procedures. J. Educ. Meas. **27**(4), 361–370 (1990)
22. Van de Ven, W.P., Van Praag, B.M.: The demand for deductibles in private health insurance: a probit model with sample selection. J. Econom. **17**(2), 229–252 (1981)
23. Westgaard, S., Van der Wijst, N.: Default probabilities in a corporate bank portfolio: a logistic model approach. Eur. J. Oper. Res. **135**(2), 338–349 (2001)
24. Zmijewski, M.E.: Methodological issues related to the estimation of financial distress prediction models. J. Account. Res. **22**, 59–82 (1984)

What If We Do Not Know Correlations?

Michael Beer[1,2], Zitong Gong[3], Ingo Neumann[4], Songsak Sriboonchitta[5], and Vladik Kreinovich[6(✉)]

[1] Institute for Risk and Reliability, Leibniz University of Hannover,
Callinstraße 34, 30167 Hannover, Germany
beer@irz.uni-hannover.de
[2] Institute for Risk and Uncertainty, University of Liverpool,
Liverpool L69 3BX, UK
[3] School of Engineering, Institute for Risk and Uncertainty, University of Liverpool,
Liverpool L69 3BX, UK
Zitong.Gong@liverpool.ac.uk
[4] Geodetic Institute, Leibniz University of Hannover,
Nienburger Strasse 1, 30167 Hannover, Germany
neumann@gih.uni-hannover.de
[5] Faculty of Economics, Chiang Mai University, Chiang Mai 50200, Thailand
songsakecon@gmail.com
[6] University of Texas at El Paso, 500 W. University, El Paso, TX 79968, USA
vladik@utep.edu

Abstract. It is well know how to estimate the uncertainty of the result y of data processing if we know the correlations between all the inputs. Sometimes, however, we have no information about the correlations. In this case, instead of a single value σ of the standard deviation of the result, we get a range $[\underline{\sigma}, \overline{\sigma}]$ of possible values. In this paper, we show how to compute this range.

1 Formulation of the Problem

Need for data processing. In many real-life situations, we are interesting in quantities y which are difficult (or even impossible) to measure directly. For example, we may be interested in the distance to a faraway star or in the amount of oil in a given oil field. Since we cannot measure y directly, a natural idea is to measure it *indirectly*, i.e., to find easier-to-measure quantities $x_1, \ldots, x_n$ which are connected to y by a known algorithm $y = f(x_1, \ldots, x_n)$, and use the results $\widetilde{x}_i$ of measuring x_i to estimate y as $\widetilde{y} = f(\widetilde{x}_1, \ldots, \widetilde{x}_n)$; see, e.g., [4].

What is the accuracy of the resulting estimate? The results $\widetilde{x}_i$ of measuring x_i are, in general, different from the actual values of the measured quantities. In other words, there is a usually a measurement error $\Delta x_i \stackrel{\text{def}}{=} \widetilde{x}_i - x_i$, so that $x_i = \widetilde{x}_i - \Delta x_i$.

As a result, the estimate $\widetilde{y} = f(\widetilde{x}_1, \ldots, \widetilde{x}_n)$ is also, in general, different from the actual value $y = f(x_1, \ldots, x_n) = f(\widetilde{x}_1 - \Delta x_1, \ldots, \widetilde{x}_n - \Delta x_n)$. It is therefore desirable to estimate the error $\Delta y \stackrel{\text{def}}{=} \widetilde{y} - y$ of the indirect measurement.

© Springer International Publishing AG 2018
L. H. Anh et al. (eds.), *Econometrics for Financial Applications*, Studies in Computational Intelligence 760, https://doi.org/10.1007/978-3-319-73150-6_5

Measurement errors are usually relatively small. In most real-life situations, the measurement errors are relatively small. As a result, we can safely ignore terms which are quadratic (or of higher order) with respect to Δx_i. For example, if the measurement error is 10%, its square is 1%, which is much smaller.

So, we can expand the expression

$$\Delta y = \widetilde{y} - y = f(\widetilde{x}_1, \ldots, \widetilde{x}_n) - f(\widetilde{x}_1 - \Delta x_1, \ldots, \widetilde{x}_n - \Delta x_n)$$

in Taylor series in Δx_i and keep only linear terms in this expansion. As a result, we get a formula

$$\Delta y = \sum_{i=1}^{n} c_i \cdot \Delta x_i, \tag{1}$$

where $c_i \stackrel{\text{def}}{=} \dfrac{\partial f}{\partial x_i}_{|(\widetilde{x}_1, \ldots, \widetilde{x}_n)}$.

What do we know about Δx_i. In the ideal case, for each measuring instrument, we know the first two moments of the measurement errors, i.e., equivalently, we know the mean value μ_i of the corresponding measurement error Δx_i, and we know the standard deviation σ_i.

If we know the exact mean, then we can re-calibrate the i-th measuring instrument by subtracting μ_i from all the measurement results. In this case, we get the mean value equal to 0.

Sometimes, we only know the mean and the standard deviation with some uncertainty, i.e., we only know the bounds $\underline{\mu}_i \le \mu_i \le \overline{\mu}_i$ and $\underline{\sigma}_i \le \sigma_i \le \overline{\sigma}_i$; see, e.g., [1–3].

Based on this information, we can estimate the mean value μ of Δy. Based on this information, we can estimate the mean μ of the desired measurement error. Namely, from (1), it follows that

$$\mu = \sum_{i=1}^{n} c_i \cdot \mu_i. \tag{2}$$

So, if we know the exact values of means μ_i, we can use the formula (2) to find μ.

If μ_i are only known with interval uncertainty, then we can represent the interval $[\underline{\mu}_i, \overline{\mu}_i]$ in the centered form $[\widetilde{\mu}_i - \Delta_i, \widetilde{\mu}_i + \Delta_i]$, where $\widetilde{\mu}_i \stackrel{\text{def}}{=} \dfrac{\underline{\mu}_i + \overline{\mu}_i}{2}$ and $\Delta_i \stackrel{\text{def}}{=} \dfrac{\overline{\mu}_i - \underline{\mu}_i}{2}$. In this representation, each value $\mu_i \in [\underline{\mu}_i, \overline{\mu}_i] = [\widetilde{\mu}_i - \Delta_i, \widetilde{\mu}_i + \Delta_i]$ can be represented as $\widetilde{\mu}_i + \Delta\mu_i$, where $\Delta\mu_i \stackrel{\text{def}}{=} \mu_i - \widetilde{\mu}_i$ takes values from the interval $[-\Delta_i, \Delta_i]$. Substituting the expression $\mu_i = \widetilde{\mu}_i + \Delta\mu_i$ into the formula (2), we conclude that $\mu = \widetilde{\mu} + \Delta\mu$, where

$$\widetilde{\mu} \stackrel{\text{def}}{=} \sum_{i=1}^{n} c_i \cdot \widetilde{\mu}_i \tag{3}$$

and

$$\Delta\mu \overset{\text{def}}{=} \sum_{i=1}^{n} c_i \cdot \Delta\mu_i.$$

The largest value of $\Delta\mu$ is attained when each of the terms $c_i \cdot \Delta\mu_i$ is the largest. For $c_i > 0$, this happens when $\Delta\mu_i$ is the largest, i.e., when $\Delta\mu_i = \Delta_i$. For $c_i \leq 0$, this happens when $\Delta\mu_i$ is the smallest, i.e., when $\Delta\mu_i = -\Delta_i$. In both cases, the largest value of $c_i \cdot \Delta\mu_i$ is equal to $|c_i| \cdot \Delta_i$. Similarly, the smallest possible value of $c_i \cdot \Delta\mu_i$ is equal to $-|c_i| \cdot \Delta_i$. Thus, we conclude that

$$\mu \in [\tilde{\mu} - \Delta, \tilde{\mu} + \Delta], \tag{4}$$

where

$$\Delta \overset{\text{def}}{=} \sum_{i=1}^{n} |c_i| \cdot \Delta_i. \tag{5}$$

What is the standard deviation σ of Δy: case when we know the correlations. To complete our description of the uncertainty Δy, we need to also estimate its standard deviation σ, i.e., equivalently, the variance $V = \sigma^2 = E[(\delta y)^2]$, where we denoted

$$\delta y \overset{\text{def}}{=} \Delta y - E[\Delta y] = \Delta y - \mu.$$

Subtracting (2) from (1), we conclude that

$$\delta y = \sum_{i=1}^{n} c_i \cdot \delta x_i, \tag{6}$$

where we denoted $\delta x_i \overset{\text{def}}{=} \Delta x_i - E[\Delta x_i] = \Delta x_i - \mu_i$. Substituting the expression (6) into the formula for the variance $\sigma^2 = E[(\delta y)^2]$ and taking into account that the mean of the linear combination is equal to the linear combination of the means, we conclude that

$$E[(\delta y)^2] = \sum_{i=1}^{n} \sum_{j=1}^{n} c_i \cdot c_j \cdot E[\delta x_i \cdot \delta x_j]. \tag{7}$$

For $i = j$, we get $E[(\delta x_i)^2] = \sigma_i^2$. For $i \neq j$, by definition of the correlation ρ_{ij}, we have $\rho_{ij} = \dfrac{E[\delta x_i \cdot \delta x_j]}{\sigma_i \cdot \sigma_j}$, thus $E[\delta x_i \cdot \delta x_j] = \rho_{ij} \cdot \sigma_i \cdot \sigma_j$, and the formula (7) takes the form

$$\sigma^2 = \sum_{i=1}^{n} c_i^2 \cdot \sigma_i^2 + \sum_{i \neq j} \rho_{ij} \cdot c_i \cdot c_j \cdot \sigma_i \cdot \sigma_j. \tag{8}$$

So, if we know all the correlations ρ_{ij}, we can use the formula (8) to estimate the desired standard deviation σ of the result y of data processing [4,5].

But what if we do not know the correlations? In some practical situations, however, we do not know the correlations. In this case, depending on the actual values of the correlations, we get different values σ. What is the range of possible values σ? This is the question that we answer in this paper.

2   Main Result: Formulation and Proofs

First result. Our first result is that if we know the exact values of the standard deviations σ_i, but we have no information about the correlations, then the range of possible values of σ is equal to $[\underline{\sigma}, \overline{\sigma}]$, where

$$\overline{\sigma} = \sum_{i=1}^{n} |c_i| \cdot \sigma_i, \tag{9}$$

and

$$\underline{\sigma} = \max\left(0, |c_{i_0}| \cdot \sigma_{i_0} - \sum_{i \neq i_0}^{n} |c_i| \cdot \sigma_i\right), \tag{10}$$

where i_0 is the index for which

$$|c_{i_0}| \cdot \sigma_{i_0} = \max_i |c_i| \cdot \sigma_i.$$

Comment. It should be noticed that the formula (9) for the upper bound $\overline{\sigma}$ of the standard deviation is, somewhat surprisingly, very similar to the formula (5) for the upper bound on the mean μ.

Proof

$1°$. It is well known that for every two random variables a and b, we have

$$\sigma^2[a + b] = v^2[u] + v^2[b] + \rho_{ab} \cdot \sigma[u] \cdot \sigma[b].$$

Since the correlation coefficient ρ_{ab} is always bounded by 1, we conclude that

$$\sigma^2[a + b] \leq \sigma^2[a] + \sigma^2[b] + 2\sigma[a] \cdot \sigma[b].$$

The right-hand side of this inequality is $(\sigma[a] + \sigma[b])^2$, thus we conclude that

$$\sigma[a + b] \leq \sigma[a] + \sigma[b].$$

In particular, for $a - b$ and b, we thus get $\sigma[a] \leq \sigma[a - b] + \sigma[b]$, hence

$$\sigma[a - b] \geq \sigma[a] - \sigma[b].$$

Let us apply these inequalities to our case.

$2°$. The overall random component $\delta y = \Delta y - E[\Delta y]$ of the measurement error Δy is the sum of n terms $c_i \cdot \delta x_i$. For each term $c_i \cdot \delta x_i$, the standard deviation is $|c_i| \cdot \sigma_i$. Thus, we can conclude that the standard deviation σ of the sum δy of these terms does not exceed the sum of standard deviations, i.e., that $\sigma \leq \sum_{i=1}^{n} |c_i| \cdot \sigma_i$.

Alternatively, we can represent δy as the difference $\delta y = c_{i_0} \cdot \delta x_{i_0} - s$, where $s \overset{\text{def}}{=} \sum_{i \neq i_0} (-c_i) \cdot \delta x_i$. Thus, by using the formula for the standard deviation of the difference, we get $\sigma \geq |c_{i_0}| \cdot \sigma[s]$. By using the formula for the standard deviation of the sum, we conclude that $\sigma[s] \leq \sum_{i \neq i_0} |c_i| \cdot \sigma_i$. Thus, we have

$$\sigma \geq |c_{i_0}| \cdot \sigma_{i_0} - \sum_{i \neq i_0} |c_i| \cdot \sigma_i.$$

Clearly also $\sigma \geq 0$, so

$$\sigma \geq \max \left(|c_{i_0}| \cdot \sigma_{i_0} - \sum_{i \neq i_0} |c_i| \cdot \sigma_i \right).$$

So, we proved that for the above expressions (9) and (10) for $\underline{\sigma}$ and $\overline{\sigma}$, we always have

$$\underline{\sigma} \leq \sigma \leq \overline{\sigma}.$$

To complete our proof, it is now sufficient to prove that the values $\underline{\sigma}$ and $\overline{\sigma}$ (described by the formulas (1) and (9)) are attainable for some random variables with given values σ_i.

$3°$. Let us first prove that the upper bound $\overline{\sigma}$ is attainable. Indeed, let η be a standard normally distributed random variable, with 0 mean and standard deviation 1. Then, we can take $\delta x_i = \text{sign}(c_i) \cdot \sigma_i \cdot \eta$, where $\text{sign}(x) \overset{\text{def}}{=} 1$ for $x \geq 0$ and $\text{sign}(x) \overset{\text{def}}{=} -1$ for $x < 0$. Due to this definition, we have $\text{sign}(x) \cdot x = |x|$ for all x.

For this selection, we have

$$\delta y = \sum_{i=1}^{n} c_i \cdot \delta_i = \sum_{i=1}^{n} c_i \cdot \text{sign}(c_i) \cdot \sigma_i \cdot \eta = \sum_{i=1}^{n} |c_i| \cdot \sigma_i \cdot \eta = \left(\sum_{i=1}^{n} |c_i| \cdot \sigma_i \right) \cdot \eta.$$

This sum has the desired standard deviation $\sum_{i=1}^{n} |c_i| \cdot \sigma_i$.

$4°$. Let us now prove that the lower bound is also attainable. We will first prove it for the case when the difference $d \overset{\text{def}}{=} |c_{i_0}| \cdot \sigma_{i_0} - \sum_{i \neq i_0} |c_i| \cdot \sigma_i$ is positive. In this case,

$$\underline{\sigma} = d.$$

To find a sum with this standard deviation, let us take $\delta x_{i_0} = \text{sign}(c_{i_0}) \cdot \sigma_{i_0} \cdot \eta$ and $\delta x_i = -\text{sign}(c_i) \cdot \sigma_i \cdot \eta$ for all $i \neq i_0$. In this case,

$$\delta y = c_{i_0} \cdot \delta x_{i_0} + \sum_{i \neq i_0} c_i \cdot \delta x_i = |c_{i_0}| \cdot \sigma_{i_0} \cdot \eta - \sum_{i \neq i_0} |c_i| \cdot \sigma_i \cdot \eta$$

$$= \left(|c_{i_0}| \cdot \sigma_{i_0} \eta - \sum_{i \neq i_0} |c_i| \cdot \sigma_i \right) \cdot \eta = d \cdot \eta.$$

Since $d > 0$, this sum has standard deviation $d = \underline{\sigma}$.

5°. To finalize the proof, we need to show that when $d < 0$, the sum Δy can have zero standard deviation.

5.1°. To prove this fact, let us prove, by induction over m, the following auxiliary result: when $a_1 \leq \ldots \leq a_m$, then for every number a from $\max\left(0, a_m - \sum_{i=1}^{m-1} a_i\right)$ and $\sum_{i=1}^{m} a_i$, there exist planar vectors A_i for which $|A_i| = a_i$ for all i and $\left|\sum_{i=1}^{m} A_i\right| = a$.

The base case $m = 2$ is straightforward. Indeed, in this case, the desired inequality takes the form $a_2 - a_1 \leq a \leq a_2 + a_1$. To get a vector A with $|A| = a_1 + a_2$, we simply take A_1 and A_2 parallel and going in the same direction. To get a vector A with $|A| = a_2 - a_1$, we take A_1 and A_2 parallel but going in different directions. By a continuous transformation of one configuration into another, we get cases with all intermediate values a.

Let us now describe the induction step. Suppose that we have already proved this result for m, we want to prove it for $m+1$. The value $a = a_1 + \ldots + a_m + a_{m+1}$ is easy to obtain: it is sufficient to take vectors A_i all parallel and all going in the same direction. If $a_{m+1} > a_1 + \ldots + a_m$, then the value $a = a_{m+1} - \sum_{i=1}^{m} a_i$ is also easy to obtain: we take all the vector parallel, the first m vectors $A_1, \ldots, A_m$ go in one direction, and the vector A_{m+1} goes in the opposite direction.

To complete the proof of induction step, we need to consider the case when $a_{m+1} < a_1 + \ldots + a_m$. In this case, we want to find the vectors for which the sum is 0. By induction assumption, for the sum $A_1 + \ldots + A_m$, any length from

$$\max(0, a_m - (a_1 + \ldots + a_{m-1}))$$

to $a_1 + \ldots + a_m$ is possible. Here, $a_{m+1} < a_1 + \ldots + a_m$, since this is the case that we are considering. Also, $a_{m+1} \geq 0$ and $a_{m+1} \geq a_m$ hence $a_{m+1} \geq a_m - \sum_{i=1}^{m-1} a_i$ and thus $a_{m+1} \geq \max\left(0, a_m - \sum_{i=1}^{m-1} a_i\right)$. So, by induction assumption, there exist vectors $A_1, \ldots, A_m$ for which $|A_1 + \ldots + A_m| = a_{m+1}$. Now, if we take

$$A_{m+1} = -(A_1 + \ldots + A_m),$$

we get $|A_{m+1}| = a_{m+1}$ and $A_1 + \ldots + A_m + A_{m+1} = 0$. The auxiliary statement is proven.

5.2°. The above statement implies that when a_{i_0} is larger than or equal to all the values a_i and $a_{i_0} \le \sum\limits_{i \neq i_0} a_i$, then there exist planar vectors A_i of lengths $|A_i| = a_i$ for which $\sum\limits_i A_i = 0$.

Let us take such vectors A_i corresponding to $a_i = |c_i| \cdot \sigma_i$. Let us select two independent standard normally distributed random variables η' and η'', with 0 mean and standard deviation 1, and assign, to each planar vector A with coordinates $A = (A', A'')$, a random variable $\eta_A \overset{\text{def}}{=} A' \cdot \eta' + A'' \cdot \eta''$. One can easily check that the variance of the resulting random variable is equal to $(A')^2 + (A'')^2$, i.e., to the square of the length of the original vector A. Thus, the standard deviation of the random variable η_A is equal to the length $|A|$ of the vector A.

It is also easy to check that the transformation $A \to \eta_A$ from vectors to random variables is linear: $\eta_{c_A \cdot A + \ldots + c_B \cdot B} = c_A \cdot \eta_A + \ldots + c_B \cdot \eta_B$ for all vectors $A, \ldots, B$ and for all values $c_A, \ldots, c_B$.

We can then take for each i, as δx_i, the random variable corresponding to the vector $\dfrac{A_i}{c_i}$. This variable has standard deviation $\left|\dfrac{A_i}{c_i}\right| = \dfrac{|A_i|}{|c_i|} = \dfrac{|c_i| \cdot \sigma_i}{|c_i|} = \sigma_i$.

Here, $c_i \cdot \delta x_i = \eta_{A_i}$. Thus, for the sum $\delta y = \sum\limits_{i=1}^{n} c_i \cdot \delta x_i$, we have

$$\delta y = \sum_{i=1}^{n} c_i \cdot \delta x_i = \sum_{i=1}^{n} \eta_{A_i} = \eta_{\sum\limits_{i=1}^{n} A_i} = \eta_0 = 0.$$

The statement is proven, and so is our first result.

Second result. If we only know the bounds $\underline{\sigma}_i$ and $\overline{\sigma}_i$ on the standard deviations, then the range of possible values of σ is equal to $[\underline{\sigma}, \overline{\sigma}]$, where

$$\overline{\sigma} = \sum_{i=1}^{n} |c_i| \cdot \overline{\sigma}_i, \tag{11}$$

and

$$\underline{\sigma} = \max\left(0, |c_{i_0}| \cdot \underline{\sigma}_{i_0} - \sum_{i \neq i_0} |c_i| \cdot \overline{\sigma}_i\right), \tag{12}$$

where i_0 is the index for which the product $|c_{i_0}| \cdot \underline{\sigma}_{i_0}$ is the largest; if there are several such indices i_0, then we select the one for which the product $|c_{i_0}| \cdot \overline{\sigma}_{i_0}$ is the smallest.

Proof is straightforward: e.g., for the upper bound, from the fact that for all possible values σ_i, we get $\sigma \le \sum\limits_{i=1}^{n} |c_i| \cdot \sigma_i$ and that $\sigma_i \le \overline{\sigma}_i$, we conclude that $\sigma \le \sum\limits_{i=1}^{n} |c_i| \cdot \overline{\sigma}_i$. Vice versa, by taking $\sigma_i = \overline{\sigma}_i$ in the example from the proof of the previous result, we get an example when σ is equal to the upper bound $\sum_{i=1}^{n} |c_i| \cdot \sigma_i$.

To get a similar example for the lower bound, we should take $\sigma_{i_0} = \underline{\sigma}_{i_0}$ and $\sigma_i = \overline{\sigma}_i$ for all $i \neq i_0$.

Acknowledgments. We acknowledge the partial support of the Center of Excellence in Econometrics, Faculty of Economics, Chiang Mai University, Thailand. This work was performed when Vladik was a visiting researcher with the Geodetic Institute of the Leibniz University of Hannover, a visit supported by the German Science Foundation.

This work was also supported in part by the US National Science Foundation grant HRD-1242122.

References

1. Jaulin, L., Kiefer, M., Dicrit, O., Walter, E.: Applied Interval Analysis. Springer, London (2001)
2. Moore, R.E., Kearfott, R.B., Cloud, M.J.: Introduction to Interval Analysis. SIAM, Philadelphia (2009)
3. Nguyen, H.T., Kreinovich, V., Wu, B., Xiang, G.: Computing Statistics Under Interval and Fuzzy Uncertainty. Springer Verlag, Heidelberg (2012)
4. Rabinovich, S.G.: Measurement Errors and Uncertainty: Theory and Practice. Springer Verlag, Heidelberg (2005)
5. Sheskin, D.J.: Handbook of Parametric and Nonparametric Statistical Procedures. Chapman and Hall/CRC, Boca Raton (2011)

Markowitz Portfolio Theory Helps Decrease Medicines' Side Effect and Speed up Machine Learning

Thongchai Dumrongpokaphan[1] and Vladik Kreinovich[2(⊠)]

[1] Faculty of Science, Chiang Mai University, Chiang Mai, Thailand
tcd43@hotmail.com
[2] Department of Computer Science, University of Texas at El Paso,
500 W. University, El Paso, TX 79968, USA
vladik@utep.edu

Abstract. In this paper, we show that, similarly to the fact that distributing the investment between several independent financial instruments decreases the investment risk, using a combination of several medicines can decrease the medicines' side effects. Moreover, the formulas for optimal combinations of medicine are the same as the formulas for the optimal portfolio, formulas first derived by the Nobel-prize winning economist H. M. Markowitz. A similar application to machine learning explains a recent success of a modified neural network in which the input neurons are also directly connected to the output ones.

1 Markowitz Portfolio Theory: A Brief Reminder

The main idea behind Markowitz portfolio theory. In his Nobel-prize winning paper [5], Markowitz proposed a method for selecting an optimal portfolio of financial investments.

To be explain the main ideas behind his method, let us start with a simple case when we have n independent financial instrument, each with a known expected return-on-investment μ_i and a known standard deviation σ_i. In principle, we can combine these portfolios, by allocating the part w_i of our investment amount to the i-th instrument. Here, we have $w_i \geq 0$ and $\sum_{i=1}^{n} w_i = 1$.

For each of these portfolios, we can determine the expected return on investment μ and the standard deviation σ from the formulas

$$\mu = \sum_{i=1}^{n} w_i \cdot \mu_i \text{ and } \sigma^2 = \sum_{i=1}^{n} w_i^2 \cdot \sigma_i^2.$$

Some of such portfolios are less risky – i.e., have smaller standard deviation – but have a smaller expected return on investment. Other portfolios have a larger expected return on investment but are more risky.

We can therefore formulate two possible problems:

© Springer International Publishing AG 2018
L. H. Anh et al. (eds.), *Econometrics for Financial Applications*, Studies in Computational Intelligence 760, https://doi.org/10.1007/978-3-319-73150-6_6

- The first problem is when we want to achieve a certain expected return on investment μ. Out of all possible portfolios that provide such expected return on investment, we want to find the portfolio for which the risk σ is the smallest possible.
- The second problem is when we know the maximum amount of risk σ that we can tolerate. There are several different portfolios that provide the allowed of risk. Out of all such portfolios, we would like to select the one that provides the largest possible return on investment.

Example. Let us consider the simplest case, when all n instruments have the same expected return on investment $\mu_1 = \ldots = \mu_n$ and the same standard deviation $\sigma_1 = \ldots = \sigma_n$. In this case, the problem is completely symmetric with respect to permutations, and thus, the optimal portfolio should be symmetric too. Therefore, all the parts must be the same: $w_1 = \ldots = w_n$. Since $\sum_{i=1}^{n} w_i = 1$, this implies that $w_1 = \ldots = w_n = \dfrac{1}{n}$. For these values w_i, the expected return on investment is equal to the same value as for each instrument $\mu = \mu_1$, but the risk decreases:

$$\sigma^2 = \sum_{i=1}^{n} w_i^2 \cdot \sigma_i^2 = n \cdot \frac{1}{n^2} \cdot \sigma_1^2 = \frac{1}{n} \cdot \sigma_1^2,$$

hence $\sigma = \dfrac{\sigma_1}{\sqrt{n}}$.

What we can conclude from this example. A natural conclusion is that if we diversify our portfolio, i.e., if we divide our investment amount between different independent financial instruments, then we can drastically decrease the corresponding risk.

A similar idea works well in measurement. If we have n results $x_1, \ldots, x_n$ of measuring the same quantity x, with measurement error $x_i - x$ with mean 0 and standard deviation σ_i, and if the measurement errors corresponding to different measurements are independent, then we can decrease the estimation error if,

- instead of the original estimates x_i for the quantity x,
- we use their weighted average $\widetilde{x} = \sum_{i=1}^{n} w_i \cdot x_i$, for some weights $w_i \geq 0$ for which

$$\sum_{i=1}^{n} w_i = 1;$$

see, e.g., [6].

In this case, the standard deviation of the estimate $\widetilde{x}$ is equal to

$$\sigma^2 = \sum_{i=1}^{n} w_i^2 \cdot \sigma_i^2.$$

We want to find the weights w_i that minimize σ^2 under the given constraint $\sum_{i=1}^{n} w_i = 1$. By using the Lagrange multiplier method, we can reduce this constraint optimization problem to the following unconstrained optimization problem:

$$\sum_{i=1}^{n} w_i^2 \cdot \sigma_i^2 + \lambda \cdot \left(\sum_{i=1}^{n} w_i - 1 \right) \to \min_i .$$

Differentiating the resulting objective function with respect to w_i and equating the derivative to 0, we conclude that $2w_i \cdot \sigma_i^2 + \lambda = 0$, thus, $w_i = c \cdot \sigma_i^{-1}$, for some constant $c \overset{\text{def}}{=} -\dfrac{\lambda}{2}$. This constant c can be found from the condition that $\sum_{i=1}^{n} w_i = 1$: we get $c = \dfrac{1}{\sum_{j=1}^{n} \sigma_j^{-2}}$ and thus,

$$w_i = \frac{\sigma_i^{-2}}{\sum_{j=1}^{n} \sigma_j^{-2}} .$$

For these weights, we get

$$\sigma^2 = \sum_{i=1}^{n} w_i^2 \cdot \sigma_i^2 = \sum_{i=1}^{n} \frac{\sigma_i^{-4}}{\left(\sum_{j=1}^{n} \sigma_j^{-2} \right)^2} \cdot \sigma_i^2 = \sum_{i=1}^{n} \frac{\sigma_i^{-2}}{\left(\sum_{j=1}^{n} \sigma_j^{-2} \right)^2}$$

$$= \frac{\sum_{i=1}^{n} \sigma_j^{-2}}{\left(\sum_{j=1}^{n} \sigma_j^{-2} \right)^2} = \frac{1}{\sum_{j=1}^{n} \sigma_j^{-2}} .$$

The sum $\sum_{j=1}^{n} \sigma_j^{-2}$ is larger than each of its terms σ_j^{-2}, and thus, the inverse σ^2 of this sum is smaller than each of the inverses σ_j^2. So, combining measurement results indeed decreases the approximation error.

In particular, when all measurements are equally accurate, i.e., when $\sigma_1 = \ldots = \sigma_n$, we get $\sigma = \dfrac{\sigma}{\sqrt{n}}$.

Optimal portfolio when different instruments are independent. In the previous text, we considered the case when different financial instruments are independent and identical. Let us now consider a more general case, when we still assume that the financial instruments are independent, but we take into account that these instrument are, in general, different, i.e., they have individual values μ_i and σ_i.

In this case, the first portfolio optimization problem takes the following form: minimize

$$\sum_{i=1}^{n} w_i^2 \cdot \sigma_i^2$$

under the constraints

$$\sum_{i=1}^{n} w_i \cdot \mu_i = \mu \text{ and } \sum_{i=1}^{n} w_i = 1.$$

For this problem, Lagrange multiplier methods leads to minimizing the expression

$$\sum_{i=1}^{n} w_i^2 \cdot \sigma_i^2 + \lambda \cdot \left(\sum_{i=1}^{n} w_i \cdot \mu_i - \mu \right) + \lambda' \cdot \left(\sum_{i=1}^{n} w_i - 1 \right).$$

Differentiating this expression with respect to w_i and equating the derivative to 0, we conclude that

$$2 w_i \cdot \sigma_i^2 + \lambda \cdot \mu_i + \lambda' = 0,$$

i.e., that

$$w_i = a \cdot (\mu_i \cdot \sigma_i^{-2}) + b \cdot \sigma_i^{-2},$$

where $a \overset{\text{def}}{=} -\dfrac{\lambda}{2}$ and $b \overset{\text{def}}{=} -\dfrac{\lambda'}{2}$. For these values w_i, the constraints $\sum_{i=1}^{n} w_i \cdot \mu_i = \mu$ and $\sum_{i=1}^{n} w_i = 1$ take the form

$$a \cdot \Sigma_2 + b \cdot \Sigma_1 = \mu; \text{ and } a \cdot \Sigma_1 + b \cdot \Sigma_0 = 1,$$

where we denoted $\Sigma_k \overset{\text{def}}{=} \sum_{i=1}^{n} (\mu_i)^k \cdot \sigma_i^{-2}$. Thus,

$$a = \frac{\Sigma_1 - \mu \cdot \Sigma_0}{\Sigma_1^2 - \Sigma_0 \cdot \Sigma_2} \text{ and } b = \frac{\mu \cdot \Sigma_1 - \Sigma_2}{\Sigma_1^2 - \Sigma_0 \cdot \Sigma_2}.$$

General case. In general, we may have correlations ρ_{ij} between different financial instruments. In this case, the standard deviation of the weighted combination has the form

$$\sum_{i=1}^{n} w_i^2 \cdot \sigma_i^2 + \sum_{i \neq j} \rho_{ij} \cdot w_i \cdot w_j \cdot \sigma_i \cdot \sigma_j.$$

This is a quadratic function, thus the Lagrange multiplier form is also quadratic, and after differentiating it and equating the derivatives to 0 we get an easy-to-solve system of linear equations.

2 How Markowitz Portfolio Theory Can Be Applied To Medicine

Formulation of the problem in informal terms. In medicine, usually, for each disease, we have several possible medicines. All these medicines are usually reasonable effective – otherwise they would not have been approved by the corresponding regulatory agency – but all of them usually have some undesirable side effects. How can we decrease these side effects?

A natural idea. The example of portfolio optimization prompts a natural idea: instead of applying individual medicines, try a combination of several medicines.

To see whether this approach will indeed work, let us reformulate our problem in precise terms.

Let us reformulate this problem in precise terms. We want to change the state of the patient: to bring the patient from a sick state to the healthy state. Each state can be described by the values of all the parameters that characterize this state: body temperature, blood pressure, etc.

We want to move the patient from the current sick state $s = (s_1, \ldots, s_d)$ to the desired healthy state $h = (h_1, \ldots, h_d)$.

We want to describe the joint effect of taking several medicines. Let us measure the dose w_i of each medicine i by considering the proportion to the actual dose to usually prescribed dose. In these units, the usually prescribed dose is $w_i = 1$. Let us describe the state of a patient after taking the doses $w = (w_1, \ldots, w_n)$ of different medicines by $f(w)$.

When no medicines are applied, i.e., when $w_i = 0$ for all i, then the patient remains sick, in the state s: $f(0) = s$. Doses of medicine are usually reasonable small, to avoid harmful side effects – we are not talking about life-and-death situations where strong measures are applied and side effects (like crushed ribs during the heart massage) are a price everyone is willing to pay to stay alive. Since the doses are small, we can expand the dependence $f(w)$ of the state on the doses w_i in Taylor series and keep only linear terms in this dependence; taking into account that $f(0) = s$, we conclude that

$$f(w) = s + \sum_{i=1}^{n} w_i \cdot a_i$$

for some vectors a_i.

We can use this formula to find the resulting state in situation when we apply the full usual dose of the i-th medicine, i.e., when we take $w_i = 1$ for this i and $w_{i'} = 0$ for all $i' = i$. In this situation, the resulting state is equal to $s + a_i$. In the ideal world, we should get the state h, i.e., we should have $a_i = h - s$, but in reality, we have side effects, i.e., deviations from this state: $\Delta a_i \overset{\text{def}}{=} a_i - (h - s) \neq 0$.

Let σ_i^2 denote the mean square values of this deviation Δa_i. Substituting the expression $a_i = (h - s) + \Delta a_i$ into the formula for the resulting state $f(w)$, we conclude that the joint effect of several medicine is equal to

$$f(w) = s + \sum_{i=1}^{n} w_i \cdot (h - s) + \sum_{i=1}^{n} w_i \cdot \Delta a_i.$$

We want to make sure that, modulo side effects, we get into the healthy state h, i.e., that $s + (h - s) \cdot \sum_{i=1}^{n} w_i = h$. This condition is equivalent to $\sum_{i=1}^{n} w_i = 1$.

Under this condition, we want to minimize the overall side effect, i.e., we want to minimize its mean squared value. When all medicines are different, side effects are independent, and thus, for the mean square error σ of the overall side effect, we have the formula $\sum_{i=1}^{n} w_i^2 \cdot \sigma_i^2$.

Thus, to get the optimal combination of medicines, we must find, among all the values w_i for which $\sum_{i=1}^{n} w_i = 1$, the combination that minimizes the sum $\sum_{i=1}^{n} w_i^2 \cdot \sigma_i^2$.

This is exactly Markowitz formula. The above optimization problem is exactly the Markowitz problem – with $\mu_i = 1$. This is also the exact same problem as we encountered when combining different independent measurement results. Thus, we conclude that we should take $w_i = \dfrac{\sigma_i^{-2}}{\sum_{j=1}^{n} \sigma_j^{-2}}$. This will enable us to decrease the side effects to the level $\sigma^2 = \dfrac{1}{\sum_{j=1}^{n} \sigma_j^{-2}}$.

In particular, in situations when all the medicines are of approximate the same quality, i.e., when all side effects are of the same strength $\sigma_1 = \ldots = \sigma_n$, we should take all the medicines with equal weight $w_1 = \ldots = w_n = \dfrac{1}{n}$. This will enable us to decrease the side effects to the level $\sigma = \dfrac{\sigma_1}{n}$.

What if side effects are correlated. The above analysis assumes that all side effects are independent. In reality, side effects may be correlated. It is therefore desirable to take this correlation into account.

In the symmetric case, when $\sigma_1 = \ldots = \sigma_n$, even if we allow the possibility of correlations - but assume that correlation is approximately the same for all pairs of medicines $\rho_{ij} \approx \rho$ – due to symmetry, we will still get the optimal combination in which each medicine is taken in the same dose $w_1 = \ldots = w_n = \dfrac{1}{n}$. The only difference is that if there is a correlation, the decrease in side effects will be not as drastic as in the independent case. Namely, we will have

$$\sigma^2 = \sum_{i=1}^{n} w_i^2 \cdot \sigma_i^2 + \sum_{i \neq j} \rho_{ij} \cdot w_i \cdot w_j \sigma_i \cdot \sigma_j$$

$$= n \cdot \frac{1}{n^2} \cdot \sigma_1^2 + n \cdot (n-1) \cdot \rho \cdot \frac{1}{n^2} \cdot \sigma_1^2 = \sigma_1^2 \cdot \left(\rho + \frac{1-\rho}{n} \right).$$

This decrease in side effects has actually been experimentally observed. Recent analysis of experimental data shows that for hypertension, a combination of quarter-doses of four different medicines indeed drastically decreases the corresponding side effect [1,4] – so this is real!

3   Applications to Machine Learning

Description of the problem. In many cases, when the inputs are small, we can use linear models – just as we did in medical applications. When the inputs are large, linear models often no longer work, and we often do not know what type of non-linear dependence we have. To describe such dependencies, we can use machine learning techniques that allow us to approximate any possible non-linear dependencies; see, e.g., [2].

   In the intermediate case, we can use both models:

- we can use a linear model, and
- we can also use machine learning techniques – such as neural networks.

Both models are not perfect: linear models are not very accurate while machine learning models are much more accurate but require a lot of time to train. Can we combine the advantages of these models?

Markowitz-motivated idea. Instead of considering the estimate $f_{NN}(x)$ generated by a neural network and a linear model $f_{lin}(x) = a_0 + \sum_{i=1}^{n} a_i \cdot x_i$, let us consider the weighted combinations of these models, i.e., functions of the type

$$f(x) = w_{NN} \cdot f_{NN}(x) + b_0 + \sum_{i=1}^{n} b_i \cdot x_i,$$

where we denoted $b_i = w_{lin} \cdot a_i = (1 - w_{NN}) \cdot a_i$.

This idea also works! It turns out that this idea can indeed drastically speed up the neural networks, see [3].

   Interestingly, the addition of linear terms did not even require big changes in the training algorithm. Indeed, usually, neural networks have:

- an intermediate layer, where the input signals $x_1, \ldots, x_n$ undergo some non-linear transformations into values $z_k = f_k(x_1, \ldots, x_k)$, followed by
- the output layer, where linear neurons transforms the values z_k coming from the intermediate layer into the final outputs $y = f_{NN}(x) = \sum_k W_k \cdot z_k - W_0$.

To incorporate additional linear terms $b_i \cdot x_i$, all we need to do is to add direct connections from the input layer to the output layer. This way, the signal produced by the output neuron is a linear combination of the signals z_k from the intermediate layer *and* the inputs x_i and thus, has the form

$$y = \sum_k W_k \cdot z_k - W_0 + \sum_{i=1}^{n} b_i \cdot x_i,$$

i.e., the desired form $y = f_{NN}(x) + \sum_{i=1}^{n} b_i \cdot x_i$, where we denoted

$$f_{NN}(x) \stackrel{\text{def}}{=} \sum_k W_k \cdot z_k - W_0 = \sum_k W_k \cdot f_k(x_1, \ldots, x_n) - W_0.$$

This minor change in the structure of a neural network still allows us to use practically the same standard computationally backpropagation algorithm (see, e.g., [2]) for training – after a very small and computationally insignificant modification.

Acknowledgments. This work was supported by Chiang Mai University, Thailand. It was also supported in part by NSF grant HRD-1242122.

One of the authors (VK) is greatly thankful to Philip Chen for valuable discussions.

References

1. Bennett, A., Chow, C.K., Chou, M., Dehbi, H.-M., Webster, R., Salam, A., Patel, A., Neal, B., Peiris, D., Thakkar, J., Chalmers, J., Nelson, M., Reid, C., Hillis, G.S., Woodward, M., Hilmer, S., Usherwood, T., Thom, S., Rodgers, A.: Efficacy and safety of quarter-dose blood pressure-lowering agents: a systematic review and meta-analysis of randomized controlled trials. Hypertension **69**(6), 85–93 (2017)
2. Bishop, C.M.: Pattern Recognition and Machine Learning. Springer, New York (2006)
3. Feng, S., Chen, C.L.P.: A fuzzy restricted Boltzmann machine: novel learning algorithms based on crisp possibilistic mean value of fuzzy numbers. IEEE Trans. Fuzzy Syst. (2017, to appear)
4. Grassi, G., Mancia, G.: Quarter dose combination therapy: good news for blood pressure control. Hypertension **69**(6), 32–34 (2017)
5. Markowitz, H.M.: Portfolio selection. J. Finan. **7**(1), 77–91 (1952)
6. Rabinovich, S.G.: Measurement Errors and Uncertainty: Theory and Practice. Springer Verlag, Berlin (2005)

A Method for Optimal Solution of Intuitionistic Fuzzy Transportation Problems via Centroid

Darunee Hunwisai[1,2(✉)], Poom Kumam[2], and Wiyada Kumam[3]

[1] Department of Mathematics and Statistics, Faculty of Science and Technology,
Valaya Alongkorn Rajabhat University under the Royal Patronage,
Bangkok, Thailand
[2] KMUTT-Fixed Point Theory and Applications Research Group (KMUTT-FPTA),
Theoretical and Computational Science Center (TaCS), Science Laboratory Building,
Faculty of Science, King Mongkut's University of Technology Thonburi (KMUTT),
126 Pracha-Uthit Road, Bang Mod, Thrung Khru,
Bangkok 10140, Thailand
{darunee.3790,poom.kumam}@mail.kmutt.ac.th
[3] Program in Applied Statistics, Department of Mathematics and Computer Science,
Faculty of Science and Technology, Rajamangala University of Technology
Thanyaburi (RMUTT), Thanyaburi 12110, Pathumthani, Thailand
wiyada.kum@rmutt.ac.th

Abstract. In this work, we introduce the method for solving intuitionistic fuzzy transportation problem (IFTP) in which supplies and availability are crisp numbers and cost is intuitionistic fuzzy number (IFN). We are using centroid of IFN for the representative value of the intuitionistic fuzzy cost. In addition we are using allocation table method (ATM) to find an initial basic feasible solution (IBFS) for the IFTP. Moreover, this method is also good optimal solution in the literature and illustrated with numerical examples.

Keywords: Transportation problem · Intuitionistic fuzzy number
Centroid

1 Introduction

The intuitionistic fuzzy set (IFS) was suggested first by Atanssov [1], which is a generalization of an ordinary Zadeh's fuzzy set [2]. In the intuitionistic fuzzy set the degree of membership and the degree of non-membership are defined simultaneously such that sum of both values is less than or equal to one. The intuitionistic fuzzy set had more abundant and flexible than the fuzzy set with uncertain information. Many researchers have also used fuzzy and intuitionistic fuzzy set for solving real word optimization problems such as scheduling, planning, transportation problems, etc. (see e.g., [3–9]).

The transportation problem is a special kind of optimization problem. The basic transportation model was first developed by Hitchcock in 1941 [10].

© Springer International Publishing AG 2018
L. H. Anh et al. (eds.), *Econometrics for Financial Applications*, Studies in Computational Intelligence 760, https://doi.org/10.1007/978-3-319-73150-6_7

In 1951, Dantzig [13] applied linear programming to solve the transportation problem. Several authors have carried out an examination about IFTP. Dinagar and Palanivel [14] investigated the transportation problem in fuzzy environment. In 2010, Pandian and Natarajan [15] introduced a new algorithm to find a fuzzy optimal solution for FTP. Hussain and Kumar [6] investigate the transportation problem with the aid of triangular intuitionistic fuzzy numbers (TIFN). In 2013, Shanmugasundari and Ganesan [16] studied a novel approach for the fuzzy optimal solution of FTP. Kaur and Kumar [17] presented a new approach for solving FTP using generalized trapezoidal fuzzy numbers. Srinivas and Ganesan [18] obtained the optimal solution for intuitionistic fuzzy transportation problem via revised distribution method. Antony et al. [8] developed method for solving intuitionistic fuzzy transportation problem of type-2. Singh and Yadav [19] suggested a new approach for solving intuitionistic fuzzy transportation problem of type-2. Ahmed et al. [20] introduced the allocation table method to finding an IBFS for the transpotation problem. Hunwisai and Kumam [9] presented a method for solving a FTP via robust ranking technique and allocation table method (ATM).

In this paper, we introduce the method for solving IFTP in which supplies and availability are crisp numbers and cost is intuitionistic fuzzy number (IFN). We are using centroid of IFN for the representative value of the intuitionistic fuzzy cost. In addition we are using ATM by Ahmed et al. [20] to find an IBFS for the IFTP and improve basic feasible solution (BFS) by modified distribution method (MODIM) to find optimal solution. Moreover, this method is also good optimal solution in the literature and illustrated with numerical examples.

The rest of this paper is organized as follows: In Sect. 2, deals with some definitions and operations IFN from literature. In Sect. 3, a method to find IBFS and optimal solution for IFTP. In Sect. 4, we give examples to illustrate to finding IBFS and the optimal solution for the IFTP. Finally, Sect. 5 contains the conclusion.

2   Preliminaries

In this section, we summarize some basic concepts of fuzzy set, intuitionistic fuzzy set, notation, definitions and operations of triangular intuitionistic fuzzy number (TIFN) which are used throughout the paper.

2.1   The Definitions and Operations of Intuitionistic Fuzzy Number

Definition 1 [2]. Let X be an arbitrary nonempty set of the universe. A *fuzzy set* A in X denoted by a set $\widetilde{A} = \{\langle x, \mu_{\widetilde{A}}(x)\rangle | x \in X\}$, where $\mu_{\widetilde{A}} : X \to [0,1]$ called $\mu_{\widetilde{A}}(x)$ is the degree of membership of element x in fuzzy set $\widetilde{A}$.

Definition 2. A fuzzy set $\widetilde{A}$ is called a *fuzzy number* if the following hold:

(i) subset of the real line.

(ii) normal, that is there exist at least one $x \in \mathbb{R}$ such that $\mu_{\widetilde{A}}(x) = 1$
(iii) convex, that is for any $x, y \in \mathbb{R}$ and $\lambda \in [0, 1]$

$$\mu_{\widetilde{A}}(\lambda x + (1 - \lambda)y) \geq \min\left(\mu_{\widetilde{A}}(x), \mu_{\widetilde{A}}(y)\right).$$

(iv) $\mu_{\widetilde{A}} : \mathbb{R} \to [0, 1]$ is piecewise continuous function, where

$$\mu_{\widetilde{A}}(x) = \begin{cases} l(x) & ; x \in (-\infty, c), \\ 1 & ; x = c, \\ r(x) & ; (c, \infty), \end{cases}$$

here, l is a picewise continuous and strictly increasing function in $(-\infty, c)$; r is a picewise continuous and strictly decreasing function in (c, ∞).

Definition 3 [1]. Let X be a arbitrary nonempty set of the universe. An *intuitionistic fuzzy set* (IFS) A in X denoted by a set $\widetilde{A}^* = \{\langle x; \mu_{\widetilde{A}^*}(x), \nu_{\widetilde{A}^*}(x)\rangle | x \in X\}$, where

$$\mu_{\widetilde{A}^*} : X \to [0, 1]$$

and

$$\nu_{\widetilde{A}^*} : X \to [0, 1]$$

the $\mu_{\widetilde{A}^*}(x)$ and $\nu_{\widetilde{A}^*}(x)$ are called the membership and non-membership degree of an element x belonging to $\widetilde{A}^* \subseteq X$, respectively. And for every $x \in X$, where $0 \leq \mu_{\widetilde{A}^*}(x) + \nu_{\widetilde{A}^*}(x) \leq 1$.

Definition 4. Let $\widetilde{A}^* = \langle(a_l, a_c, a_r); (a'_l, a_c, a'_r)\rangle$ be a TIFN in $\mathbb{R}$, whose *membership* and *non-membership function* are defined as follow:

$$\mu_{\widetilde{A}^*}(x) = \begin{cases} \dfrac{x - a_l}{a_c - a_l} & ; a_l \leq x \leq a_c \\ 1 & ; x = a_c \\ \dfrac{a_r - x}{a_r - a_c} & ; a_c \leq x \leq a_r \\ 0 & ; \text{otherwise} \end{cases} \qquad (1)$$

and

$$\nu_{\widetilde{A}^*}(x) = \begin{cases} \dfrac{a_c - x}{a_c - a'_l} & ; a'_l \leq x \leq a_c \\ 0 & ; x = a_c \\ \dfrac{x - a_c}{a'_r - a_c} & ; a_c \leq x \leq a'_r \\ 1 & ; \text{otherwise} \end{cases} \qquad (2)$$

respectively, where $a'_l \leq a_l \leq a_c \leq a_r \leq a'_r$.

Remark. From Definition 4 A TIFN $\widetilde{A}^* = \langle(a_l, a_c, a_r); (a'_l, a_c, a'_r)\rangle$ if $a_l = a'_l, a_r = a'_r$ then $\widetilde{A}^* = \langle a_l, a_c, a_r\rangle$ is a triangular fuzzy number (TFN), which is particular case of TIFN (Fig. 1).

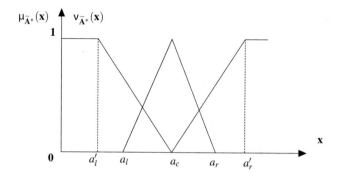

Fig. 1. A TIFN $\widetilde{A}^* = \langle (a_l, a_c, a_r); (a'_l, a_c, a'_r) \rangle$

Likewise to algebraic operation of TFN and TIFN are defined as follows.

Definition 5. Let $\widetilde{A}^* = \langle (a_l, a_c, a_r); (a'_l, a_c, a'_r) \rangle$ and $\widetilde{B}^* = \langle (b_l, b_c, b_r); (b'_l, b_c, b'_r) \rangle$ be two TIFNs with $\gamma \neq 0$ be any real number. Then, the arithmetic operations of TIFNs are defined as follows:

Addition: $\quad \widetilde{A}^* \oplus \widetilde{B}^* = \langle (a_l + b_l, a_c + b_c, a_r + b_r); (a'_l + b'_l, a_c + b_c, a'_r + b'_r) \rangle$,

Subtraction: $\quad \widetilde{A}^* \ominus \widetilde{B}^* = \langle (a_l - b_r, a_c - b_c, a_r - b_l); (a'_l - b'_r, a_c - b_c, a'_r - b'_l) \rangle$,

Multiplication: $\quad \widetilde{A}^* \otimes \widetilde{B}^* = \langle (m_l, m_c, m_r); (m'_l, m'_c, m'_r) \rangle$, where

$$m_l = \min\{a_l b_l, a_l b_r, a_r b_l, a_r b_r\} \qquad m'_l = \min\{a'_l b'_l, a'_l b'_r, a'_r b'_l, a'_r b'_r\}$$

$$m_r = \max\{a_l b_l, a_l b_r, a_r b_l, a_r b_r\} \qquad m'_r = \max\{a'_l b'_l, a'_l b'_r, a'_r b'_l, a'_r b'_r\}$$

$$m_c = a_c b_c$$

and

Scalar multiplication: $\quad \gamma \widetilde{A}^* = \begin{cases} \langle (\gamma a_l, \gamma a_c, \gamma a_r); (\gamma a'_l, \gamma a_c, \gamma a'_r) \rangle & ; \gamma > 0 \\ \langle (\gamma a_r, \gamma a_c, \gamma a_l); (\gamma a'_r, \gamma a_c, \gamma a'_l) \rangle & ; \gamma < 0 \end{cases}$

Definition 6. The (α, λ)-cut set of TIFN $\widetilde{A}^* = \langle (a_l, a_c, a_r); (a'_l, a_c, a'_r) \rangle$ is a crisp subset of $\mathbb{R}$, which is defined as follow:

$$\widetilde{A}^{*\lambda}_\alpha = \{x \in X | \mu_{\widetilde{A}^*}(x) \geq \alpha, \nu_{\widetilde{A}^*}(x) \leq \lambda\},$$

where $\alpha, \lambda \in [0, 1]$.

2.2   Centroid of Intuitionistic Fuzzy Number

Definition 7 [11]. Let $m : \mathbb{R} \to [0, 1]$ be defined by

$$m(x) = \frac{(\mu(x) - \nu(x)) + 1}{2}$$

where μ and ν are the membership and non-membership function of IFN $\tilde{A}^*$.

Lemma. m is a fuzzy number.

Proof. From Definition 7 $m(x) = \dfrac{(\mu(x) - \nu(x)) + 1}{2}$

Range of $\mu(x) - \nu(x) \in [-1, 1]$.

Let $g : \mathbb{R} \to [0, 1], h : \mathbb{R} \to [0, 1], f : \mathbb{R} \to [0, 1]$ and $k : \mathbb{R} \to [0, 1]$, where f and g are strictly increasing and h and k are strictly decreasing, we get

$$m(x) = \begin{cases} 0 & ; \ x < a'_l \\ \dfrac{-f(x) + 1}{2} & ; \ a'_l \le x < a_l \\ \dfrac{g(x) - f(x) + 1}{2} & ; \ a_l \le x < a_c \\ \dfrac{h(x) - k(x) + 1}{2} & ; \ a_c \le x < a_r \\ \dfrac{-k(x) + 1}{2} & ; \ a_r \le x < a'_r \\ 0 & ; \ x \ge a'_r \end{cases}$$

$f(a_c) = 1$. Therefore

$$m(x) = \begin{cases} 0 & ; \ x < a'_l \\ \dfrac{-f(x) + 1}{2} & ; \ a'_l \le x < a_l \\ \dfrac{g(x) - f(x) + 1}{2} & ; \ a_l \le x < a_c \\ 1 & ; \ x = a_c \\ \dfrac{h(x) - k(x) + 1}{2} & ; \ a_c < x \le a_r \\ \dfrac{-k(x) + 1}{2} & ; \ a_r < x \le a'_r \\ 0 & ; \ x > a'_r \end{cases}$$

functions f and g are piecewise continuous and strictly increasing. Function h and k are piecewise continuous and strictly decreasing. Hence m is a fuzzy number.

Next, we will find the centroid of fuzzy number.

Centroid of fuzzy number A is its geometric center and is given by the formula $\dfrac{\int_{-\infty}^{\infty} xA(x)dx}{\int_{-\infty}^{\infty} A(x)dx}$ [12]. Centroid of fuzzy number m is $\dfrac{D}{N}$, where

$$D = \int_{a_l'}^{a_l} \frac{-f(x)+1}{2} x\,dx + \int_{a_l}^{a_c} \frac{g(x)-f(x)+1}{2} x\,dx + \int_{a_c}^{a_r} \frac{h(x)-k(x)+1}{2} x\,dx$$

$$+ \int_{a_r}^{a_r'} \frac{-k(x)+1}{2} x\,dx$$

and

$$N = \int_{a_l'}^{a_l} \frac{-f(x)+1}{2} dx + \int_{a_l}^{a_c} \frac{g(x)-f(x)+1}{2} dx + \int_{a_c}^{a_r} \frac{h(x)-k(x)+1}{2} dx$$

$$+ \int_{a_r}^{a_r'} \frac{-k(x)+1}{2}$$

Definition 8. Let $\widetilde{A}^*$ be an IFN *the centroid* of an IFN $\widetilde{A}^*$ is $\dfrac{D}{N}$, where

$$D = \int_{a_l'}^{a_l} \frac{-f(x)+1}{2} x\,dx + \int_{a_l}^{a_c} \frac{g(x)-f(x)+1}{2} x\,dx + \int_{a_c}^{a_r} \frac{h(x)-k(x)+1}{2} x\,dx$$

$$+ \int_{a_r}^{a_r'} \frac{-k(x)+1}{2} x\,dx$$

and

$$N = \int_{a_l'}^{a_l} \frac{-f(x)+1}{2} dx \; | \; \int_{a_l}^{a_c} \frac{g(x)-f(x)+1}{2} dx + \int_{a_c}^{a_r} \frac{h(x)-k(x)+1}{2} dx$$

$$+ \int_{a_r}^{a_r'} \frac{-k(x)+1}{2}$$

Remark. From Definition 8 let $\widetilde{A}^* = \langle (a_l, a_c, a_r); (a_l', a_c, a_r') \rangle$ be a TIFN. By substituting the values in $\dfrac{D}{N}$ the centroid of $\widetilde{A}^*$ is

$$\frac{1}{3} \left[\frac{(a_r' - a_l')(a_c - 2a_r' - 2a_l') + (a_r - a_l)(a_l + a_c + a_r) + 3(a_r'^2 - a_l'^2)}{a_r' - a_l' + a_r - a_l} \right]. \quad (3)$$

Remark. From a TIFN $\widetilde{A}^* = \langle (a_l, a_c, a_r); (a_l', a_c, a_r') \rangle$ if $a_l = a_l'$ and $a_r = a_r'$. Then a TIFN $\widetilde{A}^*$ will become to the TFN. That is $\widetilde{A} = \langle a_l, a_c, a_r \rangle$. By substituting the values in Eq. (3) the centroid of a TFN $\widetilde{A}$ is $\dfrac{a_l + a_c + a_r}{3}$.

In this paper, centroid denoted by $R(\widetilde{A}^*)$. And $R(\widetilde{A}^*)$ to be the representative value of the IFN $(\widetilde{A}^*)$. We give index $R(\widetilde{c}_{ij}^*)$ to be the representative value of the intuitionistic fuzzy cost $(\widetilde{c}_{ij}^*)$.

3   Intuitionistic Fuzzy Transportation Problem

The formulation of the IFTP is of the following form:

$$(\text{IFTP}) \quad \text{Minimize } \widetilde{\psi} = \sum_{i=1}^{m}\sum_{j=1}^{n}\widetilde{c}_{ij}^*x_{ij}$$

$$\text{subject to} \quad \sum_{j=1}^{n}x_{ij} \leq a_i \quad, i = 1, 2, \ldots, m$$

$$\sum_{i=1}^{m}x_{ij} \geq b_j \quad, j = 1, 2, \ldots, n$$

$$x_{ij} \geq 0 \quad \text{for all } i \text{ and } j,$$

where

- $\widetilde{c}_{ij}^*$ is the intuitionistic fuzzy cost of transportation of one unit of the goods from i^{th} source to the jth destination.
- x_{ij} is the quantity of transportation from i^{th} source to the jth destination.
- a_i is the total availability of the goods at i^{th} source.
- b_j is the total demand of the goods at j^{th} destination.
- $\sum_{i=1}^{m}\sum_{j=1}^{n}\widetilde{c}_{ij}^*x_{ij}$, is the total fuzzy transportation cost.
- If $\sum_{i=1}^{m}a_i = \sum_{j=1}^{n}b_j$, then IFTP is said to be balanced.
- If $\sum_{i=1}^{m}a_i \neq \sum_{j=1}^{n}b_j$, then IFTP is said to be unbalanced.

A balanced transportation problem has total supply equal to total demand which can be expressed as

$$\sum_{i=1}^{m}a_i = \sum_{j=1}^{n}b_j \tag{4}$$

A consequence of this is that the problem is defined by $n + m - 1$ availability and demand variables since, if $a_i, i = 2, 3, 4, \ldots, m$ and $b_j, j = 1, 2, 3, \ldots, n$ are specified, then a_1 can be found from Eq. (4). This means that one of the constraint equations is not required. Thus, a balanced transportation model has $n + m - 1$ independent constraint equations (Table 1).

Table 1. The intuitionistic fuzzy transportation table

	1	2	...	N	a_i
1	$\widetilde{c}_{11}^*$	$\widetilde{c}_{12}^*$	...	$\widetilde{c}_{1n}^*$	a_1
2	$\widetilde{c}_{21}^*$	$\widetilde{c}_{22}^*$	...	$\widetilde{c}_{2n}^*$	a_2
⋮	⋮	⋮	⋮	⋮	⋮
b_j	b_1	b_2	...	b_n	$\sum_{i=1}^{m} a_i = \sum_{j=1}^{n} b_j$

Since the number of basic variables in a basic solution is the same as the number of constraints, solutions of this problem should have $n + m - 1$ basic variables which are non-zero and all the remaining variables will be non-basic and thus have the value zero.

3.1   Algorithm to Find an Initial Basic Feasible Solution (IBFS)

In this section, we use an allocation table set up to find the solution for IFTP. This method is called allocation table method (ATM) and the algorithm is illustrated as follows:

Step-1: Establish the formulated fuzzy linear programming problem into the tabular form known as intuitionistic fuzzy transportation table (IFTT). Calculate $R(\widetilde{c}_{ij}^*)$ for all the intuitionistic fuzzy cost of transportation to be put on the table allocation.

Step-2: Examine that the IFTP is balanced or unbalanced.

case (i). If the problem is balanced i.e., $\sum_{i=1}^{m} a_i = \sum_{j=1}^{n} b_j$, then go to Step-3.

case (ii). If $\sum_{i=1}^{m} a_i \neq \sum_{j=1}^{n} b_j$, then convert the unbalanced problem into balanced problem as follows:

case (iia). If $\sum_{i=1}^{m} a_i < \sum_{j=1}^{n} b_j$ then introduce a dummy source (row) with availability. Assume the fuzzy cost for transporting one unit quantity of the product from the introduced dummy source to all destinations as zero fuzzy number. Go to Step-3.

case (iib). If $\sum_{i=1}^{m} a_i > \sum_{j=1}^{n} b_j$ then introduce a dummy destination (column) with demand. Assume the fuzzy cost for transporting one unit quantity of the product from all sources to the introduced dummy destination as zero fuzzy number. Go to Step-3.

Step-3: Choose minimum odd cost (cost for transporting one unit quantity of the product which is an odd number) from every cost cells of IFTT. If there is no odd cost in cost cells of the IFTT, continue by dividing every cost cells by 2 until get at least an odd value in cost cells.

Step-4: Form a new table which revises is to be known as allocation table by keeping the minimum odd cost in the respective cost cell/cells as it was/were, and subtract selected minimum odd cost only from each of the odd cost valued cells of the IFTT. Now every the cell values are to be called as allocation cell value in allocation table.

Step-5: At first start with selected minimum odd cost in allocation table in Step-4. Delete the row (availability) or column (demand) that has been allocated to complete.

Step-6: Now specify the minimum allocation cell value and allocate minimum of availability/demand at the place of selected allocation cell value in the allocation table. In the event of same allocation cell values, select the allocation cell value where minimum allocation can be made. Afresh in the event of same allocation in the allocation cell values, choose the minimum cost cell which is corresponding to the cost cells of IFTT formed in Step-1. Afresh if the allocations and the cost cells are equal, in such case choose the nearer cell to the minimum of demand/availability which is to be allocated. Now Delete the row(availability) or column(demand) that has been allocated to complete.

Step-7: Repeat Step-6 as far as the demand and availability are depleted.

Step-8: Finally, from the IFTT, we compute the total fuzzy transportation cost.

3.2   Modified Distribution Method (MODIM) for Finding Optimal Solution

In this section, we find the best solution for IFTP by using a modified method of distribution. Algorithm of modified distribution method is illustrated as follow:

Step-1: Find IBFS by proposed ATM.
Step-2: Let

- R_i is the value assigned to row i
- K_j is the value assigned to column j
- C_{ij} is the cost in square ij (cost of shipping from source i to destination j).

Compute the values for each row R_i and column K_j, set $R_i + K_j = C_{ij}$. For example, if the square at the intersection of row 3 and column 2 is occupied, we set $R_3 + K_2 = C_{32}$. After all equations have been written, set $K_1 = 0$ (or $R_1 = 0$).

Step-3: Calculate the improvement index value for unoccupied cells by the equation $E_{ij} = C_{ij} - R_i - K_j$.

Considering only these dispatch and reception costs, it would cost $R_i + K_j$ to send 1 unit from source i to destination j. For (i, j) not corresponding to a basic variable (whenever x_{ij} is a basic variable), it will often be the case that $R_i + K_j \neq C_{ij}$. In particular, if $R_i + K_j > C_{ij}$ for a particular (i, j) not corresponding to a basic variable, then there would be a benefit from sending more goods that way.

So let $E_{ij} = C_{ij} - R_i - K_j$. The E_{ij} values are entered in the top right of the cells. Then E_{ij} is the change in cost due to allocating 1 extra unit to cell (i, j). If any E_{ij} is negative (so that $R_i + K_j > C_{ij}$), then the total cost can be reduced by allocating as many units as possible to cell (i, j). However, if *all* the E_{ij} are positive then it will be more expensive to change any of the allocations and so we have found a minimum cost.

Step-4: Consider valued of E_{ij}.

case (i). IBFS is intutionistic fuzzy optimal solution, if $E_{ij} \geq 0$ for every unoccupied cells.

case (ii). IBFS is not intutionistic fuzzy optimal solution, for at least one $E_{ij} < 0$. Go to step 5.

Step-5: Choose the unoccupied cell for the most negative value of E_{ij}.

Step-6: We construct the closed loop below.

At first, start the closed loop with choose the empty cell and move vertically and horizontally with corner cells occupied and come back to choose the empty cell to complete the loop. Use sign "+" and "−" at the corners of the closed loop, by assigning the "+" sign to the selected empty cell first.

Step-7: Look for the minimum allocation value from the cells which have "−" sign. After that, allocate this value to the choose empty cell and subtract it to the other occupied cell having "−" sign and add it to the other occupied cells having "+" sign.

Step-8: Allocation in step-7 will result an improved basic feasible solution (BFS).

Step-9: Test the optimality condition for improved BFS. The process is complete when $E_{ij} \geq 0$ for all the empty cell.

4   Numerical Example

Example 1. Consider IFTP with three sources S_1, S_2, S_3 and three destinations D_1, D_2, D_3. The cost of transporting one unit of the goods from i^{th} source to the jth destination given whose elements are TIFNs, shown in the Table 2. Find out the minimum cost of total intuitionistic fuzzy transportation.

Table 2. Data of the Example 1

Source	D_1	D_2	D_3	Availability (a_i)
S_1	$\langle(3,6,10);(2,6,11)\rangle$	$\langle(6,8,13);(4,8,15)\rangle$	$\langle(2,5,7);(1,5,11)\rangle$	30
S_2	$\langle(3,5,9);(2,5,12)\rangle$	$\langle(4,6,10);(3,6,12)\rangle$	$\langle(5,7,9);(3,7,13)\rangle$	45
S_3	$\langle(6,8,9);(4,8,11)\rangle$	$\langle(7,8,10);(5,8,14)\rangle$	$\langle(7,9,12);(4,9,15)\rangle$	40
Demand (b_j)	35	60	20	115

Since $\sum_{i=1}^{3} a_i = \sum_{j=1}^{3} b_j = 115$, the IFTP is balanced.

From Table 2 the cost of transporting one unit of the goods from i^{th} source to the jth destination are TIFNs. We use Eq. (3) for calculating the value of intuitionistic fuzzy cost.

Let us consider Table 2 where the element $\langle (3, 6, 10); (4, 6, 11) \rangle$ is TIFN. From Eq. (3)

$$\frac{1}{3} \left[\frac{(a'_r - a'_l)(a_c - 2a'_r - 2a'_l) + (a_r - a_l)(a_l + a_c + a_r) + 3(a'^2_r - a'^2_l)}{a'_r - a'_l + a_r - a_l} \right].$$

Therefore, we obtain

$$R(\widetilde{c}^*_{11}) = R\langle (3, 6, 10); (2, 6, 11) \rangle$$
$$= \frac{1}{3} \left[\frac{(11 - 2)(6 - 2(11) - 2(2)) + (10 - 3)(3 + 6 + 10) + 3(11^2 - 2^2)}{11 - 2 + 10 - 3} \right]$$
$$= 6.33.$$

Similarly, the value of the intuitionistic fuzzy costs $\widetilde{c}^*_{ij}$ are calculated as: $R(\widetilde{c}^*_{12}) = 9, R(\widetilde{c}^*_{13}) = 5.33, R(\widetilde{c}^*_{21}) = 6.08, R(\widetilde{c}^*_{22}) = 6.87, R(\widetilde{c}^*_{23}) = 7.48, R(\widetilde{c}^*_{31}) = 7.67, R(\widetilde{c}^*_{32}) = 8.83, R(\widetilde{c}^*_{33}) = 9.33$. We put all $R(\widetilde{c}^*_{ij})$ in Table 3.

From Table 3, it is found that the IFTP is balanced. Thus, move to step-3.

Step-3, minimum odd cost is 5.33 in cost cell $(1, 3)$ among all the cost cells of Table 3.

Table 3. The IFTT after calculated $R(\widetilde{c}^*_{ij})$

Source	D_1	D_2	D_3	Availability (a_i)
S_1	6.33	9	5.33	30
S_2	6.08	6.87	7.48	45
S_3	7.67	8.83	9.33	40
Demand (b_j)	35	60	20	115

According to step-5, minimum of availability/demand is 20 that is allocation in cell $(1, 3)$. After allocating this value it is found that the demand is satisfied. For which D_3 column is to be exhausted.

After step-5, only D_1 and D_2 column are to be considered. Where 6.08 is the lowest cell value in cells $(1, 1), (2, 1)$ and $(2, 2)$. Among these three cells 10 is the lowest allocation can be made in cells $(1, 1)$. For which S_1 row is to be exhausted.

Next, consider cells $(2, 1)$ and $(3, 1)$ we found that 25 is the lowest allocation of cost cells $(2, 1)$ and $(3, 1)$. Thus, select cell $(2, 1)$ because 6.08 is the lowest cell value in cells $(2, 1)$ and $(3, 1)$. For which D_1 column is to be exhausted.

Now complete the allocation by allocating 20 and 40 to the cell $(2, 2)$ and $(3, 2)$, respectively. All these allocations are made according to step-6 and step-7 of the proposed algorithm.

After that, transfer this allocation to the IFTT. The first allocation is shown in Table 4 and the final allocation is shown in Table 5.

Table 4. The first iteration of allocation cell

Source	D_1	D_2	D_3		Availability (a_i)
S_1	6.33	9	5.33	**20**	30
S_2	6.08	6.87	7.48		45
S_3	7.67	8.83	9.33		40
Demand (b_j)	35	60	20		115

Table 5. Finally, allocation of various cells are in the allocation table

Source	D_1		D_2		D_3		Availability (a_i)
S_1	6.33	**10**	9		5.33	**20**	30
S_2	6.08	**25**	6.87	**20**	7.48		45
S_3	7.67		8.83	**40**	9.33		40
Demand (b_j)	35		60		20		115

Therefore, IBFS is $x_{11} = 10, x_{13} = 20, x_{21} = 25, x_{22} = 20, x_{32} = 40$.

Finally, total intuitionistic fuzzy transportation cost is $(6.33 \times 10) + (5.33 \times 20) + (6.08 \times 25) + (6.87 \times 20) + (8.83 \times 40) = 812.5$.

Now, we apply modified distribution method (MODIM) to compute the optimal solution. Algorithm of modified distribution method as shown in Sect. 3.2.

Firstly, we compute dual variables R_i and K_j for each row and column respectively, satisfying $R_i + K_j = C_{ij}$ for each occupied cell. Therefor, let $K_1 = 0$. For each occupied cell, $C_{ij} = R_i + K_j$, we have

$$C_{11} = R_1 + K_1; \qquad R_1 = 6.33$$
$$C_{13} = R_1 + K_3; \qquad K_3 = -1$$
$$C_{21} = R_2 + K_1; \qquad K_2 = 6.08$$
$$C_{22} = R_2 + K_2; \qquad K_2 = 0.79$$
$$C_{32} = R_3 + K_2; \qquad R_3 = 8.04$$

Hence, we obtain

$$E_{12} = C_{12} - (R_1 + K_2); \qquad = 1.88$$
$$E_{23} = C_{23} - (R_2 + K_3); \qquad = 2.4$$
$$E_{31} = C_{31} - (R_3 + K_1); \qquad = -0.37$$
$$E_{33} = C_{33} - (R_3 + K_3); \qquad = 2.29$$

From above, we found that the value of E_{31} is negative, therefor IBFS is not intuitionistic fuzzy optimal.

Construction of loop
In Table 6, use sign "+" in cells $(3,1)$ and $(2,2)$. And use sign "−" in cells $(2,1)$ and $(3,2)$.

Table 6. Construction of loop

Source	D_1		D_2		D_3		Availability (a_i)
S_1	6.33	**10**	9		5.33	**20**	30
S_2	6.08	**25**	6.87	**20**	7.48		45
S_3		7.67 →	8.83	**40**	9.33		40
Demand (b_j)	35		60		20		115

Check E_{ij} again, if $E_{ij} \geq 0$ for all unoccupied cells, then the solution is intuitionistic fuzzy optimal solution. If $E_{ij} < 0$, go to step 5.

Improved basic feasible solution (Table 7)
Let $K_1 = 0$
For each occupied cell, $C_{ij} = R_i + K_j$

$$C_{11} = R_1 + K_1; \qquad R_1 = 6.33.$$
$$C_{13} = R_1 + K_3; \qquad K_3 = -1.$$
$$C_{22} = R_2 + K_2; \qquad R_2 = 5.71.$$
$$C_{31} = R_3 + K_1; \qquad R_3 = 7.67.$$
$$C_{32} = R_3 + K_2; \qquad K_2 = 1.16.$$

Table 7. Improved basic feasible solution

Source	D_1		D_2		D_3		Availability (a_i)
S_1	6.33	**10**	9		5.33	**20**	30
S_2	6.08		6.87	**45**	7.48		45
S_3	7.67	**25**	8.83	**15**	9.33		40
Demand (b_j)	35		60		20		115

Hence, we observe that

$$E_{12} = C_{12} - R_1 - K_2 = 1.15.$$
$$E_{21} = C_{21} - R_2 - K_1 = 0.37.$$
$$E_{23} = C_{23} - R_2 - K_3 = 2.77.$$
$$E_{33} = C_{33} - R_3 - K_3 = 2.66.$$

From above, we found that the value of $E_{ij} \geq 0$ for all unoccupied cells, intuitionistic fuzzy optimal solution is $x_{11} = 10, x_{13} = 20, x_{22} = 45, x_{31} = 25, x_{32} = 15$, and the minimum transportation intuitionistic fuzzy cost is $\psi = (6.33 \times 10) + (5.33 \times 20) + (6.87 \times 45) + (7.67 \times 25) + (8.83 \times 15) = 803.25$.

Example 2. Consider FTP with three sources S_1, S_2, S_3 and four destinations D_1, D_2, D_3, D_4. The cost of transporting one unit of the goods from i^{th} source to the jth destination given whose elements are trapezoidal fuzzy numbers, shown in the Table 8 below. Find out the minimum cost of total intuitionistic fuzzy transportation.

Since $\sum_{i=1}^{3} a_i = \sum_{j=1}^{4} b_j = 250$, the IFTP is balanced.

From Table 8 the cost of transporting one unit of the goods from i^{th} source to the jth destination are TIFNs. We use Eq. (3) for calculating the value of intuitionistic fuzzy cost.

Let us consider the element in Table 8 and from Eq. (3)

$$\frac{1}{3}\left[\frac{(a'_r - a'_l)(a_c - 2a'_r - 2a'_l) + (a_r - a_l)(a_l + a_c + a_r) + 3(a'^2_r - a'^2_l)}{a'_r - a'_l + a_r - a_l}\right].$$

The value of the intuitionistic fuzzy costs $\widetilde{c}^*_{ij}$ are calculated as: $R(\widetilde{c}^*_{11}) = 14.4, R(\widetilde{c}^*_{12}) = 21.25, R(\widetilde{c}^*_{13}) = 38.07, R(\widetilde{c}^*_{14}) = 40, R(\widetilde{c}^*_{21}) = 55, R(\widetilde{c}^*_{22}) = 37.52, R(\widetilde{c}^*_{23}) = 47.39, R(\widetilde{c}^*_{24}) = 24.36, R(\widetilde{c}^*_{31}) = 23.85, R(\widetilde{c}^*_{32}) = 50, R(\widetilde{c}^*_{33}) = 46.67, R(\widetilde{c}^*_{34}) = 25.18$. We put all $R(\widetilde{c}^*_{ij})$ in Table 9.

Table 8. Data of the Example 2

Source	D_1	D_2	D_3	D_4	Availability (a_i)
S_1	$\langle(10, 15, 20);$ $(6, 15, 21)\rangle$	$\langle(12, 20, 30);$ $(10, 20, 35)\rangle$	$\langle(30, 38, 45);$ $(27, 38, 50)\rangle$	$\langle(30, 40, 50);$ $(25, 40, 55)\rangle$	80
S_2	$\langle(50, 55, 60);$ $(47, 55, 63)\rangle$	$\langle(30, 35, 47);$ $(28, 35, 50)\rangle$	$\langle(40, 45, 56);$ $(38, 45, 60)\rangle$	$\langle(10, 25, 37);$ $(9, 25, 40)\rangle$	75
S_3	$\langle(12, 24, 35);$ $(10, 24, 35)\rangle$	$\langle(40, 50, 60);$ $(30, 50, 65)\rangle$	$\langle(35, 45, 60);$ $(30, 45, 65)\rangle$	$\langle(10, 25, 40);$ $(8, 25, 43)\rangle$	95
Demand (b_j)	50	90	60	50	250

Table 9. The IFTT after calculated $R(\widetilde{c}_{ij}^*)$

Source	D_1	D_2	D_3	D_4	Availability (a_i)
S_1	14.4	21.25	38.07	40	80
S_2	55	37.52	47.39	24.36	75
S_3	23.85	50	46.67	25.18	95
Demand (b_j)	50	90	60	50	250

From Table 9, it is found that the IFTP is balanced. Thus, move to step-3.

Step-3, minimum odd cost is 21.25 in cost cell $(2,2)$ among all the cost cells of Table 9.

According to step-5, minimum of availability/demand is 80 that is allocation in cell $(2,2)$. After allocating this value it is found that the demand is satisfied. For which S_1 row is to be exhausted.

After step-5, only S_2 and S_3 column are to be considered. Where 23.85 is the lowest cell value in cells $(2,2)$, $(2,4)$, $(3,1)$ and $(3,4)$. Among these four cells 10 is the lowest allocation can be made in cells $(2,2)$. For which S_1 column D_2 is to be exhausted.

Next, consider cells $(2,1)$ and $(3,1)$ we found that 23.85 is the lowest allocation of cost cells $(2,1)$ and $(3,1)$. Thus, select cell $(3,1)$. For which D_1 column is to be exhausted

Consider cells $(3,3)$ and $(3,4)$ we found that minimum of availability/demand is 45 and 25.18 is the lowest allocation of cost cells $(3,3)$ and $(3,4)$. Thus, select cell $(3,4)$. For which S_3 row is to be exhausted.

Now complete the allocation by allocating 5 and 60 to the cell $(2,4)$ and $(2,3)$, respectively. All these allocations are made according to step-6 and step-7 of the proposed algorithm.

After that, transfer this allocation to the IFTT. The first allocation is shown in Table 10 and the final allocation is shown in Table 11.

Table 10. The first iteration of allocation cell

Source	D_1	D_2		D_3	D_4	Availability (a_i)
S_1	14.4	21.25	**80**	38.07	40	80
S_2	55	37.52		47.39	24.36	75
S_3	23.85	50		46.67	25.18	95
Demand (b_j)	50	90		60	50	250

Therefore, IBFS is $x_{12} = 80, x_{22} = 10, x_{23} = 60, x_{24} = 5, x_{31} = 50, x_{34} = 45$.

Finally, total intuitionistic fuzzy transportation cost is $(21.25 \times 80) + (37.52 \times 10) + (47.39 \times 60) + (24.36 \times 5) + (23.85 \times 50) + (25.18 \times 45) = 7366$.

Table 11. Finally, allocation of various cells are in the allocation table

Source	D_1		D_2		D_3		D_4		Availability
S_1	14.4		21.25	**80**	38.07		40		80
S_2	55		37.52	**10**	47.39	**60**	24.36	**5**	75
S_3	23.85	**50**	50		46.67		25.18	**45**	95
Demand	50		90		60		50		250

Now, we apply MODIM to compute the optimal solution. Algorithm of MODIM as shown in Sect. 3.2.

Firstly, we compute dual variables R_i and K_j for each row and column respectively, satisfying $R_i + K_j = C_{ij}$ for each occupied cell. Therefor, let $R_1 = 0$.

For each occupied cell, $C_{ij} = R_i + K_j$, we have

$$
\begin{aligned}
C_{12} &= R_1 + K_2; & K_2 &= 21.25 \\
C_{22} &= R_2 + K_2; & R_2 &= 16.27 \\
C_{23} &= R_2 + K_3; & K_3 &= 31.12 \\
C_{24} &= R_2 + K_4; & K_4 &= 8.09 \\
C_{31} &= R_3 + K_1; & K_1 &= 6.76 \\
C_{34} &= R_3 + K_4; & R_3 &= 17.09
\end{aligned}
$$

Hence, we obtain

$$
\begin{aligned}
E_{11} &= C_{11} - (R_1 + K_1); & &= 7.64 \\
E_{13} &= C_{13} - (R_1 + K_3); & &= 6.95 \\
E_{14} &= C_{14} - (R_1 + K_4); & &= 31.91 \\
E_{21} &= C_{21} - (R_2 + K_1); & &= 31.97 \\
E_{32} &= C_{32} - (R_3 + K_2); & &= 16.64 \\
E_{33} &= C_{33} - (R_3 + K_3); & &= -1.54
\end{aligned}
$$

From above, we found that the value of E_{33} is negative, therefor IBFS is not intuitionistic fuzzy optimal.

Construction of loop

In Table 12, use sign "+" in cells $(3,3)$ and $(2,4)$. And use sign "−" in cells $(3,4)$ and $(2,3)$.

Check E_{ij} again, if $E_{ij} \geq 0$ for all unoccupied cells, then the solution is intuitionistic fuzzy optimal solution. If $\widetilde{E}_{ij} < 0$, go to step 5.

Table 12. Construction of loop

Source	D_1	D_2	D_3	D_4	Availability (a_i)
S_1	14.4	21.25 **80**	38.07	40	80
S_2	55	37.52 **10**	47.39 **60** ← 24.36 **5**		75
S_3	23.85 **50**	50	46.67 → 25.18 **45**		95
Demand (b_j)	50	90	60	50	250

Improved basic feasible solution (Table 13)

Let $K_1 = 0$. For each occupied cell, $C_{ij} = R_i + K_j$

$$C_{12} = R_1 + K_2; \qquad R_1 = 8.3.$$
$$C_{22} = R_2 + K_2; \qquad K_2 = 12.95.$$
$$C_{23} = R_2 + K_3; \qquad R_2 = 24.57.$$
$$C_{24} = R_2 + K_4; \qquad K_4 = -0.21.$$
$$C_{31} = R_3 + K_1; \qquad R_3 = 23.85.$$
$$C_{33} = R_3 + K_3; \qquad K_3 = 22.82.$$

Hence, we observe that

$$E_{11} = C_{11} - (R_1 + K_1) = 6.1.$$
$$E_{13} = C_{13} - (R_1 + K_3) = 6.95.$$
$$E_{14} = C_{14} - (R_1 + K_4) = 31.91.$$
$$E_{21} = C_{21} - (R_2 + K_1) = 30.436.$$
$$E_{32} = C_{32} - (R_3 + K_2) = 13.2.$$
$$E_{34} = C_{34} - (R_3 + K_4) = 1.54.$$

From above, we found that the value of $E_{ij} \geq 0$ for all unoccupied cells, intuitionistic fuzzy optimal solution is $x_{12} = 80, x_{22} = 10, x_{23} = 15, x_{24} = 50, x_{31} = 50, x_{33} = 45$ and the minimum transportation intuitionistic fuzzy cost is $\tilde{\psi} = (21.25 \times 80) + (37.52 \times 10) + (47.39 \times 15) + (24.36 \times 50) + (23.85 \times 50) + (46.67 \times 45) = 7296.70$.

Example 3 (Unbalanced intuitionistic fuzzy transportation problem). Consider the following problem with three factories and four warehouses.

Table 13. Improved basic feasible solution

Source	D_1		D_2		D_3		D_4		Availability (a_i)
S_1	14.4		21.25	**80**	38.07		40		80
S_2	55		37.52	**10**	47.39	**15**	24.36	**50**	75
S_3	23.85	**50**	50		46.67	**45**	25.18		95
Demand (b_j)	50		90		60		50		250

Table 14. Data of the Example 3

Source	W_1	W_2	W_3	W_4	Supply (a_i)
F_1	$\langle(2,7,15);$ $(0,7,18)\rangle$	$\langle(4,10,12);$ $(3,10,17)\rangle$	$\langle(3,7,14);$ $(2,7,18)\rangle$	$\langle(3,5,10);$ $(1,5,16)\rangle$	35
F_2	$\langle(2,4,7);$ $(1,4,10)\rangle$	$\langle(5,6,10);$ $(2,6,13)\rangle$	$\langle(6,7,12);$ $(4,7,12)\rangle$	$\langle(8,9,13);$ $(8,9,18)\rangle$	40
F_3	$\langle(5,8,10);$ $(4,8,11)\rangle$	$\langle(7,15,20);$ $(7,15,24)\rangle$	$\langle(3,9,15);$ $(2,9,16)\rangle$	$\langle(4,5,6);$ $(3,5,9)\rangle$	45
Deman (b_j)	30	25	60	35	

From Table 14, $\sum_{i=1}^{4} a_i = 120, \sum_{j=1}^{4} b_j = 150$, the IFTP is an unbalanced intuitionistic fuzzy transportation problem. To solve the problem, we introduce a dummy source $(F4)$ which has a capacity of 30. The amount shipped from this dummy source to a destination represents the shortage quantity at that destination.

From Table 15, $\sum_{i-1}^{4} a_i = \sum_{j-1}^{4} b_j = 150$, the IFTP is balanced.

Next, consider the element in Table 15 and from Eq. (3) we calculate the value of the intuitionistic fuzzy costs $\tilde{c}_{ij}^*$ shown in the Table 16.

From Table 16, we are using ATM to find an IBFS for the IFTP, shown in the Table 17.

From Table 17, we get IBFS is $x_{13} = 35, x_{22} = 15, x_{23} = 25, x_{31} = 10, x_{34} = 35, x_{41} = 20, x_{42} = 10$.

Finally, total fuzzy transportation cost is $(8.59 \times 35) + (7 \times 15) + (7.95 \times 25) + (7.67 \times 10) + (5 \times 35) + (0 \times 20) + (0 \times 10) = 856.1$.

Now we apply MODIM to compute the optimal solution. Algorithm of MODIM is shown in Sect. 3.2. We will show only the results of improved basic feasible solution is shown in Table 18 and calculation minimum fuzzy transportation cost is given below.

Hence, optimal solution is $x_{12} = 15, x_{13} = 20, x_{21} = 30, x_{22} = 10, x_{33} = 10, x_{34} = 30, x_{43} = 30$ and minimum fuzzy transportation cost is $\tilde{\psi} = (9.52 \times 15) + (8.59 \times 20) + (4.76 \times 30) + (7 \times 10) + (9 \times 10) + (5 \times 30) + (0 \times 30) = 767.4$

Table 15. Adding a dummy source to Example 3

Source	W_1	W_2	W_3	W_4	Supply (a_i)
F_1	$\langle(2,7,15);$ $(0,7,18)\rangle$	$\langle(4,10,12);$ $(3,10,17)\rangle$	$\langle(3,7,14);$ $(2,7,18)\rangle$	$\langle(3,5,10);$ $(1,5,16)\rangle$	35
F_2	$\langle(2,4,7);$ $(1,4,10)\rangle$	$\langle(5,6,10);$ $(2,6,13)\rangle$	$\langle(6,7,12);$ $(4,7,12)\rangle$	$\langle(8,9,13);$ $(8,9,18)\rangle$	40
F_3	$\langle(5,8,10);$ $(4,8,11)\rangle$	$\langle(7,15,20);$ $(7,15,24)\rangle$	$\langle(3,9,15);$ $(2,9,16)\rangle$	$\langle(4,5,6);$ $(3,5,9)\rangle$	45
F_4	$\langle(0,0,0);$ $(0,0,0)\rangle$	$\langle(0,0,0);$ $(0,0,0)\rangle$	$\langle(0,0,0);$ $(0,0,0)\rangle$	$\langle(0,0,0);$ $(0,0,0)\rangle$	30
Demand (b_j)	30	25	60	35	150

Table 16. The IFTT after calculated $R(\widetilde{c}_{ij}^*)$

Source	W_1	W_2	W_3	W_4	Supply (a_i)
F_1	8.19	9.52	8.59	6.91	35
F_2	4.76	7	7.95	11.11	40
F_3	7.67	14.76	9	5	45
F_4	0	0	0	0	30
Demand (b_j)	50	90	60	50	150

Table 17. Finally, allocation of various cells in the allocation table

Source	W_1		W_2		W_3		W_4		Supply (a_i)
F_1	8.19		9.52		8.59	**35**	6.91		35
F_2	4.76		7	**15**	7.95	**25**	11.11		40
F_3	7.67	**10**	14.76		9		5	**35**	45
F_4	0	**20**	0	**10**	0		0		30
Demand (b_j)	50		90		60		50		150

Table 18. Improved basic feasible solution

Source	W_1		W_2		W_3		W_4		Supply (a_i)
F_1	8.19		9.52	**15**	8.59	**20**	6.91		35
F_2	4.76	**30**	7	**10**	7.95		11.11		40
F_3	7.67		14.76		9	**10**	5	**30**	45
F_4	0		0		0	**30**	0		30
Demand (b_j)	50		90		60		50		150

5   Conclusion

In this work, the ATM is used to find an initial basic feasible solution. Using centroid of TIFN for the representative value of the IFN based on the both availability and demand are real numbers. Moreover the cost is TIFN. In addition we improve the basic feasible solution by MODIM to find the optimal solution. Finally, it can be claimed that the proposed method can be used to solve TP and FTP which is TFN and TIFN. Therefore, this method can be applied to solve the real life transportation problem.

Acknowledgements. This project was supported by the Theoretical and Computational Science (TaCS) Center (Project Grant No.TaCS2560-1).

References

1. Atanassov, K.T.: Intuitionistic fuzzy sets. Fuzzy Set Syst. **20**(1), 87–96 (1986)
2. Zadeh, L.A.: Fuzzy sets. Inf. Control **8**(3), 338–353 (1965)
3. Gani, A.N., Razak, K.A.: Two stage fuzzy transportation problem. J. Phys. Sci. **10**, 63–69 (2006)
4. Li, L., Huang, Z., Da, Q., Hu, J.: A new method based on goal programming for solving transportation problem with fuzzy cost. In: International Symposiums on Information Processing, pp. 3–8 (2008)
5. Nagarajan, R., Solairaju, A.: Computing improved fuzzy optimal Hungarian assignment problem with fuzzy costs under Robust ranking techniques. Int. J. Comput. Appl. **6**(4), 6–13 (2010)
6. Hussain, R.J., Kumar, P.S.: Algorithmic approach for solving intuitionistic fuzzy transportation problem. Appl. Math. Sci. **6**(80), 3981–3989 (2012)
7. Pramila, K., Uthra, G.: Optimal solution of an intuitionistic fuzzy transportation problem. Ann. Pure Appl. Math. **8**(2), 67–73 (2014)
8. Antony, R.J.P., Savarimuthu, S.J., Pathinathan, T.: Method for solving the transportation problem using triangular intuitionistic fuzzy number. Int. J. Comput. Algorithm **3**, 590–605 (2014)
9. Hunwisai, D., Kumam, P.: A method for solving a fuzzy transportation problem via robust ranking technique and ATM. Cogent Math. **4**, 1–11 (2017)
10. Hitchcock, F.L.: The distribution of a product several sources to numerous localities. J. Math. Phys. **20**, 224–230 (1941)
11. Varghese, A., Kuriakose, S.: Centroid of an intuitionistic fuzzy number. Notes Intuitionistic Fuzzy Set **18**(1), 19–24 (2012)
12. Wang, Y.M., Yang, J.B., Xu, D.L., Chin, K.S.: On the centroids of fuzzy numbers. Fuzzy Sets Syst. **157**, 919–926 (2006)
13. Dantzig, G.B.: Application of the simplex method to a transportation problem. In: Koopmans, T.C. (ed.) Activity Analysis of Production and Allocation, pp. 359–373. Wiley, New York (1951)
14. Dinagar, D.S., Palanivel, K.: The transportation problem in fuzzy environment. Int. J. Algorithms Comput. Math. **2**(3), 65–71 (2009)
15. Pandian, P., Natarajan, G.: A new algorithm for finding a fuzzy optimal solution for fuzzy transportation problems. Appl. Math. Sci. **4**(20), 79–90 (2010)
16. Shanmugasundari, M., Ganesan, K.: A novel approach for the fuzzy optimal solution of fuzzy transportation problem. Int. J. Eng. Res. Appl. **3**(1), 416–1421 (2013)

17. Kaur, A., Kumar, A.: A new approach for solving fuzzy transportation problems using generalized trapezoidal fuzzy numbers. Appl. Soft Comput. **12**, 1201–1213 (2012)
18. Srinivas, B., Ganesan, G.: Optimal solution for intuitionistic fuzzy transportation problem via revised distribution method. Int. J. Math. Tends Technol. **19**(2), 150–161 (2015)
19. Singh, S.K., Yadav, S.P.: A new approach for solving intuitionistic fuzzy transportation problem of type-2. Ann. Oper. Res. **243**, 349–363 (2016)
20. Ahmed, M.M., Khan, A.R., Uddin, M.S., Ahmed, F.: A new approach to solve transportation problems. Open J. Optim. **5**(1), 22–30 (2016)

The Generalized Diffie-Hellman Key Exchange Protocol on Groups

Wachirapong Jirakitpuwapat[1] and Poom Kumam[2,3(✉)]

[1] Department of Mathematics, Faculty of Science, King Mongkuts University
of Technology Thonburi (KMUTT), 126 Pracha-Uthit Road, Bang Mod,
Thrung Khru, Bangkok 10140, Thailand
wachirapong.jira@hotmail.com

[2] KMUTT-Fixed Point Research Laboratory, Department of Mathematics, Room
SCL 802 Fixed Point Laboratory, Science Laboratory Building, Faculty of Science,
King Mongkuts University of Technology Thonburi (KMUTT), 126 Pracha-Uthit
Road, Bang Mod, Thrung Khru, Bangkok 10140, Thailand
poom.kum@kmutt.ac.th

[3] KMUTT-Fixed Point Theory and Applications Research Group (KMUTT-FPTA),
Theoretical and Computational Science Center (TaCS), Science Laboratory Building,
Faculty of Science, King Mongkuts University of Technology Thonburi (KMUTT),
126 Pracha-Uthit Road, Bang Mod, Thrung Khru,
Bangkok 10140, Thailand

Abstract. In this paper, we study key exchange protocol which is similar to Diffie-Hellman key exchange protocol. This key exchange protocol uses maximal abelian subgroup of automorphism of group. We give an example group is used for key exchange protocol.

Keywords: Diffie-Hellman key exchange · Cryptography · Group

Mathematics Subject Classification: Primary 94A62
Secondary 20F28

1 Introduction

In cryptography, keys exchange is a method to send a key between a sender and a recipient. The problems of the key exchange are how they send the message so that nobody else can understand the message except for the sender and the recipient. The procedure is one of the first public key cryptographic protocols used to build up a secret key between each other over insecure channel. The protocol itself is constrained to exchange of the keys for example: we are making a key together instead of sharing data while the key exchange. We implement algorithm for exchanging information over a public channel so that building up a mutual secret between two gatherings that can use for secret communication. Diffie-Hellman is suitable to use in information communication and less frequently use for information storage or archived over a long time period.

© Springer International Publishing AG 2018
L. H. Anh et al. (eds.), *Econometrics for Financial Applications*, Studies in Computational
Intelligence 760, https://doi.org/10.1007/978-3-319-73150-6_8

In cryptographic protocol has the key exchange is the first issue. For human development, people try to hide the data from other people so that composing structure. This is assumed that is the first and primitive type of encryption, but it is just only half section of cryptography. The other half is the capacity to reproduce the first message from its hidden structure. Cryptography is like a normal message but nobody except for the exact recipient will understand the message. By that time the huge majority of the cryptosystems were private of symmetric key cryptosystems. In this two clients Alice and Bob select a key, which is their private key then use the key in a private key cryptosystem to convey information over people in public channel. We investigate a public key cryptography regarding the Diffie-Hellman key Exchange Protocol, which is the most primitive thought behind a public key cryptography. In the Diffie-Hellman key exchange protocol, two clients unknown to one another can set up a private however arbitrary key for their symmetric key cryptosystem. The Diffie-Hellman key agreement protocol (1976) was the first practical method for setting up a shared secret over an insecure communication channel.

In modern cryptography, we assume that key is a only secret. Therefore if there are many keys, then the opponent hard break cryptosystem. We will generalized Diffie-Hellman key exchange protocol on groups. We choose key which is automorphism group in maximal abelian subgroup of automorphism group.

2   Preliminaries

In this section, we will introduce Diffie-Hellman Key Exchange and group.

2.1   Diffie-Hellman Key Exchange

The simple and original key exchange protocol uses the module p and $g \in \{1, \ldots, p-1\}$ where p is a prime in [1].

Example 1. Alice and Bob want to exchange key over an insecure channel.

1. Alice and Bob agree to use the module p and $g \in \{1, \ldots, p-1\}$ where p is a prime.
2. Alice choose a secret $a \in \{1, \ldots, p-1\}$. Then she sends $A = g^a \bmod p$ to Bob.
3. Bob choose a secret $b \in \{1, \ldots, p-1\}$. Then he sends $B = g^b \bmod p$ to Alice.
4. Alice compute $B^a \bmod p$.
5. Bob compute $A^b \bmod p$.
6. Alice and Bob have common secret key $A^b = B^a \bmod p$.

2.2   Groups

Definition 1. For a nonempty G, a function $\cdot : G \times G \to G$ is called a *binary operation*. Image of $(a, b) \in G$ is denoted by ab. G with binary operation is a *group* if it has properties

1. Associativity, $\forall a, b, c \in G, (a \cdot b) \cdot c = a \cdot (b \cdot c)$,
2. Identity $\exists e \in, \forall a \in G, a \cdot e = a = e \cdot a$,
3. Inverse $\forall a \in G \exists b \in G, a \cdot b = e = b \cdot a$.

Definition 2. A group G is commutative if $\forall a, b \in G, ab = ba$. A commutative group is called *abelian group*.

Definition 3. Let G and H be groups. A *homomorphism* from G to H is a map $\phi : G \to H$ which satisfy

$$\forall a, b \in G, \phi(ab) = \phi(a)\phi(b).$$

An *isomorphism* is a homomorphism which is injection and surjection. We write $G \cong H$. An *automorphism* is a isomorphism from G to G. The automorphism group of G is denoted by $Aut(G)$.

Theorem 1 [2]. *Let G be a finite abelian group. Then G is isomorphic to a product of groups of the form*

$$H_p = \mathbb{Z}_{p^{n_1}} \times \cdots \times \mathbb{Z}_{p^{n_m}},$$

in which p is a prime number and $n_1 \leq \cdots \leq n_m$ are positive integers.

Theorem 2 [2]. *Let H and K be finite groups with relatively prime orders. Then*

$$Aut(H) \times Aut(K) \cong Aut(H \times K).$$

Theorem 3 [2]. *Let $H_p = \mathbb{Z}_{p^{n_1}} \times \cdots \times \mathbb{Z}_{p^{n_m}}$ be a group which p is a prime number and $n_1 \leq \cdots \leq n_m$ are positive integers. Setting $d_k = \max\{\ell : n_\ell = n_k\}$ and $c_k = \min\{\ell : n_\ell = n_k\}$. Then*

$$|Aut(H_p)| = \prod_{k=1}^{m}(p^{d_k} - p^{k-1})\prod_{j=1}^{m}(p^{n_j^{(m-d_j)}})\prod_{i=1}^{m}(p^{n_i-1^{(m-c_i+1)}}).$$

Lemma 1. *A abelian group $G = H_{p_1} \times \cdots \times H_{p_k}$ which $p_1 < \cdots < p_k$ are prime numbers, $H_p = \mathbb{Z}_{p^{n_1}} \times \cdots \times \mathbb{Z}_{p^{n_m}}$ and $n_1 \leq \cdots \leq n_m$ are positive integers has*

$$Aut(G) = \prod_{i=1}^{k}|Aut(H_{p_i})|.$$

Proof. It's obvious by Theorems 2 and 3.

Theorem 4 [4]. *Let n be a positive integer such that $n \geq 3$ and let $k = 2n^{n-1}$. Let $G = <x, y, z, u>$ with defined by*

1. $x^{2n} = y^2 = z^2 = u^2 = [x, z] = [x, u] = [z, u] = [y, z] = 1$,
2. $yxy = x^{k+1}$,
3. $yuy = zu$.

The $Aut(G)$ is abelian group. It is isomorphic to $\mathbb{Z}_{2^6} \times \mathbb{Z}_{2^{n-2}}$. The order of G is 2^{n+3}.

3   Key Exchange Protocol

Alice and Bob want to exchange key over an insecure channel that is similar in [3].

3.1   Key Exchange Protocol I

1. Alice and Bob choose group G and an element $g \in G$ in public information. Note that G and g are public information.
2. Alice and Bob choose automorphism ϕ_A and ϕ_B from maximal abelian subgroup S of $Aut(G)$, respectively. Note that ϕ_A and ϕ_B are private information.
3. Alice and Bob compute $\phi_A(g)$ and $\phi_B(g)$ respectively and exchange them. Note that $\phi_A(g)$ and $\phi_B(g)$ are public information.
4. Both of them compute $\phi_A(\phi_B(g)) = \phi_B(\phi_A(g))$ from their private information, which is their common secret key.

In Example 1 is special case when $\phi_A(g) = g^a$, $\phi_B(g) = g^b$ and $\phi_A(\phi_B(g)) = g^{ab}$.

Remark. The opponent hard compute $\phi_A(\phi_B(g))$ from $G, g, \phi_A(g), \phi_B(g)$.

Example 2. Alice and Bob want to exchange key over an insecure channel.

1. Alice and Bob agree to use group $G = \mathbb{Z}_{p^n} \times \mathbb{Z}_{q^m}$ where p, q are prime and $g = (g_1, g_2) \in G$.
2. Alice choose $a = (a_1, a_2)$ where $gcd(a_1, n) = gcd(a_2, m) = 1$. Her automorphism is $\phi_A(g') = (g_1'^{a_1}, g_2'^{a_2})$. Bob choose $b = (b_1, b_2)$ where $gcd(b_1, n) = gcd(b_2, m) = 1$. His automorphism is $\phi_B(g') = (g_1'^{b_1}, g_2'^{b_2})$.
3. Alice and Bob compute $\phi_A(g)$ and $\phi_B(g)$ respectively and exchange them.
4. Both of them compute $\phi_A(\phi_B(g)) = \phi_B(\phi_A(g))$ from their private information, which is their common secret key.

3.2   Key Exchange Protocol II

1. Alice and Bob choose group G in public information.
2. Alice chooses automorphism ϕ_A from maximal abelian subgroup S of $Aut(G)$ and she choose an element $g \in G$. Then she sends $\phi_A(g)$ to Bob. Note that g and ϕ_A are private information but $\phi_A(g)$ is public information.
3. Bob chooses automorphism ϕ_B from maximal abelian subgroup S of $Aut(G)$. Then he send $\phi_B(\phi_A(g))$ to Alice. Note that ϕ_B is private information but $\phi_B(\phi_A(g))$ is public information.
4. Alice compute $\phi_A^{-1}(\phi_B(\phi_A(g))) = \phi_B(g)$. Next Alice choose automorphism ϕ_H from maximal abelian subgroup S of $Aut(G)$ and compute $\phi_H(g)$. Then she sends $\phi_H(\phi_B(g))$ to Bob. Note that ϕ_H is private information but $\phi_H(\phi_B(g))$ is public information.
5. Bob compute $\phi_B^{-1}(\phi_H(\phi_B(g))) = \phi_H(g)$, which is their common secret key.

Remark. The opponent hard compute $\phi_A(\phi_B(g))$ from $G, \phi_A(g), \phi_B(\phi_A(g))$, $\phi_H(\phi_B(g))$.

Example 3. Alice and Bob want to exchange key over an insecure channel.

1. Alice and Bob agree to use group $G = \mathbb{Z}_{p^n} \times \mathbb{Z}_{q^m}$ where p, q are prime.
2. Alice chooses $a = (a_1, a_2)$ where $gcd(a_1, n) = gcd(a_2, m) = 1$. Her automorphism is $\phi_A(g') = (g_1'^{a_1}, g_2'^{a_2})$. She choose an element $g \in G$. Then she sends $\phi_A(g)$ to Bob.
3. Bob chooses $b = (b_1, b_2)$ where $gcd(b_1, n) = gcd(b_2, m) = 1$. His automorphism is $\phi_B(g') = (g_1'^{b_1}, g_2'^{b_2})$. Then he send $\phi_B(\phi_A(g))$ to Alice.
4. Alice compute $\phi_A^{-1}(\phi_B(\phi_A(g))) = \phi_B(g)$. Next Alice choose $c = (c_1, c_2)$ where $gcd(c_1, n) = gcd(c_2, m) = 1$. Her automorphism is $\phi_H(g') = (g_1'^{c_1}, g_2'^{c_2})$. Then she sends $\phi_H(\phi_B(g))$ to Bob.
5. Bob compute $\phi_B^{-1}(\phi_H(\phi_B(g))) = \phi_H(g)$, which is their common secret key.

Acknowledgements. This project was supported by the Theoretical and Computational Science (TaCS) Center under Computational and Applied Science for Smart Innovation Cluster (CLASSIC), Faculty of Science, KMUTT. The third author would like to thanks the Petchra Pra Jom Klao Ph.D. Research Scholarship for financial support.

References

1. Diffie, W., Hellman, M.: New directions in cryptography. IEEE Trans. Inf. Theory **22**, 644–654 (1976)
2. Hillar, C.J., Rhea, D.L.: Automorphism of finite abelian groups. Am. Math. Mon. **114**, 917–923 (2007)
3. Mahalanobis, A.: The Diffie-Hellman Key Exchange protocal and non-abelian groups. Israel J. Math. **165**, 161–187 (2008)
4. Struik, R.R.: Some non-abelian 2-groups with abelian automorphism groups. Arch. Math. **39**, 299–302 (1982)

Combination and Composition in Probabilistic Models

Radim Jiroušek[1]([✉]) and Prakash P. Shenoy[2]

[1] Jindřichův Hradec and Institute of Information Theory and Automation,
Czech Academy of Sciences, University of Economics, Prague, Czech Republic
radim@utia.cas.cz
[2] University of Kansas, Lawrence, KS, USA
pshenoy@ku.edu

Abstract. In probability theory, as well as in other alternative uncertainty theories, the existence of efficient processes for the multidimensional model construction is a basic assumption making the application of the respective theory to practical problems possible. Most of the approaches are based on the idea that a multidimensional model is set up from a great number of smaller parts representing pieces of local knowledge. Such a process is called *knowledge integration*. In the probabilistic framework, it means that a multidimensional probability distribution is aggregated from a number of low-dimensional (possibly conditional) ones.

Historically, two different operators of aggregation were designed for this purpose: the operator of *combination*, and the operator of *composition*. This paper, using the simplest possible framework of discrete probability theory, answers some natural questions like: *What is the difference between these operators? Is there a need for both of them? Are there situations when they can be mutually interchanged?*

Keywords: Discrete probability · Aggregation of distributions · Factorization Algebraic properties · Idempotency

1 Introduction

Broad application of probability theory in artificial intelligence that took place in the last decades of the last century was facilitated by the development of new tools and models that were incorporated into the basic theoretical gear of artificial intelligence. There are many of them, though not all of them are as famous as the Bayesian networks [7], or more generally, graphical Markov models [13]. Some of them were developed not only in the theoretical framework of probability theory but also in the framework of other uncertainty theories like possibility [6] or belief functions theories [4, 17]. This holds true also for two operators of aggregation that belong among the concepts of several uncertainty theories. The goal of this paper is to make clear the difference between these two operators of aggregation and to show that both of them have their indisputable role for uncertain knowledge modeling. Namely, they are widely used in the process of knowledge integration, the process aiming at the construction of big knowledge bases of

© Springer International Publishing AG 2018
L. H. Anh et al. (eds.), *Econometrics for Financial Applications*, Studies in Computational
Intelligence 760, https://doi.org/10.1007/978-3-319-73150-6_9

intelligent systems. To make the exposition as clear as possible we restrict our consideration only to the best-known probability theory, we will study aggregation of (discrete) probability distributions (measures).

The basic idea of the knowledge integration process copies a human-like behavior. Nobody is able to express/comprehend knowledge that is too complex. Therefore, it should be formulated in small pieces of local knowledge, and the pieces of local knowledge are then aggregated to form a complete knowledge of the area of interest. Analogously, a probabilistic model of a knowledge base should be integrated from a great number of pieces of local knowledge, which are represented by small dimensional probability distributions. This way of knowledge base representation has also an additional advantage. For such models, there exist efficient computational procedures that can be applied for making inferences [1, 11, 14–16].

In probability theory there are many ways and purposes why two or more probability distributions are aggregated; see, e.g. "Aggregating Probability Distributions" by Clemen and Winkler [3]. In this paper we restrict our attention only to two ways that can be both considered as an aggregation of knowledge in AI applications: *Combination* and *Composition*.

Combination

The purpose of the combination of probability distributions can hardly be described better than it was done by Dempster in [4]: *A probability measure may be regarded as defining degrees of belief which quantify a state of partial knowledge. ... A mechanism for combining such sources of information is a virtual necessity for a theory of probability oriented to statistical inference. The mechanism adopted here assumes* independence *of the sources, a concept whose real world meaning is not so easily described as its mathematical definition.* So, by the operation of combination we understand in this paper a proper way to combine *independent* sources of information. In agreement with the Dempster's words, with the stress on the notion of independence.

As an example, consider a situation when data files are the main source of information. Let two data files describe patients from hospitals A and B, respectively. Then, a natural way of combining these two sources of information is to join the records into one file. The respective estimation of the probability distribution π corresponding to the joint data file can be got as a weighted sum of the estimations of probability distributions π_A and π_B corresponding to the data files from hospitals A and B, respectively. So, in spite of the fact that in the described situation we do not have any objections against the employment of distribution π received as a weighted sum distributions π_A and π_B (representing two sources of information), we should not consider the weighted sum of distributions a *combination* operation for probability distributions. This is because the computation of a weighted sum is not appropriate when the sources are independent. Naturally, data collected in different hospitals cannot be considered independent. They are samples from different populations, or from two disjoint parts of a population.

Composition

As the term suggests, the operation of the composition is an inverse operation to decomposition. By decomposition, we understand the result of a process that, with the goal of simplification, divides an original object into its sub-objects. Thus, for example, a problem is decomposed into two (or more) simpler sub-problems. General properties of such decomposition can be viewed on the example familiar to everybody: decomposition of a positive integer into prime numbers. In this case, an elementary decomposition is a decomposition of an integer into two factors, the product of which gives the original integer. For this example, we see that

- the result of decomposition are two objects of the same type as the decomposed object – an integer is decomposed into two integers;
- both these sub-objects are simpler (smaller) than the original object – both factors are smaller that the original integer, we do not consider $1 \times n$ to be a decomposition of n;
- not all objects can be decomposed – prime numbers cannot be decomposed;
- there exists an inverse operation (we will call it a composition) yielding the original object from its decomposed parts – the composition of two integers is their product.

It can easily be deduced from the above-presented properties that the process of a repeatedly performed decomposition of an arbitrary (finite) object into elementary sub-objects that cannot be further decomposed is always finite.

As another example, let us note that a decomposition is studied also in graph theory. A simple[1] graph $G = (V, E)$ is decomposed into two simple graphs $G_1 = (V_1, E_1)$ and $G_2 = (V_2, E_2)$ if

- $V_1 \cup V_2 = V$, $V_1 \neq V \neq V_2$,
- both G_1 and G_2 are induced subgraphs of G (i.e., $E_j = \{(u - v) \in E \; : \; \{u, v\} \subseteq V_j\}$),
- $E_1 \cup E_2 = E$.

Note that the graphs that cannot be decomposed are called *prime-graphs*.

What is a decomposition of a finite probability distribution? Consider a two-dimensional distribution $\pi(X, Y)$. Simpler sub-objects are just one-dimensional distribution $\pi(X)$ and $\pi(Y)$. Generally, the process of marginalization is unique, but, with the exception of a degenerate distribution, we cannot unambiguously reconstruct the original two-dimensional distribution from its one-dimensional marginals. To bypass this fact, we restrict the decomposition of two-dimensional distributions $\pi(X, Y)$ into their one-dimensional marginals only for the case of independence (to denote that variables X and Y are independent for distribution π we use symbol $X \perp\!\!\!\perp Y[\pi]$ – for a precise definition see the next section). In this case, $\pi(X, Y)$ can easily be reconstructed from its marginals $\pi(X)$ and $\pi(Y)$: $\pi(X, Y) = \pi(X) \cdot \pi(Y)$, where "$\cdot$" denotes pointwise multiplication, i.e., $\pi(X, Y)(x, y) = \pi(X)(x) \, \pi(Y)(y)$ for all values x of X and y of Y.

Analogously, three-dimensional distribution $\pi(X, Y, Z)$ can be decomposed into two simpler probability distributions (marginals of $\pi(X, Y, Z)$) only if either a couple of variables (say X, Y) is independent of the remaining third variable (in this case Z), or, if two variables (say X and Z) are conditionally independent given the remaining third variable (in this case Y):

- $\{X, Y\} \perp\!\!\!\perp Z[\pi]$, then $\pi(X, Y, Z)$ can be reconstructed from $\pi(X, Y)$ and $\pi(Z)$,
- $X \perp\!\!\!\perp Z | Y[\pi]$, then $\pi(X, Y, Z)$ can be reconstructed from $\pi(X, Y)$ and $\pi(Y, Z)$.

---

[1] An undirected graph containing no loops and no multiple edges.

Thus, the composition considered in this paper will be an inverse operation to the following general operation of decomposition: *Probability distribution* $\pi(X_1,\ldots,X_n)$ *can be decomposed into* $\kappa(\{X_i\}_{i\in K})$ *and* $\lambda(\{X_i\}_{i\in L})$ *if*

1. $K\cup L = \{1,2,\ldots,n\}$;
2. $K \neq \{1,2,\ldots,n\}, L \neq \{1,2,\ldots,n\}$;
3. $\pi(\{X_i\}_{i\in K\cup L})\cdot\lambda(\{X_i\}_{i\in K\cap L}) = \kappa(\{X_i\}_{i\in K})\cdot\lambda(\{X_i\}_{i\in L})$.

Notice that in this case the original distribution $\pi(X_1,X_2,\ldots,X_n)$ can be uniquely reconstructed from distributions $\kappa(\{X_i\}_{i\in K})$ and $\lambda(\{X_i\}_{i\in L})$.

The formal definitions of both combination and composition operators as well as the notation used in the paper form the content of the next section. The main part of the paper is Sect. 3 where we show what are the common properties of the studied operators (Sect. 3.1) and in what way they differ from each other (Sect. 3.2). The last section concludes the paper referring to the relation of the presented results with other uncertainty theories.

2   Basic Definitions and Notation

In this text we deal with finite-valued random variables denoted by upper case characters of Latin alphabet X,Y,Z, with possible indices. The respective finite (nonempty) sets of values of variables X,Y,Z will be denoted by $\mathbb{X},\mathbb{Y},\mathbb{Z}$, respectively. Therefore, the values of (X,Y) are from the Cartesian product $\mathbb{X}\times\mathbb{Y}$. In the case of a subset of variables $\{X_i\}_{i\in K} \subset \{X_1,X_2,\ldots,X_n\}$, we will use a simplified notation:

$$\mathbb{X}_K = \times_{i\in K}\mathbb{X}_i$$

to denote the set of values of $\{X_i\}_{i\in K}$.

Distributions of subsets of variables will be denoted by lower-case Greek alphabets $\pi,\kappa,\lambda,\ \mu,\delta$ (again with possible indices). Thus, $\pi(X_1,X_2,\ldots,X_n)$ denotes an n-dimensional probability distribution. It's *marginal distribution* for $K \subseteq \{1,\ldots,n\}$ will be denoted $\pi(\{X_i\}_{i\in K})$, or, more often simply $\pi^{\downarrow K}$. Analogously, for $x \in \mathbb{X}_{\{1,\ldots,n\}}$, $x^{\downarrow K}$ denote the *projection* of x into $\mathbb{X}_K$. When considering marginal distributions we do not exclude situations when $K = \emptyset$. In this case, we assume that $\mathbb{X}_\emptyset = \{\blacklozenge\}$, and naturally, $\pi^{\downarrow\emptyset}(\blacklozenge) = 1$.

In what follows we will also need a symbol for conditional probability distribution. For disjoint $L,M \subseteq K$, $\pi^{\downarrow L|M}$ denote the conditional probability distribution of variables $\{X_i\}_{i\in L}$ given variables $\{X_i\}_{i\in M}$, i.e., if the marginal $\pi^{\downarrow M}$ is positive then

$$\pi^{\downarrow L|M} = \frac{\pi^{\downarrow L\cup M}}{\pi^{\downarrow M}}.$$

In a general case, for each $x \in \mathbb{X}_M$, $\pi^{\downarrow L|M}(\{X_i\}_{i\in L}|x)$ is a probability distribution of variables $\{X_i\}_{i\in L}$ such that

$$\pi^{\downarrow L\cup M}(\{X_i\}_{i\in L},\{X_i\}_{i\in M} = x) = \pi^{\downarrow L|M}(\{X_i\}_{i\in L}|x)\cdot\pi^{\downarrow M}(x).$$

Thus, the conditional probability distribution $\pi^{\downarrow L|M}$ is always defined, though sometimes ambiguously (in case that $\pi^{\downarrow M}(x) = 0$ for some $x \in \mathbb{X}_M$).

In Sect. 1, we used the symbol $\perp\!\!\!\perp$ to denote the independence of variables. Let us, now, introduce it more formally. Consider a distribution $\pi(\{X_i\}_{i \in N})$, and three disjoint subset $K, L, M \subset N$, $K \neq \emptyset$, $L \neq \emptyset$. We say that for distribution π variables $\{X_i\}_{i \in K}$ and $\{X_i\}_{i \in L}$ are *conditionally independent given variables* $\{X_i\}_{i \in M}$, if for all $x \in \mathbb{X}_{K \cup L \cup M}$

$$\pi^{\downarrow K \cup L \cup M}(x) \cdot \pi^{\downarrow M}(x^{\downarrow M}) = \pi^{\downarrow K \cup M}(x^{\downarrow K \cup M}) \cdot \pi^{\downarrow L \cup M}(x^{\downarrow L \cup M}).$$

This independence will be denoted $\{X_i\}_{i \in K} \perp\!\!\!\perp \{X_i\}_{i \in L} | \{X_i\}_{i \in M} [\pi]$. If $M = \emptyset$ the independence simplifies to (unconditional - some authors say also marginal) independence $\{X_i\}_{i \in K} \perp\!\!\!\perp \{X_i\}_{i \in L} [\pi]$.

Suppose K and L are subsets of $\{1, \dots, n\}$. Two distributions $\kappa(\{X_i\}_{i \in K})$ and $\lambda(\{X_i\}_{i \in L})$ are said to be *consistent* if their joint marginals coincide: $\kappa^{\downarrow K \cap L} = \lambda^{\downarrow K \cap L}$. Notice that if $K \cap L = \emptyset$ then $\kappa(\{X_i\}_{i \in K})$ and $\lambda(\{X_i\}_{i \in L})$ are always consistent.

Having two distributions defined for the same set of variables $\pi(\{X_i\}_{i \in K})$ and $\kappa(\{X_i\}_{i \in K})$, we say that κ *dominates* π (in symbol $\pi \ll \kappa$) if for all $x \in \mathbb{X}_K$

$$\kappa(x) = 0 \implies \pi(x) = 0.$$

Combination

Here we adopt (and adapt to the introduced notation) the definition introduced by Dempster in [4].

Definition 1. For arbitrary two distributions $\kappa(\{X_i\}_{i \in K})$ and $\lambda(\{X_i\}_{i \in L})$ their *combination* is for each $x \in \mathbb{X}_{(L \cup K)}$ given by the following formula

$$(\kappa \oplus \lambda)(x) = Const^{-1} \kappa(x^{\downarrow K}) \lambda(x^{\downarrow L}),$$

where *Const* is the normalization constant given by:

$$Const = \sum_{x \in \mathbb{X}_{K \cup L}} \kappa(x^{\downarrow K}) \lambda(x^{\downarrow L}).$$

In the case where $Const = 0$, we say that distributions κ and λ are in *total conflict*, and, for this case, their combination is undefined.

Composition

The following definition was first introduced in [8].

Definition 2. For arbitrary two distributions $\kappa(\{X_i\}_{i \in K})$ and $\lambda(\{X_i\}_{i \in L})$, for which $\kappa^{\downarrow K \cap L} \ll \lambda^{\downarrow K \cap L}$ their *composition* is for each $x \in \mathbb{X}_{(L \cup K)}$ given by the following formula[2]

$$(\kappa \triangleright \lambda)(x) = \frac{\kappa(x^{\downarrow K}) \lambda(x^{\downarrow L})}{\lambda^{\downarrow K \cap L}(x^{\downarrow K \cap L})}.$$

In case that $\kappa^{\downarrow K \cap L} \not\ll \lambda^{\downarrow K \cap L}$ the composition remains undefined.

---

[2] Define $\frac{0 \cdot 0}{0} = 0$.

The reader certainly noticed that the presented definition slightly extends the notion of composition discussed in Sect. 1. We do not require that both K and L are proper subsets of $K \cup L$. There are two reasons for this. First, we are going to compare the two operations, and combination was basically defined for the distributions defined for the same variable sets. Second, this generalization makes the formulation of some theoretical properties simpler.

3   Properties of Combination and Composition

As already said above, the two operators were designed for different purposes, and so it is not surprising that they possess different properties. Nevertheless, from a formal point of view, they manifest some similar, or even identical, properties. And, it is the purpose of this section to show what the similarities and dissimilarities between the two operators are.

3.1   Common Properties

Theorem 1. *Suppose* $\kappa(\{X_i\}_{i\in K})$ *and* $\lambda(\{X_i\}_{i\in L})$ *are probability distributions.*

1. (Domain of combination): *If* $\kappa \oplus \lambda$ *is defined, then* $\kappa \oplus \lambda$ *is a probability distribution for* $\{X_i\}_{i\in K\cup L}$*, and*

$$\kappa \oplus \lambda = \left(\kappa^{\downarrow K\cap L} \oplus \lambda^{\downarrow K\cap L}\right) \cdot \kappa^{\downarrow K\setminus L|K\cap L} \cdot \lambda^{\downarrow L\setminus K|K\cap L}. \tag{1}$$

2. (Domain of composition): *If* $\kappa \triangleright \lambda$ *is defined, then* $\kappa \triangleright \lambda$ *is a probability distribution for* $\{X_i\}_{i\in K\cup L}$*, and*

$$\kappa \triangleright \lambda = \left(\kappa^{\downarrow K\cap L} \triangleright \lambda^{\downarrow K\cap L}\right) \cdot \kappa^{\downarrow K\setminus L|K\cap L} \cdot \lambda^{\downarrow L\setminus K|K\cap L}. \tag{2}$$

3. (Disjoint domains of arguments): *If* $K \cap L = \emptyset$*, then both* $\kappa \oplus \lambda$ *and* $\kappa \triangleright \lambda$ *are defined and* $\kappa \oplus \lambda = \kappa \triangleright \lambda$*.*

4. (Simple marginalization): *Let* $(K \cap L) \subseteq M \subseteq K \cup L$*. If* $\kappa \oplus \lambda$ *is defined, then*

$$(\kappa \oplus \lambda)^{\downarrow M} = \kappa^{\downarrow K\cap M} \oplus \lambda^{\downarrow L\cap M}.$$

If $\kappa \triangleright \lambda$ *is defined, then*

$$(\kappa \triangleright \lambda)^{\downarrow M} = \kappa^{\downarrow K\cap M} \triangleright \lambda^{\downarrow L\cap M}.$$

5. (Conditional independence): *Let* $K \setminus L \neq \emptyset \neq L \setminus K$*. If* $\kappa \oplus \lambda$ *is defined, then*

$$\{X_i\}_{i\in K\setminus L} \perp\!\!\!\perp \{X_i\}_{i\in L\setminus K} | \{X_i\}_{i\in K\cap L}[\kappa \oplus \lambda],$$

and if $\kappa \triangleright \lambda$ *is defined, then*

$$\{X_i\}_{i\in K\setminus L} \perp\!\!\!\perp \{X_i\}_{i\in L\setminus K} | \{X_i\}_{i\in K\cap L}[\kappa \triangleright \lambda],$$

Proof. **Ad. 1.** Consider probability distributions $\kappa(\{X_i\}_{i\in K})$ and $\lambda(\{X_i\}_{i\in L})$, such that their combination is well defined. Then, for each $x \in \mathbb{X}_{K\cup L}$:

$$(\kappa \oplus \lambda)(x) = Const^{-1} \ \kappa(x^{\downarrow K})\lambda(x^{\downarrow L})$$
$$= Const^{-1} \ \left(\kappa^{\downarrow K \cap L}(x^{\downarrow K \cap L}) \kappa^{\downarrow K \backslash L | K \cap L}(x^{\downarrow K \backslash L} | x^{\downarrow K \cap L}) \right)$$
$$\cdot \left(\lambda^{\downarrow K \cap L}(x^{\downarrow K \cap L}) \lambda^{\downarrow L \backslash K | K \cap L}(x^{\downarrow L \backslash K} | x^{\downarrow K \cap L}) \right)$$
$$= Const^{-1} \ \left(\kappa^{\downarrow K \cap L}(x^{\downarrow K \cap L}) \cdot \lambda^{\downarrow K \cap L}(x^{\downarrow K \cap L}) \right)$$
$$\cdot \kappa^{\downarrow K \backslash L | K \cap L}(x^{\downarrow K \backslash L} | x^{\downarrow K \cap L}) \lambda^{\downarrow L \backslash K | K \cap L}(x^{\downarrow L \backslash K} | x^{\downarrow K \cap L}),$$

which yields Eq. (1), because the constant *Const* in the definition of $\kappa \oplus \lambda$ and $\kappa^{\downarrow K \cap L} \oplus \lambda^{\downarrow K \cap L}$ is the same as shown below:

$$\sum_{x \in \mathbb{X}_{K \cup L}} \kappa(x^{\downarrow K})\lambda(x^{\downarrow L}) = \sum_{x \in \mathbb{X}_{K \cup L}} \left(\kappa^{\downarrow K \cap L}(x^{\downarrow K \cap L}) \kappa^{\downarrow K \backslash L | K \cap L}(x^{\downarrow K \backslash L} | x^{\downarrow K \cap L}) \right)$$
$$\cdot \left(\lambda^{\downarrow K \cap L}(x^{\downarrow K \cap L}) \lambda^{\downarrow L \backslash K | K \cap L}(x^{\downarrow L \backslash K} | x^{\downarrow K \cap L}) \right)$$
$$= \sum_{x \in \mathbb{X}_{K \cap L}} \kappa^{\downarrow K \cap L}(x^{\downarrow K \cap L}) \cdot \lambda^{\downarrow K \cap L}(x^{\downarrow K \cap L})$$
$$\cdot \left(\sum_{y \in \mathbb{X}_{K \backslash L}} \kappa^{\downarrow K \backslash L | K \cap L}(y | x^{\downarrow K \cap L}) \right) \left(\sum_{z \in \mathbb{X}_{L \backslash K}} \lambda^{\downarrow L \backslash K | K \cap L}(z | x^{\downarrow K \cap L}) \right)$$
$$= \sum_{x \in \mathbb{X}_{K \cap L}} \kappa^{\downarrow K \cap L}(x^{\downarrow K \cap L}) \cdot \lambda^{\downarrow K \cap L}(x^{\downarrow K \cap L}).$$

The last equality holds true because both the expressions in parentheses equal 1.

Ad. 2. Equation (2) can be proven analogously to the first part of this proof.

Ad. 3. The assertion follows immediately from Eqs. (1) and (2), because $\kappa^{\downarrow \emptyset} \oplus \lambda^{\downarrow \emptyset} = \kappa^{\downarrow \emptyset} \triangleright \lambda^{\downarrow \emptyset} = 1$.

Ad. 4. Assume $(K \cap L) \subseteq M \subseteq K \cup L$, and $\kappa \oplus \lambda$ is defined. Then

$$(\kappa \oplus \lambda)^{\downarrow M} = \left(\left(\kappa^{\downarrow K \cap L} \oplus \lambda^{\downarrow K \cap L} \right) \cdot \kappa^{\downarrow K \backslash L | K \cap L} \cdot \lambda^{\downarrow L \backslash K | K \cap L} \right)^{\downarrow M}$$
$$= \left(\kappa^{\downarrow K \cap L} \oplus \lambda^{\downarrow K \cap L} \right) \cdot \left(\kappa^{\downarrow K \backslash L | K \cap L} \cdot \lambda^{\downarrow L \backslash K | K \cap L} \right)^{\downarrow M}$$
$$= \left(\kappa^{\downarrow K \cap L} \oplus \lambda^{\downarrow K \cap L} \right) \cdot \kappa^{\downarrow (K \cap M) \backslash L | K \cap L} \cdot \lambda^{\downarrow (L \cap M) \backslash K | K \cap L} = \kappa^{\downarrow K \cap M} \oplus \lambda^{\downarrow L \cap M}.$$

The respective assertion for the operator of composition was formulated (and proven) as Lemma 5.10 in [10], however, using Eq. 2 it can easily be proven analogously to the preceding part of the proof.

Ad. 5. Assume $\kappa \oplus \lambda$ is defined. Then due to already proven Property 4 (Simple marginalization)

$$(\kappa \oplus \lambda)^{\downarrow K} = \kappa \oplus \lambda^{\downarrow K \cap L}$$
$$(\kappa \oplus \lambda)^{\downarrow L} = \kappa^{\downarrow K \cap L} \oplus \lambda$$
$$(\kappa \oplus \lambda)^{\downarrow K \cap L} = \kappa^{\downarrow K \cap L} \oplus \lambda^{\downarrow K \cap L},$$

and therefore one can easily verify the validity of the equality defining the required conditional independence: For each $x \in \mathbb{X}_{K \cup L}$

$$
\begin{aligned}
(\kappa \oplus \lambda)(x) \cdot (\kappa \oplus \lambda)^{\downarrow K \cap L}(x^{\downarrow K \cap L}) & \\
= Const^{-1} \; \kappa(x^{\downarrow K})\lambda(x^{\downarrow L}) \cdot Const^{-1} \; \kappa^{\downarrow K \cap L}(x^{\downarrow K \cap L})\lambda^{\downarrow K \cap L}(x^{\downarrow K \cap L}) & \\
= Const^{-1} \; \kappa(x^{\downarrow K})\lambda^{\downarrow K \cap L}(x^{\downarrow K \cap L}) \cdot Const^{-1} \; \kappa^{\downarrow K \cap L}(x^{\downarrow K \cap L})\lambda(x^{\downarrow L}) & \\
= (\kappa \oplus \lambda)^{\downarrow K} \cdot (\kappa \oplus \lambda)^{\downarrow L}, &
\end{aligned}
$$

which proves the required property for the operator of combination (notice that $Const = \sum_{x \in \mathbb{X}_{K \cup L}} \kappa(x^{\downarrow K})\lambda(x^{\downarrow L}) = \sum_{y \in \mathbb{X}_{K \cap L}} \kappa^{\downarrow K \cap L}(y)\lambda^{\downarrow K \cap L}(y)$).

The respective assertion for the operator of composition was formulated (and proven) as Lemma 5.2 in [10], however, it can be proven analogously to the preceding part of the proof. □

Conditioning

Consider arbitrary two distributions $\kappa(\{X_i\}_{i \in K})$ and $\lambda(\{X_i\}_{i \in L})$. From formulae (1) and (2) it immediately follows that $\kappa \oplus \lambda = \kappa \triangleright \lambda$ if and only if $\kappa^{\downarrow K \cap L} \oplus \lambda^{\downarrow K \cap L} = \kappa^{\downarrow K \cap L} \triangleright \lambda^{\downarrow K \cap L}$. As expressed in Property 3 of the previous theorem, it holds if $K \cap L = \emptyset$. However it holds also in other situations. In this paragraph we are going to show that this happens also in the case when one of these distributions is a degenerate one-dimensional distribution expressing certainty. Consider variable X_k and its value $a \in \mathbb{X}_k$. The probability distribution $\delta_a(X_k)$ expressing for certain that variable $X_k = a$ is defined for each $x \in \mathbb{X}_k$ as

$$\delta_a(x) = \begin{cases} 1, & \text{if } x = a; \\ 0, & \text{otherwise.} \end{cases}$$

Let us now show that, using the respective definitions, $\delta_a(X_k) \oplus \lambda(\{X_i\}_{i \in L}) = \delta_a(X_k) \triangleright \lambda(\{X_i\}_{i \in L})$. In case that $k \notin L$ this equality holds because of Property 3 of Theorem 1. Therefore, consider the case when $k \in L$. Then for each $x \in \mathbb{X}_L$

$$(\delta_a \oplus \lambda)(x) = Const^{-1}\delta_a(x^{\downarrow \{k\}})\lambda(x) = \begin{cases} Const^{-1}\lambda(x), & \text{if } x^{\downarrow \{k\}} = a; \\ 0, & \text{otherwise,} \end{cases} \quad (3)$$

where

$$Const = \sum_{x \in \mathbb{X}_L} \delta_a(x^{\downarrow \{k\}})\lambda(x) = \sum_{x \in \mathbb{X}_L : x^{\downarrow \{k\}} = a} \lambda(x) = \lambda^{\downarrow \{k\}}(a). \quad (4)$$

Notice that $\delta_a \oplus \lambda$ is a probability distribution when $\lambda^{\downarrow \{k\}}(a)$ is positive; otherwise δ_a and λ are in total conflict and their combination is not defined.

Using Eq. (4) we can rewrite formula (3) into the form

$$(\delta_a \oplus \lambda)(x) = \frac{\delta_a(x^{\downarrow \{k\}})\lambda(x)}{\lambda^{\downarrow \{k\}}(a)} = \begin{cases} \frac{\delta_a(x^{\downarrow \{k\}})\lambda(x)}{\lambda^{\downarrow \{k\}}(x^{\downarrow \{k\}})}, & \text{if } x^{\downarrow \{k\}} = a; \\ 0, & \text{otherwise,} \end{cases}$$

which obviously equal $(\delta_a \triangleright \lambda)(x)$ in case that $\lambda^{\downarrow\{k\}}(a)$ is positive (otherwise $\lambda^{\downarrow\{k\}} \not\gg \delta_a$ and the composition is not defined), because

$$(\delta_a \triangleright \lambda)(x) = \frac{\delta_a(x^{\downarrow\{k\}})\lambda(x)}{\lambda^{\downarrow\{k\}}(x^{\downarrow\{k\}})} = \begin{cases} \frac{\delta_a(x^{\downarrow\{k\}})\lambda(x)}{\lambda^{\downarrow\{k\}}(x^{\downarrow\{k\}})}, & \text{if } x^{\downarrow\{k\}} = a; \\ 0, & \text{otherwise.} \end{cases}$$

So, let us summarize the proven equality along with what was proven about the composition operator in Theorem 2.3 in [2].

Theorem 2. *Consider a distribution* $\lambda(\{X_i\}_{i\in L})$, *variable* $X_k \in \{X_i\}_{i\in L}$, *its value* $a \in \mathbb{X}_k$, *and* $K \subseteq L \setminus \{k\}$. *If* $\lambda^{\downarrow\{k\}}(a) > 0$, *then the corresponding conditional distribution* $\lambda(\{X_i\}_{i\in K}|X_k = a)$ *can be computed*

$$\lambda(\{X_i\}_{i\in K}|X_k = a) = (\delta_a(X_k) \oplus \lambda)^{\downarrow K} = (\delta_a(X_k) \triangleright \lambda)^{\downarrow K}.$$

It is worth mentioning that this assertion formally justifies what is often called "Dempster conditioning". Recall that, for example, Dubois and Denœux describe it in [5] as *a special case of Dempster rule of combination,* which, *widely used in evidence theory, can be viewed as a revision process, understood as a prioritized merging of a sure piece of information with an uncertain one.*

3.2 Differences

Commutativity

From Definition 1 it is obvious that the operator of combination is commutative. On the other hand, it is equally evident that, generally, the operator of composition is not commutative. To show it, it is enough to consider a pair of distributions $\kappa(\{X_i\}_{i\in K})$, $\lambda(\{X_i\}_{i\in L})$, for which $\kappa^{\downarrow K \cap L} \neq \lambda^{\downarrow K \cap L}$. Let us express the respective property precisely in the following assertion, the proof of which can be found in [10].

Theorem 3. *For arbitrary two probability distributions* $\kappa(\{X_i\}_{i\in K})$ *and* $\lambda(\{X_i\}_{i\in L})$, *for which either* $\lambda^{\downarrow K \cap L} \ll \kappa^{\downarrow K \cap L}$, *or* $\lambda^{\downarrow K \cap L} \gg \kappa^{\downarrow K \cap L}$, *it holds that* κ *and* λ *are consistent if and only if* $\kappa \triangleright \lambda = \lambda \triangleright \kappa$.

Associativity

The associativity of the operator of combination is again obvious from the definition. The corresponding properties of the operator of composition are expressed in the following assertion.

Theorem 4. *Consider three probability distributions* $\kappa(\{X_i\}_{i\in K})$, $\lambda(\{X_i\}_{i\in L})$ *and* $\mu(\{X_i\}_{i\in M})$.

1. (Non-associativity): *In general,* $(\kappa \triangleright \lambda) \triangleright \mu \neq \kappa \triangleright (\lambda \triangleright \mu)$.
2. (Associativity under RIP): *Let* $\kappa \triangleright (\lambda \triangleright \mu)$ *be defined. If* $K \supset (L \cap M)$, *or* $L \supset (K \cap M)$, *then* $(\kappa \triangleright \lambda) \triangleright \mu = \kappa \triangleright (\lambda \triangleright \mu)$.
3. (Exchangeability): *If* $K \supset (L \cap M)$, *then* $(\kappa \triangleright \lambda) \triangleright \mu = (\kappa \triangleright \mu) \triangleright \lambda$.

Proof. **Ad. 1.** To show non-associativity, it is enough to consider $\kappa(X), \lambda(Y)$, and $\mu(X,Y)$, such that $X \not\perp\!\!\!\perp Y[\mu]$. Then, it is easy to show that $X \perp\!\!\!\perp Y[(\kappa \triangleright \lambda) \triangleright \mu]$ (this is because it follows from the definition that $(\kappa(X) \triangleright \lambda(Y)) \triangleright \mu(X,Y) = \kappa(X) \triangleright \lambda(Y))$, and $X \not\perp\!\!\!\perp Y[\kappa \triangleright (\lambda \triangleright \mu)]$. The latter relation follows from the fact that for the given choice of distributions

$$X \perp\!\!\!\perp Y[\kappa \triangleright (\lambda \triangleright \mu)] \iff X \perp\!\!\!\perp Y[\lambda \triangleright \mu] \iff X \perp\!\!\!\perp Y[\mu].$$

Ad. 2. This property was proven in [10] as Theorems 7.2. and 7.3.

Ad. 3. This property was proven in [10] as Lemma 5.7. $\qquad\square$

Notice that from the commutativity and associativity of the operator of combination it follows that $(\kappa \oplus \lambda) \oplus \mu = (\kappa \oplus \mu) \oplus \lambda$ holds always true. Thus, the exchangeability property holds for the combination operator trivially.

Idempotency of composition

The following assertion summarizes the basic properties of the operator of composition, neither of which, generally, hold for the operator of combination. The respective proofs can be found in [9, 10].

Theorem 5. *Suppose* $\kappa(\{X_i\}_{i \in K})$ *an* $\lambda(\{X_i\}_{i \in L})$ *are probability distributions such that* $\kappa^{\downarrow K \cap L} \ll \lambda^{\downarrow K \cap L}$. *Then the following statements hold true:*

1. (Extension): *If* $M \subseteq K$ *then,* $\kappa^{\downarrow M} \triangleright \kappa = \kappa$.
2. (Composition preserves first marginal): $(\kappa \triangleright \lambda)^{\downarrow K} = \kappa$.
3. (Reduction): *If* $L \subseteq K$ *then,* $\kappa \triangleright \lambda = \kappa$.
4. (Perfectization): $\kappa \triangleright \lambda = \kappa \triangleright (\kappa \triangleright \lambda)^{\downarrow L}$.
5. (Stepwise composition): *If* $(K \cap L) \subseteq M \subseteq L$ *then,* $(\kappa \triangleright \lambda^{\downarrow M}) \triangleright \lambda = \kappa \triangleright \lambda$.

All these properties are, in a way, connected with the fact that the operator of composition is *idempotent*. This fact supports the explanation of the difference between the combination and the composition.

The composition assembles (composes) pieces of knowledge that are supposed to have their origin by decomposition of global knowledge. So it corresponds, for example, to the reconstruction of a picture that was torn into pieces. Having one piece of the picture twice does not help us to reconstruct the picture better than if we have this very piece of picture only once.

In contrast, the combination operator combines pieces of knowledge from *independent* sources. So, in this case, one cannot have the same piece of knowledge twice. Though the two pieces of knowledge can (formally) be expressed in the same way, being from independent sources they are not (they cannot be) identical. Learning from two independent physicians that I am healthy makes me feel better than when I hear this message just from one of them. Hearing the same message repeatedly from independent sources decreases my uncertainty, and this is exactly the property, which is expressed in the following assertion. Using Shannon entropy of a probability distribution $\kappa(X)$

$$H(\kappa) = -\sum_{x \in \mathbb{X}} \kappa(x) \log_2 \kappa(x)$$

as a measure of uncertainty connected with the distribution κ, the following assertion says that getting the same amount of information from another independent source decreases our uncertainty. This, in a way, corresponds to what is often understood by the Latin proverb "Repetitio est mater studiorum".

Theorem 6. *For an arbitrary probability distribution π*

$$H(\pi \oplus \pi) \leq H(\pi). \tag{5}$$

Proof. In the proof, we will use an obvious property of Shannon entropy: For two probability distributions $\kappa(X)$ and $\lambda(X)$, such that $\kappa(x) = \lambda(x)$ for all $x \in \mathbb{X} \setminus \{a,b\}$, $\kappa(a) - \lambda(a) = \lambda(b) - \kappa(b) > 0$, and $\kappa(a) \leq \kappa(b)$ it holds that

$$H(\kappa) > H(\lambda). \tag{6}$$

This property can be proven by the following simple consideration. Denote $\varepsilon = \kappa(a) - \lambda(a)$, and

$$f(\varepsilon) = H(\kappa) - H(\lambda) = -\kappa(a)\log_2(\kappa(a)) - \kappa(b)\log_2(\kappa(b))$$
$$+ (\kappa(a) - \varepsilon)\log_2(\kappa(a) - \varepsilon) + (\kappa(b) + \varepsilon)\log_2(\kappa(b) + \varepsilon).$$

Since $f(0) = 0$, and

$$f'(\alpha) = \log_2\left(\frac{\kappa(b) + \alpha}{\kappa(a) - \alpha}\right) \cdot (\ln(2))^{-1}$$

is nonnegative for all $\alpha \in [0, \varepsilon]$, it is clear that $f(\varepsilon) > 0$, and therefore strict inequality (6) holds true.

To prove inequality (5) for $\pi(X)$ notice that

$$(\pi \oplus \pi)(x) = (\pi(x))^2 \cdot Const^{-1},$$

where $Const = \sum_{x \in \mathbb{X}} (\pi(x))^2$, and therefore

$$\begin{aligned}(\pi \oplus \pi)(x) < \pi(x) \ \text{ iff } \ \pi(x) < Const, \\ (\pi \oplus \pi)(x) > \pi(x) \ \text{ iff } \ \pi(x) > Const.\end{aligned} \tag{7}$$

To finish the proof we will construct a finite sequence of probability distributions, such that $\pi = \pi_0, \pi_1, \pi_2, \ldots, \pi_k = \pi \oplus \pi$, and $H(\pi_i) < H(\pi_{i-1})$ for all $i = 1, 2, \ldots, k$.

Consider π_i (starting with π_0), and denote $\mathbb{A}_i = \{x \in \mathbb{X} : \pi_i(\{x\}) \neq (\pi \oplus \pi)(x)\}$. Let a be the element of $\mathbb{A}_i$, for which the difference between $\pi_i(x)$ and $(\pi \oplus \pi)(x)$ is minimal, i.e.,

$$|\pi_i(a) - (\pi \oplus \pi)(a)| \leq |\pi_i(x) - (\pi \oplus \pi)(x)| \ \forall x \in \mathbb{A}_i. \tag{8}$$

Naturally, there must exist $b \in \mathbb{A}_i$ such that

$$\text{sign}(\pi_i(b) - (\pi \oplus \pi)(b)) = -\text{sign}(\pi_i(a) - (\pi \oplus \pi)(a)), \tag{9}$$

and, because of (8), $|\pi_i(b) - (\pi \oplus \pi)(b)| \geq |\pi_i(a) - (\pi \oplus \pi)(a)|$. Therefore we can define distribution π_{i+1}:

$$\pi_{i+1}(a) = (\pi \oplus \pi)(a),$$

$$\pi_{i+1}(b) = \pi_i(b) + (\pi_i(a) - (\pi \oplus \pi)(a)),$$
$$\pi_{i+1}(x) = \pi_i(x) \text{ for all } x \in \mathbf{X} \setminus \{a,b\}.$$

We immediately see that $|\mathbb{A}_{i+1}| > |\mathbb{A}_i|$, and therefore the sequence $\pi = \pi_0, \pi_1$, $\pi_2, \ldots, \pi_k = \pi \oplus \pi$ must be finite. We also can see that, because of inequalities (7), each pair π_i and π_{i+1} meets the assumptions of the property presented at the beginning of this proof. Therefore $H(\pi_{i+1}) < H(\pi_i)$, which completes the proof. $\square$

Factorization/Decomposition

Property 5 of Theorem 1 says that a relation of conditional independence holds for distributions that are created as a combination or composition of two probability distributions $\kappa(\{X_i\}_{i \in K})$ and $\lambda(\{X_i\}_{i \in L})$. The following assertion expresses the fact that for the operator of composition the assertion may be strengthen. For this operator it can be formulated in a form of equivalence that was proven as Corollary 5.3. in [10].

Theorem 7. *For arbitrary probability distribution* $\mu(\{X_i\}_{i \in M})$ *and* $K, L \subset M$ *such that* $K \setminus L \neq \emptyset \neq L \setminus K$

$$\{X_i\}_{i \in K \setminus L} \perp\!\!\!\perp \{X_i\}_{i \in L \setminus K} | \{X_i\}_{i \in K \cap L}[\mu]$$

if and only if $\mu^{\downarrow K \cup L} = \mu^{\downarrow K} \triangleright \mu^{\downarrow L}$.

Let us conclude this section by saying that combining Property 5 of Theorem 1 with Theorem 7 we get the following assertion, which casts a new viewpoint to the relation between the two studied operators.

Corollary. *Assume that* $K \setminus L \neq \emptyset \neq L \setminus K$. *For any two probability distribution* $\kappa(\{X_i\}_{i \in K})$ *and* $\lambda(\{X_i\}_{i \in L})$, *which are not in total conflict, their combination* $\kappa \oplus \lambda$ *can be expressed in the following way*

$$\kappa \oplus \lambda = (\kappa \oplus \lambda)^{\downarrow K} \triangleright (\kappa \oplus \lambda)^{\downarrow L}.$$

4 Summary and Conclusions

In this paper, we studied the properties of the operators of combination and composition. Though these operators were designed to solve different tasks, both of them may be used in the process of knowledge integration, and both of them were defined not only in classical probability theory but also in some alternative uncertainty theories like the possibility and belief functions theories. Since both these operators were introduced in Shenoy's valuation-based system framework [12, 18], it means that they can directly be applied in other uncertainty theories such as Spohn's epistemic belief theory [20], Dempster-Shafer theory [17], and others.

The purpose, for which the two operators were designed, is explained in Sect. 1. The remainder of the paper studies the formal (algebraic) properties of these operators stressing their common features and differences. To make the presentation as simple as possible, we restricted the exposition to classical probability theory. The readers familiar with alternative uncertainty theories should keep in mind that analogous results can be formulated in all the theories that meet the axioms of Shenoy's valuation-based systems [18, 19], and also in a general possibility theory [21]. It concerns also the important formulae (1) and (2), which help us to reveal surprisingly many similarities of the studied operators, the similarities that may explain why some researchers do not properly distinguish between the combination and composition. From the formal point of view, an interesting result is formulated as a corollary in the last section. It says that if a distribution is a combination of two independent distributions then it may be expressed also as a composition of its marginals.

Acknowledgements. This work has been supported in part by funds from grant GAČR 15-00215S to the first author, and from the Ronald G. Harper Distinguished Professorship at the University of Kansas to the second author.

References

1. Bína, V., Jiroušek, R.: Marginalization in multidimensional compositional models. Kybernetika **42**(4), 405–422 (2006)
2. Bína, V., Jiroušek, R.: On computations with causal compositional models. Kybernetika **51**(3), 525–539 (2015)
3. Clemen, R.T., Winkler, R.L.: Aggregating probability distributions. In: Edwards, W., Miles, J.R., Ralph, F., von Winterfeldt, D. (eds.) Advances in Decision Analysis: From Foundations to Applications, pp. 154–176. Cambridge University Press (2007)
4. Dempster, A.P.: Upper and lower probabilities induced by a multivalued mapping. Ann. Math. Stat. **38**(2), 325–339 (1967)
5. Dubois, D., Denœux, T.: Conditioning in Dempster-Shafer theory: prediction vs. revision. In: Belief Functions, pp. 385–392 (2012)
6. Dubois, D., Prade, H.: Possibility Theory: An Approach to Computerized Processing of Uncertainty. Plenum Press, New York (1988)
7. Jensen, F.V.: Bayesian Networks and Decision Graphs. IEEE Computer Society Press, New York (2001)
8. Jiroušek, R.: Composition of probability measures on finite spaces. In: Geiger, D., Shenoy, P.P. (eds.) Proceedings of the 13th Conference on Uncertainty in Artificial Intelligence (UAI 1997), pp. 274–281. Morgan Kaufmann (1997)
9. Jiroušek, R.: Decomposition of multidimensional distributions represented by perfect sequences. Ann. Math. Artif. Intell. **35**(1–4), 215–226 (2002)
10. Jiroušek, R.: Foundations of compositional model theory. Int. J. Gen. Syst. **40**(6), 623–678 (2011)
11. Jiroušek, R.: Local computations in Dempster-Shafer theory of evidence. Int. J. Approx. Reason. **53**(8), 1155–1167 (2012)
12. Jiroušek, R., Shenoy, P.P.: Compositional models in valuation-based systems. Int. J. Approx. Reason. **55**(1), 277–293 (2014)
13. Lauritzen, S.L.: Graphical Models. Oxford University Press, Oxford (1996)
14. Lauritzen, S.L., Spiegelhalter, D.: Local computation with probabilities on graphical structures and their application to expert systems. J. Roy. Stat. Soc. B **50**, 157–224 (1988)
15. Malvestuto, F.M.: Equivalence of compositional expressions and independence relations in compositional models. Kybernetika **50**(3), 322–362 (2014)
16. Malvestuto, F.M.: Marginalization in models generated by compositional expressions. Kybernetika **51**(4), 541–570 (2015)

17. Shafer, G.: A Mathematical Theory of Evidence. Princeton University Press, Princeton (1976)
18. Shenoy, P.P.: A valuation-based language for expert systems. Int. J. Approx. Reason. **3**(5), 383–411 (1989)
19. Shenoy, P.P., Shafer, G.: Axioms for probability and belief-function propagation. In: Classic Works of the Dempster-Shafer Theory of Belief Functions, pp. 499–528. Springer, Heidelberg (2008)
20. Spohn, W.: A general non-probabilistic theory of inductive reasoning. In: Shachter, R.D., Levitt, T.S., Lemmer, J.F., Kanal, L.N. (eds.) Uncertainty in Artificial Intelligence 4 (UAI 1990), pp. 274–281. North Holland (1990)
21. Vejnarová, J.: Composition of possibility measures on finite spaces: preliminary results. In: Bouchon-Meunier, B., Yager, R.R. (eds.) Proceedings of 7th International Conference on Information Processing and Management of Uncertainty in Knowledge-Based Systems IPMU 1998, pp. 25–30. Editions E.D.K., Paris (1998)

Efficient Parameter-Estimating Algorithms for Symmetry-Motivated Models: Econometrics and Beyond

Vladik Kreinovich[1(✉)], Anh H. Ly[2], Olga Kosheleva[1],
and Songsak Sriboonchitta[3]

[1] University of Texas at El Paso, 500 W. University, El Paso, TX 79968, USA
{vladik,olgak}@utep.edu
[2] Banking University of Ho Chi Minh City,
56 Hoang Dieu 2, Quan Thu Duc, Thu Duc, Ho Ch Minh City, Vietnam
[3] Faculty of Economics, Chiang Mai University, Chiang Mai 50200, Thailand
songsakecon@gmail.com

Abstract. It is known that symmetry ideas can explain the empirical success of many non-linear models. This explanation makes these models theoretically justified and thus, more reliable. However, the models remain non-linear and thus, identification or the model's parameters based on the observations remains a computationally expensive nonlinear optimization problem. In this paper, we show that symmetry ideas can not only help to select and justify a nonlinear model, they can also help us design computationally efficient almost-linear algorithms for identifying the model's parameters.

1 Formulation of the Problem

Need for prediction. In many real-life situations, we have a quantity x that changes with time t, and we want to use the previous values of this quantity to predict its future values. For example, we know how the stock price has changed with time, and we want to use this information to predict future stock prices.

In many cases, such a prediction is possible. For example, when weather records show clear yearly cycles, it is reasonable to predict that a similar yearly cycle will be observed in the future as well.

How can we predict: main idea. A usual approach to prediction is that we select some *model*, i.e., some parametric family of functions $f(t, c_1, \ldots, c_\ell)$. Based on the available observations, we find the parameters $\widetilde{c}_i$ which provide the best fit, and then we use these values $\widetilde{c}_j$ to predict the future values of the quantity x as

$$x(t) \approx f(t, \widetilde{c}_1, \ldots, \widetilde{c}_\ell).$$

Examples of models. In some cases, the dependence of the quantity x on time t is polynomial, in which case

$$f(t, c_1, \ldots, c_\ell) = c_1 + c_2 \cdot t + c_3 \cdot t^2 + \ldots + c_\ell \cdot t^{\ell-1}.$$

© Springer International Publishing AG 2018
L. H. Anh et al. (eds.), *Econometrics for Financial Applications*, Studies in Computational Intelligence 760, https://doi.org/10.1007/978-3-319-73150-6_10

For a simple periodic process, the dependence of the quantity x on time is described by a sinusoid, in which case

$$f(t, c_1, c_2, c_3) = c_1 \cdot \sin(c_2 \cdot t + c_3).$$

To get a more realistic description of a periodic process, we need to take into account higher harmonics, i.e., assume that

$$f(t, c_1, c_2, \ldots) = c_1 \cdot \sin(c_2 \cdot t + c_3) + c_4 \cdot \sin(2c_2 \cdot t + c_5) + \ldots$$

For a simple radioactive decay, the amount of radioactive material decreases exponentially:

$$f(t, c_1, c_2) = c_1 \cdot \exp(-c_2 \cdot t).$$

A more realistic model takes into account that often, a radioactive material is a mixture of several different isotopes, with different half-lives. In this case,

$$f(t, c_1, c_2, \ldots) = c_1 \cdot \exp(-c_2 \cdot t) + c_3 \cdot \exp(-c_4 \cdot t) + \ldots$$

Other models include *log-periodic model*

$$f(t, c_1, c_2, \ldots, c_7) = c_1 + c_2 \cdot (c_3 - t)^{c_4} + c_5 \cdot (c_3 - t)^{c_4} \cdot \cos(c_6 \cdot \ln(c_3 - t) + c_7)$$

which is used to predict economic crashes [2–5, 7–12, 14, 21–26], or a model

$$f(t, c_1, c_2, c_3) = c_1 \cdot \ln(t - c_2) + c_3$$

that describes, for some software packages, the dependence of the number of uncovered faults on time t; see, e.g., [15, 16].

A more complex example is a neural network, when c_j are the corresponding weights; see, e.g., [1, 6].

How do we estimate the parameters? Usually, the Least Squares method is used to estimate the values of the parameters $c_1, \ldots, c_\ell$.

In other words, based on the values $x(t_i)$ observed at different moments of time t_i, $1 \le i \le n$, we find the values c_j for which the mean square approximation error is the smallest possible, i.e., for which the following expression is minimized:

$$\sum_{i=1}^{n} (x_i - f(t_i, c_1, \ldots, c_\ell))^2. \tag{1}$$

Identifying the model's parameters is often computationally intensive. In some cases – e.g., for the polynomial dependence – the model $f(x, c_1, \ldots, c_\ell)$ linearly depends on the values of the parameters c_j. In this case, the minimized expression (1) is quadratic in c_j.

We can find the minimum of a function of several variables by equating all its partial derivatives to 0. For a quadratic objective function (1), all the partial derivatives are linear functions of c_j. Thus, by equating them all to 0, we get a system of linear equations for the unknowns c_j. For solving systems of linear equations, there are many efficient algorithms, so in this case, the problem of identifying the model's parameters is computationally easy.

On the other hand, in general, the dependence of the model on the parameters c_j is non-linear. Thus, the objective function (1) is more complex than quadratic. It is known that, in general, optimization is computationally intensive – for example, it has been proven that optimization is an NP-hard problem, meaning that it as complex as a computational problem can be; see, e.g., [13, 17, 18].

It is therefore desirable to select models for which identification is easier. This bring us to a question of how we select models in the first place.

How are models selected in the first place? Sometimes, we have an good understanding of the processes that cause the quantity x to change. In such situations, we have a theoretically justified model.

In most cases, however, the model is selected empirically. We try different models, and we select the one for which, for the same number of parameters, the approximation error is the smallest.

In many cases, the empirical efficiency of selected models can be explained by symmetry ideas. In an empirical choice, we only compare a few possible models. As a result, the fact that the selected model turned out to be better than others does not necessarily mean that this model is indeed the best for a given phenomenon: there are, in principle, many other models that we did not consider in our empirical comparison.

Good news is that in many cases, the empirical selection can be confirmed by a theoretical analysis. For example, often, it turns out that the empirically successful model can be derived from the natural symmetry requirements; see, e.g., [16]. This theoretical justification compares the selected model not just with a few others, but with *all* possible models – thus, it makes us more confident that the selected model is indeed the best.

But the model remains computationally intensive. The fact that the empirically selected model is theoretically justified does not change its formulas. So, if the dependence of this model on the corresponding parameters c_j is non-linear, the problem of identifying parameters of this model remains computationally intensive.

What we do in this paper: we show that symmetries can help in parameter identification too. In this paper, we show that symmetries are not only helpful in selecting a model, they can also help design computationally efficient algorithms for identifying parameters of the selected model.

Structure of this paper. In Sect. 2, we briefly recall what symmetries are used to derive the corresponding models, how exactly these models are derived, and what are the resulting models. In Sect. 3, we analyze the problem of determining parameters of these models, and we show how to make this identification computationally easier.

2 How Symmetries Justify Models: A Brief Reminder

Preliminaries. In some practical cases, the changes in the quantity x come from a single and simple process – this is the situation, e.g., with most oscillations. In most practical cases, however, many different factors lead to changes in x. Some of these changes are independent, and may have different intensity. Thus, the resulting value of the quantity x can be represented as a linear combination of the dependencies corresponding to different factors.

In precise terms, this means that we consider models of the type

$$C_1 \cdot e_1(t) + \ldots + C_m \cdot e_m(t) \tag{2}$$

for some functions $e_j(t)$ (which may depend on other parameters as well).

- This is the case for polynomials, when $e_1(t) = 1$, $e_2(t) = t$, $e_3(t) = t^2$, etc.
- This is the case for periodic processes, when $e_1(t)$ is the main sinusoid, $e_2(t)$ is the sinusoid corresponding to double frequency, $e_3(t)$ is the sinusoid corresponding to triple frequency, etc.
- This is the case for radioactive decay, where $e_j(t)$ are exponential functions with different hall-life.

In all these cases, the functions $e_j(t)$ are differentiable (smooth). So, without losing generality, we can assume that these functions are smooth.

In these terms, selecting a model means selecting the corresponding functions

$$e_1(t), \ldots, e_m(t).$$

What natural symmetries should we consider? Many physical processes – such as radioactive decay – do not have a starting point, their general properties do not change whether we consider the piece of a radioactive material now or in a hundred years. The exact amount of the material will decrease, but its properties – and its rate of decay – will remain the same. In such situations, the observed value $x(t)$ changes with time, but the whole family of functions (2) should not change if we simply start counting time from a different starting point.

If we start to count time from a starting point which is t_0 moments in the future, then moment t in the new scale corresponds to moment $t + t_0$ in the original scale. Thus, if in the new scale, the set of functions has the form (2), then these same functions in the original time scale have the form

$$C_1 \cdot e_1(t + t_0) + \ldots + C_m \cdot e_m(t + t_0). \tag{3}$$

The above natural requirement then says that the families (2) and (3) must coincide – i.e., that:

- every function of type (2) can be expressed in the form (3) (with, of course, different constants C_j), and
- vice versa, every function of type (3) can be expressed in the form (2).

In other cases, there *is* a natural starting (or ending) point t_0, but there is no preferred time unit. In such cases, it is reasonable to require that if we use a different unit for measuring time, nothing will change – in particular, the class (2) of possible dependencies should not change.

If we keep t_0 as the starting point, and choose a measuring unit which is λ times smaller, then we get a new numerical value $t' = t_0 + \lambda \cdot (t - t_0)$. It is therefore reasonable to require that if we make this change, the family of approximating functions remains the same, i.e., that the family

$$C_1 \cdot e_1(t_0 + \lambda \cdot (t - t_0)) + \ldots + C_m \cdot e_m(t_0 + \lambda \cdot (t - t_0)) \tag{4}$$

coincides with the original family (2).

What can we conclude from these symmetry requirements. Let us consider the two cases separately:

- first, the case (3) of shift-invariance, and
- then, the case (4) of scale-invariance.

Case of shift-invariance. In the shift-invariant case, every function from the family (3) also belongs to the family (2).

In particular, for every j and t_0, the function $e_j(t + t_0)$ belongs to the family (3): it corresponds to the case when $C_j = 1$ and $C_{j'} = 0$ for all $j' \neq j$. Thus, we conclude that the function $e_j(t + t_0)$ belongs to the family (2), i.e., that

$$e_j(t + t_0) = C_{1j}(t_0) \cdot e_1(t) + \ldots + C_{mj}(t_0) \cdot e_m(t) \tag{5}$$

for some coefficients $C_{j'j}(t_0)$ depending on the shift t_0.

For each t, if we consider the Eq. (5) at m different moments of time $t = t_1, \ldots, t_m$, then we get the following system of m linear equations with m linear unknowns $C_{1j}(t_0), \ldots, C_{mj}(t_0)$:

$$e_j(t_1 + t_0) = C_{1j}(t_0) \cdot e_1(t_1) + \ldots + C_{mj}(t_0) \cdot e_m(t_1),$$
$$e_j(t_2 + t_0) = C_{1j}(t_0) \cdot e_1(t_2) + \ldots + C_{mj}(t_0) \cdot e_m(t_2),$$
$$\ldots$$
$$e_j(t_m + t_0) = C_{1j}(t_0) \cdot e_1(t_m) + \ldots + C_{mj}(t_0) \cdot e_m(t_m). \tag{6}$$

The solution to a linear system can be explicitly described by the Cramer's rule (see. e.g., [19]), according to which this solution is a ratio of two determinants – i.e., a differentiable function of the right-hand sides and of the coefficients at the unknowns. Since the functions $e_j(t)$ are smooth, the right-hand sides and the coefficients are also smooth, and thus, thus the solution $C_{j'j}(t_0)$ is a differentiable function of differentiable functions – thus, a smooth function itself.

Since the functions $e_{j'}(t)$ and $C_{j'j}(t_0)$ are all differentiable, we can differentiate both sides of Eq. (5) by t_0 and take $t_0 = 0$. As a result, for each j, we get the following differential equation:

$$e_j'(t) = c_{1j} \cdot e_1 + \ldots + c_{mj} \cdot e_m, \tag{7}$$

where e'_j, as usual, denotes the derivatives, and $c_{j'j} \overset{\text{def}}{=} C'_{j'j}(0)$.

Thus, m functions $e_1(t), \ldots, e_m(t)$ satisfy a system of m linear differential equations (7) with constant coefficients. A general solution to this system of equations is well known: it is a linear combination of functions of the type $t^k \cdot \exp(\lambda \cdot t)$, where λ are eigenvalues of the matrix $c_{j'j}$ and factors $t, t^2, \ldots, t^q$ appear if the corresponding eigenvalue is multiple, with multiplicity q; see, e.g., [20]. Please note that the eigenvalues are, in general, complex numbers $\lambda = a + b \cdot i$, in which case

$$\exp(\lambda \cdot t) = \exp(a \cdot t) \cdot (\cos(b \cdot t) + i \cdot \sin(b \cdot t)).$$

In real-valued terms, each function $e_j(t)$ is thus a linear combination of functions of the type

$$t^k \cdot \exp(a \cdot t) \cdot (\cos(b \cdot t) + i \cdot \sin(b \cdot t)).$$

Case of scale-invariance. Let us now consider the case of scale-invariance with respect to the special point t_0. To simplify our analysis, let us consider, instead of time, an auxiliary variable $\tau \overset{\text{def}}{=} \ln(t - t_0)$. In terms of this auxiliary variable, we have $t = t_0 + \exp(\tau)$, and the original functions $e_i(t)$ take the form $E_i(\tau) = e_i(t_0 + \exp(\tau))$.

In terms of the new variable τ, the scaling transformation takes the form $\tau \to \tau + \tau_0$, where $\tau_0 \overset{\text{def}}{=} \ln(\lambda)$. Thus, for the new functions $E_j(\tau)$, scale-invariance means that the original class of functions

$$C_1 \cdot E_1(\tau) + \ldots + C_m \cdot E_m(\tau)$$

coincides with the transformed family

$$C_1 \cdot E_1(\tau + \tau_0) + \ldots + C_m \cdot E_m(\tau + \tau_0).$$

We already know what this condition implies: that each function $E_j(\tau)$ is a linear combination of functions

$$\tau^k \cdot \exp(\lambda \cdot \tau) - \tau^k \cdot \exp(a \cdot \tau) \cdot (\cos(b \cdot \tau) + i \cdot \sin(b \cdot \tau)).$$

Substituting τ's definition $\tau = \ln(t - t_0)$ into this formula, and taking into account that $\exp(\tau) = \exp(\ln(t - t_0)) = t - t_0$ and thus, $\exp(a \cdot \tau) = (\exp(\tau))^a = (t - t_0)^a$, we conclude that each function $e_j(t) = E_j(\tau) = E_j(\ln(t - t_0))$ is a linear combination of functions of the type

$$(\ln(t - t_0))^k \cdot (t - t_0)^\lambda =$$
$$(\ln(t - t_0))^k \cdot (t - t_0)^a \cdot (\cos(b \cdot \ln(t - t_0) + i \cdot \sin(b \cdot \ln(t - t_0))).$$

Comments.

- While it is good that we get expressions similar to what we have empirically observed, be it in case of predicting economic crashes or the case of predicting the number of discovered software faults, the dependence of these expressions on the corresponding parameters t_0, a, and b is highly nonlinear. So, it is computationally difficult to identify the parameters of these models from observations.

- What if we have both shift- and scale-invariance? In this cases, the expression should be both a linear combination of the terms $t^k \cdot \exp(\lambda \cdot t)$ and a combination of the terms of the type $(\ln(t - t_0))^k \cdot (t - t_0)^\lambda$. The need for the second interpretation excludes exponential terms, so such functions should be linear combinations of terms x^k, i.e., polynomials, with C_j as the only parameters. This is the only case when the dependence on the parameters is linear and so, identification of these parameters is computationally easy.

What we plan to do now. Now that we have described the symmetry-motivated models, let us described how to make identification of the parameters of these models easy.

3   Analysis of the Problem and Resulting Computationally Efficient Parameter Identification

Main idea. What we would like to do is come up with a linear differential equation with linear coefficients that describes all linear combinations of symmetry-motivated models. To describe such an equation, let us denote the differentiation operation by D, so that $(Df)(t) \overset{\text{def}}{=} f'(t)$.

Shift-invariant case: analysis of the problem. Let us start with describing shift-invariant models in these terms. In these models, every function $e_j(t)$ is a linear combination of functions of the type $x^k \cdot \exp(\lambda \cdot t)$.

To find an appropriate differential equation for these functions, let us start with the case $k = 1$, when this function takes the form $\exp(\lambda \cdot t)$. For the function

$$\exp(\lambda \cdot t),$$

we have $D \exp(\lambda \cdot t) = \lambda \cdot \exp(\lambda \cdot t)$, thus $(D - \lambda) \exp(\lambda \cdot t) = 0$.

For the next ($k = 1$) function $e(t) = t \cdot \exp(\lambda t)$, we have

$$(De)(t) = \exp(\lambda \cdot t) + \lambda \cdot \exp(\lambda \cdot t),$$

thus $((D - \lambda)e)(t) = \exp(\lambda \cdot t)$. We already know that

$$(D - \lambda) \exp(\lambda \cdot t) = 0,$$

thus we have $((D - \lambda)^2 e)(t) = 0$.

Similarly, for the function $e(t) = t^k \cdot \exp(\lambda \cdot t)$, we have

$$(De)(t) = k \cdot t^{k-1} \cdot \exp(\lambda \cdot t) + \lambda \cdot t^k \cdot \exp(\lambda \cdot t),$$

thus

$$((D - \lambda)e)(t) = k \cdot t^{k-1} \cdot \exp(\lambda \cdot t).$$

So, by induction, we can prove that for this function $e(t)$, we have $(D - \lambda)^k e = 0$.

Different expressions forming $e_j(t)$ correspond to different eigenvalues λ_ℓ, so each of them annihilated by a corresponding differential operation $D - \lambda_\ell$, or, if this eigenvalue if multiple with multiplicity q_ℓ, by an operator $(D - \lambda_\ell)^{q_\ell}$. Thus, if we apply all these operators one after another, all the terms in $e_j(t)$ will be annihilated and thus, we will have a differential operator

$$\widetilde{D} \overset{\text{def}}{=} (D - \lambda_1)^{q_1}(D - \lambda_2)^{q_2} \ldots (D - \lambda_m)^{q_m}$$

for which $\widetilde{D}e_j = 0$ for all j. Since each model $x(t)$ is a linear combination of the functions $e_j(t)$, the function $x(t)$ also satisfies the equation $\widetilde{D}x = 0$.

If we open the parentheses, we conclude that $\widetilde{D}$ is a polynomial of m-th order in terms of D, i.e., that it has the form

$$\widetilde{D} = D^m + a_1 \cdot D^{m-1} + a_2 \cdot D^{m-2} + \ldots + a_m.$$

Thus, the equation $(\widetilde{D}x)(t) = 0$ takes the form

$$\frac{d^m x}{dt^m} + a_1 \cdot \frac{d^{m-1} x}{dt^{m-1}} + a_2 \cdot \frac{d^{m-2} x}{dt^{m-2}} + \ldots + a_m \cdot x = 0. \tag{8}$$

This is the desired differential equation with constant coefficients.

Examples. For a polynomial of order $\leq m - 1$, all eigenvalues are zeros, so $\widetilde{D} = D^m$, and the corresponding differential equation has the form

$$\frac{d^m x}{dt^m} = 0.$$

One can see that solutions to this differential equation are indeed exactly polynomials of order $\leq m - 1$.

For a simple sinusoidal signal $x(t) = A \cdot \cos(\omega \cdot t + \varphi)$, we get a second order differential equation with constant coefficients

$$\frac{d^2 x}{dt^2} + a_1 \cdot \frac{dx}{dt} + a_2 \cdot x = 0.$$

To be more precise, the sinusoid correspond to the case when $a_1 = 0$ and $a_2 > 0$; other cases correspond to exponential functions or functions of the type

$$A \cdot \exp(-a \cdot t) \cdot \cos(\omega \cdot t + \varphi).$$

How can we easily identify a model: towards an algorithm. Instead of the original parameters of the model – parameters on which depends highly non-linearly – we can instead identify the parameters $a_1, \ldots, a_m$ of the corresponding differential equation (8).

Of course, we have to approximate each derivative by a finite difference, so that if we start with a sequence of values $x_1, \ldots, x_i, \ldots$ corresponding to moments of time

$$t_1, \quad t_2 = t_1 + \Delta t, \quad t_3 = t_1 + 2\Delta t, \ldots, t_i = t_1 + (i - 1) \cdot \Delta t,$$

then we form finite difference $(\Delta x)_i \overset{\text{def}}{=} \dfrac{x_i - x_{i-1}}{\Delta t}$. Then, instead of the second derivatives, we will use the values

$$(\Delta^2 x)_i \overset{\text{def}}{=} (\Delta(\Delta x))_i = \frac{(\Delta x)_i - (\Delta x)_{i-1}}{\Delta t} = \frac{x_i - 2x_{i-1} + x_{i-2}}{(\Delta t)^2}.$$

Similarly, in the general case, we have

$$(\Delta^k x)_i = (\Delta(\Delta^{k-1} x))_i = \frac{x_i - k \cdot x_{t-1} + C_2^k \cdot t_{i-1} - C_3^k \cdot t_{i-2} + \ldots + (-1)^k \cdot t_{i-k}}{(\Delta t)^k}.$$

So, instead of Eq. (8), we have an approximate equation

$$(\Delta^m x)_i + a_1 \cdot (\Delta^{m-1} x)_i + a_2 \cdot (\Delta^{m-2} x)_i + \ldots + x_i = 0. \tag{9}$$

The values $(\Delta^k x)_i$ are computed based on the observations x_i, so we get an (over-determined) system of linear equations from which we can easily find the unknowns $a_1, \ldots, a_m$ by using the Least Squares method.

Shift-invariant case: resulting algorithm. Based on the sequence of observations $x_i = x(t_i)$, we compute the sequence of values $(\Delta x)_i = \dfrac{x_i - x_{i-1}}{\Delta t}$, then the sequence $(\Delta^2 x)_i = (\Delta(\Delta x))_i$, etc., until we have computed $(\Delta^m x)_i$. Based on thus computed sequences, we find the parameters a_j by applying the Least Squares Method to the Eq. (9).

Important comments.

- No problem if observations are not equally spaced in time: just take $(\Delta x)_i = \dfrac{x_i - x_{i-1}}{\Delta t_i}$, where we denoted $\Delta t_i \overset{\text{def}}{=} t_i - t_{i-1}$.
- It should be mentioned that even when the measurements of $x_i = x(t_i)$ at different moments of time are uncorrelated, their linear combinations (as in the left-hand side of formula (9)) are correlated, since the expressions for i and for $i - 1$ now depend on the same value x_i. Thus, we need to use the Least Squares in the presence of this easy-to-compute correlation. This does not affect the computational easiness – the expression is still quadratic and equating its derivatives to 0 still leads to a system of linear equations.
- If needed, we can convert the new parameters $a_1, \ldots, a_m$ into the more traditional ones. All we need for this is to compute the derivatives of the original expressions $f(t, c_1, \ldots, c_\ell)$ and find the values a_j for which the linear combinations of these derivatives are 0s. Then, we get expressions describing a_j in terms of c_j: $a_j = f_j(c_1, \ldots, c_\ell)$. Once we know a_j, we can solve the corresponding system of equations $f_j(c_1, \ldots, c_\ell) = a_j$. This system is non-linear, but when the number of parameters is small, it is not that difficult to solve.

Scale-invariant case: analysis of the problem. As we have shown earlier, the scale-invariant case reduces to the shift-invariant case if we introduce an auxiliary variable $\tau = \ln(t - t_0)$. Thus, similarly to the above-described shift-invariant case, with respect to this new variable τ, we get a differential equation

$$\frac{d^m x}{d\tau^m} + a_1 \cdot \frac{d^{m-1} x}{d\tau^{m-1}} + \ldots + a_m \cdot x = 0. \tag{10}$$

Differentiating the relation between τ and t, we conclude that $d\tau = \dfrac{dt}{t - t_0}$. Thus, $\dfrac{d}{d\tau} = (t - t_0) \cdot \dfrac{d}{dt}$, and the Eq. (1) takes the following form:

$$(t - t_0)^m \cdot \frac{d^m x}{dt^m} + a_1 \cdot (t - t_0)^{m-1} \cdot \frac{d^{m-1} x}{dt^{m-1}} + \ldots + a_m \cdot x = 0. \tag{11}$$

There are two possibilities:

- it may be that we know t_0, or
- it may be that we need to determine t_0 from observations.

In the first subcase, all we need is to find the values a_j.

In the second subcase, to make the problem linear, we expand all the polynomials

$$(t - t_0)^j = x^j + (-j \cdot t_0) \cdot t^{j-1} + \ldots,$$

then each term $a_j \cdot (t - t_0)^{m-j} \cdot \dfrac{d^{m-j} x}{dt^{m-j}}$ becomes a linear combination of the following terms:

$$t^{m-j} \cdot \frac{d^{m-j} x}{dt^{m-j}}, \quad t^{m-j-1} \cdot \frac{d^{m-j} x}{dt^{m-j}}, \quad \ldots, \quad \frac{d^{m-j} x}{dt^{m-j}}.$$

Let us denote the coefficients at $t^{m-j-k} \cdot \dfrac{dx^{m-j}}{dt^{m-j}}$ by a_{jk}. Then, the formula (11) takes the following form:

$$t^m \cdot \frac{dx^m}{dt^m} + a_{01} \cdot t^{m-1} \cdot \frac{dx^m}{dt^m} + \ldots + a_{0m} \cdot \frac{dx^m}{dt^m}$$
$$+ a_{10} \cdot t^{m-1} \cdot \frac{dx^{m-1}}{dt^{m-1}} + a_{11} \cdot t^{m-2} \cdot \frac{dx^{m-1}}{dt^{m-1}} + \ldots + a_{1,m-1} \cdot \frac{dx^{m-1}}{dt^{m-1}}$$
$$+ \ldots$$
$$+ a_{m0} \cdot x = 0. \tag{12}$$

Thus, depending on whether we know t_0 or we don't, we arrive at the following linear algorithms.

Scale-invariant case: resulting algorithms. Based on the original sequence of observations $x_i = x(t_i)$, we compute the finite differences $(\Delta^k x)_i$ for all possible values $k \leq m$.

Then, if we know the value t_0, we compute the parameters $a_1, \ldots, a_m$ of the corresponding model by applying the Least Squares method to the following system of linear equations:

$$(t_i - t_0)^m \cdot (\Delta^m x)_i + a_1 \cdot (t_i - t_0)^{m-1} \cdot (\Delta^{m-1} x)_i + \ldots + a_m \cdot x_i = 0. \tag{13}$$

When we do not know the value t_0, then we need to find the parameters a_{jk} of the model by applying the Least Squares method to the following system of equations:

$$t_i^m \cdot (\Delta^m x)_i + a_{01} \cdot t_i^{m-1} \cdot (\Delta^m x)_i + \ldots + a_{0m} \cdot (\Delta^m x)_i$$
$$+ a_{10} \cdot t_i^{m-1} \cdot (\Delta^{m-1} x)_i + a_{11} \cdot t^{m-2} \cdot (\Delta^{m-1} x)_i + \ldots + a_{1,m-1} \cdot (\Delta^{m-1} x)_i$$
$$+ \ldots$$
$$+ a_{m0} \cdot x = 0. \tag{14}$$

Acknowledgments. We acknowledge the partial support of the Center of Excellence in Econometrics, Faculty of Economics, Chiang Mai University, Thailand. This work was also supported in part by the National Science Foundation grant HRD-1242122 (Cyber-ShARE Center of Excellence).

References

1. Bishop, C.M.: Pattern Recognition and Machine Learning. Springer, New York (2006)
2. Feigenbaum, J.A.: A statistical analyses of log-periodic precursors to financial crashes. Quant. Financ. **1**(5), 527–532 (2001)
3. Feigenbaum, J.A., Freund, P.: Discrete scaling in stock markets before crashes. Int. J. Mod. Phys. **12**, 57–60 (1996)
4. Gazola, L., Fenandez, C., Pizzinga, A., Riera, R.: The log-periodic-AR (1)-GARCH (1,1) model for financial crashes. Eur. Phys. J. B **61**(3), 355–362 (2008)
5. Geraskin, P., Fantazzinin, D.: Everything you always wanted to know about log periodic power laws for bubble modelling but were afraid to ask. Eur. J. Financ. **19**(5), 366–391 (2013)
6. Goodfellow, I., Bengio, Y., Courville, A.: Deep Leaning. MIT Press, Cambridge (2016)
7. Jiang, Z.Q., Zhou, W.H., Sornette, D., Woodard, R., Bastiaensen, K., Cauwels, P.: Bubble diagnosis and prediction of the 2005–2007 and 2008–2009 Chinese stock market bubbles. J. Econ. Behav. Organ. **74**, 149–162 (2010)
8. Johansen, A.: Characterization of large price variations in financial markets. Phys. A **324**, 157–166 (2003)
9. Johansen, A., Ledoit, O., Sornette, D.: Crashes as critical points. Int. J. Theor. Appl. Financ. **3**(2), 219–255 (2000)
10. Johansen, A., Sornette, D.: Financial anti-bubbles: log-periodicity in Gold and Nikkei collapses. Int. J. Mod. Phys. C **10**(4), 563–575 (1999)
11. Johansen, A., Sornette, D.: Large stock market price drawdowns are outliers. J. Risk **4**(2), 69–110 (2002)
12. Johansen, A., Sornette, D.: Endogenous versus exogenous crashes in financial markets. In: Contemporary Issues in International Finance. Nova Science Publishers (2004). Reprinted as a special issue of Brussels Economic Review, vol. 49, No. 3/4 2006
13. Kreinovich, V., Lakeyev, A., Rohn, J., Kahl, P.: Computational Complexity and Feasibility of Data Processing and Interval Computations. Kluwer, Dordrecht (1998)

14. Kreinovich, V., Nguyen, H.T., Sriboonchitta, S.: Log-periodic power law as a predictor of catastrophic events: a new mathematical justification. In: Proceedings of the International Conference on Risk Analysis in Meteorological Disasters, RAMD 2014, Nanjing, China, 12–13 October 2014 (2014)
15. Kreinovich, V., Swenson, T., Elentukh, A.: Interval approach to testing software. Interval Comput. **2**, 90–109 (1994)
16. Nguyen, H.T., Kreinovich, V.: Applications of Continuous Mathematics to Computer Science. Kluwer, Dordrecht (1997)
17. Papadimitriou, C.H.: Computational Complexity. Pearson, Boston (1993)
18. Pardalos, P.: Complexity in Numerical Optimization. World Scientific, Singapore (1993)
19. Poole, D.: Linear Algebra: A Modern Introduction. Cengage Learning, Independence (2014)
20. Robinson, J.C.: An Introduction to Ordinary Differential Equations. Cambridge University Press, Cambridge (2004)
21. Sornette, D.: Critical market crashes. Phys. Rep. **378**(1), 1–98 (2003)
22. Sornette, D.: Why Stock Markets Crash: Critical Events in Compelx Financial Systems. Princeton University Press, Princeton (2003)
23. Sornette, D., Johansen, A.: Significance of log-periodic precursors to financial crashes. Quant. Financ. **1**(4), 452–471 (2001)
24. Sornette, D., Zhou, W.Z.: The US 2000–2002 market descent: how much longer and deeper? Quant. Financ. **2**(6), 468–481 (2002)
25. Weatherall, J.M.: The Physics of Wall Street: The History of Predicting the Unpredictable. Houghton Mifflin Harcourt, New York (2013)
26. Zhou, W.Z., Sornette, D.: Evidence of a worldwide stock market log-periodic antibubble since mid-2000. Phys. A **330**(3), 543–583 (2003)

Quantum Ideas in Economics Beyond Quantum Econometrics

Vladik Kreinovich[1]([✉]), Hung T. Nguyen[2,3], and Songsak Sriboonchitta[3]

[1] Department of Computer Science, University of Texas at El Paso,
500 W. University, El Paso, TX 79968, USA
vladik@utep.edu
[2] Department of Mathematical Sciences, New Mexico State University,
Las Cruces, NM 88003, USA
hunguyen@nmsu.edu
[3] Faculty of Economics, Chiang Mai University, Chiang Mai 50200, Thailand
songsakecon@gmail.com

Abstract. It is known that computational methods developed for solving equations of quantum physics can be successfully applied to solve economic problems; there is a whole related research area called *quantum econometrics*. Current quantum econometrics techniques are based on a purely mathematical similarity between the corresponding equations, without any attempt to relate the underlying ideas. We believe that the fact that quantum equations can be successfully applied in economics indicates that there is a deeper relation between these areas, beyond a mathematical similarity. In this paper, we show that there is indeed a deep relation between the main ideas of quantum physics and the main ideas behind econometrics.

1 Quantum Ideas in Economics: Why and What Is Known

Why quantum ideas in economics. In most practical problems, once we have a candidate for a solution, we can feasibly check whether this candidate is indeed a solution.

For example, in mathematics, it is often difficult to find a proof of a statement or of its negation. However, once someone produces what intends to be a detailed proof, it is feasible for a referee (or even for a computer-based system) to check that all the steps in this text are indeed correct and thus, that the text does indeed constitute a proof.

Similarly, in physics, it is often difficult to find a formula that described the observed phenomena, but once such a formula is proposed, one can feasibly check whether all observations indeed satisfy this formula.

In engineering, it is often difficult to come up with a design that satisfies all the given specifications, but once a design is produced, we can use software packages to check that this design indeed satisfies the specifications. For example,

© Springer International Publishing AG 2018
L. H. Anh et al. (eds.), *Econometrics for Financial Applications*, Studies in Computational Intelligence 760, https://doi.org/10.1007/978-3-319-73150-6_11

we can check that the designed airplane is indeed stable under allowable winds, that the corresponding stresses do not exceed the prescribed level, etc.

Problems for which we can feasibly check whether a candidate is indeed a solution are known as *problems from the class NP*; see, e.g., [3,4]. The abbreviation NP stands for *Non-deterministic Polynomial*, where:

- "non-deterministic" means that we are allowed to guess, and
- "polynomial" means that once a guess is produced, the computation time needed to check whether a given guess is a solution should not exceed a polynomial of the length of the input (such polynomial bounds are a formal description of feasibility).

Not all practical problems belong to the class NP:

- For example, if we want to find an *optimal* design, then, in general, it is not easy to check that a given guess is optimal: for that, we would need to compare it with an unfeasible number of all possible designs.
- Similarly, in multi-step conflict situations, it is not easy to check whether a given move is winning or not – checking it would require going over all possible counter-moves of the opposite side.

However, many practical problem are indeed problems from the class NP.

It is still not known whether we can solve all problems from the class NP is feasible (polynomial) time: this is the famous open problem of whether the class NP is equal to the class P of all the problems that can be solved feasibly (i.e., in polynomial time). Most computer scientists believe that NP is different from P.

The fact that we do not know whether NP is different from P means that there is no problem from the class NP for which we have proven that this problem cannot be solved in polynomial time. What *is* proven is that there are problems from the class NP which are as hard as possible within this class, in the sense that every other problem from the class NP can be feasibly reduced to this problem. Such problems are known as *NP-complete*. Many problems of solving non-linear equations (and many other problems) have been proven to be NP-complete.

Historically the first problem for which NP-completeness was proven was the following *propositional satisfiability problem (SAT)*:

- given a *propositional formula F*, i.e., a formula obtained from propositional ("yes"-"no") variables v_i by using propositional connectives & (and), ∨ (or), and ¬ (not),
- find the values of the variables v_i that make the formula F true.

As an illustrative example, we can take $F = (v_1 \lor v_2 \lor \neg v_3) \& (\neg v_1 \lor v_2)$.

Here, a reduction of a problem A to problem B means that for every instance a of the problem A, we can feasibly compute an appropriate instance b of the problem B for which, once we have a solution to the instance b, we can feasibly transform this solution into a solution to the original instance a.

Let us give a simple example of reduction. The problem of solving an equation $p \cdot x^4 + q \cdot x + r = 0$ can be reduced to the problem of solving a quadratic equation

$p \cdot y^2 + q \cdot y + r = 0$. Once we have found a solution y to the quadratic equation, we can find the solutions to the original fourth order equation by computing $x = \pm\sqrt{y}$.

So, once we know that a problem is NP-complete, then any good algorithm for solving this problem automatically becomes a good algorithm for solving all other problems from the class NP. This is not just a theoretical possibility – efficient tools for solving the propositional satisfiability problem (known as *SAT-solvers*) are now used to solve many problems from different application areas.

From this viewpoint, econometrics has many complex problems. Sometimes, we do not have efficient algorithms for solving these problems. In this case, due to the above reduction, it is reasonable to look for other complex (NP-complete) problem, and see if known algorithms for solving these other problems can be used to solve economics-related problems as well.

Where can we find such other problems? Most of the practical problems deal with the physical world. Thus, it is reasonable to look into physics for examples of other complex problems for which efficient algorithms are known.

It is known that adding quantum effects makes problems more complex. Thus, if we look for complex problems in physics, it is reasonable to look for problems of quantum physics. So, we arrive at the idea of trying to see if we can apply known algorithms for solving complex problem of quantum physics to solve complex economics-related problems.

Quantum econometrics: what is known. The idea of using quantum techniques – i.e., techniques for solving quantum equations – to solve economics problems has been successfully implemented. The corresponding techniques are known as *quantum econometrics*. These techniques and their numerous applications are described, e.g., in the seminal book [1].

This book emphasizes that quantum econometrics is based on a *mathematical* similarity of equations, *not* on any similarity between physical ideas of quantum physics and economics ideas.

Our idea and what we do in this paper. The fact that quantum ideas have been very successful in econometric applications makes us think that there may be deeper reasons for the mathematical similarity between the corresponding equations, i.e., that there is indeed some relation between physical ideas of quantum physics and ideas from economics.

In this paper, we show that there is indeed such a relation.

2   Main Ideas Behind Quantum Physics: A Brief Reminder

Need for a reminder. To describe a relation between the main ideas of quantum physics and the main ideas behind econometrics – and to convince the readers that this relation is indeed fundamental, not just a mathematical similarity – let us recall the main ideas behind quantum physics (for more details, see, e.g., [2]).

Quantum physics as physics of micro-world. The main objective of physics is to learn the state of the physical world and to predict its future state. The information about the current state of the physical world comes from measurements. To get the most information about the world, we want to make the measurements as accurate as possible. This means, in particular, that the measurements should disturb the measured object as little as possible – since each such disturbance changes the state of the object.

Traditional physics is the physics of *macro-world*, the physics of objects of macro-size. For such objects, it is usually possible to measure them while disturbing them as little as possible. For example:

- We can measure the distance to an object by sending an ultrasound signal towards the object and measure the time it takes for this signal to get to the object, get reflected, and come back to the sensor.
- We can also perform a similar measurement by sending a laser beam.

In both cases, we can use relatively weak signals, so that the measured object is not affected by this signal.

However, as we study smaller and smaller objects, this becomes more and more complicated. When we send a measuring signal to a body consisting of $\approx 10^{23}$ particles, we can have a relatively very weak signal whose effect on the multi-particle body of interest is small. However, the situation drastically changes if we consider *micro-objects*.

To measure the location of an elementary particle, we need to send another particle – e.g., a photon – to interact with the particle of interest. In this case, the signal that we send is of approximately of the same size as the object itself, and there is thus no way that we can ignore the effect of this signal on the measured object.

In other words, in the micro-world, when we perform a measurement on an object, we change this object. This is one of the main features of the micro-world – known as a the quantum world – that no matter how much we try, we cannot avoid changing the state: whenever we measure the state, we change it.

There is a similar idea in economics. At first glance, economics is a macro-object: when we measure GDP or unemployment, we do not change it, the value remains very accurate. However, econometrics is *not* about measuring different parameters of economics, econometrics is about *discovering new dependencies* that describe the economic data.

From this viewpoint, econometrics has exactly the same effect as quantum physics: once we discover a new dependence, the situation changes.

Indeed, let us consider a simplified example. Suppose that a researcher finds out a better way to predict the price $x(t + 2)$ of a certain financial instrument two days from today based on the prices $x(t)$, $y(t)$, ..., $x(t-1)$, $y(t-1)$, ... of this stock and related stocks today and in the previous days.

It is known that the stock values sometimes change drastically. For such change days, based on the newly discovered dependence, we can potentially predict, at day t_0, that the stock value will drastically increase in 2 days, to the level

$$x(t_0 + 2) \gg x(t_0), x(t_0 + 1).$$

When we did not know the dependence, this could indeed be a valid prediction:

- the value $x(t_0 + 1)$ would have equal to what the model predicts based on the prices $x(t_0 - 1)$, $y(t_0 - 1)$, $x(t_0 - 2)$, $y(t_0 - 2)$, ..., and
- the value $x(t_0 + 2)$ would have been equal to what the model predicts based on the prices $x(t_0)$, $y(t_0)$, ..., $x(t - 1)$, $y(t_0 - 1)$, ...

However, since we now know the dependence, the traders in the stock exchange know that the price will rise and therefore, will start buying this stock – until its price rises, in day $t_0 + 1$, to the level $\rho \cdot x(t_0 + 2)$, where ρ is a discount that takes into account one day difference (i.e., that takes into account the interest rate that you get in one day by a safe investment like Treasury bonds or bank deposits). This will change the next day's stock price $x(t_0 + 1)$ from the previously predicted value $x(t_0 + 1) \ll x(t_0 + 2)$ to a new value $x(t_0 + 1) \approx x(t_0 + 2)$.

So, while the model worked perfectly well until it was discovered, once it is discovered, it longer provides correct predictions – because the stock traders take this model into account when trading and thus, change the dynamics of the system and consequently, modify stock prices.

Similarly, in situations in which the model originally predicted drastic decreases in stock prices, once the model becomes known, it no longer provides accurate predictions; see, e.g., [5].

This is an exact analog of the quantum physics phenomenon:

- In quantum physics, once you learn the value of a quantity describing the object, the actual value of this quantity changes, and the known value is no longer a perfect description of the current state of the particle.
- Similarly, in economics, once we discover the previously unknown dependence between economic quantities, this changes the dynamics of trade and thus, the dependence – which worked well in the past – stops working, at least stops being accurate.

This fundamental similarity may be the reason why techniques for solving quantum equations are so helpful in the economic realm.

Comment. In both cases, the size of the effect depends on the relative size of the object:

- In quantum physics, the effect of measurement on a micro-size body can be minuscule, while for micro-size body, the effect is very drastic.

- Similarly, in economics, if only one person knows the dependence and uses it to buy and sell small amounts of stock, the effect on the stock market will be small. However, nowadays, with financial companies actively investing in data analytics, a dependence uncovered by one researcher cannot be kept secret for long: it will inevitably (and very soon) be discovered by others as well. Once this happens, the effect on the stock market will become large – and it will invalidate the original dependence.

Acknowledgments. We acknowledge the partial support of the Center of Excellence in Econometrics, Faculty of Economics, Chiang Mai University, Thailand. This work was also supported in part by the US National Science Foundation grant HRD-1242122.

References

1. Baaquie, B.E.: Quantum Finance: Path Integrals and Hamiltonians for Options and Interest Rates. Camridge University Press, New York (2004)
2. Feynman, R., Leighton, R., Sands, M.: The Feynman Lectures on Physics. Addison Wesley, Boston (2005)
3. Kreinovich, V., Lakeyev, A., Rohn, J., Kahl, P.: Computational Complexity and Feasibility of Data Processing and Interval Computations. Kluwer, Dordrecht (1998)
4. Papadimitriou, C.H.: Computational Complexity. Pearson, Boston (1993)
5. Thamotharan, S.: Prediction Paradox: Neural Network – a Possible Application to the Economic Prediction, Master's thesis, Department of Computer Science, University of Texas at El Paso (1993)

An Ancient Bankruptcy Solution Makes Economic Sense

Anh H. Ly[1], Michael Zakharevich[2], Olga Kosheleva[3], and Vladik Kreinovich[3(✉)]

[1] Banking University of Ho Chi Minh City, 56 Hoang Dieu 2,
Quan Thu Duc, Thu Duc, Ho Chi Minh City, Vietnam
[2] SeeCure Systems, Inc., 1040 Continentals Way # 12, Belmont, CA 94002, USA
michael@seecure360.com
[3] University of Texas at El Paso, 500 W. University, El Paso, TX 79968, USA
{olgak,vladik}@utep.edu

Abstract. While econometrics is a reasonable recent discipline, quantitative solutions to economic problem have been proposed since the ancient times. In particular, solutions have been proposed for the bankruptcy problem: how to divide the assets between the claimants? One of the challenges of analyzing ancient solutions to economics problems is that these solutions are often presented not as a general algorithm, but as a sequence of examples. When there are only a few such example, it is often difficult to convincingly extract a general algorithm from them. This was the case, for example, for the supposedly fairness-motivated Talmudic solution to the bankruptcy problem: only in the mid 1980s, the Nobelist Robert Aumann succeeded in coming up with a convincing general algorithm explaining the original examples. What remained not so clear in Aumann's explanation is why namely this algorithm best reflects the corresponding idea of fairness. In this paper, we find a simple economic explanation for this algorithm.

1 The Bankruptcy Problem and Its Ancient Solution: An Introduction

The bankruptcy problem: reminder. When a person or a company cannot pay all its obligation, a bankruptcy is declared, and the available funds are distributed among the claimants. Since there is not enough money to give, to each claimant, what he/she is owed, claimants will get less than what they are owed. How much less? What is a fair way to divide the available funds between the claimants?

An ancient solution. The bankruptcy problem is known for many millennia, since money became available and people starting lending money to each other. Solutions to this problem have also been proposed for many millennia. One such ancient solution is described in the Talmud, an ancient commentary on the Jewish Bible [2]. Specifically, this solution is described in the Babylonian Talmud, in Ketubot 93a, Bava Metzia 2a, and Yevamot 38a. (This solution is actually about a more general problem of several contracts which cannot be all fully fulfilled).

Like many ancient texts containing mathematics, the Talmud does not contain an explicit algorithm. Instead, it contains four examples illustrating the main idea. In the first three examples, the three parties are owed the following amounts:

© Springer International Publishing AG 2018
L. H. Anh et al. (eds.), *Econometrics for Financial Applications*, Studies in Computational Intelligence 760, https://doi.org/10.1007/978-3-319-73150-6_12

- the first person is owed $d_1 = 100$ monetary units,
- the second person is owed $d_2 = 200$ monetary units, and
- the third person is owed $d_3 = 300$ monetary units:

$$d_1 = 100, \quad d_2 = 200, \quad d_3 = 300.$$

For three different available amounts E, the text describes the amounts e_1, e_2, and e_3 that each of the three person will get:

E	$d_1 = 100$	$d_2 = 200$	$d_3 = 300$
	e_1	e_2	e_3
100	$33\frac{1}{3}$	$33\frac{1}{3}$	$33\frac{1}{3}$
200	50	75	75
300	50	100	150

There is also a fourth example, formulated in a slightly different way – as the question of dividing a disputed garment. In the bankruptcy terms, it can be described as follows: the owed amounts are:

$$d_1 = 50, \quad d_2 = 100.$$

The available amount E and the recommended division (e_1, e_2) are as follows:

E	$d_1 = 50$	$d_2 = 100$
	e_1	e_2
100	25	75

Example are here, but what is a general solution? There has been, historically, a big problem with this solution: in contract to many other ancient mathematical texts, where the general algorithm is very clear from the examples, in this particular case, the general algorithm was unknown until 1985. Actually, many researchers came up with algorithms that explained *some* of these examples – while claiming that the original ancient text must have contained some mistakes.

Mystery solved, algorithm is reconstructed. This problem intrigued Robert Aumann, later the Nobel Prize winner in Economics (2005). In his 1985 paper [1], Professor Aumann came up with a reasonable general algorithm that explains the ancient solution; see also [4, 8].

To explain this algorithm, we need to first start with the the case of two claimants. Without losing generality, let us assume that the first claimant has a smaller claim $d_1 \leq d_2$.

Then, if the overall amount E is small – to be precise, smaller that d_1 – then this amount E is distributed equally between the claimants, so that each gets

$$e_1 = e_2 = \frac{E}{2}.$$

When the available amount E is between d_1 and d_2, i.e., when $d_1 \leq E \leq d_2$, then the first claimant receives $e_1 = \frac{d_1}{2}$, and the second claimant receives the remaining amount $e_2 = E - e_1$.

This policy continues until we reach the amount $E = d_2$, at which moment the first claimant receives the amount $d_1 = \frac{d_1}{2}$ and the second claimant received the amount $e_2 = d_2 - \frac{d_1}{2}$. At this moment, after receiving the money, both claimants lose the same amount of money: $d_1 - e_1 = d_2 - e_2 = \frac{d_1}{2}$.

Finally, when the overall amount is larger than d_2 (but smaller than the overall amount of debt $d_1 + d_2$), the money is distributed in such a way that the losses remain equal, i.e., that $d_1 - e_1 = d_2 - e_2$ and $e_1 + e_2 = E$. From these two conditions, we can find the corresponding claims:

$$e_1 = \frac{E + d_1 - d_2}{2}, \quad e_2 = \frac{E - d_1 + d_2}{2}.$$

The division between three (or more) claimants is then explained as the one for which for every two claimants, the amounts given to them is distributed according to the above algorithm. This can be easily checked if we select, for each pair (i, j) only the overall amount $E_{ij} = e_i + e_j$ allocated to claimants from this pair. As a result, for the pairs $(1, 2)$, $(2, 3)$, and $(1, 3)$, we get the following tables:

E_{12}	$d_1 = 100$	$d_2 = 200$
	e_1	e_2
$66\frac{2}{3}$	$33\frac{1}{3}$	$33\frac{1}{3}$
125	50	75
150	50	100

E_{23}	$d_2 = 200$	$d_3 = 300$
	e_2	e_3
$66\frac{2}{3}$	$33\frac{1}{3}$	$33\frac{1}{3}$
150	75	75
250	100	150

E_{13}	$d_1 = 100$	$d_3 = 300$
	e_1	e_3
100	$66\dfrac{2}{3}$	$33\dfrac{1}{3}$
125	50	75
200	50	150

Remaining problem. That the ancient algorithm has been reconstructed, great. We now know *what* the ancients proposed. However, based on the above description, it is still not clear *why* this solution to the bankruptcy problem was proposed.

The above solution sounds rather arbitrary. To be more precise, both idea of dividing the amount equally and dividing the losses equally make sense, but how do we combine these two ideas? And why in the region between $E = \min(d_1, d_2)$ and $E = \max(d_1, d_2)$ the claimant with the smallest claim always gets half of his/her claim while the second claimant gets more and more? How dow that fit with the Talmud's claim that the proposed division represents fairness?

What we do in this paper. In this paper, we propose an economics-based explanation for the above solution.

2   Analysis of the Problem

What is fair is not clear. At first glance, it may look like fairness means dividing the amount either equally. If everyone is equal, why should someone gets more than others?

However, this is not necessarily a fair division. Suppose that two folks start with an equal amount of 400 dollars. They both decided to invest some money in the biomedical company that promised to use this money to develop a new drug curing up-to-now un-curable disease. The first person invested $200, the second invested $300. After this, the first person has $200 left and the second person has $100 left.

The company went bankrupt, and only $300 remains in its account. If we divide this mount equally, both investors will get back the same amount of $150. As a result:

- the first person will have $350 instead of the original $400, while
- the second person will have $250 instead of the original $400.

So, the first person loses only $50, while the second person loses three times more: $150. So, the first person, who selfishly kept money to himself, gets more than the altruistic second person who invested more in a noble case: how is this fair?

How we understand fairness: let us divide equally, but with respect to what status quo point? If two people jointly find an amount of money, then fairness means that this amount should be divided equally. If two people jointly contributed to some expenses, fairness means that they should split the expenses equally.

In both cases, we have a natural status quo point $(\tilde{e}_1, \tilde{e}_2)$:

- in the first case, we take $(\tilde{e}_1, \tilde{e}_2) = (0, 0)$, and
- in the second case, we take $(\tilde{e}_1, \tilde{e}_2) = (d_1, d_2)$.

Any change from the status quo should be divided equally, i.e., we should have $e_1 - \tilde{e}_1 = e_2 - \tilde{e}_2$. So, to apply this idea to the bankruptcy problem, we need to decide what is the status quo point here.

Comment. The idea that the difference between the actual amount and the status quo point should be divided equally is not only natural and fair, it actually comes from the game-theoretic notion of bargaining solution proposed by another Nobelist John Nash; see, e.g., [6,7].

What are possible ranges for the status quo point: example. Let us consider one of the above cases, when the first person is owed $d_1 = 100$ monetary units, the second person is owed $d_2 = 200$ units, and we have an amount $E_{12} = 125$ units to distribute between these two claimants.

Depending on how we distribute this amount, the first person may get different amounts. The best possible case for the first claimant is when he get all the money he is owed, i.e., $\bar{e}_1 = 100$ monetary units. The worst possible case for the first claimant is when all the money goes to the second person, and the first person gets nothing: $\underline{e}_1 = 0$. Thus, the status quo point for the first person is somewhere in the interval

$$[\underline{e}_1, \bar{e}_1] = [0, 100].$$

Similarly, the best possible case for the second person is when the second person gets all the money, i.e., when $\bar{e}_2 = 125$. The worst possible case for the second person is when the first claimant gets everything he is owed – i.e., all 100 units, and the second person gets the remaining amount of $\underline{e}_2 = 125 - 100 - 25$ units. Thus, the status quo point for the second person is somewhere in the interval

$$[\underline{e}_2, \bar{e}_2] = [25, 125].$$

Let us perform the same analysis in the general case.

What are possible ranges for the status quo point: general case. Without losing generality, let us assume that the 1st person is the one who is owed less, i.e., that $d_1 \leq d_1$. We will consider three different cases:

- when the available amount E_{12} does not exceed d_1: $E_{12} \leq d_1$;
- when the available amount E_{12} is between d_1 and d_2: $d_1 \leq E_{12} \leq E_2$, and
- when the available amount E_{12} exceeds d_2, i.e., $d_2 \leq E_{12} \leq d_1 + d_2$.

Let us consider these three cases one by one.

Case when the overall amount does not exceed the smallest claim. Let us first consider the case when $E_{12} \leq d_1 \leq d_2$. In this case, for the first person, the best possible case is when this person gets all the amount E_{12}: $\bar{e}_1 = E_{12}$. The worst possible case is when all the available money goes to the second claimant and the first person gets nothing: $\underline{e}_1 = 0$. So, for the first person, the range of possible gains is $[\underline{e}_1, \bar{e}_1] = [0, E_{12}]$.

For the second person, the best possible case is when this person gets all the amount E_{12}: $\bar{e}_2 = E_{12}$. The worst possible case is when all the available money goes to the first claimant and the second person gets nothing: $\underline{e}_2 = 0$. So, for the second person, the range of possible gains is $[\underline{e}_2, \bar{e}_2] = [0, E_{12}]$.

Case when the overall amount is in the between the smaller and the larger claims.
Let us now consider the case when $d_1 \leq E_{12} \leq d_2$. In this case, for the first person,
the best possible case is when this person gets all the amount it is owed: $\bar{e}_1 = d_1$. The
worst possible case is when all the available money goes to the second claimant and the
first person gets nothing: $\underline{e}_1 = 0$. So, for the first person, the range of possible gains is
$[\underline{e}_1, \bar{e}_1] = [0, d_1]$.

For the second person, the best possible case is when this person gets all the amount
E_{12}: $\bar{e}_2 = E_{12}$. The worst possible case is when the first claimant gets all the money he
is owed (i.e., the amount d_1), and the second person only gets the remaining amount
$\underline{e}_2 = E_{12} - d_1$. So, for the second person, the range of possible gains is $[\underline{e}_2, \bar{e}_2] = [E_{12} -
d_1, E_{12}]$.

Case when the overall amount is larger than both claims. Let us now consider the
case when $d_1 \leq d_2 \leq E_{12}$. In this case, for the first person, the best possible case is
when this person gets all the amount it is owed: $\bar{e}_1 = d_1$. The worst possible case is
when the second person gets all the money it is owed, and the first person only gets the
remaining amount $\underline{e}_1 = E_{12} - d_2$. So, for the first person, the range of possible gains is
$[\underline{e}_1, \bar{e}_1] = [E_{12} - d_2, d_1]$.

For the second person, the best possible case is when this person gets all the amount
it is owed: $\bar{e}_2 = d_2$. The worst possible case is when the first claimant gets all the
money he is owed (i.e., the amount d_1), and the second person only gets the remaining
amount $\underline{e}_2 = E_{12} - d_1$. So, for the second person, the range of possible gains is $[\underline{e}_2, \bar{e}_2] =
[E_{12} - d_1, d_2]$.

Which points of the corresponding intervals should we select? In all three cases, for
both claimants, we have an *interval* of possible values of the resulting gain. On each of
these intervals, we need to select a status quo point that corresponds to the equivalent
cost of this interval uncertainty.

The problem of what is the fair cost $\bar{e}$ in the case of interval uncertainty $[\underline{e}, \bar{e}]$ has
been handled by yet another Nobelist, Leo Hurwicz; see, e.g., [3,5,6]. Namely, he pro-
posed to select the value

$$\tilde{e} = \alpha \cdot \bar{e} + (1 - \alpha) \cdot \underline{e},$$

where the coefficient $\alpha \in [0,1]$ describes the decision-maker's degree of optimism-
pessimism:

- the value $\alpha = 1$ describes a perfect optimist, when the decision maker only takes
  into account the most optimistic (best possible) scenario;
- the value $\alpha = 0$ describes a complete pessimist, when the decision maker only takes
  into account the worst possible scenario; and
- the values α strictly between 0 and 1 describe a realistic decision maker, who takes
  into account both the best-case and the worst-case possibilities.

Let us see what will happen if we take one of these solutions as a status-quo point and
consider a division fair if the differences between the gains e_i and the status quo are
equal: $e_1 - \tilde{e}_1 = e_2 - \tilde{e}_2$.

3 No Matter What Our Level of Optimism, We Get Exactly the Ancient Solution

Three cases: reminder. We will now show that in all the cases, we get exactly the ancient solution – so we have a good economic explanation for this solution. To show this, let us consider all three possible cases:

- case when $E_{12} \le d_1 \le d_2$,
- case when $d_1 \le E_{12} \le d_2$, and
- case when $d_1 \le d_2 \le E_{12}$.

Case when the overall amount does not exceed the smallest claim: general formulas. In this case,

$$\tilde{e}_1 = \alpha \cdot \bar{e}_1 + (1-\alpha) \cdot \underline{e}_1 = \alpha \cdot E_{12} + (1-\alpha) \cdot 0 = \alpha \cdot E_{12}$$

and similarly,

$$\tilde{e}_2 = \alpha \cdot \bar{e}_2 + (1-\alpha) \cdot \underline{e}_2 = \alpha \cdot E_{12} + (1-\alpha) \cdot 0 = \alpha \cdot E_{12}.$$

Thus, the fairness condition $e_1 - \tilde{e}_1 = e_2 - \tilde{e}_2$ takes the form $e_1 - \alpha \cdot E_{12} = e_2 - \alpha \cdot E_{12}$, i.e., the form $e_1 = e_2$.

So, in this case, no matter what is the optimism-pessimism value α, we divide the available amount E_{12} equally between the claimants:

$$e_1 = e_2 = \frac{E_{12}}{2}.$$

This is exactly what the ancient solution recommends in this case.

Case when the overall amount does not exceed the smallest claim: example. Let us consider one of the above examples, when $d_1 = 100$, $d_2 = 200$, and $E_{12} = 66\frac{2}{3}$. In this case, the above formulas recommend a solution in which $e_1 = e_2 = 33\frac{1}{3}$.

For the optimistic case $\alpha = 1$, the status quo point is $\tilde{e}_1 = \bar{e}_1 = 66\frac{2}{3}$ and $\tilde{e}_2 = \bar{e}_1 = 66\frac{2}{3}$. Thus, the condition of fairness with respect to this optimistic status quo point is indeed satisfied: $e_1 - \tilde{e}_1 = e_2 - \tilde{e}_2 = -33\frac{1}{3}$.

Case when the overall amount is in the between the smaller and the larger claims: general formulas. In this case,

$$\tilde{e}_1 = \alpha \cdot \bar{e}_1 + (1-\alpha) \cdot \underline{e}_1 = \alpha \cdot d_1 + (1-\alpha) \cdot 0 = \alpha \cdot d_1$$

and

$$\tilde{e}_2 = \alpha \cdot \bar{e}_2 + (1-\alpha) \cdot \underline{e}_2 = \alpha \cdot E_{12} + (1-\alpha) \cdot (E_{12} - d_1) = E_{12} - (1-\alpha) \cdot d_1.$$

Thus, the fairness condition $e_1 - \tilde{e}_1 = e_2 - \tilde{e}_2$ takes the form

$$e_1 - \alpha \cdot d_1 = e_2 - E_{12} + (1 - \alpha) \cdot d_1 = e_2 - E_{12} + d_1 - \alpha \cdot d_1.$$

Canceling the common term $-\alpha \cdot d_1$ on both sides, we get $e_1 = e_2 - E_{12} + d_1$. Substituting $e_2 = E - e_1$ into this formula, we conclude that $e_1 = E_{12} - e_1 - E_{12} + d_1$, i.e., $e_1 = -e_1 + d_1$. Moving the term $-e_1$ to the left-hand side, we get $2e_1 = d_1$ and $e_1 = \dfrac{d_1}{2}$.

The second person gets the remaining amount $e_2 = E_{12} - \dfrac{d_1}{2}$.

This is also exactly what the ancient solution recommends in this case.

Case when the overall amount is in the between the smaller and the larger claims: example. Let us consider one of the above examples, when $d_1 = 100$, $d_2 = 200$, and $E_{12} = 125$. In this case, the above formulas recommend a solution in which $e_1 = \dfrac{100}{2} = 50$ and $e_2 = E_{12} - e_1 = 125 - 50 = 75$.

Here, the optimistic status quo point is $\tilde{e}_1 = d_1 = 100$ and $\tilde{e}_2 = E_{12} = 125$. Thus, the condition of fairness with respect to this optimistic status quo point is indeed satisfied: $e_1 - \tilde{e}_1 = 50 - 100 = -50$ and $e_2 - \tilde{e}_2 = 75 - 125 = -50$.

Case when the overall amount is larger than both claims: general formulas. In this case,

$$\tilde{e}_1 = \alpha \cdot \bar{e}_1 + (1 - \alpha) \cdot \underline{e}_1 = \alpha \cdot d_1 + (1 - \alpha) \cdot (E_{12} - d_2)$$
$$= \alpha \cdot d_1 + (1 - \alpha) \cdot E_{12} - (1 - \alpha) \cdot d_2$$

and

$$\tilde{e}_2 = \alpha \cdot \bar{e}_2 + (1 - \alpha) \cdot \underline{e}_2 = \alpha \cdot d_2 + (1 - \alpha) \cdot (E_{12} - d_1)$$
$$= \alpha \cdot d_2 + (1 - \alpha) \cdot E_{12} - (1 - \alpha) \cdot d_1.$$

Thus, the fairness condition $e_1 - \tilde{e}_1 = e_2 - \tilde{e}_2$ takes the form

$$e_1 - \alpha \cdot d_1 - (1 - \alpha) \cdot E_{12} + (1 - \alpha) \cdot d_2$$
$$= e_2 - \alpha \cdot d_2 - (1 - \alpha) \cdot E_{12} + (1 - \alpha) \cdot d_1.$$

Canceling the comon term $-(1 - \alpha) \cdot E_{12}$ in both sides, we get

$$e_1 - \alpha \cdot d_1 + (1 - \alpha) \cdot d_2 = e_2 - \alpha \cdot d_2 + (1 - \alpha) \cdot d_1.$$

Moving terms containing d_1 and d_2 to the right-hand side, we conclude that $e_1 = e_2 + d_1 - d_2$. Substituting $e_2 = E_{12} - e_1$ into this formula, we get $e_1 = E_{12} - e_1 + d_1 - e_2$. Moving the term $-e_1$ to the left-hand side, we get $2e_1 = E_{12} + d_1 - e_2$ and $e_1 = \dfrac{E_{12} + d_1 - d_2}{2}$. The second person gets the remaining amount

$$e_2 = E_{12} - \frac{E_{12} + d_1 - d_2}{2} = \frac{E_{12} - d_1 + d_2}{2}.$$

This too is exactly what the ancient solution recommends in this case.

Case when the overall amount is larger than both claims: example. Let us consider one of the above examples, when $d_1 = 50$, $d_2 = 100$, and $E_{12} = 100$. In this case, the above formulas recommend a solution in which

$$e_1 = \frac{100 + 50 - 100}{2} = 25 \text{ and } e_2 = \frac{100 - 50 + 100}{2} = 75.$$

Here, the optimistic status quo point is $\widetilde{e}_1 = d_1 = 50$ and $\widetilde{e}_2 = d_2 = 100$. Thus, the condition of fairness with respect to this optimistic status quo point is indeed satisfied: $e_1 - \widetilde{e}_1 = 25 - 50 = -25$ and $e_2 - \widetilde{e}_2 = 75 - 100 = -25$.

Acknowledgments. This work was supported in part by the National Science Foundation grant HRD-1242122 (Cyber-ShARE Center of Excellence).

References

1. Aumann, R.J., Machler, M.: Game theoretic analysis of a banruptcy problem from the telmud. J. Econ. Theor. **36**, 195–213 (1985)
2. Epstein, L. (ed.): The Babylonian Talmud. Soncino, London (1935)
3. Hurwicz, L.: Optimality Criteria for Decision Making Under Ignorance, Cowles Commission Discussion Paper, Statistics, No. 370 (1951)
4. Kaminsky, M.: 'Hydraulic' Rationing. Math. Soc. Sci. **40**, 131–155 (2000)
5. Kreinovich, V.: Decision making under interval uncertainty (and beyond). In: Guo, P., Pedrycz, W. (eds.) Human-Centric Decision-Making Models for Social Sciences. Springer Verlag, pp. 163–193 (2014)
6. Luce, R.D., Raiffa, R.: Games and Decisions: Introduction and Critical Survey. Dover, New York (1989)
7. Nash, J.: Two-person cooperative games. Econometrica **21**, 128–140 (1953)
8. Shechter, S.: How the talmud divides an estate among creditors. In: Bridging Mathematics, Statistics, Engineering, and Technology: Controbutions from the Seminar on Mathematical Sciences and Applications. Springer Verlag, Berlin, Heidelberg, New York (2012)

Confidence Intervals for the Ratio of Means of Delta-Lognormal Distribution

Patcharee Maneerat$^{(\boxtimes)}$, Sa-Aat Niwitpong, and Suparat Niwitpong

Department of Applied Statistics, Faculty of Applied Science, King Mongkut's
University of Technology North Bangkok, Bangkok 10800, Thailand
m.patcharee@uru.ac.th, sa-aat.n@sci.kmutnb.ac.th, suparatn@kmutnb.ac.th

Abstract. This paper investigates confidence intervals for the ratio of means in the delta-lognormal distribution. The method of variance estimates recovery (MOVER) based on the variance stabilizing transformation, Wilson score method and Jeffreys method were proposed to establish confidence intervals for the ratio of delta-lonormal means. These confidence intervals were compared with the existing confidence interval based on the generalized confidence interval (GCI). The coverage probabilities and average lengths were the performance of these proposed confidence intervals which were evaluated via Monte Carlo simulation. The simulation results showed that the three MOVERs' performance is similar to the GCI in terms of coverage probability for all sample sizes except when the probability δ of having zero is close to zero and the coefficient of variation gets large. However, the MOVER based on Jeffreys provides the minimal average lengths when the coefficient of variation are small for all sample sizes. Finally, two data sets are used to illustrate examples of using the proposed confidence intervals.

1 Introduction

The ratio of means is one of the interesting parameters that indicate the comparison of the two quantitative variables measured in different units. In addition, the ratio of coefficients of variation is also included in the parameters of interest of data distribution. Mean and coefficient of variation are the statistical measures which collect information from the whole population. These parameters have been utilized in many applications such as public health, agriculture, medicine and environment. For instance, they are used to examine medical costs for patients with type I diabetes and patients for diabetic ketoacidosis [6], to investigate the percentage of fine gravel in the surface of soil types [8], to analyze the relative carboxyhemoglobin level for two large groups of nonsmokers and cigarette smokers [14], and to measure relation in millimeters between the length of frogs from top to tail and the snout-vent length of lizards [26].

In probability and statistics, the delta-lognormal distribution describes the occurrence in which data are composed of many zeros with a given probability $\delta > 0$ and lognormal data with the remaining probability $1 - \delta$, see e.g. [1,4,18,21]. This distribution, first introduced by Aitchison [1], is applied in several fields, including

© Springer International Publishing AG 2018
L. H. Anh et al. (eds.), *Econometrics for Financial Applications*, Studies in Computational
Intelligence 760, https://doi.org/10.1007/978-3-319-73150-6_13

economics, environment, fisheries survey, biology and medicine. For example, it was utilized to study the measurement of air contaminants where zeros corresponded to the case of the number of measurement that did not detect the concentration of airborne chlorine [16], the densities of fish where zeros corresponded to the case of empty trawls [15, 17, 19, 20], the certain species in different geographic areas where zeros corresponded to areas unsuitable for these species [4] and the urinary output (UO) of patients, zeros corresponded to the case when UO was not found in the critically ill patients [21].

These situations show that this distribution is used in many research areas, the achieved estimation of its parameter is critical and interval estimation gives more information on interesting parameters than point estimation. Consequently, many researchers have concentrated on the construction of confidence interval for the parameters in this distribution. For instance, Tian and Wu [21] established confidence intervals for the mean of lognormal with excess zeros based on the adjusted signed log-likelihood ratio statistic. Fletcher [11] also constructed the proposed confidence intervals of the mean derived from a profile-likelihood interval. Their results revealed that the profile likelihood performed poorly in cases of small sample size when the level of skewness was moderate to high. Wu and Hsieh [27] proposed generalized confidence interval to create confidence intervals for the mean of delta-lognormal distribution. The simulation study showed that generalized confidence interval was satisfactory in terms of coverage probabilities, expected interval lengths and reasonable relative biases. Chen and Zhou [5] proposed generalized confidence intervals for the ratio of two means for lognormal populations with zeros. They found that the approximate generalized pivotal approach outperforms all other methods, even in a small sample. However, these studies mostly examined confidence intervals for the parameter function which is mean in delta-lognormal distribution. Therefore, it is necessary to find a better confidence interval for the ratio of two means as the confidence intervals for δ performed well in many previous studies. These confidence intervals for δ were constructed by the variance stabilizing transformation that was presented by Dasgupta [7], Wu and Hsieh [27], Wilson score and Jeffreys method were recommended by Donner and Zou [9].

The purpose of this research is to look for methods for constructing the new confidence interval for the ratio means of delta-lognormal distribution. Four methods were applied: the method of variance estimate recovery (described by Donner and Zou [10]) based on the variance stabilizing transformation, Wilson score and Jeffreys method, and the generalized confidence interval by Weerahandi [23]. This article is organized as follows: The theory and method are detailed in Sect. 2. The simulation studies are described in Sect. 3 to assess the performance of all methods. All confidence interval are applied the real data in Sect. 4. Finally, the discussion and conclusions are contained in Sect. 5.

2 Confidence Intervals for the Ratio of Means of Delta-Lognormal Distribution

Let $W_i = (W_{i1}, W_{i2}, ..., W_{in_i})$ be a positive random variables of lognormal distribution, denoted as $LN(\mu_i, \sigma_i^2)$ and $Y_{ij} = \ln(W_{ij}) \sim N(\mu_i, \sigma_i^2)$ where μ_i and σ_i^2 are the mean and variance of Y_{ij}; $i = 1, 2$ $j = 1, 2, ..., n_i$, respectively. The probability density function of W_{ij} is given by

$$f\left(w_{ij}; \mu_i, \sigma_i^2\right) = \begin{cases} \frac{1}{w_{ij}\sqrt{2\pi\sigma_i^2}} \exp\left(-\frac{1}{2\sigma_i^2}\left(\ln(w_{ij}) - \mu_i\right)^2\right) & ; w_{ij} > 0 \\ 0 & ; \text{otherwise} \end{cases} \tag{1}$$

Suppose that $X_i = (X_{i1}, X_{i2}, ..., X_{in_i})$ is a non-negative random sample from delta-lognormal distribution, denoted as $\Delta(\mu_i, \sigma_i^2, \delta_i)$ where δ_i is the probabilities of having zero observations. The number of zero observations have the binomial distribution, $n_{i(0)} \sim B(n_i, \delta_i)$. Then, Tian and Wu [21] have noted that the distribution function of X_{ij} is defined as

$$G\left(x_{ij}; \mu_i, \sigma_i^2, \delta_i\right) = \begin{cases} \delta_i & ; x_{ij} = 0 \\ \delta_i + (1 - \delta_i) F(x_{ij}; \mu_i, \sigma_i^2) & ; x_{ij} > 0 \end{cases} \tag{2}$$

where $F(x_{ij}; \mu_i, \sigma_i^2)$ is the cumulative distribution function of lognormal distribution. The maximum likelihood estimator of parameter μ_i, σ_i^2 and δ_i are $\hat{\mu}_i = \frac{1}{n_{i(1)}} \sum_{j=1}^{n_{i(1)}} \ln(x_{ij})$, $\hat{\sigma}_i^2 = \frac{1}{n_{i(1)}} \sum_{j=1}^{n_{i(1)}} (\ln(x_{ij}) - \hat{\mu}_i)^2$ and $\hat{\delta}_i = \frac{n_{i(0)}}{n_i}$; $n_{i(0)} + n_{i(1)} = n_i$ where $n_{i(0)}$ and $n_{i(1)}$ are the number of zero and positive observed values from population i^{th}, respectively. The population mean and variance of X_i can be written as

$$E(X_i) = \vartheta_i = (1 - \delta_i) \exp\left(\mu_i + \frac{\sigma_i^2}{2}\right) \tag{3}$$

$$Var(X_i) = (1 - \delta_i) \exp\left(2\mu_i + \sigma_i^2\right) \left[\exp\left(\sigma_i^2\right) + \delta_i - 1\right] \tag{4}$$

and the coefficient of variation is

$$CV(X_i) = \phi_i = \sqrt{\frac{\exp\left(\sigma_i^2\right) + \delta_i - 1}{1 - \delta_i}} \tag{5}$$

For the ratio of two means is given by

$$\gamma = \frac{\vartheta_1}{\vartheta_2} = \frac{(1 - \delta_1) \exp\left(\mu_1 + \frac{\sigma_1^2}{2}\right)}{(1 - \delta_2) \exp\left(\mu_2 + \frac{\sigma_2^2}{2}\right)} \tag{6}$$

The confidence intervals for the parameter γ can be established by the following methods.

2.1   The Method of Variance Estimates Recovery for the Ratio of Means

In this case, focus is on the confidence interval for a ratio of parameters ϑ_i. Suppose that X_{ij} are random variables with $\Delta(\mu_i, \sigma_i^2, \delta_i)$. There are three parameters $\mu_i, \sigma_i^2, \delta_i$ in this distribution. First, constructing confidence interval for δ_i. In the past, researchers studied intensively the interval estimator for binomial proportion. Now, methods are suggested which consist of the variance stabilizing transformation (VST), Wilson score method and Jeffreys method. These methods are as follows:

The Variance Stabilizing Transformation
This method was presented by Dasgupta [7]. After that Wu and Hsieh [27] used the variance stabilizing transformation to apply in their study. Since $n_{i(0)} \sim B(n_i, \delta_i)$. The expected Fisher's Information for δ_i is given by

$$I_{n_i}(\delta_i) = \frac{n_i}{\delta_i(1 - \delta_i)} \tag{7}$$

Then $Var(\delta_i) = I_{n_i}^{-1}$. Apply delta theorem [7], we get that $\sqrt{n_i}\left(\hat{\delta}_i - \delta_i\right) \sim N(0, \delta_i(1 - \delta_i))$. As a result, the VST is defined as

$$g(\delta_i) = \int \frac{1/2}{\sqrt{\delta_i(1 - \delta_i)}} d\delta_i = \arcsin\sqrt{\delta_i} \tag{8}$$

Thus, $g(n_{i(0)}) = \arcsin\sqrt{\frac{n_{i(0)}}{n_i}}$ is the VST for binomial proportion. In fact, $\sqrt{n_i}\left(\arcsin\sqrt{\hat{\delta}_i} - \arcsin\sqrt{\delta_i}\right) \sim N(0, 1/4)$ so that

$$Z_{i1} = 2\sqrt{n_i}\left(\arcsin\sqrt{\hat{\delta}_i} - \arcsin\sqrt{\delta_i}\right) \tag{9}$$

where Z_{i1} converges in distribution to the standard normal distribution as $n \to \infty$. Then, the $100(1 - \alpha)\%$ asymptotically confidence interval of δ_i is

$$CI_{\delta_i.v} = [l_{v_i}, u_{v_i}] \tag{10}$$

where

$$l_{v_i} = sin^2\left(\arcsin\sqrt{\hat{\delta}_i} - Z_{i1\left(1 - \frac{\alpha}{2}\right)}\frac{1}{2\sqrt{n_i}}\right)$$

$$u_{v_i} = sin^2\left(\arcsin\sqrt{\hat{\delta}_i} + Z_{i1\left(1 - \frac{\alpha}{2}\right)}\frac{1}{2\sqrt{n_i}}\right)$$

The Wilson Score Method
The approach was proposed by Wilson [25]. After that Wilks [24] used it to construct the confidence interval. As a result, it is called the Wilson score metod.

Donner and Zou [9] found that this method perform well in small to moderate sample sizes when establishing confidence interval of proportion. Thus, the $100(1 - \alpha)\%$ confidence interval for δ_i is

$$CI_{\delta.w_i} = [l_{w_i}, u_{w_i}] \tag{11}$$

where

$$l_{w_i} = \frac{n_{i(0)} + Z_{\frac{\alpha}{2}}^2/2}{n_i + Z_{\frac{\alpha}{2}}^2} - \left(\frac{Z_{1-\frac{\alpha}{2}}}{n_i + Z_{\frac{\alpha}{2}}^2} * \sqrt{\frac{n_{i(0)}\left(n_{i(1)}\right)}{n_i} + \frac{Z_{\frac{\alpha}{2}}^2}{4}} \right)$$

$$u_{w_i} = \frac{n_{i(0)} + Z_{\frac{\alpha}{2}}^2/2}{n_i + Z_{\frac{\alpha}{2}}^2} + \left(\frac{Z_{1-\frac{\alpha}{2}}}{n_i + Z_{\frac{\alpha}{2}}^2} * \sqrt{\frac{n_{i(0)}\left(n_{i(1)}\right)}{n_i} + \frac{Z_{\frac{\alpha}{2}}^2}{4}} \right)$$

The Jeffreys Method

The Jeffreys method was developed from Brown et al. [3] to use beta priors in inference on δ [2]. Let the prior and posterior distribution of δ_i are $Beta(b_1, b_2)$ and $Beta(n_{i(0)} + b_1, n_{i(1)} + b_2)$, respectively. In this study, Jeffreys prior has the distribution Beta(1/2,1/2) so that the $100(1 - \alpha)\%$ Jeffreys prior confidence interval for δ_i is

$$CI_{\delta.J_i} = [l_{J_i}, u_{J_i}] \tag{12}$$

where

$$l_{J_i} = Beta\left(\frac{\alpha}{2}; n_{i(0)} + 1/2, n_{i(1)} + 1/2 \right)$$

$$u_{J_i} = Beta\left(1 - \frac{\alpha}{2}; n_{i(0)} + 1/2, n_{i(1)} + 1/2 \right)$$

Next, confidence intervals of σ_i^2 were considered. The unbiased estimator for σ_i^2 is

$$\hat{\sigma}_i^2 = \frac{1}{n_{i(1)} - 1} \sum_{j=1}^{n_{i(1)}} (\ln(X_{ij}) - \hat{\mu}_i)^2 \tag{13}$$

and also

$$U_i = \frac{\left(n_{i(1)} - 1\right)\hat{\sigma}_i^2}{\sigma_i^2} \sim \chi_{n_{i(1)}-1}^2 \tag{14}$$

where $\hat{\sigma}_i^2$ denoted as the sample variance for log-transformed data of non-zeros and $\chi_{n_{i(1)}-1}^2$ is chi-square distribution with $n_{i(1)} - 1$ degrees of freedom. To estimate the variance of normal distribution at a significant level, defined by α. The coverage probability for $\chi_{n_{i(1)}-1}^2$ is given by

$$P\left(\chi_{\frac{\alpha}{2},n_{i(1)}-1}^2 \leq \chi_{n_{i(1)}-1}^2 \leq \chi_{1-\frac{\alpha}{2},n_{i(1)}-1}^2 \right) = 1 - \alpha \tag{15}$$

Therefore, the $100(1 - \alpha)\%$ confidence interval for σ_i^2 is

$$CI_{\sigma_i^2} = [l_{\sigma_i^2}, u_{\sigma_i^2}] = \left[\frac{(n_{i(1)} - 1)\,\hat{\sigma}_i^2}{\chi_{1-\frac{\alpha}{2},n_{i(1)}-1}^2}, \frac{(n_{i(1)} - 1)\,\hat{\sigma}_i^2}{\chi_{\frac{\alpha}{2},n_{i(1)}-1}^2} \right] \qquad (16)$$

Then the confidence interval for γ can be substituted by $\hat{\mu}_i$, $\hat{\sigma}_i^2$, $\hat{\delta}_i$ in its parameter so that $\hat{\vartheta}_1 = (1 - \hat{\delta}_1)\exp(\hat{\mu}_1 + \frac{\hat{\sigma}_1^2}{2})$ and $\hat{\vartheta}_2 = (1 - \hat{\delta}_2)\exp(\hat{\mu}_2 + \frac{\hat{\sigma}_2^2}{2})$. Hence, the $100(1 - \alpha)\%$ two-sided confidence interval for γ based on MOVER approach is given by

$$CI_{rm} = [L_{rm}, U_{rm}] \qquad (17)$$

where

$$L_{rm} = \frac{\left(\hat{\vartheta}_1\hat{\vartheta}_2\right) - \sqrt{\left(\hat{\vartheta}_1\hat{\vartheta}_2\right)^2 - l_1 u_2 \left(2\hat{\vartheta}_1 - l_1\right)\left(2\hat{\vartheta}_2 - u_2\right)}}{u_2\left(2\hat{\vartheta}_2 - u_2\right)}$$

$$U_{rm} = \frac{\left(\hat{\vartheta}_1\hat{\vartheta}_2\right) + \sqrt{\left(\hat{\vartheta}_1\hat{\vartheta}_2\right)^2 - u_1 l_2 \left(2\hat{\vartheta}_1 - u_1\right)\left(2\hat{\vartheta}_2 - l_2\right)}}{l_2\left(2\hat{\vartheta}_2 - l_2\right)}$$

By using l_i and u_i derive from three previous methods, we can establish confidence intervals for γ.

2.1.1 The Method of Variance Estimate Recovery Based on the Variance Stabilizing Transformation

Setting $l_{i.v} = (1 - u_{v_i})\exp(\hat{\mu}_i + \frac{l_{\sigma_i^2}}{2})$ and $u_{i.v} = (1 - l_{v_i})\exp(\hat{\mu}_i + \frac{u_{\sigma_i^2}}{2})$. The new confidence interval for γ based on VST is given by

$$CI_{rm.v} = [L_{rm.v}, U_{rm.v}] \qquad (18)$$

where

$$L_{rm.v} = \frac{(\hat{\vartheta}_1\hat{\vartheta}_2) - \sqrt{(\hat{\vartheta}_1\hat{\vartheta}_2)^2 - l_{1.v}u_{2.v}(2\hat{\vartheta}_1 - l_{1.v})(2\hat{\vartheta}_2 - u_{2.v})}}{u_{2.v}(2\hat{\vartheta}_2 - u_{2.v})}$$

$$U_{rm.v} = \frac{(\hat{\vartheta}_1\hat{\vartheta}_2) + \sqrt{(\hat{\vartheta}_1\hat{\vartheta}_2)^2 - u_{1.v}l_{2.v}(2\hat{\vartheta}_1 - u_{1.v})(2\hat{\vartheta}_2 - l_{2.v})}}{l_{2.v}(2\hat{\vartheta}_2 - l_{2.v})}$$

2.1.2 The Method of Variance Estimate Recovery Based on Wilson Score

Given $l_{i.w} = (1 - u_{w_i})\exp(\hat{\mu}_i + \frac{l_{\sigma_i^2}}{2})$ and $u_{i.w} = (1 - l_{w_i})\exp(\hat{\mu}_i + \frac{u_{\sigma_i^2}}{2})$. The new confidence interval for γ based on Wilson is given by

$$CI_{rm.w} = [L_{rm.w}, U_{rm.w}] \tag{19}$$

where

$$L_{rm.w} = \frac{(\hat{\vartheta}_1\hat{\vartheta}_2) - \sqrt{(\hat{\vartheta}_1\hat{\vartheta}_2)^2 - l_{1.w}u_{2.w}(2\hat{\vartheta}_1 - l_{1.w})(2\hat{\vartheta}_2 - u_{2.w})}}{u_{2.w}(2\hat{\vartheta}_2 - u_{2.w})}$$

$$U_{rm.w} = \frac{(\hat{\vartheta}_1\hat{\vartheta}_2) + \sqrt{(\hat{\vartheta}_1\hat{\vartheta}_2)^2 - u_{1.w}l_{2.w}(2\hat{\vartheta}_1 - u_{1.w})(2\hat{\vartheta}_2 - l_{2.w})}}{l_{2.w}(2\hat{\vartheta}_2 - l_{2.w})}$$

2.1.3 The Method of Variance Estimate Recovery Based on Jeffreys

Let $l_{i.J} = (1 - u_{J_i})\exp(\hat{\mu}_i + \frac{l_{\sigma_i^2}}{2})$ and $u_{i.J} = (1 - l_{J_i})\exp(\hat{\mu}_i + \frac{u_{\sigma_i^2}}{2})$. The new confidence interval for γ based on Jeffreys is given by

$$CI_{rm.J} = [L_{rm.J}, U_{rm.J}] \tag{20}$$

where

$$L_{rm.J*} = \frac{(\hat{\vartheta}_1\hat{\vartheta}_2) - \sqrt{(\hat{\vartheta}_1\hat{\vartheta}_2)^2 - l_{1.J}u_{2.J}(2\hat{\vartheta}_1 - l_{1.J})(2\hat{\vartheta}_2 - u_{2.J})}}{u_{2.J}(2\hat{\vartheta}_2 - u_{2.J})}$$

$$U_{rm.J} = \frac{(\hat{\vartheta}_1\hat{\vartheta}_2) + \sqrt{(\hat{\vartheta}_1\hat{\vartheta}_2)^2 - u_{1.J}l_{2.J}(2\hat{\vartheta}_1 - u_{1.J})(2\hat{\vartheta}_2 - l_{2.J})}}{l_{2.J}(2\hat{\vartheta}_2 - l_{2.J})}$$

2.2 The Generalized Confidence Interval

The common method to construct the confidence interval is the generalized confidence interval (GCI) method which has been introduced by Weerahandi [23]. The GCI is based on the concept following of generalized pivotal quantity.

Let $X_{ij} = (X_{i1}, X_{i2}, ..., X_{in_i})$ be random variables with the probability density function $f_X(x_{ij}; \eta_i)$ where $\eta_i = (\mu_i, \sigma_i^2, \delta_i)$ is the vector of unknown parameters. Let $x_{ij} = (x_{i1}, x_{i2}, ..., x_{in_i})$ be observed values of X_{ij}. The generalized pivotal quantity $R(X_{ij}; x_{ij}, \eta_i)$ is required to satisfy the following conditions:

(i) Given X_{ij}, the distribution of $R(X_{ij}; x_{ij}, \eta_i)$ is free of all unknown parameters.
(ii) The observed value of $R(X_{ij}; x_{ij}, \eta_i)$, denoted by $r(x_{ij}; x_{ij}, \eta_i)$, does not depend on the nuisance parameters.

$R(X_{ij}; x_{ij}, \eta_i)$ satisfies conditions (i) and (ii) so that the $100(1 - \alpha)\%$ two-sided GCI confidence interval for η_i is $(R_{\alpha/2}, R_{1-\alpha/2})$ where R_α is the αth percentile of $R(X_{ij}; x_{ij}, \eta_i)$. Now, suppose that X_{ij} be a random variables distributed according to $\Delta(\mu_i, \sigma_i^2, \delta_i)$. The pivotal quantity for δ_i based on Dasgupta [7], Wu and Hsieh [27] is defined as

$$R_{\delta_i} = \sin^2\left[\arcsin\sqrt{\hat{\delta}_{i.0}} - \frac{Z_{i1}}{2\sqrt{n_i}}\right] \tag{21}$$

Krishnamoorthy and Mathew [12] showed that another pivotal quantity for μ_i and σ_i^2 given by

$$R_{\mu_i} = \hat{\mu}_{i.0} - Z_{i2}\sqrt{\frac{(n_{i(1)} - 1)\hat{\sigma}_{i.0}^2}{n_{i(1)}U_i}} \text{ and } R_{\sigma_i^2} = \frac{(n_{i(1)} - 1)\hat{\sigma}_{i.0}^2}{U_i} \tag{22}$$

where $Z_{i2} = (\hat{\mu}_i - \mu_i)/\sqrt{\frac{(n_{i(1)}-1)\hat{\sigma}_i^2}{n_{i(1)}U_i}}$ denoted as the standard normal distribution and $U_i = \frac{(n_{i(1)}-1)\hat{\sigma}_i^2}{\sigma_i^2}$ is the chi-square distribution with $n_{i(1)} - 1$ degrees of freedom. By the information of three pivots, the pivot for γ is defined as follows:

$$R_\gamma = \frac{R_{\vartheta_1}}{R_{\vartheta_2}} = \frac{(1 - R_{\delta_1})\exp(R_{\mu_1} + \frac{R_{\sigma_1^2}}{2})}{(1 - R_{\delta_2})\exp(R_{\mu_2} + \frac{R_{\sigma_2^2}}{2})} \tag{23}$$

Therefore, the generalized pivotal quantity of γ is

$$R_\gamma = \frac{(1 - \sin^2\left[\arcsin\sqrt{\hat{\delta}_{1.0}} - \frac{Z_{11}}{2\sqrt{n_1}}\right])\exp\left[\hat{\mu}_{1.0} - Z_{12}\sqrt{\frac{n_{1(1)}-1)\hat{\sigma}_{1.0}^2}{n_{1(1)}U_1}} + \frac{(n_{1(1)}-1)\hat{\sigma}_{1.0}^2}{2U_1}\right]}{(1 - \sin^2\left[\arcsin\sqrt{\hat{\delta}_{2.0}} - \frac{Z_{21}}{2\sqrt{n_2}}\right])\exp\left[\hat{\mu}_{2.0} - Z_{22}\sqrt{\frac{n_{2(1)}-1)\hat{\sigma}_{2.0}^2}{n_{2(1)}U_2}} + \frac{(n_{2(1)}-1)\hat{\sigma}_{2.0}^2}{2U_2}\right]} \tag{24}$$

where Z_{i1}, Z_{i2}, U_i are independent random variables and $\hat{\delta}_{i.0}$, $\hat{\mu}_{i.0}$, $\hat{\sigma}_{i.0}^2$ are the observed values of $\hat{\delta}_i$, $\hat{\mu}_i$, $\hat{\sigma}_i^2$, respectively. The expression (24) satisfies the two conditions for being a general pivotal quantity. Consequently, the $100(1 - \alpha)\%$ two-sided confidence interval for the ratio of means (γ) based on GCI is

$$CI_{rgci} = [L_{rgci}, U_{rgci}] = [R_\gamma(\alpha/2), R_\gamma(1 - \alpha/2)] \tag{25}$$

where $R_\gamma(\alpha/2)$ denotes the $100(\alpha/2)\%$ percentile of R_γ of R_γ. Additionally, the $100(1 - \alpha)\%$ GCI for γ can be investigated by the following algorithm.

Algorithm 1

begin;

for $j = 1$ *to* M **do**

Generate dataset $x_{i1}, x_{i2}, ..., x_{1n_i}$ from $\Delta(\mu_i, \sigma_i^2, \delta_i)$;

Calculate $\hat{\delta}_i$, $\hat{\mu}_i$, $\hat{\sigma}_i^2$;

for $k = 1$ *to* m **do**

Generate Z_{i1}, Z_{i2} from standard normal distribution;

Generate U_i from chi-square distribution with $n_{i(1)} - 1$ degrees of freedom;

Calculate R_{δ_i};

Calculate R_{μ_i};

Calculate $R_{\sigma_i^2}$;

Calculate R_γ;

end

Calculate the $(\alpha/2)100$th percentile of R_γ;

Calculate the $(1 - \alpha/2)100$th percentile of R_γ;

end

3   Simulation Studies

In this part, all confidence intervals are evaluated based on the coverage probabilities and average lengths which are compared via Monte Carlo simulation. These methods are investigated by using R statistical programming language [22]. Each simulation calculate the nominal confidence level of 0.95 which is utilized based on 10,000 replications and 5,000 pivotal quantities for the GCI method. The generalized confidence interval is defined as CI_{rgci}, the method of variance estimate recovery based on the variance stabilizing transformation, Wilson score and Jeffreys are defined as $CI_{rm.v}$, $CI_{rm.w}$ and $CI_{rm.J}$, respectively.

In simulation, four confidence intervals for parameter γ, are the comparison between n_1, n_2, δ_1, δ_2, ϕ_1 and ϕ_2 where the mean ϑ_i is fixed to 0; $n_1 = n_2 = 20, 50, 100$; $\delta_1 = \delta_2 = 0.2, 0.5, 0.8$ and $\phi_1 = \phi_2 = 0.2, 0.5, 1.0, 2.0$. In the study by Fletcher [11], if the number of non-zero values $E(n_{i(1)})$ is less than 10, then the results would not perform well in terms of the coverage probabilities and average lengths. For Wu and Hsieh [27], this was true also except for the cases of $n_1 = n_2 = 20$, $\delta_1 = \delta_2 = 0.8$ and $\phi_1 = \phi_2 = 0.2, 0.5, 1.0, 2.0$ because $E(n_{i(1)})$ is below 10. As a result, these combinations are excluded in this study. As the best method for computing the confidence interval, we selected method with the smallest nominal confidence level and the shortest average length.

In Tables 1 and 2, the coverage probabilities and the average lengths of confidence intervals for the ratio of means (γ) of delta-lognormal distribution are displayed. The result of this study show that the CI_{rgci} has the coverage probabilities greater than the nominal confidence level in all cases but its average lengths are rather wide and go back to narrow as sample size increases. The $CI_{rm.v}$, $CI_{rm.w}$ and $CI_{rm.J}$ also have coverage probabilities which are greater than the nominal level for all samples sizes except for cases of $\delta = 0.2$ and $\phi = 1, 2$. The average lengths of $CI_{rm.J}$ are shorter than other methods except for $\delta = 0.8$ and $\phi = 1, 2$.

Table 1. The coverage probabilities of 95% two-sided confidence intervals for the ratio of means (γ) of delta-lognormal distribution

n_1	n_2	δ_1	δ_2	ϕ_1	ϕ_2	$CI_{r.gci}$	$CI_{rm.v}$	$CI_{rm.w}$	$CI_{rm.J}$
20	20	0.2	0.2	0.2	0.2	0.984	0.971	0.988	0.969
				0.5	0.5	0.976	0.948	0.962	0.941
				1.0	1.0	0.965	0.933	0.941	0.930
				2.0	2.0	0.956	0.930	0.934	0.929
		0.5	0.5	0.2	0.2	0.993	0.994	0.994	0.985
				0.5	0.5	0.990	0.990	0.989	0.981
				1.0	1.0	0.979	0.981	0.978	0.973
				2.0	2.0	0.965	0.970	0.967	0.963
50	50	0.2	0.2	0.2	0.2	0.988	0.984	0.988	0.981
				0.5	0.5	0.972	0.947	0.952	0.941
				1.0	1.0	0.960	0.924	0.927	0.919
				2.0	2.0	0.954	0.928	0.929	0.925
		0.5	0.5	0.2	0.2	0.993	0.993	0.993	0.989
				0.5	0.5	0.987	0.988	0.987	0.983
				1.0	1.0	0.975	0.976	0.975	0.969
				2.0	2.0	0.961	0.962	0.962	0.960
		0.8	0.8	0.2	0.2	0.998	0.998	0.998	0.996
				0.5	0.5	0.994	0.996	0.996	0.994
				1.0	1.0	0.984	0.990	0.989	0.986
				2.0	2.0	0.971	0.982	0.980	0.978
100	100	0.2	0.2	0.2	0.2	0.991	0.987	0.989	0.986
				0.5	0.5	0.978	0.954	0.957	0.952
				1.0	1.0	0.962	0.928	0.929	0.927
				2.0	2.0	0.953	0.926	0.926	0.924
		0.5	0.5	0.2	0.2	0.996	0.996	0.996	0.996
				0.5	0.5	0.987	0.986	0.986	0.984
				1.0	1.0	0.975	0.975	0.974	0.973
				2.0	2.0	0.963	0.963	0.963	0.961
		0.8	0.8	0.2	0.2	0.996	0.997	0.997	0.996
				0.5	0.5	0.993	0.995	0.995	0.993
				1.0	1.0	0.982	0.989	0.988	0.987
				2.0	2.0	0.965	0.978	0.977	0.975

Table 2. The average lengths of 95% two-sided confidence intervals for the ratio of means (γ) of delta-lognormal distribution

n_1	n_2	δ_1	δ_2	ϕ_1	ϕ_2	$CI_{r.gci}$	$CI_{rm.v}$	$CI_{rm.w}$	$CI_{rm.J}$
20	20	0.2	0.2	0.2	0.2	0.627	0.596	0.651	0.529
				0.5	0.5	0.968	0.836	0.875	0.764
				1.0	1.0	1.954	1.692	1.712	1.600
				2.0	2.0	5.815	5.368	5.364	5.164
		0.5	0.5	0.2	0.2	1.286	1.288	1.222	1.065
				0.5	0.5	1.718	1.746	1.669	1.498
				1.0	1.0	3.567	3.827	3.701	3.440
				2.0	2.0	24.828	30.308	29.479	28.034
50	50	0.2	0.2	0.2	0.2	0.397	0.381	0.395	0.358
				0.5	0.5	0.578	0.507	0.516	0.483
				1.0	1.0	1.005	0.868	0.871	0.840
				2.0	2.0	1.956	1.751	1.748	1.714
		0.5	0.5	0.2	0.2	0.782	0.782	0.766	0.719
				0.5	0.5	0.963	0.965	0.947	0.897
				1.0	1.0	1.506	1.526	1.505	1.450
				2.0	2.0	3.087	3.208	3.175	3.095
		0.8	0.8	0.2	0.2	1.713	1.734	1.618	1.537
				0.5	0.5	2.090	2.232	2.125	2.023
				1.0	1.0	3.794	4.470	4.360	4.165
				2.0	2.0	22.643	31.628	31.387	30.657
100	100	0.2	0.2	0.2	0.2	0.285	0.274	0.279	0.265
				0.5	0.5	0.405	0.358	0.362	0.349
				1.0	1.0	0.678	0.587	0.588	0.576
				2.0	2.0	1.209	1.080	1.079	1.067
		0.5	0.5	0.2	0.2	0.552	0.552	0.546	0.528
				0.5	0.5	0.666	0.666	0.660	0.642
				1.0	1.0	0.981	0.985	0.978	0.959
				2.0	2.0	1.722	1.746	1.737	1.713
		0.8	0.8	0.2	0.2	1.141	1.151	1.114	1.085
				0.5	0.5	1.323	1.382	1.349	1.315
				1.0	1.0	1.946	2.135	2.108	2.061
				2.0	2.0	4.060	4.687	4.664	4.576

4   An Empirical Application

To demonstrate the calculation of all confidence intervals, the data taken from a nutrition analysis of every menu item on the United State McDonald's menu provided by Larion [13] were utilized. In this study, we are interesting in the calories (kcal) of two categories, including 27 menus of beverages and 95 menus of coffee plus tea are of interest. There were 9 and 7 menus with no calorie for beverages and coffee plus tea, respectively. Figure 1 shows the histogram plots of non-zero calories in two categories: 18 menus of beverages and 88 menus of coffee plus tea.

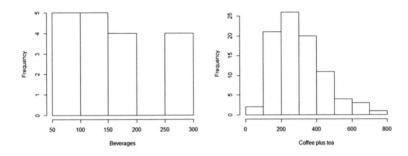

Fig. 1. The histogram plots of calories for beverages and coffee plus tea

Table 3. Point estimation and 95% confidence interval for the ratio of mean calories each method.

Methods	Point estimation of mean calories		Ratio of mean	Confidence interval		Average length
	Beverages	Coffee & tea		Lower	Upper	
GCI	114.31	285.58	0.400	0.272	0.568	0.297
MOVER based on						
VST				0.276	0.560	0.284
Wilson score				0.275	0.553	0.278
Jeffreys				0.286	0.528	0.241

Next, the distribution of these data was investigated. The Shapiro–Wilk normality test have p-values of 0.7768 and 0.1287 for the log transformation of beverages and coffee plus tea at the 5% significance level. As a result, both categories fit the delta-lognormal distribution. The summary statistics are $\hat{\mu}_1 = 5.06$, $\hat{\mu}_2 = 5.62$, $\hat{\sigma}_1^2 = 0.17$, $\hat{\sigma}_2^2 = 0.22$, $\hat{\delta}_1 = 0.33$, $\hat{\delta}_2 = 0.07$ and $n_1 = 27$, $n_2 = 95$.

From Table 3, the 95% confidence intervals for γ were computed, the MOVER based on the variance stabilizing transformation $CI_{rm.v} = [0.276, 0.560]$ with the length interval of 0.284, Wilson score $CI_{rm.w} = [0.275, 0.553]$ with the length interval of 0.278 and Jeffreys $CI_{rm.J} = [0.286, 0.528]$ with the length interval of 0.241. In comparison, the generalized confidence interval $CI_{rgci} = [0.272, 0.568]$ with the length interval of 0.297.

5   Discussion and Conclusions

In this article, the aim was to establish a new confidence interval for the ratio of means of delta-lognormal distribution based on MOVER which is based on the variance stabilizing transformation, Wilson score method and Jeffreys method, then compare with the GCI. All of the approaches were assessed via Monte Carlo simulation. The finding can be summarized as follows: the MOVER based on the variance stabilizing transformation $CI_{rm.v}$, Wilson score $CI_{rm.w}$ and Jeffreys $CI_{rm.J}$ perform similarly in terms of the coverage probability as the probability of having zero is close to one. The Jeffreys $CI_{rm.J}$ gave the narrowest average length in cases of the coefficient of variation was small in all sample sizes. Hence, the Jeffreys $CI_{rm.J}$ is recommended as the interval estimator of the ratio of means when proportion of zero is close to one and the coefficient of variation is small for each sample size. The generalized confidence interval CI_{rgci} is recommended in cases when the probability of zero outcome and the coefficients of variation are big for large sample sizes because its coverage probability would be greater than nominal coverage level and average length would also be narrower than all methods.

Furthemore, Lee and Lin [14] found that the generalized confidence intervals performed reasonably well in the case of large sample sizes for the ratio of two means of normal populations which is consistent with the finds in this research.

Acknowledgments. We would like to acknowledge referees for their important comments and recommendations that assisted with the improvements of this paper.

References

1. Aitchison, J.: On the distribution of a positive random variable having a discrete probability mass at the origin. J. Am. Stat. Assoc. **50**(271), 901–908 (1955)
2. Berger, J.O.: Statistical Decision Theory and Bayesian Analysis, 2nd edn. Springer, New York (1985)
3. Brown, L.D., Cai, T.T., Dasgupta, A.: Interval estimation for a binomial proportion. Stat. Sci. **16**(2), 101–117 (2001)
4. Buntao, N., Niwitpong, S.-A., Kreinovich, V.: Estimating statistical characteristics of lognormal and delta-lognormal distributions under Interval uncertainty: algorithms and computational complexity. Departmental Technical Report, Department of Computer Science, University of Texas at El Paso (2012)
5. Chen, Y.-H., Zhou, X.-H.: Generalized confidence intervals for the ratio or difference of two means for lognormal populations with zeros. UW Biostatistics Working Paper Series (2006)
6. Chen, Y.-H., Zhou, X.-H.: Interval estimates for the ratio and difference of two lognormal means. Stat. Med. **25**(23), 4099–4113 (2006)

7. Dasgupta, A.: Asymptotic Theory of Statistics and Probability. Springer Texts in Statistics. Springer, New York (2008)
8. Díaz-Francés, E., Sprott, D.A.: Inference for the ratio of two normal means with unspecified variances. Biom. J. **46**(1), 83–89 (2004)
9. Donner, A., Zou, G.Y.: Estimating simultaneous confidence intervals for multiple contrasts of proportions by the method of variance estimates recovery. Stat. Biopharm. Res. **3**(2), 320–335 (2011)
10. Donner, A., Zou, G.Y.: Closed-form confidence intervals for functions of the normal mean and standard deviation. Stat. Methods Med. Res. **21**(4), 347–359 (2012)
11. Fletcher, D.: Confidence intervals for the mean of the delta-lognormal distribution. Environ. Ecol. Stat. **15**(2), 175–189 (2008)
12. Krishnamoorthy, K., Mathew, T.: Inferences on the means of lognormal distributions using generalized p-values and generalized confidence intervals. J. Stat. Plan. Inference **115**(1), 103–121 (2003)
13. Larion, A.: Nutrition facts for McDonald's menu (2017). https://www.kaggle.com/mcdonalds/nutrition-facts
14. Lee, J.C., Lin, S.-H.: Generalized confidence intervals for the ratio of means of two normal populations. J. Stat. Plann. Inference **123**(1), 49–60 (2004)
15. Lo, N.C.-H., Jacobson, L.D., Squire, J.L.: Indices of relative abundance from fish spotter data based on delta-lognornial models. Can. J. Fish. Aquat. Sci. **49**(12), 2515–2526 (1992)
16. Owen, W.J., DeRouen, T.A.: Estimation of the mean for lognormal data containing zeroes and left-censored values, with applications to the measurement of worker exposure to air contaminants. Biometrics **36**(4), 707–719 (1980)
17. Pennington, M.: Efficient estimators of abundance, for fish and plankton surveys. Biometrics **39**(1), 281–286 (1983)
18. Rosales, M.: The robustness of confidence intervals for the mean of delta distribution. Dissertations (2009)
19. Smith, S.J.: Evaluating the efficiency of the $\triangle$- distribution mean estimator. Biometrics **44**(2), 485–493 (1988)
20. Smith, S.J.: Use of statistical models for the estimation of abundance from groundfish trawl survey data. Can. J. Fish. Aquat. Sci. **47**(5), 894–903 (1990)
21. Tian, L., Wu, J.: Confidence intervals for the mean of lognormal data with excess zeros. Biom. J. Biometrische Zeitschrift **48**(1), 149–156 (2006)
22. Venables, W., Smith, D.: R Development Core Team: An Introduction to R: a programming environment for data analysis and graphics. Network Theory (2015). http://cran.r-project.org/
23. Weerahandi, S.: Generalized confidence intervals. J. Am. Stat. Assoc. **88**(423), 899–905 (1993)
24. Wilks, S.S.: Shortest average confidence intervals from large samples. Ann. Math. Stat. **9**(3), 166–175 (1938)
25. Wilson, E.B.: Probable inference, the Law of succession, and statistical inference. J. Am. Stat. Assoc. **22**(158), 209–212 (1927)
26. Wongkhao, A., Niwitpong, S.-A., Niwitpong, S.: Confidence intervals for the ratio of two independent coefficients of variation of normal distribution. Far East J. Math. Sci. (FJMS) **98**(6), 741–757 (2015)
27. Wu, W.-H., Hsieh, H.-N.: Generalized confidence interval estimation for the mean of delta-lognormal distribution: an application to New Zealand trawl survey data. J. Appl. Stat. **41**(7), 1471–1485 (2014)

Modeling and Simulation of Financial Risks

Akira Namatame[(✉)]

National Defense Academy, Yokosuka, Japan
akiranamatame@gmail.com

Abstract. Networked systems offer multiple benefits and new opportunities. However, there are also disadvantages to certain networks, especially networks of interdependent systems. This paper investigates how the complexity of a networked system contributes to the emergence of systemic instability. Systemic instability is said to exist in a networked system if its integrity is compromised regardless of whether each component functions properly. An agent-based model, which seeks to explain how the behavior of individual agents can affect the outcomes of complex systems, can make an important contribution to our understanding of potential vulnerabilities and the way in which risk propagates across networked systems. Furthermore, we discuss various methodological issues related to managing risk in networks.

1 Introduction

Risks are idiosyncratic in the sense that a risk to one system may present an opportunity in another. Networks increase the level of interdependence for its components, and this affects the integrity of the system. Over the last few decades, we have witnessed the dark side to increased interdependency. A crucial question follows: Is an interconnected world a safer place in which to live, or is it more dangerous? This is especially relevant to financial institutions and firms (or agents), where the actions of a single agent can impact all the other agents in a network. In this section, we present an overview of the concepts, ideas, and examples of systemic risks inherent in a network, that is, of "networking risks" [9].

We often use the terms risk and uncertainty interchangeably. Risk exists where the outcome is uncertain, but where it is possible to determine the probability distribution of outcomes. Therefore, we can manage risk using a probability calculus. By contrast, uncertainty exists in a situation where we do not know the outcome, or the probability cannot be determined. Most systems we deal with in the real world contain uncertainties, rather than risks. Risk can be understood as the probability of an adverse outcome with consideration to the severity of the consequences should it occur. This definition is sufficient in many cases, but it is inadequate for situations of the greatest interest-for pivotal events and large-scale disturbances. Because pivotal events are generally rare and unprecedented, statistics are powerless to give any meaningful probability to the occurrence of such events.

© Springer International Publishing AG 2018
L. H. Anh et al. (eds.), *Econometrics for Financial Applications*, Studies in Computational Intelligence 760, https://doi.org/10.1007/978-3-319-73150-6_14

Most natural and social systems are continually subjected to external disturbances that can vary widely in degree. Scientific research has focused on natural
disasters, such as earthquakes, or on failures to engineered systems, such as electrical blackouts. However, many major disasters affecting human societies relate
to the internal structure of networked systems. Risk can be categorized as either
exogenous or endogenous. When we explore the sources of risk, it is important
to distinguish between these types. Exogenous risk comes from outside a system,
and is basically a condition imposed on it. For instance, disaster preparedness
involves considering exogenous risk. Other examples include the threat of avian
flu, terrorism, and hurricanes. Endogenous risk, on the other hand, refers to
the risk from disturbances that are generated and amplified within the system.
The amplification of risk proceeds from its systemic nature, and challenges the
integrity of the system. As systems increase their interdependency on other systems, we must consider the risk inherent to such interdependency. As a disturbance propagates through a network, it might encounter components known as
amplifiers, that is, components that increase the risk to other components in the
network. Amplification occurs when such an interaction results in a vicious cycle,
reinforcing the effects of amplifiers. Amplifiers are the mechanisms that boost
the scale of a disturbance in a particular system, and also the means by which
hazards are spread and intensified throughout the system. This phenomenon is
especially prevalent in interdependent systems.

Allen and Gale introduced network theory as a means for enriching our understanding of networked risks [1]. They explored critical issues in their study of
networked risk by answering fundamental questions, such as how resilient financial networks are to contagion, and how financial institutions form connections
when exposed to the risk of contagion. They showed that by increasing the
connections between financial institutions, the overall risk of contagion can be
reduced because the risk is shared. Empirical evidence shows that, whereas a
network's performance improves as its connectivity increases, there is also a corresponding increase to the risk of contagion. Failures to networked components
are significantly more egregious than the failure of any single node or component. Network interdependency is a main reason for a series of failures. Examples
include disease epidemics, traffic congestion, and electrical blackouts. These phenomena are known as cascading failures, and more commonly as chain reactions
or domino effects. The definitive feature to cascading failures is that a local failure results in a global failure on a larger scale. The result is that networked risk,
which can lead to the failure of the networked system as a whole, is not related
in any simple way to the risk profiles for each component. Such networked risks
are common with non-linear interactions, which are ubiquitous in networked
systems.

Networked risks are idiosyncratic, and they are an inherent characteristic of
interdependent networks. In a networked society, the risks faced by any one agent
depend not only on the actions of that agent, but also on those of other agents.
The fact that the risk a single agent faces is often determined in part by the
activities of other agents gives a unique and complex structure to the incentives

that agents face as they attempt to reduce their exposure to networked risks. The term "interdependent risk" refers to situations in which multiple agents decide separately whether to adopt protective management strategies, such as protection against risk, and where each agent has less of an incentive to adopt protection if others fail to do so. Protective management strategies can reduce the risk of a direct loss to any agent, but there is still a chance of suffering damage from other agents who do not adopt similar strategies. The fact that the risk is often determined in part by the actions of other agents imposes independent risk structures for the incentives that agents require to reduce risk by investing in risk-mitigation measures.

A variety of schemes exist that are designed to mitigate networked risks, but the majority of these schemes depend on centralized control and full knowledge of the system. Furthermore, centralized designs are frequently more susceptible to limited situational awareness, making them inadequate and resulting in increased vulnerability and disastrous consequences. Understanding the mechanisms and determinants of risk can only help to describe what is going on, and these mechanisms have limited predictive power. Bi-directional causal relationships are an essential component in the study of networked risk. Understanding the relationship between the different levels at which macroscopic phenomena can be observed as networked risk is possible with the tools and insights generated by combining agent-based and network models. Agent based modeling combined with network theory can explain certain networked risks. Instead of looking at the details of particular failures of nodes or agents, this approach involves investigating a series of failures caused by the dependencies among agents. A detailed discussion of different approaches can be found in [2,3,7,10,12].

2   Systemic Risks

The concept of systemic risk pertains to something undesirable happening in a system that is significantly larger and worse than the failure of any one node or component. Global financial instability and economic crises are typical examples of systemic risks. The scope and speed of the diffusion of risk in recent financial crises have stimulated an analysis of the conditions under which financial contagion can actually arise. Systemic risk describes a situation in which financial institutions fail as a result of a common shock or a contagion process. A contagion process refers to the systemic risk that the failure of one financial institution will lead to defaults in other financial institutions through a domino effect to the interbank market.

Network interdependencies are apparent in financial networks, where the actions of a single agent (e.g., a financial institution) in an interconnected network can impact other agents in the network. Increased globalization and financial innovation have prompted a sudden increase in the creation of financial linkages and trade relationships between agents. In financial systems, therefore, there is a tendency for crises to spread from one agent to another, and this tendency can lead to systemic failures on a significantly large scale.

In this section, we analyze the mechanisms to systemic risk using agent-based modeling and network theory. In an interbank market, banks facing liquidity shortages may borrow liquidity from other banks that have liquidity surpluses. This system of liquidity swapping provides the interbank market with enhanced liquidity sharing, and it decreases the risk of contagion among agents when unexpected problems arise. However, solvency and liquidity problems faced by a single agent can also travel through the interbank market to other agents and cause systemic failures.

May and Arinaminpathy provided the models to analyze financial instability and the contagion process based on mean-field analysis [9]. The assumption in their approach is that each agent in the network is identical. External or internal shocks may lead to the collapse of the entire system. If a single agent is disrupted and this causes a failure, this initial failure can lead to a cascade of failures. Several critical constellations determine whether this initial failure remains local or grows to point where it affects the system leading to systemic risk.

One of the main objectives in research on systemic risk is to identify agents that are systemically important to the contagion process. When these agents default, they influence other agents through the interconnectivity of the net-worked system. Systemic risk is not a risk of failure caused by the fundamental weakness of a particular agent. Because failed agents are not able to honour their commitments in the interbank market, other agents are likely to be influenced to default as well, which can affect more agents and cause further contagious defaults. For an agent, maintaining interconnections with other agents always implies a trade-off between risk sharing and the risk of contagion. Indeed, the more interconnected a balance sheet is, the more easily a negative shock, say a liquidity shock, can be dissipated and absorbed when an agent has multiple counter-parties with whom to discharge the negative hit. Therefore, studying the role of the level and form of connectivity in the interbank network is crucial to understanding how direct contagion works, i.e. how an idiosyncratic shock may travel through the network of agents.

In the cascade model, each agent has a threshold, which determines its capac-ity to sustain a shock. If agents fail, they redistribute their debt to neighboring agents in a network. Once we can specify the threshold distribution of the agents, we can analytically derive the size of the default cascade. Systemic risk depends much more on factors such as the network's topology.

Gai and Kapadia were the first to analyze the mechanism of systemic risk with a threshold-based cascade model [6]. They developed an analytical model to study the potential impact of contagion influenced by idiosyncratic shocks. Their model also explains the "robust-yet-fragile" tendency exhibited by finan-cial systems. This property explains a phase transition in contagion occurring when connectivity and other properties of the network vary. In their model, every agent in the network is identical, i.e., all agents have the same number of debtors and creditors. They investigated how the system responds when a single agent defaults, and, in particular, when this results in contagion events in which a finite number of default as a result. The agent network was developed using a random graph where each agent had an identical threshold. The main results were the following: (i) as a function of the average connectivity, the network displayed a window of connectivity for which the probability of contagion was finite; (ii) increasing the net worth of agents reduced the probability of contagion; (iii) when the network was well connected (i.e., when the average degree was high), the network was robust-yet-fragile (i.e., the probability of contagion was very low, but in those instances where contagion occurred, the entire network was shut down).

May and Arinaminpathy modeled systemic risk in financial networks with heterogeneous agents and discovered that the agents with the most capital were more resilient to contagious defaults [9]. They also modeled part of the tiered structure by classifying the agents in the network into large and small agents. They found that tiered structures are not necessarily more prone to systemic risk, and that whether they are depends on the degree centrality, that is, the number of connections to the central agent. As the degree centrality increases, contagious defaults increase at first, but then they begin to decrease as the number of connections to the central agent leads the dissipation of the shock.

3   Agent-Based Modeling of Systemic Risks

We portray N agents (viz., banks), randomly linked together in a weighted directed network where the weighted links represent interbank liabilities. The financial state of each agent is described by the balance sheet. The balance sheets of agents are modeled according to their assets and liabilities. Agent i's assets (denoted by A_i) include interbank loans (denoted by I_i) and external assets (denoted by E_i). Liabilities (denoted by L_i) consist of interbank borrowings (denoted by B_i), deposits (denoted by D_i) and the net worth (denoted by C_i), as shown in Fig. 1. Agent vulnerability depends on the net worth, which is defined as follows:

$$C_i \equiv I_i + E_i - B_i - D_i \tag{3.1}$$

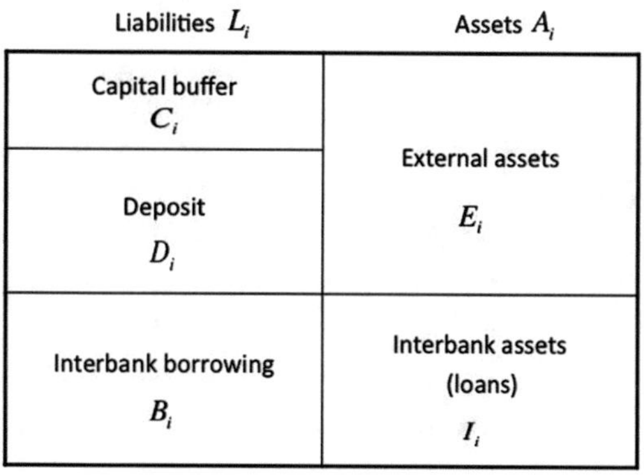

Liabilities L_i	Assets A_i
Capital buffer C_i	External assets E_i
Deposit D_i	
Interbank borrowing B_i	Interbank assets (loans) I_i

Fig. 1. Balance sheet for a bank i.

As an additional assumption, the total interbank asset positions are assumed to be evenly distributed among all incoming links, which represent loans to the other agents. The defaulting condition is as follows:

$$Agent\ i\ defaults\ if\quad (1-p)A_i\ -\ qE_i\ -\ B_i\ -\ D_i\quad <\quad 0 \qquad (3.2)$$

where p is the number of agents with obligations to agent i that have defaulted, and q is the resale price of the illiquid external assets-which takes a value between 0 and 1. When an agent fails, "zero recovery" is assumed and all of that agent's assets are lost.

The contagion process begins by selecting one defaulting agent randomly. Then, we observe whether there is a chain of defaults in the interbank network. Initially, all agents are solvent, and defaults can spread only if the neighboring agents of a defaulted agent are vulnerable. By definition, an agent is vulnerable whenever the default of one of neighboring agents with a credit relation causes a loss to the balance sheet such that it meets the defaulting condition from (3.2). The defaulting condition from (3.2) can thus be rewritten as follows:

$$p > \frac{C_i - (1-q)E_i}{L_i} \qquad (3.3)$$

By setting q = 1, we can derive the following solvency condition:

$$Agent\ i\ defaults:\ \ if\quad p > \frac{C_i}{L_i} \equiv \phi \qquad (3.4)$$

Agent vulnerability depends on the threshold ϕ, which is the ratio of the net worth (C_i) to liability (L_i).

Leverage generally refers to using credit to buy assets. It is commonly used in the context of financial markets. However, the concept of leverage covers a range of techniques from personal investments to the activities in financial markets on a national scale. One of widely used measures of financial leverage is the debt-to-net-worth ratio. This is the ratio of the liabilities of an agent with respect to the net worth. The inverse of the threshold ϕ in (3.4) represents the debt-to-net-worth ratio of an agent.

The process of contagion gains momentum and suddenly spreads to a large number of agents after the failure of some critical agents. Therefore, finding these critical agents is crucial to preventing a cascade of defaults. The criticality of an agent does not directly depend on the size of liabilities. Rather, it is determined by the debt-to-net-worth ratio. Another crucial factor is the location of an agent within the interbank network. It is on this latter point that the agent-based model may yield information useful for identifying critical agents.

4   Modeling of Financial Networks

Real-world financial networks often have fat-tailed degree distributions [4]. Many authors also note that there is some form of community structure to the network.

For example, Nier et al. [11] showed that the so-called core-periphery networks include a tightly connected core of money-center banks to which all other banks connect, and that the interbank market is tiered to a certain extent. In their study of tiered banking systems, Freixas et al. demonstrated that a tiered system of money-center banks-where banks on the periphery are linked to the center but not to each other-may also be susceptible to contagion.

Fricke and Lux conducted an empirical study as the starting point for investigating interbank networks [5]. They studied the network derived from the credit extended via the electronic market for interbank deposits (e-MID) trading platform for overnight loans between 1999 and 2010. e-MID is a privately owned Italian company and currently the only electronic brokerage market for interbank deposits. They showed that the Italian interbank market has a hierarchical core-periphery structure. This set of highly connected core banks tends to lend money to other core banks and a large number of loosely connected peripheral banks. These banks in turn tend to lend money to a small number of selected core banks, but they appear to have relatively little trade among themselves.

A core-periphery network structure is a division of the nodes into a densely connected core and a sparsely connected periphery. The nodes in the core should also be reasonably well connected to the nodes in the periphery, but the nodes in the periphery are not well connected to the core. Hence, a node belongs to the core if and only if it is well connected both to other nodes in the core and to nodes assigned to the periphery. Thus, a core structure to a network is not merely densely connected; it also tends to be central to the network in terms of short paths through the network. The latter feature also helps to distinguish a core-periphery structure from a community structure. However, many networks can have a perfect core-periphery structure as well as a community structure, so it is desirable to develop measurements that allow us to examine the various types of core-periphery structures.

Understanding the structure of a financial network is the key to understanding its function. Structural features exist at both the microscopic level, resulting from differences in the properties of single nodes, and the mesoscopic level, resulting from properties shared by groups of nodes. In general, financial networks contain unique structures, such as a core-periphery structure, by which a densely connected subset of core nodes and a subset of sparsely connected peripheral nodes coexist, or a modular structure with highly clustered sub-graphs. With a core-periphery structure, the cores consist of the important financial institutions playing a critical role in the interactions within the financial network. In a modular network, the connecting nodes (i.e., bridge nodes) of multiple modules play a key role in network contagion. Therefore, identifying the intermediate-scale structure at the mesoscopic level allows us to discover features that cannot be found by analyzing the network at the global or local scale.

We used the different characteristics of network organizations to identify the influential nodes in some typical networks as a benchmark model. We decided to use particular local measures, based either on the network's connectivity at a microscopic scale or on its community structure at a mesoscopic scale [8].

We used a mesoscopic approach to contagion models for describing risk propagation in a network, and we investigated the manner by which the network's topology impacts risk propagation. An agent fails depending on the states of its neighboring agents, and systemic risk is measured according to the proportion of failed agents. Failures are usually are absorbed by neighboring agents provided that each agent is connected to other agents, even if failures occur relatively frequently with some agents. Such failures may, however, propagate to other agents under certain conditions. When that happens, things worsen throughout the chain and the network can experience widespread failure.

We distinguish between amplification, whereby specific agents cause failures in other agents by contagion, and vulnerability, determined by the number of failures brought down by the failure of other agents. Thus, we use two indices to quantify systemic risk:

1. The *vulnerability index* to quantify the ratio of agent defaults when each agent is selected in turn as an initial defaulting agent.
2. The *amplification index* to quantify the ratio of defaulting agents caused by an agent.

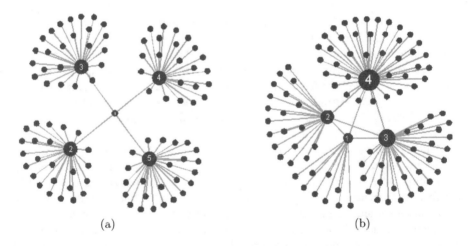

(a)   (b)

Fig. 2. Two mock networks: (a) Four symmetrical star networks are connected via one bridge agent (#1) at the center. (b) Four asymmetrical star networks of different sizes are connected through four hub agents (#1–#4). Each mock network consists of 100 agents.

We considered two mock networks, each network consisting of the N = 100 agents, as seen in Fig. 2, and we identified the influential agents in these networks. Figure 2(a) shows four symmetric star networks connected via one bridge agent in the center. Figure 2(b) shows four asymmetrical star networks of different sizes with the hub agents connected in the center of each sub-network. These four hub agents play the role of the bridge agent.

5   Simulation Results

The parameter θ denotes the ratio of total interbank loans $I = \sum_i I_i$ to total interbank assets $A = \sum_i A_i$, assuming that the ratio θ for every agent is the same across the interbank market. Given this assumption, the interbank loans I_i for agent i can be calculated from the assets A_i or the external assets E_i as follows:

$$L_i = qA_i = \frac{q}{1-q}E_i \qquad (5.1)$$

Assuming that the unit amount of interbank loans (denoted by w between any two agents is the same (i.e., the amount of interbank loans represented by one link is fixed to a certain value ω), the interbank loans I_i can be calculated by $k_{out,i}$, denoting the out-degree of agent i (i.e., the number of agents who borrow from agent i). The assets A_i and the external assets E_i of agent i can be computed using the relation in (5.1). The interbank borrowings B_i is $\omega k_{in,i}$ where $k_{in,i}$ denotes the in-degree for agent i (i.e., the number of agents who lend to agent i). If we assume an undirected network to represent interbank transactions, we have $k_{out,i} = k_{in,i} = k_i$. In the simulation study, we set these parameters at $\omega = 1$, $\theta = 0.45$ and $\phi = 0.03$.

We selected each agent in turn as an initial defaulting agent. We set the initial shock to a selected agent about to default, by stipulating a loss of 50% of its external assets. The propagation of default contagion begins with the loss of half of the external assets E_i owned by the agent i that failed initially. If the loss of the external assets E_i cannot be absorbed by the net worth (i.e., by the agent's capital buffer) C_i the remaining interbank borrowings B_i will be absorbed by each neighboring agent j, and agent j loses a portion of the interbank loans I_j. If the loss of assets A_j from agent j cannot be absorbed by the net worth C_j, the interbank borrowings B_j will be used. This series of debt propagation continues until all losses are finally absorbed in the interbank network.

We conducted 100 simulations by selecting each agent in turn as an initial defaulting agent. We derived the vulnerability index and the amplification index for each of these. Figure 3 shows the vulnerability and amplification indices for each agent in the network seen in Fig. 2(a). The vulnerability index of the bridge agent at the center (#1) is the highest. Its amplification index is the lowest, however, and the same as those of the other peripheral agents (#6–#100). The respective vulnerability indices for the four hub agents (#2–#5) were very low but their respective amplification indices are the highest, implying that when one of these hub agents is selected to default, it causes a large-scale cascade of defaults.

Figure 4 shows the vulnerability and amplification indices for each agent in the network depicted in Fig. 2(b). In this case, the vulnerability of the hub agent (#1) in the smallest star network is extremely high, whereas those of the other three hub agents (#2, #3, #4) are lower than those of the peripheral agents (#5–#100). The amplification index for each hub agent (#1–#4) in each star network increases in the proportion with the degree, implying that the amplification for a cascade of defaults is proportional to the degree of the hub agent.

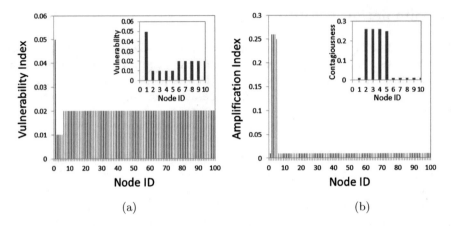

(a)   (b)

Fig. 3. The ratio of defaults when the cascade is initiated, by selecting each agent from Fig. 2(a) to default in turn: (a) vulnerability index for each agent; (b) amplification index for each agent. The inserts in the figures highlight these respective indices for the hub agents.

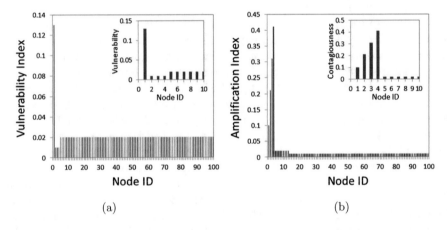

(a)   (b)

Fig. 4. The ratio of defaults when a cascade is initiated by selecting each agent in Fig. 2(b) to default in turn: (a) vulnerability index for each agent; (b) amplification index for each agent. The inserts in the figures highlight these respective indices for the hub agents.

In summary, we found strong correlations between the position of an agent in the interbank network and the likelihood that it either causes contagion or will be affected by contagion.

6   Conclusion

Agent-based models allow us to estimate the probability of economic risk-i.e. the large-scale collective behavior based on the individual behavior of interactive agents. The primary advantage to doing so is that an agent-based model can generate data where there is none or little available in the financial system. The collective effects can be studied by running multiple simulations. Then, estimations can be made about the effect that changing the rules for interactions between agents will have on the collective outcome in the financial system. For example, in this work, the likelihood of a firm or bank defaulting was shown to depend on particular regulations, such as the bailout procedure. With agent models, we can demonstrate that such regulatory measures will have specific effects; these results would be difficult, if not impossible, to surmise otherwise. The origin, development, and unfolding of economic crises is tied to the model's parameters. In reality, the relevant data is never completely available. If it were, we could more easily identify the crucial parameters. However, salient parameters can indeed be identified through a modeling and simulation process.

References

1. Allen, F., Gale, D.: Financial contagion. J. Polit. Econ. **108**, 1–33 (2000)
2. Battiston, S., et al.: DebtRank: too central to fail? Financial networks, the FED and systemic risk. Sci. Rep. **2** (2012). Nature Publishing, Paper 541
3. Eisenberg, L., Noe, H.: Systemic risk in financial systems. Manag. Sci. **47**(2), 236–249 (2001)
4. Freixas, X., Parigi, B.M., Rochet, J.-C.: Systemic risk, interbank relations and liquidity provision by the central bank. J. Money Credit Bank. **32**(3), 611–638 (2000). Part 2: What Should Central Banks Do?
5. Fricke, D., Lux,T.: Core-periphery structure in the overnight money market: evidence from the e-mid trading platform. Kiel Working Paper (2012)
6. Gai, P., Kapadia, S.: Contagion in financial networks. Proc. R. Soc. A **466**, 2401–2423 (2010)
7. Haldane, A., May, R.: Systemic risk in banking ecosystems. Nature **469**, 351–355 (2011)
8. Ide, K., Namatame, A.: A mesoscopic approach to modeling and simulation of systemic risks. In: IEEE CIFEr (2014)
9. May, R., Arinaminpathy, N.: Systemic risk: the dynamics of model banking system. J. R. Soc. Interface **7**, 823–838 (2010)
10. Namatame, A., Chen, S.: Agent-Based Models and the Network Dynamics. Oxford University Press, Oxford (2015)
11. Nier, E., Yang, J., Yorulmazer, T., Alentorn, A.: Network models and financial stability. J. Econ. Dyn. Control **31**(6), 2033–2060 (2007)
12. Soramäki, K., Bech, M.L., Arnold, J., Glass, R.J., Beyeler, W.E.: The topology of interbank payment flows. Phys. A **379**, 317–333 (2007)

Maximum Entropy Beyond Selecting Probability Distributions

Thach N. Nguyen[1], Olga Kosheleva[2], and Vladik Kreinovich[2(✉)]

[1] Banking University of Ho Chi Minh City,
56 Hoang Dieu 2, Quan Thu Duc, Ho Chi Minh City, Vietnam
Thachnn@buh.edu.vn
[2] University of Texas at El Paso, 500 W. University,
El Paso, TX 79968, USA
{olgak,vladik}@utep.edu

Abstract. Traditionally, the Maximum Entropy technique is used to select a probability distribution in situations when several different probability distributions are consistent with our knowledge. In this paper, we show that this technique can be extended beyond selecting probability distributions, to explain facts, numerical values, and even types of functional dependence.

1 How Maximum Entropy Technique is Currently Used

Need to select a distribution: formulation of a problem. Many data processing techniques assume that we know the probability distribution – e.g., the probability distributions of measurement errors, and/or probability distributions of the signals; see, e.g., [6,7].

Often, however, we have only partial information about a probability distribution. In such cases, there are several different probability distributions which are consistent with the available knowledge. To apply to this situation a data processing algorithm which is based on the assumption that the probability distribution is known, we must select a single probability distribution out of all distributions which are consistent with our knowledge. How can we select such a distribution?

Main idea. By selecting a single distribution out of several, we inevitably decrease uncertainty. It is reasonable to select a distribution for which this decrease in uncertainty is as small as possible.

How to describe this idea as a precise optimization problem. A natural way to measure uncertainty is by the average number of binary ("yes"-"no") questions that we need to ask to uniquely determine the corresponding random value (or, in the case of continuous variables, to determine the random value with a given accuracy ε).

One can show that for a probability distribution with a given probability density function $\rho(x)$, this average number of binary questions is asymptotically

© Springer International Publishing AG 2018
L. H. Anh et al. (eds.), *Econometrics for Financial Applications*, Studies in Computational Intelligence 760, https://doi.org/10.1007/978-3-319-73150-6_15

(when $\varepsilon \to 0$) proportional to the *entropy* $S(\rho) \overset{\text{def}}{=} - \int \rho(x) \cdot \ln(\rho(x)) \, dx$ of this probability distribution; see, e.g., [5] and references therein.

For a class F of distributions, the average number of binary question is asymptotically proportional to $\max\limits_{\rho \in F} S(\rho)$. We want select a single distribution ρ_0 from the class F for which the decrease in uncertainty is the smallest possible, i.e., for which the difference $\max\limits_{\rho \in F} S(\rho) - S(\rho_0)$ is the smallest possible.

How to solve the corresponding optimization problem: enter maximum Entropy technique. There is a natural solution to this optimization problem: select a distribution ρ_0 for which the entropy is the largest possible, i.e., for which $S(\rho_0) = \max\limits_{\rho \in F} S(\rho)$. In this case. the desired difference is 0 – and so the decrease in uncertainty is asymptotically negligible.

This is the main idea behind the *Maximum Entropy techniques:* when we need to select a single distribution for the class of all possible distributions, we select the distribution ρ for which the entropy $S(\rho)$ attains the largest possible value.

Simple examples of using the Maximum Entropy techniques. In some cases, all we know is that the random variable is located somewhere on a given interval $[a, b]$, but we have no information about the probability of it being in different parts of this interval. Which probability distribution would we then select to represent this situation?

If we use the Maximum Entropy approach, then we need to maximize the expression $- \int_a^b \rho(x) \cdot \ln(\rho(x)) \, dx$ under the condition that the function $\rho(x) \geq 0$ is a probability density function, i.e., that $\int_a^b \rho(x) \, dx = 1$.

Thus, we get a *constraint optimization problem:* optimize the entropy under the constraint $\int_a^b \rho(x) \, dx = 1$. To solve this constraint optimization problem, we can use the Lagrange multiplier method and reduce to the following unconstrained optimization problem of maximizing the following expression:

$$- \int_a^b \rho(x) \cdot \ln(\rho(x)) \, dx + \lambda \cdot \left(\int_a^b \rho(x) \, dx - 1 \right),$$

where λ is the *Lagrange multiplier* – a constant that needs to be determined so that the original constraint will be satisfied.

We want to find the function ρ, i.e., we want to find the values $\rho(x)$ corresponding to different inputs x. Thus, the unknowns in this optimization problem are the values $\rho(x)$ corresponding to different inputs x. To solve the resulting unconstrained optimization problem, we can simply differentiate the above expression by each of the unknowns $\rho(x)$ and equate the resulting derivative to 0. As a result, we conclude that $- \ln(\rho(x)) - 1 + \lambda = 0$, hence $\ln(\rho(x))$ is a constant not depending on x (and equal to $\lambda - 1$). Therefore, the probability density function $\rho(x)$ itself is a constant. Thus, in this case, the Maximum Entropy technique leads to a *uniform* distribution on the interval $[a, b]$.

This conclusion makes perfect sense: if we have no information about which values from the interval $[a, b]$ are more probable and which are less probable, it is reasonable to conclude that all these values are equally probable, i.e., that $\rho(x) = \text{const}$. (This idea goes back to Laplace and is known as the *Laplace Indeterminacy Principle*).

In other situations, the only information that we have about the probability distribution on a real line is its first two moments $\int x \cdot \rho(x) \, dx = \mu$ and

$$\int (x - \mu)^2 \cdot \rho(x) \, dx = \sigma^2.$$

In this case, the Maximum Entropy technique means selecting a distribution for which the entropy is the largest under the above two constraints and the constraint that $\int \rho(x) \, dx = 1$. For this problem, the Lagrange multiplier methods leads to the following unconstrained optimization problem, in which λ_i are Lagrange multipliers:

$$\text{Maximize} - \int \rho(x) \cdot \ln(\rho(x)) \, dx + \lambda_1 \cdot \left(\int x \cdot \rho(x) \, dx - \mu \right)$$
$$+ \lambda_2 \cdot \left(\int (x - \mu)^2 \cdot \rho(x) \, dx - \sigma^2 \right) + \lambda_3 \cdot \left(\int_a^b \rho(x) \, dx - 1 \right).$$

Differentiating the maximized expression with respect to each unknown $\rho(x)$ and equating the resulting derivative to 0, we conclude that

$$- \ln(\rho(x)) - 1 + \lambda_1 \cdot x + \lambda_2 \cdot (x - \mu)^2 + \lambda_3 = 0,$$

i.e., we conclude that $\ln(\rho(x))$ is a quadratic function of x and thus, that $\rho(x) = \exp(\ln(\rho(x)))$ is a Gaussian distribution.

This conclusion is also in good accordance with common sense. Indeed:

- in many case, e.g., the measurement error results from many independent small effects and,
- according to the Central Limit Theorem, the distribution of the sum of a large number of independent small random variables is close to Gaussian.

There are many other examples of a successful use of the Maximum Entropy technique; see, e.g., [4].

A natural question. Since the Maximum Entropy technique works so well for selecting a distribution, can we extend it solving other problems – e.g., explaining a fact, finding the unknown value of a quantity, or finding the formula for a functional dependence?

What we do in this paper. In this paper, we show, on several examples, that such an extension is indeed possible. We will show it on case studies that cover all three types of possible problems: explaining a fact, finding the number, and finding the functional dependence.

2 First Case Study: How Maximum Entropy Techniques Can Be Used to Explain a Fact

Fact to be explained. This fact comes from a recent study [1], and it is related to the uncertainty of expert estimates.

Experts' estimates are imprecise – just like measuring instruments are imprecise. Moreover, when we ask the same expert after some time to estimate the same quantity, he/she will, in general, give a slightly different estimate – just like when we repeatedly measure the same quantity with the same measuring instrument, we, in general, get slightly different results. We can describe the expert's estimates x_i of a quantity x as $x_i = x + \Delta x_i$, where $\Delta x_i \overset{\text{def}}{=} x_i - x$ is the estimation error.

A reasonable way to gauge the expert's accuracy is to compute the mean square value of the expert's estimation error, i.e., the value $\sigma_x \overset{\text{def}}{=} \sqrt{\dfrac{1}{N} \cdot \sum_{i=1}^{n} (\Delta x_i)^2}$, where N is the overall number of estimates performed by this expert. This quantity describes the *intra-expert* variation of the expert estimate.

We can also compare the estimates $x_i = x + \Delta x_i$ and $y_i = x + \Delta y_i$ of two (or more) different experts and compute the standard deviation

$$\sigma_{xy} \overset{\text{def}}{=} \sqrt{\frac{1}{N} \cdot \sum_{i=1}^{n} (x_i - y_i)^2} = \sqrt{\frac{1}{N} \cdot \sum_{i=1}^{n} (\Delta x_i - \Delta y_i)^2}$$

that describes the *inter-expert variation* of expert estimates.

An interesting empirical fact is that in many situations, the intra-expert and inter-expert variations are practically equal: the difference between the two variations is about 3% [1].

Why does this fact need explanation. At first glance, it may seem that the above fact is very natural and does not need any sophisticated explanation. However, as we show, a deeper analysis makes this fact truly puzzling.

Indeed, the above estimates seem to be informally based on a simple probabilistic model, in which the differences Δx_i are instances of a random variable Δx with 0 mean. The above expression for the intra-expert variance is simply a sample-based estimation of this random variable's standard deviation: $\sigma_x \approx \sigma[\Delta x]$ and thus, $\sigma_x^2 \approx \sigma^2[\Delta x] = E[(\Delta x)^2]$, where, as usual, $E[\eta]$ denotes the expected value of a random quantity η, and $\sigma[\eta]$ denotes its standard deviation.

Similarly, the inter-expert variation is approximately equal to the standard deviation of the difference $\Delta x - \Delta y$ between the random variables Δx and Δy corresponding to two experts: $\sigma_{xy} \approx \sigma[\Delta x - \Delta y]$, i.e., $\sigma_{xy}^2 \approx E[(\Delta x - \Delta y)^2]$.

Thus, the fact that the intra-expert and the inter-expert variations coincide means that $E[(\Delta x - \Delta y)^2] \approx E[(\Delta x)^2] \approx E[(\Delta y)^2]$.

If experts were fully independent, then we would have $E[(\Delta x - \Delta y)^2] = E[(\Delta x)^2] + E[(\Delta y)^2]$, so we would have $\sigma_{xy}^2 \approx 2\sigma_x^2$ and $\sigma_{xy} \approx \sqrt{2} \cdot \sigma_x$, and the inter-expert variation would be at least 40% larger than the intra-expert one.

This we do not observe. It means that there *is* a correlation between the experts. If there was the perfect correlation, we would have $\Delta x_i = \Delta y_i$, and the inter-expert variation would be exactly 0.

In situations of *partial* correlation, we would get all possible values of σ_{xy} ranging from 0 to $\sqrt{2} \cdot \sigma_x$. So why, out of all possible values from interval $[0, \sqrt{2} \cdot \sigma_x]$, the value σ_x corresponds to the average inter-expert variation?

Maximum Entropy technique can help us explain this fact. To provide our explanation, let us express the inter-expert variation in terms of the (Pearson) correlation coefficient $r \stackrel{\text{def}}{=} \dfrac{E[\Delta x \cdot \Delta y]}{\sigma[\Delta x] \cdot \sigma[\Delta y]}$.

By definition of the inter-expert correlation, we have

$$\sigma_{xy}^2 = E[(\Delta x - \Delta y)^2] = E[(\Delta x)^2] + E((\Delta y)^2) - 2E(\Delta x \cdot \Delta y).$$

Here, $E(\Delta x)^2] = E(\Delta y)^2] = \sigma_x^2$, and, by definition of the correlation coefficient, $E[\Delta x \cdot \Delta y] = r \cdot \sigma[\Delta x] \cdot \sigma[\Delta y] = r \cdot \sigma_x^2$. Thus, the above formula for the inter-expert variation takes the form

$$\sigma_{xy}^2 = 2\sigma_x^2 - 2r \cdot \sigma_x^2 = 2 \cdot (1 - r) \cdot \sigma_x^2.$$

In general, the correlation r can take any value from -1 to 1, but in this case, since we assume that all experts are indeed experts, it is reasonable to assume that their estimates are non-negatively correlated, i.e., that $r \geq 0$. Thus, in this example, the set of possible value of the correlation r is the interval $[0, 1]$.

In different situations, we may have different values of the correlation coefficient: some experts may be independent, other pairs of experts may have the same background and thus, have strongly correlated estimates. So, in real life, there will be some probability distribution on the set $[0, 1]$ of all possible values of the correlation coefficient that reflects the frequency of different pairs of experts. We would like to estimate the average value $E[r]$ of r over this distribution. Then, by averaging over r, we will get the desired relation between the intra- and inter-expert variations:

$$\sigma_{xy}^2 = 2 \cdot (1 - E[r]) \cdot \sigma_x^2.$$

We do not have any information about which values r are more probable (i.e., more frequent) and which values r are less probable. In other words, in principle, all probability distributions on the interval $[0, 1]$ are possible. To perform the above estimation, we need to select a single distribution form this class.

It is reasonable to apply the Maximum Entropy technique to select such a distribution. As we have mentioned, in this case, the Maximum Entropy technique selects a uniform distribution on the interval $[0, 1]$. For the uniform distribution on the interval $[0, 1]$, the probability density is equal to 1, and the mean value is 0.5:

$$E[r] = \int_0^1 x \cdot \rho(x) \, dx = \int_0^1 x \, dx = \frac{x^2}{2} \Big|_0^1 = \frac{1^2}{2} - \frac{0^2}{2} = 0.5.$$

Substituting the value $E[r] = 0.5$ into the above formula $\sigma_{xy}^2 = 2 \cdot (1 - E[r]) \cdot \sigma_x^2$, we conclude that $\sigma_{xy}^2 = \sigma_x^2$, which is exactly the fact that we try to explain.

3   Second Case Study: How Maximum Entropy Techniques Can Be Used to Find a Numerical Value

Empirical fact. It has been observed that when people make crude estimates, their estimates differ by half-order of magnitude; see, e.g., [2]. For example, when people estimate the size of a crowd, they normally give answers like 100, 300, 1000, but it is much more difficult for them to distinguish, e.g., between 100 and 200. Similarly, when describing income, people talk about low six figures, high six figures, etc., – which is exactly half-orders of magnitude.

So, what is so special about the ratio 3 corresponding to half-order of magnitude? Why not 2 or 4?

There are explanations for this fact, but can we have a simpler one? There are explanations for the above fact; see, e.g., [3]. However, these explanations are somewhat complicated.

For a simple fact about commonsense reasoning, it is desirable to have a simpler, more intuitive explanation.

What we do in this section. In this section, we show that the Maximum Entropy technique can be used to provide a simpler explanation for this empirical fact.

Let us formulate this problem in precise terms. Let us assume that we have two quantities a and b, and a is smaller than b. For example, a and b are the salaries of two employees on the two layers of the company's hierarchy. If all we know is that $a < b$, what can we conclude about the relation between a and b?

Applying Maximum Entropy technique: first attempt. Let us try to apply the Maximum Entropy techniques to answer this question. For this purpose, it may sound reasonable to come up with some probability distribution on the set of all possible values of a and on the set of possible values of b. Here, we do not have any bound on a and b. In this case, similar to the case of interval bounds, the Maximum Entropy technique implies that $\rho(x) = \text{const}$ for all possible real numbers x – and thus, since we want $\rho(x) > 0$, we get $\int_0^\infty \rho(x)\,dx = \infty > 1$.

Applying Maximum Entropy technique: second attempt and the resulting explanation. To be able to meaningfully apply the Maximum Entropy idea, we need to consider *bounded* quantities. One such possibility is to consider, instead of the original salary a, the *fraction* of the overall salary $a + b$ that goes to a, i.e., the ratio

$$r \stackrel{\text{def}}{=} \frac{a}{a+b}.$$

We know that $a < b$, so this ratio takes all possible values from 0 to 0.5, where 0.5 corresponds to the ideal case when the salaries a and b are equal. By using the Maximum Entropy technique, we can conclude that the variable r is uniformly distributed on the interval $[0, 0.5)$. Thus, the average value of this variable is at the midpoint of this interval, when $r = 0.25$. So, on average, the salary a of the first person takes $1/4$ of the overall amount $a + b$, and thus, the average salary b of the second person is equal to the remaining amount $1 - 1/4 = 3/4$. Thus, the ratio of the two salaries is exactly $\dfrac{b}{a} = \dfrac{3/4}{1/4} = 3$.

This corresponds exactly to the half-order of magnitude ratio that we are trying to explain. Thus, the Maximum Entropy technique indeed explains this empirical ratio.

4   Third Case Study: How Maximum Entropy Techniques Can Be Used to Find a Functional Dependence

Often, we need to find a functional dependence. In many practical situations, we know that the value of a quantity x uniquely determines the values of the quantity y, i.e., that $y = f(x)$ for some function $f(x)$.

- In some practical situations, this dependence is known, but
- in other situations, we need to find this dependence.

How the Maximum Entropy technique can help: the main idea. For each physical quantity, we usually know its bounds. Thus, we can safely assume that we know that:

- all possible values of the quantity x are in a known interval $[\underline{x}, \overline{x}]$, and
- all possible values of the quantity y are in a known interval $[\underline{y}, \overline{y}]$.

If we apply the Maximum Entropy technique to the quantity x, we conclude that x is uniformly distributed on the interval $[\underline{x}, \overline{x}]$. Similarly, if we apply the Maximum Entropy technique to the quantity y, we conclude that x is uniformly distributed on the interval $[\underline{y}, \overline{y}]$.

It is therefore reasonable to select a function $f(x)$ for which,

- when x is uniformly distributed on the interval $[\underline{x}, \overline{x}]$,
- the quantity $y = f(x)$ is uniformly distributed on the interval $[\underline{y}, \overline{y}]$.

What are the resulting functional dependencies? For a uniform distribution, the probability to be in an interval is proportional to its length. In particular, for a small interval $[x, x + \Delta]$ of width Δx, the probability to be in this interval is equal to $\rho_x \cdot \Delta x$.

The corresponding y-interval $[f(x), f(x + \Delta x)]$ has width

$$\Delta y = |f(x + \Delta x) - f(x)|.$$

For small Δx, we have

$$\frac{f(x + \Delta x) - f(x)}{\Delta x} \approx \lim_{h \to 0} \frac{f(x + h) - f(x)}{h} = f'(x).$$

Thus, for small Δx, we have $f(x + \Delta x) - f(x) \approx f'(x) \cdot \Delta x$ and therefore, $\Delta y \approx |f'(x)| \cdot \Delta x$. Since the variable y is also uniformly distributed, the probability for y to be in this interval is equal to $\rho_y \cdot \Delta y = \rho_y \cdot |f'(x)| \cdot \Delta x$.

Comparing this expression with the original formula $\rho_x \cdot \Delta x$ for the same probability, we conclude that $\rho_y \cdot |f'(x)| \cdot \Delta x = \rho_x \cdot \Delta x$, hence $|f'(x)| = \dfrac{\rho_x}{\rho_y}$, i.e., $|f'(x)| = \text{const}$. So, we conclude that *the function $f(x)$ should be linear*.

What is our result and why it is interesting. Our conclusion is that if we have no information about the functional dependence, it is reasonable to assume that this dependence is linear.

This fits well with the usual engineering practice, where indeed the first idea is usually to try a linear dependence. However, the usual motivation for using a linear dependence first is that such a dependence is the easiest to analyze – and why would nature care which dependencies are easier for us to analyze? The Maximum Entropy argument seems more convincing, since it relies on the general ideas about uncertainty itself – and not on our ability to deal with this uncertainty.

Need for nonlinear dependencies. That we came up with an explanation for a linear dependence may be nice, but in practice, linear dependence is usually only the first approximation to the true non-linear dependence. Once we know that the a linear dependence is only an approximation, we would like to find a more adequate nonlinear model.

The Maximum Entropy technique can help beyond linear dependencies. It turns out that the Maximum Entropy technique can also help in finding such a nonlinear dependence – just like for probability distributions:

- once we have an additional information which is not consistent with the assumption that the actual distribution is uniform,
- we can add this information to the corresponding Maximum Entropy problem and get a non-uniform distribution consistent with this information.

We will actually describe two alternative ideas on in which the Maximum Entropy technique can help.

The Maximum Entropy technique can help beyond linear dependencies: first idea. The first, more direct, idea is to take into account that often, not only the quantity y, but also its derivative $z \stackrel{\text{def}}{=} \dfrac{dy}{dx}$ (and sometimes, its second derivative as well) is also an observable quantity. For example, when y is a distance and x is time, then the first derivative $v \stackrel{\text{def}}{=} \dfrac{dy}{dx}$ is velocity and the second derivative $a \stackrel{\text{def}}{=} \dfrac{dv}{dx} = \dfrac{d^2y}{dx^2}$ is acceleration – both perfectly observable quantities.

If we apply the Maximum Entropy techniques to the dependence of velocity v on time x, we conclude that the velocity linearly depends on time – in which case, by integrating this dependence, we conclude that the distance is a quadratic function of time. Similarly, if we apply the Maximum Entropy technique to the dependence of acceleration a on time, then we conclude that the velocity is a quadratic function of time, and thus, that the distance is a cubic function of time.

The Maximum Entropy technique can help beyond linear dependencies: second idea. The second, less direct idea, is to take into account that when the dependence $y = f(x)$ is non-linear, then, even when the probability distribution for x is uniform, with density $\rho_x(x) = \rho_x = \text{const}$, the corresponding probability distribution $\rho_y(y)$ for the quantity y is, in general, *not* uniform.

How can we describe the dependence $\rho_y(y)$ of the probability density on y? To describe this auxiliary dependence, we can use the Maximum Entropy technique and conclude that this dependence is linear, i.e., that $\rho_y(y) = a + b \cdot y$. Now that we know the distributions for x and y, we can look for functions $f(x)$ for which:

- once x is uniformly distributed,
- the quantity $y = f(x)$ is distributed with the probability density $\rho_y(y) = a + b \cdot y$.

Similarly to the above case when both x- and y-distributions were uniform, the probability of being in the x-interval of width Δx is equal to $\rho_x \cdot \Delta x$, and on the other hand, it is equal to $\rho_y(y) \cdot |f'(x)| \cdot \Delta x = (a + b \cdot f(x)) \cdot |f'(x)| \cdot \Delta x$. By comparing these two expressions for the same probability, we conclude that

$$|f'(x)| \cdot (a + b \cdot f(x)) = \text{const},$$

i.e., that $\dfrac{df}{dx} \cdot (a + b \cdot f) = \text{const}$. By moving all the terms containing f to one side and all the terms containing x to another sides, we conclude that $\dfrac{df}{a + b \cdot f} = \text{const} \cdot x$. So, for $g \overset{\text{def}}{=} f + \dfrac{a}{b}$, we get $\dfrac{dg}{g} = c \cdot dx$. Integration leads to $\ln(g) = c \cdot x + C$ for some integration constant C, thus, $g = A \cdot \exp(c \dot{x})$, and $f = A \cdot \exp(c \cdot x) + \text{const}$.

By assuming that y is uniformly distributed, we get the inverse (logarithmic) dependence. By assuming that the dependence $\rho_y(y)$ on y is not linear but is described by one of these nonlinear formulas, we can get an even more complex dependence.

Thus, we can indeed use the Maximum Entropy technique to describe non-linear dependencies as well.

Acknowledgments. This work was supported in part by the National Science Foundation grant HRD-1242122 (Cyber-ShARE Center of Excellence).

References

1. Garibaldi, J.: Type-2 Beyond the Centroid. In: Proceedings of the IEEE International Confreence on Fuzzy Systems FUZZ-IEEE 2017, Naples, Italy, 8–12 July 2017
2. Hobbs, J.R.: Half orders of magnitude. In: Obrst, L., Mani, I. (eds.) Proceeding of the Workshop on Semantic Approximation, Granularity, and Vagueness, A Workshop of the Seventh International Conference on Principles of Knowledge Representation and Reasoning KR 2000, Breckenridge, Colorado, pp. 28–38, 11 April 2000
3. Hobbs, J., Kreinovich, V.: Optimal choice of granularity in commonsense estimation: why half-orders of magnitude. Int. J. Intell. Syst. **21**(8), 843–855 (2006)
4. Jaynes, E.T., Bretthorst, G.L.: Probability Theory: The Logic of Science. Cambridge University Press, Cambridge (2003)
5. Nguyen, H.T., Kreinovich, V., Wu, B., Xiang, G.: Computing Statistics under Interval and Fuzzy Uncertainty. Springer, Heidelberg (2012)
6. Rabinovich, S.G.: Measurement Errors and Uncertainty: Theory and Practice. Springer, Berlin (2005)
7. Sheskin, D.J.: Handbook of Parametric and Nonparametric Statistical Procedures. Chapman and Hall/CRC, Boca Raton (2011)

Confidence Intervals for Functions of Signal-to-Noise Ratios of Normal Distributions

Sa-Aat Niwitpong[(✉)]

Department of Applied Statistics, Faculty of Applied Science,
King Mongkut's University of Technology North Bangkok, Bangkok 10800, Thailand
sa-aat.n@sci.kmutnb.ac.th

Abstract. This paper examines confidence intervals for the single signal-to-noise ratio (SNR), the difference of signal-to-noise ratios (SNRs), and the common SNR of normal distributions, using the generalized confidence interval (GCI) approach, large-sample approach, and method of variance estimates recovery (MOVER) approach. The coverage probability and average width of these confidence intervals were evaluated by Monte Carlo simulation. The simulation results indicated that the MOVER approach performs better than the other approaches in terms of coverage probability for the single SNR and the difference of SNRs. Furthermore, the coverage probabilities of the adjusted MOVER approach are satisfactory for the common SNR. We also illustrate our confidence intervals using three examples.

Keywords: Normal distribution · Coefficient of variation
Signal-to-noise ratio · GCI approach · MOVER approach

1   Introduction

The coefficient of variation is a measure of variability relative to the mean. It is used to compare several populations in different units. The coefficient of variation is defined as a ratio of the standard deviation (σ) to the mean (μ). The coefficient of variation is widely used in science, medicine, and economics. That is because the coefficient of variation is a unit-free measure and can be used to compare the variation of two or more different measurement methods. The reciprocal of the coefficient of variation is called signal-to-noise ratio (SNR). The SNR is defined as the ratio of the signal mean (μ) to the standard deviation of the noise (σ). McGibney and Smith [10] described that the SNR is a measure used to quantify how much a signal has been corrupted by noise.

The issue of SNR is a serious problem in econometrics, in particular in estimation problem. For example, in the standard simple regression, Swann [13] showed that the SNR of the regression problem is a ratio of the full variance of an explanatory variable (X) to the partial variance of an explanatory variable (X). If this ratio is small, then the serious bias of the Ordinary Least Squares (OLS)

© Springer International Publishing AG 2018
L. H. Anh et al. (eds.), *Econometrics for Financial Applications*, Studies in Computational Intelligence 760, https://doi.org/10.1007/978-3-319-73150-6_16

estimator has occurred and otherwise, if this ratio is large, then any bias of the OLS is very small. The same is true for multivariate models. Hence, the estimation of SNR is one of the main problems in econometrics. In this paper, we construct a confidence interval for the SNR problems.

Confidence intervals associated with point estimates provide more information about the population characteristics than the p-values in the hypothesis test (Visintainer and Tejani [18]). The confidence interval for the SNR has been studied in the literature. For example, Sharma and Krishna [12] presented the asymptotic sampling distribution and confidence interval for the SNR. George and Kibria [6] proposed the confidence intervals for the SNR of Poisson distribution. George and Kibria [7] compared several confidence intervals estimate for the SNR by inverting confidence intervals for the coefficient of variation. Albatineh et al. [2] revisited the asymptotic sampling distribution of the SNR. Recently, Albatineh et al. [1] derived the asymptotic sampling distribution of the SNR and introduced a new confidence interval estimator for the SNR.

It is of interest to develop procedures for confidence interval estimation of the SNR. The objective of this paper is to propose the confidence intervals for functions of SNR of normal distribution. The functions are the single SNR, the difference between SNRs, and the common SNR. First, the confidence intervals for the single SNR of normal distribution are constructed based on the generalized confidence interval (GCI) approach, the large-sample approach, and the method of variance estimates recovery (MOVER) approach. Second, the GCI, large sample, and MOVER approaches are used to find the confidence interval estimates for the difference of SNRs of normal distributions. Finally, the concepts of the GCI, large sample and MOVER approaches are extended to k populations. Hence, the confidence intervals for the common SNR of several normal distributions are provided using the GCI approach, the large-sample approach, and the adjusted MOVER approach.

This paper is organized as follows. In Sect. 2, the confidence intervals for the single SNR are introduced. In Sect. 3, the confidence intervals for the difference between SNRs are provided. In Sect. 4, the confidence intervals for the common SNR are presented. In Sect. 5, the simulation study and simulation results are described. In Sect. 6, the proposed approaches are illustrated with three examples. In Sect. 7, a conclusion is presented.

2 Confidence Intervals for the Signal-to-Noise Ratio of Normal Distribution

Let $X = (X_1, X_2, \ldots, X_n)$ be a random sample of size n from the normal distribution with mean μ and variance σ^2. The SNR is defined by $\theta = \mu/\sigma$.

Let $\bar{X}$ and S^2 be sample mean and sample variance for X, respectively. Also, let $\bar{x}$ and s^2 be the observed values of $\bar{X}$ and S^2, respectively. The maximum likelihood estimators of μ and σ are $\bar{X}$ and S, respectively. Hence, the estimated SNR for θ is given by $\hat{\theta} = \bar{X}/S$.

2.1 The Generalized Confidence Interval for the Signal-to-Noise Ratio

Definition: Let $X = (X_1, X_2, \ldots, X_n)$ be a random sample from a distribution $F(x|\delta)$, where $\delta = (\mu, \sigma^2)$ is the vector of unknown parameters, and $x = (x_1, x_2, \ldots, x_n)$ is an observed value of X. Weerahandi [19] defines a generalized pivot $R = R(X, x, \mu, \sigma^2)$ for confidence interval estimation as a random variable having the following two properties:

(i) R has a probability distribution that is free of unknown parameters.
(ii) The observed value of R, $R = R(x, x, \mu, \sigma^2)$, is the parameter of interest θ.

Let $R(\alpha)$ be the 100α-th percentile of R. Hence, $(R(\alpha/2), R(1 - \alpha/2))$ becomes a $100(1 - \alpha)\%$ two-sided generalized confidence interval for the parameter of interest θ.

Following Weerahandi [20] and Tian [15], the generalized pivotal quantity for $\theta = \mu/\sigma$ is

$$R_\theta = \frac{\bar{x}}{s}\frac{S}{\sigma} - \frac{\bar{X} - \mu}{\sigma} = \frac{\bar{x}}{s}\sqrt{\frac{U}{n-1}} - \frac{Z}{\sqrt{n}}, \tag{1}$$

where U denotes the chi-square distribution with degrees of freedom $n - 1$ and Z denotes the standard normal distribution.

Therefore, the $100(1-\alpha)\%$ two-sided confidence interval for the SNR θ based on the GCI approach is

$$CI_{S.GCI} = (L_{S.GCI}, U_{S.GCI}) = (R_\theta(\alpha/2), R_\theta(1 - \alpha/2)), \tag{2}$$

where R_θ is defined in Eq. (1) and $R_\theta(p)$ denotes the p-th quantile of R_θ.

2.2 The Large-Sample Confidence Interval for the Signal-to-Noise Ratio

Albatineh et al. [1] provided the variance of $\hat{\theta} = \bar{X}/S$. The variance is given by

$$Var(\hat{\theta}) = \frac{1}{n}\left(1 - \frac{\mu}{\sigma^4}\mu_3 + \frac{\mu^2}{4\sigma^6}\mu_4 - \frac{\mu^2}{4\sigma^2}\right), \tag{3}$$

where $\mu_3 = E(X_j - \mu)^3 = \frac{1}{n}\sum_{j=1}^{n}(X_j - \bar{X})^3$ and $\mu_4 = E(X_j - \mu)^4 = \frac{1}{n}\sum_{j=1}^{n}(X_j - \bar{X})^4$.

Therefore, the $100(1-\alpha)\%$ two-sided confidence interval for the SNR θ based on the large-sample approach is defined by

$$CI_{S.LS} = (L_{S.LS}, U_{S.LS}) = (\hat{\theta} - z_{1-\alpha/2}\sqrt{Var(\hat{\theta})}, \hat{\theta} + z_{1-\alpha/2}\sqrt{Var(\hat{\theta})}), \tag{4}$$

where $z_{1-\alpha/2}$ denotes the $(1 - \alpha/2)$-th quantile of the standard normal distribution and $Var(\hat{\theta})$ is an estimate of $Var(\hat{\theta})$ in Eq. (3) with μ and σ replaced by $\bar{x}$ and s, respectively.

2.3   The Method of Variance Estimates Recovery (MOVER) Confidence Interval for the Signal-to-Noise Ratio

Krishnamoorthy [9] presented following the confidence intervals for μ and σ

$$CI_\mu = (l_\mu, u_\mu) = \left(\hat{\mu} - t_{n-1,1-\frac{\alpha}{2}} \frac{s}{\sqrt{n}}, \hat{\mu} + t_{n-1,1-\frac{\alpha}{2}} \frac{s}{\sqrt{n}} \right) \tag{5}$$

and

$$CI_\sigma = (l_\sigma, u_\sigma) = \left(\sqrt{\frac{(n-1)s^2}{\chi^2_{n-1,1-\frac{\alpha}{2}}}}, \sqrt{\frac{(n-1)s^2}{\chi^2_{n-1,\frac{\alpha}{2}}}} \right), \tag{6}$$

where $t_{n-1,p}$ and $\chi^2_{n-1,p}$ denote the p-th quantiles of the t-distribution and chi-square distribution, respectively.

Donner and Zou [4] introduced the MOVER approach to construct confidence interval for a ratio. Hence, the MOVER approach is applied to construct confidence interval for $\theta = \mu/\sigma$. The lower bound ($L_{S.MOVER}$) and upper bound ($U_{S.MOVER}$) of θ are obtained by

$$L_{S.MOVER} = \frac{\hat{\mu}\hat{\sigma} - \sqrt{(\hat{\mu}\hat{\sigma})^2 - l_\mu u_\sigma (2\hat{\mu} - l_\mu)(2\hat{\sigma} - u_\sigma)}}{u_\sigma(2\hat{\sigma} - u_\sigma)} \tag{7}$$

and

$$U_{S.MOVER} = \frac{\hat{\mu}\hat{\sigma} + \sqrt{(\hat{\mu}\hat{\sigma})^2 - u_\mu l_\sigma (2\hat{\mu} - u_\mu)(2\hat{\sigma} - l_\sigma)}}{l_\sigma(2\hat{\sigma} - l_\sigma)}, \tag{8}$$

where l_μ and u_μ are defined in Eq. (5) and l_σ and u_σ are defined in Eq. (6).

Therefore, the $100(1-\alpha)\%$ two-sided confidence interval for the SNR θ based on the MOVER approach is defined by

$$CI_{S.MOVER} = (L_{S.MOVER}, U_{S.MOVER}), \tag{9}$$

where $L_{S.MOVER}$ and $U_{S.MOVER}$ are defined in Eqs. (7) and (8), respectively.

3   Confidence Intervals for the Difference of Signal-to-Noise Ratios of Normal Distributions

Suppose that $X_1 = (X_{11}, X_{12}, \ldots, X_{1n_1})$ is a random sample from a normal distribution with mean μ_1 and variance σ_1^2. Let $X_2 = (X_{21}, X_{22}, \ldots, X_{2n_2})$ be a random sample from a normal distribution with mean μ_2 and variance σ_2^2. Also, X_1 and X_2 are independent. Let $\bar{X}_1$ and S_1^2 be sample mean and sample variance for X_1, respectively. Also, let $\bar{x}_1$ and s_1^2 be the observed values of $\bar{X}_1$ and S_1^2, respectively. Similarly, let $\bar{X}_2$ and S_2^2 be sample mean and sample variance for X_2, respectively. Also, let $\bar{x}_2$ and s_2^2 be the observed values of $\bar{X}_2$ and S_2^2, respectively.

In this section, the GCI, large-sample, and MOVER approaches are applied to establish the confidence intervals for the difference of signal to noise ratios (SNRs). The difference of SNRs is given by

$$\delta = \theta_1 - \theta_2 = \frac{\mu_1}{\sigma_1} - \frac{\mu_2}{\sigma_2}. \tag{10}$$

The estimated difference of SNRs for δ is given by

$$\hat{\delta} = \hat{\theta}_1 - \hat{\theta}_2 = \frac{\bar{X}_1}{S_1} - \frac{\bar{X}_2}{S_2}. \tag{11}$$

3.1   The Generalized Confidence Interval for the Difference of Signal-to-Noise Ratios

Using the generalized pivotal quantity from Eq. (1), the generalized pivotal quantity of δ is obtained by

$$R_\delta = \left(\frac{\bar{x}_1}{s_1} \sqrt{\frac{U_1}{n_1 - 1}} - \frac{Z_1}{\sqrt{n_1}} \right) - \left(\frac{\bar{x}_2}{s_2} \sqrt{\frac{U_2}{n_2 - 1}} - \frac{Z_2}{\sqrt{n_2}} \right), \tag{12}$$

where U_1 and U_2 denote the chi-square distribution with degrees of freedom $n_1 - 1$ and $n_2 - 1$, respectively, and Z_1 and Z_2 denote the standard normal distribution.

Therefore, the $100(1 - \alpha)\%$ two-sided confidence interval for the difference of SNRs δ based on the GCI approach is defined by

$$CI_{D.GCI} = (L_{D.GCI}, U_{D.GCI}) = (R_\delta(\alpha/2), R_\delta(1 - \alpha/2)), \tag{13}$$

where R_δ is defined in Eq. (12) and $R_\delta(p)$ denotes the p-th quantile of R_δ.

3.2   The Large-Sample Confidence Interval for the Difference of Signal-to-Noise Ratios

Firstly, the variance of $\hat{\delta}$ is considered. From

$$Var(\hat{\delta}) = Var\left(\frac{\bar{X}_1}{S_1} \right) + Var\left(\frac{\bar{X}_2}{S_2} \right).$$

From Eq. (3), the variance of $\hat{\delta}$ is obtained by

$$Var(\hat{\delta}) = \frac{1}{n_1} \left(1 - \frac{\mu_1}{\sigma_1^4} \mu_{13} + \frac{\mu_1^2}{4\sigma_1^6} \mu_{14} - \frac{\mu_1^2}{4\sigma_1^2} \right)$$
$$+ \frac{1}{n_2} \left(1 - \frac{\mu_2}{\sigma_2^4} \mu_{23} + \frac{\mu_2^2}{4\sigma_2^6} \mu_{24} - \frac{\mu_2^2}{4\sigma_2^2} \right), \tag{14}$$

where $\mu_{l3} = \frac{1}{n_l} \sum_{j=1}^{n_l} (X_{lj} - \bar{X}_l)^3$ and $\mu_{l4} = \frac{1}{n_l} \sum_{j=1}^{n_l} (X_{lj} - \bar{X}_l)^4$; $l = 1, 2$.

Therefore, the $100(1-\alpha)\%$ two-sided confidence interval for the difference of SNRs δ based on the large-sample approach is defined by

$$CI_{D.LS} = (L_{D.LS}, U_{D.LS}) = (\hat{\delta} - z_{1-\alpha/2}\sqrt{Var(\hat{\delta})}, \hat{\delta} + z_{1-\alpha/2}\sqrt{Var(\hat{\delta})}), \quad (15)$$

where $z_{1-\alpha/2}$ denotes the $(1-\alpha/2)$-th quantile of the standard normal distribution and $Var(\hat{\delta})$ is an estimate of $Var(\hat{\delta})$ in Eq. (14) with μ_l and σ_l replaced by $\bar{x}_l$ and s_l, respectively.

3.3   The Method of Variance Estimates Recovery (MOVER) Confidence Interval for the Difference of Signal-to-Noise Ratios

Following the confidence interval for θ based on the MOVER approach in Eq. (9), then

$$l_1 = \frac{\hat{\mu}_1\hat{\sigma}_1 - \sqrt{(\hat{\mu}_1\hat{\sigma}_1)^2 - l_{11}u_{12}(2\hat{\mu}_1 - l_{11})(2\hat{\sigma}_1 - u_{12})}}{u_{12}(2\hat{\sigma}_1 - u_{12})} \quad (16)$$

$$u_1 = \frac{\hat{\mu}_1\hat{\sigma}_1 + \sqrt{(\hat{\mu}_1\hat{\sigma}_1)^2 - u_{11}l_{12}(2\hat{\mu}_1 - u_{11})(2\hat{\sigma}_1 - l_{12})}}{l_{12}(2\hat{\sigma}_1 - l_{12})} \quad (17)$$

$$l_2 = \frac{\hat{\mu}_2\hat{\sigma}_2 - \sqrt{(\hat{\mu}_2\hat{\sigma}_2)^2 - l_{21}u_{22}(2\hat{\mu}_2 - l_{21})(2\hat{\sigma}_2 - u_{22})}}{u_{22}(2\hat{\sigma}_2 - u_{22})} \quad (18)$$

and

$$u_2 = \frac{\hat{\mu}_2\hat{\sigma}_2 + \sqrt{(\hat{\mu}_2\hat{\sigma}_2)^2 - u_{21}l_{22}(2\hat{\mu}_2 - u_{21})(2\hat{\sigma}_2 - l_{22})}}{l_{22}(2\hat{\sigma}_2 - l_{22})}, \quad (19)$$

where

$$(l_{i1}, u_{i1}) = (\hat{\mu}_i - t_{n_i-1,1-\frac{\alpha}{2}}\frac{s_i}{\sqrt{n_i}}, \hat{\mu}_i + t_{n_i-1,1-\frac{\alpha}{2}}\frac{s_i}{\sqrt{n_i}})$$

and

$$(l_{i2}, u_{i2}) = \left(\sqrt{\frac{(n_i-1)s_i^2}{\chi^2_{n_i-1,1-\frac{\alpha}{2}}}}, \sqrt{\frac{(n_i-1)s_i^2}{\chi^2_{n_i-1,\frac{\alpha}{2}}}} \right).$$

Donner and Zou [4] proposed the confidence interval for the difference of two parameters based on the MOVER approach. Hence, the confidence limits for $\delta = \theta_1 - \theta_2$ is obtained by

$$L_{D.MOVER} = \hat{\theta}_1 - \hat{\theta}_2 - \sqrt{(\hat{\theta}_1 - l_1)^2 + (u_2 - \hat{\theta}_2)^2} \quad (20)$$

and

$$U_{D.MOVER} = \hat{\theta}_1 - \hat{\theta}_2 + \sqrt{(u_1 - \hat{\theta}_1)^2 + (\hat{\theta}_2 - l_2)^2}, \quad (21)$$

where l_i and u_i are defined in Eqs. (16)–(19).

Therefore, the $100(1-\alpha)\%$ two-sided confidence interval for the difference of SNRs δ based on the MOVER approach is defined by

$$CI_{D.MOVER} = (L_{D.MOVER}, U_{D.MOVER}), \quad (22)$$

where $L_{D.MOVER}$ and $U_{D.MOVER}$ are defined in Eqs. (20) and (21), respectively.

4   Confidence Intervals for the Common Signa-to-Noise Ratio of Normal Distributions

Again, let $X_i = (X_{i1}, X_{i2}, \ldots, X_{in_i})$ be the i-th random sample from a normal distribution with mean μ_i and variance σ_i^2 for $i = 1, 2, \ldots, k$. Hence, the signal to noise ratio based on the i-th sample is defined by $\theta_i = \mu_i / \sigma_i$, $i = 1, 2, \ldots, k$. Let $\bar{X}_i$ and S_i^2 be sample mean and sample variance for X_i, respectively. Also, let $\bar{x}_i$ and s_i^2 be the observed values of $\bar{X}_i$ and S_i^2, respectively. The maximum likelihood estimators of μ_i and σ_i are $\bar{X}_i$ and S_i^2, respectively. Therefore, the estimated SNR for θ_i is given by $\hat{\theta}_i = \bar{X}_i / S_i$.

In this section, the confidence intervals for the common SNR based on the GCI approach, large-sample approach, and the adjusted MOVER approach are presented. The common parameter, introduced by Graybill and Deal [5], was applied. Therefore, the common SNR is obtained by

$$\hat{\eta} = \sum_{i=1}^{k} \frac{\hat{\theta}_i}{Var(\hat{\theta}_i)} \Big/ \sum_{i=1}^{k} \frac{1}{Var(\hat{\theta}_i)}, \tag{23}$$

where $\hat{\theta}_i$ is an estimator based on the i-th sample.

4.1   The Generalized Confidence Interval for the Common Signal-to-Noise Ratio

Following Tian and Wu [16], the generalized pivotal quantities for estimating μ_i and σ_i^2 based on the i-th sample are given by

$$R_{\mu_i} = \bar{x}_i - \frac{Z_i}{\sqrt{U_i}} \sqrt{\frac{(n_i - 1)s_i^2}{n_i}} \tag{24}$$

and

$$R_{\sigma_i^2} = \frac{(n_i - 1)s_i^2}{V_i}, \tag{25}$$

where Z_i denotes standard normal distribution, U_i and V_i denote chi-square distribution with degrees of freedom $n_i - 1$.

Using the generalized pivotal quantity in Eq. (1), then the generalized pivotal quantity for θ_i based on the i-th sample is defined by

$$R_{\theta_i} = \frac{\bar{x}_i}{s_i} \sqrt{\frac{U_i}{n_i - 1}} - \frac{Z_i}{\sqrt{n_i}}; i = 1, 2, \ldots, k. \tag{26}$$

The generalized pivotal quantity for the common parameter, introduced by Tian and Wu [16], is a weighted average of the generalized pivotal quantity R_{θ_i} based on k individual samples. Hence, the generalized pivotal quantity for the common SNR R_η is given by

$$R_\eta = \sum_{i=1}^{k} \frac{R_{\theta_i}}{R_{Var(\hat{\theta}_i)}} \Big/ \sum_{i=1}^{k} \frac{1}{R_{Var(\hat{\theta}_i)}}, \tag{27}$$

where R_{θ_i} is defined in Eq. (26) and $R_{Var(\hat{\theta}_i)}$ can be obtained from Eq. (3) which is obtained by

$$R_{Var(\hat{\theta}_i)} = \frac{1}{n_i}\left(1 - \frac{R_{\mu_i}}{R_{\sigma_i^2}^2}\mu_{i3} + \frac{R_{\mu_i}^2}{4R_{\sigma_i^2}^3}\mu_{i4} - \frac{R_{\mu_i}^2}{4R_{\sigma_i^2}^2}\right), \qquad (28)$$

where $\mu_{i3} = \frac{1}{n_i}\sum_{j=1}^{n_i}(X_{ij} - \bar{X}_i)^3$ and $\mu_{i4} = \frac{1}{n_i}\sum_{j=1}^{n_i}(X_{ij} - \bar{X}_i)^4$.

Therefore, the $100(1 - \alpha)\%$ two-sided confidence interval for the common SNR η based on the GCI approach is defined by

$$CI_{C.GCI} = (L_{C.GCI}, U_{C.GCI}) = (R_\eta(\alpha/2), R_\eta(1 - \alpha/2)), \qquad (29)$$

where R_η is defined in Eq. (27) and $R_\eta(p)$ denotes the p-th quantile of R_η.

4.2 The Large-Sample Confidence Interval for the Common Signal-to-Noise Ratio

Graybill and Deal [5] and Tian and Wu [16] proposed the large-sample estimator of the common parameter of interest. The large-sample estimate of the SNR, is a pooled estimated estimator, is given by

$$\hat{\eta} = \sum_{i=1}^{k}\frac{\hat{\theta}_i}{Var(\hat{\theta}_i)} \Big/ \sum_{i=1}^{k}\frac{1}{Var(\hat{\theta}_i)}, \qquad (30)$$

where $\hat{\theta}_i = \bar{X}_i/S_i$ and $Var(\hat{\theta}_i)$ is an estimate of

$$Var(\hat{\theta}_i) = \frac{1}{n_i}\left(1 - \frac{\mu_i}{\sigma_i^4}\mu_{i3} + \frac{\mu_i^2}{4\sigma_i^6}\mu_{i4} - \frac{\mu_i^2}{4\sigma_i^2}\right), \qquad (31)$$

where $\mu_{i3} = \frac{1}{n_i}\sum_{j=1}^{n_i}(X_{ij} - \bar{X}_i)^3$ and $\mu_{i4} = \frac{1}{n_i}\sum_{j=1}^{n_i}(X_{ij} - \bar{X}_i)^4$ which μ_i and σ_i^2 replaced by $\bar{x}_i$ and s_i^2, respectively.

Therefore, the $100(1 - \alpha)\%$ two-sided confidence interval for the common SNR η based on the large-sample approach is defined by

$$CI_{C.LS} = (L_{C.LS}, U_{C.LS}) \qquad (32)$$

$$= (\hat{\eta} - z_{1-\alpha/2}\sqrt{1\Big/\sum_{i=1}^{k}\frac{1}{Var(\hat{\theta}_i)}}, \hat{\eta} + z_{1-\alpha/2}\sqrt{1\Big/\sum_{i=1}^{k}\frac{1}{Var(\hat{\theta}_i)}}),$$

where $z_{1-\alpha/2}$ denotes the $(1 - \alpha/2)$-th quantile of the standard normal distribution.

4.3 The Adjusted Method of Variance Estimates Recovery (Adjusted MOVER) Confidence Interval for the Common Signal-to-Noise Ratio

Using Eqs. (16)–(19), the lower and upper limits for θ_i based on the i-th sample for $i = 1, 2, \ldots, k$ are given by

$$l_i = \frac{\hat{\mu}_i\hat{\sigma}_i - \sqrt{(\hat{\mu}_i\hat{\sigma}_i)^2 - l_{i1}u_{i2}(2\hat{\mu}_i - l_{i1})(2\hat{\sigma}_i - u_{i2})}}{u_{i2}(2\hat{\sigma}_i - u_{i2})} \tag{33}$$

and

$$u_i = \frac{\hat{\mu}_i\hat{\sigma}_i + \sqrt{(\hat{\mu}_i\hat{\sigma}_i)^2 - u_{i1}l_{i2}(2\hat{\mu}_i - u_{i1})(2\hat{\sigma}_i - l_{i2})}}{l_{i2}(2\hat{\sigma}_i - l_{i2})}, \tag{34}$$

where

$$(l_{i1}, u_{i1}) = (\hat{\mu}_i - t_{n_i-1,1-\frac{\alpha}{2}}\frac{s_i}{\sqrt{n_i}}, \hat{\mu}_i + t_{n_i-1,1-\frac{\alpha}{2}}\frac{s_i}{\sqrt{n_i}})$$

and

$$(l_{i2}, u_{i2}) = \left(\sqrt{\frac{(n_i-1)s_i^2}{\chi^2_{n_i-1,1-\frac{\alpha}{2}}}}, \sqrt{\frac{(n_i-1)s_i^2}{\chi^2_{n_i-1,\frac{\alpha}{2}}}}\right).$$

Thangjai and Niwitpong [14] introduced the adjusted MOVER approach to construct the confidence interval for the common parameter. This approach uses the concepts of the large-sample approach and MOVER approach. Hence, the common SNR η is a weighted average of $\hat{\theta}_i$ based on k individual samples. Then

$$\hat{\eta} = \sum_{i=1}^{k} \frac{\hat{\theta}_i}{Var(\hat{\theta}_i)} \bigg/ \sum_{i=1}^{k} \frac{1}{Var(\hat{\theta}_i)}, \tag{35}$$

where the variance estimate for $\hat{\theta}_i$ at $\theta_i = l_i$ and $\theta_i = u_i$ is the average variance between these two variances and given by

$$Var(\hat{\theta}_i) = \frac{1}{2}\left(\frac{(\hat{\theta}_i - l_i)^2}{z^2_{\alpha/2}} + \frac{(u_i - \hat{\theta}_i)^2}{z^2_{\alpha/2}}\right); i =, 2, \ldots, k. \tag{36}$$

Therefore, the $100(1 - \alpha)\%$ two-sided confidence interval for the common SNR η based on the adjusted MOVER approach is defined by

$$CI_{C.AM} = (L_{C.AM}, U_{C.AM}) \tag{37}$$

$$= (\hat{\eta} - z_{1-\alpha/2}\sqrt{1\bigg/\sum_{i=1}^{k}\frac{z^2_{\alpha/2}}{(\hat{\theta}_i - l_i)^2}}, \hat{\eta} + z_{1-\alpha/2}\sqrt{1\bigg/\sum_{i=1}^{k}\frac{z^2_{\alpha/2}}{(u_i - \hat{\theta}_i)^2}}),$$

where z_p denotes the p-th quantile of the standard normal distribution, $\hat{\eta}$ is defined in Eq. (35), and l_i and u_i are defined in Eqs. (33)–(34).

5   Simulation Studies

In this section, a Monte Carlo simulation was used to evaluate the performance of all confidence intervals. First, a simulation is performed to compare the given approaches in Sect. 2. Second, simulation is performed to compare the approaches in Sect. 3. Finally, the simulation is performed to compare the approaches in Sect. 4.

For single SNR, the sample sizes are $n = 15$, 25, 50, 100, and 200. Following Albatineh et al. [1], the mean $\mu = 10$ and the standard deviation $\sigma = 10$, 5, 3, 1 are set. The SNR is computed by $\theta = \mu/\sigma$ as $\theta = 1.00$, 2.00, 3.33, and 10.00. For each parameters and sample size combination, 5000 random samples are generated. For each of these 5000 random samples, 2500 R_θ's are simulated. The coverage probabilities and average widths of 95% two-sided confidence intervals for θ are reported in Table 1. The average widths of all approaches increase when the SNR value increases, but the averages widths decrease when the sample size increases. For averages widths, the simulation results are similar to those of

Table 1. The coverage probabilities (CP) and average widths (AW) of 95% two-sided confidence intervals for signal to noise of normal distribution

n	θ	$CI_{S.GCI}$		$CI_{S.LS}$		$CI_{S.MOVER}$	
		CP	AW	CP	AW	CP	AW
15	1.00	0.9512	1.2939	0.9046	1.1667	0.9546	1.3656
	2.00	0.9496	1.8673	0.8694	1.5499	0.9574	1.9127
	3.33	0.9476	2.7851	0.8468	2.2054	0.9530	2.8168
	10.00	0.9462	7.8517	0.8310	5.9761	0.9480	7.8627
25	1.00	0.9452	0.9837	0.9182	0.9290	0.9494	1.0143
	2.00	0.9512	1.4042	0.9062	1.2579	0.9576	1.4243
	3.33	0.9420	2.0929	0.8960	1.8179	0.9480	2.1057
	10.00	0.9462	5.8540	0.8784	4.9726	0.9460	5.8584
50	1.00	0.9454	0.6865	0.9360	0.6689	0.9508	0.6967
	2.00	0.9508	0.9777	0.9266	0.9220	0.9538	0.9842
	3.33	0.9448	1.4452	0.9166	1.3517	0.9488	1.4496
	10.00	0.9478	4.0361	0.9180	3.7185	0.9484	4.0364
100	1.00	0.9484	0.4826	0.9416	0.4758	0.9532	0.4858
	2.00	0.9470	0.6862	0.9372	0.6691	0.9486	0.6885
	3.33	0.9490	1.0123	0.9350	0.9763	0.9496	1.0140
	10.00	0.9488	2.8276	0.9374	2.7217	0.9492	2.8277
200	1.00	0.9492	0.3407	0.9448	0.3383	0.9482	0.3419
	2.00	0.9540	0.4821	0.9482	0.4752	0.9532	0.4827
	3.33	0.9398	0.7145	0.9322	0.7010	0.9420	0.7143
	10.00	0.9466	1.9940	0.9406	1.9515	0.9462	1.9930

Table 2. The coverage probabilities (CP) and average widths (AW) of 95% two-sided confidence intervals for the difference of signal to noise ratios of normal distributions

(n_1, n_2)	(θ_1, θ_2)	$CI_{D.GCI}$		$CI_{D.LS}$		$CI_{D.MOVER}$	
		CP	AW	CP	AW	CP	AW
(15, 15)	(10.00, 1.00)	0.9530	7.9249	0.8454	6.1028	0.9554	7.9432
	(10.00, 2.00)	0.9516	8.0476	0.8444	6.2165	0.9528	8.0631
	(10.00, 3.33)	0.9478	8.3541	0.8446	6.3921	0.9492	8.3654
	(10.00, 10.00)	0.9474	11.1580	0.8458	8.6109	0.9482	11.1512
(15, 25)	(10.00, 1.00)	0.9514	7.9195	0.8444	6.0756	0.9512	7.9313
	(10.00, 2.00)	0.9428	7.9349	0.8416	6.1089	0.9430	7.9480
	(10.00, 3.33)	0.9530	8.1045	0.8430	6.2892	0.9536	8.1120
	(10.00, 10.00)	0.9480	9.8182	0.8612	7.8710	0.9490	9.8064
(25, 25)	(10.00, 1.00)	0.9538	5.9545	0.8998	5.1092	0.9544	5.9619
	(10.00, 2.00)	0.9494	6.0569	0.8894	5.1864	0.9504	6.0641
	(10.00, 3.33)	0.9438	6.2400	0.8932	5.3868	0.9434	6.2445
	(10.00, 10.00)	0.9438	8.3374	0.8896	7.1549	0.9454	8.3288
(25, 50)	(10.00, 1.00)	0.9538	5.8785	0.8960	5.0379	0.9530	5.8850
	(10.00, 2.00)	0.9520	5.9615	0.8936	5.1121	0.9520	5.9666
	(10.00, 3.33)	0.9482	6.0506	0.8886	5.2180	0.9496	6.0517
	(10.00, 10.00)	0.9532	7.1508	0.9036	6.2866	0.9528	7.1458
(50, 50)	(10.00, 1.00)	0.9478	4.1099	0.9206	3.7970	0.9484	4.1100
	(10.00, 2.00)	0.9508	4.1687	0.9220	3.8579	0.9484	4.1710
	(10.00, 3.33)	0.9484	4.3088	0.9196	3.9905	0.9486	4.3077
	(10.00, 10.00)	0.9502	5.7434	0.9232	5.3105	0.9500	5.7378
(50, 100)	(10.00, 1.00)	0.9514	4.0841	0.9226	3.7794	0.9518	4.0844
	(10.00, 2.00)	0.9484	4.1043	0.9194	3.7964	0.9490	4.1056
	(10.00, 3.33)	0.9498	4.1691	0.9248	3.8622	0.9508	4.1702
	(10.00, 10.00)	0.9496	4.9561	0.9290	4.6531	0.9494	4.9532
(100, 100)	(10.00, 1.00)	0.9472	2.8698	0.9322	2.7598	0.9478	2.8708
	(10.00, 2.00)	0.9484	2.9136	0.9308	2.8028	0.9494	2.9138
	(10.00, 3.33)	0.9472	3.0105	0.9358	2.8962	0.9478	3.0116
	(10.00, 10.00)	0.9534	4.0128	0.9424	3.8707	0.9552	4.0105
(100, 200)	(10.00, 1.00)	0.9484	2.8554	0.9338	2.7484	0.9492	2.8554
	(10.00, 2.00)	0.9494	2.8769	0.9372	2.7655	0.9500	2.8762
	(10.00, 3.33)	0.9498	2.9215	0.9338	2.8110	0.9514	2.9215
	(10.00, 10.00)	0.9500	3.4666	0.9384	3.3591	0.9500	3.4658
(200, 200)	(10.00, 1.00)	0.9520	2.0203	0.9430	1.9830	0.9522	2.0194
	(10.00, 2.00)	0.9528	2.0520	0.9432	2.0074	0.9520	2.0507
	(10.00, 3.33)	0.9490	2.1161	0.9398	2.0724	0.9472	2.1154
	(10.00, 10.00)	0.9508	2.8184	0.9472	2.7678	0.9516	2.8169

Table 3. The coverage probabilities (CP) and average widths (AW) of 95% of two-sided confidence intervals for the common signal to noise ratio of normal distributions: 3 sample cases

(n_1, n_2, n_3)	η	$CI_{C.GCI}$		$CI_{C.LS}$		$CI_{C.AM}$	
		CP	AW	CP	AW	CP	AW
(15, 15, 15)	1.00	0.9666	0.8533	0.8990	0.6503	0.9596	0.7786
	2.00	0.9852	1.6012	0.8546	0.8329	0.9576	1.0775
	3.33	0.9966	3.9533	0.8222	1.1637	0.9560	1.5722
	10.00	1.0000	18.5896	0.8098	3.1348	0.9516	4.3531
(25, 25, 25)	1.00	0.9628	0.6225	0.9198	0.5199	0.9576	0.5810
	2.00	0.9814	1.0292	0.8928	0.6929	0.9546	0.8115
	3.33	0.9936	1.8716	0.8792	0.9960	0.9506	1.1918
	10.00	0.9982	8.2291	0.8740	2.7277	0.9504	3.3116
(50, 50, 50)	1.00	0.9586	0.4198	0.9352	0.3786	0.9544	0.4010
	2.00	0.9748	0.6568	0.9246	0.5179	0.9540	0.5639
	3.33	0.9838	1.0596	0.9102	0.7553	0.9506	0.8304
	10.00	0.9942	3.4041	0.9184	2.0888	0.9534	2.3123
(15, 25, 50)	1.00	0.9608	0.5643	0.9204	0.4800	0.9530	0.5264
	2.00	0.9802	1.0680	0.9004	0.6423	0.9582	0.7360
	3.33	0.9948	2.6839	0.8862	0.9271	0.9528	1.0826
	10.00	0.9996	13.8921	0.8688	2.5386	0.9482	3.0125
(100, 100, 100)	1.00	0.9522	0.2880	0.9422	0.2719	0.9510	0.2803
	2.00	0.9654	0.4330	0.9390	0.3790	0.9518	0.3952
	3.33	0.9696	0.6715	0.9290	0.5557	0.9502	0.5830
	10.00	0.9832	1.9730	0.9392	1.5442	0.9544	1.6256
(25, 50, 100)	1.00	0.9598	0.3861	0.9382	0.3525	0.9536	0.3699
	2.00	0.9720	0.6192	0.9222	0.4840	0.9510	0.5211
	3.33	0.9866	1.1764	0.9238	0.7063	0.9570	0.7677
	10.00	0.9958	5.6740	0.9100	1.9517	0.9484	2.1347
(50, 100, 200)	1.00	0.9548	0.2652	0.9482	0.2525	0.9564	0.2590
	2.00	0.9638	0.4011	0.9378	0.3527	0.9514	0.3656
	3.33	0.9726	0.6423	0.9354	0.5173	0.9532	0.5389
	10.00	0.9842	2.0848	0.9328	1.4368	0.9492	1.5034
(200, 200, 200)	1.00	0.9458	0.2000	0.9416	0.1940	0.9458	0.1970
	2.00	0.9622	0.2930	0.9438	0.2721	0.9536	0.2783
	3.33	0.9656	0.4446	0.9364	0.4011	0.9498	0.4108
	10.00	0.9712	1.2725	0.9426	1.1147	0.9532	1.1461

Table 4. The coverage probabilities (CP) and average widths (AW) of 95% of two-sided confidence intervals for the common signal to noise ratio of normal distributions: 6 sample cases

$(n_1, n_2, n_3, n_4, n_5, n_6)$	η	$CI_{C.GCI}$		$CI_{C.LS}$		$CI_{C.AM}$	
		CP	AW	CP	AW	CP	AW
$(15, 15, 15, 15, 15, 15)$	1.00	0.9690	0.6220	0.8904	0.4553	0.9632	0.5497
	2.00	0.9924	1.6096	0.8514	0.5776	0.9652	0.7570
	3.33	0.9994	6.7473	0.8204	0.8054	0.9578	1.1025
	10.00	1.0000	35.5802	0.7834	2.1551	0.9440	3.0478
$(25, 25, 25, 25, 25, 25)$	1.00	0.9616	0.4487	0.9214	0.3658	0.9600	0.4105
	2.00	0.9852	0.8036	0.8892	0.4831	0.9534	0.5726
	3.33	0.9970	2.1502	0.8734	0.6925	0.9508	0.8393
	10.00	1.0000	18.6178	0.8592	1.8946	0.9452	2.3275
$(50, 50, 50, 50, 50, 50)$	1.00	0.9590	0.2997	0.9320	0.2667	0.9554	0.2833
	2.00	0.9784	0.4863	0.9182	0.3643	0.9532	0.3981
	3.33	0.9922	0.8554	0.9070	0.5306	0.9512	0.5862
	10.00	0.9994	3.7716	0.9060	1.4653	0.9436	1.6312
$(15, 15, 25, 25, 50, 50)$	1.00	0.9580	0.4050	0.9238	0.3369	0.9530	0.3714
	2.00	0.9862	0.8990	0.8968	0.4507	0.9544	0.5195
	3.33	0.9982	3.3789	0.8826	0.6473	0.9550	0.7629
	10.00	1.0000	22.4223	0.8606	1.7675	0.9430	2.1186
$(100, 100, 100, 100, 100, 100)$	1.00	0.9588	0.2052	0.9464	0.1919	0.9564	0.1981
	2.00	0.9702	0.3149	0.9380	0.2671	0.9530	0.2793
	3.33	0.9798	0.5034	0.9342	0.3915	0.9538	0.4118
	10.00	0.9876	1.5616	0.9306	1.0879	0.9512	1.1480
$(25, 25, 50, 50, 100, 100)$	1.00	0.9610	0.2755	0.9394	0.2478	0.9586	0.2613
	2.00	0.9770	0.4615	0.9274	0.3413	0.9550	0.3679
	3.33	0.9926	1.1577	0.9144	0.4963	0.9472	0.5417
	10.00	0.9998	8.9890	0.9056	1.3740	0.9476	1.5096
$(50, 50, 100, 100, 200, 200)$	1.00	0.9612	0.1887	0.9506	0.1784	0.9580	0.1831
	2.00	0.9704	0.2904	0.9388	0.2483	0.9550	0.2583
	3.33	0.9778	0.4870	0.9334	0.3640	0.9500	0.3810
	10.00	0.9948	1.9740	0.9288	1.0133	0.9510	1.0632
$(200, 200, 200, 200, 200, 200)$	1.00	0.9506	0.1421	0.9466	0.1371	0.9510	0.1393
	2.00	0.9614	0.2102	0.9424	0.1922	0.9502	0.1967
	3.33	0.9666	0.3226	0.9446	0.2829	0.9514	0.2904
	10.00	0.9730	0.9370	0.9394	0.7869	0.9478	0.8096

Albatineh et al. [1]. The coverage probabilities of the GCI and large-sample approaches are less than nominal confidence level of 0.95, whereas the coverage probabilities of the MOVER approach are close to nominal confidence level of 0.95. Therefore, the confidence interval based on the MOVER approach performs the best confidence interval in terms of the coverage probability.

For the difference of SNRs, the sample sizes are $(n_1, n_2) = (15, 15)$, $(15, 25)$, $(25, 25)$, $(25, 50)$, $(50, 50)$, $(50, 100)$, $(100, 100)$, $(100, 200)$, and $(200, 200)$. The population means are $(\mu_1, \mu_2) = (10, 10)$ and the population standard deviation are $\sigma_1 = 1$, $\sigma_2 = 10, 5, 3, 1$. Hence, the SNRs $(\theta_1, \theta_2) = (10.00, 1.00)$, $(10.00, 2.00)$, $(10.00, 3.33)$, and $(10.00, 10.00)$ are obtained. For each parameter setting, 5000 random samples are generated, and within each of the 5000 random samples, 2500 R_δ's are obtained. The coverage probabilities and average widths of 95% two-sided confidence intervals for δ are presented in Table 2. The average widths of all the proposed approaches decrease when the sample size increases. Overall, the MOVER approach is better than the other approaches in terms of coverage probability.

For the common SNR, the sample cases are $k = 3$ and $k = 6$. The sample sizes are given in the following tables. For $i = 1, 2, \ldots, k$, the population means are $\mu_1 = \mu_2 = \ldots = \mu_k = \mu = 10$ and the population standard deviations are computed by $\sigma_i = \mu_i / \eta_i$, where the common SNR $\eta = 1.00, 2.00, 3.33$, and 10.00. For each parameter setting, 5000 random samples are generated and thus 2500 R_η's are obtained for each of the random samples. Tables 3 and 4 present the coverage probabilities and average widths for $k = 3$ and $k = 6$ sample cases, respectively. For $\eta = 1.00$, the coverage probabilities of the GCI approach perform as well as those of the adjusted MOVER approach, but the average widths of the adjusted MOVER approach are shorter than that of the GCI approach. For $\eta \geq 2.00$, the coverage probabilities of the GCI approach are in the range from 0.96 to 1.00. Hence, the generalized confidence interval is a conservative confidence interval when $\eta \geq 2.00$. The coverage probabilities of the large-sample approach are less than nominal confidence level of 0.95 for all cases. The coverage probabilities of the adjusted MOVER approach are greater than a nominal confidence level of 0.95 in almost all cases. Therefore, the adjusted MOVER approach can be used to estimate the confidence interval for the common SNR of normal distributions.

6   An Empirical Application

In this section, three examples are given to illustrate the proposed approaches.

Example 1: The data reported by Albatineh et al. [1] are about the heights of 50 nano-pillars in nano-meters. The summary statistics are $n = 50$, $\bar{x} = 305.58$, $s = 36.97$, and $\hat{\theta} = 8.27$. The 95% two-sided confidence interval for the SNR based on the GCI approach is $(6.6415, 9.9170)$ with confidence width 3.2755. The 95% two-sided confidence interval for the SNR based on the large-sample approach is $(6.6932, 9.8376)$ with confidence width 3.1444. The 95% two-sided confidence interval for the SNR based on the MOVER approach is $(6.6131, 9.9241)$ with confidence width 3.3110. It can be seen that the confidence interval of all the proposed approaches contains the true SNR. The confidence width of the MOVER approach is greater than the confidence widths of the large-sample and GCI approaches, while the large-sample approach provides the width shorter than the GCI approach. Therefore, these results confirm the simulation results in Table 1.

Example 2: The data considered by Devore [3] and Niwitpong and Wongkhao [11]. The data are dry density of cyclic strength using Pitcher sampling method and Block sampling method, see Devore [3]. The Pitcher sampling yields $n_1 = 24$, $\bar{x}_1 = 103.6583$, $s_1 = 3.7376$, and $\hat{\theta}_1 = 27.7339$. The Block sampling yields $n_2 = 11$, $\bar{x}_2 = 101.1091$, $s_2 = 3.6035$, and $\hat{\theta}_2 = 28.0586$. The difference of SNRs is $\hat{\delta} = \hat{\theta}_1 - \hat{\theta}_2 = -0.3247$. The 95% two-sided confidence interval for the difference of SNRs based on the GCI approach is $(-14.5837, 13.5241)$ with confidence width 28.1078. The 95% two-sided confidence interval for the difference of SNRs based on the large-sample approach is $(-7.3549, 6.7053)$ with confidence width 14.0602. The 95% two-sided confidence interval for the difference of SNRs based on the MOVER approach is $(-14.8357, 14.1452)$ with confidence width 28.9809. The confidence width of the large-sample approach is shorter than the confidence widths of the GCI approach and the MOVER approach. Hence, the results from this example support the simulation results in Table 2.

Example 3: The data collected by Tsou [17] and previously considered by Gokpinar and Gokpinar [8]. The data consist of 156 observations of numbers of birth in 1978 in $k = 3$ groups: Monday, Thursday, and Saturday, with $n_1 = n_2 = n_3 = 52$. A summary of the data are $\bar{x}_1 = 9350.3460$, $\bar{x}_2 = 9471.4620$, $\bar{x}_3 = 8309.3270$, $s_1 = 613.2140$, $s_2 = 554.8795$, $s_3 = 390.2555$, $\hat{\theta}_1 = 15.2481$, $\hat{\theta}_2 = 17.0694$, and $\hat{\theta}_3 = 21.2920$. The 95% two-sided confidence interval for the common SNR based on the GCI approach is $(15.0284, 22.9194)$ with confidence width 7.8910. The 95% two-sided confidence interval for the common SNR based on the large-sample approach is $(16.2237, 19.9769)$ with confidence width 3.7532. The 95% two-sided confidence interval for the common SNR based on the adjusted MOVER approach is $(15.2742, 19.1691)$ with confidence width 3.8949. The confidence width of the GCI approach is greater than those of the confidence widths of the large-sample and the adjusted MOVER approaches. Therefore, the results confirm the simulation results in Table 3.

7   Discussion and Conclusions

In this paper, the confidence intervals for SNR of normal distribution were constructed using the GCI approach, the large-sample approach, and the MOVER approach. Moreover, the GCI approach, the large-sample approach, and the MOVER approach were used to construct the confidence intervals for the difference between two SNRs of normal distribution. Furthermore, the confidence interval estimation for the common SNR of several normal distributions based on the GCI approach, the large-sample approach, and the adjusted MOVER approach were presented. The performances of all the proposed confidence intervals were evaluated via Monte Carlo simulations.

For single SNR, the results are similar to the paper by Albatineh et al. [1] in terms of average width. The coverage probabilities of the MOVER approach were satisfactorily stable around 0.95. Therefore, the MOVER approach is recommended to construct confidence interval for the SNR of normal distribution. The coverage probabilities of the GCI approach and the large-sample approach are less than nominal confidence level. Hence, the GCI approach and the large-sample approach are not recommended to construct the confidence intervals for the SNR of normal distribution.

For difference of SNRs, the coverage probabilities of the MOVER approach were satisfactorily stable around 0.95. Therefore, the MOVER approach is recommended to construct confidence interval for the difference between two SNRs of normal distributions. Moreover, the GCI approach can be used as an alternative for constructing the confidence interval.

For common SNR, the generalized confidence interval is a conservative confidence interval when $\eta \geq 2.00$. Hence, the GCI approach can be used to estimate confidence interval for common SNR of normal distributions when $\eta = 1.00$, but it is not recommended to estimate the confidence interval when $\eta \geq 2.00$. The large-sample approach is not recommended to construct the confidence interval for the common SNR because the coverage probabilities are less than nominal confidence level. The coverage probabilities of the adjusted MOVER approach are greater than a nominal confidence level 0.95. Therefore, the adjusted MOVER approach is recommended to estimate the confidence interval for the common SNR of normal distributions.

References

1. Albatineh, A.N., Boubakari, I., Kibria, B.M.G.: New confidence interval estimator of the signal-to-noise ratio based on asymptotic sampling distribution. Commun. Stat. Theory Methods **46**, 574–590 (2017)
2. Albatineh, A.N., Kibria, B.M.G., Zogheib, B.: Asymptotic sampling distribution of inverse coefficient of variation and its applications: revisited. Int. J. Adv. Stat. Probab. **2**, 15–20 (2014)
3. Devore, J.L.: Probability and statistics for engineering and the sciences, Books/Cole (2012)
4. Donner, A., Zou, G.Y.: Closed-form confidence intervals for function of the normal mean and standard deviation. Stat. Methods Med. Res. **21**, 347–359 (2010)
5. Graybill, F.A., Deal, R.B.: Combining unbiased estimators. Biometrics **15**, 543–550 (1959)
6. George, F., Kibria, B.M.G.: Confidence intervals for signal to noise ratio of a poisson distribution. Amer. J. Biost. **2**, 44–55 (2011)
7. George, F., Kibria, B.M.G.: Confidence intervals for estimating the population signal-to-noise ratio: a simulation study. J. Appl. Stat. **39**, 1225–1240 (2012)
8. Gokpinar, F., Gokpinar, E.: A computational approach for testing equality of coefficients of variation in k normal populations. Hacet. J. Math. Stat. **5**, 1197–1213 (2015)
9. Krishnamoorthy, K.: Handbook of Statistical Distributions with Applications. Chapman and Hall/CRC, Boca Raton (2006)

10. McGibney, G., Smith, M.R.: An unbiased signal-to-noise ratio measure for magnetic resonance images. Med. Phys. **20**, 1077–1078 (1993)

11. Niwitpong, S., Wongkhao, A.: Confidence intervals for the difference between inverse of normal means. Adv. Appl. Stat. **48**, 337–347 (2016)

12. Sharma, K.K., Krishna, H.: Asymptotic sampling distribution of inverse coefficient of variation and its applications. IEEE Trans. Reliability **43**, 630–633 (1994)

13. Swann, G.M.P.: Putting Econometrics in its Place: A New Direction in Applied Economics Edward Elgar, Cheltenham (2006)

14. Thangjai, W., Niwitpong, S.: Confidence intervals for the weighted coefficients of variation of two-parameter exponential distributions. Cogent Math. **4**, 1–16 (2017)

15. Tian, L.: Inferences on the common coefficient of variation. Stat. Med. **24**, 2213–2220 (2005)

16. Tian, L., Wu, J.: Inferences on the common mean of several log-normal populations: the generalized variable approach. Biometrical J. **49**, 944–951 (2007)

17. Tsou, T.S.: A robust score test for testing several coefficients of variation with unknown underlying distributions. Commun. Stat. Theory Methods **38**, 1350–1360 (2009)

18. Visintainer, P.F., Tejani, N.: Understanding and using confidence intervals in clinical research. J. Matern. Fetal Med. **7**, 201–202 (1998)

19. Weerahandi, S.: Generalized confidence intervals. J. Am. Stat. Assoc. **88**, 899–905 (1993)

20. Weerahandi, S.: Exact Statistical Methods for Data Analysis. Springer, New York (1995)

Fuzzy vs. Probabilistic Techniques in Time Series Analysis

Vilém Novák[✉]

Institute for Research and Applications of Fuzzy Modeling, NSC IT4Innovations,
University of Ostrava, 30. dubna 22, 701 03 Ostrava 1, Czech Republic
Vilem.Novak@osu.cz

Abstract. In this paper, we discuss the difference between probabilistic and fuzzy techniques used in time series analysis. First, we focus on the fundamental difference between vaguenes and uncertainty phenomena. Then we briefly describe probabilistic view of time series. In the main part, we demonstrate how special fuzzy techniques, namely the fuzzy natural logic and fuzzy transform can be applied in the analysis of time series and what is their outcome in comparison with the probabilistic approach. We argue that fuzzy techniques enable to obtain knowledge that is either more difficult or impossible to obtain using probabilistic techniques.

Keywords: Vagueness · Fuzzy transform · F-transform
Fuzzy natural logic · Components of time series

1 Introduction

In this section, we will demonstrate that probabilistic and fuzzy techniques are based on modeling of different phenomena, namely vagueness and uncertainty. Both phenomena are usually present and require different mathematical principles. Hence, in the reality both kinds of techniques are complementary rather than competitive.

2 Uncertainty and Vagueness

Two phenomena whose importance in science raised especially in 20th century are *uncertainty* and *vagueness* (cf. [2,28]). Both of them characterize situations in which the amount, character and extent of knowledge we have at disposal is essential. It is important to stress that both uncertainty as well as vagueness form two complementary facets of a more general phenomenon called *indeterminacy*[1]. In the reality, we often meet indeterminacy *with both its facets* present, i.e., *vague* phenomena can be at the same time also *uncertain*.

---

[1] This phenomenon is sometimes called "uncertainty in wider sense".

© Springer International Publishing AG 2018
L. H. Anh et al. (eds.), *Econometrics for Financial Applications*, Studies in Computational Intelligence 760, https://doi.org/10.1007/978-3-319-73150-6_17

2.1   Potentiality and Uncertainty

When observing the surrounding world, we encounter events of two kinds: those that already occurred and *potential* ones that can, but need not, occur. For example, consider a company producing tires. We know that today it produced, say 300 of them. But the number of tires produced the next day is not known. We may expect production of, e.g., 350 tires but the concrete number is uncertain because, for example, technical or personal problems on the production line may appear. From it follows that the *uncertainty phenomenon* emerges when there is a *lack of knowledge* about *occurrence* of some *event* (e.g., the production of tires). In general, we may state that uncertainty is encountered when a certain kind of experiment (process, test, etc.) is to proceed, the result of which is not known to us. It may refer to variety of potential outcomes, ways of solution, choices, etc.

Specific form of uncertainty is *randomness* which is uncertainty raising in connection with time. There is no randomness (uncertainty) after the experiment was realized (the event has occurred) and the result is known to us. Note that it is connected with the question whether a given event may be regarded within some time period, or not. This becomes apparent on the typical example with tossing a player's cube. The phenomenon to occur is *the number of dots on the cube* and it occurs after the experiment (i.e. tossing the cube one times) has been realized. Thus, we refer here to the future, to events that are *potential*; not yet existing.

Let us remark, however, that the variety of potential events may raise even a more abstract uncertainty that is less dependent on time. We may, for example, analyze uncertainty in potentiality (that is, lack of knowledge) without necessary reference to time, or with reference to the past (such as a posterior Bayesian probability).

The mathematical model (i.e. quantified characterization) of the uncertainty phenomenon is provided especially by the *probability theory*. In everyday terminology, probability can be thought of as a numerical measure of the likelihood that a particular event will occur. There are also other mathematical theories addressing the mentioned abstract uncertainty, for example possibility theory, belief measures and others.

2.2   Vagueness and Actuality

The *vagueness phenomenon* raises when we try to *group* together all objects that have a certain property φ. The result is a grouping of objects

$$X = \{o \mid o \text{ is an object having the property } \varphi\}. \tag{1}$$

We see the grouping X as one object consisting of objects o that all are at our disposal at once because we have already grouped them together. We say that X is *actualized*.

In general, however, X *cannot be taken as a set* since the property φ may be of such a character that when checking that a given object $\hat{o}$ has the property φ, we hardly obtain a definite answer. For example, consider the property $\varphi =$ 'to be expensive' and let the total amount of money we have at disposal for all our expenses be 50,000 \$. Let o_1 be a car for 20,000 \$, o_2 a car for 48,000 \$, and o_3 a car for 35,000 \$. Then o_1 is not expensive at all, i.e., $\varphi(o_1)$ is false and $\varphi(o_2)$ is true. But what about $\varphi(o_3)$? This car is not really expensive but also not too cheap. Hence, we cannot say that the grouping X in (1) is a set because a set is formed only of objects that we *unambiguously* know that they have the property φ. Hence, we say that φ is *vague*. There can exist *borderline* elements o for which it is unclear whether they have the property φ (and thus, whether they belong to X), or not. On the other hand, it is always possible to characterize, at least some *typical objects* (prototypes), i.e. objects having typically the property in concern. For example, everybody can show a "blue sweater" or "huge building", "expensive car" but it is impossible to show "all expensive cars".

Vagueness is opposite to exactness and we argue that it cannot be avoided in the human way of regarding the world. Any attempt to explain an extensive detailed description necessarily leads to using vague concepts since precise description can contain such abundant number of details that we will be lost when learning all of them. To understand it, we must group them together — and this can hardly be done precisely. This idea was formulated by Zadeh in [30] as the *incompatibility principle*. The problem consists in the way how people regard the phenomena around them. This would be impossible without presence of vagueness.

The (so far) best mathematical concept that can be used to model vague groupings is that of a *fuzzy set*. Formally, a fuzzy set A is a function

$$A : U \longrightarrow L$$

where U is some universal set containing all the elements (objects) that may be considered to fall into the considered vague grouping and L is a set of *membership degrees* which is a special lattice. The function A is also called the membership function. Note that the fuzzy set is *identified* with its membership function. Sometimes we use the symbol $A \subseteq U$ to emphasize that A is a fuzzy set in the universe U. The value $A(x) \in L$ for any $x \in U$ is called the *membership degree* of the element x in A.

2.3 Actuality vs. Potentiality

In the discussion above we touched two phenomena: actuality and potentiality. A classical set is always understood as being *actual*[2], i.e. we take all its elements as already existing and at our disposal in one moment. Therefore, our reasoning about any set stems from the assumption that it is at our disposal as a whole. Of course, when a set is infinite then only God is able to see it as a whole while we can see only a part of it. It should be emphasized that the set theory (and so, the modern mathematics) can deal with *actualized sets* only!

---

[2] Cf. the analysis by Vopěnka in [29].

On the other hand, most events around us are only *potential*, i.e. they may, but need not, occur or happen. Thus, to create a grouping of objects, we may have only a method how a new element can be created but all of them will never exist together. For example, if a machine has on its input one piece of metal, then it can produce various products of it but only one will actually be finished. It is even impossible to imagine all products produced by the machine from one piece of metal. Note that the same we observe at the company producing tires considered above. In one day it produces from the given amount of material only one number of tires.

As already mentioned, there are two kinds of events: those that already happened and those that have not yet happened. We know the first ones because they are at our disposal and we know that they have a given property φ. However, we do not know the second ones and we even do not know whether some new events having the property φ will occur or not. We encounter uncertainty; we speculate about the whole X (1), but only part of it indeed exists. But as noted, mathematical description of X is possible only if it is actualized. The only solution thus is to imagine all (or, at least some) still not existing elements of X as existing. The "added" part may be, or may be not, possible to happen but we search for methods providing us with the estimation of the information about their possible occurrence.

For example, we can imagine all dots on a dice that can be tossed, i.e., we imagine the tossed numbers $X = \{1, \ldots, 6\}$ as already existing (though they cannot be tossed all of them together). For example, let the numbers $\{1, 3, 5\}$ be already tossed. Then they already exist (this is the actualized part of X) and now we may try to guess whether another number will indeed be tossed (i.e., whether the given element of X will indeed occur). The measure of information about such possibility is modeled using the *probability theory*. As a mathematical theory, however, it works with the whole X, i.e., the problem that X is not yet created is disregarded.

Note that the vagueness phenomenon is not related to occurrence of whatever event. It concerns the question *how* is the given grouping X formed, i.e., what is the character of the property φ in (1) determining it. If for any object, either $\varphi(o)$ holds or not then φ is sharp. If it allows borderline cases then it is vague. Vagueness applies to an *actualized* non-sharply delineated grouping. Once an actualized (i.e. already existing) grouping of objects X is at our disposal, we may speak about *truth* of the fact that an object o has the property φ; that is, we *know the truth* of $o \in X$.

In probability theory we introduce the concept of a probabilistic space $\langle \Omega, \mathscr{A}, P \rangle$, where Ω is a set of elementary random events, $\mathscr{A}$ is a σ-algebra of subsets of Ω and $P : \mathscr{A} \longrightarrow [0, 1]$ is a probabilistic measure. With respect to the discussion above, Ω is a sharp grouping of objects that is *actualized*. Moreover, we deal with the actualized set (σ-algebra) $\mathscr{A}$ of subsets of Ω. Any element $Y \in \mathscr{A}$ is a mathematical model of an event that may, or may not occur. From the mathematical point of view, in fact, Y already exists but we pretend that

$\mathscr{A}$ it is only potential and take P as the measure of information about possible occurrence of Y.

3   Probabilistic View on Time Series

The mathematical model of time series is based on the assumption that a probabilistic space $\langle \Omega, \mathscr{A}, P \rangle$ is given. A time series is then a stochastic process (see [1,7])

$$X : \mathbb{T} \times \Omega \longrightarrow \mathbb{R} \qquad (2)$$

where $\mathbb{T}$ is a set of time moments. In general, it can be $\mathbb{T} = [a,b] \subset \mathbb{R}$ but in economy and elsewhere we usually take $\mathbb{T} = \{1, \ldots, p\} \subset \mathbb{N}$ being a finite set of natural numbers. These are usually construed as hours, days, weeks, months, or years. Instead of the general form (2) we usually write time series as a system of random variables

$$\{X(t) \mid t \in \mathbb{T}\} \qquad (3)$$

where each $X(t)$ is a random variable $X(t) : \Omega \longrightarrow \mathbb{R}$, $t \in \mathbb{T}$, i.e., it is a measurable function w.r.t. Borel sets on $\mathbb{R}$ and $\mathscr{A}$. This enables us to define a function

$$F_t(x) = P\{\omega \in \Omega \mid X(t)(\omega) < x\}.$$

called the *distribution function*, which characterizes the probability distribution of values of the random variable $X(t)$. More generally, we may consider a multidimensional distribution function

$$F_{t_1,\ldots,t_n}(x_1, \ldots, x_n) = P\{\omega \in \Omega \mid X(t_1)(\omega) < x_1, \ldots, X(t_n)(\omega) < x_n\}, \qquad (4)$$

where $t_1, \ldots, t_n \in \mathbb{T}$. When speaking about time series, we will usually write it simply as X without marking the time variable t.

This model assumes existence of a distribution function of each $X(t)$, $t \in \mathbb{T}$, or a joint distribution function (4) of a finite set of them. Let us realize that by this model, the time series is considered to be a sequence of values being measurements of outcomes of some real process that proceeds in time. We do not know which outcome really occurs but we assume to have information about probability of its occurrence. Such information, however, is very rough and does not enable us to penetrate into the substance of the considered process.

Something more we can learn from the following characteristics.

(a) *Mean value* of the time series:

$$\mathbf{E}(X(t)) = \int_{\mathbb{R}} x \, dF_t(x). \qquad (5)$$

(b) *Covariance function* of the time series:

$$R(s,t) = \mathbf{E}((X(s) - \mathbf{E}(X(s)))(X(t) - \mathbf{E}(X(t)))). \qquad (6)$$

Additional used characteristics is *variance*

$$\mathbf{D}(X(t)) = \int_{\mathbb{R}} [x - \mathbf{E}(X(t))]^2 \, dF_t(x).$$

The behavior of these characteristics gives rise to specific kinds of time series. We say that the time series is *strictly stationary* if

$$F_{t_1+h,\ldots,t_n+h}(x_1,\ldots,x_n) = F_{t_1,\ldots,t_n}(x_1,\ldots,x_n) \tag{7}$$

holds for all $t_1,\ldots,t_n \in \mathbb{T}$ and $h \in \mathbb{R}$ such that $t_1+h,\ldots,t_n+h \in \mathbb{T}$. This means that the joint probability distribution does not depend on time. Such time series behaves in a dully uniform way.

We say that the time series is *weak-sense stationary* if the following holds for all $t, s \in \mathbb{T}$:

(i) $\mathbf{E}(X(t)) = \mu$,
(ii) $R(s, t) = R(t - s)$.

This means that the mean value remains the same independently on time and the covariance function is determined by the distance between time moments but not on the position in time.

It is important to emphasize that if we fix $\omega \in \Omega$ then the time series (1) becomes an ordinary function $X : \mathbb{T} \longrightarrow \mathbb{R}$. We call it *realization* of the time series. Note that in practice, we always have *one* realization at disposal only. This fact, however, makes the assumption (3) not fully sound. In extreme case it means that we derive conclusions about time series in a given time moment on one measurement only. But this contradicts the basic assumptions of the probability theory, especially the mass scale, i.e., that its predictions are the more reliable the more measurements of a given random variable are at disposal. We are thus implicitly forced to assume that the real process does not (significantly) change during the time, i.e., whenever we measure its outcome, we measure the same random variable more or less independently on time.

Probabilistic methods, however, led to amazingly well working methods for analysis and prediction of time series. The best known is the *autoregressive moving-average model* ARMA(p, q) (also referred to as Box-Jenkins model) whose general formula is the following:

$$X(t) = \alpha_1 X(t-1) + \cdots + \alpha_p X(t-p) + Z(t) + \beta_1 Z(t-1) + \cdots + \beta_q Z(t-q) \tag{8}$$

where $\{Z(t) \mid t \in \mathbb{T}\}$ is a simple strictly stationary time series with zero mean value and bounded variance. The α_i are autoregressive coefficients and β_j are moving-average coefficients. This model, however, assumes that the time series is stationary, which is rarely the case. In practice, trends and periodicity exists in many datasets, so there is a need to remove these effects before applying such models. This is the fertile ground for application of fuzzy techniques to the analysis of time series.

Let us mention one more important concept, namely the *periodogram*. This is a function of frequencies

$$I(\lambda) = \frac{1}{2\pi N} \left| \sum_{t=1}^{N} X(t) e^{-it\lambda} \right|^2, \qquad -\pi \le \lambda \le \pi. \tag{9}$$

This function makes it possible to identify distinguished frequencies contained in the time series X. Using the well known formula $T = \frac{2\pi}{\lambda}$ we can compute characteristic periodicities in X.

4   Fuzzy Techniques for Time Series Analysis

In this section we will describe basic techniques that are based on the concept of a fuzzy set and that turned out to be very useful in the analysis and prediction of time series. We will very briefly describe the main concepts. More details can be found in the book [24] and the other cited literature.

4.1   Fuzzy Transform

The fuzzy (F-)transform is a universal technique introduced by Perfilieva in [26,27] that has many kinds of applications. Its fundamental idea is to map a bounded continuous function $f : [a, b] \longrightarrow \mathbb{R}$ to a finite vector of numbers and then to transform it back. The former is called a *direct F-transform* and the latter an *inverse one*. The result of the inverse F-transform is a function $\hat{f}$ that *approximates* the original function f. The advantage of this approach consists in the possibility to set the parameters of the F-transform in such a way that the approximating function $\hat{f}$ has desired properties.

The power of the F-transform stems from its approximation abilities, from its ability to filter out high frequencies and from the ability to reduce noise [14,15,25]. Another outcome is the ability to estimate values of first and second derivatives in an area given approximately (cf. [11]).

4.1.1   Fuzzy Partition

The first step of the F-transform procedure is to form a *fuzzy partition* of the domain $[a, b]$. It consists of a finite set of fuzzy sets

$$\mathscr{A} = \{A_0, \dots, A_n\}, \qquad n \ge 2, \tag{10}$$

defined over nodes

$$a = c_0, \dots, c_n = b. \tag{11}$$

The properties of the fuzzy sets from $\mathscr{A}$ are specified by five axioms, namely: *normality, locality, continuity, unimodality,* and *orthogonality* that is formally defined by

$$\sum_{i=0}^{n} A_i(x) = 1, \qquad x \in [a, b.] \tag{12}$$

(Equation (12) is sometimes called *Ruspini condition*).

A fuzzy partition $\mathscr{A}$ is called *h-uniform* if the nodes $c_0, \ldots, c_n$ are *h-equidistant*, i.e., for all $k = 0, \ldots, n-1$, $c_{k+1} = c_k + h$, where $h = (b-a)/n$ and the fuzzy sets $A_1, \ldots, A_{n-1}$ are shifted copies of a *generating function* $A : [-1, 1] \longrightarrow [0, 1]$ such that for all $k = 1, \ldots, n-1$

$$A_k(x) = A\left(\frac{x - x_k}{h}\right), \qquad x \in [c_{k-1}, c_{k+1}]$$

(for $k = 0$ and $k = n$ we consider only half of the function A, i.e. restricted to the interval $[0, 1]$ and $[-1, 0]$, respectively). The membership functions $A_0, \ldots, A_n$ of fuzzy sets forming the fuzzy partition $\mathscr{A}$ are usually called *basic functions*.

Let us emphasize that the concept of fuzzy partition is crucial for the F-transform. Moreover, it is a typical concept used in many fuzzy techniques. Its main advantage for applications consists in the possibility that the neighboring fuzzy sets can overlap, which is not the case of the classical partition of a set.

4.1.2   Zero Degree F-transform

Once the fuzzy partition $A_0, \ldots, A_n \in \mathscr{A}$ is determined, we define a *direct F-transform* of a continuous function f as a vector $\mathbf{F}[f] = (F_0[f], \ldots, F_n[f])$, where each k-th *component* $F_k[f]$ is equal to

$$F_k[f] = \frac{\int_a^b f(x) A_k(x)\, dx}{\int_a^b A_k(x)\, dx}, \qquad k = 0, \ldots, n. \tag{13}$$

Clearly, the $F_k[f]$ component is a *weighted average* of the functional values $f(x)$ where weights are the membership degrees $A_k(x)$. The *inverse F-transform* of f with respect to $\mathbf{F}[f]$ is a continuous function[3] $\hat{f} : [a, b] \longrightarrow \mathbb{R}$ such that

$$\hat{f}(x) = \sum_{k=0}^{n} F_k[f] \cdot A_k(x), \qquad x \in [a, b].$$

Theorem 1. *The inverse F-transform $\hat{f}$ has the following properties:*

(a) *The sequence of inverse F-transforms $\{\hat{f}_n\}$ determined by a sequence of uniform fuzzy partitions based on uniformly distributed nodes with $h = (b-a)/n$ uniformly converges to f for $n \to \infty$.*

(b) *The F-transform is linear, i.e., if $f(x) = \alpha u(x) + \beta v(x)$ then $\hat{f}(x) = \alpha \hat{u}(x) + \beta \hat{v}(x)$ for all $x \in [a, b]$.*

All the details and full proofs can be found in [26, 27].

---

[3] By abuse of language, we call by direct as well as inverse F-transform both the procedure as well as its respective results $\mathbf{F}[f] = (F_0[f], \ldots, F_n[f])$ and $\hat{f}$.

4.1.3   Higher Degree F-transform

The F-transform introduced above is F^0-transform (i.e., zero-degree F-transform). Its components are real numbers. If we replace them by polynomials of arbitrary degree $m \geq 0$, we arrive at the higher degree F^m transform. This generalization has been in detail described in [27]. Let us remark that the F^1 transform enables to estimate also derivatives of the given function f as weighted average values over a vaguely specified area.

The direct F^1-*transform* of f with respect to $A_1, \ldots, A_{n-1}$ is a vector $F^1[f] = (F^1_1[f], \ldots, F^1_{n-1}[f])$ where the components $F^1_k[f]$, $k = 1, \ldots, n-1$ are linear functions

$$F^1_k[f](x) = \beta^0_k + \beta^1_k(x - c_k) \tag{14}$$

with the coefficients β^0_k, β^1_k given by

$$\beta^0_k = \frac{\int_{c_{k-1}}^{c_{k+1}} f(x) A_k(x) dx}{\int_{c_{k-1}}^{c_{k+1}} A_k(x) dx}, \tag{15}$$

$$\beta^1_k = \frac{\int_{x_{k-1}}^{x_{k+1}} f(x)(x - c_k) A_k(x) dx}{\int_{c_{k-1}}^{c_{k+1}} (x - c_k)^2 A_k(x) dx}. \tag{16}$$

Note that $\beta^0_k = F_k[f]$, i.e. the coefficients β^0_k are just the components of the F^0 transform given in (13). The F^1 transform has also the properties stated in Theorem 1 (see [27]).

We will also use the F^2 transform. Its components are the functions

$$F^2_k[f](x) = \beta^0_k + \beta^1_k(x - c_k) + \left(\beta^2_k(x - c_k)^2 - \frac{h^2}{6} \right)$$

(provided that the basic functions are triangles).

Theorem 2 ([11]). *If f is four-times continuously differentiable on $[a, b]$ then for each $k = 1, \ldots, n-1$,*

$$\beta^0_k = f(c_k) + O(h^2), \tag{17}$$

$$\beta^1_k = f'(c_k) + O(h^2). \tag{18}$$

$$\beta^2_k = \frac{f''(c_k)}{2} + O(h^2). \tag{19}$$

Thus, the F-transform components provide a weighted average of values of the function f in the area around the node c_k (17), and also a weighted average of slopes (27) of f and that of its second derivatives (19) in the same area.

Remark 1 (important). It should be noted that only the nodes $c_1, \ldots, c_{n-1}$ should be considered when dealing with the F-transform and the edge nodes c_0, c_n should be omitted. The reason is that the areas $[c_0, c_1]$ and $[c_{n-1}, c_n]$ are covered by halves of the basic functions A_0, A_n, respectively and so, the approximation of f in these areas is subject to too large error. Hence, we should consider the function $\hat{f}$ on the interval $[c_1, c_{n-1}]$ only.

4.2   Fuzzy Natural Logic

This is a special formal logical theory whose goal is to model the reasoning of people for which it is specific to use natural language. So far, it is not a unified theory but a bunch of the following theories:

(a) A formal theory of evaluative linguistic expressions explained in detail in [18] (see also [17,24]).
(b) A formal theory of fuzzy IF-THEN rules and approximate reasoning [16,22–24].
(c) A formal theory of intermediate and generalized fuzzy quantifiers [5,13,19] and elsewhere.

4.2.1   Evaluative Linguistic Expressions

The central role in all these theories is played by the theory of *evaluative linguistic expressions*. These are expressions with the general form

$$\langle \text{linguistic modifier} \rangle \langle \text{TE-adjective} \rangle \tag{20}$$

where $\langle \text{TE-adjective} \rangle^4$ is one of the adjectives "small, medium, big" (and possibly other specific adjectives, especially the so called gradable or evaluative ones), or "zero" as well as arbitrary symmetric fuzzy number. The $\langle \text{linguistic modifier} \rangle$ is a special expression that belongs to a wider linguistic phenomenon called *hedging* and that specifies more closely the topic of utterance. In our case, the linguistic modifier makes the meaning of the $\langle \text{TE-adjective} \rangle$ more specific. Quite often it is represented by an intensifying adverb such as "very, roughly, approximately, significantly", etc. The linguistic modifiers can have narrowing ("extremely, significantly, very, typically") and widening effect ("more or less, roughly, quite roughly, very roughly") on the meaning of the $\langle \text{TE-adjective} \rangle$.

If $\langle \text{linguistic hedge} \rangle$ is not present (expressions such as "weak, large", etc.) then we take it as presence of *empty linguistic hedge*. Thus, all the simple evaluative expressions have the same form (20). Since they characterize values on an ordered scale, we may consider also scales divided into two parts that are usually interpreted as *positive* and *negative*. Hence, the evaluative expressions may have also a sign, namely "positive" or "negative".

Simple evaluative expressions of the form (20) can also be combined using logical connectives (usually "and" and "or") to obtain *compound* ones. A limited usage of the particle "not" is also possible. Let us emphasize, however, that syntactic and semantic limitations of natural language *prevent the compound evaluative expressions to form a boolean algebra!*

We distinguish abstract evaluative expressions from more specific *evaluative predications*. The latter are expressions of natural language of the form 'X is $\mathscr{A}$' where $\mathscr{A}$ is an evaluative expression and X is a variable which stands for objects, for example "degrees of temperature, height, length, speed", etc. Examples are "temperature is high", "speed is extremely low", "quality is very high", etc.

---

[4] The "TE" is a short for "trichotomic evaluative".

In general, the variable X represents certain features of objects such as "size, volume, force, strength," etc. and so, its values are often real numbers (Fig. 1).

Important notion is that of *linguistic context*. In our theory it is an interval $w = [v_L, v_S] \cup [v_S, v_R]$ determined by a triple of (real) numbers $w = \langle v_L, v_S, v_R \rangle$ where v_L is the leftmost typically small value, v_S is typically medium value and v_R is the rightmost typically big value. For example, when speaking about temperature of water, we may set $v_L = 15\,°\mathrm{C}$, $v_S = 50\,°\mathrm{C}$ and $v_R = 100\,°\mathrm{C}$. In the sequel, we will consider a set of all linguistic contexts

$$W = \{w = \langle v_L, v_S, v_R \rangle \mid v_L, v_S, v_R \in \mathbb{R}, v_L < v_S < v_R\}. \qquad (21)$$

The element x belongs to a context $w \in W$ if $x \in [v_L, v_R]$. Then we write $x \in w$.

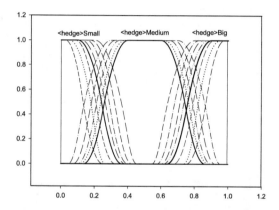

Fig. 1. Shapes of extensions of some evaluative expressions in the context $\langle 0, 0.5, 1 \rangle$. The hedges are {Extremely, Significantly, Very, empty hedge} for "small" and "big" and {More-or-Less, Roughly, Quite Roughly, Very Roughly} for "small", "medium", and "big".

The meaning of an evaluative linguistic expression $\mathscr{A}$ (as well as of a predication) is represented by its *intension*

$$\mathrm{Int}(X \text{ is } \mathscr{A}) : W \longrightarrow \mathscr{F}(\mathbb{R}) \qquad (22)$$

where $\mathscr{F}(\mathbb{R})$ is a set of all fuzzy sets on $\mathbb{R}$. For each context $w \in W$, the *extension* $\mathrm{Ext}_w(X \text{ is } \mathscr{A})$ is a specific fuzzy set on $\mathbb{R}$. Example of extensions of several evaluative linguistic expressions is in Fig. 7. Let us emphasize that their shapes have been established on the basis of logical analysis of the meaning of the corresponding evaluative expressions (for the details, see [18]).

4.2.2   Linguistic Description

The evaluative linguistic predications are basic constituents of fuzzy/linguistic IF-THEN rules that are special conditional clauses of natural language. A set of such rules is called a *linguistic description*, that is, a finite set of fuzzy/linguistic IF-THEN rules

$$\mathscr{R}_1 = \text{IF } X \text{ is } \mathscr{A}_1 \text{ THEN } Y \text{ is } \mathscr{B}_1,$$

$$\dots\dots\dots\dots\dots\dots\dots\dots\dots\dots \qquad (23)$$

$$\mathscr{R}_m = \text{IF } X \text{ is } \mathscr{A}_m \text{ THEN } Y \text{ is } \mathscr{B}_m$$

where "X is $\mathscr{A}_j$", "Y is $\mathscr{B}_j$", $j = 1, \dots, m$ are evaluative linguistic predications. The linguistic description can be understood as a specific kind of a (structured) text that can be used for description of various situations and processes.

4.2.3   Perception-Based Logical Deduction

Linguistic description taken as a special text requires a special inference method, namely the *Perception-based Logical Deduction* (PbLD). This inference method works with genuine evaluative linguistic expressions and it is based on formal properties of mathematical fuzzy logic (see [16, 17, 23]). The method is based on local properties of the linguistic description, so that we distinguish the rules as such but at the same time deal with them as vague expressions of natural language. The PbLD has nothing in common with the classical Mamdani's inference (cf., e.g., [10]) (Fig. 2).

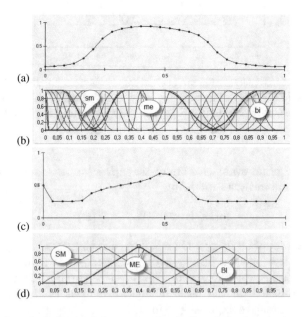

Fig. 2. (a) A function obtained from the simple linguistic description (24) using the PbLD method with smooth DEE defuzzification. (b) Extensions of the used evaluative expressions "small–medium–big" in the context $\langle 0, 0.4, 1 \rangle$. (c) A function obtained using Mamdani's-COG method from linguistic description of the form (24) interpreted as fuzzy relation constructed using triangular membership functions depicted in (d).

The PbLD requires a defuzzification method called DEE (Defuzzification of Evaluative Expressions). Its variant realized using the F-transform is called *smooth DEE* (see [23]).

To demonstrate PbLD, let us consider the following linguistic description:

$$\mathscr{R}_1 = \text{IF } X \text{ is } small \text{ THEN } Y \text{ is } small,$$
$$\mathscr{R}_2 = \text{IF } X \text{ is } medium \text{ THEN } Y \text{ is } big, \tag{24}$$
$$\mathscr{R}_3 = \text{IF } X \text{ is } big \text{ THEN } Y \text{ is } small.$$

This description characterizes linguistically a function that has small functional values on the left and right side of the graphs and big ones in the middle. The result using PbLD method is depicted in part (a) of Fig. 3. In part (b) are extensions of the used evaluative expressions in the context $\langle 0, 0.4, 1 \rangle$.

To see that PbLD method cannot be replaced by the Mamdani's method that is often used in various kinds of applications, we depicted in Fig. 3(c) and (d) the result obtained from (24) using it the basis of triangular fuzzy sets often (incorrectly) considered in literature as extensions of evaluative expressions. The reason why Mamdani's method does not work in this case is the fact that it provides very good approximation of a function, but it is not logical inference suitable for manipulation with linguistic expressions.

4.2.4   Learning of Linguistic Description

In applications of the methods describe above, very important is the possibility to use a learning procedure developed in FNL (cf. [24]). If the data and a context w are given, we can learn linguistic description of the form (23) that linguistically characterizes the data. Using the PbLD inference method, we can obtain various kinds of specific information.

The learning procedure is realized by implementing a function of *local perception*

$$LPerc(x, w) = \mathscr{A} \tag{25}$$

where $w \in W$ is a given context and $x \in w$ is a given value. The evaluative linguistic expression $\mathscr{A}$ characterizes the value x in the given context w. For example, the value $x = 0.15$ in a context $w = \langle 0, 4, 10 \rangle$ is evaluated by the evaluative expression "very small".

Using this simple idea, we can transform data in the form

$$\begin{bmatrix} u_{11} & u_{12} & \dots & u_{1c} & v_1 \\ u_{21} & u_{22} & \dots & u_{2c} & v_2 \\ \vdots & \vdots & \vdots & \vdots & \vdots \\ u_{m1} & u_{m2} & \dots & u_{mc} & v_m \end{bmatrix}$$

into a linguistic description consisting of m fuzzy/linguistic IF-THEN rules of the form IF X_1 is $\mathscr{A}_1$ AND $\cdots$ AND X_c is $\mathscr{A}_c$ THEN Y is $\mathscr{B}$.

The outcome of this procedure is twofold: first, it provides us with the succinct information understandable to people about the content of the data. Second, we can obtain answers to many "what if" questions and, on the basis of that, make proper decisions.

5   Analysis and Forecasting of Time Series Using Fuzzy Techniques

As discussed above, the fuzzy set theory (and fuzzy logic) is the mathematical model of vaguely determined actualized groupings of objects. No occurrence of any event is considered. In this section we will describe how fuzzy techniques can be applied when dealing with time series. This requires a slightly different view of time series. We will show that these techniques are able to compete with the probabilistic ones in forecasting not only future values of time series, but also to fit well the idea of their trend or trend cycle. But even more, the fuzzy models have the potential to bring new hints for analysis of time series that are not possible in the probabilistic approach. We have in mind especially applications of the model of the semantics of natural language using which we can obtain additional information about the behavior of time series which is, moreover, well understandable to people.

5.1   Decomposition of Time Series

In the probabilistic model, a time series is a sequence of random variables $\{X(t), t \in \mathbb{T}\}$ without considering their structure. A more apt model is the following: the time series is decomposed into several components

$$X(t) = Tr(t) + C(t) + S(t) + R(t), \qquad t \in \mathbb{T}, \tag{26}$$

where Tr is the *trend*, C is a *cyclic* component, S is a *seasonal* component that is a mixture of periodic functions and R is a random *noise*, i.e., a sequence of independent random variables $R(t)$ such that for each $t \in \mathbb{T}$, the $R(t)$ has zero mean and finite variance.

The seasonal component S in (26) is assumed to be a sum of periodic functions

$$S(t) = \sum_{j=1}^{r} P_j \sin(\lambda_j t + \varphi_j), \qquad t \in \mathbb{T}, \tag{27}$$

for some finite r where λ_j are frequencies, φ_j are phase shifts and P_j are amplitudes[5].

In the practice, it is often difficult to distinguish trend and cycle. Therefore, these two components are often joined into one component called *trend-cycle*

$$TC(t) = Tr(t) + C(t), \qquad t \in \mathbb{T}.$$

---

[5] Because $\cos x = \sin(x + \pi/2)$, it is sufficient to consider only sin.

Hence, we will replace the decomposition (26) by the simpler one

$$X(t) = TC(t) + S(t) + R(t), \qquad t \in \mathbb{T}.$$

The difference between trend and trend-cycle was informally summarized by the following OECD definitions.

The *trend* is a component of a time series that represents variations of low frequency in a time series, the high and medium frequency fluctuations having been filtered out.

The *trend-cycle* is a component that represents variations of low and medium frequency in a time series, the high frequency fluctuations having been filtered out. This component can be viewed as those variations with a period longer than a chosen threshold (usually 1 year is considered as the minimum length of the business cycle). Form the mathematical point of view, we assume that both trend as well as trend-cycle are (continuous) functions with *small modulus of continuity*[6].

Note that the decomposition model keeps the idea that the time series is a sequence of random variables but randomness is present only at the noise which is an unpredictable random component with specific properties. The rest are non-random components with clear interpretation. We argue, that fuzzy techniques provide more powerful means for extracting these components and, moreover, they make it possible to extract also additional information about time series. This information is usually vaguely specified, provided often in natural language and, therefore, it that cannot be obtained using the probabilistic methods.

The following theorem assures us that we can find a fuzzy partition enabling us to estimate either the trend Tr or the trend cycle TC with high fidelity.

Theorem 3. *Let $\{X(t) \mid t \in \mathbb{T}\}$ be a continuous realization of the stochastic process*

$$X(t) = Tr(t) + \sum_{j=1}^{r} P_j \sin(\lambda_j t + \varphi_j) + R(t), \qquad t \in \mathbb{T}$$

where $\mathbb{T} = [0, b]$, $Tr : \mathbb{T} \longrightarrow \mathbb{R}$ is a function with small modulus of continuity, $\lambda_1 \leq \ldots \leq \lambda_r$ are frequencies and R is the noise from (26).

Let us construct an h-uniform fuzzy partition $\mathscr{P}$ over nodes $c_0, \ldots, c_n$ with $h = d\,\bar{T}_1$, where $\bar{T}_1 = \frac{2\pi}{\lambda_1}$ and $d \geq 1$ is a real number. Let us compute the direct F-transform $F[X]$. Then there exists a number $D(d)$ such that $D(d) = 0$ for $d \to \infty$ and

$$|\hat{X}(t) - Tr(t)| \leq D(d), \qquad t \in [c_1, c_{n-1}]$$

where $\hat{X}$ is the corresponding inverse F-transform of X.

This theorem holds both for triangular as well as raised cosine fuzzy partition. The precise expressions for D in both cases and the proof of this theorem can be found in [14, 25]. It can also be proved that $D(d)$ is minimal if $d \in \mathbb{N}$.

---

[6] Let $f : [a, b] \longrightarrow \mathbb{R}$ be a continuous function. Then $\omega(h, f) = \max_{\substack{|x-y|<h \\ x,y \in [a,b]}} |f(x) - f(y)|$ is the modulus of continuity of f.

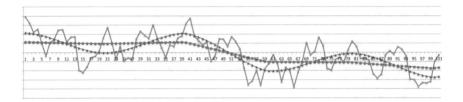

Fig. 3. Real trend and trend-cycle of the artificial time series.

Corollary 1. *Let* $\{X(t) \mid t \in \mathbb{T}\}$ *be a continuous realization of the stochastic process (26), Tr its trend and TC its trend-cycle. Then there exist numbers $D_1(d), D_2(d)$ such that $D_k(d) = 0$ for $d \to \infty$, $k = 1, 2$ and*

(a) $|\hat{X}(t) - Tr(t)| \le D_1(d)$,
(b) $|\hat{X}(t) - TC(t)| \le D_2(d)$

for corresponding inverse F-transform $\hat{X}$ of X and all $t \in [c_1, c_{n-1}]$.

It follows from this corollary that we can form the h-uniform fuzzy partition $\mathscr{P}$ with h corresponding to the largest periodicity of a periodic constituent occurring either in the cyclic or the seasonal component of $S(t)$. Then all the subcomponents with shorter periodicities (i.e., higher frequencies) are almost "wiped down", and also, the noise is significantly reduced. In other words, either the components C, S and R in (26) are almost completely removed and we obtain estimation of the trend

$$Tr(t) \approx \hat{X}(t), \qquad (28)$$

or the components S and R are removed and we obtain estimation of the trend-cycle

$$TC(t) \approx \hat{X}(t), \qquad (29)$$

$t \in [c_1, c_{n-1}]$.

To demonstrate the above outlined theory for estimation of trend Tr and trend-cycle TC of a time series, we constructed an artificial time series using the following formula:

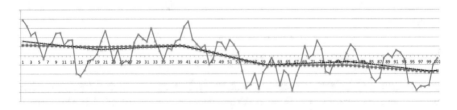

Fig. 4. Real trend (dotted line) of the artificial time series and its estimation (crossed line) using the F-transform with $h = 40$.

$$X(t) = \underbrace{Tr(t) + \sin(0.157t + 1.5)}_{TC} + \sin(0.283t + 0.34) + \sin(0.628t + 1.12)$$

$$+ \sin(1.57t + 0.79) + R(t) \qquad (30)$$

where R is a random noise. The frequencies ω in this time series correspond to the following periodicities T, respectively: 40, 22.2, 10, 4. The trend $Tr(t)$ is determined explicitly by data and has no predefined shape (Fig. 4).

Using Periodogram, we found in the artificial time series (30) the following periodicities T: 36.9, 22.7, 16.6, 14.2, 9.9, 4. Note that Periodogram found two more not existing periodicities and also, that estimation of the periodicity $T = 40$ is not too precise (Fig. 5).

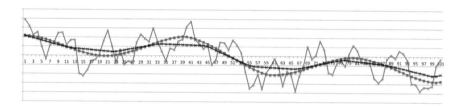

Fig. 5. Real trend-cycle (dotted line) of the artificial time series and its estimation (crossed line) using the F-transform with $h = 22$.

Finally, let us remark that the method is very robust towards missing values and outliers, i.e., there is no visible change of the trend or trend-cycle if omit some values of the time series.

5.2   Forecasting Future Course of Time Series

The linguistic description and PbLD inference method mentioned in Sect. 4.2.2 can be applied to forecasting of the trend Tr or trend-cycle TC. The method was described in detail in [24], and so, we will only briefly review its main ideas. Let $\bar{\mathbb{T}} \subset \mathbb{T}$. Then by $X|\bar{\mathbb{T}}'$ we denote the restriction of X to $\bar{\mathbb{T}}$. For the consistency of notation, we will write the time series (26) as $X|\mathbb{T}$.

Let $\mathbb{T}' \supset \mathbb{T}$ be a new time domain. Our task is to extrapolate values of X to $X|(\mathbb{T}' \setminus \mathbb{T})$ on the basis of the known values of $X|\mathbb{T}$. The method for finding the former is called *forecasting*. As noted above, there are many forecasting methods mostly formulated using probability theory (cf. [3,7,9]). In this section, we present methods based on fuzzy techniques.

Recall that trend or trend-cycle are obtained using the F-transform on the basis of an h-uniform fuzzy partition $\mathscr{P}$. The result of the direct F-transform is a vector of F-transform components

$$\mathbf{F}[X] = (F_1[X], \ldots, F_{n-1}[X]), \qquad (31)$$

where each component $F_i[X]$ represents a weighted average of values of $X(t)$ in the area of width $2h$. The components (31) can be used as data for learning of a linguistic description. Then, using it and the PbLD method, we can forecast future F-transform components

$$F_n[X], \dots, F_{n+l}[X] \tag{32}$$

and from them, we can compute estimation of the future development either of trend or trend cycle using the inverse F-transform $\hat{X}$.

The learned linguistic description consists of fuzzy/linguistic rules of the form, for example,

IF $\Delta^2 X_{i-1}$ is $\mathscr{A}_{\Delta^2 i-1}$ AND ΔX_{i-1} is $\mathscr{A}_{\Delta i-1}$ AND X_i is $\mathscr{A}_i$ THEN X_{i+1} is $\mathscr{B}_{i+1}$. (33)

where

$$\Delta F_i[X] = F_i[X] - F_{i-1}[X], \qquad\qquad i = 1, \dots, n-1 \tag{34}$$
$$\Delta^2 F_i[X] = \Delta F_i[X] - \Delta F_{i-1}[X], \qquad\qquad i = 2, \dots, n-1 \tag{35}$$

are the first and second differences, respectively. Let us remark that in practice, all kinds of combinations of the F-transform components and their first and second differences can occur both in the antecedent as well as in the consequent of (33). Example of such description is the following:

Rule	$F_i[X]$	$\Delta F_i[X]$	$\Rightarrow F_{i+1}[X]$
1	ex bi	ra me	qr bi
2	ro bi	-ml me	qr bi
3	ro bi	-ex sm	vr sm
4	ze	-ex bi	vr sm
5	si sm	si sm	ra me
⋮			

(the used shorts: ze-zero, sm-small, me-medium, bi-big, ex-extremely, ro-roughly, qr-quite roughly, vr-very roughly, ra-rather, si-significantly, ml-more or less).

Note that forecasting of the future values, the learned linguistic description provides us also with information in linguistic form (i.e., understandable to people) explaining how the forecast was obtained, i.e., what are the inner characteristics of the time series that led to the forecast. The differences (34) and (35) characterize dynamics of the time series as well as logical dependencies of the trend-cycle changes (hidden cycle influences).

5.3   Mining Knowledge on Time Series

5.3.1   Linguistic Evaluation of the Local Trend

If a certain time interval is given, it may be interesting to learn what trend (tendency) of the time series can be recognized in it. Surprisingly, recognition of trend in by no means is a trivial task even when watching the graph. Recall that by Theorem 23, the F^1-transform provides estimation of the average slope (tangent). Therefore, it is a convenient tool for estimation of the course of trend of the given time series. Such estimation can be, moreover, expressed in natural language. For example, we can say *"fairly large decrease (huge increase) of trend"*, *"the trend is stagnating (negligibly increasing)"*, etc. These expressions characterize trend (tendency) of the time series in an area specified by the user. It is quite important achievement of the fuzzy techniques that it provides algorithms using which it is possible to generate automatically this kinds of linguistic evaluations. The method is based on the theoretical results in fuzzy natural logic and was described in more detail in [20,21]. Its idea is outlined below.

First, we must specify, what does it mean "extreme increase (decrease)". In practice, it can be determined as the largest acceptable difference of time series values with respect to a given (basic) time interval (for example 12 months, 31 days) that is, a minimal and maximal tangent. In practice, we set only the largest tangent v_R while the smallest one is usually $v_L = 0$. The typically medium value v_S is determined analogously as v_R. The result is the context $w_{tg} = \langle v_L, v_S, v_R \rangle$. Furthermore, we must specify the time interval $I \subset \mathbb{T}$ interesting for the inspection. The next step is to compute a basic function A with the support I (cf. Subsect. 4.1.1) and compute the coefficient β^1 using formula (16).

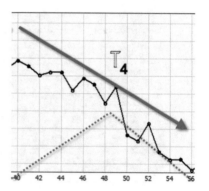

Fig. 6. The principle of linguistic of evaluation of direction of trend: *clear decrease*. The necessary parameter is the context w_{tg} specifying the lowest, typically medium and the largest value of the tangent. The triangle above the x-axis is the basic function of the F^1-transform.

Finally, we generate a linguistic evaluation of the trend of the time series X in the area characterized by A with respect to the context w_{tg}. The required evaluative expression $\mathscr{A}$ is obtained using the function of local perception

$$\mathscr{A} = LPerc(\beta^1, w_{tg}). \tag{36}$$

Demonstration of the principle of evaluation is in Fig. 6[7].

5.3.2   Mining More Kinds of Knowledge

Fuzzy techniques suggest more methods for mining knowledge from time series. One of them is finding *perceptionally important points* (PIP). According to [6], these are points where the time series essentially changes its course. Because

---

[7] The results were obtained using the experimental software LFL Forecaster (see http://irafm.osu.cz/en/c110_lfl-forecaster/) which implements the described method. Its author is Viktor Pavliska.

of the complicated character following from the presence of various frequencies and noise, we cannot expect that this is just one isolated time point but better a certain area that cannot be precisely determined. Therefore, a very suitable method is based on the higher-degree F-transform because it makes it possible to estimate the first and second derivatives of a function with complicated course in a vaguely specified area. Namely, this can be done by looking for small values of the β^1 coefficient (27).

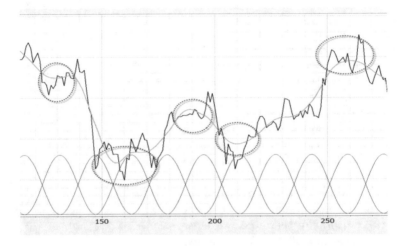

Fig. 7. Time series with marked perceptionally important points and F^1-approximation of its course. Along the x-axis is also depicted the fuzzy partition.

Demonstration of the result of searching PIP in a part of the Monthly Closing of Dow-Jones index is in Fig. 7[8]. The points are found in areas covered by the corresponding basic functions of the fuzzy partition. They correspond to values of β^1 close to zero[9]. To find the points we must shift the fuzzy partition to localize β^1 with minimal values. Note that it can be equal to zero only in case of ideal line parallel with x-axis.

Other interesting possibility is to find time intervals in which the trend of the time series X exhibits monotonous behavior which is also characterized linguistically. This means that we must decompose the time domain $\mathbb{T}$ into a set of time intervals $\mathbb{T}_i \subseteq \mathbb{T}$, $i = 1, \ldots, s$, with monotonous trend of X (increasing, decreasing, stagnating). Each interval $\mathbb{T}_i$ is a union of one or more adjacent intervals $\bar{\mathbb{T}}_j$. As a final result, direction of the trend of $X|\mathbb{T}_i$ is linguistically evaluated similarly as is outlined above. The detailed algorithm can be found in [20].

An area that is becoming still more attractive is automatic summarization of knowledge about time series. This task is addressed by several authors (see,

---

[8] The points were obtained using the experimental software FT-Studio whose author is Radek Valášek.

[9] The actual values range between 0.07 and 1.4.

e.g., [4,8]). The fuzzy natural logic suggests sophisticated formal theory of *intermediate quantifiers*. The summarized information may address either one time series or a set of time series. The theory includes a formalism on the basis of which we can develop a model of the meaning of linguistic statements containing quantified information, as is usual in natural language, but also human-like syllogistic reasoning that is based on the formal model of *generalized Aristotle's syllogisms*. For more details, see [12,20].

6 Conclusion

In this paper, we discussed the difference between vagueness and uncertainty phenomena and their role in the fuzzy and probabilistic techniques applied to time series analysis. While probabilistic techniques assume that the time series is a stochastic process consisting of random variables, fuzzy techniques stem from the decomposition of the time series into deterministic components, assuming that only noise is random. We argue that, because both techniques have at disposal one realization of the time series only, statistically relevant processing is possible only under quite strong assumptions on the origins of the time series. While such assumptions are in the case of noise natural, for the whole time series they are too strong.

As fuzzy techniques are based on the model of the vagueness phenomenon, they are robust which means that they are little sensitive to changes of the data. Moreover, they enable to obtain information that cannot be obtained using probabilistic techniques. This concerns especially the area of mining knowledge from time series. On one hand, this knowledge can be obtained in an easier and straightforward way than classical methods (e.g., finding perceptionally important points). On the other hand, the knowledge can be often obtained directly in expressions or even sentences of natural language, which is the form well understandable to people.

References

1. Anděl, J.: Statistical Analysis of Time Series. SNTL, Praha (1976). (in Czech)
2. Black, M.: Vagueness: an exercise in logical analysis. Philos. Sci. **4**, 427–455 (1937). reprinted in Int. J. Gen. Syst. **17**, 107–128 (1990)
3. Bovas, A., Ledolter, J.: Statistical Methods for Forecasting. Wiley, New York (2003)
4. Castillo-Ortega, R., Marín, N., Sánchez, D.: A fuzzy approach to the linguistic summarization of time series. Multiple-Valued Logic Soft Comput. **17**(2–3), 157–182 (2011)
5. Dvořák, A., Holčapek, M.: L-fuzzy quantifiers of the type ⟨1⟩ determined by measures. Fuzzy Sets Syst. **160**, 3425–3452 (2009)
6. Fu, T.-C.: A review on time series data mining. Eng. Appl. Artif. Intell. **24**, 164–181 (2011)
7. Hamilton, J.: Time Series Analysis. Princeton University Press, Princeton (1994)
8. Kacprzyk, J., Wilbik, A., Zadrożny, S.: Linguistic summarization of time series using a fuzzy quantifier driven aggregation. Fuzzy Sets Syst. **159**, 1485–1499 (2008)

9. Kedem, B., Fokianos, K.: Regression Models for Time Series Analysis. Wiley, New York (2002)
10. Klir, G., Bo, Y.: Fuzzy Set Theory: Foundations and Applications. Prentice Hall, Upper Saddle River (1995)
11. Kreinovich, V., Perfilieva, I.: Fuzzy transforms of higher order approximate derivatives: a theorem. Fuzzy Sets Syst. **180**, 55–68 (2011)
12. Murinová, P., Novák, V.: A formal theory of generalized intermediate syllogisms. Fuzzy Sets Syst. **186**, 47–80 (2012)
13. Murinová, P., Novák, V.: The structure of generalized intermediate syllogisms. Fuzzy Sets Syst. **247**, 18–37 (2014)
14. Nguyen, L., Novák, V.: Filtering out high frequencies in time series using F-transform with respect to raised cosine generalized uniform fuzzy partition. In: Proceedings of International Conference FUZZ-IEEE 2015. IEEE Computer Society, CPS, Istanbul (2015)
15. Nguyen, L., Novák, V.: Trend-cycle forecasting based on new fuzzy techniques. In: Proceedings of the International Conference FUZZ-IEEE 2017, Naples, Italy (2017)
16. Novák, V.: Perception-based logical deduction. In: Reusch, B. (ed.) Computational Intelligence, Theory and Applications, pp. 237–250. Springer, Berlin (2005)
17. Novák, V.: Mathematical fuzzy logic in modeling of natural language semantics. In: Wang, P., Ruan, D., Kerre, E. (eds.) Fuzzy Logic - A Spectrum of Theoretical & Practical Issues, pp. 145–182. Elsevier, Berlin (2007)
18. Novák, V.: A comprehensive theory of trichotomous evaluative linguistic expressions. Fuzzy Sets Syst. **159**(22), 2939–2969 (2008)
19. Novák, V.: A formal theory of intermediate quantifiers. Fuzzy Sets Syst. **159**(10), 1229–1246 (2008)
20. Novák, V.: Linguistic characterization of time series. Fuzzy Sets Syst. **285**, 52–72 (2016)
21. Novák, V.: Mining information from time series in the form of sentences of natural language. Int. J. Approx. Reason. **78**, 192–209 (2016)
22. Novák, V., Lehmke, S.: Logical structure of fuzzy IF-THEN rules. Fuzzy Sets Syst. **157**, 2003–2029 (2006)
23. Novák, V., Perfilieva, I.: On the semantics of perception-based fuzzy logic deduction. Int. J. Intell. Syst. **19**, 1007–1031 (2004)
24. Novák, V., Perfilieva, I., Dvořák, A.: Insight into Fuzzy Modeling. Wiley, Hoboken (2016)
25. Novák, V., Perfilieva, I., Holčapek, M., Kreinovich, V.: Filtering out high frequencies in time series using F-transform. Inf. Sci. **274**, 192–209 (2014)
26. Perfilieva, I.: Fuzzy transforms: theory and applications. Fuzzy Sets Syst. **157**, 993–1023 (2006)
27. Perfilieva, I., Daňková, M., Bede, B.: Towards a higher degree F-transform. Fuzzy Sets Syst. **180**, 3–19 (2011)
28. Russell, B.: Vagueness. Aust. J. Phi. **1**, 84–92 (1923)
29. Vopěnka, P.: Mathematics in the Alternative Set Theory. Teubner, Leipzig (1979)
30. Zadeh, L.A.: Outline of a new approach to the analysis of complex systems and decision processes. IEEE Trans. Syst. Man Cybern. **SMC–3**, 28–44 (1973)

Is It Legitimate Statistics or Is It Sexism: Why Discrimination Is Not Rational

Martha Osegueda Escobar[1], Vladik Kreinovich[1(✉)], and Thach N. Nguyen[2]

[1] Department of Computer Science, University of Texas at El Paso,
500 W. University, El Paso, TX 79968, USA
mcoseguedaescobar@miners.utep.edu, vladik@utep.edu
[2] Banking University of Ho Chi Minh City, 56 Hoang Dieu 2, Quan Thu Duc,
Thu Duc, Ho Chi Minh City, Vietnam
Thachnn@buh.edu.vn

Abstract. While in the ideal world, everyone should have the same chance to succeed in a given profession, in reality, often the probability of success is different for people of different gender and/or ethnicity. For example, in the US, the probability of a female undergraduate student in computer science to get a PhD is lower than a similar probability for a male student. At first glance, it may seem that in such a situation, if we try to maximize our gain and we have a limited amount of resources, it is reasonable to concentrate on students with the higher probability of success – i.e., on males, and only moral considerations prevent us from pursuing this seemingly economically optimal discriminatory strategy. In this paper, we show that this first impression is wrong: the discriminatory strategy is not only morally wrong, it is also not optimal – and the morally preferable inclusive strategy is actually also economically better.

1 Is It Legitimate Statistics or Is It Sexism?

There are statistical differences. For different reasons, people of different gender and/or different ethnicity have different success rates in different disciplines.

For example, while there are many highly successful female computer scientists, in the US, the percentage of undergraduate female computer science students who go to eventually defend a PhD is lower than for similar male students – while in some other disciplines (and in some other countries), the difference is reverse.

Similarly with ethnicity: for example, the corresponding percentage is higher among Asian-American students than among white students.

An important and difficult challenge. The very fact that the percentage of successful females strongly varies from country to country – even for countries with similar ethnicity – shows that the reasons for the statistical differences are not biological. We need to learn from the success of other countries and other disciplines to make sure that everyone has an equal chance to succeed. This idea

L. H. Anh et al. (eds.), *Econometrics for Financial Applications*, Studies in Computational Intelligence 760, https://doi.org/10.1007/978-3-319-73150-6_18

may sound straightforward, but in reality, how to do it is an important and difficult challenge, way beyond the scope of this paper.

The problem that we deal in this paper. In this paper, we deal with a more mundane problem: what is the best strategy in the current situation?

A seemingly rational strategy. The situation is very simple and straightforward. We want to graduate a certain number of PhDs. We have limited resources. So, at first glance, it seems that a rational strategy is to concentrate on undergraduate students for whom the probability of success is higher – i.e., on male students – and ignore the female students (since for them, the probability of success is lower).

This argument has nothing to do with prejudice against females: if in a few years, the situation reverses, and the probability of a female student succeeding becomes higher than for male ones, a person following this rational will start concentrating on promising female students only and ignore male students completely.

A similar argument can be applied to hiring: female applicants tend to have a higher probability of retiring early because of their family obligations, so should we stop hiring them? Should we just ignore resumes coming from female applicants and only hire males?

The resulting discriminatory strategy may sound rational, but is it moral? The usual argument against the above hypothetical strategy is that while it may sound rational, it goes against the basic moral principles. Everyone should get a chance to succeed, we should judge every person based on his/her individuality, not based on their gender, race, ethnicity, etc.

This is an explanation many people give; see, e.g., [1].

What we show in this paper. In this paper, a detailed analysis reveals that discriminatory strategies are not just immoral, they are actually *not rational*.

2   Why Discrimination Is Not Rational

Let us start analyzing the problem. Without losing generality, let us consider the problem of hiring. The same argument can be used for selecting the most promising students to "groom" them for graduate school, etc.

For simplicity, let us assume that the candidates belong to two possible groups:

- a group for which the probability of success p is higher; for simplicity, we will call this group *majority* (we say "for simplicity", since in computer science, males are actually a majority, but, e.g., Asian-Americans are not), and
- a group for which the probability of success p' is somewhat lower: $p' < p$; for simplicity, we will call this group *minority*.

The probabilities p and p' can be estimated as the proportion of those who succeeded in each group.

We will compare two strategies. Let us consider two possible strategies:

- a *discriminatory* strategy, in which we ignore all minority applicants and only consider more-probable-to-succeed majority applicants, and
- an *inclusive* strategy, in which we consider all applicants.

We plan to analyze these two strategies from a purely rational, purely economic viewpoint: which one brings more benefit to the company.

From this viewpoint, each of these two strategies has its gains and its losses:

- in the discriminatory strategy, we save some money on analyzing minority applicants, but we miss potential gains that we could have if we hired good female employees;
- in the inclusive strategy, we lose some money on checking the applications of all minority applicants, but we may gain by hiring good female employees.

The question is: if we combine these gains and losses, which of the two strategies will turn out to be the most beneficial?

Let us prepare to evaluate gains and losses. The cost of analyzing an application is approximately the same for all candidates. Let us denote this cost by a.

There is also a cost of training a person and supporting this person through the probation period. Let us denote this cost by t.

What can be drastically different is the gain. Like many other things, potential gains are distributed according the *Zipf law* (see, e.g., [2]): if we denote the lifetime gain from hiring the best possible candidate by G, then:

- the gain from hiring the second best candidate is $\dfrac{G}{2}$,
- the gain from hiring the third best candidate is $\dfrac{G}{3}$,
- and, in general, the gain from hiring the i-th best candidate is $\dfrac{G}{i}$.

Case of inclusive strategy. Let us first consider the profit in the case of the inclusive strategy. Let us first count expenses.

The easiest to evaluate are the expenses related to reviewing applications. In the inclusive strategy, we review all $N + N'$ applications. Reviewing each application requires amount a, so overall, we spend the amount $a \cdot (N + N')$ on these reviews.

The next expense item is training. Let us assume that we have k positions that we want to be eventually filled – e.g., in the case of a university, we have k tenured positions.

Since some of the people we hire will not succeed after a probation period, we hire more people to make sure that at the end, we have k successful folks. In general, from N majority candidates, $p \cdot N$ will succeed if hired. From N' minority candidates, $p' \cdot N'$ will succeed if hired. Overall, if we could hire all of them, we would end up with $p \cdot N + p' \cdot N'$ successful folks. Out of these folks, we select k best. Out of successful folks, the probability of being among the k best is the same, whether it is a successful majority or a successful minority. Thus, out of k best, we will have proportionally many majority and minority folks:

- $k_0 \overset{\text{def}}{=} k \cdot \dfrac{p \cdot N}{p \cdot N + p' \cdot N'}$ majority folks and

- $k_0' \overset{\text{def}}{=} k \cdot \dfrac{p' \cdot N'}{p \cdot N + p' \cdot N'}$ minority folks.

For a majority applicant, the probability of success is p. Thus, to make sure that at the end, we have $k \cdot \dfrac{p \cdot N}{p \cdot N + p' \cdot N'}$ majority employees, we need to hire

$$n_0 \overset{\text{def}}{=} \frac{k_0}{p} = k \cdot \frac{N}{p \cdot N + p' \cdot N'}$$

majority applicants.

For a minority applicant, the probability of success is p'. Thus, to make sure that at the end, we have $k \cdot \dfrac{p' \cdot N'}{p \cdot N + p' \cdot N'}$ minority employees, we need to hire

$$n_0' \overset{\text{def}}{=} \frac{k_0'}{p'} = k \cdot \frac{N'}{p \cdot N + p' \cdot N'}$$

majority applicants.

Overall, we need to hire $n_0 + n_0'$ applicants. Training one hire costs the amount t, so the overall expenses on training are equal to

$$t \cdot (n + n_0) = t \cdot \frac{k \cdot (N + N')}{p \cdot N + p' \cdot N'}.$$

Let us now count the gains. Since we considered all the applicants, we are sure that the k folks that remain after the probation period are the k best ones. The best of these folks brings the gain G, the second best brings the gain $\dfrac{G}{2}$, the third best the gain $\dfrac{G}{3}$, etc., until we reach the k-th person who contributes the gain $\dfrac{G}{k}$. The overall gain from all these folks is

$$G + \frac{G}{2} + \frac{G}{3} + \ldots + \frac{G}{k} = G \cdot \left(1 + \frac{1}{2} + \frac{1}{3} + \ldots + \frac{1}{k}\right).$$

The sum in parentheses is an integral sum for the interval

$$\int_0^k \frac{1}{x}\, dx = \ln(x)|_1^k = \ln(k),$$

so the above sum is approximately equal to this integral. Thus, the gain is equal to $G \cdot \ln(k)$.

Subtracting the expenses from this gain, we conclude that if we use the inclusive hiring strategy, our overall profit is equal to

$$G \cdot \ln(k) - a \cdot (N + N') - t \cdot \frac{k \cdot (N + N')}{p \cdot N + p' \cdot N'}. \tag{1}$$

Case of discriminatory strategy. Let us now consider the case of the discriminatory strategy.

In this case, we only screen N majority candidates, so the amount we spend on screening is $a \cdot N$ (smaller amount that for the inclusive strategy).

We want to end up with k candidates. We only hire majority folks, for whom the probability of success is p. Thus, to end with k employees after the probation period, we need to hire $\frac{k}{p}$ folks. The cost of training all these hires is equal to $t \cdot \frac{k}{p}$.

What is the gain of all these hires? Out of all $p \cdot N + p' \cdot N'$ potentially successful folks, we hired only the majority persons, i.e., our hiring pool consisted of $p \cdot N$ folks out of $p \cdot N + p' \cdot N'$. The probability p_b that the best of the $p \cdot N + p' \cdot N'$ folks is a majority is thus equal to the proportion of successful majority folks among all successful folks: $p_b = \frac{p \cdot N}{p \cdot N + p \cdot N'}$. So, in the formula for the expected gain, the contribution of the best person is not G (as in the case of the inclusive strategy), but rather the product $p_b \cdot G = G \cdot \frac{p \cdot N}{p \cdot N + p \cdot N'}$.

Similarly, the probability that the second best person is in the majority (and thus, among the hires) is also equal to p_b. Thus, the contribution of this second best person into the formula for the expected gain is not $\frac{G}{2}$, but $p_b \cdot \frac{G}{2}$. Same with the 3rd best person, etc.

We need to be careful now as we count further. We end up with k employees, but they are not k best folks, they are k best out of *majority* folks. Overall, there are $p \cdot N$ potentially successful majority folks out of the total amount of $p \cdot N + p' \cdot N'$ successful folks. Thus, when we select k top majority top, there are overall

$$K \stackrel{\text{def}}{=} k \cdot \frac{p \cdot N + p' \cdot N'}{p \cdot N}$$

folks of similar quality. Here, $K = 1 + \frac{p' \cdot N'}{p \cdot N}$.

So, in counting down in quality, we have to go down to the K-th person. As a result, the overall gain for this strategy is equal to

$$p_b \cdot G + p_b \cdot \frac{G}{2} + p_b \cdot \frac{G}{3} + \ldots + p_b \cdot \frac{G}{K} = p_b \cdot G \cdot \left(1 + \frac{1}{2} + \frac{1}{3} + \ldots + \frac{1}{K}\right).$$

Here, as before,

$$1 + \frac{1}{2} + \frac{1}{3} + \ldots + \frac{1}{K} \approx \ln(K),$$

where

$$\ln(K) = \ln\left(k \cdot \left(1 + \frac{p' \cdot N'}{p \cdot N}\right)\right) = \ln(k) + \ln\left(1 + \frac{p' \cdot N'}{p \cdot N}\right).$$

Thus, for the discriminatory strategy, the gain is equal to

$$\frac{p \cdot N}{p \cdot N + p \cdot N'} \cdot G \cdot \left(\ln(k) + \ln\left(1 + \frac{p' \cdot N'}{p \cdot N}\right)\right).$$

By subtracting the expenses from this gain, we conclude that the overall profit of using this strategy is equal to

$$\frac{p \cdot N}{p \cdot N + p \cdot N'} \cdot G \cdot \left(\ln(k) + \ln\left(1 + \frac{p' \cdot N'}{p \cdot N}\right)\right) - a \cdot N - t \cdot \frac{k}{p}. \tag{2}$$

So which of the two strategies is the most profitable? In several realistic numerical examples that we tried, the profit (1) from the inclusive strategy exceeds the profit (2) from the discriminatory strategy.

To have a general result, let us consider the case when what we called "minority" is really a minority, i.e., when the ratio $m \stackrel{\text{def}}{=} \dfrac{N'}{N}$ is small (so that we can ignore terms which are quadratic or of higher order in terms of m). Let us denote $r \stackrel{\text{def}}{=} \dfrac{p'}{p} < 1$.

In this case, dividing both numerator and denominator of the training-expenses term in the formula (1), we get

$$t \cdot \frac{k \cdot (N + N')}{p \cdot N + p' \cdot N'} = t \cdot \frac{k(1 + m)}{p + p' \cdot m}.$$

Here, $p + p' \cdot m = p \cdot (1 + r \cdot m)$, and

$$\frac{1}{1 + r \cdot m} = 1 - r \cdot m + o(m).$$

Thus, the training-expenses term takes the form

$$t \cdot \frac{k}{p} + t \cdot \frac{k}{p} \cdot (1 - r) \cdot m.$$

So, the overall profit from using the inclusive strategy has the form

$$G \cdot \ln(k) - a \cdot N - t \cdot \frac{k}{p} - a \cdot N \cdot m - t \cdot \frac{k}{p} \cdot (1 - r) \cdot m. \tag{1'}$$

Similarly, terms in the formula (2) take the following form:

$$\frac{p \cdot N}{p \cdot N + p' \cdot N'} = \frac{1}{1 + m \cdot r} \approx 1 - m \cdot r$$

and

$$\ln\left(1 + \frac{p' \cdot N'}{p \cdot N}\right) = \ln(1 + r \cdot m) \approx r \cdot m.$$

Thus, the formula (2) takes the following form:

$$G \cdot \ln(k) - a \cdot N - t \cdot \frac{k}{p} - G \cdot \ln(k) \cdot r \cdot m + G \cdot r \cdot m. \tag{2'}$$

So, in comparison with the identical expressions corresponding to $m = 0$, we lose the following amounts proportional to m:

- in the discriminatory case, we lose the amount proportional to $G \cdot (\ln(k) - 1)$, while
- in the inclusive case, we lose the amount proportional to $a \cdot N + t \cdot \frac{k}{p} \cdot (1 - r)$.

To compare these losses, we need to take into account that even for the weakest of the k hires, for whom the gain is equal to $\frac{G}{k}$, this gain is still much larger than all the expenses on selection and training – otherwise, the company would not be hiring this person in the first place. The expenses of selecting a person are equal to a, the expenses of training $\frac{1}{p}$ persons (needed for one person to succeed) are $t \cdot \frac{1}{p}$. Thus, we conclude that

$$\frac{G}{k} \gg a + t \cdot \frac{1}{p},$$

hence

$$G \gg a \cdot k + t \cdot \frac{k}{p}$$

thus

$$G \gg a \cdot k + t \cdot \frac{k}{p} \cdot (1 - r).$$

Thus, to make sure that the discriminatory-strategy loss is larger than the inclusive-strategy loss, it is sufficient to make sure that

$$G \cdot ((\ln(k) - 1) \cdot r - 1) \geq a \cdot (N - k):$$

then, by adding the last inequality, we would get the desired one.

This last inequality is definitely true: even the gain $\frac{G}{k}$ of the least productive hire is of the same order as this person's lifetime salary, i.e., in the US, several million dollars, while the cost of scanning all N candidates, even if there is a thousand of them, does not exceed $\$100 \times 1000$, i.e., 100 000 dollars, which is much smaller than G.

Thus, *the inclusive strategy is indeed economically preferable.*

Comment. Of course, from the purely mathematical viewpoint, there are cases when the discriminatory strategy is more profitable: e.g., when the probability p' of the minority hire's success is close to 0, there is no gain in hiring them, only additional expenses in screening and training. However, in practice, the ratio p'/p is not 0: it can be 0.5, even somewhat less – but still sufficiently positive to make sure that the inclusive strategy is economically preferable.

Acknowledgments. This work was supported in part by the National Science Foundation grant HRD-1242122 (Cyber-ShARE Center of Excellence).

References

1. Carmichael, L., Stalla-Bourdillon, S., Taab, S.: Data mining and automated discrimination: a mixed legal/technical perspective. IEEE Intell. Syst. **31**(6), 51–55 (2016)
2. Saichev, A.I., Malevergne, Y., Sornette, D.: Theory of Zipf's Law and Beyond. Springer, Heidelberg (2010)

Dimensionality Reduction by Fuzzy Transforms with Applications to Mathematical Finance

Irina Perfilieva[✉]

Institute for Research and Applications of Fuzzy Modeling, NSC IT4Innovations,
University of Ostrava, 30. dubna 22, 701 03 Ostrava 1, Czech Republic
Irina.Perfilieva@osu.cz

Abstract. Two distinguished properties of the F-transform: the best approximation in a local sense and the reduction in dimension imply the fact that the F-transform has many successful applications. We show that the technique of F-transform fully agrees with the technique of dimensionality reduction, based on Laplacian eigenmaps. To justify this claim, we characterize the processed by the F-transform data in terms of the adjacency graph that reflects their (data) intrinsic geometry.

In the second part, we give an overview of the F-transform applications to mathematical finance: we discussed the estimation of market volatility and the numerical scheme and solution to the one-factor Black-Scholes partial differential equation.

Keywords: F-transform · Dimensionality reduction
Laplacian eigenmaps · Fuzzy partition · Market volatility
Black-Scholes equation

1 Introduction

Modeling and processing large data bases (texts, images, video signals, cash flows, etc.) motivate applying machine learning theory. "Big data" mining is based on extracting structured knowledge from spatio-temporally correlated data. The first step is a certain granulation of data, which means developing a low dimensional representation of data that arises from sampling a complex high dimensional data. The generic problem of dimensionality reduction is to find a set of points $\mathbf{y}_1, \ldots, \mathbf{y}_k$ in the space $\mathbb{R}^m$ such that $\mathbf{y}_i$ "represents" a point $\mathbf{x}_i$ from the given dataset $\mathbf{x}_1, \ldots, \mathbf{x}_k$ that belongs to the space $\mathbb{R}^l$ with the substantially larger dimension so that $m \ll l$.

Let us give some remarks regarding the history of the problem of dimensionality reduction, see [1]. Classical approaches include principal components analysis (PCA) and multidimensional scaling. Various methods that generate nonlinear maps have also been considered. Most of them, such as self-organizing maps and other neural network-based approaches (e.g., [2]), set up a nonlinear optimization problem whose solution is typically obtained by gradient descent that is guaranteed only to produce a local optimum; global optima are difficult

© Springer International Publishing AG 2018
L. H. Anh et al. (eds.), *Econometrics for Financial Applications*, Studies in Computational Intelligence 760, https://doi.org/10.1007/978-3-319-73150-6_19

to attain by efficient means. To our knowledge, the approach of generalizing the PCA through kernel-based techniques in [14] does not have this shortcoming. However, most of methods in [14] do not explicitly consider the structure of the manifold (space $\mathbb{R}^l$) on which the data may possibly lie.

In [1], an approach that builds a graph incorporating neighborhood information of the data set is proposed. Using the notion of the Laplacian of the graph, a low-dimensional representation of the data set that optimally preserves local neighborhood information is computed. The representation map generated by the proposed algorithm may be viewed as a discrete approximation to a continuous map that naturally arises from the geometry of the manifold. The most important feature of the solution consists in reflecting the intrinsic geometric structure of the manifold. The latter is approximated by the adjacency graph computed from the data points.

In the new theoretical part of this contribution (Sect. 3), we show that the technique of F-transforms ("F" stands for fuzzy) fully agrees with the technique in [1]. The theory of fuzzy (F)-transforms relates to a modern mathematical modeling. It provides a (dimensionally) reduced and robust representation of original data. It is based on a granulation of a domain (fuzzy partition) and gives a tractable image of an original data. The main characteristics with respect to input data: size reduction, noise removal, invariance to geometrical transformations, knowledge transfer from conventional mathematics, fast computation. The F-transform has been applied to: image processing, computer vision, on-line pattern recognition in big data bases, time series analysis and forecasting, mathematical finance, numerical methods for differential equations, deep learning neural networks.

To justify the main claim, we characterize the processed by the F-transform data in terms of the adjacency graph that reflects their intrinsic geometry. In the second part (Sect. 4), we give an overview of the F-transform applications to mathematical finance.

2   Preliminaries

2.1   Laplacian Eigenmaps for Dimensionality Reduction

In this part, we shortly remind details of the technique proposed in [1]. Following [1], we give an algorithm that constructs representatives $\mathbf{y}_i$'s from the space $\mathbb{R}^m$ for the set of points $\mathbf{x}_1, \ldots, \mathbf{x}_k$ from the space $\mathbb{R}^l$ where $m \ll l$. The result is in the form of embedding map and is provided by computing the eigenvectors of the graph Laplacian.

We start with the construction of a weighted graph $G = (V, E)$ with k nodes, one for each point $\mathbf{x}_i$, where the edges from E connect neighboring points. We assume that the graph is connected and that the resulting embedding $(\mathbf{y}_1, \ldots, \mathbf{y}_k)^T$ maps vertices of G to the real line $\mathbb{R}$ in such a way that connected points stay as close as possible.

Step 1 (constructing the adjacency graph). We put an edge between nodes i and j, if $\mathbf{x}_i$ and $\mathbf{x}_j$ are "close".

Step 2 (choosing the weights). In [1], two variants for weighting the edges were considered:

(a) *Heat kernel.* If vertices i and j are connected, then

$$W_{ij} = \exp(-\frac{\|\mathbf{x}_i - \mathbf{x}_j\|^2}{t}),$$

otherwise $W_{ij} = 0$. Parameter $t \in \mathbb{R}$ should be specified beforehand.

(b) *Simple-minded assignment.* $W_{ij} = 1$, if vertices i and j are connected by an edge and $W_{ij} = 0$, otherwise.

Step 3 (eigenmaps). Compute eigenvalues and eigenvectors for the generalized eigenvector problem

$$\mathbf{L}\mathbf{f} = \lambda \mathbf{D}\mathbf{f}, \tag{1}$$

where D is the diagonal weight matrix such that

$$D_{ii} = \sum_{j=1}^{k} W_{ij},$$

and $L = D - W$ is the Laplacian matrix. Below, we will show that matrix L is symmetric and positive semidefinite.

Let $\mathbf{f}_0, \ldots, \mathbf{f}_{k-1}$ be the solutions of equation (1), ordered according to their eigenvalues. If we leave out the eigenvector $\mathbf{f}_0$ corresponding to the eigenvalue 0 and use the next m eigenvectors, then the embedding in m-dimensional Euclidean space is given by

$$\mathbf{x}_i \to (\mathbf{f}_1(i), \ldots, \mathbf{f}_m(i)).$$

Let us stress that the proposed solution reflects the intrinsic geometric structure of the manifold. It is reported in [1] that the above given Laplacian eigenmap algorithm is relative insensitive to outliers and noise. This is due to locality-preserving character of the adjacency graph. Moreover, by trying to preserve local information in the embedding, the algorithm implicitly emphasizes the natural clusters in the data.

In order to justify the proposed algorithm, we show that matrix L is symmetric and positive semidefinite. The first property is obvious, and to prove the second one, we choose an arbitrary vector $\mathbf{y} = (y_1, \ldots, y_k)$ and show that

$$\mathbf{y}^T L \mathbf{y} = \frac{1}{2} \sum_{i,j=1}^{k} (y_i - y_j)^2 W_{ij}. \tag{2}$$

Indeed,

$$\sum_{i,j=1}^{k} (y_i - y_j)^2 W_{ij} = \sum_{i,j=1}^{k} (y_i^2 + y_j^2 - 2y_i y_j)^2 W_{ij}$$

$$= \sum_{i=1}^{k} y_i^2 D_{ii} + \sum_{j=1}^{k} y_j^2 D_{jj} - 2 \sum_{i,j=1}^{k} y_i y_j^2 W_{ij} = 2\mathbf{y}^T L \mathbf{y}.$$

It is known that vector $\mathbf{y}$ that minimizes $\mathbf{y}^T L \mathbf{y}$ is given by the minimum eigenvalue solution to the generalized eigenvalue problem (1). By (2), this minimum eigenvalue solution minimizes the following objective function

$$\sum_{i,j=1}^{k} (y_i - y_j)^2 W_{ij}. \tag{3}$$

Moreover, if the graph G is connected, then $\mathbf{y} = \mathbf{1}$ is the only eigenvector for $\lambda = 0$.

In the following sections, we show that the F-transform technique can be reinterpreted within the framework of dimensionality-reduction based on intrinsic geometry of data. Moreover, we connect the F-transform with the generalized kernel-based PCA with the discrete Laplacian kernel.

2.2   F-transforms

We recall some definitions from [11].

Fuzzy Partition. Let $[a, b]$ be an interval on the real line $\mathbb{R}$. Fuzzy sets on $[a, b]$ are identified by their membership functions; i.e., they are mappings from $[a, b]$ into $[0, 1]$.

Definition 1. *Let $[a, b]$ be an interval on $\mathbb{R}$, $n \geq 2$, and let $x_0, x_1, \ldots, x_n, x_{n+1}$ be nodes such that $a = x_0 \leq x_1 < \ldots < x_n \leq x_{n+1} = b$. We say that fuzzy sets $A_1, \ldots, A_n : [a, b] \to [0, 1]$, which are identified with their membership functions, constitute a* fuzzy partition *of $[a, b]$ if for $k = 1, \ldots, n$, if they fulfill the following conditions:*

1. *(locality) - $A_k(x) = 0$ if $x \in [a, x_{k-1}] \cup [x_{k+1}, b]$,*
2. *(continuity) - $A_k(x)$ is continuous,*
3. *(covering) - $A_k(x) > 0$ if $x \in (x_{k-1}, x_{k+1})$.*

The membership functions $A_1, \ldots, A_n$ are called basic functions.

We say that the fuzzy partition $A_1, \ldots, A_n$, $n \geq 2$, is *h-uniform* if nodes $x_0, \ldots, x_{n+1}$ are *h-equidistant*; i.e., for all $k = 1, \ldots, n+1$, $x_k = x_{k-1} + h$, where $h = (b - a)/(n + 1)$ and the following three additional properties are fulfilled:

4. for all $k = 1, \ldots, n$, $A_k(x)$ strictly increases on $[x_{k-1}, x_k]$ and strictly decreases on $[x_k, x_{k+1}]$,
5. for all $k = 1, \ldots, n$, and for all $x \in [0, h]$, $A_k(x_k - x) = A_k(x_k + x)$,
6. for all $k = 2, \ldots, n$, and for all $x \in [x_{k-1}, x_{k+1}]$, $A_k(x) = A_{k-1}(x - h)$.

It can be easily shown that for an h-uniform fuzzy partition $A_1, \ldots, A_n$, of $[a, b]$, there exists a continuous and even function $A_0 : [-1, 1] \to [0, 1]$ such that it vanishes on boundaries and for all $k = 1, \ldots, n$,

$$A_k(x) = A_0 \left(\frac{x - x_k}{h} \right), \quad x \in [x_{k-1}, x_{k+1}]. \tag{4}$$

We call A_0 a *generating function* of an h-uniform fuzzy partition.

Hilbert Space with Weighted Inner Product. Let us fix $[a, b]$ and its fuzzy partition $A_1, \ldots, A_n$, $n \geq 2$. Let k be a fixed integer from $\{1, \ldots, n\}$, and let $L_2(A_k)$ be a set of square-integrable functions on $[x_{k-1}, x_{k+1}]$, and $L_2([a, b])$ a set of square-integrable functions on $[a, b]$. Let us denote

$$s_k = \int_{x_{k-1}}^{x_{k+1}} A_k(x)dx,$$

and consider $\frac{1}{s_k}\int_{x_{k-1}}^{x_{k+1}} f(x)g(x)A_k(x)dx$ as a Lebesgue integral $\int_{x_{k-1}}^{x_{k+1}} f(x)$ $g(x)d\mu_k$, where $d\mu = A_k(x)dx/s_k$ and the measure μ_k on $[x_{k-1}, x_{k+1}]$ is defined as follows:

$$\mu_k(E) = \frac{\int_E A_k(x)dx}{\int_{x_{k-1}}^{x_{k+1}} A_k(x)dx}.$$

The functions $f, g \in L_2(A_k)$ are *orthogonal* in $L_2(A_k)$ if $\langle f, g \rangle_k = 0$. The function $f \in L_2(A_k)$ is orthogonal to a subspace B of $L_2(A_k)$ if $\langle f, g \rangle_k = 0$ for all $g \in B$.

Let us denote by $L_2^m(A_k)$ a linear subspace of $L_2(A_k)$ with the basis given by orthogonal functions $P_k^0, P_k^1, P_k^2 \ldots, P_k^m$.

The following lemma gives analytic representation of the orthogonal projection on the subspace $L_2^m(A_k)$.

Lemma 1 *([11]). Let function F_k^m be the orthogonal projection of $f \in L_2(A_k)$ on $L_2^m(A_k)$. Then,*

$$F_k^m = c_{k,0}P_k^0 + c_{k,1}P_k^1 + \cdots + c_{k,m}P_k^m, \tag{5}$$

where for all $i = 0, 1, \ldots, m$,

$$c_{k,i} = \frac{\langle f, P_k^i \rangle_k}{\langle P_k^i, P_k^i \rangle_k} = \frac{\int_{x_{k-1}}^{x_{k+1}} f(x)P_k^i(x)A_k(x)dx}{\int_{x_{k-1}}^{x_{k+1}} P_k^i(x)P_k^i(x)A_k(x)dx}. \tag{6}$$

The n-tuple $(F_1^m, \ldots, F_n^m)$ is an F^m-transform of f with respect to $A_1, \ldots, A_n$, or formally,

$$F^m[f] = (F_1^m, \ldots, F_n^m).$$

F_k^m is called the kth F^m-transform component of f.

In particular, let us consider the case where the basis of $L_2^m(A_k)$ is given by orthogonal polynomials $P_k^0, P_k^1, P_k^2 \ldots, P_k^m$ and P_k^0 is a constant function with the value 1. Then, the F^0-transform of f or simply, the F-transform of f with respect to the partition $A_1, \ldots, A_n$ is given by the n-tuple $(c_{1,0}, \ldots, c_{n,0})$ of constant functions (0-degree polynomials) where for $k = 1, \ldots, n$,

$$c_{k,0} = \frac{\langle f, 1 \rangle_k}{\langle 1, 1 \rangle_k} = \frac{\int_{x_{k-1}}^{x_{k+1}} f(x)A_k(x)dx}{\int_{x_{k-1}}^{x_{k+1}} A_k(x)dx}. \tag{7}$$

The F^1-transform of f with respect to $A_1, \ldots, A_n$ is given by the n-tuple $(c_{1,0} + c_{1,1}(x-x_1), \ldots, c_{n,0} + c_{n,1}(x-x_n))$ of linear functions (1-degree polynomials). The latter are fully represented by their 2D coefficients $((c_{1,0}, c_{1,1}), \ldots, (c_{n,0}, c_{n,1}))$, which in addition to (7), have the following particular representation:

$$c_{k,1} = \frac{\langle f, x - x_k \rangle_k}{\langle (x - x_k), (x - x_k) \rangle_k} = \frac{\int_{x_{k-1}}^{x_{k+1}} f(x)(x - x_k) A_k(x) dx}{\int_{x_{k-1}}^{x_{k+1}} (x - x_k)^2 A_k(x) dx}. \tag{8}$$

The *inverse F^m-transform* of function f with respect to the partition $A_1, \ldots, A_n$ is a function represented by the following *inversion formula*:

$$f_{F,n}^m(x) = \sum_{k=1}^{n} F_k^m A_k(x). \tag{9}$$

The following results demonstrate approximation properties of the direct and inverse F^m-transforms.

Lemma 2 *([11]). Let $m \geq 0$, and let functions F_k^m and F_k^{m+1} be the k-th F^m- and F^{m+1}- transform components of f, respectively. Then,*

$$\|f - F_k^{m+1}\| \leq \|f - F_k^m\|.$$

Theorem 1 *([10, 11]). Let $A_1, \ldots, A_n$, $n \geq 2$, be an h-uniform fuzzy partition of $[a, b]$, let functions f and A_k, $k = 1, \ldots, n$, be four times continuously differentiable on $[a, b]$, and let $F^1[f] = (c_{1,0} + c_{1,1}(x - x_1), \ldots, c_{n,0} + c_{n,1}(x - x_n))$ be the F^1-transform of f with respect to $A_1, \ldots, A_n$. Then, for every $k = 1, \ldots, n$, the following estimation holds true:*

$$c_{k,0} = f(x_k) + O(h^2),$$
$$c_{k,1} = f'(x_k) + O(h^2)$$

Theorem 2 *([10]). Let f be a continuous function on $[a, b]$. For every $\varepsilon > 0$, there exist an integer n_ε and the related fuzzy partition $A_1, \ldots, A_{n_\varepsilon}$ of $[a, b]$ such that for all $x \in [a, b]$,*

$$|f(x) - f_{F,n_\varepsilon}(x)| < \varepsilon,$$

where f_{F,n_ε} is the inverse F-transform of f with respect to $A_1, \ldots, A_{n_\varepsilon}$.

Theorem 3 *([11]). Let $A_1, \ldots, A_n$, $n \geq 2$, be an h-uniform fuzzy partition of $[a, b]$ that fulfills the Ruspini condition on $[a + h, b - h]$. Let functions f and A_k, $k = 1, \ldots, n$, be four times continuously differentiable on $[a, b]$, and let $f_{F,n}^m$ be inverse F^m-transform of f where $m \geq 1$. Then,*

$$\int_{a+h}^{b-h} |f_{F,n}^m(x) - f(x)| dx \leq O(h^2).$$

The discrete F-transforms were introduced in [10] and then further elaborated in a number of papers, see e.g., [8].

3   Dimensionality Reduction by the F-transform

In this section, we will connect the above considered techniques. In more detail, we will show that the F^0- and the F^1-transform are solutions of equation (1), provided that the Laplacian matrix L is properly constructed. This fact confirms that these F-transforms provide with a low-dimensional representation of a given dataset.

At first, we specify the dataset that will be characterized by their F-transform-based low-dimensional representation in the form of embedding maps and finally, by components. This dataset will be connected with a discrete representation of a function, say f on some domain. This function can be a signal, time series, image, etc. For simplicity, we assume that the domain is an interval $[a, b]$ of the real line, and the function f is given on a discrete set, say P of points where $P \subseteq [a, b]$. Then, we assume that $[a, b]$ is partitioned into the collection of fuzzy sets $A_1, \ldots, A_n$ (not necessarily uniform), as it is described in Definition 1. Moreover, we assume that for every k, $1 \le k \le n$, there is one point $x_k \in P$ (we call it *node*) such that $A_k(x_k) = 1$ and $A_j(x_k) = 0$, $j \ne k$.

The chosen partition determines geometry of the set P given by a directed weighted graph $D = (U, A)$. Each vertex from U corresponds to one point in P, and if a point $p_i \in P$ is *covered by basic function* A_k, i.e. $A_k(p_i) > 0$, then the corresponding to it vertex i is connected by the directed edge (p_i, x_k) with the vertex k, corresponding to the node x_k. The weight of the edge (p_i, x_k) is equal to $A_k(p_i)$. It is easy to see that graph D is split into n connected components $D_1, \ldots, D_n$ (connectedness respects the orientations of edges in D), each has a "star" shape. We will continue with each connected component separately, and construct the low-dimensional representation of the corresponding to it dataset.

Let us choose and fix one basic function A_k, $1 \le k \le n$, and consider a finite set of points $\{p_1, \cdots, p_{l_k}\}$ that are covered by A_k. Let us remark that the node x_k is among these points. Let $f_i = f(p_i)$, $1 \le i \le l_k$. The set X_k of data points allocated for the low-dimensional representation is $\{(p_1, f_1), \ldots, (p_{l_k}, f_{l_k})\}$.

The next step is to construct an ordinary weighted graph $G_k = (V_k, E_k)$ with l_k vertices, one for each data point (p_i, f_i). This graph corresponds to the respective connected component D_k with the same edges, but without orientation. Therefore, every vertex i, $1 \le i \le l_k$, is only connected with the vertex k. In the corresponding to G_k weight matrix W, each edge, connecting i and k, is represented by $W_{ik} = W_{ki} = A_k(p_i)$. This means that W is a symmetrical matrix. Then, we proceed with *Step 3* of the algorithm described in Sect. 2.

The minimum eigenvalue solution to the generalized eigenvalue problem (1) is a constant vector that corresponds to the zero eigenvalue. This solution minimizes the objective function $\mathbf{y}^T L \mathbf{y}$ (see explanations in Sect. 2.1). If we put the additional constraint to this minimization problem and consider

$$\mathbf{f}_0 = \operatorname*{argmin}_{(W\mathbf{f})_k = (D\mathbf{y})_k} \mathbf{y}^T L \mathbf{y}, \tag{10}$$

where $\mathbf{f} = (f_1, \ldots, f_{l_k})$ and $(\cdot)_k$ is the k-th vector component, then $\mathbf{f}_0$ is the constant vector, whose k-th component is specified by the constraint. This constraint agrees with the discrete version of the k-th F^0-transform component in (7). The solution of (10) can be interpreted as a weighted projection of $\mathbf{f}$ on the constant vector $\mathbf{1}$.

To obtain a non-constant vector solution to the minimization of $\mathbf{y}^T L \mathbf{y}$, let us put an additional constraint of orthogonality and consider

$$\mathbf{y}_1 = \operatorname*{argmin}_{\langle \mathbf{f}_0, \mathbf{y} \rangle_k = 0} \mathbf{y}^T L \mathbf{y}, \tag{11}$$

where $\langle \cdot, \cdot \rangle_k$ is the weighted inner product in $L_2(A_k)$ (discrete case). Then the k-th F^1-transform component $\mathbf{f}_1$ has the following representation

$$\mathbf{f}_1 = \mathbf{f}_0 + \frac{\langle \mathbf{y}_1, \mathbf{f} \rangle_k}{\langle \mathbf{y}_1, \mathbf{y}_1 \rangle_k} \mathbf{y}_1.$$

This representation agrees with (5) and (6), because the orthogonality constraint in (11) does not automatically guarantee that $\mathbf{y}_1$ has a polynomial form.

To conclude, the vectors $\mathbf{f}_0$ and $\mathbf{f}_1$ with the corresponding F-transform components provide with a low-dimensional representation of the considered data set X_k. In particular, every (x_i, f_i) from X_k is represented by $(\mathbf{f}_{0,i}, \mathbf{f}_{1,i})$ or even better, the whole set X_k is represented by the two numbers $c_{k,0} = \mathbf{f}_{0,i}$ and $c_{k,1} = \frac{\langle \mathbf{y}_1, \mathbf{f} \rangle_k}{\langle \mathbf{y}_1, \mathbf{y}_1 \rangle_k}$.

Our final remark is about the role of the Laplacian matrix L in the above given construction. We remind that the problem of a low-dimensional representation of the considered data set X_k was reduced to find a mapping of the weighted graph G (weights correspond to degrees of closeness between vertices) to the real line ($\mathbb{R}$) so that connected points stay as close together as possible. If $\mathbf{y}$ is such a map, then it should minimize the following objective function

$$\sum_{i,j} (y_i - y_j)^2 W_{ij},$$

where W_{ij} is a weight of the edge, connecting corresponding vertices. It turned out (see (2)) that for any $\mathbf{y}$,

$$\frac{1}{2} \sum_{i,j} (y_i - y_j)^2 W_{ij} = y^T L y,$$

where $L = D - W$ and D is a diagonal weight matrix. Thus, the problem of minimization has been reformulated in terms of matrix L. The latter reflects the intrinsic geometric structure of the considered data.

4   F-transforms in Mathematical Finance

Two distinguished properties of the F-transform: the best approximation in a local sense and the reduction in dimension imply the fact that the F-transform has many successful applications. Let us mention some of them: image processing, computer vision, on-line pattern recognition in big data bases, time series analysis and forecasting, mathematical finance, numerical methods for differential equations, deep learning neural networks. In this section, we will give a short overview of applications in mathematical finance.

4.1 Estimation of Market Volatility

At first, we will be focused on estimations of volatility. Informally, volatility refers to a degree of variation of a trading price series. It is distinguished between a *historic volatility* (derived from time series of past market prices) and *implied volatility* (derived from the market price of a market traded derivative). Implied volatility returns a value of the current market price of the option, when input is an option pricing model (such as Black-Scholes). It is important to realize that the values of implied volatilities depend on the model used to calculate them. In [3], the estimation of the implied volatility surface for pricing of illiquid options is proposed. In general, a smooth model of discrete observations of implied volatilities (the implied volatility surface) should respect no-arbitrage conditions. In [3], the smooth model of discrete observations of implied volatilities was obtained applying the novel nonparametric regression approach based on a higher degree F-transform. The proposed technique allows easy calculation of partial derivatives of the estimated implied volatility and this is important for the calculation of several useful measures such as a state price density.

In [15], the F-transform and its inverse are used as alternative models of historic volatility. A common measure of historical volatility is defined as standard deviation of returns. This measure (under certain conditions) is widely accepted as a measure of risk. Two important axiomatic properties of a risk measure are known as the Luce axioms [7]. First, a risk measure is proportional to any change of scale of returns and second, it is represented with the help of an integral of a modified density of returns distribution. Below, we give some details of how the model of the measure of volatility was proposed in [15].

At first, a time horizon T is chosen and a uniform fuzzy partition of time with T as a distance between nodes is established. Then, the direct and inverse F-transform of daily returns r_t is computed. The latter is denoted b_t and referred to as a *baseline*. It plays the same role as the mean in computing the standard deviation. Similarly, the direct and inverse F-transform of absolute daily returns h_t is computed. Finally, the measure of volatility d_t is proposed in the form $d_t = h_t - b_t$. It has been shown in [15], that the obtained measure of volatility d_t is compatible with the definition of a risk measure given by Luce. In other words, it fulfils the Luce axioms.

For testing the quality of the proposed model, the authors of [15] considered the application to NIFTY 50 that is the benchmark index used by National Stock Exchange of India for the equity market. The period used for the analysis is between the 20th September 2000 and the 9th February 2017, entailing 4040 days. The authors concluded that volatility measured by means of F-transform looks smoother, if compared to volatility computed as a standard deviation. Although smoother, the F-transform based volatility follows the trend given by the series of daily returns. Thus, it is related to values given by the STD volatility. This aspect was further investigated on the used data, and the high correlation was shown by both volatility measures: 0.9167 on the yearly basis, 0.8755 on the monthly basis and 0.7523 on the weekly basis (Pearson's correlation coefficients). Moreover, calculations on the basis of the NIFTY 50 stock index show that the F-transform based volatility offers a more regular/stable and better centered measure of volatility than that based on the standard deviation. Both are able to mitigate the effect of extreme returns.

4.2   *F*-transform Method for Option Pricing

The option pricing is an important part of market practice. A standard approach to option pricing is based on Black-Scholes type models utilizing the no-arbitrage argument of complete markets. In literature, one can find other models and methodologies for option pricing, e.g., Heston or Lévy models. It is well-known that no analytic solution can be found for complex models. Therefore, instead of exact solutions one seeks approximate solutions by means of classical numerical methods such as finite difference, finite elements, the method of Monte Carlo, or the wavelet method. In [4,5], the focus was made on the one-factor Black-Scholes partial differential equation (PDE) describing a single plain vanilla option pricing problem. The numerical (approximate) solution with the help of the F-transform technique was proposed.

Historically the first publication on the F-transform based solution of ordinary differential equations appeared in [9]. In this paper, the numeric method, which generalizes the Euler method, has been proposed for the ordinary initial value problem (IVP). This method (as well as its classical prototype) has the same accuracy and belongs to the class the order one methods. In [6], three new numeric schemes for solving the IVP have been presented. All these schemes are based on the F-transform of various degrees and are comparable with the corresponding Runge-Kutta methods. The F-transform-based approach for boundary value problems (BVPs) was successfully applied to linear second order ordinary differential equations (ODEs) and to nonlinear second-order ODEs, see [12,13] and references therein.

Theoretical scheme and numerical solution to the one-factor Black-Scholes partial differential equation (PDE) have been discussed in [4,5]. Usually, the Black-Scholes PDE with boundary (Dirichlet) and terminal conditions is used to be solved numerically. In certain cases (European call), it is possible to solve this problem analytically. In [4,5], after the discretization of the time domain by the Crank-Nicolson method the problem was reduced to the BVP with respect to a two-variable function. Then, the F-transform based method (similar to that discussed in [5]) was successfully applied. In [13], the comparison with the analytically expressed exact solution was done. The error of the F-transform based numerical solution on the domain $(0, 1/3) \times (0, 16000)$ at time $t = 1/3$ for the triangle 2D fuzzy partition with $M = 256$ (x axis) and $L = 32$ (t axis) basic functions is equal to 4.58572e–05.

5   Conclusions

In this contribution, we showed that the technique of F-transforms fully agrees with the technique of dimensionality reduction, based on Laplacian eigenmaps. To justify this claim, we characterized the processed by the F-transform data in terms of the adjacency graph that reflects their intrinsic geometry. Moreover, we showed that the F- and the F^1-transform of a function f are solutions of the generalized eigenvector problem where the Laplacian matrix L is properly constructed. This fact confirms that the F-transforms provide with a low-dimensional representation of a given data set.

In the second part, we gave an overview of the F-transform applications to mathematical finance: we discussed the estimation of market volatility and the numerical scheme and solution to the one-factor Black-Scholes partial differential equation.

Acknowledgment. This work was supported by the project LQ1602 IT4Innovations excellence in science. The additional support was also provided by the Czech Science Foundation (GAČR) through the project of No. 16-09541S.

References

1. Belkin, M., Niyogi, P.: Laplacian eigenmaps for dimensionality reduction and data representation. Neural Comput. **15**(6), 1373–1396 (2003)
2. Haykin, S.: Neural Networks: A Comprehensive Foundation. Prentice Hall, Upper Saddle River (1999)
3. Holčapek, M., Tichý, T.: Nonparametric regression via higher degree F-transform for implied volatility surface estimation. In: Proceedings of Mathematical Methods in Economics, Liberec, pp. 277–282 (2016)
4. Holčapek, M., Števuliáková, P., Perfilieva, I.: F-transform method for option pricing. In: Proceedings of the 8th International Scientific Conference Managing and Modelling of Financial Risks, Ostrava, Czech Republic, VŠB (2016)
5. Holčapek, M., Valášek, R.: Numerical solution of partial differential equations with the help of fuzzy transform technique. In: Proceedings of the IEEE International Conference on Fuzzy Systems, FUZZ-IEEE, Naples, Italy (2017)
6. Khastan, A., Perfilieva, I., Alijani, Z.: A new fuzzy approximation method to Cauchy problems by fuzzy transform. Fuzzy Sets Syst. **288**, 75–89 (2016)
7. Luce, R.D.: Several possible measures of risk. Theor. Dec. **12**(3), 217–228 (1980)
8. Novák, V., Perfilieva, I., Dvořák, A.: Insight into Fuzzy Modeling. Wiley, Hoboken (2016)
9. Perfilieva, I.: Fuzzy transform: application to the Reef growth problem. In: Demicco, R.V., Klir, G.J. (eds.) Fuzzy Logic in Geology, pp. 275–300. Academic Press, Amsterdam (2003)
10. Perfilieva, I.: Fuzzy transform: theory and application. Fuzzy Sets Syst. **157**, 993–1023 (2006)
11. Perfilieva, I., Danková, M., Bede, B.: Towards a higher degree F-transform. Fuzzy Sets Syst. **180**, 3–19 (2011)
12. Perfilieva, I., Števuliáková, P., Valášek, R.: F-transform-based shooting method for nonlinear boundary value problems. Soft Comput. **21**(13), 3493–3502 (2017)

13. Perfilieva, I., Števuliáková, P., Valášek, R.: F-transform for numerical solution of two-point boundary value problem. Iran. J. Fuzzy Syst. (2017, in press)
14. Scholkopf, B., Smola, A., Mulller, K.-R.: Nonlinear component analysis as a kernel eigenvalue problem. Neural Comput. **10**(5), 1299–1319 (1998)
15. Troiano, L., Mejuto, E., Kriplani, P.: An alternative estimation of market volatility based on fuzzy transform. In: Proceedings of IFSA-SCIS 2017, Otsu, Shiga, Japan (2017)

Confidence Intervals for the Signal to Noise Ratio of Two-Parameter Exponential Distribution

Luckhana Saothayanun[1] and Warisa Thangjai[2(✉)]

[1] University of the Thai Chamber of Commerce, Bangkok 10400, Thailand
luckhana_sao@utcc.ac.th
[2] Department of Applied Statistics, Faculty of Applied Science,
King Mongkut's University of Technology North Bangkok, Bangkok 10800, Thailand
wthangjai@yahoo.com

Abstract. This paper investigates the performance of confidence intervals for signal to noise ratio (SNR) of two-parameter exponential distribution. The confidence intervals were constructed using generalized confidence interval (GCI) approach, large sample (LS) approach, and method of variance estimates recovery (MOVER) approach. The coverage probability and average length of the confidence intervals were evaluated by a Monte Carlo simulation. The results found that the confidence intervals based on the GCI approach provide the best coverage probabilities for all cases. The approaches are illustrated using an example.

Keywords: GCI approach · LS approach · MOVER approach

1 Introduction

Two-parameter exponential distribution is often used to represent the time to failure. Lawless [12] examined lifetime data using two-parameter exponential distribution and obtained a prediction interval for a future observation from this distribution. Several researchers investigated the problem concerning confidence interval estimation for the parameter in two-parameter exponential distribution. For example, Chiou [5] proposed a method to construct confidence intervals for scale parameters of two-parameter exponential distributions. Roy and Mathew [15] constructed an exact lower confidence limit for the reliability function of two-parameter exponential distribution using GCI approach. For more details about confidence interval estimation for the parameter in two-parameter exponential distribution, see the research papers of Li and Zhang [14], Kharrati-Kopaei et al. [9], Singh and Singh [20], Li et al. [13], Sangnawakij et al. [16], Sangnawakij and Niwitpong [17], and Thangjai and Niwitpong [21].

The coefficient of variation is defined as a ratio of the standard deviation to the mean; see Kelley [10]. It is a unit-free measure of variability relative to the mean. The coefficient of variation is often used to compare two or more

© Springer International Publishing AG 2018
L. H. Anh et al. (eds.), *Econometrics for Financial Applications*, Studies in Computational Intelligence 760, https://doi.org/10.1007/978-3-319-73150-6_20

distributions measured on different units. The reciprocal of the coefficient of variation is called signal to noise ratio (SNR). The SNR is defined as a ratio of the mean to the standard deviation.

In econometrics, the parameters are estimated by using an indirect inference and direct description. The indirect inference is used to estimate model parameters from data on all model variables. In a multivariate model, regression techniques can only estimate model parameters. The part of the variance of each regressor, is independent of all other explanatory variables, is used. The partial variance is much smaller than the full variance when the explanatory variables are closely related to each other. Therefore, the signal is low. It is not practical to include all possible explanatory variables in an econometric model. The direct description do evidence on the nature and magnitude of relationships. It can use both the partial variance and the full variance of each variable. Therefore, the noise, is the error of measurement in the explanatory variables of interest, is low. Hence, it is practical to take account of a much larger number of possible explanatory variables than in an econometric model.

The SNR is of importance in applications. For example, the SNR is used to consider a quality of hi-fi system. A good quality SNR for the hi-fi system is probably in the region of 60 dB or higher, whereas the system is of poor quality when the SNR is less than 60 dB. The SNR of 60 dB means that the signal is 10^6 times stronger than the noise. Therefore, the economist is interested in finding the high SNR. This is because a high SNR denotes a good radio broadcast and good estimate of the econometric parameter, whereas a low SNR denotes a poor radio signal and inaccurate measure of the econometric parameter; see Swann [19].

The confidence interval provides information respecting the population value of the quantity much more than the point estimate; see Casella and Berger [4]. Several researchers have studied a problem concerning confidence interval estimation for the SNR; for example, see Sharma and Krishna [18], George and Kibria [7], Albatineh et al. [2], and Albatineh et al. [1]. To our knowledge, there is no previous work on constructing confidence interval for the SNR of two-parameter exponential distribution. Therefore, this paper proposes the new confidence intervals for the SNR of two-parameter exponential distribution based on generalized confidence interval, large sample confidence interval, and method of variance estimates recovery confidence interval. A Monte Carlo simulation is used to evaluate the performance of the proposed confidence intervals.

The remainder of this paper is organized as follows. Section 2, the proposed approaches and computational procedures to construct confidence intervals for the SNR of two-parameter exponential distribution are presented. Section 3, simulation results are presented to evaluate the performances of the GCI approach, covering the LS approach, and the MOVER approach on coverage probabilities and average lengths. Section 4, the three approaches are illustrated with an example. And finally, Sect. 5 summarizes this paper.

2   Confidence Intervals for Signal to Noise Ratio of Two-Parameter Exponential Distribution

Suppose that random sample $X = (X_1, X_2, \ldots, X_n)$ follows a two-parameter exponential distribution with probability density function

$$f(x; \beta, \lambda) = \frac{1}{\lambda} \exp\left(-\frac{x - \beta}{\lambda}\right) I_{[\beta, \infty)}(x), \tag{1}$$

where $\beta \in (-\infty, +\infty)$ and $\lambda \in (0, +\infty)$ are location and scale parameters, respectively. The mean and variance of X are $E(X) = \lambda + \beta$ and $Var(X) = \lambda^2$, respectively. The SNR is defined as follows

$$\theta = \frac{E(X)}{\sqrt{Var(X)}} = \frac{\lambda + \beta}{\lambda}. \tag{2}$$

It is well known that the following likelihood estimators of β and λ

$$\hat{\beta} = X_{(1)} = \min(X_1, X_2, \ldots, X_n) \quad \text{and} \quad \hat{\lambda} = \bar{X} - X_{(1)}, \tag{3}$$

where $X_{(1)}$ and $\bar{X}$ denote the first order statistic and sample mean of X, respectively.

The following estimate of θ

$$\hat{\theta} = \frac{\bar{X} - X_{(1)} + X_{(1)}}{\bar{X} - X_{(1)}} = \frac{\bar{X}}{\bar{X} - X_{(1)}}. \tag{4}$$

2.1   Generalized Confidence Interval

Definition. *Let $X = (X_1, X_2, \ldots, X_n)$ be a random sample from a distribution $F(x|\delta)$ and let x be the observed value of X. Let $\delta = (\theta, \nu)$ be a vector of unknown parameters where θ is a parameter of interest and ν is a vector of nuisance parameters. Let $R(X; x, \delta)$ be a function of X, x, and δ. Let $R(x; x', \delta)$ be a function of interest, where x and x' are n-dimensional tuples and δ is a pair. The function $R(x; x', \delta)$ is called be generalized pivotal quantity (GPQ) if the following two conditions are satisfied; see Weerahandi [22]:*

(i) For any given x', the distribution of the random variable $R(X; x', \delta)$ is the same for all δ. The distribution depends only on x'.
(ii) For all x and δ, the observed value $R(x; x, \delta)$ is the parameter of interest.

Let $R(\alpha/2)$ and $R(1 - \alpha/2)$ be the $100(\alpha/2)$-th and the $100(1 - \alpha/2)$-th percentiles of $R(x; x', \delta)$, respectively. Hence, $(R(\alpha/2), R(1 - \alpha/2))$ becomes a $100(1 - \alpha)\%$ two-sided generalized confidence interval for the parameter of interest.

The pivots of Roy and Mathew [15] have the following form

$$W_1 = \frac{2n(\hat{\beta} - \beta)}{\lambda} \sim \chi_2^2 \quad \text{and} \quad W_2 = \frac{2n\hat{\lambda}}{\lambda} \sim \chi_{2n-2}^2, \tag{5}$$

where χ_p^2 denotes a chi-square distribution with p degrees of freedom. The SNR can be defined as follows

$$\theta = \left(\frac{2n\hat{\lambda}}{W_2} + \left(\hat{\beta} - \frac{W_1\lambda}{2n} \right) \right) \left(\frac{2n\hat{\lambda}}{W_2} \right)^{-1} = 1 + \frac{1}{2n} \left(\frac{\hat{\beta}W_2}{\hat{\lambda}} - W_1 \right). \qquad (6)$$

Therefore, the generalized pivotal quantity for θ is defined as follows

$$R_\theta = 1 + \frac{1}{2n} \left(\frac{\hat{\beta}W_2}{\hat{\lambda}} - W_1 \right), \qquad (7)$$

where $\hat{\beta}$ and $\hat{\lambda}$ are defined in Eq. (3). The generalized pivotal quantity for θ satisfies above two conditions. Therefore, the $100(1 - \alpha)\%$ two-sided confidence interval for the SNR based on the GCI approach is defined as follows

$$CI_{GCI} = (L_{GCI}, U_{GCI}) = (R_\theta(\alpha/2), R_\theta(1 - \alpha/2)), \qquad (8)$$

where $R_\theta(p)$ denotes the $100(p)$-th percentile of R_θ. Algorithm 1 is used to construct the confidence interval based on the GCI approach.

Algorithm 1
input : $\bar{x}, x_{(1)}$
output: $R_\theta(\alpha/2), R_\theta(1 - \alpha/2)$
begin
    for $g = 1$ **to** m **do**
        Generate $W_1 \sim \chi_2^2$ and $W_2 \sim \chi_{2n-2}^2$;
        Compute R_θ from Eq. (7);
    end
    Obtain an array of R_θ;
    Rank the array of R_θ from small to large;
    Compute $R_\theta(\alpha/2)$ and $R_\theta(1 - \alpha/2)$;
end

2.2   Large Sample Confidence Interval

Theorem 1. *Let* $X = (X_1, X_2, \ldots, X_n)$ *be a random sample from two-parameter exponential distribution with location parameter* β *and scale parameter* λ. *Let* $\theta = (\lambda + \beta)/\lambda$ *be the SNR. Also, let* $\hat{\theta} = \bar{X}/(\bar{X} - X_{(1)})$ *be an estimator of* θ. *The following mean and variance of* $\hat{\theta}$

$$E(\hat{\theta}) = \frac{n^2(\lambda + \beta)}{(n-1)^2\lambda} \quad and \quad Var(\hat{\theta}) = \frac{2n^2\lambda^2 - n\lambda^2 + 2n^2\lambda\beta + n^2\beta^2}{(n-1)^3\lambda^2}. \qquad (9)$$

Proof. Let $X_{(1)}$ *be the first order statistic. The mean and variance of* $X_{(1)}$ *are* $E(X_{(1)}) = \beta + (\lambda/n)$ *and* $Var(X_{(1)}) = \lambda^2/n^2$, *respectively. Let* $\bar{X}$ *be the sample mean. The mean and variance of* $\bar{X}$ *are* $E(\bar{X}) = \lambda + \beta$ *and* $Var(\bar{X}) = \lambda^2/n$, *respectively. Sangnawakij and Niwitpong [17] proposed the following covariance of $X_{(1)}$ and $\bar{X}$*

$$Cov(X_{(1)}, \bar{X}) = \frac{\lambda^2}{n^2}.$$

First, the mean and variance of $\bar{X} - X_{(1)}$ are considered. The mean of $\bar{X} - X_{(1)}$ is

$$E(\bar{X} - X_{(1)}) = E(\bar{X}) - E(X_{(1)})$$
$$= \lambda + \beta - \beta - \frac{\lambda}{n}$$
$$= \frac{(n-1)\lambda}{n}.$$

The variance of $\bar{X} - X_{(1)}$ is

$$Var(\bar{X} - X_{(1)}) = Var(\bar{X}) + Var(X_{(1)}) - 2Cov(X_{(1)}, \bar{X})$$
$$= \frac{\lambda^2}{n} + \frac{\lambda^2}{n^2} - \frac{2\lambda^2}{n^2}$$
$$= \frac{(n-1)\lambda^2}{n^2}.$$

The concepts of Blumenfeld [3] are used to conduct the mean and variance of ratio of $\bar{X}$ to $\bar{X} - X_{(1)}$. The following mean of $\bar{X}/(\bar{X} - X_{(1)})$

$$E\left(\frac{\bar{X}}{\bar{X} - X_{(1)}}\right) = \left(\frac{E(\bar{X})}{E(\bar{X} - X_{(1)})}\right)\left(1 + \frac{Var(\bar{X} - X_{(1)})}{(E(\bar{X} - X_{(1)}))^2}\right)$$
$$= \left(\frac{n(\lambda + \beta)}{(n-1)\lambda}\right)\left(1 + \frac{(n-1)n^2\lambda^2}{(n-1)^2n^2\lambda^2}\right)$$
$$- \frac{n^2(\lambda + \beta)}{(n-1)^2\lambda}.$$

The following variance of $\bar{X}/(\bar{X} - X_{(1)})$

$$Var\left(\frac{\bar{X}}{\bar{X} - X_{(1)}}\right) = \left(\frac{E(\bar{X})}{E(\bar{X} - X_{(1)})}\right)^2\left(\frac{Var(\bar{X})}{(E(\bar{X}))^2} + \frac{Var(\bar{X} - X_{(1)})}{(E(\bar{X} - X_{(1)}))^2}\right)$$
$$= \left(\frac{n(\lambda + \beta)}{(n-1)\lambda}\right)^2\left(\frac{\lambda^2}{n(\lambda + \beta)^2} + \frac{(n-1)n^2\lambda^2}{(n-1)^2n^2\lambda^2}\right)$$
$$= \frac{2n^2\lambda^2 - n\lambda^2 + 2n^2\lambda\beta + n^2\beta^2}{(n-1)^3\lambda^2}.$$

Hence, Theorem 1 is proved.

Again, let $\hat{\theta} = \bar{X}/(\bar{X} - X_{(1)})$. The variance of $\hat{\theta}$ is defined in Eq. (9) with β and λ replaced by $x_{(1)}$ and $\bar{x} - x_{(1)}$, respectively. Therefore, the $100(1 - \alpha)\%$

two-sided confidence interval for the SNR based on the LS approach is defined as follows

$$CI_{LS} = (L_{LS}, U_{LS}) = (\hat{\theta} - z_{1-\alpha/2}\sqrt{Var(\hat{\theta})}, \hat{\theta} + z_{1-\alpha/2}\sqrt{Var(\hat{\theta})}), \quad (10)$$

where $z_{1-\alpha/2}$ denotes the $100(1 - \alpha/2)$-th percentile of the standard normal distribution.

2.3   Method of Variance Estimates Recovery Confidence Interval

Sangnawakij and Niwitpong [17] proposed the following confidence interval for $\lambda + \beta$

$$CI_{\lambda+\beta} = (l_1, u_1), \quad (11)$$

where

$$l_1 = \hat{\lambda} + \hat{\beta} - \sqrt{\left(\hat{\lambda} - \frac{n\hat{\lambda}}{z_{\alpha/2}\sqrt{n-1} + (n-1)}\right)^2 + \left(\frac{\hat{\lambda}}{n}\ln\left(\frac{\alpha}{2}\right)\right)^2}$$

and

$$u_1 = \hat{\lambda} + \hat{\beta} + \sqrt{\left(\frac{n\hat{\lambda}}{-z_{\alpha/2}\sqrt{n-1} + (n-1)} - \hat{\lambda}\right)^2 + \left(\frac{\hat{\lambda}}{n}\ln\left(1 - \frac{\alpha}{2}\right)\right)^2}.$$

It is well known that the confidence interval for λ is defined as follows

$$CI_\lambda = (l_2, u_2), \quad (12)$$

where

$$l_2 = \frac{n\hat{\lambda}}{\sqrt{n-1}\left(z_{\alpha/2} + \sqrt{n-1}\right)},$$

$$u_2 = \frac{n\hat{\lambda}}{\sqrt{n-1}\left(-z_{\alpha/2} + \sqrt{n-1}\right)},$$

and $z_{1-\alpha/2}$ denotes the $100(1 - \alpha/2)$-th percentile of the standard normal distribution.

The concept of Donner and Zou [6] is used to construct a $100(1 - \alpha)\%$ two-sided confidence interval (L_{MOVER}, U_{MOVER}) for $\theta = (\lambda + \beta)/\lambda$. The lower limit and upper limit are defined as follows

$$L_{MOVER} = \frac{(\hat{\lambda} + \hat{\beta})\hat{\lambda} - \sqrt{((\hat{\lambda} + \hat{\beta})\hat{\lambda})^2 - l_1 u_2(2(\hat{\lambda} + \hat{\beta}) - l_1)(2\hat{\lambda} - u_2)}}{u_2(2\hat{\lambda} - u_2)} \quad (13)$$

and

$$U_{MOVER} = \frac{(\hat{\lambda} + \hat{\beta})\hat{\lambda} + \sqrt{((\hat{\lambda} + \hat{\beta})\hat{\lambda})^2 - u_1 l_2(2(\hat{\lambda} + \hat{\beta}) - u_1)(2\hat{\lambda} - l_2)}}{l_2(2\hat{\lambda} - l_2)}, \quad (14)$$

where (l_1, u_1) and (l_2, u_2) are defined in Eqs. (11) and (12), respectively. There-fore, the $100(1 - \alpha)\%$ two-sided confidence interval for the SNR based on the MOVER approach is defined as follows

$$CI_{MOVER} = (L_{MOVER}, U_{MOVER}),\tag{15}$$

where L_{MOVER} and U_{MOVER} are defined in Eqs. (13) and (14), respectively.

3   Simulation Studies

In this section, simulation studies are carried out to evaluate the performance of the confidence intervals for the SNR of two-parameter exponential distribution. Coverage probabilities and average lengths were estimated for the confidence intervals based on the GCI, LS, and MOVER approaches. Algorithm 2 is used to compute the coverage probabilities and average lengths of the proposed con-fidence intervals.

Algorithm 2
input : M, m, n, λ, β, θ
output: CP, AL
begin
    for $h = 1$ **to** M **do**
        Generate $X_j \sim Exp(\lambda, \beta)$; $j = 1, 2, \ldots, n$;
        Compute $\bar{x}$ and $x_{(1)}$;
        Use Algorithm 1 to construct (L_{GCI}, U_{GCI});
        Use Eq. (10) to construct (L_{LS}, U_{LS});
        Use Eq. (15) to construct (L_{MOVER}, U_{MOVER});
        if $(L_{(h)} \leq \theta \leq U_{(h)})$ **then**
            $p_{(h)} = 1$;
        else
            $p_{(h)} = 0$;
        end
        Compute $U_{(h)} - L_{(h)}$;
    end
    Compute mean of $p_{(h)}$;
    Compute mean of $U_{(h)} - L_{(h)}$;
end

In the simulation, each confidence interval was computed at the nominal confidence level of 0.95. The sample size was $n = 30, 50, 100$, and 200. The scale parameter was $\lambda = 1$ and the location parameter was computed by $\beta = \lambda(\theta - 1)$. The SNR was $\theta = -20, -10, -5, -3, -2, 2, 3, 5, 10$, and 20. The performance of the proposed confidence intervals can be obtained when using Algorithm 1 with $m = 2500$ simulations and Algorithm 2 with $M = 5000$ simulations.

Table 1 presents the coverage probabilities and average lengths of 95% two-sided confidence intervals for the signal to noise of two-parameter exponential distribution. The simulation results indicated that the coverage probabilities of

Table 1. The coverage probabilities (CP) and average lengths (AL) of 95% two-sided confidence intervals for the signal to noise of two-parameter exponential distribution

n	λ	θ	CI_{GCI}		CI_{LS}		CI_{AM}	
			CP	AL	CP	AL	CP	AL
30	1	-20	0.9548	15.7778	0.9602	16.1828	0.9428	15.1483
		-10	0.9462	8.2698	0.9396	8.1547	0.9282	7.6721
		-5	0.9486	4.4949	0.9290	4.1384	0.9280	3.9646
		-3	0.9526	2.9743	0.9054	2.5506	0.9244	2.5268
		-2	0.9508	2.2304	0.8872	1.8058	0.9236	1.8647
		2	0.9530	0.7871	0.9992	1.7507	0.9928	4.2736
		3	0.9506	1.5351	0.9980	2.5066	0.9574	6.1967
		5	0.9506	3.0367	0.9918	4.0781	0.8886	10.1248
		10	0.9498	6.8182	0.9790	8.1007	0.8414	20.1301
		20	0.9474	14.3206	0.9710	16.1364	0.8272	40.1028
50	1	-20	0.9482	11.9650	0.9466	11.9179	0.9356	11.4568
		-10	0.9550	6.2706	0.9456	5.9955	0.9366	5.7796
		-5	0.9500	3.4170	0.9190	3.0444	0.9182	2.9647
		-3	0.9516	2.2735	0.8974	1.8875	0.9066	1.8740
		-2	0.9494	1.7057	0.8698	1.3363	0.8936	1.3609
		2	0.9532	0.5891	1.0000	1.3092	0.9966	2.0289
		3	0.9482	1.1553	0.9976	1.8617	0.9674	2.8909
		5	0.9512	2.2892	0.9928	3.0115	0.9266	4.6706
		10	0.9480	5.1488	0.9756	5.9657	0.8896	9.2400
		20	0.9474	10.8452	0.9662	11.8931	0.8728	18.4128
100	1	-20	0.9520	8.3245	0.9452	8.1126	0.9386	7.9547
		-10	0.9514	4.3768	0.9334	4.0882	0.9302	4.0138
		-5	0.9448	2.3815	0.9024	2.0719	0.9006	2.0441
		-3	0.9522	1.5875	0.8932	1.2855	0.8962	1.2803
		-2	0.9522	1.1887	0.8718	0.9073	0.8830	0.9156
		2	0.9494	0.4043	1.0000	0.9006	0.9980	1.0971
		3	0.9504	0.8001	0.9978	1.2773	0.9826	1.5546
		5	0.9502	1.5943	0.9884	2.0644	0.9504	2.5076
		10	0.9432	3.5947	0.9718	4.0866	0.9146	4.9578
		20	0.9460	7.5625	0.9598	8.1276	0.9116	9.8566
200	1	-20	0.9522	5.8689	0.9456	5.6585	0.9424	5.6033
		-10	0.9460	3.0713	0.9274	2.8363	0.9240	2.8104
		-5	0.9512	1.6753	0.9098	1.4408	0.9084	1.4309
		-3	0.9574	1.1136	0.8848	0.8909	0.8870	0.8891
		-2	0.9496	0.8364	0.8562	0.6306	0.8626	0.6334
		2	0.9476	0.2814	1.0000	0.6280	0.9998	0.6902
		3	0.9522	0.5598	0.9974	0.8894	0.9874	0.9767
		5	0.9504	1.1181	0.9884	1.4361	0.9680	1.5750
		10	0.9508	2.5169	0.9736	2.8354	0.9458	3.1073
		20	0.9468	5.3064	0.9598	5.6441	0.9336	6.1840

the GCI approach close to nominal confidence level of 0.95 and are stable. The LS approach yields the coverage probabilities under nominal confidence level of 0.95 when θ is a negative value. For a positive value of θ, the LS approach provides the coverage probabilities close to 1.00 when θ is small, and close to nominal confidence level of 0.95 when θ is large. In almost all cases, the MOVER approach has the coverage probabilities under nominal confidence level of 0.95. Therefore, the confidence intervals based on the GCI approach provide the best coverage probabilities for all cases.

4   An Empirical Application

An example previously considered by Grubbs [8] and Krishnamoorthy and Thomas [11]. The data represent the failure mileages of 19 military carriers. The failure mileages are given in Table 2. Krishnamoorthy and Thomas [11] showed that the data fit a two-parameter exponential distribution.

Table 2. Failure mileages of 19 military carriers

162	200	271	302	393	508	539	629	706	777
884	1008	1101	1182	1463	1603	1984	2355	2880	

The summary statistics are $\bar{x} = 997.2105$, $x_{(1)} = 162$, $\hat{\theta} = 1.1940$. The generalized confidence interval for the SNR was $(0.9857, 1.2482)$ with the length of interval 0.2625. The large sample confidence interval was $(0.4428, 1.9451)$ with the length of interval 1.5023. In comparison, the MOVER confidence interval was $(0.2109, 31.3566)$ with the length of interval 31.1457. The numerical results showed that all the proposed confidence intervals contain the true SNR. However, the length of the generalized confidence interval is shorter than those of the other confidence intervals. Therefore, the GCI approach performs much better than the other approaches. The results confirm the simulation study in the previous section.

5   Discussion and Conclusions

The aim of this paper was to propose new confidence intervals for the SNR of two-parameter exponential distribution. The study was carried out to investigate the performance of confidence intervals based on the GCI approach, the LS approach, and the MOVER approach. The results indicated that the GCI approach provides much better confidence interval estimates than the other approaches for all cases. This is because the coverage probabilities of the confidence interval based on the GCI approach are satisfactory. Therefore, the GCI approach can be considered as an alternative to estimate the confidence interval for the SNR of two-parameter exponential distribution.

As a final note, the generalized confidence interval for the coefficient of variation of two-parameter exponential distribution, proposed by Sangnawakij and Niwitpong [17], is an exact confidence interval. Using the invariance property, the generalized confidence interval for the SNR of two-parameter exponential distribution is an exact confidence interval.

References

1. Albatineh, A.N., Boubakari, I., Kibria, B.M.G.: New confidence interval estimator of the signal to noise ratio based on asymptotic sampling distribution. Commun. Stat. Theor. Methods **46**, 574–590 (2017)
2. Albatineh, A.N., Kibria, B.M.G., Zogheib, B.: Asymptotic sampling distribution of inverse coefficient of variation and its applications. Int. J. Adv. Stat. Prob. **2**, 15–20 (2014). Revisited
3. Blumenfeld, D.: Operations Research Calculations Handbook. CRC Press, Boca Raton (2001)
4. Casella, G., Berger, R.L.: Statistical Inference. Duxbury Press, Pacific Grove (2002)
5. Chiou, P.: Interval estimation of scale parameters following a pre-test for two exponential distributions. Comput. Stat. Data Anal. **23**, 477–489 (1997)
6. Donner, A., Zou, G.Y.: Closed-form confidence intervals for function of the normal standard deviation. Stat. Methods Med. Res. **21**, 347–359 (2010)
7. George, F., Kibria, B.M.G.: Confidence intervals for estimating the population signal to noise ratio: a simulation study. J. Appl. Stat. **39**, 1225–1240 (2012)
8. Grubbs, F.E.: Approximate fiducial bounds on reliability for the two parameter negative exponential distribution. Technometrics **13**, 873–876 (1971)
9. Kharrati-Kopaei, M., Malekzadeh, A., Sadooghi-Alvandi, M.: Simultaneous fiducial generalized confidence intervals for the successive differences of exponential location parameters under heteroscedasticity. Stat. Probab. Lett. **83**, 1547–1552 (2013)
10. Kelley, K.: Sample size planning for the coefficient of variation from the accuracy in parameter estimation approach. Behav. Res. Methods **39**, 755–766 (2007)
11. Krishnamoorthy, K., Thomas, M.: Statistical Tolerance Regions: Theory Applications, and Computation. Wiley, New York (2009)
12. Lawless, J.F.: Prediction intervals for the two parameter exponential distribution. Technometrics **19**, 469–472 (1977)
13. Li, J., Song, W., Shi, J.: Parametric bootstrap simultaneous confidence intervals for differences of means from several two-parameter exponential distributions. Stat. Probab. Lett. **106**, 39–45 (2015)
14. Li, J.B., Zhang, R.Q.: Inference of parameters ratio in two-parameter exponential distribution. Chin. J. Appl. Probab. Stat. **26**, 81–88 (2010)
15. Roy, A., Mathew, T.: A generalized confidence limit for the reliability function of a two-parameter exponential distribution. J. Stat. Plan. Inference **128**, 509–517 (2005)
16. Sangnawakij, P., Niwitpong, S., Niwitpong, S.: Confidence intervals for the ratio of coefficients of variation in the two-parameter exponential distributions. Lecture Notes in Artificial Intelligence, vol. 9978, pp. 542–551 (2016)
17. Sangnawakij, P., Niwitpong, S.: Confidence intervals for coefficients of variation in two-parameter exponential distributions. Commun. Stat. Simul. Comput. **46**(8), 6618–6630 (2017)

18. Sharma, K.K., Krishna, H.: Asymptotic sampling distribution of inverse coefficient of variation and its applications. IEEE Trans. Reliab. **43**, 630–633 (1994)
19. Swann, G.M.P.: Putting Econometrics in Its Place: A New Direction in Applied Economics. Edward Elgar Publishing Ltd. (2006)
20. Singh, P., Singh, N.: Simultaneous confidence intervals for ordered pairwise differences of exponential location parameters under heteroscedasticity. Stat. Probab. Lett. **83**, 2673–2678 (2013)
21. Thangjai, W., Niwitpong, S.: Confidence intervals for the weighted coefficients of variation of two-parameter exponential distributions. Cogent Math. **4**, 1–16 (2017)
22. Weerahandi, S.: Generalized confidence intervals. J. Am. Stat. Assoc. **88**, 899–905 (1993)

Why Student Distributions? Why Matern's Covariance Model? A Symmetry-Based Explanation

Stephen Schön[1], Gael Kermarrec[1], Boris Kargoll[1], Ingo Neumann[1],
Olga Kosheleva[2], and Vladik Kreinovich[2(✉)]

[1] Leibniz Universität Hannover, Nienburger Strasse 1, 30167 Hannover, Germany
schoen@ife.uni-hannover.de, gael.kermarrec@web.de,
{kargoll,neumann}@gih.uni-hannover.de
[2] University of Texas at El Paso, 500 W. University, El Paso, TX 79968, USA
{olgak,vladik}@utep.edu

Abstract. In this paper, we show that empirical successes of Student distribution and of Matern's covariance models can be indirectly explained by a natural requirement of scale invariance – that fundamental laws should not depend on the choice of physical units. Namely, while neither the Student distributions nor Matern's covariance models are themselves scale-invariant, they are the only one which can be obtained by applying a scale-invariant combination function to scale-invariant functions.

1   Formulation of the Problem

Scale-invariance: a natural property of the physical world. Scientific laws are described in terms of numerical values of the corresponding quantities, be it physical quantities such as distance, mass, or velocity, or economic quantities such as price or cost.

These numerical values, however, depend on the choice of a measuring unit; see, e.g., [9]. If we replace the original unit by a new unit which is λ times smaller, then all the numerical values of the corresponding quantity get multiplied by λ. For example, if instead of meters, we start using centimeters – a 100 smaller unit – to describe distance, then all the distances get multiplied by 100, so that, e.g., 2 m becomes $2 \cdot 100 = 200$ cm.

It is reasonable to require that the fundamental laws describing objects from the physical world – be it material objects or human beings – do not change if we simply change the measuring unit. In other words, it is reasonable to require that the laws be invariant with respect to *scaling* $x \to \lambda \cdot x$.

Of course, if we change a measuring unit for one quantity, then we may need to also correspondingly change the measuring unit for related quantities as well. For example, in a simple motion, the distance d is equal to the product $v \cdot t$ of velocity v and time t. If we simply change the unit for time without changing the units for distance or velocity, the formula stops being true. However, the formula

© Springer International Publishing AG 2018
L. H. Anh et al. (eds.), *Econometrics for Financial Applications*, Studies in Computational Intelligence 760, https://doi.org/10.1007/978-3-319-73150-6_21

remains true if we accordingly change the unit for velocity. For example, if we started with seconds and m/sec, then, once we change seconds to hours, we should also change the measuring unit for velocity from m/sec to m/hr.

Thus, scale-invariance means that if we arbitrarily change the units of one or more fundamental quantities, then, after an appropriate re-scaling of related units, we should get, in the new units, the exact same formula as in the old units.

Heavy-tailed distributions: a situation in which we expect scale-invariance. Measurements are rarely absolutely accurate. Usually, the measurement result $\widetilde{x}$ is somewhat different from the actual (unknown) value x of the corresponding quantity. In many cases, we know the upper bound of the measurement error, so that the probability of exceeding this bound is either equal to 0 or very small (practically equal to 0).

In many other practical situations, however, the probability of having reasonably large measurement errors $\Delta x \stackrel{\text{def}}{=} \widetilde{x} - x$ is positive – and does not become negligibly small. In such cases, we talk about *heavy-tailed* distributions.

Such distributions are ubiquitous in physics, in economics, etc., and they have the same shape in different application areas; see, e.g., [8, 10]. This ubiquity seems to indicate that there is a fundamental reason for such distributions. It therefore seems reasonable to expect that for this fundamental law – just like for all other fundamental laws – we have the scale-invariance property. In other words, it is reasonable to expect that, for the corresponding probability density function $\rho(x)$, for every $\lambda > 0$, there exists a value $\mu(\lambda)$ for which

$$\rho(\lambda \cdot x) = \mu(\lambda) \cdot \rho(x). \tag{1}$$

Alas, no scale-invariant pdf is possible. At first glance, the above scale-invariance criterion sounds reasonable, but, alas, it is never satisfied.

Indeed, the pdf should have the property that the overall probability to be somewhere should be equal to 1 (i.e., $\int \rho(x)\,dx = 1$), and be measurable, and it is known (see, e.g., [1,2]) that every measurable solution of the Eq. (1) has the power law form

$$\rho(x) = c \cdot x^\alpha$$

for some c and α. For this function, the integral over the real line is always infinite:

- for $\alpha \geq -1$, it is infinite in the vicinity if 0, while
- for $\alpha \leq -1$, it is infinite for $x \to \infty$.

A simple explanation of why power laws are the only scale-invariant ones. If we additionally assume that the function $\rho(x)$ is differentiable, then the fact that power laws are the only solutions can be easily derived.

Indeed, in this case, the function $\mu(\lambda) = \dfrac{\rho(\lambda \cdot x)}{\rho(x)}$ is also differentiable, as a ratio of two differentiable functions $\rho(\lambda \cdot x)$ and $\rho(x)$. Since both functions $\rho(x)$ and $\mu(\lambda)$ are differentiable, we can differentiate both sides of the Eq. (1)

by λ and take $\lambda = 1$; we then conclude that $x \cdot \dfrac{d\rho}{dx} = \alpha \cdot \rho$, where we denoted $\alpha \stackrel{\text{def}}{=} \dfrac{d\mu}{d\lambda}_{|\lambda=1}$. By moving all the terms containing ρ to the left-hand side and the terms containing x to the right-hand side, we conclude that $\dfrac{d\rho}{\rho} = \alpha \cdot \dfrac{dx}{x}$. Integrating both sides, we get $\ln(\rho) = \alpha \cdot \ln(x) + C$, hence for $\rho = \exp(\ln(\rho))$, we get $\rho(x) = c \cdot x^{\alpha}$, where we denoted $c \stackrel{\text{def}}{=} \exp(C)$.

What is usually done. A usual idea is to abandon scale-invariance completely. For example, one of the most empirically successful ways to describe heavy-tailed distributions is to use non-scale-invariant *Student distributions*, with the probability density $\rho(x) = \text{const} \cdot (1 + a \cdot x^2)^{-\nu}$ for some coefficients const, a, and ν (see, e.g., [7,11]).

What we show in this paper. In this paper, we "rehabilitate" scale-invariance: namely, we show that while the distribution cannot be "directly" scale-invariant, it can be "indirectly" scale-invariant – namely, it can be described as a scale-invariant combination of two scale-invariant functions.

Interestingly, under a few reasonable additional conditions, we get exactly the empirically successful Student distributions – and thus, indirect scale-invariance explains their empirical success.

This line of reasoning also provides us with a reasonable next approximation (that is worth trying if we want a more accurate description): namely, a scale-invariant combination of three or more scale-invariant functions.

Multi-D case. A similar situation occurs in the multi-D case, e.g., in the analysis of spatial data. Often, spatial data is described as a homogeneous and isotropic process. To describe such processes, it is convenient to use Fourier transforms: namely, to describe, for each frequency ω, the mean value $S(\omega)$ of the square of the absolute value of the ω-Fourier component of the original multi-D data. The value $S(\omega)$ is known as the *spectral density*.

In some cases, this function $S(\omega)$ is mainly concentrated at some frequencies. However, in many other practical situations, the corresponding values do not become negligible neither for small nor for large ω. In many such cases, the shape of the spectral density is approximately the same, so it looks like we have a fundamental law of spatial dependence.

Since it is a fundamental law, it is reasonable to expect it to be scale-invariant, i.e., satisfy the condition $S(\lambda \cdot \omega) = \mu(\lambda) \cdot S(\omega)$.

We already know that every measurable solution to this functional equation has the form $S(\omega) = \text{const} \cdot \omega^{\alpha}$ for some const and α. However, for such functions, we have $\int S(\omega)\, d\omega = +\infty$, while the integral is equal to the overall energy of the spatial signal and should, therefore, be finite.

Similar to the 1-D case, a usual solution is to abandon scale-invariance and to use some non-scale-invariant function for which $\int S(\omega)\, d\omega < +\infty$. It turns out that among all such functions, Matern's function $S(\omega) = \text{const} \cdot (a_0 + a_1 \cdot \omega^2)^{-\nu}$ (for some const, a_i, and ν) is, empirically, the best; see, e.g., [3].

In this paper, we show that while this function is not directly scale-invariant, it is indirectly scale-invariant – as a result of applying a scale-invariant combination function to two scale-invariant functions $S(\omega)$. Moreover, it turns out that, under reasonable assumptions, Matern's functions are the only such combinations. Thus, scale invariance explains their empirical success.

We also provide a natural next approximation to Matern's function – a scale-invariant combination of three or more scale-invariant functions.

2 Let Us Describe Scale-Invariant Combination Functions

A combination function: reasonable requirements. By a combination function we would like to mean an operation $a * b$ that transforms two non-negative numbers into a new non-negative number. Intuitively, a combination of a and b should be the same as a combination of b and a, so the operation $*$ should be commutative: $a * b = b * a$.

Similarly, a combination of a, b, and c should not depend on the order in which we combine them, so this operation must be associative: $(a * b) * c = a * (b * c)$. It is also reasonable to require that this operation is continuous (if $a_n \to a$ and $b_n \to b$, then we should have $a_n * b_n \to a * b$) and monotonic (non-decreasing in each of its variables). So, we arrive at the following definition.

Definition 1. *By a combination function $*$ we mean a commutative associative continuous non-decreasing function from pairs of non-negative real numbers to non-negative real numbers.*

Scale-invariance. Scale-invariance means that if we have $a * b = c$, then after re-scaling all three values a, b, and c, we conclude that $(\lambda \cdot a) * (\lambda \cdot b) = \lambda \cdot c$. Substituting $c = a * b$ into this formula, we get the following definition.

Definition 2. *We say that a combination function is scale-invariant if for all a, b, and λ, we have $(\lambda \cdot a) * (\lambda \cdot b) = \lambda \cdot (a * b)$.*

Proposition. *The only scale-invariant combination functions are $a * b = \min(a, b)$, $a * b = \max(a, b)$, and $a * b = (a^\beta + b^\beta)^{1/\beta}$ for some β.*

For reader's convenience, the proof of this result is given in the Appendix.

3 Resulting Derivation of Student Distribution and Matern's Covariance Model

Derivation of Student distribution. If we use a scale-invariant combination operation to combine two scale-invariant functions $c_i \cdot x^{\alpha_i}$, we get the expressions $\min(c_1 \cdot x^{\alpha_1}, c_2 \cdot x^{\alpha_2})$, $\max(c_1 \cdot x^{\alpha_1}, c_2 \cdot x^{\alpha_2})$, and

$$((c_1 \cdot x^{\alpha_1})^\beta + (c_2 \cdot x^{\alpha_2})^\beta)^{1/\beta} = (C_1 \cdot x^{\gamma_1} + C_2 \cdot x^{\gamma_2})^\gamma,$$

where $C_i = (c_i)^\beta$, $\gamma_i = \beta \cdot \alpha_i$, and $\gamma = 1/\beta$.

It is reasonable to require:

- that the pdf if *analytical* in x – i.e., can be expanded in Taylor series – and
- that it is *monotonically decreasing with* x – since it is reasonable to require that the larger the measurement error, the less probable it is.

Analyticity excludes min and max.

For the sum, if both γ_i are different from 0, the value at 0 is either 0 or infinity. It cannot be infinite – then $\rho(x)$ would be not analytical, and it cannot be 0 – then it will not be able to monotonically decrease to 0. Thus, one of the coefficients γ_i is equal to 0, and we have

$$\rho(x) = C \cdot (1 + c \cdot x^{\gamma_2})^\gamma.$$

This expression is analytical when γ_2 is a positive integer. We cannot have $\gamma_2 = 1$, because then we would get $\rho(x) \to +\infty$ either when $x \to +\infty$ or when $x \to -\infty$. Thus, we must have $\gamma_2 \geq 2$.

Out of all possible functions of this type, the *generic* case – when both the 0-th and the second coefficient at Taylor expansion are not 0 – is when $\gamma_2 = 2$. Thus, we get exactly the Student distribution.

Derivation of Matern's covariant model. For dependence of the spectral density on ω, we similarly get exactly Matern's covariance model.

What next? If the scale-invariant combination of *two* scale-invariant functions does not work well, we can try a scale-invariant combination of three or more such functions: $f(x) = \left(\sum_{i=1}^{k} C_i \cdot x^{\gamma_i} \right)^\gamma .$

4   Alternative Symmetry-Based Explanation

How to explain normal distributions: reminder. Many practical applications assume that the distribution is Gaussian (normal). One way to derive the Gaussian distribution is to consider, among all distributions with mean 0 and known standard deviation σ, the distribution with the largest entropy $\mathscr{S}(\rho) \stackrel{\text{def}}{=} - \int \rho(x) \ln(\rho(x)) \, dx$ (see, e.g., [4]), i.e., to optimize entropy under the constraints

$$\int \rho(x) \, dx = 1, \quad \int x \cdot \rho(x) \, dx = 0, \quad \text{and} \quad \int x^2 \cdot \rho(x) \, dx = \sigma^2. \qquad (2)$$

For this constraint optimization problem, the Lagrange multiplier method reduces it to the following unconditional optimization problem

$$- \int \rho(x) \cdot \ln(\rho(x)) \, dx + \lambda_0 \cdot \left(\int \rho(x) \, dx - 1 \right) + \lambda_1 \cdot \left(\int x \cdot \rho(x) \, dx \right)$$
$$+ \lambda_2 \cdot \left(\int x^2 \cdot \rho(x) \, dx - \sigma^2 \right) \to \max .$$

Differentiating the objective function with respect to $\rho(x)$ and equating the derivative to 0, we conclude that

$$-\ln(\rho(x)) - 1 + \lambda_0 + \lambda_1 \cdot x + \lambda_2 \cdot x^2 = 0,$$

hence $\rho(x) = \exp((\lambda_0 - 1) + \lambda_1 \cdot x + \lambda_2 \cdot x^2)$. The requirement that the mean is 0 implies that $\lambda_1 = 0$, so we get the usual Gaussian distribution.

Entropy is scale-invariant. Entropy is scale-invariant in the sense that:

- if we have two distributions $\rho(x)$ and $\rho'(x)$ for which $\mathscr{S}(\rho) = \mathscr{S}(\rho')$, and
- we re-scale x and thus, transform the original distributions into the re-scaled ones $\rho_\lambda(x)$ and $\rho'_\lambda(x)$, then these re-scaled distributions will also have the same entropy $\mathscr{S}(\rho_\lambda) = \mathscr{S}(\rho'_\lambda)$.

Scale-invariant generalizations of entropy. It turns out that entropy is not the only functional with the above scale-invariance properties. All such scale-invariant functions have been described [5,6]. In addition to entropy, we can also have $\int \ln(\rho(x)) \, dx$ and $\int (\rho(x))^q \, dx$ for some q.

For scale-invariant generalizations of entropy, we get Student distribution. Optimizing $\int \ln(\rho(x)) \, dx$ under constraints (2) leads to

$$\int \ln(\rho(x)) \, dx + \lambda_0 \cdot \left(\int \rho(x) \, dx - 1 \right) + \lambda_1 \cdot \left(\int x \cdot \rho(x) \, dx \right)$$
$$+ \lambda_2 \cdot \left(\int x^2 \cdot \rho(x) \, dx - \sigma^2 \right) \to \max.$$

Differentiating the objective function with respect to $\rho(x)$ and equating the derivative to 0, we conclude that $\dfrac{1}{\rho(x)} - 1 + \lambda_0 + \lambda_1 \cdot x + \lambda_2 \cdot x^2 = 0$, hence

$$\rho(x) = \frac{1}{(1 - \lambda_0) - \lambda_1 \cdot x - \lambda_2 \cdot x^2}.$$

The requirement that the mean is 0 implies that $\lambda_1 = 0$, so we indeed get a particular case of the Student distribution.

Similarly, optimizing $\int (\rho(x))^q \, dx$ under constraints (2) leads to

$$\int (\rho(x))^q \, dx + \lambda_0 \cdot \left(\int \rho(x) \, dx - 1 \right) + \lambda_1 \cdot \left(\int x \cdot \rho(x) \, dx \right)$$
$$+ \lambda_2 \cdot \left(\int x^2 \cdot \rho(x) \, dx - \sigma^2 \right) \to \max.$$

Differentiating the objective function with respect to $\rho(x)$ and equating the derivative to 0, we conclude that $q \cdot (\rho(x))^{q-1} + \lambda_0 + \lambda_1 \cdot x + \lambda_2 \cdot x^2 = 0$, hence $\rho(x) = (a_0 + a_1 \cdot x + a_2 \cdot x^2)^{1/(q-1)}$. The requirement that the mean is 0 implies that $a_1 = 0$, so we indeed get (the generic case of) the Student distribution.

Acknowledgments. This work was performed when Olga Kosheleva and Vladik Kreinovich were visiting researchers with the Geodetic Institute of the Leibniz University of Hannover, a visit supported by the German Science Foundation. This work was also supported in part by NSF grant HRD-1242122.

A   Proof

$1°$. Depending on whether the value $1*1$ is equal to 1 or not, we have two possible cases: $1*1 = 1$ and when $1*1 \neq 1$. Let us consider these two cases one by one.

$2°$. Let us first consider the case when $1*1 = 1$. In this case, the value $0*1$ can be either equal to 0 or different from 0. Let us consider both subcases.

$2.1°$. Let us first consider the first subcase, when $0*1 = 0$.

In this case, for every $b > 0$, scale invariance with $\lambda = b$ implies that

$$(b \cdot 0) * (b \cdot 1) = (b \cdot 0),$$

i.e., that $0*b = 0$. By taking $b \to 0$ and using continuity, we also get $0*0 = 0$. Thus, $0*b = 0$ for all b.

By commutativity, we have $a*0 = 0$ for all a. So, to fully describe the operation $a*b$, it is sufficient to consider the cases when $a > 0$ and $b > 0$.

$2.1.1°$. Let us prove, by contradiction, that in this subcase, we have $1*a \leq 1$ for all a.

Indeed, let us assume that for some a, we have $b \overset{\text{def}}{=} 1*a > 1$. Then, due to associativity and $1*1 = 1$, we have $1*b = 1*(1*a) = (1*1)*a = 1*a = b$.

Due to scale-invariance with $\lambda = b$, the equality $1*b = b$ implies that $b*b^2 = b^2$. Thus, $1*b^2 = 1*(b*b^2) = (1*b)*b^2 = b*b^2 = b^2$.

Similarly, from $1*b^2 = b^2$, we conclude that for $b^4 = (b^2)^2$, we have $1*b^4 = b^4$, and, in general, that $1*b^{2^n} = b^{2^n}$ for every n.

Scale invariance with $\lambda = b^{-2^n}$ implies that $b^{-2^n}*1 = 1$. In the limit $n \to \infty$, we get $0*1 = 1$, which contradicts to our assumption that $0*1 = 0$. This contradiction shows that indeed, $1*a \leq 1$.

$2.1.2°$. For $a \geq 1$, monotonicity implies $1 = 1*1 \leq 1*a$, so $1*a \leq 1$ implies that $1*a = 1$.

Now, for any a' and b' for which $0 < a' \leq b'$, if we denote $r \overset{\text{def}}{=} \dfrac{b'}{a'} \geq 1$, then scale-invariance with $\lambda = a'$ implies that $a' \cdot (1*r) = (a' \cdot 1)*(a' \cdot r) = a'*b'$. Here, $1*r = 1$, thus $a'*b' = a' \cdot 1 = a'$, i.e., $a'*b' = \min(a', b')$. Due to commutativity, the same formula also holds when $a' \geq b'$. So, in this case, $a*b = \min(a, b)$ for all a and b.

$2.2°$. Let us now consider the second subcase of the first case, when $0*1 > 0$.

$2.2.1°$. Let us first show that in this subcase, we have $0*0 = 0$.

Indeed, scale-invariance with $\lambda = 2$ implies that from $0*0 = a$, we can conclude that

$$(2 \cdot 0) * (2 \cdot 0) = 0*0 = 2 \cdot a.$$

Thus $a = 2 \cdot a$, hence $a = 0$. The statement is proven.

2.2.2°. Let us now prove that in this subcase, $0 * 1 = 1$.

Indeed, in this case, for $a \overset{\text{def}}{=} 0 * 1$, we have, due to $0 * 0 = 0$ and associativity, that

$$0 * a = 0 * (0 * 1) = (0 * 0) * 1 = 0 * 1 = a.$$

Here, $a > 0$, so by applying scale invariance with $\lambda = a^{-1}$, we conclude that

$$0 * 1 = 1.$$

2.2.3°. Let us now prove that for every $a \leq b$, we have $a * b = b$. So, due to commutativity, we have $a * b = \max(a, b)$ for all a and b.

Indeed, from $1 * 1 = 1$ and $0 * 1 = 1$, due to scale invariance with $\lambda = b$, we conclude that $0 * b = b$ and $1 * b = b$. Due to monotonicity, $0 \leq a \leq b$ implies that $b = 0 * b \leq a * b \leq b * b = b$, thus $a * b = b$. The statement is proven.

3°. Let us now consider the remaining case when $1 * 1 \neq 1$.

3.1°. Let us denote $v(k) \overset{\text{def}}{=} 1 * \ldots * 1$ (k times). Then, for every m and n, the value $v(m \cdot n) = 1 * \ldots * 1$ ($m \cdot n$ times) can be represented as

$$(1 * \ldots * 1) * \ldots * (1 * \ldots * 1),$$

where we divide the 1 s into m groups with n 1s in each. For each group, we have $1 * \ldots * 1 = v(n)$. Thus, $v(m \cdot n) = v(n) * \ldots * v(n)$ (m times).

We know that $1 * \ldots * 1$ (m times) $= v(m)$. Thus, by using scale-invariance with $\lambda = v(n)$, we conclude that $v(m \cdot n) = v(m) \cdot v(n)$, i.e., that that function $v(n)$ is multiplicative. In particular, this means that for every number p and for every positive integer n, we have $v(p^n) = (v(p))^n$.

3.2°. If $v(2) = 1 * 1 > 1$, then by monotonicity, we get $v(3) = 1 * v(2) \geq 1 * 1 = v(2)$, and, in general, $v(n + 1) \geq v(n)$. Thus, in this case, the sequence $v(n)$ is (non-strictly) increasing.

Similarly, if $v(2) = 1 * 1 < 1$, then we get $v(3) \leq v(2)$ and, in general, $v(n + 1) \leq v(n)$, i.e., in this case, the sequence $v(n)$ is strictly decreasing.

Let us consider these two cases one by one.

3.2.1°. Let us first consider the case when the sequence $v(n)$ is increasing. In this case, for every three integers m, n, and p, if $2^m \leq p^n$, then $v(2^m) \leq v(p^n)$, i.e., $(v(2))^m \leq (v(p))^n$.

For all m, n, and p, the inequality $2^m \leq p^n$ is equivalent to $m \cdot \ln(2) \leq n \cdot \ln(p)$, i.e., to $\dfrac{m}{n} \leq \dfrac{\ln(p)}{\ln(2)}$. Similarly, the inequality $(v(2))^m \geq (v(p))^n$ is equivalent to $\dfrac{m}{n} \leq \dfrac{\ln(v(p))}{\ln(v(2))}$. Thus, the above conclusion

$$\text{if } 2^m \leq p^n, \text{ then } (v(2))^m \leq (v(p))^n$$

takes the following form:

$$\text{for every rational number } \frac{m}{n}, \text{ if } \frac{m}{n} \le \frac{\ln(p)}{\ln(2)} \text{ then } \frac{m}{n} \le \frac{\ln(v(p))}{\ln(v(2))}.$$

Similarly, for all m', n', and p, if $p^{n'} \le 2^{m'}$, then $v(p^{n'}) \le v(2^{m'})$, i.e., $(v(p))^{n'} \le (v(2))^{m'}$. The inequality $p^{n'} \le 2^{m'}$ is equivalent to $n' \cdot \ln(p) \le m' \cdot \ln(2)$, i.e., to $\dfrac{\ln(p)}{\ln(2)} \le \dfrac{m'}{n'}$. Also, the inequality $(v(p))^{n'} \le (v(2))^{m'}$ is equivalent to $\dfrac{\ln(v(p))}{\ln(v(2))} \le \dfrac{m'}{n'}$. Thus, the above conclusion

$$\text{if } p^{n'} \le 2^{m'}, \text{ then } (v(p))^{n'} \le (v(2))^{m'}$$

takes the following form:

$$\text{for every rational number } \frac{m'}{n'}, \text{ if } \frac{\ln(p)}{\ln(2)} \le \frac{m'}{n'} \text{ then } \frac{\ln(v(p))}{\ln(v(2))} \le \frac{m'}{n'}.$$

Let us denote $\alpha \overset{\text{def}}{=} \dfrac{\ln(v(2))}{\ln(2)}$ and $\beta \overset{\text{def}}{=} \dfrac{\ln(v(p))}{\ln(p)}$. For every $\varepsilon > 0$, there exist rational numbers $\dfrac{m}{n}$ and $\dfrac{m'}{n'}$ for which $\alpha - \varepsilon \le \dfrac{m}{n} \le \alpha \le \dfrac{m'}{n'} \le \alpha + \varepsilon$. For these numbers, the above two properties imply that $\dfrac{m}{n} \le \beta$ and $\beta \le \dfrac{m'}{n'}$ and thus, that $\alpha - \varepsilon \le \beta \le \alpha + \varepsilon$, i.e., that $|\alpha - \beta| \le \varepsilon$. This is true for all $\varepsilon > 0$, so we conclude that $\beta = \alpha$, i.e., that $\dfrac{\ln(v(p))}{\ln(v(2))} = \alpha$. Hence, $\ln(v(p)) = \alpha \cdot \ln(p)$ and thus, $v(p) = p^{\alpha}$ for all integers p.

3.2.2°. We can reach a similar conclusion $v(p) = p^{\alpha}$ when the sequence $v(n)$ is decreasing.

3.3°. By definition of $v(n)$, we have $v(m) * v(m') = v(m + m')$. Thus, we have

$$m^{\alpha} * (m')^{\alpha} = (m + m')^{\alpha}.$$

By using scale-invariance with $\lambda = n^{-\alpha}$, we get

$$\frac{m^{\alpha}}{n^{\alpha}} * \frac{(m')^{\alpha}}{n^{\alpha}} = \frac{(m + m')^{\alpha}}{n^{\alpha}}.$$

Thus, for $a = \dfrac{m^{\alpha}}{n^{\alpha}}$ and $b = \dfrac{(m')^{\alpha}}{n^{\alpha}}$, we get $a * b = (a^{\beta} + b^{\beta})^{1/\beta}$, where $\beta \overset{\text{def}}{=} 1/\alpha$.

Rational numbers $r = \dfrac{m}{n}$ are everywhere dense on the real line, hence the values r^{α} are also everywhere dense, i.e., every real number can be approximated, with any given accuracy, by such numbers. Thus, continuity implies that $a * b = (a^{\beta} + b^{\beta})^{1/\beta}$ for every two real numbers a and b.

The proposition is proven.

References

1. Aczél, J.: Lectures on Functional Equations and Their Applications. Dover, New York (2006)
2. Aczél, J., Dhombres, H.: Functional Equations in Several Variables. Cambridge University Press, Cambridge (1989)
3. Cressie, N.: Statistics for Spatial Data. Wiley, New York (2015)
4. Jaynes, E.T., Bretthorst, G.L.: Probability Theory: The Logic of Science. Cambridge University Press, Cambridge (2003)
5. Kosheleva, O.: Symmetry-group justification of maximum entropy method and generalized maximum entropy methods in image processing. In: Erickson, G.J., Rychert, J.T., Smith, C.R. (eds.) Maximum Entropy and Bayesian Methods, pp. 101–113. Kluwer, Dordrecht (1998)
6. Kreinovich, V., Kosheleva, O., Nguyen, H.T., Sriboonchitta, S.: Why some families of probability distributions are practically efficient: a symmetry-based explanation. In: Huynh, V.N., Kreinovich, V., Sriboonchitta, S. (eds.) Causal Inference in Econometrics, pp. 133–152. Springer Verlag, Cham, Switzerland (2016)
7. Lange, K.L., Little, R.J.A., Taylor, J.M.G.: Robust statistical modeling using the t-distribution. J. Am. Stat. Assoc. **84**(408), 881–896 (1989)
8. Mandelbrot, B.: The Fractal Geometry of Nature. Freeman, San Francisco (1983)
9. Rabinovich, S.G.: Measurement Errors and Uncertainty: Theory and Practice. Springer, Berlin (2005)
10. Resnick, S.I.: Heavy-Tail Phenomena: Probabilistic and Statistical Modeling. Springer, New York (2007)
11. Sheskin, D.J.: Handbook of Parametric and Nonparametric Statistical Procedures. Chapman and Hall/CRC, Boca Raton, Florida (2011)

Confidence Intervals for Common Mean of Lognormal Distributions

Narudee Smithpreecha$^{(\boxtimes)}$, Sa-Aat Niwitpong$^{(\boxtimes)}$, and Suparat Niwitpong$^{(\boxtimes)}$

Department of Applied Statistics, Faculty of Applied Science,
King Mongkut's University of Technology North Bangkok, Bangkok 10800, Thailand
narudee.s@rmutp.ac.th, {sa-aat.n,suparat.n}@sci.kmutnb.ac.th

Abstract. This paper presents new confidence intervals for the common mean of lognormal distributions by transforming the lognormal data. Three approaches were based on generalized confidence intervals (GCI) and adjusted method of variance estimates recovery (adjusted MOVER). A Monte Carlo simulation was used to assess the coverage probability and average length. The simulation study found that the adjusted MOVER approach based on Angus's conservative method (AM2) is appropriate and had the smallest coverage error in all of the scenarios. The generalized confidence interval approach (GCI) had the second smallest coverage error and had the smallest average lengths among the three approaches when the coverage probabilities were close to nominal level 0.95. Real data examples illustrate this approach.

1 Introduction

One of the important right skewed distributions with a long tail is lognormal distribution. It is widely used in many fields, such as environmental study, survival analysis, biostatistics and other statistical fields. The lognormal distribution has closely resembled a normal distribution. Simple to implement and easy to understand, by taking the natural logarithm of a random variable, the random variable will have a normal distribution.

Interval estimation of lognormal means for one, two and several populations have received widespread attention in papers of science and statistical literature. Statistical methods for interval estimation involving common mean for several lognormal distributions have also appeared frequently in many journals, such as biometrical journal by Tian and Wu [1] defined the concept of generalized variable and the large sample criteria to provide approach for the confidence interval estimation and hypothesis testing of the common mean of several lognormal populations. Lin and Wang [2] in a journal of applied statistics focused on making inferences on several log-normal means based on the modification of the quadratic method. There are also many other journals such as journal of statistical research by Ahmed et al. [3], journal of probability and statistical science by Baklizi and Ebrahem [4] and measurement science review by Cimermanová [5] interested in construction of the confidence intervals for common mean of several lognormal distributions.

© Springer International Publishing AG 2018
L. H. Anh et al. (eds.), *Econometrics for Financial Applications*, Studies in Computational Intelligence 760, https://doi.org/10.1007/978-3-319-73150-6_22

In this paper, the interest is to construct confidence intervals for common mean of lognormal distributions. A simple approach to construct the confidence intervals for lognormal mean by transform the lognormal data would be to log-transfer data prior to analyzing statistical. There have been several researchers used the log-transformed data to construct confidence intervals for mean of lognormal distribution. For example, the paper by Krishnamoorthy and Mathew [6] and Olsson [7]. It seems evident that the results for interval estimation of the common mean of several lognormal populations by taking the natural logarithm have not been studied. Furthermore, there has not been much discussion on methods for interval estimation of common lognormal means by taking the natural logarithm of data. Therefore, researchers have proposed new simple approaches to construct the confidence intervals for the common lognormal mean. The first approach is generalized confidence intervals (GCI) which is based on the concepts of generalized confidence interval and was introduced by Weerahandi [8]. The second and the third approaches are adjusted method of variance estimates recovery approach (adjusted MOVER) based on cox's method (AM1) and Angus's conservative method (AM2) which are based on the concepts of the method of variance of estimates recovery (MOVER) introduced by Zou and Donner [9]. The GCI approach, the MOVER approach and the adjusted MOVER approach have been successfully used to construct the confidence interval for many common parameters. As reviewed in Tian [10], Tian and Wu [1], Krishnamoorthy and Lu [11], Ye et al. [12], Donner and Zou [13], Suwan and Niwitpong [14], Li et al. [15], Wongkhao [16] and Thangjai and Niwitpong [17]. Therefore, the focus is to develop interval estimation procedures with three approaches for the common mean of lognormal distributions and then compare them to each of the situation in terms of coverage probability and average length.

This paper is organized as follows. The properties of lognormal distribution and the parameter of interest will be briefly introduced in Sect. 2. The three approaches developed and descriptions of computational procedures are presented in Sect. 3. Section 4 presents simulation results to evaluate performances of the three approaches on coverage probabilities and average lengths. Section 5 illustrates the proposed approaches with real examples. Finally, conclusions are given in Sect. 6.

2 Lognormal Distribution and the Parameter of Interest

Let $Y_1, Y_2, ..., Y_n$ be a random variable having lognormal distribution with two parameters. This means that the log-transformed variables $X_1 = \log Y_1, X_2 = \log Y_2, ..., X_n = \log Y_n$ are normally distributed, that has mean value $E(X) = \mu$ and variance $var(X) = \sigma^2$. The mean of Y is $E(Y) = \exp\left(\mu + \frac{\sigma^2}{2}\right)$, by taking the natural logarithm of a random variable we get $\log(E(Y)) = \mu + \frac{\sigma^2}{2}$.

According to Olsson [7], An estimator of $\log(E(Y))$ can be calculated from sample data as $\log\left(E(\widehat{Y})\right) = \overline{X} + \frac{S^2}{2}$ and an estimator of the variance of $\log\left(E(\widehat{Y})\right)$ is given by $var\left[\log\left(E(\widehat{Y})\right)\right] = \frac{S^2}{2} + \frac{S^4}{2(n-1)}$.

Consider k independent lognormal populations with a common mean α. Let $Y_{i1}, Y_{i2}, ..., Y_{in_i}$ be a random sample from the i-th lognormal population as follows:

$$X_{ij} = \log Y_{ij} \sim \left(\mu_i, \sigma_i^2\right), \text{ for } i = 1, 2, ..., k, j = 1, 2, ..., n_i.$$

Thus, the common mean is $\alpha = \exp\left(\mu_i + \frac{\sigma_i^2}{2}\right)$, $\log \alpha = \left(\mu_i + \frac{\sigma_i^2}{2}\right)$.

3   The Approaches of Confidence Interval Estimation

3.1   The Generalized Confidence Interval Approach

Weerahandi [8] introduced the concept of generalized confidence intervals (GCI) which is based on the generalized pivotal quantity (GPQ) for a parameter of interest θ and ν is a vector of nuisance parameters. A generalized pivot $R(X, x, \theta, \nu)$ for interval estimation, where x is an observed value of X, as a random variable having the following two properties:

1. $R(X, x, \theta, \nu)$ has a distribution free of the vector of nuisance parameters ν.
2. The value of $R(X, x, \theta, \nu)$ is θ.

Let R_α be the 100α-th percentile of R. Then R_α becomes the $100(1-\alpha)\%$ lower bound for θ and $\left(R_{\alpha/2}, R_{1-\alpha/2}\right)$ becomes a $100(1-\alpha)\%$ two-side generalized confidence interval for θ.

Consider k independent lognormal populations with a common mean α.

Thus, we have $\qquad \alpha = \exp\left(\mu_i + \frac{\sigma_i^2}{2}\right)$.

The common log-mean, $\quad \theta = \log \alpha = \left(\mu_i + \frac{\sigma_i^2}{2}\right)$.

Let $\overline{X}_i$ and S_i^2 denote the sample mean and variance for data X_{ij} for the i-th sample and let $\overline{x}_i$ and s_i^2 denote the observed sample mean and variance respectively.

Thus $\sigma_i^2 = \frac{(n_i-1)S_i^2}{V_i}$ $\quad$ where $V_i \sim \chi_{n_i-1}^2$.

where V_i is χ^2 variates with degrees of freedom and $n_i - 1$, we have the generalized pivot

$$R_{\sigma_i^2} = \frac{(n_i - 1)\, s_i^2}{V_i} \sim \frac{(n_i - 1)\, s_i^2}{\chi_{n_i-1}^2}. \tag{1}$$

The generalized pivotal quantity to estimate μ_i based on the i-th sample can be defined as

$$R_{\mu_i} = \overline{x}_i - \frac{Z_i}{\sqrt{U_i}}\sqrt{\frac{(n_i - 1)s_i^2}{n_i}}, \tag{2}$$

where Z_i and U_i denote standard normal variate and χ^2 variate with degree of freedom $n_i - 1$ respectively.

The generalized pivotal quantity for estimating θ based on the i-th sample is

$$R_\theta^{(i)} = R_{\mu_i} + \frac{R_{\sigma_i^2}}{2}. \tag{3}$$

From the i-th sample, the maximum likelihood estimator of θ is

$$\widehat{\theta}^{(i)} = \widehat{\mu}_i + \frac{\widehat{\sigma}_i^2}{2}, \quad where \ \widehat{\mu}_i = \overline{X}_i, \widehat{\sigma}_i^2 = S_i^2. \tag{4}$$

The variance for $\widehat{\theta}^{(i)}$ is

$$var\left(\widehat{\theta}^{(i)}\right) = \frac{\sigma_i^2}{n_i} + \frac{\sigma_i^4}{2(n_i - 1)}, \quad see \ Olsson \ [7]. \tag{5}$$

According to Ye et al. [12], the generalized pivotal quantity proposed for the common log-mean $\theta = \log \alpha$ is a weighted average of the generalized pivot $R_\theta^{(i)}$ based on k individual samples defined as

$$R_\theta = \frac{\sum\limits_{i=1}^{k} R_w R_\theta^{(i)}}{\sum\limits_{i=1}^{k} R_{w_i}}, \tag{6}$$

where

$$R_{w_i} = \frac{1}{R_{var\left(\widehat{\theta}^{(i)}\right)}}, \tag{7}$$

$$R_{var\left(\widehat{\theta}^{(i)}\right)} = \frac{R_{\sigma_i^2}}{n_i} + \frac{R_{\sigma_i^4}}{2(n_i - 1)}. \tag{8}$$

That is, $R_{var\left(\widehat{\theta}^{(i)}\right)}$ is $var\left(\widehat{\theta}^{(i)}\right)$ with σ_i^2 replaced by $R_{\sigma_i^2}$.

$\left[L_{Gci}, U_{Gci}\right] = \left(R_{\alpha/2}, R_{1-\alpha/2}\right)$ is the $100\left(1 - \alpha\right)\%$ two-side generalized confidence interval of the common log-mean $\theta = \log \alpha$.

$\left[\exp\left(L_{Gci}\right), \exp\left(U_{Gci}\right)\right] = \left(\exp\left(R_{\alpha/2}\right), \exp\left(R_{1-\alpha/2}\right)\right)$ is the $100\left(1 - \alpha\right)\%$ two-side generalized confidence interval of the common mean θ.

Computing algorithms

For a given data set X_{ij} for $i = 1, 2, ..., k, j = 1, 2, ..., n_i$, the generalized confidence intervals for θ can be computed by the following steps.

1. Compute $\overline{x}_i$ and s_i^2 for $i = 1, 2, ..., k$.
2. Generate $V_i \sim \chi_{n_i-1}^2$ and then calculate R_{σ^2} from (1) for $i = 1, 2, ..., k$.
4. Generate $Z_i \sim N(0, 1)$ and $U_i \sim \chi_{n_i-1}^2$ then calculate R_{μ_i} from (2) for $i = 1, 2, ..., k$.
4. Calculate $R_\theta^{(i)}$ from (3) for $i = 1, 2, ..., k$.
5. Repeat steps 2–3, calculate R_{w_i} from (7) and (8) for $i = 1, 2, ..., k$.
6. Compute R_θ following (6).
7. Repeat step 2–6 a total m times and obtain an array of R_θ's.
8. Rank this array of R_θ's from small to large. The 100α-th percentile of R_θ's, $R_\theta\left(\alpha\right)$, is an estimate of the lower bound of the one-sided $100\left(1 - \alpha\right)\%$ confidence interval and $\left(R_\theta\left(\alpha/2\right), R_\theta\left(1 - \alpha/2\right)\right)$ is a two-sided $100\left(1 - \alpha\right)\%$ confidence interval.
9. Calculate the interval length.
10. Count the number of successes in 5,000 independent generated datasets.
11. Calculate coverage probability and average length.

3.2 The Adjusted Method of Variance Estimates Recovery Approach

The concepts of the method of variance estimates recovery (the Mover approach) and the large sample method are used to create the adjusted method of variance estimates recovery (The adjusted MOVER approach).

The Mover approach was introduced by Zou and Donner [9] which considers two parameters $\theta_1 + \theta_2$ which have $100\,(1-\alpha)\%$ confidence limits (l_1, u_1) and (l_2, u_2), respectively. Under the assumption of the point estimates $\hat{\theta}_1$ and $\hat{\theta}_2$ are independence, the lower limit L and the upper limit U are given by

$$[L_1, U_1] = \left(\hat{\theta}_1 + \hat{\theta}_2\right) \pm z_{\alpha/2}\sqrt{\widehat{var}\left(\hat{\theta}_1\right) + \widehat{var}\left(\hat{\theta}_2\right)}, \tag{9}$$

where $\widehat{var}\left(\hat{\theta}_i\right) = \dfrac{\left(\hat{\theta}_i - l_i\right)^2}{z_{\alpha/2}^2}$, $\widehat{var}\left(\hat{\theta}_i\right) = \dfrac{\left(u_i - \hat{\theta}_i\right)^2}{z_{\alpha/2}^2}$.

For $i = 1, 2$. Using these estimates with from (9), two-side $100\,(1-\alpha)\%$ confidence limits for $\theta_1 + \theta_2$ given as

$$L = \left(\hat{\theta}_1 + \hat{\theta}_2\right) - \sqrt{\left(\hat{\theta}_1 - l_1\right)^2 + \left(\hat{\theta}_2 - l_2\right)^2}$$

$$U = \left(\hat{\theta}_1 + \hat{\theta}_2\right) + \sqrt{\left(U_1 + \hat{\theta}_1\right)^2 + \left(U_2 + \hat{\theta}_2\right)^2}.$$

Let $\theta^{(1)}, \theta^{(2)}, ..., \theta^{(k)}$ be k parameters of interest, where the estimates $\hat{\theta}^{(1)}, \hat{\theta}^{(2)}, ..., \hat{\theta}^{(k)}$ are independent. Use concept of Donner and Zou [13] to construct of a $100\,(1-\alpha)\%$ two-sided confidence interval (L, U) for $\theta^{(1)}, \theta^{(2)}, ..., \theta^{(k)}$.

Thus $[L, U] = \left(\hat{\theta}^{(1)} + \hat{\theta}^{(2)} + ... + \hat{\theta}^{(k)}\right) \pm z_{\alpha/2}\sqrt{\widehat{var}\left(\hat{\theta}^{(1)}\right) + ... + \widehat{var}\left(\hat{\theta}^{(k)}\right)}$.

Where the variance estimate for $\hat{\theta}^{(i)}$ at $\theta^{(i)} = l_i$ is $\widehat{var}\left(\hat{\theta}^{(i)}\right) = \dfrac{\left(\hat{\theta}^{(i)} - l_i\right)^2}{z_{\alpha/2}^2}$.

And the variance estimate at $\theta^{(i)} = u_i$ is $\widehat{var}\left(\hat{\theta}^{(i)}\right) = \dfrac{\left(u_i - \hat{\theta}^{(i)}\right)^2}{z_{\alpha/2}^2}$.

Therefore, the lower limit L and upper limit U for $\theta^{(1)}, \theta^{(2)}, ..., \theta^{(k)}$ is given by

$$L = \left(\hat{\theta}^{(1)} + \hat{\theta}^{(2)} + ... + \hat{\theta}^{(k)}\right) - \sqrt{\left(\hat{\theta}^{(1)} - l_1\right)^2 + ... + \left(\hat{\theta}^{(k)} - l_k\right)^2},$$

$$U = \left(\hat{\theta}^{(1)} + \hat{\theta}^{(2)} + ... + \hat{\theta}^{(k)}\right) + \sqrt{\left(u_1 - \hat{\theta}^{(1)}\right)^2 + ... + \left(u_k - \hat{\theta}^{(k)}\right)^2}.$$

The next step uses concept of the method of the large sample by Tain and Wu [1] for parameter with pooled estimate of the common parameter from k populations. It is defined as

$$\hat{\theta} = \dfrac{\displaystyle\sum_{i=1}^{k} \dfrac{\hat{\theta}^{(i)}}{var\left(\hat{\theta}^{(i)}\right)}}{\displaystyle\sum_{i=1}^{k} \dfrac{1}{var\left(\hat{\theta}^{(i)}\right)}}, \tag{10}$$

which gives a variance estimate for $\widehat{\theta}^{(i)}$ at $\theta^{(i)} = l_i$ and $\theta^{(i)} = u_i$ of

$$var\left(\widehat{\theta}^{(i)}\right) = \frac{1}{2}\left(\frac{\left(\widehat{\theta}^{(i)} - l_i\right)^2}{z_{\alpha/2}^2} + \frac{\left(u_i - \widehat{\theta}^{(i)}\right)^2}{z_{\alpha/2}^2}\right). \tag{11}$$

Therefore, the lower limit L for the common parameter θ is given by

$$L = \widehat{\theta} - z_{1-\alpha/2}\sqrt{\frac{1}{\sum\limits_{i=1}^{k}\frac{1}{\frac{\left(\widehat{\theta}^{(i)} - l_i\right)^2}{z_{\alpha/2}^2}}}}. \tag{12}$$

Similarly, the upper limit U for the common parameter θ is given by

$$U = \widehat{\theta} + z_{1-\alpha/2}\sqrt{\frac{1}{\sum\limits_{i=1}^{k}\frac{1}{\frac{\left(u_i - \widehat{\theta}^{(i)}\right)^2}{z_{\alpha/2}^2}}}}. \tag{13}$$

Hence, the adjusted MOVER solution for confidence interval estimation is

$$\left(\widehat{\theta} - z_{1-\alpha/2}\sqrt{1/\sum\limits_{i=1}^{k}\frac{z_{\alpha/2}^2}{\left(\widehat{\theta}^{(i)} - l_i\right)^2}}, \quad \widehat{\theta} + z_{1-\alpha/2}\sqrt{1/\sum\limits_{i=1}^{k}\frac{z_{\alpha/2}^2}{\left(u_i - \widehat{\theta}^{(i)}\right)^2}}\right).$$

From the i-th sample, where $i = 1, 2, ..., k$. The common log-mean θ is

$$\theta = \log\alpha = \left(\mu_i + \frac{\sigma_i^2}{2}\right).$$

The maximum likelihood estimator of common log-mean θ is

$$\widehat{\theta}^{(i)} = \widehat{\mu}_i + \frac{\widehat{\sigma}_i^2}{2}, \quad where \ \widehat{\mu}_i = \overline{X}_i, \widehat{\sigma}_i^2 = S_i^2.$$

Chami et al. [18] have presented Cox's method and Angus's conservative method for constructing confidence interval for log-mean of lognormal.

According to Cox's method, the confidence interval (CI_C) for $\widehat{\theta}^{(i)}$ is

$$l_{i1} = \overline{X}_i + \frac{S_i^2}{2} - Z_{1-\alpha/2}\sqrt{\frac{S_i^2}{n_i} + \frac{S_i^4}{2(n_i - 1)}}, \tag{14}$$

$$u_{i1} = \overline{X}_i + \frac{S_i^2}{2} + Z_{1-\alpha/2}\sqrt{\frac{S_i^2}{n_i} + \frac{S_i^4}{2(n_i - 1)}}, \tag{15}$$

According to Angus's conservative method, the confidence interval (CI_A) for $\widehat{\theta}^{(i)}$ is

$$l_{i2} = \overline{X}_i + \frac{S_i^2}{2} - \frac{t_{1-\alpha/2}}{\sqrt{n_i}}\sqrt{S_i^2\left(1 + \frac{S_i^2}{2}\right)}, \tag{16}$$

$$u_{i2} = \overline{X}_i + \frac{S_i^2}{2} + \frac{q_{\alpha/2}}{\sqrt{n_i}}\sqrt{S_i^2\left(1 + \frac{S_i^2}{2}\right)}, \tag{17}$$

Which $t_{1-\alpha/2}$ be $1-\alpha$ percentile of a t-distribution with n_i-1 degrees of freedom, and let $q_{\alpha/2} = \sqrt{\frac{n}{2}\left(\frac{n-1}{\chi_\alpha^2} - 1\right)}$ where χ_α^2 is the α- percentile of the chi-square distributions with $n_i - 1$ degrees of freedom.

Researchers use the method of the large sample for log-mean with pooled estimate. It is defined in Eq. (10) and variance estimate for $\widehat{\theta}^{(i)}$ in Eq. (11). Consequently, L and U are defined in Eqs. (12) and (13). One gains these two groups of confidence intervals (CI_C) for $\widehat{\theta}^{(i)}$ in Eqs. (14), (15) and (CI_A) for $\widehat{\theta}^{(i)}$ in Eqs. (16) and (17) defines AM1 and AM2.

Hence, the adjusted MOVER approach based on cox's method (AM1) for confidence interval estimation of common log-mean $\theta = \log\alpha$ is $\left[L_{AM1}, U_{AM1}\right]$

$$= \left(\widehat{\theta} - z_{1-\alpha/2}\sqrt{1/\sum_{i=1}^{k}\frac{z_{\alpha/2}^2}{\left(\widehat{\theta}^{(i)} - l_{i1}\right)^2}}, \widehat{\theta} + z_{1-\alpha/2}\sqrt{1/\sum_{i=1}^{k}\frac{z_{\alpha/2}^2}{\left(u_{i1} - \widehat{\theta}^{(i)}\right)^2}}\right). \tag{18}$$

or the confidence interval for common mean α is $\left[\exp\left(L_{AM1}\right), \exp\left(U_{AM1}\right)\right]$.

The adjusted MOVER approach based on Angus's conservative method (AM2) for confidence interval estimation of common log-mean $\theta = \log\alpha$ is $\left[L_{AM2}, U_{AM2}\right]$

$$= \left(\widehat{\theta} - z_{1-\alpha/2}\sqrt{1/\sum_{i=1}^{k}\frac{z_{\alpha/2}^2}{\left(\widehat{\theta}^{(i)} - l_{i1}\right)^2}}, \widehat{\theta} + z_{1-\alpha/2}\sqrt{1/\sum_{i=1}^{k}\frac{z_{\alpha/2}^2}{\left(u_{i1} - \widehat{\theta}^{(i)}\right)^2}}\right). \tag{19}$$

or the confidence interval for common mean α is $\left[\exp\left(L_{AM2}\right), \exp\left(U_{AM2}\right)\right]$.

Computing algorithms

For a given data set X_{ij} for $i = 1, 2, ..., k, j = 1, 2, ..., n_i$, the adjusted method of variance estimates recovery for θ can be computed by the following steps.

1. Compute $\overline{x}_i$ and s_i^2 for $i = 1, 2, ..., k$.
2. Compute $l_{i1}, u_{i1}, l_{i2}, u_{i2}$ from (14), (15), (16), (17) for $i = 1, 2, ..., k$.
3. Calculate $var((\widehat{\theta}^{(i)})$ from (11) for $i = 1, 2, ..., k$.
4. Compute $\widehat{\theta}$ following (10).
5. Calculate confidence interval estimation from (18), (19) for $i = 1, 2, ..., k$.
6. Calculate the interval length.
7. Count the number of successes in 5,000 independent generated datasets.
8. Calculate coverage probability and average length.

4   Simulation Studies

A simulation study was performed with the coverage probabilities and average lengths of the common mean of the lognormal distributions for various combinations of the number of samples $k = 2$, 3 and 10, the sample sizes $n = (n_1, , ..., n_k)$ and the population variance $\sigma^2 = (\sigma_1^2, , ..., \sigma_k^2)$, the values were different and the common $\theta = \log \alpha$ take 0 and 10. In this simulation study, 95% confidence intervals from three approaches were compared, comprising of the proposed procedure generalized confidence interval approach (GCI), the adjusted MOVER approach based on cox's method (AM1) and based on Angus's conservative method (AM2). For each parameter setting, 5000 random samples were generated, 2500 R_θ's were obtained for each of the random samples.

In Tables 1, 2 and 3, the following notation applies $n = (n_1, , ..., n_k)$ and $\sigma^2 = (\sigma_1^2, , ..., \sigma_k^2)$. For $k = 2$, we have $n_1^{(2)} = (15, 15), n_2^{(2)} = (10, 20), n_3^{(2)} = (20, 10),$ $n_4^{(2)} = (50, 50), n_5^{(2)} = (50, 100)$ and $\sigma_1^{2(2)} = (1, 1), \sigma_2^{2(2)} = (1, 9), \sigma_3^{2(2)} = (1, 25),$ $\sigma_4^{2(2)} = (1, 100)$. For $k = 3$, we have $n_1^{(3)} = (15, 15, 15), n_2^{(3)} = (10, 15, 20), n_3^{(3)} = (20, 15, 10), n_4^{(3)} = (30, 50, 100)$ and $\sigma_1^{2(3)} = (0.02, 0.2, 2), \sigma_2^{2(3)} = (1, 1, 1), \sigma_3^{2(3)} = (1, 4, 9), \sigma_4^{2(3)} = (1, 9, 100)$. For $k = 10$, we have $n_1^{(10)} = (15, 15, 15, 15, 15, 15, 15,$ $15, 15, 15), n_2^{(10)} = (15, 15, 1515, 15, 10, 10, 10, 10, 10), n_3^{(10)} = (30, 30, 30, 30, 30,$ $30, 30, 30, 30, 30), n_4^{(10)} = (50, 50, 80, 100, 100, 50, 50, 80, 100, 100)$ and $\sigma_1^{2(10)} = (1, 1, 1, 1, 1, 1, 1, 1, 1, 1), \sigma_2^{2(10)} = (1, 1, 9, 9, 25, 25, 49, 49, 81, 81), \sigma_3^{2(10)} = (0.01, 0.01, 0.1, 0.1, 1, 1, 10, 10, 100, 100)$.

Tables 1, 2, and 3 presents the coverage probabilities and average lengths for 2, 3 and 10 sample cases respectively. In sample case 2, the GCI approach overestimated the coverage probabilities for all of the scenarios. In other sample cases, the GCI approach tended to overestimate the coverage probabilities for all of the scenarios but this depends on n_i and σ_i^2, especially when the n_i is small and σ_i^2 are similar values. It is also shown that the GCI approach tends to be drop from the nominal level 0.95. In all sample cases, The AM1 approach provides the underestimate coverage probabilities and has the coverage probabilities close to the nominal level 0.95 when the number of samples went up. The AM2 approach performs very well for all of the scenarios, although it tends to produce a wider average lengths more than the GCI approach, but the average length is slightly higher than the GCI approach.

In this paper, the average lengths of all intervals are considered since the approaches provide the coverage probability above the nominal level of 0.95 for all cases. Finally, it was discovered that the AM2 approach provided much better results over the another approaches in terms of average lengths for all cases. However, the average lengths of the GCI approach were the shortest when the coverage probabilities were close to the nominal level of 0.95.

5   An Empirical Application

In this section, two real data examples are exhibited to illustrate the generalized confidence interval approach (GCI), the adjusted MOVER approach based on

Table 1. Coverage probabilities (CP) and average length (AL) of approximate 95% two – side confidence bounds for common mean α of lognormal distributions, $\mu_i = \log \alpha - \left(\sigma_i^2/2\right)$ (based on 5000 simulations, m of gci = 2500): 2 sample cases.

n	σ^2	θ	GCI		AM1		AM2	
			CP	AL	CP	AL	CP	AL
$n_1^{(2)}$	$\sigma_1^{2(2)}$	0	0.9578	1.0027	0.9076	0.8422	0.9804	1.1652
		10	0.9546	1.0033	0.8968	0.8423	0.9796	1.1654
	$\sigma_2^{2(2)}$	0	0.9558	1.4735	0.9144	1.2177	0.9812	1.6822
		10	0.9514	1.4784	0.9166	1.2204	0.9820	1.6859
	$\sigma_3^{2(2)}$	0	0.9586	1.4822	0.9224	1.2424	0.9856	1.7172
		10	0.9586	1.4795	0.9286	1.2381	0.9834	1.7113
	$\sigma_4^{2(2)}$	0	0.9600	1.4808	0.9260	1.2457	0.9866	1.7219
		10	0.9540	1.4831	0.9236	1.2468	0.9814	1.7233
$n_2^{(2)}$	$\sigma_1^{2(2)}$	0	0.9502	1.0042	0.8968	0.8361	0.9790	1.1672
		10	0.9532	1.0042	0.8990	0.8356	0.9752	1.1664
	$\sigma_2^{2(2)}$	0	0.9540	1.9166	0.8954	1.4605	0.9744	2.0687
		10	0.9550	1.9141	0.8956	1.4592	0.9770	2.0669
	$\sigma_3^{2(2)}$	0	0.9574	1.9796	0.9064	1.5040	0.9770	2.1362
		10	0.9584	1.9909	0.9144	1.5111	0.9802	2.1461
	$\sigma_4^{2(2)}$	0	0.9616	2.0357	0.9172	1.5351	0.9806	2.1811
		10	0.9642	2.0363	0.9142	1.5376	0.9830	2.1848
$n_3^{(2)}$	$\sigma_1^{2(2)}$	0	0.9532	1.0051	0.8976	0.8368	0.9770	1.1683
		10	0.9514	1.0022	0.8942	0.8342	0.9748	1.1647
	$\sigma_2^{2(2)}$	0	0.9502	1.2441	0.9112	1.0588	0.9828	1.4620
		10	0.9552	1.2474	0.9142	1.0603	0.9814	1.4641
	$\sigma_3^{2(2)}$	0	0.9524	1.2442	0.9236	1.0769	0.9838	1.4866
		10	0.9510	1.2422	0.9240	1.0777	0.9828	1.4877
	$\sigma_4^{2(2)}$	0	0.9524	1.2339	0.9346	1.0842	0.9852	1.4965
		10	0.9538	1.2276	0.9308	1.0800	0.9860	1.4908
$n_4^{(2)}$	$\sigma_1^{2(2)}$	0	0.9524	0.5026	0.9388	0.4757	0.9854	0.6928
		10	0.9504	0.5014	0.9316	0.4747	0.9866	0.6913
	$\sigma_2^{2(2)}$	0	0.9524	0.6985	0.9428	0.6678	0.9854	0.9722
		10	0.9528	0.6993	0.9380	0.6691	0.9874	0.9741
	$\sigma_3^{2(2)}$	0	0.9534	0.7089	0.9420	0.6791	0.9864	0.9889
		10	0.9538	0.7086	0.9416	0.6786	0.9892	0.9881
	$\sigma_4^{2(2)}$	0	0.9540	0.7119	0.9450	0.6798	0.9894	0.9898
		10	0.9522	0.7122	0.9418	0.6804	0.9904	0.9907
$n_5^{(2)}$	$\sigma_1^{2(2)}$	0	0.9514	0.4037	0.9408	0.3887	0.9860	0.5935
		10	0.9544	0.4038	0.9438	0.3888	0.9862	0.5934
	$\sigma_2^{2(2)}$	0	0.9558	0.6884	0.9464	0.6597	0.9866	0.9643
		10	0.9512	0.6857	0.9386	0.6577	0.9886	0.9614
	$\sigma_3^{2(2)}$	0	0.9540	0.7080	0.9452	0.6782	0.9876	0.9882
		10	0.9482	0.7058	0.9424	0.6768	0.9868	0.9861
	$\sigma_4^{2(2)}$	0	0.9508	0.7123	0.9458	0.6801	0.9880	0.9903
		10	0.9528	0.7113	0.9454	0.6791	0.9862	0.9889

Table 2. Coverage probabilities (CP) and average length (AL) of approximate 95% two − side confidence bounds for common mean α of lognormal distributions, $\mu_i = \log \alpha - \left(\sigma_i^2/2\right)$ (based on 5000 simulations, m of gci = 2500): 3 sample cases.

n_i	σ_i^2	θ	GCI		AM1		AM2	
			CP	AL	CP	AL	CP	AL
$n_1^{(3)}$	$\sigma_1^{2(3)}$	0	0.9542	0.1518	0.9248	0.1335	0.9786	0.1867
		10	0.9516	0.1511	0.9256	0.1332	0.9762	0.1862
	$\sigma_2^{2(3)}$	0	0.9512	0.8110	0.8876	0.6755	0.9774	0.9350
		10	0.9486	0.8146	0.8856	0.6777	0.9746	0.9381
	$\sigma_3^{2(3)}$	0	0.9380	1.3733	0.8818	1.1257	0.9736	1.5537
		10	0.9380	1.3662	0.8802	1.1181	0.9710	1.5432
	$\sigma_4^{2(3)}$	0	0.9562	1.4534	0.9136	1.2115	0.9814	1.6738
		10	0.9528	1.4615	0.9134	1.2125	0.9800	1.6751
$n_2^{(3)}$	$\sigma_1^{2(3)}$	0	0.9540	0.1909	0.9158	0.1572	0.9776	0.2262
		10	0.9514	0.1917	0.9100	0.1582	0.9728	0.2276
	$\sigma_2^{2(3)}$	0	0.9468	0.8142	0.8774	0.6725	0.9756	0.9364
		10	0.9440	0.8107	0.8806	0.6706	0.9746	0.9337
	$\sigma_3^{2(3)}$	0	0.9382	1.6750	0.8744	1.3048	0.9656	1.8374
		10	0.9352	1.6612	0.8628	1.2942	0.9674	1.8230
	$\sigma_4^{2(3)}$	0	0.9524	1.9390	0.8898	1.4702	0.9738	2.0834
		10	0.9564	1.9388	0.8986	1.4699	0.9758	2.0829
$n_3^{(3)}$	$\sigma_1^{2(3)}$	0	0.9544	0.1301	0.9384	0.1181	0.9804	0.1645
		10	0.9552	0.1300	0.9362	0.1181	0.9798	0.1645
	$\sigma_2^{2(3)}$	0	0.9460	0.8152	0.8796	0.6739	0.9724	0.9384
		10	0.9448	0.8141	0.8844	0.6739	0.9730	0.9383
	$\sigma_3^{2(3)}$	0	0.9346	1.1852	0.8876	0.9956	0.9750	1.3737
		10	0.9422	1.1955	0.8902	1.0011	0.9768	1.3812
	$\sigma_4^{2(3)}$	0	0.9536	1.2248	0.9220	1.0622	0.9848	1.4657
		10	0.9510	1.2147	0.9212	1.0584	0.9842	1.4604
$n_4^{(3)}$	$\sigma_1^{2(3)}$	0	0.9514	0.0981	0.9398	0.0929	0.9800	0.1316
		10	0.9534	0.0982	0.9404	0.0929	0.9812	0.1317
	$\sigma_2^{2(3)}$	0	0.9516	0.3716	0.9372	0.3528	0.9844	0.5294
		10	0.9518	0.3716	0.9376	0.3530	0.9886	0.5297
	$\sigma_3^{2(3)}$	0	0.9508	0.8136	0.9274	0.7564	0.9918	1.0740
		10	0.9516	0.8139	0.9280	0.7565	0.9902	1.0741
	$\sigma_4^{2(3)}$	0	0.9504	0.9190	0.9308	0.8545	0.9878	1.1996
		10	0.9542	0.9169	0.9342	0.8523	0.9892	1.1966

Table 3. Coverage probabilities (CP) and average length (AL) of approximate 95% two – side confidence bounds for common mean α of lognormal distributions, $\mu_i = \log \alpha - \left(\sigma_i^2/2\right)$ (based on 5000 simulations, m of gci = 2500): 10 sample cases.

n_i	σ_i^2	θ	GCI		AM1		AM2	
			CP	AL	CP	AL	CP	AL
$n_1^{(10)}$	$\sigma_1^{2(10)}$	0	0.8806	0.4456	0.8020	0.3598	0.9586	0.4983
		10	0.8724	0.4450	0.7896	0.3593	0.9566	0.4976
	$\sigma_2^{2(10)}$	0	0.9248	0.9981	0.8730	0.8216	0.9694	1.1360
		10	0.9308	0.9933	0.8704	0.8180	0.9722	1.1311
	$\sigma_3^{2(10)}$	0	0.9600	0.0763	0.9226	0.0650	0.9734	0.0909
		10	0.9548	0.0761	0.9178	0.0648	0.9738	0.0907
$n_2^{(10)}$	$\sigma_1^{2(10)}$	0	0.8648	0.5010	0.7488	0.3852	0.9412	0.5403
		10	0.8674	0.5018	0.7624	0.3857	0.9444	0.5410
	$\sigma_2^{2(10)}$	0	0.9182	1.0171	0.8598	0.8205	0.9660	1.1346
		10	0.9274	1.0147	0.8730	0.8192	0.9706	1.1329
	$\sigma_3^{2(10)}$	0	0.9566	0.0763	0.9204	0.0649	0.9690	0.0908
		10	0.9576	0.0764	0.9178	0.0650	0.9692	0.0909
$n_3^{(10)}$	$\sigma_1^{2(10)}$	0	0.9052	0.2965	0.8744	0.2660	0.9874	0.3729
		10	0.9086	0.2967	0.8736	0.2661	0.9886	0.3730
	$\sigma_2^{2(10)}$	0	0.9482	0.6563	0.9202	0.5953	0.9848	0.8342
		10	0.9444	0.6564	0.9182	0.5947	0.9830	0.8333
	$\sigma_3^{2(10)}$	0	0.9532	0.0509	0.9318	0.0471	0.9702	0.0664
		10	0.9522	0.0509	0.9318	0.0471	0.9712	0.0663
$n_4^{(10)}$	$\sigma_1^{2(10)}$	0	0.9202	0.1791	0.9118	0.1713	0.9896	0.2615
		10	0.9304	0.1793	0.9216	0.1714	0.9890	0.2618
	$\sigma_2^{2(10)}$	0	0.9510	0.4992	0.9312	0.4582	0.9714	0.6694
		10	0.9526	0.5011	0.9316	0.4599	0.9680	0.6718
	$\sigma_3^{2(10)}$	0	0.9524	0.0372	0.9424	0.0355	0.9658	0.0524
		10	0.9502	0.0372	0.9390	0.0355	0.9618	0.0524

cox's method (AM1) and Angus's conservative method (AM2). All examples have been studied and reported by Lin and Wang [2]. The first example (A) was the medical charge data divided into two groups, 119 of them were an American group and 106 of them were the white group. The second example (B) was the pharmacokinetics data equally divided into three groups, 22 of them were group 1, group 2 and group 3. The data sets are presented in Table 4 and the results of confidence interval for two data sets are presented in Table 5. It can be seen that the interval of the adjusted MOVER approach based on Angus's conservative method (AM2) has the confidence interval close to the sample mean and wider lengths more than other approaches.

Table 4. The summary statistics of the log-transformed data sets.

Data set	n_i	$\overline{X}_i$	s_i^2	$\widehat{\theta}_i$
(A) The medical charge data				
American group	119	9.067	1.825	9.979
White group	106	8.693	2.693	10.039
(B) The pharmacokinetics data				
Group 1	22	2.601	0.24	2.721
Group 2	22	2.596	0.20	2.696
Group 3	22	2.599	0.17	2.684

Table 5. The results of confidence interval for four data sets.

The approaches	Confidence interval	Length
Data set (A) the medical charge data		
The generalized confidence interval approach (GCI)	(9.724, 10.288)	0.564
The adjusted MOVER approach		
Based on cox's method (AM1)	(9.723, 10.274)	0.551
Based on Angus's conservative method (AM2)	(9.722, 10.601)	0.879
Data set (B) the pharmacokinetics data		
The generalized confidence interval approach (GCI)	(2.582, 2.825)	0.243
The adjusted MOVER approach		
Based on cox's method (AM1)	(2.585, 2.811)	0.226
Based on Angus's conservative method (AM2)	(2.578, 2.893)	0.315

6   Discussion and Conclusions

This paper has presented three simple approaches to construct confidence intervals for common mean of lognormal distributions. The proposed confidence intervals were constructed by the generalized confidence interval approach (GCI), the adjusted MOVER approach based on cox's method (AM1) and based on Angus's conservative method (AM2). By the simulation studies, coverage probabilities form the generalized confidence interval approach (GCI) was always close to the nominal confidence level at 0.95. But there were a few cases, it seems that the generalized confidence interval approach (GCI) performed well only when σ_i^2 were more various. The adjusted MOVER approach based on cox's method (AM1) provided underestimated coverage probabilities for all cases. The adjusted MOVER approach based on Angus's conservative method (AM2) yielded coverage probabilities which tended to be high compared with the nominal level of 0.95 for all almost cases and the average lengths were wide to a little bit as compared with the GCI approach. Overall, the generalized confidence interval approach

(GCI) provided coverage probabilities close to nominal level 0.95 and average lengths is shorter than other approaches for $k = 2$. The adjusted MOVER approach based on Angus's conservative method (AM2) provides stable coverage probabilities for all k. In conclusion, the adjusted MOVER approach based on Angus's conservative method (AM2) can be successfully used to estimate the common mean of lognormal distributions.

The results of the generalized confidence interval approach (GCI) for common mean ($k \geq 2$) are similar to the simulation of the generalized confidence interval approach (GCI) for single mean of lognormal distribution which is studied by Olsson [7]. However, the coverage probabilities for $k \geq 2$ is decrease when k increased. In addition, this paper is constructing the confidence intervals for common mean of lognormal distributions by transform the lognormal data prior to use the generalized confidence interval approach (GCI). It is simple to construct the confidence intervals for the common mean. The results show that, it provided coverage probabilities close to nominal level 0.95. However, the results are not good equal to results of Lin and Wang [2], but average lengths are shorter.

Acknowledgments. The first author gratefully acknowledges the financial support from Rajamangala University of Technology Phra Nakhon of Thailand.

References

1. Tian, L., Wu, J.: Inferences on the common mean of several log-normal populations: the generalized variable approach. Biometrical J. **49**, 944–951 (2007)
2. Lin, S.H., Wang, R.S.: Modified method on the means for several log-normal distributions. J. Appl. Stat. **40**(1), 194–208 (2013)
3. Ahmed, S.E., Tomkins, R.J., Volodin, A.I.: Test of homogeneity of parallel samples from lognormal populations with unequal variances. J. Stat. Res. **35**(2), 25–33 (2001)
4. Baklizi, A., Ebrahem, M.: Interval estimation of common lognormal mean of several populations. J. Probab. Stat. Sci. **3**(1), 1–16 (2005)
5. Cimermanová, K.: Estimation of confidence intervals for the log-normal means and for the ratio and the difference of log-normal means with application of breath analysis. Measur. Sci. Rev. **7**(4), 31–36 (2007)
6. Krishnamoorthy, K., Mathew, T.: Inferences on the means of lognormal distributions using generalized p-values and generalized confidence intervals. J. Stat. Plann. Infer. **115**, 103–121 (2003)
7. Olsson, U.: Confidence intervals for the mean of a log-normal distribution. J. Stat. Educ. **13**(1) (2005). Available via DIALOG. www.amstat.org/puplications/jse/v13n1olsson.html. Accessed 11 Aug 2016
8. Weerahandi, S.: Generalized confidence intervals. J. Am. Stat. Assoc. **88**, 899–905 (1993)
9. Zou, G.Y., Donner, A.: Construction of confidence limits about effect measures: a general approach. Stat. Med. **27**, 1693–1702 (2008)
10. Tian, L.: Inferences on the common coefficient of variation. Stat. Med. **24**, 2213–2220 (2005)

11. Krishnamoorthy, K., Lu, Y.: Inference on the common means of several normal populations based on the generalized variable method. Biometrics **59**, 237–247 (2003)
12. Ye, R.D., et al.: Inferences on the common mean of several inverse Gaussian populations. Comput. Stat. Data Anal. **54**, 906–915 (2010)
13. Donner, A., Zou, G.Y.: Closed-form confidence intervals for function of the normal mean and standard deviation. Stat. Meth. Med. Res. **21**(4), 347–359 (2012)
14. Suwan, S., Niwitpong, S.: Estimated variance ratio confidence interval of nonnormal distributions. Far. East J. Math. Sci. **4**, 339–350 (2013)
15. Li, H.Q., et al.: Confidence intervals for ratio of two poisson rates using the method of variance estimates recovery. Comput. Stat. **29**, 869–889 (2014)
16. Wongkhao, A.: Confidence intervals for parameters of normal distribution. Ph.D. thesis: King Mongkut's University of Technology North Bangkok (2014)
17. Thangjai, W., Niwitpong, S.: Confidence intervals for the weighted coefficients of variation of two-parameter exponential distributions. Cogent Math. **4**, 1–16 (2017)
18. Chami, P., Antoine, R., Sahai, A.: On efficient confidence intervals for the lognormal mean. J. Appl. Sci. **7**(13), 1790–1794 (2007)

Perspectives and Experiments of Hybrid Particle Swarm Optimization and Genetic Algorithms to Solve Optimization Problems

Apirak Sombat[1], Teerapol Saleewong[2], and Poom Kumam[2(✉)]

[1] KMUTT-Fixed Point Research Laboratory, Room SCL 802 Fixed Point Laboratory, Science Laboratory Building, Department of Mathematics, Faculty of Science, King Mongkut's University of Technology Thonburi (KMUTT), 126 Pracha-Uthit Road, Bang Mod, Thrung Khru, Bangkok 10140, Thailand
apirak.som@mail.kmutt.ac.th
[2] Science Laboratory Building, KMUTT-Fixed Point Theory and Applications Research Group (KMUTT-FPTA), Faculty of Science, Theoretical and Computational Science Center (TaCS), King Mongkut's University of Technology Thonburi (KMUTT), 126 Pracha-Uthit Road, Bang Mod, Thrung Khru, Bangkok 10140, Thailand
{teerapol.sal,poom.kum}@kmutt.ac.th

Abstract. Nowadays, there are many tools to solve the optimization problem. One of the popular tool is the population-based metaheuristics can be viewed as an iterative improvement in a population of solutions. Algorithms such as Particle swarm optimization (PSO) is the swarm intelligent that find the answer by global and local search with the velocity and genetic algorithm (GA) is the stochastic search procedure based on the mechanics of natural selections. Both of them belong to this class of metaheuristics. In this paper is to present the perspective and experiments of the hybrid algorithm of genetic algorithm and particle swarm optimization to solve the optimization problems.

Keywords: Particle swarm optimization · Genetic algorithm Hybrid PSO GA

1 Introduction

The optimization problem is the problem of finding the best solution from all feasible solutions. A well-know example is the traveling salesman problem in which the salesman intends to visit for example 40 cities, exactly once so as to minimize the overall distance traveled or the overall travelling cost. Optimization problem occurs in the minimization of time, cost and risk or the maximization of profit, quality and efficiency. A large number of real-life optimization problems are complex and difficult to solve. They cannot be solved in an exact manner within a reasonable amount of time. Using approximate algorithms is the main

© Springer International Publishing AG 2018
L. H. Anh et al. (eds.), *Econometrics for Financial Applications*, Studies in Computational Intelligence 760, https://doi.org/10.1007/978-3-319-73150-6_23

alternative to solve this class of problems. In order to find satisfactory solutions for these problems, metaheuristics can be used. Metaheuristics represent approximate algorithms applicable to a large variety of optimization problems. Metaheuristic algorithm are becoming very powerful in solving hard optimization problems, and they have been applied in almost all major areas of science and engineering as well as industrial applications. Metaheuristics provide "acceptable" solutions in a reasonable time for solving hard and complex problems. Metaheuristic algorithms have been developed with three main purposes: solving faster, solving large problem and obtaining robust algorithms. Metaheuristic is nature-inspire by mimic natural metaphors to solve the problems. There are 2 classifications of metaheuristic are single-solution based metaheurisctics (S-metaheuricstics) and population-based metaheuristics (P-metaheuristics). In this is paper focus Genetic algorithm and Particle swarm optimization both of them is under population-base metaheuristics.

Genetic algorithm (GA) is established by Holland [1] in 1960 and later modified by Goldberg [2,3]. He got the inspiration from Darwin's theory: survival of the fittest and the mechanism of natural selection. GA is a population-based search and optimization method which mimics the process of natural evolution. GA imitate the genetic process in which hereditary characteristics that children are received from parents. GA works with a population of points rather than a single point. The GA comprised of reproduction, crossover, and mutation. Reproduction is the process of survival of the fittest selection. Crossover is to generate new individual offspring from parent. Mutation is the occasional random inversion of bit values that generates no-recursive offspring. GA is a method for moving from one population of "chromosomes" to a new population by using a kind of "natural selection" together with the genetics operators as explained in Fig. 1.

Particle swarm optimization (PSO) is established in 1995 by Kennedy and Eberhart [4,5]. It was inspired by bird flocking and fish schooling to find a place

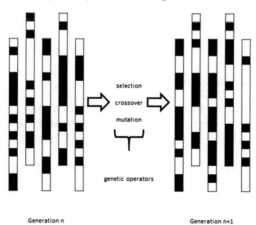

selection

crossover

mutation

genetic operators

Generation n   Generation n+1

Fig. 1. GA terminology [6]

with enough food. Indeed, in those swarm, a coordinated behavior using local movements emerges without any central control. In the basic model, a swarm consist of N particles flying around in a $D - dimensional$ search space. Each particle i is a candidate solution to the problems, and is represented by the vector x_i in the decision space. A particle has its own position and velocity, which means the flying direction and step of the particle show in Fig. 2

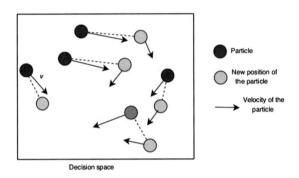

Fig. 2. PSO with their associated positions and velocities.

In this paper is divided into 4 sections including the introduction, Sect. 2 hybrid pso-ga. Section 3 consider the application of hybrid pso-ga. Section 4 conclusion.

2   Hybrid PSO-GA

Hybrid technique is a combination of the desirable properties of different approaches to using their individual advantage to solve the problems. The strengths and weaknesses of the metaheuristics have in each method. So, to use the advantage from each method to solve the optimization is to develop a hybrid technique.

Hybridization can be done in several ways:

1. Initiate the algorithm with one technique and then apply by other technique,
2. Embed the unique operators of one technique to another technique,
3. apply local search to improve the solution obtained by the global search etc.

Hybridization is the main topic of this article which is a perspective on hybrid algorithm of PSO and GA.

Robinson et al. [7] proposed two hybrid algorithms; GA-PSO and PSO-GA for solving a particular electromagnetic application of profiled corrugated horned antenna. He used GA to generate the initial population for PSO and used PSO to generate the initial population for GA repectively. His results showed that hybrid PSO-GA outperforms than PSO and GA.

Grimaldi et al. [8] proposed hybrid technique combining GA and PSO called genetical swarm optimization (GSO) for solving combinatorial optimization problems. The population is divided into two parts and is evolved with GA and PSO in every iteration. They defined a new parameter Hybridization Constant (HC) that expresses the percentage of population that is evolved with GA in every iteration $HC = 0$ implies that the procedure is pure PSO (the whole population is updated according to PSO operators) and $HC = 1$ implied that pure GA is being followed.

Juang [9] proposed hybridization strategy of PSO and GA (HGAPSO) upper half of the best performing individuals in population is regarded as elite. Before using GA operators, the algorithm is first enhanced by means of PSO, instead of being reproduced directly to the next generation. They applied HGAPSO to neural/fuzzy network design.

Shi [10] proposed variable population-size genetic algorithm (VPGA) by introducing the dying probability for the individuals and the war/disease process for the population. A novel PSO-GA based hybrid algorithm (PGHA).

Esmin [11] proposed a PSO algorithm coupled with a GA mutation operator only. They called HPSOM and used it for solving unconstrained global optimization problems.

Kim [12] proposed an improved GA called GA-PSO by using PSO and the concept of Euclidean distance on mutation procedure of GA.

Kao [13] proposed hybrid method combining GA and PSO for the global optimization of multimodal functions. This technique incorporates concepts from GA and PSO and creates individuals in a new generation.

Premalatha [14] proposed a discrete PSO with GA operators for document clustering. A reproduction operation by GA operators when the stagnation in movement of the particle is detected.

Jeong [15] proposed GA/PSO-hybrid algorithm. Half the population is updated by GA operation and the other half is updated by PSO.

Dhadwal [16] proposed particle swarm assisted genetic algorithm (PSGA) is a hybrid improved PSO and GA. PSO phase involves the enhancement of worst solutions by using the global-local best inertia weight and acceleration coefficients to increase the efficiency. GA phase is a new rank-based multi-parent crossover is used by modifying the crossover and mutation operators which favors both the local and global.

Andalib Sahnehsaraei [17] proposed hybrid algorithm makes use of the functions and operations of both algorithms such as mutation, traditional or classical crossover, multiple-crossover and the PSO formula. Selection of these operators is based on a fuzzy probability.

Garg [18] proposed hybrid technique named as a PSO-GA for solving the constrained optimization. PSO operates in the direction of improving the vector while GA has been used for modifying the decision vectors using GA operators.

Sebt [19] proposed hybridization of genetic algorithm and fully informed particle swarm for solving the multi-mode resource-constrained project scheduling problem. He show that the proposed mode improvement procedure remarkably improves the project makespan. Comparing the results of the proposed HGFA with other approaches using well-know PSPLIB benchmark sets validates the effectiveness of the proposed algorithm to solve the MRCPSP.

3   Experimental

In this section is to show the example of some hybridization techniques to solve the optimization problems.

1. Andalib Sahnehsaraei [17] multi-objective test function and single-objective test function were used to evaluate the capabilities of the hybrid algorithm. Contrasting the results of the hybrid algorithm with other algorithms demonstrates the superiority of the hybrid algorithm with regard to single and multi-objective optimization problems show in Fig. 3.

	GA (traditional crossover)	GA (multiple crossover)	PSO (standard)	Hybrid algorithm GA-PSO
Mean	$1.33 \times 10^{+2}$	$8.25 \times 10^{+1}$	$2.04 \times 10^{+1}$	5.34×10^{-1}
Standard deviation	$1.32 \times 10^{+2}$	$5.51 \times 10^{+1}$	$2.53 \times 10^{+1}$	1.38×10^{0}

Rosenbrock function

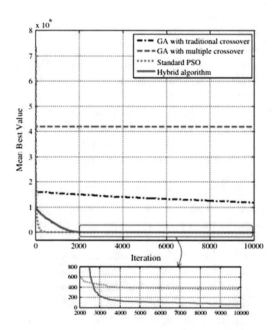

Fig. 3. The evolutionary trajectory of the single-objective optimization algorithm on the Rosenbrock test function [17]

2. Jeong et al. [15] The configurations of the two consisting methods, GA and PSO, were investigated to enhance the diversity of the former and the fast convergence of the latter simultaneously. The new hybrid algorithm was applied to two test function problems, and the results indicated that the search ability was improved by suitable tuning of the configurations.

$$Minimize \ f_1(\overrightarrow{x}) = cos(\frac{\pi}{2}x_1)cos(\frac{\pi}{2}x_2)(1 + g(\overrightarrow{x})) \tag{1}$$

$$Minimize \ f_2(\overrightarrow{x}) = cos(\frac{\pi}{2}x_1)sin(\frac{\pi}{2}x_2)(1 + g(\overrightarrow{x})) \tag{2}$$

$$Minimize \ f_2(\overrightarrow{x}) = cos(\frac{\pi}{2}x_1)sin(\frac{\pi}{2}x_2)(1 + g(\overrightarrow{x})) \tag{3}$$

$$where \ g(\overrightarrow{x}) = 100[k + \sum_{i=1}^{k}[(x_i - 0.5)^2 - cos(20\pi(x_i - 0.5))]] \tag{4}$$

	New HGAPSO	Simple HGAPSO	Pure GA	Pure PSO
Best	0.000001	0.003518	76.397080	0.068120
Worst	0.054241	3.674206	167.861800	49.161160
Average	0.003208	0.358010	134.630400	14.584250

Results of general distance

3. Shi et al. [10] The proposed PGHA synthesizes the merits in both PSO and GA. It is a simple and yet effective model to handle different kinds of continuous optimization problems. Simulated experiments for the optimization of nonlinear functions show that the PGHA algorithm is superior to the PSO and GA in both the speed of convergence and the ability of finding the global optimum.

	Success(%)	Time	F^*
PSO	100	4.812	0.000020
SGA	100	0.015	0.000000
VPGA	80	0.220	0.000000
PGHA	100	0.062	0.000000

Comparisons for Rosenbrock test function

4   Conclusions

PSO and GA are the powerful optimization techniques that have been applied to a wide range of optimization problems. This paper focus on perspective and experimental of hybrid techniques, which are quite useful and popular idea being applied to PSO and GA in order to increase efficiency and robustness.

From this present perspective and experimental of hybrid PSO and GA is more powerful techniques by combining the advantage of PSO and GA. The present study may motivate and help the researchers working in the field of metaheuristics to develop new hybrid models or to apply the existing models to new application areas.

Acknowledgements. This project was supported by the Theoretical and Computational Science (TaCS) Center under Computational and Applied Science for Smart Innovation Cluster (CLASSIC), Faculty of Science, KMUTT.

References

1. Holland, J.H.: Adaptation in Natural and Artificial Systems. University of Michigan Press, Ann Arbor (1975)
2. Goldberg, D.E.: Genetic Algorithms in Search, Optimization and Machine learning. Studies in Computational Intelligence, 1st edn. Addison-Wesley Longman Publishing Co., Inc., Boston (1989)
3. Goldberg, D.E., Deb, K.: A comparative analysis of selection schemes used in genetic algorithms. Found. Genetic Algorithms **1**, 69–93 (1991). Morgan Kaufman
4. Kennedy, J., Eberhart, R.: Particle swarm optimization. In: IEEE International Conference on Neural Networks, vol. 4, pp. 1942–1948 (1995)
5. Eberhart, R.C., Kennedy, J.: A new optimizer using particle swarm theory. In: Proceedings of the Sixth International Symposium on Micro Machine and Human Science EP'98, Piscataway, Nagoya, Japan, pp. 332–339. IEEE (1995)
6. Meenu, Verma, A.: A survey on hybrid genetic algorithm. Int. J. Adv. Res. Eng. Technol. **2**(V) (2014). www.ijaret.org, ISSN 2320-6802
7. Robinson, J., Sinton, S., Samii, Y.R.: Particle swarm, genetic algorithm, and their hybrids: optimization of a profiled corrugated horn antenna. In: Proceedings of the IEEE International Symposium in Antennas and Propagation Society 2002, pp. 314–317 (2002)
8. Gimaldi, E.A., Grimacia, F., Mussetta, M., Pirinoli, P., Zich, R.E.: A new hybrid genetical - swarm algorithm for electromagnetic optimization. In: Proceedings of International Conference on Computational Electromagnetic and its application, Beijing, China, pp. 157–160 (2004)
9. Juang, C-F.: A hybrid of genetic algorithm and particle swarm optimization for recurrent network design. IEEE Trans. Syst. Man Cybern. Part B Cybern. **34**, 997–1006 (2004)
10. Shi, X.H., Liang, Y.C., Lee, H.P., Lu, C., Wang, L.M.: An improved GA and a novel PSO-GA based hybrid algorithm. Inf. Process. Lett. **93**, 255–261 (2005)
11. Esmin, A.A., Lambert-Torres, G., Alvarenga, G.B.: Hybrid evolutionary algorithm based on PSO and GA mutation. In: Proceedings of 6th International Conference on Hybrid Intelligent Systems, pp. 57–62 (2006)

12. Kim, H.: Improvement of genetic algorithm using PSO and Euclidean data distance. Int. J. Inform. Technol. **12**, 142–148 (2006)
13. Kao, Y.-T., Zahara, E.: A hybrid genetic algorithm and particle swarm optimization for multimodal functions. Appl. Soft Comput. **8**, 849–857 (2008)
14. Premalatha, K., Natarajan, A.M.: Discrete PSO with GA operators for document clustering. Int. J. Recent Trends Eng. **1**, 20–24 (2009)
15. Jeong, S., Hasegawa, S., Shimoyama, K., Obayashi, S.: Development and investigation of efficient GA/PSO-Hybrid algorithm applicable to Real-World design optimization. IEEE Computational Intelligence (2009)
16. Dhadwal, M.K., Jung, S.N., Kim, C.J.: Advanced particle swarm assisted genetic algorithm for constrained optimization problems. Comput. Optim. Appl. **58**, 781–806 (2014)
17. Andalib Sahnehsaraei, M., Mahmoodabadi, M.J., Taherkhorsandi, M., Castillo-Villar, K.K., Mortazavi Yazdi, S.M.: A hybrid global optimization algorithm: particle swarm optimization in association with a genetic algorithm. In: Complex System Modelling and Control Through Intelliegent Soft Computations. Studies in Fuzziness and Soft Computing, vol. 319 (2015)
18. Garg, H.: A hybrid PSO-GA algorithm for constrained optimization problems. Appl. Math. Comput. **274**(2016), 292–305 (2016)
19. Sebt, M.H., Afshar, M.R., Alipouri, Y.: Hybridization of genetic algorithm and fully informed particle swarm for solving the multi-mode resource-constrained project scheduling problem. Eng. Optim. **49**(3), 513–530 (2017)

Simultaneous Confidence Intervals for All Differences of Means of Two-Parameter Exponential Distributions

Warisa Thangjai[(⊠)], Sa-Aat Niwitpong, and Suparat Niwitpong

Department of Applied Statistics, Faculty of Applied Science,
King Mongkut's University of Technology North Bangkok, Bangkok 10800, Thailand
wthangjai@yahoo.com, {sa-aat.n,suparat.n}@sci.kmutnb.ac.th

Abstract. For constructing simultaneous confidence intervals for the differences of means of several two-parameter exponential distributions, generalized confidence interval (GCI) approach and method of variance estimates recovery (MOVER) approach are proposed. The performance of the proposed approaches is compared with parametric bootstrap (PB) approach. Simulation studies showed that the MOVER approach performs better than the other approaches: its coverage probability is close to the nominal confidence level and average length is shorter than the other approaches. Three approaches are illustrated using an example.

Keywords: Parametric bootstrap · GCI · MOVER

1 Introduction

Two-parameter exponential distribution is widely used in many fields such as medicine, biology, and engineering. For example, the distribution is used to model the effective duration of a drug in dose response experiments, see Kharrati-Kopaei [4]. For more detailed applications, see Zelen [12], Johnson and Kotz [2], Lawless and Singhal [5], and Maurya et al. [7]. Recently, several researchers have investigated performance of confidence intervals of two-parameter exponential distribution; see Sangnawakij et al. [8], Sangnawakij and Niwitpong [9], and Thangjai and Niwitpong [10].

Several researchers have considered the problem of multiple comparisons of two-parameter exponential parameters; location parameter, scale parameter, and mean. For example, Kharrati-Kopaei et al. [3] proposed simultaneous fiducial generalized confidence intervals (SFGCIs) for the successive differences of exponential location parameters with unequal scale parameters. Kharrati-Kopaei [4] constructed simultaneous confidence intervals (SCIs) for the differences of location parameters of successive exponential distributions in the unbalanced case under heteroscedasticity. Li et al. [6] proposed a parametric bootstrap (PB) approach to construct SCIs for all pairwise differences of means from several two-parameter exponential distributions.

© Springer International Publishing AG 2018
L. H. Anh et al. (eds.), *Econometrics for Financial Applications*, Studies in Computational Intelligence 760, https://doi.org/10.1007/978-3-319-73150-6_24

It is of practical and theoretical importance to develop procedures for constructing SCIs of differences of means of two-parameter exponential distributions. Li et al. [6] proposed the SCIs of differences of means of two-parameter exponential distributions based on PB approach. This paper extends the paper of Li et al. [6] and constructs the SCIs based on the PB approach and the two new SCIs using generalized confidence interval (GCI) approach and method of variance estimates recovery (MOVER) approach. The GCI approach, introduced by Weerahandi [11], is based on the concept of generalized pivotal quantity. Zou and Donner [13] and Zou et al. [14] introduced the MOVER approach to construct confidence interval for the sum of two populations. Donner and Zou [1] proposed the MOVER approach to construct confidence interval for difference of two populations. This paper extends these results to k populations using the concepts of the GCI and MOVER approaches for constructing the SCIs of differences of means of two-parameter exponential distributions.

The remainder of this paper is organized as follows. The PB, GCI, and MOVER approaches to construct the SCIs for differences of means of several two-parameter exponential distributions are described in Sect. 2. Section 3, the performance of the proposed confidence intervals is evaluated via simulation studies. Section 4, the three approaches are illustrated with an example. And finally, the conclusions are presented in Sect. 5.

2 Simultaneous Confidence Intervals

Suppose that random samples are taken from k populations where the i-th population follows a two-parameter exponential distribution with probability density function

$$f(x; \beta_i, \lambda_i) = \frac{1}{\lambda_i} \exp\left(-\frac{x - \beta_i}{\lambda_i}\right) I_{[\beta_i, \infty)}(x), \tag{1}$$

where $\beta_i \in (-\infty, +\infty)$ and $\lambda_i \in (0, +\infty)$ are location and scale parameters, respectively, and $i = 1, 2, \ldots, k$.

For $i = 1, 2, \ldots, k$, let $X_i = (X_{i1}, X_{i2}, \ldots, X_{in_i})$ be an independent random sample from two-parameter exponential distribution with location parameter β_i and scale parameter λ_i. Let $\theta_i = \beta_i + \lambda_i$ be the mean of the i-th sample. Let $X_{(1)i} = \min(X_{i1}, X_{i2}, \ldots, X_{in_i})$ be the first order statistic of X_i and let $S_i = \frac{1}{n_i - 1} \sum_{j=1}^{n_i} (X_{ij} - X_{(1)i})$. Furthermore, let $x_{(1)i}$ and s_i be the observed values of $X_{(1)i}$ and S_i, respectively. The maximum likelihood estimators of β_i and λ_i are given by

$$\hat{\beta}_i = X_{(1)i} \quad \text{and} \quad \hat{\lambda}_i = S_i = \frac{1}{n_i - 1} \sum_{j=1}^{n_i} (X_{ij} - X_{(1)i}). \tag{2}$$

It is well know that $X_{(1)i}$ and S_i are independently distributed with

$$X_{(1)i} \sim Exp\left(\beta_i, \frac{\lambda_i}{n_i}\right) \quad \text{and} \quad \frac{(2n_i - 2)S_i}{\lambda_i} \sim \chi^2_{2n_i - 2}, \tag{3}$$

where $\chi^2_{2n_i-2}$ denotes a chi-square distribution with $2n_i - 2$ degrees of freedom. $X_{(1)i}$ and S_i converge to β_i and λ_i, respectively. Therefore, an estimate of θ_i is

$$\hat{\theta}_i = X_{(1)i} + S_i; i = 1, 2, \ldots, k. \tag{4}$$

The parameters of interest are $\theta_{il} = \theta_i - \theta_l$ for all $i, l = 1, 2, \ldots, k, i \neq l$. Hence, the estimate of θ_{il} is

$$\hat{\theta}_{il} = \hat{\theta}_i - \hat{\theta}_l = X_{(1)i} - X_{(1)l} + S_i - S_l; i, l = 1, 2, \ldots, k, i \neq l. \tag{5}$$

The variance of $\hat{\theta}_{il}$ proposed in Li et al. [6] is

$$Var(\hat{\theta}_{il}) = Var(\hat{\theta}_i - \hat{\theta}_l) = \frac{\lambda_i^2}{n_i^2} + \frac{\lambda_i^2}{n_i - 1} + \frac{\lambda_l^2}{n_l^2} + \frac{\lambda_l^2}{n_l - 1}; i \neq l. \tag{6}$$

The unbiased estimator of $Var(\hat{\theta}_{il})$ is

$$V_{il} = \frac{n_i - 1}{n_i^3}S_i^2 + \frac{1}{n_i}S_i^2 + \frac{n_l - 1}{n_l^3}S_l^2 + \frac{1}{n_l}S_l^2. \tag{7}$$

2.1   Parametric Bootstrap Simultaneous Confidence Intervals

The parametric bootstrap simultaneous confidence intervals for all differences of means of two-parameter exponential distributions were proposed by Li et al. [6]. Without loss of generality, all β_i's are assumed to be zeros. Let

$$X_{(1)i}^{PB} \sim Exp\left(0, \frac{s_i}{n_i}\right) \quad \text{and} \quad S_i^{PB} \sim \frac{s_i \chi^2_{2n_i-2}}{2n_i - 2}, \tag{8}$$

where $\chi^2_{2n_i-2}$ denotes a chi-square distribution with $2n_i - 2$ degrees of freedom. Define

$$D_n^{PB} = \max_{i \neq l} \left| \frac{X_{(1)i}^{PB} - X_{(1)l}^{PB} + S_i^{PB} - S_l^{PB} - (s_i - s_l)}{\sqrt{V_{il}^{PB}}} \right|, \tag{9}$$

where

$$V_{il}^{PB} = \frac{n_i - 1}{n_i^3}(S_i^{PB})^2 + \frac{1}{n_i}(S_i^{PB})^2 + \frac{n_l - 1}{n_l^3}(S_l^{PB})^2 + \frac{1}{n_l}(S_l^{PB})^2. \tag{10}$$

Therefore, the $100(1 - \alpha)\%$ two-sided parametric bootstrap simultaneous confidence intervals for all pairwise differences $\theta_i - \theta_l$ ($i \neq l$) of means of two-parameter exponential distributions are

$$SCI_{(il)PB} = (\hat{\theta}_i - \hat{\theta}_l - q_\alpha^{n.PB}\sqrt{V_{il}}, \hat{\theta}_i - \hat{\theta}_l + q_\alpha^{n.PB}\sqrt{V_{il}}), \tag{11}$$

where $q_\alpha^{n.PB}$ denotes the $(1 - \alpha)$-th quantile of the distribution of D_n^{PB} and V_{il} is defined in Eq. (7).

2.2   Simultaneous Generalized Confidence Intervals

The generalized confidence interval proposed by Weerahandi [11]. In the following concept of generalized confidence interval is described.

Definition 1: *Let $X = (X_1, X_2, \ldots, X_n)$ be a random sample from a distribution $F_X(x; \delta)$, where $\delta = (\beta, \lambda)$ is the vector of unknown parameters. Let $x = (x_1, x_2, \ldots, x_n)$ be an observed value of $X = (X_1, X_2, \ldots, X_n)$. A generalized confidence interval is computed using the percentiles of a generalized pivotal quantity $R(X; x, \delta)$. We are interested in some parameter $p = f(\delta)$. We are looking for a function $R(x; x', \delta)$, where x and x' are n-dimensional tuples, and δ is a pair, with the following two properties:*

(i) For any x', if all X_j are independent and distributed according to the distribution corresponding to the parameters δ, then the distribution of the random variable $R((X_1, X_2, \ldots, X_n); x', \delta)$ is the same for all δ. This distribution depends only on x'.

(ii) For all tuples x and for all pairs δ, we have $R(x; x, \delta) = f(\delta)$.

Let $R(\alpha)$ be the α-th quantile of $R(x; x', \delta)$. Hence, $(R(\alpha/2), R(1 - \alpha/2))$ becomes $100(1 - \alpha)\%$ two-sided generalized confidence interval for parameter of interest.

It is known that (see Li et al. [6]) $\hat{\beta}_i$ and $\hat{\lambda}_i$ are independent with

$$U_i = \frac{2n_i(\hat{\beta}_i - \beta_i)}{\lambda_i} = \frac{2n_i(X_{(1)i} - \beta_i)}{\lambda_i} \sim \chi_2^2 \tag{12}$$

and

$$W_i = \frac{(2n_i - 2)\hat{\lambda}_i}{\lambda_i} = \frac{(2n_i - 2)S_i}{\lambda_i} \sim \chi_{2n_i - 2}^2, \tag{13}$$

where χ_p^2 denotes a chi-square distribution with p degrees of freedom.

The generalized pivotal quantities of β_i and λ_i are defined as follows

$$R_{\beta_i} = X_{(1)i} - \frac{U_i R_{\lambda_i}}{2n_i} = X_{(1)i} - \frac{(n_i - 1)U_i S_i}{n_i W_i} \quad \text{and} \quad R_{\lambda_i} = \frac{(2n_i - 2)S_i}{W_i}. \tag{14}$$

Therefore, the generalized pivotal quantity of θ_i is

$$R_{\theta_i} = R_{\beta_i} + R_{\lambda_i} = X_{(1)i} - \frac{(n_i - 1)U_i S_i}{n_i W_i} + \frac{(2n_i - 2)S_i}{W_i}. \tag{15}$$

The generalized pivotal quantity of $\theta_{il} = \theta_i - \theta_l$ is

$$R_{\theta_{il}} = R_{\theta_i} - R_{\theta_l}, \tag{16}$$

where R_{θ_i} and R_{θ_l} are defined in Eq. (15), and $i, l = 1, 2, \ldots, k$, $i \neq l$.

Therefore, the $100(1 - \alpha)\%$ two-sided simultaneous generalized confidence intervals for all pairwise differences $\theta_i - \theta_l$ ($i \neq l$) of means of two-parameter exponential distributions are

$$SCI_{il(GCI)} = (R_{\theta_{il}}(\alpha/2), R_{\theta_{il}}(1 - \alpha/2)), \tag{17}$$

where $R_{\theta_{il}}(p)$ denotes the p-th quantile of $R_{\theta_{il}}$.

2.3 MOVER Simultaneous Confidence Intervals

Recall that the confidence interval for mean of two-parameter exponential distribution using central limit theorem is

$$CI_{\theta_i} = (l_i, u_i) = (\hat{\theta}_i - z_{1-\alpha/2}\sqrt{Var(\hat{\theta}_i)}, \hat{\theta}_i + z_{1-\alpha/2}\sqrt{Var(\hat{\theta}_i)}), \qquad (18)$$

where $\hat{\theta}_i$ is defined in Eq. (4) and $Var(\hat{\theta}_i)$ is an estimate of $Var(\hat{\theta}_i) = (\lambda_i^2/n_i^2) + (\lambda_i^2/(n_i - 1))$ with λ_i replaced by s_i.

For $i = 1, 2$, let θ_1 and θ_2 be two parameters of interest. Donner and Zou [1] introduced confidence interval for $\theta_1 - \theta_2$ using MOVER approach. The confidence interval is

$$CI_{12(MOVER)} = \left(L_{12(MOVER)}, U_{12(MOVER)}\right), \qquad (19)$$

where

$$L_{12(MOVER)} = \hat{\theta}_1 - \hat{\theta}_2 - \sqrt{\left(\hat{\theta}_1 - l_1\right)^2 + \left(u_2 - \hat{\theta}_2\right)^2}$$

and

$$U_{12(MOVER)} = \hat{\theta}_1 - \hat{\theta}_2 + \sqrt{\left(u_1 - \hat{\theta}_1\right)^2 + \left(\hat{\theta}_2 - l_2\right)^2},$$

where (l_i, u_i) contains the parameter value for θ_i, $i = 1, 2$.

For $i = 1, 2, \ldots, k$, the concept of Donner and Zou [1] is extended to k parameters. Therefore, the $100(1 - \alpha)\%$ two-sided MOVER simultaneous confidence intervals for all pairwise differences $\theta_i - \theta_l$ $(i \neq l)$ of means of two-parameter exponential distributions are

$$SCI_{il(MOVER)} = \left(L_{il(MOVER)}, U_{il(MOVER)}\right), \qquad (20)$$

where

$$L_{il(MOVER)} = \hat{\theta}_i - \hat{\theta}_l - \sqrt{\left(\hat{\theta}_i - l_i\right)^2 + \left(u_l - \hat{\theta}_l\right)^2}$$

and

$$U_{il(MOVER)} = \hat{\theta}_i - \hat{\theta}_l + \sqrt{\left(u_i - \hat{\theta}_i\right)^2 + \left(\hat{\theta}_l - l_l\right)^2},$$

where $\hat{\theta}_i$ is defined in Eq. (4) and l_i and u_i are defined in Eq. (18).

3 Simulation Studies

In this section, Monte Carlo simulation was carried out to evaluate the performance of GCI approach and MOVER approach. These approaches were compared with PB approach. The PB approach was proposed by Li et al. [6].

Table 1. The coverage probabilities of 95% two-sided simultaneous confidence intervals for all differences of means of two-parameter exponential distributions: 3 sample cases.

(n_1, n_2, n_3)	SCI_{PB}	SCI_{GCI}	SCI_{MOVER}
$(\beta_1, \beta_2, \beta_3) = (0,0,0)$, $(\lambda_1, \lambda_2, \lambda_3) = (0.01, 0.02, 0.08)$			
$(15, 15, 15)$	0.9728	0.9523	0.9413
$(30, 30, 30)$	0.9745	0.9499	0.9479
$(60, 60, 60)$	0.9773	0.9492	0.9497
$(15, 15, 30)$	0.9737	0.9477	0.9441
$(15, 15, 60)$	0.9779	0.9539	0.9505
$(15, 30, 30)$	0.9759	0.9493	0.9473
$(15, 60, 60)$	0.9809	0.9522	0.9516
$(30, 30, 120)$	0.9798	0.9531	0.9528
$(60, 60, 120)$	0.9774	0.9484	0.9481
$(15, 15, 120)$	0.9775	0.9499	0.9467
$(15, 30, 120)$	0.9793	0.9469	0.9493
$(15, 60, 120)$	0.9809	0.9523	0.9537
$(15, 120, 120)$	0.9789	0.9479	0.9501
$(30, 120, 120)$	0.9763	0.9497	0.9499
$(60, 120, 120)$	0.9753	0.9481	0.9473
$(120, 120, 120)$	0.9775	0.9499	0.9476
$(\beta_1, \beta_2, \beta_3) = (0,0,0)$, $(\lambda_1, \lambda_2, \lambda_3) = (1.00, 2.00, 2.00)$			
$(15, 15, 15)$	0.9809	0.9500	0.9499
$(30, 30, 30)$	0.9798	0.9481	0.9518
$(60, 60, 60)$	0.9777	0.9443	0.9467
$(15, 15, 30)$	0.9810	0.9521	0.9531
$(15, 15, 60)$	0.9757	0.9481	0.9474
$(15, 30, 30)$	0.9841	0.9483	0.9525
$(15, 60, 60)$	0.9839	0.9516	0.9564
$(30, 30, 120)$	0.9821	0.9518	0.9537
$(60, 60, 120)$	0.9807	0.9515	0.9523
$(15, 15, 120)$	0.9765	0.9519	0.9495
$(15, 30, 120)$	0.9791	0.9484	0.9510
$(15, 60, 120)$	0.9819	0.9485	0.9538
$(15, 120, 120)$	0.9815	0.9501	0.9534
$(30, 120, 120)$	0.9815	0.9457	0.9495
$(60, 120, 120)$	0.9815	0.9513	0.9523
$(120, 120, 120)$	0.9789	0.9495	0.9516

Table 2. The average lengths (standard errors) of 95% two-sided simultaneous confidence intervals for all differences of means of two-parameter exponential distributions: 3 sample cases.

(n_1, n_2, n_3)	SCI_{PB}	SCI_{GCI}	SCI_{MOVER}
$(\beta_1, \beta_2, \beta_3) = (0,0,0), (\lambda_1, \lambda_2, \lambda_3) = (0.01, 0.02, 0.08)$			
$(15, 15, 15)$	0.0840 (0.0267)	0.0725 (0.0229)	0.0646 (0.0206)
$(30, 30, 30)$	0.0550 (0.0175)	0.0475 (0.0150)	0.0450 (0.0143)
$(60, 60, 60)$	0.0376 (0.0120)	0.0325 (0.0103)	0.0316 (0.0101)
$(15, 15, 30)$	0.0599 (0.0153)	0.0520 (0.0126)	0.0483 (0.0123)
$(15, 15, 60)$	0.0452 (0.0086)	0.0397 (0.0068)	0.0372 (0.0071)
$(15, 30, 30)$	0.0553 (0.0167)	0.0486 (0.0143)	0.0458 (0.0138)
$(15, 60, 60)$	0.0390 (0.0108)	0.0344 (0.0091)	0.0330 (0.0092)
$(30, 30, 120)$	0.0313 (0.0059)	0.0270 (0.0049)	0.0262 (0.0050)
$(60, 60, 120)$	0.0282 (0.0073)	0.0242 (0.0062)	0.0238 (0.0062)
$(15, 15, 120)$	0.0362 (0.0046)	0.0323 (0.0037)	0.0301 (0.0038)
$(15, 30, 120)$	0.0318 (0.0053)	0.0281 (0.0042)	0.0269 (0.0045)
$(15, 60, 120)$	0.0298 (0.0062)	0.0263 (0.0051)	0.0253 (0.0053)
$(15, 120, 120)$	0.0289 (0.0069)	0.0253 (0.0057)	0.0244 (0.0058)
$(30, 120, 120)$	0.0271 (0.0075)	0.0236 (0.0065)	0.0232 (0.0064)
$(60, 120, 120)$	0.0264 (0.0080)	0.0228 (0.0069)	0.0225 (0.0068)
$(120, 120, 120)$	0.0262 (0.0083)	0.0225 (0.0072)	0.0222 (0.0071)
$(\beta_1, \beta_2, \beta_3) = (0,0,0), (\lambda_1, \lambda_2, \lambda_3) = (1.00, 2.00, 2.00)$			
$(15, 15, 15)$	3.1325 (0.3443)	2.9403 (0.3297)	2.5694 (0.2800)
$(30, 30, 30)$	2.1258 (0.2064)	1.9027 (0.1890)	1.7824 (0.1726)
$(60, 60, 60)$	1.4822 (0.1332)	1.2890 (0.1180)	1.2475 (0.1120)
$(15, 15, 30)$	2.6807 (0.3053)	2.5155 (0.2969)	2.2435 (0.2530)
$(15, 15, 60)$	2.4702 (0.3616)	2.3012 (0.3555)	2.0571 (0.2988)
$(15, 30, 30)$	2.2205 (0.1536)	2.0649 (0.1357)	1.8965 (0.1304)
$(15, 60, 60)$	1.7113 (0.0815)	1.5751 (0.0857)	1.4652 (0.0696)
$(30, 30, 120)$	1.7009 (0.2433)	1.5120 (0.2227)	1.4331 (0.2046)
$(60, 60, 120)$	1.2967 (0.1328)	1.1242 (0.1171)	1.0939 (0.1120)
$(15, 15, 120)$	2.3966 (0.4332)	2.1890 (0.4123)	1.9544 (0.3520)
$(15, 30, 120)$	1.8630 (0.2070)	1.6989 (0.1936)	1.5690 (0.1741)
$(15, 60, 120)$	1.5742 (0.1131)	1.4357 (0.1189)	1.3353 (0.0956)
$(15, 120, 120)$	1.4093 (0.1117)	1.2801 (0.1279)	1.1887 (0.0935)
$(30, 120, 120)$	1.2063 (0.0398)	1.0618 (0.0399)	1.0253 (0.0338)
$(60, 120, 120)$	1.0981 (0.0586)	0.9495 (0.0504)	0.9295 (0.0496)
$(120, 120, 120)$	1.0422 (0.0895)	0.8909 (0.0776)	0.8763 (0.0752)

Table 3. The coverage probabilities of 95% two-sided simultaneous confidence intervals for all differences of means of two-parameter exponential distributions: 5 sample cases.

$(n_1, n_2, n_3, n_4, n_5)$	SCI_{PB}	SCI_{GCI}	SCI_{MOVER}
$(\beta_1, \beta_2, \beta_3, \beta_4, \beta_5) = (2.00, 2.00, 2.00, 2.00, 2.00)$, $(\lambda_1, \lambda_2, \lambda_3, \lambda_4, \lambda_5) = (3.00, 3.00, 3.00, 3.00, 3.00)$			
$(15, 15, 15, 15, 15)$	0.9989	0.9501	0.9607
$(30, 30, 30, 30, 30)$	0.9958	0.9505	0.9573
$(60, 60, 60, 60, 60)$	0.9945	0.9530	0.9564
$(15, 15, 30, 30, 30)$	0.9952	0.9510	0.9569
$(15, 15, 60, 60, 60)$	0.9913	0.9503	0.9510
$(60, 60, 120, 120, 120)$	0.9928	0.9487	0.9509
$(120, 120, 120, 120, 120)$	0.9942	0.9519	0.9537
$(\beta_1, \beta_2, \beta_3, \beta_4, \beta_5) = (3.25, 3.25, 3.50, 3.50, 3.50)$, $(\lambda_1, \lambda_2, \lambda_3, \lambda_4, \lambda_5) = (0.75, 0.75, 0.50, 0.50, 0.50)$			
$(15, 15, 15, 15, 15)$	0.9966	0.9514	0.9578
$(30, 30, 30, 30, 30)$	0.9942	0.9502	0.9548
$(60, 60, 60, 60, 60)$	0.9932	0.9486	0.9515
$(15, 15, 30, 30, 30)$	0.9890	0.9472	0.9471
$(15, 15, 60, 60, 60)$	0.9898	0.9502	0.9469
$(60, 60, 120, 120, 120)$	0.9922	0.9481	0.9497
$(120, 120, 120, 120, 120)$	0.9939	0.9512	0.9530
$(\beta_1, \beta_2, \beta_3, \beta_4, \beta_5) = (0, 0, 0, 0, 0)$, $(\lambda_1, \lambda_2, \lambda_3, \lambda_4, \lambda_5) = (0.01, 0.01, 0.02, 0.08, 0.08)$			
$(15, 15, 15, 15, 15)$	0.9890	0.9520	0.9443
$(30, 30, 30, 30, 30)$	0.9896	0.9512	0.9477
$(60, 60, 60, 60, 60)$	0.9899	0.9461	0.9460
$(15, 15, 30, 30, 30)$	0.9886	0.9480	0.9453
$(15, 15, 60, 60, 60)$	0.9920	0.9494	0.9504
$(60, 60, 120, 120, 120)$	0.9909	0.9507	0.9512
$(120, 120, 120, 120, 120)$	0.9914	0.9487	0.9477

The performance of these three approaches was evaluated through the coverage probabilities, the average lengths, and the standard errors of the confidence intervals.

In the simulation, the nominal confidence level was computed at 0.95. The simulation study was performed with four configuration factors: (1) sample cases k: $k = 3$ and $k = 5$; (2) sample sizes n_i; (3) location parameters β_i; (4) scale parameters λ_i, $i = 1, 2, \ldots, k$. The specific combinations were given in the following tables. For the PB approach, 1000 values were obtained in each of 5000 simulations. Furthermore, 1000 pivotal quantities were also obtained in each of 5000 simulations for the GCI approach. The simulation results from $k = 3$ and $k = 5$ were presented in the following four tables.

It is seen from Tables 1, 2, 3 and 4 that the performance of three approaches does not depend on the values of β_i and λ_i. The GCI approach and MOVER approach perform much better than the PB approach in term of coverage probability for all sample sizes. The coverage probabilities of the GCI and MOVER approaches are close to the nominal confidence level and, by comparing the

Table 4. The average lengths (standard errors) of 95% two-sided simultaneous confidence intervals for all differences of means of two-parameter exponential distributions: 5 sample cases.

(n_1, n_2, n_3)	SCI_{PB}	SCI_{GCI}	SCI_{MOVER}
$(\beta_1, \beta_2, \beta_3, \beta_4, \beta_5) = (2.00, 2.00, 2.00, 2.00, 2.00)$, $(\lambda_1, \lambda_2, \lambda_3, \lambda_4, \lambda_5) = (3.00, 3.00, 3.00, 3.00, 3.00)$			
$(15, 15, 15, 15, 15)$	6.1777 (0.2835)	5.1999 (0.2435)	4.5120 (0.2056)
$(30, 30, 30, 30, 30)$	4.1664 (0.1325)	3.3406 (0.1109)	3.1151 (0.0989)
$(60, 60, 60, 60, 60)$	2.9383 (0.0649)	2.2516 (0.0542)	2.1733 (0.0480)
$(15, 15, 30, 30, 30)$	5.0563 (0.2814)	4.1138 (0.2544)	3.6974 (0.2044)
$(15, 15, 60, 60, 60)$	4.5402 (0.3972)	3.5300 (0.3430)	3.2026 (0.2790)
$(60, 60, 120, 120, 120)$	2.4665 (0.1022)	1.8493 (0.0812)	1.8015 (0.0746)
$(120, 120, 120, 120, 120)$	2.0904 (0.0327)	1.5542 (0.0283)	1.5272 (0.0239)
$(\beta_1, \beta_2, \beta_3, \beta_4, \beta_5) = (3.25, 3.25, 3.50, 3.50, 3.50)$, $(\lambda_1, \lambda_2, \lambda_3, \lambda_4, \lambda_5) = (0.75, 0.75, 0.50, 0.50, 0.50)$			
$(15, 15, 15, 15, 15)$	1.2704 (0.0794)	1.0478 (0.0657)	0.9109 (0.0565)
$(30, 30, 30, 30, 30)$	0.8540 (0.0450)	0.6742 (0.0358)	0.6298 (0.0331)
$(60, 60, 60, 60, 60)$	0.5992 (0.0286)	0.4554 (0.0221)	0.4401 (0.0210)
$(15, 15, 30, 30, 30)$	1.1173 (0.1034)	0.8789 (0.0866)	0.7873 (0.0725)
$(15, 15, 60, 60, 60)$	1.0546 (0.1322)	0.7936 (0.1061)	0.7164 (0.0896)
$(60, 60, 120, 120, 120)$	0.5322 (0.0441)	0.3957 (0.0336)	0.3852 (0.0319)
$(120, 120, 120, 120, 120)$	0.4239 (0.0191)	0.3146 (0.0145)	0.3092 (0.0140)
$(\beta_1, \beta_2, \beta_3, \beta_4, \beta_5) = (0, 0, 0, 0, 0)$, $(\lambda_1, \lambda_2, \lambda_3, \lambda_4, \lambda_5) = (0.01, 0.01, 0.02, 0.08, 0.08)$			
$(15, 15, 15, 15, 15)$	0.1086 (0.0185)	0.0779 (0.0133)	0.0691 (0.0118)
$(30, 30, 30, 30, 30)$	0.0692 (0.0115)	0.0509 (0.0085)	0.0481 (0.0080)
$(60, 60, 60, 60, 60)$	0.0467 (0.0076)	0.0347 (0.0057)	0.0337 (0.0055)
$(15, 15, 30, 30, 30)$	0.0703 (0.0109)	0.0522 (0.0080)	0.0490 (0.0076)
$(15, 15, 60, 60, 60)$	0.0491 (0.0067)	0.0371 (0.0049)	0.0355 (0.0049)
$(60, 60, 120, 120, 120)$	0.0328 (0.0050)	0.0245 (0.0037)	0.0242 (0.0037)
$(120, 120, 120, 120, 120)$	0.0322 (0.0052)	0.0240 (0.0039)	0.0237 (0.0038)

average lengths, it is seen that the MOVER approach performs shorter than the GCI approach. Therefore, the MOVER approach can be recommended to construct SCIs for all pairwise differences of means of two-parameter exponential distributions.

4   An Empirical Application

In this section, an example previously considered by Maurya et al. [7] and Kharrati-Kopaei et al. [3] are presented. The data consist of 36 observations of survival days of patients with inoperable lung cancer in $k = 4$ groups of tumor type: squamous (group 1), small (group 2), adeno (group 3), and large (group 4). The summary statistics are as follows: $n_1 = 9$, $\bar{x}_1 = 51.00$, $x_{(1)1} = 8$, $s_1 = 48.38$, $n_2 = 9$, $\bar{x}_2 = 22.11$, $x_{(1)2} = 13$, $s_2 = 10.25$, $n_3 = 9$, $\bar{x}_3 = 72.89$, $x_{(1)3} = 3$, $s_3 = 78.63$, $n_4 = 9$, $\bar{x}_4 = 197.89$, $x_{(1)4} = 103$, and $s_4 = 106.75$. The differences of means are $\theta_2 - \theta_1 = -33.13$, $\theta_3 - \theta_1 = 25.25$, $\theta_4 - \theta_1 = 153.37$, $\theta_3 - \theta_2 = 58.38$, $\theta_4 - \theta_2 = 186.50$, and $\theta_4 - \theta_3 = 128.12$.

Table 5. The 95% simultaneous confidence intervals for all pairwise differences of means of two-parameter exponential distributions.

Parameters	CI_{PB}		CI_{GCI}		CI_{MOVER}	
	Lower	Upper	Lower	Upper	Lower	Upper
$\theta_2 - \theta_1$	−94.4476	28.1976	−83.5414	−3.6280	−66.9889	0.7389
$\theta_3 - \theta_1$	−89.2313	139.7313	−45.4777	120.2537	−37.9695	88.4695
$\theta_4 - \theta_1$	8.0339	298.7161	74.2676	279.9462	73.1140	233.6360
$\theta_3 - \theta_2$	−39.9543	156.7043	14.1379	143.5728	4.0750	112.6750
$\theta_4 - \theta_2$	53.5086	319.4914	125.3317	310.5229	113.0588	259.9412
$\theta_4 - \theta_3$	−36.2898	292.5398	21.8491	257.0133	37.3310	218.9190

For illustration, 95% simultaneous confidence intervals for the six differences of means, $\theta_i - \theta_l$ ($i \neq l$), obtained by using the three approaches, were given in Table 5. It is seen that the MOVER approach provides shorter intervals than the other approaches. Therefore, these results confirm the simulation study in the previous section.

5   Discussion and Conclusions

This paper extends the research paper of Li et al. [6]. The PB approach, introduced by Li et al. [6], was presented. Also, the GCI and MOVER approaches were proposed to construct simultaneous confidence intervals for differences of means of two-parameter exponential distributions. The simulation studies showed that the proposed approaches perform better than the PB approach in term of coverage probability. Furthermore, the MOVER approach provides shorter average lengths than the GCI approach. Therefore, the MOVER approach is suggested to construct simultaneous confidence intervals for all pairwise differences of mean from several two-parameter exponential distributions.

References

1. Donner, A., Zou, G.Y.: Closed-form confidence intervals for function of the normal mean and standard deviation. Stat. Methods Med. Res. **21**, 347–359 (2010)
2. Johnson, N.L., Kotz, S.: Continuous Univariate Distributions. Wiley, New York (1970)
3. Kharrati-Kopaei, M., Malekadeh, A., Sadooghi-Alvandi, M.: Simultaneous fiducial generalized confidence intervals for the successive differences of exponential location parameters under heteroscedasticity. Stat. Probab. Lett. **83**, 1547–1552 (2013)
4. Kharrati-Kopaei, M.: A note on the simultaneous confidence intervals for the successive differences of exponential location parameters under heteroscedasticity. Stat. Methodol. **22**, 1–7 (2015)

5. Lawless, J.F., Singhal, K.: Analysis of data from life test experiments under an exponential model. Naval Res. Logist. Q. **27**, 323–334 (1980)
6. Li, J., Song, W., Shi, J.: Parametric bootstrap simultaneous confidence intervals for differences of means from several two-parameter exponential distributions. Stat. Probab. Lett. **106**, 39–45 (2015)
7. Maurya, V., Goyal, A., Gill, A.N.: Simultaneous testing for the successive differences of exponential location parameters under heteroscedasticity. Stat. Probab. Lett. **8**, 1507–1517 (2011)
8. Sangnawakij, P., Niwitpong, S., Niwitpong, S.: Confidence intervals for the ratio of coefficients of variation in the two-parameter exponential distributions. In: Huynh, V.N., Inuiguchi, M., Le, B., Le, B., Denoeux, T. (eds.) Integrated Uncertainty in Knowledge Modelling and Decision Making. Lecture Notes in Artificial Intelligence, vol. 9978, pp. 542–551. Springer, Cham (2016)
9. Sangnawakij, P., Niwitpong, S.: Confidence intervals for coefficients of variation in two-parameter exponential distributions. Commun. Stat. Simul. Comput. **46**(8), 6618–6630 (2017)
10. Thangjai, W., Niwitpong, S.: Confidence intervals for the weighted coefficients of variation of two-parameter exponential distributions. Cogent Math. **4**, 1–16 (2017)
11. Weerahandi, S.: Generalized confidence intervals. J. Am. Stat. Assoc. **88**, 899–905 (1993)
12. Zelen, M.: Application of exponential models problems in cancer research. J. Roy. Stat. Soc. **129**, 368–398 (1966)
13. Zou, G.Y., Donner, A.: Construction of confidence limits about effect measures: a general approach. Stat. Med. **27**, 1693–1702 (2008)
14. Zou, G.Y., Taleban, J., Hao, C.Y.: Confidence interval estimation for lognormal data with application to health economics. Comput. Stat. Data Anal. **53**, 3755–3764 (2009)

The Skew-t Option Pricing Model

Claudia Yeap, S. T. Boris Choy$^{(\boxtimes)}$, and S. Simon Kwok

University of Sydney, Sydney, Australia
claudia.yeap@gmail.com, {boris.choy,simon.kwok}@sydney.edu.au

Abstract. This paper presents a skew-t option pricing model. It is constructed analogously to the variance gamma option pricing model proposed by [14]. This proposed skew-t model inherits the variance gamma model's three parameters and their respective interpretations. In addition, it also has a fat-tailed, skewed distribution and infinite-activity (pure jump) stock dynamics, which is achieved through modelling the length of time intervals as stochastic. This paper has three main insights. From a theoretical perspective, a result is obtained for the correlation between the variance gamma model's logarithm returns and its gamma stochastic variance. This result holds for the skew-t model as well, which has reciprocal gamma variance, and it provides a new way to quantify the leverage effect under each model. The focus then shifts to the numerical procedures required for estimating the skew-t model's parameters. Finally, an empirical comparison between the skew-t, variance gamma and Black-Scholes models is conducted. The discussion links four pieces of analysis - pricing errors, pricing biases, the higher moments of the distributions and the market's implied volatility.

Keywords: Black-Scholes · Mean-variance mixture
Variance gamma · Reciprocal gamma

1 Introduction

For almost twenty years, the variance gamma (VG) option pricing model [14] has risen to prominence as the pure-jump, skewed and fat-tailed alternative to the Black-Scholes option pricing model [3]. While the Black-Scholes model assumes normality on the underlying stock's logarithm returns, the VG model assumes them to be skew-VG distributed. In this paper, the VG option pricing model is referred to as the skew-VG model to remain cognisant of this definition. The merits of the skew-VG model lie in its enhanced ability to model extreme returns in the market and to capture shocks around announcement periods. In 2007, it achieved the milestone of becoming a built-in pricing function on Bloomberg terminals [4] for traders' everyday use.

And yet, for most practitioners, students and academics, the most well-known fat-tailed alternative to the normal distribution is the Student-t distribution. It might reasonably be expected then that an option pricing model based on a

© Springer International Publishing AG 2018
L. H. Anh et al. (eds.), *Econometrics for Financial Applications*, Studies in Computational Intelligence 760, https://doi.org/10.1007/978-3-319-73150-6_25

skewed t-distribution would also be an available alternative to the skew-VG option pricing model in the extant literature. This was not found to be the case.

This paper fills the gap in the literature for a skew-t option pricing model that is comparable to the skew-VG option pricing model. Along side the contribution of the new, stand alone option pricing model, comes fresh empirical insights into whether the skew-VG's documented pricing accuracy improvements over Black-Scholes is attributable to the option market's particular preference for the skew-VG distribution or rather a more generic preference for a skewed, fat-tailed distribution.

The theoretical conduit for developing the skew-t model has been the skew-t and skew-VG distributions' location and scale mixture of normals representation, as it unveils a striking parallel between the two distributions. Where the skew-VG uses a gamma mixing density, the skew-t uses a reciprocal (or more commonly known as inverse) gamma mixing density. It is due to this ability to reduce the two distributions' differences to a singular theoretical point, that after switching the mixing densities, we can use all other techniques implemented by Madan et al. [14] and the result is an original skew-t model.

The skew-t model inherits all the same properties as the skew-VG model, including being a pure-jump model, having a time-deformed Geometric Brownian motion interpretation, as well as the skew-VG's three parameters with exactly the same interpretation. The skew-VG's three parameters are the Brownian motion diffusion parameters, σ (which is comparable to Black-Scholes' sole parameter of volatility), a skewness parameter, θ, that governs how much the latent mixing variable impacts the drift, and ν, the variance of the mixing density, which drives the kurtosis of the mixture distributions.

After the model is specified, this paper has three further sections. The first is the comparison of the skew-t model to other option pricing models. This paper offers a quantifiable relationship via moments matching to the skew-VG model and parameter matching to Heston's stochastic volatility model [10]. The second section concerns numerical strategies for estimating the parameters of the skew-t model using options data, that is under the risk-neutral measure. Gradient-free algorithms were employed and the surface of the objective function and convergence are portrayed in this section in preparation for the empirical test that follows. The third part is an empirical test that examines three years of data commencing with the Global Financial Crisis. Both a cross-sectional and time-series treatment of the pricing results are provided, and we investigate the predictability of the pricing errors to measure model misspecification.

In pursuit of a closed-form solution, the proposed option pricing model is applicable to European call and put options. The resulting model is semi-analytic one. For pricing American options using this model we would need to appeal to simulation methods. Using the S&P500 index as the underlying asset of the data set, there was no shortage of European options data available. Over three years, 39,667 data points are accounted for, each for puts and calls.

Section 2 specifies the properties of the skew-t option pricing model. Section 3 covers parameter estimation. Section 4 encompasses the empirical results. Section 5 concludes this paper.

2 The Skew-t Option Pricing Model

The skew-t option pricing model is the central contribution of this paper. It has been derived using the same principles as Madan, Carr and Chang's skew-VG option pricing model. This section shows the steps used to derive the skew-t model. Once the model has been defined, we investigate how the proposed model compares to other option pricing models. In particular we compare it with other t-distributions, the skew-VG model and Heston's stochastic volatility model.

2.1 Skew-t Model Specification

In parallel with the skew-VG option pricing model, we disaggregate the derivation of the skew-t model into three tiers: the distribution of the stationary increment, the dynamic process for log stock price and the option pricing model.

2.1.1 Step 1: The Skew-t Distribution

The skew-t distribution that we use in this paper is constructed via the normal mean-variance mixture representation. Introducing the latent scale mixture variable λ, the probability density function (pdf) of the skew-t distribution with location μ, scale σ, skewness θ and degrees of freedom ν is given by

$$f(x) = \int_0^\infty N\left(x \mid \mu + \theta\lambda, \sigma^2\lambda\right) R\Gamma\left(\lambda \,\middle|\, \alpha, \beta\right) d\lambda \qquad (1)$$

where $R\Gamma$ is the reciprocal gamma distribution with pdf given by

$$R\Gamma\left(\lambda \mid \alpha, \beta\right) = \frac{\beta^\alpha}{\Gamma(\alpha)} \lambda^{-(\alpha+1)} \exp\left(-\frac{\beta}{\lambda}\right), \quad \lambda, \alpha, \beta > 0. \qquad (2)$$

The objective of our skew-t option pricing model is to be able to compare it against the skew-VG option pricing model. As such we select parameters of the mixing density so that our model's parameters have the exact same interpretation as skew-VG's parameters. We set the mean of the reciprocal gamma density to be 1 (so that the σ still remains comparable with the Brownian motion diffusion parameter as the standard deviation of log returns over a time unit), and then we further choose the parameters so that the variance of the time change density is ν, for $\nu > 0$. Based on the mean and variance for the reciprocal gamma density, the resulting parameters to use are

$$\text{For } R\Gamma(\alpha, \beta): \quad E[\lambda] = \frac{\beta}{\alpha - 1} = 1 \quad \text{and} \quad V[\lambda] = \frac{\beta^2}{(\alpha - 1)^2(\alpha - 2)} = \nu.$$

The solution is

$$\alpha = \frac{1}{\nu} + 2 \quad \text{and} \quad \beta = \frac{1}{\nu} + 1.$$

Finlay and Seneta [8] also parameterise their skew-t distribution as a normal mean-variance mixture form with the reciprocal gamma distribution having the unit mean. In addition, it has the degrees of freedom equal to 2α or $\frac{2}{\nu} + 4$. In this paper, the skew-t distribution is reparameterised to have variance equal to ν.

Substituting the parameters into the reciprocal gamma density provided above, and taking the normal density function, the resulting conditional density for the proposed skew-t distribution is:

$$
\begin{aligned}
f(x) &= \int_0^\infty N\left(x \mid \mu + \theta\lambda, \sigma^2\lambda\right) R\Gamma\left(\lambda \mid \frac{1}{\nu} + 2, \frac{1}{\nu} + 1\right) d\lambda \\
&= \int_0^\infty \frac{1}{\sqrt{2\pi\sigma^2\lambda}} \exp\left(-\frac{(x - (\mu + \theta\lambda))^2}{2\sigma^2\lambda}\right) \\
&\quad \frac{(1+\nu)^{(\frac{1}{\nu}+2)}\lambda^{-\frac{1}{\nu}-3} \exp\left(\frac{-(1+\nu)}{\nu\lambda}\right)}{\nu^{(\frac{1}{\nu}+2)}\Gamma(\frac{1}{\nu}+2)} d\lambda.
\end{aligned}
\tag{3}
$$

Finlay and Seneta [8] provide the marginal density for the skew-t distribution. After reparameterising with $R\Gamma(\frac{1}{\nu} + 2, \frac{1}{\nu} + 1)$, we obtain the marginal pdf of the skew-t distribution as

$$
f(x) =
\begin{cases}
\dfrac{\Gamma(\frac{1}{\nu}+\frac{5}{2})\sqrt{\nu}}{\sqrt{2\sigma^2(1+\nu)}\sqrt{\pi}\Gamma(\frac{1}{\nu}+2)}\left(1 + \left(\dfrac{(x-\mu)\sqrt{\nu}}{\sqrt{2\sigma^2(1+\nu)}}\right)^2\right)^{-(\frac{1}{\nu}+\frac{5}{2})}, & \text{for } \theta = 0 \\[2em]
\sqrt{\dfrac{2}{\pi}}\dfrac{(1+\nu)^{(\frac{1}{\nu}+2)} \exp\left(\frac{(x-\mu)\theta}{\sigma^2}\right)}{\nu^{(\frac{1}{\nu}+2)}\sigma\Gamma(\frac{1}{\nu}+2)}\left(\dfrac{\theta^2\nu}{2\sigma^2(1+\nu)+(x-\mu)^2\nu}\right)^{\frac{1}{2}(\frac{1}{\nu}+\frac{5}{2})} \\[1em]
\quad K_{\frac{1}{\nu}+\frac{5}{2}}\left(\dfrac{|\theta|\sqrt{2\sigma^2(1+\nu)+(x-\mu)^2\nu}}{\sigma^2\sqrt{\nu}}\right), & \text{for } \theta \neq 0
\end{cases}
\tag{4}
$$

where $K(\cdot)$ is a modified Bessel function of the second kind.

2.1.2   Step 2: The Skew-t Process for Log Stock Prices

The second step is to verify whether we can apply the Lévy process theory to make a continuous, dynamic process out of the skew-t distribution. We saw above that Lévy processes require the distribution to be infinitely divisible. Since the t-distribution is a special case of the GH distribution and the GH distribution and all its special cases are infinitely divisible (Barndorff-Nielsen and Shephard [2]), we can invoke Lévy process theory here, the sum of independent and identically distributed increments. Using X_τ to denote the log returns over period, τ, the skew-t process is defined as follows.

$$
\begin{aligned}
X_1 &= C_j \\
C_j &\sim \text{skew-}t \\
X_\tau &= \sum_{j=1}^\tau C_j.
\end{aligned}
\tag{5}
$$

2.1.3   Step 3: Risk-Neutral Pricing and the Skew-*t* Option Pricing Model

The third step is to extend the log stock process into an option pricing context by first fixing the drift μ to enforce the martingale condition. By deriving the characteristic functions for a normal and a reciprocal gamma distributions, it can be shown that μ is given by

$$\mu = (r-q) - \ln \left(\frac{2\left(-\frac{1+\nu}{\nu}(\theta + \frac{\sigma^2}{2})\right)^{\frac{(\frac{1}{\nu}+2)}{2}}}{\Gamma(\frac{1}{\nu}+2)} K_{\frac{1}{\nu}+2}\left(\sqrt{-4\frac{1+\nu}{\nu}(\theta + \frac{\sigma^2}{2})}\right)\right). \quad (6)$$

Since $\nu > 0$, the real drift constraint for the skew-*t* model is

$$\theta + \frac{\sigma^2}{2} < 0. \quad (7)$$

Moreover, the constraint that $\sigma > 0$ forces θ to be negative. However under the risk-neutral pricing framework, we expect any skewness parameter to be negative as it corresponds to investors' risk-averse behaviour anyway. This is also supported empirically, by the work of Konikov and Madan [13] on the skew-VG option pricing model for S&P500 index options (which is the data used in this paper), all estimates for θ were negative.

The final step is to provide the option pricing model itself. The *t*-distribution is not closed in convolution density and so we use the semi-analytic option pricing model from Carr and Madan [5]. Now the skew-*t* option pricing model is presented as follows.

$$C_0 = e^{q\tau} S_0 \Pi_1 - e^{-r\tau} X \Pi_2$$
$$P_0 = e^{-r\tau} X(1 - \Pi_2) - e^{q\tau} S_0(1 - \Pi_1)$$
$$\Pi_1 = \frac{1}{2} + \frac{1}{\pi} \int_0^\infty Re\left(\frac{e^{-iux}\phi_{s_\tau}(u-i)}{iu\phi_{s_\tau}(-i)}\right) du \quad (8)$$
$$\Pi_2 = \frac{1}{2} + \frac{1}{\pi} \int_0^\infty Re\left(\frac{e^{-iux}\phi_{s_\tau}(u)}{iu}\right) du$$

$\phi_{s_\tau}(u)$ = Characteristic function of the log stock price at time, τ

For skew-$t_{\frac{2}{\nu}+4}$:

$$\phi_{s_\tau}(u)^{\{T\}} = e^{iu(s_0+\mu\tau)}$$

$$\left[\frac{2\left(-\frac{1+\nu}{\nu}\left(iu\theta - \frac{\sigma^2 u^2}{2}\right)\right)^{\frac{\frac{1}{\nu}+2}{2}}}{\Gamma\left(\frac{1}{\nu}+2\right)} K_{\frac{1}{\nu}+2}\left(\sqrt{-4\frac{1+\nu}{\nu}\left(iu\theta - \frac{\sigma^2 u^2}{2}\right)}\right)\right]^\tau \quad (9)$$

where X is the strike price not the log returns, C_0 is the current call price, P_0 is the current put price, the relationship between the call and put price is defined by Put-Call parity. τ is the time to maturity in years, r is the annual risk-free rate of return, q is the annual dividend yield, Π_1 and Π_2 are risk-neutral probabilities, $s_\tau = \ln S_\tau$, the log of the spot price, $x = \ln X$, the log of the strike price and $i = \sqrt{-1}$. μ is the martingale corrected mean given in equation (6). The degrees of freedom is $\frac{2}{\nu} + 4$. $\sigma > 0$, $\nu > 0$, and $\theta + \frac{\sigma^2}{2} < 0$.

2.1.4   The Skew-t, Skew-VG and Comparison with Heston's Stochastic Volatility Model

From the literature review, we saw that an unresolved rivalry exists between the skew-VG and Heston's models (Kim and Kim [12]). It is interesting therefore to directly compare these two models and the new skew-t model theoretically. This section provides new insights into how we can match the parameters of Heston's model with skew-VG and skew-t.

Under Heston's model [10], we model the log returns to be normally distributed with a fixed drift but a stochastic variance. The pair of formulas is provided below. That variance process, V_t, is mean-reverting by rate, κ, to the long run mean variance, μ_V, and with a Gaussian innovation term and volatility of stochastic variance parameter, σ.

As a result of the time varying variance, the initial V_0 becomes another parameter to estimate for Heston's model. It then has a fifth parameter for the correlation between its two processes, the log returns process and variance process to be estimated through ρ. In the expression below, log returns are denoted with X_t (τ notation is done away with in this section as it was used previously to be consistent with the option time to maturity, τ but the insights offered here are on a stock dynamics level).

$$\text{Heston's model:} \quad X_t = \mu + \sqrt{V_t}z_t^{(1)}$$
$$V_t = V_{t-1} + \kappa(\mu_V - V_{t-1}) + \sigma\sqrt{V_{t-1}}z_t^{(2)} \tag{10}$$
$$z_t^{(1)} \sim N(0,1), \ z_t^{(2)} \sim N(0,1)$$

For the skew-VG and skew-t models, through the random time change interpretation to the mixing variable, λ, the interpretation described so far has been that log returns are distributed normally, with fixed mean and variance, *conditional* upon a *random* length of the time increment. Another way to interpret this is that, modelling the process over fixed time intervals, the mean and the variance of the log returns are changing each period, as controlled by the λ. The interpretation of changing drift and variance for skew-VG and skew-t is more conducive to comparison with Heston's model.

We can subscript λ as λ_t without loss of generality, where it can be interpreted as the latent variable's disturbance to the average drift $(\mu + \theta)$ and variance (σ^2) of the skew-VG process. The result is that, where Heston's model has two processes, the log returns and stochastic variance, the skew-VG and skew-t both have three processes, with the addition of stochastic behaviour in the drift $(r_t = \mu + \theta\lambda_t)$.

The skew-VG and skew-t model can be written in terms of the three processes as such:

$$
\begin{aligned}
X_t &= \mu + \theta \lambda_t + \sigma \sqrt{\lambda_t} z_t \\
r_t &= \mu + \theta \lambda_t \\
V_t &= \sigma^2 \lambda_t \\
z_t &\sim N(0,1)
\end{aligned}
\tag{11}
$$

$\lambda_t \sim$ Reciprocal gamma for skew-t,

Gamma for skew-VG, mean $= 1$ and variance $= \nu$

Table 1 provides a side-by-side comparison between Heston and the skew-t (and VG) models. It also expresses the processes as distributions to aid further comparison of the means and variances. This analysis reveals to us a close-knit relationship between the skew-t (and VG) and Heston models. Specifically, we are able to retrieve the skew-t (VG) model from Heston's model by:

1. Setting Heston's mean reversion parameter, $\kappa = 0$, such that the expected variance is constant over time. Under Heston's model, as a result, $\mu_V = V_0$, the long run mean variance is the same as the initial variance,
2. Instead of the stochastic variance following a normal distribution, we ascribe it a gamma distribution,
3. We then introduce a stochastic component to the drift of the returns process which is driven by the same gamma innovation in the stochastic variance process from (2) and scaled by a parameter, θ,
4. We do not directly estimate the correlation ρ between the log returns and variance process.

In net, the VG process sets a restriction on two of Heston's give parameters, $\kappa = 0$ and $V_0 = \mu_V$, does not estimate another of its parameters, the correlation, ρ and introduces its own drift skewness parameter, θ, to come to its total of three parameters. $\kappa = 0$ for skew-t and skew-VG means that the models will likely perform poorly during times of volatility clustering such as crisis periods.

Two further remarks on the comparison—Heston's stochastic variance is normally distributed and so requires a further condition to make it positive, but the skew-t (VG)'s stochastic variance will always be positive as it is driven by non-negative distributions, the gamma or reciprocal gamma. Indeed this would be the case for any time subordinator model (time cannot be negative). Another note, is that Heston's σ is comparable to skew-t (VG)'s ν in that they both drive the volatility of the stochastic variance, and in turn kurtosis.

From this section, the similarities that have been formalised between the skew-t (VG) and Heston model shed light on why there is the rivalry between the skew-VG and Heston model in the literature to date—*both* models have stochastic variance processes and can model the leverage effect. The deficiency however of the skew-VG and skew-t model is that restricts Heston's $\kappa = 0$.

Table 1. Comparison between Heston's model and the skew-t and skew-VG models. Conditional drifts and variances are shown.

	Heston's model	Skew-t and Skew-VG model *Replace reciprocal gamma with gamma for skew-VG results*
Log returns process	$X_t = \mu + \sqrt{V_t} z_t^{(1)}$	$X_t = \mu + \theta \lambda_t + \sigma \sqrt{\lambda_t} z_t$
Drift of log returns process	$r_t = \mu$ (constant)	$r_t = \mu + \theta \lambda_t$
Variance of log returns process	$V_t = V_{t-1} + \kappa(\mu_V - V_{t-1}) + \sigma \sqrt{V_{t-1}} z_t^{(2)}$	$V_t = (\sigma \sqrt{\lambda_t})^2 = \sigma^2 \lambda_t$
Innovation terms	$z_t^{(1)} \sim N(0,1)$, $z_t^{(2)} \sim N(0,1)$	$z_t \sim N(0,1)$ $\lambda_t \sim$ Reciprocal gamma mean = 1 var = v, $v > 0$
Distribution of drift	r is constant over time	$r_t \sim$ Reciprocal gamma mean = $\mu + \theta$ var = $\theta^2 v$
Distribution of variance	$V_t \sim$ Normal mean = $V_{t-1} + \kappa(\mu_V - V_{t-1})$ var = $\sigma^2 V_{t-1}$	$V_t \sim$ Reciprocal gamma mean = σ^2 var = $\sigma^4 v$
Correlation between log returns, X_t, and variance, V_t	$-1 \leq \rho \leq 1$	$\rho = \dfrac{\theta v}{\sqrt{v(\theta^2 v + \sigma^2)}}$ $1 < \rho < 1$, since $\sigma > 0$
Correlation between drift, r_t, and variance, V_t	0, drift is deterministic	$\mathrm{corr}(r_t, V_t) = \begin{cases} -1 & \text{for } \theta < 0 \\ 1 & \text{for } \theta > 0 \\ 0 & \text{for } \theta = 0 \end{cases}$

3 Parameter Estimation Procedures

This section discusses how the parameters of the option pricing models were calibrated. It has two focuses. First, I formulate the optimisation problem and then test the calibration methods on a simulated data set without noise and then with noise introduced. The purpose of this section is to verify that we are able to recover estimates close to the true parameter values in preparation for the empirical section that follows.

3.1   Formulating the Optimisation Problem

3.1.1   The Objective Function

The objective function is the root mean square percentage pricing errors (RMSPE), where the percentage errors are found through taking the log difference. This was used by Madan et al. [14], the percentage errors used due to the range of call prices across different strikes and maturities. The authors also give the result that the optimisation problem is asymptotically equivalent to the maximum likelihood method.

The minimisation of square percentage errors has since been adopted in other empirical tests on the skew-VG model (see Kim and Kim [12] where instead of using log difference for percentage errors, a discrete percentage error measure was used, $([C - C^*]/C)$ where C^* is the call price estimated by the model and C is the observed call price. In this paper, with the chosen optimisation algorithm, outlined below, it was found that the log-difference aided the calibration compared to using the discrete measure from Kim and Kim [12]. As such the objective function used in this paper, from Madan, Carr and Chang (their Equation (30)) is:

$$\text{Objective function} = \sqrt{\frac{1}{n}\sum_{i=1}^{n}(\ln(C^*)\ln(C))^2}. \tag{12}$$

where n is the number of call prices, and ln is the natural logarithm.

3.1.2   Constraints

We impose the positivity constraints on the Black-Scholes σ and the skew-VG and skew-*t*'s, ν and σ through log transformations. For the constraints required to ensure that the drift, μ is real in the skew-VG and skew-*t* model, we use a penalty function that adds an arbitrarily large quantity to the minimisation objective value in the case of violation.

Recalling from above these constraints, for skew-VG: $\nu\left(\theta + \frac{\sigma^2}{2}\right) < 1$ and for skew-*t*: $\theta + \frac{\sigma^2}{2} < 0$, skew-*t*'s θ is restricted to negative values but skew-VG's θ is not. For our optimisation, this restricted domain on skew-*t*'s θ was an advantage for obtaining more accurate estimates compared to the skew-VG. It was therefore decided that a negativity constraint would be imposed on skew-VG's θ as well. This firstly, makes for a fairer competition between the models but also is theoretically sound based on the arguments raised above, namely investors' risk-aversion and previous empirical studies on the skew-VG model estimating all negative θs.

To ensure the negativity of skew-VG and skew-*t*'s θs, where we directly estimate the $-\log(\theta)$ and retrieve θ through taking the exponential and then negative of that value. Together, the log transformations and penalty functions allow us to then use an unconstrained optimisation procedure.

3.1.3   The Optimisation Algorithm

The optimisation algorithm used is the Nelder-mead simplex algorithm, instructed by Matlab's *fminsearch* function. *fminsearch* is Matlab's unconstrained, non-linear optimiser where a simplex (a 2-dimensional line for the Black-Scholes with only one parameter and tetrahedron (a simplex with $3 + 1$ faces) for the skew-VG and skew-t models each with three parameters) descends the surface of the objective function. At each step, the simplex can be reflected, expanded, contracted or shrunk. Its changing step size makes it a preferred algorithm for noisy objectives, such as that for the option pricing models with numerical integration (Gilli and Schumann [9]).

In other papers on the calibration of such option pricing models using Matlab, the *lsqnonlin* function (non-linear least squares) has also been implemented (Moodley [15]). However *lsqnonlin*, *fminunc* and *fmincon* performed poorly on calibrating our models. It was observed that the global optimiser, *patternsearch*, could recover the true parameter values as well as *fminsearch*, but it took an unworkable about of time. *fminsearch* on the other hand took only eight hours to estimate the parameters for all 161 weeks (close to 40,000 prices) using parallel computing and four workers.

3.1.4   Initial Values, Stopping Criterion and Convergence

The initial values were set based on the average parameter values estimated in Madan, Carr and Chang's empirical testing of the skew-VG model [14]. The initial values for the skew-VG and skew-t model were selected to be the same. Respectively, for θ, ν and σ, they are -0.2, 0.2 and 0.15. For Black-Scholes the initial value for σ is 0.2, set higher than the initial σ in the other two models because it does not have the availability of the extra parameters to further increase the variance of the distribution.

There are two operative stopping criteria in our algorithm. The first is a minimum tolerance on gains made on the objective value, $1e-04$, and the second is a minimum tolerance on movement in the variables, $1e-04$. These criteria are adequate to reach objective values that are small enough to allow us to observe convergence in the parameters estimated in the empirical study.

3.1.5   Obtaining Standard Errors

We encountered the issue that, unlike the gradient-based optimisation functions in Matlab, the *fminsearch* function does not output the Jacobian or Hessian from which we can calculate the standard errors. In their calibration study, Gilli and Schumann [9] did not provide standard errors.

Calculating the standard errors required first obtaining the Jacobian matrix ($n \times p$, where p is the number of parameters), which was found through evaluating the slope of the objective function for each individual call price and for each parameter. The interval over which the slope was evaluated was set to a small value, $1e-03$. Then it was also required that we convert the standard errors obtained on the log-transformed parameters to the standard errors for the parameters in their original form, the exponential of the log-transformed parameters. To so this, we invoked the Delta method and so multiplied the standard errors by the absolute value of the first derivative of the exponential (or negative exponential for θ) function of the log-parameters.

4 Empirical Study

We cast the following empirical hypotheses. Given how different the skew-VG and skew-t distributions are with the same parameter values, it is likely that in their attempt to fit a mutual distribution they will return rather differing parameter estimates. Another empirical observation we are likely to see is that during times of volatility clustering or financial crisis, the skew-VG and skew-t may perform worst.

It should be noted that one of the key advantages of focusing on the comparison between two models with the same number of parameters is that no advantage is given to one of them in in-sample fitting. Black-Scholes however, with only one parameter will be at a disadvantage for the in-sample fitting, which along with the practicality of being interested forecasting prices, is the motivation for conducting out-of-sample tests as well.

This empirical section is based on the methodology employed by Bakshi et al. [1] for comparing option pricing models based on two main criteria: minimising pricing error and minimising the predictability of the pricing errors by moneyness (strike:spot ratio) and maturity. Subsequent empirical tests such as Madan et al. [14], Kim and Kim [12] and Eberlein et al. [6] use similar test designs. Unlike the aforementioned tests, this paper has a distributional focus and provides additional analysis of the higher moments, tail-fitting and quantile plots.

4.1 The Data

The data used are the daily prices of European call and put options on the S&P500 index. In order to cull highly illiquid options data we restrict the data to options with moneyness between 0.97 and 1.03 and maturities between 1 month away and 1 year away (Kim and Kim [12]). All option data was obtained from Option Metrics. The risk-free rate is the rate of return on a one month US treasury bill and the dividend is that on the S&P500 index.

In order to account for frequently changing parameter values over time, the training period was only one week long. The size of the data available caters for the use of short training samples. There are 245 data points per week on average. For out-of-sample we tested the model on the data one day out of the sample.

Table 2 summarises the data used. There are 39,667 data points each for calls and puts, spanning 162 weeks (1 to 161 for training and 2 to 162 for testing) starting on the 23 February 2009. It was desired that the three years encompassed bust, boom and quiet periods but otherwise the particular start date for the sample is arbitrary. This period captures part of the GFC, the Euro-debt crisis,

Table 2. Sample properties of the S&P500 Index European call options and put option in parenthesis. Categorisation is in terms of moneyness (strike to spot price ratio) and time to maturity. Dates span from 23 February 2009 29 March 2012, 162 weeks (161 training samples).

Call (Put)	Time to maturity			{Sample size}
Moneyness	<3 months	3–6 months	>6 months	Subtotal
In (Out of) 0.97–0.99	$50.68 ($30.59) {7226}	$68.60 ($52.15) {3197}	$96.65 ($88.12) {2684}	$64.47 ($47.63) {13107}
At 0.99–1.01	$36.00 ($39.48) {7340}	$54.59 ($61.52) {3300}	$82.92 ($97.70) {2760}	$50.24 ($56.90) {13400}
Out of (In) 1.01–1.03	$24.04 ($51.08) {7344}	$41.98 ($72.42) {3314}	$71.02 ($109.40) {2502}	$37.49 ($67.54) {13160}
Subtotal	$36.83 ($40.44) {21910}	$54.90 ($62.15) {9811}	$83.81 ($98.15) {7946}	$50.71 ($57.37) {39667}
Number of weeks				162
Average number of data points in a training period (a week)				245

the US debt-ceiling crisis and some recovery periods in between. As expected, we can see that for calls, the price of the options increases with the time left until maturity and the deeper the option is in-the-money.

Table 2 is also useful for highlighting how differing the option prices are within our data set. The range on the averages for each category is from $36.83 to $96.65. This motivates the use of percentage pricing errors rather than pricing errors in dollar terms. The main body of this paper therefore will only include the percentage pricing errors (mean, mean absolute and root mean square percentage errors).

4.2   Empirical Parameter Estimation and In-Sample Fitting

There are a few notes to make regarding the summary of parameters given in Table 3. The skew-VG and skew-t use different parameter values to fit a common distribution. For θ, on average, the skew-VG will fit a more negative skewness parameter than skew-t and in turn, skew-t appears to model a higher σ over skew-VG. For ν, looking just at the average value estimated, there is not much between the two models, skew-VG at 0.234 and skew-t at 0.238. In the context of the skew-t distribution, this is on average 12.4 degrees of freedom ($d.f. = \frac{2}{\nu} + 4$ from above).

Although the average ν's are similar, it is noticeable that skew-t's maximum value for ν is extremely high. Specifically we observed four kurtoses for skew-t that exceeded 10 (11.06, 40.30, 324.67 and 733.52). This means that the distribution has very fat tails and is very peaked at these points. Theoretically, the kurtosis is able to be infinite. While the skew-t can have very large values for ν, we notice that that the skew-VG can take on very large (negative) values of θ, -2.781.

Table 3. The parameters of the risk-neutral densities of log annual S&P500 Index returns based on the Black-Scholes option pricing model, skew-VG and skew-t option pricing models. There are 161 training samples. Standard errors have been omitted for neatness and because, just as was seen in the simulation, they were generally close to zero. The objective function is the root mean square percentage pricing error (RMSPE). Results are for calls. Puts results were similar and so are not provided to save space.

Model	BSM	Skew VG			Skew t		
Parameter	σ	θ	ν	σ	θ	ν	σ
Mean	0.220	−0.397	0.234	0.206	−0.174	0.238	0.239
Standard deviation	0.059	0.463	0.143	0.072	0.067	0.587	0.058
Minimum	0.134	−2.781	0.000	0.025	−0.697	0.025	0.152
Maximum	0.412	0.000	0.558	0.412	−0.064	5.520	0.434
Optimised objective value	0.114	0.0547			0.0695		

This is further evidence of differing behaviour across skew-VG and skew-t when it comes to data fitting. (By way of reference for the values on skew-VG's θ, given the length of our sample (161 weeks) our estimates seem reasonable, compared to previous empirical tests. Konikov and Madan's ([13] empirical fitting of the skew-VG model to S&P500 data using only five training periods had a minimum $\theta = -1.789$).

The objective value (root mean square percentage pricing error) is used as an yardstick for in-sample fitting (Bakshi et al. [1]). On average, we can see that the skew-VG achieved the best fit of 5.47%, skew-t of 6.95% and both are vast improvements on Black-Scholes at 11.39%.

4.2.1 Moments of the Fitted Distributions

Following on from the parameter values, the moments for underlying log stock returns distribution are next provided (Table 4). For Black-Scholes, the variance is just the square of the σ parameter. For skew-VG and skew-t, all three parameters contribute to all three of the moments, that is, it is not the case that the skewness is solely determined by, but increases in magnitude with, the skewness parameter, θ for example. The moments were obtained through simulating the distributions based on the parameter values (1,000,000 simulations) instead of relying on the closed-form expressions for the moments as they cannot be used around $\nu = 0.5$ or 1 for the skew-t model.

As we would expect from the larger θ for skew-VG, on average the distribution fit under the skew-VG model has greater skewness (−0.596) compared to the skew-t (−0.494). Due to the outlying kurtoses (such as the two > 300), we also provide the median kurtosis (in addition to the mean kurtosis). It is interesting to note that while the skew-t's maximum value for ν and largest kurtosis estimated is a lot higher than for skew-VG, the median kurtosis fit by the skew-t is lower than the median kurtosis for the skew-VG. This suggests that the skew-VG's

Table 4. The average moments for the underlying log returns distributions for calls. Puts were similar and so not displayed in the interest of space.

Model	BSM	Skew-VG	Skew-t
Mean	−0.047	−0.054	−0.053
Variance	0.052	0.068	0.066
Skewness	0	−0.596	−0.494
Kurtosis	3	3.989	10.778
Median kurtosis	3	3.930	3.898

parameter values together, very likely driven by its higher θ, are overall fitting just as fat-tailed distributions as the skew-t.

4.3   Out-of-Sample Pricing Performance

This section presents three different perspectives on pricing performance. The first is cross-sectionally, that is how to the different models perform for options of varying moneyness ratios and maturities, second time-series analysis is provided and a comparison made between above average market volatility levels and below average. Third, we return to a more in depth understanding of the moments under high and low volatility to understand how the two distributions respond to the different financial climates.

4.3.1   Cross-Sectional Analysis and Tail-Fitting

In relation to Table 5, as a measure of overall fit, we can compare the root mean squared percentage errors (RMSPE) for all option types (bottom right corner of table), to those in the in-sample fitting. It appears that the out-of-sample fitting (7.39%, 4.10% and 4.86% for Black-Scholes, skew-VG and skew-t) is better than the in-sample fitting (11.39%, 5.47% and 6.95% respectively). However, this is a misleading comparison since the out-of-sample fit is on data only for one day (49 data points on average), where as the in-sampling fitting tried to fit one set of parameters to a whole week of data (245 data points on average). The results would therefore, instead, show signs of the variation that exists within a week.

Turning to the cross-sectional analysis and MPEs, (log predicted - log observed). Across moneyness, all models overprice OTM calls and underprice ITM calls. Black-Scholes has the greatest error differential between the moneyness categories and skew-VG has the least. Across maturities, the MPEs indicate a lower average error for short-term contracts compared to long-term, but checking the MAPE and RSMPE, we can see that the short term contracts (less than 3 months) actually have the highest absolute and root squared errors, especially OTM calls. We can also note that while for Black-Scholes the most accurately priced maturities are mid-term (3–6 months), for skew-VG and skew-t, the fit is best in the long-term contracts (6–12 months).

Table 5. 1 day out-of-sample percentage pricing errors using a week long training period and parameters estimated on option data for all maturities and moneyness (strike to spot price) levels. The percentage error is found by taking the log difference. A corresponding table of results for non-percentage errors is provided in the appendix. 161 training samples.

Calls			Months to maturity			
			<3	3–6	>6	All
	Strike/Spot		**A. Mean percentage errors (%)**			
Black-Scholes	ITM	0.97–0.99	−1.51	−4.62	−7.71	−3.53
	ATM	0.99–1.01	1.99	−3.00	−7.10	−1.12
	OTM	1.01–1.03	8.45	−0.18	−5.87	3.61
		All	2.99	−2.59	−7.00	−0.44
Skew-VG	ITM	0.97–0.99	0.97	−0.25	−1.73	0.12
	ATM	0.99–1.01	0.82	0.54	−0.88	0.38
	OTM	1.01–1.03	1.01	1.72	0.27	1.11
		All	0.92	0.68	−0.82	0.50
Skew-*t*	ITM	0.97–0.99	−1.33	−1.03	−2.42	−1.46
	ATM	0.99–1.01	−1.40	0.64	−1.22	−0.89
	OTM	1.01–1.03	2.92	3.62	0.51	2.68
		All	0.07	1.08	−1.10	0.07
	Strike/Spot		**B. Mean absolute percentage errors (%)**			
Black-Scholes	ITM	0.97–0.99	3.16	4.95	8.03	4.56
	ATM	0.99–1.01	4.83	4.20	7.72	5.23
	OTM	1.01–1.03	10.19	4.20	7.05	8.03
		All	6.07	4.42	7.69	5.92
Skew-VG	ITM	0.97–0.99	3.19	2.22	2.51	2.78
	ATM	0.99–1.01	3.97	2.80	2.50	3.36
	OTM	1.01–1.03	5.54	3.75	2.73	4.55
		All	4.22	2.92	2.59	3.53
Skew-*t*	ITM	0.97–0.99	2.89	2.43	3.26	2.81
	ATM	0.99–1.01	4.26	2.99	2.86	3.63
	OTM	1.01–1.03	6.64	4.99	2.92	5.51
		All	4.59	3.46	3.06	3.95
	Strike/Spot		**C. Root mean squared percentage errors (%)**			
Black-Scholes	ITM	0.97–0.99	3.42	5.06	8.08	5.33
	ATM	0.99–1.01	5.39	4.40	7.78	6.14
	OTM	1.01–1.03	11.15	4.53	7.16	9.47
		All	7.73	4.87	7.85	7.39
Skew-VG	ITM	0.97–0.99	3.27	2.30	2.57	3.08
	ATM	0.99–1.01	4.15	2.86	2.57	3.68
	OTM	1.01–1.03	6.03	3.83	2.81	5.15
		All	4.72	3.16	2.80	4.10
Skew-*t*	ITM	0.97–0.99	3.07	2.50	3.33	3.11
	ATM	0.99–1.01	4.68	3.09	2.98	4.16
	OTM	1.01–1.03	7.25	5.18	3.07	6.26
		All	5.55	4.02	3.34	4.86

There is one category where skew-*t* offers the best model by MAPE and RSMPE: in-the-money, short-term. However, this is not a compelling result, rather it seems an anomaly cross-sectionally. Indeed for puts - skew-VG is superior in all categories. Further, it is the OTM rather than the ITM options fitting that yields the most direct interpretation. Due to the limited downside for options, their pricing performance only reveals information about the fit of the stock price distribution to the right (left) of the strike price for calls (puts). Therefore, in-the-money options show more of a general fitting of the distribution (all but one tail), where as OTM options carry *only* the tail information. Table 6 gives the results for OTM puts and OTM calls to compare the left and right tail fitting of the skew-VG and skew-*t* distributions.

Table 6. Relative fit of the tails by the skew-VG and skew-*t* distributions

	Left tail fitting of the stock price distribution *(Reflects the probability that the spot price will fall below a strike price that is less than 99% of the current spot price)*	Right tail fitting of the stock price distribution *(Reflects the probability that the spot price will rise above a strike price that is more than 101% of the current spot price)*
	Out-of-the-money puts	Out-of-the-money calls
Mean absolute percentage error		
Skew-VG	3.62%	4.55%
Skew-*t*	3.77%	5.51%
Root mean square percentage error		
Skew-VG	4.06%	5.15%
Skew-*t*	4.27%	6.26%

From the out-of-the-money options fitting, it is evident that both models fit the left tail of the stock price distribution better than the right tail and that the skew-*t*'s performance is more competitive with skew-VG's performance in the left tail fitting as well (0.15% greater MAPE for OTM puts, where as 0.96% greater MAPE for OTM calls).

We hypothesised in Sect. 2.1.4 that during crisis periods, the skew-*t* and skew-VG models would perform the worst. Based on the instances when neither the skew-VG or skew-*t* are able to improve upon the Black-Scholes model, these indeed occur when the volatility is at its highest. Particularly, these periods are at the start of our sample, 23 February 2009 during the Global Financial Crisis, and from the 25th July 2011, coinciding with the US debt-ceiling crisis. It is found that 4% of the time (161 weeks), neither the skew-VG or the skew-*t* could improve upon Black-Scholes. Outside this 4%, we are interested in the specific comparison between skew-*t* and skew-VG. Overall, the skew-VG outperforms skew-*t* more often than not and this is consistent with the cross-sectional results. The break-down between the percentage of times the skew-VG versus the skew-*t* outperforms is shown in Table 7.

Table 7. The percentage of weeks each model is the best performer across low versus high volatility for call options. The data sample begins on the 23 February 2009 (during the Global Financial Crisis) to 29 March 2012

Superior model	Skew-VG	Skew-t	Black-Scholes	Total
All levels of volatility	65%	30%	4%	161
Above average volatility	57%	34%	9%	70
Below average volatility	71%	27%	1%	91

To conclude on the pricing analysis for calls, we see that the skew-VG outperforms skew-t in almost all cross-sectional categories when the results are averaged across time. But to look at the results from a time-series perspective, it is apparent that while in below average volatility, the skew-VG is the superior model, in above average volatility we observe skew-t's performance becomes more competitive with the skew-VGs.

5   Conclusion

The resounding conclusion is that without any information about the market volatility, the skew-VG option pricing model is more likely to outperform the skew-t than otherwise. However, if we use the S&P500 volatility index (VIX) or Black-Scholes implied volatility, then under more volatile conditions, the skew-VG's performance starts to deteriorate and the skew-t model becomes more competitive with the skew-VG. A core contribution of this paper has therefore been that, for the skew-VG, we now have evidence that its superior performance in past empirical studies showed a more powerful result than merely the option market's preference for a fat-tailed, skewed distribution and/or infinite activity, pure jump model. Now that we can compare it against another model that bears all these traits as well, (and with the same few number of parameters,) we can confirm that empirically the skew-VG distribution performs well due to some its idiosyncracies that distinguish it from the skew-t. Considering the variability in the relatively performance of the skew-VG and skew-t models, a generalised model which nests both as special cases, enables pricing fitting and predicting that is at least as good as the superior model of the skew-VG or skew-t. Such a generalised model is proposed in Yeap et al. [7], and assumes the underlying log return distribution is generalised hyperbolic.

The final model that we gained deeper insight into, only in the theory section, was Heston's Stochastic Volatility model. The five parameters of Heston's model were matched to skew-VG and skew-t who each only have three. The result was that we can obtain the symmetric VG and symmetric t from Heston's model through setting the rate of mean reversion, κ, to 0, and by having a gamma stochastic variance process rather than normal. This paper also provided the skew-t and skew-VG models' equivalent values for Heston's correlation coefficient, ρ, such that the leverage effect can now be quantified within their framework as well.

Further research would be aimed at conducting empirical tests between the skew-t and skew-VG with models that can model time-dependency. This paper nominated three such models: Heston's model, Heyde and Gay's dependent t-option pricing model [11] and a two-state Markov model such as Konikov and Madan's [13]. The motivations for these extensions were discussed as each arose.

References

1. Bakshi, G., Cao, C., Chen, Z.: Empirical performance of alternative option pricing models. J. Finance **52**, 2003–2049 (1997)
2. Barndorff-Nielsen, O.E., Shephard, N.: Basics of Lévy processes. Draft chapter of Lévy Driven Volatility Models (2012)
3. Black, F., Scholes, M.: The pricing of options and corporate liabilities. J. Polit. Econ. **81**, 637–654 (1973)
4. Carr, P., Hogan, A., Stein, H.: The for a change: the variance gamma model and option pricing. Working paper (2007)
5. Carr, P., Madan, D.B.: Option valuation using the fast fourier transform. J. Comput. Finance **2**, 61–73 (1998)
6. Eberlein, E., Keller, U., Prause, K.: New insights into smile, mispricing and value at risk: the hyperbolic model. J. Bus. **71**(3), 371–405 (1998)
7. Yeap, C., Choy, S.T., Kwok, S.: A flexible generalised hyperbolic option pricing model and its special cases. J. Financ. Econometrics (2017, to appear)
8. Finlay, R., Seneta, E.: Stationary-increment variance gamma and t models: simulation and parameter estimation. Int. Stat. Rev. **76**(2), 167–186 (2008)
9. Gilli, M., Schumann, E.: Calibrating optionpricing models with heuristics. Natural Comput. Comput. Finance **380**, 9–37 (2012)
10. Heston, S.L.: A closed-form solution fo roptions with stochastic volatility with applications to bond and currency options. Rev. Financ. Stud. **6**(2), 327–343 (1993)
11. Heyde, C.C., Gay, R.: Fractals and contingent claims. Australian National University, Preprint (2002)
12. Kim, I.J., Kim, S.: Empirical comparison of alternative stochastic volatility option pricing models: evidence from Korean KOSPI 200 index options market. Pacific-Basin Financ. J. **12**, 117–142 (2004)
13. Konikov, M., Madan, D.B.: Option pricing using variance gamma Markov chains. Rev. Deriv. Res. **5**, 81–115 (2002)
14. Madan, D.B., Carr, P.P., Chang, E.C.: The variance gamma process and option pricing. Eur. Finance Rev. **2**, 79–105 (1998)
15. Moodley, N.: The Heston model: A practical approach. Faculty of Science, University of Witwatersrand, Johannsesburg, Honours Project (2005)

Confidence Intervals for the Coefficient of Variation of the Delta-Lognormal Distribution

Noppadon Yosboonruang[✉], Sa-Aat Niwitpong, and Suparat Niwitpong

Department of Applied Statistics, Faculty of Applied Science, King Mongkut's
University of Technology North Bangkok, Bangkok 10800, Thailand
noppadonyos@gmail.com, sa-aat.n@sci.kmutnb.ac.th

Abstract. This paper proposes confidence intervals for the coefficient of variation of the delta-lognormal distribution which are based on the generalized confidence interval (GCI) method, it is compared with the modified Fletcher method. The coverage probabilities and the expected lengths are used for assessing the performance of these confidence interval. Simulation results show that the GCI method perform best in term of the coverage probability and expected length.

1 Introduction

The coefficient of variation is used to measure of the dispersion of data, and it is useful for comparing the degree of variation from one data series to another when different units are used in different series. It is commonly used in many fields such as engineering, medical, economic and science. Several researchers have proposed confidence intervals for the coefficient of variation. Wong and Wu [23] proposed small sample asymptotic method to obtain approximate confidence intervals for the coefficient of variation for both normal and nonnormal models. Tian [19] considered the problem of making inference about the common population coefficient of variation when it is a priori suspected that several independent samples are from populations with a common coefficient of variation. Mahmoudvand and Hassani [11] introduced an approximately unbiased estimator to construct two new confidence intervals for the population coefficient of variation in a normal distribution. Sangnawakij et al. [15] proposed two new confidence intervals for the ratio of coefficients of variation in the gamma distributions based on the method of variance of estimates recovery with the methods of Score and Wald intervals. Wongkhao et al. [24] proposed two new confidence intervals for the ratio of two independent coefficients of variation of normal distribution based on the concept of the general confidence interval (GCI) and the method of variance estimates recovery (MOVER). Sangnawakij and Niwitpong [14] constructed the confidence intervals for the single coefficient of variation and the difference of coefficients of variation in the two-parameter exponential distributions, using the method of variance of estimates recovery (MOVER), the generalized confidence interval (GCI), and the asymptotic confidence interval (ACI). Thangjai and Niwitpong [18] presented a new confidence intervals for

© Springer International Publishing AG 2018
L. H. Anh et al. (eds.), *Econometrics for Financial Applications*, Studies in Computational Intelligence 760, https://doi.org/10.1007/978-3-319-73150-6_26

the weighted coefficients of variation of two-parameter exponential distributions based on the adjusted method of variance estimates recover method (adjusted MOVER).

In various situations, the data are highly skewed and include many zero numbers, it is discussed in Aitchison [1]. The distribution for the non-negative data, is a lognormal distribution, and containing the zero values is the delta-lognormal [2,6] which is widely used, e.g., to fishery data in fisheries research; see, e.g., [10,12,13,16,17]. The delta-lognormal distribution was also found in insurance, reliability, and meteorology [9]. Recently, Fletcher [7] considered three methods including Aitchison's method, a modification of Cox's method and a profile-likelihood interval to construct the 95% confidence interval for the mean of the delta-lognormal distribution. Buntao and Niwitpong [4] presented new confidence intervals for the difference of the coefficients of variation by using the generalized pivotal approach (GPA) and the closed form method of variance estimation (CFM) for the lognormal distributions and the delta-lognormal distribution. Kvanli et al. [9] constructed a confidence interval based on the likelihood ratio test approach for a population mean when there are a very large number of zeros. Buntao and Niwitpong [5] used the concept of the generalized variable approach (GPA) and the method of variance estimates recovery (MOVER) to established the confidence intervals for the ratio of coefficients of variation of delta-lognormal distribution. Wu and Hsieh [25] proposed a heuristic method an estimation based on asymptotic generalized pivotal quantity to construct the generalized confidence interval for the mean of the delta-lognormal distribution. In this paper, we are interested in establishing the confidence intervals for the coefficient of variation of the delta-lognormal distribution.

The aim of this paper is to propose the new confidence interval for the coefficient of variation of the delta-lognormal distribution based on the generalized confidence interval, as presented by Weerahandi [22], and compare it with the modified Fletcher method from Fletcher [7]. In the next section, we present the research method including details of theory and method used in this study. Simulation results and example will be given in Sects. 3 and 4, respectively. Finally, discussion and conclusions are summarized in Sect. 5.

2   Confidence Intervals for the Coefficient of Variation in Delta-Lognormal Distribution

Let $\mathbf{X} = (X_1, X_2, ..., X_n)$ be a non-negative random sample from a delta-lognormal distribution, denoted by $\Delta\left(\mu, \sigma^2, \delta\right)$, where $\delta = P\left(X_i > 0\right), i = 1, 2, ..., n$. The number of zeros and non-zeros observed value are denoted by n_0 and n_1, respectively, where $n = n_0 + n_1$ the number of zero observations has a binomial distribution, $n_0 \sim Bin\left(n, 1 - \delta\right)$. Aitchison and Brown [2] and Tian and Wu [20] presented the distribution function of the delta-lognormal population as

$$G\left(x, \mu, \sigma^2, \delta\right) = \begin{cases} 1 - \delta & ; \quad x = 0 \\ (1 - \delta) + \delta F\left(x, \mu, \sigma^2\right) & ; \quad x > 0 \end{cases} \qquad (1)$$

where $F\left(x, \mu, \sigma^2\right)$ is the lognormal cumulative distribution function. Let $\mathbf{S} = \left(S_1, S_2, ..., S_n\right)$ be a positive random variable having a lognormal distribution, denoted as $LN\left(\mu, \sigma^2\right)$, and $Y = \ln\left(S\right) \sim N\left(\mu, \sigma^2\right)$ where μ and σ^2 denoted the mean and variance of Y, respectively. Then, the probability distribution function of lognormal distribution is

$$f\left(x, \mu, \sigma^2\right) = \begin{cases} \frac{1}{x\sigma\sqrt{2\pi}} \exp\left\{-\frac{1}{2\sigma^2}\left(\ln\left(x\right) - \mu\right)^2\right\} & ; \quad x > 0 \\ 0 & ; \quad \text{otherwise.} \end{cases} \tag{2}$$

Aitchison [1] describes the mean and the variance of the delta-lognormal distribution

$$E\left(X\right) = \mu_X = \delta \exp\left(\mu + \frac{\sigma^2}{2}\right) \tag{3}$$

and

$$Var\left(X\right) = \sigma_X^2 = \delta \exp\left(2\mu + \sigma^2\right)\left[\exp\left(\sigma^2\right) - \delta\right]. \tag{4}$$

Thus, the coefficient of variation of X is given by

$$\begin{aligned} CV\left(X\right) = \eta &= \frac{\sqrt{Var\left(X\right)}}{E\left(X\right)} \\ &= \frac{\sqrt{\delta \exp\left(2\mu + \sigma^2\right)\left[\exp\left(\sigma^2\right) - \delta\right]}}{\delta \exp\left(\mu + \frac{\sigma^2}{2}\right)} \\ &= \sqrt{\frac{\exp\left(\sigma^2\right) - \delta}{\delta}}. \end{aligned} \tag{5}$$

In the following, we propose the methods for establishing the confidence interval for the parameter η.

2.1 The Generalized Confidence Interval for the Coefficient of Variation

Weerahandi [22] proposed the generalized confidence interval (GCI) for constructing confidence interval. This method is based on a generalized pivotal quantity (GPQ) which is defined as follows.

Let $\mathbf{X} = \left(X_1, X_2, ..., X_n\right)$ be a vector of random variables and $f\left(\mathbf{X}; \theta\right)$, where $\theta = \left(\mu, \sigma^2, \delta\right)$, is the probability density function of $\mathbf{X}$. The σ^2 and δ are the parameters of interest and μ is a nuisance parameter. Let $\mathbf{x} = \left(x_1, x_2, ..., x_n\right)$ be the observed value of random variables $\mathbf{X}$. Let $R = R\left(\mathbf{X}; \mathbf{x}, \theta\right)$ be a function of $\mathbf{X}$, $\mathbf{x}$ and θ. Then R is called a generalized pivotal quantity if the following two conditions hold.

(i) The distribution of R is free of all unknown parameters.
(ii) The observed value of R, $r\left(\mathbf{x}; \mathbf{x}, \theta\right)$, does not depend on the nuisance parameter.

Then, if $R\left(\mathbf{X}; \mathbf{x}, \theta\right)$ satisfies the conditions (i) and (ii), the quantiles of R form a $(1 - \alpha)$ confidence interval for θ. Now, let $R(\alpha)$ be the αth quantile of $R\left(\mathbf{X}; \mathbf{x}, \theta\right)$. Subsequently, the $100\left(1 - \alpha\right)\%$ two-sided generalized confidence intervals for parameter θ is $\left(R_{\alpha/2}, R_{1-\alpha/2}\right)$. Also, for constructing the GCI for the coefficient of variation of the delta-lognormal distribution, the GPQs for δ and σ^2 are required. The GPQ for δ proposed by Anirban [3] and Wu and Hsieh [25] is as follows:

$$R_\delta = \sin^2\left[\arcsin\sqrt{\hat{\delta}} - \frac{1}{2\sqrt{n}}Z\right] \tag{6}$$

where $Z = 2\sqrt{n}\left(\arcsin\sqrt{\hat{\delta}} - \arcsin\sqrt{\delta}\right) \xrightarrow{D} N\left(0, 1\right)$, as $n \to \infty$. Wu and Hsieh [25] defined the GPQ for σ^2 by using the concept of Krishnamoorthy and Mathew [8], it is given by

$$R_{\sigma^2} = \frac{(n_1 - 1)\hat{\sigma}^2}{U} \tag{7}$$

where $U = (n_1 - 1)\hat{\sigma}^2/\sigma^2$ is the Chi-square distribution with $n_1 - 1$ degrees of freedom. By Eqs. (6) and (7), R_δ and R_{σ^2} satisfy the conditions (i) and (ii) for being GPQs for δ and σ^2, respectively. Hence, the GPQ for η can be defined as

$$R_\eta = \sqrt{\frac{\exp\left(R_{\sigma^2}\right) - R_\delta}{R_\delta}}. \tag{8}$$

Since R_η is a function which only depend on R_{σ^2} and R_δ that are obviously independent, thus, R_η is a GPQ. Therefore, a $100(1 - \alpha)\%$ confidence interval for the coefficient of variation (η) based on GCI is simply

$$CI_\eta^{(gci)} = \left[R_{\eta,(l)}, R_{\eta,(u)}\right] = \left[R_\eta\left(\alpha/2\right), R_\eta\left(1 - \alpha/2\right)\right] \tag{9}$$

where $R_\eta\left(\alpha/2\right)$ is the $(\alpha/2)\,100^{th}$ percentile of $R\left(\mathbf{X}; \mathbf{x}, (\sigma^2, \delta)\right)$. For estimating the $R_\eta\left(\alpha/2\right)$ and $R_\eta\left(1 - \alpha/2\right)$, we can use the following algorithm.

Algorithm
(For $i = 1$ to K)
    Generate x_i, $i = 1, 2, ..., n$ from $\Delta\left(\mu, \sigma^2, \delta\right)$;
    Compute $\hat{\delta}$ and $\hat{\sigma}^2$;
    (For $j = 1$ to k)
        Generate Z and U;
        Compute R_δ, R_{σ^2}, and R_η;
    (End j loop)
    Compute the $(\alpha/2)\,100^{th}$ percentile of R_η;
    Compute the $(1 - \alpha/2)\,100^{th}$ percentile of R_η;
(End i loop)

2.2   The Modified Fletcher Method for the Coefficient of Variation

Fletcher [7] presented a confidence interval for $\theta = \ln(\mu_y)$ of the delta-lognormal distribution. For the coefficient of variation of the delta-lognormal distribution, we adjust the method as follows. First, the confidence interval for $\xi = \ln(\eta)$ is constructed, and then we back-transform the endpoints. From Eq. (5), we have

$$\xi = \ln(\eta) = \ln\left[\frac{\exp(\sigma^2) - \delta}{\delta}\right]^{\frac{1}{2}} = \frac{1}{2}\ln\left[\frac{\exp(\sigma^2) - \delta}{\delta}\right] \tag{10}$$

The estimators for ξ are given by $\hat{\sigma}^2$ and $\hat{\delta}$, which are sufficient for σ^2 and δ. Then

$$\hat{\xi} = \frac{1}{2}\ln\left[\frac{\exp(\hat{\sigma}^2) - \hat{\delta}}{\hat{\delta}}\right]. \tag{11}$$

An estimate of the variance of $\hat{\xi}$ is defined as follows (see Appendix A):

$$\hat{V}\left(\hat{\xi}\right) \approx \frac{\left(\hat{b} - \hat{a}\right)\left(1 - \hat{a}\hat{b}\right) - n_1\left(1 - \hat{a}\right)^2}{4n_1\left(1 - \hat{a}^2\right)} + \frac{\hat{\sigma}^4}{2\left(n_1 - 1\right)} \tag{12}$$

where $\hat{a} = \left(1 - \hat{\delta}\right)^{n-1}$ and $\hat{b} = 1 + (n - 1)\hat{\delta}$. Assuming that $\hat{\xi}$ is approximately normal, a back-transformed $100(1 - \alpha)\%$ confidence interval for the coefficient of variation (η) based on the modified Fletcher method is given by

$$CI_\eta^{(F)} = \exp\left\{\hat{\xi} \pm 2\sqrt{\hat{V}\left(\hat{\xi}\right)}\right\} \tag{13}$$

3   Simulation Results

In this section, the performance of the confidence intervals is gauged by the coverage probabilities and the expected lengths via Monte Carlo simulation by using R statistical programming language [21]. For each simulation, the nominal value of 0.95 is calculated based on 15,000 replications and 5,000 pivotal quantities are used for GCI method. In this study, the generalized confidence interval is defined as $CI_\eta^{(gci)}$ and the confidence interval obtained by modified Fletcher method is denoted by $CI_\eta^{(F)}$. The expected lengths of the GCI method and the modified Fletcher method are defined as $EL^{(gci)}$ and $EL^{(F)}$, respectively. The confidence intervals obtained by the simulation for the coefficient of variation, η, are compared for different combinations of n, δ and CV. These include $n = 20, 50, 100$; $\delta = 0.2, 0.5, 0.8$ and CV $= 0.2, 0.5, 1.0, 2.0$. In the simulation study we do not consider the cases of the combinations between $n = 20$, $\delta = 0.2$ and CV $= 0.2, 0.5, 1.0, 2.0$ because in these cases $E(x_1)$ is less than 10, following Fletcher [7] and Wu and Hsieh [25]. For each combination, we chose the best

Table 1. The coverage probability and the expected length of 95% two-sided confidence intervals for the coefficient of variation of the delta-lognormal distribution

n	δ	CV	$CI_\eta^{(gci)}$	$EL^{(gci)}$	$CI_\eta^{(F)}$	$EL^{(F)}$
20	0.5	0.2	0.963	0.820	0.626	0.361
		0.5	0.972	1.074	0.729	0.497
		1.0	0.971	3.306	0.776	1.320
		2.0	0.959	43.080	0.820	6.483
	0.8	0.2	0.917	0.474	0.377	0.089
		0.5	0.971	0.588	0.517	0.172
		1.0	0.958	1.529	0.711	0.665
		2.0	0.953	7.783	0.826	3.316
50	0.2	0.2	0.963	1.299	0.870	0.956
		0.5	0.974	1.688	0.879	1.168
		1.0	0.974	4.852	0.869	2.414
		2.0	0.961	58.744	0.853	10.120
	0.5	0.2	0.951	0.518	0.678	0.245
		0.5	0.966	0.603	0.739	0.328
		1.0	0.963	1.260	0.807	0.788
		2.0	0.955	4.746	0.865	3.035
	0.8	0.2	0.939	0.311	0.394	0.062
		0.5	0.961	0.347	0.515	0.113
		1.0	0.960	0.749	0.740	0.403
		2.0	0.953	2.542	0.872	1.772
100	0.2	0.2	0.964	0.915	0.871	0.706
		0.5	0.968	1.089	0.887	0.847
		1.0	0.968	2.279	0.884	1.654
		2.0	0.956	9.552	0.885	5.773
	0.5	0.2	0.953	0.371	0.679	0.182
		0.5	0.961	0.421	0.746	0.240
		1.0	0.959	0.801	0.821	0.554
		2.0	0.952	2.598	0.890	2.029
	0.8	0.2	0.950	0.224	0.395	0.047
		0.5	0.961	0.243	0.524	0.083
		1.0	0.954	0.499	0.750	0.285
		2.0	0.950	1.541	0.883	1.197

confidence interval from the coverage probability that larger than or the closest to the nominal confidence level; and that has the shortest expected length.

The results in Table 1 show that the coverage probabilities of the GCI method are closer to the nominal level of 0.95 than that of the modified Fletcher method. However, the expected lengths of the GCI method are wide when CV = 2.0 and tend to narrow down if the sample size is large.

4   Example

For illustrate the computation of confidence interval in this article, we used the data from a trawl survey carried out by the National Institute of Water & Atmospheric Research in New Zealand. The sample size is 67 trawls, it is focused on the density (kg/km^2) of red cod (*Pseudophycis bachus*) there were 54 trawls with red cod. Fletcher [7] and Wu and Hsieh [25] indicated that the red cod densities in this study follow the delta-lognormal distribution. For non-zero densities, the coefficient of variation is 1.7 and the probability of non-zero value is 0.806. Hence, the 95% confidence interval for η by using the GCI method is given by (1.626, 3.307).

5   Discussion and Conclusions

This paper presents the new confidence intervals for the coefficient of variation of the delta-lognormal distribution based on the GCI method and the modified Fletcher method. The results obtained from the coverage probabilities and the expected lengths through simulation studies, presented in the previous section, indicate that the GCI method is better than that of the modified Fletcher method in term of the coverage probability and expected length. Hence, the GCI method would be recommended for establishing the confidence intervals for the coefficient of variation of the delta-lognormal distribution. For the modified Fletcher method, it is not recommended for constructing the confidence interval since its coverage probability is smaller than the nominal level in all cases.

Acknowledgements. The author would like to thank referees for their important comments and recommendations, leading to many improvements in this article.

A   Appendix: Variance of $\hat{\xi}$

By Eq. (11), we have

$$\hat{\xi} = \frac{1}{2} \ln \left[\frac{\exp\left(\hat{\sigma}^2\right) - \hat{\delta}}{\hat{\delta}} \right]$$

$$= \frac{1}{2} \left[\ln \left(\exp\left(\hat{\sigma}^2\right) - \hat{\delta} \right) - \ln \left(\hat{\delta} \right) \right]$$

$$= \frac{1}{2} \left\{ \hat{\sigma}^2 + \ln \left[1 - \frac{\hat{\delta}}{\exp\left(\hat{\sigma}^2\right)} \right] - \ln \left(\hat{\delta} \right) \right\}.$$

Since $\hat{\delta} \in [0, 1]$ and $\hat{\sigma}^2 > 0$, we can compute the limit

$$\lim_{\exp(\hat{\sigma}^2) \to \infty} \left[\frac{1}{2} \left\{ \hat{\sigma}^2 + \ln \left[1 - \frac{\hat{\delta}}{\exp(\hat{\sigma}^2)} \right] - \ln \left(\hat{\delta} \right) \right\} \right] = \frac{1}{2} \left(\hat{\sigma}^2 - \ln \left(\hat{\delta} \right) \right).$$

Now, we will find the variance of $\hat{\xi}$. Let $\mathbf{T} = T_1, T_2, ..., T_n$ be the random variables defined as

$$T_i = \begin{cases} 1 & ; X_i > 0 \\ 0 & ; X_i = 0. \end{cases}$$

Let n_1 have a truncated binomial distribution, $n_1 = \sum_{i=1}^{n} T_i$, and we consider repeated sampling under the constraint $n_1 > 1$, as described in [7]. Then,

$$\begin{aligned} P(n_1 = y) &= \frac{1}{1 - (1 - \delta)^n - n\delta (1 - \delta)^{n-1}} \binom{n}{y} \delta^y (1 - \delta)^{n-y} \\ &= \frac{1}{1 - \left\{ (1 - \delta)^{n-1} [1 + (n-1)\delta] \right\}} \binom{n}{y} \delta^y (1 - \delta)^{n-y} \\ &= \frac{1}{1 - ab} \binom{n}{y} \delta^y (1 - \delta)^{n-y} \quad ; y = 2, 3, ..., n \end{aligned}$$

where $a = (1 - \delta)^{n-1}$ and $b = 1 + (n - 1)\delta$. The variance of $\hat{\xi}$ is given by

$$V\left(\hat{\xi} \right) = V_{\mathbf{T}} \left[E\left(\hat{\xi} \mid \mathbf{T} \right) \right] + E_{\mathbf{T}} \left[V\left(\hat{\xi} \mid \mathbf{T} \right) \right].$$

Consider,

$$\begin{aligned} E\left(\hat{\xi} \mid \mathbf{T} \right) &= E\left[\frac{1}{2} \left(\hat{\sigma}^2 - \ln \left(\hat{\delta} \right) \right) \mid \mathbf{T} \right] \\ &= \frac{1}{2} \left[E\left(\hat{\sigma}^2 \mid \mathbf{T} \right) - E\left(\ln \left(\hat{\delta} \right) \mid \mathbf{T} \right) \right] \\ &= \frac{1}{2} \left(\sigma^2 - \ln \left(\hat{\delta} \right) \right) \end{aligned}$$

and

$$\begin{aligned} V\left(\hat{\xi} \mid \mathbf{T} \right) &= V\left[\frac{1}{2} \left(\hat{\sigma}^2 - \ln \left(\hat{\delta} \right) \right) \mid \mathbf{T} \right] \\ &= \frac{1}{4} \left[V\left(\hat{\sigma}^2 \mid \mathbf{T} \right) + V\left(\ln \left(\hat{\delta} \right) \mid \mathbf{T} \right) \right] \\ &= \frac{\sigma^4}{2(n_1 - 1)}. \end{aligned}$$

Thus, by delta method we have

$$
\begin{aligned}
V\left(\hat{\xi}\right) &= V_{\mathbf{T}}\left[\frac{1}{2}\left(\sigma^2 - \ln\left(\hat{\delta}\right)\right)\right] + E_{\mathbf{T}}\left[\frac{\sigma^4}{2\left(n_1 - 1\right)}\right] \\
&= \frac{1}{4}\left[V_{\mathbf{T}}\left(\sigma^2\right) + V_{\mathbf{T}}\left(\ln\left(\hat{\delta}\right)\right)\right] + \frac{1}{2}E_{\mathbf{T}}\left[\frac{\sigma^4}{n_1 - 1}\right] \\
&= \frac{1}{4}V_{\mathbf{T}}\left(\ln\left(\hat{\delta}\right)\right) + \frac{\sigma^4}{2}E_{\mathbf{T}}\left[\frac{1}{n_1 - 1}\right] \\
&\approx \frac{1}{4}\left\{\frac{V_{\mathbf{T}}\left(\hat{\delta}\right)}{\left[E_{\mathbf{T}}\left(\hat{\delta}\right)\right]^2}\right\} + \frac{\sigma^4}{2}E_{\mathbf{T}}\left(\frac{1}{n_1 - 1}\right) \\
&= \frac{1}{4}\left\{\frac{V_{\mathbf{T}}\left(n_1\right)}{\left[E_{\mathbf{T}}\left(n_1\right)\right]^2}\right\} + \frac{\sigma^4}{2}E_{\mathbf{T}}\left(\frac{1}{n_1 - 1}\right) \\
&= \frac{1}{4}\left\{\frac{E_{\mathbf{T}}\left(n_1^2\right) - \left[E_{\mathbf{T}}\left(n_1\right)\right]^2}{\left[E_{\mathbf{T}}\left(n_1\right)\right]^2}\right\} + \frac{\sigma^4}{2}E_{\mathbf{T}}\left(\frac{1}{n_1 - 1}\right) \\
&= \frac{1}{4}\left\{\frac{E_{\mathbf{T}}\left(n_1^2\right)}{\left[E_{\mathbf{T}}\left(n_1\right)\right]^2} - 1\right\} + \frac{\sigma^4}{2}E_{\mathbf{T}}\left(\frac{1}{n_1 - 1}\right).
\end{aligned}
$$

Now, we find

$$
\begin{aligned}
E_{\mathbf{T}}\left(n_1\right) &= \frac{1}{1 - ab}\sum_{y=2}^{n} y\binom{n}{y}\delta^y\left(1 - \delta\right)^{n-y} \\
&= \frac{1}{1 - ab}\left\{\sum_{y=0}^{n} y\binom{n}{y}\delta^y\left(1 - \delta\right)^{n-y} - \sum_{y=0}^{1} y\binom{n}{y}\delta^y\left(1 - \delta\right)^{n-y}\right\} \\
&= \frac{n\delta\left(1 - a\right)}{1 - ab}
\end{aligned}
$$

and

$$
\begin{aligned}
E_{\mathbf{T}}\left(n_1^2\right) &= \frac{1}{1 - ab}\sum_{y=2}^{n} y^2\binom{n}{y}\delta^y\left(1 - \delta\right)^{n-y} \\
&= \frac{1}{1 - ab}\left\{\sum_{y=0}^{n} y^2\binom{n}{y}\delta^y\left(1 - \delta\right)^{n-y} - \sum_{y=0}^{1} y^2\binom{n}{y}\delta^y\left(1 - \delta\right)^{n-y}\right\} \\
&= \frac{n\delta\left(b - a\right)}{1 - ab}.
\end{aligned}
$$

Hence we have

$$
V\left(\hat{\xi}\right) = \frac{1}{4}\left\{ \frac{\frac{n\delta(b-a)}{1-ab}}{\left[\frac{n\delta(1-a)}{1-ab}\right]^2} - 1\right\} + \frac{\sigma^4}{2}E_{\mathbf{T}}\left(\frac{1}{n_1 - 1}\right)
$$

$$
= \frac{1}{4}\left\{ \frac{(b-a)(1-ab)}{n\delta(1-a)^2} - 1\right\} + \frac{\sigma^4}{2}E_{\mathbf{T}}\left(\frac{1}{n_1 - 1}\right).
$$

By normal theory results for the sample variance, we can obtain unbiased estimate for $\sigma^4 E_{\mathbf{T}}\left(\frac{1}{n_1 - 1}\right)$,

$$
E\left(\frac{\hat{\sigma}^4}{n_1 - 1}\right) = E_{\mathbf{T}}\left\{ E\left(\frac{\hat{\sigma}^4}{n_1 - 1} \mid \mathbf{T}\right)\right\} = \sigma^4 E_{\mathbf{T}}\left(\frac{1}{n_1 - 1}\right).
$$

Therefore, an approximately unbiased estimate of $V\left(\hat{\xi}\right)$ is

$$
\hat{V}\left(\hat{\xi}\right) \approx \frac{\left(\hat{b} - \hat{a}\right)\left(1 - \hat{a}\hat{b}\right) - n_1\left(1 - \hat{a}\right)^2}{4n_1\left(1 - \hat{a}\right)^2} + \frac{\hat{\sigma}^4}{2(n_1 - 1)}
$$

where $\hat{a} = \left(1 - \hat{\delta}\right)^{n-1}$ and $\hat{b} = 1 + (n-1)\hat{\delta}$.

References

1. Aitchison, J.: On the distribution of a positive random variable having a discrete probability and mass at the origin. J. Am. Stat. Assoc. **50**(271), 901–908 (1955)
2. Aitchison, J., Brown, J.A.C.: The Lognormal Distribution. Cambridge University Press, Toronto (1957)
3. Anirban, D.: Asymptotic Theory of Statistics and Probability Theory of Statistics and Probability. Springer, New York (2008)
4. Buntao, N., Niwitpong, S.: Confidence intervals for the difference of coefficients of variation for lognormal distributions and delta-lognormal distributions. Appl. Math. Sci. **6**(134), 6691–6704 (2012)
5. Buntao, N., Niwitpong, S.: Confidence intervals for the ratio of coefficients of variation of delta-lognormal distribution. Appl. Math. Sci. **7**(77), 3811–3818 (2013)
6. Crow, E.L., Shimizu, K.: Lognormal Distribulations: Theory and Applicatins. CRC Press, New York (1987)
7. Fletcher, D.: Confidence intervals for the mean of the delta-lognormal distribution. Environ. Ecol. Stat. **15**(2), 175–189 (2008)
8. Krishnamoorthy, K., Mathew, T.: Inferences on the means of lognormal distributions using generalized p-values and generalized confidence intervals. J. Stat. Plan. Infer. **115**(1), 103–121 (2003)
9. Kvanli, A.H., Shen, Y.K., Deng, L.Y.: Construction of confidence intervals for the mean of a population containing many zero values. J. Bus. Econ. Stat. **16**(3), 362–368 (2012)
10. Lo, N.C.-H., Jacobson, L.D., Squire, J.L.: Indices of relative abundance from fish spotter data based on delta-lognormal models. Can. J. Fish. Aquat. Sci. **49**(12), 2515–2526 (1992)

11. Mahmoudvand, R., Hassani, H.: Two new confidence intervals for the coefficient of variation in a normal distribution. J. Appl. Stat. **36**(4), 429–442 (2009)
12. Myers, R.A., Pepin, P.: The robustness of lognormal-based estimators of abundance. Biometrics **46**(4), 1185–1192 (1990)
13. Pennington, M.: Efficient estimators of abundance, for fish and plankton surveys. Biometrics **39**(1), 281–286 (1983)
14. Sangnawakij, P., Niwitpong, S.: Confidence intervals for coefficients of variation in two-parameter exponential distributions. Commun. Stat. Simul. Comput. **46**(8), 6618–6630 (2017)
15. Sangnawakij, P., Niwitpong, S., Niwitpong, S.: Confidence intervals for the ratio of coefficients of variation of the gamma distribution, Lecture Notes in Computer Science, pp. 193–203. Springer, Cham (2015)
16. Smith, S.J.: Evaluating the efficiency of the Δ-distribution mean estimator. Biometrics **44**(2), 485–493 (1988)
17. Smith, S.J.: Use of statistical models for the estimation of abundance from groundfish trawl survey data. Can. J. Fish. Aquat. Sci. **47**(5), 894–903 (1990)
18. Thangjai, W., Niwitpong, S.: Confidence intervals for the weighted coefficients of variation of two-parameter exponential distributions. Cogent Math. **4**, 1315880 (2017). https://doi.org/10.1080/23311835.2017.1315880
19. Tian, L.: Inferences on the common coeffieient of variation. Stat. Med. **24**(14), 2213–2220 (2005)
20. Tian, L., Wu, J.: Confidence intervals for the mean of lognormal data with excess zeros. Biom. J. Biom. Z. **48**(1), 149–156 (2006)
21. Venables, W.N., Smith, D.M.: An Introduction to R, 2nd edn. Network Theory Ltd., Bristol (2009)
22. Weerahandi, S.: Generalized confidence intervals. J. Am. Stat. Assoc. **88**(423), 899–905 (1993)
23. Wong, A.C.M., Wu, J.: Small sample asymptotic inference for the coefficient of variation: normal and nonnormal models. J. Stat. Plan. Infer. **104**(1), 73–82 (2002)
24. Wongkhao, A., Niwitpong, S., Niwitpong, S.: Confidence intervals for the raio of two independent coefficients of variation of normal distribution. Far East J. Math. Sci. (FJMS) **98**(6), 741–757 (2015)
25. Wu, W.-H., Hsieh, H.-N.: Generalized confidence interval estimation for the mean of delta-lognormal distribution: an application to New Zealand trawl survey data. J. Appl. Stat. **41**(7), 1471–1485 (2014)

Fixed-Point Theory

Best Proximity Point Theorems for Generalized α-ψ-Proximal Contractions

Muhammad Usman Ali[1], Arslan Hojat Ansari[2], Konrawut Khammahawong[3,4], and Poom Kumam[3,4(✉)]

[1] Department of Mathematics, COMSATS Institute of Information Technology, Attock, Pakistan
muh_usman_ali@yahoo.com
[2] Department of Mathematics, Karaj Branch, Islamic Azad University, Karaj, Iran
analsisamirmath2@gmail.com
[3] KMUTT-Fixed Point Research Laboratory, Department of Mathematics, Room SCL 802 Fixed Point Laboratory, Science Laboratory Building, Faculty of Science, King Mongkut's University of Technology Thonburi (KMUTT), 126 Pracha-Uthit Road, Bang Mod, Thrung Khru, Bangkok 10140, Thailand
k.konrawut@gmail.com
[4] KMUTT-Fixed Point Theory and Applications Research Group (KMUTT-FPTA), Theoretical and Computational Science Center (TaCS), Science Laboratory Building, Faculty of Science, King Mongkut's University of Technology Thonburi (KMUTT), 126 Pracha-Uthit Road, Bang Mod, Thrung Khru, Bangkok 10140, Thailand
poom.kum@kmutt.ac.th

Abstract. In this paper, we introduce a new generalization of multi-valued α-ψ-proximal contraction and prove some best proximity point theorems for such mappings on complete metric spaces. An example is also constructed to show the generality of our results.

Keywords: Best proximity point · α-ψ-proximal contraction
$(h, F, \alpha, \mu, \psi)$-proximal contraction
$(h, F, \alpha, \beta, \mu, \psi)$-proximal contraction

Mathematics Subject Classification: Primary 47H10
Secondary 54H25.

1 Introduction and Mathematical Preliminaries

Jleli and Samet [1] introduced the notion of α-proximal admissible and α-ψ-proximal contractive type mappings and proved some best proximity point theorems. The earlier generalization of this α-ψ-proximal contractive type mapping was also given by Jleli et al. [2]. Ali et al. [3] and Choudhurya et al. [4], independently, gave the multivalued generalization of these notions and proved some best proximity point theorems for such mappings. The literature of best proximity point is very rich and there are some significant results some of them are given in [1–27].

© Springer International Publishing AG 2018
L. H. Anh et al. (eds.), *Econometrics for Financial Applications*, Studies in Computational Intelligence 760, https://doi.org/10.1007/978-3-319-73150-6_27

Ansari and Shukla *et al.* [28], introduced a new class of implicit type functions and by using these functions, they introduced the (F, h) pair of type I and type II. With the help of these pairs they obtained the contraction condition which successfully generalized many existing contraction condition.

In this paper, by combining the above ideas we generalize the results of Ali *et al.* [3] and those contained therein. Now, we recollect some basis notions, definitions, and results, which we require subsequently. Readers interested in details of Ali *et al.* results are referred to the paper [3].

Let (X, d) be a metric space. For $A, B \subset X$, we use following notions:

$$dist(A, B) = \inf\{d(a, b) : a \in A, \ b \in B\},$$
$$D(x, B) = \inf\{d(x, b) : b \in B\},$$
$$A_0(A, B) = \{a \in A : d(a, b) = dist(A, B) \ for \ some \ b \in B\},$$
$$A_0(B, A) = \{b \in B : d(a, b) = dist(A, B) \ for \ some \ a \in A\}.$$

We denote by Ψ a family of strictly increasing functions $\psi : [0, \infty) \rightarrow [0, \infty)$ such that $\sum_{n=1}^{\infty} \psi^n(t) < \infty$ for each $t > 0$, where ψ^n is the n-th iteration of ψ. By $CL(X)$, we denote the set of all nonempty closed subsets of X. For every $A, B \in CL(X)$, let

$$H(A, B) = \begin{cases} \max\{\sup_{x \in A} D(x, B), \sup_{y \in B} D(y, A)\} & \text{if the maximum exists;} \\ \infty & \text{otherwise.} \end{cases}$$

Such a map H is called Hausdorff metric induced by d.

Definition 1 [9]. Let (A, B) be a pair of nonempty subsets of a metric space (X, d) with $A_0(A, B) \neq \emptyset$. Then the pair (A, B) is said to have the *weak P-property* if and only if for any $x_1, x_2 \in A$ and $y_1, y_2 \in B$,

$$\begin{cases} d(x_1, y_1) = dist(A, B) \\ d(x_2, y_2) = dist(A, B) \end{cases} \implies d(x_1, x_2) \leq d(y_1, y_2).$$

Definition 2 [12]. Let A and B be non-empty subsets of a metric space X and $T : A \rightarrow 2^B$ be a multivalued mapping. A point $x^* \in A$ is said to be a *best proximity point* of a multivalued mapping T if it satisfies the following condition

$$D(x^*, Tx^*) = dist(A, B).$$

Ali *et al.* [3] generalized the results of Jleli and Samet [1] in the following way:

Definition 3 [3]. Let A and B be non-empty subsets of a metric space (X, d) and $\alpha : A \times A \rightarrow [0, \infty)$ be a function. A mapping $T : A \rightarrow 2^B$ is said to be α-*proximal admissible* if

$$\begin{cases} \alpha(x_1, x_2) \geq 1 \\ d(u_1, y_1) = d(A, B) \\ d(u_2, y_2) = d(A, B) \end{cases} \implies \alpha(u_1, u_2) \geq 1$$

for all $x_1, x_2, u_1, u_2 \in A$, $y_1 \in Tx_1 \in B$ and $y_2 \in Tx_2$.

Definition 4 [3]. A mapping $T : A \to CL(B)$ is said to be an α-ψ-*proximal contraction*, if

$$\alpha(x,y)H(Tx,Ty) \leq \psi(d(x,y)) \quad \forall x,y \in A,$$

where $\alpha : A \times A \to [0,\infty)$ and $\psi \in \Psi$.

Following is the main result of [3].

Theorem 1 [3]. *Let A and B be two nonempty closed subsets of a complete metric space (X,d) such that $A_0(A,B)$ is nonempty. Suppose that $T : A \to CL(B)$ is an α-ψ-proximal contraction satisfying the following conditions:*

(i) $Tx \subset A_0(B,A)$ for each $x \in A_0(A,B)$ and (A,B) satisfies the weak P-property;

(ii) T is α-proximal admissible;

(iii) there exist elements $x_0, x_1 \in A_0(A,B)$ such that

$$D(x_1,Tx_0) = dist(A,B) \quad and \quad \alpha(x_0,x_1) \geq 1;$$

(iv) either T is a continuous, or, if $\{x_n\}$ is a sequence in A such that $\alpha(x_n, x_{n+1}) \geq 1$ for all n and $x_n \to x \in A$ as $n \to \infty$, then there exists a subsequence $\{x_{n_k}\}$ of $\{x_n\}$ such that $\alpha(x_{n_k}, x) \geq 1$ for all k.

Then there exists an element $x^* \in A_0(A,B)$ such that $D(x^*,Tx^*) = dist(A,B)$.

We have the following notions due to [28]:

Definition 5 [28]. We say that the function $h\colon \mathbb{R}^+ \times \mathbb{R}^+ \to \mathbb{R}$ is a *function of subclass type I*, if $x \geq 1 \implies h(1,y) \leq h(x,y)$ for all $x,y \in \mathbb{R}^+$.

Definition 6 [28]. Let $h, F\colon \mathbb{R}^+ \times \mathbb{R}^+ \to \mathbb{R}$, then we say that the pair (F,h) is an *upper class of type I* if h is a function of subclass type I and: (i) $0 \leq s \leq 1 \implies F(s,t) \leq F(1,t)$, (ii) for all $t,y \in \mathbb{R}^+$, $h(1,y) \leq F(1,t) \implies y \leq t$.

Definition 7 [28]. We say that the function $h : \mathbb{R}^+ \times \mathbb{R}^+ \times \mathbb{R}^+ \to \mathbb{R}$ is a *function of subclass type II*, if $x,y \geq 1 \implies h(1,1,z) \leq h(x,y,z)$ for all $z \in \mathbb{R}^+$.

Definition 8 [28]. Let $h\colon \mathbb{R}^+ \times \mathbb{R}^+ \times \mathbb{R}^! \to \mathbb{R}$ and $F\colon \mathbb{R}^+ \times \mathbb{R}^+ \to \mathbb{R}$, then we say that the pair (F,h) is an *upper class type II* if h is a subclass of type II and: (i) $0 \leq s \leq 1 \implies F(s,t) \leq F(1,t)$, (ii) for all $s,t,z \in \mathbb{R}^+$, $h(1,1,z) \leq F(s,t) \implies z \leq st$.

We need the following definition and property in our results.

Definition 9. Let A and B be non-empty subsets of a metric space (X,d) and $\mu : A \times A \to [0,\infty)$ be a function. A mapping $T : A \to 2^B$ is said to be μ-*proximal subadmissible* if

$$\begin{cases} \mu(x_1,x_2) \leq 1 \\ d(u_1,y_1) = d(A,B) \\ d(u_2,y_2) = d(A,B) \end{cases} \implies \mu(u_1,u_2) \leq 1$$

for all $x_1, x_2, u_1, u_2 \in A$, $y_1 \in Tx_1 \in B$ and $y_2 \in Tx_2$.

(C,α,μ): If $\{x_n\}$ is a sequence in A such that $\alpha(x_n,x_{n+1}) \geq 1$, $\mu(x_n,x_{n+1}) \leq 1$ for all n and $x_n \to x \in A$, then $\alpha(x_n,x) \geq 1$, $\mu(x_n,x) \leq 1$ for all n.

2 Main Result

We begin this section by introducing the following definition.

Definition 10. Let A and B be two nonempty subsets of a metric space (X, d). A mapping $T : A \to CL(B)$ is said to be $(h, F, \alpha, \mu, \psi)$ -*proximal contraction*, if there exist functions $\alpha, \mu : A \times A \to [0, \infty)$ and pair (F, h) of upper class of type I, such that

$$h(\alpha(x, y), H(Tx, Ty)) \leq F(\mu(x, y), \psi(\max\{d(x, y), D(x, Tx) - dist(A, B),$$
$$D(y, Ty) - dist(A, B)\})), \quad \forall x, y \in A.$$

Lemma 1 [29]. *Let (X, d) be a metric space and $B \in CL(X)$. Then $x \in X$ with $D(x, B) > 0$ and for each $q > 1$, there exists an element $b \in B$ such that*

$$d(x, b) < qD(x, B). \tag{1}$$

Now, we state and prove our first main result.

Theorem 2. *Let A and B be two nonempty closed subsets of a complete metric space (X, d) such that $A_0(A, B)$ is nonempty. Suppose that $T : A \to CL(B)$ is a continuous $(h, F, \alpha, \mu, \psi)$-proximal contraction mapping and satisfies the following conditions:*

(i) *$Tx \subseteq A_0(B, A)$ for each $x \in A_0(A, B)$ and (A, B) satisfies the weak P-property;*
(ii) *T is α-proximal admissible and μ-proximal subadmissible;*
(iii) *there exist $x_0, x_1 \in A_0(A, B)$ and $y_1 \in Tx_0$ such that*

$$\alpha(x_0, x_1) \geq 1, \mu(x_0, x_1) \leq 1 \quad and \quad d(x_1, y_1) = dist(A, B).$$

Then there exists an element $x^* \in A_0(A, B)$ such that $D(x^*, Tx^*) = dist(A, B)$.

Proof. From condition (iii), there exist $x_0, x_1 \in A_0(A, B)$ and $y_1 \in Tx_0$ such that

$$d(x_1, y_1) = dist(A, B) \quad and \quad \alpha(x_0, x_1) \geq 1, \mu(x_0, x_1) \leq 1. \tag{2}$$

Assume that $y_1 \notin Tx_1$, for otherwise x_1 is a best proximal point. By hypothesis, we have

$$h(1, H(Tx_0, Tx_1)) \leq h(\alpha(x_0, x_1), H(Tx_0, Tx_1))$$
$$\leq F(\mu(x_0, x_1), \psi(\max\{d(x_0, x_1), D(x_0, Tx_0) - dist(A, B),$$
$$D(x_1, Tx_1) - dist(A, B)\}))$$
$$\leq F(1, \psi(\max\{d(x_0, x_1), D(x_0, Tx_0) - dist(A, B),$$
$$D(x_1, Tx_1) - dist(A, B)\})). \tag{3}$$

This implies that

$$\begin{aligned}
D(y_1, Tx_1) \leq H(Tx_0, Tx_1) &\leq \alpha(x_0, x_1) H(Tx_0, Tx_1) \\
&\leq \psi(\max\{d(x_0, x_1), D(x_0, Tx_0) - dist(A, B), \\
&\quad D(x_1, Tx_1) - dist(A, B)\}) \\
&\leq \psi(d(x_0, x_1)),
\end{aligned} \tag{4}$$

since,

$$D(x_0, Tx_0) \leq d(x_0, x_1) + d(x_1, y_1) + D(y_1, Tx_0) \leq d(x_0, x_1) + dist(A, B)$$

and

$$D(x_1, Tx_1) \leq d(x_1, y_1) + D(y_1, Tx_1) = D(y_1, Tx_1) + dist(A, B).$$

For $q > 1$, it follows from Lemma 1 that there exists $y_2 \in Tx_1$ such that

$$0 < d(y_1, y_2) < qD(y_1, Tx_1). \tag{5}$$

From (4) and (5), we have

$$0 < d(y_1, y_2) < qD(y_1, Tx_1) \leq q\psi(d(x_0, x_1)). \tag{6}$$

As $y_2 \in Tx_1 \subseteq A_0(B, A)$, there exists $x_2 \neq x_1 \in A_0(A, B)$ such that

$$d(x_2, y_2) = dist(A, B), \tag{7}$$

for otherwise x_1 is a best proximity point. As (A, B) satisfies the weak P-property. From (2) and (7), we have

$$0 < d(x_1, x_2) \leq d(y_1, y_2). \tag{8}$$

From (6) and (8), we have

$$0 < d(x_1, x_2) < qD(y_1, Tx_1) \leq q\psi(d(x_0, x_1)). \tag{9}$$

Since ψ is strictly increasing, we have $\psi(d(x_1, x_2)) < \psi(q\psi(d(x_0, x_1)))$. Put $q_1 = \frac{\psi(q\psi(d(x_0, x_1)))}{\psi(d(x_1, x_2))}$. Also, we have $\alpha(x_0, x_1) \geq 1$, $\mu(x_0, x_1) \leq 1$ and $d(x_1, y_1) = dist(A, B)$, $d(x_2, y_2) = dist(A, B)$. Since T is α-proximal admissible and μ-proximal subadmissible, then $\alpha(x_1, x_2) \geq 1, \mu(x_1, x_2) \leq 1$. Thus we have

$$\alpha(x_1, x_2) \geq 1, \mu(x_1, x_2) \leq 1 \quad \text{and} \quad d(x_2, y_2) = dist(A, B). \tag{10}$$

Assume that $y_2 \notin Tx_2$, for otherwise x_2 is a best proximity point. From hypothesis, we have

$$\begin{aligned}
h(1, H(Tx_1, Tx_2)) &\leq h(\alpha(x_1, x_2), H(Tx_1, Tx_2)) \\
&\leq F(\mu(x_1, x_2), \psi(\max\{d(x_1, x_2), D(x_1, Tx_1) - dist(A, B), \\
&\quad D(x_2, Tx_2) - dist(A, B)\})) \\
&\leq F(1, \psi(\max\{d(x_1, x_2), D(x_1, Tx_1) - dist(A, B), \\
&\quad D(x_2, Tx_2) - dist(A, B)\})).
\end{aligned}$$

This implies that

$$0 < D(y_2, Tx_2) \le H(Tx_1, Tx_2) \le \alpha(x_1, x_2)H(Tx_1, Tx_2)$$
$$\le \psi(\max\{d(x_1, x_2), D(x_1, Tx_1) - dist(A, B),$$
$$D(x_2, Tx_2) - dist(A, B)\}))$$
$$\le \psi(d(x_1, x_2)), \tag{11}$$

since,

$$D(x_1, Tx_1) \le d(x_1, x_2) + d(x_2, y_2) + D(y_2, Tx_1) \le d(x_1, x_2) + dist(A, B)$$

and

$$D(x_2, Tx_2) \le d(x_2, y_2) + D(y_2, Tx_2) = D(y_2, Tx_2) + dist(A, B).$$

For $q_1 > 1$, it follows from Lemma 1 that there exists $y_3 \in Tx_2$ such that

$$0 < d(y_2, y_3) < q_1 D(y_2, Tx_2). \tag{12}$$

From (11) and (12), we have

$$0 < d(y_2, y_3) < q_1 D(y_2, Tx_2) \le q_1 \psi(d(x_1, x_2)) = \psi(q\psi(d(x_0, x_1))). \tag{13}$$

As $y_3 \in Tx_2 \subseteq A_0(B, A)$, there exists $x_3 \ne x_2 \in A_0(A, B)$ such that

$$d(x_3, y_3) = dist(A, B), \tag{14}$$

for otherwise x_2 is a best proximity point. As (A, B) satisfies the weak P-property. From (10) and (14), we have

$$0 < d(x_2, x_3) \le d(y_2, y_3). \tag{15}$$

From (13) and (15), we have

$$0 < d(x_2, x_3) < q_1 D(y_2, Tx_2) \le q_1 \psi(d(x_1, x_2)) = \psi(q\psi(d(x_0, x_1))). \tag{16}$$

Since ψ is a strictly increasing, we have $\psi(d(x_2, x_3)) < \psi^2(q\psi(d(x_0, x_1)))$. Put $q_2 = \frac{\psi^2(q\psi(d(x_0, x_1)))}{\psi(d(x_2, x_3))}$. Also, we have $\alpha(x_1, x_2) \ge 1$, $\mu(x_1, x_2) \le 1$, $d(x_2, y_2) = dist(A, B)$ and $d(x_3, y_3) = dist(A, B)$. Since T is an α-proximal admissible and μ-proximal subadmissible, then $\alpha(x_2, x_3) \ge 1$, $\mu(x_2, x_3) \le 1$. Thus, we have

$$\alpha(x_2, x_3) \ge 1, \mu(x_2, x_3) \le 1 \text{ and } d(x_3, y_3) = dist(A, B). \tag{17}$$

Continuing in the same way, we get sequences $\{x_n\}$ in $A_0(A, B)$ and $\{y_n\}$ in $A_0(B, A)$, such that $y_n \in Tx_{n-1}$ for each $n \in \mathbb{N}$,

$$\alpha(x_n, x_{n+1}) \ge 1, \mu(x_n, x_{n+1}) \le 1 \text{ and } d(x_{n+1}, y_{n+1}) = dist(A, B). \tag{18}$$

Further, we have

$$d(y_{n+1}, y_{n+2}) < \psi^n(q\psi(d(x_0, x_1))). \tag{19}$$

As $y_{n+2} \in Tx_{n+1} \subseteq A_0(B, A)$, there exists $x_{n+2} \neq x_{n+1} \in A_0(A, B)$ such that

$$d(x_{n+2}, y_{n+2}) = dist(A, B). \tag{20}$$

Since (A, B) satisfies the weak P-property form (18) and (20), we have $d(x_{n+1}, x_{n+2}) \leq d(y_{n+1}, y_{n+2})$. Then from (19), we have

$$d(x_{n+1}, x_{n+2}) < \psi^n(q\psi(d(x_0, x_1))). \tag{21}$$

For $m > n$ we have

$$d(x_n, x_m) \leq \sum_{i=n}^{m-1} d(x_i, x_{i+1}) < \sum_{i=n}^{m-1} \psi^{i-1}(q\psi(d(x_0, x_1))). \tag{22}$$

Hence $\{x_n\}$ is a Cauchy sequence in A. Similarly, we show that $\{y_n\}$ is a Cauchy sequence in B. Since A and B are closed subsets of a complete metric space, there exist x^* in A and y^* in B such that $x_n \to x^*$ and $y_n \to y^*$ as $n \to \infty$. By the (20), we conclude that $d(x^*, y^*) = dist(A, B)$. Since T is continuous and $y_n \in Tx_{n-1}$, we have $y^* \in Tx^*$. Hence $dist(A, B) \leq D(x^*, Tx^*) \leq d(x^*, y^*) = dist(A, B)$. Therefore x^* is a best proximity point of the mapping T.

Theorem 3. *Let A and B be two nonempty closed subsets of a complete metric space (X, d) such that $A_0(A, B)$ is nonempty. Suppose that $T : A \to CL(B)$ is a $(h, F, \alpha, \mu, \psi)$-proximal contraction mapping with continuous ψ and satisfies the following conditions:*

(i) $Tx \subseteq A_0(B, A)$ for each $x \in A_0(A, B)$ and (A, B) satisfies the weak P-property;

(ii) T is α-proximal admissible and μ-proximal subadmissible;

(iii) there exist $x_0, x_1 \in A_0(A, B)$ and $y_1 \in Tx_0$ such that

$$\alpha(x_0, x_1) \geq 1, \mu(x_0, x_1) \leq 1 \quad and \quad d(x_1, y_1) = dist(A, B);$$

(iv) Property (C, α, μ) holds.

Then there exists an element $x^* \in A_0(A, B)$ such that $D(x^*, Tx^*) = dist(A, B)$.

Proof. Following the proof of Theorem 2, there exist Cauchy sequences $\{x_n\}$ in A and $\{y_n\}$ in B such that (18) holds and $x_n \to x^* \in A$, $y_n \to y^* \in B$. From hypothesis, we have $\alpha(x_n, x^*) \geq 1$ and $\mu(x_n, x^*) \leq 1$ for all n. We claim that $D(y^*, Tx^*) = 0$. Suppose on contrary it is wrong. As T is $(h, F, \alpha, \mu, \psi)$-proximal contraction, we have

$$\begin{aligned}
h(1, H(Tx_n, Tx^*)) &\leq h(\alpha(x_n, x^*), H(Tx_n, Tx^*)) \\
&\leq F(\mu(x_n, x^*), \psi(\max\{d(x_n, x^*), D(x_n, Tx_n) - dist(A, B), \\
&\quad D(x^*, Tx^*) - dist(A, B)\})) \\
&\leq F(1, \psi(\max\{d(x_n, x^*), D(x_n, Tx_n) - dist(A, B), \\
&\quad D(x^*, Tx^*) - dist(A, B)\})), \forall n.
\end{aligned}$$

This implies that

$$D(y_n, Tx^*) \leq H(Tx_n, Tx^*)$$
$$\leq \psi(\max\{d(x_n, x^*), D(x_n, Tx_n) - dist(A, B),$$
$$D(x^*, Tx^*) - dist(A, B)\})$$
$$\leq \psi(\max\{d(x_n, x^*), d(x_n, x_{n+1}) + d(x_{n+1}, y_{n+1}) + D(y_{n+1}, Tx_n) - dist(A, B),$$
$$d(x^*, x_n) + d(x_n, y_n) + D(y_n, Tx^*) - dist(A, B)\})$$
$$= \psi(\max\{d(x_n, x^*), d(x_n, x_{n+1}), d(x^*, x_n) + D(y_n, Tx^*)\}).$$

Letting $n \to \infty$ in above inequality, we get

$$D(y^*, Tx^*) \leq \psi(D(y^*, Tx^*)).$$

This is only possible if $D(y^*, Tx^*) = 0$. By continuity of the metric d, we have

$$d(x^*, y^*) = \lim_{n \to \infty} d(x_{n+1}, y_{n+1}) = dist(A, B). \tag{23}$$

Hence $dist(A, B) \leq D(x^*, Tx^*) \leq d(x^*, y^*) = dist(A, B)$. Therefore x^* is a best proximity point of the mapping T.

Definition 11. Let A and B be two nonempty subsets of a metric space (X, d). A mapping $T : A \to 2^B \setminus \emptyset$ is called (α, β)-*proximal admissible* if there exist two functions $\alpha, \beta : A \to [0, \infty)$ such that

$$\begin{cases} \alpha(x_1) \geq 1, \beta(x_2) \geq 1 \\ d(u_1, y_1) = dist(A, B) \quad \implies \alpha(u_1) \geq 1, \beta(u_2) \geq 1. \\ d(u_2, y_2) = dist(A, B) \end{cases}$$

where $x_1, x_2, u_1, u_2 \in A$ and $y_1 \in Tx_1, y_2 \in Tx_2$.

Definition 12. Let A and B be two nonempty subsets of a metric space (X, d). A mapping $T : A \to CL(B)$ is said to be a $(h, F, \alpha, \beta, \mu, \psi)$-*proximal contraction*, if there exist four functions $\alpha, \beta : A \to [0, \infty)$ and $\mu : A \times A \to [0, \infty)$, $\psi \in \Psi$ and pair (F, h) is of upper class of type II such that

$$h(\alpha(x), \beta(y), H(Tx, Ty)) \leq F(\mu(x, y), \psi(\max\{d(x, y), D(x, Tx) - dist(A, B),$$
$$D(y, Ty) - dist(A, B)\})), \quad \forall x, y \in A.$$

Theorem 4. *Let A and B be two nonempty closed subsets of a complete metric space (X, d) such that $A_0(A, B)$ is nonempty. Suppose that $T : A \to CL(B)$ is a continuous $(h, F, \alpha, \beta, \mu, \psi)$-proximal contraction mapping satisfying the following conditions:*

(i) *$Tx \subseteq A_0(B, A)$ for each $x \in A_0(A, B)$ and (A, B) satisfies the weak P-property;*
(ii) *T is (α, β)-proximal admissible and μ-proximal subadmissible;*
(iii) *there exist $x_0, x_1 \in A_0(A, B)$ and $y_1 \in Tx_0$ such that*

$$\alpha(x_0) \geq 1, \beta(x_1) \geq 1 \quad and \quad \mu(x_0, x_1) \leq 1 \quad and \quad d(x_1, y_1) = dist(A, B).$$

Then there exists an element $x^* \in A_0(A, B)$ such that $D(x^*, Tx^*) = dist(A, B)$.

Theorem 5. *Let A and B be two nonempty closed subsets of a complete metric space (X, d) such that $A_0(A, B)$ is nonempty. Suppose $T : A \rightarrow CL(B)$ is a $(h, F, \alpha, \beta, \mu, \psi)$-proximal contraction mapping with continuous ψ and satisfies the following conditions:*

(i) $Tx \subseteq A_0(B, A)$ for each $x \in A_0(A, B)$ and (A, B) satisfies the weak P-property;

(ii) T is (α, β)-proximal admissible and μ-proximal subadmissible;

(iii) there exist $x_0, x_1 \in A_0(A, B)$ and $y_1 \in Tx_0$ such that

$$\alpha(x_0) \geq 1, \beta(x_1) \geq 1 \quad and \quad \mu(x_0, x_1) \leq 1 \quad and \quad d(x_1, y_1) = dist(A, B).$$

(iv) If $\{x_n\}$ is a sequence in A such that $\beta(x_{n+1}) \geq 1$, $\mu(x_n, x_{n+1}) \leq 1$ for all n and $x_n \rightarrow x \in A$, then $\beta(x) \geq 1$ and $\mu(x_n, x) \leq 1$ for all n.

Then there exists an element $x^* \in A_0(A, B)$ such that $D(x^*, Tx^*) = dist(A, B)$.

Corollary 1. *Let A and B be two nonempty closed subsets of a complete metric space (X, d) such that $A_0(A, B)$ is nonempty. Let $\alpha, \mu : A \times A \rightarrow [0, \infty)$, $\psi \in \Psi$ and $T : A \rightarrow CL(B)$ is a mapping satisfying the following conditions:*

(i) $Tx \subseteq A_0(B, A)$ for each $x \in A_0(A, B)$ and (A, B) satisfies the weak P-property;

(ii) T is α-proximal admissible and μ-proximal subadmissible;

(iii) there exist $x_0, x_1 \in A_0(A, B)$ and $y_1 \in Tx_0$ such that

$$\alpha(x_0, x_1) \geq 1, \mu(x_0, x_1) \leq 1 \quad and \quad d(x_1, y_1) = dist(A, B);$$

(iv) for each $x, y \in A$, we have

$$\alpha(x, y)H(Tx, Ty) \leq \mu(x, y)\psi(\max\{d(x, y), D(x, Tx) - dist(A, B),$$
$$D(y, Ty) - dist(A, B)\});$$

(v) either T is continuous, or, property (C, α, μ) holds and ψ is continuous.

Then there exists an element $x^* \in A_0(A, B)$ such that $D(x^*, Tx^*) = dist(A, B)$.

If we define $h : \mathbb{R}^+ \times \mathbb{R}^+ \rightarrow \mathbb{R}$ by

$$h(x, y) = \begin{cases} y & \text{if } x \geq 1 \\ 0 & \text{otherwise} \end{cases}$$

and $F : \mathbb{R}^+ \times \mathbb{R}^+ \rightarrow \mathbb{R}$ by $F(x, y) = y$, and $\mu : A \times A \rightarrow [0, \infty)$ by $\mu(x, y) = 1$. Then, by Theorem 3, we get the following result:

Corollary 2. *Let A and B be two nonempty closed subsets of a complete metric space (X, d) such that $A_0(A, B)$ is nonempty. Let $\alpha : A \times A \rightarrow [0, \infty)$, $\psi \in \Psi$ and $T : A \rightarrow CL(B)$ is a mapping satisfying the following conditions:*

(i) $Tx \subseteq A_0(B, A)$ for each $x \in A_0(A, B)$ and (A, B) satisfies the weak
    P-property;
(ii) T is α-proximal admissible;
(iii) there exist $x_0, x_1 \in A_0(A, B)$ and $y_1 \in Tx_0$ such that

$$d(x_1, y_1) = dist(A, B) \quad and \quad \alpha(x_0, x_1) \geq 1;$$

(iv) for each $x, y \in A$ with $\alpha(x, y) \geq 1$, we have

$$H(Tx, Ty) \leq \psi(\max\{d(x, y), D(x, Tx) - dist(A, B), D(y, Ty) - dist(A, B)\}); \quad (24)$$

(v) either T is continuous, or, ψ is continuous, further, for each sequence $\{x_n\}$
    in A with $\alpha(x_n, x_{n+1}) \geq 1$ for all n and $x_n \to x \in A$, we have $\alpha(x_n, x) \geq 1$
    for all n.

Then there exists an element $x^* \in A_0(A, B)$ such that $D(x^*, Tx^*) = dist(A, B)$.

Example 1. Let $X = [0, \infty) \times [0, \infty)$ be endowed with the usual metric d. Suppose
that $A = \{(\frac{1}{4}, x) : 0 \leq x < \infty\}$ and $B = \{(0, x) : 0 \leq x < \infty\}$. Define
$T : A \to CL(B)$ by

$$T(\frac{1}{4}, a) = \begin{cases} \{(0, \frac{x}{4}) : 0 \leq x \leq a\} & if \ a \leq 1 \\ \{(0, \sqrt{x}) : x \geq a\} & if \ a > 1, \end{cases}$$

and $\alpha : A \times A \to [0, \infty)$ by

$$\alpha(x, y) = \begin{cases} 1 & if \ x, y \in \{(\frac{1}{4}, a) : 0 \leq a \leq 1\} \\ 0 & otherwise. \end{cases}$$

Let $\psi(t) = \frac{t}{2}$ for all $t \geq 0$. Notice that that $A_0(A, B) = A$, $A_0(B, A) = B$ and
$Tx \subseteq A_0(B, A)$ for each $x \in A_0(A, B)$. Also, the pair (A, B) satisfies the weak
P-property. Let $x_0, x_1 \in \{(\frac{1}{4}, x) : 0 \leq x \leq 1\}$, then $Tx_0, Tx_1 \subseteq \{(0, \frac{x}{4}) : 0 \leq x \leq 1\}$. Consider $y_1 \in Tx_0$, $y_2 \in Tx_1$ and $u_1, u_2 \in A$ such that $d(u_1, y_1) = dist(A, B)$
and $d(u_2, y_2) = dist(A, B)$. Then we have $u_1, u_2 \in \{(\frac{1}{4}, x) : 0 \leq x \leq \frac{1}{4}\}$. Hence
T is an α-proximal admissible map. For $x_0 = (\frac{1}{4}, 1) \in A_0(A, B)$ and $y_1 = (0, \frac{1}{4}) \in Tx_0$ in $A_0(B, A)$, we have $x_1 = (\frac{1}{4}, \frac{1}{4}) \in A_0(A, B)$ such that $d(x_1, y_1) = dist(A, B)$ and $\alpha(x_0, x_1) = 1$. If $\alpha(x, y) = 1$, then $x, y \in \{(\frac{1}{4}, a) : 0 \leq a \leq 1\}$,
thus, we have

$$H(Tx, Ty) = \frac{|x - y|}{4} \leq \frac{1}{2}d(x, y).$$

Hence T satisfies (24). Moreover, if $\{x_n\}$ is a sequence in A such that
$\alpha(x_n, x_{n+1}) = 1$ for all n and $x_n \to x \in A$ as $n \to \infty$, then $\alpha(x_n, x) = 1$
for all n. Therefore all the conditions of Corollary 2 (In other wards Theorem 3)
hold and T has a best proximity point. Note that the Theorem 1 is not applicable
on this example.

Acknowledgements. This project was supported by the Theoretical and Computational Science (TaCS) Center under Computational and Applied Science for Smart
Innovation Cluster (CLASSIC), Faculty of Science, KMUTT. The third author would
like to thank the Research Professional Development Project Under the Science
Achievement Scholarship of Thailand (SAST) for financial support.

References

1. Jleli, M., Samet, B.: Best proximity point for α-ψ-proximal contraction type mappings and applications. Bull. Sci. Math. **137**, 977–995 (2013)
2. Jleli, M., Karapinar, E., Samet, B.: Best proximity points for generalized α-ψ-proximal contractive type mappings. J. Appl. Math. **2013**, 10 pages (2013). Article ID 534127
3. Ali, M.U., Kamran, T., Shahzad, N.: Best proximity point for α-ψ-proximal contractive multimaps. Abstr. Appl. Anal. **2014**, 6 pages (2014). Article ID 181598
4. Choudhurya, B.S., Maitya, P., Metiya, N.: Best proximity point results in set-valued analysis. Nonlinear Anal. Modell. Contr. **21**, 293–305 (2016)
5. AlThagafi, M.A., Shahzad, N.: Best proximity pairs and equilibrium pairs for Kakutani multimaps. Nonlinear Anal. **70**(3), 1209–1216 (2009)
6. AlThagafi, M.A., Shahzad, N.: Convergence and existence results for best proximity points. Nonlinear Anal. **70**, 3665–3671 (2009)
7. Bari, C.D., Suzuki, T., Vetro, C.: Best proximity point for cyclic Meir-Keeler contraction. Nonlinear Anal. **69**, 3790–3794 (2008)
8. Eldred, A., Veeramani, P.: Existence and convergence of best proximity points. J. Math. Anal. Appl. **323**, 1001–1006 (2006)
9. Zhang, J., Su, Y., Cheng, Q.: A note on A best proximity point theorem for Gerathy-contractions. Fixed Point Theory Appl. **2013**, 4 pages (2013). Article ID 83
10. Abkar, A., Gbeleh, M.: Best proximity points for asymptotic cyclic contraction mappings. Nonlinear Anal. **74**, 7261–7268 (2011)
11. Abkar, A., Gbeleh, M.: Best proximity points for cyclic mappings in ordered metric spaces. J. Optim. Theory Appl. **151**, 418–424 (2011)
12. Abkar, A., Gbeleh, M.: The existence of best proximity points for multivalued non-self mappings. RACSAM **107**(2), 319–325 (2012)
13. Alghamdi, M.A., Alghamdi, M.A., Shahzad, N.: Best proximity point results in geodesic metric spaces. Fixed Point Theory Appl. **2012**, 12 pages (2012). Article ID 234
14. Al-Thagafi, M.A., Shahzad, N.: Best proximity sets and equilibrium pairs for a finite family of multimaps. Fixed Point Theory Appl. **2008**, 10 pages (2008). Article ID 457069
15. Derafshpour, M., Rezapour, S., Shahzad, N.: Best proximity points of cyclic φ-contractions in ordered metric spaces. Topol. Meth. Nonlin. Anal. **37**, 193–202 (2011)
16. Bari, C.D., Suzuki, T., Vetro, C.: Best proximity point for cyclic Meir-Keeler contraction. Nonlinear Anal. **69**, 3790–3794 (2008)
17. Markin, J., Shahzad, N.: Best proximity points for relatively u-continuous mappings in Banach and hyperconvex spaces. Abstr. Appl. Anal. **2013**, 5 pages (2013). Article ID 680186
18. Rezapour, S., Derafshpour, M., Shahzad, N.: Best proximity points of cyclic ϕ-contractions on reflexive Banach spaces. Fixed Point Theory Appl. **2010**, 7 pages (2010). Article ID 946178
19. Basha, S.S., Shahzad, N., Jeyaraj, R.: Best proximity point theorems for reckoning optimal approximate solutions. Fixed Point Theory Appl. **2012**, 9 pages (2012). Article ID 202
20. Vetro, C.: Best proximity points: convergence and existence theorems for p-cyclic mappings. Nonlin. Anal. **73**(7), 2283–2291 (2010)

21. Mongkolkeha, C., Kumam, P.: Best proximity point Theorems for generalized cyclic contractions in ordered metric Spaces. J. Optim. Theory Appl. **155**, 215–226 (2012)

22. Sintunavarat, W., Kumam, P.: Coupled best proximity point theorem in metric spaces. Fixed Point Theory Appl. **2012**, 16 pages (2012). Article ID 93

23. Nashine, H.K., Vetro, C., Kumam, P.: Best proximity point theorems for rational proximal contractions. Fixed Point Theory Appl. **2013**, 11 pages (2013). Article ID 95

24. Cho, Y.J., Gupta, A., Karapinar, E., Kumam, P., Sintunavarat, W.: Tripled best proximity point theorem in metric space. Math. Inequal. Appl. **4**, 1197–1216 (2013)

25. Mongkolkeha, C., Kongban, C., Kumam, P.: The existence and uniqueness of best proximity point theorems for generalized almost contraction. Abstr. Appl. Anal. **2014** (2014). Article ID 813614, 11 pages

26. Kumam, P., Salimi, P., Vetro, C.: Best proximity point results for modified α-proximal C-contraction mappings. Fixed Point Theory Appl. **2014**, 16 pages (2014). Article ID 99

27. Pragadeeswarar, V., Marudai, M., Kumam, P., Sitthithakerngkiet, K.: The existence and uniqueness of coupled best proximity point for proximally coupled contraction in a complete ordered metric space. Abstr. Appl. Anal. **2014**, 7 pages (2014). Article ID 274062

28. Ansari, A.H., Shukla, S.: Some fixed point theorems for ordered F-$(\mathcal{F}, h)$-contraction and subcontractions in 0-f-orbitally complete partial metric spaces. J. Adv. Math. Stud. **9**, 37–53 (2016)

29. Ali, M.U., Kamran, T.: On (α^*, ψ)-contractive multi-valued mappings. Fixed Point Theory Appl. **2013**, 7 pages (2013). Article ID 137

Zeroes and Fixed Points of Different Functions via Contraction Type Conditions

Muhammad Usman Ali[1], Khanitin Muangchoo-in[2,3], and Poom Kumam[2,3(✉)]

[1] Department of Mathematics, COMSATS Institute of Information Technology,
Attock, Pakistan
muh_usman_ali@yahoo.com

[2] KMUTT-Fixed Point Research Laboratory, Department of Mathematics,
Room SCL 802 Fixed Point Laboratory, Science Laboratory Building,
Faculty of Science, King Mongkut's University of Technology Thonburi (KMUTT),
126 Pracha-Uthit Road, Bang Mod, Thrung Khru, Bangkok 10140, Thailand
khanitin.math@mail.kmutt.ac.th, poom.kum@kmutt.ac.th

[3] KMUTT-Fixed Point Theory and Applications Research Group (KMUTT-FPTA),
Theoretical and Computational Science Center (TaCS), Science Laboratory Building,
Faculty of Science, King Mongkut's University of Technology Thonburi (KMUTT),
126 Pracha-Uthit Road, Bang Mod, Thrung Khru, Bangkok 10140, Thailand

Abstract. The purpose of this paper is to introduce some results which help us to ensure the existence of fixed points and zero points of three different functions satisfying a single contraction-type condition. We also provide an example to support our result.

Keywords: ϕ-fixed points $\cdot$ (ϕ, ψ)-fixed points $\cdot$ Zero points

Mathematics Subject Classification: Primary 47H10
Secondary 54H25

1 Introduction and Mathematical Preliminaries

The notion of ϕ-fixed point of a self mapping T was introduced by Jleli et al. [1]. They said that an element x of X is ϕ-fixed point of $T : X \to X$ and $\phi : X \to [0, \infty)$ if $x \in F_T \cap Z_\phi$, where $F_T = \{x \in X : x = Tx\}$ and $Z_\phi = \{x \in X : \phi(x) = 0\}$. They proved some theorems for the existence of ϕ-fixed points of a self mapping T. The contraction type conditions used in the results of Jleli et al. [1] are based on the following family of functions: A family of functions $\mathfrak{F}$ contain the functions
$F : [0, \infty) \times [0, \infty) \times [0, \infty) \to [0, \infty)$ such that

(i) $\max\{a, b\} \leq F(a, b, c)$ for each $a, b, c \geq 0$;
(ii) $F(0, 0, 0) = 0$;
(iii) F is continuous.

Following we list the most significant result of Jleli et al. [1].

© Springer International Publishing AG 2018
L. H. Anh et al. (eds.), *Econometrics for Financial Applications*, Studies in Computational Intelligence 760, https://doi.org/10.1007/978-3-319-73150-6_28

Theorem 1 *[1]. Let (X, d) be a complete metric space and $\phi : X \to [0, \infty)$ be lower semi continuous. Let $T : X \to X$ be a mapping such that*

$$F(d(Tx, Ty), \phi(Tx), \phi(Ty)) \leq k \cdot (F(d(x, y), \phi(x), \phi(y)))$$

for each $x, y \in X$, where $0 \leq k < 1$ and $F \in \mathfrak{F}$. Then T has a ϕ-fixed point.

Wardowski [2] gave a nice generalization of Banach contraction principle. For this purpose, he used a family of specific type of functions. The conditions imposed on these functions seem to be very strong, but this family has many interesting examples, as mentioned in [2]. Following we list the work of Wardowski [2], briefly.

Definition 1 [2]. Let $\mathbb{F}$ be the class of all functions $F : (0, \infty) \to \mathbb{R}$ satisfying the following three axioms:

(i) F is strictly increasing, that is, for each $a_1, a_2 \in (0, \infty)$ with $a_1 < a_2$, we have $F(a_1) < F(a_2)$.
(ii) For each sequence $\{a_n\}$ of positive real numbers we have $\lim_{n \to \infty} a_n = 0$ if and only if $\lim_{n \to \infty} F(a_n) = -\infty$.
(iii) There exists $k \in (0, 1)$ such that $\lim_{a \to 0+} a^k F(a) = 0$.

Theorem 2 *[2]. Let (X, d) be a complete metric space and let $T : X \to X$ be an F-contraction, that is, there exists $F \in \mathbb{F}$ and $\tau > 0$ such that for each $x, y \in X$ with $d(Tx, Ty) > 0$, we have*

$$\tau + F(d(Tx, Ty)) \leq F(d(x, y)).$$

Then T has a unique fixed point.

This result has been extended by several authors, see for example [3–11].

The purpose of this paper is to introduce the notion of (ϕ, ψ)-fixed points of self mappings and establish some results for the existence of (ϕ, ψ)-fixed points. We have established our contraction type condition in such a way that we get many other contractions from our notion. Hence many results are direct consequence of our main result.

2   Main Result

We begin this section by introducing the notion of (ϕ, ψ)-fixed points.

Let us denote the set of all fixed points of $T : X \to X$ is denoted by $F_T = \{x \in X : x = Tx\}$ and the set of all zeros of $\phi : X \to [0, \infty)$ and $\psi : X \to [0, \infty)$ by $Z_\phi = \{x \in X : \phi(x) = 0\}$ and $Z_\psi = \{x \in X : \psi(x) = 0\}$, respectively. Further, $Z_{(\phi, \psi)} = Z_\phi \cap Z_\psi$. An element x of X is (ϕ, ψ)-fixed point of T if $x \in F_T \cap Z_{(\phi, \psi)}$. The following family of functions is required in order to define our contraction condition.

Definition 2. Let $\mathfrak{M}$ be the class of functions $M : (0, \infty) \times [0, \infty) \times [0, \infty) \to \mathbb{R}$ satisfying the following two axioms:

(M_1) For sequences $\{a_n : a_n > 0 \ \forall n \in \mathbb{N}\}, \{b_n : b_n \geq 0 \ \forall n \in \mathbb{N}\}$ and $\{c_n : c_n \geq 0 \ \forall n \in \mathbb{N}\}$ there exists $\lim_{n \to \infty} a_n = \lim_{n \to \infty} b_n = \lim_{n \to \infty} c_n = 0$ if and only if $\lim_{n \to \infty} M(a_n, b_n, c_n) = -\infty$.

(M_2) For sequences $\{a_n : a_n > 0 \ \forall n \in \mathbb{N}\}, \{b_n : b_n \geq 0 \ \forall n \in \mathbb{N}\}$ and $\{c_n : c_n \geq 0 \ \forall n \in \mathbb{N}\}$ such that $\lim_{n \to \infty} a_n = \lim_{n \to \infty} b_n = \lim_{n \to \infty} c_n = 0$, there exists $k \in (0,1)$ such that $\lim_{n \to \infty} a_n^k M(a_n, b_n, c_n) = 0$.

Example 1. Consider the mapping $M_i : (0, \infty) \times [0, \infty) \times [0, \infty) \to \mathbb{R}$ for $i = 1, 2$ as defined below:

(i) $M_1(a, b, c) = \ln(a + b + c)$ for each $a > 0$ and $b, c \geq 0$.
(ii) $M_2(a, b, c) = (a + b + c) + \ln(a + b + c)$ for each $a > 0$ and $b, c \geq 0$.

Then one can observe that $M_i \in \mathfrak{M}$ for $i = 1, 2$.

In the following definition we introduce the notion of $M_{(\phi, \psi)}$-contraction mappings.

Definition 3. Let (X, d) be a metric space and $\phi, \psi : X \to [0, \infty)$ are functions. A mapping $T : X \to X$ is $M_{(\phi, \psi)}$-contraction if there exists the function $M \in \mathfrak{M}$ and a constant $\tau > 0$ such that for each $x, y \in X$ with $d(Tx, Ty) > 0$, we have

$$\tau + M(d(Tx, Ty), \phi(Tx), \psi(Ty)) \leq M(d(x, y), \phi(x), \psi(y)). \tag{1}$$

Theorem 3. *Let (X, d) be a complete metric space and let $T : X \to X$ be a continuous $M_{(\phi, \psi)}$-contraction such that the mappings $\phi, \psi : X \to [0, \infty)$ are lower semi continuous. Then T has a (ϕ, ψ)-fixed point.*

Proof. Let $x_0 \in X$, we can construct a sequence $\{x_n\}$ such that $x_n = Tx_{n-1}$ and $x_n \neq x_{n-1}$ for each $n \in \mathbb{N}$. From (1), we have

$$\tau + M(d(x_1, x_2), \phi(x_1), \psi(x_2)) = \tau + M(d(Tx_0, Tx_1), \phi(Tx_0), \psi(Tx_1))$$
$$\leq M(d(x_0, x_1), \phi(x_0), \psi(x_1)). \tag{2}$$

Again from (1), we have

$$\tau + M(d(x_2, x_3), \phi(x_2), \psi(x_3)) = \tau + M(d(Tx_1, Tx_2), \phi(Tx_1), \psi(Tx_2))$$
$$\leq M(d(x_1, x_2), \phi(x_1), \psi(x_2)). \tag{3}$$

From (2) and (3), we have

$$M(d(x_2, x_3), \phi(x_2), \psi(x_3)) \leq M(d(x_0, x_1), \phi(x_0), \psi(x_1)) - 2\tau.$$

Continuing in this way we get the following inequality

$$M(d(x_n, x_{n+1}), \phi(x_n), \psi(x_{n+1})) \leq M(d(x_0, x_1), \phi(x_0), \psi(x_1)) - n\tau \tag{4}$$

for each $n \in \mathbb{N}$.

Letting $n \to \infty$ in (4), we get $\lim_{n \to \infty} M(d(x_n, x_{n+1}), \phi(x_n), \psi(x_{n+1})) = -\infty$. Thus, by property (M_1), we have $\lim_{n \to \infty} d(x_n, x_{n+1}) = \lim_{n \to \infty} \phi(x_n) = \lim_{n \to \infty} \psi(x_{n+1}) = 0$. Let $d_n = d(x_n, x_{n+1})$, $\phi_n = \phi(x_n)$ and $\psi_n = \psi(x_{n+1})$ for each $n \in \mathbb{N}$. From (M_2) there exists $k \in (0, 1)$ such that

$$\lim_{n \to \infty} d_n^k M(d_n, \phi_n, \psi_n) = 0.$$

From (4) we have

$$d_n^k M(d_n, \phi_n, \psi_n) - d_n^k M(d_0, \phi_0, \psi_0) \leq -d_n^k n\tau \leq 0 \quad \text{for each } n \in \mathbb{N}. \tag{5}$$

Letting $n \to \infty$ in (5), we get

$$\lim_{n \to \infty} n d_n^k = 0. \tag{6}$$

This implies that there exists $n_1 \in \mathbb{N}$ such that $n d_n^k \leq 1$ for each $n \geq n_1$. Thus, we have

$$d_n \leq \frac{1}{n^{1/k}}, \quad \text{for each } n \geq n_1. \tag{7}$$

To prove that $\{x_n\}$ is a Cauchy sequence. Consider $m, n \in \mathbb{N}$ with $m > n > n_1$. By using the triangular inequality and (7), we have

$$d(x_n, x_m) \leq d(x_n, x_{n+1}) + d(x_{n+1}, x_{n+2}) + \cdots + d(x_{m-1}, x_m)$$
$$= \sum_{i=n}^{m-1} d_i \leq \sum_{i=n}^{\infty} d_i \leq \sum_{i=n}^{\infty} \frac{1}{i^{1/k}}.$$

Since $\sum_{i=1}^{\infty} \frac{1}{i^{1/k}}$ is convergent series. Thus, $\lim_{n,m \to \infty} d(x_n, x_m) = 0$. This implies that $\{x_n\}$ is a Cauchy sequence. As (X, d) is complete, there exists $x^* \in X$ such that $x_n \to x^*$ as $n \to \infty$. Since $x_n \to x^*$ and ϕ, ψ are lower semi continuous functions with $\lim_{n \to \infty} \phi(x_n) = \lim_{n \to \infty} \psi(x_n) = 0$. Thus we have $\phi(x^*) = \psi(x^*) = 0$. By continuity of T, we have $x_{n+1} = Tx_n \to Tx^*$. Thus, we have $x^* = Tx^*$ and $\phi(x^*) = \psi(x^*) = 0$. Hence, x^* is a (ϕ, ψ)-fixed point of T.

Corollary 1. *Let (X, d) be a complete metric space and $\phi, \psi : X \to [0, \infty)$ are lower semi continuous functions. Let $T : X \to X$ be a continuous mapping such that*
$$d(Tx, Ty) + \phi(Tx) + \psi(Ty) \leq k(d(x, y) + \phi(x) + \psi(y))$$
for each $x, y \in X$, where $0 \leq k < 1$. Then T has a (ϕ, ψ)-fixed point.

The conclusion of this corollary can be obtained from Theorem 3, by considering $M(a, b, c) = M_1(a, b, c)$ for each $a > 0$ and $b, c \geq 0$.

Example 2. Let $X = [0, \infty)$ be endowed with the usual metric $d(x, y) = |x - y|$. Define the functions $T : X \to X$, $\phi : X \to [0, \infty)$ and $\psi : X \to [0, \infty)$ by

$$Tx = \frac{x}{2(x+1)}, \quad \phi(x) = x, \quad \text{and} \quad \psi(x) = \frac{x}{2}.$$

Then one can prove the following inequality

$$d(Tx, Ty) + \phi(Tx) + \psi(Ty) = \left| \frac{x}{2(x+1)} - \frac{y}{2(y+1)} \right| + \frac{x}{2(x+1)} + \frac{1}{2} \left(\frac{y}{2(y+1)} \right)$$
$$\leq \frac{1}{2}|x - y| + \frac{x}{2} + \frac{y}{4}$$
$$= \frac{1}{2}(d(x, y) + \phi(x) + \psi(y)).$$

Thus, by Corollary 1 (or Theorem 3) we reach the conclusion that T has a $\left(x, \frac{x}{2}\right)$-fixed point, that is (ϕ, ψ)-fixed point.

In the following text, we introduced the notion of $\mathfrak{Z}$ family. By using this family we shall state a new (ϕ, ψ)-fixed point theorem, as an interesting consequence of our result.

Let $\mathfrak{Z}$ denotes the family of functions $Z : (0, \infty) \times [0, \infty) \times [0, \infty) \to (0, \infty)$ such that

(i) $\max\{a, b\} \leq Z(a, b, c)$ for each $a > 0$ and $b, c \geq 0$
(ii) For sequences $\{a_n\}, \{b_n\}$ and $\{c_n\}$ of real numbers we have $\lim_{n \to \infty} a_n = a$, $\lim_{n \to \infty} b_n = \lim_{n \to \infty} c_n = 0$ if and only if $\lim_{n \to \infty} Z(a_n, b_n, c_n) = a$. In particular, $Z(a, 0, 0) = a$.

Theorem 4. *Let (X, d) be a complete metric space and $\phi, \psi : X \to [0, \infty)$ are lower semi continuous functions. Let $T : X \to X$ be a continuous mapping such that*

$$Z(d(Tx, Ty), \phi(Tx), \psi(Ty)) \leq k \cdot (Z(d(x, y), \phi(x), \psi(y)))$$

for each $x, y \in X$ whenever $d(Tx, Ty) > 0$, where $0 \leq k < 1$ and $Z \in \mathfrak{Z}$. Then T has a (ϕ, ψ)-fixed point.

Proof. The conclusion of this theorem can be obtained from Theorem 3, by considering $M(a, b, c) = \ln(Z(a, b, c))$ for each $a > 0$ and $b, c \geq 0$, where $Z \in \mathfrak{Z}$.

Open Problem: Can we extend Theorem 1 for the existence of (ϕ, ψ)-fixed points of T under the influence of the function $F \in \mathfrak{F}$?

The following theorem follows from Theorem 3 by considering $\psi(x) = \phi(x)$ for each $x \in X$.

Theorem 5. *Let (X, d) be a complete metric space and let $T : X \to X$ be a continuous mapping such that there exists the function $M \in \mathfrak{M}$ and a constant $\tau > 0$ satisfying the following condition*

$$\tau + M(d(Tx, Ty), \phi(Tx), \phi(Ty)) \leq M(d(x, y), \phi(x), \phi(y)) \tag{8}$$

for each $x, y \in X$ with $d(Tx, Ty) > 0$, where $\phi : X \to [0, \infty)$ is lower semi continuous. Then T has a ϕ-fixed point.

The following corollary follows from Theorem 4 by taking $\psi(x) = \phi(x)$ for each $x \in X$.

Corollary 2. *Let (X, d) be a complete metric space and $\phi : X \to [0, \infty)$ is lower semi continuous function. Let $T : X \to X$ be a continuous mapping such that*

$$Z(d(Tx, Ty), \phi(Tx), \phi(Ty)) \leq k \cdot (Z(d(x, y), \phi(x), \phi(y)))$$

for each $x, y \in X$ whenever $d(Tx, Ty) > 0$, where $0 \leq k < 1$ and $Z \in \mathfrak{Z}$. Then T has a ϕ-fixed point.

Now, we are going to discuss our main results for Graphic $M_{(\phi, \psi)}$-contractions. In the following definition we introduce the notion of graphic contraction mappings.

Definition 4. *Let (X, d) be a metric space. A mapping $T : X \to X$ is graphic contraction if there exists $\alpha \in [0, 1)$ such that*

$$d(T(x), T^2(x)) \leq \alpha d(x, T(x)), \tag{9}$$

for all $x \in X$.

Theorem 6. *Let (X, d) be a complete metric space and $\phi, \psi : X \to [0, \infty)$ are lower semi continuous functions. Let $T : X \to X$ be a continuous graphic $M_{(\phi, \psi)}$-contraction, that is, there exists the function $M \in \mathfrak{M}$ and a constant $\tau > 0$ such that for each $x \in X$ with $d(Tx, T^2x) > 0$, we have*

$$\tau + M(d(Tx, T^2x), \phi(Tx), \psi(T^2x)) \leq M(d(x, Tx), \phi(x), \psi(Tx)). \tag{10}$$

Then T has a (ϕ, ψ)-fixed point.

Proof. Let $x_0 \in X$, we can construct a sequence $\{x_n\}$ such that $x_n = T^n x_0$ and $x_n \neq x_{n-1}$ for each $n \in \mathbb{N}$. From (10), we have

$$\tau + M(d(Tx_0, T^2x_0), \phi(Tx_0), \psi(T^2x_0)) \leq M(d(x_0, Tx_0), \phi(x_0), \psi(Tx_0)). \tag{11}$$

Again from (10), we have

$$\tau + M(d(T^2x_0, T^3x_0), \phi(T^2x_0), \psi(T^3x_0)) \leq M(d(Tx_0, T^2x_0), \phi(Tx_0), \psi(T^2x_0)). \tag{12}$$

From (11) and (12), we have

$$M(d(T^2x_0, T^3x_0), \phi(T^2x_0), \psi(T^3x_0)) \leq M(d(x_0, Tx_0), \phi(x_0), \psi(Tx_0)) - 2\tau.$$

Continuing in this way and by using the fact that $x_n = T^n x_0$ for each $n \in \mathbb{N}$, we get the following inequality

$$M(d(x_n, x_{n+1}), \phi(x_n), \psi(x_{n+1})) \leq M(d(x_0, x_1), \phi(x_0), \psi(x_1)) - n\tau \tag{13}$$

for each $n \in \mathbb{N}$.

Rest of the proof is analogous to the proof of Theorem 3. Thus, we conclude that T has a (ϕ, ψ)-fixed point.

The following theorem follows from above result by considering $\psi(x) = \phi(x)$ for each $x \in X$.

Theorem 7. *Let (X, d) be a complete metric space and let $T : X \to X$ be a continuous mapping such that there exists the function $M \in \mathfrak{M}$ and a constant $\tau > 0$ satisfying the following condition*

$$\tau + M(d(Tx, T^2x), \phi(Tx), \phi(T^2x)) \le M(d(x, Tx), \phi(x), \phi(Tx)) \qquad (14)$$

for each $x \in X$ with $d(Tx, T^2x) > 0$, where $\phi : X \to [0, \infty)$ is lower semi continuous. Then T has a ϕ-fixed point.

Acknowledgements. This project was supported by the Theoretical and Computational Science (TaCS) Center under Computational and Applied Science for Smart Innovation Cluster (CLASSIC), Faculty of Science, KMUTT. The third author would like to tank the Research Professional Development Project Under the Science Achievement Scholarship of Thailand (SAST) for financial support.

References

1. Jleli, M., Samet, B., Vetro, C.: Fixed point theory in partial metric spaces via ϕ-fixed point's concept in metric spaces. J. Inequal. Appl. **2014**, 426 (2014)
2. Wardowski, D.: Fixed points of a new type of contractive mappings in complete metric spaces. Fixed Point Theory Appl. **2012**, 94 (2012)
3. Kumrod, P., Sintunavarat, W.: A new contractive condition approach to ϕ-fixed point results in metric spaces and its applications. J. Comput. Appl. Math. **311**, 194–204 (2017)
4. Ali, M.U., Kamran, T.: Multivalued F-contraction and related fixed point theorems with application. Filomat **30**, 3779–3793 (2016)
5. Kamran, T., Postolache, M., Ali, M.U., Kiran, Q.: Feng and Liu type Fcontraction in b-metric spaces with application to integral equations. J. Math. Anal. **7**, 18–27 (2016)
6. Ali, M.U., Kamran, T., Postolache, M.: Solution of volterra integral inclusion in b-metric spaces via a new fixed point theorem. Nonlinear Anal. Model. Control **22**, 17–30 (2017)
7. Cosentino, M., Vetro, P.: Fixed point results for F-contractive mappings of Hardy-Rogers-type. Filomat **28**(4), 715–722 (2014)
8. Minak, G., Helvac, A., Altun, I.: Ciric type generalized F-contractions on complete metric spaces and fixed point results, Filomat (in Press)
9. Sgroi, M., Vetro, C.: Multi-valued F-contractions and the solution of certain functional and integral equations. Filomat **27**(7), 1259–1268 (2013)
10. Paesano, D., Vetro, C.: Multi-valued F-contractions in 0-complete partial metric spaces with application to Volterra type integral equation. RACSAM **108**, 1005–1020 (2013)
11. Piri, H., Kumam, P.: Some fixed point theorems concerning F-contraction in complete metric spaces. Fixed Point Theory Appl. **2014**, 210 (2014)

A Globally Stable Fixed Point in an Ordered Partial Metric Space

Umar Yusuf Batsari[1,2,3] and Poom Kumam[2,3(✉)]

[1] Department of Mathematics and Statistics, College of Science and Technology,
Hassan Usman Katsina Polytechnic, Katsina, Katsina State, Nigeria
uyub2k@yahoo.com
[2] Department of Mathematics, Faculty of Science,
King Mongkut's University of Technology Thonburi (KMUTT),
126 Pracha-Uthit Road, Bang Mod, Thung Khru, Bangkok 10140, Thailand
poom.kumam@mail.kmutt.ac.th
[3] Theoretical and Computational Science Center (TaCS),
Science Laboratory Building,
King Mongkut's University of Technology Thonburi (KMUTT), Bangkok, Thailand
umar.batsari@mail.kmutt.ac.th

Abstract. The research in this paper was motivated by Kamihigashi and Stachurski [11], and Matthews [14]. The application of the research made by Kamihigashi and Stachurski [11] does not cover some set of real life problems; as some problems are not compatible with normal metric (d) they used. In a move to cover such problems, a partial metric was used and an analogue operator of asymptotic contraction in a partial metric space was introduced, the existence of a globally stable fixed point in an ordered partial metric space was established. The results we obtained extend and improve the applicability of many existing results in the literature. In particular, the results we obtained covered the research of Kamihigashi and Stachurski [11] and can have a real life application in computer semantics as earlier shown by Matthews [14].

Keywords: Partial metric · Transitive order
Regularity of order in partial metric
Identifying property in partial metric
Fixed point of a self mapping · Globally stable fixed point
Asymptotic mapping in partial metric and order preserving property of a mapping

1 Introduction

The area of fixed point theory received much attention from mathematicians due to its vast scope of applications (see [7,21]). Many applications of fixed point theory can be found in [2,5,10,11,13,14,18,21]. Furthermore, spaces and mappings are very important when discussing the theory of fixed points i.e.

© Springer International Publishing AG 2018
L. H. Anh et al. (eds.), *Econometrics for Financial Applications*, Studies in Computational Intelligence 760, https://doi.org/10.1007/978-3-319-73150-6_29

metric spaces, partial metric spaces, fuzzy spaces, smooth spaces, contractive mappings, monotone mappings and so on, see [1,3,6,7,10,19–21].

In 1992, Matthews [15] introduced the definition and concept of a partial metric (*pmetric*) due to the failure of a metric on his work in computer studies, more explanation can be found in [14]. After introducing partial metric, Matthews [14] also proved the partial metric version of Banach fixed point theorem; this makes it useful in fixed point theory. In 1999, Heckmann [9] established some results using a generalization of partial metric called a *weak partial metric*. In 2004, Oltra and Valero [17] also generalized the Matthews's fixed point theorem using a complete partial metric space in the sense of O'Neill. In 2013, Shukla et al. [20] introduced the notion of asymptotically regular mappings in a partial metric space and established the fixed point results of those mappings. Very recently, Onsod et al. [18] established some fixed point results using a complete partial metric space endowed with a graph.

Another important part of fixed point theory applications is the advent of fixed point theorems in ordered spaces like that of Tarski's in 1955 [2], many and important results were established as improvements or generalization of Tarski's fixed point theorem [4,5,11–13,16]. In this direction, Heikkila [10] proved some fixed point results of increasing operators in a partially ordered set and show some applications in partially ordered Polish space. In 2011, Hassen [8] established some fixed point results in an ordered partial metric space that guarantee the existence of a fixed point for a nondecreasing mapping. In 2012, Kamihigashi and Stachurski [12,13] established an important criteria for investigating stability of a chain (linear order) in an order theoretic sense. In 2013, Kamihigashi and Stachurski [11] established some existence and uniqueness of a fixed point results with order preserving mapping and show some application on probability distribution functions.

In this paper, motivated by Kamihigashi and Stachurski [11] and Matthews [14,15], we generalize the work of Kamihigashi and Stachurski [11] from a metric space to a partial metric space so as to have wider coverage in applications.

2   Preliminaries/Definitions

Let X and B be non empty sets, $\mathbb{R}_+$ be the set of non negative real numbers. The following definitions can be found in [11,14,15] unless otherwise stated.

Let $\preceq$ be a binary relation on the set X, then the relation $\preceq$ is

1. **Reflexive** if $x \preceq x$, $\forall x \in X$.
2. **Antisymmetric** if $x \preceq y$ and $y \preceq x \implies x = y$, $\forall x, y \in X$.
3. **Transitive** if $x \preceq y$ and $y \preceq z \implies x \preceq z$, $\forall x, y, z \in X$.

The binary relation $\preceq$ is called a **partial order** if it satisfies all of the above conditions (1–3), we call the pair (X, $\preceq$) a partial ordered set.

A partial ordered set $(X, \preceq)$ is called a **lattice** if for any $a, b \in X$ there exist

1. a least upper bound called "**join**" or simply sup$\{a, b\}$ and denoted by $a \cup b$.
2. a greatest lower bound called "**meet**" or simply inf$\{a, b\}$ and denoted by $a \cap b$.

A lattice $(X, \preceq)$ is **complete** if every non empty subset A of X has a least upper bound $\cup A$ and greatest lower bound $\cap A$.

In view of Kamihigashi and Stachurski [11], a function $\Psi_b : X \times X \to \mathbb{R}_+$ for $b \in B$ is

1. **Identifying** if $\Psi_b(y, y) = \Psi_b(y, x) = \Psi_b(x, x) \Rightarrow x = y$, $\forall x, y \in X$ and $b \in B$.
2. **One dimensional** if $\Psi_b(x, y)$ is independent of $b \in B$, i.e. $\Psi_b(x, y) = \Psi_{b'}(x, y)$, $\forall b, b' \in B$.
3. **Regular** if whenever $x \preceq y \preceq z$ then $\max\{\Psi_b(x, y), \Psi_b(y, z)\} \leq \Psi_b(x, z)$, $\forall x, y, z \in X$, where $(X, \preceq)$ is an ordered space.

A function $f : X \to \mathbb{R}$ with $X \subseteq \mathbb{R}$ is increasing if $f(x) \leq f(y)$ whenever $x \preceq y$, $\forall x, y \in X$.

In an ordered space $(X, \preceq)$, **a sequence $\{x_n\} \subseteq X$ is increasing** if $x_n \preceq x_{n+1}$, $\forall n \in \mathbb{N}$.

A **partial metric or** *pmetric* on the set X is a function $p : X \times X \to \mathbb{R}_+$ such that
(P1) $x = y \Longleftrightarrow p(x, x) = p(x, y) = p(y, y)$, $\forall x, y \in X$.
(P2) $p(x, x) \leq p(x, y)$, $\forall x, y \in X$.
(P3) $p(x, y) = p(y, x)$, $\forall x, y \in X$.
(P4) $p(x, z) \leq p(x, y) + p(y, z) - p(y, y)$, $\forall x, y, z \in X$.
Every metric is a partial metric with $p(x, x) = 0$, $\forall x \in X$.

For any partial metric p on X, there exists an **induced metric** $d_p : X \times X \to \mathbb{R}_+$ defined by $d_p(x, y) = 2p(x, y) - p(x, x) - p(y, y)$ [14].

Example 1: Define a mapping $p : X \times X \to \mathbb{R}_+$ by $p(x, y) = \min\{x, y\}$, for $X \subset \mathbb{R}$. Clearly, (P2) fails if $x > y$, thus p is not a partial metric.

Example 2: Define a mapping $\Psi_b : \mathbb{R} \times \mathbb{R} \to \mathbb{R}_+$ by $\Psi_b(x, y) = |bx - by|$, $b \in B \subset (0, 1]$. So, Ψ_b is a metric which implies partial metric but, not one dimensional.

Example 3: Let $B = (-15, \infty)$, define a mapping $\Psi_b : F \times F \to [0, 1]$ by

$$\Psi_b(k, h) = \max_{(-15, b]} \{\gamma_{k,b}, \gamma_{h,b}\} \tag{1}$$

where F is a family of some continuous and integrable real valued functions from $\mathbb{R}$ to $[0, 1]$, $\gamma_{k,b}$ is the minimum value of the function $k \in F$ over the interval $(-15, b] \subset B$. So, Ψ_b is not a metric but a partial metric which is not one dimensional (a partial metric for each b).

Example 4: Let $B \subseteq \mathbb{R}$, define a mapping $\Psi_b : \mathbb{R} \times \mathbb{R} \to \mathbb{R}_+$ by $\Psi_b(x_1, x_2) = \max\{x_1, x_2\}$, $x_1, x_2 \in \mathbb{R}$, $b \in [0, 1]$. Ψ_b is not a metric but, it is a one dimensional partial metric.

An **open ball** for a partial metric $p : X \times X \to \mathbb{R}_+$ is a set of the form $B_\varepsilon^p := \{y \in X : p(x, y) < \varepsilon\}$ for each $\varepsilon > 0$ and $x \in X$. Matthews [14] established that, every partial metric is a T_0 **topology** (τ_p); τ_p is the topology induced by B_ε^p.

Every partial metric defines a partial order with an induced order $(\preceq_p)$ defined by $x \preceq_p y \iff p(x, x) = p(x, y)$, $\forall x, y \in X$ [14].

A sequence $\{x_n\}$ in (X, p) **converges with respect to the topology** τ_p to a point x in (X, p) iff

$$\lim_{n \to \infty} p(x_n, x) = p(x, x), \tag{2}$$

it is **Cauchy** if the below limit exists and is finite

$$\lim_{n, m \to \infty} p(x_n, x_m). \tag{3}$$

Also, the analogue of contraction mapping theorem in a complete partial metric space (X, p) was established in [14].

A partial metric space (X, p) is complete if every Cauchy sequence $\{x_n\}$ in (X, p) converges with respect to the topology τ_p (topology generated by p) to a point $x \in X$ such that

$$\lim_{n, m \to \infty} p(x_n, x_m) = p(x, x).$$

Let (X, p) be a partial metric space and (X, d_p) be the induced metric on X, then

1. $\{x_n\}$ is a Cauchy sequence in (X, p) iff $\{x_n\}$ is a Cauchy in (X, d_p), (see [17]).
2. (X, p) is complete iff (X, d_p) is complete, (see [17]).
3. $\lim_{n \to \infty} d_p(x_n, x) = 0$ iff $p(x, x) = \lim_{n \to \infty} p(x_n, x) = \lim_{n \to \infty} p(x_n, x_m)$, (see [17]).

Let $U : X \to X$ be a mapping, (X, p) be a partial metric space and $\preceq$ be the induced order on X, then

1. U is **order preserving** if $x \preceq y \implies Ux \preceq Uy$, $\forall x, y \in X$ (see [11]).
2. $x \in X$ is a **fixed point** of U if $U(x) = x$, (see [11]).

Definition 1: For $x, y \in X$, U is $[x, y]$-**order preserving** if whenever $U^n x \preceq U^m y$ then $U(U^n x) \preceq U(U^m y)$, $\forall m, n \in \{0, 1, 2 \cdots\}$.

Definition 2: For $x \in X$, U is x-**order preserving** if whenever $U^n x \preceq U^m x$ then $U(U^n x) \preceq U(U^m x)$, $\forall m, n \in \{0, 1, 2 \cdots\}$. Note that, x-order preserving is the same as $[x, x]$-order preserving.

In view of asymptotically contractive mapping used by Kamihigashi and Stachurski [11], a mapping $U : X \to X$ is **asymptotically contractive in a partial metric space** (X, p) if

$$p(U^n x, U^n y) \to 0, \; \forall x, y \in X \tag{4}$$

Also, **a fixed point** $x \in X$ **is a globally stable** fixed point of $U : X \to X$ in a partial metric space (X, p) if $\forall y \in X$, $p(x, U^i y,) \to 0$.

Example 5: Define a mapping $\Psi : F \times F \to \mathbb{R}_+$ by

$$\Psi(k, h) = \max\{A_k, A_h\}, \tag{5}$$

where F is a finite set of some continuous functions over $[0,10]$ with the property that graphs of these functions do not intersect and A_k is the area of the region bounded by $k \in F$ and the horizontal axis. Define a mapping $U : F \to F$ by

$$U(k) = g_k, \tag{6}$$

where $g_k \in F$ is the function with shortest distance to k over the interval $[0,10]$. Using (5) and (6) we can check and see that

$$\Psi(U^n f, U^n g) \to 0.$$

Let $\mathfrak{P}_S$ be the space of all probability measures (distributions) on $(S, \mathfrak{B})$, where S is a topological space equipped with borel sets $\mathfrak{B}$, a sequence $\{\mu_n\} \subset \mathfrak{P}_S$ is called tight if, for all $\varepsilon > 0$, there exists a compact $K \subset S$ such that $\mu_n(S \backslash K) < \varepsilon$ for all n, see [13].

3   Main Results

Let X be nonempty, $\preceq$ an order relation defined on X, $U : X \to X$ a self mapping and $p : X \times X \to \mathbb{R}_+$. We also assume the followings

Assumption 3.1: $\preceq$ is transitive.

Assumption 3.2: p is identifying.

Assumption 3.3: p is regular.

Assumption 3.4: If $x \in X$ is a fixed point of U then $p(y, x) = p(x, y), \forall y \in X$.

Theorem 3.1. *Suppose $\preceq$ is reflexive, if there exist $s, t \in X$ such that U is both t-order preserving and $[s, t]$-order preserving, and the conditions*

$$p(U^i s, U^i t) \to 0, \tag{7}$$
$$U^i s \preceq t, \; \forall \, i \in \mathbb{N}, \tag{8}$$
$$t \preceq Ut, \tag{9}$$

are satisfied, then U has a fixed point.

Proof. Let the Eqs. (7)–(9) hold, for $\preceq$ reflexive and U t-order preserving we have $Ut \preceq Ut \preceq U^i t$, using Assumption 3.1 together with (8) and (9) we have

$$U^i s \preceq t \preceq U^i t.$$

By regularity of p we have

$$0 \leq p(t,t)$$
$$\leq p(t, Ut)$$
$$\leq p(t, U^i t)$$
$$\leq p(U^i s, U^i t)$$
$$\to 0.$$

Similarly, the above inequality is true if we replace $p(t,t)$ with $p(Ut, Ut)$, thus we have $p(Ut, Ut) = p(t, Ut) = p(t,t)$, by identifying property of p we have $t = Ut$ and hence t is a fixed point of U. $\qquad\square$

Lemma 3.2. *Suppose p satisfies (P2), U is asymptotically contractive map on X and has a fixed point t, then t is unique and globally stable.*

Proof. Let t be a fixed point of U and $y \in X$, then

$$p(t, U^i y) = p(U^i t, U^i y)$$
$$\to 0,$$

therefore t is a globally stable fixed point of U. Also, for fixed points s and t of U we have

$$0 \leq p(t,t)$$
$$\leq p(s,t)$$
$$= p(s, U^i t)$$
$$= p(U^i s, U^i t)$$
$$\to 0.$$

Similarly, the above inequality is true if we replace $p(t,t)$ with $p(s,s)$, thus we have $p(t,t) = p(s,t) = p(s,s)$, by identifying property of p we have $t = s$. $\quad\square$

Theorem 3.3. *If $\preceq$ is reflexive, p satisfy (P2) and (P4), then U has a unique globally stable fixed point iff U is asymptotically contractive and there exist s, t satisfying (8) and (9).*

Proof. In view of Theorem 3.1 and Lemma 3.2 it suffices to show that, if p satisfies (P4) and U has a globally stable fixed point, then U is asymptotically contractive. Let $x, y, t \in X$ such that t is a globally stable fixed point of U then

$$p(U^i x, U^i y) \leq p(U^i x, t) + p(t, U^i y) - p(t,t)$$
$$\leq p(t, U^i x) + p(t, U^i y)$$
$$\to 0,$$

thus U is asymptotically contractive. $\qquad\square$

4   Complete Space Case

Below are some additional assumptions.

Assumption 4.1: (X, p) is a complete partial metric space and $\preceq$ is reflexive.

Assumption 4.2: For any increasing sequence $\{x_i\}_{i \in \mathbb{N}} \subset X$ converging to some $x \in X$ we have $x_i \preceq x$ and $x_i \preceq Ux$, $\forall i \in \mathbb{N}$.

Assumption 4.3: For any increasing sequence $\{x_i\}_{i \in \mathbb{N}} \subset X$ converging to some $x \in X$, if there exist $y \in X$ such that $x_i \preceq y$, $\forall i \in \mathbb{N}$ then, $x \preceq y$ and $x_i \preceq Ux, \forall i \in \mathbb{N}$.

Theorem 4.1. *Suppose for any $x, y \in X$ we have*

$$x \preceq y \Longrightarrow p(U^i x, U^i y) \to 0. \tag{10}$$

Suppose also there exist $t, s \in X$ with U both t-order preserving and $[s,t]$-order preserving such that

$$t \preceq Ut, \tag{11}$$
$$U^i t \preceq s, \quad \forall i \in \mathbb{N}, \tag{12}$$

then U has a fixed point.

Proof. For a fixed point to exists, it is enough to show the existence of $\hat{x}$ and t satisfying conditions (7)–(9).

Now, let $x_i = U^i t$, $\forall i \in \mathbb{N}$. It follows from (11) and t-order preserving condition that $\{x_i\}_{i \in \mathbb{N}}$ is increasing. We next need to show $\{x_i\}$ is Cauchy by using (10)–(12) and regularity of p. Let $\varepsilon > 0$, from (10)–(12) there exists $m \in \mathbb{N}$ such that $p(U^m t, U^m s) < \varepsilon$. Let $j, k \in \mathbb{N}$ such that $k > j > m$ and $N = k - m$. Clearly, $x_m \preceq x_j \preceq x_k$ so that

$$\begin{aligned}
p(x_j, x_k) &\leq p(x_m, x_k) \\
&= p(U^m t, U^k t) \\
&= p(U^m t, U^m U^N t) \\
&\leq p(U^m t, U^m s) < \varepsilon.
\end{aligned}$$

Hence $\lim_{j,k \to \infty} p(x_j, x_k) = 0$ and thus $\{x_i\}_{i \in \mathbb{N}}$ is Cauchy. By completeness of (X, p) there exists $\hat{x} \in X$ such that $x_i \to \hat{x}$, i.e. $\lim_{n \to \infty} p(x_i, \hat{x}) = p(\hat{x}, \hat{x})$. Now, using Assumption 4.2, (11) and t-order preserving condition of U, we have

$$t \preceq U^i t \preceq \hat{x}, \quad \forall i \in \mathbb{N}, \tag{13}$$

Equation (13) implies Eq. (8). Equation (7) follows from (10) and (13). From Assumptions 4.2 and 4.3 we have $\hat{x} \preceq U\hat{x}$, hence (9) holds. From the proof of Theorem 3.1 we can conclude that $\hat{x}$ is a fixed point of U. $\square$

5 Conclusion

The results in this paper

1. extend the work of Kamihigashi and Stachurski [11] from a metric space to a partial metric space.
2. improves the work of Altun and Erduran [3] by providing more simpler proving technique and do not use any continuity property on U if compared our Theorem 3.1 and their Theorem 2.1.
3. our Theorem 3.1 improves Matthews's [14] Theorem 5.3; as our theorem requires no completeness property.
4. can be used in programming language semantic; for our research being in partial metric [14] and if restricted to a one dimensional partial metric.

Acknowledgment. The authors acknowledge the financial support provided by King Mongkut's University of Technology Thonburi through the "KMUTT 55th Anniversary Commemorative Fund". Umar Yusuf Batsari was supported by the Petchra Pra Jom Klao Doctoral Academic Scholarship for Ph.D. Program at KMUTT. Moreover, the second author was supported by Theoretical and Computational Science (TaCS) Center, under Computational and Applied Science for Smart Innovation Cluster (CLASSIC), Faculty of Science, KMUTT.

References

1. Agarwal, R., El-Gebeily, M.A., O'Regan, D.: Generalized contractions in partially ordered metric spaces. Appl. Anal. **87**, 109–116 (2008)
2. Alfred, T.: A Lattice-theoretical fixed point theorem and its applications. Pac. J. Math. **5**(2), 285–309 (1955)
3. Altun, I., Erduran, A.: Fixed point theorems for monotone mappings on partial metric spaces. Fixed Point Theory Appl. **2011** (2011). https://doi.org/10.1155/2011/508730
4. Altun, I., Simsek, H.: Some fixed point theorems on ordered metric spaces and application. Fixed Point Theory Appl. **2010** (2010). Article ID 621492
5. Amanda, G.T.: Convegence of Markov processes near a saddle fixed point. Ann. Probab. **35**(3), 1141–1171 (2007)
6. Charalambos, D.A., Kim, C.B.: Infinite Dimensional Analysis: A Hitchhiker's Guide, 3rd edn. Springer, Berlin (2006)
7. Granas, A., Dugundji, J.: Fixed Point Theory. Springer, New York (2003)
8. Hassen, A.: Fixed point theorems for generalized weakly contractive condition in ordered partial metric spaces. J. Nonlinear Anal. Optim. **2**(2), 269–284 (2011)
9. Heckmann, R.: Approximation of metric spaces by partial metric spaces. Appl. Categ. Struct. **7**(1–2), 71–83 (1999)
10. HeikkilÃd', S.: Fixed point results and their applications to Markov processes. Fixed Point Theory Appl. **2005**(3), 307–320 (2005)
11. Kamihigashi, T., Stachurski, J.: Simple fixed point results for order-preserving self-maps and applications to nonlinear Markov operators. Fixed Point Theory Appl. **2013**, 351 (2013)
12. Kamihigashi, T., Stachurski, J.: An order-theoretic mixing condition for monotone Markov chains. Stat. Probab. Lett. **82**, 262–267 (2012)

13. Kamihigashi, T., Stachurski, J.: Stochastic stability in monotone economies. Theor. Econ. **9**, 383–407 (2014)
14. Matthews, S.G.: Partial metric topology. In: Proceedings of the 8th Summer Conference on General Topology and Applications. Annals of the New York Academy of Sciences, vol. 728, pp. 183–197 (1994)
15. Matthews, S.G.: Partial metric spaces, 8th British Colloquium for Theoretical Computer Science. In: Research report 212, Dept. of Computer Science, University of Warwick, March 1992
16. Nieto, J.J., Rodríguez-López, R.: Existence and uniqueness of fixed point in partially ordered sets and applications to ordinary differential equations. Acta Math. Sin. Engl. Ser. **23**, 2205–2212 (2007)
17. Oltra, S., Valero, O.: Banach's fixed point theorem for partial metric spaces. Rend. Istit. Mat. Univ. Trieste **XXXVI**, 17–26 (2004)
18. Onsod, W., Kumam, P., Cho, Y.J.: Fixed points of α-Θ-geraghty type and Θ-geraghty graphic type contractions. Appl. General Topol. **18**(1), 153–171 (2017). https://doi.org/10.4995/agt.2017.6694
19. Penot, J.P.: A fixed-point theorem for asymptotically contractive mappings. Proc. Am. Math. Soc. **131**, 2371–2377 (2003)
20. Shukla, S., Altun, I., Sen, R.: Fixed point theorems and assymptotically regular mappings in partial metric spaces. Comput. Math. **2013** (2013). https://doi.org/10.1155/2013/602579
21. Stokey, N.L., Lucas, E.R., Prescott, C.E.: Recursive Methods in Economic Dynamics. Havard University Press, Massachusetts (1999)

An (α, ϑ)-admissibility and Theorems for Fixed Points of Self-maps

Aziz Khan[1], Kamal Shah[2], Poom Kumam[3,4]([⊠]), and Wudthichai Onsod[3,4]

[1] Department of Mathematics, University of Peshawar,
P.O. Box 25000,
Khybar Pakhtunkhwa, Pakistan
azizkhan927@yahoo.com

[2] Department of Mathematics, University of Malakand,
P.O. Box 18000,
Khybar Pakhtunkhwa, Pakistan
kamalshah408@gmail.com

[3] KMUTT-Fixed Point Research Laboratory, Department of Mathematics, Room SCL 802 Fixed Point Laboratory, Science Laboratory Building, Faculty of Science, King Mongkut's University of Technology Thonburi (KMUTT), 126 Pracha-Uthit Road, Bang Mod, Thrung Khru, Bangkok 10140, Thailand

[4] KMUTT-Fixed Point Theory and Applications Research Group (KMUTT-FPTA), Theoretical and Computational Science Center (TaCS), Science Laboratory Building, Faculty of Science, King Mongkut's University of Technology Thonburi (KMUTT), 126 Pracha-Uthit Road, Bang Mod, Thrung Khru, Bangkok 10140, Thailand
poom.kum@kmutt.ac.th, wudthichai.ons@mail.kmutt.ac.th

Abstract. We introduce (α, ϑ)-admissibility and an $(\Upsilon, \wp)$-integral-type contraction with applications to new fixed point theorems for the admissible and continuous mapping $F : X \to X$ on a complete metric space (X, d). For the application, an interesting example is added which demonstrate our results.

Keywords: Fixed point theorems · Integral-type contractions
Self continuous mappings

Mathematics Subject Classification: Primary 47H09
Secondary 54H25

1 Introduction and Preliminaries

Fixed point theory have a lot of applications in different disciplines of pure and applied mathematics, image processing, engineering, nonlinear functional analysis, computer science, economics, dynamical system etc. [11–13]. In 1922, Banach [6] provided an outstanding theorem which has led to many follow-up results have been proven. In this area, one can observe a large number of new fixed point theorems which modify the pre-existing theorems. For instance, Agarwal et al.

© Springer International Publishing AG 2018
L. H. Anh et al. (eds.), *Econometrics for Financial Applications*, Studies in Computational Intelligence 760, https://doi.org/10.1007/978-3-319-73150-6_30

[1] pointed out many important consequences of the new fixed point theorems in multiplicative metric space. Samet and Vetro [5] provided the $(\alpha\text{-}\psi)$-contractive type mappings and produced new fixed point theorems and provided some interesting applications of their results. Shatanawi and Rawashdeh [4] proved new fixed point theorems in order metric space by (ψ, ϕ)-contractive-type mapping and applied their theorems to some functional equations. Hussian et al. [3] established new fixed point theorems by using α-admissible contraction in complete metric space and gave some interesting instructive example in the applications of their work. Chandok [2] produced (α, β)-admissible Geraghty-type contractive mappings in metric space and for the usability he provided some explanatory examples. Altun et al. [16] produced new fixed point theorems for weakly compatible mappings sustaining integral-type contractions and provided helpful examples. Rhoades [17] published two fixed point theorems for mappings by the use of integral-type contractions and gave some instructive example. Farajzadeh et al. [14] introduced a new $(\alpha, \eta, \psi, \xi)$ contraction for multi-valued mappings and added interesting instructive example for the applications of their fixed point theorems. Bota et al. [10] introduced $(\alpha\text{-}\psi)$-Ciric-type contraction for the multi valued operator and proved fixed point theorems in b-metric space.

In this paper, we introduce an (α, ϑ)-admissibility and an $(\Upsilon, \wp)$-integral-type contraction for a self-continuous mapping $F : X \rightarrow X$ on a complete metric space (X, d) and produce new fixed point theorems. We also provide an interesting instructive example which demonstrate an application of our results.

The following definitions are given from the available literature.

Definition 1 [8]. Let (X, d) be a metric space and $F : X \rightarrow X$ be a mapping and $\alpha : X \times X \rightarrow [0, \infty)$. A mapping F is called α-admissible mapping if the following condition holds:

$$x, y \in X \text{ with } \alpha(x, y) \geq 1 \Rightarrow \alpha(Fx, Fy) \geq 1. \tag{1}$$

Definition 2. Let $F : X \rightarrow X$ and $\alpha : X \times X \rightarrow [0, \infty)$ be a given mappings. A mappings $F : X \rightarrow X$ is called a triangular α-admissible if

$(\mathfrak{C}_1)$ F is α-admissible;
$(\mathfrak{C}_2)$ $\alpha(x, y) \geq 1$ and $\alpha(y, z) \geq 1 \Rightarrow \alpha(x, z) \geq 1$, $x, y, z \in X$.

By Ψ we mean a class of functions $\wp : [0, \infty) \rightarrow [0, \infty)$ satisfying the following assumptions:

(1) $\wp$ is non decreasing function;
(2) $\sum_{n=1}^{\infty} \wp^n(t) < \infty$ for all $t > 0$, where $\wp^n$ is the nth iteration of $\wp$;
(3) $\lim_{n \to \infty} \wp^n(t) = 0$ for all $t > 0$;
(4) $\wp(t) < t$ for each $t > 0$;
(5) $\wp(0) = 0$.

The Ψ is known as Bianchini-Grandolf gauge function.

Now the Ξ denotes the family of functions $\Upsilon : [0, \infty) \rightarrow [0, \infty)$ sustain the following assumptions:

(1) Υ is continuous;
(2) Υ is nondecreasing on $[0, \infty)$;
(3) $\Upsilon(t) = 0$ if and only if $t = 0$;
(4) $\Upsilon(t) > 0$ for all $t \in (0, \infty)$;
(4) Υ is subadditive.

We denote by $CL(X)$ the class of all nonempty closed subsets of X.

Lemma 1 *[14]. Let (X, d) be a metric space. $\Upsilon \in \Xi$ and $\mathscr{B} \in CL(X)$. If there exist $x \in X$ such that $\Upsilon(d(x, \mathscr{B})) \geq 0$, then there exists $y \in \mathscr{B}$, such that*

$$\Upsilon(d(x, y)) < q\Upsilon(d(x, \mathscr{B})), \tag{2}$$

where $q > 1$.

2   Main results

Here we give definition of (α, ϑ)-admissibility.

Definition 3. Let (X, d) be a metric space and $F : X \to X$ be a self mapping and $\alpha, \vartheta : X \times X \to [0, \infty)$. A mapping F is said to be (α, ϑ)-admissible mapping if the following condition holds; for all $x, y \in X$

$$\alpha(x, y)\vartheta(x, y) \geq 1 \text{ implies } \alpha(Fx, Fy) \geq 1 \text{ and } \vartheta(Fx, Fy) \geq 1.$$

If for all $x, y \in X$, we have $\vartheta(x, y) = 1$ then (α, ϑ)-admissibility becomes α-admissibility given in Definition 1.

Example 1. Let $(X = [0, \infty), d)$ be a metric space. Define

$$Fx = \begin{cases} x + 2, & \text{if } x \neq 0; \\ 0, & \text{if } x = 0, \end{cases} \tag{3}$$

and

$$\alpha(x, y) = \begin{cases} 1.2, & \text{if } x, y \neq 0; \\ 0, & \text{if } x = 0 \text{ or } y = 0, \end{cases} \qquad \vartheta(x, y) = \begin{cases} 1.5, & \text{if } x, y \neq 0; \\ 0, & \text{if } x = 0 \text{ or } y = 0. \end{cases} \tag{4}$$

We discuss two cases.

Case I. When x or $y = 0 \Rightarrow \alpha(x, y) = 0 < 1$, so we omit the case.

Case II. When $x, y \neq 0$, $\alpha(x, y) = 1.2$. Now we have to check whether $\alpha(Fx, Fy) \geq 1$, or not. Since

$$Fx = x + 2, \quad Fy = y + 2, \tag{5}$$

for $x \neq 0$, $y \neq 0$, then we have

$$\alpha(Fx, Fy) = \alpha(x + 2, y + 2) = 1.2 \geq 1.$$

Now we check whether $\vartheta(Fx, Fy) \geq 1$ or not. Where $Fx = x + 2$, $Fy = y + 2$, which implies

$$\vartheta(x + 2, y + 2) = 1.5 \geq 1. \tag{6}$$

Thus the self-mapping F is an (α, ϑ)-admissible map.

Definition 4. Let (X, d) be a complete metric space and $F : X \to X$ be an (α, ϑ)-admissible mapping. The mapping F satisfies $(\Upsilon, \wp)$-integral-type contraction if there exist $\alpha, \vartheta : X \times X \to [0, \infty)$ such that for all $x, y \in X$, such that $\alpha(x, y)\vartheta(x, y) \geq 1$ implies

$$\Upsilon\left(\int_0^{d(y, Fy)} \zeta(j)dj\right) \leq \wp\left(\Upsilon \max\left\{\int_0^{M_1(x,y)} \zeta(j)dj, \int_0^{M_2(x,y)} \zeta(j)dj, \int_0^{M_3(x,y)} \zeta(j)dj\right\}\right), \tag{7}$$

where $M_i(x, y)$ for $i = 1, 2, 3$ are

$$M_1(x, y) = \max\left\{d(x, y), d(x, Fx), d(y, Fy), \frac{d(x, Fy) + d(y, Fx)}{2}\right\},$$

$$M_2(x, y) = \max\left\{d(x, y), d(x, Fx), d(y, Fy)\right\},$$

$$M_3(x, y) = \max\left\{d(x, y), d(y, Fy)\right\},$$

$$M(x, y) = \max\left\{M_1(x, y), M_2(x, y), M_3(x, y)\right\},$$

for $\wp \in \Psi$, $\Upsilon \in \Xi$ and $\mathfrak{P} \geq 1$, $\zeta : \mathbb{R}^+ \to \mathbb{R}^+$ Lebesgue integrable with finite integral such that $\int_0^\epsilon \zeta(j)dj > 0$, for each $\epsilon > 0$.

Theorem 1. *Let $F : X \to X$ be a self-mapping on a complete metric space (X, d) and F satisfies $(\Upsilon, \wp)$-integral-contraction with the following assumptions:*

$(\mathfrak{C}_1)$ *F is (α, ϑ)-admissible mapping;*
$(\mathfrak{C}_2)$ *there exist $x_0, y_0 \in X$ such that $\alpha(x_0, y_0) \geq 1$;*
$(\mathfrak{C}_3)$ *F satisfies $(\Upsilon, \wp)$-integral-type contraction.*

Then F has a unique fixed point in (X, d).

Proof. Since F is (α, ϑ)-admissible self mapping. Then there exist $x_0, y_0 \in X$ such that $\alpha(x_0, y_0) \geq 1$ which implies $\alpha(Fx_0, Fy_0) \geq 1$ implies $\vartheta(Fx_0, Fy_0) \geq 1$. Since $F : X \to X$, again there exist some $x_1 \in X$ such that $Fx_0 = x_1$, similarly $Fy_0 = y_1$ for some $y_1 \in X$, thus $\alpha(x_1, y_1) \geq 1$. By (C_1), $\alpha(Fx_1, Fy_1) \geq 1$ which implies $\vartheta(Fx_1, Fy_1) \geq 1$. By continuing this process and using mathematical induction, we may have $\alpha(x_n, y_n) \geq 1$ implies $\alpha(Fx_n, Fy_n) \geq 1$ which give us $\vartheta(Fx_n, Fy_n) \geq 1$. Ultimately, we have

$$\alpha(x_n, y_n)\vartheta(x_n, y_n) \geq 1. \tag{8}$$

Now by $(\mathfrak{C}_3)$, we may use the inequality (7). By putting $x = x_0$ and $y = x_1$, in the inequality (7), we have

$$\Upsilon\left(\int_0^{d(x_1, Fx_1)} \zeta(j)dj\right) \leq \wp\left(\Upsilon \max\left\{\int_0^{M_1(x_0, x_1)} \zeta(j)dj, \int_0^{M_2(x_0, x_1)} \zeta(j)dj, \int_0^{M_3(x_0, x_1)} \zeta(j)dj\right\}\right), \tag{9}$$

where $M_i(x_1, x_2)$ for $i = 1, 2, 3$

$$M_1(x_0, x_1) = \max \Big\{ d(x_0, x_1), d(x_0, Fx_0), d(x_1, Fx_1),$$
$$\frac{d(x_0, Fx_1) + d(x_1, Fx_0)}{2} \Big\}$$
$$\leq \max \Big\{ d(x_0, x_1), d(x_0, x_1), d(x_1, Fx_1),$$
$$\frac{d(x_0, Fx_1) + d(x_1, x_1)}{2} \Big\}$$
$$\leq \max \Big\{ d(x_0, x_1), d(x_1, Fx_1), \frac{d(x_0, x_1) + d(x_1, x_1)}{2} \Big\}$$
$$\leq \max \Big\{ d(x_0, x_1), d(x_1, Fx_1) \Big\},$$

$$M_2(x_0, x_1) = \max \Big\{ d(x_0, x_1), d(x_0, Fx_0), d(x_1, Fx_1) \Big\}$$
$$\leq \max \Big\{ d(x_0, x_1), d(x_0, x_1), d(x_1, Fx_1) \Big\}$$
$$\leq \max \Big\{ d(x_0, x_1), d(x_1, Fx_1) \Big\},$$

$$M_3(x_0, x_1) = \max \Big\{ d(x_0, x_1), d(x_1, Fx_1) \Big\}$$
$$\leq \max \Big\{ d(x_0, x_1), d(x_1, Fx_1) \Big\}.$$

Now if $\max \Big\{ d(x_0, x_1), d(x_1, Fx_1) \Big\} = d(x_1, Fx_1)$. From (9), we proceed

$$0 < \Upsilon \Big(\int_0^{d(x_1, Fx_1)} \zeta(j) dj \Big)$$
$$\leq \wp \Big(\Upsilon \max \Big\{ \int_0^{d(x_1, Fx_1)} \zeta(j) dj, \int_0^{d(x_1, Fx_1)} \zeta(j) dj, \int_0^{d(x_1, Fx_1)} \zeta(j) dj \Big\} \Big)$$
$$\leq \wp \Big(\Upsilon \Big(\int_0^{d(x_1, Fx_1)} \zeta(j) dj \Big) \Big)$$
$$< \Upsilon \Big(\int_0^{d(x_1, Fx_1)} \zeta(j) dj \Big).$$

This is contradiction. If we consider $\max \Big\{ d(x_0, x_1), d(x_1, Fx_1) \Big\} = d(x_0, x_1)$, then we have

$$0 < \Upsilon \Big(\int_0^{d(x_1, Fx_1)} \zeta(j) dj \Big) \leq \wp \Big(\Upsilon \Big(\int_0^{d(x_0, x_1)} \zeta(j) dj \Big) \Big). \tag{10}$$

From Lemma 1, we have

$$\Upsilon \Big(\int_0^{d(x_1, x_2)} \zeta(j) dj \Big) < q \Upsilon \Big(\int_0^{d(x_1, Fx_1)} \zeta(j) dj \Big),$$

for some $x_2 = Fx_1$ and $q > 1$. If $x_2 = Fx_2$, then x_2 is the fixed point of F, we assume $x_2 \neq Fx_2$ then from Eqs. (10) and (11), we have

$$0 < \Upsilon\left(\int_0^{d(x_1,x_2)} \zeta(j)dj\right) < q\wp\left(\Upsilon\left(\int_0^{d(x_0,x_1)} \zeta(j)dj\right)\right). \qquad (11)$$

Applying $\wp$ is non-decreasing to the inequality (11), we obtain

$$0 < \wp\left(\Upsilon\left(\int_0^{d(x_1,x_2)} \zeta(j)dj\right)\right) < \wp\left(q\wp\left(\Upsilon\left(\int_0^{d(x_0,x_1)} \zeta(j)dj\right)\right)\right).$$

This implies

$$q_1 = \frac{\wp\left(q\wp\left(\Upsilon\left(\int_0^{d(x_0,x_1)} \zeta(j)dj\right)\right)\right)}{\wp\left(\Upsilon\left(\int_0^{d(x_1,x_2)} \zeta(j)dj\right)\right)} > 1. \qquad (12)$$

Next, by putting $x = x_2, y = x_2$ in the inequality (7), we have

$$0 < \Upsilon\left(\int_0^{d(x_2,Fx_2)} \zeta(j)dj\right)$$
$$\leq \wp\left(\Upsilon \max\left\{\int_0^{M_1(x_1,x_2)} \zeta(j)dj, \int_0^{M_2(x_1,x_2)} \zeta(j)dj, \int_0^{M_3(x_1,x_2)} \zeta(j)dj\right\}\right), \qquad (13)$$

where $\phi(M_i(x_1,x_2))$ for $i = 1, 2, 3$

$$M_1(x_1,x_2) = \max\left\{d(x_1,x_2), d(x_1,Fx_1), d(x_2,Fx_2),\right.$$
$$\left.\frac{d(x_1,Fx_2) + d(x_2,Fx_1)}{2}\right\},$$
$$\leq \max\left\{d(x_1,x_2), d(x_1,x_1), d(x_2,Fx_2),\right.$$
$$\left.\frac{d(x_1,Fx_2) + d(x_2,x_1)}{2}\right\},$$
$$\leq \max\left\{d(x_1,x_2), d(x_2,Fx_2), \frac{d(x_1,x_2) + d(x_2,x_2)}{2}\right\},$$
$$\leq \max\left\{d(x_1,x_2), d(x_2,Fx_2)\right\},$$

$$M_2(x_1,x_2) = \max\left\{d(x_1,x_2), d(x_1,Fx_1), d(x_2,Fx_2)\right\},$$
$$\leq \max\left\{d(x_1,x_2), d(x_1,x_2), d(x_2,Fx_2)\right\},$$
$$\leq \max\left\{d(x_1,x_2), d(x_2,Fx_2)\right\},$$

$$M_3(x_1,x_2) = \max\left\{d(x_1,x_2), d(x_2,Fx_2)\right\},$$
$$\leq \max\left\{d(x_1,x_2), d(x_2,Fx_2)\right\}.$$

Now if $\max\{d(x_1, x_2), d(x_2, Fx_2)\} = d(x_1, x_2)$. Then from (13) and $\wp(t) < t$, we have

$$0 < \Upsilon\left(\int_0^{d(x_2, Fx_2)} \zeta(j)dj\right) \leq \wp\left(\Upsilon \max\left\{\int_0^{d(x_1, x_2)} \zeta(j)dj, \int_0^{d(x_1, x_2)} \zeta(j)dj, \right.\right.$$

$$\left.\left.\int_0^{d(x_1, x_2)} \zeta(j)dj\right\}\right) \tag{14}$$

$$\leq \wp\left(\Upsilon\left(\int_0^{d(x_1, x_2)} \zeta(j)dj\right)\right) < \Upsilon\left(\int_0^{d(x_1, x_2)} \zeta(j)dj\right).$$

Therefore, we get

$$0 < \Upsilon\left(\int_0^{d(x_2, Fx_2)} \zeta(j)dj\right) \leq \wp\left(\Upsilon\left(\int_0^{d(x_1, x_2)} \zeta(j)dj\right)\right). \tag{15}$$

As $q_1 > 1$ from the Lemma 1 and there exists some $x_3 \in X$, such that $x_3 = Fx_2$, which gives us

$$\Upsilon\left(\int_0^{d(x_2, x_3)} \zeta(j)dj\right) < q_1\Upsilon\left(\int_0^{d(x_2, Fx_2)} \zeta(j)dj\right). \tag{16}$$

From (13), (15) and (16), we have

$$0 < \Upsilon\left(\int_0^{d(x_2, x_3)} \zeta(j)dj\right) \leq q_1\wp\left(\Upsilon\left(\int_0^{d(x_1, x_2)} \zeta(j)dj\right)\right)$$

$$= \wp\left(q\wp\left(\Upsilon\left(\int_0^{d(x_0, x_1)} \zeta(j)dj\right)\right)\right). \tag{17}$$

Applying $\wp$ on (17), we have

$$0 < \wp\left(\Upsilon\left(\int_0^{d(x_2, x_3)} \zeta(j)dj\right)\right) < \wp^2\left(q\wp\left(\Upsilon\left(\int_0^{d(x_0, x_1)} \zeta(j)dj\right)\right)\right). \tag{18}$$

Continuing the same process upto x_n with the assumption that $x_n \neq x_{n+1} = Fx_n$, we have

$$0 < \Upsilon\left(\int_0^{d(x_{n+1}, x_{n+2})} \zeta(j)dj\right) < \wp^n\left(q\wp\left(\Upsilon\left(\int_0^{d(x_0, x_1)} \zeta(j)dj\right)\right)\right),$$

for all $n \in N_0$. Now we show that $\{x_n\}$ in X is a Cauchy sequence. For this, let $m, n \in N$ such that $m > n$, and triangle inequality then we have

$$0 < \Upsilon\left(\int_0^{d(x_m, x_n)} \zeta(j)dj\right) \leq \sum_{i=n}^{m-1} \Upsilon\left(\int_0^{d(x_i, x_{i+1})} \zeta(j)dj\right) \tag{19}$$

$$< \sum_{i=n}^{m-1} \wp^{i-1}\left(q\wp\left(\Upsilon\left(\int_0^{d(x_0, x_1)} \zeta(j)dj\right)\right)\right).$$

Applying $\lim_{n,m\to\infty}$ to (19), and $\wp \to 0$ as $n \to \infty$, therefore, we have

$$\lim_{n,m\to\infty} \Upsilon\left(\int_0^{d(x_m,x_n)} \zeta(j)dj\right) = 0. \tag{20}$$

By the continuity of Υ, we get

$$\lim_{n,m\to\infty} \int_0^{d(x_m,x_n)} \zeta(j)dj = 0.$$

Thus $\{x_n\}$ is a Cauchy sequence in X. Since (X,d) is complete, therefore there exists $x^* \in X$ such that $x_n \to x^*$ as $n \to \infty$, thus we have $\lim_{n,m\to\infty} \int_0^{d(x_n,x^*)} \zeta(j)dj = 0$ from the continuity of F we have

$$\lim_{n,m\to\infty} \int_0^{d(Fx_n,Fx^*)} \zeta(j)dj = 0,$$

and

$$\int_0^{d(x^*,Fx^*)} \zeta(j)dj = \lim_{n\to\infty} \int_0^{d(x_{n+1},Fx^*)} \zeta(j)dj = \lim_{n\to\infty} \int_0^{d(Fx_n,Fx^*)} \zeta(j)dj = 0.$$

This implies $d(x^*, Fx^*) = 0$ or $x^* = Fx^*$ and therefore, x^* is a fixed point of F in (X,d).

Theorem 2. *Let $F : X \to X$ be a self-mapping on a complete metric soace (X,d) and F satisfies $(\Upsilon, \wp)$-integral-contraction with the following assumptions:*

$(\mathfrak{C}_1^*)$ F is (α, ϑ)-admissible self mapping;
$(\mathfrak{C}_2^*)$ there exist $x_0, x_1 \in X$ such that $\alpha(x_0, x_1) \geq 1$ and $\alpha(x_0, x_1)\vartheta(x_0, x_1) \geq 1$;
$(\mathfrak{C}_3^*)$ $f\{x_n\}$ is a sequence in X with $x_{n+1} \in Fx_n$, $x_n \to x \in X$ as $n \to \infty$ and $\alpha(x_n, x_{n+1}) \geq 1$ for all $n \in N_0$.

Then we have

$$\Upsilon\left(\int_0^{d(x_{n+1},Fx)} \zeta(j)dj\right) \leq \wp\left(\Upsilon \max\left(\int_0^{M1(x_n,x)} \zeta(j)dj, \int_0^{M2(x_n,x)} \zeta(j)dj, \int_0^{M3(x_n,x)} \zeta(j)dj\right)\right),$$

for all $n \in N_0$. Then F has a fixed point in X.

Proof. Let $\{x_n\}$ be a Cauchy sequence in X, such that $x_n \to x^*$ as $n \to \infty$, then

$$\alpha(x_n, x_{n+1}) \geq 1,$$

and

$$\alpha(x_n, x_{n+1})\vartheta(x_n, x_{n+1}) \geq 1,$$

for all $n \in N$. Then from $(\mathfrak{C}_3^*)$, we have

$$\gamma\left(\int_0^{d(x_{n+1}, Fx^*)} \zeta(j)dj\right) \leq \wp\left(\gamma \max\left(\int_0^{M1(x_n, x^*)} \zeta(j)dj, \int_0^{M2(x_n, x^*)} \zeta(j)dj, \int_0^{M3(x_n, x^*)} \zeta(j)dj\right)\right),$$

(21)

where M_i for $i = 1, 2, 3$, are

$$M_1(x_n, x^*) = \max\left\{d(x_n, x^*), d(x_n, Fx_n), d(x^*, Fx^*),\right.$$
$$\left.\frac{d(x_n, Fx^*) + d(x^*, Fx_n)}{2}\right\},$$

$$M_2(x_n, x^*) = \max\left\{d(x_n, x^*), d(x_n, Fx_n), d(x^*, Fx^*)\right\},$$

$$M_3(x_n, x^*) = \max\left\{d(x_n, x^*), d(x^*, Fx^*)\right\},$$

for all $n \in N$. Here, we assume that $d(x^*, Fx^*) > 0$ and let $\epsilon := \frac{d(x_n, Fx^*)}{2}$. Since $x_n \to x^*$ as $n \to \infty$, so we can find $N_1 \in N_0$ such that

$$d(x^*, Fx_n) < \frac{d(x^*, Fx^*)}{2},$$

for all $n \geq N_1$. Furthermore, we obtain

$$d(x^*, Fx_n) \leq d(x^*, x_{n+1}) \leq \frac{d(x_n, Fx^*)}{2},$$

Since $\{x_n\}$ is Cauchy sequence, therefore there exists $N_2 \in N_0$ such that

$$d(x_n, Fx_n) \leq d(x_n, x_{n+1}) \leq \frac{d(x^*, Fx^*)}{2},$$

for all $n \geq N_2$. Since $d(x_n, Fx^*) \to d(x^*, Fx^*)$ as $n \to \infty$, it follows that there exists $N_3 \in N_0$ such that

$$d(x_n, Fx^*) < \frac{3d(x^*, Fx^*)}{2},$$

for all $n \geq N_3$, and we get

$$M_1(x_n, x^*) = \max\left\{d(x_n, x^*), d(x_n, Fx_n), d(x^*, Fx^*),\right.$$
$$\left.\frac{d(x_n, Fx^*) + d(x^*, Fx_n)}{2}\right\}$$
$$= d(x^*, Fx^*),$$

$$M_2(x_n, x^*) = \max\left\{d(x_n, x^*), d(x_n, Fx_n), d(x^*, Fx^*)\right\} = d(x^*, Fx^*),$$

$$M_3(x_n, x^*) = \max\left\{d(x_n, x^*), d(x^*, Fx^*)\right\} = d(x^*, Fx^*),$$

for all $n \geq N := \max\{N_1, N_2, N_3\}$. For each $n \geq N$, from (21) by using the triangular inequality we have

$$\Upsilon\left(\int_0^{d(x^*,Fx^*)} \zeta(j)dj\right) \leq \Upsilon\left(\int_0^{d(x^*,x_{n+1})} \zeta(j)dj\right) + \Upsilon\left(\int_0^{d(x_{n+1},Fx^*)} \zeta(j)dj\right)$$

$$\leq \Upsilon\left(\int_0^{d(x^*,x_{n+1})} \zeta(j)dj\right) + \wp\left(\Upsilon \max\left\{\int_0^{M_1(x_n,x)} \zeta(j)dj,\right.\right.$$

$$\int_0^{M_2(x_n,x)} \zeta(j)dj, \int_0^{M_3(x_n,x)} \zeta(j)dj\right\}\right)$$

$$= \Upsilon\left(\int_0^{d(x^*,x_{n+1})} \zeta(j)dj\right) + \wp\left(\Upsilon\left(\int_0^{d(x^*,Fx^*)} \zeta(j)dj\right)\right).$$

Letting $n \to \infty$ in the above inequality we obtain

$$\Upsilon\left(\int_0^{d(x^*,Fx^*)} \zeta(j)dj\right) \leq \wp\left(\Upsilon\left(\int_0^{d(x^*,Fx^*)} \zeta(j)dj\right)\right).$$

This is a contradiction of $\wp(t) \leq t$. This implies $\Upsilon\left(\int_0^{d(x^*,Fx^*)} \zeta(j)dj\right) = 0$, which further implies that $\int_0^{d(x^*,Fx^*)} \zeta(j)dj = 0$. Consequently, we have $x^* = Fx^*$.

3   Applications

Example 2. Let $(X = [0,10], d)$ be a metric space with $d(x,y) = |x - y|$, for $x, y \in X$. Defining $F : X \to X$ and $\alpha, \vartheta : X \times X \to [0, \infty)$ as

$$F(x) = \begin{cases} 2.5x, & \text{if } x \in [0,5); \\ \frac{x}{64}, & \text{if } x \in [5,10], \end{cases}$$

and

$$\alpha(x,y) = \begin{cases} 6, & \text{if } x \in [5,10]; \\ 2.5, & \text{otherwise}, \end{cases} \qquad \vartheta(x,y) = \begin{cases} 7, & \text{if } x \in [5,10]; \\ 3, & \text{otherwise}. \end{cases} \tag{22}$$

Let $\wp, \Upsilon : [0, \infty) \to [0, \infty)$ by $\wp(t) = \frac{t}{8}$, $\Upsilon(t) = 5\sqrt{t}$, $\zeta(t) = 1$ and $\mathfrak{P} = 1$. Since $\wp \in \Psi$ and $\Upsilon \in \Xi$. To show that F is a $(\Upsilon, \wp)$-integral-type contraction. For this, let $x, y \in X$, then we have

$$\alpha(x,y)\vartheta(x,y) \geq 1.$$

Let $x, y \in [5, 10]$, then we have

$$\Upsilon\left(\int_0^{d(y, Fy)} \zeta(j)dj\right) = 5\sqrt{\left(\int_0^{d(y, Fy)} \zeta(j)dj\right)}$$

$$= 5\sqrt{\left(\int_0^{d(Fx, Fy)} \zeta(j)dj\right)}$$

$$= 5\sqrt{\left(\int_0^{\frac{|x-y|}{64}} \zeta(j)dj\right)}$$

$$= \frac{5}{8}\sqrt{|x-y|}$$

$$\leq \frac{5}{8}\sqrt{\int_0^{\phi(M(x,y))} \zeta(j)dj}$$

$$= \wp\left(\Upsilon \max\left(\int_0^{M1(x,y)} \zeta(j)dj, \int_0^{M2(x,y)} \zeta(j)dj, \int_0^{M3(x,y)} \zeta(j)dj\right)\right).$$

Therefore F is a $(\Upsilon, \wp)$-integral-type contractive mapping. Now for the condition $(\mathfrak{C}_3^*)$, we assume a sequence $\{x_n\}$ in X with $x_{n+1} = Fx_n$ where $x_n \to x \in X$ as $n \to \infty$ and $\alpha(x_n, x_{n+1}) \geq 1$ which implies $\vartheta(x_n, x_{n+1}) \geq 1$ for all $n \in N$, then

$$\Upsilon\left(\int_0^{d(x_{n+1}, Fx)} \zeta(j)dj\right) = 5\sqrt{\left(\int_0^{d(x_{n+1}, Fx)} \zeta(j)dj\right)}$$

$$= 5\sqrt{\left(\int_0^{d(Fx_n, Fx)} \zeta(j)dj\right)}$$

$$= 5\sqrt{\left(\int_0^{\frac{|x_n - x|}{64}} \zeta(j)dj\right)}$$

$$= \frac{5}{8}\sqrt{|x_n - x|}$$

$$\leq \frac{5}{8}\sqrt{\int_0^{\phi M(x_n, x)} \zeta(j)dj}$$

$$= \wp\left(\Upsilon \max\left(\int_0^{M1(x_n,x)} \zeta(j)dj, \int_0^{M2(x_n,x)} \zeta(j)dj, \int_0^{M3(x_n,x)} \zeta(j)dj\right)\right).$$

Thus $(\mathfrak{C}_3^*)$ of Theorem 2 holds. Therefore, we conclude that 0 is a fixed point of the self-mapping F.

4   Conclusion

In this paper, the notion of (α, ϑ)-admissibility, $(\Upsilon, \wp)$-integral-type contraction are given and new fixed point theorems are proved for the admissible and continuous self-mapping $F : X \to X$ on a complete metric space (X, d). For the fixed point of $F : X \to X$, we further assume that F be an (α, ϑ)-admissible mapping, there exist $x_0, y_0 \in X$ such that $\alpha(x_0, y_0) \geq 1$ and the map F sustaining $(\Upsilon, \wp)$-integral-type contraction. For the application, we have presented an interesting example which fulfill all the conditions of our Theorem 2 and the self-map X has a fixed point 0.

Acknowledgements. This project was supported by the Theoretical and Computational Science (TaCS) Center under Computational and Applied Science for Smart Innovation Cluster (CLASSIC), Faculty of Science, KMUTT. The third author would like to tank the Research Professional Development Project Under the Science Achievement Scholarship of Thailand (SAST) for financial support.

References

1. Agarwal, R.P., Karapinar, E., Samet, B.: An essential remark on fixed point results on multiplicative metric spaces. Fixed Point Theor. Appl. **2016**, 21 (2016)
2. Chandok, S.: Some fixed point theorems for (α, β)-admissible Geraghty type contractive mappings and related results. J. Math. Sci. **9**, 127–135 (2015)
3. Hussian, N., Karapinar, E., Salimi, P., Akbar, F.: α-admissibility mappings and related fixed point theorems. J. Inequal. Appl. **2013**, 114 (2013)
4. Shatanawi, W., Al-Rawashdeh, A.: Common fixed point of almost generalized (ψ,ϕ)-contractive mappings in ordered metric spaces. Fixed Point Theor. Appl. **2012**, 80 (2012)
5. Samet, B., Vetro, C.: Fixed point theorem for $(\alpha\text{-}\psi)$-contractive type mappings. Nonlinear Anal. **75**, 2154–2165 (2012)
6. Banach, S.: Surles operations dans les ensembles abstrats et leur application aux equations itegrales. Fundam. Math. **3**, 133–181 (1922)
7. Felhi, A., Aydi, H., Zhang, D.: Fixed point for α-admissible contractive mappings via simulation functions. J. Nonlinear Sci. Appl. **9**, 5544–5560 (2016)
8. Kutbi, M.A., Sintunavarat, W.: On the wealky (α, ψ, ξ)-contractive condition for multi-valued operator in metric spaces and related fixed point result. Open Math. **14**, 167–180 (2016)
9. Farajzadeh, A., Chuadchawna, P., Kaewcharoen, A.: Fixed point theorem for $(\alpha, \eta, \psi, \xi)$-contractive multi-valued mappings on α-η-complete partial metric spaces. J. Nonlinear Sci. App. **19**, 1977–1990 (2016)
10. Bota, M.F., Chifu, C., Karapinar, E.: Fixed point theorems for generalied $(\alpha_*\text{-}\psi)$-Ciric-type contractive multivalued operators in b-metric spaces. J. Nonlinear Sci. Appl. **9**, 1165–1177 (2016)
11. Baleanu, D., Khan, H., Jafari, H., Khan, R.A., Alipour, M.: On existence results for solutions of a coupled system of hybrid boundary value problems with hybrid conditions. Adv. Differ. Equ. **2015**, 318 (2015)
12. Jafari, H., Baleanu, D., Khan, H., Khan, R.A., Khan, A.: Existence criterion for the solutions of fractional order p-Laplacian boundary value problems. Bound. Value Probl. **2015**, 164 (2015)
13. Baleanu, D., Agarwal, R.P., Khan, H., Khan, R.A., Jafari, H.: On the existence of solution for fractional differential equations of order $3 < \delta_1 \leq 4$. Adv. Differ. Equ. **2015**, 362 (2015)
14. Ali, M.U., Kamran, T., Karapinar, E.: $(\alpha, \wp, \xi)$-contractive multi-valued mappings. Fixed Point Theor. Appl. **2014**, 7 (2014)
15. Hussain, N., Isik, H., Abbas, M.: Common fixed point results of generalized almost rational contraction mappings with an application. J. Nonlinear Sci. Appl. **9**, 2273–2288 (2016)
16. Altun, I., Turkoglu, D., Rhoades, B.E.: Fixed point of weakly compatible maps satisfying a general contractive condition of integral type. Fixed Point Theor. Appl. **9**, 1155–17301 (2007)
17. Rhoades, B.E.: Two fixed-point theorems for mappings satisfying a genral contractive condition of integral type. Int J. Math. Math. Sci. **63**, 4007–4013 (2003)

The Modified Multi-step Iteration Process for Pairwise Generalized Nonexpansive Mappings in CAT(0) Spaces

Nuttapol Pakkaranang[1], Phanuphan Kewdee[1], Poom Kumam[1,2(✉)], and Piyachat Borisut[1]

[1] KMUTT-Fixed Point Research Laboratory, Department of Mathematics, Room SCL 802 Fixed Point Laboratory, Science Laboratory Building, Faculty of Science, King Mongkuts University of Technology Thonburi (KMUTT), 126 Pracha-Uthit Road, Bang Mod, Thung Khru, Bangkok 10140, Thailand
{nuttapol.pak,phanuphan.1}@mail.kmutt.ac.th, poom.kum@kmutt.ac.th, piyachat.b@hotmail.com
[2] KMUTT-Fixed Point Theory and Applications Research Group (KMUTT-FPTA), Theoretical and Computational Science Center (TaCS), Science Laboratory Building, Faculty of Science, King Mongkut's University of Technology Thonburi (KMUTT), 126 Pracha-Uthit Road, Bang Mod, Thrung Khru, Bangkok 10140, Thailand

Abstract. In this paper, we modified multi-step procedure to find approximation fixed point of pairwise generalized nonexpansive mappings in CAT(0) spaces. We also prove both strong and Δ-convergence theorems for such a mapping with under mild conditions.

Keywords: CAT(0) spaces · Convergence theorems
Generalized nonexpansive mappings · Common fixed point

Mathematics Subject Classification 2010: Primary 47H09
Secondary 47H10

1 Introduction

Let C be a nonempty subset of a metric space (X, d). A mapping $T : C \to C$ is said to be *nonexpansive* if $d(Tx, Ty) \leq d(x, y)$ for all $x, y \in C$.

A point $x \in C$ is called a *fixed point* of T if $x = Tx$ and $F(T)$ denotes the set of fixed points of the mapping T. A sequence $\{x_n\}$ in C is called *approximate fixed point sequence* if

$$\lim_{n \to \infty} d(x_n, Tx_n) = 0.$$

© Springer International Publishing AG 2018
L. H. Anh et al. (eds.), *Econometrics for Financial Applications*, Studies in Computational Intelligence 760, https://doi.org/10.1007/978-3-319-73150-6_31

In 2008, Suzuki [20] introduced a class of generalized nonexpansive mappings as follows.

Definition 1. Let C be a nonempty subset of a metric space (X, d). We say that a mapping $T : C \to C$ satisfies condition (C) if

$$\frac{1}{2}d(x, Tx) \leq d(x, y) \text{ implies } d(Tx, Ty) \leq d(x, y).$$

for all $x, y \in C$.

He also proved the existence of fixed points of such mappings with condition (C).

Later on, in 2011, García-Falset et al. [10] introduced two classes of generalized nonexpansive mappings which in turn include Suzuki's generalized nonexpansive mappings as follows.

Definition 2. Let C be a nonempty subset of a metric space (X, d). For $\lambda \in (0, 1)$ we say that a mapping $T : C \to C$ satisfies condition (C_λ) for all $x, y \in C$

$$\lambda d(x, Tx) \leq d(x, y) \text{ implies } d(Tx, Ty) \leq d(x, y)$$

Definition 3. Let C be a nonempty subset of a metric space (X, d). For $\mu \geq 1$ we say that a mapping $T : C \to C$ satisfies condition (E_μ) on C if there exists $\mu \geq 1$ such that for all $x, y \in C$,

$$d(x, Ty) \leq \mu d(x, Tx) + d(x, y).$$

In 2017, Zhang et al. [22] introduced new classes of generalized nonexpansive mappings for pairwise generalized nonexpansive mappings which extended two classes generalized nonexpansive mappings of condition (C_λ) and (E_μ) in metric spaces as follows.

Definition 4. Let C be a nonempty subset of a metric space (X, d). Let $T, S : C \to C$ be two mappings and we say that T and S are pairwise generalized nonexpansive mappings if

$$d(Tx, Sy) \leq d(x, y).$$

for all $x, y \in C$.

Definition 5. Let C be a nonempty subset of a metric space (X, d). Let $T, S : C \to C$ be two mappings and $\lambda \in (0, 1)$ and we say that T and S are pairwise generalized nonexpansive mappings satisfying condition (PC_λ) if

$$\lambda d(x, Tx) \leq d(x, y) \text{ and } \lambda d(x, Sx) \leq d(x, y)$$

implies

$$d(Tx, Sy) \leq d(x, y). \tag{1}$$

for all $x, y \in C$.

Obviously, when $T = S$, the mapping T(or S) satisfies condition (C_λ).

Definition 6. Let C be a nonempty subset of a metric space (X, d). Let $T, S :$ $C \to C$ be two mappings and $\mu \geq 1$ and we say T and S are pairwise generalized nonexpansive mappings satisfying condition (PE_μ) on C if

$$d(x, Ty) \leq \mu d(x, Sx) + d(x, y) \tag{2}$$

for all $x, y \in C$.

Obviously, when $T = S$, the mapping T(or S) satisfies condition (E_μ).

On the other hand, in 1953, Mann [16] introduced the following iteration process to approximate fixed points of a nonexpansive mapping T in a Hilbert space H: a sequence $\{x_n\}$ is defined by $x_1 \in C$ and

$$x_{n+1} = (1 - \alpha_n)x_n + \alpha_n T x_n$$

for all $n \geq 1$, where $\{\alpha_n\}$ is a real sequence in $(0, 1)$.

In 1974, Ishikawa [12] introduced the following iteration process to approximate fixed points of a nonexpansive mapping T in a Hilbert space H: a sequence $\{x_n\}$ is defined by for $x_1 \in C$ and

$$\begin{aligned} y_n &= (1 - \alpha_n)x_n + \alpha_n T x_n, \\ x_{n+1} &= (1 - \beta_n)x_n + \beta_n T y_n \end{aligned}$$

for all $n \geq 1$, where $\{\alpha_n\}$ and $\{\beta_n\}$ are real sequences in $(0, 1)$. This Ishikawa iteration reduces to the Mann iteration when $\beta_n = 0$ for all $n \in \mathbb{N}$.

In 2000, Noor [17] introduced the following iteration process to approximate fixed points of a nonexpansive mapping T in a Hilbert space H: a sequence $\{x_n\}$ is defined by for $x_1 \in C$ and

$$\begin{aligned} z_n &= (1 - \alpha_n)x_n + \alpha_n T x_n, \\ y_n &= (1 - \beta_n)x_n + \beta_n T z_n, \\ x_{n+1} &= (1 - \gamma_n)x_n + \gamma_n T y_n \end{aligned}$$

for all $n \geq 1$, where $\{\alpha_n\}, \{\beta_n\}$ and $\{\gamma_n\}$ are real sequences in $(0, 1)$. This Noor iteration reduces to the Ishikawa iteration when $\beta_n = 0$ for all $n \in \mathbb{N}$.

In 2007, Agarwal et al. [2] introduced the following so-called S-iteration to approximate fixed points of a nearly asymptotically nonexpansive mapping T in a Banach space E: a sequence $\{x_n\}$ is defined by for $x_1 \in C$ and

$$\begin{aligned} y_n &= (1 - \alpha_n)x_n + \alpha_n T x_n, \\ x_{n+1} &= (1 - \beta_n)T x_n + \beta_n T y_n \end{aligned}$$

for all $n \geq 1$, where $\{\alpha_n\}$ and $\{\beta_n\}$ are real sequences in $(0, 1)$. Note that this iteration can not reduce which is independent and converges faster than both of the Ishikawa and Mann iterations.

Later on, in 2014, Abbas and Nazir [1] introduced the following iteration process, a sequence $\{x_n\}$ is defined by for $x_1 \in C$ and

$$\begin{aligned} z_n &= (1 - \alpha_n)x_n + \alpha_n T x_n, \\ y_n &= (1 - \beta_n)T x_n + \beta_n T z_n, \\ x_{n+1} &= (1 - \gamma_n)T y_n + \gamma_n T z_n, \ n \in \mathbb{N}, \end{aligned}$$

for all $n \geq 1$, where $\{\alpha_n\}, \{\beta_n\}$ and $\{\gamma_n\}$ are real sequences in $(0, 1)$.

Recently, Thakur et al. [21] introduced the following multi-step iteration process for approximating a fixed point of a nonexpansive mapping T in a Banach space E, a sequence $\{x_n\}$ is defined by for $x_1 \in C$ and

$$
\begin{aligned}
z_n &= (1 - \alpha_n)x_n + \alpha_n T x_n, \\
y_n &= (1 - \beta_n)T z_n + \beta_n T z_n, \\
x_{n+1} &= (1 - \gamma_n)T z_n + \gamma_n T y_n, \ n \in \mathbb{N},
\end{aligned}
$$

where $\{\alpha_n\}, \{\beta_n\}$ and $\{\gamma_n\}$ are real sequences in $(0, 1)$ with some conditions. They showed that this process converges faster than Abbas and Nazir's iteration process [1], S-iteration process [2], Mann iteration [16], Ishikawa iteration [12] and Noor iteration [17].

Inspired and motivated by above results, in this paper, we introduce modified multi-step procedure to find approximation fixed point of pairwise generalized nonexpansive mappings in CAT(0) spaces. We also prove both strong and Δ-convergence theorems for such a mapping under mild conditions.

2   Preliminaries

Let (X, d) be a metric space. A geodesic path joining x to y is a isometric map c from a closed interval $[0, l] \subset \mathbb{R}$ to X for $x, y \in X$ such that $c(0) = x$, $c(l) = y$ and $d(x, y) = l$. The image of c is called a *geodesic* (or metric) segment joining x and y denoted by $[x, y]$ whenever it is unique. The spaces (X, d) is said to be a (uniquely) geodesic space if every two points of X are joined by (exactly) one geodesic segment. A geodesic triangle $\Delta(x_1, x_2, x_3)$ in a geodesic space X consists of three points x_1, x_2, x_3 of X and three geodesic segments joining each pair of vertices. A comparison triangle of a geodesic triangle $\Delta(x_1, x_2, x_3)$ is the triangle $\bar{\Delta}(x_1, x_2, x_3) := \Delta(\bar{x}_1, \bar{x}_2, \bar{x}_3)$ in the Euclidean space $\mathbb{E}^2$ such that

$$
d(x_i, x_j) = d_{\mathbb{E}^2}(\bar{x}_i, \bar{x}_j), \text{ for } i, j = \{1, 2, 3\}.
$$

A geodesic space is called a *CAT(0) space*, if for each geodesic triangle $\Delta(x_1, x_2, x_3)$ in X and its comparison triangle $\bar{\Delta} := \Delta(\bar{x}_1, \bar{x}_2, \bar{x}_3)$ in $\mathbb{E}^2$, the CAT(0) inequality

$$
d(x, y) \leq d_{\mathbb{E}^2}(\bar{x}, \bar{y})
$$

for all $x, y \in \Delta$ and $\bar{x}, \bar{y} \in \bar{\Delta}$ holds.

A thorough discussions of these spaces are given in [4, 18].

Lemma 1 *[9]. Let (X, d) be a CAT(0) space.*
(I) For each $x, y \in X$ and $\alpha \in [0, 1]$, there exists a unique point $z \in [x, y]$ such that

$$
d(z, x) = \alpha d(x, y) \quad and \quad d(z, y) = (1 - \alpha)d(x, y).
$$

Note that $z = (1 - \alpha)x \oplus \alpha y$.
(II) For each $x, y, z \in X$ and $\alpha \in [0, 1]$, we have

$$
d((1 - \alpha)x \oplus \alpha y, z) \leq (1 - \alpha)d(x, z) + \alpha d(y, z) \tag{3}
$$

(III) For all $t \in [0,1]$ and $x, y, z \in X$,

$$d^2((1-t)x \oplus ty, z) \leq (1-t)d^2(x, z) + td^2(y, z) - t(1-t)d^2(x, y) \qquad (4)$$

The inequality (4) is called *(CN) inequality*. A geodesic space X is a CAT(0) space if and only if *(CN)* inequality [6] holds.

CAT(0) spaces have a remarkably nice geometric structure. One can see almost immediately from Lemma 1 that in such spaces angles exist satisfied geometric property and also the distance function is convex, one has both uniform convexity and orthogonal projection onto convex subsets, etc. Also, because of their generality, CAT(0) spaces arise in a wide variety of contexts. Some examples of CAT(0) spaces are pre-Hilbert spaces (see [4]), R-trees (see [13]), Euclidean buildings (see [5]), the complex Hilbert ball with a hyperbolic metric (see [11]), Hadamard manifolds (see [3]) and many others.

We now give the notion of Δ-convergence and list some of its basic properties. Let $\{x_n\}$ be a bounded sequence in a CAT(0) space X. For $z \in X$, we set

$$r(z, \{x_n\}) = \limsup_{n \to \infty} d(z, x_n).$$

The *asymptotic radius* $r(\{x_n\})$ of $\{x_n\}$ is given by

$$r(\{x_n\}) = \inf\{r(z, \{x_n\}) : z \in X\}.$$

The *asymptotic radius* $r_D(\{x_n\})$ of $\{x_n\}$ with respect to $D \subseteq X$ is given by

$$r_D(\{x_n\}) = \inf\{(z, \{x_n\}) : z \in D\}.$$

The *asymptotic center* $A(\{x_n\})$ of $\{x_n\}$ is the set

$$A(\{x_n\}) = \{z \in X : r(z, \{x_n\}) = r(\{x_n\})\}.$$

And the *asymptotic center* $A_D(\{x_n\})$ of $\{x_n\}$ with respect to $D \subseteq X$ is the set

$$A_D(\{x_n\}) = \{z \in D : r(z, \{x_n\}) = r(\{x_n\})\}.$$

It is well known [8] that $A(\{x_n\})$ consists of exactly one point in a framework of CAT(0) space. In 1976, Lim [15] introduced the concept of Δ-convergence in a general metric space. In 2008, Kirk and Panyanak [9] brought in Δ-convergence to CAT(0) spaces and proved that there is an analogy between Δ-convergence and weak convergence.

Definition 7 [14]. A sequence $\{x_n\}$ in a CAT(0) space X is said to $\Delta-$converge to $x \in X$ if x is the unique asymptotic center of $\{a_n\}$ for every subsequence $\{a_n\}$ of $\{x_n\}$. In this case, we write $\Delta - \lim_{n \to \infty} x_n = x$ and call x the Δ-limit of $\{x_n\}$

Lemma 2 [14]. *If C is a closed convex subset of a complete CAT(0) space and if $\{x_n\}$ is a bounded sequence in C, then the asymptotic center of $\{x_n\}$ is in C.*

Lemma 3 [14]. *Every bounded sequence in a complete CAT(0) space always has a Δ- convergent subsequence.*

Lemma 4 *[9]. If $\{x_n\}$ is a bounded sequence in a complete CAT(0) space with $A(\{x_n\}) = \{p\}$, $\{u_n\}$ is a subsequence of $\{x_n\}$ with $A(\{u_n\}) = \{u\}$, and the sequence $\{d(x_n, u)\}$ converges, then $p = u$.*

Lemma 5 *[7]. Let X be a CAT(0) space, $x \in X$ be a given point and $\{t_n\}$ be a sequence in $[a, b]$ with $a, b \in (0, 1)$. Let $\{x_n\}$ and $\{y_n\}$ be any sequence in X such that*

$$\limsup_{n \to \infty} d(x_n, x) \le c, \ \limsup_{n \to \infty} d(y_n, x) \le c,$$
$$and \ \lim_{n \to \infty} d((1 - t_n)x_n \oplus t_n y_n, x) = c$$

for some $c \ge 0$. Then $\lim_{n \to \infty} d(x_n, y_n) = 0$.

3   Main results

We now modify multi-step iteration process introduced by Thakur et al. [21] to CAT(0) spaces as follows:

Let C be a nonempty, closed and convex subset of a CAT(0) space (X, d). Let T, S be two mappings of C into C. For $x_1 \in C$, be a sequence $\{x_n\}$ be generated by

$$\begin{aligned} w_n &= (1 - \alpha_n)x_n \oplus \alpha_n S x_n, \\ y_n &= (1 - \beta_n)w_n \oplus \beta_n T w_n, \\ x_{n+1} &= (1 - \gamma_n)T w_n \oplus \gamma_n S y_n, \end{aligned} \quad (5)$$

for all $n \ge 1$, where α_n, β_n and γ_n are real sequences in $[a, b]$ for some $a, b \in (0, 1)$.

Lemma 6. *Let C be a nonempty closed and convex subset of a complete CAT(0) space X. Suppose that $T, S : C \to C$ are pairwise generalized nonexpansive mapping satisfying conditions (PC_λ) for some $\lambda \in (0, 1)$ and (PE_μ) for some $\mu \ge 1$. If $\{x_n\}$ is a sequence defined by (5), then the common fixed point set of T and S is nonempty if and only if $\lim_{n \to \infty} d(x_n, S x_n) = 0$ and $\lim_{n \to \infty} d(w_n, T w_n) = 0$.*

Proof. Suppose that the common fixed point set of T and S is nonempty. We will show that $\lim_{n \to \infty} d(x_n, S x_n) = 0$ and $\lim_{n \to \infty} d(w_n, T w_n) = 0$

(I) First we prove that $\lim_{n \to \infty} d(x_n, p)$ exists.
Taking arbitrary common fixed point p of T and S, we have

$$\begin{aligned} \lambda d(p, Tp) = 0 \le d(p, x_n), \\ \lambda d(p, Sp) = 0 \le d(p, x_n), \end{aligned}$$

for all $n \in \mathbb{N}$. From (1) we know that there exists some $\lambda \in (0, 1)$ such that

$$d(S x_n, p) \le d(x_n, p). \quad (6)$$

Since T and S are pairwise mappings satisfying condition (PE_μ), we know that there exists $\mu \ge 1$ such that

$$d(T w_n, p) \le \mu d(Sp, p) + d(w_n, p),$$

which implies that

$$d(Tw_n, p) \leq d(w_n, p), \tag{7}$$

and

$$d(p, Sy_n) \leq \mu d(Tp, p) + d(y_n, p),$$

which implies that

$$d(Sy_n, p) \leq d(y_n, p). \tag{8}$$

By Lemma 1 and from (6) and (7), we get

$$\begin{aligned}
d(w_n, p) &= d((1 - \alpha_n)x_n \oplus \alpha_n Sx_n, p) \\
&\leq (1 - \alpha_n)d(x_n, p) + \alpha_n d(Sx_n, p) \\
&\leq (1 - \alpha_n)d(x_n, p) + \alpha_n d(x_n, p) \\
&= d(x_n, p),
\end{aligned}$$

which implies that

$$d(w_n, p) \leq d(x_n, p). \tag{9}$$

And we have

$$\begin{aligned}
d(y_n, p) &= d((1 - \beta_n)w_n \oplus \beta_n Tw_n, p) \\
&\leq (1 - \beta_n)d(w_n, p) + \beta_n d(Tw_n, p) \\
&\leq (1 - \beta_n)d(w_n, p) + \beta_n d(w_n, p) \\
&= d(w_n, p),
\end{aligned}$$

which implies that

$$d(y_n, p) \leq d(x_n, p). \tag{10}$$

Again by Lemma 1 and from (9) and (10), we have

$$\begin{aligned}
d(x_{n+1}, p) &= d((1 - \gamma_n)Tw_n \oplus \gamma_n Sy_n, p) \\
&\leq (1 - \gamma_n)d(Tw_n, p) + \gamma_n d(Sy_n, p) \\
&\leq (1 - \gamma_n)d(w_n, p) + \gamma_n d(y_n, p) \\
&\leq (1 - \gamma_n)d(x_n, p) + \gamma_n d(x_n, p) \\
&= d(x_n, p),
\end{aligned}$$

that is

$$d(x_{n+1}, p) \leq d(x_n, p). \tag{11}$$

So $\{d(p, x_n)\}$ is bounded and decreasing for each common fixed point p of T and S. Therefore, $\lim_{n \to \infty} d(x_n, p)$ exists.

(II) Next we prove that

$$\lim_{n \to \infty} d(Sy_n, Tw_n) = 0.$$

Since $\lim_{n \to \infty} d(p, x_n)$ exists, we can assume that

$$\lim_{n \to \infty} d(x_n, p) = c \geqslant 0. \tag{12}$$

Since

$$\begin{aligned}
d(Sx_n, p) &\leq d(x_n, p), \\
d(Sy_n, p) &\leq d(x_n, p), \\
d(Tw_n, p) &\leq d(x_n, p).
\end{aligned}$$

Taking the limit supremum on both sides of the above inequalities, we get

$$\limsup_{n\to\infty} d(Sx_n, p) \leq c, \tag{13}$$

$$\limsup_{n\to\infty} d(Sy_n, p) \leq c, \tag{14}$$

$$\limsup_{n\to\infty} d(Tw_n, p) \leq c. \tag{15}$$

On the other hand, it follows from (5) and (12) that

$$\lim_{n\to\infty} d(x_{n+1}, p) = \lim_{n\to\infty} d((1-\gamma_n)Tw_n \oplus \gamma_n Sy_n, p) \leq c.$$

This implies that

$$\lim_{n\to\infty} d((1-\gamma_n)Tw_n \oplus \gamma_n Sy_n, p) = c. \tag{16}$$

From (14), (15), (16) and Lemma 5 we get

$$\lim_{n\to\infty} d(Sy_n, Tw_n) = 0. \tag{17}$$

(III) Next, we prove that

$$\lim_{n\to\infty} d(x_n, Sx_n) = 0 \quad and \quad \lim_{n\to\infty} d(w_n, Tw_n) = 0.$$

Since

$$\begin{aligned}
d(x_{n+1}, Sy_n) &= d((1-\gamma_n)Tw_n \oplus \gamma_n Sy_n, Sy_n) \\
&\leq (1-\gamma_n)d(Tw_n, Sy_n) + \gamma_n d(Sy_n, Sy_n) \\
&\leq (1-\gamma_n)d(Tw_n, Sy_n) \\
&\leq bd(Tw_n, Sy_n) \\
&\to 0
\end{aligned}$$

as $n \to \infty$, and

$$\begin{aligned}
d(x_{n+1}, Tw_n) &= d((1-\gamma_n)Tw_n \oplus \gamma_n Sy_n, Sy_n) \\
&\leq (1-\gamma_n)d(Tw_n, Tw_n) + \gamma_n d(Sy_n, Tw_n) \\
&\leq \gamma_n d(Tw_n, Sy_n) \\
&\leq bd(Tw_n, Sy_n) \\
&\to 0
\end{aligned}$$

as $n \to \infty$. So we have

$$\begin{aligned}
d(x_{n+1}, p) &= d(x_{n+1}, Tw_n) + d(Tw_n, p) \\
&\leq d(x_{n+1}, Tw_n) + d(w_n, p)
\end{aligned}$$

we can obtain that

$$c \leq \liminf_{n\to\infty} d(w_n, p).$$

On the other hand, from (9) and (12), we get

$$\limsup_{n\to\infty} d(w_n, p) \leq \limsup_{n\to\infty} d(x_n, p) = c.$$

Hence

$$c = \lim_{n \to \infty} d(w_n, p) = \lim_{n \to \infty} d((1 - \alpha_n)x_n \oplus \alpha_n Sx_n, p). \qquad (18)$$

From (12), (13), (18) and Lemma 5, we get

$$\lim_{n \to \infty} d(x_n, Sx_n) = 0, \qquad (19)$$

and also

$$
\begin{aligned}
d(w_n, Sx_n) &= d((1 - \alpha_n)x_n \oplus \alpha_n Sx_n, Sx_n) \\
&\leq (1 - \alpha_n)d(x_n, Sx_n) \\
&\to 0
\end{aligned}
\qquad (20)
$$

as $n \to \infty$. So we have

$$
\begin{aligned}
d(x_{n+1}, p) &= d(x_{n+1}, Sy_n) + d(Sy_n, p) \\
&\leq d(x_{n+1}, Sy_n) + d(y_n, p),
\end{aligned}
$$

implies that

$$c \leq \lim_{n \to \infty} \inf d(y_n, p).$$

On the other hand, from (9) and (12), we get

$$\limsup_{n \to \infty} d(y_n, p) \leq \limsup_{n \to \infty} d(x_n, p) = c.$$

Hence

$$c = \lim_{n \to \infty} d(y_n, p) = \lim_{n \to \infty} d((1 - \beta_n)w_n \oplus \beta_n Tw_n, p). \qquad (21)$$

From (12), (13), (18) and Lemma 5, we obtain that

$$\lim_{n \to \infty} d(w_n, Tw_n) = 0. \qquad (22)$$

Conversely, suppose that $\lim_{n \to \infty} d(x_n, Sx_n) = 0$ and $\lim_{n \to \infty} d(w_n, Tw_n) = 0$. We next show that the common fixed point set of T and S is nonempty. Since C is bounded, closed and convex, $\{x_n\}$ and $\{w_n\}$ are bounded. Let $A(\{x_n\}) = x$ and $A(\{w_n\}) = w$. Then, from Lemma 2, we know that $x, w \in C$

(IV) Next, we prove that $x = w$.
It follows from (19) and (20), we get

$$
\begin{aligned}
d(x_n, w_n) &\leq d(x_n, Sx_n) + d(Sx_n, w_n) \\
&\to 0
\end{aligned}
\qquad (23)
$$

as $n \to \infty$.
If $x \neq w$, then by the uniqueness of asymptotic centers,

$$
\begin{aligned}
\limsup_{n \to \infty} d(x_n, x) &< \limsup_{n \to \infty} d(x_n, w) \\
&\leq \limsup_{n \to \infty} [d(x_n, w_n) + d(w_n, w)] \\
&= \limsup_{n \to \infty} d(w_n, w) \\
&< \limsup_{n \to \infty} d(w_n, x) \\
&\leq \limsup_{n \to \infty} [d(x_n, w_n) + d(x_n, x)] \\
&= \limsup_{n \to \infty} d(x_n, x),
\end{aligned}
$$

which is a contradiction. That means,

$$x = w. \tag{24}$$

(V) Finally, we prove that x is a common fixed point of T and S.
Since T and T are pairwise mappings satisfying condition (PE_μ), from (2) we know that there exists $\mu \geqslant 1$ such that

$$d(x_n, Tx) \leq \mu d(x_n, Sx_n) + d(x_n, x),$$
$$d(w_n, Sx) \leq \mu d(w_n, Tw_n) + d(w_n, x).$$

Taking the limit supremum on both sides of the above inequalities, we get

$$\limsup_{n \to \infty} d(x_n, Tx) \leq \limsup_{n \to \infty} d(x_n, x),$$
$$\limsup_{n \to \infty} d(w_n, Sx) \leq \limsup_{n \to \infty} d(w_n, x).$$

By the uniqueness of asymptotic centers, we have $Tx = x = Sx$. Therefore, x is a common fixed point of T and S.

Next, we ready to prove Δ- convergence for pairwise mappings satisfying condition (PC_λ) and (PE_μ).

Theorem 1. *Let C be a nonempty closed and convex subset of a complete $CAT(0)$ space X. Suppose that $T, S : C \to C$ are pairwise generalized nonexpansive mapping satisfying conditions (PC_λ) for some $\lambda \in (0,1)$ and (PE_μ) for some $\mu \geq 1$ on C with a nonempty common fixed point set. Then the sequence $\{x_n\}$ defined by (5) Δ-convergent to a common fixed point of T and S.*

Proof. Since $\{x_n\}$ is bounded, from Lemma 3 we know that $\{x_n\}$ has a Δ- convergent subsequence. We now prove that every Δ-convergent subsequence of $\{x_n\}$ has unique Δ-limit in common fixed point set of T and S. For this, let a be a Δ-limit of the subsequence $\{a_n\}$ of $\{x_n\}$. We claim that a is a common fixed point of T and S. From (19) and (22) in the proof of Lemma 6, we know that there exists a sequence $\{b_n\}$ such that

$$\lim_{n \to \infty} d(a_n, Sa_n) = 0 \text{ and } \lim_{n \to \infty} d(b_n, Tb_n) = 0.$$

As T and S are pairwise generalized nonexpansive mappings satisfying condition (PE_μ), there exists $\mu \geq 1$ such that

$$d(a_n, Ta) \leq \mu d(a_n, Sa_n) + d(a_n, a),$$
$$d(b_n, Sa) \leq \mu d(b_n, Tb_n) + d(b_n, a).$$

Taking the limit supremum on both sides of the above inequalities, we get

$$\limsup_{n \to \infty} d(a_n, Ta) \leq \limsup_{n \to \infty} d(a_n, Sa_n) + d(a_n, a),$$
$$\limsup_{n \to \infty} d(b_n, Sa) \leq \limsup_{n \to \infty} d(b_n, Tb_n) + d(b_n, a).$$

By the uniqueness of asymptotic centers, we get $Ta = a = Sa$. Thus, a is a common fixed point of T and S. Assume that a' is a Δ-limit of the subsequence $\{a'_n\}$ of $\{x_n\}$. Similarly, we can prove that $Ta' = a' = Sa'$. Next we will show that $a = a'$. If $a \neq a'$, by the uniqueness of asymptotic centers,

$$
\begin{aligned}
\limsup_{n\to\infty} d(x_n, a) &= \limsup_{n\to\infty} d(a_n, a) \\
&< \limsup_{n\to\infty} d(a_n, a') \\
&= \limsup_{n\to\infty} d(x_n, a') \\
&= \limsup_{n\to\infty} d(a'_n, a') \\
&< \limsup_{n\to\infty} d(a'_n, a) \\
&= \limsup_{n\to\infty} d(x_n, a),
\end{aligned}
$$

which is a contradiction. Hence, $a = a'$.

Finally, under mild conditions, we prove a strong convergence theorem for pairwise generalized nonexpansive mappings satisfying conditions (PC_λ) and (PE_μ).

Theorem 2. *Let C be a nonempty closed and convex subset of a complete $CAT(0)$ space (X, d), and suppose also that C is a compact subset of X. Suppose that $T, S : C \to C$ are pairwise generalized nonexpansive mappings satisfying condition (PC_λ) for some $\lambda \in (0, 1)$ and (PE_μ) for some $\mu \geq 1$ on C with a nonempty common fixed point set. Then, the sequence $\{x_n\}$ defined by (5) converges strongly to a common fixed point of T and S.*

Proof. From (19) and (22) in the proof of Lemma 6, we have

$$
\lim_{n\to\infty} d(x_n, Sx_n) = 0 \quad and \quad \lim_{n\to\infty} d(w_n, Tw_n) = 0.
$$

Since C is compact, there exists a subsequence $\{x_{n_k}\}$ of $\{x_n\}$ such that $x_{n_k} \to p$ for some $p \in C$. From (23) in the proof of Lemma 6, we know that there exists a subsequence $\{w_{n_k}\}$ of $\{w_n\}$ such that $d(x_{n_k}, w_{n_k}) \to 0$ as $k \to \infty$. This implies that $w_{n_k} \to p$. As T and S are pairwise mappings satisfying condition (PE_μ), there exists $\mu \geq 1$ such that

$$
\begin{aligned}
d(x_{n_k}, Tp) &\leq \mu d(x_{n_k}, Sx_{n_k}) + d(x_{n_k}, p), \\
d(w_{n_k}, Sp) &\leq \mu d(w_{n_k}, Tw_{n_k}) + d(w_{n_k}, p).
\end{aligned}
$$

for all $k \in \mathbb{N}$. Setting $k \to \infty$, we conclude that $\{x_{n_k}\}$ converges to Tp and $\{w_{n_k}\}$ converges to Sp. This implies that $Tp = p = Sp$, that is, p is a common fixed point of T and S. It follows from (11) in the proof of Lemma 6, we have $\lim_{n\to\infty} d(x_n, p)$ exists, therefore p is the strong limit of the sequence $\{x_n\}$ into itself.

Recall that self-mapping T, S are said to satisfy condition (I) [19] if there exists a nondecreasing function $f : [0, \infty) \to [0, \infty)$ with $f(0) = 0$ and $f(r) > 0$

for all $r > 0$ such that $d(x, Sx) \geq f(d(x, F(S)))$ and $d(w, Tw) \geq f(d(w, F(T)))$ for all $x, w \in C$, where $d(x, F(S)) = \inf_{p \in F(S)} d(x, p)$ and $d(w, F(T)) = \inf_{p \in F(T)} d(w, p)$.

Theorem 3. *Let C be a nonempty closed and convex subset of a complete $CAT(0)$ space (X, d). Suppose that $T, S : C \to C$ are pairwise generalized nonexpansive mapping satisfying conditions (PC_λ) for some $\lambda \in (0, 1)$ and (PE_μ) for some $\mu \geq 1$ on C with a nonempty common fixed point set. If T and S satisfies condition (I) then, the sequence $\{x_n\}$ defined by (5) converges strongly to a common fixed point of T and S.*

Proof. By T and S satisfy condition (I), we have

$$f(d(x_n, F(S))) \leq d(x_n, Sx_n),$$
$$f(d(w_n, F(T))) \leq d(w_n, Tw_n),$$

for all $n \in \mathbb{N}$. It follows from (19) and (22), that

$$\lim_{n \to \infty} d(x_n, F(S)) = \lim_{n \to \infty} d(w_n, F(T)) = 0.$$

Similarly, by Lemma 6 we know that $\lim_{n \to \infty} d(x_n, w_n) = 0$. Consequently, it follows from (24) we have $x = w$.

We can choose a subsequence $\{x_{n_k}\}$ of $\{x_n\}$ and sequence $\{p_k\}$ in $F(S) \cap F(T)$ such that

$$d(x_{n_k}, p_k) < \frac{1}{3^k} \tag{25}$$

for all $k \in \mathbb{N}$.

It follows from (11), that

$$d(x_{n_{k+1}}, p_k) \leq d(x_{n_k}, p_k) < \frac{1}{3^k}.$$

So

$$
\begin{aligned}
d(p_{k+1}, p_k) &\leq d(p_{k+1}, x_{n_{k+1}}) + d(x_{n_{k+1}}, p_k) \\
&\leq \frac{1}{3^{k+1}} + \frac{1}{3^k} \\
&< \frac{1}{3^{k-1}} \\
&\to 0
\end{aligned}
$$

as $k \to \infty$.

This shows that $\{p_k\}$ is a Cauchy sequence in $F(S) \cap F(T)$. Since $F(S) \cap F(T)$ is closed in X, there exists a point $p \in F(S) \cap F(T)$ such that $\lim_{k \to \infty} p_k = p$. It follows from (23) and (25) that $\lim_{k \to \infty} x_{n_k} = \lim_{k \to \infty} w_{n_k} = p$. By Lemma 6 $\lim_{k \to \infty} d(x_n, p)$ exists, we ensures be the case that $\lim_{k \to \infty} d(x_n, p) = 0$. Therefore we obtain the desired result.

Acknowledgements. The first author would like to thank the Research Professional Development Project Under the Science Achievement Scholarship of Thailand (SAST) for financial support. Also, this project was supported by the Theoretical and Computation Science (TaCS) Center under Computational and Applied Science for Smart Innovation Cluster (CLASSIC), Faculty of Science, KMUTT.

References

1. Abbas, M., Nazir, T.: A new faster iteration process applied to constrained minimization and feasibility problems. Mat. Vesnik **66**(2), 223–234 (2014)
2. Agarwal, R.P., O'Regan, D., Sahu, D.R.: Iterative construction of fixed points of nearly asymptotically nonexpansive mappings. J. Nonlinear Convex Anal. **8**(1), 61–79 (2007)
3. Borisenko, A.A., Gallego, E., Reventós, A.: Relation between area and volume for λ-convex sets in Hadamard manifolds. Differential Geom. Appl. **14**(3), 267–280 (2001)
4. Bridson, M.R., Haefliger, A.: Metric spaces of non-positive curvature. Grundlehren der Mathematischen Wissenschaften [Fundamental Principles of Mathematical Sciences], vol. 319. Springer, Heidelberg (1999)
5. Brown, K.S.: Buildings. Springer, New York (1989)
6. Bruhat, F., Tits, J.: Groupes réductifs sur un corps local. Inst. Hautes Études Sci. Publ. Math. **41**, 5–251 (1972)
7. Chang, S.S., Wang, L., Lee, H.W.J., Chan, C.K., Yang, L.: Demiclosed principle and Δ-convergence theorems for total asymptotically nonexpansive mappings in CAT(0) spaces. Appl. Math. Comput. **219**(5), 2611–2617 (2012)
8. Dhompongsa, S., Kirk, W.A., Sims, B.: Fixed points of uniformly Lipschitzian mappings. Nonlinear Anal. **65**(4), 762–772 (2006)
9. Dhompongsa, S., Panyanak, B.: On Δ-convergence theorems in CAT(0) spaces. Comput. Math. Appl. **56**(10), 2572–2579 (2008)
10. García Falset, J., Llorens-Fuster, E., Suzuki, T.: Fixed point theory for a class of generalized nonexpansive mappings. J. Math. Anal. Appl. **375**(1), 185–195 (2011)
11. Goebel, K., Reich, S.: Uniform convexity, hyperbolic geometry, and nonexpansive mappings. Monographs and Textbooks in Pure and Applied Mathematics, vol. 83. Marcel Dekker Inc., New York (1984)
12. Ishikawa, S.: Fixed points by a new iteration method. Proc. Amer. Math. Soc. **44**, 147–150 (1974)
13. Kirk, W.A.: Fixed point theorems in CAT(0) spaces and R-trees. Fixed Point Theor. Appl. **2004**(4), 309–316 (2004)
14. Kirk, W.A., Panyanak, B.: A concept of convergence in geodesic spaces. Nonlinear Anal. **68**(12), 3689–3696 (2008)
15. Lim, T.C.: Remarks on some fixed point theorems. Proc. Amer. Math. Soc. **60**, 179–182 (1977). 1976
16. Mann, W.R.: Mean value methods in iteration. Proc. Amer. Math. Soc. **4**, 506–510 (1953)
17. Noor, M.A.: New approximation schemes for general variational inequalities. J. Math. Anal. Appl. **251**(1), 217–229 (2000)
18. Pakkaranang, N., Ngiamsunthorn, P.S., Kumam, P., Cho, Y.J.: Convergence theorems of the modified S-type iterative method for (α, β)-generalized hybrid mappings in CAT(0) spaces. J. Math. Anal. **8**(1), 103–112 (2017)
19. Senter, H.F., Dotson Jr., W.G.: Approximating fixed points of nonexpansive mappings. Proc. Amer. Math. Soc. **44**, 375–380 (1974)
20. Suzuki, T.: Fixed point theorems and convergence theorems for some generalized nonexpansive mappings. J. Math. Anal. Appl. **340**(2), 1088–1095 (2008)
21. Thakur, B.S., Thakur, D., Postolache, M.: A new iteration scheme for approximating fixed points of nonexpansive mappings. Filomat **30**(10), 2711–2720 (2016)
22. Zhang, S., Zhou, J., Ciu, Y.: Common fixed point properties for pairwise generalized nonexpansive type mappings in geodesic spaces. J. Nonlinear Sci. Appl. (preprint)

Applications

Interbank Contagion: An Agent-Based Model for Vietnam Banking System

Anh T. M. Vu$^{(\boxtimes)}$, Thong P. Le, Thanh D. X. Duong, and Tai T. Nguyen

John von Neumann Institute, Vietnam National University,
Ho Chi Minh City 70000, Vietnam
anhmaivu88@gmail.com, lpthong90@gmail.com,
{thanh.duong,tai.nguyen.qcf}@jvn.edu.vn

Abstract. Network connectivity and credit contagion have drawn a great concern after financial crises with collapses of too-big-to-fail institutions and their consequences. The matter in question here is the impact and mechanism of contagion, which means how collapses of one or several institutions can trigger subsequent failures and in turn affect the whole system. This article proposes an agent-based approach to construct an interactive inter-bank system simulating the decisions of 19 Vietnamese banks and their balance sheets. A stress-testing mechanism is also provided to test the effects of idiosyncratic and systemic shocks of different magnitudes on the system. Initial results suggest that while idiosyncratic shocks don't substantially damage the banking network, systemic impairment could devastate the system, particularly in case it stimulates contagion of bank defaults.

Keywords: Agent-based · Credit risk · Contagion
Inter-bank network · Vietnam banking

1 Introduction

Agent-based method is a rather new approach in economics and finance, as we reach the point where computation capacity and behavioral economics have developed strongly enough to be used in practical applications. This method has a clear advantage in building better models to represent our realistic, finite, imperfect decision-making human-beings society, since agents in such models are not required to behave perfectly rationally as in traditional research. Better insight of fundamental patterns in the market is achieved by observing results of interaction between numerous agents.

Related to inter-bank system, there are a few studies using an agent-based approach. However, most of studies of Vietnamese banking network are qualitative and have not provided interaction mechanism. This article, therefore, aims to offer an advanced approach which combines financial knowledge, mathematical modeling and computation to build a full-scale agent-based model (ABM) to represent Vietnam's banking system. The network is comprised of 19 banks

© Springer International Publishing AG 2018
L. H. Anh et al. (eds.), *Econometrics for Financial Applications*, Studies in Computational Intelligence 760, https://doi.org/10.1007/978-3-319-73150-6_32

whose available data is adequate to construct such a model. Rules for lending and borrowing between banks are agent-driven and based on banks' historical balance sheets and general behavior pattern from practical findings.

Unfortunately, due to data insufficiency, it is not the intention of this article to construct a real Vietnamese market, but it is to make a tool to simulate and study the development of inter-bank network under various circumstances, to gain insight into the questions: how different kinds of shock affect the whole system? More specifically, how default of a particular bank affects other banks and the system; how default of multiple banks affects the system; and how system shock affects the system? Whether the topology of banking system affects its resilience and how?

The first contribution of this article is to introduce the mechanism of inter-bank activities based on banks' balance sheets, in which banks can form new relationship and reform the network. Several approaches to build the full network are also provided. The article makes a second contribution by testing the network under a number of shocks to examine its resilience and role of inter-bank network on contagion effect. The third contribution is to provide a tool for the regulators to interact with the model, change parameters and test effects of new policies.

The article is structured as follows. Section 2 reviews current literature related to inter-bank network, associated risks, contagions, and agent based approach. Section 3 discusses methods to extrapolate inter-bank matrix - one of the most important inputs of the model from raw data. Section 4 provides model experiments. Section 5 presents the results. Finally, the article concludes in Sect. 6 by assessing the results and the methodology's contribution.

2   Literature Review

This section provides overview on three key aspects of modeling interconnections in a banking system: (1) credit risk and contagion (2) inter-bank network structure (3) agent-based modeling applied in inter-bank market.

2.1   Credit Risk and Contagion

Credit risk can be defined as a risk of changes in the value that is associated with unexpected changes in the credit quality of other counter-parties; see, e.g., [35]. Banks are particularly vulnerable to this kind of risk, as their main function is to receive money and make loans to other parties.

Two approaches that are often applied in early studies are structural and reduced-form models. The former explicitly modeled assets of banks as dynamic time series and assumed that default occurs when asset value decreased below a threshold [32]. The latter modeled a company's time to default as a stochastic process with parameters acquired from historical data.

However, despite their complexity, these models were criticized for not being able to predict financial turmoils. A major critique was that they failed to evaluate the market as a complex system with large-number of heterogeneous entities interacting and affecting each other. This is particularly true for inter-bank market where banks interlace each others through links created from their inter-bank exposures.

Research on spread of contagion (see, e.g., [1,2,5,25]) had identified the role of connectivity and network topology in systemic risk. Increasing connectivity made network less affected by systemic risk thanks to risk sharing mechanism. However, too high connectivity can lead to a crisis especially in the presence of high magnitude shocks as the trigger. In this case, linkages between banks became propagation channels for contagion and a source of systemic risk. This could in turn significantly froze the economy, as firms (even firms with high credibility) could not obtain capital to finance their projects. Currently, there are not many studies on the formation of banking network, as well as changes of this network after shocks like occurrences of defaults or introduction of new policies happen. Fortunately this trend is attracting more researchers and some progress can be seen in this direction. Some early research focused on agents' decisions to balance costs and benefits of maintaining and forming links with other agents (see, e.g., [4]). More recent works have paid attention to the impact of network formation to systemic risk (see, e.g., [1]), whereas [21] used features of agent trading decisions to form a network.

2.2 Inter-bank Network Structure

The first challenge to be overcome before testing the resilience of the network using agent-based approach is to reproduce full inter-bank exposures. Contagions and system risks spread through the inter-bank network, which is represented by lending and borrowing matrix of all banks in the network. This matrix is unobserved since banks do not public detailed information of inter-bank loans. Available data is just the total of inter-bank lending and borrowing of a particular bank in the network. Therefore, several methods to estimate the network has been developed, that can be named: Maximum Entropy (ME) method, Minimum Density (MD) method and agent-based simulation method.

The Maximum Entropy is the most popular one, which was based on the assumption that banks diversified their exposures by spreading borrowing and lending across all other banks in the network; see, e.g., [38]. This approach created a complete dense network; see, e.g., [11].

The Minimum Density adopted an opposite approach in that the network is built minimizing the number of linkages in while satisfying other constraints; see, e.g., [3]. It was based on the real fact that banks tended to lend or borrow other banks it had relationship with. This was to minimize the cost of inter-bank linkage. This approach created a sparse network.

Other approaches tried to create a mechanism to simulate banks' business and assumed banks follow rules, such as determining their lending-borrowing structure by maximizing profit and minimizing risk associated. For example, [10] offered Nash equilibrium of networks maximizing the investors' payoff which depended on the inter-bank neighborhood of banks in which an investor allocated their fund. [22] proposed a portfolio optimization model, whereby banks allocated their inter-bank exposures while balancing the return and risk of counterparty default risk, from that new links were created and networks were formed in equilibrium.

2.3 Agent-Based Approach for Inter-bank Network

Agent-based modelling (ABM) is defined as a simulation framework which comprised autonomous agents with interacting behaviors, connections between agents, and an exogenous environment; see, e.g., [31]. ABM is a strong tool to replicate real social phenomena, to represent adaptive behaviors and information diffusion of agents.

Recently, agent-based method has become a new trend in financial research, thanks to plenty of advantages and development in computation capacity and behavioral economics. Comparing to traditional network theoretic-based models, agent based modelling promises flexibility and enhances fidelity to observed data.

In case of inter-bank network, ABMs was utilized to analyze contagion risk through inter-bank channels; see, e.g., [19,29]. Moreover, different kinds of inter-bank loans (short-term, long-term and overnight) were also considered in extended research; see, e.g., [40]. Furthermore, [28] studied a multi-layer relationship which represented three different kinds of dependencies (long-term, short-term exposures and sharing of common asset portfolio) among banks. For other approaches and results, see [6,8,14,15,18,20,23,27,30,36] and references therein.

3   Inter-bank Network Extrapolation

3.1 Inter-bank Network Problem

The inter-bank network consists of N banks. The matrix $X \in [0,\infty)^{N \times N}$ represents gross inter-bank positions, where x_{ij} represents the loan bank i lending to bank j. For each bank i, the row sum of X shows total inter-bank assets, while the column sum of X shows total inter-bank liabilities.

- Total inter-bank assets: $A_i = \sum_{j=1}^{N} x_{ij}$.
- Total inter-bank liabilities: $L_i = \sum_{j=1}^{N} x_{ji}$.

The value of x_{ij} is unobservable since banks do not public this kind of information. However, we have values of A_i and L_i from banks' balance sheets. Since each value of bilateral positions is crucial for further modeling, we need to resort to a method to fill the inter-bank matrix, given the marginals A_i, L_i.

3.2 Maximum Entropy Approach

The first and most popular approach is Maximum Entropy; see, e.g., [17,37]. The fundamental assumption of this approach is that we don't have any information, so the matrix is filled in the way of "spreading exposures as evenly as possible given the assets and liabilities reported in the balance sheets of all other banks". The rationale is that when there is no information about parameters, uniform distribution is used. Such assumption ensures that no information which would affect the estimates is offered. Bilateral exposures x_{ij} minimizing the relative entropy function subjects to constraints of total assets and liabilities is the solution of this approach. Translated to the current setting, maximizing entropy of the matrix X means no any structure is imposed besides the information in the balance sheets.

Maximizing entropy of a matrix was first applied to contagion issue by [34]. It was then developed by [16,38] to solve the zeros diagonal of the matrix.

The method works as follows:

1. Finding matrix X as the solution of maximizing entropy without the constraint of zero diagonal. a_i and l_i with appropriate standardization can be consider as realizations of marginal distributions $f(a)$ and $f(b)$ and x_{ij} is realization of joint distribution. If $f(a)$ and $f(b)$ are independent, then $x_{ij} = a_i l_j$.

   However, this can lead to non-zero diagonal, which is not consistent with the fact that bank does not lend to itself.

2. Finding X^*, the most similar matrix to X and satisfy the zero-diagonal constraint. First, we must modify the independence assumption by setting $x_{ij} = 0$ for $i = j$. Then we need to minimize the relative entropy of a matrix X^* with respect to the previous maximum entropy matrix X:

$$\min_{x*} x^{*'} \ln \frac{x^*}{x} = \min \sum_{k=1}^{N^2-N} x_k^* \ln \frac{x_k^*}{x_k}, \tag{1}$$

such that $x \geq 0$ and $Ax^* = [a', l]'$,

where x^* and x are $(N^2 - N) \times 1$ vectors containing the off-diagonal elements of X^* and X, A is a matrix containing the adding-up restrictions $a_i = \sum_j x_{ij}$ and $l_j = \sum_i x_{ij}$.

The RAS algorithm can be used to solve the problem, since the objective function is strictly concave; see, e.g., [38].

The Maximum entropy approach is criticized for delivering an unrealistic network structure. The created network is a complete network in which linkages of banks are dense. Such a network tends to underestimate contagion in stress tests; see, e.g., [2]. Simulations show that these networks may reduce the probability of crisis, but may also raise its severity when a crisis happens; see, e.g., [33].

3.3   Minimum Density Approach

Minimum Density Approach is based on the premise of high cost in establishing and maintaining network linkages. These costs includes information processing and hedging against credit risks. Actually, banks often lend and borrow banks having relationship with them previously and inter-bank network is sparse in reality. [7,12] show that the number of active linkages is less than 1% of potential bilateral linkages. The network is also disassortative (or negatively assortative) in the sense that agents with dissimilar characteristics tend to have relationship. In the case of an inter-bank network, less-connected banks are more likely to trade with well-connected banks than with other less-connected banks; see, e.g., [7, 26].

The Minimum Density adopts an opposite approach to Maximum Entropy in that it determines a pattern of linkages for allocating inter-bank positions that is efficient in minimizing these costs; see, e.g., [3]. This means the matrix X should minimize the total number of linkages necessary for allocating inter-bank positions, subject to the total lending and borrowing constraints:

$$\min_{X} c \sum_{i=1}^{N} \sum_{j=1}^{N} \mathbf{1}_{[x_{ij}>0]}, \tag{2}$$

such that

$$\sum_{j=1}^{N} x_{ij} = A_i \forall i = 1, 2, ...N,$$

$$\sum_{i=1}^{N} x_{ij} = L_j \forall j = 1, 2, ...N,$$

$$x_{ij} \geq 0 \ \forall i, j.$$

This problem is very similar to the optimal network design problem in transportation and communication network. They are known to be non-deterministic polynomial-time hard except in very special cases; see, e.g., [9]. To create a more realistic network, in this paper, we will loosen some constraints of the problem, as well as adding elements in the objective function.

First, we soften the constrains by assigning penalties for deviations from the marginals:

$$AD_i = (A_i - \sum_{j} x_{ij}), \tag{3}$$

$$LD_i = (L_i - \sum_{j} x_{ji}). \tag{4}$$

LD_i measures bank i's current deficit. Defining these deviations helps to make the optimization problem smooth, so that the only non-smooth part lies in the cost of links; see, e.g., [3]. We want to create the sparse network, but also want to minimize the deviations from marginals. Thus, the objective function is to maximize

$$V(X) = -c \sum_{i=1}^{N} \sum_{j=1}^{N} \mathbf{1}_{[x_{ij}>0]} - \sum_{i=1}^{N} [\alpha_i AD_i^2 + \sigma_i LD_i^2]. \tag{5}$$

We also want to cover the disassortative feature of the network. This feature reflects the reality that small banks tend to seek relationship (lending or borrowing) with larger banks. The probability that bank i lends to bank j increases if either bank

i or bank j is a small bank and the other is a large one. We denote this feature by Q_{ij}:

$$Q_{ij} = \max\left(\frac{AD_i}{LD_j}, \frac{LD_i}{AD_j}\right). \qquad (6)$$

Sparse network will have high value of $V(X)$, while the usage of Q_{ij} ensure relationships between large and small banks. [39] proposed a method to cover this trade-off. Defining $P(X)$ as the probability distribution over network configurations, our target now is to maximize the objective function as sum of two terms. The first term favors sparse network and less deviation from marginals. The second term is to create a disassortative network in which small and large banks tend to make relationship. The optimization problem now reads

$$\max_{P} \sum_{X} P(X)V(X) + \theta R(P\|Q), \qquad (7)$$

in which $P(X)$ is the probability distribution over network configurations, θ is a scaling parameter, and R is the relative entropy between Q and the optimal distribution. We note that relative entropy measures the 'distance' between two probability distributions. The interpretation here is that we guess Q as the true probability distribution, but only trust it partly (reflected by the value of θ). With this belief, we consider many other plausible probabilities P with plausibility diminishing proportionally to their 'distance' from Q. [24] provides the solution, and the first-order conditions is

$$P(X) \propto Q(X)e^{\theta V(X)}. \qquad (8)$$

If all network configurations can be obtained and sorted based on their value of $P(X)$, we can get the optimal solution. However, due to complexity, heuristic approaches are usually applied. [3] provides a heuristic procedure to allocate links.

Result of this approach is a sparse network. This result and the maximum entropy result are at opposite extremes of the spectrum, which is far from reality. However, with the introduction of the parameter $\lambda < 1$, we could scale the loadings for selected links as $x_{ij} = \lambda \min(AD_i, LD_j)$. This will produce the effect of creating more linkages between banks. The advantage here is that we are free to choose the parameter λ and create networks of different structures.

3.4   Network Analysis

This section provides several criteria to evaluate the connectivity of a network, so that we can analyze the inter-bank networks obtained from the above methods. In-degree ($d_{in}(i)$) of node i represents the number of links terminating at i, and out-degree ($d_{out}(i)$) represents the number of links originating from i.

3.4.1   Network Density

The density of a graph is the ratio of the number of edges and the number of possible edges. For directed graphs, the formula for graph density is

$$D = \frac{|E|}{|n|(|n| - 1)},\tag{9}$$

where E is the number of edges and n is the number of nodes.

3.4.2   Network Centrality

Centrality is the fundamental concept to measure a network's connectivity. Here, we will present four kinds of centrality to evaluate the importance of a bank in the network, as well as the level of connectivity of a particular network.

- **Degree centrality**
  Degree centrality of a node is its degree. The relative degree centrality of node (bank) i is

$$C_D(i) = \frac{c_D(i)}{(n - 1)}.\tag{10}$$

  Degree centralization of a network is the variation in the degrees of nodes divided by the maximum degree variation which is possible in a network of the same size (see, e.g., [13]):

$$C_D(G) = \frac{\sum_{i=1}^{n}[c_D(n^*) - c_D(n_i)]}{n^2 - 3n + 2},\tag{11}$$

  where $c_D(n^*)$ is the maximum value in the network.

- **Closeness centrality**
  Closeness centrality of a node measures how close a bank is to other banks.

$$c_C(i) = \frac{1}{\sum_{j \in G} d(i, j)},\tag{12}$$

  in which $d(i.j)$ is the network theoretic distance between banks i and j and G is the set of all banks in the network. Relative closeness centrality is defined as

$$C_C(i) = (n - 1)c_C(i).\tag{13}$$

  Closeness centralizaton of a network is the variation in the closeness centrality of nodes divided by the maximum variation in closeness centrality scores possible in a network of the same size.

$$C_C(G) = \frac{\sum_{i=1}^{n}[C_C(n^*) - C_C(n_i)]}{(n^2 - 3n + 2)(2n - 3)}.\tag{14}$$

- **Betweeness centrality**
  The betweenness centrality of a bank reflects the amount of control that this bank exerts over the interactions of other banks in the network. Betweenness value of node i with respect to node pair j, k is the ratio:

$$b_{jk}(i) = \frac{g_{jk}(i)}{g_{jk}}, \tag{15}$$

in which g_{jk} is the number of j, k shortest paths, and $g_{jk}(i)$ is the number of j, k shortest paths that contains i.
Betweenness centrality of a node i is

$$c_B(i) = \sum_{k=1}^{n} \sum_{j=1}^{k-1} b_{jk}(i). \tag{16}$$

Relative betweeness centrality of node i is

$$C_B(i) = \frac{2c_B(i)}{n^2 - 3n + 2}. \tag{17}$$

Betweeness centrality of a network is

$$C_B(G) = \frac{\sum_{i=1}^{n} [C_B(n^*) - C_B(n_i)]}{n - 1}. \tag{18}$$

- **Eigen-vector Centrality**
  Eigen-vector Centrality can be said to be enhancement of degree centrality. In-degree centrality counts every link a node participate. However, not all nodes are relevant, some nodes are more important than others, so links from a more relevant node should be awarded. The importance of a node depends on the value of its eigenvector centrality.
  Let $A = (a_{i,j})$ be the adjacency matrix of a graph, i.e., $a_{i,j} = 1$ if node i is linked to node j and $a_{i,j} = 0$ otherwise, then the eigenvector centrality of node i is:

$$x(i) = \frac{1}{\lambda} \sum_{j \in G} a_{i,j} \, x(j), \tag{19}$$

where $\lambda \neq 0$ is a constant.
This can be written in the form of an eigenvector equation as follows

$$Ax = \lambda x. \tag{20}$$

4   Model

4.1   Network and Banks

A network is a simple simulation of a banking system and comprises by only one kind of agent, which is bank. It is assumed that a network is comprised of N banks.

4.1.1   Bank Agent and Attributes

There are 3 kinds of attributes associated with bank agent:

- Balance sheet attributes: cash, inter-bank lending, and external asset comprising Bank's asset; deposit, inter-bank borrowing, equity comprising Bank's liability (see Table 1) in which:
    - Cash, external asset, deposit and equity are numbers, reflecting amount of money in VND.
    - Inter-bank lending IB_i^l of bank i is a vector containing lending of a bank i to other banks, i.e., $IB_i^l = [X_{i1}, X_{i2}, ..., X_{iN}]$.
    - Inter-bank borrowing IB_i^b of bank i is a vector containing borrowing of a bank i from other banks, i.e., $IB_i^b = [X_{1i}, X_{2i}, ..., X_{Ni}]$.

Table 1. Bank's balance sheet

Total Asset (A)	Total Liability (L)
Cash (C)	Deposit (D)
Inter-bank Lending (IB^l)	Inter-bank Borrowing (IB^b)
External Asset (EA)	Equity (E)

- Performance parameter features, which includes:
    - State, which can be one of the values: activating, bankrupt, large, small;
    - Deposit growth rate of a bank, which is assumed to follow log-normal distribution, with mean and standard variation based on historical data;
    - Return on Equity (ROE) of a bank, which is assumed to follow normal distribution with mean and standard variation based on historical data; and
    - Ratio between short-term inter-bank lending (borrowing) and total inter-bank lending (borrowing) are assumed to be fixed for simplicity's sake.
- Decision parameter features. Banks decide their decision on lending/borrowing by setting targets of lending, borrowing which is a ratio on total asset, which includes:
    - Ratio between deposit and total liability,
    - Ratio between total inter-bank borrowing and total liability,
    - Ratio between external asset and total asset, and
    - Ratio between total inter-bank lending and total asset.

4.2   Network and Attributes

- Number of agents includes: total number of banks, number of bankrupt banks, number of affected banks, and numbers of large and small banks.
- Network asset features:
    - Network total Assets, in million VND,
    - Defaulted Assets in a particular period, in million VND, and
    - Proportion of defaulted asset over total assets in a particular period, in %.

- Network equity features:
  - Network total equity,
  - Value of defaulted equity, in million VND, and
  - Proportion of defaulted equity over network total equity, in %.
- Network inter-bank matrix: the matrix X of inter-bank lending and borrowing. From this network, we can calculate related coefficients, including:
  - Degree Centrality C_D,
  - Closeness Centrality C_C,
  - Betweeness Centrality C_B, and
  - Eigen-vector Centrality C_E.

4.3 Bank Activities and Behaviors

Bank activities include three rounds, whose details are given below.

4.3.1 Round 1: Update Deposit, Set Liability and Asset Targets

At the beginning of each step, banks will update Deposit - mimicking activities relating to deposit collecting of banks. Then, with this updated deposit, bank calculates targeted total asset, and allocates to external assets and inter-bank lending. Bank also calculates targeted inter-bank borrowing.

Update Deposit

Deposit (D) is assumed to follow log-normal distribution and calculated from historical data. Deposit gain or loss will be recognized as Cash in Asset. New deposit D^n is calculated as follows

$$D^n = D \times \exp(g_d), \tag{21}$$

in which deposit growth g_d follows normal distribution.

Set Liability and Asset allocation Target

Bank sets target of its Total Liability based on its new deposit and its preferred proportion of deposit to Total Liability. Let target deposit over total liability as $\frac{D}{L} := R_{tg}^d$, then targeted total asset (A_{tg}) and targeted total liability (L_{tg}) are

$$A_{tg} = L_{tg} = \frac{D^n}{R_{tg}^d}. \tag{22}$$

Bank also calculates its inter-bank borrowing target (IB_{tg}^b) based on targeted inter-bank borrowing over total asset ratio R_{tg}^b as

$$IB_{tg}^b = R_{tg}^b \times A_{tg}. \tag{23}$$

From this target, bank set its new borrowing targets (IB_{tg}^{nb}) the difference between target and old lending (in case this value is positive) as

$$IB_{tg}^{nb} = \min(IB_{tg}^b - IB^l, 0). \tag{24}$$

Next, bank set target of inter-bank lending and External Asset as fraction of targeted Total Asset. Let $\frac{IB^l_{tg}}{A} := R^l_{tg}$ and $\frac{EA}{A} := R^{ea}_{tg}$, then

$$IB^l_{tg} = R^l_{tg} \times A_{tg}, \tag{25}$$

$$EA_{tg} = R^{ea}_{tg} \times A_{tg}. \tag{26}$$

Bank also calculates it available IB lending it wants to make as the difference between target and old lending (in case this value is positive):

$$IB^{nl}_{tg} = \min(IB^l_{tg} - IB^l, 0). \tag{27}$$

4.3.2   Round 2: Lending-Borrowing Mechanism

Request for loan. Based on new loan target IB^{nl}_{tg} computed in Round 1, bank asks other banks for a loan. The amount of asking loan is the remaining of loans a bank need to fulfill its new borrowing target. A bank will ask until one of these condition met: the target is fulfilled unless there are no other banks available to borrow.

Response mechanism. The response mechanism is built with pretty simple rules which ensures that larger banks are more likely to lend and banks are more likely to lend other banks having relationship with them previously:

- Probability that a large bank lends to a bank it has relationship previously ranges from 0.6 to 1,
- Probability that a large bank lends to a bank it does not have relationship previously ranges from 0.2 to 0.8,
- Probability that a small bank lends to a bank it has relationship previously ranges from 0.4 to 0.9, and
- Probability that a small bank lends to a bank it does not have relationship previously ranges from 0 to 0.7.

Amount of loan. The amount of loan a bank wants to lend is a fraction of its available resource. This fraction follows a uniform distribution. Between the amount that lending bank want to give and the amount that borrowing bank want to borrow, the lower one will be the actual value of the loan. After the amount of a loan is set, the inter-bank lending and borrowing accounts of banks will be added. Also, cash account of lending bank will be subtracted, while cash account of borrowing bank will be added.

Also, each new loan is divided to short-term and long-term loans for convenience in the repayment step of the next round.

The order of this process is random, which is suitable to reflect the reality that the time for a loan request depends on many factors and is unpredictable. This request-response process will be repeated until borrowing schedule or lending schedule is emptied.

4.3.3   Round 3: Update Equity and Repay Inter-bank Loans

Update Equity. Bank income will be assumed to depend on its equity and derived return-on-equity ratio (ROE), that is,

$$\text{Net Income} = ROE \times E. \tag{28}$$

ROE is assumed to follow normal distribution with mean and variance calculated from historical data. This net income will be recognized as Cash in the Asset and as additional equity in Liability.

Inter-bank Borrowing Repayment. Banks make payment on their debts P_{IB} and receive debt payment R_{IB}. There are 2 types of debts: short-term and long-term debts. Most of short-term loans are repaid within three months, and majority of long-term loans are repaid in less than one year (four periods). Therefore, 25% of long-term loan is paid and 75% continue to exist and will be paid gradually in next 3 periods.

We will not simulate the activities (lending, borrowing, repayment) on Deposit and External Asset since this paper aims at analyzing inter-bank activities. For simplicity's sake, these activities will be combined in updating activity.

4.3.4   Bank Default

Definition: A bank goes to default if the value of its equity (E) becomes negative, i.e.,

$$E_i(t) < 0. \tag{29}$$

A bank who meets the default condition will be transmitted to Bankrupting process list. This list contains banks who are in the progress of settling its assets and liabilities before becoming bankrupt and removed from the network. When bank i fails at time t, it will default on its inter-bank borrowing. Becoming default forces banks to sell its illiquid assets and leads to loss on these assets' value. The model assumes that banks will recover from 50% to 80% of the value of its external assets, that is,

$$EA_i^r = U(50\%, 80\%) \times EA_i. \tag{30}$$

The amount of money from selling these assets will be added to Cash account, while external assets will be set back to 0. Bank recognizes the loss in its equity.

After selling external assets, bank will ask other banks that borrow from it to buy back the loan. The asking amount will be smaller than the actual loan and have recovery rate randomly selected from 50% to 80%:

$$IB_{i,j}^a = U(50\%, 80\%) \times IB_{i,j}. \tag{31}$$

Then, there will be 2 cases. If bank asked (bank j) is able to repay the loan (repaying the loan doesn't force j to bankrupt condition), asking bank take the money to Cash, set its lending to j as 0, recognize the loss to equity. While j

takes the money from its Cash, sets its borrowing from i as 0, recognizes the gain to equity.

In case bank asked meets bankrupt condition, it pays the amount of money it can and then is moved to bankrupting list to settle later:

$$IB_{j,i}^p = \min(C_j, IB_{i,j}^a), \tag{32}$$

$$WD_{i,j} = IB_{i,j}^l - IB_{j,i}^p. \tag{33}$$

This is recognized as all of the money bank i can take from j. It adds the money collected to its Cash, set its lending to j as 0, recognize the loss to equity. Whereas, j set its borrowing from i as 0, subtract its Cash and recognizes the gain to equity.

After selling all the assets, the bankrupt bank will use the money collected to pay its creditors. The allocation is based on the proportion of the loan. Bank i's creditor get the repayment and transfer to its Cash account, set lending to i as 0 and recognize the lost. It then checks the bankrupt condition and will be moved to bankrupting list if the conditions are met.

$$IB_{j,i}^{rc} = C_i \times \frac{IB_{j,i}^l}{D_i + IB_i^b}, \tag{34}$$

$$WD_{j,i} = IB_{j,i}^l - IB_{j,i}^{rc}. \tag{35}$$

5   Experiment and Results

5.1   Data Description

The data is collected from public financial statements source, including 19 banks' quarter balance sheets, which reflect banks' business and lending-borrowing practices. These balance sheets are simplified to create inter-bank system. To be more specific, simplified balance sheet cover these below six accounts:

- Cash: most liquid asset of banks, including cash and cash equivalent.
- Inter-bank Lending: loans a bank giving to other banks in the inter-bank network.
- External Asset: covers all other kinds of banks' asset, including stocks, bonds, lending to enterprises, personal loans and other kinds of external asset.
- Deposit: money bank receives from depositors.
- Inter-bank Borrowing: the money a bank borrows from other banks in the inter-bank network.
- Equity: capital of banks from shareholders.

The data sample covers 7 years from 2010 to 2016, although for some banks the data is not sufficient for all 7 years due to public data shortage. For initial balance sheet position, data as of June 30, 2016 is used for the purpose of maximizing the number of banks. A cumulative initial value of banks' assets is 3,805,256 billions VND.

5.2   Interbank-Network Results

5.2.1   Maximum Entropy Network

As expected, the Maximum entropy method creates a network with every node connecting to others, so every bank has equal and highest degree centrality (18) and the density of the network is maximum (1). Degree centrality of the network is 0 in both in-degree and out-degree cases for the fact that no bank dominates the network. Also, the betweenness centrality and closeness centralization are both 0 as every node connects directly to each other.

5.2.2   Minimum Density - Highly Dense Network ($\lambda = 0.2$)

The most outstanding characteristic of the network is high level of independence. Every bank in the network can connect to each other, although the network is less dense than Maximum Entropy approach network since not every node connects directly.

VCB and CTG are two most important nodes in the network. They lend and borrow from all other banks in the network. Their betweenness centralities are also highest, 39.5 and 37.1, respectively. This means they are frequently present between linkages of other nodes. CTG is the biggest banks in the system with total asset of 850,209 billion VND, while VCB is the second-biggest banks in the system with total asset of 678,274 billion VND. Either of these two banks going bankrupt may devastate the whole network.

5.2.3   Minimum Entropy - Slightly Dense Network ($\lambda = 0.5$)

With $\lambda = 0.5$ this network is less dense than when the density is set to be 0.2. Two most important nodes are still CTG and VCB, but their roles are not as significant as in the Highly Dense Network. The network has high out-centrality (out-degree centrality: 0.798, out-close centrality: 0.583) and low in-centrality (in-degree centrality: 0.298, in-close centrality: 0.196). This means the in-centrality of the network is pretty equal.

CTG and VCB are most important nodes and frequently appear in shortest paths between two other nodes, with betweenness centrality of 101.3 and 129, respectively.

5.2.4   Minimum Entropy - Slightly Sparse Network ($\lambda = 0.7$)

This network is special comparing to other networks as CTG and VCB do not take as so significant role as in other network. The number of links created for CTG is just 16, and for VCB is just 13. However, while they do not create many direct links, the presence of them in indirect link are still significant (168.5 for CTG and 86 for VCB). The linkages are spread over the network. Some banks (BACA, MBB) have high numbers of connections but low eigen-vector centrality since they only have relationship with small banks. Meanwhile, several banks (ACB, VIB) have low connections but high eigen-vector centrality as they have relationship with large bank.

5.2.5   Minimum Entropy - Highly Sparse Network ($\lambda = 1$)

This network is the most sparse one among analyzed networks, with the number of linkages are minimized. The network can be relatively considered as an island network with two islands. CTG and VCB are centers of these two islands, which makes them have outstanding high betweenness centrality (238.5 for CTG and 179.5 for VCB). Other banks have a few connections to other banks. This can make the weigh of edge large, and bankruptcy of a particular bank could seriously affect the bank lend to it. However, the contagion will be limited in this connection only.

5.3   Simulations and Result

We simulate on 5 above networks. For each network, we first run the system without shocks for 10 periods (quarters) and observe changes in total assets and capitals of each banks and the whole system.

Then, for the purpose of stress-testing, we apply different kinds of shock for each repetition. There are two kinds of shocks: idiosyncratic shock and systemic shock.

- **Idiosyncratic Shock** represents default of several individual banks. Default of a bank will affect banks that lend money by the manner loss to inter-bank asset, then directly write this loss to capital of the affected bank. Three different scenarios are employed:
    - Defaults of individual banks (50 observations).
    - Simultaneous defaults of three random banks (500 observations).
    - Simultaneous defaults of five random banks (500 observations).
  Each bank chosen to be bankrupt will suffered loss from 50% of its external asset. It will ignite the bankrupt mechanism as explained above.
- **Systemic Shock** represents a sudden drop in the value of external asset, such as mortgage loans. In this scenario, there are two forces: direct effects of a shock which reduce bank's mortgage loan assets, and the contagion effects which reduce bank's inter-bank assets. The combination of 2 effects makes a systemic shock more damaging.

    We start with a shock (drop in external asset) of 1%. Then we repeat the experiment 19 times more, so that default rate ranges from 1% to 20% with an increment of 1%. Affected by shocks, some weak banks in the system may get into the bankrupting conditions. In this case, they are moved to bankrupting list and will be settled in succession as described in the bankrupting mechanism.

For every scenario, we will observe the number of defaulted banks, suffered banks, the value of assets losses, capital losses to compare impacts of shocks.

5.3.1   Networks Without Shocks

In normal condition, asset and liability accounts of banks grow gradually as expected. The most interesting thing happening here is the development of inter-bank matrix. Whatever the initial condition of the matrix, the matrix tends to

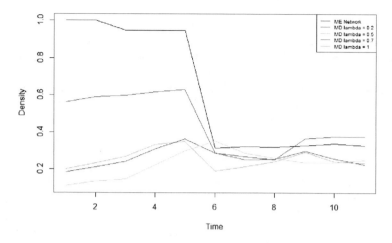

Fig. 1. Density of 5 networks over time of without-shock simulation

develop into a modestly sparse matrix. Figure 1 presents the changes in density of matrices over periods. After five periods, the density of the networks converges to values from 0.2–0.4 and fluctuates modestly then. The reason here is that after 5 periods, banks repay all initial inter-bank debts and rules of lending-borrowing start to dominate the trend. All the networks from this period tend to be modestly sparse. This fact shows some insight about inter-bank matrix formation. Since banks in the system will follow particular rules, the network will converge to a particular pattern. Then, combining experience in banking domain and agent-based simulation could be a potential approach to reproduce actual formation of the inter-bank market.

5.3.2   Network Under Systemic Shock

A system shock in the model is presented by a sudden drop in external asset. In reality, the shock can come from events that affect the whole economy system such as drop in housing or oil's prices. It provokes two sorts of negative impact. The first one is loss caused by fall on external asset value, which is recognized as equity loss and drives bank into bankrupt. The second is unfavorable post-event in the market like small drop in income or loss as the result of bankrupts of partners. Banks surviving from the first impact with low equity are highly vulnerable to such unfavorable events.

As analyzed in the previous section, since initial equity of bank is pretty thin, most of banks (17 out of 19) go bankrupt at shock of 11%. Two banks survive are SaiGonBank and PGbank who have highest equity over asset ratio.

With shocks less than 4%, no bank goes bankrupt. SCB - the weakest node in the system goes bankrupt under shock of 5% magnitude. The ME network and MD network ($\lambda = 0.2$) have most affected banks (19 and 17 respectively) as they are two most dense network. In other networks, number of affected banks ranges from 5 to 7.

Under a shock of 7% magnitude we notice differences between networks. In the most dense ME network and the most sparse MD network ($\lambda = 1$), the number of bankrupt bank are least (3 for each), while the 3 remaining networks observes bankruptcy of 5 banks. 3 bank goes bankrupt as the result of first impact are: ACB, LienVietPostBank, SCB. Banks go bankrupt under the contagion of unfavorable conditions are CTG (in MD Network $\lambda = 0.2$), SeaBank (in MD networks $\lambda = 0.2$ and $\lambda = 0.5$), VCB (in MD Network $\lambda = 0.7$) and STB (in MD Network $\lambda = 0.7$). The explanation is that the highly connected networks spread the impact over the network, while in case of a sparse network, bankruptcy of a particular bank only affects a limited number of other banks.

Under shock of 8% magnitude, the contagion effects fade and the numbers of bankrupt bank in networks are pretty even, ranges from 9 to 11. This number increases to 12–13 under shock of 9% magnitude. As shock magnitude rises from then, the behavior of networks are pretty alike and there are few signs of credit contagion effect. At 11%, 17 out of 19 banks go bankrupt and at 19% the whole network collapses in all five structures.

Dynamic of Defaulted Asset and Equity is pretty similar among networks and share the same pattern to dynamic of bank defaulted number.

5.3.3   Network Under Idiosyncratic Shock

Analyzing the network under idiosyncratic shock, we can see that the contagion effect is insignificant. Bankrupts of one, three, or five banks only lead to bankruptcy of two more banks at most. However, the dense network is most resilient since even in the case of 5 bankrupt banks, there is no other banks beside banks shocked at initial stage going bankrupt. In this case, Highly Dense Network under shocks of idiosyncratic acts as risk-sharing mechanism and spread the impact to the whole network. The Highly Sparse Network is also more resilient than the slightly sparse and slightly dense network.

As expected, one bank going bankrupt affects all the other banks in ME network, but only impact an average of 4 banks in MD network with $\lambda = 1$. However, the level of impact is least significant in the ME network with no more banks going bankrupt. Generally, highly connected networks (ME and MD with $\lambda = 0.2$) are least impacted, while moderately connected networks (MD with $\lambda = 0.5$ and MD with $\lambda = 0.7$) are most impacted. However, when the network is particularly sparse, the impact tend to be lessened. On average, the most connected network (ME) loses 5.13% total asset and 7.36% total equity, while MD network with $\lambda = 0.7$ loses 6.62% of total asset and 9.93% of total equity. In the worst case, ME network only loses 25.15% of asset and 37.54% of equity, while MD network with $\lambda = 0.7$ loses 32.04% of asset and 43.85% of equity.

In cases of three and five bankrupt banks, the average loss in asset and equity are pretty similar over networks (about 19% for asset loss and 26% for equity loss in case of three banks going bankrupt, 32% for asset loss and 42% for equity loss in case of five banks going bankrupt).

Detail statistics of network features under idiosyncratic shock are represented in Tables 2, 3, 4, 5 and 6 below.

Table 2. ME Network - Selected end-of-period statistics after a default of a single bank, three random banks and five random banks.

	Def	Affected	DA	% DA	DE	% DE
One bank bankrupt						
Mean	1.00	19.00	195,044.00	5.13	20,449.00	7.36
Min	1.00	19.00	17,770.00	0.47	3,766.00	1.36
Max	1.00	19.00	983,448.00	25.84	105,616.00	38.02
Std. Dev.	0	0	216,495.00	5.69	21,610.00	7.78
Three banks bankrupt						
Mean	3.00	19.00	724,277.00	19.03	71,988.00	25.92
Min	3.00	19.00	92,085.00	2.42	14,615.00	5.26
Max	3.00	19.00	2,142,920.00	56.31	171,988.00	61.92
Std. Dev.	0	0	421,110.00	11.07	37,434.00	13.47
Five banks bankrupt						
Mean	5.00	19.00	1,202,378.00	31.60	114,548.00	41.24
Min	5.00	19.00	286,963.00	7.54	35,290.00	12.70
Max	5.00	19.00	2,697,124.00	70.88	207,638.00	74.75
Std. Dev.	0	0	498,611.00	13.10	38,230.00	13.76

Note: DA, DE in billions VND; %DA, %DE in %

Table 3. MD Network ($\lambda = 0.2$) - Selected end-of-period statistics after a default of a single bank, three random banks and five random banks

	Def	Affected	DA	% DA	DE	% DE
One bank bankrupt						
Mean	1.00	15.40	228,342.00	6.00	23,398.00	8.42
Min	1.00	11.00	26,086.00	0.68	5,155.00	1.86
Max	1.00	19.00	1,160,273.00	30.49	114,484.00	41.21
Std. Dev.	0	2.86	243,620.00	6.40	24,699.00	8.89
Three banks bankrupt						
Mean	3.00	18.60	740,241.00	19.45	72,536.00	26.11
Min	3.00	15.00	89,370.00	2.35	14,459.00	5.20
Max	3.00	19.00	2,267,858.00	59.60	174,609.00	62.85
Std. Dev.	0	0.73	450,051.00	11.83	37,910.00	13.64
Five banks bankrupted						
Mean	5.00	18.96	1,218,086.00	32.00	115,114.00	41.44
Min	5.00	17.00	229,271.00	6.03	37,883.00	13.64
Max	5.00	19.00	2,814,111.00	73.95	203,258.00	73.17
Std. Dev.	0	0.21	527,609.00	13.86	40,695.00	14.65

Note: DA, DE in billions VND; %DA, %DE in %

Table 4. MD Network ($\lambda = 0.5$) - Selected end-of-period statistics after a default of a single bank, three random banks and five random banks

	Def	Affected	DA	% DA	DE	% DE
One bank bankrupt						
Mean	1.08	7.18	258,378.00	6.79	25,749.00	9.27
Min	1.00	4.00	27,835.00	0.73	5,290.00	1.90
Max	2.00	19.00	956,974.00	25.15	104,288.00	37.54
Std. Dev.	0.27	3.90	243,065.00	6.39	22,538.00	8.11
Three banks bankrupted						
Mean	3.29	13.77	740,676.00	19.46	72,965.00	26.27
Min	3.00	8.00	102,556.00	2.69	15,797.00	5.69
Max	5.00	19.00	2,155,437.00	56.64	172,284.00	62.02
Std. Dev.	0.49	3.59	413,399.00	10.86	36,704.00	13.21
Five banks bankrupt						
Mean	5.40	16.65	1,242,814.00	32.66	117,598.00	42.33
Min	5.00	11.00	222,059.00	5.84	32,545.00	11.71
Max	7.00	19.00	2,793,161.00	73.40	203,610.00	73.30
Std. Dev.	0.59	2.24	525,731.00	13.81	40,773.00	14.68

Note: DA, DE in billions VND; %DA, %DE in %

Table 5. MD Network ($\lambda = 0.7$) - Selected end-of-period statistics after a default of a single bank, three random banks and five random banks

	Def	Affected	DA	% DA	DE	% DE
One bank bankrupted						
Mean	1.20	7.76	251,966.00	6.62	27,580.00	9.93
Min	1.00	4.00	20,495.00	0.54	3,819.00	1.37
Max	3.00	17.00	1,219,146.00	32.04	121,803.00	43.85
Std. Dev.	0.49	3.95	311,997.00	8.20	30,546.00	10.99
Three banks bankrupt						
Mean	3.27	13.82	736,149.00	19.35	72,643.00	26.15
Min	3.00	7.00	93,124.00	2.45	14,675.00	5.28
Max	5.00	19.00	2,435,421.00	64.00	181,685.00	65.40
Std. Dev.	0.50	3.21	439,711.00	11.55	37,329.00	13.44
Five banks bankrupt						
Mean	5.35	16.80	1,295,765.00	34.05	121,322.00	43.67
Min	5.00	11.00	275,059.00	7.23	41,179.00	14.82
Max	7.00	19.00	2,822,117.00	74.16	208,913.00	75.21
Std. Dev.	0.54	2.00	543,785.00	14.29	41,350.00	14.88

Note: DA, DE in billions VND; %DA, %DE in %

Table 6. MD Network ($\lambda = 1$) - Selected end-of-period statistics after a default of a single bank, three random banks and five random banks

	Def	Affected	DA	% DA	DE	% DE
One bank bankrupt						
Mean	1.08	4.12	232250.00	6.10	23462.00	8.45
Min	1.00	2.00	19659.00	0.52	3802.00	1.37
Max	2.00	13.00	1086269.00	28.55	114557.00	41.24
Std. Dev.	0.27	3.10	263283.00	6.92	25556.00	9.20
One bank bankrupt						
Mean	3.26	9.55	722068.00	18.98	71702.00	25.81
Min	3.00	5.00	84280.00	2.21	14276.00	5.14
Max	5.00	18.00	2045732.00	53.76	177325.00	63.84
Std. Dev.	0.44	3.48	428628.00	11.26	38766.00	13.96
One bank bankrupt						
Mean	5.30	12.97	1231765.00	32.37	115694.00	41.65
Min	5.00	7.00	317864.00	8.35	36832.00	13.26
Max	7.00	19.00	2822595.00	74.17	209910.00	75.57
Std. Dev.	0.48	3.13	523855.00	13.77	39791.00	14.32

Note: DA, DE in billions VND; %DA, %DE in %

6   Conclusion

This article builds an agent-based model presenting a banking system of 19 Vietnamese banks, in which banks interact with each others via their balance sheets. Different network structures are evaluated to see the potential effects of credit contagion under various circumstances. These networks are acquired using two most well-established inter-bank network reconstruction methods: Maximum Entropy and Minimum Density. There are several outstanding points which can be inferred from the model:

- Under normal condition, the network structures tend to converge to a slightly dense network whatever the initial condition is. This result suggests simulation method with proper lending-borrowing behavior maybe a good approach to build actual inter-bank network structure.
- The equity buffers of Vietnamese banks are pretty thin, that make them become more vulnerable to system shocks. Even a system shocks of 11% magnitude on external asset can make 17 out of 19 banks going bankrupt and the whole system collapses at shock of 19% magnitude.
- Generally, under both system shock and idiosyncratic shock, contagion effect for networks simulated from Vietnam's banking data is not significant. One reason here is the proportion of inter-bank asset and liability over total asset is small, while ratio of equity over total liability is low and does not vary much.

- In case of system shock, under shocks of low magnitude, highly connected network and highly sparse network tend to be most resilient. The reason is highly connected networks spread the impact over the network and act as a risk-sharing mechanism, while in case of sparse network, bankrupt of particular bank only affects limited number of other banks and limit the impact of crisis. However, when the magnitude of shock increases, the impact of shocks on different network are pretty equal since most banks meet bankrupting condition. We can only observe the difference in bankrupting behavior of networks under shocks magnitude ranges from 5 to 10%.

- In case of idiosyncratic shock, the contagion effect is insignificant. Bankrupts of one, three, or five banks only lead to bankruptcy of two more banks at most. However, the dense network is more resilient since even in the case of 5 banks shocked, there are no other banks beside banks shocked at initial stage going bankrupt. The Highly Sparse Network is also more resilient than the slightly sparse and slightly dense network. However, when the number of banks shocked increase, the impact over all the networks tend to converge.

This article succeeds in building a complete network and gains some insights about inter-bank network and behavior. However, there are a lot of works to improve and make the model more realistic. First, other inter-bank network reconstruction methods could be implemented. Second, lending-borrowing rules in the model could be refined to reflect the reality more accurately. Third, the model could be expanded to include not only banks but also other agents like depositors, customers, traders, government. Forth, other kinds of external force such as changes in policy or liquidity shocks could be included in the model.

All in all, agent-based method is a promising approach to understand interaction in the inter-bank network. It can be a tool for government to perform stress test under different kinds of shocks, as well as experiment new regulations to see whether their targets can be met.

References

1. Acemoglu, D., Ozdaglar, A., Tahbaz-Salehi, A.: Systemic risk and stability in financial networks. Am. Econ. Rev. **105**(2), 564–608 (2015)
2. Allen, F., Gale, D.: Financial contagion. J. Polit. Econ. **108**(1), 1–33 (2000)
3. Anand, K., Craig, B., Von Peter, G.: Filling in the blanks: network structure and interbank contagion. Quant. Financ. **15**(4), 625–636 (2015)
4. Bala, V., Goyal, S.: A noncooperative model of network formation. Econometrica **68**(5), 1181–1229 (2000)
5. Battiston, S., Gatti, D.D., Gallegati, M., Greenwald, B., Stiglitz, J.E.: Liaison dangerous: Increasing connectivity, risk sharing, and systemic risk. J. Econ. Dyn. Control **36**(8), 1121–1141 (2012)
6. Battiston, S., Puliga, M., Kaushik, R., Tasca, P., Caldarelli, G.: Debtrank: too central to fail? financial networks, the fed and systemic risk, Scientific reports, vol. 2, Article 541 (2012)
7. Bech, M.L., Atalay, E.: The topology of the federal funds market. Phys. A **389**(22), 5223–5246 (2010)

8. Bookstaber, R., Paddrik, M., Tivnan, B.: An agent-based model for financial vulnerability. J. Econ. Interact. Coord. 1–34 (2017)
9. Campbell, J.F., O'Kelly, M.E.: Twenty-five years of hub location research. Transp. Sci. **46**(2), 153–169 (2012)
10. Castiglionesi, F., Navarro, N.: Optimal fragile financial networks. SSRN Working Paper Series. https://papers.ssrn.com/sol3/papers.cfm?abstract_id=1089357
11. Cocco, J.F., Gomes, F.J., Martins, N.C.: Lending relationships in the interbank market. J. Financ. Intermediation **18**(1), 24–48 (2009)
12. Craig, B., Von Peter, G.: Interbank tiering and money center banks. J. Financ. Intermediation **23**(3), 322–347 (2014)
13. De Nooy, W., Mrvar, A., Batgelj, V.: Exploratory Social Network Analysis with Pajek, vol. 27. Cambridge University Press, Cambridge (2011)
14. Duffie, D., Singleton, K.J.: Modeling term structures of defaultable bonds. Rev. Financ. Stud. **12**(4), 687–720 (1999)
15. Eisenberg, L., Noe, T.H.: Systemic risk in financial systems. Manage. Sci. **47**(2), 236–249 (2001)
16. Elsinger, H., Lehar, A., Summer, M.: Using market information for banking system risk assessment. Int. J. Central Bank. **2**(1), 137–165 (2006)
17. Elsinger, H., Lehar, A., Summer, M.: Network models and systemic risk assessment. In: Handbook on Systemic Risk, vol. 1, pp. 287–305 (2013)
18. Furfine, C.: Interbank exposures: quantifying the risk of contagion. J. Money Credit Bank. **35**(1), 111–128 (2003)
19. Georg, C.P.: The effect of the interbank network structure on contagion and common shocks. J. Bank. Financ. **37**(7), 2216–2228 (2013)
20. Glasserman, P., Young, H.P.: How likely is contagion in financial networks? J. Bank. Financ. **50**, 383–399 (2015)
21. Gofman, M.: Efficiency and stability of a financial architecture with too-interconnected-to fail institutions. J. Financ. Econ. **124**(1), 113–146 (2017)
22. Hałaj, G., Kok, C.: Assessing interbank contagion using simulated networks. CMS **10**(2–3), 157–186 (2013)
23. Hałaj, G., Kok, C.: Modeling the emergence of the interbank networks. Quantit. Finan. **15**(4), 653–671 (2015)
24. Hansen, L., Sargent, T.: Robust control and model uncertainty. Am. Econ. Rev. **9**(2), 60–66 (2001)
25. Iori, G., Jafarey, S., Padilla, F.G.: Systemic risk on the interbank market. J. Econ. Behav. Organ. **61**(4), 525–542 (2006)
26. Iori, G., De Masi, G., Precup, O.V., Gabbi, G., Caldarelli, G.: A network analysis of the Italian overnight money market. J. Econ. Dynamics Control **32**(1), 259–278 (2008)
27. Jarrow, R.A., Turnbull, S.M.: Pricing derivatives on financial securities subject to credit risk. J. Financ. **50**(1), 53–85 (1995)
28. Kok, C., Montagna, M.: Multi-layered interbank model for assessing systemic risk. Kiel Working Paper No 1873 (2013)
29. Ladley, D.: Contagion and risk-sharing on the inter-bank market. J. Econ. Dynamics Control **37**(7), 1384–1400 (2013)
30. Lelyveld, I., Liedorp, F.: Interbank contagion in the dutch banking sector: a sensitivity analysis. Int. J. Central Bank. **2**(2), 99–133 (2006)
31. Macal, C.M., North, M.J.: Tutorial on agent-based modeling and simulation. J. Simul. **4**(3), 151–162 (2010)
32. Merton, R.C.: On the pricing of corporate debt: the risk structure of interest rates. J. Financ. **29**(2), 449–470 (1974)

33. Nier, E., Yang, J., Yorulmazer, T., Alentorn, A.: Network models and financial stability. J. Econ. Dynamics Control **31**(6), 2033–2060 (2007)
34. Sheldon, G., Maurer, M., et al.: Interbank lending and systemic risk: an empirical analysis for Switzerland. Swiss J. Econ. Stat. (SJES) **134**, 685–704 (1998)
35. Steinbacher, M.: Simulating portfolios by using models of social networks. Ph.D. Dissertation, University of Ljubljana, Ljubljana (2012)
36. Streit, R.E., Borenstein, D.: An agent-based simulation model for analyzing the governance of the Brazilian financial system. Expert Syst. Appl. **36**(9), 11489–11501 (2009)
37. Upper, C.: Simulation methods to assess the danger of contagion in interbank markets. J. Financ. Stab. **7**(3), 111–125 (2011)
38. Upper, C., Worms, A.: Estimating bilateral exposures in the German interbank market: Is there a danger of contagion? Europ. Econ. Rev. **48**(4), 827–849 (2004)
39. Weisbuch, G., Kirman, A., Herreiner, D.: Market organization and trading relationships. Econ. J. **110**, 411–436 (2000)
40. Yang, S.Y., Liu, A., Zhang, X., Paddrik, M.E.: Interbank contagion: an ABM approach to endogenously formed networks. OFR Working Paper (2016). https://papers.ssrn.com/sol3/papers.cfm?abstract_id=2777507

Assessment of the Should be Effects of Corruption Perception Index on Foreign Direct Investment in ASEAN Countries by Spatial Regression Method

Bui Hoang Ngoc[1,2(✉)], Dang Bac Hai[1,3], and Truong Hoang Chinh[1,2]

[1] HCM City of Open University, Ho Chi Minh City, Vietnam
buihoangngoc.ulsa@gmail.com
[2] Faculty of Administration, University of Labour
and Social Affairs, Ho Chi Minh City, Vietnam
[3] Environmental and Natural Resources Economics Faculty, HCM University
of Natural Resources and Environment, Ho Chi Minh City, Vietnam

Abstract. Studies of corruption and its relationship with foreign direct investment (FDI) have yielded mixed results. Some have found that corruption deters FDI but others have found the opposite. This study applies Spatial Regression in combination with dynamic panel data to investigate how different corruption perceptions index (CPI) impact upon one of the fundamental decisions made by foreign investors, the choice of FDI location within the selected host country. For a sample of Asean countries and the time period 2005–2015, we find a clear positive relationship between CPI and FDI.

Keywords: FDI · Corruption · Spatial regression · Asean

1 Introduction

Corruption is usually defined narrowly as the abuse of public office for personal gain. This definition is reflected in reported measures of the perceptions index of national corruption levels. Such public corruption may have a corrosive effect on the integrity of a national's entire system: it may reduce operational efficiency, distort public policy, slow the dissemination of information, negatively impact upon income distribution, and increase the poverty of an entire nation.

In international business, alternative conceptions of distance (for example, cultural, psychic, institutional) are often more pertinent and have been put forward as explanations for various aspects of multinational enterprise (MNEs) behavior. MNEs may use care when choosing host countries for their foreign subsidiaries because of their concern for the additional uncertainly and operational costs associated with corruption. Corruption has, consequently, been considered a deterrent to FDI. Corruption varies widely across different locations in its scope in an economy as well as in the level of uncertainty it creates. Also, not all MNEs perceive and respond to corruption in the same manner. Besides the

© Springer International Publishing AG 2018
L. H. Anh et al. (eds.), *Econometrics for Financial Applications*, Studies in Computational Intelligence 760, https://doi.org/10.1007/978-3-319-73150-6_33

direct impact of host country corruption on inward FDI, formal institutions in the host country may interact with institutions in the home country, which may themselves interact with informal institutions and therefore affect the behavior of foreign investors. In that sense, the degree of uncertainty and the costs associated with corruption may vary depending on the country of origin of the foreign investors. For this reason, recent studies have concluded that MNEs located in countries with low levels of corruption avoid investing in highly corrupt countries. With little knowledge and skills for dealing with this phenomenon at home, they are more likely to be deterred by high levels of corruption as well as their unfamiliarity with it abroad. On the other hand, rms which originated in highly corrupt environments may not be as sensitive to high corruption levels abroad; they may be attracted by the environment and even take advantage of corrupt activities.

Based on the premise that the relative differences between CPI in home and host countries may influence FDI, the understanding of corruption and its effects on FDI can be extended by replicating earlier studies within the unique context of Asean. We argue that not all foreign investors are affected equally by corruption in the host country and specically, that rms based in highly corrupt countries are not excessively affected by high levels of corruption abroad or by corruption distance. The next section addresses these research questions in relation to corruption and FDI, by reviewing the theoretical literature on corruption. Subsequent sections detail hypotheses, methodology, results, and conclusions.

2   Theoretical Background and Literature Review

The study of FDI has generally focused on efficiency based on transaction cost analysis. The transaction cost theory (TCT) utilizes transactions as its basic unit of analysis. According to Williamson (1985) a transaction "occurs when a good or service is transferred across a technologically separable interface". Therefore, the organization of economic activity is understood in transaction cost terms. In this sense, TCT is concerned with the costs of integrating an operation within the firm as apposed to the costs of using an external market to act for the firm in an overseas market.

Building on TCT, Dunning developed his Ownership - Location - Internalization paradigm (OLI) by using of transaction cost theory to analyze FDI activities. The OLI paradigm argues that a firm's international activities are determined by three factors: ownership (O) advantages, location (L) advantages, and internalization (I) advantages. The main premise of the paradigm is that MNEs develop competitive O advantages at their home country and then transfer them abroad to countries where they can be exploited (based on L advantages) through FDI, which allows the multinational enterprise (MNEs) to internalize such O advantages (Dunning 1981).

Analyzed through the TCT lens, corruption in a host location can be seen in a cost/benefit manner that will deter foreign investors if the costs of the potential deal exceed its benefits (Rose-Ackerman 2008). This might suggest that while some firms with no experience in dealing with corruption at home might be at a disadvantage when operating in highly corrupt foreign countries, the same might not be true for those firms familiar with operating in highly corrupt home countries. MNEs with knowledge of dealing with corrupt environments at home may be encouraged by their location-bound-ownership advantages and willing to invest in similar locations. Thus, when analyzing how corruption affects FDI, it is important to know if strategic knowledge of coping with corruption may be acquired at home by some firms and redeployed abroad without incurring high costs.

Corruption is an important part of a country's institutions (Wei 2000). Therefore, corruption (or its absence) lies at the core of any national environment. Institutions are seen as consciously designed, man-made and tangible features, including "structures of codified and explicit rules and standards" (Holmes et al. 2012). One compelling perspective, according to North (1990), on the entire national environment proposes the co-evolution of informal and formal institutions, whereby customs, habits and social norms become codified and institutionalized. This co-evolutionary view is echoed by Holmes et al. (2012), who note that "formal institutions reflect, embody and reinforce the country's culture across the population". Likewise, Dunning and Lundan (2008) following North (1990) insist that "anything that is likely to influence individual decision making, such as education, social mores and belief systems, is also likely to affect the choice of institutions" of any location.

Habib and Zurawicki (2002) assess corruption in two manners - the level of corruption of the host country and the difference between levels of corruption of the home and host countries, pointing out that high levels corruption in the host country deters FDI. Habib and Zurawicki (2002) state that countries with different levels of corruption avoid trading with each other and that "foreign firms are unwilling to deal with the planning and operational pitfalls related to an environment with a different corruption level". Harms and Ursprung (2002) provide evidence that FDI inflows per capita positively depend on political rights and civil liberties. Further illustrate that FDI increases as a country's degree of political risk decreases. Egger and Winner (2003) find a positive impact of the viability of contracts on FDI inflows. A high level of corruption is usually associated with an unfavorable institutional environment. Accordingly, one would expect FDI to decline with corruption: Corruption is expressed as a "grabbing hand" that increases costs of multinational firms, so lowering the incentives to invest abroad. On the other hand, in the presence of regulations and other administrative controls, corruption can act as a "helping hand" to foster FDI, as proposed by Leff (1964).

Despite these studies showing corruption as a deterrent of FDI, some empirical studies have found no relationship between the two variables (Henisz 2000; Wheeler and Mody 1992). Furthermore, other authors have actually found that corruption can be positive as it facilitates transactions in countries with too many regulations (Egger and Winner 2005; Huntington 1968). For example, when studying location decisions for US MNEs, Wheeler and Mody (1992) used a combination of transaction costs and institutional variables including corruption. They found that corruption, political risk, and short-term incentives have little effect on the attraction of US FDI in developing economies; investors preferred good infrastructure development, specialized suppliers, and a growing market.

3   Data and Methods

In this paper, we draw upon economic geography to investigate how difference CPI impact upon one of fundamental decisions made by foreign investors, namely the choice of FDI location within the selected host country. First, we outline the basic characteristics of the data, describe how the variables have been measured, explain the methodology for the estimation of the spatial panel model, and consider alternative specifications for the elements to be used in the spatial weights matrix.

The method of Spatial Regression has been used all over the world, so in this section, we want to explain clearly the research model, how the variables have been used and measured and how the basic spatial models have been used. There are many different disciplines of economics that use quantitative analysis in their research, we focus on the core relations of the research problem, i.e. on the ideal that in the model, the number of explanatory variables are as smaller as suitable. Thus, inherited previous studies by Godinez and Liu (2014), Blanc-Brude et al. (2013), Egger and Winner (2003), Iloie (2015), we use the following the research model about the corruption perceptions index (CPI) affect to FDI attraction in ASEAN countries as follows:

$$LnFDI_{it} = (\beta_0 + v_i) + \beta_1 LnCPI_{it} + \beta_2 LnGNI_{it} + \beta_3 LnDES_{it} + e_{it}. \quad \text{(Model 1)}$$

In this model as presented above

$i = 1, 2, .., 11$ indicates a country: $i = 1$ is Brunei, $i = 2$ is Cambodia, $i = 3$ is Indonesia, $i = 4$ is Laos, $i = 5$ is Malaysia, $i = 6$ is Myanmar, $i = 7$ is Philippines, $i = 8$ is Singapore, $i = 9$ is Thailand, $i = 10$ is Timor-Leste, $i = 11$ is Vietnam.

v_i is the individual characteristics of each country in the study (Table 1).

t is the time (from 2005 to 2015).

Table 1. List of variables, their measures and source

Variable	Measure	Unit	Source
FDI	The annual FDI inflows per capita	USD/Person	UNCTAD
CPI	The Corruption Perception Index	Score	Transparency International
GNI	Per capita income (calculated at constant 2010 prices)	USD	World Bank
DES	Population density	Person/km^2	World Bank

Model (1) helps to look at individual characteristics of individual countries, but it disregards the spatial relationship between countries. Peracchi and Meliciani (2001) noted that there is a strong correlation economic growth among neighboring countries. The countries in the same area often strongly interact with each other in economic terms through the flow of capital, labor force, import-export turnover, etc. The countries have similarities in geography, climate conditions, and natural resources they often imitate between good economic policies, so there will appear spillover effect on economic policy among neighboring countries, that includes attracting foreign direct investment policy.

According to Anselin and Bera (1998) in the case of spatial dependence, the assumptions of the OLS regression method are no longer guaranteed. If the spatial correlation coefficient differs from 0, the OLS estimators will result be biased and not robust. If the spatial correlation coefficient is not 0, the OLS estimation will make the regression coefficients bias and un robustness.

There are three basic spatial regression models: Spatial Error Model (SEM); Spatial Autoregression Model (SAR); Spatial Durbin Model (SDM) model. Three models can be illustrated in matrixes as follows:

Spatial Error Model (SEM)

$$Y = \beta X + U,$$
$$U = \lambda W_u + \epsilon \qquad \text{(Model 2)}$$

Y is a dependent variable; Z is the vector of explanatory variables, U is the regression vector of spatially correlated spatial dimensions; λ is the spatial autocorrelation coefficient; W_u is the space-weight matrix

Spatial Autoregression Model (SAR)

In SAR, the correlation in terms of spatial are put directly into the regression model through the space of late variables depend. Meanwhile, performances of the SAR models in the form of a matrix of the form

$$Y = \rho WY + \beta Z + \epsilon, \qquad \text{(Model 3)}$$

where ρ autoregression regression coefficient of dependent variable.

Spatial Durbin Model (SDM)

$$Y = \rho WY + \beta Z + \gamma WZ + \epsilon, \qquad \text{(Model 4)}$$

The most difficulty in applying spatial correlation methods is to determine the spatial weight matrix(W). There are two basic ways: First is to use the adjacent weight matrix, The countries sharing the land border line are given a value of 1, with no boundary value 0. Second is to use the actual distance matrix that measure the actual air distance between the capitals of the countries. This paper uses actual distance matrices, the results are shown in Sect. 4.

4   Empirical Results

The Corruption Perceptions Index (CPI) published annually by Transparency International is based on perceptions of business people and public sector corruption experts. The overall score is 100, which has a high score indicating that the national administration is transparent and low scores mean that corruption in the public sector is still present and has not improved. In 2016, ranking 176 countries and territories, the Asean region is ranked with average transparency compared to the world. Only Singapore is in the top 10 most transparent countries in the world, in the ASEAN region six countries increased its position and five countries dropped out its position. Vietnam, Malaysia increased one step, Indonesia increased two steps, Cambodia and the Philippines increased six steps, but the most impressive to mention Thailand increased by 25 steps (Table 2).

Table 2. CPI 2016 score and rank CPI of Asean countries.

Countries	CPI score 2016	Rank 2016	Rank 2015	Rank difference
Thailand	35	101	76	25
Singapore	84	7	8	−1
Malaysia	49	55	54	1
Cambodia	21	156	150	6
Philippines	35	101	95	6
Indonesia	37	90	88	2
Vietnam	33	113	112	1
Laos	30	123	139	−16
Myanmar	28	136	147	−11
Timor-Leste	35	101	123	−22
Brunei	58	41	N/A	N/A

In the attraction of foreign direct investment (FDI), the ASEAN region is still an attractive destination for investors from the US, Europe, Japan, Korea and Taiwan. Despite the severe effects of the global economic crisis in the global economy for the period 2008–2013, FDI flows into the ASEAN region are still rising. Attracting the best FDI in the region is Singapore thanks to high technology and financial services. Indonesia has a large population to claim second place in attracting FDI, Viet Nam has made remarkable efforts in improving the investment environment, holding the number three position in 2015. The authors use Stata 14 software to process the data. For model (1), the paper is estimated by 3 methods Pooled, FEM, REM, comparative test results with FEM & Pooled; FEM & REM showed that the FEM model is best suited to the data sample. Further test of heteroscedasticity, regression results of model 1 are shown in column 1 of Table 3. As a result, only the average GNI variable is statistically significant and positively impacted on attracting FDI, CPI variable has positive signal but not significant so there is not enough basis to conclude that CPI has an impact on FDI. According to Le Gallo et al. (2003), Blanc-Brude et al. (2013), when measuring economic relationships that ignore spatial correlations, can lead to biased and unreliable estimates. However, as mentioned above, model (1) ignores spatial relationships.

In addition, according to Anis Omri et al. (2014), FDI data is usually persistent time series data, the amount of FDI attracted in the following years often have very strong relationship with the amount of FDI of previous years. This paper continues to test the impact of CPI on FDI in the form of model 4 - the Spatial Durbin Model. Test results at column 2 in Table 3 shows that spatial correlation in the regression model (Global Moran MI coefficient = 0.1578) was statistically significant at 1%. Test of additional spatial correlation of dependent

Table 3. Empirical results

Dependent variable: Ln_FDI	Model 1		Model 4	
	β coefficient	Prob	β coefficient	Prob
Ln_CPI	0.68715	0.774	0.20741	0.029
Ln_GNI	2.37711	0.000	1.72496	0.000
Ln_DES	0.21031	0.905	0.06831	0.462
Intercept	−18.4166	0.011	−2.04925	0.007
W1xLn_CPI			6.26e-06	0.681
W1xLn_GNI			0.00006	0.051
W1xLn_DES			−0.00012	0.007
No.Obs	121		121	
F_test	19.73***			
FEM & OLS test	13.04***			
Hausman test	35.65***			
Wald test	32.47			
Wooldridge test	2.835**			
Test	**Z**			
Error Spatial AutoCorrelation Tests				
GLOBAL Moran MI	0.1578***			
GLOBAL Geary GC	0.1299			
LM Error (Robust)	49.0302***			
Spatial Lagged Dependent Variable Spatial AutoCorrelation Tests				
LM Lag (Anselin)	18.6498***			
LM Lag (Robust)	59.4752***			
Panel Heteroscedasticity Tests				
Engle LM ARCH Test	3.1732**			

Note ***, ** & * indicate 1%; 5% and 10% level of significance.

variables by two criteria LM Lag (Anselin) and LM Error (Robust) are statistically significant. This is evidence that after controlling the dynamic nature of the model and correlation factors of space is actually CPI has a direct impact on the attraction of foreign direct investment in the countries of Southeast Asia. In particular, as with other conditions do not change while the corruption perception index risees by 1%, then the attraction FDI per capita increases by 0.207%.

5   Conclusions and Policy Implications

In the process of development, there is a need for capital to invest in infrastructure, social security, education, health care, defense, etc. is always necessary. Pressure to maintain a positive growth rate along with higher national incomes compels Governments to consider FDI as a priority choice. By spatial regression method, with data from 11 Southeast Asian countries in the period 2005–2015, we conclude that improving the transparency of the executive branch of government (i.e., reducing corruption in the public sector) will have a positive impact on the attraction of foreign direct investment. In addition, from the research results the authors suggest some practical applications as follows:

Firstly, there has the spatial correlation between ASEAN countries, regression models constructed to study socio-economic factors in this region. Therefore, we should be very careful to avoid abandonment spatial dependence leads to unreliable research results.

Secondly, the spatial correlation between the ASEAN countries is positive, while the ASEAN countries are integrating deeply and broadly within the bloc as well as with other countries in the world. So, when a country plans its economic policies, it also needs to take into account the impact of these policy on the surrounding countries and vice versa the impact of changing the policies of its neighbors to its own country.

References

Rose-Ackerman, S.: Corruption and government. J. Int. Peace Keep. **15**, 328–343 (2008). (Special issue on post-conflict peacebuilding and corruption)

Wei, S.J.: How taxing is corruption on international investor? Rev. Econ. Stat. **82**, 1–11 (2000)

Holmes, R., Miller, T., Hitt, M., Salmador, M.: The interrelationships among informal institutions, formal institutions, and inward foreign direct investment (2012)

North, D.: Institutions, Institutional Change and Economic Performance. Cambridge University Press, Cambridge (1990)

Habib, M., Zurawicki, L.: Corruption and foreign direct investment. J. Int. Bus. Rev. **33**(2), 291–307 (2002)

Harms, P., Ursprung, H.W.: Do civil and political repression really boost foreign direct investment? Econ. Inq. **40**, 651–663 (2002)

Godinez, J.R., Liu, L.: Corruption distance and FDI inflows into Latin America. J. Int. Bus. Rev. **24**(1), 33–42 (2014)

Egger, P., Winner, H.: Evidence on corruption as an incentive for foreign direct investment. Europ. J. Polit. Econ. **21**, 932–952 (2005)

Wheeler, D., Mody, A.: International investment location decisions: the case of US firms. J. Int. Econ. **33**, 57–76 (1992)

Henisz, W.: The institutional environment for multinational investment. J. Law Econ. Organ. **16**, 334–364 (2000)

Blanc-Brude, F., Cookson, G., Piesse, J., Strange, R.: The FDI location decision: Distance and the effects of spatial dependence. J. Int. Bus. Rev. **23**, 797–810 (2013)

Iloie, R.E.: Connections between FDI, corruption index and country risk assessments in Central and Eastern Europe. J. Procedia Econ. Financ. **32**, 626–633 (2015)

Peracchi, F., Meliciani, V.: Convergence in per capita GDP across European regions a reappraisal (2001)

Williamson, O.: The Economic Institutions of Capitalism: Firms, Markets, Relational Contracting. Free Press, New York (1985)

Dunning, J.: Toward an eclectic theory of international production: some empirical test. J. Int. Bus. Stud. **11**, 9–31 (1981)

Dunning, J., Lundan, S.: Institutions and the OLI paradigm of the multinational enterprise. Asia Pac. J. Manag. **25**(4), 573–593 (2008)

Egger, P., Winner, H.: Does contract risk impede foreign direct investment? Swiss J. Econ. Stat. **139**, 155–172 (2003)

Huntington, S.: Political Order in Changing Societies. Yale University Press, New Heaven (1968)

Leff, N.: Economic development throught bureaucratic corruption. Am. Behav. Sci. **8**, 8–14 (1964)

Anselin, L., Bera, A.: Spatial dependence in linear regression models with an introduction to spatial econometrics. In: Ullah, A., Giles, D.E.A. (eds.) Handbook of Applied Economic Statistics, pp. 237–289. Marcel Dekker, New York (1998)

Le Gallo, J., Ertur, C., Baumont, C.: A spatial econometric analysis of convergence across European regions, 1980-1995. In: Fingleton, B. (ed.) European regional growth. Springer, Heidelberg (2003). https://doi.org/10.1007/978-3-662-07136-6_4

Omri, A., Nguyen, D.K., Rault, C.: Causal interactions between CO2 emissions, FDI, and economic growth: evidence from dynamic simultaneous-equation models. Econ. Model. **42**, 382–389 (2014)

Using SmartPLS 3.0 to Analyse Internet Service Quality in Vietnam

Bui Huy Khoi[1] and Ngo Van Tuan[2(✉)]

[1] Industrial University of Ho Chi Minh City,
12 Nguyen Van Bao, Govap District, Ho Chi Minh City, Vietnam
buihuykhoi@iuh.edu.vn
[2] Banking University of Ho Chi Minh City,
36 Ton That Dam, District 1, Ho Chi Minh City, Vietnam
tuannv@buh.edu.vn

Abstract. The aim of this research is to explore relationship both customer satisfaction and loyalty of internet users in Vietnam to stay with their current provider among existing Internet Service Providers (ISP) in the context of Vietnam telecommunication sector. Survey data was collected from 200 people in HCM City. The research model is proposed from the study of customer satisfaction and loyalty of some authors in abroad. The reliability and validity of the scale were tested by Cronbach's Alpha, Average Variance Extracted (Pvc) and Composite Reliability (Pc). The analysis results of structural equation model (SEM) showed that the customer satisfaction and loyalty have a relationship with each other.

1 Introduction

Customer loyalty are believed to be the most effective weapons for providers to complete with other rivals because these factors are the primary variables that impact customers' intention to stay with their current providers (Khatibi et al. 2002). The providers which satisfy customers can expect higher market shares and greater profitability and, in turn, encourage their customers to become loyalty (Khatibi et al. 2002). Aaker (1991) has discussed the role of loyalty in the brand equity process and has specifically noted that brand loyalty leads to certain marketing advantages such as reduced marketing costs, more new customers and greater trade leverage. Expanding loyal customers can assist in generating a business's profitability because the cost of retaining existing customers is less than that of acquiring new ones (Reichheld and Teal 1996). Therefore, understanding how and why a sense of loyalty develops in customers remains one of the crucial management issues of our day. By accessing the Internet service users in Ho Chi Minh city, this research will give useful information of a range of factors impact on Vietnamese internet users' perceptions of their current ISPs, from which factors effecting customer loyalty, especially in telecommunication sector, will be derived. In other words, this research provides ISPs the direction and valuable understandings of what factors should be to become successful and have more loyal customers, which ultimately result in more profitability and leads to their strong competitiveness and stable growth in the future.

© Springer International Publishing AG 2018
L. H. Anh et al. (eds.), *Econometrics for Financial Applications*, Studies in Computational Intelligence 760, https://doi.org/10.1007/978-3-319-73150-6_34

2   Literature Review

2.1   Network Quality and Customer Satisfaction

Network quality is one of the most important drivers of overall service quality which leads to customer satisfaction in the context of telecommunication (Chun and Hahn 2007; Wang et al. 2004). Previous literature suggested that stability, transmission speed and network coverage are the core attributes of network quality (Woo and Fock 1999; Yaacob 2011). The stability and transmission speed of internet service were important to users and they will consider changing to other providers due to stability and speed factor (Yaacob 2011). Moreover, the uptime of service was tested to have impact on customer satisfaction as well as customer loyalty (Wang et al. 2004). The results of Wang et al. (2004) study suggested that service providers to be successfully competitive, they must try to improve the availability of service. We, therefore, hypothesize that:

H1: Network quality has a positive impact on customer satisfaction.

2.2   Price Perception and Customer Satisfaction

With regards to price perception, although Internet broadband users are willing to pay more for better service, they will consider changing to another provider because of the price factor (Yaacob 2011). Therefore, we believe that customers of Internet services are sensitive to price and the higher price level could lead to low demand, accordingly. Ranaweera and Neely (2003) showed that price perception has a direct linear relationship with customer satisfaction in telecommunications sector. We therefore believe that such relationship may be more explicit in the fierce price competitive in telecommunication market like Vietnam. Hence, we formulate the following hypothesis:

H2: Service price is positively related to customer satisfaction.

2.3   Perceived Quality and Customer Satisfaction

Perceived can be defined as the customers' overall assessment of the utility of products or services based on perceptions of what is received and what is sacrificed (Monroe 1991; Zeithaml et al. 1996). Simply stated, perceived value is trade-off between benefits and sacrifices. The perceived sacrifice includes all the costs the customer faces when they are in the process of purchasing such as purchase price, acquisition costs, installation, repair and maintenance, risk of failure or poor performance (Monroe 1991; Parasuraman et al. 1996). The perceived benefits include physical and service attributes, technical support, and other indicators of perceived service quality. Monroe 1991, Parasuraman et al. (1996) point out that perceived quality is subjective and individual, and therefore varies among customers. Research evidence suggests that customers who perceive that they receive value for money are more satisfied and tend to become loyal customers more often than ones who do not (Zeithaml et al. 1996). Customers who remain with a service provider for a longer because they are pleased with the perceived quality and value of the services tend to buy more additional services and spread favorable word-of-mouth messages and have a willingness to recommend the firm to others (Zeithaml et al. 1996). Perceived quality, from the service providers' standpoint, should be viewed as a major competitive weapon and can be used to create and improve their sustainable competitive advantage (Khatibi et al. 2002). By adding more perceived quality to the service, customer satisfaction can be increased. We, therefore, hypothesize that:

H3: Perceived value has a positive effect on customer satisfaction.

2.4   Trust and Customer Satisfaction

Trust can be viewed as an antecedent of customer loyalty (Dick and Basu 1994). To create customer loyalty toward a brand, building long-term relationships with customers to win their trust is considered advantageous (Morgan and Hunt 1994). When customers have trust in the products or services of their preferred brand, they are willing to purchase more of the brands without the need to be heavily persuaded, and this can result in lower advertising and other related costs (Murphy 2001). Trust can be viewed as being complementary to satisfaction, in ensuring long-term relationships, and strengthening customer satisfaction (Hart and Johnson 1999). In addition, trust is a willingness to rely on other parties in whom one has confidence, when the buyers trust the sellers, it affects their purchasing behavior (Chow and Holden 1997). This leads to the following hypothesis:

H4: Trust positively influences customer satisfaction.

2.5   Customer Satisfaction and Customer Loyalty

Many studies have provided empirical evidence to support the statement that customer satisfaction has positive relationship on repurchase intention and customer loyalty (Aksoy 2014; Amin et al. 2013; Kashif et al. 2015). For example, Ramseook-Munhurrun and Naidoo (2011) found that there is a significant relationship between customer satisfaction and customer loyalty in internet banking (Amin 2016).

H5: Customer sastisfation influences customer loyalty.

2.6   Controlling Factors (AIJS)

Choudrie (2005) find there are strong influences of demographic variables such as gender, income on customer loyalty. In this research, the relationship between gender and income also were tested. Besides, the age and job should be examined its impact on customer loyalty (Fig. 1).

H6: Controlling factors (AIJS) positively influences loyalty.

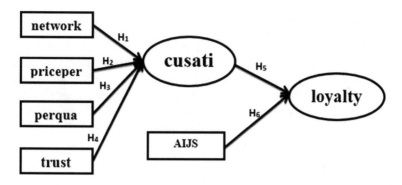

Fig. 1. Research model (**network**: Network quality, **priceper**: Price perception, **perqua**: Perceived quality, **cusati**: Customer satisfaction, **AIJS**: Age, Income, Job, Sex, **trust**: Trust, **loyalty**: Customer loyalty) **Source:** *Author*

3   Research Method

Research methodology is implemented through two steps: qualitative research and quantitative research. Qualitative research was conducted with a sample of 30 people. First period 1 is tested on a small sample to discover the flaws of the questionnaire. Second period of the official research was carried out as soon as the question was edited from the test results. Respondents were selected by convenient methods with a sample size of 200 people. There were 104 males and 96 females in this survey. Their ages were from 21 to 42 years old. Their jobs and income are as follows (Table 1):

Table 1. Job and income

Job	Amount	Percent (%)		VND (1000,000)	Amount	Percent (%)
Businessman	32	16.0		<5	58	29.0
Technician	49	24.5		5 – <7	81	40.5
Officer	39	19.5	Income	7 – <9	45	22.5
Bank officer	51	25.5		>=9	16	8.0
Other	29	14.5		**Total**	**200**	**100.0**
Total	**200**	**100.0**				

Source: *Author*

The questionnaire answered by respondents is the main tool to collect data. The questionnaire contained questions about the position of the customer satisfaction and loyalty and their personal information.

The survey was conducted on May 03, 2017. Data processing and statistical analysis software is used by SmartPLS 3.0. The reliability and validity of the scale were tested by Cronbach's Alpha, Average Variance Extracted (Pvc) and Composite Reliability (Pc). Followed by a linear structural model SEM was used to test the research hypotheses.

3.1   Results and Discussion

Structural Equation Modeling (SEM) is used on the theoretical framework. Partial Least Square method can handle many independent variables, even when multicollinearity exists. PLS can be implemented as a regression model, predicting one or more dependent variables from a set of one or more independent variables or it can be implemented as a path model. Partial Least Square (PLS) method can associate with the set of independent variables to multiple dependent variables.

3.2   Consistency and Reliability

In this reflective model convergent validity is tested through composite reliability or Cronbach's alpha. Composite reliability is the measure of reliability since Cronbach's alpha sometimes underestimates the scale reliability. Table 2 shows that composite reliability varies from 0.876 to 0.900 which is above preferred value of 0.5. This proves that model is internally consistent. To check whether the indicators for variables display convergent validity, Cronbach's alpha is used. From Table 2, it can be observed that all the factors are reliable (>0.60) and Pvc > 0.5.

Table 2. Cronbach's alpha, composite reliability (Pc) and AVE values (Pvc)

Factor	Cronbach's alpha	Average Variance Extracted (Pvc)	Composite reliability (Pc)	P	Findings
cusatis	0.857	0.776	0.912	0.000	Supported
loyalty	0.844	0.689	0.897	0.000	Supported
network	0.812	0.640	0.876	0.000	Supported
perqua	0.861	0.644	0.900	0.000	Supported

Source: *Author*

3.3   Structural Equation Modeling (SEM) in the First

SEM results in the first showed that the model is compatible with data research. The customer satisfaction is affected by network and perceive quality. The customer loyalty is affected by customer satisfaction and AIJS. The price perception and trust are not relative with customer satisfaction as Table 3 (Fig. 2).

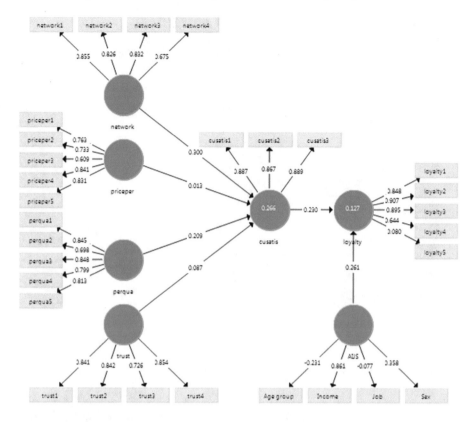

Fig. 2. Structural Equation Modeling (SEM) in the first **Source:** *Author*

Table 3. Structural Equation Modeling (SEM)

Relation	Beta	SE	T-value	P	Findings
AIJS->loyalty	0.261	0.087	2.994	0.003	Supported
cusatis->loyalty	0.230	0.079	2.922	0.004	Supported
network->cusatis	0.300	0.086	3.476	0.001	Supported
perqua->cusatis	0.209	0.087	2.403	0.017	Supported
priceper->cusatis	0.013	0.065	0.203	0.839	Unsupported
trust->cusatis	0.87	0.088	0.994	0.321	Unsupported

Source: *Author*

3.4 Structural Equation Modeling (SEM) in the Last

SEM results showed that the model is compatible with data research: SRMR, d_ULS and d_G has P-value = 0.000 (<0.05) (Henseler et al. 2016) (Fig. 3 and Table 4).

Table 4. Standard of model SEM

Standard	Beta	SE	T-value	P	Findings
SRMR	0.073	0.004	16.966	0.000	Supported
d_ULS	0.811	0.068	11.918	0.000	Supported
d_G	0.494	0.056	8.864	0.000	Supported

Source: *Author*

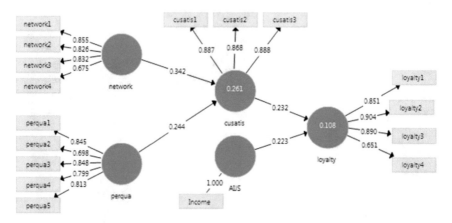

Fig. 3. Structural Equation Modeling (SEM) in the last **Source:** *Author*

In bootstrapping, resampling methods are used to compute the significance of PLS coefficients. Output of significance levels can be retrieved from boot-strapping option. Table 5 shows the results of hypotheses testing; all the t values above 1.96 are significant at the .05 level. Hypotheses H1, H3, H5 and H1 were supported. The results indicated a positive ($\beta = 0.232$) and significant (p < 0.05) association between customer satisfaction and loyalty (Table 5).

Observations have Outer Loadings greater than 0.5 (P < 0.05) so they are significant (Wong 2013) (Table 6).

Observations have Outer Weights lower than 0.05 (P < 0.05) so they are supported (Table 7).

Table 5. Structural Equation Modeling (SEM)

Relation	Beta	SE	T-value	P	Findings
AIJS → loyalty	0.223	0.067	3.328	0.001	Supported
cusatis → loyalty	0.232	0.079	2.950	0.003	Supported
network → cusatis	0.342	0.082	4.184	0.000	Supported
perqua → cusatis	0.244	0.081	3.015	0.003	Supported

Source: *Author*

Table 6. Outer loadings

Relation	Beta	SE	T-value	P	Findings
cusatis1 ← cusatis	0.887	0.022	40.380	0.000	Supported
cusatis2 ← cusatis	0.868	0.036	24.341	0.000	Supported
cusatis3 ← cusatis	0.888	0.022	39.733	0.000	Supported
loyalty1 ← loyalty	0.851	0.036	23.912	0.000	Supported
loyalty2 ← loyalty	0.904	0.028	32.053	0.000	Supported
loyalty3 ← loyalty	0.890	0.031	29.172	0.000	Supported
loyalty4 ← loyalty	0.651	0.072	9.074	0.000	Supported
network1 ← network	0.855	0.027	32.158	0.000	Supported
network2 ← network	0.826	0.036	22.998	0.000	Supported
network3 ← network	0.832	0.037	22.782	0.000	Supported
network4 ← network	0.675	0.064	10.516	0.000	Supported
perqua1 ← perqua	0.845	0.021	39.360	0.000	Supported
perqua2 ← perqua	0.698	0.057	12.333	0.000	Supported
perqua3 ← perqua	0.848	0.027	31.391	0.000	Supported
perqua4 ← perqua	0.799	0.039	20.484	0.000	Supported
perqua5 ← perqua	0.813	0.033	24.495	0.000	Supported

Source: *Author*

Table 7. Outer loadings

Relation	Beta	SE	T-value	P	Findings
cusatis1 ← cusatis	0.349	0.030	11.749	0.000	Supported
cusatis2 ← cusatis	0.356	0.033	10.690	0.000	Supported
cusatis3 ← cusatis	0.429	0.039	10.986	0.000	Supported
loyalty1 ← loyalty	0.269	0.039	10.986	0.000	Supported
loyalty2 ← loyalty	0.329	0.032	10.189	0.000	Supported
loyalty3 ← loyalty	0.332	0.036	9.124	0.000	Supported
loyalty4 ← loyalty	0.273	0.064	4.279	0.000	Supported
network1 ← network	0.400	0.04	8.988	0.000	Supported
network2 ← network	0.311	0.034	9.117	0.000	Supported
network3 ← network	0.292	0.037	7.969	0.000	Supported
network4 ← network	0.235	0.051	4.596	0.000	Supported
perqua1 ← perqua	0.276	0.031	8.873	0.000	Supported
perqua2 ← perqua	0.206	0.045	4.581	0.000	Supported
perqua3 ← perqua	0.269	0.034	7.955	0.000	Supported
perqua4 ← perqua	0.208	0.037	5.638	0.000	Supported
perqua5 ← perqua	0.281	0.032	8.896	0.000	Supported

Source: *Author*

4 Conclusion

As the result of data analysis in results and discussion, four hypotheses were accepted. Hypothesis H5 was proved to be true regardless of income. The first hypothesis was about the impact of network quality on customer satisfaction which was proven that there was a positive effect of network quality on customer satisfaction. With the Beta equals to 0.342, network quality was considered the first one that had the most positive on customer satisfaction. Perceived quality had Beta of 0.244 - the second value. Finally, customer satisfaction and Income was the factors had positive impact on customer loyalty.

4.1 Limitations and Suggestions for Further Research

Besides the implications, this research has some limitations. The scope of this research has some limitations which apply to how broadly its outcomes can be applied. These limitations must be acknowledged to add context to the conclusions from the research results. Firstly, this questionnaire used in this research was answered by respondents living in Ho Chi Minh city only. According to the result of this research and the limitations above, further research may consider the following points:

This research's result is applied only to Vietnam. The same findings may not occur in other countries because of culture differences. Further research could apply the findings of this thesis as an example to the Internet market in other countries. In addition, to get a clear picture and better understanding of customer loyalty, further research may investigate other factors such as the impact of cultural differences on different countries. Future research may also study different and additional variables that could influence customer loyalty.

In brief, the findings of this research provide direction for ISP to determine which attributes to focus on enhancing overall customer loyalty. The research has value because it proposed and justify a theoretical model to explain customer loyalty in the context of Vietnamese telecommunication.

References

Aaker, D.A.: Managing Brand Equity: Capitalizing on the Value of a Brand Name. The Free Press, New York (1991)

Aksoy, L.: Linking satisfaction to share of deposits: an application of the wallet allocation rule. Int. J. Bank Mark. 32(1), 28–42 (2014)

Amin, M., Isa, Z., Fontaine, R.: Islamic banks: contrasting the drivers of customer satisfaction on image, trust, and loyalty of Muslim and non-Muslim customers in Malaysia. Int. J. Bank Mark. 31(2), 79–97 (2013)

Amin, M.: Internet banking service quality and its implication on e-customer satisfaction and e-customer loyalty. Int. J. Bank Mark. 34(3), 280–306 (2016)

Choudrie, J.: The demographics of broadband residential consumers in a British local community: the London borough of Hillingdon. J. Comput. Inf. Syst. 1(3), 93–101 (2005)

Chow, S., Holden, R.: Toward an understanding of trust. J. Manag. Issues **9**(3), 257–298 (1997)

Chun, Y.S., Hahn, M.: Network externality and future usage of internet services. Internet Res. **17**(2), 156–168 (2007)

Dick, A.S., Basu, K.: Customer loyalty: toward an integrated conceptual framework. Acad. Mark. Sci. **22**(2), 99–113 (1994)

Henseler, J., Hubona, G., Ray, P.A.: Using PLS path modeling in new technology research: updated guidelines. Ind. Manag. Data Syst. **116**(1), 2–20 (2016)

Khatibi, A.A., Ismail, S., Thyagarajan, V.: What drives customer loyalty: an analysis from the telecommunication industry. J. Target. Meas. Anal. Mark. **11**(1), 34–44 (2002)

Kashif, M., Wan Shukran, S.S., Rehman, M.A., Sarifuddin, S., Estelami, H., Heinonen, K.: Customer satisfaction and loyalty in Malaysian Islamic banks: a PAKSERV investigation. Int. J. Bank Mark. **33**(1), 23–40 (2015)

Hart, C.W., Johnson, M.D.: Growing the trust relationship. Mark. Manag. **8**(1), 8–19 (1999)

Yaacob, M.R.B.: Determinants of customer satisfaction towards broadband services in Malaysian, pp. 123–134 (2011)

Monroe, K.B.: Pricing - Making Profitable Decisions. McGraw-Hill, New York (1991)

Morgan, R.M., Hunt, S.D.: The commitment-trust theory of relationship marketing. J. Mark. **58**(3), 20–38 (1994)

Murphy, J.A.: The Lifebelt: The Definition Guide to Managing Customer Retention. Wiley, Chichester (2001)

Ranaweera, C., Neely, A.: Some moderating effects on the service quality-customer retention link. Int. J. Oper. Prod. Manag. **23**, 230–248 (2003)

Ramseook-Munhurrun, P., Naidoo, P.: Customers' perspectives of service quality in internet banking. Serv. Mark. Q. **32**(4), 247–264 (2011)

Reichheld, F.F., Teal, T.: The Loyalty Effect. Havard Business School Press, Boston (1996)

Zeithaml, V.A., Berry, L.L., Parasuraman, A.: The behavioral consequences of service quality. J. Mark. **60**(2), 31–46 (1996)

Wang, Y., Lo, H., Yang, Y.: An integrated framework for service quality, customer value, satisfaction: evidence from China's telecommunication industry. Inf. Syst. Front. **6**(4), 325–340 (2004)

Wong, K.: Partial least squares structural equation modeling (PLS-SEM) techniques using SmartPLS. Mark. Bull. **24**(1), 1–32 (2013)

Woo, K.S., Fock, H.K.Y.: Customer satisfaction in the Hong Kong mobile phone industry. J. Serv. Ind. **19**(3), 162–174 (1999)

A Convex Combination Method for Quantile Regression with Interval Data

Somsak Chanaim[1,2(✉)], Chatchai Khiewngamdee[1,3], Songsak Sriboonchitta[1,2], and Chongkolnee Rungruang[4]

[1] Center of Excellence in Econometrics,
Chiang Mai University, Chiang Mai 50200, Thailand
somsak_ch@cmu.ac.th
[2] Faculty of Economics, Chiang Mai University, Chiang Mai 50200, Thailand
songsakecon@gmail.com
[3] Department of Agricultural Economy and Development,
Faculty of Agriculture, Chiang Mai University, Chiang Mai 50200, Thailand
getliecon@gmail.com
[4] Faculty of Commerce and Management, Prince of Songkla University,
Trang Campus, Trang 92000, Thailand
chongkolnee.r@psu.ac.th

Abstract. This paper studies a quantile regression under asymmetric Laplace distribution (semi-parametric model) with interval valued data. Generally, the center point of the interval data has been used to represent the sample data for estimated parameter of the model. This paper uses the convex combination method to find the best point to estimate parameter in the quantile regression model. We apply the quantile capital asset pricing model (quantile CAPM) to present the result.

1 Introduction

Since introduced by Koenker and Bassett [1] in 1978, quantile regression has been a useful tool in economics and statistics. Unlike least squares estimation of conditional mean models, quantile regression estimates the differential effects of covariates on various quantiles in the conditional distribution of a response variable. It becomes an appropriate model when the distribution of the response variable appears to be unknown or conditional expectation does not exist. More importantly, quantile regression provides a framework for robust inference.

Recently, many studies have developed quantile regression model by combining with other technique in order to expand the usage of quantile regression model and gain more accurate results. For instance, Xie et al. [2] applied varying-coefficient approach to quantile regression modeling under random data censoring. They found that varying-coefficient method can be further improved when implemented within a composite quantile regression framework. Frumento and Bottai [3] introduced an estimation equation for censored, truncated quantile regression. This method allows estimation of quantile regression models under

© Springer International Publishing AG 2018
L. H. Anh et al. (eds.), *Econometrics for Financial Applications*, Studies in Computational Intelligence 760, https://doi.org/10.1007/978-3-319-73150-6_35

random, covariate-dependent truncation and censoring. Furthermore, Das and Ghosal [4] proposed a Bayesian semi-parametric approach for fitting simultaneous quantile regression using quadratic and cubic B-spline basis function. Wu and Yao [6] studied a semi-parametric mixture of quantile regression model and Galvao and Kato [5] studied fixed effects estimation of smoothed quantile regression models for panel data.

Interval data analysis has been applied in several frameworks of regression models (Billard and Diday [11], Neto and de Carvalho [15] and Fagundes et al. [7]). Similarly, quantile regression modeling has also applied interval-valued data. In Fagundes et al. [18], a quantile regression model for interval data was proposed. In this study, range and center of the input data represent the interval variable and a smooth function between vectors composed by interval variables is defined. However, the parameter estimated from center method does not ensure the minimal error of the regression model, as shown in Chanaim et al. [22], the Akaike information criterion (AIC) value from center method is greater than one obtained from convex combination method. Therefore, in this paper, we propose a convex combination approach for quantile regression modeling in order to generalize the obtained values of parameter from the center method to the solution set of convex combination technique. In doing so, we apply the application of quantile capital asset pricing model (quantile CAPM).

The rest of the paper is organized as follows. Section 2 presents methodology. Section 2.2.2 explains the convex combination approach for quantile regression model with interval data. The empirical results are discussed in Sect. 3 and Sect. 4 concludes the paper.

2   Methodology

2.1   Operation with Interval Arithmetics

Let $P_i = [\underline{P}_i, \overline{P}_i]$, be a bounded lower and bounded higher interval data i, $i = 1, 2, \cdots, n$. We can define arithmetic operations.

For addition

$$P_i + P_j = [\underline{P}_i + \underline{P}_j, \overline{P}_i + \overline{P}_j]. \tag{1}$$

For subtraction

$$P_i - P_j = [\underline{P}_i - \overline{P}_j, \overline{P}_i - \underline{P}_j]. \tag{2}$$

For multiplication

$$P_i \cdot P_j = [\min\ B, \max\ B],\ B = \{\underline{P}_i\underline{P}_j, \underline{P}_i\overline{P}_j, \overline{P}_i\underline{P}_j, \overline{P}_i\overline{P}_j\}. \tag{3}$$

For division, we can define when $\underline{P}_i > 0$, we have

$$1/P_i = [1/\overline{P}_i, 1/\underline{P}_i] \tag{4}$$

$$P_i/P_j = P_i \cdot (1/P_j). \tag{5}$$

Addition and Multiplication by scalar

$$P_i + a = [\underline{P}_i + a, \overline{P}_i + a] \tag{6}$$

$$a \cdot P_i = \begin{cases} [a \cdot \overline{P}_i, a \cdot \underline{P}_i], & a < 0 \\ 0, & a = 0 \\ [a \cdot \underline{P}_i, a \cdot \overline{P}_i], & a > 0. \end{cases} \tag{7}$$

For logarithm function we can define, if $\underline{P}_i > 0$ then

$$\log P_i = [\log \underline{P}_i, \log \overline{P}_i]. \tag{8}$$

For more details about interval arithmetics see Moore et al. ([12], Chap. 2) or Nguyen et al. [13].

2.2 Linear Regression Models with Interval Data

In this section, we review the traditional methods for linear regression models with interval data and introduce our new method.

2.2.1 The Center Method

This method was proposed by Billard and Diday [11] with the main idea to use the center of the interval data to make a prediction using regression equation in the simplest form:

$$Y_c = X_c \beta + \varepsilon_c, \tag{9}$$

where $Y_c = \dfrac{\underline{Y} + \overline{Y}}{2}, X_c = \left(1, \dfrac{\underline{X}_1 + \overline{X}_1}{2}, \cdots, \dfrac{\underline{X}_m + \overline{X}_m}{2}\right), \beta = (\beta_0, \beta_1, \cdots,$

$\beta_m)^T$. This model can use OLS or maximum likelihood method to estimate parameter β and the coefficient of determination (R^2) is easy to define like an ordinary coefficient of determination (R^2) by

$$R_c^2 = \frac{\sum\limits_{i=1}^{n}(\hat{Y}_{c,i} - \overline{Y}_c)^2}{\sum\limits_{i=1}^{n}(Y_{c,i} - \overline{Y}_c)^2}, \quad \overline{Y}_c = \sum\limits_{i=1}^{n}\frac{Y_{c,i}}{n}. \tag{10}$$

2.2.2 Convex Combination Method for Quantile Regression Model with Interval-Valued Data

In 2016, Fagundes et al. [18] used the center method for quantile regression and Chanaim et al. [22] proposed the convex combination method for regression with interval data. This method is the generalized from of center method. In the quantile regression at quantile level $\tau \in (0,1)$ we can apply convex combination method and define it in the same way as the mean regression by

$$Y = \alpha_y(\tau)(\underline{Y} + (1-\alpha_y(\tau))\overline{Y}, \ \alpha_y(\tau) \in [0,1] \tag{11}$$

$$X_j = \alpha_j(\tau)\underline{X}_j + (1-\alpha_j(\tau))\overline{X}_j, \ \alpha_j(\tau) \in [0,1] \ \forall j = 1,2,\cdots,m \tag{12}$$

$$Y = \beta_0(\tau) + \sum_{j=1}^{m} \beta_j(\tau)X_j + \varepsilon(\tau). \tag{13}$$

It is easy to see that if we choose $\alpha_y(\tau), \alpha_j(\tau)$ equal $\dfrac{1}{2}$, this is the center method. In this method we find parameters $\alpha_y(\tau), \ \alpha_j(\tau)$ and $\beta_j(\tau)$ by minimizing the objective function

$$\min_{\alpha_y(\tau),\alpha_1(\tau),\cdots,\alpha_m(\tau),\beta_0(\tau),\beta_1(\tau),\cdots,\beta_m(\tau)} \sum_{i=1}^{n} \rho_\tau \left(Y_i - \beta_0(\tau) - \sum_{j=1}^{m} \beta_j(\tau)X_{ji} \right), \tag{14}$$

where $\rho_\tau(\cdot)$ is the check function defined by

$$\rho_\tau(x) = x(\tau - 1_{\tau<0}) = \begin{cases} x(\tau-1), & x < 0 \\ x\tau, & x \geq 0 \end{cases} \tag{15}$$

This method is equivalent to maximum likelihood methods for the following asymmetric Laplace Distribution (ALD), Autchariyapanitkul et al. [19,20] used ALD in this form

$$f(\varepsilon|\tau,\sigma) = \frac{\tau(1-\tau)}{\sigma} \exp\left[-\frac{\varepsilon(\tau-1_{\varepsilon<0})}{\sigma} \right], \quad \varepsilon \in \mathbb{R}$$

$$F(\varepsilon|\tau,\sigma) = \begin{cases} \tau \exp\left[\dfrac{(1-\tau)\varepsilon}{\sigma} \right] & , \quad \varepsilon < 0 \\ 1 + (\tau-1)\exp\left[-\dfrac{\tau\varepsilon}{\sigma} \right] & , \quad \varepsilon \geq 0 \end{cases}$$

$$F^{-1}(u|\tau,\sigma) = \varepsilon = \begin{cases} \dfrac{\sigma(\ln u - \ln \tau)}{1-\tau} & , \quad 0 < u \leq \tau \\ -\dfrac{\sigma}{\tau} \ln\left(\dfrac{1-u}{1-\tau} \right) & , \quad \tau < u < 1. \end{cases}$$

where $1_{\varepsilon<0}$ is indicator function, $f(\cdot)$ is probability density function, $F(\cdot)$ is cumulative distribution function, $F^{-1}(\cdot)$ is quantile function, u is uniform(0,1) and $\mathbb{E}(\varepsilon) = 0$. The log likelihood function for ALD with $\varepsilon_1,\cdots,\varepsilon_n$ is

$$LL = n\left(\log \tau + \log(1-\tau) - \log \sigma \right) - \sum_{i=1}^{n} \frac{\varepsilon_i(\tau - 1_{\varepsilon_i<0})}{\sigma}, \tag{16}$$

where $\varepsilon_i = y_i - \beta_0(\tau) - \sum_{j=1}^{m} \beta_j(\tau)x_{ji}, \ i = 1,\cdots,n$ and we can define the validation measure for quantile regression model by applying the goodness of fit proposed by Koenker [8] as

$$R_1(\tau) = 1 - \frac{\tau \sum_{\varepsilon_i \geq 0} |\varepsilon_i| + (1-\tau) \sum_{\varepsilon_i < 0} |\varepsilon_i|}{\tau \sum_{\delta_i \geq 0} |\delta_i| + (1-\tau) \sum_{\delta_i < 0} |\delta_i|} \tag{17}$$

where $\delta_i = y_i - y_\tau$, y_τ is equal quantile at level τ of data set $y_1, \cdots, y_n$, $y_i = \alpha_y(\tau)\underline{y}_i + (1 - \alpha_y(\tau))\overline{y}_i, i = 1, \cdots, n$.

3 Empirical Results

In this section, we apply the capital asset pricing model (CAPM). Specifically we use the quantile CAPM model to explain the relationship between the stock market and stock price. First, we investigate the relationship of the S&P500 stock index interval return and stock price interval return of Advance Micro Devices, Inc (AMD). Second, we investigate the relationship of the S&P500 stock index interval return and stock price interval return of Intel Corp. (INTC). For the data point of risk free we use 10-year US government bond yield and convert the return to interest rate per week. We use weekly data between Jan. 1, 2012–May 31, 2017. From the Sect. 2, we can define return interval by

$$\underline{r}_i = \log\left(\frac{\underline{s}_{i+1}}{\overline{s}_i}\right), \quad \overline{r}_i = \log\left(\frac{\overline{s}_{i+1}}{\underline{s}_i}\right),$$

where $\underline{r}_i$ is a lower return, $\overline{r}_i$ is a higher return, $\underline{s}_i$ and $\overline{s}_i$ are stock price or index of the stock market at lower and higher respectively at time i. The quantile CAPM model has the form

$$\alpha_{r_s}(\tau)\underline{r}_s + (1 - \alpha_{r_s}(\tau))\overline{r}_s - r_f = \beta_0(\tau) + \beta_1(\tau)\left[\alpha_{r_m}(\tau)\underline{r}_m + (1 - \alpha_{r_m}(\tau))\overline{r}_m - r_f\right] + \varepsilon(\tau),$$

where $\underline{r}_s$ is the lower return of stock, $\overline{r}_s$ is the upper return of stock, $\alpha_{r_s}(\tau)$ is the weight point of stock price at quantile τ, $\underline{r}_m$ is the lower return of stock index, $\overline{r}_m$ is the upper return of stock index, $\alpha_{r_m}(\tau)$ is the weight point of stock index return at quantile τ and r_f is the risk free point. By MLE method and ALD assumption of $\varepsilon(\tau)$ for density function, $\mathbb{E}(\varepsilon) = F_\varepsilon^{-1}(\tau) = 0$ at quantile τ (Fig. 1).

$$\mathbb{E}\left[\alpha_{r_s}(\tau)\underline{r}_s + (1 - \alpha_{r_s}(\tau))\overline{r}_s - r_f\right] = \beta_0(\tau) + \beta_1(\tau)\left[\alpha_{r_m}(\tau)\underline{r}_m + (1 - \alpha_{r_m}(\tau))\overline{r}_m - r_f\right].$$

Table 1 shows the general summary of statistics of interval return data of AMD, INTC and data point of the risk free. Then, we estimated parameters of quantile CAPM at quantile level $\tau = \{.05, .50, .95\}$ of S&P500 and AMD (as shown in Table 2), the prediction quantile regressions are

$$\mathbb{E}(r_{AMD}) = -0.0021 + 1.3210\left(0.2617\underline{r}_{SP500} + 0.7383\overline{r}_{SP500}\right), \quad \text{for } \tau = .05,$$
$$\mathbb{E}(r_{AMD}) = 0.0067 + 1.9529\left(0.6012\underline{r}_{SP500} + 0.3988\overline{r}_{SP500}\right), \quad \text{for } \tau = .50,$$
$$\mathbb{E}(r_{AMD}) = -0.0039 + 0.0214\left(0.7984\underline{r}_{SP500} + 0.2016\overline{r}_{SP500}\right), \quad \text{for } \tau = .95,$$

where $r_{AMD} = \alpha_{r_s}(\tau)\underline{r}_{AMD} + (1 - \alpha_{r_s}(\tau))\overline{r}_{AMD}$, the goodness of fit $R_1(\tau)$ equals 0.0559, 0.1059 and 0.0740, respectively.

Finally, in Table 3, we estimated parameters of quantile CAPM at quantile level $\tau = \{.05, .50, .95\}$ of S&P500 and INTC, the prediction quantile regressions are

$$\mathbb{E}(r_{INTC}) = -0.0017 + 0.8265\left(0.1792\underline{r}_{SP500} + 0.8208\overline{r}_{SP500}\right), \quad \text{for } \tau = .05,$$
$$\mathbb{E}(r_{INTC}) = -0.0007 + 1.1246\left(0.5335\underline{r}_{SP500} + 0.4665\overline{r}_{SP500}\right), \quad \text{for } \tau = .50,$$
$$\mathbb{E}(r_{INTC}) = -0.0013 + 0.7657\left(0.8373\underline{r}_{SP500} + 0.1628\overline{r}_{SP500}\right), \quad \text{for } \tau = .95,$$

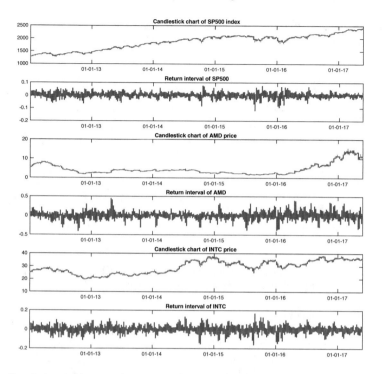

Fig. 1. Graph interval price (candlestick) and interval return of S&P500, AMD and INTC.

Table 1. Summary statistics of data

	S&P500 return		AMD return		INTC (Intel) return		Risk free
	Low	High	Low	High	Low	High	
Mean	−0.0207	0.0252	−0.1039	0.1089	−0.0419	0.0445	0.0004
Medean	−0.0165	0.0229	−0.0848	0.0896	−0.0358	0.0400	0.0004
Std	0.0172	0.0144	0.0795	0.0798	0.0292	0.0271	0.0001
Min	−0.1193	−0.0027	−0.4070	−0.0141	−0.1605	−0.0126	0.0003
Max	0.0033	0.0764	0.0334	0.4471	0.0012	0.1680	0.0006
Skewness	−1.8124	0.9491	−1.4003	1.4908	−1.3177	1.0439	0.1388
Kurtosis	8.1394	3.9772	4.9286	5.5925	4.8973	4.5475	1.9935
Obs.	282						

where $r_{INTC} = \alpha_{r_s}(\tau)\underline{r}_{INTC} + (1 - \alpha_{r_s}(\tau))\overline{r}_{INTC}$, the goodness of fit $R_1(\tau)$ equals 0.1495, 0.2136 and 0.1504 respectively.

Table 2. Parameter estimation and standard error of quantile CAPM model at quantile $\tau = \{.05, .50, .95\}$ by using MLE for $S\&P500$ and AMD

	AMD					
	$\tau = .05$		$\tau = .50$		$\tau = .95$	
	Parameter	SD	Parameter	SD	Parameter	SD
$\alpha_{r_s}(\tau)$	0.0520	0.0002	0.5079	0.0199	0.9346	0.0004
$\alpha_{r_m}(\tau)$	0.2617	0.0009	0.6012	0.0294	0.7984	0.0060
$\beta_0(\tau)$	−0.0021	0.0001	0.0067	0.0039	0.0008	0.0002
$\beta_1(\tau)$	1.3210	0.0014	1.9529	0.1470	1.2463	0.0074
σ	0.0047	0.0002	0.0214	0.0011	0.0045	0.0002
$R_1(\tau)$	0.0559		0.1059		0.0740	

Table 3. Parameter estimation and standard error of quantile CAPM model at quantile $\tau = \{.05, .50, .95\}$ by using MLE for $S\&P500$ and AMD

	INTC (Intel)					
	$\tau = .05$		$\tau = .50$		$\tau = .95$	
	Parameter	SD	Parameter	SD	Parameter	SD
$\alpha_{r_s}(\tau)$	0.0522	0.4438	0.5190	0.0029	0.9482	1.4081
$\alpha_{r_m}(\tau)$	0.1792	1.4981	0.5351	0.0014	0.8373	4.6080
$\beta_0(\tau)$	−0.0017	0.0710	−0.0007	0.0003	−0.0013	0.0045
$\beta_1(\tau)$	0.8265	1.0668	1.1246	0.0029	0.7657	11.4847
σ	0.0017	0.0001	0.0071	0.0004	0.0016	0.0001
$R_1(\tau)$	0.1495		0.2136		0.1504	

From Figs. 2 and 3, we can observe that $\alpha_{r_{AMD}}(\tau)$ and $\alpha_{r_{INTC}}(\tau)$ look like an increasing function as τ increases that means center point is not a good representative data for interval.

Figure 4 shows the quantile regression line of S&P500 and AMD on the scatter plot of data point from convex combination at $\tau = \{.05, .50, .95\}$, respectively(left side hand) and shows the quantile regression line of S&P500 and ITNC on the scatter plot of data point from convex combination at $\tau = \{.05, .50, .95\}$, respectively (right hand side).

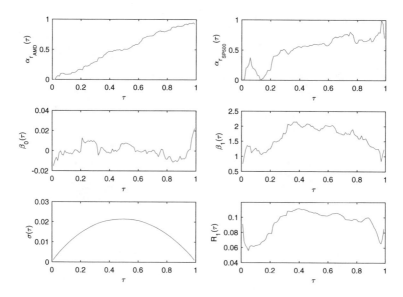

Fig. 2. Graph of all parameters in quantile CAPM of S&P500 and AMD for $\tau \in (0, 1)$.

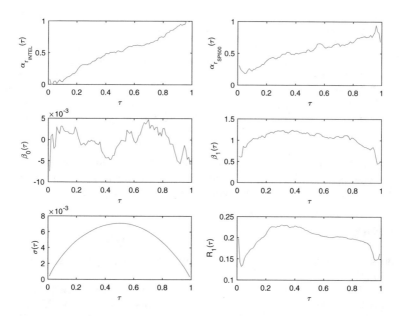

Fig. 3. Graph of all parameters in quantile CAPM of S&P500 and INTC for $\tau \in (0, 1)$.

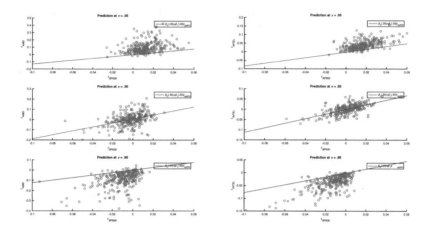

Fig. 4. Forecast of AMD return (left) and forecast of ITNC return (right) at quantile $\{\tau = .05, .50, .95\}$.

4   Conclusions

The convex combination method is the generalized from of the center method. For quantile regression, this method can work very well by using Maximum likelihood technique to estimate. From using the real example data for quantile CAPM, we can see that we should not use the center point to represent the interval data but we should find best the convex combination for each quantile because from the estimation of $\alpha_{r_{AMD}}(\tau)$ or $\alpha_{r_{ITNC}}(\tau)$ the center point is good if we use quantile regression around median. The next future research we will apply this method on financial time series interval data to forecast stock price or volatility(Garch model family) and to find the optimal portfolio for investment.

Acknowledgement. The authors thank Prof. Dr. Vladik Kreinovich for giving comments on manuscript. We are grateful for financial support from Center of Excellence in Econometrics, Faculty of Economics and Graduate School, Chiang Mai University.

References

1. Koenker, R., Bassett Jr., G.: Regression quantiles. Econom. J. Econom. Soc. **46**(1), 33–50 (1978)
2. Xie, S., Wan, A.T., Zhou, Y.: Quantile regression methods with varying-coefficient models for censored data. Comput. Stat. Data Anal. **88**, 154–172 (2015)
3. Frumento, P., Bottai, M.: An estimating equation for censored and truncated quantile regression. Comput. Stati. Data Anal. **113**, 53–63 (2016)
4. Das, P., Ghosal, S.: Bayesian quantile regression using random B-spline series prior. Comput. Stat. Data Anal. **109**, 121–143 (2017)
5. Galvao, A.F., Kato, K.: Smoothed quantile regression for panel data. J. Econom. **193**(1), 92–112 (2016)
6. Wu, Q., Yao, W.: Mixtures of quantile regressions. Comput. Stat. Data Anal. **93**, 162–176 (2016)

7. Fagundes, R.A., De Souza, R.M., Cysneiros, F.J.A.: Interval kernel regression. Neurocomputing **128**, 371–388 (2014)

8. Koenker, R., Machado, J.A.: Goodness of fit and related inference processes for quantile regression. J. Am. Stat. Assoc. **94**(448), 1296–1310 (1999)

9. Tanaka, H., Uejima, S., Asai, K.: Linear regression analysis with fuzzy model. IEEE Trans. Syst. Man Cybern. **12**(6), 903–907 (1982)

10. William, F.: Capital asset prices: a theory of market equilibrium under conditions of risk. J. Financ. **19**(3), 425–442 (1964)

11. Billard, L., Diday, E., Regression analysis for interval-valued data. In: Data Analysis, Classification, and the Related Methods, pp. 369–374. Springer (2000)

12. Moore, R.E., Kearfott, R.B., Cloud, M.J.: Introduction to Interval Analysis, pp. 7–18. SIAM, Philadelphia (2009)

13. Nguyen, H.T., Kreinovich, V., Wu, B., Xiang, G.: Computing Statistics Under Interval and Fuzzy Uncertainty. Studies in Computational Intelligence, vol. 393. Springer, Heidelberg (2012)

14. Billard, L., Diday, E.: Symbolic regression analysis. In: Classification, Clustering, and Data Analysis, pp. 281–288. Springer (2002)

15. Neto, E.A.L., Carvalho, F.A.T.: Centre and range method for fitting a linear regression model to symbolic interval data. Comput. Stat. Data Anal. **52**, 1500–1515 (2008)

16. Maia, A.L.S., de Carvalho, F.D.A., Ludermir, T.B.: Forecasting models for interval-valued time series. Neurocomputing **71**(16), 3344–3352 (2008)

17. Domingues, M.A.O., Souza, R.M.C.R., Cysneiros, F.J.A.: A robust method for linear regression of symbolic interval data. Pattern Recogn. Lett. **31**, 1991–1996 (2010)

18. Fagundes, R.A., de Souza, R.M., Soares, Y.M.: Quantile regression of interval-valued data. In: Proceeding of the 2016 23rd International IEEE Conference on Pattern Recognition, pp. 2586–2591

19. Autchariyapanitkul, K., Chanaim, S., Sriboonchitta, S.: Quantile regression under asymmetric Laplace distribution in capital asset pricing model. In: Econometrics of Risk, pp. 219–231. Springer (2015)

20. Autchariyapanitkul, K., Chanaim, S., Sriboonchitta, S.: Evaluation of portfolio returns in Fama-French model using quantile regression under asymmetric Laplace distribution. In: Econometrics of Risk, pp. 233–244. Springer (2015)

21. John, L.: The valuation of risk assets and selection of risky investments in stock. Rev. Econ. Stat. **47**(1), 13–37 (1965)

22. Chanaim, S., Sriboonchitta, S., Rungruang, C.: A convex combination method for linear regression with interval data. In: Proceeding of the 2016 5th the International Symposium on Integrated Uncertainty in Knowledge Modelling and Decision Making (IUKM2016), pp. 469-480. Springer

The Influence of Corporate Culture on Employee Commitment

Do Huu Hai[1], Nguyen Minh Hai[2], and Nguyen Van Tien[2(✉)]

[1] Ho Chi Minh City University of Food Industry, Ho Chi Minh City, Vietnam
haidh1975@gmail.com
[2] Banking University of Ho Chi Minh City, Ho Chi Minh City, Vietnam
minhhai.nguyen77@gmail.com, lananh1903@gmail.com

Abstract. In this paper, we study the influence of corporate culture on employee commitment. This study is based on a survey of 289 employees, managers, and company leaders in Hanoi and Vinh City, Vietnam. We used Cronbach's alpha and EFA for preliminary assessments, CFA for testing whether the actual data fit a hypothesized measurement model, and SEM for checking the theoretical model. The collected data have been analyzed with SPSS 23 and AMOS 23 to assess the suitability of the research model, to test the relationships, and to assess the reliability and validity of measurement scales for evaluating the influence of the corporate culture on all aspects of employee commitment: affective commitment, normative commitment, and continuance commitment. The results show that our model is in very good accordance with the data.

1 Introduction

To be efficient, a company, in addition to capital, employees, technology, and management skills, must also have a favorable corporate culture, a culture ensuring employee satisfaction and their desire to help the company succeed. Many studies show that corporate culture has a considerable influence on employee commitment, and thus, on the company's success. Therefore, it is in the best interest of each company to develop a favorable corporate culture.

2 Research Overview

Several studies analyzed how corporate culture affects employee commitment. Some of these studies are based on data from Vietnam, some on data from other countries.

2.1 International Research on the Employee Commitment

Many research papers study the relationship between employees and their workplace (Allen and Meyer 1990, p. 1; Hult 2005, p. 249; Lok and Crawford 2004,

L. H. Anh et al. (eds.), *Econometrics for Financial Applications*, Studies in Computational Intelligence 760, https://doi.org/10.1007/978-3-319-73150-6_36

p. 321; Meyer et al. 1993, p. 538; Mowday 1998, p. 387; Mowday et al. 1979, p. 224; Porter et al. 1974, p. 604; Rashid et al. 2003, p. 713). In particular, several papers show that such phenomena as taking on extra tasks, refusing to take on extra tasks, absences, etc., affect the employee's work performance (Allen and Meyer 1990, p. 1; Cohen 2007, p. 34; Hogg and Terry 2001, p. 110; Mathieu and Zajac 1990, p. 171; Porter et al. 1974, p. 604; Rashid et al. 2003, p. 713; Wasti 2003, p. 303; Yu and Egri 2005, p. 336).

In particular, Silverthorne (2004) studied employee commitment to work for the company. On the one hand, employees with such a commitment work more productively. Vice versa, Greenberg and Baron (2003, p. 163) show that, based on the employee's behavior and attitude to work, one can predict his/her future commitment. Companies need employee commitment to minimize the personnel replacement and to avoid ineffectiveness of the existing employees (Mathieu and Zajac 1990, p. 171). For this purpose, individuals showing their commitment of staying may be respected and appreciated, and obtain a lot of benefits and incentives by their workplace (Mathieu and Zajac 1990, p. 171). The employee commitment benefits the society as a whole since it enhances the overall productivity and work performance (Mathieu and Zajac 1990, p. 171).

McKinnon et al. (2003, p. 28) showed that individuals are strongly affected by core values, respects, beliefs, and visions of their workplace. Like McKinnon et al. (2003, p. 28), Martin (2001, p. 263) has demonstrated that the employee commitment makes them willing to, if needed, carry out extra tasks which are not a part of their official job requirements. In general, it has been shown that the performance employees with high level of commitment usually exceed expectations (Martin 2001, p. 623). Similarly, Porter et al. (1974, p. 604; Brooks and Wallace 2006, p. 233) showed that employees with organizational commitment are more cheerful and more willing to volunteer to make contributions into the organizations. This is not a minor issue, having employees that are willing to take on extra tasks often provides an organization with a competitive edge, that makes it a winner in the marketplace competition; see Mathieu and Zajac (1990, p. 171).

Silverthorne (2004, pp. 592–593) and Ogaard et al. (2005, p. 25) show that the employee's suitability for the organization significantly affects their commitment. Nazir (2005, pp. 47–48) shares a similar view to Silverthorne (2004, pp. 592–593) that employee's satisfaction and organizational commitment largely result from the employee's suitability. Hult (2005, p. 250) has pointed out that if an employee is suitable for the working environment, her or his organizational commitment is high. To increase the employee commitment, it is therefore necessary to identify corporate cultures which are favorable for nurturing the employee commitment.

Becker (1960) was one of the first researchers to propose a theoretical framework for definitions of organizational commitment. In his opinion, employees exchange their loyalty and commitment for advantages and benefits such as good working conditions, efficient performance of familiar works in the organization. This theoretical framework is based on definitions of organizational commitment as the extent to which an employee determines his or her worth in the organization.

The study of *"Antecedents and Outcomes of Organizational commitment among Malaysia engineers"* by Muthuveloo and Rose (2005) demonstrates that organizational commitment has a considerable influence on the organization's operation. In order to achieve organizational commitment of employees, it is therefore imperative for managers and leaders to build and improve good relationship between employees and the organization, and to motivate employee's loyalty and dedication to the organization. The more appreciated an employees is, the longer he/she will stay.

In general, employee commitment is an important factor deciding the success of an organization (Chen 2004, p. 438). A committed employee is a valuable resource for the organization: such employees remain with their organization regardless of ups and downs of the organization, and this enhances the organization's competitiveness (Allen and Meyer 1990, p. 1; Rashid et al. 2003, p. 713; Yu and Egri 2005, p. 336). According to Buchanan (1974, p. 339) and Mathieu and Zajac (1990, p. 171), organizations have various methods to retain their talented employees and obtain their commitment and loyalty. Rashid et al. (2003, p. 713) consider the employee commitment as an honorable reward for the organization.

Among a variety of different models of organizational commitment, we focused on two models: by Mowday et al. (1979) and by Allen and Meyer (1990). After a thorough analysis of these models, we selected the model of Allen and Meyer (1990) for our analysis.

The paper by Allen and Meyer (1990) recommends a three-component model of commitment, incorporating affective, normative, and continuance commitment (Fig. 1).

- Affective commitment is defined as the individual's emotional attachment to – and involvement and identification with – the organization (Allen and Meyer 1990, p. 2; Lee et al. 2001, p. 597; Mowday 1998, p. 390; Rashid et al. 2003, p. 714; Wasti 2003, p. 303). Buchanan (1974, p. 533) clarifies that affective commitment is the attachment to objectives and values of the organization and a clear understanding of the employees's roles ion regards to these objectives and values. Greenberg and Baron (2003, p. 162) show that the employee active engagement and long-term work for an organization originated from the deep affinity with main objectives and values of the organization. In other words, when an employee have strong affective commitment, he or she tends to actively engage with the organization, and stay with it for a long time (Clugston et al. 2000, p. 7; Meyer et al. 1993, p. 539). According to Wasti (2003, p. 304), many empirical studies in commitment have shown there is a reliable evidence of the strong relation between emotional aspects and commitment – stronger than with other job aspects.
- Continuance commitment is a commitment related to the perceived or real cost of leaving (Greenberg and Baron 2003, p. 161; Lee et al. 2001, p. 597; Rashid et al. 2003, p. 714; Wasti 2003, p. 303). In term of this commitment, the fewer job alternatives for an individual to work at other organizations,

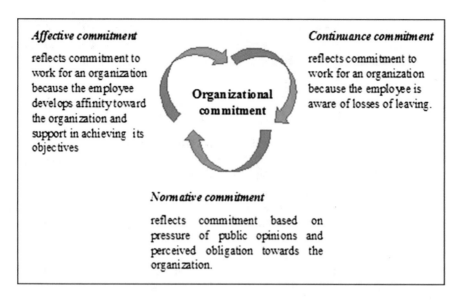

Affective commitment

reflects commitment to work for an organization because the employee develops affinity toward the organization and support in achieving its objectives

Organizational commitment

Continuance commitment

reflects commitment to work for an organization because the employee is aware of losses of leaving.

Normative commitment

reflects commitment based on pressure of public opinions and perceived obligation towards the organization.

Fig. 1. The three-component model of commitment by Allen and Meyer (1990).

the stronger the continuance commitment to the organization (Rashid et al. 2003, p. 714).

Employees tend to develop continuance commitment when they perceive that the cost of leaving the organization is too high (Clugston et al. 2000, p. 7; Meyer et al. 1993, p. 539).

- Normative commitment is the employee's perceived obligation to continue employment with their organization (Lee et al. 2001, p. 597; Rashid et al. 2003, p. 714; Wasti 2003, p. 303). Employees who work for the organization for a long time usually show normative commitment (Rashid et al. 2003, p. 714). A possible explanation is that organizational socialization takes place in the same way as societal socialization: social norms put employees under the tremendous pressure towards working for the organization (Clugston et al. 2000, p. 5). Individuals continue their employment with the organization because they feel that they should; for example, because their organisational loyalty is respected at the organisation or in the society (Clugston et al. 2000, p. 5; Meyer et al. 1993, p. 539; Rashid et al. 2003, p. 714). Greenberg and Baron (2003, p. 163) also show that employees remain in the organization because they are worried about their colleagues' disapproval if they decide yo leave.

In general, according to Meyer and Allen (1990), organizational commitment is a psychological state of the employee toward the organization related to the decision to remain a member of the organization. An employee's organizational commitment demonstrates their responsibility for work, their loyalty, and their belief in organizational values (O'Reilly 1986).

In other words, organizational commitment is earnestness and trust in the organization, as *"Employee commitment is the extent to which an employee gives their earnestness and trust in the organization's objectives and desires to stay with the organization".*

2.2   Vietnamese Research on Employee Commitment

Vietnamese studies on employee commitment have mainly focused on benefits, incentives, and other favorable conditions for employees to actively engage in the organization. Here are some typical studies.

The study *"Assessment of human resource management of travel businesses in Ho Chi Minh City"* (Dung 1999) collected the data from 86 businesses and found four major influences on employee's engagement to the organization including (1) assigned jobs, (2) training and promotion opportunities, (3) working conditions and environment and (4) income. Of these form, working conditions and environment is the most influential factor affecting the employee's engagement, and the effect of training and promotion opportunities is the smallest.

The study *"Evaluation on employee's organizational commitment based on job satisfaction in Vietnam"* (Dung and Abraham 2005) explored the relationship between job satisfaction and organizational commitment. The study shows three components of organization commitment – pride, effort, and loyalty – are affected by the following five main factors: work, payment, colleagues, supervision, and promotion. This study was based on the survey of 396 full-time employees from Ho Chi Minh City. The results of this study mean that an employee works very hard and efficiently and takes pride in the organization when he/she likes his/her job and gets on well with colleagues and the boss. It should be mentioned that this study focuses mostly on the organizational commitment – whether the employee will stay or not, and does not specifically analyze factors influencing the employee's degree of engagement with the organization.

The study *"Evaluation on employee's job satisfaction at Long Shin Company Limited"* (Lam 2009) gives some recommendations to enhance employees' job satisfaction by strengthening (1) senior-junior relationship, (2) sympathy for personal affairs, (3) payment and benefits, (4) working conditions and environment, (5) career development and prospects, (6) prospects of company development, and (7) efficiency of training. The author emphasized that the most important way to promote employees' satisfaction is to improve the senior-junior relationship.

3   Research Methodology

In this paper, we have used expert opinions, document analysis, and, more generally, qualitative research methods to decide how to gauge the corresponding quantities. Once the scales for all these quantities have been selected, we can apply quantitative methods to study how all four aspects of the corporate culture – mission, adaptability, consistency and participation – influence three components of

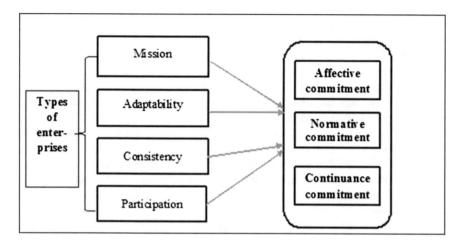

Fig. 2. The research model.

employee commitment: affective commitment, continuance commitment, and normative commitment; see Fig. 2.

This study uses the results of the survey of full-time employees, managers, and enterprise leaders in Hanoi and Vinh City. The questionnaire used for data collection consists of two parts: (1) Part A includes personal information and enterprise's information: gender, age, qualification, position in the company, the type, scale and main business line of the company. (2) Part B presents statements corresponding to mission, adaptability, consistency, and participation and to three components of commitment: affective commitment, continuance commitment and normative commitment. For each of these statements, a respondent selects one of the 5 possible Likert-scale-type answers ranging from (1) "Strongly disagree" to (5) "Strongly agree". Pronoun "I" is used in the questionnaires.

The statement are as follows:

Corporate culture

- Mission

    M1: The information about decisions made by managers is mostly correct.

    M2: Leaders and managers always "do what they say".

    M3: The company's development strategies force other companies to change their strategies to compete in a sector.

    M4: The company has a clear mission and my performance is mission-oriented and meaningful.

    M5: Everyone knows what to do to achieve the sustainable success.

    M6: The vision of the company motivates and promotes the employees.
- Adaptability

    A1: Creation and innovation are encouraged in the company.

    A2: The company is always willing to carry out new ideas and methods.

    A3: New ideas are always supported for development.

A4: We could adjust decisions according to customers.

A5: Failures are valuable lessons.

A6: We respond well to competitors and to other changes in the business environment.

- Consistency

  C1: It is easy for us to get unanimous resolution even when facing the most difficult problems.

  C2: We have a clear agreement on how to work properly and which ways are wrong.

  C3: Policies of the company are consistent and well-planned.

  C4: Objectives of management at different levels are consistent.

  C5: A person who ignores core values will face difficulties.

- Participation

  P1: The company always encourages the cooperation between/among its departments.

  P2: The ability of the staff regularly improves.

  P3: Everyone believes that they have a positive impact in the organization.

  P4: Tasks are assigned so that individuals see the connection between their work and the organization's objectives.

  P5: People work as members in a team.

  P6: Business plans are continuously made and everyone is involved in this process; emotional commitment.

Commitment

- Affective commitment

  AC1: I would be very happy to build my career in my current organization.

  AC2: I like discussing my organization with others.

  AC3: I do not see that "I am a member" in my organization.

  AC4: I do not feel engaged to my organization.

  AC5: I do not like to mention my organization.

  AC6: I am proud of my job and of my position in the organization.

- Continuance commitment

  CC1: It is difficult for me to leave the organization right now, even if that's what I want.

  CC2: I continue to work for my current organization because I have made a significant sacrifice here, and another organization may not be in line with my interests.

  CC3: If I did not have many personal relations in this organization, I could consider working elsewhere.

  CC4: Too many changes in my life will happen if I leave my organization now.

  CC5: I'm not afraid of what might happen if I quit my job without a substitute staff.

  CC6: I believe that those who have been trained in my profession must be responsible for the profession.

- Normative commitment

  NC1: I believe that frequent job changes are normal.

  NC2: One of the main reasons why I continue to work for this organization is that I believe that my loyalty is important and I have a moral obligation to work here.

  NC3: If I get an offer for a better job elsewhere, I feel that it is not nice to leave my organization immediately.

  NC4: I do not believe that a person should always be loyal to his/her organization.

  NC5: I do not feel any obligation to the leader when I leave the organization.

  NC6: I really feel that the problems of the organization are also my problems.

4 Research Results

4.1 Statistics of Sample Description

We sent the survey to 1000 folks. Of these folks, 315 filled in the questionnaires. 26 of them provided insufficient information. As a result, 289 valid surveys were used as data for the research.

4.2 Testing of Scale

4.2.1 Testing Reliability of the Scale

For each of eight characteristics – five characteristics describing the corporate culture and three characteristics describing employee commitment – we asked several questions which are, in our opinion, reflecting this characteristic. Our expectation is that, since all these questions describe the same characteristic, the corresponding answers will be strongly correlated.

To check to what extent they are indeed strongly correlated, we used the Cronbach's Alpha technique. For this purpose, we estimate the correlation among all the variables (which describes the reliability of the corresponding multi-statement scale), as well as the correlation between the combined total score and each of the variables. Only variables which are strongly correlation with the total score are retained, while variables with low correlation should be deleted from the scale. In general, variables are accepted with the reliability coefficient at or over 0.6 and the correlation coefficient between variables and the total score over 0.3. The Cronbach's alpha coefficients of variables are as follows.

Scale of Mission: The Cronbach's alpha coefficient is 0.807, and the correlation between each of the observed variables and the total variable is > 0.3; therefore, all six observed variables are retained in this scale.

Scale of Adaptability: The Cronbach's alpha coefficient is 0.896, and the correlation between each of the observed variables and the total variable is > 0.3; therefore, all six observed variables are retained in this scale.

Scale of Participation: The Cronbach's alpha coefficient is 0.816, and the correlation between each of the observed variables and the total variable is > 0.3; therefore, all six observed variables are retained in this scale.

Scale of Consistency: The Cronbach's alpha coefficient is 0.881, and the correlation between each of the observed variables and the total variable is > 0.3; therefore, all five observed variables are retained in this scale.

Scale of Affective commitment: The Cronbach's alpha coefficient is 0.812, and the correlation between each of the observed variables and the total variable is > 0.3; therefore, all six observed variables are retained in this scale.

Scale of Continuance commitment: The Cronbach's alpha coefficient is 0.832, and the correlation between each of the observed variables and the total variable is > 0.3; therefore, all six observed variables are retained in this scale.

Scale of Normative commitment: The Cronbach's alpha coefficient is 0.878, and the correlation between each of the observed variables and the total variable is > 0.3; therefore, all six observed variables are retained in this scale.

Overall, all 41 observed variables are retained.

4.2.2   Results of Factor Analysis

How consistent are different aspects of corporate culture? different aspects of employee commitment? To answer these questions, we used the KMO and Bartlett's Test is applied with the KMO (Kaiser-Meyer-Olkin) index in the range from 0.5 to 1 to assess whether the exploratory factor analysis (EFA) is indeed suitable in this case or not. The coefficient extraction method uses the Varimax rotation, and stops when extracting factors with the eigenvalue of 1. The scale is acceptable when the total variance of extracted factors is 50% or more. The results of applying this technique to the scales of the corporate culture and commitment are presented below.

Scale of the corporate culture: The KMO and Bartlett's Test in the analysis of 23 observed variables of the corporation culture with the high KMO index (0.845) and the significance level of 0 (sig = 0.000) demonstrates that the exploratory factor analysis is appropriate in this scale. With 23 variables introduced into the factor analysis, the results are extracted into four groups as shown in Table 1. The total variance is 59.58%. Thus, the factor analysis results of the corporate culture extract the number of groups as expected.

Scale of commitment: The KMO and Bartlett's Test in the analysis of 23 observed variables of the corporation culture with the high KMO index (0.833) and the significance level of 0 (sig = 0.000) demonstrates that the exploratory factor analysis is appropriate in this scale. With 18 variables introduced into the factor analysis, the results are extracted into three groups as shown in Table 2. The total variance is 56.64%. Thus, the factor analysis results of commitment extract the number of groups as expected.

Table 1. The exploratory factor analysis of the corporate culture.

No.	Group of factors	Variables
1	M- Mission	M1, M2, M3, M4, M5 and M6
2	A-Adaptability	A1,A2, A3, A4, A5 and A6
3	P - Participation	P1, P2, P3, P4, P5 and P6
4	C-Consistency	C1, C2, C3, C4 and C5

Kaiser-Meyer-Olkin Measure of Sampling Adequacy: 0.845
Bartlett's Test of Sphericity: Sig. = 0.000
Total variance is 59.58% >50%

Table 2. The exploratory factor analysis of the Commitment.

No.	Group of factors	Variables
1	AC - Affective commitment	AC1, AC2, AC3, AC4, AC5 and AC6
2	CC - Continuance commitment	CC1, CC2, CC3, CC4, CC5 and CC6
3	NC - Normative commitment	NC1, NC2, NC3, NC4, NC5 and NC6

Kaiser-Meyer-Olkin Measure of Sampling Adequacy: 0.833
Bartlett's Test of Sphericity: Sig. = 0.000
Total variance is 56.64% > 50%

4.3 Testing of the Scale by the Confirmatory Factor Analysis (CFA)

4.3.1 Testing of the Research Model

Testing of the formal theoretical model: The results of the SEM theoretical model (as illustrated on Fig. 2) are shown in Fig. 3. Here, Chi-square / df = 1.593, GFI = 0.838, TLI = 0.925, CFI = 0.931, and RMSEA = 0.050, which shows that the theoretical models are appropriate for our data.

The positive (+) and statistically significant ($p \leq 0.05$) estimated results of the weights in Table 3 demonstrate that: Mission (M), Adaptability (A), Consistency (C) and Participation (P) impact three components of organizational commitment including Affective commitment (AC), Continuance commitment (CC) and Normative commitment (NC). It means that "Each measurement is related to the scales as theoretically expected". In other words, the scales of the concepts in the model meet the standards of the value related to the theory.

In addition, the results of standardized estimates in Fig. 3 and Table 3 show that:

- Mission (MS) has the most powerful impact (0.675) on the employee's continuance commitment. In other words, the mission is the most important factor affecting the employee's continuance commitment in Vietnamese enterprises.
- The second most powerful factor is consistency (0.661) that impacts employee's normative commitment;
- The third is mission (0.628) influencing employee's normative commitment;
- The fourth is adaptability (0.572) impacting employee's continuance commitment;
- The fifth is adaptability (0.478) affecting employee's affective commitment;

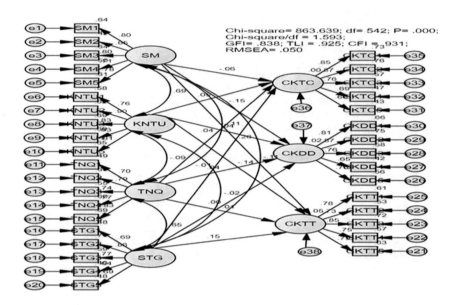

Fig. 3. Official SEM theoretical model results (standardized).

Table 3. Results of testing the causal relationship among the concepts of formal theoretical model (standardized).

Relationship			Estimate	S.E	C.R	P	Label
AC	<—	M	0.318	0.061	11.12	***	
NC	<—	M	0.628	0.050	7.39	***	
CC	<—	M	0.675	0.048	6.81	***	
AC	<—	A	0.478	0.057	9.19	***	
NC	<—	A	0.276	0.062	11.65	***	
CC	<—	A	0.572	0.053	8.07	***	
AC	<—	C	0.278	0.062	11.62	***	
AC	<—	P	0.133	0.053	1.962	0.050	
NC	<—	P	0.149	0.044	2.202	0.028	
CC	<—	P	0.454	0.058	9.47	***	
NC	<—	C	0.661	0.049	6.98	***	
CC	<—	C	0.165	0.059	2.105	0.035	

Of which Estimate: Average value estimate; SE: standard error; CR: Criticalratio; P: probability value; $***$: $p < 0,001$.

- The sixth is participation (0.454) impacting the employee's affective commitment;
- The seventh is mission (0.318) affecting employee's affective commitment;
- The next is consistency (0.278) affecting employee's affective commitment;
- The ninth is adaptability (0.276) affecting the normative employee commitment;
- The tenth is consistency (0.165) effecting employee's continuance commitment;
- The eleventh is participation (0.149) affecting the normative employee commitment, and
- Finally, the least important factor is participation (0.133) affecting affective commitment.

4.3.2   Testing of Research Hypotheses

Our results show that all 12 research hypotheses: H1, H2, H3, H4, H5, H6, ..., H12 described in Fig. 2 have been experimentally confirmed.

Specifically, Table 3 shows that all the weights are positive (+) and statistically significant ($p \leq 0.05$), thus M (Mission), Adaptability (A), Consistency (C), and Participation (P) all positively and statistically significantly impact all aspects of commitment:

- mission is positively correlated with affective commitment;
- consistency is positively correlated with affective commitment;
- adaptability is positively correlated with affective commitment;
- participation is positively correlated with affective commitment;
- mission is positively correlated with normative commitment;
- consistency is positively correlated with normative commitment;
- adaptability is positively correlated with normative commitment;
- participation is positively correlated with normative commitment;
- mission is positively correlated with continuance commitment;
- consistency is positively correlated with continuance commitment;
- adaptability is positively correlated with continuance commitment;
- participation is positively correlated with continuance commitment.

5   Conclusion

It is known that the employee commitment benefits both the employee – by making him or her more satisfied – and the company – by making it more efficient and more productive. It is therefore important to maintain employee commitment. This commitment is known to be affected by the corporate culture. To maintain employee commitment, it is therefore desirable to study how different components of corporate culture affect different aspects of employee commitment. In this study, we analyze the relation between corporate culture and employee commitment in Vietnam based on the results of a survey of 289 employees, managers, and enterprise leaders from Hanoi and Vinh City. Our conclusion is that each component of corporate culture has a statistically significant effect on all aspects of employee commitment.

Our conclusion is that to enhance the employee commitment to the organization, it is really necessary to develop all aspects of favorable corporate culture.

An improved corporate culture benefits employees, benefits corporations, and benefits the country as a whole – by increasing its economic productivity. It is therefore desirable that this improvement be a joint effort of the corporations' leaders and managers, corporations' employees, and the state as a whole.

References

Allen, N.J., Meyer, J.P.: The measurement and antecedents of affective, continuance and normative commitment to the organization. J. Occup. Psychol. **63**, 1–18 (1990)

Al-Rasheedi, S.: Infuence of National Culture on Employee Commitment Forms: A case study of Saudi-Western IJVs vs Saudi Domestic companies. University of Warwick, WMG (2012)

Aviv, S., Maria, S.A., Minoo, F.: Hofstede's dimensions of culture in international marketing studies. J. Bus. Res. **60**(3), 277–284 (2007)

Becker, H.S.: Notes on the concept of commitment. Am. J. Sociol. **66**, 32–40 (1960)

Bennis, W.: Leaders and visions: orchestrating the corporate culture. In: Berman, M.A. (ed.) Corporate Culture and Change. The Conference Board Inc., New York (1986)

Boje, M., Fedor, B., Rowland, M.: Myth making: a qualitative step in OD interventions. J. Appl. Behav. Sci. **18**, 17–28 (1982)

Brooks, G.R., Wallace, J.P.: A discursive examination of the nature, determinants and impact of organisational commitment. Asia Pac. J. Hum. Resour. **44**(2), 222–239 (2006)

Buchanan, B.: Building organisational commitment: the socialisation of managers in work organisations. Adm. Sci. Q. **19**(4), 533–546 (1974)

Calder, B., Phillips, L., Tybout, A.: Designing research for application. J. Consum. Res. **8**, 197–207 (1981)

Cameron, K.S., Quinn, R.E.: Diagnosing and Changing Organizational Culture. Addison-Wesley, New York (1999)

Chen, Y.S.: Green organizational identity: sources and consequence. Department of Business Administration, National Taipei University, Taipei, Taiwan (2011)

Clugston, M., Howell, J.P., Dorfman, P.W.: Does cultural socialization predict multiple bases and foci of commitment? J. Manag. **26**(1), 5–30 (2000)

Cohen, A.: An examination of the relationship between commitments and culture among five cultural groups of Israeli teachers. J. Cross-Cult. Psychol. **38**(1), 34–49 (2007)

Davis, S.M.: Managing Corporate Culture. Ballinger, New York (1984)

Deal, P., Kennedy, A.: Corporate Cultures. Addison-Wesley, Reading (1982)

Denison, D.R.: Corporate culture and organizational effectiveness. Wiley, New York (1990)

Hai, D.H.: Corporate culture - The pinnacle of wisdom. Monographs, Transport and Communications Publishing House (2016)

Ghani, R.A., Nordin, F., Mamat, L.: Organizational commitment among the academic staff in the distance education program. Int. J. Educ. Dev. **1**, 29–43 (2004)

Greenberg, J., Baron, R.A.: Behaviour in Organisations: Understanding and Managing the Human Side of Work, 8th edn. Pearson Education Inc., Upper Saddle River (2003)

Hult, C.: Organisational commitment and person-environment fit in six Western countries. Organ. Stud. **26**(2), 249–270 (2005)

Hogg, M.A., Terry, D.J.: Social identity process in organisational contexts. Sheridan Books, Michigan (2001)

Kotter, J.P., Heskett, J.L.: Corporate Culture and Performance. Free Press, New York (1992)

Lee, K., Allen, N.J., Meyer, J.P., Rhee, K.-Y.: The three-component model of organisational commitment: an application to South Korea. Appl. Psychol. Int. Rev. **50**(4), 596–614 (2001)

Lim, B.: Examining the organizational culture and organizational performance link. Leadersh. Organ. Dev. J. **16**(5), 16–21 (1995)

Lok, P., Crawford, J.: Antecedents of organisational commitment and the mediating role of job satisfaction. J. Manag. Psychol. **16**(8), 594–613 (2001)

Lok, P., Crawford, J.: The effects of organisational culture and leadership style on job satisfaction and organisational commitment. J. Manag. Dev. **23**(4), 321–338 (2004)

Louis, M.R.: Surprise and sense making: what newcomers experience in entering organizational settings. Adm. Sci. Q. **25**, 226–251 (1980)

Lund, D.B.: Organizational culture and job satisfaction. J. Bus. Ind. Mark. **18**, 219–236 (2003)

Martin, J.: Organisational Behaviour, 2nd edn. Thomson Learning, London (2001)

Mathieu, J.E., Zajac, D.M.: A review and meta-analysis of the antecedents, correlates and consequences of organisational commitment. Psychol. Bull. **108**(2), 171–194 (1990)

McKinnon, J.L., Harrison, G.L., Chow, C.W., Wu, A.: Organisational culture: association with commitment, job satisfaction, propensity to remain, and information sharing in Taiwan. Int. J. Bus. Stud. **11**(1), 25–44 (2003)

Meyer, J.P., Allen, N.J.: A three-component conceptualization of organizational commitment. Hum. Resour. Manag. Rev. **1**, 61–89 (1991)

Meyer, J.P., Allen, N.J., Smith, C.A.: Commitment to organisations and occupations: extension and test of the three-component conceptualisation. J. Appl. Psychol. **78**(4), 538–551 (1993)

Mitroff, I., Kilmann, H.: On organizational stories: an approach to the design and of organizations through myths and stories. In: Kilmann, H., Pondy, R., Slevin, P. (eds.) The Management of Organization Design, pp. 189–207. Elsevier-North Holland, New York (1976)

Mowday, R.T.: Reflections on the study and relevance of organisational commitment. Hum. Resour. Manag. Rev. **8**(4), 387–401 (1998)

Mowday, R.T., Steers, R.M., Porter, L.W.: The measurement of organizational commitment. Vocat. Behav. **14**, 224–247 (1979)

Muthuveloo, R., Rose, R.C.: Antecedents and outcomes of organizational commitment among Malaysia engineers. Am. J. Appl. Sci. **2**, 1095–1100 (2005)

Nazir, N.A.: Person-culture fit and employee commitment in banks. Vikalpa **30**(3), 39–51 (2005)

Tinh, N.T. et al.: Analysis of the factors affecting young employees' long-term engagement to the enterprise. J. Econ. Integr. **27**(7) (2012)

O'Reilly, C., Chatman, J.: Organizational commitment and psychological attachment: the effects of compliance, identification and internalization on prosocial behavior. J. Appl. Psychol. **71**, 492–499 (1986)

Nongol, E.S., Ikyanyon, D.N.: The influence of corporate culture on employee commitment to the organization. MBA Master Thesis, University Benue (2012)

Ogaard, T., Larsen, S., Marnburg, E.: Organisational culture and performance: evidence from the fast food restaurant industry. Food Serv. Technol. **5**(1), 23–34 (2005)

Ooi, K.B.: The influence of corporate culture on organizational commitment: case study of semiconductor organizations in Malaysia. MBA Master Thesis, University Teknology Malaysia (2009)

Porter, L.W., Steers, R.M., Mowday, R.T., Boulian, P.V.: Organisational commitment, job satisfaction, and turnover among psychiatric technicians. J. Appl. Psychol. **59**(5), 603–609 (1974)

Quinn, R.E., McGrath, M.R.: The transformation of organizational culture: a competing values. Paper presented at the Conference of Organizational Culture and Meaning of Life in the Workplace, Vancouver (1984)

Rashid, Z.A., Sambasivan, M., Johari, J.: The influence of corporate culture and organisational commitment on performance. J. Manag. Dev. **22**(8), 708–728 (2003)

Recardo, R., Jolly, J.: Organizational culture and teams. S.A.M Adv. Manag. J. **62**, 4 (1997)

Robbins, S., Judge, T.: Organizational Behavior. Prentice Hall, Upper Saddle River (2009)

Rose, R.M., Che, R.: Antecedents and outcomes of organizational commitment among Malaysian engineers. Am. J. Appl. Sci. **2**, 1095–1100 (2005). Science Publications

Saeed, M., Hassan, A.: Organizational culture and work outcomes: evidence from some Malaysian organizations. Malays. Manag. Rev. **35**, 54–59 (2000)

Schein, E.H.: Organizational Culture and Leadership. Jossey-Bass, San Francisco (1985)

Sekaran, U.: Research Methods for Business: A Skill-Building Approach, 3rd edn. Wiley, Chichester (2000)

Siehl, C., Martin, J.: Learning Organizational Culture. Working Paper, Graduate School of Business, Stanford University (1981)

Silverthorne, C.: The impact of organisational culture and person-organisation fit on organisational commitment and job satisfaction in Taiwan. Leadersh. Organ. Dev. J. **25**(7), 522–599 (2004)

Smircich, L.: Concepts of culture and organizational analysis. Sci. Q. **28**, 339–358 (1983)

Steers, R.M.: Antecedents and outcomes of organisational commitment. Adm. Sci. Q. **22**(1), 46–56 (1977)

Syauta, J.H. et al.: The influence of organizational culture, organizational commitment to job satisfaction and employee performance: study at municipal waterworks of Jayapura, Papua Indonesia. MBA Master Thesis, Cendrawasih University (2012)

Nguyen, T.V.: The traditional ideology of senior management: obstacles to entrepreneurship and ninovation in the reform of state-owned enterprises in Vietnam. Int. J. Entrep. Innov. Manag. **5**, 227–250 (2005)

Tharp, M.B.: Four Organizational Culture Types. Organizational Culture white paper, Haworth (2009)

Thompson, K.R., Luthans, F.: Organizational Culture: A Behavioral Perspective. Organizational climate and Culture. Jossey-Bass, San Francisco (1990)

Trice, H.M., Beyer, J.M.: The Culture of Work Organizations. Prentice Hall, Englewood Cliff (1993)

Wasti, S.A.: Organizational commitment, turnover intentions and the influence of cultural values. J. Occup. Organ. Psychol. **76**(3), 303–321 (2003)

Wilkins, A., Martin, J.: Organizational Legends. Working Paper, Graduate School of Business, Stanford University (1980)

Wilkins, A.L., Ouchi, W.G.: Efficient cultures: exploring the relationship between culture and organizational performance. Adm. Sci. Q. **28**, 468–481 (1983)

William, H.M., Wang, L., Fang, K.: Measuring and developing organizational culture. Havard Bus. Rev. (China) **3**, 128–139 (2005)

Yu, B.B., Egri, C.P.: Human resource management practices and affective organizational commitment: a comparison of Chinese employees in a state-owned enterprise and a joint venture. Asia Pac. J. Hum. Resour. **43**(3), 332–360 (2005)

http://tapchibcvt.gov.vn/nghien-cuu-moi-quan-he-giua-van-hoa-doanh-nghiep-doi-voi-su-gan-bo-cua-nhan-vien-vnpt-tai-khu-vuc-dbscl-2329-bcvt.htm

http://www.tc-consulting.com.vn/vi/thu-vien/bai-viet-chuyen-gia/ly-thuyet-ve-su-cam-ket-va-y-nghia-doi-voi-lanh-dao-su-thay-doi

On a New Calibrated Mixture Model for a Density Forecast of the VN30 Index

Dung Tien Nguyen, Son Phuc Nguyen[(✉)], Thien Dinh Nguyen, and Uyen Hoang Pham

University of Economics and Law, VNU-HCM,
Khu pho 3, phuong Linh Xuan, quan Thu Duc,
Ho Chi Minh City, Viet Nam
{dungnt,sonnp,thiennd,uyenph}@uel.edu.vn

Abstract. Regarding predictions of business and financial quantities, seldom has a consensus been reached among experts which, in certain cases, creates insurmountable difficulties for decision-makers to reach a final decision. In this paper, motivated by the quest for a reliable forecast of the VN 30 index, we introduce a novel method to mix and calibrate a number of stocks to better predict the index. Treating each stock as one "expert's opinion", we construct an integrated forecast by applying a beta calibration to a mixture model of stock historical data to derive a combined and calibrated density function for the VN 30 index. Since all the computations are carried out within the framework of bayesian statistics, our new technique is part of the bayesian semi-parametric methods.

1 Introduction

There are a variety of methods to combine experts' opinions to obtain a single consensus forecast. In the statistical sense, it by and large boils down to combining experts' predictive cumulative distribution functions. There have been numerous work on this topics; to be more specific, the most popular approach is to utilize a method of pooling predictive distributions like the linear pooling etc. These pooling techniques have been quite successful in physics, engineering and other fields where the data of interest are not sensitive to human behavior. Unfortunately, business and financial data do not enjoy this insensitivity. This makes it a lot harder for researchers to build a trustworthy forecast.

In this paper, we propose using a beta distribution function to calibrate a linear pooling of stocks in VN 30 to construct a density forecast for the index. The techniques use the Bayesian approach to estimate the model parameters. All Markov chain Monte Carlo simulations are carried out using the probabilistic programming language Stan in Python.

The rest of the paper is as follows: Sect. 2 contains a brief literature review. Section 3 introduces the models to predict individual stocks as well as the notion of calibration of a linear mixture of stocks. Section 4 discusses the Bayesian inference for the calibrated models. Section 5 presents the results with the VN 30 index data. Finally, the paper ends with a conclusion in Sect. 6.

L. H. Anh et al. (eds.), *Econometrics for Financial Applications*, Studies in Computational Intelligence 760, https://doi.org/10.1007/978-3-319-73150-6_37

2 Literature Review

Volatility has always been a main concern of every financial business throughout history. Therefore, a wide range of techniques have been devised to predict volatility of stocks and market indices. One of the earlier work is the ARCH model by Engle [6] which allows heteroskedasticity to be taken into account. Since then there is a whole line of research to continue to extend the ARCH model in various directions. The most notable ones are the GARCH model by Bollerselev [1] for a flexible lag structure, the power GARCH to test the long-memory characteristic of a time series, or the GJR-GARCH model to capture the impacts of stocks on the conditional variance and so on. Based on these remarkable theoretical achievements, a lot of practitioners have been successful in applying the family of ARCH/GARCH models to build time series forecasts.

The following table consists of a few notable papers on predicting stock prices and indices worthwide; see also [2–5,7,9,10,14–17].

Author(s)	Year	Model	Data	Results
Nam, Pyun and Arize [12]	2002	Nonlinear GARCH-M	US market indices Jan 1926–Dec 1997	Negative return, on average, reverted more quickly in the long term than positive returns
Sun and Zhou [18]	2014	IGARCH	S&P 500 index, 12 S&P equity sector indices, Jan, 1995–Dec, 2010	GARCH with student t innovation gives better results than GARCH
Abdalla	2012	EGARCH(1,1)	Jan 2000–Nov 2011	Negative shocks caused higher volatility than positive shocks
Dritsaki [13]	2017	EGARCH(1,1) and GJR-GARCH(1,1)	Stockhom Sept 1986–May, 2016	Negative shocks have larger impacts than positive ones
Horvarth and Sopov [19]	2016	GARCH, EGARCH and GJR-GARCH	S&P 500 returns 1995–2014	EGARCH with student t innovation captures the fatness of the tail of distribution well
Laurent et al. [20]	2016	ARMA-GARCH(1,1), ARMA-GJR GARCH(1,1)	Exchange rates (USD/JPY, EUR/USD, GBP/USD) Jan, 2005–May, 2011	The standard GARCH estimated on a filtered returns outperforms other models

In this paper, we take up a different path. Instead of calling upon a complex version of GARCH model, we will mix, then calibrate the GARCH(1,1) model for a variety of stock prices to create a density prediction for the VN30 Index.

3   Statistical Models

3.1   Generalized Autoregressive Conditional Heteroskedasticity (GARCH)

The linear regression model is the crucial tool in a great number of applications in business and finance. However, the model has a few assumptions which are not always satisfied in reality. One of the common violation of the model assumption is heteroskedasticity i.e. the variances of the error terms are not the same; to be more precise, the error variances may reasonably be expected to be larger for some points or ranges of the data than for others. In the presence of heteroskedasticity, the regression coefficients for an ordinary least squares regression are still unbiased, but the standard errors and confidence intervals estimated by conventional procedures will be too narrow, giving a false sense of precision. Instead of considering this as a problem to be corrected, GARCH models treat heteroskedasticity as a variance to be modeled. As a result, not only are the deficiencies of least squares corrected, but a prediction is computed for the variance of each error term. This turns out often to be of interest particularly in finance.

In the present paper, we are going to use GARCH(1,1) to model log returns for each of the stocks in the group VN 30. Let (r_t) be some log return time series. Then, the GARCH(1,1) for (r_t) is as follows

$$r_t = \mu + \sigma_t \epsilon_t$$
$$\sigma_t^2 = \alpha_0 + \alpha_1 \sigma_{t-1}^2 + \beta_1 \epsilon_{t-1}^2 \tag{1}$$

where $\epsilon_t \sim N(0,1)$ (normal GARCH) or $\epsilon_t \sim T(\nu)$ (student t-GARCH).

Here, μ, α_0, α_1, β_1 for each stock are estimated by MCMC algorithms, see [8].

3.2   Mixture Models

Mixture models were invented to address the situations where data do not seem to follow a single pattern but rather a mixture of several patterns. In this case, researchers can design one model for each pattern and mix them together to capture the whole phenomenon. This method is notably powerful in clustering problem, for instance. In addition, since any continuous distribution can be approximated arbitrarily well by a finite mixture of normal densities with common variance, mixture models provide a convenient semiparametric framework to model unknown distributional shapes whether the objective is a density estimation or a construction of Bayesian priors.

In this paper, we make use of a linear mixture models of all the Bayesian GARCH(1,1) estimations of individual stocks to construct a density estimation for the log returns of VN30 index. Suppose every stock i has a cumulative distribution function F_i, $1 \le i \le K$. Then, the linear mixture distribution function is defined as

$$H = \omega_1 F_1 + \cdots + \omega_K F_K, \quad \sum_{i=1}^{K} \omega_i = 1 \tag{2}$$

and the corresponding density function is

$$h = \omega_1 f_1 + \cdots + \omega_K f_K, \quad \sum_{i=1}^{K} \omega_i = 1 \tag{3}$$

The idea is that the VN30 index behaves like stock i with probability ω_i.

It should be emphasized that we have experimented with several more complex methods of mixture like the logarithmic or harmonic ones, but the results are not significantly better. Hence, we decided to proceed with the simple linear mixture model only for the sake of computational complexity.

3.3 Beta Calibration

In risk management, distortion functions are widely popular. They provide important tools to calibrate risk measures in applications. In details, a distortion function $g : [0, 1] \longrightarrow [0, 1]$ is a non-decreasing function such that $g(0) = 0$ and $g(1) = 1$.

In particular, the beta cumulative distribution function (beta cdf) is a highly useful distortion function which is applied, in our work, to the linear mixture model to calibrate our density forecast for the VN30 index. It has two parameters α and β which will be estimated by an MCMC algorithm.

So, the final predictive cumulative distribution function has the form:

$$\begin{aligned} G &= B_{\alpha, \beta}(H) \\ &= B_{\alpha, \beta}(\omega_1 F_1 + \cdots + \omega_K F_K) \end{aligned} \tag{4}$$

where $B_{\alpha,\beta}$ is the beta cumulative distribution function. Thus, the corresponding density function is:

$$\begin{aligned} g &= b_{\alpha, \beta}(H) \cdot (\omega_1 f_1 + \cdots + \omega_K f_K) \\ &= b_{\alpha, \beta}(H) \cdot h \end{aligned} \tag{5}$$

4   Bayesian Inference of the Beta Calibration of Mixture Models

Bayesian approaches to mixture modelling have attracted great interest among researchers and practitioners alike. The Bayesian paradigm allows for probability statements to be made directly about the unknown parameters. Moreover, prior information or expert opinions can be included in the analysis of data. When necessary, hierarchical descriptions of both local and global features of the model can be constructed. This framework also allows the complicated structure of a mixture model to be decomposed into a set of simpler structures through the use of hidden or latent variables. When the number of components is unknown, it can well be argued that the Bayesian paradigm is the only sensible approach to its estimation, see [11].

Despite all the advantages, the Bayesian framework used to be impractical due to the prohibitively high computational complexity. Modern algorithms, especially the Markov Chain Monte Carlo family, together with powerful computers allow researchers to build and estimate a lot of Bayesian predictive models.

In this paper, using the probabilistic programming language Stan implemented in Python, we manage to estimate all parameters in our beta calibrated mixture model and generate good density forecast for the VN 30 index log return. Below are the details of the model

Step 1: Train the GARCH(1,1) model for each time series of the 30 stocks in the VN30. The results of this step are 8000 samples for each parameter in the tuples

$$(\mu_i, \alpha_{0i}, \alpha_{1i}, \beta_{1i}), \ 1 \leq i \leq 30$$

For the details about the GARCH(1,1) parameters, please refer to formula (1).

Step 2: Use the VN30-index time series to find the weights of the 30 stocks in the linear mixture model. In other words, find $(\omega_1, \ldots, \omega_{30})$, $\sum_{i=1}^{30} \omega_i = 1$ which best fits the VN30-index. The results are the cumulative distribution H and the density h as follows

$$H = \omega_1 F_1 + \cdots + \omega_{30} F_{30}$$
$$h = \omega_1 f_1 + \cdots + \omega_{30} f_{30}$$

Please refer to formulas (2) and (3) for more details.

Step 3: The final step is to estimate the parameters (α, β) for the beta calibration. So, we complete our distribution and density forecast

$$G = B_{\alpha, \beta}(\omega_1 F_1 + \cdots + \omega_K F_K)$$
$$g = b_{\alpha, \beta}(\omega_1 F_1 + \cdots + \omega_K F_K) \cdot (\omega_1 f_1 + \cdots + \omega_K f_K) \tag{6}$$

All computations are done on our personal laptops and it take several hours to train and infer the parameters.

5   The Density Forecast for the VN30 Index

The data in this paper are the time series of all stocks in the VN30 list together with the index. The time frame is from July 20th, 2015 to February 13th, 2017. The start and end dates are based on the availability of the dataset which we can access. It is not related to the model; we believe the model will work well with other time frame as well.

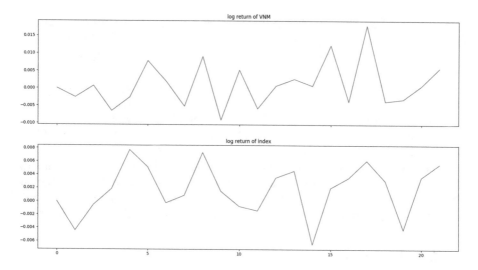

Fig. 1. Vinamilk and index log returns

We divide the dataset into two sets: a training set consisting of 473 days of close-price log returns of the stocks and VN30 Index, and a test set consisting of the newest 22 days in our data. The picture below shows the log return of Vinamilk (VNM) and log return of the VN index 30 for the 22 days in our test set. Please refer to Fig. 1.

The out-of-sample test set is one month of trading. The training set is used to train the beta calibration model to obtain estimations for all the model

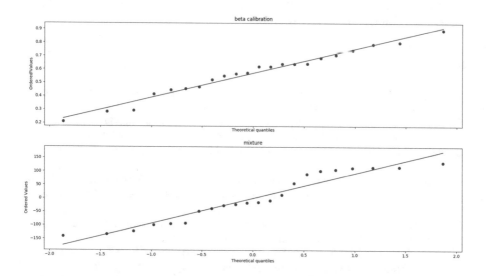

Fig. 2. Calibration and mixture density forecasts

parameters. Using the cumulative distribution function H with maximum likelihood estimators for the parameters (see formula (6)), we apply the probability integral transform to the out-of-sample test set to obtain the probability plot as follows

In Fig. 2, the upper plot is the result of the beta calibration model. The red line is the theoretical unknown distribution of the log return index and the blue dots are the integral transforms of the 22 days out-of-sample test set. In the ideal situation, when the model is "correct", all blue dots will lie on the red line. Our results in this paper is very close to the ideal case.

The lower plot is drawn on the same scale for comparison. It is the integral transforms of the well-known linear mixture model. Compared to the beta calibration model, the blue dots of the linear mixture model are farther away from the red line which indicates a lower accuracy in predicting the density function.

6   Conclusion

In this paper, we introduce a novel method of calibration to study the prediction problem in financial data. This can be considered a new practical sequential analysis in analyzing stock market data. Starting with as many as 30 stocks in the VN 30 groups, we apply the Bayesian paradigm to estimate all GARCH parameters for each stocks. Then, we mix the stocks together and calibrate the combination by a beta distribution function. The reason we choose the linear mixture is its simplicity and its comparable performance compared to more complicated mixture techniques. In our opinions, the density forecast is satisfactory when applied to the VN 30 Index data.

In subsequent work, we will utilize our method in a wide variety of data. Moreover, we also plans to create meaningful point estimates and interval estimates based on the predictive density.

References

1. Bollerslev, T.: Generalised autoregressive conditional heteroskedasticity. J. Econometrics **31**, 307–327 (1986)
2. Bassetti, F., Casarin, R., Ravazzolo, F.: Bayesian nonparametric calibration and combination of predictive distributions. J. Am. Stat. Assoc. (2016, accepted)
3. Bates, J.M., Granger, C.W.J.: Combination of forecasts. Oper. Res. Q. **20**, 451–468 (1969)
4. Casarin, R., Grassi, S., Ravazzolo, F., Van Dijk, H.K.: Dynamic predictive density combinations for large data sets in economics and finance, Norges Bank Research (2015)
5. Casarin, R., Mantoan, G., Ravazzolo, F.: Bayesian calibration of generalized pools of predictive distributions. Econometrics **4**(1), 17 (2016)
6. Engle, R.: Autoregressive conditional heteroskedasticity with estimates of the variance of United Kingdom inflation. Econometrica **5**, 987–1008 (1982)
7. Hall, S.G., Mitchell, J.: Combining density forecasts. Int. J. Forecast. **23**, 1–13 (2007)

8. Hoffman, M.D., Gelman, A.: The No-U-turn sampler: adaptively setting path lengths in Hamiltonian Monte Carlo. J. Mach. Learn. Res. **15**(1), 1593–1623 (2014)
9. Jore, A.S., Mitchell, J., Vahey, S.P.: Combining forecast densities from VARs with uncertain instabilities. J. Appl. Econ. **25**, 621–634 (2010)
10. Kling, J.L., Bessler, D.A.: Calibration-based predictive distributions: an application of prequential analysis to interest rates, money, prices, and output. J. Bus. **62**, 477–499 (1989)
11. Richardson, S., Green, P.: On Bayesian analysis of mixtures with an unknown number of components (with discussion). J. Royal Statist. Soc. Ser. B **59**, 731–792 (1997)
12. Nam, K., Pyun, C.S., Arize, C.A.: Asymmetric mean-reversion and contrarian profits: ANSTGARCH approach. J. Empirical Finan. **9**, 563–588 (2002)
13. Dritsaki, C.: An empirical evaluation in GARCH volatility modeling: evidence from the stockholm stock exchange. J. Math. Finan. **2017**(7), 366–390 (2017)
14. Higgins, M.L., Bera, A.K.: A class of nonlinear ARCH models. Int. Econ. Rev. **33**(1), 137–158 (1992)
15. Nelson, D.B.: Conditional heteroscedasticity in asset returns: a new approach. Econometrica **59**(2), 347–370 (1991)
16. Zakoian, J.M.: Threshold heteroscedasticity models. J. Econ. Dyn. Control **18**(5), 931–955 (1994)
17. Glosten, L.R., Jagannathan, R., Runkle, D.E.: Relationship between the expected value and the volatility of the nominal excess return on stocks. J. Finan. **48**(5), 1779–1801 (1993)
18. Sun, P., Zhou, C.: Diagnosing the distribution of GARCH innovations. J. Empirical Finan. **29**, 287–303 (2014)
19. Horváth, R., Sopov, B.: GARCH models, tail indexes and error distributions: an empirical investigation. N. Am. J. Econ. Finan. **37**(2016), 1–15 (2016)
20. Laurent, S., Lecourt, C., Palm, F.C.: Testing for jumps in conditionally Gaussian ARMA-GARCH models, a robust approach. Comput. Stat. Data Anal. **100**, 383–400 (2016)

An Improved Fuzzy Time Series Forecasting Model

Ha Che-Ngoc[1], Tai Vo-Van[2], Quoc-Chanh Huynh-Le[1], Vu Ho[3],
Thao Nguyen-Trang[1,4(✉)], and Minh-Tuyet Chu-Thi[1,2,3,4]

[1] Faculty of Mathematics and Statistics, Ton Duc Thang University,
Ho Chi Minh City, Vietnam
nguyentrangthao@tdt.edu.vn
[2] Department of Mathematics, Can Tho University, Can Tho, Vietnam
[3] Faculty of Mathematical Economics, Banking University of Ho Chi Minh City,
Ho Chi Minh City, Vietnam
[4] Division of Computational Mathematics and Engineering, Institute
for Computational Science, Ton Duc Thang University, Ho Chi Minh City, Vietnam

Abstract. This model is developed from the model of Abbasov and
Mamedova (2003) in which the parameters are investigated by meth-
ods and algorithm to obtain the most suitable values for each data set.
The experiments on Azerbaijan's population, Vietnam's population and
Vietnam's rice production demonstrate the feasibility and applicability
of the proposed methods.

Keywords: Fuzzy time series · Abbasov-Mamedova · Population
GDP · Vietnam

1 Introduction

Forecasting is the process of making prediction for the future based on summing
experiences, assembling knowledge and analyzing related problems. It is considered as the basis process, the first step for organizations as well as governments to
build their policies and objectives. Because of its important role in many fields,
forecasting has received much attention from scientists. Despite several discussions in the literature, the problems of forecasting have not yet been completely
solved. Based on the historical data, looking for principles and rules to establish
a suitable forecasting model is the major method of statistics. Time series and
regression models have important roles in forecasting using statistical methods,
but they have many disadvantages in practice. A regression model (Galton (1888);
Pearson (1896)) requires a number of assumptions that are unsatisfactory, whereas
a time series model, like ARIMA (Box and Jenkins (1976)), performs poorly when
there are abnormal changes or the time series is nonstationary. To overcome the
disadvantages of these two models, various models have recommended by many
researches, such as (Zecchin et al. (2011); Wang and Fu (2006); Wang et al. (2001);
Ren et al. (2016); Gupta and Wang (2010); Zhu and Wang (2010); Park (2010);

© Springer International Publishing AG 2018
L. H. Anh et al. (eds.), *Econometrics for Financial Applications*, Studies in Computational
Intelligence 760, https://doi.org/10.1007/978-3-319-73150-6_38

Teo et al. (2001); Ghazali et al. (2009)). These proposals are the important contributions for forecasting problem because they have given good results in considered data sets. However, we could not obtain optimum for all cases. Some other models like Artificial Neuron Network, Supported Vector Regression (Cortes and Vapnik (1995)), Multivariable Adaptive Regression Spline (Friedman (1991)), Adaptive Spline Threshold Autoregressive (Lewis and Stevens (1991)), Autoregressive conditional heteroscedasticity (Engle (1982)) or hybrid models (Zhang (2003)); (de Oliveira and Ludermir (2014)) were also proposed; however, most of them still have many disadvantages in real forecasting applications.

Based on the fuzzy theory of Zadeh (Zadeh (1965)), fuzzy time series (FTS) introduced by Song (Song and Chissom (1993)) can solve the gap mentioned above. FTS has been then interested to research and have been shown to be more efficient than traditional statistical techniques (Song and Chissom (1993)); (Tseng and Tzeng (2002)). Among them, Abbasov and Manedova (AM) proposed the model where the variations of data are represented by language level to forecast the population of Azerbaijan (Abbasov and Mamedova (2003)). Because of its better performance for some kinds of forecasting problems, AM model has been applied in many applications; for instance, Sasu utilized the AM model to forecast the population of Romanian (Sasu (2010)). Some other important studies of FTS can be listed as the models in (Chen (2004); Huarng (2001); Singh (2008)). Nonetheless, all of the above methods use only historical fuzziness data without forecasting. Moreover, the parameters in the models are not properly investigated to find the optimal values for each data set. One model is only rated as better than the others in some specific cases. As a result, there is no model that is considered optimal in all situations.

To overcome the gap mentioned above, this article proposes the methods to identify the suitable parameters in the AM model. Specifically, w, the number of elements in the data set used as prior information to forecast the data is chosen through the value of partial autocorrelation function (PACF), the number of fuzzy sets n is selected through an index that can evaluating the compactness of the divided intervals. After determining suitable w and n, the optimal choice of C is searched via an efficient algorithm so that the forecasting error is the smallest. The numerical examples illustrate the proposed theories in detail and prove that this method can improve the performance in term of forecasting accuracy.

The remainder of this paper is organized as follows. Section 2 reviews the AM model and some the related definitions. Section 3 proposes the modifications for AM model, in which the suitable parameters are determined by new methods and algorithm. The numerical examples are presented in Sects. 4 and 5 is the conclusion.

2   Related Definitions and Abbasov-Mamadova Model

2.1   Related Definitions

Definition 1. Let U be a universe (domain), with a generic element of U denoted by u. A fuzzy set A on universe U is a set defined by the membership function $\mu_A(u)$ which is a mapping from the universe U into the unit interval:

$$\mu_A(u) : U \to [0, 1]$$

The value of the membership function of a specific u, $\mu_A(u)$ is called as the membership degree or grade of membership. If the grade of membership equals one, u belongs completely to the fuzzy set. If the grade of membership equals zero, u does not belong to the set. If the grade of membership is between 0 and 1, u is a partial member of the fuzzy set.

$$\mu_A(u) \begin{cases} = 1 & \text{u is a full member of A} \\ \in (0,1) & \text{u is a partial member of A} \\ = 0 & \text{u is not member of A} \end{cases}$$

In above definition, a fuzzy set A in U is characterized by a membership function $\mu_A(u)$. There are several ways to define the membership function $\mu_A(u)$. Some of forms of membership functions which are often used such as trapezoidal membership function, triangular membership function, Gaussian membership function, etc. An example of some membership functions with different shapes is presented in Fig. 1. We next examine a special case of Definition 1 where the universe is a time series and introduce the definition proposed by (Song and Chissom (1993)) as follows.

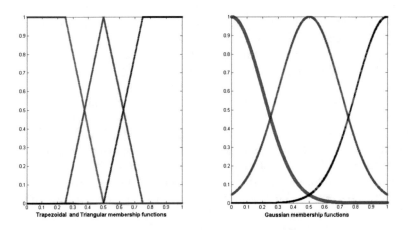

Fig. 1. Algorithm to determine C.

Definition 2. Let $Y(t) \in R$, $t = 0, 1, 2, \ldots$ be a time series, with a generic element denoted by y_t. If $\mu_A(y_t)$ is the membership function which is a mapping from the universe $Y(t)$ into $[0, 1]$ and $F(t) = \{\mu_A(y_0), \mu_A(y_1), \mu_A(y_2), \ldots\}$ is a collection of $\mu_A(y_t)$ then $F(t)$ is called a fuzzy time series.

Definition 3. Given the actual data $\{X_i\}$ and predictive value $\left\{\hat{X}_i\right\}$, $i = 1, 2, \ldots, m$, respectively, we have the popular indexes to evaluate established model as follows:

Mean absolute error:

$$MAE = \frac{1}{m} \sum_{i=1}^{m} \left|\hat{X}_i - X_i\right|. \tag{1}$$

Mean squared error:

$$MSE = \frac{1}{m} \sum_{i=1}^{m} \left(\hat{X}_i - X_i\right)^2. \tag{2}$$

Mean absolute percentage error:

$$MAPE = \frac{1}{m} \sum_{i=1}^{m} \left(\frac{\left|\hat{X}_i - X_i\right|}{X_i}.100\right). \tag{3}$$

2.2 Abbasov-Mamedova Model

Given the historical data X_t corresponded to year $t = 1, 2, ..., m$. The AM model consists of the following six steps.

- **Step 1:** Compute the variation V_t between every next and previous year by Formula (4). Then define the universal set U by Formula (5).

$$V_t = X_t - X_{t-1} \tag{4}$$

$$U = [V_{\min} - D_1, V_{\max} + D_2] \tag{5}$$

  where $V_{\min}$ is the smallest variation, $V_{\max}$ is the greatest variation, D_1 and D_2 are positive numbers.

  Step 2: Divide the universal set U into n equal-length intervals u_i, $i = 1, 2, \ldots, n$, such that each interval u_i contains at least one variation value. Then find the middle points u_m^i of each interval.
- **Step 3:** Define the fuzzy set A_i, $i = 1, 2, \ldots, n$, on the universal set U by the following formula:

$$\mu_{A_i}(u) = \frac{1}{1 + [C \times (u - u_m^i)]^2}, \tag{6}$$

  where u is a generic element of universal set U, u_m^i is the middle point of the corresponding interval u_i, $(i = 1, 2, \ldots, n)$ and C is a constant.
- **Step 4:** Convert the input data, time-point variations, into fuzzy values by Formula 6.
- **Step 5:** Select an integer w, $1 < w < l$, where l is the number of years, prior to the current year included in experimental evaluation. Based on the chosen w and Mamdani inference system, we establish an operation matrix $O^w(t)$ of size $i \times j$ (here i is the number of rows, which conforms to the sequence of

years $t-2, t-3, \ldots, t-w$, j is the number of columns, which conforms to the number of variation intervals) and a criteria matrix $K(t)$ of size $1 \times j$ (a row matrix corresponding to fuzzy variation in total population for the year $t-1$). After that, the relationship matrix $R(t)$ is calculated as follows.

$$R(t)[i, j] = O^w[i, j] \cap K(t)[j],$$

or

$$R(t) = O^w(t) \otimes K(t) = \begin{bmatrix} R_{11} & R_{12} & \ldots & R_{1j} \\ R_{21} & R_{22} & \ldots & R_{2j} \\ \ldots & \ldots & \ldots & \ldots \\ R_{i1} & R_{i2} & \ldots & R_{ij} \end{bmatrix},$$

where $O^w(t)$ is the operation matrix, $K(t)$ is the criteria matrix, $\otimes$ is the min operator ($\cap$).

Define $F(t)$, the fuzzy forecasting of variations for the year t, in a fuzzy form as follows.

$$F(t) = [\max(R_{11}, \ldots, R_{i1}), \ldots, \max(R_{1j}, \ldots, R_{ij})]$$
$$= [\mu_{A_1}(V_t), \mu_{A_2}(V_t), \ldots, \mu_{A_m}(V_t)].$$

- **Step 6:** Defuzzify the obtained results of the 5-*th* step according to the Formula 7.

$$V(t) = \frac{\sum_{i=1}^{m} \mu_{A_i}(V_t) \times u_m^i}{\sum_{i=1}^{m} \mu_{A_i}(V_t)}, \qquad (7)$$

where $\mu_{A_i}(V_t)$ is the value of membership function of the forecast variation in interval i, $V(t)$ is the defuzzified forecast variation.

In orders to estimate the forecast value $X(t)$ for year t, the following formula is utilized:

$$X(t) = X(t-1) + V(t), \qquad (8)$$

where $X(t-1)$ is the forecast value for year $t-1$, $V(t)$ is the variation for year t.

3   The Proposed Method

In AM model, there are three parameters including the number of equal-length intervals n, the positive integer w and the constant C have effects on the forecasting result. However, in the studies of (Abbasov and Mamedova (2003); Sasu (2010)), these parameters were only identified according to the experiences. Hence, this method is not suitable when dealing with various types of time series. For w, Song and Chissom conducted a survey on the specific data and pointed out that $w = 2$ is the best. They also concluded that the forecasting result is better if we utilize a less complex model, with smaller w (Song and Chissom (1993)). However, this conclusion is only drawn from a few specific surveys, so lose generality. In fact, w is number of previous times that have a strong influence on current value of time series (it is similar to the partial autocorrelation p in autoregressive integrated moving average, ARIMA). Therefore, it is not reasonable if we utilize a model, with a fixed value of w, for all type of time series. For instance, when dealing with a monthly or quarterly data, $w = 7$ is consider as an unreasonable parameter. According to above remarks, it is certain that the forecasting performance of the AM model can be significantly improved if its parameters are determined in reasonable ways. To overcome the limitations mentioned above, this section proposes a method called MAM (Modified Abbasov-Mamedova model), which can identify the parameters n, w and C in a reasonable way. Details of the proposed method are presented as follows.

3.1   Determine the Number of Interval n

In the fuzzification step, the middle point u_0^i is used as the representative element of ith interval. Therefore, if the data in each interval are well-represented by u_0^i, the forecasting performance can be improved. In general, we can evaluate whether data are well-represented by u_0^i or not according to the compact measure between this middle point and elements in the interval i. Figure 2 illustrates a few cases of representative elements. It can be seen that the universal set U is defined as the interval $(0, 3)$ and divided into three equal-length intervals. The distance between middle points (red points) and elements belonging to the intervals $(0, 1)$ and $(1, 2)$ are really large; therefore, using u_0^1 and u_0^2 as the representative elements can lead to a low forecasting performance. Conversely, u_0^3 is close to the elements in $3rd$ interval, and it can lead to a good measure of compactness as well as a high forecasting performance. Therefore, it is important to point out the number of intervals n so that the measure of compactness is optimized. According to mentioned idea, this paper proposes a measure denoted as MMSE to evaluate the compactness of algorithm. MMSE is computed by (9):

$$MMSE = \frac{1}{n} \sum_{i=1}^{n} \sum_{v_t \in u_i} \frac{\left(v_t - u_0^i\right)^2}{n_i}, \qquad (9)$$

where n is the number of equal-intervals, v_t is the variation, u_0^i and n_i is the middle point and the number of elements belonging to the interval i, respectively.

Clearly, the smaller distances between middle point u_0^i and variations v_t are, the smaller of numerator as well as the MMSE criterion. Therefore, MMSE can be used to evaluate the compactness of time series model with different number of intervals n. In addition, when the number of intervals n is extended up to a specific number, the empty intervals containing no variation values are created. Meanwhile, the denominator in (9) is equal to 0, and MMSE converges to infinity. Hence, the choice of n in this case is not suitable. This is also entirely consistent with the constraint mentioned in (Abbasov and Mamedova (2003)).

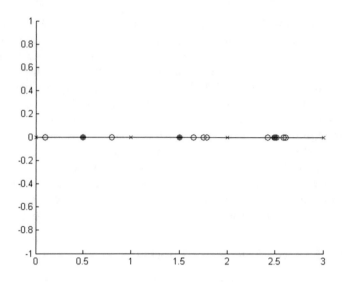

Fig. 2. Illustrate for representing elements.

3.2   Determine w

In literature, the parameter w is chosen according to experiences. Although Song and Chissom conducted a survey on specific datasets and pointed out that $w = 2$ is the best, this conclusion is only drawn from a few specific surveys and does lose generality. Here, an effective method that can determine w is proposed, based on the partial autocorrelation function (PACF) of the time series.

Let ϕ_{kk} be PACF at lag k $(k = 1, 2, \ldots)$, ϕ_{kk} can be obtained according to the recursive formula (Durbin (1960)) as follows:

$$\phi_{p+1,j} = \phi_{p,j} - \phi_{p+1,p+1}\phi_{p,p-j+1}, \tag{10}$$

$$\phi_{p+1,p+1} = \frac{r_{p+1} - \sum_{j=1}^{p} \phi_{p,j} r_{p+1-j}}{1 - \sum_{j=1}^{p} \phi_{p,j} r_j}, \tag{11}$$

where r_k is the autocorrelation function (ACF) at lag k, $\phi_{1,1} = r_1$.

The PACF at lag k considers only the direct correlation between v_t and v_{t-k}, with the linear dependences between the intermediate variables are removed. Therefore, it can accurately reflect the number of the previous years, on which the current year depends. Based on this result, we can determine the appropriate w.

Note that, when PACF presents the largest value at lag 1, w is considered as 2 so that the AM model conditions are fitted. In practice, based on statistical programs including R, Matlab, etc., it is possible to calculate the PACF and determine the reasonable w for fuzzy time series model.

3.3 Determine the Constant C

Given AM model with specific parameters w and n, the value of $\mu_{A_i}(u_i)$ as well as the forecasting result is strongly affected by the constant C. However, the previous studies of (Abbasov and Mamedova (2003); Sasu (2010)) did not offer guidance on how to determine the reasonable C for each specific dataset. This subsection proposes an algorithm to determine optimum C (DOC), using the following five steps:

Step 1. Initialize an integer k ($k > 499$), a very small positive number ϵ, where k is the number points which divided for each iteration and ϵ is the error of C.

Step 2. When $t = 0$, assign values: $a^{(0)} = 0$, $b^{(0)} = 1$, $\Delta C^{(0)} = \frac{1}{2}$, $n^{(0)} = 1$.

Step 3. When $t = i$, $i \geq 1$, calculate the values

$$a^{(t)} = a^{(t-1)} + \left[n^{(t-1)} - 1\right] \Delta C^{(t-1)}$$

$$b^{(t)} = a^{(t-1)} + \left[n^{(t-1)} + 1\right] \Delta C^{(t-1)}, \Delta C^{(t)} = \frac{b^{(t)} - a^{(t)}}{k}$$

If $a = 0$ and $b = 1$, then $C_i^{(t)} = a^{(t)} + i\Delta C^{(t)}$, $i = 1, 2, \ldots, k - 1$.
If $a = 0$ and $b \neq 1$, then $C_i^{(t)} = a^{(t)} + i\Delta C^{(t)}$, $i = 1, 2, \ldots, k$.
If $a \neq 0$ and $b = 1$, then $C_i^{(t)} = a^{(t)} + i\Delta C^{(t)}$, $i = 1, 2, \ldots, k - 1$.
If $a \neq 0$ and $b \neq 1$, then $C_i^{(t)} = a^{(t)} + i\Delta C^{(t)}$, $i = 1, 2, \ldots, k$.

Step 4. Run the Abbasov-Mamedova model with all the values $C_i^{(t)}$ in Step 3. Find $C_n^{(t)}$ at which the criterion CEF is the current best.

Step 5. With the new n, repeat the Step 3 and Step 4 to find $C = C_n^{(m)} = a^{(m)} + n\Delta C^{(m)}$ until $b^{(m)} - a^{(m)} < \varepsilon$.
   Note that,

(i) In each iteration, $(k+1)$ values of C are considered. In the numerical examples in this paper, $k = 1000$ is chosen.
(ii) ϵ is a very small number and is chosen arbitrarily. The smaller ϵ is, the more iterations and computer time it are required. In fact, the optimum value of C can be determined with an acceptable error depending on the value of ϵ. In numerical examples, $\epsilon = 10^{-6}$ is chosen.
(iii) There are a many criterions considered to evaluate the forecasting model (CEF). In this article, we use MAE, MAPE and MSE presented in Subsect. 2.1 to compare established models.

The DOC algorithm is illustrated by Fig. 3.

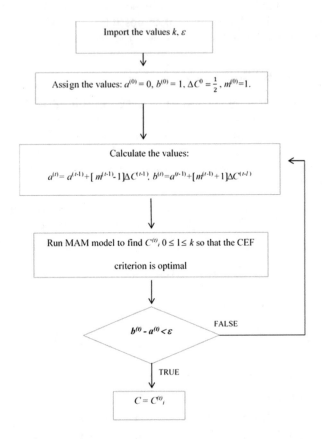

Fig. 3. Algorithm to determine C.

4   Numerical Examples

Section 3 proposes the methods of determining w, n and C in order to improve the forecasting performance of AM model. In Sect. 4, this paper presents two examples to illustrate and test the forecasting performance of proposed method. Specifically, Example Sect. 4.1 presents in detail the proposed method when dealing with the well-known data, Azerbaijan's population. This example, in addition to clarifying the proposed algorithm, can test its forecasting performance. In Example Sect. 4.2, the new method is applied to forecast the GDP per capita in Vietnam. In each example, we compare the forecasting results of proposed method with those of the AM model (Abbasov and Mamedova (2003)), the Chen model (Chen (1996)) and the Huarng model (Huarng (2001)). Furthermore, to present that the proposed method is more efficient in predicting time series than the traditional statistical methods, MAM is also compared with the auto-regressive model AR(p) where p is choose based on AIC criterion.

4.1   Example 1

This example forecasts the annual population (thousand persons) of Azerbaijan from 1980 to 2001 to clarify the proposed method and test its performance. This is a well-known dataset presented in (Abbasov and Mamedova (2003)). The detailed procedure is presented by six following steps.

Step 1. Table 1 presents the annual populations over 1980–2001 and variations in all given years. Variation for the current year is the difference between the population values in current year and previous year. For example, variation for 1990 is equal to $7131900 - 7021200 = 1110700$. To define the universal set U, first of all, the smallest and greatest variation values must be found over the interval $[1980, 2001]$, later, to ensure the smoothness of boundaries of the interval, adequate non-negative numbers D_1, D_2 are selected. After that, the universal set U can be defined as $U : U = [V_{min} - D_1, V_{max} + D_2]$, where $V_{min} = 62800$ is the smallest variation, $V_{max} = 115900$ is the greatest variation, $D_1 = 0$, $D_2 = 0$. Thus, the universal set U is defined as: $U = [62800, 115900]$. Based on the variation in Table 1, MMSE and PACF that are computed and presented in Fig. 4 are considered as the criterion to find the suitable n and w. From Fig. 4, it can be seen that MMSE is inversely correlated with the number of equal intervals n. It proves that the more intervals we have, the better representation the middle points make. However, the split of intervals must stop at a specific level. Specifically, in Fig. 4, the empty intervals, which is associated with n, are created when n is greater or equal to 10. As a result, MMSE is unspecified, and the method is considered to be unreasonable in those cases, which stands for the suitable number of equal intervals is 9. For w, according to Fig. 4, it can be observed that PACF reaches the maximum value at lag 1. As mentioned in Sect. 3, w is chosen as 2 in this case. It is also suitable with the survey of (Song and Chissom (1993)). In summary, based on MMSE and PACF, $n = 9$ and $w = 2$ are utilized in the AM model. With $n = 9$ and $w = 2$, performing the DOC algorithm, the optimum value of C is reach at 0.3197.

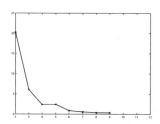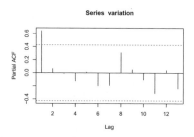

Fig. 4. MMSE according to n and the PACF.

Step 2. The universal set U must be divided into 9 equal intervals:

$$u_1 = [62800, 68700], u_2 = [68700, 74600]$$
$$u_3 = [74600, 80500], u_4 = [80500, 86400]$$
$$u_5 = [86400, 92300], u_6 = [92300, 98200]$$
$$u_7 = [98200, 104100], u_8 = [104100, 110000]$$
$$u_9 = [110000, 115900]$$

The middle points of the intervals are determined as follows: $u_m^1 = 65750$, $u_m^2 = 71650$, $u_m^3 = 77550$, $u_m^4 = 83450$, $u_m^5 = 89350$, $u_m^6 = 95250$, $u_m^7 = 101150$, $u_m^8 = 107050$, $u_m^9 = 112950$.

Steps 3 and 4. Define the fuzzy sets $A_1, A_2, \ldots, A_9$ on the universal set U and convert the input data into fuzzy values by formula 6. An exemplary growth of the continuous membership functions of fuzzy sets A_i is shown in Fig. 5. For the sake of briefly, the results of fuzzification for all the given years with last two digits are shown in Table 1.

Table 1. Population of Azerbaijan

T	N_t	V_t	Fuzzy time series F_t
1980	6114.3	0	
1981	6206.7	92.4	0.00 0.02 0.04 0.11 0.51 0.55 0.11 0.04 0.02
1982	6308.8	102.1	0.01 0.01 0.02 0.03 0.06 0.17 0.92 0.29 0.08
1983	6406.3	97.5	0.01 0.01 0.02 0.05 0.13 0.66 0.42 0.10 0.04
1984	6513.3	107.0	0.01 0.01 0.01 0.02 0.03 0.07 0.22 1.00 0.22
1985	6622.4	109.1	0.01 0.01 0.01 0.01 0.02 0.05 0.13 0.70 0.40
1986	6717.9	95.5	0.01 0.02 0.03 0.06 0.21 0.99 0.23 0.07 0.03
1987	6822.7	104.8	0.01 0.01 0.01 0.02 0.04 0.10 0.42 0.66 0.13
1988	6928.0	105.3	0.01 0.01 0.01 0.02 0.04 0.09 0.36 0.76 0.14
1989	7021.2	93.2	0.01 0.02 0.04 0.09 0.40 0.70 0.13 0.05 0.02
1990	7131.9	110.7	0.00 0.01 0.01 0.01 0.02 0.04 0.10 0.42 0.66
1991	7218.5	86.6	0.02 0.04 0.11 0.50 0.56 0.12 0.04 0.02 0.01
1992	7324.1	105.6	0.01 0.01 0.01 0.02 0.04 0.08 0.33 0.82 0.15
1993	7440.0	115.9	0.00 0.00 0.01 0.01 0.01 0.02 0.04 0.11 0.53
1994	7549.6	109.6	0.01 0.01 0.01 0.01 0.02 0.05 0.12 0.60 0.47
1995	7643.5	93.9	0.01 0.02 0.04 0.08 0.32 0.84 0.16 0.05 0.03
1996	7726.2	82.7	0.03 0.07 0.27 0.95 0.18 0.06 0.03 0.02 0.01
1997	7799.8	73.6	0.14 0.72 0.39 0.09 0.04 0.02 0.01 0.01 0.01
1998	7879.7	79.9	0.05 0.13 0.64 0.44 0.10 0.04 0.02 0.01 0.01
1999	7953.4	73.7	0.13 0.70 0.40 0.09 0.04 0.02 0.01 0.01 0.01
2000	8016.2	62.8	0.53 0.11 0.04 0.02 0.01 0.01 0.01 0.00 0.00
2001	8081.0	64.8	0.92 0.17 0.06 0.03 0.02 0.01 0.01 0.01 0.00

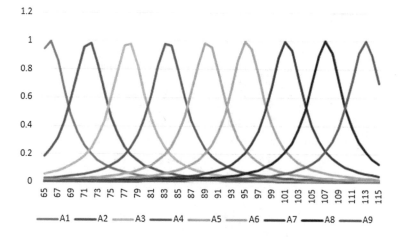

Fig. 5. Membership functions of 9 fuzzy sets.

Step 5. Apply the $\min - \max$ operator to forecast the population in 1990, we have the results:

$$O^2(1990) = [0.01 \ 0.01 \ 0.01 \ 0.02 \ 0.04 \ 0.09 \ 0.36 \ 0.76 \ 0.14]$$
$$K(1900) = [0.01 \ 0.02 \ 0.04 \ 0.09 \ 0.40 \ 0.70 \ 0.13 \ 0.05 \ 0.02]$$
$$R(1990) = [0.01 \ 0.01 \ 0.01 \ 0.02 \ 0.04 \ 0.09 \ 0.13 \ 0.05 \ 0.02]$$

Hence, the fuzzy forecasting of the variation for the year 1990, $F(1990)$, is

$$[0.01 \ 0.01 \ 0.01 \ 0.02 \ 0.04 \ 0.09 \ 0.13 \ 0.05 \ 0.02]$$

Step 6. Finally, compute the variations in 1990 by the Formula 7.

$$V(1990) = \frac{0.01 * 65750 + \ldots + 0.02 * 112950}{0.01 + \ldots + 0.02} = 97181$$

Hence, the forecasting population in 1990 is:

$$X(1990) = X(1989) + V(1990)$$
$$= 7021200 + 97181 = 7118381$$

Perform in a similar way for the remainder, the forecasting results are presented in Table 2 and Fig. 6. As shown in Table 3, in addition to the proposed method, the performance of models presented in (Abbasov and Mamedova (2003)); Chen (1996); Huarng (2001)) are examined for comparison purpose.

It can be observed that MAM model outperforms others in term of accuracy for all cases of criterion. The result verifies that the proposed method is suitable at first and need to be retested in actual application as follows.

Table 2. Actual and forecast population (thousand person)

Years	Actual		Forecasted	
	Total	Variation	Total	Variation
1988	6928.0	105.3	6921.096	98.396
1989	7021.2	93.2	7031.659	103.659
1990	7131.9	110.7	7118.381	97.181
1991	7218.5	86.6	7230.382	98.482
1992	7324.1	105.6	7314.037	95.537
1993	7440.0	115.9	7418.289	94.189
1994	7549.6	109.6	7545.407	105.407
1995	7643.5	93.9	7658.468	108.868
1996	7726.2	82.7	7742.128	98.628
1997	7799.8	73.6	7814.826	88.626
1998	7879.7	79.9	7879.700	79.900
1999	7953.4	73.7	7958.389	78.689
2000	8016.2	62.8	8032.098	78.698
2001	8081.0	64.8	8089.909	73.709

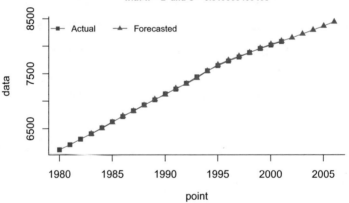

Actual series vs forecated series by Abbasov–Mamedova model of 9 fuzzy set with w = 2 and C = 0.319363409408

Fig. 6. The forecasting result of proposed model.

Table 3. The performance of comparative methods

	MAM	AM	Chen (1996)	Huarng (2001)
MAE	**9.989**	15.007	77.756	77.756
MAPE	**0.136**	0.197	1.099	1.099
MSE	**127.751**	290.459	8835.054	8835.054

4.2   Example 2

GDP is gross domestic product converted to international dollars using purchasing power parity rates. It is one of the major measures of nation's economic health. Therefore, GDP forecasting has an essential role for countries all over the world. In this example, the new method is applied to forecast the GDP per capita in Vietnam. In particular, the GDP per capita (USD) from 1990 to 2015 are collected (http://data.worldbank.org). Similar to Example 1, the performance of the proposed method are compared with those of (Abbasov and Mamedova (2003); Chen (1996); Huarng (2001)) (Fig. 7 and Table 4). In addition, the forecasts for the 5-year period 2016–2020 are presented (Table 5). Table 4 and Fig. 7 show that the proposed method has good forecasting results. The established model fits well almost all the actual data, with the mean of absolute error is less than 26USD% per year and the mean of absolute percentage error is less than 8% in comparison with actual data. In addition, Table 4 and Fig. 7 demonstrate

Table 4. Forecasting results of comparative methods

GDP per capita (USD)				
	MAM model	AM model	The Chen mode	The Huarng model
MAE	**25.669**	42.942	201.552	201.552
MAPE	**0.744**	1.224	7.86	7.86
MSE	**1117.06**	2834.763	62332.37	62332.37

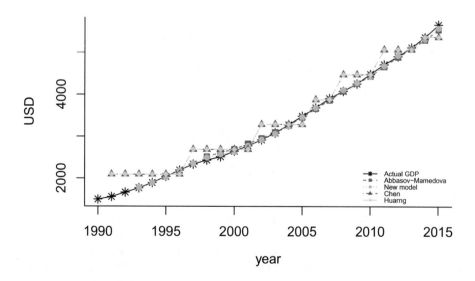

Actual GDP vs forecated GDP by models

Fig. 7. Vietnam GDP per capita forecasting results of comparative models

Table 5. Out-of-sample forecasting results

GDP	
Year	Forecast
2016	5908.166
2017	6148.920
2018	6372.591
2019	6588.328
2020	6796.071

the superiority of proposed method over comparative models, when it always shows the best results. For out-of-sample forecasting, it can be seen from Table 5 that Vietnam's GDP continues to increase steadily, with average growth rates of over 200 USD each year. GDP will reach over 6500 USD per capita by 2019.

4.3 On the Comparison Between MAM and the Traditional Statistical Method

As mentioned earlier, we resolve two above experiments in which MAM is compared with the auto-regressive model $AR(p)$, (auto-regressive order p is choose based on AIC criterion). The brief summary of results in Table 6 present that the MAM model outperforms the AR model. Based on the above examples, at first, we can be see that MAM is a good and competitive model in comparison with traditional statistical method as well as other fuzzy time series model. It is feasible and capable of practical problems, particularly of population and GDP forecasting.

Table 6. The results of MAM and AR models for Example 1 and Example 2

	Example 1		Example 2	
	MAM	AR	MAM	AR
MAE	9.989	93.908	25.669	165.612
MSE	127.751	8995.331	1117.06	30881.76
MAPE	0.136	1.329	0.744	5.323

5 Conclusion

This study proposes an improved fuzzy time series forecasting model based on the methods to determine the suitable parameters for each data set in the AM model. The numerical examples prove that the proposed method is more feasible and capable of practical problems. In future, a program will be written in the R statistical software to apply the proposed model in many different practice problems.

References

Abbasov, A.M., Mamedova, M.H.: Application of fuzzy time series to population forecasting. Vienna Univ. Technol. **12**, 545–552 (2003)

Box, G.E.P., Jenkins, G.M.: Time series analysis: forecasting and control. Holden-Day Series in Time Series Analysis, Revised edn. Holden-Day, San Francisco (1976)

Chen, S.M.: Forecasting enrollments based on fuzzy time series. Fuzzy Sets Syst. **81**(3), 311–319 (1996)

Chen, S.M., Hsu, C.C.: A new method to forecast enrollments using fuzzy time series. Int. J. Appl. Sci. Eng. **2**(3), 234–244 (2004)

Cortes, C., Vapnik, V.: Support-vector networks. Mach. Learn. **20**(3), 273–297 (1995)

Durbin, J.: The fitting of time-series models. Revue de l'Institut International de Statistique **28**(3), 233–244 (1960)

Engle, R.F.: Autoregressive conditional heteroscedasticity with estimates of the variance of United Kingdom inflation. Econometrica **50**, 987–1007 (1982). Journal of the Econometric Society

Friedman, J.H.: Multivariate adaptive regression splines. Ann. Stat. **19**, 1–67 (1991)

Galton, F.: Co-relations and their measurement, chiefly from anthropometric data. Proc. Roy. Soc. Lond. **45**(273–279), 135–145 (1888)

Ghazali, R., Hussain, A.J., Al-Jumeily, D., Lisboa, P.: Time series prediction using dynamic ridge polynomial neural networks. In: 2009 Second International Conference on Developments in eSystems Engineering (DESE), pp. 354–363. IEEE (2009)

Gupta, S., Wang, L.P.: Stock forecasting with feedforward neural networks and gradual data sub-sampling. Aust. J. Intell. Inf. Process. Syst. **11**(4), 14–17 (2010)

Huarng, K.: Heuristic models of fuzzy time series for forecasting. Fuzzy Sets Syst. **123**(3), 369–386 (2001)

Lewis, P.A.W., Stevens, J.G.: Nonlinear modeling of time series using multivariate adaptive regression splines (MARS). J. Am. Stat. Assoc. **86**(416), 864–877 (1991)

de Oliveira, J.F.L., Ludermir, T.B.: A distributed PSO-ARIMA-SVR hybrid system for time series forecasting. In: 2014 IEEE International Conference on Systems, Man and Cybernetics (SMC), pp. 3867–3872. IEEE (2014)

Park, D.C.: A time series data prediction scheme using bilinear recurrent neural network. In: 2010 International Conference on Information Science and Applications (ICISA), pp 1–7. IEEE (2010)

Pearson, K.: Mathematical contributions to the theory of evolution. III. Regression, heredity, and panmixia. Philos. Trans. Roy. Soc. Lond. Ser. A Contain. Pap. Math. Phys. Character **187**, 253–318 (1896)

Ren, Y., Suganthan, P.N., Srikanth, N., Amaratunga, G.: Random vector functional link network for short-term electricity load demand forecasting. Inf. Sci. **367**, 1078–1093 (2016)

Sasu, A.: An application of fuzzy time series to the Romanian population. Bulletin Transilv. Univ. Brasov **3**, 52 (2010)

Singh, S.R.: A computational method of forecasting based on fuzzy time series. Math. Comput. Simul. **79**(3), 539–554 (2008). https://doi.org/10.1016/j.matcom.2008.02.026

Song, Q., Chissom, B.S.: Forecasting enrollments with fuzzy time series part I. Fuzzy Sets Syst. **54**(1), 1–9 (1993)

Teo, K., Wang, L., Lin, Z.: Wavelet packet multi-layer perceptron for chaotic time series prediction: effects of weight initialization. In: Computational Science-ICCS 2001, pp 310–317 (2001)

Tseng, F.M., Tzeng, G.H.: A fuzzy seasonal ARIMA model for forecasting. Fuzzy Sets Syst. **126**(3), 367–376 (2002). https://doi.org/10.1016/S0165-0114(01)00047-1. http://www.sciencedirect.com/science/article/pii/S0165011401000471

Wang, L., Fu, X.: Data Mining with Computational Intelligence. Springer Science & Business Media, New York (2006)

Wang, L., Teo, K.K., Lin, Z.: Predicting time series with wavelet packet neural networks. In: Proceedings of the International Joint Conference on Neural Networks, IJCNN 2001. vol 3, pp 1593–1597. IEEE (2001)

Zadeh, L.A.: Fuzzy sets. Inf. Control **8**(3), 338–353 (1965)

Zecchin, C., Facchinetti, A., Sparacino, G., De Nicolao, G., Cobelli, C.: A new neural network approach for short-term glucose prediction using continuous glucose monitoring time-series and meal information. In: 2011 Annual International Conference of the IEEE Engineering in Medicine and Biology Society, EMBC, pp 5653–5656. IEEE (2011)

Zhang, G.P.: Time series forecasting using a hybrid ARIMA and neural network model. Neurocomputing **50**, 159–175 (2003)

Zhu, M., Wang, L.: Intelligent trading using support vector regression and multilayer perceptrons optimized with genetic algorithms. In: The 2010 International Joint Conference on Neural Networks (IJCNN), pp. 1–5. IEEE (2010)

Testing J-Curve Phenomenon in Vietnam: An Autoregressive Distributed Lag (ARDL) Approach

Le Hoang Phong[1(✉)], Ho Hoang Gia Bao[1], and Dang Thi Bach Van[2]

[1] School of Management, Ho Chi Minh City University of Law,
02 Nguyen Tat Thanh, District 4, Ho Chi Minh City, Vietnam
{lhphong,hhgbao}@hcmulaw.edu.vn
[2] School of Public Finance, University of Economics, Ho Chi Minh City,
59C Nguyen Dinh Chieu, District 3, Ho Chi Minh City, Vietnam
bachvan@ueh.edu.vn

Abstract. The assessment of factors having impacts on trade balance, especially exchange rate, is very important for effective implementation of macroeconomic policy. This paper investigates the existence of the short-run and long-run impacts of real exchange rate on trade balance in Vietnam by employing an Autoregressive Distributed Lag (ARDL) approach using quarterly data ranging from 2000Q1 to 2015Q4. Our results present the evidence of both short-term and long-term effects. In the short-run, Error Correction Model (ECM) based on ARDL approach indicates that real exchange rate has negative impact on trade balance. Also, impulse response functions based on ECM exhibit J-curve pattern of trade balance. In the long-run, real exchange rate has positive impact on trade balance. The stability of the long-run trade balance equations is also checked through CUSUM and CUSUMSQ stability tests.

1 Introduction and Literature Review

Devaluation is normally considered to have negative effect on trade balance in the short-run as well as positive one in the long-run, which refers to the term "J-curve" that has been widely studied since its first introduction in 1973; see, e.g., [24]. J-curve effect can be explained by the lagged adjustment of quantities to the changes in prices right after the devaluation occurred, which worsens trade balance in the short-run due to the increasing import value and the decreasing export value; see, e.g., [21,24]. In the long-run, however, the quantities can be appropriately adjusted to the prices, which fosters trade balance. Consequently, policy makers try to exploit the long-run favorable impact of devaluation to improve trade balance. The well-known Marshall-Lerner condition states that the absolute sum of long-run export and import elasticities of demand should be greater than one so as to make such devaluation effective.

Many researchers have tested the J-curve phenomenon in various countries. Although they apply different econometric methods, their studies can be

© Springer International Publishing AG 2018
L. H. Anh et al. (eds.), *Econometrics for Financial Applications*, Studies in Computational Intelligence 760, https://doi.org/10.1007/978-3-319-73150-6_39

classified into two categories based on the types of data: aggregate and disaggregated ones. Studies with aggregate data normally tried to test the existence of J-curve in a country by treating all her main trading partners as a whole and building aggregate variables which can represent incomes and exchange rates of all her trading partners. Studies with disaggregated data, on the other hand, tried to find the occurrence of J-curve in the relationship between a country with each of her main trading partners.

Regarding studies employing aggregate data, the findings are mixed. An intensive research by Bahmani-Oskooee and Alse testing J-curve phenomenon in 41 countries found the evidence of J-curve in Costa Rica, Ireland, Netherlands and Turkey; see, e.g., [3]. Also, Halicioglu showed that J-curve existed in Turkey; see, e.g., [19]. Bahmani-Oskooee and Kutan, however, did not detect J-curve in Turkey as well as 7 East European countries (Cyprus, Czech Republic, Hungary, Poland, Romania, Slovakia, and Ukraine); see, e.g., [10]. Rather, they concluded that Croatia, Bulgaria and Russia witnessed J-curve effect, which supported the result of Stučka in terms of the evidence of J-curve in Croatia; see, e.g., [10,34]. Anju and Uma found the proof of J-curve in Japan; see, e.g., [1]. Nevertheless, using Johansens cointegration technique and impulse response function when examining 7 countries including Indonesia, Japan, Korea, Malaysia, Philippines, Singapore and Thailand, Lal and Lowinger reported J-curve phenomenon in all countries except for Japan; see, e.g., [23]. Trinh indicated that J-curve effect happened in Vietnam; see, e.g., [35]. Similarly, Kyophilavong, Shahbaz and Uddin confirmed the occurrence of J-curve in Laos; see, e.g., [22]. Besides, Felmingham found no sign of J-curve in Australia; see, e.g., [16]. Bahmani-Oskooee and Gelan detected no trace of J-curve in 9 African countries including Burundi, Egypt, Kenya, Mauritius, Morocco, Nigeria, Sierra Leone, South Africa and Tanzania, which went in line with the findings of Ziramba and Chifamba about the case of South Africa; see, e.g., [6,38].

Regarding studies employing disaggregated data, the results also varied. As researchers faced bias with the application of aggregate data, they have utilized disaggregated data at 2 levels: bilateral and industry ones. Firstly, at bilateral level, Rose and Yellen found no evidence of J-curve for the case of USA from 1960 to 1985; see, e.g., [32]. Bahmani-Oskooee and Brooks did not find J-curve proof in USA, either; see, e.g., [4]. Bahmani-Oskooee and Kantipong analyzed the trade between Thailand and her 5 main partners (alphabetically listed as Germany, Japan, Singapore, UK and USA), reporting that J-curve appeared only in the trade with USA and Japan; see, e.g., [9]. Wilson examined bilateral data of Korea, Malaysia and Singapore in trading with Japan and USA and noticed J-curve in the pair Korea-USA only; see, e.g., [37]. For China, Bahmani-Oskooee and Wang discovered no J-curve; see, e.g., [11]. Halicioglu reported no indication of J-curve in Turkey; see, e.g., [18]. Bahmani-Oskooee and Harvey studied data between Singapore and her 13 largest partners (alphabetically listed as Australia, Canada, China, Hong Kong, India, Japan, Korea, Malaysia, Philippines, Saudi Arabia, Thailand, UK and USA) and witnessed J-curve effect in the trade between Singapore and Canada, Philippines, Saudi Arabia and USA; see, e.g., [8]. Dash researched the trade between India and her 4 major partners (alphabetically listed as Germany, Japan, UK and USA) and demonstrated J-curve effect in India-Germany and India-Japan cases; see, e.g., [14]. Secondly, at industry level, Baek tested J-curve hypothesis of 5 forest products (softwood lumber, hardwood lumber, panel/plywood product, logs and chips, and other wood product) in the trade between USA and Canada, and little evidence was found; see, e.g., [2]. Bahmani-Oskooee and Mitra detected J-curve phenomenon in 8 out of 38 industries in India-USA trade; see, e.g., [5]. Bahmani-Oskooee and Hajilee observed the proof of J-curve in 23 out of 87 industries in the trade between Sweden and USA; see, e.g., [7]. Šimáková and Stavárek, analyzed 10 product groups in Czech Republic's trade relations with Germany, Slovakia, Poland, France, Italy and Austria and reported no sign of J-curve; see, e.g., [33]. Vural employed disaggregated data at industry level for the case of Turkey (in relationship with her major trading partner Germany) and revealed J-curve appearance in 20 out of 96 industries; see, e.g., [36].

2   Theoretical Framework

We start by reviewing the standard "two-country" model of trade which was used by Rose and Yellen and Stučka; see, e.g., [32,34]. The requirement of this model is that imported and exported goods cannot totally substitute for domestic goods. Hence, the majority of traded goods' demand and supply elasticities can be estimated; see, e.g., [34]. The volume of imports demanded domestically and the volume of imports demanded by the rest of the world are displayed in Eqs. (1) and (2) respectively:

$$D_m = D_m(Y, p_m), \frac{\partial D_m}{\partial Y} > 0, \frac{\partial D_m}{\partial p_m} < 0. \tag{1}$$

$$D_m^* = D_m^*(Y^*, p_m^*), \frac{\partial D_m^*}{\partial Y^*} > 0, \frac{\partial D_m^*}{\partial p_m^*} < 0. \tag{2}$$

where $D_m(D_m^*)$ is the quantity of goods imported by home country (the rest of the world); $Y(Y^*)$ is the real income of home country (the rest of the world); p_m is the relative price of imported goods of home country, calculated as the ratio between the price of imported goods (measured in home currency) and the overall price level of home country; p_m^* is the relative price of imported goods of the rest of the world, calculated as the ratio between the price of imported goods (measured in the rest of the world currency) and the overall price level of the rest of the world. The real incomes have positive correlations with the volumes of goods imported while the relative prices have negative ones.

Under perfect competition assumption, the volume of goods exported by home country and the volume of goods exported by the rest of the world are given in Eqs. (3) and (4) respectively:

$$S_x = S_x(p_x), \frac{\partial S_x}{\partial p_x} > 0. \tag{3}$$

$$S_x^* = S_x^*(p_x^*), \frac{\partial S_x^*}{\partial p_x^*} > 0. \tag{4}$$

where $S_x(S_x^*)$ is the quantity of goods exported by home country (the rest of the world); p_x is the relative price of exported goods of home country, calculated as the ratio between the price of exported goods (measured in home currency) and the overall price level of home country; p_x^* is the relative price of exported goods of the rest of the world, calculated as the ratio between the price of exported goods (measured in the rest of the world currency) and the overall price level of the rest of the world. The relative prices have positive correlations with the quantities of goods exported.

Call E the nominal exchange rate between home country currency and the rest of the world currency (1 unit of the rest of the world currency = E units of home country currency). Call e the real exchange rate between home country currency and the rest of the world currency. Call P_m the price of imported goods of home country (measured in home currency). Call P_x^* the price of exported goods of the rest of the world (measured in the rest of the world currency). Call P the overall price level of home country. Call P^* the overall price level of the rest of the world. The relationship between p_m and p_x^* is demonstrated as:

$$p_m = \frac{P_m}{P} = \frac{E \cdot P_x^*}{P} = \frac{E \cdot P^*}{P} \cdot \frac{P_x^*}{P^*} = e \cdot p_x^*. \tag{5}$$

Similarly, the relationship between p_m^* and p_x is as follows:

$$p_m^* = \frac{p_x}{e}. \tag{6}$$

In equilibrium, the volume of goods imported by home country equals the volume of goods exported by the rest of the world and the volume of goods imported by the rest of the world equals the volume of goods exported by home country:

$$D_m = S_x^*. \tag{7}$$

$$D_m^* = S_x. \tag{8}$$

Call B the real value of trade balance of home country, defined as the real value of export minus the real value of import:

$$B = p_x \cdot S_x - p_m \cdot D_m = p_x \cdot D_m^* - e \cdot p_x^* \cdot D_m. \tag{9}$$

Equations from (1) to (8) can be solved for D_m, D_m^*, p_x, p_x^* as functions of e, Y and Y^*. Consequently, B can be expressed as a "partial reduced form":

$$B = B(e, Y, Y^*), \frac{\partial B}{\partial e} > 0, \frac{\partial B}{\partial Y} < 0, \frac{\partial B}{\partial Y^*} > 0. \tag{10}$$

In Eq. (10), real exchange rate (e) and real income of the rest of the world (Y^*) have positive correlations with trade balance (B) while real income of home country (Y) has negative one.

The trade balance as the difference between export value and import value faces a weakness which is sensitive to the unit of measurement; see, e.g., [4]. Accordingly, we apply the definition provided by Ziramba and Chifamba which defines trade balance as the ratio between export value and import value; see, e.g., [38]. The natural logarithm form of trade balance is as follows:

$$\ln B_t = \pi \cdot \ln e_t + \delta \cdot \ln Y_t + \alpha \cdot \ln Y_t^*. \tag{11}$$

From Eq. (11), π is expected to be positive, and if so, the Marshall-Lerner condition is satisfied; see, e.g., [38]. δ is expected to be negative and α is expected to be positive.

3   Estimation Methodology, Sample, Data and Results

Estimation Methodology. Our theoretical framework follows the work of Rose and Yellen and Stučka, while our empirical methodology follows that of Ziramba and Chifamba; see, e.g., [32,34,38]. The empirical equation is modeled as following:

$$TB_t = \alpha + \beta_1 \cdot REER_t + \beta_2 \cdot GDP_t + \beta_3 \cdot GDP_t^* + \varepsilon_t. \tag{12}$$

where $TB, REER, GDP, GDP^*$ represent trade balance, real effective exchange rate, domestic output and foreign output respectively. We use $REER$ to represent real exchange rate since it can measure the value of VND in relation to the currencies of major trading partners of Vietnam.

Many methods could be used for analyzing the cointegration of the trade balance's function. Among them, there are 2 common methods including the residual-based one introduced by Engle and Granger and the maximum likelihood-based one introduced by Johansen and Juselius. Both of these methods require that the variables have the same order of integration; see, e.g., [15,20]. In case this requirement is not met, Autoregressive Distributed Lag (ARDL) is an optimal alternative as it does not need the classification of I(0) and I(1) variables; see, e.g., [30,31]. Thus, unit root test is not necessary when employing ARDL approach for cointegration test. Besides, ARDL approach has several more advantages. First, in terms of the required sample size, ARDL approach needs a smaller size than the Johansen cointegration technique; see, e.g., [17]. Second, if some regressors in the model are endogenous, the ARDL approach can provide unbiased long-run estimates associated with valid t statistics; see, e.g., [25,26]. Third, it can measure the short-term and long-term effects of independent variables on the dependent variable in only one equation without the need of solving a set of equations; see, e.g., [13].

Due to the above advantages, we apply ARDL approach, which is widely used in empirical study, in this paper to analyze the trade balanceof Vietnam.

The specified trade balance function of Eq. (12) can be written as unrestricted error correction version of ARDL model as the following:

$$\Delta TB_t = \alpha + \sum_{i=1}^{p_1}(\beta_{1,i} \cdot \Delta TB_{t-i}) + \sum_{j=0}^{p_2}(\beta_{2,j} \cdot \Delta REER_{t-j})$$

$$+ \sum_{k=0}^{p_3}(\beta_{3,k} \cdot \Delta GDP_{t-k}) + \sum_{l=0}^{p_4}(\beta_{4,l} \cdot \Delta GDP^*_{t-l})$$

$$+ \lambda_1 \cdot TB_{t-1} + \lambda_2 \cdot REER_{t-1} + \lambda_3 \cdot GDP_{t-1} + \lambda_4 \cdot GDP^*_{t-1} + \varepsilon_t. \quad (13)$$

The ARDL procedure begins by doing the check for existence of the long-run relation among the variables in the function by utilizing "bound tests". The null hypothesis of no cointegration (or no long-run relationship) is shown by H0: $\lambda_1 = \lambda_2 = \lambda_3 = \lambda_4 = 0$. The alternative hypothesis is H1: $\lambda_1 \neq \lambda_2 \neq \lambda_3 \neq \lambda_4 \neq 0$. These hypotheses are tested by computing the F-statistics and compare it with the two sets of critical values (one set is calculated under I(0) variable assumption and the other is calculated under I(1) assumption) of the F-statistics; see, e.g., [29,31]. If the F-statistics is higher than the upper bound of the critical value, we reject H0. In contrast, if the F-statistics is below the lower bound, we cannot reject H0. In case the F-statistics is between the lower and upper bounds, the result is inconclusive.

The next step is that, after checking the occurrence of cointegration among variables, we choose the optimal lag orders of the variables using Schwarz Bayesian Criteria (SBC); see, e.g., [29]. The appropriate selection of the lag orders of variables is very important because it can help identify the true dynamics of the models. In order to verify the performance of the model, we conduct the diagnostic tests to check some issues including the serial correlation, functional form, normality, and heteroscedasticity; see, e.g., [28]. Furthermore, we also apply Cumulative Sum (CUSUM) and cumulative sum of squares (CUSUMSQ) of recursive residuals tests to assess the stability of the model; see, e.g., [12]. In addition, we use the impulse response function based on obtained ECM to examine the response of the trade balance to the real exchange rate.

Estimation Sample and Data. We use quarterly data from International Financial Statistics (IFS) and Direction of Trade Statistics (DOTS) of IMF in this paper. The time range is from 2000(1) to 2015(4). The trade balance (TB) is defined as the ratio of Vietnam's export value to import value. GDP is Vietnam's real GDP index. GDP^* is the foreign real GDP index and it is the sum total of real GDP indices of 22 trading partners after each of the index is multiplied by their respective percentage of export and import volume in trading with Vietnam. $REER$ is the real effective exchange rate. REER is the geometric mean of real exchange rates between VND and 22 trading partners' currencies associated with the respective percentage of export and import volume of each partner in trading with Vietnam. These 22 partners occupy approximately 90% of trade volume of Vietnam. All variables are in index forms with base period 2000(1) = 100 and converted into natural logarithms.

The empirical results. In order for F-test is valid for cointegration "bounds test", we need to use unit root test to confirm that the variables are not integrated at I(2); see, e.g. [27].

We apply the Augmented Dickey-Fuller (ADF) test and the Phillips-Perron (PP) test for our study. Table 1 reports the test results for the four variables.

All variables are in natural logarithms, D is the first difference operator.

From Table 1, we can see that TB, GDP, and GDP^* are stationary at their level forms, meaning that the order of integration is I(0). $REER$ is non-stationary at its level form, but stationary in first-difference, meaning that the order of integration of REER is I(1).

Table 1. ADF and PP tests results for non-stationarity of variables.

Variable	ADF test statistic	PP test statistic
TB	−6.003339***	−5.904744***
$REER$	−1.982632	−2.270239
GDP	−8.307229***	−8.315656***
GDP^*	−7.549758***	−7.680978***
$D(TB)$	−17.52242***	−28.91540***
$D(REER)$	−5.395644***	−5.364851***
$D(GDP)$	−15.70245***	−26.84714***
$D(GDP^*)$	−14.11703***	−24.50858***

Note: ***, ** and * are respectively the 1%, 5% and 10% significance level.

We display the F-statistics results according to various lag orders (shown in Table 2) to check the existence of the cointegration among variables in the model.

Table 2. F-statistics of bound tests, 10% CV [2.711, 3.800], 5% CV [3.219, 4.378], 1% CV [4.385, 5.615].

Lag order	1	2	3	4	5	6
F-statistic	4.285*	2.709	5.080**	1.768	1.295	1.587

Note: ** and * are respectively the 5% and 10% significance level.

The results of Table 2 show that some of the values of F-statistics are above the upper bounds of the critical values (CV) of standard signicance levels; see, e.g., [29]. Hence, we reject the hypothesis H0, showing that there is cointegration or long-run relationship between trade balance and its determinants in case of Vietnam.

Next, we estimate Eq. (13) and use Schwartz Bayesian Criterion (SBC) to select the optimal lag length. The maximum lag order is 6, which allows saving the degree of freedom. From Table 3, ARDL(1,2,2,0) is obtained based on SBC.

Our model can account for about 60% of trade balance performance, meaning that real exchange rate, domestic income and world income can explain for 60%

Table 3. Autoregressive Distributed Lag Estimation Results (Autoregressive Distributed Lag Estimates ARDL(1,2,2,0) selected based on Schwarz Bayesian Criterion).

Dependent variable: TB		
Variable	Coefficient	t-statistic
TB_{t-1}	0.33421***	3.4368
$REER_t$	0.35542	0.61417
$REER_{t-1}$	−1.3553	−1.5345
$REER_{t-2}$	1.5016***	2.7032
GDP_t	−0.36290***	−4.9789
GDP_{t-1}	−0.036870	−0.45102
GDP_{t-2}	−0.44896***	−6.1911
GDP_t^*	−0.15673**	−2.1285
$constant$	5.5894***	3.7369
$Adj - R^2 = 0.60054$		
$DW - statistics = 1.8766$		
$SE\ of\ Regression = 0.080581$		
Diagnostic tests	A: Serial Correlation	$F(1, 52) = 0.32779$ [0.569]
	B: Functional Form	$F(1, 52) = 0.33912$ [0.563]
	C: Normality	$ChiSQ(2) = 3.8046$ [0.149]
	D: Heteroscedasticity	$F(1, 60) = 0.51002$ [0.478]

Note: ***, ** and * are respectively the 1%, 5% and 10% significance level.
A: Lagrange multiplier test of residual serial correlation
B: Ramsey's RESET test using the square of the fitted values
C: Based on a test of skewness and kurtosis of residuals
D: Based on the regression of squared residuals on squared fitted values.

of Vietnam's trade balance, which is higher than that of the prior study in Vietnam; see, e.g., [35]. The reason is that our study includes the independent variable GDP^* (an important variable affecting trade balance); see, e.g., [32,38]. Also, we employ longer range time series. Besides, the real effective exchange rate used in our study represents the exchange rates between VND and the currencies of 22 countries occupying nearly 90% the trade volume of Vietnam, which is better than the prior study and reinforces the existence of J-curve in Vietnam; see, e.g., [35].

The diagnostic tests are also provided in Table 3, which suggests that there is no issue with this model. Thus, short-run and long-run impacts derived from the model are reliable.

The estimation of short-run coefficients of the ARDL model is shown in Table 4. In short-run, our results indicate that Vietnam's real income lagged 1 quarter has a positive and significant impact on trade balance at the current quarter. Vietnam's real income, foreign income and real exchange rate lagged

Table 4. The Error Correction Representation for the Selected ARDL model (Error Correction Representation for the Selected ARDL Model ARDL(1,2,2,0) selected based on Schwarz Bayesian Criterion).

Dependent variable: ΔTB		
Variable	Coefficient	t-statistic
$\Delta REER_t$	0.35542	0.61417
$\Delta REER_{t-1}$	−1.5016***	−2.7032
ΔGDP_t	−0.36290***	−4.9789
ΔGDP_{t-1}	0.44896***	6.1911
ΔGDP_t^*	−0.15673**	−2.1285
$constant$	5.5894***	3.7369
EC_{t-1}	−0.66579***	−6.8465
$Adj - R^2 = 0.73222$		

Note: ***, ** and * are respectively the 1%, 5% and 10% significance level.

1 quarter have negative and significant impacts on trade balance in the case of Vietnam. This indicates the negative impact of devaluation on trade balance in the short-run.

$$EC_{t-1} = TB_{t-1} - 0.75364 \cdot REER_{t-1} + 1.2748 \cdot GDP_{t-1}$$
$$+ 0.23541 \cdot GDP_{t-1}^* - 8.3951 \cdot constant. \tag{14}$$

From the table, it is obvious that the error correction term (EC_{t-1}) has negative sign and is statistically significant. This result once again provides the evidence of cointegration among variables in the model. This shows the speed of adjustment from short-run towards long-run. Specifically, the estimated value of EC_{t-1} is −0.66579 (about 66%) which indicates the speed of adjustment to equilibrium following short-run shocks.

The impulse response function based on obtained ECM allows us to examine the evolution of the trade balance over time subsequent to a real devaluation of VND. The results reported in Fig. 1 shows that the trade balance declines and hits bottom in the third quarter. After that, the trade balance

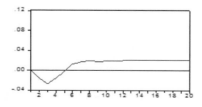

Fig. 1. Evolution of trade balance following real depreciation. Source: Authors' calculation based on response of TB to Cholesky One S.D. REER Innovation.

Table 5. Long-run estimation results (Estimated Long-Run Coefficients using the ARDL Approach ARDL(1,2,2,0) selected based on Schwarz Bayesian Criterion).

Dependent variable: TB		
Variable	Coefficient	t-statistic
$REER_t$	0.75364**	1.7973
GDP_t	−1.2748***	−4.9127
GDP_t^*	−0.23541**	−2.1733
$constant$	8.3951***	4.4704

Note: ***, ** and * are respectively the 1%, 5% and 10% significance level.

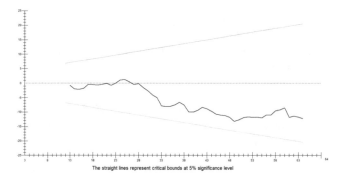

Fig. 2. Plot of cumulative sum of recursive residuals (CUSUM).

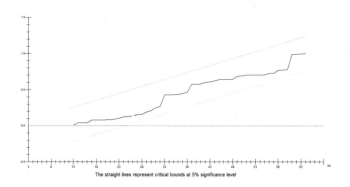

Fig. 3. Plot of cumulative sum of squares of recursive residuals (CUSUMSQ).

starts to improve. 6 quarters after devaluation, the new equilibrium is stable. Thus, impulse response result shows the occurrence of J-curve phenomenon in the relationship between real exchange rate and trade balance of Vietnam.

The results of long-run impacts are presented in Table 5. They indicate that the signs of the estimated coefficients of GDP and GDP^* are negative. The positive sign of real exchange rate's estimated coefficient implies that the devaluation of VND makes Vietnam's trade balance improve in the long-run.

To verify the stability of the estimated model, the tests of CUSUM and CUSUMSQ are employed in this study. Figures 2 and 3 respectively provide the graphs of CUSUM and CUSUMSQ statistics staying within the critical bounds indicating the stability of trade balance equation.

4   Conclusion

In this paper, by using aggregate data and employing ARDL approach, we find that the real effective exchange rate (between VND and a basket of 22 main trading partners' currencies) has statistically significant impacts on Vietnam's trade balance in both short-run and long-run.

Both results from analyzing ARDL model and impulse response function indicate the occurrence of J-curve phenomenon in Vietnam.

In short-run, a devaluation of real exchange rate causes considerably negative impact on Vietnam's trade balance. Nevertheless, this negative effect of real exchange rate on trade balance only lasts for 5 quarters. Trade balance will be improved from the fifth quarter after the devaluation. In the long-run, the increase of real exchange rate helps improve the trade balance.

Our study employs aggregate data to test the overall existence of J-curve in Vietnam by treating main trading partners of Vietnam as a whole. Consequently, we cannot examine the J-curve phenomenon in the trade between Vietnam and each of her major trading partners. Future researches could use disaggregated data at bilateral level between Vietnam and her largest trading countries or even at industry level for more detailed examination.

References

1. Anju, G.K., Uma, R.: Is there a J-curve? a new estimation for Japan. Int. Econ. J. **13**(4), 71–79 (1999)
2. Baek, J.: The J-curve effect and the US–Canada forest products trade. J. For. Econ. **13**(4), 245–258 (2007)
3. Bahmani-Oskooee, M., Alse, J.: Short-run versus long-run effects of devaluation: error-correction modeling and cointegration. East. Econ. J. **20**(4), 453–464 (1994)
4. Bahmani-Oskooee, M., Brooks, T.J.: Bilateral J-curve between U.S. and her trading partners. Weltwirtschaftliches Archiv **135**(1), 156–165 (1999)
5. Bahmani-Oskooee, M., Mitra, R.: The J-curve at industry level: evidence from U.S-India trade. Econ. Bull. **29**(2), 1520–1529 (2009)
6. Bahmani-Oskooee, M., Gelan, A.: Is there a J-curve in Africa? Int. Rev. Appl. Econ. **26**(1), 73–81 (2012)
7. Bahmani-Oskooee, M., Hajilee, M.: The J-curve at industry level: evidence from Sweden-US trade. Econ. Syst. **33**(1), 83–92 (2009)

8. Bahmani-Oskooee, M., Harvey, H.: J-curve: Singapore versus her major trading partners. Econ. Pap. **31**(4), 515–522 (2012)
9. Bahmani-Oskooee, M., Kantipong, T.: Bilateral J-curve between Thailand and her trading partners. J. Econ. Dev. **26**(2), 107–117 (2001)
10. Bahmani-Oskooee, M., Kutan, A.M.: The J-curve in the emerging economies of Eastern Europe. Appl. Econ. **41**(20), 2523–2532 (2009)
11. Bahmani-Oskooee, M., Wang, Y.: The J-curve: China versus her trading partners. Bull. Econ. Res. **58**(4), 323–343 (2006)
12. Brown, R.L., Durbin, J., Evans, M.: Techniques for testing the constancy of regression relations over time. J. Roy. Stat. Soc. **37**(2), 149–192 (1975)
13. Bentzen, J., Engsted, T.: A revival of the autoregressive distributed lag model in estimating energy demand relationship. Energy **26**(1), 45–55 (2001)
14. Dash, A.: Bilateral J-curve between India and her trading partners: a quantitative perspective. Econ. Anal. Policy **43**(3), 315–338 (2013)
15. Engle, R.F., Granger, C.W.J.: Co-integration and error correction: representation, estimation, and testing. Econometrica **55**(2), 251–276 (1987)
16. Felmingham, B.: Where is the Australian J-curve? Bull. Econ. Res. **40**(1), 43–56 (1988)
17. Ghatak, S., Siddiki, J.: The use of the ARDL approach in estimating virtual exchange rate in India. J. Appl. Stat. **28**(5), 573–583 (2001)
18. Halicioglu, F.: The bilateral J-curve: Turkey versus her 13 trading partners. J. Asian Econ. **19**(3), 236–243 (2008)
19. Halicioglu, F.: The J-curve dynamics of Turkey: an application of ARDL model. Appl. Econ. **40**(18), 2423–2429 (2008)
20. Johansen, S., Juselius, K.: Maximum likelihood estimation and inference on cointegration with application to the demand for money. Oxford Bull. Econ. Stat. **52**(2), 169–210 (1990)
21. Junz, H.B., Rhomberg, R.R.: Price competitiveness in export trade among industrial countries. Am. Econ. Rev. **63**(2), 412–418 (1973)
22. Kyophilavong, P., Shahbaz, M., Uddin, G.S.: Does J-curve phenomenon exist in case of Laos? an ARDL approach. Econ. Model. **35**, 833–839 (2013)
23. Lal, A.K., Lowinger, T.C.: The J-curve: evidence from East Asia. J. Econ. Integr. **17**(2), 397–415 (2002)
24. Magee, S.P.: Currency contracts, pass-through and devaluation. Brook. Pap. Econ. Activity **4**(1), 303–325 (1973)
25. Narayan, P.K.: The saving and investment nexus for China: evidence from cointegration tests. Appl. Econ. **37**(17), 1979–1990 (2005)
26. Odhiambo, N.M.: Energy consumption and economic growth nexus in Tanzania: an ARDL bounds testing approach. Energy Policy **37**(2), 617–622 (2009)
27. Ouattara, B.: Modelling the long run determinants of private investment in Senegal, The School of Economics Discussion Paper Series 0413, Economics. The University of Manchester (2004)
28. Pesaran, B., Pesaran, M.H.: Time Series Econometrics Using Microt 5.0. Oxford University Press, Oxford (2009)
29. Pesaran, M.H., Pesaran, B.: Microfit 4.0 (Window Version), Oxford University Press, Oxford (1997)
30. Pesaran, M.H., Shin, Y., Smith, R.J.: Bounds testing approaches to the analysis of level relationships. DEA Working Paper 9622, Department of Applied Economics, University of Cambridge (1996)
31. Pesaran, M.H., Shin, Y., Smith, R.J.: Bounds testing approaches to the analysis of level relationships. J. Appl. Econ. **16**(3), 289–326 (2001)

32. Rose, A., Yellen, J.: Is there a J-curve? J. Monet. Econ. **24**(1), 53–68 (1989)
33. Šimáková, J., Stavárek, D.: Exchange-rate impact on the industry-level trade flows in the Czech Republic. Procedia Econ. Financ. **12**, 679–686 (2014)
34. Stučka, T.: The Effects of Exchange Rate Change on the Trade Balance in Croatia, IMF Working paper (2004)
35. Trinh, P.T.T.: The impact of exchange rate fluctuation on trade balance in short and long run. Depocen Working Paper Series No. 2012/23 (2012)
36. Vural, B.T.: Effect of real exchange rate on trade balance: commodity level evidence from turkish bilateral trade data. Procedia Econ. Financ. **38**, 499–507 (2016)
37. Wilson, P.: Exchange rates and the trade balance for dynamic Asian economies: Does the J-curve exist for Singapore, Malaysia and Korea? Open Econ. Rev. **12**(4), 389–413 (2001)
38. Ziramba, E., Chifamba, R.: The J-curve dynamics of South African trade: evidence from the ARDL approach. Europ. Sci. J. **10**(19), 346–358 (2014)

An Analysis of Eigenvectors of a Stock Market Cross-Correlation Matrix

Hieu T. Nguyen$^{(\boxtimes)}$, Phuong N. U. Tran, and Quang Nguyen

John von Neumann Institute, Vietnam National University,
Ho Chi Minh City, Vietnam
htnguyen.satz@gmail.com, uyenphuong.tran@gmail.com,
quang.nguyen@jvn.edu.vn

Abstract. Random matrix theory (RMT) has been used to great effect in analysing the structure of the stock return cross-correlation matrix. Common results have been found in various markets: only a few eigenmodes (or portfolios) appear significant and the rest resemble that of a random matrix. Specifically, the eigenvalues spectrum consists of an outstanding large eigenvalue representing the whole market mode, followed by several other large ones, plus the bulk which mostly agrees with the theoretical spectrum distribution predicted by the RMT. The body of work on eigenvector components is not as abundant however. In this work, we analyse in detail the components of the eigenvectors and found that the market mode components depend linearly on what we called the correlation weights of the stocks. Therefore, the corresponding market portfolio must be viewed as a market-correlation-portfolio rather than a market-cap-portfolio. Other informative eigenvectors also show important structures. Those results could be very meaningful in analysing the structure of financial markets and their applications.

1 Introduction

The study of cross-correlations between stocks is of immense importance not only for understanding the complex structure of the stock market, but also for assets allocation and risk management. Several established statistical methods employed from physics have been used to study the cross-correlation matrix including the minimal spanning tree (MST) (first applied in [1]) and random matrix theory (RMT) (first applied in [2,3]). The MST method constructs a tree that connects all stocks in the market. Based on this tree, one can analyse the market topology, which is usually represented by sectors. On the other hand, the RMT method can serve as a powerful noise filter that reveals important correlation modes in the market.

Under the lenses of RMT, the eigenvalues and eigenvectors of the cross-correlation matrix are calculated and compared to those of a random matrix, generated from randomly generated uncorrelated time-series (referred to as RMT eigenvalues and eigenvectors). It was found in [2–12] that the majority of eigenvalues from a stock returns cross-correlation matrix overlap with the bulk of RMT eigenvalues, except:

© Springer International Publishing AG 2018
L. H. Anh et al. (eds.), *Econometrics for Financial Applications*, Studies in Computational Intelligence 760, https://doi.org/10.1007/978-3-319-73150-6_40

- The largest eigenvalue λ_1
- Several large eigenvalues beyond the theoretical upper bound λ_+, though considerably lower compared to the top eigenvalue.
- Some very small eigenvalues smaller than the theoretical lower bound λ_-

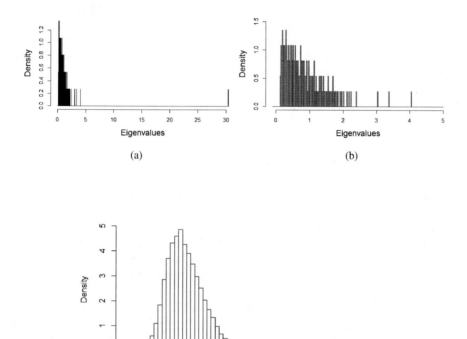

(a)   (b)

(c)

Fig. 1. (a): Distribution of all empirical eigenvalues (b): Distribution of all empirical eigenvalues without the largest eigenvalue (c): Distribution of Correlation Coefficients of the Correlation Matrix

One can easily observe the largest eigenvalue λ_1 is substantially higher than subsequent eigenvalues. This implies that most stocks follow a dominant correlation mode: there is a tendency of a positive mutual correlation of all stocks in the market. From a financial point of view, it is intuitive to understand because all listed companies operate in the same economy and therefore, are influenced by many (macro) common factors. This view is also coherent with the CAPM theory and the co-movement of the market as a whole is called the market mode.

From Fig. 1(c) it is seen the correlation coefficients are distributed in a positive narrow range and mostly positive. In our previous work [13] we modeled the correlation matrix by a constant correlation ρ_1 (one-factor correlation matrix model) and indeed found the appearance of the largest eigenvalue λ_1. Furthermore, the value of λ_1 can be approximated by the product of ρ_1 and N, where N is the total number of stocks.

While there are ample studies focusing on dynamics of the eigenvalues, fewer focus on the eigenvectors. Moreover, most studies suggest that the eigenvector components associated with the largest eigenvalue indicate a uniform distribution of weights among stocks. However, we re-examined this distribution (Fig. 3(a)) and attempt to instead relate it to the correlation degree of each stock. We define the correlation weight by:

$$ w_i = \frac{1}{N-1} \cdot \left(\sum_{j=1}^{N} \rho_{ij} - 1 \right) \tag{1} $$

and found that a_i^1, the component i of the eigenvector corresponding to the largest eigenvalue λ_1, is linearly dependent on w_i, as demonstrated in Fig. 3(b). To our knowledge this result is not addressed yet in studies using RMT for financial markets. It is then reasonable to interpret this market mode portfolio as a correlation-weight-portfolio, unlike other definitions of market-portfolio (usually market-capitalization weighted or uniformly weighted). For example, in the Vietnamese market, the heaviest-weighted stock in the market mode is SSI, which is a mid-cap securities firm. The risk and behaviour of such portfolio is a subject for our future work.

We also study the components of eigenvectors associated with other large eigenvalues lying beyond the bulk predicted by RMT. Considering that the majority of components of the market-correlation-portfolio are negative except for a few positive ones, the components of other eigenvectors are expected to be distributed around zero in order to satisfy the orthogonality condition.

In addition, eigenvectors with significant positive and negative components values may indicate groups of stocks belonging to specific financial sectors. Other studies on developed markets such as the US, Japan found that they represent industry sectors (see [3–5]) while some studies on emerging markets such as China and India found a mix of industry sectors and other so-called "specific treatment" stock groups, as in [9–12]. In the Vietnamese stock market, we show that one of the major groups is the speculative group. In fact, it depends on the property of the market and their investors behaviour to create such structure. We also found that within each mode, the *long* and the *short* groups need to be positively correlated within each group and possibly negatively correlated between groups.

The paper is organized as follows: Sect. 2 explains the data and method, Sect. 3 presents our main results and in Sect. 4 we discuss, conclude and suggest some future work.

2   Data and Methods

2.1   Data

The data is collected from the two Vietnamese stock exchanges – the Hanoi Stock Exchange and the Ho Chi Minh Stock Exchange – for the period from 1/1/2015 to 31/5/2017. We selected active stocks that were listed prior to 2015. Moreover, to ensure some degree of liquidity, the stocks must have, on daily average, trading volume exceeding 50,000 units and transaction value exceeding one billion Vietnam Dong. The prices on days with no trades, due to technical reasons such as a change of stock exchange, are assumed to be constant with the last trading day. In the end, our dataset consists of $N = 186$ stocks with daily prices spanning over a total of $T = 600$ days.

We consider the standardized stock returns as follows: (i) we first obtain the log-returns $R_i(t)$ from the daily closing prices $R_i(t) = \log(S_i(t+1)) - \log(S_i(t))$ (ii) we then remove the (empirical) mean of each stock from its returns: $R_i^*(t) = R_i(t) - (R_i)$ where $(.)$ denotes the sample average; (iii) we finally normalize each stock returns by its volatility which, in this work, is simply estimated by the standard deviation: $r_i(t) = \frac{R_i^*(t)}{\sigma(R_i)}$. From this, we obtain an estimation of the $N \times T$ correlation matrix C of returns using the Pearson estimator for the correlation co-efficients C_{ij}:

$$C_{ij} = (r_i(t)r_j(t))$$

2.2   Methods

Once the eigenvalues and (normalised) eigenvectors are obtained from the estimated correlation matrix C, the eigenvalues are plotted and compared against a theoretical eigenvalues distribution given by Random Matrix Theory. Deviations of the empirical eigenvalues from the theoretical distribution indicate presence of true informative eigenvalues, separated from noise-corrupted eigenvalues.

In a more general setting, let N and T be respectively the number of stocks and the number of realizations of data for each stock. We consider a random matrix called the Wilshart matrix of the form $W = \frac{1}{T}YY^*$ where Y is a $N \times T$ matrix of T i.i.d vectors of size N. We note that in this context, the Wilshart matrix is in fact the sample covariance matrix (or correlation matrix, after normalization). From the seminal paper [14], it is known that: in the limits of T and N to infinity such that $Q := \frac{T}{N} > 1$ and under the null hypothesis of a purely random matrix, the limiting spectral density – more famously known as the *Marchenko-Pastur* (MP) density – reads:

$$\rho_{MP}(\lambda) = \frac{Q}{2\pi\sigma^2} \cdot \frac{\sqrt{(\lambda_+ - \lambda)(\lambda - \lambda_-)}}{\lambda}$$

where λ_+, λ_- are the upper and lower bounds of the bulk component containing all eigenvalues:

$$\lambda_\pm = \sigma^2 \cdot \left(1 \pm \frac{1}{\sqrt{Q}}\right)^2$$

and σ^2 is equal to the variance of the elements, which is 1 given the normalisation of stock returns.

In order to highlight the non-randomness in the components of informative eigenvectors, as in [2], we may also compare these components (normalized such that the eigenvector has length $\sqrt{N}$) against the entropy-maximizing Porter-Thomas distribution – in this case simply the standard normal distribution:

$$P(u) = \frac{1}{\sqrt{2\pi}} \cdot \exp\left(-\frac{u^2}{2}\right)$$

If there is no information in an eigenvector, we expect the components to obey the distribution.

3   Results

3.1   Overview

As expected, there is one outstanding eigenvalue at $\lambda_1 = 30.4$ well separated from the rest, a bulk containing the majority of eigenvalues and a few others outside this bulk but lower in value than the largest one.

Figure 1(a) shows the correlation coefficients distributed in a narrow range, indicating that there exists a coherent movement of all the stocks – the market mode.

In order to separate the informative from the noisy, we first compare the empirical eigenvalues distribution with the MP-density where $Q = \frac{T}{N} = \frac{599}{186} \approx 3.22$ and $\sigma^2 = 1$ as per normalization.

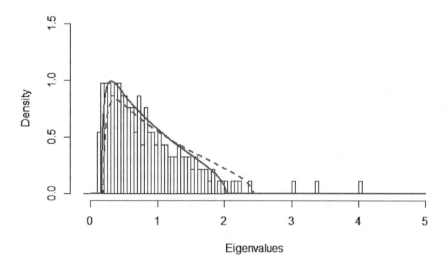

Fig. 2. Distribution of the eigenvalues, excluding the top eigenvalue. Also included is a plot of the MP density with $\sigma^2 = 1$ (dotted line) and with $\sigma^2 = 0.836$ (solid line)

As in the work of Laloux et al. [3], a fit of the MP - density can also be constructed with a modified $\sigma^2 = 1 - \frac{\lambda_1}{N} = 0.836$. This corresponds to the observation that the top eigenvalue and eigenvector is clearly not random and so it is reasonable to subtract the contribution of the market eigenvalue from the variance of the random part. A fit of the corresponding plot is also included in Fig. 2.

We note that a better fit can be made still, considering three reasons: (i) larger eigenvalues are prone to overestimation (ii) outlying eigenvalues containing true information may effectively reduce the variance σ^2 of the random part of the matrix (iii) volatility correlations may cause Q to deviate from the effective Q for the MP density.

Still the current fit is convincing, with 90% of the empirical eigenvalues lying within the bulk. In particular, the eigenvalues 1 to 4 (in order of magnitude) are likely to contain information.

3.2   On the Top Eigenvector

We continue with a study of the top eigenvector associated with the largest eigenvalue, often referred to as the market eigenvector, market mode or the market portfolio. Figure 3 shows the components of the top eigenvector, where all the components are negative (or of the same sign). As eigenvectors are unique up to multiplication by scalar, we may interpret the portfolio as short-only or more economically realistically, long-only. In this work, we opt for the latter interpretation. Figure 3(b) shows the linear relationship between the components and the correlation weights defined in 1. This indicates a correlation-based structure to the market eigenvector, instead of a market capitalization-based weighting or a uniform weighting structure. Indeed, for our data, the stocks holding the most weight include the tickers SSI, HCM, VND, PVT etc. These are in fact not among the highest-cap stocks in the market. Whether this structure is stable or can be proved theoretically is a subject for future work.

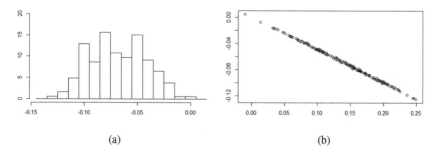

(a)   (b)

Fig. 3. (a): Distribution of market mode components (b): Plot of market mode components against corresponding correlation weights of the stocks

3.3 On Subsequent Eigenvectors

We make the first observation that: since the market portfolio is long-only, to ensure orthogonality, all others eigenvectors must have both negative and positive components. Say differently, all other portfolios are some sort of long-short portfolio, as illustrated in Fig. 4, referred to from here on as *long* and *short* *sub-portfolios*. It has been documented in literature the *localization effect* displayed in the eigenvector components, that is, each eigenvector is dominated by a group of stocks,. This is particularly evident for developed markets [4] where the eigenvectors exhibit strong presence of financial sectors: the blue-chips group, the technology sector, the gold sector, to name a few. The case with emerging markets is less apparent where the top eigenvectors are dominated by fewer and less traditional sectors: for China (see [9,10]), the sectors include blue-chip companies, the so-called ST stocks (the stocks with Special Treatment defined in [9]) and Shanghai-listed real estate businesses; still the eigenvectors exhibit localization in the component values. Figure 4 however shows that for the Vietnamese stock market, this localization effect is not as clear-cut among different eigenvectors. Specifically, except the eigen-portfolio 4 and the short-only sub-eigen-portfolio 3 where there are groups of dominating stocks, the weights are spread diffusely.

We shall look at the physical components of these eigenvectors to further understand this phenomenon. For each eigenvector, we consider the top 12 contributing stocks in each of the long and short sub-portfolios. The stocks are listed in Table 1, together with a general description.

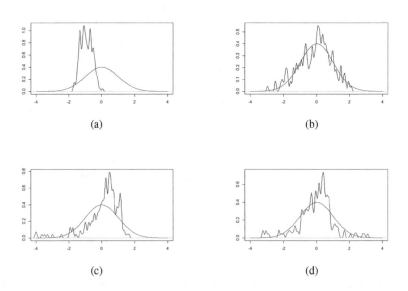

Fig. 4. Smoothed densities of the components of the eigenvectors 1, 2, 3 and 4 (plot a, b, c and d respectively) against the Porter-Thomas distribution, plotted with a red thick line.

Table 1. Notable physical components of the Eigen-portfolios 2, 3 and 4

Portfolio	Long Sub-Portfolio	Short Sub-Portfolio
Eigen-portfolio 2	BMP, PNJ, VSC, VCS, CAV, SVC, PAC, DHG, VNS, KSB, DHA, C32	ITA, HAI, FLC, VHG, KSA, KLF, HQC, FIT, S99, DLG, ASM, HAR
	Mid and low-cap, strong upward trend overall, high EPS and ROE, low PE	Downward trend, slump in profits, EPS < 1 while still maintaining solid liquidity
Eigen-portfolio 3	EVE, DLG, TIG, FLC, TDH, KSA, TTF, VPH, HAD, DRH, TSC, FIT	PVD, PVS, PVS, GAS, PXS, PVB, PGS, CTG, BID, PVT, BVH, VCB
	Speculative mid-cap, on a downward trend, despite high income levels for some of the stocks	Stocks in the Gas-Oil Sector and in the Banking Sector
Eigen-portfolio 4	PVI, SHB, PGI, VNM, BVH, BMI, BIC, MBB, VCB, ACB, BID, CTG	PVC, PVD, PXS, PVB, PVS, GAS, VHG, C32, PET, HTI, CTI, TRC
	VNM, the highest-cap stock in Vietnam, and others from the Banking Sector	Stocks in the Gas-Oil Sector

Curiously, the second eigen-portfolio is composed of mid-cap stocks whose performance has significantly slumped (forming the short sub-portfolio) or improved (forming the long sub-portfolio). That the portfolio is composed of mid and low-caps instead of blue-chip stocks (as in developed markets and to a lesser extent, emerging markets) is surprising. A similar observation can be made for the long sub-portfolio yielded by the third eigenvector, though the contributing stocks are of a different nature: they are mostly mid-cap and speculative with volatile price trajectories. It should be pointed out that these stocks are not entirely similar to the "ST" stocks defined in [9], which are stocks with an abnormal financial situation, with profits sharply decreasing over the data period.

The above observation leads to an interpretation that the eigenvectors with significant eigenvalues do not (only) reflect the dominating sectors within the economy, rather the investing appetites of the stock market participants. This view is particularly understandable for Vietnam where the presence of speculative individual investors, informed or not, is significant relative to that of financial institutions. It would be interesting to see whether for a shorter time-period, perhaps with intra-day data, it is possible to study the short-term investing behaviour of the market.

Meanwhile, the subsequent eigenvectors displays a more sectoral structure often seen in more developed markets, where the two largest sectors in Vietnam, Oil & Gas and Banking, dominate. Considering that few other sectors in Vietnam boast enough listed companies to form a meaningful group, the modest sectoral behaviour is expected. Still, it is surprising the rather unremarkable presence of the blue-chip stocks as a group within the eigen-portfolios.

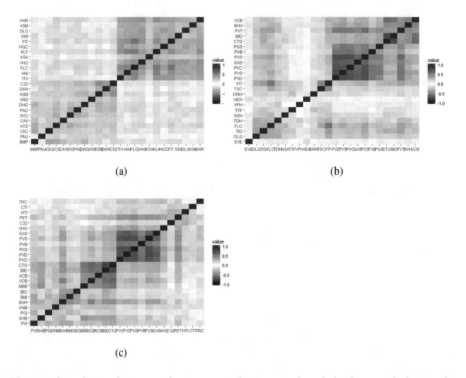

(a)   (b)

(c)

Fig. 5. Correlation between the top contributing stocks of the long and short sub-portfolios within the eigen-portfolio 2, 3 and 4 (picture a, b and c respectively)

Finally, we observe the correlation structure of the eigen-portfolios. As indicated in the Fig. 5, there is strong correlation within each of the sub-portfolios and weak correlation between sub-portfolios.

4   Conclusion

In this work, we have analysed the components of the important eigenvectors of a cross-correlation matrix in the Vietnamese stock market. We found the following results: firstly, the component i of the top eigenvector associated with λ_1 is linearly dependent on the correlation weight w_i of each stock. This finding is important because as λ_1 is far higher than other eigenvalue, its corresponding portfolio has a great contribution of variance (or risk) to any suitably diversified portfolio. We redefined this portfolio as a correlation-weighted market portfolio and will study its role in future work, as well as the relation above. Secondly, as found in literature, all markets have this common correlation-weighted market mode, we suppose that all stocks in the market have a tendency to be mutually positively correlated and the correlation weight w_i of each stock is positive. In consequence, the components of the market portfolio are positive and those of other eigen-portfolios are equally distributed around zero. In other words, if the correlation-weighted market portfolio is the common market mode, all other eigen-portfolios are formed by long and short sub-portfolios. Thirdly, we also highlight some important groups of contributing stocks in some modes and found mixed results on the nature of these groups compared to previous work on developed and emerging markets. In conclusion, this work focuses on the eigenvector components of cross-correlation matrix using RMT. We have found some results that can be used in other risk management study, especially for a diversified portfolio. Further work is needed to complement these initial findings and get a better view about the stock market topology.

Acknowledgements. This research is funded by Vietnam National University Ho Chi Minh City (VNU-HCM) under grant number B2017-42-01.

References

1. Mantegna, R.N.: Hierarchical structure in financial markets. Eur. Phys. J. B **11**, 193 (1999)
2. Laloux, L., Cizeau, P., Bouchaud, J.P., Potters, M.: Noise dressing of financial correlation matrices. Phys. Rev. Lett. **83**(7), 1467 (1999)
3. Laloux, L., Cizeau, P., Potters, M., Bouchaud, J.P.: Random matrix theory and financial correlations. Int. J. Theoret. Appl. Finance **3**(03), 391–397 (2000)
4. Plerou, V., Gopikrishnan, P., Rosenow, B., Amaral, L.A.N., Guhr, T., Stanley, H.E.: Random matrix approach to cross correlations in financial data. Phys. Rev. E **65**(6), 066126 (2002)
5. Utsugi, A., Ino, K., Oshikawa, M.: Random matrix theory analysis of cross correlations in financial markets. Phys. Rev. E **70**, 026110 (2004)
6. Wilcox, D., Gebbie, T.: On the analysis of cross-correlations in South African market data. Phys. A **344**, 294 (2004)
7. Oh, G., Eom, C., Wang, F., Jung, W.S., Stanley, H.E., Kim, S.: Statistical properties of crosscorrelation in the Korean stock market. Eur. Phys. J. B **79**, 55 (2011)
8. Cukur, S., Eryigit, M., Eryigit, R.: Cross-correlation in an emerging market financial data. Physics A **376**, 555 (2007)
9. Shen, J., Zheng, B.: Cross-correlation in financial dynamics. Europhys. Lett. **86**, 48005 (2009)
10. Jiang, X.F., Zheng, B.: Anti-correlation and subsector structure in financial systems. EPL **97**, 48006 (2012)
11. Pan, R.K., Sinha, S.: Collective behavior of stock price movements in an emerging market. Phys. Rev. E **76**, 046116 (2007)
12. Kulkarni, V., Deo, N.: Volatility of an Indian stock market: a random matrix approach, e-print arXiv:physics/0512169 (2005)
13. Nguyen, Q.: One-factor model for cross-correlation matrix in the Vietnamese stock market. Phys. A **392**, 2915–2923 (2013)
14. Marchenko, V.A., Pastur, L.A.: Distribution of eigenvalues for some sets of random matrices. Matematicheskii Sbornik **114**(4), 507–536 (1967)

Factors Impacting Tax Revenue of Southeast Asian Countries

Ly Hoang Anh and Tran Quoc Thinh[(✉)]

Banking University of Ho Chi Minh City, Ho Chi Minh City, Vietnam
`thinhtq@buh.edu.vn`

Abstract. Tax revenue is important for the country to ensure important government spending. This is even more important for a developing country. By analyzing panel data, we find factors that affect tax revenues of Southeast Asian countries during 10 years, for the period 2006–2015. The results show that there are four positive factors for tax revenues: GDP per capita, Trade volume, Agricultural sector, Industry sector. This result is in accordance with the characteristics and general conditions of these countries.

1 Introduction

Taxes account for a high proportion of total revenues in most countries in the world. Taxation is one of the best tools to boost the performance of the public sector, to fund social security, and to pay for public debt. A sound national financial system must rely primarily on internal revenues.

It is therefore important to find which factors affect the tax revenue. Several authors have studied this problem; see, e.g., (Piancastelli 2001; Eltony 2002; Pesino and Fenochietto 2010; Dioda 2012). Most of these papers study the immediate effect of different factors on tax revenue; some papers show that there is also a delayed effect; see, e.g., (Castro and Ramirez 2014). Most of these paper use table data technique, since this technique enables us to take into account individual characteristics and also to take into account how things change with time.

Some of the corresponding papers study tax revenue in developing countries; see, e.g., Gupta (2007), Bird et al. (2008), Ajaz and Ahmad (2010). Piancastelli (2001) studied both developed and developing countries, and showed that the factors affecting tax revenue are somewhat different in these two groups of countries. Several studies focus on regional groups such as Arab countries (Eltony 2002), Latin America and the Caribbean (Dioda 2012), and Middle East countries in general (Imam and Jacobs 2014). In all these cases, there were regional specifics, and taking these specifics into account made the resulting models more accurate. It is therefore desirable to conduct a study of Southeast Asian countries, to get more accurate models of tax revenue in these countries.

It is reasonable to group these countries together: they have not only geographical similarity, they also are at similar levels of economic development, and

© Springer International Publishing AG 2018
L. H. Anh et al. (eds.), *Econometrics for Financial Applications*, Studies in Computational Intelligence 760, https://doi.org/10.1007/978-3-319-73150-6_41

they are actively trying to integrate their economies: e.g., recently, the ASEAN Economic Community (AEC) was formed. The analysis of tax revenue in these countries is the main objective of this research.

This paper consists of 5 sections. Section 2 presents theoretical background and the results of prior studies. Section 3 describe the methodology of this study. Section 4 contains the result of our research and related discussion. Section 5 contains conclusions.

2   Theoretical Foundations of Tax Revenue Analysis and the Results of Previous Studies

2.1   Theoretical Foundations of Tax Revenue Analysis

In economics, the study of tax revenues is based of several related theories: the theory of service costs, benefit theory, and social and political theories (Ojong et al. 2016).

According to the theory of service costs, service costs should be proportional to the benefits that people get from these services. Taxes can be viewed as a particular case of the services, since the state uses taxes to support as health care, education, and other activities.

With respect to taxes, the conclusions of this theory are similar to the conclusions of the benefit theory, according to which people should be required to pay taxes in proportion to the benefits that they receive from government services.

The social and political theory of taxes emphasizes the social and political goals as key factors in choosing taxes. This theory emphasizes the benefits to the society as a whole as opposed to the benefits for each individual.

2.2   Overview of Previous Studies

Several researchers analyzed factors affecting tax revenues. Piancastelli (2001) examined the factors affecting the tax revenues of 75 developing and developing countries during the period of 1985–1995. The study concluded that the ratio of trade to GDP-per-capita has a positive impact on tax revenue, while the agriculture sector has a negative impact on tax revenue.

Eltony (2002) studied the factors affecting tax revenues of the Arab nations, namely, a sample of 10 Arab countries that do not produce oil, during the period from 1994 to 2000. The result was that per-capita GDP, export and import ratio, the mineral sector, and the amount of the external debt positively affected the tax revenue, while the agricultural sector had the opposite effect.

Eltony (2002) also studied 6 Arab countries that produce oil. For these countries, GDP per capita was positively associated with tax revenue while the minerals industry was inversely related to tax revenue.

Gupta (2007) studied the factors affecting the tax revenues of 105 developing countries form 1980 to 2004. This study showed that GDP per capita, trade, foreign aid, and economic stability indicators have a positive relationship with tax revenues, while the agriculture sector and the corruption perception index were inversely related to tax revenue.

Bird et al. (2008) investigated the relationship between tax revenues of 110 developing countries in period of 1990–1999. Their analysis showed that GDP per capita, population growth rate, and net import-export ratio were inversely related to tax revenue, while the non-agricultural sectors and the indicator of government accountability were positively correlated with tax revenue.

Ajaz and Ahmad (2010) examined the factors affecting the tax revenues of 25 developing countries in 1990–2005. They concluded that the industry sector and good government governance positively affected the tax revenue, while corruption had the opposite effect on tax revenues.

Pessino and Fenochietto (2010) analyzed factors affecting tax revenues of 96 countries in 1991–2006. They concluded that GDP per capita, openness of the economy, and ratio of public investment in education were positively correlated with tax revenue. The study also pointed to the negative relationship between corruption, inflation, the value-added ratio of agriculture to tax revenue.

Dioda (2012) used panel data to analyze 32 Latin American and Caribbean countries for the period 1990–2009. The results show that civil liberties, the number of female workers, age structure, political stability, educational attainment, and population density have a positive impact on tax revenue. In contrast, the agriculture sector and the size of the underground economy had a negative impact on tax revenue.

Castro and Ramirez (2014) studied the factors affecting tax revenues for 34 OECD member countries for the period 2001–2011. They show that GDP per capita and industry sector have a positive impact on tax revenue, while the rate of foreign direct investment, agricultural sector, indicators of civil liberties and life expectancy were negatively affecting tax revenues.

Imam and Jacobs (2014) studied factors affecting the tax revenues of the 12 Middle East countries in 1990–2003. The study concluded that inflation positively impacts the tax revenue, while GDP per capita has the opposite effect on tax revenue.

Ayenew (2016) examined the factors affecting the tax revenues of Ethiopia for the period 1975–2013. The results show that the industrial sector, GDP growth, and foreign aid have a positive impact on tax revenue, while inflation rate has a negative impact on tax revenues.

3   Methodology of This Study

3.1   Research Methods

To estimate the regression model according to the table data, three approaches are commonly used: Pooled Ordinary Least Square (POLS), fixed effects model (FEM), and random effect model (REM).

POLS is a regression model in which each coefficient is the same for all the countries and for all moments of time. For the problem of estimating tax revenues, the POLS model has the form

$$Y_{it} = \beta_0 + \beta_1 X_{1it} + \ldots + \beta_n X_{nit} + \varepsilon_{it},$$

where:

- Y_{it} is the dependent variable – tax revenue of country i in year t;
- X_{it} are independent variables – factors affecting tax revenues,
- β_0 is a constant coefficient, and
- ε_{it} is the model's inaccuracy.

POLS is the simplest regression model, a model that does not take into account the difference between different countries. As a result, for this model, however, the Durbin-Watson coefficient is usually quite small (less than 1), and there is often a positive autocorrelation. If we take into account differences between the two countries, we get more accurate models. In this paper, we consider the Fixed Effects Model (FEM) and the Random Effects Model (REM).

In both these models, we still assume that the coefficients β_1 through β_n are the same for all the countries, but we take into account that the value of a constant term β_0 may differ from one country to another. As a result, we arrive at the following models: the FEM model

$$Y_{it} = \beta_{0i} + \beta_1 X_{1it} + \ldots + \beta_n X_{nit} + \varepsilon_{it}.$$

and the REM model

$$Y_{it} = \beta_{0i} + \beta_1 X_{1it} + \ldots + \beta_n X_{nit} + u_i + \varepsilon_{it}.$$

The difference between the two models is that in REM, we assume that X_{it} and β_{0i} are not correlated, while in the FEM, we allow the possibility of such a correlation.

To select between POLS and FEM, the authors used the F-test to decide between the following two hypotheses:

- H_0: $u_i = 0$ (POLS)
- H_1: $u_i \neq 0$ (FEM)

To select between the FEM model and the REM model, the authors use the Hausman (1978) test to decide between the following two hypotheses:

- H_0: $cov(u_i, x_{it}) = 0$ (REM)
- H_1: $cov(u_i, x_{it}) \neq 0$ (FEM)

Once the model is selected, we use regression analysis to find out how exactly the national tax revenue depends on the selected factors.

3.2 Describing the Details of the Study

There are 10 Southeast Asian countries, but since the information related to Myanmar is not published, we conducted research based on the data from 9 countries: Indonesia, Thailand, Vietnam, Malaysia, Philippines, Laos, Cambodia, Brunei Darussalam, and Singapore. This research is based on the 10-year period 2006–2015. Countries' data is taken from the World Bank (2017) database.

3.3   Research Models

Dependent variable: tax revenue. Tax revenue is the percentage of the GDP that goes to the state budget through taxes. It is computed by dividing the overall amount of taxes by the country's GDP.

Independent Variables

GDP per capita. Most studies show that GDP per capita positively affects tax revenues such as Eltony (2002), Gupta (2007), Penssino and Fenochetto (2010), Castro and Remize (2014), Ayenew (2016), while the results of the research by Imam and Jacobs (2014) suggest that GDP per capita has the opposite effect on tax revenues. Because of this, the authors propose the following hypothesis:

Hypothesis H_1: GDP per capita positively affects tax revenue.

Trade volume. Trade volume reflects the openness of a country to international trade. It is an important factor to determining tax revenue, since taxes on international trade are one of the major sources of revenue in many countries. Studies on the relationship between trade volume and tax revenues provide different results. Studies by Eltony (2002), Gupta (2007), Pessino and Fenochietto (2010) show that the increase in trade volume increases tax revenues, while Bird et al. (2008) come up with the opposite conclusion. Research by Ajaz and Ahmad (2010), Castro and Ramirez (2014), Imam and Jacobs (2014) did not find any link between trade volume and tax revenue. Since more papers found a positive relation that the negative one, we propose the following hypothesis:

Hypothesis H_2: The higher the trade volume, the higher the tax revenue.

Foreign direct investment (FDI). FDI occurs when an investor from one country (investor country) acquires a property in another country (country attracting investment), along with the right to manage the property. Research by Gupta (2007) and Ayenew (2016) shows that FDI has a positive impact on tax revenue, while the study by Castro and Ramirez (2014) shows a negative relationship between FDI and tax revenue. The authors believe that FDI inflow increases national competitiveness and thus, increases the tax base. Thus, we propose the following hypothesis:

Hypothesis H_3: Foreign direct investment positively affects tax revenue.

Agricultural sector. The studies of Piancastelli (2001), Gupta (2007), Pessino and Fenochietto (2010), Dioda (2012), and Castro and Ramirez (2014) found the agricultural sector has a negative correlation with tax revenue, while the studies by Ajaz and Ahmad (2010), Imam and Jacobs (2014), and Ayenew (2016) suggest that the agricultural sector does not have an impact on tax revenue. The majority of previous studies have shown that the agricultural sector has a negative impact on tax revenue. Therefore, we propose the following hypothesis:

Hypothesis H_4: The agricultural sector negatively affects tax revenue.

Industry sector. Bird et al. (2008), Ajaz and Ahmad (2010), Castro and Ramirez (2014), and Ayenew (2016) show that there is a positive relationship between industry sector and tax revenues. So the authors propose the following hypothesis:

Hypothesis H_5: The industry sector has a positive impact on tax revenue.

Civil liberties. There are different results on how civil liberties such as the right to life, liberty and security of individual, freedom of movement and residence, affect tax revenues. Diona (2012) found out that citizen freedoms have a positive impact on tax revenue, while the empirical study by Castro and Ramirez (2014) found the opposite effect. We propose the following hypothesis:

Hypothesis H_6: Civil liberties positively affect tax revenue.

Political freedom. Political rights include the right to participate in state administration, freedom of thought, freedom of speech, the right to form associations, and the right for peaceful assembly. Research by Castro and Ramirez (2014) did not find any statistically significant link between political freedom and tax revenue. Our opinion, however, is that there is such an effect:

Hypothesis H_7: The larger the political freedom, the greater the tax revenue.

Education level. Dioda (2012) found out that the average level of education of the people has a positive impact on tax revenue. This makes sense: a country with a high level of intellectual level will have quality human resources with sophisticated production methods that increase production efficiency and thus, increase tax revenues. So our hypothesis is:

Hypothesis H_8: The educational level has a positive impact on tax revenues.

Life expectancy. Castro and Ramirez (2014) show the adverse effects of longevity on tax revenue – the higher average age of the population, the higher proportion of retired people, and thus, the lower tax revenues. So the authors propose the following hypothesis:

Hypothesis H_9: The life expectancy has negative impact on tax revenues.

Infant mortality rate. High-income countries tend to have low infant mortality rates. Thus, it is reasonable to expect that the lower the infant mortality rates, the higher tax revenues. So, although Castro and Ramirez (2014) did not find any statistically significant relation between infant mortality rate and tax revenue, we still propose the following hypothesis:

Hypothesis H_{10}: Infant mortality has a negative impact on tax revenue.

Research models. In this study, we use a model from Castro and Ramirez (2014), with 10 independent variables:

$$TAXREV_{it} = \alpha + \beta_1 GDP_{it} + \beta_2 TRADE_{it} + \beta_3 FDI_{it} + \beta_4 AGR_{it} + \beta_5 IND_{it} + \beta_6 CIVILB_{it}$$
$$+ \beta_7 POLRIG_{it} + \beta_8 SCHTER_{it} + \beta_9 LIFEEXP_{it} + \beta_{10} INFMOR_{it} + u_i + \epsilon_{it},$$

where

- i represent a country and t represents a year.
- $TAXREV$ is tax revenue, as measured by the percentage of the year's tax revenue to GDP.
- GDP is GDP per capita (actually, natural logarithm of GDP per capita).
- $TRADE$ is trade volume: the percentage of total exports to GDP.
- FDI is foreign direct investment, measured by the percentage of FDI inflows to GDP.
- AGR is the percentage of Agricultural sector value-added to GDP.
- IND is the percentage of industrial value-added to GDP.
- $CIVLIB$ is Civil Liberties Index (Likert scale from 1 to 7).
- $POLRIG$ is Political Freedom Index (Likert scale from 1 to 7).
- $SCHTER$ is Educational level as measured by the university enrollment rate.
- $LIFEEXP$ is life expectancy (in terms of years of life).
- $INFMOR$ is infant mortality rate: number of children deaths per 1,000 births.

4   Results of Our Study

The authors used the F test to choose between the POLS and FEM models, with the hypothesis:

- H_0: $u_i = 0$ (POLS)
- H_1: $u_i \neq 0$ (FEM)

The results presented in Fig. 1 show that $Prob > F = 0.0052$. So, we reject the null hypothesis, thus selecting the FEM technique.

Similarly, to select between the FEM and the REM models, the authors use the Hausman (1978) test to test the following hypothesis:

- $H_0 : cov(u_i, x_{it}) = 0$ (REM)
- $H_1 : cov(u_i, x_{it}) \neq 0$ (FEM)

The results presented in Fig. 2 show that $Prob > \chi^2 = 0.0011$, so we reject the null hypothesis, thus selecting the FEM technique.

The regression models corresponding to all three techniques are described in Table 1; additional material is given in Figs. 3, 4 and 5 which are placed in the Appendix.

According to research results, R-squared equals 0.3347, meaning that independent variables in the model have accounted for 33.47% of the dependent variable.

After processing the variance of change and self-correlation in FEM, the aggregated results are shown in Table 2 (the details are shown in Fig. 6 from the appendix). The results show that there are four factors that affect the tax revenue: GDP, TRADE, AGR, and IND.

```
Fixed-effects (within) regression        Number of obs      =        68
Group variable: ID                       Number of groups   =         9

R-sq:  within  = 0.3347                  Obs per group: min =         6
       between = 0.6930                                 avg =       7.6
       overall = 0.6259                                 max =         9

                                         F(10,49)           =      2.47
corr(u_i, Xb)  = -0.9236                 Prob > F           =    0.0178
```

TAXREV	Coef.	Std. Err.	t	P>|t|	[95% Conf. Interval]	
GDP	6.396665	2.956949	2.16	0.035	.4544538	12.33888
TRADE	.0284548	.0290609	0.98	0.332	-.0299453	.0868549
FDI	-.0162932	.1447833	-0.11	0.911	-.3072462	.2746598
AGR	.587125	.3216964	1.83	0.074	-.0593481	1.233598
IND	.7173956	.1876165	3.82	0.000	.3403662	1.094425
CIVLIB	1.468334	2.170242	0.68	0.502	-2.892931	5.829598
POLRIG	.3234508	.8633965	0.37	0.710	-1.411609	2.058511
SCHTER	-.0693186	.1340197	-0.52	0.607	-.3386412	.200004
LIFEEXP	1.49181	1.308346	1.14	0.260	-1.137411	4.12103
INFMOR	.3199768	.2630439	1.22	0.230	-.2086297	.8485834
_cons	-195.8969	91.68984	-2.14	0.038	-380.1545	-11.63926
sigma_u	10.512017					
sigma_e	2.6347753					
rho	.94089091	(fraction of variance due to u_i)				

```
F test that all u_i=0:     F(8, 49) =    3.21              Prob > F = 0.0052
```

Fig. 1. F Test.

The results of this study are similar to most previous studies, especially Castro and Ramirez (2014), the only difference is that according to our study, AGR is positively correlated with TAXRE.

For GDP and TRADE factors that have a positive impact on TAXREV with a 10% significance level. Thus, we can conclude that GDP and TRADE increase TAXREV. For GDP, the economic growth comes from the development of the businesses, so when the business grows, the corresponding tax revenue increases. For TRADE – i.e., for trade volume with foreign countries – when the

Table 1. Synthesis of regression techniques of POLS, FEM and REM.

Variables	TAXREV		
	(1)	(2)	(3)
	POLS	FEM	REM
GDP	0.957	6.397**	0.957
	(0.698)	(2.957)	(0.698)
TRADE	−0.0257*	0.0285	−0.0257*
	(0.0136)	(0.0291)	(0.0136)
FDI	−0.0684	−0.0163	−0.0684
	(0.141)	(0.145)	(0.141)
AGR	−0.0180	0.587*	−0.0180
	(0.150)	(0.322)	(0.150)
IND	0.187*	0.717***	0.187**
	(0.0952)	(0.188)	(0.0952)
CIVLIB	−0.538	1.468	−0.538
	(1.254)	(2.170)	(1.254)
POLRIG	1.540***	0.323	1.540***
	(0.569)	(0.863)	(0.569)
SCHTER	−0.121**	−0.0693	−0.121**
	(0.0568)	(0.134)	(0.0568)
LIFEEXP	0.990**	1.492	0.990**
	(0.417)	(1.308)	(0.417)
INFMOR	0.0568	0.320	0.0568
	(0.0959)	(0.263)	(0.0959)
Constant	−68.74**	−195.9**	−68.74**
	(27.97)	(91.69)	(27.97)
Observations	68	68	68
R-squared	0.844	0.335	
Number of ID		9	9

Standard errors in parentheses, ***$p < 0.01$, **$p < 0.05$, *$p < 0.1$.
Source: Synthesis of the authors.

import-export business increases, businesses have more opportunities for development and expansion, so the ability to collect taxes from these units increases.

For AGR factors, there is a positive effect on TAXREV with a 5% significance level and this also increases TAXREV. This result is different from previous studies. This difference many be explained by the fact that most ASEAN countries are developing countries, many – such as Thailand, Vietnam, and Laos – with a major agricultural sector. It therefore makes sense that the contribution of this sector positively affects the tax revenue.

Table 2. FEM regression (processed variable variance and self-correlation)

Variables	TAXREV
GDP	6.397*
	(3.236)
TRADE	0.0285*
	(0.0149)
FDI	−0.0163
	(0.0670)
AGR	0.587**
	(0.205)
IND	0.717***
	(0.209)
CIVLIB	1.468
	(0.988)
POLRIG	0.323
	(0.456)
SCHTER	−0.0693
	(0.0573)
LIFEEXP	1.492
	(2.114)
INFMOR	0.320
	(0.368)
Constant	−195.9
	(138.6)
Observations	68
Number of groups	9

Standard errors in parentheses, ***$p < 0.01$, **$p < 0.05$, *$p < 0.1$.
Source: Synthesis of the authors.

For the IND factor, there is a positive impact on TAXREV with a 1% confidence level. For the development of the country, industry is one of the key industries – this is even more true for developing countries such as Southeast Asian countries – thus contributing to the increase in tax revenue.

Therefore, to increase the tax revenue, Southeast Asian countries should try their best to increase all four factors: they should increase GDP, they should actively increase their exports, and they should have strategies for developing both the industrial sector and the high-tech agriculture sector.

5   Conclusions

Tax revenue is an important contribution to the national budget, needed to ensure government spending. Therefore, to increase tax revenue, we need to study factors affecting it. Based on the data from Southeast Asian countries, we show that four factors affect the tax revenue: GDP per capita, trade volume, agricultural sector, and industry sector. The results of this study are similar to most previous studies, particularly to those of Castro and Ramirze (2014). The difference from previous studies is that, in contrast to the previous studies, our research shows that the agriculture sector variable has a positive impact on tax revenue.

Therefore, to increase the tax revenue, Southeast Asian countries should try their best to increase all four factors: they should increase GDP, they should actively increase their exports, and they should have strategies for developing both the industrial sector and the high-tech agriculture sector.

In the future, it is desirable to extend this study to other countries, and also to take into account other possible factors, such as educational level and life expectancy.

Appendix

| | — Coefficients — | | | |
| | (b) | (B) | (b-B) | sqrt(diag(V_b-V_B)) |
	FEM	REM	Difference	S.E.
GDP	6.396665	.9571024	5.439563	2.873443
TRADE	.0284548	-.0257437	.0541984	.025702
FDI	-.0162932	-.0683787	.0520855	.034822
AGR	.587125	-.0179674	.6050923	.2845496
IND	.7173956	.1868843	.5305113	.1616722
CIVLIB	1.468334	-.5378942	2.006228	1.771022
POLRIG	.3234508	1.539811	-1.21636	.6494679
SCHTER	-.0693186	-.1214535	.0521349	.1213792
LIFEEXP	1.49181	.9896903	.5021193	1.240123
INFMOR	.3199768	.0567878	.263189	.2449552

b = consistent under Ho and Ha; obtained from xtreg

B = inconsistent under Ha, efficient under Ho; obtained from xtreg

Test: Ho: difference in coefficients not systematic

$$chi2(10) = (b-B)'[(V\_b-V\_B)^(-1)](b-B)$$

$$= 29.45$$

Prob>chi2 =   0.0011

(V_b-V_B is not positive definite)

Fig. 2. Hausman test.

Source	SS	df	MS		
Model	2803.58423	10	280.358423		
Residual	518.398008	57	9.09470189		
Total	3321.98223	67	49.5818244		

```
Number of obs =      68
F( 10,    57) =   30.83
Prob > F      =  0.0000
R-squared     =  0.8439
Adj R-squared =  0.8166
Root MSE      =  3.0157
```

TAXREV	Coef.	Std. Err.	t	P>|t|	[95% Conf. Interval]	
GDP	.9571024	.6977625	1.37	0.176	-.4401428	2.354348
TRADE	-.0257437	.0135626	-1.90	0.063	-.0529023	.001415
FDI	-.0683787	.1405334	-0.49	0.628	-.349792	.2130346
AGR	-.0179674	.1500669	-0.12	0.905	-.3184711	.2825364
IND	.1868843	.0951948	1.96	0.055	-.00374	.3775085
CIVLIB	-.5378942	1.254365	-0.43	0.670	-3.049716	1.973927
POLRIG	1.539811	.568898	2.71	0.009	.4006121	2.679009
SCHTER	-.1214535	.0568188	-2.14	0.037	-.2352311	-.007676
LIFEEXP	.9896903	.4169708	2.37	0.021	.1547207	1.82466
INFMOR	.0567878	.0958596	0.59	0.556	-.1351678	.2487433
_cons	-68.74306	27.97216	-2.46	0.017	-124.7564	-12.72977

Fig. 3. POLS

```
Fixed-effects (within) regression          Number of obs     =      68
Group variable: ID                         Number of groups  =       9

R-sq:  within  = 0.3347                     Obs per group: min =      6
       between = 0.6930                                     avg =    7.6
       overall = 0.6259                                     max =      9

                                           F(10,49)          =    2.47
corr(u_i, Xb)  = -0.9236                    Prob > F          =  0.0178
```

TAXREV	Coef.	Std. Err.	t	P>|t|	[95% Conf. Interval]	
GDP	6.396665	2.956949	2.16	0.035	.4544538	12.33888
TRADE	.0284548	.0290609	0.98	0.332	-.0299453	.0868549
FDI	-.0162932	.1447833	-0.11	0.911	-.3072462	.2746598
AGR	.587125	.3216964	1.83	0.074	-.0593481	1.233598
IND	.7173956	.1876165	3.82	0.000	.3403662	1.094425
CIVLIB	1.468334	2.170242	0.68	0.502	-2.892931	5.829598
POLRIG	.3234508	.8633965	0.37	0.710	-1.411609	2.058511
SCHTER	-.0693186	.1340197	-0.52	0.607	-.3386412	.200004
LIFEEXP	1.49181	1.308346	1.14	0.260	-1.137411	4.12103
INFMOR	.3199768	.2630439	1.22	0.230	-.2086297	.8485834
_cons	-195.8969	91.68984	-2.14	0.038	-380.1545	-11.63926
sigma_u	10.512017					
sigma_e	2.6347753					
rho	.94089091	(fraction of variance due to u_i)				

```
F test that all u_i=0:    F(8, 49) =   3.21              Prob > F = 0.0052
```

Fig. 4. FEM (self-correlated autocorrelation and variance)

```
Random-effects GLS regression              Number of obs      =       68
Group variable: ID                         Number of groups   =        9

R-sq:  within  = 0.0789                     Obs per group: min =        6
       between = 0.9856                                      avg =      7.6
       overall = 0.8439                                      max =        9

                                            Wald chi2(10)      =   308.27
corr(u_i, X)   = 0 (assumed)                Prob > chi2        =   0.0000
```

TAXREV	Coef.	Std. Err.	z	P>|z|	[95% Conf. Interval]	
GDP	.9571024	.6977625	1.37	0.170	-.4104869	2.324692
TRADE	-.0257437	.0135626	-1.90	0.058	-.0523259	.0008385
FDI	-.0683787	.1405334	-0.49	0.627	-.3438192	.2070617
AGR	-.0179674	.1500669	-0.12	0.905	-.312093	.2761583
IND	.1868843	.0951948	1.96	0.050	.0003059	.3734626
CIVLIB	-.5378942	1.254365	-0.43	0.668	-2.996404	1.920615
POLRIG	1.539811	.568898	2.71	0.007	.4247911	2.65483
SCHTER	-.1214535	.0568188	-2.14	0.033	-.2328163	-.0100908
LIFEEXP	.9896903	.4169708	2.37	0.018	.1724426	1.806938
INFMOR	.0567878	.0958596	0.59	0.554	-.1310936	.2446691
_cons	-68.74306	27.97216	-2.46	0.014	-123.5675	-13.91863

sigma_u	0					
sigma_e	2.6347753					
rho	0	(fraction of variance due to u_i)				

Fig. 5. REM

```
Regression with Driscoll-Kraay standard errors   Number of obs      =       68
Method: Fixed-effects regression                 Number of groups   =        9
Group variable (i): ID                           F( 10,    8)       =   128.11
maximum lag: 1                                   Prob > F           =   0.0000
                                                 within R-squared   =   0.3347
```

TAXREV	Coef.	Drisc/Kraay Std. Err.	t	P>|t|	[95% Conf. Interval]	
GDP	6.396665	3.236457	1.98	0.084	-1.066617	13.85995
TRADE	.0284548	.0148522	1.92	0.092	-.0057944	.062704
FDI	-.0162932	.0669776	-0.24	0.814	-.1707438	.1381574
AGR	.587125	.2054908	2.86	0.021	.1132624	1.060988
IND	.7173956	.2089592	3.43	0.009	.2355347	1.199256
CIVLIB	1.468334	.9877302	1.49	0.175	-.8093763	3.746044
POLRIG	.3234508	.4559348	0.71	0.498	-.7279367	1.374838
SCHTER	-.0693186	.0573021	-1.21	0.261	-.2014574	.0628202
LIFEEXP	1.49181	2.114069	0.71	0.500	-3.383241	6.36686
INFMOR	.3199768	.3677148	0.87	0.410	-.5279751	1.167929
_cons	-195.8969	138.6019	-1.41	0.195	-515.5134	123.7196

Fig. 6. FEM (processed variance variation and self-correlation)

References

Ajaz, T., Ahmad, E.: The effect of corruption and governance on tax revenues. The Pakistan Dev. Rev. **49**(4), 405–417 (2010). Part II

Ayenew, W.: Determinants of tax revenue in Ethiopia (Johansen Co-Integration Approach). Int. J. Bus. Econ. Manag. **3**(6), 69–84 (2016)

Bird, R.M., Martinez-Vazquez, J., Torgler, B.: Tax effort in developing countries and high income countries: the impact of corruption voice and accountability. Econ. Anal. Policy **38**(1), 55–71 (2008)

Castro, A., Ramirez, D.: Determinants of tax revenue in OECD countries over the period 2001–2011. Contadura y Administracia: Revista Internacional **59**(3), 35–60 (2014)

Dioda, L.: Structural determinants of tax revenue in Latin America and the Caribbean, 1990–2009. Economic Commission for Latin America and the Caribbean. Press in United Nations Mexico, 2012–041, pp. 1–41 (2012)

Eltony, M.N.: Measuring tax effort in Arab countries. Economic Research Forum for the Arab Countries, Iran and Turkey. Working Paper Series 0229, pp. 1–20 (2002)

Imam, P.A., Jacobs, D.: Effect of Corruption on Tax Revenues in the Middle East. Rev. Middle East Econ. Fin., vol. 10, no. 1, pp. 1–24 (2014)

Gupta, A.S.: Determinants of tax revenue efforts in developing countries. IMF Working paper 07(184), pp. 1–39 (2007)

Hausman, J.A.: Spacification tests in econometrics. Econométrica **46**(6), 1251–1271 (1978)

Ojong, C.M., Anthony, O., Arikpo, O.F.: The impact of tax revenue on economic growth: evidence from Nigeria. J. Econ. Finan. **7**(1), 32–38 (2016)

Pessino, C., Fenochietto, R.: Determining countries' tax effort. Hacienda Publica Espanola/Revista de Economica Publica **195**, 65–87 (2010)

Piancastelli, M.: Measuring the tax effort of developed and developing Countries. Cross country Panel data analysis 1985/95. Instituto de Pesquisa Economica Applicada, pp. 1–18 (2001)

World Bank: World Developing Indicators (2017). http://databank.worldbank.org/data/download/archive/WDI_excel_2017_01.zip

Mixed-Copulas Approach in Examining the Relationship Between Oil Prices and ASEAN's Stock Markets

Paravee Maneejuk[1,2(✉)], Woraphon Yamaka[1,2], and Songsak Sriboonchitta[1,2]

[1] Faculty of Economics, Chiang Mai University, Chiang Mai, Thailand
mparavee@gmail.com
[2] Centre of Excellence in Econometrics, Chiang Mai University,
Chiang Mai, Thailand

Abstract. This study aims to examine the relationship between oil prices and stock markets in five ASEAN countries: Thailand, Indonesia, Malaysia, Singapore, and the Philippines. Copula approach is used for modelling dependence structure between variables. In essence, this study considers four classes of copula, namely Archimedean copulas, Elliptical copulas, extreme value copulas, and mixed copulas, to examine the dependency between oil prices and stock market prices. We found that Thai, Malaysian, and Indonesian stock markets are likely to boom when crude oil prices increase while the Singaporean stock market as well as the Philippines's stock market tend to move in the opposite direction to crude oil prices. However, the results show that these relationships are not strong.

Keywords: Copulas · Mixed copulas · Extreme value copulas
Co-movement · Dependency

1 Introduction

Different perspectives on the relationship between the price of crude oil and stock market prices are presented. The conventional wisdom holds that oil prices and performance of stock market are negatively correlated. An increase in oil prices will cause most businesses higher input costs as well as higher transportation costs, which in turn brings about a reduction in corporate earnings. Additionally, more expensive fuel can also affect consumer behaviour. When oil prices rise, consumers are forced to spend more money on fuel or gasoline and have to reduce discretionary spending. As a result, conversely, lower oil prices should be good news for both consumers and corporates. Economic activities should increase and stock markets should boom due to the lower cost. In contrast, some economists, financialists, or researchers often draw a positive line between them. As it can be seen through the real data of crude oil prices and stock indexes they tend to move together, decline in oil prices is then cited as one major cause of weaknesses in the stock markets. One explanation for this positive

L. H. Anh et al. (eds.), *Econometrics for Financial Applications*, Studies in Computational Intelligence 760, https://doi.org/10.1007/978-3-319-73150-6_42

relationship is that both just reflect a common thing, reduction of the global aggregate demand, which can hurt corporate profits and also demand for crude oil. Many attempts have been made to discover this relationship between crude oil prices and stock market performances, but agreeable conclusions seem to hardly exist. For instance, Nandha and Faff [8] provided evidence that increasing oil price has negative impact on the global stock market. But Nguyen and Bhatti [9]; Hamma et al. [3]; Aloui et al. [1], and Pastpipatkul et al. [4], just found a positive dependence between oil prices and stock market prices through various cases of countries and regions; see also [2].

Understanding this relationship is crucial because, as we show, the reaction of the stock returns to changes in the oil prices differs across countries and the presence of the oil assets in a portfolio of stocks can permit improvement of the portfolio's risk-return characteristics. Accordingly, this study will examine the relationship between crude oil prices and stock market prices for ASEAN countries, especially, for Thailand, Singapore, Indonesia, the Philippines, and Malaysia, where plenty of capital is flown into but only few studies have been conducted on this area. Therefore, this study is conducted with an attempt to address this issue and to provide information for investors as well as policy makers. However, examining the dependency among the prices of crude oil and stock market is what this study actually focuses on and aims to improve.

This study considers one of the most well-known and efficient methods for modeling the dependency between these two variables that is copula approach. Recently, Nguyen et al. [10] list several advantages of copula. First, it allows us to use different marginal distributions to model the dependence structure through various copula functions, thus constructing a complicate non-normal marginal distributions that can capture dependencies, possibly exhibiting co-movements among financial variables. Second, the results from mixed copula model significantly improve the degree and the structure of the dependence implying that the co-movement of the variables during a market downturn and/or upturn. Third, copula dependence structure will not change under any types of data transformation.

Literature review regarding the relationship between prices of crude oil and stock market prices shows that copula had an ability to capture nonlinearity and asymmetry. As we can see, this technique has been commonly used in many studies, for example, Nguyen and Bhatti [9], Aloui et al. [1], and Pastpipatkul et al. [4]. However, only the basic copulas were used. This study also considers the advantage of copula to model the dependency but branch out into a various classes of copula-based ARMA-GARCH model. In essence, this study considers both parametric and nonparametric methods of four main classes of copulas, namely (i) Archimedean copulas, (ii) Extreme value copulas, (iii) Elliptical copulas, and (iv) Mixed copulas. This will be the first study that uses extreme value copulas and mixed copulas in examining the dependence structure between oil prices and ASEAN's stock market returns. This allows us to have a better flexibility of capturing almost all possible dependence structures between oil prices and stock returns, which have not been shown in the related literature.

The rest of this paper is structured as follows. In Sect. 2, we will explain methodology especially the considered classes of copula functions. Section 3 contains data description and graphs illustrating historical prices of crude oil and the stock markets. Section 4 presents estimated results and some discussions related to the findings. And Sect. 5 contains conclusions.

2   Methodology

2.1   GARCH Models for Univariate Distributions

To model the marginal distribution of each random variable, we employ a univariate ARMA(p, q)-GARCH(m, n) specification which can be described as

$$y_t = \phi_0 + \sum_{i=1}^{p} \phi_i y_{t-i} + \sum_{j=1}^{q} \theta_j \varepsilon_{t-j} + \varepsilon_t, \tag{1}$$

$$\varepsilon_t = h\eta_t, \tag{2}$$

$$h_t^2 = \alpha_0 + \sum_{i=1}^{m} \alpha_i \alpha_{t-i}^2 + \sum_{j=1}^{n} \beta_j h_{t-j}^2, \tag{3}$$

where Eqs. (1) and (3) are the conditional mean and variance equation, respectively. ε_t is the residual term which consists of the standard variance, h_t, and the standardized residual, η_t, which is assumed to have a normal distribution, a Student-t distribution, and a skewed-t distribution, skewed normal distribution, generalized error distribution and skewed generalized error distribution. The best-fit ARMA(p, q)-GARCH(m, n) will give the standardized residuals which are transformed into a uniform distribution in [0,1].

2.2   Basic Concepts of Copula

According to Sklar's theorem [5], copula is described as a multivariate distribution function joining marginal probability distribution of random variables. Sklar's theorem states that for any joint distribution $F(x_1, ..., x_n)$ with marginal distribution $F_1(x_1), ..., F_n(x_n)$, it can be represented as follows:

$$F(x_1, ..., x_n) = C(F_1(x_1), ..., F_n(x_n)) = C(u_1, ..., u_n), \tag{4}$$

where $x_1, ..., x_n$ is the random variable vector and C is copula distribution function of a n dimensional random variable with uniform margin $[0, 1]$. The copula density c is obtained by differentiating Eq. (4); thus, we get

$$c(u_1, .., u_n) = \frac{\partial^n C(F_1(x_1), ..., F_n(x_n))}{\partial u_1 .... \partial u_n}, \tag{5}$$

where c is the multivariate copula density function.

2.3   Bivariate Copula Function

As we mentioned earlier, we attempt to measure the dependence structure of oil prices and ASEAN's stock markets by using copula approach. Thus, in this subsection, we will briefly describe on three classes of bivariate copula density functions namely elliptical copulas, Archimedean copulas, and extreme value copulas.

Elliptical copulas

(1) Gaussian copula
Considering the case of two-dimensional type, the Gaussian copula can be defined by Schepsmeier and Stöber [6]. The density of the Gaussian copula can be written as

$$c_G(u, v \,|\theta_G) = \int\limits_{-\infty}^{\Phi^{-1}(u}\int\limits_{-\infty}^{\Phi^{-1}(v))} \frac{1}{2\pi\sqrt{1-\theta_G}} \exp\left(\frac{x_1^2 + x_2^2 - 2\theta_G x_1 x_2}{2(1-\theta_G^2)}\right) dx_1 x_2, \quad (6)$$

where Φ is bivariate standard normal cumulative distribution and θ_G is the correlation copula parameter on the interval $[-1, 1]$.

(2) Student's t copula
Student's t copula is a copula function which has a fat tail. θ_T is the parameter of the copula on the interval $[-1, 1]$ and v is a degree of freedom. In the bivariate copula, we can define the probability density function of Student's t copula as

$$c_T(u, v \,|\theta_T) =$$
$$\int\limits_{-\infty}^{t_v^{-1}(u)}\int\limits_{-\infty}^{t_v^{-1}(v)} \frac{1}{2\pi\sqrt{1-\theta_T}} \exp\left(\frac{x_1^2 + x_2^2 - 2\theta_T x_1 x_2}{v(1-\theta_T^2)}\right)^{-(v+2)/2} dx_1 x_2, \quad (7)$$

where $t_d(v, 0, \theta_T)$ is the standard univariate student's t distribution with v degree of freedom, mean 0 and variance $(v + 2)/2$.

Archimedean copulas

In Archimedean copula class, the copula function allows us to capture the asymmetric nature of the data. In the estimation, this class of copula allows the modeling of the dependence with only one parameter. Let's consider the probability density function of the Archimedean copulas consisting of Clayton, Gumbel, Frank, and Joe.

(1) Clayton copula
The Clayton copula is an asymmetric copula exhibiting greater dependence in the negative tail than in the positive one. This copula is given by

$$c_C(u, v \,|\theta_C) = \left(1 + (u^{-\theta_C} - v^{-\theta_C} - 1)\right)^{-1/\theta_C}, \quad (8)$$

where θ_C is a degree of dependence on the value $0 < \theta_C < \infty$. If $\theta_C \to \infty$, the Clayton copula converges to the monotonicity copula with positive dependence. But if $\theta = 0$, it corresponds to independence.

(2) Gumbel copula

The density function of Gumbel copula can be defined by

$$c_G(u, v \,|\theta_G) = \exp\left(-\left((-\ln(u))^{\theta_G} + (-\ln(v))^{\theta_G}\right)^{1/\theta_G}\right), \qquad (9)$$

where the parameter θ_G is a degree of dependence on the value $1 < \theta_G < \infty$.

(3) Frank copula

The density function of Frank copula can be defined by

$$c_F(u, v \,|\theta_F) = -\theta_F^{-1} \log\left(1 + \frac{(e^{-\theta_F(u)} - 1)(e^{-\theta_F(v)} - 1)}{e^{-\theta_F} - 1}\right), \qquad (10)$$

where θ_F is a degree of dependence on the value $-\infty < \theta_F < \infty$.

(4) Joe copula

The density function of Joe copula is defined by

$$c_J(u, v \,|\theta_J) = 1 - \left((1 - u)^{\theta_J} + (1 - v)^{\theta_J} - (1 - u)^{\theta_J}(1 - v)^{\theta_J}\right)^{1/\theta_J}, \qquad (11)$$

where the parameter θ_J is a degree of dependence on the value $1 < \theta_J < \infty$.

Extreme value copulas

The theory of multivariate maxima in Extreme Value Theory can be expressed in terms of copulas. Extreme value copulas belong to a class of copulas that emerge as the natural limiting dependence structures for multivariate maxima. The following are three EV copulas commonly used in this study (see, [7])

(1) E-Gumbel copula

$$c_{EG}(u, v \,|\theta_{EG}) = \exp(-[(-\ln u)^{\theta_{EG}} + (-\ln v)^{\theta_{EG}}]^{1/\theta_{EG}}), \qquad (12)$$

where the parameter θ_{EG} is a degree of dependence on the value $1 < \theta_{EG} < \infty$.

(2) Galambos copula

$$c_{Gal}(u, v \,|\theta_{Gal}) = uv \exp([(-\ln u)^{-\theta_{Gal}} + (-\ln v)^{-\theta_{Gal}}]^{-1/\theta_{Gal}}), \qquad (13)$$

where the parameter θ_{Gal} is a degree of dependence on the value $0 < \theta_{EG} < \infty$.

(3) Husler-Reiss copula

$$c_H(u, v \,|\theta_H) = \exp\left\{-\tilde{u}\Phi\left(\frac{1}{\theta_H} + \frac{1}{2}\theta_H \ln(\frac{\tilde{u}}{\tilde{v}})\right) - \tilde{v}\Phi\left(\frac{1}{\theta_H} + \frac{1}{2}\theta_H \ln(\frac{\tilde{u}}{\tilde{v}})\right)\right\}, \qquad (14)$$

where the parameter θ_H is a degree of dependence on the value $0 < \theta_H < \infty$. $\tilde{u} = -\ln(u)$, $\tilde{v} = -\ln(v)$ and Φ is standardized normal distribution.

2.4   Mixed Copulas

The conventional copulas may lead to the misspecification problem of dependence measure since each family has a strong restriction on its dependence parameter. Mixed copulas class is proposed to allow better flexibility of capturing almost all possible dependence structures between variables. By using convex combination approach, the mixed copula can be computed by

$$c_{Mix}(u, v \,|\theta_1, \theta_2) = wc_{\theta_1}(u, v \,|\theta_1) + (1 - w)c_{\theta_2}(u, v \,|\theta_2), \quad w \in [0, 1], \qquad (15)$$

where w is called the weight parameter of $[0, 1]$. The advantage of this method lies in the flexibility to assign weights in calculating the appropriate value between two copula functions.

3   Data

This paper uses daily time series of five stock indexes in ASEAN namely the Stock Exchange of Thailand (SET), Singapore Exchange (SGX), Kuala Lumpur Stock Exchange (KLSE), Indonesia Stock Exchange (IDX), and the Philippines Stock Exchange (PSI) together with crude oil Brent price (BRENT). The oil price is measured in US dollars per barrel (unit: USD/bbl.). All data series are collected from January 2, 2008 to June 30, 2017, totally 2,320 observations. Stock indexes are then transformed into log-return before performing the experiment. Table 1 presents descriptive statistics of these variables.

Table 1. Descriptive statistics

	SET	SGX	KLSE	IDX	PSI	BRENT
Mean	0.0001	0	0.0001	0.0001	0.0002	−0.0001
Median	0.0003	0.0001	0.0001	0.0005	0.0001	0
Maximum	0.0328	0.0327	0.0176	0.0331	0.0306	0.0483
Minimum	−0.0482	−0.0378	−0.0433	−0.0476	−0.0568	−0.0475
Std. Dev	0.0055	0.005	0.0033	0.0059	0.0061	0.0094
Skewness	−0.6962	−0.1718	−1.2491	−0.649	−0.84	−0.0245
Kurtosis	11.0753	10.1581	19.8238	11.2608	9.6702	6.7412
Jarque-Bera	6488.32	4962.35	24106.87	6756.55	4571.65	1352.65
Prob	0	0	0	0	0	0
ADF-test	−45.66***	−46.45***	−40.19***	−43.22***	−42.73***	−51.19***

Source: Calculation.

Note: "***" denotes a significance at 1% level.

The following graphs (Fig. 1) show the levels of the stock markets and oil prices, from January 2, 2008 to June 30, 2017. The main purpose of this figure is to let the data tell a story and allow us to observe the movements of oil prices and stocks in each market. From this figure, we can see that the pattern of co-movement between oil prices and stocks is not stable over this period. We expect that there are three possible patterns of their movements: opposite direction, same direction, and no discernable connection. The first pattern happened before 2009. Both stocks and oil prices went down but a sharp decrease could be seen more clearly in the movement of oil prices. The second pattern started from around 2009 to 2014, in which both stocks and oil prices changed in the same direction. The temporally deviated movements could be seen in this period of time; however, overall data showed that their correlation was pretty strong during this co-movement. The third pattern started from mid-2014 or the beginning of 2015 for some countries- until the end. We can see that the divergences between the stock markets and crude oil prices became more obvious and were strong in that period of time; the oil prices went down rapidly while the stock prices tended to increase.

Fig. 1. Levels of ASEAN's stock markets and crude oil prices (unit: USD/bbl.); SET (top left), SGX (top right), KLSE (middle left), IDX (middle right), and PSI (bottom)

4   Estimated Results

Four classes of copula, i.e., Archimedean copulas, extreme value copulas, Elliptical copulas, and mixed copulas, as mentioned earlier, are considered here for modeling the dependency or relationship between oil prices and the stock markets in ASEAN, particularly SET, SGX, KLSE, IDX, and PSI. Given a set of copulas, the Akaike information criterion (AIC) is used to choose the best fitting copula among these candidates. The model with minimum AIC will be chosen. Table 2 presents AIC values for the comparison of models. We find that the conventional classes of copula, Archimedean and Elliptical copulas, are selected for every pair of the stock indexes and oil prices. For the pair of SET with oil prices, Frank copula is the best-fit copula as it has the smallest AIC, −4.9057. For the pair of SGX with oil prices, Gaussian copula has the minimum value of AIC, 1.2500. Therefore, the best copula for this pair is Gaussian. The Clayton and Frank copulas are chosen for the pairs of IDX and PSI indexes with oil prices, respectively, since they have minimum values of AIC, −1.4170 and 0.0944. In the pair of KLSE index with oil prices, the Clayton copula is chosen with the lowest AIC, −1.6102.

Table 2. Model comparison through AIC values

Copula	SET-BRENT	SGX-BRENT	KLSE-BRENT	IDX-BRENT	PSI-BRENT
Gaussian	−3.9759	**1.25**	1.9395	1.9245	0.6672
Student-t	−1.1522	3.4498	2.3512	2.4908	1.0829
Gumbel	−1.0199	5.251	5.0606	3.0054	1.0199
Clayton	−0.5995	1.9314	**−1.6102**	**−1.417**	1.9975
Frank	**−4.9057**	1.2867	1.9969	1.9907	**0.0944**
Joe	0.6364	4.0248	5.0663	0.6364	4.1632
E-Gumbel	−1.0199	1.53	2.1124	1.637	1.5114
Galambos	1.1254	2.0125	2.0001	2.0141	2.2589
Husler-Reiss	−1.0154	2.0111	2.0002	2.0125	2.1114
C-G	2.4569	6.0068	2.3897	2.5829	6.3414
C-F	−0.9015	3.7115	6.5337	−0.3741	1.4414
F-G	−0.9406	4.3332	5.9882	4.254	0.7076
C-J	−0.2615	6.0068	2.39	2.5829	6.3414
G-J	0.0904	8.3084	9.0663	8.4199	8.1632

Source: Calculation.
Note: C = Clayotn, G = Gumbel, F = Frank, J = Joe.

Table 3 shows the estimated values of copula parameter, Kendall's tau, and upper and lower tail dependence. From this table, it can be seen that the values of Kendall's tau are positive for the pairs of the stock markets of Thailand, Malaysia, and Indonesia with crude oil prices, implying the stock markets are likely to boom (or crash) with oil prices, and conversely. On the other hand,

Table 3. Estimated parameters

Parameter	SET-BRENT	SGX-BRENT	KLSE-BRENT	IDX-BRENT	PSI-BRENT
Copula parameter (θ_1)	0.3253	−0.0179	0.0334	0.0356	−0.1718
Kendall's Tau	0.0361	−0.0114	0.0164	0.0175	−0.0191
Upper tail	0	0	0.00001	0.00001	0
Lower tail	0	0	0	0	0

Source: Calculation.

the values of tau are negative for the pairs of the stock markets of Singapore
and Philippines with oil prices, which means that stocks and oil prices tend to
move in the opposite direction. Moreover, they also exhibit left tail dependence
in the pairs of Malaysia's and Indonesia's stock markets with oil prices. This
implies that these stock markets may go up with the oil price during a booming
economy. This finding corresponds to what the real data just displayed in Fig. 1.
The stocks and oil prices tended to move together but this co-movement was
not so clear during that time. Until the mid-2014, some countries, especially
the Philippines, just showed an obvious divergence among these two variables'
movements. And in this section, the estimated parameter just supports this idea.
The tau value of Philippines-oil price pair is negative, indicating that PSI is likely
to move in the opposite direction to fuel costs.

Overall results just indicate that there are the co-movements between
ASEAN's stocks and oil prices in either the opposite or the same directions,
but even then, the relationships are weak. This empirical result may not surprise us, and here are some reasons. First of all, the economy is more complex
than that, so it might be imprudent to expect only one commodity to drive the
stock markets significantly. Second, there are lots of factor in the real economic
system that can affect the stock prices besides energy costs. Even though it
does, businesses and corporations in these days have become more sophisticated
at reading future situations and better prepared themselves for an increase in
input prices, c.g., buying a futures contract. On the other hand, there are also
several factors in the economy that can offset changes in oil prices besides the
stocks, such as, interest rates, technology development, industrial metals and the
government policy. Third, changes in crude oil prices depend on the supply and
demand for petroleum, for instance, higher oil prices can be caused by increased
consumption or decreased production, which occurs independently of the stock
prices or certain corporates.

5   Conclusions

In this study, we used four classes of copulas to examine the relationships
between oil prices and stock markets in ASEAN, particularly Thailand, Singapore,
Indonesia, Malaysia, and the Philippines. Here, the four classes of copulas consist
of Archimedean copulas, Elliptical copulas, extreme value copulas, and mixed copulas. The results show that Thailand's, Malaysia's, and Indonesia's stock markets

are likely to boom with oil prices, where the Frank copula is the best dependence structure for the case of Thailand while Clayton is the best copula for the cases of Malaysia and Indonesia. On the other hand, it is found that the Singaporean stock market tends to move in the opposite direction to the crude oil prices, and the same for Philippines's stock market. These dependence structures are best represented by Gaussian copula (for Singapore) and Frank copula (for Philippines).

Interestingly, the results show that these relationships are weak. Our finding corresponds to what researchers at many well-known financial and economic institutions just highlight; that is, there is no certain pattern can describe the relationship between them throughout the time. Instead, those weakness like to explain this relationship on a given day after they realize that this relationship can have varying patterns. According to our finding, we would conclude that the relationship between oil prices and stock markets in ASEAN are weak. Even though we can obtain some evidences about direction of the relationships, we cannot really predict the way stock markets react to changes in crude oil prices. Hence, it might be better to compare results with other methods and rely on others such as news or media. A certain value (or estimated value) may not well explain what really happens in the markets, but with some knowledge from plot, graphs, or chart sets, it will bring about better understanding of this.

Acknowledgements. The authors are grateful to Puay Ungphakorn Centre of Excellence in Econometrics, Faculty of Economics, Chiang Mai University for the financial support.

References

1. Aloui, R., Hammoudeh, S., Nguyen, D.K.: A time-varying copula approach to oil and stock market dependence: the case of transition economies. Energy Econ. **39**, 208–221 (2013)
2. Arouri, M.E.H., Nguyen, D.K.: Oil prices, stock markets and portfolio investment: evidence from sector analysis in Europe over the last decade. Energy Policy **38**(8), 4528–4539 (2010)
3. Hamma, W., Jarboui, A., Ghorbel, A.: Effect of oil price volatility on Tunisian stock market at sector-level and effectiveness of hedging strategy. Procedia Econ. Financ. **13**, 109–127 (2014)
4. Pastpipatkul, P., Yamaka, W., Sriboonchitta, S.: Co-movement and dependency between New York stock exchange, London stock exchange, Tokyo stock exchange, oil price, and gold price. In: International Symposium on Integrated Uncertainty in Knowledge Modelling and Decision Making, pp. 362–373. Springer, Cham (2015)
5. Sklar, M.: Fonctions de repartition n-dimensions et. leurs marges. Publ. Inst. Stat. Univ. Paris **8**, 229–231 (1959)
6. Schepsmeier, U., Stöber, J.: Derivatives and Fisher information of bivariate copulas. Stat. Pap. **55**(2), 525–542 (2014)

7. Lu, J., Tian, W.J., Zhang, P.: The extreme value copulas analysis of the risk dependence for the foreign exchange data. In: 4th International Conference on Wireless Communications, Networking and Mobile Computing, WiCOM 2008, pp. 1–6. IEEE (2008)
8. Nandha, M., Faff, R.: Does oil move equity prices? a global view. Energy Econ. **30**(3), 986–997 (2008)
9. Nguyen, C.C., Bhatti, M.I.: Copula model dependency between oil prices and stock markets: evidence from China and Vietnam. J. Int. Financ. Mark. Inst. Money **22**(4), 758–773 (2012)
10. Nguyen, C., Bhatti, M.I., Komorníková, M., Komorník, J.: Gold price and stock markets nexus under mixed-copulas. Econ. Model. **58**, 283–292 (2016)

Forecasting Credit-to-GDP

Kobpongkit Navapan[1], Jianxu Liu[1,2(✉)], and Songsak Sriboonchitta[1,2]

[1] Faculty of Economics, Chiang Mai University,
Chiang Mai 52000, Thailand
`kobpongkit.nav@gmail.com, songsakecon@gmail.com`
[2] Center of Excellence in Econometrics, Chiang Mai University,
Chiang Mai 52000, Thailand
`liujianxu1984@163.com`

Abstract. After the Global Financial Crisis in 2008, a great attempt has been placed on studying of early warning indicators (EWIs) in order to forecast possible future crises. EWIs have played a crucial role not only in explaining which macroprudential policies should be involved and put into effect, but also indicating when it is an appropriate timing for implementation of the policies. Accurate prediction of EWIs therefore has become a big issue. The paper aims to forecast a credit-to-GDP gap, by using three different models: linear, Markov switching, quantile models with some selected macroeconomic variables; set index, exchange rate and export. The empirical results show that the quantile 25th model performs the most accurate forecasting ability based on RMSE and MAPE. Furthermore, the forecast results indicates that there is a slight downturn of the predicted values during 2006 to 2007.

Keywords: Early warning indicators · Linear model
Markov switching model · Quantile model

1 Introduction

Since the global crisis in 2008, literature on early warning indicators (EWIs) of financial crisis has recently attached considerable attention, especially in banking sector. Negative consequences caused by the crisis to the banking sector can vary depending on the length of time. For example, for the short period some consequences are that banks losing money on mortgage defaults, or interbank transactions becoming to freeze. For the longer period one of the consequences is the situation that generate new regulatory actions are internationally generated through Basel III. It is quite clear that the crisis has multiple impacts.

In the literature, credit-to-GDP has been regularly used, not only to provide valuable message as EWIS, but also help policy makers to understand a financial environment of a nations economy. The credit-to-GDP ratio is defined in financial literature as the measurement of the relative size of the outstanding debt of non-financial private sector with respect to GDP. It is known that a high level of debt has significantly played an important role on bank performance. In case of facing

© Springer International Publishing AG 2018
L. H. Anh et al. (eds.), *Econometrics for Financial Applications*, Studies in Computational Intelligence 760, https://doi.org/10.1007/978-3-319-73150-6_43

too many liabilities coming due with not enough cash to satisfy those liabilities, banks might confront liquidity risk. Moreover, when the bank problems spread all over a country, they can expand into other countries through the banking network. This types of crisis is called systemic banking crisis.

The after of one of well-known international indicators, Basel III uses the gap between the credit-to-GDP ratio and its long-term trend as a guide for setting countercyclical capital buffers – an extension to the regulatory capital framework used by policymakers for adjusting in a time-varying way for banks (Drehmann and Tsatsaronis [2]). Furthermore, it is recommended by the Basel Committee on Banking Supervision (BCBS) when a nations credit-to-GDP ratio is higher than its long run trend by two percentage points. Then, policymakers can use this information for taking a decision with respect to capital buffer along with other indicators.

Many studies showed effects in the same direction. When the credit-to-GDP ratio continues to climb up, relatively financing costs keep rising (Mackenzie [5]). This empirical result is consistent with the finding of Kaminsky et al. [4] which shows that credit-to-GDP gap performs as the best signal for long-term prediction.

Many studies also demonstrated that the debt level also has a very important role to economic growth. One of them is the study of Randveer et al. [7], who have related the growth performance of countries in the study with the countries debt level. His finding was that a higher level of debt is correlated with a smaller economic growth. Meanwhile, Panizza and Presbitero [6] also provide a very strong support to that with a sample of OECD countries, which shows a negative correlation between debt and growth, Alessi and Detken [1] also presented that excess credit growth can provide significant information as an advance signal of financial crises.

However, some economists have criticized the credit-to-GDPs properties. Edge and Meisenzahl [3] show that the credit to GDP gap measures can be unreliable in real time as the gap can yield false-positives by signaling excessively high level of credit. As a result, it causes capital shortfalls in the banking sector and unnecessary lending restraint. Sarma [8] also cited that in an economy the credit-to-GDP tends to have its no optimum desirable level.

In spite of these weakness, a lot of policymakers still find the credit-to-GDP ratio useful rather than ignoring it. However, it always comes available in the time series. As known in real practice, we often meet time series data that exhibit strong non-linear features. This could be a reason why linear models are not suitable for use. The study of Yuan [10] regarding exchange rate demonstrated that fundamentals-based linear models generally fail to detect the persistence in it. Consequently the models are incapable of out-forecasting the random walk. In addition, Timmermann [9] cited that Markov-switching models are successfully used for fitting a lot of economic and financial time series, whereas the quantile regression approach allows us to directly capture the impact of different magnitudes of shocks that hits the real exchange rate condition on its past history and can capture asymmetric, dynamic adjustment of the real exchange rate towards its long run equilibrium.

To our knowledge there has been few papers which relate the credit-to-GDP performance with other macroeconomics variables and also compare the quantile regression approach with Markov-switching models. The primary objective of this paper is to fulfill this gap by forecasting credit-to-GDP values with three different model, linear model, Markov switching model, quantile model.

The remainder of the paper is organized as follows: Sect. 2 outlines the methodologies of linear, Markov switching and quantile regression (QR) models. Section 3 discusses data. Section 4 shows the empirical results along with model selection regarding to root mean square error approach (RMSE), and mean absolute percentage error approach (MAPE). Section 5 concludes.

2   Methodology

2.1   Linear Model

This section begins with the classical linear regression model. The most significant assumption for linear analysis is that the coefficients parameter are significantly assumed to be constant over time. While, in general time series data can be different over time. The linear equation estimated by OLS method can be expressed as

$$CTG_t = \beta_0 + \beta_1 EXP_t + \beta_2 EXC_t + \beta_3 SET_t + \varepsilon_t, \tag{1}$$

where CTG_t denotes to the credit to GDP gap. EXP denotes to the value of Thailands export. EXC denotes to exchange rate for United States Dollar to Thailand Baht. SET denotes the Stock Exchange of Thailand SET Index. The OLS estimators β_0, β_1, β_2 and β_3 obtained will be best linear unbiased estimators of β_0, β_1, β_2 and β_3 respectively.

2.2   Markov Switching Model

It is also known that the regime switching model is very popular in modeling the nonlinear time series data. A Markov switching model is to investigate the different dynamic patterns throughout different time periods because it comprises multiple structure. Therefore, it can characterize the behavior the time series through different regimes. It also allows to switch between these structures which can capture the complex dynamic patterns. The mechanism of Markov switching model is controlled by an unobservable state variable following a first-order Markov chain. The first-order Markov chain refers to the current value of the state variable state variable only depend on its last previous value. The original Markov switching model concentrates on the mean behavior of variables which is broadly applied in economic and finance area (Schaller and van Norden [9]); see also [11].

We begin with the random variable y_t, that follows a process depending on the value of an unobserved discrete state variable s_t with a possibility of M regimes in state or regime m in period t when $s_t = m$, for $m = 1, ..., M$.

A different regression model associated with each regime is technically assumed along with the assumption that regressors X_t and Z_t are given, the conditional mean of in the regimes is the linear specification:

$$u_t(m) = X_t'\beta_m + Z_t'\gamma, \tag{2}$$

where both y and β_m are vectors of coefficients k_x and k_z, whereas the coefficients y which are associated with Z_t denote regime invariant. The coefficients β_m for X_t are indexed by regime. Finally the regression errors are assumed to be normally distributed with variance depending on the regime. The equation is relatively expressed as

$$y_t = \mu_t(m) + \sigma(m)\varepsilon_t, \tag{3}$$

where ε_t iid denotes a standard normal distribution with respect to $s_t = m$. Then σ the standard deviation may be regime dependent, where $\sigma(m) = \sigma_m$.

For a given observation, the likelihood contribution is formed by weighting the density function in each of the regimes with the one-step ahead probability in that regime:

$$L_t(\beta, \gamma, \sigma, \delta) = \sum_{m=1}^{M} \frac{1}{\sigma_m} \phi\left(\frac{y_t - \mu_t(m)}{\sigma(m)}\right).P(S_t = m|\mathfrak{S}_{t-1}, \delta), \tag{4}$$

where $\sigma = (\sigma_1, ..., \sigma_M), \delta, \beta = (\beta_1, ..., \beta_M)$. Parameters are presented by δ to verify the regime probabilities. Whereas $\phi(.)$ refers to the standard normal density. $\mathfrak{S}_{t-1}$ denotes the information set in Period $t-1$. For a simple scenario, the regime probabilities is presented by the δ. The equation of the full log likelihood is expressed as

$$l(\beta, \gamma, \sigma, \delta) = \sum_{m=1}^{M} \log\left\{\frac{1}{\sigma_m} \phi\left(\frac{y_t - \mu_t(m)}{\sigma(m)}\right).P(S_t = m|\mathfrak{S}_{t-1}, \delta)\right\}, \tag{5}$$

which may be maximized with respect to $(\beta, \gamma, \sigma, \delta)$.

A simple switching model is introduced in term of featuring independent regime probabilities. Then the evaluation of likelihood and estimation of the probabilities are described by focusing on the specification of the regime probabilities. They are simply treated as additional parameters in the likelihood, since the probabilities are constant values. While in general, to allow for varying probabilities, we assume that it is a function of vectors of exogenous observables G_{t-1} and coefficients δ parameterized using a multinomial logit specification:

$$P(S_t = m|\mathfrak{S}_{t-1}, \delta) \equiv p_m(G_{t-1}, \delta) = \frac{\left(G_{t-1}'\delta_m\right)}{\sum\limits_{j=1}^{M}\left(G_{t-1}'\delta_j\right)}, \tag{6}$$

where $\delta = (\delta_1, \delta_1, .., \delta_M)$ with the identifying normalization $\delta_M = 0$. The special case of constant probabilities is handled by choosing G_{t-1} to be identically equal to 1.

2.3   Quantile Regression (QR)

Given a set of conditioning variables, QR provides an ability to model the quantiles of the dependent variables. It also provide an linear estimation describing the relationship between a specified quantile of the dependent variable and regressors. It can be used to supplement the OLS estimates to gain a more detailed and complete picture of the relationship between the dependent variable and explanatory variables. As it allows a more complete description of the conditional distribution than conditional mean analysis alone, this provide a ability describing how the median of the response variable can be impacted by regressors variables at the 10th or 95th as an example. That is, a determinant that is beneficial for the credit-to-GDP gap, on average, as revealed by the OLS method may become unimportant or even detrimental for firms located at extremely high or low quantiles of the credit-to-GDP. By estimating only the conditional mean relationship between the dependent and explanatory variables, the OLS method masks the heterogeneity in the estimated relationship.

We begin a random variable with probability distribution function

So that for $0 < \tau < 1$, the τ-th quantile of Y can be defined as the smallest y to satisfy $F(y) \geq \tau$;

$$Q(\tau) = \inf\{y : F(y) \geq \tau\} \tag{7}$$

Providing a set of N observations on Y, and the traditional empirical distribution function can be expressed by

$$F_n(y) = \sum_k I(Y_i \leq y), \tag{8}$$

where $I(Y_i \leq y)$ denotes an indicator function, and it takes the value of 0 if the argument Z is untrue and the value of 1 if the argument is true. The associated empirical quantile can be provided by

$$Q_n(\tau) = \inf\{y : F_n(y) \geq \tau\} \tag{9}$$

or equivalently in the form of a simple optimization problem

$$Q_n(\tau) = \arg\min_\xi \left\{ \sum_{i:Y_i \geq \xi} \tau |Y_i - \xi| + \sum_{i:Y_i < \xi} (1 - \tau) |Y_i - \xi| \right\} \tag{10}$$

$$= \arg\min_\xi \left\{ \sum_i \rho_\tau (Y_i - \xi) \right\},$$

where $\rho_\tau(u) = u(\tau - 1(u < 0))$ which weights positive and negative values asymmetrically. Quantile regression extends this simple formulation to allow for regressors X. It assume a linear specification for the conditional quantile of the response variable Y given values for the p-vector of explanatory variables X:

$$Q(\tau | X_i, \beta(\tau)) = X_i' \beta(\tau), \tag{11}$$

where $\beta(\tau)$ associated with the τ-th quantile is the vector of coefficients. Then the analog to the unconditional quantile minimization becomes the conditional quantile regression estimator:

$$\hat{\beta}_n(\tau) = \min_{\beta(\tau)} \left\{ \sum_i \rho_\tau \left(Y_i - X_i' \beta(\tau) \right) \right\} \tag{12}$$

2.4 Testing for Forecasting Ability

For testing for robustness for forecasting ability, we focus on two popular approaches. The first one is root mean square error (RMSE) and the second one is mean absolute percentage error (MAPE). RMSE measures the sample standard deviation. In other words, it presents the values calculated from differences comparing actual values with predicted values. The residuals are the differences of these individual when the calculations are performed over the data sample that was used for estimation and RMSE's equation is written as,

$$RMSE = \sqrt{\frac{1}{n} \sum_{i=1}^{n} (A_t - F_t)^2}, \tag{13}$$

where F_t represents the predicted value, and A_t represents the original value. Whereas MAPE is a measure that captures accuracy as a percentage and MAPE's equation is written as

$$MAPE = \frac{1}{n} \sum_{t=1}^{n} \left| \frac{A_t - F_t'}{A_t} \right| \tag{14}$$

3 Data

Time series data is obtained from two main sources, Bank for International Settlements (BIS) database and data stream database. The data of export, exchange rate and set index of Thailand are from BOT database, whereas credit-to-GDP gap data is data stream database. The data is the quarterly time series during the period from 2006 quarter 1 to 2015 quarter 4. There are 40 observations in each variable, and 160 observations in total.

4 Empirical Results

We initially related credit-to-GDP gap with the other variables without data transformation. Figure 1 presents three paired graphs; credit-to-GDP gap and exchange rate, credit-to-GDP gap and export and credit-to-GDP gap and set index. The graphs show both downward and upward trends. The only first paired graph has a upward trend, whereas the other two graphs have an opposite trend.

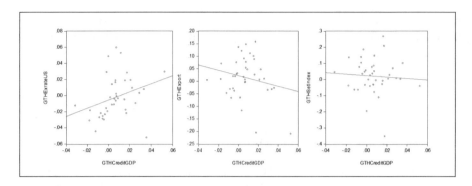

Fig. 1. The paired graphs between credit-to-GDP gap and export, exchange rate and set index respectively

Later, unit root test is employed to check that the data is stationary. The results are non stationary. Consequently all observations have been transferred into the form of growth and applied with the unit root test, which consists of three different methods as follows a unit root, a unit root with growth, and a unit root with growth and deterministic time trend. The results in Table 1 indicate that the growth of credit to GDP, export, exchange rate and set index are stationary.

Table 1. Results of unit root test

Observation	ADF			PP		
	Intercept	Trend and intercept	None	Intercept	Trend and intercept	None
GcreditGDP	0.0003	0.0009	0.0288	00003	0.0008	0.0002
Gtheexport	0.0000	0.0002	0.0000	0.0000	0.0000	0.0000
Gthexrates	0.0031	0.0029	0.0001	0.0032	0.0029	0.0001
Gtheindex	0.0005	0.0036	0.0000	0.0128	0.0623	0.0008

Notably state dependence in the transition probabilities considering with higher probability of remaining in the origin regime is 0.38 for the low output state and 0.62 for high output state. Meanwhile, it is about 2.66 and 1.60 quarters for the corresponding expected durations in regime.

Table 2 shows estimated parameters of the three selected models. For Quantile regression, we considers three quantiles; 25th, 50th, and 75th indicating low, median, and high credit-to-GDP gap respectively, whereas for Markov switching model, we considers two regimes; upper and lower regime. The regime 1 is considered as a market lower regime, meanwhile regime 2 is referred to market upper regime.

We can see that Thai exports growth is negatively related to credit-to-GDP gap, which shows statistically a negatively significant values of 0.089 at the 10% level for 75th quantile and 0.116 at the 1% level for Markov switching regime 2.

Table 2. Estimated parameters of the three models

Variable	Linear	Quantile			Markov switching	
		25th	50th	75th	Regime 1	Regime 2
Constant	0.009	−0.004	0010	0.016	0.003	0.0260
	(0.0037)***	(0.423)	(0.017)**	(0.000)***	(0.39)	(0.000)***
Gtheexport	−0.059	0.0124	−0.0536	−0.0887	0.017	−0.116
	(0.1052)	(0.8711)	(0.346)	(0.049)*	(0.735)	(0.0008)***
Gthexrates	0.3823	0.4085	0.509	0.626	0.403	−0.155
	(0.013)***	(0.0493)**	(0.027)***	(0.001)***	(0.049)***	(0.4892)
Gtheindex	0.051	0.0657	0.0492	0.0967	−0.005	0.0011
	(0.121)	(0.255)	(0.264)	(0.009)**	(0.863)	(0.9825)

Note: ***significant at the 1% level, **significant at the 5% level, *significant at the 10% level.

This indicates that Thailands exports are negatively correlated with high credit-to-GDP gap for quantile regression which accords to the result of Markov switching.

This seems to contract to the other two variables. Both exchange rate growth and set index are positively with credit-to-GDP. However, only 75th quantile shows a positively significant value of 0.097 at 10% level for the set index growth. All of the models shows that exchange rate has a positive relationship with credit-to-GDP gap, 0.382, 0.509, 0.626, 0.403 significant at 1% level for linear, 50th quantile, 75th quantile and Markov switching regime 1 respectively and 0.409 at 5% level for 25th quantile except Markov switching regime 2. This indicates that no matter of low or high credit-to-GDP gap, exchange rate positively influences credit-to-GDP gap.

For testing forecasting ability regarding MAPE and RMSE approaches, the data is separated into two groups. The first group is called in sample and the second group is called out of sample. In sample consists of the first 96 observations during 2006 quarter 1 to 2013 quarter 4 and the rest are in out of sample for the period 2014 quarter 1to 2015 quarter 4. We begin the prediction calculation by using the in sample data to generated new predicted values and compare with the existing values in out of sample period.

Table 3. Forecasting criteria for the best selected model

Models	Linear	Markov switching	Quantile		
			0.25th	0.50th	0.75th
RMSE	0.009	0.06	0.007***	0.014	0.019
MAPE	57.33	523.63	55.72***	103.14	99.90

Note: ***denotes the lowest values.

In order to forecast credit-to-GDP gap with export, exchange rate, and set index, it is shown that the results of the three models, linear, quantile, and switching models in Table 3 are varying. It is also found that quantile model at

0.25th has the smallest errors in both MAPE and RMSE followed by linear model and quantile model at 0.75th respectively. Moreover, the MAPE and RMSE values generated by the three models are remarkably different. Meanwhile, the MAPE and RMSE values of the Markov Switching model are 0.06 and 523.63 respectively. Overall the quantile model at 0.25th performs the most forecasting ability than the other models.

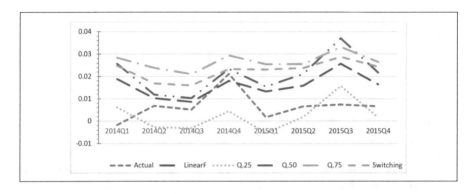

Fig. 2. Plotting actual variables with predicted values of all the models

Figure 2 shows a set of graphs generated by the different models. It is a clear evidence that the line generated by quantile model at 0.25th moves closer to the actual line during 2014 quarter 1 to 2015 quarter 4 than the other lines. This result is also consistent by the empirical results as demonstrated in Table 3.

Another interesting point is that there are three intercepts between both the actual line and the quantile 25 line in the periods of 2014 quarter 1 to 2014 quarter 2, 2015 quarter 2 to 2015 quarter 3 and 2015 quarter 3 to 2015quarter 4 followed by two intercepts between the linear line and the actual line during the period of 2014 quarter 3 to2015 quarter 1. Meanwhile the other lines generated by quantile 50, quantile 75, and Markov switching tend to be overestimate during the whole period. However, Fig. 3 also shows that the predicted values generated by quantile 25 are overestimated the actual values only in 2014 quarter 1 and 2015 quarter 3, while the other periods the predicted values are underestimated.

The next step is to begin to forecast the credit-to-GPD gap for next two year, which starts from 2016 quarter 1 to 2017 quarter 4 with the quantile 25 model by using the data for the whole period. The foretasted values of the credit to GPD gap in 2016 quarter 1, 2, 3, 4 and 2017 quarter 1, 2, 3, 4 are 123.42, 123.22. 122.96, 122.68, 122.40, 122.12, 122.84, and 121.56 respectively. Figure 3 plots the actual values during the period 2006 quarter 1 to 2015 quarter 4 followed by the predicted values during the period 2016 quarter 1 to 2017 quarter 4. The graph shown in Fig. 3 presents a slight downturn during 2016 quarter 1 to 2017 quarter 4.

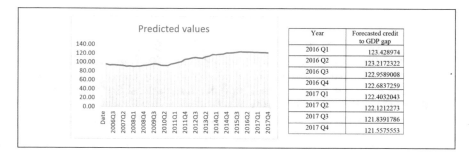

Fig. 3. Plotting the predicted values of the quantile 25th model during 2016 Q1–2017 Q4

5   Conclusion

To forecast credit-to-GDP gap with three selected variables namely export, exchange rate and set index with the three different models, namely linear, quantile and Markov switching models indicates that the 25th quantile model trends to be the most accurate method followed by linear model and switching model respectively. However, the ability of prediction during the out of sample period shows that the 25th quantile model produces three intercepts between both the actual line and its line. The first intercept occurs during the periods of 2014 quarter 1 to 2014 quarter 2, the second one is during 2015 quarter 2 to 2015 quarter 3 and the last one is during 2015 quarter 3 to 2015 quarter 4. Meanwhile linear model has only two intercepts which both take place during the period of 2014 quarter 3 to 2015 quarter 1. Meanwhile the other lines generated by quantile 50, quantile 75, and Markov switching tend to be overestimate during the whole period without intercept. The empirical results generated by using the quantile 25th to forecast the credit-to-GDP gap during the period of 2016 quarter 1 to 2017 quarter 4 also show that predicted credit-to-GDP gap is slightly dropping from 123.42 at 2016 quarter 1 to 121.56 at 2017 quarter 4.

The study draws more attention to policymakers on the need to observe the credit-to-GDP gap, because the forecast values indicates a slight downward trend during 2016 to 2017. Even though this credit indicator is recognized as one of the efficient tools, there are many economic indicators causing significant affects in economy, for example rapid devaluation in foreign currency loans can lead to a remarkable increase in the credit-to-GDP gap. If it is used as a single reference without considering the others, it can produce misinterpretation. Furthermore, banking crisis indicators are significantly associated with banks' performance, Consequently financial indicators, such as, leverage ratios, types of loans, and changes in property prices should be monitored as importantly as the credit-to-GDP gap.

References

1. Alessi, L., Detken, C.: 'Real time' early warning indicators for costly asset price boom/bust cycles: a role for global liquidity (2009). ECB Working Paper No. 1039
2. Drehmann, M., Tsatsaronis, K.: The credit-to-GDP gap and countercyclical capital buffers: questions and answers. Bank for International Settlements (BIS), it publishes Quarterly Reviews (2014)
3. Edge, R.M., Meisenzahl, R.R.: The unreliability of credit-to-GDP ratio gaps in real-time: implications for countercyclical capital buffers. Int. J. Cent. Bank. **7**(4), 261–298 (2011)
4. Kaminsky, G., Lizondo, S., Reinhart, C.M.: Leading indicators of currency crises. Staff Pap. **45**(1), 1–48 (1998)
5. Mackenzie, K.: Chinas credit-to-GDP ratio, updated (and why it matters), 17 April 2013. https://ftalphaville.ft.com/2013/04/17/1463992/chinas-credit-to-gdp-ratio-updated-and-why-it-matters/
6. Panizza, U., Presbitero, A.F.: Public debt and economic growth: is there a causal effect? J. Macroecon. **41**, 21–41 (2014)
7. Randveer, M., Kulu, L., Uuskla, L.: The impact of private debt on economic growth. Working Paper Series, No. 20, Eesti Pank (2011). Bank of Estonia Eesti Pank
8. Sarma, M.: Index of Financial Inclusion. Indian Council for Research on International Economics Relations, New Delhi (2008)
9. Schaller, H., Norden, S.V.: Regime switching in stock market returns. Appl. Finan. Econ. **7**(2), 177–191 (1997)
10. Timmermann, A.: Moments of Markov switching models. J. Econ. **96**(1), 75–111 (2000)
11. Yuan, C.: Forecasting exchange rates: the multi-state Markov-switching model with smoothing. Int. Rev. Econ. Finan. **20**(2), 342–362 (2011)

Resilience of Stock Cross-Correlation Network to Random Breakdown and Intentional Attack

Ngoc Kim Khanh Nguyen[1] and Quang Nguyen[2(✉)]

[1] Faculty of Basic Sciences, Van Lang University,
45 Nguyen Khac Nhu, Co Giang Ward, District 1,
Ho Chi Minh City, Vietnam
nguyenngockimkhanh@vanlanguni.edu.vn
[2] John von Neumann Institute, Vietnam National University Ho Chi Minh City,
Linh Trung Ward, Thu Duc District, Ho Chi Minh City, Vietnam
quangnguyen@jvn.edu.vn

Abstract. We study the network constructed by the correlation coefficients between stocks which are listed in the Vietnamese stock exchanges. Network edges between nodes (stocks) are established if the correlations between stocks are higher than certain value (0.25). We found that this network is scale-free, having connectivity distribution $P(k) \sim k^{-\gamma}$ (where k is the node connectivity) with a relatively low power exponent of $\gamma \sim 1.0$. This result accords with the highly co-movement of listed stocks in a market found previously. The low power-law distribution exponent coefficient corresponding to a dense connectivity makes it robust even under intentional attacks: its critical fraction is of range $30.77\% - 50.36\%$. Finally, we compare different intentional attack strategies and find that: if we want to fully break apart the network, recalculated degree distribution-based attack is the most efficient; if we only want to break the network to a certain level (for example, to break half of the size of the largest network component), the recalculated betweenness centrality-based attack is more efficient. This result may be used to enhance the structure design of some real-life network systems.

1 Introduction

The research of complex network has been recently of interest [1–9] because many real-life systems, ranging from biology to medicine, sociology, financial and engineering, can be modeled as networks. Some of the most well known networks are the Internet [1], the World Wide Web [2], the metabolic networks [3]. Most of those networks, though constructed from objects of different natures, share a common property of having a power-law degree distribution $P(k) \sim k^{-\gamma}$ (with an exponent γ ranges in the scope of 1.5–4) and are called scaled-free. Pioneer work of Albert et al. [1] has shown that scaled-free network with exponent ≤ 3 is extremely resilient to random failure of its nodes. If some proportion of nodes are randomly removed, the network still maintains its structure (the existence of a spanning cluster a cluster whose size is proportional to the size of the network).

© Springer International Publishing AG 2018
L. H. Anh et al. (eds.), *Econometrics for Financial Applications*, Studies in Computational Intelligence 760, https://doi.org/10.1007/978-3-319-73150-6_44

It will only be disrupted when nearly all of its nodes are removed. This fact is particularly important to our daily life as most networks are safe if some of their components fail by accident: the Internet still functions if some routers at somewhere in the globe are down, the economy still runs if some corporates file for bankruptcy due to some management mistakes... However, the authors also found that such networks are fragile to intentional attacks. A typical scale-free network will be completely broken apart if less than 10% of their most connected nodes are removed [10, 11].

In reality, some networks such as actors collaboration and science citations may have high critical threshold even under attack. This better resilience is thanks to either a high density/connectivity or more complicated distribution (although still scaled-free) [12]. A better understanding about the structure of such networks may help to design a robust network under both random failure and intentional attack.

While many works have done on physical or social networks, few are found on economic systems. The banking system, the economy, the corporates leader network, to name a few, can all be studied using network framework [13]. After the recent financial crisis, the systemic risk, the risk that the whole system does not function properly, is a particularly important research subject. Such a risk can be considered as a network percolation phenomena and be analyzed by the same study on other networks.

In this work, we study the resilience of a financial network constructed from the correlation coefficient between stocks which are listed in the exchanges. The nodes represent stocks while edges represent the possibility that the correlation coefficient between 2 stock returns is higher than a certain level. Such network is the subject of some previous studies [14, 15]. Most of those researches use the

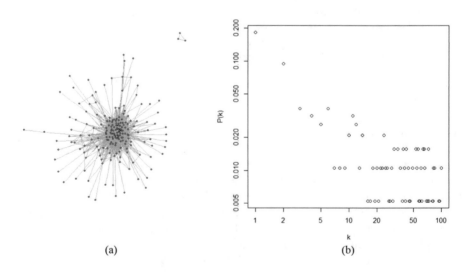

(a)   (b)

Fig. 1. Stock correlation network (a) and its degree distribution (b)

cross-correlation matrix to generate the network. Here, we use approach similar to [15] but fix a level of correlation coefficient above which an edge between 2 stocks is defined. Apparently, an edge represents a strong correlation between 2 stocks which can be the result of several factors: common business environment, same industry, common shareholders, common investors... The topology of our network is shown in Fig. 1.

2   Material and Method

We calculate the cross-correlation matrix C from stocks trading in Vietnam from Jan 01st 2015 to May 19th 2017 using the method described in [16,17]. In the analysis of principal component of this matrix, Vietnamese stock market data has been found to be coherent with other markets in the world [18]. In this work, we select stocks that have more than 250 active trading days and an average daily volume of more than 20,000 shares/day. There are in total 347 stocks in our dataset. A network adjacency matrix A is constructed directly from the cross-correlation matrix, where $A_{ij}(i \neq j)$ is 1 if the element C_{ij} is higher than 0.25 and 0 otherwise. We select a level of 0.25 as it is considered significant correlation level in financial market. By this construction, there are 191 stocks that have at least one connection and the network of those stocks is selected.

According to the percolation theory, a network is considered robust if there exists a giant component, i.e. a spanning cluster. It has been shown in [19–21] that random uncorrelated network with degree distribution $P(k)$ (k is the site connectivity) will lose its global connectivity (giant component) when

$$\kappa = \frac{\langle k^2 \rangle}{\langle k \rangle} < 2, \tag{1}$$

where k is the degree of nodes and the angular brackets are the average over all nodes of the network.

Different methods have been used to break the network:

- Random breakdown: At each step, one node is selected randomly and is deleted from the network together with its edges.
- Initial degree distribution (ID): Firstly, the network's node degree is calculated and ranked. Then we delete nodes and their edges according to this ranking from largest to smallest degree.
- Initial betweenness centrality (IB): Nodes are deleted similarly to ID but the ranking was done by the betweenness centrality of initial network.
- Recalculated degree distribution (RD): At each step, node degree of the current network is calculated, then the largest degree node is deleted from the network together with its edges.
- Recalculated degree distribution (RB): At each step, node betweenness centrality of the current network is calculated, then the largest betweenness centrality node is deleted from the network together with its edges.

We use Monte-Carlo simulation to demonstrate the random breakdown [22] and estimate the threshold q_c, the fraction of node that need to removed in order to break the network using formula (1). The number of nodes i removed up to that point is recorded and the threshold threshold q_c is calculated simply by $q_c = i/N$. In fact, we run several simulations and estimate q_c as the average $\langle i \rangle / N$, where N is the number of network nodes (191 in our case). For intentional attack, we will continuously remove nodes and their corresponding edges following attacking rules and recalculate κ until we meet the condition (1). In this work, we attack the network based on its structural information [4,23,24]: the initial degree distribution, the initial betweenness centrality, the recalculated degree distribution and the recalculated betweenness centrality. It has been shown that the resilience of the network, the effectiveness of the attack strategy, depends greatly on the attacking strategies [12]. With each strategy, we firstly estimate the threshold q_c by removing a fraction q of nodes whose degree/betweenness centrality are largest until the giant component is considered as missing according to criterion (1), and secondly measure the ratio of the largest component' size as a function of the fraction of removed nodes.

3   Result

Figure 1(a) shows the network topology of 191 stocks which represent nodes. Edges between nodes are created if the correlation coefficient of stock pairs are higher than 0.25 as previously described. Apparently, one found that there are many nodes which have a high number of connection, similar to many other real-life networks. Figure 1(b) shows the network degree distribution which has a power-law with an exponent of $\gamma \sim 1.3$.[1] This demonstrated that the cross-correlation network is also scaled-free like many real-life networks. However, its exponent γ of 1.3 is relatively small in comparison to other scale-free networks such as film actors [25,26], sexual contacts [27,28], word co-occurrence [29,30], Internet [2,31], peer-to-peer network [32,33], metabolic network [3], stock correlation network [34], etc. which typically range from 1.8 to 3.2. The low exponent corresponds to a high connectivity between nodes as we can see from Fig. 1(a).

As mention previously, scaled-free network is extremely resilient to random failure of its nodes as we found a critical thresholds of 95% approximately. In other words, one needs to destroy over 95% of nodes of this correlation network before the spanning cluster collapses. In contrast, when using intentional attack strategies, the thresholds are much lower.

It was found for random scaled-free network that q_c is about 0.1 with γ near 2 [11]. Our critical thresholds are considerably higher. This is probably due to the smaller exponent found above which resulting a much highly dense network [35].

We also found that the attacks with recalculated strategies are more useful than with initial strategies, similarly to the result of [36]. This is reasonable

---

[1] For other level of 0.2 (more dense network) and 0.4 (less dense network), the exponents are 1.1 and 1.7 respectively.

Remove by order of	Critical threshold for percolation
Initial degree distribution	49.74%
Initial nodes betweenness centrality	45.55%
Recalculated degree distribution	36.65%
Recalculated betweenness centrality	43.98%

because when removing nodes, the network structure (including degree distribution and betweenness centrality) changes. The attacks using initial network information (ID and IB) will not be as efficient as the updated strategies (RD and RB). Figure 2 confirms this remark by comparing the relative size of giant component after removing a fraction q of nodes by above strategies and random breakdown strategy whose removed nodes are chosen randomly. We can clearly see the sharper decrease of RD and RB compare to that of ID and IB strategy curve, respectively.

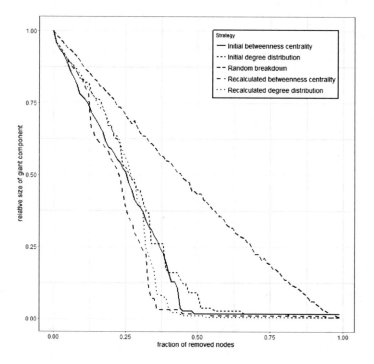

Fig. 2. Fraction of nodes belonging to the giant component P_∞ as a function of the fraction of removed nodes under different attack strategies.

When comparing between the removing strategies by degree distribution and betweenness distribution, we found interesting results. Using initial information, the IB strategy is more efficient than the ID strategy. This could be due to the fact that, when removing the nodes with highest degrees, we also delete many edges and change the remaining network degree distribution significantly. In consequence, the initial degree distribution no longer reflects the new structure and the attack strategy based on that outdated information becomes less efficient. In contrast, the path between nodes in a network are multiple. When removing the nodes with highest betweenness, we remove the paths that go through this node but other paths still exist and the overall centrality (betweenness) ranking is less affected. This out-of-date problem is solved when using recalculated strategies and we found improvement in both strategies especially the recalculated degree one.

Finally, we compare the recalculated degree distribution and recalculated betweenness strategies. Despite the fact that the critical threshold of the former is smaller than that of the later (36.65% compares to 43.98%), we found that in fact, the RB strategy is quite efficient at the beginning of the removing process. Figure 3 illustrates the network structure after removing 25%, 34%, 36% and 41% of nodes by each strategy, respectively. The RB strategy breaks the network into many sub-networks thus the maximum giant component's relative size decreases faster than in the RD strategy which removes important nodes inside of the initial giant cluster. However, near the critical point, the RD strategy gets stuck in a highly connected cluster[2], which has many nodes but low betweenness, therefore the RB strategy removes nodes in other clusters and the maximum

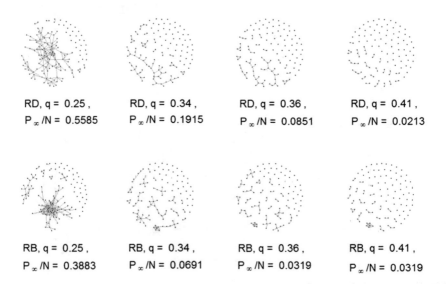

| RD, q = 0.25, | RD, q = 0.34, | RD, q = 0.36, | RD, q = 0.41, |
| P_∞/N = 0.5585 | P_∞/N = 0.1915 | P_∞/N = 0.0851 | P_∞/N = 0.0213 |

| RB, q = 0.25, | RB, q = 0.34, | RB, q = 0.36, | RB, q = 0.41, |
| P_∞/N = 0.3883 | P_∞/N = 0.0691 | P_∞/N = 0.0319 | P_∞/N = 0.0319 |

Fig. 3. Vietnamese stocks correlation network corresponding to fraction of removed nodes q under recalculated degree distribution-based attack (RD) and recalculated betweenness centrality-based attack (RB), where P_∞ is the giant component' size and N is the number of stocks in the initial network, i.e. $N = 191$.

---

[2] For a dense network, the existence of such sub-network is very likely.

size remains unchanged, as can be seen in Fig. 3. The RB strategy finally breaks the network only when q deposes the critical threshold of 43.98%.

It shows that one strategy can be more efficient than other in some situations and less efficient in another one. In case where we need to completely break the network (validated by the criterion (1)), RD is the most efficient strategy. By contrast, in case where we only need to decrease the size of the largest cluster to a certain level (for example 5% or 20% of the initial cluster size), RB could be a better choice.

4   Discuss and Conclusion

We have analyzed the network of highly correlated stock in the Vietnamese stock exchanges and found that it has scaled-free property, a power-law degree distribution, like many other real-life networks. In addition, its exponent is about 1.3, relatively smaller than that of other real scaled-free networks, corresponding to a dense structure. Using Monte-Carlo simulation, we demonstrate that it is extremely robust against random failure. We also compare different intentional attack strategies and found that the network is less resilient to the recalculated degree strategy, even though the recalculated betweenness is also effective in breaking down the network to a certain level. In other words, the attacking strategy needs to be chosen based on the structure of the network and the ultimate objective of the attack. Last but not least, this work may help to improve the design of some networks in order to be protected from attack.

Acknowledgments. This research is funded by Vietnam National University - Ho Chi Minh City (VNU-HCM) under grant number B2017-42-01.

References

1. Albert, R., Jeong, H., Barabasi, A.-L.: Internet: diameter of the world-wide web. Nature **401**, 130–131 (1999)
2. Faloutsos, M., Faloutsos, P., Faloutsos, C.: On power-law relationships of the Internet topology. Comput. Commun. Rev. **29**, 251–262 (1999)
3. Jeong, H., Tombor, B., Albert, R., Oltvai, Z.N., Barabasi, A.-L.: The large-scale organization of metabolic networks. Nature **407**, 651–654 (2000)
4. Albert, R., Barabasi, A.-L.: Statistical mechanics of complex networks. Rev. Mod. Phys. **74**, 47–97 (2002)
5. Barrat, A., Barthelemy, M., Vespignani, A.: Dynamical Processes on Complex Networks. Cambridge University Press, Cambridge, New York (2008)
6. Boccaletti, S., Latora, V., Moreno, Y., Chavez, M., Hwang, D.-U.: Complex networks: Structure and dynamics. Phys. Rep. **424**, 175–308 (2006)
7. Dorogovtsev, S.N., Mendes, J.F.F.: Evolution of Networks: From Biological Nets to the Internet and WWW. Oxford University Press, Oxford, New York (2003)
8. Newman, M.E.J.: The structure and function of complex networks. SIAM Rev. **45**, 167–256 (2002)

9. Pastor-Satorras, R., Vespignani, A.: Evolution and Structure of the Internet: a Statistical Physics Approach. Cambridge University Press, Cambridge, New York (2003)

10. Albert, R., Jeong, H., Barabasi, A.L.: Error and attack tolerance of complex networks. Nature **406**, 378–382 (2000)

11. Cohen, R., Erez, K., ben-Avraham, D., Havlin, S.: Breakdown of the internet under intentional attack. Phys. Rev. Lett. **86**, 3682–3685 (2001)

12. Gallos, L.K., Cohen, R., Liljeros, F., Argyrakis, P., Bunde, A., Havlin, S.: Attack strategies on complex networks. In: Workshop on Networks: Structure and Dynamics Conference, ICCS. Lecture Notes in Computer Science, vol. 3993, pp. 1048–1055 (2006)

13. Bonanno, G., Caldarelli, G., Lillo, F., Micciche, S., Vandewalle, N., Mantegna, R.N.: Networks of equities in financial markets. Eur. Phys. J. B **38**, 363–371 (2004)

14. Onnela, J.-P., Kaski, K., Kertsz, J.: Clustering and information in correlation based financial networks. Eur. Phys. J. B **38**, 353–362 (2004)

15. Garas, A., Argyrakis, P., Havlin, S.: The structural role of weak and strong links in a financial market network. Eur. Phys. J. B **63**, 265–271 (2008)

16. Plerou, V., Gopikrishnan, P., Rosenow, B., Amaral, L.A.N., Stanley, H.E.: Universal and nonuniversal properties of cross correlations in financial time series. Phys. Rev. Lett. **83**, 1471–1474 (1999)

17. Laloux, L., Cizeau, P., Bouchad, J.-P., Potters, M.: Noise dressing of financial correlation matrices. Phys. Rev. Lett. **83**, 1467–1470 (1999)

18. Nguyen, Q.: One-factor model for the cross-correlation matrix in the Vietnamese stock market. Phys. A **392**, 2915–2923 (2013)

19. Molley, M., Reed, B.: A critical point for random graphs with a given degree sequence. Random Struct. Algorithms **6**, 161–180 (1995)

20. Cohen, R., Erez, K., ben-Avraham, D., Havlin, S.: Resilience of the internet to random breakdowns. Phys. Rev. Lett. **85**, 4626–4628 (2000)

21. Callaway, D.S., Newman, M.E.J., Strogatz, S.H., Watts, D.J.: Network robustness and fragility: percolation on random graphs. Phys. Rev. Lett. **85**, 5468–5471 (2000)

22. Paul, G., Screenivasan, S., Stanley, H.E.: Resilience of complex networks to random breakdown. Phys. Rev. E **72**, 056130 (2005)

23. Lichtenwalter, R. Chawla, N.V.: DisNet: a framework for distributed graph computation. In: Proceedings of the International Conference on Advances in Social Networks Analysis and Mining, Notre Dame (2011)

24. Crucitti, P., Latora, V., Marchiori, M., Rapisarda, A.: Error and attack tolerance of complex networks. Phys. A **340**, 388–394 (2004)

25. Amaral, L.A.N., Scala, A., Barthelemy, M., Stanley, H.E.: Classes of small-word networks. Proc. Natl. Acad. Sci. USA **97**, 11149–11152 (2000)

26. Watts, D.J., Strogatz, S.H.: Collective dynamics of "small word" networks. Nature **393**, 440–442 (1998)

27. Liljeros, F., Edling, C.R., Amaral, L.A.N., Stanley, H.E., Aberg, Y.: The web of human sexual contacts. Nature **411**, 907–908 (2001)

28. Liljeros, F., Edling, C.R., Amaral, L.A.N.: Sexual networks: implications for the transmission of sexually transmitted infections. Microbes Infect. **5**, 189–196 (2003)

29. Cancho, R.F.I., Sole, R.V.: The small word of human language. Proc. Roy. Soc. Lond. Ser. B Biol. Sci. **268**, 2261–2262 (2001)

30. Dorogovtsev, S.N., Mendes, J.F.F.: Language as an evolving word web. Proc. Roy. Soc. Lond. Ser. B **268**, 2603–2606 (2001)

31. Chen, Q., Chang, H., Govindan, R., Jamin, S.: The origin of power laws in internet topologies revisited. In: Proceedings of the Twenty-First Annual Joint Conference of the IEEE Computer and Communications Societies, INFOCOM 2002, vol. 2, pp. 608–617 (2002)
32. Adamic, L.A., Lukose, R.M., Puniyani, A.R., Huberman, B.A.: Search in power law networks. Phys. Rev. E **64**, 046135 (2001)
33. Ripeanu, M., Foster, I., Iamnitchi, A.: Mapping the Gnutella network: properties of large-scale peer-to-peer systems and implications for system design. IEEE Internet Comput. **6**, 50–57 (2002)
34. Kim, H.J., Lee, Y., Kim, I.-M., Kahng, B.: Scale-free networks in financial correlations, cond-mat/0107449 (2001)
35. Nguyen, Q., Nguyen, N.K.K.: in preparation
36. Nie, T., Guo, Z., Zhao, K., Lu, Z.-M.: New attack strategies for complex networks. Phys. A **424**, 248–253 (2015)

Constructing a Financial Stress Index for Vietnam: An Application of Autoregressive Conditional Heteroskedastic Models

Nguyen Chi Duc[1(✉)] and Ho Thuy Ai[1,2]

[1] Banking University of Ho Chi Minh City,
36 Ton That Dam, District 1, Ho Chi Minh City, Vietnam
{ducnc,aiht}@buh.edu.vn
[2] Lingnan University, 8 Castle Peak Road,
Tuen Mun, New Territories, Hong Kong
thuyaiho@ln.hk

Abstract. This paper constructs an index to measure financial stress for Vietnam with monthly data from April 2007 to December 2016. Various measures of stress are selected based on literature and Vietnam's practice. An important stress measure, the volatility of stock market, bond market, money market and banking sector, is estimated by variants of the general autoregressive conditional heteroskedasticity (GARCH) model. Individual stress variables are combined together to make an aggregate index using equal variance weighting scheme. The constructed index is a useful tool for policy makers to monitor the riskiness of domestic financial system as well as academics to conduct further research about financial crisis.

1 Introduction

The onset of the Global Financial Crisis (GFC) 2008–2009, which resulted in a severe economic downturn worldwide in the past few years, has again challenged both economists and policymakers about the validity and reliability in forecasting such a crisis. In particular, it poses a question whether we can develop an early warning system and systemic risk indicators for future crises so that we can have timing measures to prevent the crises from happening.

One of many great efforts to address the problem is the creation of a contemporaneous index namely Financial Stress Index (FSI) to capture the systemic risk of the financial system by combining a variety of stress indicators [13, 25–27]. The index's unusual upper values are considered as signs of stress while its extreme values corresponding to some kind of crisis. Since the pioneering work in 2003 of two economists from the Bank of Canada, Mark Illing and Ying Liu, other researchers have constructed a single index for a specific country [9, 22, 35, 45] and a composite index for a group of countries [4, 8, 25, 41]. Such indices are created in different ways to cover various elements of the financial system. The researchers have proved that their indices adequately capture stress episodes in some countries at some periods of time. These indices can also be used as an early warning indicator for financial distress and therefore, help policymakers

© Springer International Publishing AG 2018
L. H. Anh et al. (eds.), *Econometrics for Financial Applications*, Studies in Computational Intelligence 760, https://doi.org/10.1007/978-3-319-73150-6_45

diagnose accurately the current health of the financial system and take timely measures to protect the system from instability and crisis.

Despite FSI's usage and popularity in developed and emerging countries, limited research has been paid to the establishment of early warning indicators such as the FSIs for small opened dynamic economies in Southeast Asia, especially Vietnam. Therefore, our study is to contribute to literature by constructing a continuum indicator measuring financial stress level for Vietnam's financial system. As volatility is an important measure of risk, we employ the autoregressive conditional heteroskedasticity (ARCH) model proposed by Engle [16] and its variant, the general autoregressive conditional heteroskedasticity (GARCH) model introduced by Bollerslev [5] to estimate the volatility of financial time series. All the stress variables are combined using the most popular weighting method, equal variance weighting scheme. The index can be used not only as a tool to support the macro-prudential policy but also as a continuum variable for studies about financial crisis.

This paper is organized as follows. Section 2 explains the concepts of financial stability and measures of financial stability, then reviews literature on FSI. Section 3 presents steps to construct an FSI for Vietnam with the application of GARCH models. The estimated variables and constructed index are shown in Sect. 4. Section 5 concludes with some recommendations for further studies.

2 Literature Review

2.1 Definition of Financial (In)stability

Unlike monetary stability or macroeconomic stability whose analytical framework has been long developed and received wide consensus, there has not existed a generally accepted analytical framework or standard measures for financial stability [2,3,11,19,21,43]. Whereas one may agree that price stability is measurable and observable, financial stability is characterized by sophisticated interaction among different aspects of the financial system and cannot be measured by a single indicator. Literature on defining financial stability presents two courses of approach: directly interpretation and its negative counterpart, financial instability.

Some researchers and practitioners prefer the financial stability approach rather than its absence because they believe that it is likely to envisage policies and avoid bias policy decisions, analyses, and analytical frameworks towards sacrificing both private and social benefits of finance [43]. Crockett [11] and Laker [34] claim that financial stability requires that the key financial institutions and the key markets are stable while Schinasi [43] emphasizes the continuum aspect of financial stability. He argues that financial stability is not implied the ability to return to a single and sustainable position or time path after a shock but rather a range or a continuum. This continuum has multidimensional nature occurring across a set of measurable variables which forms a set of tolerable boundaries. Observable states within these boundaries indicate the financial system performs its key functions well while observations outside these boundaries are considered instability.

The second approach to define financial stability is the demonstration of its absence. The reason behind is that financial stability is not completely observable. There are cases where the economy looks stable but its structure could amplify certain shocks and such an appearance of stability may last quite long before a shock occurs. For that reason, specifying features characterizing an episode of financial instability would be preferred and financial stability can just be inferred as the condition where an episode of financial stability is unlikely to occur [3]. In defining financial instability Mishkin [36] focuses on information asymmetry problem of the financial system while Ferguson [18] considers the distortion of asset prices as an essential manifestation of financial instability. This point is not, however, totally agreed by Davis [12] who interprets financial instability and systemic risks as a heightened risk of a financial crisis and excludes the asset price volatility and misalignment from the definition of financial instability. In addition, Allen and Wood [3] argue that asset price bubbles should not always be considered constitution of financial instability. This is because asset prices express expectations about the future while expectations could also volatile by a large degree, e.g. technology changes. Only in case asset price bubbles seriously weaken financial condition of financial institutions, we could say it constitutes an episode of financial stability. This point of view is also supported by Borio and Lowe [6] who claim that the more relevant issue for policymakers is not a rapid rise in asset prices or a rapid rise in credit growth but rather the combination of events in the financial and real sectors which exposes the financial system to a materially increased level of risk.

Despite the lack of global consensus about what is financial stability or instability, some similarities among definitions could be found. First of all, majority of authors agree that the financial system is stable when its basic functions are performed properly to facilitate the effective allocation of economic resources. Two main functions commonly highlighted include channeling savings to worthy investments and providing payment services. Second, financial stability is characterized by its ability to resist unforeseen shocks and to restore its functioning on its own without major disruptions. In other words, the system should have capacity to dampen the shocks rather than amplify them. Thirdly, in episodes of financial instability, a great number of parties in the economy are seriously negatively affected. Also, most of researchers and practitioners consent that a stable financial system requires the stability of all its components, financial institutions, markets and infrastructure.

2.2 Measure of Financial (In)stability - Financial Stress Index

Challenge to the current literature is not only about the consensus on the definition of financial stability but also about how to measure it. Because the financial system consists of many key sectors and complicated interactions between them, using individual indicators of a certain sector cannot assess the overall condition of the financial system [19]. Over the past few decades, studies on measures of financial stability have paid more attention to diagnosing the health of the financial system as a whole rather than its individual components. From previous literature on financial crises and early warning indicators, microeconomic and macroeconomic variables have been chosen and combined to make a set of indicators or aggregate indices to better capture the dynamic behaviour and the potential build-up of unstable condition of the financial system.

For cross-country comparison, the International Monetary Fund (IMF) proposed a set of financial soundness indicators to measure the strength and vulnerability of its state members, whereas the European Central Bank (ECB) initiated the so-called Macro-Prudential Indicators (MPIs) to monitor the financial soundness of the European banking sectors [37]. Although MPIs and FSIs have similarities in terms of containing indicators to measure the health of the financial system, compared to FSIs, MPIs include far more indicators and focus more on banking sector with a consolidated basis in compiling and publishing [20].

Financial crises occurring more frequently since the last two decades of the 20th century have also urged numerous researchers to investigate indicators which would better summarize the financial system's conditions and predict the onset of financial distress. At first, some attempt to discern the fluctuations in a certain part of the financial system, for instance the foreign exchange market and banking sector. Lately, more endeavours have been made to identify stress conditions of the complete financial system rather than its individual parts.

Among very first scholars who have tried to construct a stress measure for banking sector, a vital element of the financial system, Kibritcioglu [31] proposes a monthly banking sector fragility index by averaging standardized value of three leading sectoral indicators of banking crises: bank deposits, bank claims on domestic private sector and foreign liabilities of banks. Drops in value of the index are interpreted as increases in fragility of the banking sector and vice versus. Gersl and Hermanek [20] also create an index to measure the stability of the Czech Republic's banking sector with similar interpretation but include more balance sheet data variables and assign arbitrarily different weights to the partial indicators of the index. Hanschel and Monnin [23] develop a stress index on yearly basis for the Swiss banking sector by combining different types of variables: market price data, balance sheet data, non-public data of the supervisory authorities, and other structural variables. Applying the technique proposed by Borio and Lowe [6], the authors use deviations from long-term trend of the variables rather than their level or growth rate. Increases in the index level indicate the stress is higher.

To gauge conditions in the financial system as a whole, Hawkins and Klau [24] calculate three composite indices to measure potential vulnerabilities in 24 emerging market economies: exchange market pressure index (EMPI), external vulnerability index and banking system vulnerability index. Each index is a weighted sum of scores of individual indicators in such a way that its maximum value is ten and increase in the index shows a rise in risk. Van den End [44] creates a financial stability condition index (FSCI), which is a combination between the traditional monetary condition index and the financial condition index, for the Netherlands and some other OECD countries. FSCI is bounded within a corridor where the system is considered functioning well. Any movements exceeding the upper or lower boundary signal the imbalance condition or instability, respectively, of the financial system.

Regarded as pioneers in constructing a single continuous index to capture developments of the financial system's conditions, Illing and Liu [27] propose a financial stress index for the Canadian financial system based on variables selected from banking sector, foreign exchange market, debt market and equity market. The authors apply different weighting methods and compare the computed index with experts' assessment in evaluating the index's performance. Their idea has been absorbed by other researchers, particularly after the GFC 2007–2008, with some adjustments, for instance Cardarelli et al. [8] with financial stress index for each of 17 developed economies using real-time and high frequent market-based data in banking, securities and foreign exchange markets, Yiu et al. [45] with a financial stress index for Hong Kong SAR characterized by using option-implied volatility of the exchange rate rather than its level or foreign reserves, Jakubik and Slacik [29] with a financial instability index for some Central, Eastern and Southeastern Europe, Aboura and van Roye [1] with a financial stress index for France.

To measure systemic stress in the European financial system Hollo et al. [25] introduce the Composite Indicator of Systemic Stress (CISS) by applying basic portfolio theory to the aggregation of five market-specific sub-indices created from 15 individual financial stress measures. This method is then adopted by Louzis and Vouldis [35] to calculate a financial systemic stress index for Greece, Braga, Pereira, and Reis [7] to create a composite indicator of financial stress for Portugal, Huotari [26] to construct a financial stress index for Finland, and Kota and Saqe [33] to compile a financial systemic stress index for Albania.

The Federal Reserve Board has also collected a wide range of indicators, including both market-based and balance sheet data, quantitative and qualitative data, to assess the condition of the financial system. Besides measures based on interest rates and asset prices, mortgage market indicators, measures of individual institutions' condition, they construct an index of financial fragility to estimate the probability that the U.S. financial system is currently under severe stress [39]. The index is created by first, reducing 12 market-based individual variables to 3 summary statistics that capture their level, their rate of change and their correlation, and second, combining the 3 summary statistics into a single measure of fragility by a logit model. Also, Hakkio and Keeton [22] introduce the Kansas City Financial Stress Index which focuses more on investor uncertainty, and Kliesen and Smith [32] use the method similar to Hakkio and Keeton's to construct a financial stress index for the Federal Reserve Bank of St. Louis based on 18 weekly data series.

It could be seen from literature that a useful tool for financial stability analysis and macro-prudential policy as FSI has been vastly growing across developed countries but still limited in emerging economies. In this paper we construct an FSI for Vietnamese financial system with monthly frequency. By applying

GARCH models to extract the volatility in different segments of the system and using other stress variables, our index can signal episodes when the system is undergoing severe stress and likely to suffer a crisis.

3 Methodology

3.1 Criteria for Choosing a Stress Measure

We select variables to measure stress level of Vietnam's financial system based on four criteria as follows.

First, variables must cover main components of the financial system. Similar to other countries Vietnam's financial system consists of three primary components: financial markets, institutions and infrastructure. Whereas stress in financial markets and institutions can be observed directly, it is not totally obvious in the case of financial infrastructure due to limited data. Therefore, we pay much attention to choosing stress variables which cover as much as possible the evolution of domestic financial markets and institutions. The selected variables are necessary to measure stress in five segments of Vietnam's financial system, namely stock market, bond market, money market, foreign exchange market, and banking sector.

Second, variables must express one phenomenon or more of financial stress. Balakrishnan et al. [4] define financial stress as a situation that the financial system is under strain and cannot fulfill intermediation function smoothly. Hakkio and Keeton [22] characterize some features of financial stress including increased uncertainty, increased information asymmetry, decreased willingness to hold risky assets and/or illiquid assets. Following literature, we focus on variables which express financial stress phenomena by measuring the volatility of asset prices, changes in asset returns, risk or uncertainty of financial system and the health of the banking sector.

Third, data for variables has to be available in terms of high frequency and accessibility. Because the main purpose of FSI is to signal contemporaneous severity of financial stress, it is vital to have indicators with frequency as high as possible. Most of relating studies calculate monthly FSI while a few produce weekly, or daily or even real time index (e.g. KCFSI of the Federal Reserve Bank of Kansas City). Moreover, for the index to be widely used and easily tested, data on underlying variables need to be obtained with no considerate difficulty, meaning accessible to the public, not confidential data.

Finally, stress variables have to show its connection with the real economy. In fact, there are a large number of variables expressing financial tension in Vietnam. Selected ones must be related to the real economy because eventually, financial stress is of particular concern when it adversely affects production, business activities, and social welfare. Based on this criterion we select variables related to the real economy by calculating their correlation coefficients with industrial production index (IPI), consumer price index (CPI) and gross domestic product (GDP) growth. This criterion is also emphasized by Huotari [26], Islami and Kurz-Kim [28] in constructing FSI for European countries.

To satisfy the aforementioned criteria we choose ten variables to represent stress in five components of Vietnam's financial system. Data on all the variables covers the period of April 2007–December 2016. This period is picked because we want to make a consistent time frame for all the variables. Note that the principal data to calculate stress measures for Vietnam's banking sector, the banking stock index, is available from April 2007 only, whereas the others exist some years before that time. This period also covers the GFC 2008–2009, therefore, reveals an important event to validate the ability to capture financial stress of the constructed index. Details on the meaning and calculation of each variable, sub-indices and the aggregate index are demonstrated in Sect. 3.3.

3.2   GARCH Models for Stress Indicators

Most of FSI studies agree that volatility is an important measure of financial stress. Modeling volatility possess a long history since the seminal work of Engle [16], who shared the 2003 Nobel Prize in Economics with Clive W.J. Granger for the invention of ARCH model. The model has speedily and powerfully influenced research in finance discipline and been modified with the exploration of numerous variants to better capture the time-varying characteristic of volatility as well as the persistency of time series variance. Among those influential contributions are the generalized ARCH (or GARCH) model of Bollerslev [5], the exponential GARCH (EGARCH) of Nelson [38], and the integrated GARCH (or IGARCH) of Engle and Bollerslev [17].

To measure the volatility of Vietnam's financial markets and banking sector, we utilize the aforementioned classical conditional heteroskedastic models for stock market returns, banking stock returns and money market interest rates. It is worth emphasizing that our ultimate purpose is not to forecast the volatility but to extract the volatility in those segments and provide inputs for Vietnam's FSI computation. The models for volatility are selected based on model selection criteria such as AIC and SBC given the satisfaction of model assumptions.

The basic heteroskedatic model for financial time series is the ARCH of Engle [16]. The main idea of this model is that the mean and variance of a stationary random process y_t can be modeled using its own past information. Suppose the first order autoregression of y_t is

$$y_t = \phi_0 + \phi_1 y_{t-1} + u_t \tag{1}$$

where u_t is a white noise with variance σ^2. Given the stability of y_t ($|\phi_1| < 1$),

the unconditional mean of y_t is $E y_t = \phi_0/(1 - \phi_1)$,

the conditional mean of y_t is $E y_t = \phi_0 + \phi_1 y_{t-1}$,

the unconditional variance of y_t is $Var(y_t) = \sigma^2/(1 - \phi_1^2)$, and

the conditional variance y_t is $Var(y_t|y_{t-1}) = E_{t-1}(u_t)^2 = \sigma^2$.

If the conditional variance is not constant, it can be forecasted by using an autoregressive of order q AR(q) model as follows:

$$\hat{u}_t^2 = \alpha_0 + \alpha_1 \hat{u}_{t-1}^2 + ... + \alpha_q \hat{u}_{t-q}^2 + e_t \tag{2}$$

where $\hat{u}_{t-j}^2 (j = 1, ..., q)$ is the squares of estimated residuals from Eq. (1); e_t is a white noise process with mean zero and constant variance.

If all the coefficients of autoregressive terms in (2) are zero, the conditional variance becomes a constant α_0. Otherwise, the conditional variance changes over time depending on the estimation of AR(q) model in (2). This is the formation of the so-called ARCH(q) model.

Engle generalizes his model of conditional variance in a multiplicative form instead of the additional form in (2) so that he can apply maximum likelihood (ML) estimation method. The model for conditional variance with order one (ARCH(1)) is rewritten as

$$u_t = e_t \sqrt{h_t}, e_t \; IID(0, 1) \tag{3}$$

$$h_t = \alpha_0 + \alpha_1 u_{t-1}^2, \alpha_0 > 0, 0 < \alpha_1 < 1 \tag{4}$$

Obviously, by this setup the process u_t has zero mean, constant variance and is uncorrelated. It is easy to show that

the unconditional mean of u_t is

$$Eu_t = E[v_t(\alpha_0 + \alpha_1 u_{t-1}^2)^{1/2}] = Ev_t E(\alpha_0 + \alpha_1 u_{t-1}^2)^{1/2} = 0,$$

the conditional mean of u_t is

$$E_{t-1}u_t = E_{t-1}[v_t(\alpha_0 + \alpha_1 u_{t-1}^2)^{1/2}] = E_{t-1}v_t E_{t-1}(\alpha_0 + \alpha_1 u_{t-1}^2)^{1/2} = 0$$

the autocorrelation of u_t is

$$Eu_t u_{t-1} = E(e_t \sqrt{h_t})E(e_{t-i}\sqrt{h_{t-i}}) = 0, i \neq 0,$$

and the unconditional variance of u_t is

$$Eu_t^2 = E[e_t^2(\alpha_0 + \alpha_1 u_{t-1}^2)] = Ee_t^2 E(\alpha_0 + \alpha_1 u_{t-1}^2) = \alpha_0/(1 - \alpha_1).$$

However, the conditional variance of u_t depends on the realization of u_{t-1}^2 as

$$E_{t-1}u_t^2 = E_{t-1}[e_t^2(\alpha_0 + \alpha_1 u_{t-1}^2)] = E_{t-1}e_t^2 E_{t-1}(\alpha_0 + \alpha_1 u_{t-1}^2) = \alpha_0 + \alpha_1 u_{t-1}^2.$$

In short, the ARCH model can estimate the evolution of conditional variance of a stationary time series based on the past realization of the error process. How persistent the conditional variance is depends on the magnitude of α_1. When u_{t-1} departs widely from 0, the estimated u_t also tends to depart from 0 due to the effect of $(\alpha_1 u_{t-1}^2)$. This in turn results in the larger change in y_t. Therefore, the ARCH model can capture the serial correlation of changes in financial asset prices where high volatile periods seem to occur together in group and so do low volatile periods. Engle extends the basic ARCH(1) by considering higher order of the conditional variance such that all shocks from u_{t-1} to u_{t-q} impact directly on u_t:

$$u_t = e_t \sqrt{h_t} = e_t \sqrt{\alpha_0 + \alpha_1 u_{t-1}^2 + \alpha_2 u_{t-2}^2 + ... + \alpha_q u_{t-q}^2} = e_t \sqrt{\alpha_0 + \sum_{i=1}^{q} \alpha_i u_{t-i}^2}$$

(5)

where $\alpha_0 > 0, \alpha_i \geq 0, i = 1, ..., q$.

Bollerslev [5] generalizes the ARCH model by allowing not only the past errors but also the past conditional variances to affect the current conditional variance. His generalized ARCH (or GARCH) model in the simplest form is GARCH(1,1) as follows:

$$u_t = e_t \sqrt{h_t}, e_t \ IID(0,1)$$

$$h_t = \alpha_0 + \alpha_1 u_{t-1}^2 + \beta_1 h_{t-1}, \alpha_0 > 0, \alpha_1 \geq 0, \beta_1 \geq 0, (\alpha_1 + \beta_1) < 1 \qquad (6)$$

The more general form of GARCH is GARCH(p, q) where q is the number of lags of the squared errors and p is the number of lags of the conditional variances introduced in (6):

$$h_t = \alpha_0 + \sum_{i=1}^{q} \alpha_i u_{t-1}^2 + \sum_{j=1}^{p} \beta_j h_{t-j} \qquad (7)$$

where

$$p \geq 0, \qquad\qquad q \geq 0$$
$$\alpha_0 > 0, \qquad\qquad \alpha_i \geq 0, i = 1, ..., q$$
$$\beta_j \geq 0, j = 1, ..., p$$

The stability condition now is $\sum_{i=1}^{q} \alpha_i + \sum_{j=1}^{p} \beta_j < 1$.

The GARCH model is considered more beneficial than the ARCH because it allows for both longer memory and a more flexible lag structure but achieves similar properties. While the high-order ARCH process can bring trouble in estimation because many coefficients and restrictions involve, a more parsimonious GARCH model is much easier to identify and estimate. Consider the GARCH(1,1), by including h_{t-1} the higher orders of the ARCH process have actually been accounted for. If $\beta_1 = 0$, the GARCH model is nothing but the ARCH. For this reason, the GARCH model has become more popular. Empirical works have also shown that GARCH(p, q) with $p, q \leq 2$ is sufficient to cover dynamics of the conditional variance. Especially, GARCH(1,1) specification is the most popular form for modeling time-varying volatility of financial time series [15].

Other variants of the initial ARCH model have been developed by relaxing some constraints of the original version. For instance, Engle and Bollerslev [17] analyze the case where the GARCH process exhibits a unit root rather than a stationarity. If it is the case, we cannot reject the null hypothesis of $(\alpha_1 + \beta_1) = 1$ with a conventional Wald type test for the GARCH(1,1) process. The integrated GARCH(1,1) is then written as:

$$h_t = \alpha_0 + \alpha_1 u_{t-1}^2 + (1 - \alpha_1)h_{t-1}, \alpha_0 > 0, 0 \leq \alpha_1 < 1 \qquad (8)$$

The j-step-ahead forecast of h_t is $E_t h_{t+j} = j\alpha_0 + h_t$.

Equation (8) is called the IGARCH(1,1) model with a trend. If $\alpha_0 = 0$, the model becomes very closely the traditional random walk without drift. The IGARCH model possesses interesting properties because it shows that shocks to the conditional variance will not be forgotten but have permanent impact.

Another extension of the standard GARCH model is the exponential GARCH (or EGARCH) model. Nelson [39] relaxes the nonnegativity constraints of the standard form by introduce the log-linear equation for the conditional variance. His EGARCH model also allows for asymmetric effect of shocks on the conditional variance which is usually the case in financial time series that "volatility tends to rise in response to "bad news" (excess returns lower than expected) and to fall in response to "good news" (excess returns higher than expected)". The EGARCH specification takes the form:

$$\ln(h_t) = \alpha_0 + \alpha_1 (u_{t-1})/\sqrt{h_{t-1}} + \lambda_1 |u_{t-1}/\sqrt{ht - 1}| + \beta_1 \ln(h_{t-1}) \qquad (9)$$

The aforementioned part summarizes some popular variants of the standard GARCH model. For the purpose of FSI construction, GARCH(1,1) is often used in most of the relating studies [4, 8, 27, 41, 45]. They apply the model directly without seriously conducting a comparison between models to choose the best fit. We are instead more cautious by checking the model assumptions and comparing different specifications to choose the one which can provide most of the information on volatility and highest numbers of observation.

Our strategy in applying the GARCH models to capture the time-varying volatility is as follows:

(i) construct a mean equation for the time series of interest (e.g. stock return, interest rate);
(ii) test for the serially correlation property of the residuals and the existence of the heteroskedasticity (or ARCH effect);
(iii) estimate the GARCH specifications for the conditional variance and choose the best fitting one based on model selecting criteria.

It is possible that each time series corresponds with a specific class of the GARCH model. That depends on the characteristics of each series realizations. The estimated results are presented in Sect. 4.

3.3 Data Description and Construction of FSI

Given the stated requirements for selecting stress measures, we take into account the following variables for constructing Vietnam's FSI (VNFSI):

3.3.1 Equity Market

Extremely high level of stress in the equity market may result in a crisis and usually manifest with a sharp fall in stock market indices. The growing stress is also measured by higher volatility of stock prices, when investors face increasing uncertainty and risks, greater expected loss and are more doubtful about future profits of firms. Three variables used to measure stress in Vietnam's equity market are

- $CMAX$ for stock market price index (StockCMAX): this is a common measure used in stock market crisis literature [7,25,27,35,40,42]. The $CMAX$ shows how sharp the decline in stock prices is corresponding to previous time period. StockCMAX for Vietnam is calculated as follows:

$$CMAX = \frac{x_t}{\max[x \in (x_{t-j}|j = 0, 1, ..., T)]} \tag{10}$$

  where x_t is VNindex at month t, T is 12-month time window. Data on VNindex is from Datastream. Daily frequency is converted to monthly by using the month-end values. Because $CMAX$ is always less than or equal to 1, it is tranformed into $(1-CMAX)$ when computing the VNFSI.
- Volatility of stock price (StockVol): the volatility of VNindex is estimated with a GARCH model. The monthly stock returns is the time series of interest, calculated by taking the log difference of VNindex at current month and the previous month. Month-end data is used to approximate the monthly index. Higher volatility is interpreted higher stress.
- Stock market return (StockRet): facing high stress, investors tend to switch their portfolio from risky to risk-free assets (flight to quality) and/or increase holding cash instead of illiquid assets (flight to liquidity). This phenomenon results in a sharp decrease in stock price, and consequently, a great fall in stock return. We calculate the stock return variable by taking the log difference of VNindex between current and previous month. Month-end data is used and collected from Datastream as usual.

3.3.2 Debt Market

Debt markets are usually classified into two types: corporate debt versus government debt. It is common in emerging countries that corporate debt markets are still underdeveloped and their trading data is very limited to the public. For that reason, our study restricts stress measures of Vietnamese debt market to the government debt only. Two variables are taken into account:

- Sovereign debt spread (BondSpread): the difference in returns between domestic government bonds and a benchmark (e.g. U.S. Treasury bonds) are often considered as a measure of stress in the bond market. Because of limited data on Vietnamese bond market, we use the Vietnam EMBIG spread of JP Morgan as a proxy for Vietnamese sovereign debt spread. The spread is a weighted average of (USD denominated) Vietnamese government bond yields minus the 10-year U.S. Treasury yield. This measure is also widely used in literature on emerging market debt crisis [4,9,10]. Daily Vietnam EMBIG spread is collected from Datastream, then converted to monthly frequency by using month-end values.

- Volatility of sovereign debt index (BondVol): as in the case of stock market, we employ the GARCH model to extract the time-varying volatility of Vietnamese sovereign debt (total return) index which is part of JP Morgan's emerging market bond indices. The data is collected from Datastream. Monthly data is drawn from month-end values.

3.3.3 Money Market

The money market plays a vital role in the financial system because it facilitates short-term funding flows. Any disruption in the market can represent the increasing worry of market participants for short-term financing. The stress in money market is often measured by TED spread (spread between 3-month interbank interest rate and Treasury bill of the same maturity). The growing anxiety in the market can also be expressed by increasing volatility of short-term interest rates. Due to limited data on Vietnam's money market our study uses the volatility of 1-month interest rate in Vietnam's interbank market (VN1M) as a stress variable. The time-varying volatility of the interbank interest rate (MoneyVol) is estimated by the GARCH model. Daily data on VN1M is collected from Datastream and converted to monthly frequency by using month-end value.

3.3.4 Foreign Exchange Market

Episodes of currency crisis often happen with a sharp decline in exchange rate (domestic currency depreciation). There are cases that central banks react to speculative attacks by depleting foreign reserves or raising interest rates to maintain exchange rates. Therefore, the exchange market pressure index (EMPI) proposed by Eichengreen et al. [14] is widely used as a measure of foreign exchange market stress in emerging countries. We also construct an EMPI for Vietnam to capture the depreciation of Vietnam dong against US dollar and the fall in foreign reserves as follows:

$$EMPI_t = \frac{(\Delta e_t - \mu_{\Delta e})}{\sigma_{\Delta e}} - \frac{(\Delta RES_t - \mu_{\Delta RES})}{\sigma_{\Delta RES}} \tag{11}$$

where Δe and ΔRES are month-over-month percentage changes in the exchange rate and foreign reserves of Vietnam, μ and σ are mean and standard deviation, respectively. Data on the monthly average exchange rate, which expresses the value of one US dollar in terms of Vietnam dong, and foreign reserves are collected from CEIC. Higher value of EMPI shows increasing stress in the currency market.

3.3.5 Banking Sector

The possibly most important financing channel of Vietnam's financial system is the banking sector. Stress in the banking sector may occur with dramatic bank runs leading to bank closures, merges and takeovers and large-scale government assistance to save important banks [30]. Studies on banking crises usually investigate indicators of soundness and fragility of the banking sector such as capital adequacy ratio, profitability, non-performing loans. Stress in this sector can also manifest in a tremendous drop in banking stock price. Literature on FSI, however, pays less attention to those indicators because they are usually released at low frequency (financial data on Vietnamese banks is announced quarterly or even annually). On the other hand, the indicators tend to highly correlated to the economic conditions. Illing and Liu [27] argue that stress measures for the banking sector should be isolated from the overall market stress and reflect the characteristics of the banking sector only. Based on this idea, three variables considered as measures of stress in Vietnamese banking sectors are:

- Banking sector beta (BankBeta): beta of Vietnamese banking sector is calculated from the capital asset pricing model as follows:

$$\beta_t = \frac{cov(r_t, m_t)}{var(m_t)} \tag{12}$$

where r_t and m_t are month-over-month banking and market returns, respectively. Banking sector index and VNindex from Datastream are used to calculate the returns. Monthly returns are the log differences between current and previous respective index. Daily index is converted to monthly frequency by using month-end values. A beta greater than 1 indicates that banking stocks more volatile than the overall market and implies higher risk. As in Balakrishnan et al. [4]; Illing and Liu [27]; Park and Mercado [41], two restrictions are imposed in the beta series: the banking sector is deemed stressful when beta greater than 1 and banking returns lower than stock market returns. If those two conditions are not met, beta will take value of 0. With those restrictions banking beta can capture better idiosyncratic risk in the banking sector rather than reflecting its comovement with the overall market.
- *CMAX* for the banking sector (BankCMAX): calculation of BankCMAX is similar to StockCMAX in (10) but VNindex is replaced by Vietnamese banking sector index from Datastream.
- Volatility of banking stock price (BankVol): we utilize the GARCH model to calculate the banking stock volatility. Banking stock return is the monthly return as explained in BankBeta calculation.

After obtaining value of all stress variables, we construct a sub-index of stress for each component of the financial system, namely BankFSI, StockFSI, BondFSI, MoneyFSI and ForexFSI. Five sub-indices are then combined together to make an aggregate FSI for Vietnam, VNFSI. Weighting scheme for all the combination is the popular equal-variance weight. According to this method, all individual variables are normalized by standardization, where each variable, first subtracts its mean, and then is divided by its standard deviation. By transforming all variables into a standardized form with zero mean and unit variance, we assign equal weight to each variable when combining them together.

4   Results

4.1   Estimation of GARCH Models for Volatility

4.1.1   Volatility of Stock Market

Before estimating the GARCH model we check the stationarity of the time series of interest (e.g. stock return in this case) and the serial correlation in the mean equation to choose the appropriate model.

Literature on financial time series often gives evidence of the stationarity of stock returns. Vietnam's stock market return is not an exception. The Augmented Dickey-Fuller unit root test with 12 lags for the variable StockRet confirms its stationary property with t-statistic of -8.675. The null hypothesis that StockRet has a unit root is therefore rejected at 1% significance level. Correlogram of stock market return shows evidence of autocorrelation with 5% significance at lag 1 and 10% level for some other lags. We conclude that an AR(1) process could capture the dynamics of the stock return. Therefore, our initial mean equation for stock return (Mean-Model 1) is

$$y_t = \phi_0 + \phi_1 y_{t-1} + u_t \tag{13}$$

where y_t is the series of Vietnam's stock market return.

To check whether Mean-Model 1 exposes to any problem of serial correlation we use the Q-statistics with 12 lags for the residuals. None of the autocorrelations (AC) or partial autocorrelations (PCA) is significant at all conventional levels. However, based on the Breusch-Godfrey LM test for 12 lags we reject the null hypothesis of no serial correlation at 10% level although it cannot be rejected at 5% significance.

To be more cautious in the model specification we consider the second mean equation (Mean-Model 2) which excludes the serial correlations of the residuals but minimizes loss of degree of freedom as follows:

$$y_t = \phi_0 + \phi_1 y_{t-1} + \phi_2 y_{t-2} + u_t \tag{14}$$

Both Q-statistics and LM test support the absence of serial correlation in Mean-Model 2. However, we still consider Mean-Model 1 because the presence of serial correlation is marginal while less number of coefficients to be estimated. Moreover, the results estimated by OLS for the mean equation may be greatly different from the ones estimated by ML in the GARCH model. Heteroskedasticity tests for both two models confirms the existence of ARCH effect. Our next step is, therefore, to estimate the GARCH models for both (13) and (14), then compare them together.

Estimation of Mean-Model 1:

(i) GARCH(1,1)

Mean equation:
$$\hat{y}_t = \underset{(0.004)}{0.002} + \underset{(0.000)}{0.172} y_{t-1} \tag{15}$$

Variance equation:

$$\hat{h}_t = -8.93E - 5 - \underset{(2.98E-5)}{} \underset{(-0.820)}{0.016}\, u^2_{t-1} + \underset{(46.39)}{1.01}\, h_{t-1} \tag{16}$$

Equations (15) and (16) present the estimated coefficients and their standard error (in parentheses) of the GARCH(1,1) model. The result shows violation of coefficient restriction of the GARCH model with negative α_1 (-0.016). Therefore, we apply the EGACH model where the nonnegativity constraints are relaxed.

(ii) EGARCH

Mean equation:

$$\hat{y}_t = \underset{(0.004)}{-0.003} + \underset{(0.084)}{0.142}\, y_{t-1} \tag{17}$$

Variance equation:

$$\widehat{\ln(h_t)} = \underset{(0.057)}{-0.06} - \underset{(-0.060)}{0.165}\left(\frac{u_{t-1}}{\sqrt{h_{t-1}}}\right) - \underset{(0.065)}{0.152}\left|\frac{u_{t-1}}{\sqrt{h_{t-1}}}\right| + \underset{(0.000)}{0.979}\ln(h_{t-1}) \tag{18}$$

The estimation result of the EGARCH model confirms the asymmetric effect of news on the conditional variance with negative α_1 (-0.165) in (18). All the coefficients in the variance equation are highly significant. Residual diagnostic tests prove no serial correlation in standardized residuals and no ARCH effect with 4 lags. Also, we cannot reject the null hypothesis of normal distribution of the residuals with the Jarque-Berra statistic of 4.238. Consequently, this model is taken into account to compare with the Mean-Model 2 in the following part.

Estimation of Mean-Model 2:

(i) GARCH(1,1)

Mean equation:

$$\hat{y}_t = \underset{(0.006)}{0.003} + \underset{(0.102)}{0.108}\, y_{t-1} + \underset{(0.096)}{0.005}\, y_{t-2} \tag{19}$$

Variance equation:

$$\hat{h}_t = \underset{(0.000)}{2.06E - 5} + \underset{(0.059)}{0.104}\, u^2_{t-1} + \underset{(0.050)}{0.879}\, h_{t-1} \tag{20}$$

In the estimated variance Eq. (20) both the news (u^2_{t-1}) and the past realizations of the conditional variance (h_{t-1}) are significant. The size of β_1 much larger than that of α_1 indicates a strong persistency of the volatility. Results from all the residual diagnostic tests for serial correlation, heteroskedasticity and normal distribution strongly support this model specification. Hence, in the following step we compare this model with the EGARCH model above.

Table 1. Comparison between EGARCH model and GARCH model for stock return

Criteria	Mean-Model 1 (EGARCH)	Mean-Model 2 (GARCH)
SSR	0.8252	0.8050
AIC	−2.4291	−2.3396
SBC	−2.2867	−2.1964
HQC	−2.3713	−2.2815

Table 1 summarizes some goodness-of-fit measures of the two competing models. Mean-Model 1 has greater SSR but lower AIC, SBC and HQC than Mean-Model 2. Accordingly, we choose the conditional variance estimated in Mean-Model 1 as the measure of volatility of Vietnam's stock market.

In the following sections we carry out the same procedure as the one in this section to estimate volatility of debt market, money market and banking sector. For brevity we report the estimation results and final conclusion only.

4.1.2   Volatility of Sovereign Debt Index

The augmented Dickey Fuller test results in the stationarity property of government bond return while the correlogram with 12 lags shows that autocorrelation is not serious. Accordingly, mean equation of bond return is estimated as follows:

$$y_t = \phi_0 + u_t \tag{21}$$

where y_t is the series of bond return.

Residual diagnostic tests show that there is no evidence on autocorrelation but exists ARCH effect. However, modelling the conditional variance of bond return poses more challenge than the case of stock return. Using the basis GARCH(1,1) model we come up with the results violating the nonnegativity and stability constraints of the model. Estimating the EGARCH model, all the coefficients are highly significant but residual diagnostic test reveals the presence of ARCH effect which signals a misspecification problem. Considering the possibility of non-normally distributed residuals, we re-estimate the GARCH(1,1) model with assumption of student's distribution error rather than the normal distribution. Coefficient negativity of the variance equation disappears but we cannot reject the null hypothesis of $(\alpha_1 + \beta_1) = 1$. Finally, we decide to employ the IGARCH model with the t-distribution or generalized error distribution (GED) rather than normality. Both the models result in no serial correlation of residuals or squared residuals. They are compared together (see Table 2) and the IGARCH (GED) is chosen based on model selection criteria.

4.1.3   Volatility of Interbank Interest Rate

Because the short-term interest rate time series is a unit root process, we estimate the GARCH model for its first difference, namely DVN1M, instead. We add one

Table 2. Comparison of GARCH models for bond return

Criteria	IGARCH (t-distribution)	IGARCH (GED)
SSR	0.1670	0.1671
AIC	−4.3793	−4.3802
SBC	−4.3085	−4.3094
HQC	−4.3506	−4.3515

lag of DVN1M to the mean equation to capture serial correlation of the residuals. Heteroskedasticity test reveals an obvious presence of ARCH effect while both Q-statistics and LM test for 12 lags give no evidence on serial correlation. The estimated GARCH(1,1) model provides the high significance of all variance equation coefficients and satisfying the non-negativity as well as stability restrictions. However, the residuals are not normally distributed although there is no serial correlation or heteroskedasticity. We then come up with a GARCH(1,1) model with t-distribution degree of freedom fixed at 10 (after trying many forms of the GARCH model). The estimated results are as follows

Mean equation:

$$\hat{y}_t = \underset{(0.079)}{0.014} + \underset{(0.104)}{0.094}\, y_{t-1} \tag{22}$$

Variance equation:

$$\hat{h}_t = \underset{(0.114)}{0.364} + \underset{(0.284)}{0.789}\, u_{t-1}^2 + \underset{(0.109)}{0.176}\, h_{t-1} \tag{23}$$

4.1.4   Volatility of Banking Stock Price

Taking into consideration all the concerning assumptions and constraints, we use the IGARCH model (t-distribution with fixed degree of freedom 10) for modeling the conditional variance of banking stock return with results as follows

Mean equation:

$$\hat{y}_t = \underset{(0.007)}{-0.002} + \underset{(0.091)}{0.16}\, y_{t-1} \tag{24}$$

Variance equation:

$$\hat{h}_t = \underset{(0.039)}{0.14}\, u_{t-1}^2 + \underset{(0.039)}{0.86}\, h_{t-1} \tag{25}$$

4.2   Vietnam Financial Stress Index and Its Components

The aggregate FSI for Vietnam is calculated by combining all the stress variables by equal variance weighting method. As mentioned in Sect. 3.3 we create stress index for each of the five financial system segments, StockFSI, BondFSI, MoneyFSI, ForexFSI, and BankFSI. The aggregate FSI (VNFSI) is constructed by

VNFSI

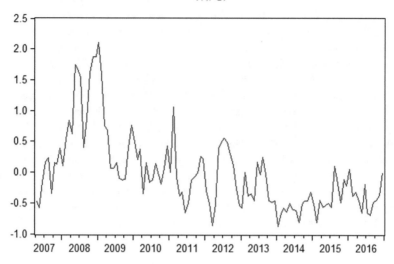

Fig. 1. Vietnam's Financial Stress Index

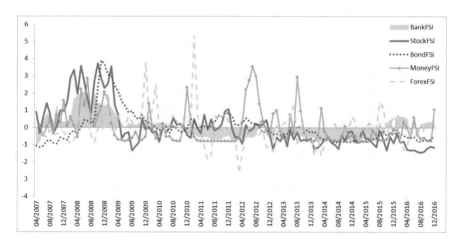

Fig. 2. Components of Vietnam's Financial Stress Index

taking average of five sub-indices. By separating the calculation into two steps we can track the source of stress, whether it comes from a certain segment of the system or is a result of high stress level of some or all the segments at the same time. Following the literature, we define an episode of financial stress is the period that the index is greater than its mean by 1.5 standard deviations [4].

Figure 1 demonstrates the developments of VNFSI from April 2007 to December 2016. The index clearly shows extreme values in the mid-2008 and last until the early 2009 despite a little drop in some months between. Unarguably, those peaked values of VNFSI coincide with the happening of the 2008–2009 GFC. In the final quarter of 2008 the global financial system was dramatically shaken by the collapse of financial institutions and large-scale bail-outs in developed countries. Vietnam was also adversely affected by the crisis with inverted capital flows, increased trade deficit, extremely high inflation and sharp drop in GDP growth. Looking at components of the VNFSI (see Fig. 2) we can see how the severe financial stress was built up. StockFSI and BankFSI started to rise since early 2008 and reached their highs in autumn of the year. ForexFSI and MoneyFSI evolved nearly in the same pattern after the first half of 2008 with tremendous upsurge. Stress in the bond market came a bit later but they all interacted and resulted in the extreme stress level in the beginning 2009. The stress index reserved its high until the second quarter and reverted to its mean (zero) after that. The next period of time is characterized by some fluctuation of VNFSI, however, the aggregate index never deviates greatly from its average. It actually possesses some stable condition since 2013.

5   Conclusions

This paper presents a construction of financial stress index for Vietnam to measure the severity of stress for the whole system. All the stress measures are selected based on criteria of market coverage, data availability, stress representation ability and connection with the real economy. Different from other studies on FSI, we cautiously take into account various forms of the GARCH model to estimate time-varying volatility, an important indicator of stress, for the financial markets. The aggregate FSI appears to capture well the most serious financial stress in the sample period and can explain the stress build-up by its component evolution.

One may argue that VNFSI reaches extreme level in the second half 2008 and first half 2009 because the stress is too obvious, it may fail to capture other stress episodes. We would emphasize that the ultimate purpose of this index is to signal the symmetric risk of the financial system. From macro-prudential perspective, it is much more important to give a special care when all financial markets and institutions experience high stress at the same time. There could be the time when certain part of the system poses some anxiety but the others remain stable, then that should not be considered financial stress but swings only. By construction, that phenomenon can be observed by investigating the developments of sub-indices as well.

There are some limitations of this paper which can be improved in future research. First of all, literature provides numerous way to construct an index but we choose the most popular and simplest one, equal variance weight. Other studies may use the principal component analysis, credit weight, cumulative distribution function, or portfolio theory weight to construct alternative VNFSI and compare the performance of different methods. Secondly, it is possible to use VNFSI to weather financial crisis for Vietnam. Some forecasting models can be developed by combining this FSI with other financial and economic indicators.

Thirdly, we can use VNFSI and FSI of other countries to study the interconnectivity between Vietnam and these countries, how financial stress is transmitted and in what channels. Last but not least, VNFSI can be used to empirically analyze the connection between the financial system and the real economy which is an important topic but still not completely understood so far.

References

1. Aboura, S., van Roye, B.: Financial stress and economic dynamics: an application to France, Kiel Working Paper, No. 1834, March 2013
2. Alawode, A.A., Al Sadek, M.: What is financial stability? Financial Stability Paper Series (2008)
3. Allen, W.A., Wood, G.: Defining and achieving financial stability. J. Finan. Stab. **2**(2), 152–172 (2006)
4. Balakrishnan, R., Danninger, S., Elekdag, S., Tytell, I.: The transmission of financial stress from advanced to emerging economies, IMF Working Paper, WP/09/133 (2009)
5. Bollerslev, T.: Generalized autoregressive conditional heteroskedasticity. J. Econ. **31**(3), 307–327 (1986)
6. Borio, C., Lowe, P.W.: Asset prices, financial and monetary stability: exploring the nexus, BIS Working Paper, No. 114 (2002)
7. Braga, J.P., Pereira, I., Reis, T.B.: Composite indicator of financial stress for Portugal, Economic Bulletin and Financial Stability Report Articles (2014)
8. Cardarelli, R., Elekdag, S.A., Lall, S.: Financial stress, downturns, and recoveries, IMF Working Papers, WP/09/100 (2009)
9. Cevik, E.I., Dibooglu, S., Kenc, T.: Measuring financial stress in Turkey. J. Policy Model. **35**(2), 370–383 (2013)
10. Cevik, E.I., Dibooglu, S., Kutan, A.M.: Measuring financial stress in transition economies. J. Finan. Stab. **9**(4), 597–611 (2013)
11. Crockett, A.: Why is financial stability a goal of public policy? Econ. Rev. (Kansas City) **82**(4), 5 (1997)
12. Davis, P.E.: Towards a Typology for Systemic Financial Instability. Economics and Finance Section, School of Social Sciences, Brunel University, London (2003)
13. Duca, M.L., Peltonen, T.A.: Assessing systemic risks and predicting systemic events. J. Bank. Finan. **37**(7), 2183–2195 (2013)
14. Eichengreen, B., Rose, A.K., Wyplosz, C.: Contagious currency crises, National Bureau of Economic Research (1996)
15. Enders, W.: Applied econometric time series. Technometrics **46**(2), 264 (2004)
16. Engle, R.F.: Autoregressive conditional heteroscedasticity with estimates of the variance of United Kingdom inflation. Econometrica J. Econ. Soc. **50**, 987–1007 (1982)
17. Engle, R.F., Bollerslev, T.: Modelling the persistence of conditional variances. Econ. Rev. **5**(1), 1–50 (1986)
18. Ferguson, R.W.: Should financial stability be an explicit central bank objective. Challenges to Central Banking from Globalized Financial Systems, International Monetary Fund, Washington DC, pp. 208–223 (2003)
19. Gadanecz, B., Jayaram, K.: Measures of financial stability - a review, Irving Fisher Committee Bulletin, No. 31, pp. 365–383 (2008)

20. Gersl, A., Hermanek, J.: Financial stability indicators: advantages and disadvantages of their use in the assessment of financial system stability, Occasional Publications-Chapters in Edited Volumes, pp. 69–79 (2007)
21. Goodhart, C.: A framework for assessing financial stability. J. Bank. Finan. **30**(12), 3415–3422 (2006)
22. Hakkio, C.S., Keeton, W.R.: Financial stress: what is it, how can it be measured, and why does it matter? Econ. Rev. - Federal Reserve Bank of Kansas City **94**(2), 5–50 (2009)
23. Hanschel, E., Monnin, P.: Measuring and forecasting stress in the banking sector: evidence from Switzerland. BIS papers, vol. 22, pp. 431–449 (2005)
24. Hawkins, J., Klau, J.: Measuring potential vulnerabilities in emerging market economies, BIS Working Papers, No. 91, October 2000
25. Hollo, D., Kremer, M., Duca, M.L.: CISS-a composite indicator of systemic stress in the financial system, ECB Working Paper Series, No. 1426, March 2012
26. Huotari, J.: Measuring financial stress – a country-specific stress index for Finland, Bank of Finland Research Discussion Paper 7 (2015)
27. Illing, M., Liu, Y.: An index of financial stress for Canada, Bank of Canada Working Paper, No. 14, June 2003
28. Islami, M., Kurz-Kim, J.R.: A single composite financial stress indicator and its real impact in the Euro area. Int. J. Finan. Econ. **19**(3), 204–211 (2014)
29. Jakubik, P., Slacik, T.: Measuring financial (in)stability in emerging Europe: a new index-based approach. Finan. Stab. Rep. **25**, 102–117 (2013)
30. Kaminsky, G.L., Reinhart, C.M.: The twin crises: the causes of banking and balance-of-payments problems. Am. Econ. Rev. **89**, 473–500 (1999)
31. Kibritcioglu, A.: Excessive risk-taking, banking sector fragility, and banking crises, University of Illinois at Urbana-Champaign, College of Commerce and Business Administration, Working Paper (02–0114) (2002)
32. Kliesen, K.L., Smith, D.C.: Measuring financial market stress, Federal Reserve Bank of St. Louis-Economic Synopses, No. 2 (2010)
33. Kota, V., Saqe, A.: A financial systemic stress index for Albania, Bank of Albania Working Paper, Series 03 (42) (2013)
34. Laker, J.: Monitoring financial system stability, Reserve Bank of Australia Bulletin, October 1999
35. Louzis, D.P., Vouldis, A.T.: A methodology for constructing a financial systemic stress index: an application to Greece. Econ. Model. **29**(4), 1228–1241 (2012)
36. Mishkin, F.S.: Global financial instability: framework, events, issues. J. Econ. Perspect. **13**, 3–20 (1999)
37. Mörttinen, M.L., Poloni, P., Sandars, P., Vesala, J.M.: Analysing banking sector conditions: how to use macro-prudential indicators, ECB occasional paper, No. 26 (2005)
38. Nelson, D.B.: Conditional heteroskedasticity in asset returns: a new approach. Econometrica J. Econ. Soc. **59**, 347–370 (1991)
39. Nelson, W.R., Perli, R.: Selected indicators of financial stability. In: Risk Measurement and Systemic Risk, vol. 4, pp. 343–372 (2007)
40. Oet, M.V., Dooley, J.M., Ong, S.J.: The financial stress index: identification of systemic risk conditions. Risks **3**(3), 420–444 (2015)
41. Park, C.Y., Mercado, R.V.: Determinants of financial stress in emerging market economies. J. Bank. Finan. **45**, 199–224 (2014)
42. Patel, S.A., Sarkar, A.: Crises in developed and emerging stock markets. Finan. Anal. J. **54**(6), 50–61 (1998)

43. Schinasi, M.G.J.: Defining financial stability, IMF Working Papers, WP/04/187 (2004)
44. Van den End, J.W.: Indicator and boundaries of financial stability, DNB Working Paper, No. 97, March 2006
45. Yiu, M.S., Ho, W.Y.A., Jin, L.: A measure of financial stress in Hong Kong financial market? The financial stress index, Hong Kong Monetary Authority Research Note, 2 (2010)

Bank Competition and Financial Stability: Empirical Evidence in Vietnam

Tuyen L. Nguyen[1], Anh H. Le[2(✉)], and Dao M. Tran[1]

[1] Banking University of Ho Chi Minh City, 36 Ton That Dam,
Nguyen Thai Binh Ward, District 1, Ho Chi Minh City, Vietnam
nguyenluutuyen1983@gmail.com, tranminhdaosg@gmail.com
[2] HCMC University of Food Industry, 140 Le Trong Tan, Tay Thanh Ward,
Tan Phu District, Ho Chi Minh City, Vietnam
anhlh@cntp.edu.vn

Abstract. The relationship between competition and stability of the commercial banking system has been at the heart of scholarly and policy debates over the past two decades, especially since the 2008 financial crisis. In this study, we focus on analysing the relationship between competition and stability with 24 commercial banks in Vietnam for the period 2008–2016. Our study results show that increasing competition helps Vietnam's banking system become more stable. However, the relationship between competition and stability is nonlinear. The results of our study also show the impact of competition on the stability of Vietnam's banking system in a crisis situation. In the context of financial crisis, the instability and NPL of Vietnam commercial banks are increased. At the same time, increased competition can cause instability for the commercial banking system in Vietnam.

1 Introduction

In recent years there have been several debates regarding the relationship between competition and the stability of the banking system (Beck 2008; Carletti 2003). The debates on this relationship have formed two contradictory views: "competition-fragility" and "competition-stability". From the standpoint of "competition-fragility", increasing bank competition reduces the market power, bank's profit margins and consequently decreases the franchise value of the bank (Berger et al. 2009). This encourages banks to take more risks to seek profits, causing instability in the banking system (Marcus 1984; Keeley 1990; Carletti and Hartmann 2003).

In contrast, the "competition-stability" view holds that there is a positive relationship between banking competition and the stability of the banking system. Increased competition will lead to the stability of the banking system and vice versa ((Maggie) Fu et al. 2014). In a market where competition between banks is low, it can be more risky when large banks are often considered too important to fail and thus, when faced with difficulties in operating, those banks usually receive support from the government (Mishkin 1999). In addition, in a low-competition

© Springer International Publishing AG 2018
L. H. Anh et al. (eds.), *Econometrics for Financial Applications*, Studies in Computational Intelligence 760, https://doi.org/10.1007/978-3-319-73150-6_46

market, large market-power banks will offer higher lending rates, which will cause difficulties for borrowers in repayment capacity and increase the risk exposure of the bank ((Maggie) Fu et al. 2014). In contrast, in a market where competition among banks is high, lending rates are low; problems of "too big to fail" receive less attention, and therefore positively impact on the stability of the banking system (Boyd and De Nicolo 2005; Beck et al. 2006; Schaeck 2006; Turk-Ariss 2010).

Studies supporting these two points suggest that the effects of competition on the stability of the banking system are inconsistent across countries. In addition, very few studies examine the relationship between competition and stability before and after the financial crisis. To strengthen the theoretical foundation of the relationship between competition and stability, our study aims to assess the relationship between competition and stability of the commercial banking system in Vietnam. The study also looks at this relationship in the context of the financial crisis of 2008 and 2010.

This study is designed in 6 sections. Following the introduction of the research in Sects. 1 and 2 will present theoretical background and empirical evidences about the relationship between competition and stability of the banking system. Section 3 will show the research methodology and data. The empirical results and discussion are presented in Sect. 5. Based on the results of the study, Sect. 6 will draw conclusions and policy implications.

2   Theoretical Background and Related Studies

2.1   Theoretical Background of Competition and Stability of the Banking System

From the traditional viewpoint of a "competition-fragility" relationship, a more competition or less centralized banking system increase instability. This is explained by the franchise value theory studied by Marcus (1984) and Keeley (1990), suggesting that competition motivates banks to pursue more risky strategies. These studies show that less competition or a more exclusive monopoly of some banks will lead to higher franchise value of these banks, and may prevent excessive risky decisions of the bank's executives. Because when the Franchise value is higher, the opportunity cost of bankruptcy is higher, leading to the reluctance of bank executives and bank shareholders to participate in dangerous decisions, thereby improving the quality of bank assets.

Boot and Greenbaum (1993) and Allen and Gale (2000, 2004) show that in a competition environment, banks receive less information from their relationships with borrowers, making it difficult to check credit records and increase the risk and instability.

Boyd et al. (2004) argue that banks with a higher level of presence or higher monopolies in a centralized banking system can increase profits and thereby reduce financial breach ability by providing "Buffer Capital" to protect the system against macroeconomic shocks and external liquidity problem.

From a "competition-stability" standpoint, a more competitive or less monopoly banking system will be more stable, in other words, less competitive or more monopoly banking system will be more unstable. This can be explained by the "too big to fail" theory proposed by Mishkin (1999) indicating that policymakers will be more concerned about the collapse of the bank when there are so few banks in the banking system. Thus, large banks are more likely to receive government guarantees or grants, which can create moral hazard problems, encourage dangerous decisions and increase instability of the banking system. Moreover, the spreading risk may increase in the centralized banking system with large banks.

Caminal and Matutes (2002) argue that less competition can lead to easier credit granting and larger loans, which increases the probability of bank collapse. Boyd and De Nicolo (2005) argue that high monopoly banking systems allow banks to charge higher interest rates, and may encourage borrowers to take greater risks. Therefore, the amount of non-performing loans can increase, resulting in higher probability of bank's bankruptcy. However, Martinez-Miera and Repullo (2010) suggests that higher lending rates also bring higher interest income to banks. This offset effect can create a U-shaped relationship between bank competition and stability.

2.2   Empirical Evidence of Competition and Stability of the Banking System

There is considerable debate relating to the impact of bank competition on the stability from the literature, especially in the context of financial crisis. As observed from the recent financial crisis, instability can be spread widely to the entire economy through the banking system. Vulnerabilities are mainly due to the collapse of the interbank lending and payment markets, the reduction in credit supply, and the freezing of deposits (Berger et al. 2008). There have been a number of studies showing that the greater the bank competition is, the more likely it is that financial instability will be triggered by a decline in market power, which in turn will reduce profits and lower franchise value. These studies support the "competition-fragility" hypothesis. From this view, banks are encouraged to take more risks to increase profitability and deteriorate the quality of their loan portfolio (Marcus 1984; Keeley 1990 and Carletti and Hartmaan 2003). There have been various empirical studies supporting this relationship, such as Keeley (1990), finding that increased bank competition and the deregulation in the US during the 1990s reduced the monopoly and contributed to bank failures. Hellmann et al. (2000) concluded that removing the interest rate ceiling, and thereby creating higher price competition, reduced franchise value and encouraged more moral hazard behavior in banks. Jimenez et al. (2007) study the banking sector in Spain and show that the greater the bank competition is, the higher the risk of the loan portfolio (higher non performing loans). Berger et al. (2008) study 23 developed countries and came to the conclusion in favor of the "competition-fragility" view, in which the higher the market power or lower competition would reduce the risk of the bank. However, this study also shows that higher market power increases the risk of loan portfolio, which can be interpreted as evidence of a "competition-stability" view. Vives (2010) evaluate theories and empirical studies on the "competition-stability" relationship and argue that although competition is not a determinant of instability but could aggravate more instability issues.

However, many recent studies have advocated a "competition-stability" view. Beck et al. (2006) study a group of 69 countries and find that countries with low market concentration or high competition are less likely to suffer financial crisis. Boyd and De Nicolo (2005) argue that the greater market power or less competition in lending markets increases the risk for banks because higher interest rates make it more difficult for customers to repay. This can exacerbate moral risk and at the same time, higher interest rates will attract higher risk borrowers. Moreover, in highly centralized monopolies, financial institutions may believe that they are "too big to fail" and this can lead to riskier investments (Berger et al. 2008). There are some recent empirical studies supporting this hypothesis. Studies of Boyd et al. (2006), De Nicolo and Loukoianova (2006) find the opposite relationship between high levels of market monopoly concentration or low competition and stability of the banking system. These studies suggest that the risk of bank failures increases in higher monopoly markets. Financial stability is estimated using the Z-score and the level of market monopoly concentration is measured by the Herfindahl-Hirschman index (HHI). Schaeck et al. (2006) study the banking sector of a group of countries by applying logit model and time analysis. Furthermore, the Rosse-Panzar index (H-Statistics) is used to measure the level of competition. The key finding of these studies show that the higher level of bank competition, the lower likelihood of banks failure or being more stable than the monopoly banking system.

Other recent studies adopt the Lerner Index and measure bank stability through the Z-score to examine the relationship between competition and stability of the banking system. Berger et al. (2008) study the sampling of more than 8,000 banks in 23 countries using the Generalized Methods of Moments (GMM) data table. The main results of this study indicate that banks with higher market power or less competition have less overall risk, favoring "competition-fragility" views. On the other hand, this study also find evidence of the positive relationship between competition and stability, implying that higher market power increases the credit risk. Turk-Ariss (2010) studies the impact of market power on banking efficiency and financial stability in the banking sector of a group of emerging economies. This study applies three different techniques of Lerner's competitive index and uses Z-score to represent the stability of the banking system. Research results indicate that increasing market power leads to more bank stability, despite significant losses in cost-effectiveness. Liu et al. (2010) analyzes competitive conditions in 11 EU countries for the period 2000–2008 to examine "competition-stability" relationships in the banking sector. This study uses Lerner index and Z-score to respectively represent the level of bank competition and bank stability. The results show the non-linear relationship between competition and stability in the banking sector in Europe. More specifically, they see the shift of risk in highly concentrated markets, where the increase in bank competition reduces net interest margins (higher deposit rates and lower lending rates) and increase bank stability. However, they also recognize that the marginal effect exists in highly competition markets, where increased competition reduces interest payments and bad debt provisions.

3   Research Methodology and Data

In this study, we seek evidence of the impact of competition on the stability of the Vietnam's commercial banks under normal and crisis conditions. The study uses data from 24 commercial banks in Vietnam for the period 2008–2016. To overcome the potential endogenous problem in the model, we use the GMM estimation technique.

3.1   Measure the Stability of Commercial Banks in Vietnam

There have been many studies that developed the methods of measuring commercial bank stability, most of which use Z-scores. In this study, we follow the studies of Boyd and Graham (1986), Hannan and Hanweck (1988), and Boyd et al. (1993) to use Z-scores, which are calculated as follows:

$$Zscore_{it} = \frac{ROA_{it} - EQTA_{it}}{ROA_{ip}}.$$

Where: $Zscore_{it}$ is the Z-score measures the bank i's financial stability in year t.

ROA_{it} is the return on total assets of bank i in year t, calculated as the after-tax profit divided by total assets.

$EQTA_{it}$ is the ratio of equity to total assets of bank i in year t, calculated by the average equity divided by total assets.

ROA_{ip} is the standard deviation of the bank's ROA in the study period p.

According to the above formula, the lower the Z-score, the lower the financial stability of the bank. In contrast, the higher the Z-score, the higher the financial stability of the bank.

3.2   Measure the Level of Competition of Commercial Banks

To proxy the degree of competition of commercial banks, we use the Lerner index which has been used by Berger et al. (2008), Fernández de Guevara et al. (2005), Maudos and Solis (2009), (Maggie) Fu et al. (2014). The Lerner index for banks is calculated as follows:

$$Lerner = \frac{P_{it} - MC_{it}}{P_{it}}.$$

where P_{it} is the output price of bank i in year t, calculated by the ratio of total income to total assets. MC_{it} is the marginal cost of bank i in year t. However, marginal cost can not be observed directly, so it is estimated based on the function of the total bank cost (Ariss 2010; Fernández de Guevara et al. 2005; (Maggie) Fu et al. 2013). The total cost function is as follows:

$$LnTC = \alpha_0 + \sum_{j=1}^{3} \alpha_j \times w_{it}^j + \frac{1}{2}\sum_{j=1}^{3}\sum_{k=1}^{3} \alpha_{jk} \ln w_{it}^j \ln w_{it}^k + \beta_1 \ln Y_{it} + \frac{1}{2}\beta_2(\ln Y_{it})^2$$

$$+ \sum_{j=1}^{3} \beta_j' \ln Y_{it} \ln w_{it}^j + \varphi_{1t} \times T + \frac{1}{2}\varphi_{2t}T^2 + \sum_{j=1}^{3} \varphi_{3t}T \ln w_{it}^j + \varphi_{4t}T \ln Y_{it} + \mu_t + \varepsilon_{it}$$

where TC is the total cost, w is the price of the three inputs (personnel expenses/total assets, interest expenses/total deposits, and other operating expenses/fixed assets), Y is total asset, T is the time trend reflecting the effect of technical progress, μ captures the individual fixed effects, and ϵ is the error term.

Total bank cost functions are estimated using fixed effects with robust standard error. After estimating the total cost TC, the MC marginal cost is determined by taking the first derivative of the total cost function, as follow:

$$MC = \frac{TC}{Y}\left(\beta_1 + \beta_2 \ln Y_{it} + \sum_{j=1}^{3} \beta_j' \ln w_{it}^j + \varphi_{4t}T\right).$$

3.3   Empirical Model

To search for empirical evidence for "competition-stability" and "competition-fragility" views, we use dynamic model proposed by Fernández and Garza-García (2015). According to Gambacorta (2005) and Gunji and Yuan (2010), a dynamic model is also constructed with the latency of the dependent variable as a independent variable. Because the sample size is relatively small, only the first latency is considered, so the research model is as follows (Table 1):

$$Ln(Z_{it}) = \alpha_{it} + \delta ln(Z_{it-1}) + \beta_1 lerner_{it} + \beta_2(lerner_{it})^2$$
$$+ \beta_3 banksize_{it} + \beta_4 loanta_{it} + \beta_5 own_{it} + \varepsilon_{it} \tag{1}$$

$$NPL_{it} = \alpha_{it} + \delta NPL_{it-1} + \beta_1 lerner_{it} + \beta_2(lerner_{it})^2$$
$$+ \beta_3 banksize_{it} + \beta_4 loanta_{it} + \beta_5 own_{it} + \varepsilon_{it} \tag{2}$$

To proxy risk, we use 2 variables Z-score and NPL. In these two models, $ln(Z_{it})$ is the natural logarithm of Z-score, which measures the stability of commercial banks. The degree of fragility of commercial banks is measured by Non performing loans in terms of total loans (NPLs) (Fernández and Garza-García 2015). The level of competition of commercial banks is measured by the Lerner index. According to Liu et al. (2010), there is a non-linear relationship between competition and stability existing in banking system. So we use $lerner^2$ as the squared measure of the Lerner index to test this non-linear relationship. Also, according to studies by Schaeck and Cihak (2008), Laeven and Levine (2009) and Uhde and Heimeshoff (2009),

Table 1. Variables definition & data source

Variables	Measures	Data sources
Dependent variables		
lnZ	ln $\frac{ROA_{it} - EQTA_{it}}{ROA_{ip}}$	Financial statements of 24 commercial banks of Vietnam
NPL	(Non performing loans)/(Total loans)	Financial statements of 24 commercial banks of Vietnam
Independent variables		
Lerner	$Lerner = \frac{P_{it} - MC_{it}}{P_{it}}$	Financial statements of 24 commercial banks of Vietnamfirst derivative of the total cost function
banksize	Ln(total assets)	Financial statements of 24 commercial banks of Vietnam
loanta	(Total loans)/(Total assets)	Financial statements of 24 commercial banks of Vietnam
own	Dummy variable with value1 reflecting foreign ownership; and value 0 reflecting non-foreign ownership in bank capital	Financial statements of 24 commercial banks of Vietnam
crisis	Dummy variable with value 1 in crisis period of 2008, 2010 and value 0 in the remaining years	

Source: author's summary.

Table 2. Descriptive statistics of sample.

Variable	Obs	Mean	Std. dev.	Min	Max
Z-score	216	24.69598	12.18316	1.32173	62.19548
banksize	216	17.97906	1.256402	14.69872	20.72988
Loanta	216	.5113043	.1564076	.0046616	.8516832
NPL	216	.0232807	.0153248	0.001	.1260667
Lerner	216	.2957634	.0849353	.0214135	.6085381

Source: Calculating result from Stata 12.0.

we also include a series of characteristic variables for each bank: Banksize is defined as the logarithm of total assets, the ratio of total loans to total assets (loanta), and the own dummy variable reflecting foreign ownership in bank capital. In addition, in order to find evidence of the impact of competition on the stability of commercial banks under crisis conditions, we add the crisis dummy variable representing period of financial crisis. The dummy variable value is 1 in 2008, 2010 and is set to 0 in the remaining years. Specific models are as follows:

$$Ln(Z_{it}) = \alpha_{it} + \delta ln(Z_{it-1}) + \beta_1 lerner_{it} + \beta_2 (lerner_{it})^2$$
$$+ \beta_3 crisis_t + \beta_4 banksize_{it} + \beta_5 loanta_{it} + \beta_6 own_{it} + \varepsilon_{it} \qquad (3)$$

$$NPL_{it} = \alpha_{it} + \delta NPL_{it-1} + \beta_1 lerner_{it} + \beta_2(lerner_{it})^2$$
$$+ \beta_3 crisis_t + \beta_4 banksize_{it} + \beta_5 loanta_{it} + \beta_6 own_{it} + \varepsilon_{it} \qquad (4)$$

The dual impact of competition in crisis conditions on the stability of commercial banks is assessed by the lerner x crisis, as follows:

$$Ln(Z_{it}) = \alpha_{it} + \delta ln(Z_{it-1}) + \beta_1 lerner_{it} + \beta_2 lerner_{it} \times crisis_t + \beta_3 banksize_{it}$$
$$+ \beta_4 loanta_{it} + \beta_5 own_{it} + \varepsilon_{it}. \qquad (5)$$

$$NPL_{it} = \alpha_{it} + \delta NPL_{it-1} + \beta_1 lerner_{it} + \beta_2 lerner_{it} \times crisis_t + \beta_3 banksize_{it}$$
$$+ \beta_4 loanta_{it} + \beta_5 own_{it} + \varepsilon_{it}. \qquad (6)$$

This study uses the Difference GMM (DGMM) method of Arellano and Bond (1991). This method is commonly used in dynamic panel data estimation. This estimator is designed for situations with "small T, large N" panels, meaning few time periods and many individuals; with independent variables that are not strictly exogenous, meaning correlated with past and possibly current realizations of the error; with fixed effects; and with heteroskedasticity and auto-correlation within individuals. In these cases, the classical linear estimates of panel data such as FE (fixed effects), RE (random effects), LSDV (least squares dummy variable) are no longer effective and reliable. The DGMM method is appropriate for this study because the panel data has small T (8 years), large N (25 banks), which means few time periods and many individuals.

4   Research Data

This study uses a sample of 24 commercial banks in Vietnam for the period from 2008 to 2016. This is a balanced panel, consisting of 216 observations. The data is derived from the annual financial statements of commercial banks. Information needed for research collected from audited financial statements, annual reports and public disclosures of commercial banks. The descriptive statistics for the variables used in this study are shown in Table 2.

Table 2 shows that the sample of banks studied had an average Z-score of 24.696. Of which the most stable bank has a relatively high Z-score of 62.195. The bank with the lowest stability has Z-score of 1.322. Banks in the sample had an average NPL ratio of 2.33% at safe levels compared to the regulations of the State Bank of Vietnam. The Lerner index of banks in the sample has an average value of 0.296 indicating that the degree of monopoly in the banking market in Vietnam is relatively low.

5   Empirical Results

Table 3 presents the empirical results using the DGMM estimation. Two dependent variables Z-score and NPL representing stability and degree of fragility are used in the model to assess the impact of competition on the stability of Vietnam commercial banks. For each dependent variable, we estimate three models described in Sect. 3.3. Estimates for each dependent variable are conducted in the following order: (i) consider the impact of competition on the stability of the commercial banks in Vietnam; (ii) consider the impact of the financial crisis on the stability of the Vietnam banking system; (iii) consider the dual effects of competition in crisis conditions on the stability of the Vietnamese banking system.

The following model reliability tests have been performed:

Verification of self-correlation of the residual: According to Arellano and Bond (1991), the GMM estimate requires a first degree correlation and no second correlation of the residual. Thus, when testing the hypothesis H_0: there is no first ordered correlation (test AR (1)) and no second ordered correlation of residual (test AR (2)), we reject H_0 at the AR test (1) and acceptance of H_0 at the AR test (2), the model satisfies the requirements.

Similar to other models, F test will test statistical significance for the coefficient of estimation of the explanatory variable with the hypothesis H_0: that all coefficients in the equation are equal to 0, therefore, in order for the model to conform, we must reject the hypothesis H_0. In addition, the Sargan/Hansen test is also used to test the validity of instrument variable based on the hypothesis H_0: the instrumental variables are consistent.

Estimated results in Table 3 show that all six models have a p-value of AR (1) test less than 5% and have a p-value of AR (2) test greater than meaning level of 5%. Therefore, the model has a first ordered correlation but no second ordered correlation of the residual. At the same time, the Hansen test in all six models has a p-value greater than 5%, meaning that the instrument variables used in the model are appropriate. On the other hand, the p-value of the F test is also less than the 5% significance level, indicating that the estimated modelfit the panel data reasonably well. With Z-score variable, the results in Table 3 show that the regression coefficients of lerner and $lerner^2$ in (1) and (3) are statistically significant at 1%. In addition, the regression coefficient of lerner is positive while the regression coefficient of the $lerner^2$ is negative, indicating that there exists an inverse U-shape nonlinear relationship between the two variables lerner and Z-score. Specifically, the higher the lerner index the higher Z-score. In other word, increased banking competition results in greater stability of Vietnam banking system. This result is consistent with studies by Boyd et al. (2006), De Nicolo and Loukoianova (2006) supporting for a "competition-stability" hypothesis. However, when the lerner index exceeds a certain limit, the Z-score will decrease which means increase the instability of Vietnam banking system. This inverse U-shaped nonlinear relationship has also been shown in studies by Ariss (2010), Liu et al. (2010).

The results of the model (3) also show evidence of financial crisis impact on the stability of the commercial banking system in Vietnam. Specifically, the regression coefficient of the crisis variable is statistically significant at 1% and negative. This shows that when the financial crisis occurs, there will be a negative

Table 3. Estimation result

	Dependant variable: Z-score			Dependant variable: NPL		
	(1)	(3)	(5)	(2)	(4)	(6)
lnZ_{it-1}	−0.0899	0.0528	−0.0647			
NPL_{it-1}				0.4014***	0.1637***	0.2193***
$lerner_{it}$	4.4571***	3.6892***	0.7913*	0.5479***	0.2452***	−0.1976
$lerner_{it}2$	−5.0216***	−4.8862***		−0.9924***	−0.6843***	
$banksize_{it}$	−0.2870***	−0.2706***	−0.3042***	−0.0117***	−0.0210***	−0.0229***
$loanta_{it}$	−0.1392**	−0.1177	−0.1538	−0.0023	−0.0309**	−0.0154
own_{it}	0.2601**	0.0189	0.0627	0.0125	0.0039	−0.0030
$crisis_t$		−0.0348***			0.0051***	
$lerner_it \times crisis_t$			−0.0677**			0.0156***
AR (1) p-value	0.036	0.041	0.005	0.028	0.023	0.014
AR (1) p-value	0.932	0.424	0.424	0.447	0.783	0.766
Hansen p-value	0.198	0.143	0.178	0.183	0.193	0.245
Number of groups	24	24	24	24	24	24
Number of instruments	22	23	22	22	21	20
F-test p-value	0.000	0.000	0.000	0.000	0.000	0.000

Source: Calculating result from Stata 12.0.

The estimation results of models examining the effects of competition on the stability of the banking system in Vietnam are made using the DGMM method. Z-score variable represents the stability used inmedel (1), (3), (5). NPL dependent variable represents the degree of fragility used in (2), (4), (6). Lerner variable represents the level of competition. Other independent variables in the model are the banksize, total loans over total asset ratio (loanta), the dummy variable representing the foreign ownership (own), and the dummy variable representing the Financial crisis period of 2008, 2010 (crisis). AR (1), AR (2) p-value is the p-value of the first and second correlation tests of the residual. Hansen p-value is the p-value value of the Hansen test for the suitability of tool variables in the model. The second stage F-test p-value is the p-value of the F test for conformance of the model.

*** indicate statistically significance at 1%; ** indicate statistically significance at 5%; * indicate statistically significance at 10%.

impact on the stability of Vietnam commercial banks. In addition, the results of model (5) also show evidence of the dual effects of competition in the context of the financial crisis on the stability of the banking system. The regression coefficient of the lernerxcrisis variable was statistically significant at 5% and negative. This indicates that when the financial crisis occurs, competition will have a negative impact on the stability of Vietnam commercial banks.

With the NPL dependent variable, the results in Table 3 show that the regression coefficients of the variables lerner and $lerner^2$ in (2) and (4) are statistically significant at 1%. In addition, the regression coefficient of lerner is positive while the regression coefficient of $lerner^2$ is negative, indicating that there is an inverse U-shaped nonlinear relationship between the lerner variables and the NPL. This is consistent with the results from Z-score model. In particular, as the lerner index increases the NPLs increased, i.e. increased bank competition results in increased fragility of the banking system in Vietnam. Under competition pressure, commercial banks may loosen their lending conditions and lead to an increase in NPL. This result is consistent with studies by Marcus (1984), Keeley (1990), Carletti and Hartmaan (2003), Berger et al. (2008) supporting a "competition-fragility" hypothesis. The results of the model (4) also show evidence of financial crisis impact on the fragility of the commercial banking system in Vietnam. In particular, the regression coefficient of the crisis variable is statistically significant at 1%

and has a positive value. This shows that when the financial crisis occurs, there will be a positive impact on the non-performing loans of Vietnam commercial banks. In addition, the results of the model (6) also show evidence of the dual effects of competition in the context of the financial crisis on the fragility of the banking system. The regression coefficient of the lernerxcrisis variable is statistically significant at 1% and is positive. This shows that when the financial crisis occurs, competition will have a negative impact on the non performing loans of Vietnam commercial banks.

In all 3 models (2), (4) and (6), the ratio of non performing loans over total loans of the previous period also had an impact on that of the current period. Specifically, the regression coefficient of the $NPL_{(t-1)}$ delayed variable was statistically significant at 1% and positive.

6   Conclusions and Policy Implications

6.1   Conclusions

Competition is an important factor not only for non-financial businesses, but also for commercial banks especially in the context of higher challenges and risks in banking and financial industry of Vietnam. In this study, we examine the impact of bank competition on the bank stability in order to test "competition-stability" and "competition-fragility" views for the commercial banking system of Vietnam. The Lerner Index is used to measure the level of competition among banks, while the degree of bank stability and fragility are measured through the Z-score and ratio of non performing loans over total outstanding loans. The results from DGMM estimation support the "competition-stability" hypothesis. In particular, increased competition results in Vietnam's banking system to be more stable. However, if the level of competition exceeds a certain threshold, it will cause instability. This implies that the relationship between competition and stability of the commercial banking system in Vietnam is an inversed U-shaped nonlinear relationship. In addition, the results of our study also show that as competition intensifies, NPL ratio increased as banks under competitive pressure may be forced to loosen their lending conditions. This result partly shows that the "competition-fragility" view can happen in medium and long term.

In addition, the results of our study also show that in the context of financial crisis, the instability and NPL of Vietnam commercial banks are also increased. At the same time, increased competition can cause instability for the commercial banking system in Vietnam.

6.2   Policy Implications

Based on the results of our study, we draw some policy implications as follow:

Firstly, in managing bank's operations, the executive officers need to well manage the bank's operating expenses and incomes. This will help banks to improve their competitiveness thus contributing to the stability of the bank.

Second, under increasing and intensive competition in Vietnam's banking market, bank executives need to have action plans to improve the quality of their products and services, to take advantages of modern technology to maximize the satisfaction of customers needs. Banks should be alert and avoid loosening their lending conditions to compete for market share and customers.

Third, the Government and the State Bank of Vietnam should take actions to encourage and promote healthy and transparent competition in the banking system.

Fourth, in parallel with the creation of an environment that encourages healthy competition, the Government and the State Bank of Vietnam need to increase the inspection and supervision of credit quality at commercial banks to limit excessive risk exposure.

References

Allen, F., Gale, D.: Financial contagion. J. Polit. Econ. **108**, 1–33 (2000)

Allen, F., Gale, D.: Competition and financial stability. J. Money Credit Bank. **36**, 453–480 (2004)

Arellano, M., Bond, S.: Some tests of specification for panel data: Monte Carlo evidence and an application to employment equations. Rev. Econ. Stud. **58**(2), 277–297 (1991)

Beck, T.: Bank Competition and Financial Stability: Friends or Foes? Policy Research Working Paper No. 4656, World Bank (2008)

Beck, T., Demirguc-Kunt, A., Levine, R.: Bank concentration, competition, and crises: first results. J. Bank. Financ. **30**, 1581–1603 (2006)

Berger, A., Klapper, L., Turk-Ariss, R.: Bank competition and financial stability. J. Financ. Serv. Res. **35**, 99–118 (2009)

Berger, A., Klapper, L., Turk-Ariss, R.: Bank Competition and Financial Stability, World Bank Policy Research Working Paper 4696 (2008)

Boot, A., Greenbaum, S.: Bank regulation, reputation and rents: theory and policy implications. In: Mayer, C., Vives, X. (eds.) Capital Markets and Financial Intermediation, pp. 262–285. Cambridge University Press, Cambridge (1993)

Boyd, J.H., De Nicolo, G., Smith, B.D.: Crises in competitive versus monopolistic banking systems. J. Money Credit Bank. **36**, 487–506 (2004)

Boyd, J.H., De Nicolo, G.: The theory of bank risk-taking and competition revisited. J. Financ. **60**, 1329–1343 (2005)

Boyd, J.H., De Nicolo, G., Jalal, A.M.: Bank risk taking and competition revisited: New theory and evidence. IMF working paper, WP/06/297 (2006)

Boyd, J.H., Graham, S.L.: Risk, regulation, and bank holding company expansion into nonbanking. Res. Depart. Fed. Reserv. Bank Minneap. **10**(2), 2–17 (1986)

Boyd, J.H., Graham, S.L., Hewitt, R.S.: Bank holding company mergers with nonbank financial firms: effects on the risk of failure. J. Bank. Financ. **17**(1), 43–63 (1993)

Caminal, R., Matutes, C.: Market power and bank failures. Int. J. Ind. Organ. **20**, 1341–1361 (2002)

Carletti, E., Hartmann, P.: Competition and financial stability: what's special about banking? In: Mizen, P. (ed.) Monetary History, Exchange Rates and Financial Markets: Essays in Honour of Charles Goodhart, vol. 2. Edward Elgar, Cheltenham (2003)

De Nicolo, G., Loukoianova, E.: Bank Ownership, Market Structure, and Risk. IMF Working paper, WP/07/215 (2006)

Carletti, E.: Competition and regulation in banking. In: Boot, A.W.A., Thakor, A. (eds.) Handbook of Financial Intermediation and Banking. Elsevier, Amsterdam (2008)

Fernández de Guevara, J., Maudos, J., Perez, F.: Market power in European banking sectors. J. Financ. Serv. Res. **27**, 109–137 (2005)

(Maggie) Fu, X., (Rebecca) Lin, Y., Molyneux, P.: Bank competition and financial stability in Asia Pacific. J. Bank. Finan. **38**(1), 64–77. https://doi.org/10.1016/j.jbankfin.2013.09.012 (2013)

(Maggie) Fu, X., (Rebecca) Lin, Y., Molyneux, P.: Bank competition and financial stability in Asia Pacific. J. Bank. Financ. **38**(issue C), 64–77 (2014)

Gambacorta, L.: Inside the bank lending channel. Eur. Econ. Rev. **49**(7), 1737–1759 (2005)

Gunji, H., Yuan, Y.: Bank profitability and the bank lending channel: evidence from China. J. Asian Econ. **21**(2), 129–144 (2010)

Hannan, T.H., Hanweck, G.A.: Bank insolvency risk and the market for large certificates of deposit. J. Money Credit Bank. **20**(2), 203–211 (1988)

Hellmann, T., Murdock, K., Stiglitz, J.: Liberalization, moral hazard in banking, and prudential regulation: are capital requirements enough? Am. Econ. Rev. **90**, 147–165 (2000)

Jimenez, G., Lopez, J., Saurina, J.: How does Competition Impact Bank Risk Taking? Banco de Espana Working Papers 1005 (2007)

Keeley, M.: Deposit insurance, risk, and market power in banking. Am. Econ. Rev. **80**, 1183–1200 (1990)

Laeven, L., Levine, R.: Bank governance, regulation and risk taking. J. Finan. Econ. **93**, 259–275 (2009)

Liu, H., Molyneux, P., Wilson, J.: Competition and Stability in European Banking: A Regional Analysis. Working Paper No. BBSWP/10/019. School of Management, University of St. Andrews (2010)

Marcus, A.J.: Deregulation and bank financial policy. J. Bank. Financ. **8**, 557–565 (1984)

Martinez-Miera, D., Repullo, R.: Does competition reduce the risk of bank failure? Rev. Financ. Stud. **23**, 3638–3664 (2010)

Maudos, J., Solis, L.: The determinants of net interest income in the Mexican banking system: an integrated model. J. Bank. Financ. **33**, 1920–1931 (2009)

Mishkin, F.S.: Financial consolidation: dangers and opportunities. J. Bank. Financ. **23**, 675–691 (1999)

Fernández, R.O., Garza-García, J.G.: The Relationship between Bank Competition and Financial Stability: A Case Study of the Mexican Banking Industry. Working Paper 03.12 (2015)

Schaeck, K., Cihak, M., Wolfe, S.: Are More Competitive Banking Systems More Stable? IMF Working Paper WP/06/143 (2006)

Schaeck, K., Cihak, M.: How does competition affect efficiency and soundness in banking? New empirical evidence. Working Paper No. 932, European Central Bank (2008)

Turk-Ariss, R.: On the implications of market power in banking: evidence from developing countries. J. Bank. Financ. **34**(4), 765–775 (2010)

Vives, X.: Competition and Stability in Banking. IESE Working Paper 852 (2010)

Uhde, A., Heimeshoff, U.: Consolidation in banking and financial stability in Europe: empirical evidence. J. Bank. Finan. **33**(7), 1299–1311 (2009)

Analysing the Effects of the Exporting on Economic Growth in Vietnam

Nguyen Minh Hai[1]([✉]), Do Huu Hai[2], and Nguyen Manh Hung[3]

[1] Faculty of Mathematical Economics, Banking University of Ho Chi Minh City,
Ho Chi Minh City, Vietnam
minhhai.nguyen77@gmail.com
[2] Ho Chi Minh City University of Food Industry, Ho Chi Minh City, Vietnam
[3] Banking University of Ho Chi Minh City, Ho Chi Minh City, Vietnam

Abstract. The main objective of this study is to verify the relation between exports and economic growth. At the same time, considering whether export-led economic growth is really the right choice of the Vietnamese economy. The results show that exporting activities of Vietnam in the period 2000–2016 have a positive impact on the ability to maintain the economic growth in the long run. This has supported Vietnam's export-oriented economic growth in the past time as an appropriate option. Based on the analysis results, the paper proposes some recommendations to promote the positive effects of exports towards the sustainable economic growth.

Keywords: The structure of exporting · Economic growth · Vietnam

JEL classification: C51 · E41

1 Introduction

The export-oriented economic growth has been considerably highlighted in many studies in Vietnam. Most studies have mainly emphasized on the improvement of policies to promote the export rather than effects of the export on the economic growth. The export-led growth is currently an economic strategy used by some developing countries to achieve the rapid, stable and sustainable economic growth. Korea and Thailand are two of typical successful countries in the export-driven strategy for economic growth. It can be seen that the export-led growth contributes to make Korea become a powerful economy from the backwardness and increase Thailands economic growth rate in many years and make Thailand become one of the highlights of Asia. However, many countries such as South Asian and Latin America countries fail to make advantages of the export for the economic growth. The target of the socio-economic development strategy up to 2020 is "To encourage exports to promote industrialization and modernization, create jobs, to shift the export structure towards value-added products, increase to export high-tech and intellectual products, and to expand and diversify the

© Springer International Publishing AG 2018
L. H. Anh et al. (eds.), *Econometrics for Financial Applications*, Studies in Computational Intelligence 760, https://doi.org/10.1007/978-3-319-73150-6_47

market and business methods". As a result, it is necessary to answer whether export-led economic growth is really the right choice of the Vietnamese economy. This paper has focused on effects of the export on the economic growth in Vietnam in recent years and has been expected to be an useful reference for orienting the future sustainable development policies.

2   Overview and Analytical Framework

2.1   Overview

The classical theory of international trade, modern theory of trade, new theory of growth and theoretical researches with demand-side approaches such as ones of Balassa (1985), de Pineres et al. (1997), Awokuse (2008) argued that export enhancement would provide foreign exchange to fund import of intermediate goods and increase capital formation, which contributes actively to expand domestic production and promote growth. On the contrary to the demand-side approaches, other researches with supply-side approaches argued that export just only creates "level effects". It means that it just creates effects to change the pace of economic growth and its long-term trend. With factor surplus and accumulated technology arguments, endogenous growth theory argued that export has an effect on economic growth in not only short term and medium term but also long term, of which emphasizing on the roles of productivity, improvement and labour collectively called Total Factor Productivity (TFP) as well as the roles of shifting production structure to sustainable growth.

Along with theoretical researches, experiments on the impact of exports on economic growth have also been conducted. Significantly, the study of Michalopoulos and Jay (1973), using cross data involving 39 developing countries in Organization for Economic Co-operation and Development (OECD) in the period 1960–1973, indicated that rapid growth in export sector had an effect on bettering general economic growth because it originated from increased specialization and competition to the extend of likely exploit economics of scale form a huge market. Also using this method, with different samples, Balassa (1985), Tyler (1981) and Kavoussi (1984) all showed results supporting the trend of export-led economic growth. They affirmed that the countries prioritizing export often achieve higher economic growth than other countries. However, those results didn't reflect the characteristics of many developing countries because technology level of each country in model is deemed to be the same and there is no difference in structure of each economy.

To fix this down-side, experts researched time series data and more realistic results are shown in researches of Erfani (1999) for Asia developing countries and Latin America in the period of 1965–1995, Mayer and Wood (2001) for Malaixia in the period of 1959–2000, Kaushik and Klein (2008) for India in the period of 1971–2005. However, all studies using time series data and cross data indicated that export enhancement would affect growth in short term and long term. It resulted from positive impact of export on Total Factor Productivity (TFP).

However, no small number of other researches indicated that exports did not have impact or negligible or even negative impact on economic growth. Study of Jansen (2004) when applying VAR model to OECD countries is typical evidence of unclear impact of exports on economic growth. The study of Reppas and Christopoulos (2005) for 22 underdeveloped countries in Asia and Africa indicated that export enhancement had a negative impact on economic growth because it led to the over-investment in a certain number of export-oriented industries. In this context, countries would be trapped in production of goods while its benefits were gradually depleted.

Besides, studies on the effects of export commodity on economic growth are also considered. Significant studies such as Erfani (1999), Levin (1997) for more than 30 developing countries argued that the countries export-intensive countries were always more profitable than countries exporting raw and semi-processed commodities. Affirmatively, export of processed goods strongly influenced growth while export of semi-processed goods had a small effect on short-term growth and negative impact on long-term. Accordingly, studies on the impact of specialization and diversification of exports on growth were also undertaken. Love (1986), Jansen (2004) for a huge number of countries confirmed that the more diversified export basket of the countries was, the likely higher countries economic growth was in the long term. Other related publications include Anh (2008), Lee and Huang (1985), Keong et al. (2001), Konya (2004), Lim and Saborowski (2011), McCombie (1998), Taylor and Francis (2003), Thirlwall (2000) and Thuy (2014).

In conclusion, different results of studies on efforts of export on economic growth interpreted in various ways depending on the specific factors of the nations, in each period of development and accuracy of the data. Therefore, to exploit the essence of export-growth relation, we need to deeply research on micro-structure platform of each country. It required a clear separation of effects of exported goods on economic growth. This study will focus on the impact of exports on Vietnam's economic growth in the period of 2000–2016.

2.2   Analytical Framework

Recent studies showed that, for example Bahamni-Oskooee et al. (1991), Sharma et al. (1991), the relationship between export and economic growth was measured by common measures such as productivity (GDP), resources (K), export (E) and labour (L). Therefore, the use of the growth rate of G_GDP representing the quantity growth of GDP, growth rate of labour (G_L) representing labour (L), physical resources (K) measured by the ratio of investment I GDP and export (E) measured by the ratio of exportGDP is a common choice to calculate easily based on data of General Statistics Office of Vietnam. According to Mayer and Wood (2001), the ratio of exportGDP analyzed into 3 main components:

$$E = E1 + E2 + E3 \qquad (1)$$

Of which $E1$-the proportion of raw and semi-processed exports, $E2$-the proportion of labor-intensive processed export/GDP, $E3$-the proportion of skill-intensive processed export/GDP.

In addition, the quality of the export basket is assessed on two criteria: the degree of specialization and stability of the basket.

$$HI_t = \sum_{i=1}^{n} P_{it}^2 \qquad (2)$$

Of which P_{it} is the proportion of commodity groups in total commodity structure, H_{it} belonging to $(0, 1)$. The closer to 1 the H_{it} is, the higher the degree of specialization is and the lower the level of diversification is.

To assess the level of diversification of exports, the TE variable based on the Theil Entropy index proposed by Ibrahim and Amin (2003) is also included. The Ibrahim and Amin (2003) showed that the degree of export diversification was influenced by two different levels: level of diversity in width (TW-by sector) and level of diversification in depth (TD-internal group). Thus, the TE variable is rewritten as follows:

$$TE = TW + TD \qquad (3)$$

On the other hand, to assess the stability of Vietnamese exports during the study period, the variable reflecting the level of adaptability of commodity exports in relation to economic growth (ECC) is also included:

$$ECC_t = \sum_{i=1}^{n} \min(P_{it}, P_{it-1}) \qquad (4)$$

Of which, P_k is the proportion of commodity group k in total export turnover, k is the degree of diversification in same-type goods. The combination of Eqs. (1)–(4), we have an analytical framework for the factors that affect economic growth in the period 2000–2015 as follows:

3   Research Methodology

3.1   Research Model

The starting point of the empirical regression model in this paper is the extended Cobb-Duglas model with technological advances in Hicks neutral, endogenous factors and unchanging yields on scale (Fig. 1):

$$Y_t = TFP_t \cdot K_t^{\alpha} \cdot L_t^{\beta} \qquad (5)$$

Of which, Y_t represents for the total productivity of the whole economy in the period of t; TFP_t is the total factor productivity; K_t and L_t are accumulation of resources and labour respectively; α and β are constants belonging to $(0, 1)$ about the contribution of resources and labour into the total productivity.

The model is assumed that export has an import on the TFP. All K, L, α, β are directly measured while the TFP is directly measured by

$$TFP_t = \exp(\log Y_t - \alpha \cdot \log K_t - \beta \cdot \log L_t) \qquad (6)$$

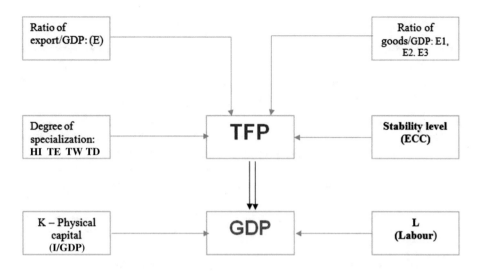

Fig. 1. Source: Author synthesized from previous studies.

According to the analytical framework, the total factor productivity TFP can be rewritten in a function with the export and other exogenous factors (C_t), which are assumed to be unrelated to the export:

$$TFP_t = \gamma_1 + \gamma_2 \cdot E_t + \gamma_3 \cdot E1_t + \gamma_4 \cdot E2_t + \gamma_5 \cdot E3_t$$
$$+ \gamma_6 \cdot T_t + \gamma_7 TD_t + \gamma_8 \cdot TW_t + \gamma_9 \cdot HI_t + \gamma_{10} \cdot ECC_t + \omega_t \quad (7)$$

Because the variables in the model have late effects, fluctuations in export in a particular quarter not only affects the growth of such quarter but also affects the growth of subsequent quarters. To assess both short and long term effects, a dynamic model is used to estimate this effects with the accompanying hypotheses:

Hypotheses	Description of H_0	Expected result
Hypothesis 1	E has positive impact on economic growth	+
Hypothesis 2	E1 has negative impact on economic growth	−
Hypothesis 3	E2 has negative impact on economic growth	−
Hypothesis 4	E3 has positive impact on economic growth	+
Hypothesis 5	HI, TE has positive impacts on economic growth	+
Hypothesis 6	TD has positive impact on economic growth	+
Hypothesis 7	TW has impact on economic growth	+
Hypothesis 8	ECC has positive impact on economic growth	+

3.2   Research Processes

Research data: For model estimation, the research data were quarterly collected from the General Statistics Office (GSO) from the first quarter of 2000 to the fourth quarter of 2016 and the total number of observed variables is 60. It is the most effective set of data quarterly collected.

Estimation: First of all, it is necessary to linearize the model (2) based on log so that it can be estimated by using OLS method. Next, the data strings need to be log-packed to be smoother, and eliminate other characteristics of the data. Then, the use of difference removes the trend components in the series. Finally, it is necessary to check the stationary of all series. Data series will be performed according to the ADF test with the maximum latency of 8 and AIC automatically selects the appropriate latency. The test results show that the first difference of the characteristic variables all have the stationary at the significance level of 5% (Table 1).

According to the above analysis, the regression model evaluates that the impacts of exports on economic growth is decomposed into equations according to components of exports that affect economic growth, namely:

Equation 1: Impact of the export/GDP on economic growth

$$
DGDP = \alpha_1 + \sum_{i=1}^{k} \alpha_2 DGDP_{t-1} + \sum_{i=1}^{k} \alpha_3 D(\Delta K_{t-1}) + \sum_{i=1}^{k} \alpha_4 \Delta L
$$
$$
+ \sum_{i=1}^{k} \alpha_5 D(E) + \sum_{i=1}^{k} \alpha_6 D(HI) + \sum_{i=1}^{k} \alpha_7 D(ECC)
$$
$$
+ \alpha_8 D2007 + \alpha_9 E \cdot D2007 + u_1. \tag{8}
$$

Table 1. Test results of data series after the adjustment

Variable	ADF value (latency)	Critical value (level of significance 5%)	Conclusion
DGDP	$ADF(1) = -3.6389$	-3.4746	Stationary
DK	$ADF(2) = -6.8574$	-3.4738	Stationary
DE	$ADF(3) = -6.745$	-3.4098	Stationary
DE1	$ADF(3) = -5.7240$	-3.4044	Stationary
DE2	$ADF(3) = -6.7316$	-3.4050	Stationary
DE3	$ADF(2) = -6.3826$	-3.4553	Stationary
DECC	$ADF(4) = -5.6389$	-3.5746	Stationary
DHI	$ADF(2) = -14.857$	-3.5784	Stationary
DTE	$ADF(4) = -12.745$	-3.5744	Stationary
DTD	$ADF(2) = -11.724$	-3.5091	Stationary
DTW	$ADF(4) = -5.6389$	-3.5081	Stationary

Source calculated by the writer.

Equation 2: Impact of the export of goods/GDP on economic growth

$$D(GDP) = \beta_1 + \sum_{i=1}^{k} \beta_2 D(GDP_{t-1}) + \sum_{i=1}^{k} \beta_3 D(\Delta K_{t-1}) + \sum_{i=1}^{k} \beta_4 \Delta L$$

$$+ \sum_{i=1}^{k} \beta_5 D(E1) + \sum_{i=1}^{k} \beta_6 D(E2) + \sum_{i=1}^{k} \beta_7 D(E3) + \sum_{i=1}^{k} \beta_8 D(HI)$$

$$+ \sum_{i=1}^{k} \beta_9 D(ECC) + u_2. \tag{9}$$

Equation 3: Impact of the degree of specification/diversification on economic growth

$$D(GDP) = \lambda_1 + \sum_{i=1}^{k} \lambda_2 D(GDP_{t-1}) + \sum_{i=1}^{k} \lambda_3 D(\Delta K_{t-1})$$

$$+ \sum_{i=1}^{k} \lambda_4 \Delta L + \sum_{i=1}^{k} \lambda_5 D(E) + \sum_{i=1}^{k} \lambda_6 D(TE) + \sum_{i=1}^{k} \lambda_7 D(ECC)$$

$$+ \lambda_8 D2007 + \lambda_9 E \cdot D2007 + u_3. \tag{10}$$

Equation 4: Impact of the degree of specification/diversification among groups of goods and same-type goods on economic growth

$$D(GDP) = \mu_1 + \sum_{i=1}^{k} \mu_2 D(GDP_{t-1}) + \sum_{i=1}^{k} \mu_3 D(\Delta K_{t-1}) + \sum_{i=1}^{k} \mu_4 \Delta L$$

$$+ \sum_{i=1}^{k} \mu_5 D(E) + \sum_{i=1}^{k} \mu_6 D(TB) + \sum_{i=1}^{k} \mu_7 D(TW) + \sum_{i=1}^{k} \mu_8 D(ECC)$$

$$+ \mu_9 D2007 + \mu_{10} E \cdot D2007 + u_4. \tag{11}$$

It is noticed that Vietnam's economic structure was changed a lot after it joined the WTO (in 2007). The Dummy D2007 and D2007 · E variables are applied to compare differences in economic growth between the pre and post-WTO period. Therefore, two variables of D2007 and D2007 · E are included in the analytical model.

4 Result and Discussion

Equations are estimated to have maximum latency of 8. According to OLS method, parameters without statistical significance are phased out. In addition, the variables latency in each equation must be large enough to ensure that there are no autocorrelation error and no variance in changeable error. The estimated results of equations from Eviews 8.0 as follows:

Equation 1:

$$DGDP_t = -0.001 + 0.2753 \, DGDP_{t-2} - 0.026 \, D\left(\Delta K_{t-7}\right)$$
$$ {\scriptstyle(-0.8696)} \quad {\scriptstyle(2.5501)} \qquad\qquad {\scriptstyle(-3.100)}$$

$$- \, 0.034 \, D\left(\Delta K_{t-8}\right) + 0.011 \, D(E_{t-8}) - 0.063 \, D(HI_{t-8})$$
$$ {\scriptstyle(-4.178)} \qquad\qquad {\scriptstyle(3.483)} \qquad\qquad {\scriptstyle(-2.296)}$$

$$+ \, 0.019 \, D(ECC_{t-4}) + 0.020 \, D(ECC_{t-6}) + 0.024 \, D(ECC_{t-7})$$
$$ {\scriptstyle(2.4068)} \qquad\qquad {\scriptstyle(2.533)} \qquad\qquad {\scriptstyle(3.033)}$$

$$+ \, 0.019 \, D(ECC_{t-8}) + \hat{u}_1$$
$$ {\scriptstyle(2.185)}$$

$$\bar{R}^2 = 0.69; \; p(ARCH) = 0.44; \; p(LM) = 0.75$$

Equation 2:

$$DGDP_t = -0.001 + 0.183 \, DGDP_{t-2} - 0.013 \, D(E1_{t-1})$$
$$ {\scriptstyle(-2.445)} \quad {\scriptstyle(1.788)} \qquad\qquad {\scriptstyle(-3.7391)}$$

$$+ \, 0.097 \, D(E2_{t-1}) - 0.057 \, D(E2_{t-6}) - 0.081 \, D(E2_{t-7})$$
$$ {\scriptstyle(6.272)} \qquad\qquad {\scriptstyle(-3.314)} \qquad\qquad {\scriptstyle(-4.9844)}$$

$$+ \, 0.122 \, D(E3_{t-3}) - 0.073 \, D(HI_{t-8}) + 0.023 \, D(ECC_{t-4})$$
$$ {\scriptstyle(2.617)} \qquad\qquad {\scriptstyle(-2.9469)} \qquad\qquad {\scriptstyle(3.2431)}$$

$$+ \, 0.035 \, D(ECC_{t-7}) + \hat{u}_2$$
$$ {\scriptstyle(4.735)}$$

$$\bar{R}^2 = 0.674; \; p(ARCH) = 0.57; \; p(LM) = 0.62.$$

Equation 3:

$$DGDP_t = -0.002 + 0.198 \, DGDP_{t-2} - 0.022 \, D(\Delta K_{t-7}) - 0.04 \, D(\Delta K_{t-8})$$
$$ {\scriptstyle(-0.366)} \quad {\scriptstyle(1.913)} \qquad\qquad {\scriptstyle(-2.678)} \qquad\qquad {\scriptstyle(-4.776)}$$

$$+ \, 0.011 \, D(E_{t-8}) - 0.009 \, D(TE_{t-3}) + 0.012 \, D(TE_{t-8})$$
$$ {\scriptstyle(3.644)} \qquad\qquad {\scriptstyle(-1.984)} \qquad\qquad {\scriptstyle(2.053)}$$

$$+ \, 0.025 \, D(ECC_{t-4}) + 0.02 \, D(ECC_{t-6}) + 0.023 \, D(ECC_{t-8}) + \hat{u}_3$$
$$ {\scriptstyle(3.084)} \qquad\qquad {\scriptstyle(2.4946)} \qquad\qquad {\scriptstyle(3.036)}$$

$$\bar{R}^2 = 0.617; \; p(ARCH) = 0.54; \; p(LM) = 0.93.$$

Equation 4:

$$DGDP_t = -0.000 + 0.176 \, DGDP_{t-2} - 0.024 \, D(\Delta K_{t-7})$$
$$ {\scriptstyle(-0.446)} \quad {\scriptstyle(1.810)} \qquad\qquad {\scriptstyle(-2.969)}$$

$$- \, 0.036 \, D(\Delta K_{t-8}) - 0.021 \, D(TW_{t-3}) + 0.017 \, D(TW_{t-7})$$
$$ {\scriptstyle(-4.988)} \qquad\qquad {\scriptstyle(-2.763)} \qquad\qquad {\scriptstyle(2.092)}$$

$$+ \, 0.024 \, D(TW_{t-8}) + 0.024 \, D(ECC_{t-4}) + 0.024 \, D(ECC_{t-6})$$
$$ {\scriptstyle(3.062)} \qquad\qquad {\scriptstyle(3.158)} \qquad\qquad {\scriptstyle(3.134)}$$

$$+ \, 0.027 \, D(ECC_{t-7}) + 0.019 \, D(ECC_{t-8}) + \hat{u}_4$$
$$ {\scriptstyle(3.687)} \qquad\qquad {\scriptstyle(2.428)}$$

$$\bar{R}^2 = 0.62; \; p(ARCH) = 0.44; \; p(LM) = 0.69.$$

Besides, the Granger causality test is used to check whether the causal relationship between export and economic development exists in the long run.

To understand the causal relationship between export (E) and economic growth (GDP), the pairs of variables expressing the export characteristics are respectively replaced in the equation.

$$GDP_t = a + \sum_{i=1}^{M} \alpha_i GDP_{t-1} + \sum_{i=1}^{N} \beta_i E_{t-1} + e_t$$

$$E_t = b + \sum_{i=1}^{K} \gamma_i GDP_{t-1} + \sum_{i=1}^{L} \lambda_i E_{t-1} + u_t \tag{12}$$

Of which, GPD_t is the economic growth rate, E_t is the export replaced by variables expressing the export characteristics. The Granger test results in the Var equation for each pair of variables between growth and export are given in the following Table 2:

Table 2. The results of Granger causality test in the Var equation

Equation	Variables	P-value	Conclusion
Var 1	DGDP ↔ DE	0.081	Two-way causality
Var 2	DGDP ↔, DE1	0.072	Two-way causality
Var 3	DGDP ↔, DE2	0.031	E2 causally affects DGDP
Var 4	DGDP ↔ DE3	0.061	Two-way causality
Var 5	DGDP ↔, DHI	0.031	DHI causally affects DGDP
Var 6	DGDP ↔ DTE	0.030	DTE causally affects DGDP
Var 7	DGDP ↔ DTD	0.000	DTD causally affects DGDP
Var 8	DGDP ↔ DTW	0.009	DTW causally affects DGDP
Var 9	DGDP ↔ DECC	0.019	Two-way causality

Source: Calculated by Granger test.

According to the results obtained from the multivariate regression estimation and causal verification of each variable pair, variables which determine the dynamics of current growth include: economic growth, export scale, quality of export structure, specialization, and export stability in the past. In particular, there is no evidence of any difference in the impact of goods export on economic growth before and after Vietnams WTO accession (Table 3).

Hypothesis H1: Export (E) has positive effect on economic growth

The elastic coefficient of D (E) in equation 1 equals 0.011, which shows that if export/GDP ratio increases by 1%, current economic growth will increase by 0.011% after 8 quarters in case other factors are unchanged. This indicates that export has effect on TFP, with small impact coefficient and large latency. The signal supports the theory that boosting export has positive impact on maintaining long-term economic growth. Moreover, according to the result from Var 1 equation, export and economic growth has a 2-way causal relationship.

Table 3. Test result of quantitative research hypotheses.

Hypotheses	Description	Conclusion
H1	E has positive impact on economic growth	Positive impact
H2	E1 has negative impact on economic growth	Negative impact
H3	E2 has negative impact on economic growth	Negative impact
H4	E3 has positive impact on economic growth	Positive impact
H5	HI, TE has positive impacts on economic growth	Positive impact
H6	TD has positive impact on economic growth	Positive impact
H7	TW has impact on economic growth	No evidence
H8	ECC has positive impact on economic growth	Positive impact

Source: Synthesized from the estimation equation

Firstly, boosting export promotes economic growth, and in return, economic growth promotes export to increase... Continuing this process, boosting export has effect not only in short-term, but in mid-term and long-term.

Hypothesis H2: Increasing export of raw and semi-processed goods (E1) negatively affects economic growth

From equation 2, variable impact coefficient DE1 at the latency of 3 is −0.013, which shows that if the export proportion of raw and semi-processed goods increases by 1%, economic growth will decrease 0.013% after 3 quarters in case other factors are unchanged. This is entirely appropriate with the current condition of Vietnam, the continuous increase in the export of raw materials has negative effect on environment as well as society, and the benefit of exporting raw materials cannot offset the long-term negative effects, leading to negative impacts of raw and semi-processed export on TFP. This affirmation shows that economic growth may increase if the proportion of raw and semi-processed export decreases.

Hypothesis H3: Increasing export of the labour-intensive processing goods (E2) negatively affects economic growth

According to the estimated result of equation 2, the labour-intensive export proportion (E2) affects economic growth at the latency of 1, 6, 7 with coefficients respectively: 0.097, −0.057, −0.081. Initially, increasing export proportion of the labour-intensive processing goods has positive effect on economic growth by 0.097%. This positive effect is suppressed and becomes negative effect since the 6th quarter. Thus, in the long term, if the proportion of labour-intensive processing goods increases by 1%, economic growth will decrease 0.041%.

Hypothesis H4: Increasing export of skill-intensive goods (E3) positively affects economic growth

From equation 2, the coefficient of DE3 is 0.012 at latency of 3. This implies that if export proportion of skill-intensive goods increases by 1%, economic growth will increase 0.122 %. Obviously, this is the strongest factor affecting economic growth. The causal assessment of Var 4 equation shows that there is a causal relationship between economic growth and export of skill-intensive goods (E3). This result is perfectly consistent with the roadmap of economic development in countries around the world. Initially, export of skill-intensive goods will create productivity as well as added value, and promote long-term economic growth... Economic growth will tend to shift the structure of export goods, to focus on the export of skill-intensive goods in the modern way, to emphasis on skill-intensive, high intellectual content... This effect sustains for a long term.

Hypothesis H5: Diversification of export has positive impact on economic growth

Estimated results from equations 1, 2 show that increasing specification (reducing diversification) has negative impact on economic growth at the latency of 8. In the long term, all three equations 1, 2, 3 reflect that increasing diversification (reducing specification) has positive impact on economic growth. In comparison with the experimental rule, with static comparative advantage of resources, and cheap labor in Vietnam, continuing specification by over-exploitation of resources cannot offset cost as well as negative consequences to the environment, which destroys other resources. Meanwhile, diversification helps to stabilize income and expand scale as well as added value of export goods, creating the necessary precondition for the refocusing of goods with advantages of high-level and dynamic comparison which may affect to growth.

Hypothesis H6: Diversifying in width has positive impact on economic growth

From equation 4, diversification level harmonizing within groups of goods (width) affects negatively economic growth at the latency of 3, positively at the latency of 7, 8. In the long term, diversifying in width has positively effect (0.02%). This result supports export expansion, stimulates new industries and manufacturing sectors to develop, diversifies and develops market.

Hypothesis H7: Diversification in depth affects economic growth

There is no evidence to confirm that diversification in depth affects economic growth. This shows that the process of diversification in depth of Vietnam has not been really properly implemented. Value content does not increase and shift to higher stages in the global value chain, but only focuses mainly on goods processing which requires large investment with lowest surplus in the value chain. The causality test from equations Var 5, 6 shows that specification level has causal effect on economic growth, but there is no evidence of opposite effect. Because the export policy and strategy are not really effective, diversification target is not properly valued.

Hypothesis H8: Export stability has positive effect on economic growth

All four equations show that ECC variable is statistically significant, which means that export stability has positive impact on economic growth. Equation

Var 9 shows that there is a 2-way relationship between export stability and economic growth. Moreover, in the impact of goods export after Vietnams WTO accession, there is no evidence of the impact from economic structural change and structural change to economic growth.

5   Conclusion and Policy Recommendations

Based on the result of above analysis, as well as current Vietnamese economic context, this research article proposes a number of views and solutions to improve the goods export performance, to promote the positive of goods export to Vietnamese economic growth in upcoming years as follows:

Firstly, the quality of goods export is an important factor affecting economic growth. Consequently, the persistent orientation of industrialization towards export should be thoroughly understood in the strategic planning and development policy of Vietnam in the new period. In the current context, Vietnam should choose an export-oriented economy combining at a certain level with import substitution. This strategy will maximally allow comparative advantages to expand export of processed goods, to overcome obstacles especially for businesses, and step by step to implement trade liberalization and integration.

Secondly, for fuel and minerals, it is necessary to gradually reduce and minimize export raw goods and roughly processed goods. For goods which have advantages and competing capability with low added value, it is necessary to increase productivity, quality and added value, giving priority for export goods with application of advanced science and technology. Especially, for goods with potential development and needed in global market, it is necessary to develop supporting technology, to raise domestic value ratio, and to reduce dependence on imported raw materials.

Thirdly, it is necessary to develop export sustainably and rationally between width and depth, between quantity and quality, to expand export scale as well as to pay attention to increasing added value, quality of export goods structure, to self-control in the research and development of designs, materials, accessories... to form a network of production and business to build the value chain of each production branch which Vietnam businesses dominate from goods sources to the direct distribution network in the main export markets.

Finally, it is necessary to create breakthroughs that change export quality, optimize resources, promote export growth and sustainable growth. It is necessary to specific solutions and roadmap to realize the development strategy with each goods industries, good groups, goods items, to harmonize between economic and social objectives, between short-term and long-term benefits, to create real value to the economy, and to contribute to improving peoples lives.

6   Limitations and Directions for Further Studies

The limitation of the research is that it focuses only on the impact of export structure on economic growth, ignores the factor of service export because this factor accounts for a very small proportion while the mechanism of influence is quite more complicated than structure of analytical goods. Therefore, this research will be a prerequisite for subsequent study on economic growth.

References

Anh, M.P.: Can Vietnams Economic Growth can be Explained Bay Investment or Export, Vietnam Development Forum (2008)

Awokuse, T.O.: Trade openness and economic growth: is growth export-led or import-led? Appl. Econ. **40**, 161–173 (2008)

Bahamni-Oskooee, M., Mohtadi, H., Shabsign, G.: Exports, growth and causality in LDCs: a reexamination. J. Dev. Econ. **36**, 405–415 (1991)

Balassa, B.: Exports, policy choices and economic growth in developing countries after the 1973 oil shock. J. Dev. Econ. **18**, 23–35 (1985)

Lee, C.-H., Huang, B.-N.: The relationship between exports and economic growth in east countries: a multivariate threhold autoregressive approach. J. Econ. Dev. **27**(2), 45–67 (2002)

Erfani, G.R.: Export and economic growth in developing countries. Int. Adv. Econ. Res. **5**(1), 112–123 (1999)

de Pineres, G., Amin, S., Ferrantini, M.: Export diversification and structural dynamics in the growth process: the case of Chile. J. Dev. Econ. **51**, 375–391 (1997)

Mayer, J., Wood, A.: South Asias export structure in a comparative perspective. Oxf. Dev. Stud. **29**, 6–29 (2001)

Jansen, M.: Income Volatility in small and developing economies: export concentration matters, WTO Discussion Papers N 03. World Trade Orgnaization, Geneva (2004)

Kavoussi, R.M.: Export expansion and economic growth. J. Dev. Econ. **14**, 241–250 (1984)

Kaushik, K.K., Klein, K.K.: Export growth, export instability, investment and economic growth in India: a time series analysis. J. Dev. Areas. **41**, 155–170 (2008)

Keong, C.C., Yusop, Z., Liew, V.K.: Export-led growth hypothesis in Malaysia: an application of two-stage least square technique. Appl. Econ. **30**, 1055–1065 (2001)

Konya, L.: Export-led growth, growth-driven export, both or none? Granger causality analysis on OECD countries. Appl. Econ. Int. Dev. AEEADE 4–1, 73–94 (2004)

Lim, J.J., Saborowski, C.: Export diversification in a transitioning economy: the case of Syria, Policy Research Working, paper 5811 (2011)

Levin, A.: Complementarities between exports and human capital in economic growth: evidence from the semi-industrialized countries. Econ. Dev. Cult. Change **46**(1), 155–174 (1997)

Love, J.: Commodity concentration and export earnings instability - a shift from cross-section to time series analysis. J. Dev. Econ. **24**(2), 787–793 (1986)

McCombie, J.: Increasing Returns and Manufacturing Industries: Some Emprical Issues. Manchester School (1998)

Michalopoulos, C., Jay, K.: Growth of exports and income in the developing world: a neoclassical view, Washington, D.C., US Agency of International Development AID Discussion Paper, No. 28, pp. 47–65 (1973)

Ibrahim, M.H., Amin, R.M.: Export expansion, export structure and economic performance in Malaysia. Asia Pac. J. Econ. Bus. **7**(2), 89–104 (2003)

Reppas, P.A., Christopoulos, D.K.: The export-output growth nexus: evidence from African and Asian countries. J. Policy Model. **27**(8), 929–940 (2005)

Sharma, S.C., Norris, M., Cheung, D.W.: Exports and economic growth in industrialized countries. Appl. Econ. **23**, 697–708 (1991)

Taylor, T.G., Francis, B.: Agricultural export diversification in latin America and Caribbean. J. Agric. Appl. Econ. **35**, 77–78 (2003)

Thirlwall, A.P.: Trade, trade liberalisation and economic growth: theory and evidence, Economic Research Papers No. 63, The Afirican Development Bank (2000)

Thuy, T.T.N.: Impact of Exporting on Vietnam Economic Growth, Thesis of National Economic University, Vietnam (2014)

Tyler, W.: Growth and export expansion in developing countries: some empirical evidence. J. Dev. Econ. **9**, 121–130 (1981)

The Impact of Supermoon on Stock Market Returns in Vietnam

Nguyen Ngoc Thach[1](✉) and Nguyen Van Diep[2](✉)

[1] Banking University of Ho Chi Minh City, 36 Ton That Dam Street, District 1,
Ho Chi Minh City, Vietnam
thachnn@buh.edu.vn
[2] Ho Chi Minh City Open University, Ho Chi Minh City, Vietnam
vandiep1302@gmail.com

Abstract. The objective of the research is to analyze effects of the supermoon phenomenon on stock market returns in Vietnam. Data were obtained from daily series the VN-Index collected from HCMC Stock Exchange (HOSE) from 13/3/2002 to 31/12/2015, using estimation models including GARCH(1,1), GARCH-M(1,1) and TGARCH(1,1). The analysis result shows that GARCH-M(1,1) model proved to be effective in describing daily stock returns features. The findings shows that supermoon phenomenon has a significantly negative impact on the stock returns. This empirical evidence implies that the supermoon phenomenon has effects on behavior of investors, thus affecting financial decisions.

1 Introduction

The idea that the moon has influence on health and behavior of humans began since the Romans' day. For thousands of years, psychics believed there is a strong emotional connection between humans and the moon. Meanwhile, effects of the moon on disconcerting human behavior and psychology is called "The Transylvania Effect" (Geller and Shannon 1976; from Radin and Rebmam 1994).

Frijda (1988) shows that mood affects human behavior. Hirshleifer and Shumway (2003) research on behavioral finance pointing out that mood of investors has effects on asset returns. Some research shows that the weather and other geographical issues also affect asset valuation (Floros 2008; Floros and Tan 2013; Borowski 2015). These evidences point out that investors' financial decisions may base on mood fluctuation.

If the moon (full moon) affects mood of investors, then it can also affect asset prices. Therefore, asset returns during full moon phases may be different from those during new moon phases (Yuan et al. 2006; Floros 2008; Floros and Tan 2013; Borowski 2015).

If full moon phases affect human mood, then the supermoon phenomenon also affects the human mood, since the supermoon only happens when the full moon is close to the earth, and this will affect investors' decisions. The result is that the supermoon phenomenon may have effects on asset valuation. Therefore, asset

© Springer International Publishing AG 2018
L. H. Anh et al. (eds.), *Econometrics for Financial Applications*, Studies in Computational Intelligence 760, https://doi.org/10.1007/978-3-319-73150-6_48

returns during full moon phases may be different (negative or lower) compared to those not in the supemoon phases. From the ideas given on effects of the moon on mood and behavior of human, the research also focuses on analyzing effects of the supermoon phenomenon on stock market returns in Vietnam.

2   Review of Literature

2.1   Effects of the Moon on Human Behavior

Moon phase is when the surface of the moon appears, lightened by the sun in observations from the earth. Full moon is one of the phases of the moon. Full moon occurs when the moon is completely illuminated as seen from the Earth (Seidelmann 2005).

The supermoon (also called perigee full moon) is named after the perigee phenomenon between the Earth and the Moon. The supermoon is a phenomenon when the moon comes extremely close to the earth on the day of the full moon (Nolle 2007). When the supermoon occurs, the moon size can increase by 12–14%, becoming 30% brighter than usual.

Biological evidences have shown that the moon can affect the human body and behavior. More specifically, psychologist Wiseman from University of Hertfordshire discovered that during a full moon, humans might have frequent "crazy and strange" dreams (University of Herfordshire 2014). After running a brain scan on 33 volunteers, Cajochen et al. (2013) discovered that during the full moon, the brain's activities lowered during a state of deep sleep. On average, the volunteers get 20 min less of sleep compared to the whole month and the quality of their sleep also deteriorated. Rotten and Kelly (1985) concluded that the supermoon is related to mental illness and suicide. The supermoon also stimulates seizures in humans (Bendadis et al. 2004).

Additionally, once the supermoon occurs, the tidal waves rise 18% and other natural phenomena may occur due to gravitation of the supermoon, this can be seen in eruption of volcanoes, earthquakes, and in prolonged harsh weather (Hamilton 2013).

There arc two explanations for the effects of the supermoon on the mood and behavior of humans:

First, scientists from University of California (USA) discovered effects of the moonlight on human health. They explained that the moonlight affected human sleep, which caused insomnia during the moon phenomenon and affected our nervous system, resulting in strange, unusual behaviors and actions (Thach and Van Diep 2014).

Second, according to the theory of "Biological Tides" (Crimson Tide) and Zimecki (2006) comprehensive research, if the gravity of the moon can create tidal waves on the earth, then it can affect the human blood and endocrine glands, as 90% of the human body is made up of water. This is the reason why blood circulation in the brain changes, causing imbalance and loss of brain's behavior control system, which leads to irrational behaviors that can cause unexpected accidents and other serious effects.

2.2 Efficient Market Hypothesis

Efficient market hypothesis in financial sector comes from the concept of "perfect market" in economics. In the general research on economics and specifically in finance, market is considered to be perfect (efficient) if it proves efficiency in all three basic aspects: distribution, organization work and information (Shiller 2003).

Market is efficient in terms of distribution when such market has the chance to bring scarce resources to customers, so that based on given resources, customers can create a maximum output, meaning they can optimize the use of resources. This also means that only the highest bidder might obtain the right to use these resources and utilize them to the full potential (Shiller 2003).

Market is efficient in terms of organization when such market has transaction costs in rights-to-use resources in the market determined by supply and demand relationship. In other words, market is deemed to be efficient in organization when it is capable of minimizing transaction costs and bringing these costs slowly back to zero through a competitive system of creation and operation between participants in the market (Shiller 2003).

Market is efficient in terms of information when the value of goods exchanged in the market fully and instantly reflects related information. More specifically, on the aspect of information in an efficient market, investors have little chance to surpass each other because all investors have the same of information access opportunities. According to this hypothesis, stocks are exchanged equal values on stock markets, so that there is no case of price-inflated stocks or forced purchase (Shiller 2003).

In brief, efficient market is a market which is capable of best allocating resources to users with prices fully reflecting related information and minimizing transaction costs.

According to Fama (1970), foundation of the efficient market is based on the following hypotheses: (i) efficient market requires a large number of opponents competing in the market. With the goal of maximizing profits, these investors will conduct analysis and valuation of types of stocks independently; (ii) new information on stocks will be published randomly and automatically in the market, decision on the publishing time will also be independent; (iii) investors will find ways to adjust stock values accordingly to reflect the precise influence of information. The rapid regulation of stock prices is due to the competition between investors in maximizing profits. Since stock prices are adjusted in accordance with new information, so they reflect given information in a true light and are widely published at all time, including the risk of maintaining stocks.

With the above hypotheses, in an efficient market, no investors can crush the market and maintain long-term returns. In addition, the expected returns in the existing price of a stock reflect the degree of risks that the stock may contain.

2.3 Abnormal Financial Market and Theory of Behavioral Finance

Efficient market hypothesis has explained financial market using the assumption that people joining the market are rational individuals. However, subsequent research shows that the overall stock market and behaviors of individuals cannot be explained by the theory of behavioral finance, even Fama – a researcher supporting efficient market – also agrees that the market cannot at all times operate 100% efficiently. It takes some time for stock values to react to new information. Thus, the absolute efficiency of the market is impossible and abnormalities can appear in the financial market.

According to Levy and Post (2005), the circumstances that cause returns of stocks or a group of stocks to go off from assumptions of the efficient market hypothesis, and movements or events that cannot be explained by the efficient market hypothesis are called the financial market's abnormalities. Evidence of these abnormalities has proven the market inefficiency, because investors can foresee its impact on assets and collect high stock returns. Additionally, the abnormalities of the financial market can be a result of an ineffective flow of information in the financial market and that is a violation on a basic assumption of the efficient market hypothesis. The two typical examples are: (i) the curse of the victorious person, where big achievements from the auction tend to exceed the intrinsic value of purchased material. This is mostly due to imbalance of information and human emotion that lead to a higher value being placed (Thaler 1988); (ii) the calendar effect, an abnormality in the financial market when stock returns depend on the exact time of the calendar year (Rozeff and Kinney 1976).

On the calendar effect, the evidence that stands out the most from these effects are the "Monday effect". More specifically, returns on Monday are lower than those on remaining transaction days of the week (Condoyanni et al. 1987; Rystrom and Benson 1989). The Monday effect is a type of inefficient market when returns on Monday are affected by the stock returns from previous Friday. This phenomenon is relatable to the investor's financial behavior and psychology. More specifically, Rystrom and Benson (1989) explained that if investors sense more negativity on Monday than on any other day then they would sell stocks or devaluate their prices. Vice versa, if they feel more positive on Friday, they should increase stock buys. The research by Pettengill (2003) on investors' decision between risk assets and risk-free assets shows similar results (Rystrom and Benson 1989). This means all rational investors must choose more risk-free assets on Friday.

Behavioral finance is an approach to analyzing of financial market that takes emotions into account. Taking into account the influence on the investor's emotions on investment decisions seems to resolve all theoretical difficulties that the traditional finance has to face (Shiller 2003). This theory assumes that some phenomena happening in the financial market can be understood better under the assumption that investors are not completely rational. The models of behavioral finance show that not all of investors' decisions are made based on reason, they are also based on emotions and on other types of not perfectly rational behavior. For example, Tversky and Kahnerman (1992) explained the evidence of abnormal behaviors in their "prospect theory". According to the prospect theory, psychology can also have an impact on the extent of loss due to investors keeping devaluated stocks for too long and sell high-rise stocks too soon.

Normally, they should keep high-rise stocks to increase the value of profits and sell devaluated stocks to minimize the loss. Overconfidence is also an investor's abnormal behavior. Overconfidence can be defined as too much self-assurance, meaning, "Investors tend to think they are better than themselves". Investors tend to overrate their abilities, considering what they analyzed to be of great potential. This can lead to inaccurate evaluation of returns. Meanwhile, investors constantly find evidence to support their arguments and focus on the positive, while forgetting the negative side. The combination of overconfidence and positivity can be the cause of over-evaluated understanding ability, underestimated risks, and the investors over-assuming that they have the ability to manage all situations. This creates a market bubble. Many researchers showed that investors are overconfident in their ability (Cheng 2007; Deaves et al. 2010). Meanwhile, the herd mentality can also greatly influence abnormal behaviors. According to Asch (1951), 35% of the subjects followed the herd mentality even though they understood that it is false. This is because they feel uncomfortable in becoming the numbered few against the powerful majority. In stock market activities, these issues still occur.

3   Data, Model and Methodology

3.1   Data

The paper uses daily returns data of VN-Index string during the time from 13/3/2002 to 31/12/2015. Daily information collected on the VN-Index database from the web of Ho Chi Minh City Stock Exchange. Daily returns of VN-Index are calculated according to the formula:

$$R_t = \frac{P_t - P_{t-1}}{P_{t-1}} \times 100,$$

where:

- P_t is the index value of VN-Index at the end of the transaction day t, and
- R_t is the returns on the day t.

Meanwhile, data on the supermoon from 13/3/2002 to 31/12/2015 was recovered from AsstroPixels.com (provided by Fred Espenak). Table 1 provides a summarized description on returns taken from VN-Index. More specifically, returns in this period include 3,445 observations with the daily average returns of 0.0330%. The highest daily returns are 8.5795% and the lowest daily returns are −8.5481%. The assessment of returns in both the supermoon and the normal phase, the preliminary statistics result shows that the average returns are negative (−0.0498%) during the supermoon and positive (0.0360%) in the phase without the supermoon.

Table 1. Basic statistics for Vietnam's stock market returns. (Source: Authors' own.)

Statistics	R_t	R_t (days with no supermoon)	R_t (days with supermoon)
Mean	0.0330	0.0360	−0.0498
Median	−0.0086	−0.0057	−0.0787
Standard deviation	1.7106	1.7122	1.6719
Minimum	−8.5481	−8.5481	−5.0409
Maximum	8.5797	8.5797	6.7168
Skewness	0.0049	−0.0154	0.6144
Kurtosis	3.1648	3.1846	2.8402
Obs.	3,445	3,325	120

3.2   Research Model

To analyze the supermoon effects on stock returns in Vietnam, the author established a window frame of the day's cycle according to the occurrence of the supermoon. Specifically, this window frame is established as follows: one day before occurrence of the supermoon + day of the supermoon + one day after the supermoon. The research model has the pattern bellow:

$$R_t = \beta_0 + \beta_1 Supermoon + u_t, \tag{1}$$

where:

- R_t is VN-Index's daily returns;
- Supermoon is a dummy variable used to analyze effects of the supermoon on stock returns (Supermoon holds the value of 1 in the time of supermoon occurrence and 0 in ordinary time);
- u_t is the model error on day t.

Apart from examining effects of the supermoon, the research also takes into consideration the calendar effect on fluctuation of returns in Vietnam stock market. With a daily sequence of VN-Index stock returns, the research examines the weekday effect, especially the Monday effect. The Monday effect is a phenomenon on which the average stock returns on this day are lower than those of any other day. The result is negative (meaning the average rate of profitable shares on Monday is devaluating). The Monday effect is evidence against the efficient market hypothesis when returns on Monday are affected by returns on other days, especially in the condition of previous Friday. Therefore, model (1) becomes:

$$R_t = \beta_0 + \beta_1 Supermoon + \beta_2 Monday + \beta_3 Tuesday + \beta_4 Wednesday$$
$$+ \beta_5 Thursday + u_t, \tag{2}$$

where Monday, Tuesday, Wednesday and Thursday are the dummy variables for Monday, Tuesday, Wednesday and Thursday respectively. Specifically, each of these variables receives a value of 1 if it happen on the same specified day and if not, then it receives a value of 0. For example, Monday receives a value of 1 if the day is Monday, if not then it receive a value of 0. Tuesday receives a value of 1 if the day is Tuesday, if not then it receive a value of 0, etc.

3.3   Research Method

Previous research on stock market returns in Vietnam (Thach and Van Diep 2014; Tram et al. 2015) used the ordinary least squares (OSL) to evaluate the models. However, Floros and Tan (2013) show that the OLS method fails to accurately capture the stock returns in the financial market, because this method assumes that the variance does not change with time. In practice, however, the data series in finance and economics have low and high oscillation phases, when the variance is different.

In this research, the GARCH model (Generalized Auto-Regressive Conditional Heteroscedasticity) is used. At present, the GARCH model is the most commonly used model. It was introduced in Bollerslev (1986) to analyze how the data change with time. Specifically, GARCH(1,1) is the most widely used model in experimental research. For example, Gokcan (2000) concluded that this model is most suitable for emerging stock markets. This GARCH(1,1) model is demonstrated below:

$$R_t = \beta_0 + \beta_1 Supermoon + \beta_2 Monday + \beta_3 Tuesday + \beta_4 Wednesday$$
$$+\beta_5 Thursday + u_t, \tag{3}$$

with $u_t \sim N(0, h_t)$ and $h_t = \gamma_0 + \delta_1 h_{t-1} + \gamma_1 u_{t-1}^2$.

The variance h_t exist depends on the past values of the shock (this dependence is represented by the squared approximation error) and on the past values of h_t (represented by h_{t-1}). The vital condition for the GARCH(1,1) model to have meaning is when both the values of δ_1 and γ_1 are positive and statistically significant.

In finance, the stock returns depend on the risk that comes with it. If investors are intimidated by risk, they have a tendency to demand a risk compensation as the insurance for holding a risky asset. Therefore, the risk compensation is a function of risk: the higher the risk, the higher the risk compensation. When the risk measured by the standard deviation, then taking this risk into account transforms the original GARCH model into GARCH-in-Mean model (GARCH-M) that has the following form:

$$R_t = \beta_0 + \beta_1 Supermoon + \beta_2 Monday + \beta_3 Tuesday + \beta_4 Wednesday$$
$$+\beta_5 Thursday + \theta \sqrt{h_t} + u_t. \tag{4}$$

with $u_t \sim N(0, h_t)$ and $h_t = \gamma_0 + \delta_1 h_{t-1} + \gamma_1 u_{t-1}^2$. The parameter θ is called the risk compensation. If $\theta > 0$, then, when the risk increases, the share price increases.

In finance, the unpleasant news (in this case, the supermoon phenomenon) can impact and influence more than the positive news, because negative news can paralyze the investors, making them wait passively for market signs. To take this difference into account, scholars developed the TGARCH model. The goal of the TGARCH model is to examine the imbalance characteristics between positive and negative information. This model uses dummy variables d_t, where d_t has a value equal to 1 if $u_t < 0$ and a value of 0 if $u_t > 0$. The resulting equation for the variance has a form $h_t = \gamma_0 + \delta_1 h_{t-1} + \gamma_1 u_{t-1}^2 + \vartheta_1 u_{t-1}^2 d_{t-1}$. When ϑ_1 is statistically significantly different from 0, the positive and negative information have different effect.

In this paper, we use all three models GARCH(1,1), GARCH-M(1,1), and TGARCH(1,1). To select the best of these three models, we use the AIC criteria: the model with the smallest AIC value is selected as the most accurate one.

4   Result and Discussion

4.1   Verification of Stationarity

To check whether we really need to apply each of these three models GARCH(1,1), GARCH-M(1,1), and TGARCH(1,1), we need to make sure that the sequence R_t satisfies the stationarity condition. There are many methods to verify stationarity on time series. In this paper, we used the ADF stationarity test (by Dickey and Fuller) and the PP test (by Phillips-Perron).

Each of these tests checks the following three hypotheses: (i) the sequence R_t is a purely stationary random sequence, with 0 mean, (ii) the sequence R_t is a stationary random sequence with a non-zero mean, and (iii) the sequence R_t is obtained by adding a stationary random sequence to a linear trend. According to Table 2, according to both tests, the sequence R_t fails all three stationarity tests with significance level below 1%. Thus, we need to use more sophisticated models that take non-stationarity into account.

Table 2. ADF and PP unit root tests. (Source: Authors' own.)

Test statistic	ADF	PP
None	−40.0331***	−49.0472***
Intercept	−40.0447***	−49.0347***
Intercept and trend	−40.0457***	−49.0257***

*, **, *** significant at 10%, 5% and 1% levels, respectively.

4.2   Model Estimation Results

The values R_t are described in Fig. 1. In addition to the above test, we also explicitly tested whether the squares of the daily changes R_t can be described

RT

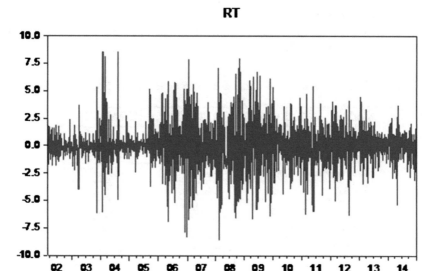

Fig. 1. The trend graph of VN-Index. (Source: Authors' own.)

by a stationary process – or they have periods of low and high fluctuations. The results – given in Table 3 – show that stationarity is rejected at 1% confidence level. Therefore, it is reasonable to use the GARCH(1,1), GARCH-M(1,1), and TGARCH(1,1) models.

The results of fitting the GARCH(1,1), GARCH-M(1,1), and TGARCH(1,1) models to the data are presented in Table 4.

Table 3. Test for ARCH effect. (Source: Authors' own.)

	F-Statistic	$n * R^2$
R_t	276,026***	254,2284***

*, **, *** significant at 10%, 5% and 1% levels, respectively.

When we estimate the parameters of the GARCH(1,1) model based on our data, we conclude that the coefficients at h_{t-1} ($\delta_1 = 0.7837$) and at u_{t-1}^2 ($\gamma_1 = 0.2322$) estimation are positive at the confidence level 1%. Thus, we indeed need the additional terms that form the GARCH(1,1) model.

Similarly, for the GARCH-M(1,1) model, the estimated coefficients at h_{t-1} ($\delta_1 = 0.7826$) and at u_{t-1}^2 ($\gamma_1 = 0.2335$), and the parameter θ ($\theta = 0.0923$) are all positive with significance level 1%. Thus, we indeed need the additional terms that form the GARCH-M(1,1) model.

For the TGARCH(1,1) model, however, while the coefficients at h_{t-1} ($\delta_1 = 0.7831$) and at u_{t-1}^2 ($\gamma_1 = 0.2204$) are statistically significantly positive, the coefficient at the asymmetric term $u_{t-1}^2 d_{t-1}$ ($\vartheta_1 = 0.0263$) is not statistically significantly different from 0. Therefore, the conclusion is that there is no observable difference between positive and negative news. In other words, in our case, the influence of information is symmetrical. This indicate that the TGARCH(1,1) model is not suitable for this situation.

Table 4. Coefficient estimation and model test for supermoon. (Source: Authors' own.)

Models	GARCH(1,1)	GARCH-M(1,1)	TGARCH(1,1)
Mean equation			
Constant	0,0418	−0,0452	0,0348
Supermoon	−0,1831***	−0,1800**	−0,1828***
Monday	−0,0301	−0,0309	−0,0271
Tuesday	−0,1361***	−0,1357***	−0,1338***
Wednesday	−0,0737	−0,0723	−0,0729
Thursday	−0,0084	−0,0086	−0,0075
$\sqrt{h_t}$	−	0,0923***	−
Variance equation			
h_{t-1}	0,7837***	0,7826***	0,7831***
u_{t-1}^2	0,2322***	0,2335***	0,2204***
$u_{t-1}^2 d_{t-1}$	−	−	0,0263
AIC	3,4298	3,4281	3,4299

*, **, *** significant at 10%, 5% and 1% levels, respectively.

In summary, only the GARCH(1,1) and the GARCH-M(1,1) are adequate. Out of these two models, the better model is the model with the lower AIC value. In our case, for the GARCH(1,1) model, the AIC value is 3.4298, while for the GARCH-M(1,1) model, the AIC value is 3.4281. This indicated that the GARCH-M(1,1) model is the most plausible model in this situation. Thus, the upcoming analysis based on the estimation result of the GARCH-M(1,1) model.

In the resulting GARCH-M(1,1) model, the coefficient β_1 describing the influence of the supermoon is negative: $\beta_1 = -0.1800$. This means that during the supermoon, the rate with which the stock index increases is, on average, 0.18% lower than the average rate of increase on all other days. This effect is statistically significant with level of confidence 5%. Thus, the hypothesis that the supermoon has an effect on stock returns is confirmed.

A reasonable possible explanation is that the supermoon affects the Vietnam's stock market through its influence on the investors' emotions. The supermoon phenomenon can affect sensitive investors, change their mood, thus affecting their asset evaluations and resulting stock-related decisions. This example explicitly shows that psychology influences the investor's behavior and actions, and can this affect their financial decisions and lead to changes in the stock market.

With the effects related to the days of the week, our results shown that while the coefficients describing the effects of Monday, Wednesday, and Thursday are negative, they are not statistically significantly different from 0. In particular, there is no evidence about the existence of the Monday effect in Vietnam's stock market. Only for Tuesday, the corresponding coefficient $\beta_3 = -0.1357$ is negative with confidence level 1%.

The explanation of this phenomenon may come from Condoyanni et al. (1987), who showed that while the Monday effect affects huge International Stock Markets such as the ones in the United States, Europe and Japan, this effect created an impact on the rate of profitable shares on Tuesday in the markets that open after that, including the Vietnam's market.

Therefore, the only conclusion we can make here is that Tuesdays have a different rate of shares price increase than Fridays, while there is not enough evidence to check whether rate on the other days of the week are different from the Friday rates. Specifically, the average Tuesday growth is lower than on Fridays by 0.1357%/day.

Finally, the fact that $\theta = 0.0923$ is statistically significantly positive, at the confidence level 1%, shows that investors demand a risk compensation for the strong fluctuations of VN-Index. The reason for this demand is explained by the theory on financial behavior. Specifically, based on the prospect theory, Benartzi and Thaler (1995) showed the investor's fear of devaluation of their investments makes them treat stocks with more fluctuating values and more high-risk, they requiring more compensation.

5 Conclusion

Research analyzes effects of the supermoon on stock returns in Vietnam from 13/3/2002 to 31/12/2015, relying on string of daily returns from VN-Index. In conclusion, experiment shows that the GARCH-M(1,1) is the best model to explain the negative effects of the supermoon on stock returns.

There is no evidence to confirm the Monday effect exist in Vietnam stock market, but there is evidence that the Tuesday effect exists. Interestingly, the influence of the supermoon on stock returns is stronger than the influence of the Tuesday effect.

The negative regression figures imply that during the supermoon, the shares growth rate rate decreases. This can be explained by the supermoon's influence on human emotion and thus, on their decisions. This means emotions and mood may lead investors to be less rational. Thus, the efficient market hypothesis is not very accurate, and its predictions may greatly deviate from the actual stock values.

Finally, the research presents implications for investors, financial managers, and policy makers. The supermoon effect causes mood fluctuation, leading to abnormal behaviors in the process of making decisions. Therefore, in financial transactions, individual investors must control all emotional aspects to enhance financial decisions more efficiently. Financial managers must be cautious about making decisions regarding price evaluation, dividend policy, mergers, acquisition activities or publishing information during the supermoon. For the policy makers, the supermoon may provide assistance in acknowledging and analyzing investors' essential behavior to establish management policies and financial market surveillance.

References

Asch, S.E.: Effects of group pressure on the modification and distortion of judgements. In: Guetzkow, H. (ed.) Groups, Leadership and Men, pp. 177–190. PA Carnegie Press, Pittsburgh (1951)

Benartzi, S., Thaler, R.H.: Myopic loss aversion and the equity premium puzzle. Q. J. Econ. **110**(1), 73–92 (1995)

Benbadis, S.R., Chang, S., Hunter, J., Wang, W.: The influence of the full moon on seizure frequency: myth or reality? Epilepsy Behav. **5**, 596–597 (2004)

Bollerslev, T.: Generalized autoregressive conditional heteroskedasticity. J. Econom. **31**(3), 307–327 (1986)

Borowski, K.: Moon phases and rates of return of WIG index on the Warsaw stock exchange. Int. J. Econ. Financ. **7**(8), 256–264 (2015)

Cajochen, C., Altanay-Ekici, S., Munch, M., Frey, S., Knoblauch, V., Wirz-Justice, A.: Evidence that the lunar cycle influences human sleep. Current Biol. **23**(15), 1485–1488 (2013)

Cheng, P.Y.K.: The trader interaction effect on the impact of overconfidence on trading performance: an empirical study. J. Behav. Financ. **8**(2), 59–69 (2007)

Condoyanni, L., O'hanlon, J., Ward, C.W.: Day of the week effects on stock returns: international evidence. J. Bus. Financ. Account. **14**(2), 159–174 (1987)

Deaves, R., Laders, E., Schroder, M.: The dynamics of overconfidence: evidence from stock market forecasters. J. Econ. Behav. Organ. **75**(3), 402–412 (2010)

Fama, E.F.: Efficient capital markets: a review of theory and empirical work. J. Financ. **25**(2), 383–417 (1970)

Floros, C.: Stock market returns and the temperature effect: new evidence from Europe. Appl. Financ. Econ. Lett. **4**(6), 461–467 (2008)

Floros, C., Tan, Y.: Moon phases, mood and stock market returns: international evidence. J. Emerg. Mark. Financ. **12**(1), 107–127 (2013)

Frijda, N.: The laws of emotion. Am. Psychol. **43**(5), 249–358 (1988)

Geller, S.H., Shannon, H.W.: The moon, weather and mental hospital contacts: confirmation and explanation of the Transylvania effect. J. Psychiatr. Nurs. **14**, 13–17 (1976)

Gokcan, S.: Forecasting volatility of emerging stock markets: linear versus non-linear GARCH models. J. Forecast. **19**(6), 499–504 (2000)

Hamilton, L.: Supermoons: how can they affect you? http://www.psychicsuniverse.com/articles/astrology/supermoons-how-can-they-affect-you. Accessed 24 Sept 2016

Hirshleifer, H., Shumway, T.: Good day sunshine: stock returns and the weather. J. Financ. **58**(3), 1009–1032 (2003)

Levy, H., Post, T.: Investments. Pearson Education, London (2005)

Thach, N.N., Van Diep, N.: Friday the thirteenth and the stock market in Vietnam. Vietnam Bank. Technol. Rev. **104**, 36–44 (2014)

Nolle, R.: The Super Moon and Other Lunar Extremes. The Mountain Astrologer, 21–24 October/November 2007

Pettengill, G.N.: A survey of the monday effect literature. Q. J. Bus. Econ. **42**(3/4), 3–27 (2003)

Radin, D.I., Rebmam, J.M.: Lunar correlates of normal, abnormal and anomalous human behavior. Subtle Energ. Energy Med. J. Arch. **5**(3), 209–238 (1994)

Rotton, J., Kelly, I.W.: Much ado about the full moon: a meta-analysis of lunar-lunacy research. Psychol. Bull. **97**(2), 286–306 (1985)

Rozeff, M.S., Kinney, W.R.: Capital market seasonality: the case of stock returns. J. Financ. Econ. **3**, 379–402 (1976)

Rystrom, D.S., Benson, E.D.: Investor psychology and the day-of-the-week effect. Financ. Anal. J. **45**(5), 75–78 (1989)

Seidelmann, P.K.: Explanatory Supplement to the Astronomical Almanac. University Science Books, Mill Valley (2005)

Shiller, R.J.: From efficient markets theory to behavioral finance. J. Econ. Perspect. **17**(1), 83–104 (2003)

Thaler, R.H.: Anomalies: the ultimatum game. J. Econ. Perspect. **2**(4), 195–206 (1988)

Tram, T.X.H., Vo, X.V., Nguyen, P.C.: Monday effect on the Vietnamese stock exchange pre-crisis, and post-crisis. Vietnam J. Dev. Integr. **20**(30), 55–60 (2015)

Tversky, A., Kahneman, D.: Advances in prospect theory: cumulative representation of uncertainty. J. Risk Uncertain. **5**(4), 297–323 (1992)

University of Hertfordshire: Mass participation experiment reveals how to create the perfect dream (2014). www.sciencedaily.com/releases/2014/03/140326212710.htm. Accessed 24 May 2016

Yuan, K., Zheng, L., Zhu, Q.: Are investors moonstruck? Lunar phases and stock return. J. Empir. Financ. **13**(1), 1–23 (2006)

Zimecki, M.: The lunar cycle: effects on human and animal behavior and physiology. Postepy Hig. Med. Dosw. **60**, 1–7 (2006)

Capital Structure of the Firms in Vietnam During Economic Recession and Economic Recovery: Panel Vector Auto-regression (PVER) Approach

Nguyen Ngoc Thach[1] and Tran Thi Kim Oanh[2(✉)]

[1] International Economic Faculty, Banking University of Ho Chi Minh City,
36 Ton That Dam Street, District 1, Ho Chi Minh City, Vietnam
thachnn@buh.edu.vn
[2] Ho Chi Minh City Technical and Economic College, Ho Chi Minh City, Vietnam
kimoanhtdnh@gmail.com

Abstract. The paper examines the differences between capital structure of enterprises in the circumstances of macro variables's fluctuations due to the economic recession, the instability of economic recovery and the impact of macro variables to capital structure. The authors analyze the data of 82 companies listed on Vietnam Stock Exchange, Quarter 1/2007–Quarter 2/2016, by using PVAR. The results show that the deviation of average capital structure during economic recession versus economic recovery is 0.0142%. There is a relationship between macro variables such as economic growth, credit market, bond market, stock market and capital structure. This relationship explains 4% of the change in these variables. In addition, micro variables such as profitability, business risk, liquidity negatively impact and asset structure, and growth rate positively impact on the capital structure.

1 Introduction

The business target of any enterprise is to maximize its profits and values. As a result, the managers should make the right decisions in the process of selecting investment opportunities as well as organize and manage them comprehensively. In particular, capital structure policy is one of the most important tasks.

The world's economic recession in 2007–2010 has seriously affected the Vietnam's economy. According to the General Statistics Office, 47,000 enterprises have been dissolved or suspended in 2010, and this situation has continued for the following years even when the world's economy began to recover. From 2011 to 2015, approximately 63,520 enterprises are disbanded and suspended annually – a 76.65% increase. One of the causes of this problem comes from the fluctuation of macro variables leading to difficulties of enterprises, especially in financial issues. However, in spite of such economic conditions, most Vietnam's

L. H. Anh et al. (eds.), *Econometrics for Financial Applications*, Studies in Computational Intelligence 760, https://doi.org/10.1007/978-3-319-73150-6_49

firms do not have a specific long-term capital restructuring plan, they just primarily base their decisions on subjective views without taking the economic cycle into consideration.

In recent years, much of the literature on capital structure has been published and most studies analyze the impact of micro and macro variables on capital structure in different countries. These studies use different approaches, most of them are: the Pooled OLS model, the fixed-effects model (FEM), the random-effects model (REM) and the General Method of Moments (GMM). In this paper, the authors use PVAR to investigate the differences between capital structure before and after the economic recession, and to investigate the mechanism that drives the impact of macro variables on the capital structure of Vietnamese enterprises in the context of the unsustainable economic recovery. Then, the article suggests that the managers should adjust capital structure to help the firm overcome the difficult period.

2   Literature Review

2.1   Theoretical Review

About the recession and recovery phases of the economic cycle. In economics, there are many perspectives on the structure of the economic cycle. Karl Marx, the famous 19th century economist, first divided the industrial cycle into four main phases: (1) Crisis; (2) Recession; (3) Recovery; (4) Growth (Industrial Growth) (Retrieved from Zamalnova (1996)). According to Haberler (1958), the economic cycle goes through four phases but its structure is different: (1) Growth (domination, expanse); (2) Climax (moving from domination to decline); (3) Recession (degradation); (4) Bottom (moving from recession to growth). In this paper, the authors choose the approach of Samuelson et al. Economic recession and recovery are two out of three phases (recession, recovery and growth) of the economic cycle (Samuelson and Nordhalls 2007). This approach differs from previous views in the naming and non-considering the bottom and the peak as independent phases. It differs from Haberler's approach since in the modern economy, the economic bottom-line crisis – the stagnant production, widespread unemployment, bankruptcy – rarely happens due to government interventions to mitigate the consequences. Thus, the authors consider only three phases: recession, recovery and growth.

The economic recession is usually defined a decline in real economic growth over two or more continuous quarters (Samuelson and Nordhalls 2007). However, this definition is not universally accepted. For example, according to the National Bureau of Economic Research (NBER) of the United States, economic recession is a decline in economic activities across the country that lasts for months (NBER, 2010). The economic recession may be related to the simultaneous decline of economic indicators such as employment, investment, corporate profits, and may be associated with inflation or deflation. A severe and prolonged recession is called an economic crisis. Economic recovery is the period when GDP begins to rise again and ends at the time when it reaches the same level of GDP before a recession occurs. The post-recession recovery process is determined by the fact that the economy has been growing from the bottom for three consecutive quarters (Samuelson and Nordhalls 2007).

The economic recession and cycle of the market economy are inevitable and periodic phenomena, causing serious consequences for the socio-economic development of many countries. Hence, nations pay a lot of attention to the research on economic recessions. There are different views explaining the causes of the economic recession; they can be summarized into three main causes:

- supply shock – due to technological change, climate, natural disasters, input costs – which affects the production and business activities of the economy;
- demand shock – as a result of changes in expenditure, consumption or investment; and
- policy shock – the consequences of economic policies which have impact on the aggregate demand (Thach and Anh 2014).

By identifying the causes of the economic recession, countries would implement appropriate micro and macro policies to mitigate the losses caused by the economic recession, particularly the policy of economic restructure, including enterprise restructure.

About capital structure. Most of the research on capital structure focuses on the following theories:

MM theory. The theory proposed by Modigliani and Miller (1958) is based on the assumption of perfect market and the tax-free environment, concluding that the firm's value and the weighted average cost of capital (WACC) are independent from capital structure. This theory continues to be investigated in the tax environment (1963), leading to the conclusion that firm's value increases when firms use debt due to shields interest from taxes. WACC of firms using debt are lower than firms that do not use debt.

The theory of trade-off (TOT). The static TOT theory, initiated by Kraus and Litzenberger (1973), continues to evolve into a dynamic TOT theory (Myers and Majluf 1984). Although these studies differ in their views, there is a common approach based on the trade-off between costs and benefits to achieve optimal capital structure, maximizing firm's value.

Pecking Order Theory (POT). This theory states that capital structure decisions follow this funding order: internal capital from retained earnings, then debt, and finally new equity issuance.

Agency Theory. According to this theory, information asymmetry between managers and business owners gives rise to agency costs. So, in order to reduce agency costs, enterprises tend to increase the use of debt.

Signalling Theory. The theory states that the issuance of additional debt is a good sign of business prospects; as a result, stock prices rise. On the contrary, when the enterprise issues more new shares, that is a bad signal and stock prices fall.

Market Timing Theory. This theory suggests that the difference between market value and book value is the decisive factor affecting the decision of the capital structure. If the ratio of market value on the book value (P/B) is high, the enterprise will issues more new shares when capital is needed; if the ratio is low, the enterprise usually uses the debt.

Despite their different point of views, the above theories complement each other to better explain the manager's decision to choose the source of capital funding.

Relationship between economic recession and capital structure. According to Keynesian theory and herd hypothesis, with the information about the economic recession, the investors are pessimistic about their investment activities. As a result, they tend to sell off their financial assets, and hold cash or pay back debt instead. On the other hand, according to Cetorelli and Goldberg (2011), in this period, enterprises seem to shrink the business scale as they are worried about the risk on the reduction of expected profit; furthermore, they would be more cautious in using debt and actively exploit the source of internal capital. This view is in line with the implication of TOT theory that during the period when the cost of financial exhaustion increases, firms should use more equity instead of debt. This, in turn, will lead to a deeper economic recession, which can push the economy far away from its potential target and make it difficult to recover to the economic utilization unless there is the government's support.

On the other hand, according to the POT theory, during the economic recession, the demand decreases, enterprises' profits declined and in order to maintain the demand of business capital, enterprises would use more debt. However, this may increase the risk of bankruptcy, negatively impacting the economy, especially in the period of economic recession, making the economy easily hit bottom, which may lead to the economic crisis (Randveer et al. 2011).

2.2 Previous Related Studies

Although theoretical and empirical studies of capital structure are different in views and methodologies, in general, the researches mainly exploit the following aspects: investigating the impact of micro-variables on capital structure, for which Trinh and Thao (2015), Vătavu (2015), Jong et al. (2008), Nor et al. (2011), Khanna et al. (2015) combine micro and macro variables in their studies. Recently, some studies have examined the impact of capital structure on corporate value or the optimal capital structure threshold, such as the work of Ahmad and Abdullah (2012), Wang and Zhu (2014). Most of these papers, however, study the capital structure of enterprises over a long period of time without linking it to the specific economic context of the economic cycle.

Since the 1970s, the economic recession has occurred with high frequency, intensity and complexity level, causing severe consequences for the nations and the world. Several papers study the effect of economic recessions such as 1997–1998, 2007–2010 on capital structure. However, there were very few such papers.

Several related oversea studies can be mentioned, such as Ariff et al. (2008), Fosberg (2013), Iqbal and Kume (2014). In Vietnam, only the paper of Trinh and Thao (2015) addresses this issue. Most of these papers use Pooled OLS, FEM, REM, and GMM to analyze the impact of the economic recession on capital structure or to merely demonstrate the impact of macro variables on capital structure. However, according to Keynesian theory and herd hypothesis, in addition to the impact of macro variables on the enterprise's financial decisions, these decisions also have impact on the macro variables and thus contribute to explaining the instability of the macro environment. On the other hand, the signalling theory suggests that the enterprise's increase or decrease in the amount of debt use would affect stock prices and the development of the stock market. From the above analysis, it is possible to identify the two-way relationship between macro variables and capital structure of enterprises. Not many studies have investigated this relationship or have analyzed the mechanism of these variables's impact on the capital structure of enterprises, especially in Vietnam.

Recently, Khanna et al. (2015) studied the impact of macro variables such as economic growth, inflation and stock indexes on enterprises' capital structure, using PVAR on panel data. For Vietnam, to the best of our knowledge, no paper has applied PVAR. This is a gap in the choice of methods of previous studies. PVAR is the structural analysis of enterprises in the context of unstable economic recovery, and the macroeconomic instability would seriously affect the financial situation of enterprises. As a result, within the scope of this paper, the authors use PVAR to study the difference between Vietnamese enterprises' capital structure adjustments in the context of macroeconomic fluctuations due to the unstable economic recession and recovery. In addition, the authors study the mechanism as well as predict the direction of macro-variables' impact on capital structure decisions of Vietnamese enterprises under the context of the unstable economy after a long period of recession.

3   Research Methodology

3.1   Data

Based on previous theories and empirical studies, the following variables affecting the capital structure were selected for the research model (Table 1).

The authors use balanced panel data, extracted from the financial statements of 82 enterprises, which are selected randomly from non-financial enterprises (operating in various fields) listed on Vietnam Stock Exchange, and have continuous operation from Quarter 1/2007 to Quarter 2/2016 ($82 \times 38 = 3,116$ observations). The data is relatively representative and is taken from the source of Viet Capital Securities. Macro variables are collected from IMF, ADB and AsianBondsOnline.

3.2   Model

The PVAR model with latency k is written as follows:

$$Y_{it} = \mu_0 + A_1 Y_{it-1} + \ldots + A_k Y_{it-k} + \beta_x X_i t + D + e_{it}, \quad i = 1, 2, \ldots, N, \ t = 1, 2, \ldots, T,$$

Table 1. Variables and measurement

Notation	Variables	Expectation	Theories	Measurement	Studies
Endogenous variables					
TDR_{it-1}	Latency of dependent variable			Total debt$_{it}$/ Total asset$_{it}$	Anh et al. (2014), Vătavu (2015)
GDP_{t-1}	Latency of economic growth rate	+	TOT, agency cost	$(GDP_t - GDP_{t-1})/ GDP_{t-1}$	Ariff (2008), Jong et al. (2008), Khanna et al. (2015)
		−	POT		
$RATE_{t-1}$	Latency of average loan's interest rare	+	MM	Average loan's interest rate	Nor et al. (2011), Zerriaa and Noubbigh (2015)
		−	POT, TOT		
$LNVNI_{t-1}$	Latency of stock market	−	Market timing	LN(VNINDEX$_t$)	Jong et al. (2008), Khanna et al. (2015)
$BOND_{t-1}$	Latency of bond market	+	TOT, agency cost	Market capitalization value$_t$/GDP$_t$	Jong et al. (2008), Nor et al. (2011)
Exogenous variables					
D	Economic recession 2007–2010	+	POT		Zarebski and Dimovski (2012), Fosberg (2013), Iqbal and Kume (2014)
		−	TOT		
$SIZE_{it}$	Enterprise's size	+	TOT, agency cost	Ln(Total asset$_{it}$)	Zarebski and Dimovski (2012), Anh et al. (2014)
		−	POT		
$TANG_{it}$	Asset structure	+	TOT, POT	Fixed asset$_{it}$/ Total asset$_{it}$	Anh et al. (2014), Vătavu (2015)
		−	Agency cost		
GRO_{it}	GDP growth	+	POT	(Total asset$_{it}$ − Total asset$_{it-1}$)/ Total asset$_{it-1}$	Jong et al. (2008), Anh et al. (2014)
		−	agency cost,		
LIQ_{it}	Short-term liquidity	+	TOT	Short-term asset$_{it}$/ Short-term loan$_{it}$	Anh et al. (2014), Vătavu (2015)
		−	POT agency cost		
ROE	Profitability	+	POT	Profit after tax/ Equity$_{it}$	Nor et al. (2011), Trinh and Phuong (2015)
		−	TOT		
VOU_{it}	Business risk	+	TOT	Standard deviation (EBIT$_{it}$/ Total asset$_{it}$)	Thuy et al. (2014), Vătavu (2015)
		−	POT		
MTR_{it}	Corporate income tax	+	TOT	Corporate income tax$_{it}$/Profit before tax$_{it}$	Jong et al. (2008), Vătavu (2015)
			MM		

where:

- $Y_{it} = (TDR_{it}, GDP_{it}, RATE_{it}, LNVNI_{it}, BOND_{it})$ is a random vector of dependent variables, of size 5;
- Y_{it-p} is a vector of dependent variables (latent) also of size 5;

- $A_1, A_2, \ldots, A_k$ are $k \times k$ matrices;
- X_{it} are exogenous vectors of size k, that include including control variables listed in Table 1;
- β_x are matrices of coefficients;
- D is the dummy variable: $D = 0$ in the period of economic recovery Q1/2011–Q2/2016, and $D = 1$ during the economic recession Q1/2007–Q4/2010;
- e_{it} are effects due to unobservable characteristics of enterprises and constant effects over time; $e_{it}|y_{it-1} \sim N(0; \sigma_i)$.

The classic VAR model is derived from the study of the macro-variable transmission mechanism of Sims (1980). Holtz-Eakin et al. (1988) further proposed the VAR model used in panel data (PVAR). However, the PVAR proposed by Holtz-Eakin et al. (1988) is based on the classical VAR, thus, there are some limitations such as the estimated parameter can be deviated from the observation due to lags. To overcome this disadvantage, the article using a version of PVAR provided by Love and Zicchino (2006). This version is based on the application of GMM. This ensures the uniformity of balance variance, autocorrelation and data conservation.

3.3   Basic Tests

To apply PVAR, the variables used must be stationary. For checking stationarity, the authors use the Augmented Dickey-Fuller (ADF) method. Table 2 shows that all variables satisfy the stationarity condition I(0).

Optimal latency test. Andrews and Lu (2001) proposed using The Moment Model Selection Criteria (MMSC) with CD-determination coefficients and J-P value statistics to determine the optimal latency. Estimation results in Table 3 shows the optimal latency of PVAR is 3.

Model stability test. Stability was checked by using the AR test. Figure 1 shows that all typical polynomial solutions of the model are in unit circle, ensuring the stability and sustainability of the model.

Table 2. PVAR unit root test results

Variables	T-statistics	Variables	T-statistics
TDR	526,1311**	TANG	291,4663**
ROE	2405,4509**	LIQ	702,1896**
VOL	1858,4298**	LNVNI	381,9070**
SIZE	493,4786**	BOND	207,7378**
MTR	1638,3719**	RATE	207,7378**
GRO	1343,1092**	GDP	561,0923**

Note: ** corresponding to the significant level of 5%, Source: Authors' calculation.

Table 3. Lags criteria results

Latency	CD	J	MBIC	MAIC	MQIC
1	0,9996	2170,706	1791,351	2074,706	1972,248
2	0,9998	1374,241	1121,338	1310,241	1241,936
3	0,9995	1050,594*	924,142*	1018,594*	985,441*

Note: * represents the selected latency corresponding to the criterion. Source: Authors' calculation.

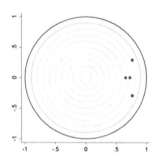

Fig. 1. AR root test results

4   Results and Discussion

4.1   Results

The regression results are shown in Table 4.

The results show that there is a difference between the capital structure of an enterprise in periods of economic recession and economic recovery. The difference in average capital structure of the economic recession over the economic recovery period is −0.0142%. This finding implies that during the economic recession, Vietnamese enterprises used less debt than in the economic recovery period, corresponding to a decrease of 0.0142%, and there is an increase in the use of equity. This result is consistent with TOT theory and with Zarebski and Dimovski (2012).

To analyze the mechanism and the direction of macro variables' impact on the capital structure of Vietnamese enterprises under the shocks, the authors analyze impulse response (Figs. 2 and 3).

Impact of the bond market on capital structure. Figure 2 shows that when BOND increased by one standard deviation (which means that the bond market improves), the company's debt increases by 0.118% debt in the first quarter and decreases in the second quarter. This result is consistent with the TOT theory and with Jong et al. (2008), Nor et al. (2011). In contrast, Fig. 3 shows that when the TDR changes by a standard deviation, its impact on the bond market is rather weak: 0.009% at a 5% significance level; this effect lasts only for 5 quarters. According to Altunbas (2009), enterprises with high debt ratio seem

Table 4. Regression results of PVAR

Variables	Coefficient of estimation
L.TDR	0,754**
	[23,02]
L.LNVNI	$-0,0193*$
	[2,10]
L.BOND	0,262**
	[4,30]
L.RATE	0,0826*
	[2,44]
L.GDP	$-0,357**$
	$[-3,14]$
ROE	$-0,0873**$
	$[-3,48]$
VOL	$-0,223*$
	[2,08]
SIZE	$-0,0084$
	$[-0,31]$
MTR	$-0,0214$
	$[-1,38]$
GRO	0,00590**
	[2,72]
TANG	0,0852**
	[2,58]
LIQ	$-0,000766**$
	$[-3,30]$
D	$-0,0142**$
	$[-2,79]$
N	3.116

Note: *, ** corresponding to significant levels of 5%, 1%; []: Value of standard error. Source: Authors' calculation.

to have difficulties to mobilize capital from the issuance of bonds. Therefore, as enterprises increase their use of debt, it may lead to a definite decline in bond market. In Vietnam, this result can be explained by two reasons:

- In general, the loan's interest rate in the study period tends to decrease and stabilize at 6.96%. Credit institutions tend to expand credit, while the liquidity in the bond market is low because of the remain of risk avoidance's behavior due to the unsustainable economic recovery.

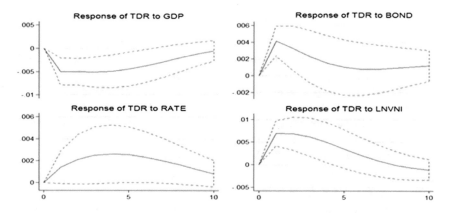

Fig. 2. Response of TDR to macro variables shock

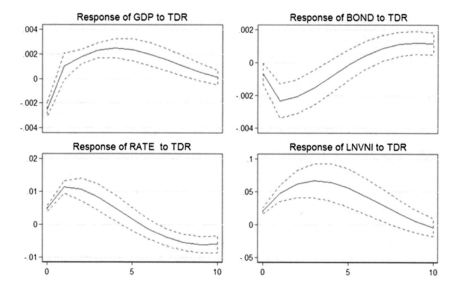

Fig. 3. Response of macro variables to TDR shock

- On the other hand, credit is the traditional channel for raising capital among Vietnamese enterprises. Therefore, although enterprises tend to increase debt using but mainly through credit channels not the bond market.

Impact of economic growth on capital structure. Figure 2 shows that when economic growth increases by one standard deviation, enterprises reduce the use of debt by 0.357%. This effect lasts for 10 quarters, which is consistent with Anh et al. (2014). In contrast, Fig. 3 shows that when enterprises' debt increases by one standard deviation, the pressure on debt repayment rises, the cost of financial exhaustion increases, which affects the business results, increasing the default risk, leading to a decrease in economic growth by 0.009% at 5% significance level and lasting for two quarters.

Impact of credit market on capital structure. Figure 2 shows that, when the credit market shrinks, credit balance drops, lending conditions increases, and RATE increases by one standard deviation, TDR increases by 0.083% at 5% significance level and diminishes in Quarter 4. This result is consistent with MM theory and with Zerriaa and Noubbigh (2015). Conversely, Fig. 3 shows that when TDR increases by 1%, the demand for credit immediately increases, and the banks raise interest rates by 0.044% in the first quarter.

Impact of the stock market on capital structure. Figure 2 shows that when the stock market rises by one standard deviation, TDR increases by 0.019%, significant at 1%. This increase lasts for seven quarters. The increase is consistent with the study of Khanna et al. (2015). In contrast, Fig. 3 shows that when TDR increases by one standard deviation, the stock market increases by 0.235% at 5% significance level. This increase lasts for nine quarters. This can be explained based on signalling theory and Baker et al. (2003).

As mentioned above, there is a relationship between capital structure and macro variables. However, TDR, apart from its dependence on macro-shock shocks, is also affected by micro variables.

The previous capital structure has a positive impact on the capital structure of the next period. This implies that an increase in past-term debt would increase the debt use by 0.754% in the following period, which is consistent with Ariff (2008), Nor et al. (2011), and Khanna et al. (2015).

Business risk has a negative impact on capital structure. The implication is that enterprises with high business risk are less able to pay due-debt, having high financial exhaustion cost and using less debt. The results are consistent with the TOT theory and with Jong et al. (2008) and Anh et al. (2014).

Profitability has a negative impact on capital structure, which can be explained by POT theory and with Nor et al. (2011) and Trinh and Thao (2015). However, the above results also show that the use of debt of Vietnamese enterprises is not effective, leading to the reduction in enterprises's profit.

Asset structure has a positive impact on capital structure. Indeed, large fixed-asset enterprises have more advantage in debt issuance and better policies. This conclusion is in line with POT theory, with TOT theory, and with Fosberg (2012), Iqbal and Kume (2014).

Growth rates have a positive impact on capital structure, which is consistent with POT theory and Nor et al. (2011). Liquidity has a negative impact on capital structure, which is consistent with POT theory and with Nor et al. (2011).

The results of TDR covariance decomposition analysis are presented in Table 5, which shows that the macro variables contribute only 4% to the change in capital structure, the rest is attributed to internal factors of enterprises. In particular, the contribution of economic growth is 1.4%, and the contributions of the bond market, stock market and credit market are 0.4%, 1.8% and 0.4% respectively. Thereby, it can be seen that the decisions on capital selection are

Table 5. Decomposition of variance

Stage	TDR	LNVNI	BOND	RATE	GDP
1	1	0	0	0	0
2	0,9853	0,0076	0,0027	0,0003	0,0040
4	0,9707	0,0152	0,0038	0,0015	0,0089
6	0,9634	0,0175	0,0038	0,0027	0,0126
8	0,9607	0,0174	0,0038	0,0036	0,0145
10	0,9599	0,0173	0,0039	0,0040	0,0149

mainly determined by enterprises' managers and internal factors, while macro variables have not been properly considered when planing capital structure.

4.2 Discussion

The results of PVAR regression show that the capital structure is affected by the economic recession. During the economic recession, Vietnamese enterprises used less debt than the economic recovery period, corresponding to a decrease of 0.0142%.

The bond market has a positive impact on the capital structure, however, this level of impact lasts for a short time and stops in the second quarter. Conversely, the capital structure has a slow and weak influence on the bond market.

Economic growth negatively affects the capital structure and lasts for ten quarters. In contrast, capital structure negatively affects economic growth but in the short term of two quarters.

The credit market have a negative impact on capital structure. When the credit market shrinks, credit balance decrease, loan conditions become more difficult and lending interest rates increases by one standard deviation, however, enterprise still increases using debt, capital structure increases. In contrast, the capital structure has a weak and negative impact on the credit market.

The stock market has a positive impact on capital structure. On the other hand, the capital structure, by increasing a standard deviation, will have a positive impact on the stock market.

In addition, the results of the study show that macro variables contribute to 4% of the change in capital structure decision. The remaining 96% is attributed to internal factors such as growth rate and asset structure, which have the positive impacts; or profitability, business risk and liquidity, which have the negative impacts on capital structure.

5 Conclusion and Policy Suggestions

5.1 Conclusion

The article uses data from 82 enterprises listed on Vietnam stock exchanges, from Quarter 1/2007 to Quarter 2/2016 and PVAR techniques to analyze the impact of the economic recession on the capital structure. The results show that the economic recession has a negative impact of -0.0142% on the capital structure compared to the economic recovery period, the macro variables have relationship and contribute 4% to the change in capital structure. In addition, micro variables such as profitability, business risk, liquidity have the negative impacts, while asset structure, growth rate have positive impacts on capital structure.

5.2   Policy Suggestions

For enterprises. First, enterprises should diversify forms of capital mobilization in the direction of reducing debt, using much more capital and increasing the use of financial instruments.

Second, the capital restructure must be linked to each stage of firms' development, growth objectives, specific financial situation, size, internal and external macro environment, especially the restructure must be linked with different phases of the economic cycle - recession and economic recovery. In particular, in the economic recession, Vietnamese enterprises should apply the TOT theory to control the proper use of debt, avoiding the increase of debt using to solve temporary financial problems. However, as the economy recovers, enterprises can increase their use of debt to meet capital needs for the expansion of their business.

Third, capital restructure should be implemented in the direction of increasing equity and self-financing capacity of enterprises. The research results indicate that the business profitability has a negative impact on the capital structure. It proves that profit is one of the important sources of capital to help businesses in keeping control, and in actively meeting capital needs.

Fourth, capital restructure must be associated with the restructure of investment portfolios, especially investment in fixed assets, but must be linked to long-term capital enhancement, avoiding the financial imbalances. On the other hand, the results also show that Vietnamese enterprises should prioritize investment in short-term and high-yield projects and assets, assuring the short-term liquidity, limiting the risks arising from the unsustainable economic recovery.

For policy makers. First, the Government should implement policies in a synchronous manner to ensure macroeconomic stability, inflation control, proper economic growth and favorable conditions for the business environment, and business support for easier access to restructuring funds.

Second, the Government should improve the regulations and policies to promote and facilitate the development of the financial markets, especially the bond market and the stock market. This is the effective channel for capital mobilization and capital withdrawal under the market mechanism. Hence, healthy and developing financial markets is important conditions to ensure a successful business capital restructure. For bond market, this is the traditional channel of enterprises' capital mobilization. As a result, the issue of enterprises' capital restructure must be associated with the development of this market. The Government should have policies to strengthen, stabilize and develop this market.

References

Ahmad, Z., Abdullah, N.M.H.: Capital structure effect on firms performance: focusing on consumers and industrials sectors on Malaysian firms. Int. Rev. Bus. Res. Pap. **8**(5), 137–155 (2012)

Andrews, D.W.K., Lu, B.: Consistent model and moment selection procedures for GMM estimation with application to dynamic panel data models. J. Econom. **101**, 123–164 (2001)

Ariff, M., Taufq, H., Shamsher, M.: How capital structure adjusts dynamically during fnancial crises. J. Fac. Bus. Bond Univ. **12**, 15–25 (2008)

Baker, M., Stein, J., Wurgler, J.: When does the market matter? Stock prices and the investment of equity-dependent firms. Q. J. Econ. **118**, 969–1005 (2003)

Cetorelli, N., Goldberg, L.S.: Global banks and international shock transmission: evidence from the crisis. IMF Econ. Rev. **59**, 41–76 (2011)

Holtz-Eakin, D., Newey, W., Rosen, H.S.: Estimating vector autoregressions with panel data. Econometrica **6**, 1371–1395 (1988)

Fosberg, R.H.: Short-term debt financing during the financial crisis. Int. J. Bus. Soc. Sci. **4**, 1–5 (2013)

Haberler, G.: Prosperity and Depression: A Theoretical Analysis of Cyclical Movements, 4th edn. George Allen and Unwin, London (1958)

Iqbal, A., Kume, O.: Impact of financial crisis on firms' capital structure in UK, France, and Germany. Multinatl. Financ. J. **18**, 249–280 (2014)

Jong, A.D., Kabir, R., Nguyen, T.T.: Capital structure around the world: the roles of of firm- and country-specific determinants. J. Bank. Financ. **32**(9), 1954–1969 (2008)

Khanna, S., Srivastava, A., Medury, Y.: The effect of macroeconomic variables on the capital structure decisions of Indian firms: a vector error correction model/vector autoregressive approach. Int. J. Econ. Financ. **5**(4), 968–978 (2015)

Kraus, A., Litzenberger, R.H.: A state-preference model of optimal financial leverage. J. Financ. **33**, 911–922 (1973)

Love, I., Zicchino, L.: Financial development and dynamic investment behavior: evidence from panel VAR. Q. Rev. Econ. Financ. **46**, 190–210 (2006)

Modigliani, F., Miller, M.H.: The cost of capital, corporation finance and the theory of investment. Am. Econ. Rev. **48**, 261–297 (1958)

Myers, S.C., Majluf, N.S.: Corporate financing and investment decisions when firms have information that investors do not have. J. Financ. Econ. **13**, 187–221 (1984)

Nor, F.M., Haron, R., Ibrahim, K., Ibrahim, I., Alias, N.: Determinants of target capital structure evidence on south east Asia countries. J. Bus. Policy Res. **6**, 39–61 (2011)

Thach, N.N., Anh, L.H.: Coordination of Monetary Policy and Fiscal Policy in the Economic Crisis. Economy of Ho Chi Minh City, Ho Chi Minh City (2014)

Randveer, M., Uuskula, L., Kulu, L.: The impact of private debt on economic growth. Working Papers of Eesti Pank (2011). https://www.eestipank.ee/en/publication/working-papers/2011/102011-martti-randveer-lenno-uuskula-and-liina-kulu-impact-private-debt-economic-growth. Accessed 01 Dec 2016

Samuelson, P., Nordhaus, W.: Economics. McGraw Hill, New York (2007)

Sims, C.: Macroeconomics and reality. Econometrica **48**(1), 1–48 (1980)

Trinh, T.H., Thao, N.P.: Determinants of capital structure of A-REITS and the global financial crisis. Pac. Rim Prop. Res. J. **18**, 3–19 (2015)

Vătavu, S.: The impact of capital structure on financial performance in Romanian listed companies. Procedia Econ. Financ. **32**, 1314–1322 (2015)

Anh, V.T.T., Ly, T.K., Anh, L.T.N., Dung, T.T.: Study the impact of macro factors on the capital structure of listed companies on Vietnam's stock market. J. Econ. Dev. **207**, 19–27 (2014)

Wang, J., Zhu, W.: The impact of capital structure on corporate performance based on panel threshold model. Comput. Model. New Technol. **18**(5), 162–167 (2014)

Zarebski, P., Dimovski, B.: Determinants of capital structure of A-REITS and the global financial crisis. Pac. Rim Prop. Res. J. **18**(1), 3–19 (2012)

Zerriaa, M., Noubbigh, H.: Determinants of capital structure: evidence from Tunisian listed firms. Int. J. Bus. Manag. **10**, 121–135 (2015)

The Efficient Sterilization of Central Bank: Suitable Estimation Method

Phung T. K. Nguyen[(✉)], Hac D. Le, and Hang T. T. Hoang

Banking University of Ho Chi Minh City,
56 Hoang Dieu 2, Thu Duc District, Ho Chi Minh City, Vietnam
{phungntk,hacld,hanghtt}@buh.edu.vn

Abstract. A country's foreign exchange reserve plays an important role for the security of its economy. However, a large accumulation of foreign exchange reserves without central bank's intervention will lead to inflation. Therefore, those must be accompanied by neutralizing its impact on the monetary market. The different econometric methods have been chosen in the previous literature to estimate the sterilization effect in many countries in the world. This study provides an overview of estimation methods for the efficiency of central bank's sterilization in foreign market. It also shows empirical evidences for the application of these methods in studies in the world. The results show that the 2SLS is the most suitable method to estimate the eficient sterilization of Central bank.

1 Introduction

When Central bank intervenes in the foreign exchange market, if there is not its sterilization, it will lead to inflation or deflation (Heller 1976; Steiner 2009) which affecting the operation of monetary policy; therefore, the central bank has to sterilize. According to Krugman et al. (2012) definition, sterilization is a policy that central bank carries out equal foreign and domestic asset transactions in opposite directions to nullify the impact of their foreign exchange operations on the domestic money supply. The researchers have investigated the central bank's sterilization in two ways: Neutralizing the money base (Sterilization in the narrow sense) or neutralizing the money supply (Sterilization in a broad sense). Most studies refer to sterilization in the narrow sense. Therefore, the scope of this study is the effectiveness of central bank's sterilization in the narrow sense. In that way, exploring what degree efficiency of central bank's sterilized policy is related to central bank variables such as Net Foreign Assets (NFA), Net Domestic Assets (NDA), Monetary Base (MB) and other macroeconomics variables such as Inflation, Money supply (Ms), interested rate, exchange rate, etc. Because these variables are related to each other, i.e. the change of this variable leads to the change of other variables and vice versa, appropriate estimation are needed to assess the actual effectiveness of the neutral intervention of Central bank.

© Springer International Publishing AG 2018
L. H. Anh et al. (eds.), *Econometrics for Financial Applications*, Studies in Computational Intelligence 760, https://doi.org/10.1007/978-3-319-73150-6_50

- **Subject of the study.** This study aims to understand the methods of estimation the effectiveness of central banks' sterilization in the world.
- **Objectives of the study:** Objectives of the study are expressed through the following questions:

  (i) What econometrics estimation methods were used to investigate the efficiency of central bank's sterilized policy?
  (ii) Which econometrics estimation method is the most popular and appropriate?
  (iii) Which econometrics estimation method is the most suitable for evaluating efficient sterilization of State Bank of Vietnam?

- **Theoretical basis and analytical framework:** The mechanism of sterilization of the central bank is reflected through the change of indicators on the Central bank's Balance sheet as follows (Table 1):
  According to Central bank's Balance sheet, Monetary Base can be expressed as:

  $$\text{Monetary Base} = \text{Net Foreign Assets} + \text{Net Domestic Assets}$$

  Or

  $$MB = NFA + NDA \tag{1}$$

  From this formula, we can give some conclusions: The increase of foreign exchange reserves (NFA) will result in the MB increase. If the central bank

Table 1. Changes in Central bank's Balance sheet after accumulation foreign exchange reserve.

Items	Non sterilization	Sterilization
Net Foreign Assets	+	+
Foreign Assets	+	+
Foreign Liabilities		
Net Domestic Assets	0	-
Monetary Base	+	0
+Currency in circulation	0	0
+Deposits of commercial banks	+	0

Source: Author's synthesis
Note: 0: No change in value;
+: Increase in value;
-: Decrease in value.

purchases foreign currencies in the foreign exchange market (without sterilization), NFA and MB will increase (+). The MB expands without a corresponding increase in production. To ease the threat of currency appreciation or inflation, the central bank often uses open market operations, that is, selling Treasury Bills or other domestic securities, to reduce the value of the monetary base. After the sale securities, NDA of central bank reduce (−) with the amount that equals decrease value of deposits of commercial banks. With this mechanism, in a successful sterilization operation, the monetary base is reduced to offset the reserve inflow, at least temporarily. And so that, to investigate the central bank's efficiency sterilization, it necessary to evaluate the change of NDA after NFA change.

- **Methodology:** In response to the research questions above, the study uses synthesis method and comparative analysis of estimation methods which empirical studies in the world to evaluate the ability of central bank's sterilization in the foreign market.

2   Results from Empirical Studies

Empirical studies of the efficiency of central bank's sterilization in the world were divided into two main approaches. The first approach primarily examined the relationship between the NDA and the NFA in the monetary policy reaction function with two main estimation methods, the Ordinary Least Square (OLS) method and the Vector Auto Regression (VAR) model. The second approach used simultaneous equations to examine the relationship between NDA and NFA and applies the Two Stage Least Squares (2SLS) for the estimation of both equations.

2.1   The First Approach

The regression model was constructed from the monetary policy reaction function which derived by Cumby and Obstfeld (1983). They supposed that the central bank should neutralize the monetary impact of accumulation international reserves by changing its NDA. Therefore, the monetary policy response function is expressed as follows:

$$\Delta NDA_t = \lambda_1(CUA_t + CAA_t) + \lambda_2 Z_1 + \varepsilon_t \tag{2}$$

where ΔNDA_t is the change in the NDA of central bank, λ_1 is the sterilization coefficient, CUA is the current account balance, CAA is the capital account balance, Z_1 is the vector of other variables that could also effect monetary policy actions.

Because the balance of payments is explained: (Disregarding errors and omissions):

$$CUA + CAA = NFA \tag{3}$$

Equation (2) can be presented as follows:

$$\Delta NDA_t = \lambda_1(\Delta NFA_t) + \lambda_2 Z_1 + \varepsilon_t \tag{4}$$

In the above equation, λ_1 is represented as the sterilization coefficient. The expected value of the sterilization coefficient is -1 if sterilization is complete and 0 if the central bank does not sterilize at all.

According to empirical studies, Eq. (4) was estimated by the OLS method or by the VAR model.

OLS Estimation Method

The OLS method was first introduced by German mathematician, Carl Friedrich Gauss, in the late 18th century. This is the most common method used to esti-mate parameters in a linear regression equation. Estimated results are used to deduce the coefficients in the population, from which there are real policy appli-cations. Empirical studies using the OLS estimation to evaluate sterilized effec-tiveness of the central bank are as follows:

Takagi and Esaka (2001) tested for the effectiveness of sterilization by esti-mating the extent to which foreign assets (FA) in the monetary base explains or predicts of monetary supplies. Both narrow money (M_1) and broad money (M_2) were used as measures of monetary supplies. They used data of East Asian countries from Q1/1987 through Q2/1997. The evidence showed that steriliza-tion was successful in limiting the growth of monetary supplies during 1987–1997 in all countries.

Cavoli and Rajan (2006) used a series of related empirical tests of the dynamic links between international capital flows, the extent to which they are sterilized in the five crisis economies (Indonesia, Korea, Malaysia, Philippines and Thailand) over the period from January 1990 to May 1997. The OLS estimation results indicated that the sterilization coefficients were ranged from -07 (Indonesia) to -1.1 (Korean). Thus, monetary sterilization was incomplete in Indonesia and almost complete sterilization on average in the other Asia economies.

Aizenman and Glick (2009) investigated the extent of sterilization within emerging market countries as they liberalized markets and integrated with the world economy (the period from the first quarter of 1996 to fourth quarter of 2007). They estimated the change in the degree of sterilization by a simple regression equation between the change in central bank's NDA and the change in central bank's NFA. Nominal GDP growth rate was also used as the control variable. Sterilization coefficient was estimated with OLS method. And they found that this coefficient had risen to varying degrees in Asia as well as in Latin America, consistent with greater concerns about the potential inflationary impact of reserve inflows.

Glick and Hutchison (2009) measured the degree to which the People's Bank of China had been able to insulate monetary base growth from the liquid-ity effects associated with rapid foreign asset accumulation. They estimated the extent of sterilization by inheriting the research model of Aizenman and Glick (2009). Data was collected in China from Q3/1985 to Q4/2007 and pro-cessed with OLS estimation. The results showed that the sterilization coefficient ranged from −0.6 in 2000 to −1.5 in the first quarter of 2006, then increased again to −0.8 in the fourth quarter of 2016 and retained its value during 2007. Thus, sterilization dropped precipitously in 2006 in the face of the ongoing massive buildup of international reserves, leading to a surge in reserve money growth.

Pham and Nguyen (2011) studied the degree of sterilization in Vietnam from the Q1/2000 to the Q3/2010 with OLS estimation. The intervention were relatively low in Vietnam during the study period (−0.24).

Dang (2015) measured the level of neutralization in Vietnam by examining the regression model according to Aizenman and Glick (2009) with data collected from 2000 to 2013. The result was that the sterilization coefficient is −0.475, so the effectiveness of sterilization policy was not effective yet.

VAR Model
The VAR model is a generalization of the one-dimensional self-regression model in predicting a set of variables, i.e. a vector of time series variables. It considers the variables in the model interacting with the corresponding lags. Thus, the VAR model is more advantage than the OLS estimate. It has a more comprehensive view of the interactions between the variables. Moreover, VAR model allows one to trace out the time path of the various shocks on the variables contained in the VAR system. If a shock from foreign assets is associated with an offsetting decrease in domestic money creation, it can be concluded that the sterilization is significant.

Empirical studies using the VAR model to evaluate the efficiency of sterilization are included:

Moreno (1996) used the VAR model with four variables (the nominal exchange rate, CPI, foreign assets and domestic credit) to find out how the monetary authorities reacted to shocks in Korea and Taiwan during January 1981 to December 1994. Analysis suggests that sterilization are an important factor in responding to shocks that alter foreign assets in both economies.

He et al. (2005) considered the effectiveness of sterilization in China. They also found out whether the domestic credit boom had been influenced by foreign inflows. The VAR model was built with four variables: NDA, NFA, domestic credit (DCR) and interest rates (IR). On the effectiveness of sterilization, dynamic analysis showed that the NDA responds to NFA fluctuations, suggesting that the NDA was used by the Chinese central bank to neutralize the impact of capital inflows on the base. Increasing NFA by one unit leaded to a decrease in NDA and most reactions occurred within one month.

2.2   The Second Approach

In the second approach, the researchers uses the simultaneous equation to examine the relationship between NDA and NFA with the 2SLS estimation method. They consider this aspect because there is a causal relationship between these variables in the model. The main issue here is that NFA and NDA change at the same time and thus should be considered endogenous. Domestic currency conditions are affected by capital inflows, whereas capital inflows are adversely affected by domestic currency conditions. The research model has the general form are as follows:

$$\Delta NDA_t = \lambda_{11}(\Delta NFA_t) + \lambda_{12}Z_1 + \varepsilon_{1t} \tag{5a}$$

$$\Delta NFA_t = \lambda_{21}(\Delta NDA_t) + \lambda_{22}Z_2 + \varepsilon_{2t} \tag{5b}$$

where Z_1 and Z_2 are the vectors of control variables in the capital - flow equation (Eq. 5a) and the monetary reaction function (Eq. 5b). The coefficient λ_{11} is represented as offset coefficient and the coefficient λ_{12} is represented as sterilization coefficient. In contrast to the monetary reaction function, the capital-flow equation built from Kouri and Porter (1974) model shows another view of the relationship between accumulation of foreign reserves and monetary policy measures. It allows estimating the effect of monetary policy on capital inflows from abroad. Accordingly, the dependent variable is represented by the change of NFA, and the independent variable is the change of NDA. The offset coefficient value was also expected in the range from -1 to 0. If the offset coefficient $\lambda_{11} = -1$, it means that the capital flows are fully mobile and sterilization was not effective because the total amount of the decline in the central bank's NDA was replaced by additional capital inflows in the same amount, and they contributed to the increase in NFA. This additional inflow then needed to be sterilized again, which results in a vicious cycle of high capital inflows and a need for further neutralization interventions. If the offset coefficient is close to zero, it means the NDA change of the central bank due to the effects of sterilization of central bank partially or wholly remains within the system, affects the total money supply. In general, the higher the volatility and the higher the level of substitution between the NDA and the NFA, the lower the degree of sterilization. The low value offset coefficient and high value sterilization coefficient indicate a relatively high degree of monetary policy effectiveness (Ouyang et al. 2008).

2SLS Estimation Method

The 2SLS estimate has been a common estimation for evaluating the concurrency of variables in a model. At the same time, 2SLS estimates also address endogenous issues. At the econometric points, the occurrence of endogenous variables would lead to cases such as abandoning variables, error in variables, or being concurrently determined by other explanatory variables. In these cases, the OLS method and the VAR model are not reliable estimators. And the general appropriate method is 2SLS estimation. Empirical studies using the 2SLS estimation method to evaluate the effectiveness of the central bank's sterilization are included:

Brissimis et al. (2002) who based on the loss function of the monetary authority constructed a simultaneous equation for estimating offset coefficient and sterilization coefficient. Empirical study was conducted in Germany from the second quarter of 1980 to the second quarter of 1992 with the 2SLS estimation method. The results showed that the offset coefficient was relatively low (-0.22) during the maintenance of the Exchange Rate Mechanism (ERM) and the sterilization coefficient was approximately -1. That proved that Germany had succeeded with the goal of both short-term and long-term interventions.

Ouyang et al. (2010) assessed the extent of sterilization and capital mobility in China using monthly data between mid-2000 and late-2008. They applied 2SLS to estimate two simultaneous equations:

$$
\begin{aligned}
\Delta NFA_t^* &= \alpha_0 + \sum_{i=0}^{n} \alpha_{1i}(\Delta NDA_{t-i}^*) + \sum_{i=0}^{n} \alpha_{2i}(\Delta mm_{t-i}) + \sum_{i=0}^{n} \alpha_{3i}(\Delta p_{t-i}) \\
&+ \sum_{i=0}^{n} \alpha_{4i}(\Delta yc_{t-i}) + \sum_{i=0}^{n} \alpha_{5i}(\Delta REER_{t-i}) \\
&+ \sum_{i=0}^{n} \alpha_{6i}(\Delta(r_{t-i+E_t e_{t+1-i}}^*)) + \varepsilon_t
\end{aligned}
\tag{6a}
$$

$$
\begin{aligned}
\Delta NDA_t^* &= \alpha_0 + \sum_{i=0}^{n} \beta_{1i}(\Delta NFA_{t-i}^*) + \sum_{i=0}^{n} \beta_{2i}(\Delta mm_{t-i}) + \sum_{i=0}^{n} \beta_{3i}(\Delta p_{t-i}) \\
&+ \sum_{i=0}^{n} \beta_{4i}(\Delta yc_{t-i}) + \sum_{i=0}^{n} \beta_{5i}(\Delta G_{t-i}) \\
&+ \sum_{i=0}^{n} \beta_{6i}(\Delta(r_{t-i+E_t e_{t+1-i}}^*)) + v_t
\end{aligned}
\tag{6b}
$$

They used government expenditure (G) as an instrument for NDA and real effective exchange rate (REER) for NFA. The result showed that China had typically sterilized around 90% of the reserve inflows.

Wang (2010) gave an empirical evaluation of the effectiveness of China's sterilization and capital mobility regulations, measured by sterilization and offset coefficients, using monthly data between mid-1999 and March 2009. Different from other studies, this paper paid more attention to the effectiveness of China's sterilizations in terms of M_2. He found that the effectiveness of China's sterilizations is almost perfect in terms of the monetary base, but not in terms of M_2, and that China's capital controls still work but are not quite effective.

Ljubaj et al. (2010) considered the degree the Croatian National Bank sterilizes capital inflows in the period from 2000 to 2009. The sterilization coefficient was estimated through the estimation of the monetary policy reaction function, while the capital-flow equation was used to estimate the offset coefficient. Econometric estimation (2SLS) was used to find the relevant explanatory variables that enabled the estimation of the mentioned coefficients. According to the estimated coefficients, it could be concluded that during the study period, the Croatian central bank implemented a strong sterilization policy but did not completely neutralize the capital inflows.

Ouyang and Rajan (2011) conducted empirical research to determine the true level of harm reduction and capital flow in Singapore and Taiwan using quarterly data from 1990 to 2008. The simultaneous equation between NDA and NFA in the Brissimis et al. (2002) were modified. They empirical results suggested that, since, 2001 both Singapore and Taiwan have a high degree of capital mobility. To date, this high-effective capital mobility has not undermined the ability of the central bank in either economies to sterilize their respective foreign exchange intervention but may make the process increasingly difficult over time.

3   Conclusion

As discussed above, empirical studies on the efficient sterilization have been assessed by two approaches. The first approach evaluates the NDA fluctuation under the NFA based on the monetary reaction function. The second approach uses a simultaneous equations based on the monetary reaction function and the capital – flow function. With the first approach, researchers ignored the simultaneous trends by assuming that capital flows are exogenous. But it is important that domestic monetary conditions are affected by changes in international capital flows and foreign exchange reserves. At the same time, international capital flows respond to changes in domestic monetary conditions. Therefore, through using the simultaneous equations towards the second approach, this similarity problem has been solve.

The main econometric methods have been used to estimate the sterilization effect: OLS method, VAR and 2SLS. The simplest estimate is OLS estimation. VAR model has the advantage of tracking the impact of any variable on others through the impulse response function. Moreover, VAR model can be used to estimate the lagged effects of NDA and NFA. However, the limitation of VAR model is that it tends to treat all endogenous variable symmetrically. Consequently, the model cannot estimate the contemporary effect of variables without restrictions. The 2SLS estimate overcomes the limitations of the VAR model, which is the most appropriate estimation method for assessing the efficiency of central bank's sterilization policy.

In Vietnam, foreign exchange reserves have tended to accumulate increasingly in recent years. From the first quarter of 2000 to the third quarter of 2016, foreign exchange reserves increased from 3.5 billion to 37.6 billion (Source: International Financial Statistic). However, in this period, inflation and money supply also change complicatedly with the same increasing trend. In particular, accumulation of foreign exchange reserves is one of the factors making money supply increase. Thus, sterilization of the SBV has not been effectively and is needed to evaluate in terms of causes and degree. As discussed above, the second approach is the most appropriate approach for assessing the efficient sterilization. Therefore, to assess the efficiency of the SBV's sterilization accurately, the appropriate estimation method is the 2SLS estimation. With this method, the simultaneous equations (Eqs. 6a and 6b) are able to clearly estimate the sterilization coefficient and offset coefficient together with the factors influenting these relationships.

References

Aizenman, J., Glick, R.: Sterilization, monetary policy, and global financial integration. Rev. Int. Econ. **17**(4), 777–801 (2009)

Brissimis, S.N., Gibson, H.D., Tsakalotos, H.D.: A unifying framework for analysing offsetting capital flows and sterilization: Germany and the ERM. Int. J. Finan. Econ. **7**(1), 63–78 (2002)

Cavoli, T., Rajan, R.S.: Capital inflows problem in selected Asian economies in the 1990s revisited: the role of monetary sterilization. Asian Econ. J. **20**(4), 409–423 (2006)

Cumby, R.E., Obstfeld, M.: Capital mobility and the scope for sterilization: Mexico in the 1970s. In: Financial Policies and the World Capital Market: The Problem of Latin American Countries, pp. 245–276. University of Chicago Press (1983)

Dang, V.D.: Measure the neutralization response to foreign currency inflows into Vietnam. J. Econ. Forecast. **16**, 74–77 (2015)

Glick, R., Hutchison, M.: Navigating the trilemma: capital flows and monetary policy in China. J. Asian Econ. **20**(3), 205–224 (2009)

He, D., Chu, C., Shu, C., Wong, A.: Monetary management in Mainland China in the face of large capital inflows, Research Memorandum - Hong Kong Monetary Authority, No. 7 (2005)

Heller, W.W.: The Economy: Old Myths and New Realities. W. W. Norton & Co., New York (1976)

Krugman, P.R., Obstfeld, M., Melitz, M.J.: International Economics: Theory and Policy, 9/E, p. 467. Pearson Education, Noida (2012)

Kouri, P.J., Porter, M.G.: International capital flows and portfolio equilibrium. J. Polit. Econ. **82**(3), 443–467 (1974)

Ljubaj, I., Martinis, A., Mrkalj, M.: Capital inflows and efficiency of sterilization-estimation of sterilization and offset coefficients, Working papers - Croatian National Bank, No. W-24 (2010)

Moreno, R.: Intervention, sterilization, and monetary control in Korea and Taiwan. Econ. Rev. Fed. Reserve Bank San Franc. **1996**(3), 23–33 (1996)

Ouyang, A.Y., Rajan, R.S., Willett, I.: Managing the monetary consequences of reserve accumulation in emerging Asia. Global Econ. Rev. **37**(2), 171–199 (2008)

Ouyang, A.Y., Rajan, R.S., Willett, I.: China as a reserve sink: the evidence from offset and sterilization coefficients. J. Int. Money Finan. **29**(5), 951–972 (2010)

Ouyang, A.Y., Rajan, R.S.: Reserve accumulation and monetary sterilization in Singapore and Taiwan. Appl. Econ. **43**(16), 2015–2031 (2011)

Pham, T.T.T., Nguyen, T.H.V.: Impact of capital inflows on money supply and the level of intervention by SBV. J. Bank. Technol. **6**(66) (2011)

Steiner, A.: Does the Accumulation of International Reserves Spur Inflation? A Panel Data Analysis. University of Osnabrueck, Osnabrueck (2009)

Takagi, S., Esaka, S.: Sterilization and the capital inflow problem in East Asia, 1987–97. In: Regional and Global Capital Flows: Macroeconomic Causes and Consequences, NBER-EASE (2001)

Wang, Y.: Effectiveness of capital controls and sterilizations in China. China World Econ. **18**(3), 106–124 (2010)

Application of Statistical Methods for Tax Inspection of Enterprises: A Case Study in Vietnam

Nguyen Thi Loan[1]([✉]), Le Dinh Hac[1], and Nguyen Viet Hong Anh[2]

[1] Department of Accounting and Auditing, Banking University
of Ho Chi Minh City, Ho Chi Minh City, Vietnam
loannt@buh.edu.vn
[2] Binh Thanh District Tax Department, Ho Chi Minh City, Vietnam

Abstract. In this study, we apply statistical methods based on Benford's Law for checking accuracy of tax reports. Specifically, instead of the usual practice of randomly selecting documents for detailed scrutiny, the proposed method selects the most suspicious document. Our experience of using this method has shown that its application has drastically increased the probability of detecting tax fraud. This method is relatively easy to use, it is based on a simple Excel-based algorithm.

1 Formulation of the Problem

According to the Oxford dictionary, tax is the amount of money that the government can use to cover public services. Accordingly, people declare and pay tax on their income, and businesses pay tax on their profits. In addition, taxes should be paid on goods and services.

The amount of tax that an enterprise needs to pay is determined based on the accounting data provided by these enterprises. To avoid tax fraud, it is therefore necessary to analyze the accounting data, so as to detect possible irregularities.

The overall amount of accounting data is huge, and it is not possible to inspect all this data in detail. As a result, only a sample of enterprises – and/or a sample of their accounting documents – undergo a detailed tax audit. At present, the selection of enterprises to audit – and the selection of documents that will be audited – are determined by the experience of a tax inspector. This often leads to a tax fraud which is not initially detected – and to inspector's time wasted on audits in which no tax fraud is revealed. To make tax audits more efficient, it is therefore desirable to supplement the tax inspectors' subjective experience with statistical techniques so as to increase the probability of fraud detection.

In this paper, we propose such techniques based on Benford's Law.

L. H. Anh et al. (eds.), *Econometrics for Financial Applications*, Studies in Computational Intelligence 760, https://doi.org/10.1007/978-3-319-73150-6_51

2   Theoretical Background and Related Studies

2.1   Benford's Law

At first glance, nature should not prefer numbers starting with 1 or with 2. It therefore seems reasonable to expect that in a big database, numbers starting with all 9 digits should be equally probable: that numbers starting with 1 should occur with the same probability 1/9 as numbers starting with 2, 3, etc. Surprisingly, his belief was shown to be wrong. In 1938, an American scientist Frank Albert Benford published a paper in which he showed that, in many databases, the frequency of numbers starting with a digit d_1 is proportional to

$$P(d = d_1) = \log_{10}\left(1 + \frac{1}{d_1}\right),$$

Benford called this The Law of Anomalous Numbers; it is now known as the first-digit law or Benford's law (Table 1).

Table 1. Frequencies based on Benford's Law and its extensions (see Nigrini 1996)

Digit	1st place	2nd place	3rd place	4th place
0		.11968	.10178	.10018
1	.30103	.11389	.10138	.10014
2	.17609	.19882	.10097	.10010
3	.12494	.10433	.10057	.10006
4	.09691	.10031	.10018	.10002
5	.07918	.09668	.09979	.09998
6	.06695	.09337	.09940	.09994
7	.05799	.00035	.09902	.09990
8	.05115	.08757	.09864	.09986
9	.04579	.08500	.09827	.09982

Similar formulas are known for the frequency of different 2-digit, 3-digit, etc. combinations starting a number; see, e.g., Table 2.

From the theoretical point, Benford's Law remains largely a mystery. More than 150 theoretical research articles have been published, and there is still no single convincing explanation for this empirical law; see, e.g., Durschi et al. (2004).

Of course, not all the data follow Benford's Law: while the raw data points follow it, the results of processing raw data often deviate from Benford's Law; see Tables 3 and 4.

Table 2. Types of data appropriate for Benford analysis

Data satisfying Benford's Law	Examples
The data obtained by applying arithmetic operations to two or more numbers	Accounts receivable (sale amount * price)
Secondary transaction data	Disbursements, taxes, costs
Big Data Set - The more observations, the better	Full transaction of the year
Items that appear reasonable – when the average value of a set of numbers is greater than the median and the positive quantities are indeed positive	Most of the accounting data sets

(Source: Durtschi et al. 2004)

Table 3. Data types not consistent with Benford's analysis

Data not satisfying Benford's Law	Examples
The dataset consists of the assigned numbers	Examination number, invoice number, zip code
Numbers are influenced by human thought	Nominal, Referral, ATM withdrawals
Accounts with large numbers of specific numbers have identified	A dedicated account is set up to record the refund
Accounts are set to the smallest or largest	The set of assets whose value must meet the specified level is recorded

(Source: Durtschi et al. 2004)

Table 4. Results of the survey on tax inspection techniques of enterprises by tax authorities.

Code	The techniques are currently being used to carry out tax inspections	People agree	Proportion
1	Manual testing only	41/187	22,16%
2	The whole test performed on a computer	63/187	34,05%
3	Combination the manual test with computerized tax examination	81/187	43,78%

2.2 Use of Statistical Methods and Benford's Law to Detect Accounting Fraud: A review of Previous Studies

To the best of our knowledge, the first application of statistical techniques to audits was done by Donald R. Cressey (1987). He used statistical methods to analyze frauds. His model – which he called fraudulent triangulation model – helped to find factors leading to fraudulent behavior: pressure, opportunity, and personal attitude. Alon and Dwyer (2010) offered a more detailed model that take into account the auditor's judgment, customer information, task division, and audit procedures.

Research of this type was also performed on the example of Vietnam tax returns, we should mention Tan (2009) who analyzed the most popular frequently used fraud techniques; Giang (2015) who analyzed the possibility of serious misconduct on the financial statements of enterprises listed on the stock market of Vietnam; Nga (2011) who performed a critical assessment of audit risk to improve the quality of audits performed by Vietnamese auditing firms; and Ba and Hung (2016) who proposed a statistical model of fraudulent accounting that enable the auditors to evaluate the truthfulness of financial reports.

Several researchers applied Benford's Law to detect accounting and tax fraud. Nigrini (1996) applied Benford's Law to examine the US income tax returns. This study was expanded in his 1999 paper. Cleary and Thibodeau (2005) applied Benford's Law to detect fraud in general accounting data. Nigrini and Miller (2009) used the distribution of the first two digits to detect accounting discrepancies.

However, all these studies did not lead to an algorithmic tool for fraud detection.

3   Research Methodology

3.1   Need for Statistical Tools

To understand whether there is indeed a need for statistics-based tools for facilitating tax inspection, we conducted a survey of 200 tax officers in charge of tax inspection and tax examination at Tax Department in Ho Chi Minh City. This survey was performed from March 2017 to May 2017. As a result of this survey, we received 187 replies.

The results of the survey in Table 5 showed that the tax officials have started to use computers in auditing. In addition to 34% of tax officers who use mostly computer-based tools in their audits, 44% of tax officers supplement manual auditing with the use of computer-based tools. Thus, the vast majority of tax inspectors are familiar with at least computer-based inspection techniques.

However, as Table 6, the existing computer-based tax inspection tools have limited scope and functionality. The most frequently used tool is the general-purpose MS Excel software, and special tax-related software tools are rarely used. It is therefore desirable to enhance the use of such software.

Table 5. Results of the survey on the level of using supporting tools in tax inspection by the tax authorities.

Code	Application of supporting tools in tax inspection	People agree	Proportion
1	TPR Risk analysis software	24/187	12,97%
2	TMS – Tax Management System	34/187	18,38%
3	iHTKK – Software support tax declaration	67/187	36,22%
4	Excel – Microsoft Excel Software	93/187	50,27%
5	SPSS, Eview, Stata – Econometric software	3/187	1,62%

Table 6. Results of error analysis for Value Added Tax (VAT).

Code	Enterprises
1	Banh Phap Trading Service Co., Ltd.
2	Marine Engineering Joint Stock Company
3	Truong An Phu Company Limited
4	Tue Tinh Electric Mechanical Construction Co., Ltd.
5	Nm Shipping Co., Ltd.
6	E3 Furniture Co., Ltd.
7	Hoang Gia Container Trading Co., Ltd.
8	Global Sun Global GNHH Company Limited
9	Lam Viet Trading & Printing Services Co., Ltd.
10	Hydraulic Construction Labor Supplying Joint Stock Company
11	First Industrial Safety Inspection Joint Stock Company
12	Thao Tung Trading Production Co., Ltd.
13	Hoa Lan Viet Co., Ltd.
14	An Huy Transportation Service Co., Ltd.
15	Dai Nam Trade Construction Consultant Co., Ltd.
16	Phuong Nghi Transportation Service Trading Co., Ltd.
17	Kim Gia Phat Transport Company Limited
18	Hung Long Tourism Company
19	Thy Khanh Linh Transportation Service Company Limited
20	Indochina Pharmaceutical Company Limited
21	Viet Thang Transport Company Limited
22	Long Giang Production and Trading Co., Ltd.
23	Minh Vy Technology Company Limited
24	Thuan Phat Chemicals Co., Ltd.
25	Huu Thien Loc Company Limited
26	Gimiko Manufacturing & Trading Co., Ltd.
27	EMC Medical Equipment Limited Company
28	Southern Fuel Company Limited
29	Thanh Tan Container Co., Ltd.
30	Anh Huy Trading - Construction - Manufacture Co., Ltd.

(Source: Calculated from enterprise datasets)

3.2 Designing a Statistics-Based Algorithm for Detecting Tax Errors at the Enterprise Level

We apply Benford's Law to the accounting data of 30 Ho Chi Minh City enterprises, data provided to the tax authorities in 2014. The objective of this method is to estimate the error level of items showing tax liability and illustrate how to identify areas of material discrepancy across the entire accounting data of the item under review to support the tax inspection and tax examination for enterprises in Vietnam.

According to Benford's Law, in a given set of data values, the numbers starting with the i-th digit should appear with probability p_i as described by Benford's formula. The actual frequency f_i if such numbers is, in general, somewhat different from p_i.

According to the mathematical statistics, with certainty α (e.g., 95%), we observed frequencies should satisfy the inequality

$$p_i - \varepsilon_i < f_i < p_i + \varepsilon_i,$$

where ε_i is the estimation accuracy

$$\varepsilon_i = t_\alpha \cdot \sqrt{\frac{p_i \cdot (1 - p_i)}{n}},$$

and n is the sample size (number of observations). Thus:

- if $|p_i - f_i| < \varepsilon_i$, there is nothing suspicious with the data,
- on the other hand, if $|p_i - f_i| \geq \varepsilon_i$, the situation is suspicious, so this part of the company's tax report should probable be edited.

For each i, we check suspiciousness based on the value of the difference $|f_i - p_i|$. As a measure of overall suspicion, it is therefore to take the average value of this difference, i.e., the expression

$$\sum_{i=1}^{9} f_i \cdot |p_i - f_i|.$$

This is the criterion that we used in this paper – that companies with the highest value of this average difference should be audited.

4   Resulting Methodology

For each set of data, we compute the frequencies f_i of numbers starting with different digits i, and then compute the above-defined average difference. Companies and parts of the reports with a high value of average difference should be audited.

The results of using this methodology are shown in Table 7.

The results show that for each company, tax officers can identify the areas of material misstatement in the Value Added Tax item as numeric digits beginning with the numbers shown in the "Critical error" column. Also, tax authorities can also estimate the error rate for each accounting items related to the tax liability of enterprises quickly and accurately, in order to sample the data area for tax inspections and tax examinations.

Table 7. Results of error analysis for Value Added Tax (VAT).

Code	Tax code	Source of data	Number of observations	The error rate of the item	Critical error area 5%
1	0305054514	NKC	1.536	0,0227	2, 3
2	0305499979	NKC	56	0,0417	1, 2, 3, 4, 6
3	0304859900	NKC	2.523	0,0331	1, 2, 3, 6
4	0311836157	NKC	45	0,0453	1, 2,3,9
5	0312491804	NKC	350	0,1269	1, 2, 3, 4, 5
6	0312542907	NKC	98	0,1464	1, 5
7	0311978930	NKC	470	0,0287	1, 2, 3
8	0303841643	NKC	619	0,0968	1, 2, 5
9	0308837947	NKC	354	0,0112	1, 2, 3, 4, 6, 7
10	0309801982	NKC	3.364	0,0731	1, 3, 4
11	0307225042	NKC	60	0,1253	1, 2, 3, 5
12	0303428732	NKC	95	0,0782	1, 2, 3
13	0303286453	NKC	622	0,0671	1, 3
14	0310930816	NKC	115	0,0389	1, 3
15	0310209033	NKC	868	0,1665	1
16	0311691328	NKC	42	0,0837	1, 5, 6, 7
17	0310000874	NKC	437	0,0411	1, 2, 3, 4, 5, 6
18	0309132717	NKC	872	0,1438	1, 2, 3
19	0308164385	NKC	108	0,1293	1, 2
20	0310566451	NKC	606	0,0215	1, 2, 3, 4, 5, 9
21	0304969283	NKC	218	0,0365	1, 2, 5, 6, 7, 8
22	0303886468	NKC	100	0,0384	1, 2, 3, 4
23	0311817034	NKC	371	0,2903	1
24	0311588539	NKC	151	0,0543	1, 2, 3
25	0309467470	NKC	26	0,0326	1, 2, 9
26	0301692805	NKC	1.249	0,0352	1, 2, 3, 4
27	0310631397	NKC	1.855	0,0082	1, 2, 3, 4, 5, 6, 8
28	0312009110	NKC	326	0,1054	1, 2, 3
29	0311795454	NKC	140	0,0454	1, 2, 3
30	0304475566	NKC	1.021	0,0363	1, 2, 3

(Source: Calculated from enterprise datasets)

5   Conclusion

This study has developed methodologies for using the Benford's Law in to select test patterns on accounting data. The results of this study may be supplemented with a system of sampling to check methods, providing tax officers with a basis for assessing and concluding the honesty of the business accounting data more accurately compared with the current test methods. Moreover, this method makes it easier for inspectors to measure the error rate of accounting items.

With the widespread application of MS Excel software in tax inspections, this study provided a tool to assist tax officials to perform faster and more efficient inspections. This will significantly reduce the workload and overall inspection time in the tax administration in Vietnam.

Desirable future work is related to the fact that other methods are known to improve tax fraud detection. For example, Sharma (2010) applied data mining technique – techniques for finding rules in data – to detect abnormal data groups in audit items. Khomurji et al. (2014) use artificial neural networks (ANNs) to detect accounting fraud. It is reasonable to combine these techniques with others, e.g., techniques that use data mining, Bayesian networks, and neural networks.

References

Alon, A., Dwyer, P.: The impact of groups and decision aid reliance on fraud risk assessment. Manag. Res. Rev. **33**(3), 240–256 (2010)

Cleary, R., Thibodeau, J.: Applying digital analysis using Benford's Law to detect fraud: the dangers of type I errors. Audit. J. Pract. Theor. **24**(1), 77–81 (2005)

Durtschi, C., Hillison, W., Pacini, C.: The effective use of Benford's Law to assist in detecting fraud in accounting data. J. Forensic Account. **5**, 17–34 (2004)

Nga, D.T.T.: Critical evaluation study and audit risk to improve the quality of operations in Vietnamese auditing firms (2011)

Khormuji, M.K., Bazrafkan, M., Sharifian, M.: Credit card fraud detection with a cascade artificial neural network and imperialist competitive algorithm. Int. J. Comput. Appl. **96**(25), 1–9 (2014)

Giang, L.T.C.: Identify potential misstatements on financial statements of companies listed on the Vietnam stock market. Law on Tax Administration No. 78/2006/QH11 dated 29/11/2006 (2015)

Nigrini, M.J.: A taxpayer compliance application of Benford's Law. J. Am. Tax. Assoc. **18**(1), 72–91 (1996)

Nigrini, M., Miller, S.: Audit. J. Pract. Theor. Data diagnostics using second-order tests of Benford's Law **28**(2), 305–324 (2009)

Hindls, R., Hronova, S.: Benford's Law and Possibilities for Its Use in Governmental Statictic. University of Economics, Prague (2015)

Tan, T.T.G.: The most common financial fraud reports and the actual situation in Vietnam (2009)

Ba, T.T., Hung, N.V.: Study the method of error checking of accounting data in support of audit of financial statements. J. Sci. Hanoi Natl. Univ. **32**(3), 61–70 (2016)

Sharma, A., Panigrahi, P.K.: A review of financial accounting fraud detection based on data mining techniques. Int. J. Comput. Appl. **39**(1), 37–47 (2012)

Detecting Corporate Income Tax Non-compliance from Financial Statements: A Case Study of Vietnam

Nguyen Thi Loan[1]([✉]), Nguyen Viet Hong Anh[2], and Pham Phu Quoc[3]

[1] Department of Accounting and Auditing,
Banking University of Ho Chi Minh City, Ho Chi Minh City, Vietnam
loannt@buh.edu.vn
[2] Binh Thanh District Tax Department, Ho Chi Minh City, Vietnam
[3] Ho Chi Minh City Finance and Investment, State-Owned Company,
Ho Chi Minh City, Vietnam

Abstract. In this paper, we combine interviews, surveys, and panel data regression to find factors affecting corporate income tax (CIT) non-compliance. This study is based on the analysis of 105 Vietnamese companies which were inspected by tax officials in 2011–2015. The results show that the following seven factors affect CIT non-compliance: the ratio of Working Capital/Total Assets, Turnover/Total Assets, Previous Loss, Inventories/Total Assets, Accounts Receivable/Turnover, size of the enterprise, and debt fines for tax administrative/tax amounts payable in the period. The article shows that the information from the financial statements can help the tax officials detect the CIT non-compliance, and suggest appropriate tax management policies for enterprises in Vietnam.

1 Formulation of the Problem

Tax is an important source of government revenue in every country. All the subjects in the economy are obliged to comply with the tax laws. The enterprises are the main taxpayers contributing to the government budget. Businesses try to minimize the amount of taxes. Sometimes, this is done legitimately, by using legal tax reductions, but sometimes, to decrease the tax obligations, companies submit financial statements with fraudulent information fraud about sales, assets, profits, liabilities, costs, and/or losses (Guenther 1994; Spathis 2002). Badertscher et al. (2006) found out that in most cases, this fraud involves decreasing the stated declared earnings. As a result, researching in the fraud of financial statements is one of the important tasks of the tax administration (Rohaya et al. 2012). Therefore, the detection of tax non-compliance is important for ensuring adequate budget revenue.

Hai and See (2011) emphasized that since tax obligations are based on the companies' financial statements, it is important to analyze these statements to detect possible non-compliance. Firms are sometimes taking advantages of different regulations between financial reporting and tax reporting to lower their

© Springer International Publishing AG 2018
L. H. Anh et al. (eds.), *Econometrics for Financial Applications*, Studies in Computational Intelligence 760, https://doi.org/10.1007/978-3-319-73150-6_52

income tax liabilities (Rohaya et al. 2008). It is therefore important to use not only financial statements submitted to the tax authorities, but also financial statements provided to investors, donors and shareholders of the company.

In Vietnam, since 2013, tax policies have been regularly amended not only to support businesses of enterprises, but also to enhance the CIT compliance. Corporate taxes form a significant proportion to the state budget. The Vietnamese Corporate Income Tax Law No. 14/2008/QH12 dated June 3, 2008 is still valid up to now; it was amended and supplemented by Laws No. 32/2013/QH13 dated June 19, 2013 and Law No. 71/2014/QH13 dated November 26, 2014. This law has created a large legal framework of regulations on CIT with which enterprises have to comply. To create a favorable business environment for enterprises, the CIT Law has adjusted the tax rate down from 28% in 2008 to 20% in 2016. As a result, the proportion of CIT in the Vietnam state budget revenue decreased from 49.63% to 43.05% in 2015. This decrease makes it important to ensure that all CITs are paid in full. In practice, however, the proportion of tax amount collected through tax inspection and tax examinations also increased a lot, which indicates that the number of tax evasions has increased.

This increase can be partly explained by the fact that, in accordance with the general trend of administrative reform and integration, the tax law of Vietnam is encouraging enterprises to fulfill their tax obligations by self-declaration, self-submission, and self-responsibility. It is therefore becoming more and more important for the tax authorities to detect tax non-complicance based on the data provided by the enterprises.

The main objective of this study is to use the current data about tax non-compliance by different companies to develop a model for detecting possible non-compliance based on the financial statements. The paper consists of five sections. Following this introductory section is Sect. 2 that provides an overview of the previous studies. Section 3 describes our research methodology. The results of this research are presented in Sect. 4. Conclusions and corresponding policy recommendations form the last Sect. 5.

2   An Overview of Previous Studies

2.1   The Theory of Corporate Income Tax and of Corporate Income Tax Non-compliance

According to the Oxford Dictionary, tax is the amount payable to the State to cover public services. People pay taxes on their income and businesses pay taxes on the profits they generate. Tax is also payable on goods and services. According to the Vietnamese dictionary, tax is a sum of money or things which must be paid to the government according to regulations. According to the Organization for Economic Co-operation and Development (1996), taxes are mandatory, limited contributions to the government. The taxes paid will be refunded respectively by the benefits brought by the government in special form.

Article 4 of Law on Tax Administration of Vietnam No. 78/2006/QH11 dated November 29, 2006 stipulates that "Taxes constitute a major revenue source of the state budget. It is an obligation and a right of all organizations and individuals to pay taxes in accordance with law. Agencies, organizations and individuals shall participate in tax administration." The subjects in the economy must be obliged to pay taxes, and for this purpose, enterprises are very important subjects.

According to Oxford Dictionary and Vietnamese Dictionary definition, CIT is the tax amount which businesses pay on their profits. According to Article 6 and Article 7 of the Law on Corporate Income Tax of Vietnam No. 14/2008/QH12 dated June 3, 2008, the basis for calculation of CIT tax is taxed income and tax rate. Taxed income in a tax period is the taxable income minus tax-exempt incomes and minus losses carried forward from previous years. In other words, taxable income is turnover minus deductible expenses for production and business activities plus other incomes, including income received outside Vietnam. Therefore, two important elements of CIT are:

Taxed income = Revenue in the period − Deductible Expenses in the period.

Jones (2009) defines that tax compliance as:

- filing and disclosure of all the information related to tax obligations of tax-payer, and
- paying the payable tax amounts in full and on time.

If taxpayers do not pay tax and/or declare the fraudulent income, thus reducing the CIT amount, this constitutes tax non-compliance. According to Kasipillai (2006); Tan and Sawyer (2003), the CIT compliance is defined as the accurate declaration of income and expenses in accordance with tax law. The companies that fail to accurately report these two elements or do not carry out tax obligations are CIT non-compliant.

Alabede et al. (2011) explain that there are two types of non-compliance. Some individuals and companies do not comply with tax law accidentally or because do not fully understand the current regulations. Others perform what is known as tax fraud − i.e., perform illegal actions that reduce the amount of tax payable. According to GAO (2012), tax non-compliance is measured by tax differences that include both accidental irregularities and fraud.

As mentioned in Sect. 1, financial statements are the legal basis for defining the tax obligations of the business. Rezaee (2002) showed that misrepresentations in the financial statements include may affect transactions, receivables, and payables. Specifically, companies try to inflate expenses or hide turnover in order to reduce CIT amount (Harris et al. 1993). This paper will analyze how we can detect such fraud.

2.2 Review of Related Studies

Many authors from all over the world analyzed tax non-compliance of enterprises. Rice (1992) and Kamdar (1997) studied the factors affecting CIT compliance in the United States. They used the Tobit regression method to find factors that affect tax compliance. For this study, they analyzed possible affecting factors such as profitability, firm size, and tax rate. Later, Hanlon et al. (2007) expanded the study on the CIT non-compliance to the data from 29,141 companies in the United States. By using Tobit regression, the results of the study have confirmed that the financial factors of enterprises affect the tax non-compliance.

Several other researchers have found a relationship between tax non-compliance and information from the financial reports. For example, Mills (1996) found that the probability of tax fraud increases with the difference between the book profits and taxable income that enterprises have declared. Also, unreasonable information in financial reports sometimes lead to their inconsistency, which is often an indication of tax non-compliance of enterprises (Frank et al. 2004).

By analyzing Malaysian firms, Rohaya et al. (2012) came up with the following regression model describing non-compliance:

$$TE = \beta_0 + \beta_1 WC + \beta_2 SAL + \beta_3 DEBT + \beta_4 ETR + \beta_5 INV + \beta_6 AR + \varepsilon,$$

where

- TE = Non-declared taxable income/Total Assets,
- WC = Working Capital/Total Assets,
- SAL = Total Revenue/Total Assets,
- $DEBT$= Total Debt/Total Assets,
- ETR = Tax paid/Net profit,
- INV = Inventory Value/Revenue,
- AR = Receivable Amount/Total Revenue.

This model was expanded in Yusof et al. (2014) that took into account other factors such as business field, enterprise size, and ownership. The resulting model is as follows:

$$\log(CTNC) = \beta_0 + \beta_1 PEN + \beta_2 MTR + \beta_3 LIQ + \beta_4 FORE + \beta_5 \log ASSETS \\ + \beta_6 MAN + \beta_7 CON + \beta_8 WSALE + \beta_9 SER + \varepsilon,$$

where

- $CTNC$ = Log (Unreported Revenue/Total Actual Revenue),
- PEN = 1 if the penalty rate for tax non-compliance is at or below 45% and 1 if it is above 45%,
- MTR = Tax Payable/Gross Profit Before Tax,
- LIQ = Working capital/Total assets,
- $FORE$ = 1 if the company is foreign-owned, 0 if it is wholly domestic-owned,
- $\log(ASSETS)$ = log(Total Assets),
- MAN = 1 if the company is in the manufacturing sector, 0 otherwise, CON = 1 if the company is in the construction sector, 0 otherwise,
- $WSALE$ = 1 if the company is in wholesale trade business, 0 otherwise,
- SER = 1 if a company in the service sector, 0 otherwise.

Several studies have been performed on tax non-compliance in Vietnam. Truong and Chien (2013) anlayzed tax evasion behaviors by Vietnamese businesses. Phan My Hanh (2003) analyzes the behavior of Value Added Tax fraud in enterprises. They concluded that the available data was not sufficient to detect which factors affect non-compliance.

In this study, we show that for CIT, there is already enough data to determine the corresponding factors.

3   Research Methodology and Model

3.1   Research Methodology

In this study, we only analyze the CIT obligations as described in the CIT Law, in the Law on Tax Administration in Vietnam, and in the legal documents amending and supplementing these laws. We focus only on non-financial companies from Vietnam. We do not analyze foreign-invested and financial enterprises, since for them, regulations are somewhat different.

In our study, we first use the expertise of tax auditors to find possible factors affecting tax non-compliance. For this purpose, we presented all the factors from the models described in Rohaya et al. (2012) and Yusof et al. (2014) to a group of 10 experts who has experience in tax auditing. The purpose of this group discussion was to find out the elements of financial statements which affect CIT liability and which may therefore be used to detect CIT non-compliance in companies.

Based on the results of group discussions with experts, we developed survey questionnaires and conduct the survey on a broader sample of 200 tax officials, to determine the factors that may influence the tax payable of the enterprise. The resulting factors combine factors from the previous research with additional factors reflecting the specifics of Vietnam.

We then apply the panel data analysis to come up with a numerical model describing the effect of all these factors. By using this model, the tax officials can meaningfully decide which enterprises should be further audited.

Analytical data is derived from the tax information system of the tax administration in Vietnam, including the financial data from the financial statements providing the basis for tax calculation. A sample of 105 enterprises was randomly selected among all Ho Chi Minh City enterprises which were inspected by tax officials from 2011 to 2015.

This selection was made because Ho Chi Minh City is home to many businesses from a wide range of industries, and thus, Ho Chi Minh City businesses are a good sample representing Vietnamese enterprises.

3.2   Research Model

3.2.1   The Experience of Tax Officials in Determining the Factors Which Affect the CIT Non-compliance

In March 2017, we have conducted a group discussion of 10 experts with a large amount of experience in the field of tax inspection of companies, experts currently working in the tax departments in Ho Chi Minh City.

According to these experts, incidents of CIT non-compliance can be classified into two main categories: (1) concealment, under-reporting of revenue, and (2) overstating expenses.

There are three typical forms of under-reporting revenue:

- Using two different accounting books.
- Performing sales without issuing invoices.
- Listing, on the invoice, the selling price which is lower than the actual selling price.

There are several typical forms of overstating expenses:

- Creating a forged labor contracts or declaring a fictitious expense.
- Declare larger depreciation expenses than is allowed by the regulations.
- Including, in the financial documentation, expenses without invoices and vouchers and/or listing invoices which are not related to the operation of the enterprise.

According to the experts, the following factors can help to determine whether a company is in compliance with CIT:

- Revenue/Total assets
- Working capital/Total assets
- Total debt/Total assets
- Value of inventory/Total assets
- Accounts receivable/Sales
- Declaration of CIT payable/Profit before tax
- Loss the previous year
- Accumulated depreciation/Total assets
- Scale of business
- Fines related to tax violations/Tax amounts payable in the period

Based on the results of this group discussion, from March to May 2017, we conducted a survey of 200 tax officials conducting tax inspection and tax examination of enterprises in the tax departments of Ho Chi Minh City. The results of this survey – as presented in Table 1 – confirm that the majority of tax inspectors agree that the above 10 factors are indeed important to determine whether a company is in compliance with CIT.

3.2.2   The Research Model and Corresponding Hypotheses

The results of our survey are consistent with previous studies, such as Rohaya et al. (2012) and Nor et al. (2012). Based on these results, we designed a quantitative research model. This model – as illustrated by Fig. 1 – has the following form:

$$nct_{it} = \beta_0 + \beta_1 wc_{it} + \beta_2 sale_{it} + \beta_3 debt_{it} + \beta_4 loss_{it} + \beta_5 inv_{it}$$
$$+ \beta_6 ar_{it} + \beta_7 depr_{it} + \beta_8 etr_{it} + \beta_9 size_{it} + \beta_{10} pen_{it} + u_{it}.$$

Table 1. Results of factors influencing CIT non-compliance in Vietnam.

Code	The factors influencing CIT non-compliance	People agree	Proportion
1	Revenue/Total assets	132	70,59%
2	Working capital/Total assets	112	59,89%
3	Total debt/Total assets	143	76,47%
4	Value of inventory/Total assets	138	73,80%
5	Accounts receivable/Sales	173	92,51%
6	Declaration of CIT payable/Profit before tax	104	55,61%
7	Loss last year adjacent	122	65,24%
8	Accumulated depreciation/Total assets	109	58,29%
9	Scale of business	154	82,35%
10	Fines related tax administrative violations/Tax amounts payable in the period	99	52,94%

(Source: Synthetic of the authors group)

Fig. 1. The proposed research model.

3.2.3 Measuring CIT Non-compliance

In order to measure CIT non-compliance, Rice (1992), Joulfaian and Rider (1998) measured non-compliance of tax by the absolute difference between the declared tax amount and the actual tax amount. The same difference was used in Rohaya (2013).

Hanlon et al. (2007) and Zainal Abidin et al. (2010) used the ratio of the tax deficiency to either total amount of assets or to annual revenue. The study by Yusof et al. (2014) measures non-compliance of taxes as the ratio of unannounced income to total real income. Hanlon et al. (2005) supposed that non-compliance should be measured by the ratio of decrease in tax liability to either total assets or to the turnover. This view is consistent with Hanlon et al. (2007) and Zainal Abidin et al. (2010).

In general, tax non-compliance means that there is a difference between the actual taxable income – as uncovered by the inspection – and the originally declared taxable income. Because of this, in this study, we measure CIT non-compliance by the ratio between this difference and the actual total income.

3.2.4 Relationship Between CIT Non-compliance and Independent Variables

Working capital/Total assets. The liquidity of a firm is measured as working capital divided by total assets. Working capital is defined as the amount of available liquid assets a company has to operate its business. In general, the high value of the working capital indicates that the company is more successful, since it can use this capital to improve its operations (Rohaya 2012). Spathis (2002) found that a low ratio of working capital to total assets has been associated with false financial statements on the public listed companies in Greece. Yusof et al. (2014) explained this observation: for low liquidity firms, it is important to getting more liquidity, and to some of them, one way to do it is to pay less taxes.

wc = (Money + Short-term investment + Short-term receivable)/Total assets.

Hypothesis 1: The smaller the working capital/total assets ratio, the larger the chance of CIT non-compliance.

Revenue/Total assets. This ratio show how much revenue a company made on its investment assets. In effect, it shows how well a company can do business. Persons (1995) and Skousen et al. (2009) suggested that the ratio represents the financial stability of firms. If this rate increases, it shows that businesses are in stable finance situation and the pressure leading to fraudulent information on financial statements is negligible. Therefore, in this view, the turnover target of total assets may have a negative relationship with the CIT non-compliance of enterprises.

sale = (Total Revenue/Total Assets).

Hypothesis 2: The more revenue generated from the total investment in a business, the smaller the chance of CIT non-compliance.

Total debt/Total assets: Financial leverage is measured by the ration of total debt to total assets. Spathis (2002) and Person (1995) found a positive relationship between financial leverage and corporate financial obligations. The studies by Rohaya et al. (2012); Yusof et al. (2014); Lisowsky (2010) show that there is a relationship between financial leverage and tax non-compliance. When companies have high financial leverage, enterprises benefit from tax shields because of high interest expenses. This helps companies reduce the amount of tax payable. Therefore, many businesses often take advantages of interest expenses to reduce CIT expenses. The more debt a company has, the bigger the pressure on the company to decrease its expenses – and for some companies, this leads to tax non-compliance.

$$\text{debt} = (\text{Total Debt/Total Assets}).$$

Hypothesis 3: The more debt a business has, the more likely that it does not comply with CIT obligation.

Loss the previous year: In addition to the factors mentioned above, business losses create a financial pressure on businesses (Lou and Wang 2011). When the pressure increases, businesses are very likely to exhibit tax non-compliance. This relation was explicitly shown in Duong (2011).

Hypothesis 4: Enterprises with losses the previous year are more likely to be non-compliant with CIT.

Value of inventory/Total assets: The large amount of end-of-period inventory decreases the company's tax obligations. Thus, it is reasonable to expect that companies with low value of inventory may inflate this value to decrease taxes; see Loebbecke et al. (1989); Rohaya et al. (2012); Summers and Sweeney (1998); Skousen et al. (2009), etc.

$$\text{inv} = \text{Value of inventory/Total assets}.$$

Hypothesis 5: The lower the inventory level, the more likely it is for the company not to comply with CIT.

Accounts receivable/sales: Schilit (2002) suggested that there is a link between revenue and turnover reporting. Rohaya et al. (2012) found that the higher the ratio of receivables to sales, the larger are the frequencies of tax concealment.

$$\text{ar} = \text{Account Receivable/Total Revenue}.$$

Hypothesis 6: The more receivables a company has, the more likely it is to fail to comply with CIT.

Accumulated depreciation/Total assets: Depreciation is an expense actually incurred in the operation of a business. This is deductible when calculating taxable income or as a non-debt tax shield for businesses. Enterprises with depreciation costs benefit from the tax shields. Mills (1996) argued that the net depreciation of assets is related to the tax non-compliance of firms. In Vietnam, the survey shows that overstating depreciation expenses in enterprises is one of the violations that, according to the experts, occurs rather frequently.

$$\text{depr} = \text{Accumulated depreciation/Total assets}.$$

Hypothesis 7: The more depreciation costs a company declares, the greater chance of CIT non-compliance.

Declaration of CIT payable/profit before tax: The above ratio is what we would call an effective tax rate. According to Rohaya et al. (2012), effective tax rates determine the ability to control the tax burden of businesses. Previous studies have provided some evidence of the relationship between tax rates and compliance behavior. Rice (1992); Joulfaian (2002); Zainal Abidin et al. (2010) and Yusof et al. (2014) have shown that tax rates have a negative relationship with on compliance behavior: the larger the tax rate, the higher the frequency of tax fraud.

$$\mathbf{etr} = \mathbf{declared\ CIT/profit\ before\ tax}.$$

Hypothesis 8: The higher the effective tax rate is, the less likely it is that the company fails to comply with CIT.

Scale of business: Hanlon et al. (2007); Rice (1992) have found a positive relationship between firm size and tax non-compliance. In addition, Juahir et al. (2010); Joulfaian (2000) suggested that the scale is inversely related to tax non-compliance. This relation was also studied in Yusof et al. (2014). In this study, following Yusof et al. (2014), we measure the scale of the business as *log* of total assets.

$$\mathbf{size} = log(\mathbf{Total\ Assets}).$$

Hypothesis 9: The smaller the size of the enterprise is, the more likely it is not to comply with high CIT.

Fines related to tax administrative violations/tax amounts payable in the period: The relationship between fines and tax non-compliance has been mentioned in many previous studies. Allingham and Sandmo (1972) found a positive relationship between fines and declared earnings. Some studies did not find a correlation between these two factors, such as Kamdar (1997); Braithwaite (2009). Yusof et al. (2014) also incorporated this variable into the study of factors affecting tax non-compliance.

If this ratio is high, this means that the company paid heavily for past violations, and it is therefore unlikely that the company will do it again.

$$\mathbf{pen} = \mathbf{tax\ fines/tax\ amount\ payable\ in\ the\ period}.$$

Hypothesis 10: Enterprises with a lower tax fines over tax amount payable in the period ratio are more likely to fail to comply with high CIT.

The proposed hypotheses and expectations associated with the relationship between the independent and dependent variables are summarized in Table 3 below:

Note: (+) means we expect a positive relationship with dependent variable; (−) means that we expect a negative relationship with the dependent variable.

4   Research Results

Table 4 presents the descriptive statistics of the variables in the proposed study model with data sets of 105 enterprises that have been audited during 2011–2015.

According to Green (1991), to test the overall fit of a model by using R^2, we need the sample size of at least $50 + 8k$, where k is the number of independent variables. In our case, $k = 10$, so we need at least 130 observations. When we want to test each factor, the minimum sample size is $104 + k$, i.e., in our case, at least 114 observations. For our analysis, we use the data from 105 enterprises in each year from the five-year study period from 2011 to 2015, so we have a total of 525 observations, thus meeting the minimum sample size requirements.

We have performed F-tests and Hausman testing to select the right model. After that, we tested for multi-collinearity, variance changes, and self-correlation. The results showed that multi-collinearity in the regression model was negligible, but the model was affected by variance changes and self-correlation. Therefore,

Table 2. Hypothesis and expectation of the sign of independent variable.

Code	Name of variables	Symbol	Hypothesis	Expectation of the sign
1	Working capital/Total assets	**wc**	H1: The lower the working capital/total assets, the greater chance of CIT non-compliance	$(-)$
2	Revenue/Total assets	**sale**	H2: The more revenue generated from the total investment in a business, the greater chance of CIT non-compliance	$(+)$
3	Total debt/Total assets	**debt**	H3: The more debt a business has, the more likely does not comply with CIT obligation	$(+)$
4	Loss last year	**loss**	H4: Enterprises with losses the previous year are more likely to be non-compliant with CIT	$(+)$
5	Value of inventory/Total assets	**inv**	H5: The lower the inventory level, the more likely it is not to comply with CIT	$(-)$
6	Accounts receivable/Sales	**ar**	H6: The more receivables a company has, the more likely they are to fail to comply with CIT	$(+)$
7	Accumulated depreciation/Total assets	**depr**	H7: The more depreciation costs a company has, the greater chance of CIT non-compliance	$(+)$
8	Declaration of CIT payable/Profit before tax	**etr**	H8: The higher the effective tax rate is, the less likely it is to fail to comply with CIT	$(-)$
9	Scale of business	**size**	H9: The smaller the size of the enterprise is, the more likely it is not to comply with high CIT	$(-)$
10	Fines related tax administrative violations/Tax amounts payable in the period	**pen**	H10: Enterprises with a lower tax fines over tax amount payable in the period ratio are more likely to fail to comply with high CIT	$(-)$

Source: Synthetic of the authors group.

Table 3. Statistics describing variables.

(1)	(2)	(3)	(4)	(5)	
VARIABLES	N	mean	sd	min	max
code	525	53	30.34	1	105
year	525	2,013	1.416	2,011	2,015
nct	525	0.345	0.476	0	5.006
wc	525	0.540	0.273	0.0128	1.290
sale	525	2.255	2.374	0	33.73
debt	525	0.594	0.264	0	1.438
loss	525	0.177	0.382	0	1
inv	525	0.258	0.246	0	0.912
ar	525	0.650	2.344	0	49.67
depr	525	0.0906	0.256	0	4
etr	525	0.170	0.170	0	2.637
size	525	7.085	0.528	5.673	8.676
pen	525	0.300	0.120	0.101	1.766

Source: Calculated from Stata 12.0

Table 4. The results of panel data regression for OLS, FEM, REM.

(1)	(2)	(3)	
VARIABLES	OLS	Fixed effects	Random effects
wc	−0.222**	0.0222	−0.192
	(0.0988)	(0.213)	(0.119)
sale	−0.0223**	−0.0160	−0.0208**
	(0.00891)	(0.0118)	(0.00950)
debt	0.0378	0.0780	0.0517
	(0.0800)	(0.165)	(0.0962)
loss	0.160***	0.0714	0.121**
	(0.0562)	(0.0663)	(0.0571)
inv	−0.188*	−0.0460	−0.164
	(0.106)	(0.269)	(0.132)
ar	0.0386***	0.0397***	0.0390***
	(0.00856)	(0.00892)	(0.00829)
depr	0.0773	−0.0485	0.0417
	(0.0776)	(0.119)	(0.0867)
etr	−0.0606	−0.121	−0.0927
	(0.123)	(0.130)	(0.120)
size	−0.193***	−0.109	−0.184***
	(0.0423)	(0.0985)	(0.0518)
pen	−0.564***	−0.591***	−0.579***
	(0.165)	(0.165)	(0.155)
Constant	2.029***	1.271*	1.948***
	(0.310)	(0.719)	(0.378)
Observations	525	525	525
R-squared	0.152	0.098	0.093
Number of code		105	105

Standard errors in parentheses *** $p < 0.01$, ** $p < 0.05$,
* $p < 0.1$
(Source: Calculated from Stata 12.0)

Table 5. The results of the generalized least squares regression model (GLS).

```
Cross-sectional time-series FGLS regression

Coefficients:   generalized least squares
Panels:         homoskedastic
Correlation:    no autocorrelation

Estimated covariances        =        1      Number of obs     =       525
Estimated autocorrelations   =        0      Number of groups  =       105
Estimated coefficients       =       11      Time periods      =         5
                                             Wald chi2(10)     =     94.27
Log likelihood               = -311.5914     Prob > chi2       =    0.0000
```

nct	Coef.	Std. Err.	z	P>|z|	[95% Conf. Interval]	
wc	-.2221683	.0977668	-2.27	0.023	-.4137877	-.0305489
sale	-.0223191	.0088122	-2.53	0.011	-.0395907	-.0050474
debt	.0378039	.0791275	0.48	0.633	-.1172831	.1928909
loss	.1602422	.0556354	2.88	0.004	.0511989	.2692855
inv	-.1883528	.1053386	-1.79	0.074	-.3948126	.018107
ar	.0386059	.0084661	4.56	0.000	.0220126	.0551992
depr	.077319	.0767448	1.01	0.314	-.0730981	.227736
etr	-.0605559	.1217829	-0.50	0.619	-.299246	.1781343
size	-.1932184	.0418764	-4.61	0.000	-.2752946	-.1111422
pen	-.5643326	.1628215	-3.47	0.001	-.8834569	-.2452083
_cons	2.029407	.307142	6.61	0.000	1.427419	2.631394

we used the generalized least squares regression model (GLS) which is appropriate in such situations. The resulting regression model is as follows:

$$nct = 2.029 - 0.222wc - 0.022sale + 0.038debt + 0.16loss - 0.188inv$$
$$+ 0.039ar + 0.077depr - 0.061etr - 0.193size - 0.564pen.$$

The corresponding signs are explained in Table 2.

According to the results, 7 independent variables in the proposed model have a significant impact on the CIT non-compliance of enterprises in Vietnam: working capital/total assets, turnover/total assets, losses in the previous year, value of inventory Depreciation/Total Assets, Accounts Receivable/Sales, Enterprise Scale, and Debt fines/Tax Payable during the period.

We can use the resulting model to predict, for each enterprise, its expected amount of tax non-compliance, and then select the enterprises with the largest values of this predicted amount for auditing (Table 5).

5   Conclusions and Policy Implications

5.1   Conclusions

In this study, we developed a linear regression model that estimates the expected CIT non-compliance of an enterprise based on the information contained in its financial statements. To develop this model, we use the data from a sample of 105

Vietnamese enterprises that have been tax inspected. Specifically, the following 7 factors can be used for this estimation:

(1) **Working Capital/Total Assets:** When enterprises have high working capital ratios, mean that they are in stable finance situation and are this able to develop their operations effectively. Our research shows that this ratio has a negative relationship with the obligation of CIT: enterprises with low ratio are more probable to be not CIT compliant. This result is consistent with findings from Rohaya (2012); Spathis (2002) and OECD (2010).

(2) **Revenue/Total assets:** Our study has found that this ratio is negatively correlated with the CIT non-compliance of enterprises. The lower the ratio, the higher the risk of noncompliance, the more likely it is that the company tries to hide a portion of sales to reduce the CIT liability. The research results is contrary to our expectations, but it is consistent with the results of Persons (1995) and Skousen et al. (2009).

(3) **Loss in the previous year:** The study found that Vietnamese enterprises often take advantage of the loss declaration to avoid the obligation to pay taxes. Our research shows that the loss variable is statistically significant and has a clear impact on CIT non-compliance of enterprises. The result is consistent with (Lou and Wang 2011) and with the current status of qualitative research in Vietnam.

(4) **Value of inventory/Total assets:** The quantitative results show that Hypothesis 4 – about the value of inventories affecting the amount of corporate tax payable – is perfectly appropriate. Enterprises with high inventories will take advantage of the provisioning to reduce the CIT taxable income. Therefore, tax officials should pay attention to this factor when examining the CIT tax for enterprises. This conclusion is consistent with Summers and Sweeney (1998); Skousen et al. (2009) and Rohaya (2012).

(5) **Accounts Receivable/Revenue:** Our research shows that in Vietnam, the recognition of untrue revenues is a common violation. For the tax inspector, it is difficult to detect this problem. This can be made easier if we use the Accounts receivable/turnover ratio as a criterion for detecting possible tax evasion. This result is consistent with the findings of Schilit (2002) and Rohaya (2012) about the positive relationship between these two variables.

(6) **Scale of business:** The smaller the company, the more likely it is that it will fail to comply with the tax. This result is consistent with the findings of Juahir et al. (2010); Joulfaian (2000) and Yusof et al. (2014).

(7) **Fines for tax administrative violations/Tax amounts payable in the period:** The more fines the company paid in the past, the more probably it is that this time, the company will be in tax compliance. This result is consistent with the findings of Allingham and Sandmo (1972).

Some studies did not find have a statistically significant relationship between this factor and non-compliance, see, e.g., Kamdar (1997); Braithwaite (2009); Yusof et al. (2014). However, in Vietnam, our study found a negative correlation.

5.2 Policy Implications

Based on our study, we can make the following recommendations for the tax administration policy related to Vietnamese enterprises:

Our main recommendation is that tax officials should pay attention to the appropriate elements of the business declaration on the financial statements when planning which enterprises to inspect. Specifically, we suggest, for each enterprise, to estimate the values of the above factors, to use our model to combined the several values into a single estimate, and then to inspect the enterprises with the largest values of this estimate.

In particular, since one of the important factors in estimating the probability of tax non-compliance is the the size of enterprise, it is important to pay special attention to small enterprises, since for them, the risk of non-compliance is higher than for the big ones.

Also, since penalties for tax non-compliance drastically improve tax compliance, it is important that these penalties be assigned more rigorously, so that businesses can see the risks and benefits when they fulfill their tax obligations.

We hope that the use of our model will lead to better detection of tax fraud and thus, to better tax compliance.

In our model, we only took into account financial data. It is desirable to try to supplement our model with other possible factors – e.g., describing the type of a business and/or the business structure. This may lead to a further improvement of our model.

References

Alabede, J.O., Ariffin, Z.B.Z., Idris, K.M.: Determinants of tax compliance behaviour: a proposed model for Nigeria. Int. Res. J. Financ. Econ. **78** (2011)

Allingham, M.G., Sandmo, A.: Income tax evasion: a theoretical analysis. J. Public Econ. **1**(3/4), 323–338 (1972)

Badertscher, B., Phillips, J., Pincus, M., Rego, S.O.: Tax Implications of Earnings Management Activities: Evidence from Restatements. Merage School of Business, University of California (2006)

Braithwaite, V.: Defiance in Taxation and Governance: Resisting and Dismissing Authority in a Democracy. Edward Elgar, Cheltenham (2009)

Duong, C.M.: How market mispricing affects investor behavior, corporate investment & real earnings management: the UK evidence. Doctoral dissertation, University of Kent (2011)

Frank, M.M., Lynch, L.J., Rego, S.O.: Does Aggressive Financial Reporting Accompany Aggressive Tax Reporting (and Vice Versa)? Working Paper, University of Virginia (2004)

GAO: Tax Gap: Sources of Noncompliance and Strategies to Reduce It. United States Government Accountability Office (2012)

Green, S.B.: How many subjects does it take to do a regression analysis. Multivar. Behav. Res. **26**(3), 499–510 (1991)

Guenther, D.A.: Earning management in response to corporate tax rate changes: evidence from the 1986 Tax Reform Act. Account. Rev. **69**(1), 230–243 (1994)

Hanlon, M., Mills, L., Slemrod, J.: An empirical examination of corporate tax noncompliance. In: Conference on Taxing Corporate Income in the 21st Century, University of Michigan and the University of California (2005)

Hanlon, M., Mills, L., Slemrod, J.: An empirical examination of corporate tax non- compliance. In: Auerbach, A., Hines, J.R., Slemrod, J. (eds.) Taxing Corporate Income in the 21st Century, pp. 171–210. Cambridge University Press, Cambridge (2007)

Harris, D., Morck, R., Slemrod, J., Yeung, B.: Income Shifting in U.S. Multinational Corporations. NBER Working Paper, 3924 (1993)

Hai, O.T., See, L.M.: Intention of Tax non compliance examine the gaps. Int. J. Bus. Soc. Sci. **2**(7), 79–83 (2011)

Joulfaian, D.: Corporate income tax evasion and managerial preferences. Rev. Econ. Stat. **82**(4), 698–701 (2002)

Juahir, M.N., Norsiah, A., Norman, M.S.: Fraudulent financial reporting and company characteristics: tax audit evidence. J. Financ. Report. Account. **8**(2), 1985–2517 (2010)

Kamdar, N.: Corporate income tax compliance: a time series analysis. Atlantic Econ. J. **25**(1), 37–49 (1997)

Kasipillai, J., Hijatullah, A.J.: Gender and ethnicity differences in tax compliance. Asian Acad. Manag. J. **11**(2), 73–88 (2006)

Lewis, A., Carrera, S., Cullis, J., Jones, P.: Individual, cognitive and cultural differences in tax compliance: UK and Italy compared. J. Econ. Psychol. **30**, 431–445 (2009)

Lisowsky, P.: Seeking shelter: empirically modeling tax shelters using financial statement information. Account. Rev. **85**, 1693–1720 (2010)

Loebbecke, J.K., Eining, M.M., Willingham, J.J.: Auditors' experience with material irregularities: frequency, nature, & detectability. Audit. J. Pract. Theory **9**(1), 1–28 (1989)

Lou, Y.-I., Wang, M.-L.: Fraud risk factor of frau triangle assessing the likelihood of fraudulent financial reporting. J. Bus. Econ. Res. (JBER) **7**(2) (2011)

Law on Tax Administration No. 78/2006 / QH11 Dated 29/11/2006

Mills, L.F.: Corporate tax compliance and financial reporting. Nat. Tax J. **49**(3), 421–435 (1996)

Organization for Economic Co-operation and Development. Definition of Taxes. International Monetary Fund and the United Nations System of National Accounts (SNA) (1996)

Organization for Economic Co-operation and Development. Forum on tax administration: Small/Medum Enterprise (SME) compliance group- Understanding and influencing taxpayers' compliance variable (2010)

Persons, O.S.: Using financial statement data to indentify factors associated with fraudulent financial reporting. J. Appl. Bus. Res. **11**(3), 38–46 (1995)

Rezaee, R.: Causes, consequences, and deterrence of financial statement fraud. Crit. Perspect. Account. **16**, 277–289 (2002)

Rice, E.M.: The Corporate Tax Gap: Evidence on Tax Compliance by Small Corporations. University of Michigan Press, Ann Arbor (1992)

Rohaya, M.N., Alizan, A.A., Nor, A.M., Norashikin, I.: Tax fraud indicators. Malays. Account. Rev. **11**(1), 43–57 (2012)

Rohaya, M.N., Nor'azam, M., Barjoyai, B.: Corporate effective tax rates: a study on Malaysian public listed companies. Malays. Account. Rev. **7**(1), 1–20 (2008)

Rohaya, M.N., Nurul, E.J., Normah, O., Rozainun, A.A.: Measuring tax gap in the service industry. In: Finance and Economics Conference, Rydges Melbourne, Australia, 5–7 May 2013

Schilit, H.: Financial Shenanigans, 2nd edn. McGraw-Hill, New York (2002)

Skousen, C.J., Smith, K.R., Wright, C.J.: Detecting & predicting financial statement fraud: the effectiveness of the fraud triangle & SAS. Adv. Financ. Econ. **13**(99), 53–81 (2009)

Spathis, C.T.: Detecting false financial statements using published data: some evidence from Greece. Manag. Audit. J. **17**(4), 179–191 (2002)

Summers, S.L., Sweeney, J.T.: Fraudulently misstated financial statements & insider trading: an empirical analysis. Account. Rev. **73**, 131–146 (1998)

Tan, L.M., Sawyer, A.J.: A synopsis of taxpayer compliance studies: overseas vis-a-vis New Zealand. N. Z. J. Taxation Law Policy **9**, 431–454 (2003)

Truong, L.X., Chien, N.D.: Identify tax fraud acts. Financ. J. (2013)

Vietnamese Enterprise Income Tax Law No. 14/2008/QH12 dated 03/06/2008

Yusof, N.A.M., Ling, L.M., Wah, Y.B.: Tax non-compliance among SMCs in Malaysia: tax audit evidence. J. Appl. Account. Res. 215–234 (2014)

Zainal Abidin, M.Y., Hasseldine, J., Paton, D.: An analysis of tax non-compliance behaviour of small and medium-sized corporations in Malaysia. In: Datt, K., Tran-Nam, B., Bain, K. (eds.) International Tax Administration: Building Bridges, pp. 9–24. CCH Australia Limited, Sydney (2010)

GARCH Models in Forecasting the Volatility of the World's Oil Prices

Nguyen Trung Hung[1], Nguyen Ngoc Thach[2(✉)], and Le Hoang Anh[3]

[1] Chiang Mai University, Chiang Mai, Thailand
[2] Institute of Science and Technology, Banking University of Ho Chi Minh City,
Ho Chi Minh City, Vietnam
thachnn@buh.edu.vn
[3] HCMC University of Food Industry, 140 Le Trong Tan, Tay Thanh Ward,
Tan Phu District, Ho Chi Minh City, Vietnam
anhlh@cntp.edu.vn

Abstract. This study was conducted to forecast the volatility of the world's oil prices. Using the daily data of the WTI spot oil price collected from the US Energy Information Administration in the period from 01/02/1986 to 25/4/2016, estimation using models such as GARCH(1,1), EGARCH(1,1), GJR-GARCH(1,1) was made under 4 different distributions: normal distribution, Student's t-distribution, generalized error distribution (GED), skewed Student's t-distribution. The results show that the EGARCH(1,1) model with Student's t-distribution provides the most accurate forecast. In addition, it is also shown that the volatility of crude oil price in the future can be predicted by the past volatility while crude oil price shock has a relatively small impact on oil price volatility.

1 Problem Set

Oil is one of the strategic sources of energy for the socio-economic development of every nation. Oil resources, also being called "black gold", serve as important inputs in most economic sectors. Oil is one of the sources used to produce electricity, it also fuels all means of transportation. In the petrochemical industry, it is also used to make plastics and many other products. Therefore, the volatility of oil prices has a significant impact on the economy. The volatility of oil prices in the past has caused high inflation and prolonged recession in some countries. Oil price volatility started with a record rise for the first time in the 1970s. After some adjustments in the following years, a strong up-trend continued in the period from April 1980 to July 2008.

Oil price started to recover and fluctuate steadily between March 2009 and May 2014. However, after that, oil prices started to decline sharply during the period from June 2014 to April 2016.

The volatility of oil price has the impact on the performance of the economy through different channels. First, as oil is an important input of the manufacturing process, its price increase causes a supply shock, i.e., a drop in output. Second, rising oil price is a sign of an increase in the scarcity of basic input

© Springer International Publishing AG 2018
L. H. Anh et al. (eds.), *Econometrics for Financial Applications*, Studies in Computational Intelligence 760, https://doi.org/10.1007/978-3-319-73150-6_53

sources of the manufacturing process, thus, businesses will cut output because of the idea that high initial costs will reduce profits. Finally, the volatility of oil prices means increasing uncertainty about future supply of energy, thus reducing the incentive to expand production. As a result, unpredictable changes in oil price have become a major concern for policy makers from the enterprise level to the state management agencies. Thus, forecasting the volatility of oil price is the basis for the risk management process to suggest appropriate development strategies.

In recent years, the Auto-Regressive Conditional Heteroskedasticity (ARCH) models and the models developed from the ARCH (Engle 1982) has become popular as a tool for measuring the volatility of time series in the financial sector that is characterized by distribution with thick tails and volatility clustering. These models use time series data of yields to model conditional variance, which represents the volatility. In forecasting the volatility of energy prices, countries, exporters and importers participating in this market need the best forecasting models. Many studies aimed at identifying the best possible models for the energy market have been conducted. Day and Lewis (1993) compare the volatility of actual oil prices with the forecasted results from the ARCH models to provide forecasts of fluctuations in the crude oil market. A similar study was conducted by Dufe and Gray (1995). These authors compare the accuracy of forecasts with the ARCH models, the Markov transformation model, and fluctuations in the markets for crude oil and natural gas. Most recently, Sadorsky (2006) compares the results of the prediction of gasoline prices from different models such as random walk pattern, historical-average model, moving-average model, linear regression model, autoregressive models, and GARCH.

Our study aims to develop models for predicting the volatility of crude oil price and determining the best forecasting models. The structure of the paper is as follows. In Sect. 1, we formulate the problem. Section 2 contains theoretical review and research methodology. The results of the study and discussion are presented in Sect. 3. Based on the results of the study, Sect. 4 draws conclusions and policy implications.

2   Theoretical Background and Related Studies

2.1   Forecasting Models for Volatility of Oil Price

Previous quantitative models have only partially explained the volatility of the dependent variable. The fluctuations that the model can not explain are expressed through random errors added to the model. For ease in estimating these models, the expected variance of the random error is assumed to be constant over time. This shows that in predicting the fluctuation of a variable, the predictability of previous quantitative models is very limited due to assumptions about error. In 1982, Robert F. Engle created a revolution in econometrics, in particular, in forecasting the volatility of time series data. The proposed model allows the expected variation of random error to change over time, thus helping to overcome the disadvantages of traditional quantitative models in which the expected variance of the random error is assumed to be constant over time. This method is based on the assumption that the variance of the random error in a given statistical model, in a given period, depends on the previous random errors referred to as Auto-Regressive Conditional Heteroskedasticity (ARCH) model. Engle also demonstrated that ARCH models can be estimated and put into empirical estimation for the hypothesis that the conditioned transformations of a random variable are constant.

According to Engle (1982)'s ARCH model, the variance of the random errors at time t depends on the squared noises at the preceding periods. The ARCH(q) model has the form:

$$Y_t = \mu_t + u_t,$$

where $u_t \sim N(0, \sigma_t^2)$,

$$\sigma_t^2 = \alpha_0 + \sum_{j=1}^{q} \alpha_j u_{t-j}^2,$$

and $\alpha_j \geq 0$.

The ARCH(q) model is used in time series prediction of price, as the main advantage is to consider time series of price data in which it allows the variance of the data series to depend on the past variance values, so that the model can be used to estimate the level of risk and to forecast the degree of volatility of the price series with high fluctuation.

However, besides advantages, this model has the fundamental disadvantage that there is no standard to determine the model's q-rank. In addition, according to Engle and McFadden (1994), one of the drawbacks of the ARCH model is that it resembles a moving-average model rather than an autoregressive model. A new idea is that we should add the latent variables of conditional variance to the equation of variance in the autoregressive form. Furthermore, if the ARCH effects have too much latency, it will affect the estimation results due to a significant reduction in the number of degrees of freedom in the model.

The GARCH model was developed by Bollerslev (1986) to overcome the disadvantages of the ARCH model. The GARCH(q,p) model allows conditional variance to depend on previous latency:

$$Y_t = \mu_t + u_t,$$

where $u_t \sim N(0, \sigma_t^2)$,

$$\sigma_t^2 = \alpha_0 + \sum_{j=1}^{q} \alpha_j u_{t-j}^2 + \sum_{i=1}^{p} \beta_i \sigma_{t-i}^2,$$

$$\sum_{t=1}^{\max\{q,p\}} (\alpha_t + \beta_t) < 1,$$

and $\alpha_j, \beta_i \geq 0$.

Parameters in the GARCH model are estimated using the Maximum Likelihood Estimation (MLE).

Although some of the disadvantages of the ARCH model are overcome, the GARCH model does not distinguish between the effects of negative shocks and positive shocks. Besides, the coefficients of the variance equation in the GARCH model are not negative. In order to overcome this disadvantage, the EGARCH model was developed. The EGARCH(q,p) model has the form:

$$Y_t = \mu_t + u_t,$$

where $u_t \sim N(0, \sigma_t^2)$ and

$$\ln(\sigma_t^2) = \alpha_0 + \sum_{j=1}^{q} \left(\alpha_j \left| \frac{u_{t-j}}{\sigma_{t-j}} \right| + \gamma_j \frac{u_{t-j}}{\sigma_{t-j}} \right) + \sum_{i=1}^{p} \beta_i \ln(\sigma_{t-i}^2).$$

GARCH models are still used to predict the volatility of financial variables. In financial markets, researchers find that when a market index decreases (or increases) then it usually increasingly (or decreasingly) fluctuates. But with the same magnitude of increase or decrease, the volatility of the decreasing index is higher than the figure of the increasing index. In order to reflect increasing and decreasing shocks, Glosten et al. (1993) developed the GJR-GARCH model by introducing a dummy variable that represent for the shocks. The GJR-GARCH(q,p) model has the form:

$$Y_t = \mu_t + u_t,$$

where $u_t \sim N(0, \sigma_t^2)$,

$$\sigma_t^2 = \alpha_0 + \sum_{j=0}^{q} \left(\alpha_j + \gamma_j d_{t-j} u_{t-j}^2 \right) + \sum_{i=1}^{p} \beta_i \sigma_{t-i}^2$$

and $d_{t-j} = 1$ if $u_{t-j} < 0$, $d_{t-j} = 0$ if $u_{t-j} \geq 0$.

2.2   Related Studies About the Volatility of Oil Price

Related studies in the world often predict the volatility of oil prices based on historical data. Forecasting models are usually developed in two orientations:

1. Time series of oil price with volatility clustering; such studies often use autoregressive models,
2. Using large numbers of variables to explain the volatility of oil price; such studies often propose complex structural models, and then forecast oil price's volatility based on predicting the value of the explanatory variables in the model.

The second approach encountered some difficulties caused by the assumption that the characteristics of random errors do not change over time. Therefore, in this study, the authors focus on the first research direction. Studies in this direction often use autoregressive models like ARCH and GARCH models.

In oil price's forecasts, econometric models examine the relationship between spot and future prices. Bopp and Lady (1991) examined the effect of future price and the lag of spot price on futures price. The authors used monthly oil price data collected from the New York Mercantile Exchange (NYMEX) between December 1980 and October 1988. Oil price is forecasted through autoregressive models. The predicted results of the autoregressive model were compared with the random walk model. The results show that autoregressive models developed by these authors have the same predictive ability. Day and Lewis (1993) compared the volatility of oil price forecasted by the GARCH(1,1) model, EGARCH(1,1), and historical simulation model, on daily oil price data from November 1986 to March 1991. They use the OLS regression to identify the volatility of oil prices in the sample, then perform a forecast and test the deviation between the predicted value and the actual value. Accuracy of forecasts from the models was compared by using the Mean Error (ME) method, the Mean Absolute Error (MAE) method, and the Root Mean Squared Error (RMSE) method. Their research shows that the GARCH(1,1), EGARCH(1,1) models have good predictive power, but the GARCH(1,1) model gives the best predictive results.

Dufe and Gray (1995) conduct a study on the volatility of crude oil, kerosene and natural gas between May 1988 and June 1992. The authors use the GARCH(1,1), EGARCH(1,1), bi-variate GARCH and historical simulations. The predicted results of these models are compared by the RMSE method. Their results show that the historical simulation model gives poorer predictions than the GARCH(1,1) models.

Sadorsky (2006) compares gasoline and oil price forecasts from different models including random walk patterns, historical average models, moving average model, linear regression models, and different types of autoregressive and GARCH models. The study used the daily WTI crude oil and kerosene futures price data for the period from February 5, 1988 to January 31, 2003. The results show that the TGARCH model gives the best forecast for WTI kerosene, while the GARCH model provides better forecast for WTI crude oil. The results also show that predicting the volatility of oil prices by GARCH models gives better results than other forecasting techniques.

This study sets up GARCH models to predict the volatility of world's crude oil price. Among the subgroups of the GARCH(p, q) model, studies of Day and Lewis (1993), Dufe and Gray (1995), Sadorsky (2006) show that the GARCH(1,1), EGARCH(1,1), GJR-GARCH(1,1) give the most accurate forecast. Following the results of previous studies, the authors also use these models in predicting the volatility of world's crude oil price.

3   Research Methodology

3.1   Research Model

If the crude oil price in day t is P_t then the relative price change is determined by the formula $r_t = \ln\left(\dfrac{P_t}{P_{t-1}}\right)$. The volatility of oil price change is determined by the variance σ_t^2 of the relative price change. The GARCH(1,1) model has the form:

$$r_t = \mu_t + u_t,$$

where $u_t \sim N(0, \sigma_t^2)$, $\sigma_t^2 = \alpha_0 + \alpha_1 u_{t-1}^2 + \beta_1 \sigma_{t-1}^2$.

The form of EGARCH(1,1) is as follows:

$$r_t = \mu_t + u_t,$$

where $u_t \sim N(0, \sigma_t^2)$ and

$$\ln(\sigma_t^2) = \alpha_0 + \alpha_1 \left| \frac{u_{t-1}}{\sigma_{t-1}} \right| + \gamma_1 \frac{u_{t-j}}{\sigma_{t-j}} + \beta_1 \ln(\sigma_{t-1}^2).$$

To reflect the rising and falling shocks of crude oil price, the GJR-GARCH(1,1) model was used. This model uses a dummy variable d_{t-1} and has the form:

$$r_t = \mu_t + u_t,$$

where $u_t \sim N(0, \sigma_t^2)$,

$$\sigma_t^2 = \alpha_0 + \left(\alpha_1 + \gamma_1 d_{-1} u_{-1}^2 \right) + \beta_1 \sigma_{-1}^2,$$

and $d_{-1} = 1$ if $u_{-1} < 0$, $d_{-1} = 0$ if $u_{-1} \geq 0$.

GARCH models are estimated under four different distributions: normal distribution, Student's t-distribution, generalized error distribution (GED), and skewed Student's t-distribution. Model selection is based on the Akaike information criterion (AIC) and the Schwarz information criterion (SIC). Accordingly, the selected model will have the smallest AIC and SIC value. In addition, to determine the best predicting model, the authors used the method developed by Day and Lewis (1993), Dufe and Gray (1995) to evaluate the difference between the actual change in crude oil prices and estimated variance from the model through the following approaches:

- The Mean Squares Error (MSE) method
- The Mean Absolute Error (MAE) method, and
- The Root Mean Squared Error (RMSE) method.

3.2   Research Data

The data used is the daily spot price of WTI crude oil, collected from the US Energy Information Administration for the period from January 2, 1986 to April 25, 2016. The research sample is divided into two parts. The first part is a sample of crude oil price data from January 2, 1986 to April 25, 2014, used to estimate GARCH models. The second part is a sample of crude oil price data from April 26, 2014 to April 25, 2016 used for checking the forecasted levels of models estimated through MSE, MAE, and RMSE methods. Statistical results describing the crude oil price data from January 2, 1986 to April 25, 2016 are displayed in Table 1.

Table 1. Descriptive statistics of the sequence of changes in crude oil price.

Index	Value
Mean	6,38E−05
Median	0,0004
Max	0,1915
Mean	−0,4064
Standard deviation	0,0255
Deviation	−0,6582
Sharpness	16,7545
Jarque-Bera Test	60903,0500***
ADF Test	−64,7679***
Observations	7.656

(***) corresponds to the significance level of 1%.

The Jarque-Bera test results in Table 1 show that the data series on the change in crude oil price r_t do not follow the standard distribution with a 1% significance level. This is consistent with the characteristics of financial data series that usually have a longer and thicker tail than the tail of the normal distribution. In addition, the results of the augmented Dickey-Fuller (ADF) test show that the series r_t is stopped with a 1% significance level, thus meeting the requirement for analysis of time series data (Table 2).

In addition, Fig. 1 shows that changes in crude oil price with large volatility tend to have closely scattered distribution, and vice versa, changes in crude oil price with small volatility are also trend to have closely scattered distribution. Thus, the series r_t has volatility clustering characteristics. This implies that the ARCH effect exists.

Table 2. Heteroskedasticity test: ARCH.

F-statistic	88,7758	Prob. F (1.7653)	0,0000
Obs*R-squared	87,7807	Prob. Chi-Square (1)	0,0000

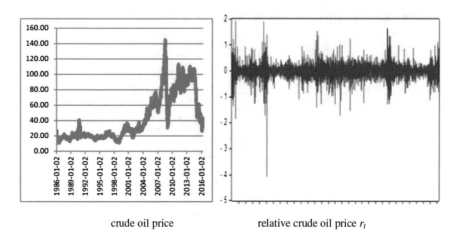

crude oil price   relative crude oil price r_t

Fig. 1. Volatility of oil price from January 2, 1986 to April 25, 2016 - Change in relative crude oil price r_t. (Source: The Unites State Energy Information Administration).

4   Estimation Results and Discussion

Next, we study the estimation of GARCH(1,1), EGARCH(1,1), GJR-GARCH(1,1) models under four different distributions: normal distribution, Student's t-distribution, generalized error distribution (GED), and skewed Student's t-distribution.

The results of the GARCH(1,1) model estimation under the four distributions are presented in Table 3. Both AIC and SIC criteria both select the GARCH(1,1) model corresponding to the Student's t-distribution. In this model, the coefficient $\beta_1 = 0.9228$ is statistically significant at 1%, i.e., 92.28% of the volatility of the oil price change σ_t^2 at any $k + 1$ is explained by the volatility of the oil price change at time k. The coefficient $\alpha_1 = 0.0684$ is statistically significant at 1%, i.e., 6,84% of the volatility of the oil price change any $k + 1$ is explained by the volatility of the oil price change at time k.

Table 3. Estimated results of the GARCH(1,1) model

Coefficient	Normal	Student-t	GED	Skewed Student-t
α_0	$5{,}77 \times 10^{-6}$	$5{,}73 \times 10^{-6}$	$5{,}70 \times 10^{-6}$	$5{,}41 \times 10^{-6}$
α_1	0,0943***	0,0684***	0,0766***	0,0689***
β_1	0,9019***	0,9228***	0,9156***	0,9189***
AIC	−4,8386	−4,9046	−4,8940	−4,8988
SIC	−4,8348	−4,8998	−4,8892	−4,8950

(***) corresponds to the significance level of 1%.

Estimated results of the EGARCH(1,1) model with four distributions are presented in Table 4. Both AIC and SIC criteria select the EGARCH(1,1) model corresponding to the Student's t-distribution. In this model, the coefficient β_1 = 0.9897 is statistically significant at 1%, i.e., 98.97% of the volatility of the oil price change at any $k+1$ is by the volatility of the oil price change at time k. The coefficient $\alpha_1 = 0.1406$ is statistically significant at 1%, i.e., 14.06% of the volatility of the oil price change at the time $k+1$ is explained by the change in oil price at the time k. The coefficient $\gamma_1 = -0.0236$ is statistically significant at 1%. This shows that suddenly rising or falling oil price shocks have affected on the volatility of oil prices at $k+1$ but the effect is small, only 2.36%.

Table 4. Estimated results of the EGARCH(1,1) model

Coefficient	Normal	Student-t	GED	Skewed Student-t
α_0	−0,2316	−0,1852	−0,2021	−0,1944
α_1	0,1938***	0,1406***	0,1594***	0,1450***
γ_0	−0,0073**	−0,0236***	−0,0162**	−0,0215***
β_1	0,9887***	0,9897***	0,9894***	0,9894***
AIC	−4,8410	−4,9102	−4,8979	−4,9044
SIC	−4,8362	−4,9044	−4,8921	−4,8996

(**), (***) corresponds to the significance level of 5%, 1%.

The results of the GJR-GARCH(1,1) model with four distributions are presented in Table 5. The AIC and SIC criteria select GJR-GARCH with the Student's t-distribution. In this model, the coefficient $\beta_1 = 0.924$ is statistically significant at 1%, i.e., 92.40% of the volatility of the oil change at any $k+1$ is explained by the volatility of the oil price change at time k. The coefficient $\alpha_1 = 0.0565$ is statistically significant at 1%, i.e., 5.65% of the volatility of the oil price change at the time $k+1$ is explained by the change in oil price σ_t^2 at the time k. The coefficient $\gamma_1 = 0.0208$ was statistically significant at 1%. This suggests that sudden suddenly rising or falling oil price shocks have affected the volatility of oil price change at $k+1$ but the effect is small, only 2.08%.

Table 5. Estimated results of the GJR-GARCH(1,1) model.

Coefficient	Normal	Student-t	GED	Skewed Student-t
α_0	5.78×10^{-6}	5.66×10^{-6}	5.62×10^{-6}	5.33×10^{-6}
α_1	0.094655***	0.056468***	0.069904***	0.058366***
γ_1	−0.000581	0.020760**	0.011320	0.018094**
β_1	0.901842***	0.923972***	0.916436***	0.920204***
AIC	−4.838345	−4.904921	−4.893919	−4.899052
SIC	−4.833539	−4.899154	−4.888152	−4.894246

(**), (***) corresponds to the significance level of 5%, 1%.

Table 6. Evaluation of the forecasting quality of the models.

Model	MSE	RMSE	MAE
GARCH(1,1) – Student-t	0,00083752	0,02894	0,021136
EGARCH(1,1) – Student-t	0,00083712	0,028933	0,021121
GJR-GARCH(1,1) – Student-t	0,00083729	0,028936	0,021128

To determine the best among the predicting model among the GARCH(1,1), EGARCH(1,1), GJR-GARCH(1,1) models with Student-t distributions, the authors use the method implemented by Day and Lewis (1993), Dufe and Gray (1995) to evaluate the difference between the actual change in crude oil price and the variance estimated from the model at the time t through the following four approaches: the Mean Squares Error (MSE) method, the Mean Absolute Error (MAE) method, and the Root Mean Squared Error (RMSE) method. The sample of crude oil price data from April 26, 2014 to April 25, 2016 is used for the purpose of examining the forecasting quality of the models. The results are shown in Table 6.

According to Table 6, all three methods MSE, RMSE, and MAE show that the prediction error of the EGARCH(1,1) model with Student-t distribution is the lowest. Thus, the EGARCH(1,1) model with the Student-t distribution can most accurately predict the volatility of crude oil prices.

5   Conclusions and Policy Implications

The data series of the change in crude oil prices by day during the period from January 2, 1986 to April 25, 2016, is stationary and has the ARCH effect. The GARCH(1,1), EGARCH(1,1), GJR-GARCH(1,1) models with four different distributions have been used by the authors to predict the volatility of crude oil prices. The AIC and SIC criteria have shown that GARCH(1,1), EGARCH(1,1), GJR-GARCH(1,1) models with the Student-t distribution are most consistent with the data.

The results show that, among the three methods, the forecast error is the smallest for the EGARCH(1,1) model with the Student-t distribution. Thus, this model has the ability to accurately forecast the volatility in crude oil price.

Studies by Day and Lewis (1993), Dufe and Gray (1995) show that the EGARCH(1,1) model is also capable of predicting oil price. However, they found that the forecasting quality of this model was inferior to the GARCH(1,1) model and this difference could be due to the choice of distribution of error in the estimation model: namely, in the studies of Day and Lewis (1993), Dufe and Gray (1995), models are assumed to be estimated according to the normal distribution.

The results of the EGARCH(1,1) model estimation with the Student-t distribution show that the volatility of future crude oil price can be predicted by the volatility and change of crude oil price in the past. Specifically, 98.97% of the volatility of future crude oil price is explained by the volatility of crude oil price in the past. This shows that the change in crude oil prices in the future reflects fully the market information in the past. In addition, the EGARCH(1,1) model with the Student-t distribution also shows the effects of good and bad information on the volatility of future crude oil price by causing rising/falling oil price shock. Specifically, 2.36% volatility of oil price changes can be explained by the good and bad news in the market.

The results of the study show that the EGARCH(1,1) model by the Student distribution is the most accurate predicting model of the volatility in world oil price. Therefore, the authors suggest that functional agencies, especially the Ministry of Finance and the Vietnam Petrolimex Group, use this model for forecasting the volatility in oil prices. In order to make effective use of the model, the authors propose two recommendations:

- First, when collecting data for modeling, it is important to ensure that data is collected over a sufficient period of time, at least one economic cycle. Also, data should be collected from highly reliable sources.
- Second, when modeling, it is recommended to divide the sample into two phases: the first phase is to build the forecasting model, the next phase is used to test the predictive accuracy of the models. The model with the highest predictive accuracy should be selected for applications.

References

Bollerslev, T.: Generalized autoregressive heteroskedasticity. J. Econ. **31**, 307–327 (1986)

Bopp, A.E., Lady, G.M.: A comparison of petroleum futures versus spot prices as predictors of prices in the future. Energy Econ. **13**, 274–282 (1991)

Day, T.E., Lewis, C.M.: Forecasting futures market volatility. J. Deriv. **1**, 33–50 (1993)

Dufe, D., Gray, S.: Volatility in Energy Prices. Managing Energy Price Risk, pp. 39–55. Risk Publications, London (1995)

Engle, R F., McFadden, D.L.. ARCH models. In: Handbook of Econometrics, vol. IV, pp. 2961–3038. Elsevier Science (1994)

Engle, R.F.: Autoregressive conditional heteroskedasticity with estimates of the variance of United Kingdom inflation. Econometrica **50**, 987–1007 (1982)

Glosten, L.R., Jagannathan, R., Runkle, D.E.: On the relation between the expected value and the volatility of the nominal excess return on stocks. J. Finan. **48**(5), 1779–1801 (1993)

Sadorsky, P.: Modeling and forecasting petroleum futures volatility. Energy Econ. **28**, 467–488 (2006)

Price Transmission Mechanism
for Natural Gas in Thailand

Natnicha Nimmonrat[1(✉)], Pathairat Pastpipatkul[1,2], Woraphon Yamaka[1,2],
and Paravee Maneejuk[1,2]

[1] Faculty of Economics, Chiang Mai University, Chiang Mai 50200, Thailand
`natnicha.nimm@gmail.com`
[2] Center of Excellence in Econometrics, Chiang Mai University,
Chiang Mai 50200, Thailand
`ppthairat@hotmail.com`, `woraphon.econ@gmail.com`, `mparavee@gmail.com`

Abstract. This study aimed to analyze natural gas price transmission in Thailand using the MS-VAR model. We focused on the data set related to Thai natural gas prices in two main groups. The first group is the price of natural gas procurement source, including price of gulf gas, natural gas price of Myanmar and price of liquefied natural gas. Second group is the prices of natural gas used for electric power generation, separation plants, and production of compressed natural gas, which are considered as natural gas consumption. The data is collected from M1/2011 to M12/2016. By using the Bayesian approach, we estimated the model with two regimes; namely high price regime and low price regime. We found that the shocks e.g., energy crisis, the shortage of natural gas sources from the gulf gas of Thailand or Myanmar stopping selling natural gas to Thailand may have a direct substantial effect on the natural gas market. Therefore, the shocks of each natural gas sources price will have implications on other natural gas sources price and thereby leading are higher consumption costs.

1 Introduction

Natural gas is important for Thailand's economic system. From playing a role as an important and indispensable factor input to several sectors. At present, the continuous expansion and growth of the economy and industry heighten the energy requirement and in future, Thailand has to import the energy in larger quantity because of the country's limited production capacity and the price volatility of natural gas attributable to the oil and natural gas prices in the global market. Natural gas is heavily used in 3 main sectors including electricity generation, fuel (Compressed Natural Gas: CNG), and gas separation plants where the highest usage ratio around 60% belongs to the electricity generation sector. The products from natural gas of each sector would have different prices because, due to different production process and transportation cost. These products are expected to affect one another though a mechanism called "Price transmission".

It is certain that the concession of Thai's natural gas in gulf will expire in 2022 as well as the contract between Thailand and Myanmar will end in 2030.

L. H. Anh et al. (eds.), *Econometrics for Financial Applications*, Studies in Computational Intelligence 760, https://doi.org/10.1007/978-3-319-73150-6_54

These will directly affect the amount of natural gas a variable in the market and also prices of its products. In other words, it can be said that the quantity of natural gas production in gulf will continuously decrease and can lead to a reduction of industrial sector. Under this scenario, the domestic supply of natural gas may not be enough for meeting the demand in Thailand. So, the price of natural gas product in all main sectors will increase. The higher price of natural gas products will lead to a higher production cost in many industries in Thailand. Moreover, as we mentioned previously, these products are expected to affect one another many papers in literature already dealt with the transmission effects among natural gas products, [see Anisie [1], Haug [2], Reungkajon [3], and Soithong [4]]. These papers mostly researched on the natural gas prices in the United States. Therefore, this paper focuses on the study of the price transmission mechanism of natural gas in Thailand. It would be great to study the effect of natural gas product prices on different sectors especially during the high price regime, in order to make an appropriate policy control in natural gas market.

This study will consider two regimes namely high price and low price regimes of the natural gas product prices. We employ the Markov-Switching Vector Autoregressive (MS-VAR) model to measure the price transmission mechanism of Thai natural gas. For this purpose, we employ a Bayesian approach as the estimator for MS-VAR model. The advantage of Bayesian estimator is that it can deal with the over-parameterization problem in the MS-VAR model which is estimated by maximum likelihood estimation (MLE) [Pastpipatkul et al. [5]].

For an overview of this paper, we begin with the importance of natural gas in Thailand and the price transmission. The following Sect. 2 provides methodology and procedures for estimating MS-BVARs. The results are reported in Sect. 3. The economic implications are presented in Sect. 4. Section 5 provides the conclusion.

2   Methodology

2.1   Markov-Switching VAR

Our motivation for estimating MS-VAR models rests on the idea that price transmission represents behavior differently in different state of economy. The model is initially estimated by using a block EM algorithm where the blocks are vector autoregressive (VAR) regression coefficients for each regime (separating for intercepts, AR coefficients, and error covariance) and the transition matrix [Brandt and Appleby [6]]. The general specification of a Markov Switching VAR(p) model can be written as

$$Y_t' = A_0(s_t) + \sum_{p=1}^{P} A_p(s_t)Y_{t-p}' + \varepsilon_t'\Gamma^{-1}(s_t), \ t = 1, \cdots, T, \tag{1}$$

where Y_t' is m-dimensional column vector of endogenous variables; A_0 is $m \times 1$ vector of constant coefficient. A_p is $m \times m$ matrix of autoregressive parameters; $\varepsilon_t' = [\varepsilon_{1t}', \cdots, \varepsilon_{mt}']$ is m dimension vector normal white noise process with a regime-dependent variance covariance Γ, $\varepsilon_t \sim NID(0, \Sigma(s_t))$; and $s_t \in [1, \cdots, h]$ state or regime. Thus, Sims et al. [7] provide a distributional assumption with densities of the MS-VAR disturbances as follows:

$$Pr(\varepsilon_t | Y_{t-p}, s_t, \omega, \Theta) = N(\varepsilon_t | 0_{m \times 1}, I_m), \tag{2}$$

and on the information set

$$Pr(Y_t | Y_{t-p}, s_t, \omega, \Theta) = N(Y_t | u_Y(s_t), \Sigma_z(s_t)), \tag{3}$$

where $Y_t = [Y_1', \cdots, Y_t']$, $s_t = [s_0', s_1', cdots, s_t']'$, $u_z(s_t)$ is the mean of each equation, and ω is the vector of probabilities which is estimated by the Markov chain, and $\Theta = [A_0(1), A_0(2), \cdots, A_0(h), A_p(1), \cdots, A_p(h)]'$. We also assume that the evolution of the regime or state variable s_t, is governed by a first-order markov chain with transition probabilities matrix. Suppose, we has h regimes, the transition matrix may take the form:

$$Q = \begin{bmatrix} p_{11} & p_{12} & \cdots & p_{1h} \\ p_{21} & p_{22} & \cdots & p_{2h} \\ \vdots & & \ddots & \vdots \\ p_{h1} & p_{h2} & \cdots & p_{hh} \end{bmatrix} \tag{4}$$

The matrix Q given in Eq. 4 contains a set of transition probability dynamics which has been estimated by the MS-VAR model. Therefore, the MS-VAR, and data could generate estimates that repeatedly move between regimes during the sample.

2.2 Bayesian Estimation

In the estimation of the MS-VAR model in Eq. 1, the study employs the Bayesian estimation technique. The model relies on the joint posterior distribution of Θ and ω. By using the Bayes's rule, the posterior of MS-BVAR model can be derived by

$$Pr(\omega, \Theta | Y_t, Y_{t-p}, \omega, \Theta) \propto Pr(Y_t | Y_{t-p}, \omega, \Theta) Pr(\omega, \Theta) \tag{5}$$

where $Pr(\omega, \Theta)$ denotes the prior of ω and Θ; and $Pr(Y_t | Y_{t-p}, \omega, \Theta)$ is the likelihood of the model which can be defined as

$$P(Y_t | \omega, \Theta) = \prod_{t=1}^{T} \left[\sum_{s_t \in H} Pr(Y_t | Y_{t-p}, \omega, \Theta) Pr(s_t, Y_{t-p}, \omega, \Theta) \right], \tag{6}$$

where $Pr(Y_t | \alpha_t, \omega, \Theta)$ is the density used to sample the probability that s_t is in regime h conditional on s_{t-1}. In this Bayesian framework, we follow the estimation of Sims et al. [7] and employ the Gibb sampling method to construct the likelihood along with the conditional densities of Θ, $Pr(\Theta|Y_{t-1}s_t, \omega)$ and ω, $Pr(\Theta|Y_{t-1}, s_t, \Theta)$, where the vector of regimes, s_T, is integrated out of the log-likelihood.

Since the Bayesian approach allow us to incorporate the prior for our unknown parameters, we consider the prior distributions for the MS-VAR coefficients that belong to the following Normal-Inverse-Wishart prior. We assume $A(s_t) \sim N(b, \Sigma)$, $Q \sim Dirichlet(\alpha)$ and $\Sigma \sim IW(\Psi, d)$, where b, Σ, Ψ, α and d each is a vector of hyper parameters. Then we employ the Markov chain Monte Carlo (MCMC) Gibb sampler in order to estimate the marginal likelihood and Bayes factor or marginal posterior distribution of interest for inference by running 2,000 steps of MCMC simulator as follows:

(1) Draw A^1 from $Pr(\Sigma^1|\Sigma^0, A^0, Q^0, Y_t)$.
(2) Draw Σ^1 from $Pr(\Sigma^1|\Sigma^0, A^1, Q^0, Y_t)$.
(3) Draw Q^1 from $Pr(C^1|\Sigma^1, A^1, Q^0, Y_t)$.

This completes a Gibb iteration and we obtain A^1, Γ^1 and Q^1. Then, using these new parameters as starting values in order to repeat the prior iteration of $A^{(i)}, \Gamma^{(i)}, C^{(i)}$ and $Q^{(i)}$ draws. Repeating the previous iterations for 10,000 times to obtain a sequence of random draws:

$$(A^1, \Sigma^1, C^1, Q^1), \cdots, (A^{10000}, \Sigma^{10000}, C^{10000}, Q^{10000})$$

Finally, we can obtain the estimated parameters from the mean of each parameter draw.

3 Empirical Result

3.1 Data

The data are monthly times series data. It consists of prices of three natural gas related groups. First group is price of natural gas procurement source, including price of gulf gas (PGG), price of Myanmar gas (PMG) and price of liquefied natural gas (PLNG). Second group is the prices of natural gas used for electric power generation (PE), separation plants (PGSP), and production of compressed natural gas (PCNG), which are considered as natural gas consumption. Finally, the last group, we consider all variables from both first and second groups. The data is collected from January 2011 to December 2016. The data are obtained from Energy statistics, Energy Policy and Planning Office (EPPO), Ministry of Energy. All data are coded Baht per Million British Thermal unit (MMBtu) (Table 1).

3.2 Transition Probabilities

In this section, the matrix of transition probability parameters is presented in Table 2. The estimated probability means the conditional probability based on the information available throughout the whole sample period at future date t.

Table 1. Descriptive statistics.

Price	Mean	Std. Dev.	Max	Min
PGG	0.0001	0.0325	0.1175	−0.1408
PMG	−0.0049	0.0408	0.0955	−0.1623
PLNG	−0.0019	0.1050	0.3174	−0.2425
PE	−0.0001	0.0349	0.0951	−0.1109
PGSP	−0.0021	0.0327	0.1117	−0.1396
PCNG	0.0053	0.0223	0.0910	−0.0562

Source: Calculation.

The result of first group shows that the probability of switching from regime 1 (low price) to regime 2 (high price) is 37%, while remaining in regime 1 is 63%, and have a duration of approximately 3 months. Similarly, the probability of switching from regime 2 (high price) to regime 1 (low price) is 25%, while remaining in regime 2 is 75%. In addition, the duration of regime 2 is approximately 4 months. Next the result of the second group shows that the probability of switching from regime 1 (high price) to regime 2 (low price) is 36%, while remaining in regime 1 is 64%, and has a duration of approximately 3 months. Similarly, the probability of switching from regime 2 (low price) to regime 1 (high price) is 25%, while remaining in regime 2 is 75%. In addition, the duration of regime 2 is approximately 4 months. And the last group, the results show that the probability of switching from regime 1 (low price) to regime 2 (high price) is 37%, while remaining in regime 1 is 63%, and has a duration of approximately 3 months. Similarly, the probability of switching from regime 2 (High price) to regime 1 (low price) is 25%, while remaining in regime 2 is 75%. In addition, the duration of regime 2 is approximately 4 months, as well.

Table 2. Estimates of transition matrices

	Price					
	Natural gas procurement source group		Natural gas products for consumption group		Natural gas procurement source and Natural gas products for consumption group	
Regime 1	0.631	0.369	0.639	0.361	0.638	0.362
Regime 2	0.249	0.751	0.253	0.747	0.250	0.750
	Duration		**Duration**		**Duration**	
Regime 1	2.709		2.773		2.761	
Regime 2	4.011		3.954		4.005	

Source: Calculation.

3.3 Estimates MS-BVAR(1): Natural Gas Procurement Source Group

Regime 1:

$$
\begin{bmatrix} PGG_t \\ PMG_t \\ PLNG_t \end{bmatrix} = \begin{bmatrix} -0.274 \\ 0.987 \\ 6.818 \end{bmatrix} + \begin{bmatrix} -3.947 & 6.174^* & -0.426 \\ -5.156 & 0.793 & -0.032 \\ -0.077 & 1.252 & -0.016 \end{bmatrix} \times \begin{bmatrix} PGG_{t-1} \\ PMG_{t-1} \\ PLNG_{t-1} \end{bmatrix} + \begin{bmatrix} \varepsilon_t^{PGG} \\ \varepsilon_t^{PMG} \\ \varepsilon_t^{PLNG} \end{bmatrix}
$$
(7)

Regime 2:

$$
\begin{bmatrix} PGG_t \\ PMG_t \\ PLNG_t \end{bmatrix} = \begin{bmatrix} 10.476^* \\ 3.727 \\ 4.947 \end{bmatrix} + \begin{bmatrix} -1.317 & 0.505 & -0.222 \\ 6.550 & 2.455 & -0.185 \\ -0.844 & -0.576 & -0.069 \end{bmatrix} \times \begin{bmatrix} PGG_{t-1} \\ PMG_{t-1} \\ PLNG_{t-1} \end{bmatrix} + \begin{bmatrix} \varepsilon_t^{PGG} \\ \varepsilon_t^{PMG} \\ \varepsilon_t^{PLNG} \end{bmatrix}
$$
(8)

$*, **$, and $* * *$ denote 10%, 5%, and 1% significant level respectively.

Equations (7) and (8) show that the estimated means (intercept) of the MS-BVAR(1) model for each of the two regime seem to have an economic interpretation. The first regime indicates that most of the values of mean in each equation are less than those of the second regime. Thus, this indicates that regime 1 is in the low price regimes of natural gas procurement source, while regime 2 is in the high price regime. Furthermore, Eq. (7) provides the results for the estimated coefficients in the first lag term. In the low price regime, only the PMG that seems to be significantly driven by the PGG and has positive effect in low price regime. This result shows that a 1% PMG change in the previous time will be followed by PGG change 6.17% at the current time in same direction. However, we cannot find any significant transmission effect of natural gas procurement source in high price regime.

3.4 Estimates MS-BVAR(1): Natural Gas Products for Consumption Group

Regime 1:

$$
\begin{bmatrix} PE_t \\ PGSP_t \\ PCNG_t \end{bmatrix} = \begin{bmatrix} 1.626 \\ 13.319 \\ 15.302 \end{bmatrix} + \begin{bmatrix} -12.884 & -20.383 & -0.433 \\ 12.634 & -2.157 & -0.123 \\ -15.327 & -3.056 & -0.527 \end{bmatrix} \times \begin{bmatrix} PE_{t-1} \\ PGSP_{t-1} \\ PCNG_{t-1} \end{bmatrix} + \begin{bmatrix} \varepsilon_t^{PE} \\ \varepsilon_t^{PGSP} \\ \varepsilon_t^{PCNG} \end{bmatrix}
$$
(9)

Regime 2:

$$
\begin{bmatrix} PE_t \\ PGSP_t \\ PCNG_t \end{bmatrix} = \begin{bmatrix} -1.436 \\ -5.285 \\ -4.919 \end{bmatrix} + \begin{bmatrix} 1.899 & -3.724 & 0.065 \\ -4.576 & -7.395 & -0.151 \\ 10.595 & -6.883 & 0.385 \end{bmatrix} \times \begin{bmatrix} PE_{t-1} \\ PGSP_{t-1} \\ PCNG_{t-1} \end{bmatrix} + \begin{bmatrix} \varepsilon_t^{PE} \\ \varepsilon_t^{PGSP} \\ \varepsilon_t^{PCNG} \end{bmatrix}
$$
(10)

$*, **$, and $* * *$ denote 10%, 5%, and 1% significant level respectively.

Equations (9) and (10) show that there are 2 regimes in the natural gas products for consumption group model. We observe that the intercept term of regime 1 seems to be larger than regime 2. This indicates that regime 1 is in high price regime, while regime 2 is in the low price regime. However, we cannot find any significant transmission effect of natural gas products in low price and high price regime.

3.5   Estimates MS-BVAR(1): Natural Gas Procurement Group and Natural Gas Products for Consumption Group

Regime 1:

$$
\begin{bmatrix} PGG_t \\ PMG_t \\ PLNG_t \\ PE_t \\ PGSP_t \\ PCNG_t \end{bmatrix} = \begin{bmatrix} -15.540 \\ 1.011 \\ 1.823 \\ -25.200^* \\ -10.120 \\ 19.610 \end{bmatrix} + \begin{bmatrix} -15.070 & -9.892^* & -10.190 & -2.273 & 9.256 & -0.643 \\ -8.424 & 1.653 & 5.892 & -0.222 & 0.481 & 0.475 \\ 0.398 & 3.562^{**} & -6.818 & 1.958 & 8.863 & -0.001 \\ -2.462 & -11.050^{**} & 0.609 & -5.259 & -3.314 & -0.273 \\ 16.110 & 3.324 & 4.992 & -4.561 & -29.350 & 0.348 \\ 6.122 & -6.906 & -15.310 & 26.470 & -12.740 & 0.753 \end{bmatrix}
$$

$$
\times \begin{bmatrix} PGG_{t-1} \\ PMG_{t-1} \\ PLNG_{t-1} \\ PE_{t-1} \\ PGSP_{t-1} \\ PCNG_{t-1} \end{bmatrix} + \begin{bmatrix} \varepsilon_t^{PGG} \\ \varepsilon_t^{PMG} \\ \varepsilon_t^{PLNG} \\ \varepsilon_t^{PE} \\ \varepsilon_t^{PGSP} \\ \varepsilon_t^{PCNG} \end{bmatrix} \tag{11}
$$

Regime 2:

$$
\begin{bmatrix} PGG_t \\ PMG_t \\ PLNG_t \\ PE_t \\ PGSP_t \\ PCNG_t \end{bmatrix} = \begin{bmatrix} -12.880 \\ -0.139 \\ 1.592 \\ 21.590^* \\ -0.425 \\ -11.010 \end{bmatrix} + \begin{bmatrix} -9.311 & 1.380 & 13.680 & -7.668 & 0.541 & -0.044 \\ 4.854 & 0.435 & -7.247 & 5.099 & -9.209 & 0.136 \\ 0.376 & -1.115 & 4.981 & -11.600^{***} & -7.046 & -0.028 \\ 5.692 & 3.816 & -4.567 & -0.748 & -10.790 & 0.024 \\ 1.430 & -9.424^{**} & -8.940 & 2.905 & 5.463 & 0.373 \\ -24.160 & 1.540 & 6.692 & -18.070 & -2.312 & 0.365 \end{bmatrix}
$$

$$
\times \begin{bmatrix} PGG_{t-1} \\ PMG_{t-1} \\ PLNG_{t-1} \\ PE_{t-1} \\ PGSP_{t-1} \\ PCNG_{t-1} \end{bmatrix} + \begin{bmatrix} \varepsilon_t^{PGG} \\ \varepsilon_t^{PMG} \\ \varepsilon_t^{PLNG} \\ \varepsilon_t^{PE} \\ \varepsilon_t^{PGSP} \\ \varepsilon_t^{PCNG} \end{bmatrix} \tag{12}
$$

$*, **$, and $***$ denote 10%, 5%, and 1% significant level respectively.

Equations (11) and (12) provide the estimated parameters of the 2 regimes between natural gas procurement source and natural gas products for consumption group. The results show that the estimated mean (intercept) of the first regime indicates that most of the values of mean in each equation are less than those of the second regime. Thus, this indicates that regime 1 is the low price regime, while regime 2 is the high price regime. Therefore, considering each equation in the low price regime, result shows that a 1% PMG change in the previous

time will be followed by PGG and PE change 9.89% and 11.05% respectively at
the current time in the opposite direction. A 1% PMG change in the previous
time will be attended by PLNG to change 3.56% at the current time in the same
direction. In the high price regime, result shows that a 1% PMG change in the
previous time will be succeeded by PGSP change 9.42% at the current time in
the opposite direction and a 1% PE change in the previous time will be followed
by PLNG change 11.60% at the current time in the opposite direction.

3.6   Regime Probabilities

Figures 1, 2 and 3 display the results on smooth probabilities for MS-VAR model
of low price regime and high price regime.

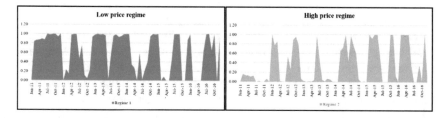

Fig. 1. Two regime probabilities: MS-BVAR (2) model natural gas procurement sources
group, 2011–2016.

Figure 1 form the result on smooth probabilities show that in 2011, PTT
engaged into natural gas sales agreements in domestic and overseas gas field
e.g., Yadana, Yetgun and Zawtika, in Myanmar of April 2013 and 2015, the natural gas energy crisis occurred. Myanmar stopped selling natural gas to Thailand
because the sea gas drilling platform collapsed in this short period. Thailand is
required to bring gas from the gulf gas of Thailand and LNG reserves to use
for domestic electricity generation. Therefore, we can see the increasing trend of
natural gas prices. In 2016, the natural gas price rose as oil prices had declined
further. Therefore, the cost of petroleum production in the gulf gas of Thailand
has increased. Especially natural gas from Bongkot and Arthit sources had problem about costs of source natural gas development and the amount of natural
gas delivered to PTT. Therefore Thailand had to increase LNG import which
contributed to higher production costs.

Figure 2 shows the smooth probability plots, whereas smooth probability is
the probability of staying in either regime 1 or regime 2. During 2011–2012
Thailand was hit by disastrous floods is was followed by an increase in the price
of natural gas in the transportation sector (NGV). The impact from floods was
also manifested by the increased cost of living as well. Later, at the end of 2012,
PTT raised the price of CNG as required by the government in order to reduce
the burden on the improvement of the natural gas supply system. By the end of

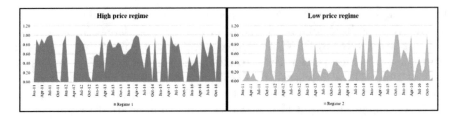

Fig. 2. Two regime probabilities: MS-BVAR (2) model natural gas products for consumption group, 2011–2016.

2014, the global crude oil price had dropped, resulting in declining gas prices. The cost of fuel used in power generation was down. And the price of gas in industry, which is based on bunker oil prices, declined, including transportation fuels. This is good for consumers to reduce the cost.

Figure 3 shows what are similar to Fig. 1. Since Thailand uses 70% of its natural gas for electricity generation, In this period Myanmar has stopped selling natural gas to Thailand which would affect the security of the Thai power system, Therefore, Thailand had to adjust by preparing measures to manage the energy by planning to increase natural gas production from the gulf source of Thailand, including alternative energy sources such as fuel oil and diesel for use in power generation. This situation in effect caused higher fuel costs which mean a greater burden on consumers. But the government issued a policy to stabilizes the energy prices to minimize the impact on consumers. The crisis did have impacts on natural gas consumption in electricity generation, fuel and gas separation plant sectors. The Electricity Generating Authority of Thailand (EGAT) and PTT Public Company Limited thus coordinated their efforts to minimized the undesirable effects.

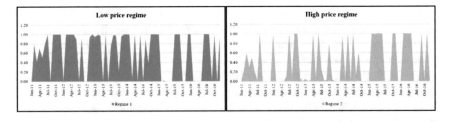

Fig. 3. Two regime probabilities: MS-BVAR (2) model natural gas procurement source and natural gas products for consumption group, 2011–2016.

4   Economic Implications

4.1   Impulse Response

Figure 4 reports the impulse responses for the natural gas procurement source group. The first panel displays the impulse response in the low price regime, while the second panel displays the high price regime. In the low price regime, it can be seem that when shocks occur in one unit standard deviation of PGG in Thailand, PGG and other observed variables, namely PMG and PLNG, respond to shocks that occur in the same direction but with varying response sizes. PGG is most likely to respond to PGG shocks and will be normalized after it has passed the 12 months. Moreover, when shocks occur in one unit standard deviation of PMG in Thailand, PMG and other observed variables, namely PGG and PLNG, respond to shocks that occur in the same direction but with varying response sizes and will be normalized after it has passed the 6 months. And when shocks occur in one unit standard deviation of PLNG in Thailand, PMG and PLNG, respond to shocks that occur in the same direction and PGG, and respond to shocks that occur in the opposite direction and PLNG shocks will be normalized after it has passed the 8 months. In the high price regime, it is shown that when shocks occur in one unit standard deviation of PGG in Thailand, PGG and other observed variables, namely PMG and PLNG, respond to shocks that occur in the same direction and will be normalized after it has passed the 6 months. In addition, when shocks occur in one unit standard deviation of PMG in Thailand, PMG and other observed variables, namely PGG and PLNG, respond to shocks

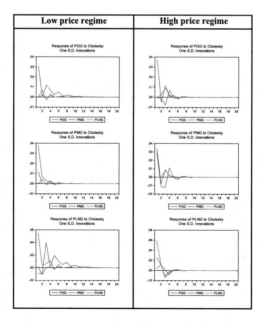

Fig. 4. The impulse response for the natural gas procurement source group model.

that occur in the same direction as well and PMG shocks will be normalized
after it has passed the 9 months. And when shocks occur in one unit standard
deviation of PLNG in Thailand, PLNG and other observed variables, namely
PGG and PMG, respond to shocks that occur in the same direction and PLNG
shocks will be normalized after it has passed the 5 months.

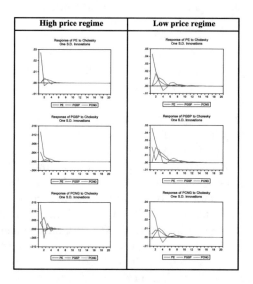

Fig. 5. The impulse response for the natural gas products for consumption group
model.

Figure 5 reports the impulse responses for the natural gas products for consumption group model. In the first panel, it displays the impulse response in
the high price regime, while the second panel of the figure displays the low
price regime. In the high price regime when shocks occur in one unit standard
deviation of PE in Thailand, PGSP and PCNG, respond to shocks that occur
in the same direction and PE, respond to shocks that occur in opposite direction and PE shocks will be normalized after it has passed the 3 months. When
shocks occur in one unit standard deviation of PGSP in Thailand, PGSP and
PE, respond to shocks that occur in the same direction and PCNG, respond
to shocks that occur in opposite direction and PGSP shocks will be normalized
after it has passed the 6 months. When shocks occur in one unit standard deviation of PCNG in Thailand, PCNG and PE, respond to shocks that occur in the
same direction and PGSP, respond to shocks that occur in opposite direction
and PCNG shocks will be normalized after it has passed the 6 months. In the
low price regime when shocks occur in one unit standard deviation of PE in
Thailand, PGSP and PCNG, respond to shocks that occur in the same direction
and PE, respond to shocks that occur in opposite direction and PE shocks will
be normalized after it has passed the 10 months. When shocks occur in one unit
standard deviation of PGSP in Thailand, PGSP and PE, respond to shocks that

occur in the same direction and PCNG, respond to shocks that occur in opposite direction and PGSP shocks will be normalized after it has passed the 12 months. When shocks occur in one unit standard deviation of PCNG in Thailand, PE and PGSP, respond to shocks that occur in the same direction and PCNG, respond to shocks that occur in opposite direction and PCNG shocks will be normalized after it has passed the 14 months.

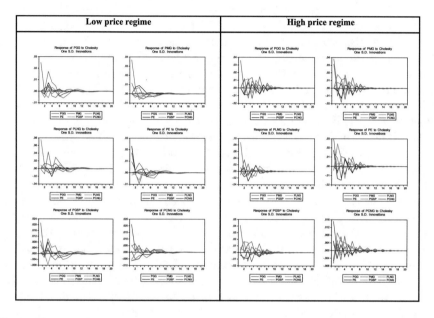

Fig. 6. The impulse response natural gas procurement source group and natural gas products for consumption group model.

Figure 6 reports the impulse responses between natural gas procurement source group and natural gas products for consumption group model. In the first panel, it displays the impulse response in the low price regime, while the second panel of the figure displays the high price regime. When consider impulse response in the low price regime, shown that when shocks occur in one unit standard deviation of PGG in Thailand, PGG, PLNG, PE and PGSP respond to shocks that occur in the same direction and PMG, PCNG respond to shocks that occur in the opposite direction and PGG will be normalized after it has passed the 14 months. When shocks occur in one unit standard deviation of PMG in Thailand, PMG, PGG, PLNG and PCNG respond to shocks that occur in the same direction and PE, PGSP respond to shocks that occur in the opposite direction and PMG will be normalized after it has passed the 13 months. When shocks occur in one unit standard deviation of PLNG in Thailand, PLNG, PMG, PE and PGSP respond to shocks that occur in the same direction and PGG, PCNG respond to shocks that occur in the opposite direction and PLNG will be normalized after it has passed the 15 months. When shocks occur in one unit standard deviation of PE in Thailand, PE, PMG, PLNG and PCNG respond to shocks that occur in the same direction and PGG, PGSP respond to shocks that occur in the opposite direction and PE will be normalized after it has passed the 13 months. When shocks occur in one unit standard deviation of PGSP in Thailand, PGSP, PGG, PMG, PLNG and PE respond to shocks that occur in the same direction and PCNG respond to shocks that occur in the opposite direction and PGSP will be normalized after it has passed the 15 months. When shocks occur in one unit standard deviation of PCNG in Thailand, PCNG and PGSP respond to shocks that occur in the same direction and PGG, PMG, PLNG and PE respond to shocks that occur in the opposite direction and PCNG will be normalized after it has passed the 12 months. In the high price regime, it is shown that when shocks occur in one unit standard deviation of PGG in Thailand, PGG, PMG, PLNG, PE and PCNG respond to shocks that occur in the same direction and PGSP respond to shocks that occur in the opposite direction and PGG will be normalized after it has passed the 12 months. When shocks occur in one unit standard deviation of PMG in Thailand, PMG, PGG, PE and PCNG respond to shocks that occur in the same direction and PLNG, PGSP respond to shocks that occur in the opposite direction and PMG will be normalized after it has passed the 13 months. When shocks occur in one unit standard deviation of PLNG in Thailand, PLNG, PMG, PGSP and PCNG respond to shocks that occur in the same direction and PGG, PE respond to shocks that occur in the opposite direction and PLNG will be normalized after it has passed the 10 months. When shocks occur in one unit standard deviation of PE in Thailand, PE, PGG, PMG and PCNG respond to shocks that occur in the same direction and PLNG, PGSP respond to shocks that occur in the opposite direction and PE will be normalized after it has passed the 13 months. When shocks occur in one unit standard deviation of PGSP in Thailand, PGSP and PGG respond to shocks that occur in the same direction and PMG, PLNG, PE and PCNG respond to shocks that occur in the opposite direction and PGSP will be normalized after it has passed the 13 months. And last when shocks occur in one unit standard deviation of PCNG in Thailand, PCNG and PMG and PE respond to shocks that occur in the same direction and PGG, PLNG and PGSP respond to shocks that occur in the opposite direction and PCNG will be normalized after it has passed the 18 months.

5   Conclusions

Based on the results, from the limited supply of natural gas in the gulf of Thailand and likely to be exhausted in the future, the government plans to reduce gas production in the gulf of Thailand and plans to import LNG to substitute natural gas. The import of LNG to replace natural gas in the gulf of Thailand will affect price of LNG to become higher than price of natural gas in the gulf of Thailand. As a consequence, consumers will face more consumption expense. In addition, the increase of LNG import will affect the electricity sector.

But the adjustment of natural gas price will not take place immediately. Moreover, if Myanmar stops to sell natural gas to Thailand, the risk of the country will increase due to the use of natural gas to generate about 70% of electricity supply. In the future, the production capacity of the gas separation plants in Thailand will decrease due to the reduction of gas resources in the gulf of Thailand and domestic gas production. As a result, the government should consider preparing natural gas consumption plan in order to suffice the domestic gas consumption by negotiating with other foreign countries in order to prepare for consumption expansion as well as providing the alternative energy enough to meet the demand of the country.

References

1. Anisie, A.: Natural gas pricing and competitiveness: the impact of natural gas prices upon the industry's dynamics. Universidad Pontificia Comillas, Master's Degree in the Electric Power Industry (MEPI)(2014)
2. Haug, J.: Seasonality in natural gas prices. Norwegian School of Economics, Financial Economics and Economic thesis degree of master (2013)
3. Reungkajon, P.: Crude oil, coal and natural gas price-return volatility estimation by ARIMA-EGARCH, ARIMA-GARCH-M and ARIMA-GARCH methods. Chiang Mai University, Economic thesis degree of master (2007)
4. Soithong, N.: Analysis of trends in Thailand's natural gas use and import quantity using ARIMA method. Chiang Mai University, Economic thesis degree of master (2014)
5. Pastpipatkul, P., Yamaka, W., Wiboonpongse, A., Sriboonchitta, S.: Spillovers of quantitative easing on financial markets of Thailand, Indonesia, and the Philippines. In: Huynh, V.-N., Inuiguchi, M., Denoeux, T. (eds.) IUKM 2015. LNCS (LNAI), vol. 9376, pp. 374–388. Springer, Cham (2015). https://doi.org/10.1007/978-3-319-25135-6_35
6. Brandt, P.T., Appleby, J.: MSBVAR: Bayesian vector autoregression models, impulse responses and forecasting. R package version 0.3.1 (2007)
7. Sims, C.A., Waggoner, D.F., Zha, T.: Methods for inference in large multiple-equation Markov-switching models. J. Econom. **146**(2), 255–274 (2008)

Portfolio Selection with Stock, Gold and Bond in Thailand Under Vine Copulas Functions

Pathairat Pastpipatkul, Woraphon Yamaka[✉], and Songsak Sriboonchitta

Faculty of Economics, Centre of Excellence in Econometrics,
Chiang Mai University, Chiang Mai, Thailand
woraphon.econ@gmail.com

Abstract. The paper aims to measure the risk and find the optimal weights of portfolio containing three instruments: Stock Exchange of Thailand, Thai Baht gold, and Treasury 10-year bond yield. The study employs the C-D vine copulas approach to construct the dependency of each pair instruments and uses the Monte Carlo simulation technique to generate the simulated data to compute Value at Risk (VaR) and Expected Shortfall (ES). Our results show that there exists a weak significant dependency between Stock Exchange of Thailand index and Thai Baht gold and dependency between Treasury 10-year bond yield and Thai Baht gold. Moreover, we find that the desired portfolio allocation is 49.8% of SET, 18.8% of Bond, and 31.4% of Gold where risk and return of the portfolio are 2.7% and 0.05%, respectively.

Keywords: Portfolio selection · CD-vine copulas · Value at risk
Thai asset

1 Introduction

Thailand's financial markets have grown significantly in both value and trading along the last decade for providing investors the opportunities to gain high returns when compared with the other emerging markets. In the first quarter of 2016, the portfolio investment of Thailand increased 2.93% from the 2015 level. This indicates that Thai financial markets remain attractive to many investors. However, the increased interdependence among the world financial markets and world financial crisis create the flow in and out of the Thai financial portfolios and result in the high volatility and uncertainty in the markets. This situation leads the investors to confront higher risk and hampers their ability to make accurate decision in investment allocation. Therefore, Thai and foreign investors, portfolio managers and financial institutions need to be cautious while making any investment decision as well as diversifying the investment in different types of stocks, bond, gold and other instruments in order to improve their gain at any level of risk.

This study focuses on three financial instruments of Thailand namely, the Stock Exchange of Thailand (SET), Bond, and gold. These instruments are considered to be key role players in the investors portfolio in Thailand. Nevertheless,

© Springer International Publishing AG 2018
L. H. Anh et al. (eds.), *Econometrics for Financial Applications*, Studies in Computational Intelligence 760, https://doi.org/10.1007/978-3-319-73150-6_55

Thai financial markets are likely to be sensitive to the negative news and events, leading to greater market fluctuations. Thus, investors have to perform portfolio diversification by investing in a variety of assets. But, the problem is how to determine the optimal proportion of the various assets in the portfolio, in such a way to improve the portfolio return. According to the Modern portfolio theory of Markowitz [11], the optimal portfolio is a tradeoff between risk and expected return to achieve the efficient portfolio. Kaura [9] suggested that optimal portfolio can be computed by the portfolio risk. To measure the portfolio risk, Value-at Risk (VaR) and Expected Shortfall (ES) will be used. In order to compute the VaR and ES of a portfolio, it is important to specify the assumption on the distribution of asset price changes and estimate the dependency or correlation between the assets. There is a large body of studies of the correlation and risk measurement among the stock, gold and bond and those studies still employed conventional methods which have a strong assumption of normal distribution in the estimation and might not provide an accurate result (see, [3, 6, 10, 14]).

Normally, the asset returns have shown that they are far from the normal distribution thus using the parametric approach which imposes the normal distribution assumption becomes inflexible in the case of financial data [7]. One way to estimate the dependence among stock, gold and bond is to use the copula approach since it enables us to model the dependence between return series which have a different marginal distribution. In the last decade, the multivariate normal copula has been used to estimate the dependency of the large number of assets in many studies. However, Kang [8] and Ortobelli et al. [12]) suggested that financial data is asymmetric and heavy tailed thus the dependence model could not be estimated by multivariate normal copula. To solve this problem, many studies employed the conventional multivariate t copula such as Kang [8], Venter et al. [18] and Autchariyapanitkul et al. [2]. Because they only take into account dependence between pairs of variables. Thus vine copulas become more flexible since they allow us to select the different bivariate copulas [13, 17]. Vine copulas have been mostly used in financial time series and firstly introduced by Aas et al. [1]. They proposed the canonical vine (C-Vine) and the drawable vine (D-vine) copulas, which give a starting point for high dimensions and allow the employment of different dependency structures between the different pair-copulas.

The first objective of this study is to model the co-movement of and the dependence among stock price, gold price and bond yield using C-vine and D-vine copula approach. The second aim is to measure the volatility of portfolio of stock, gold and bond by the concept of VaR and ES. Third, we extend the obtained expected return and ES to find the optimal weights of the assets in this portfolio and efficient frontier.

The remainder of this study is constructed as follows. In Sect. 2, we describe the C-vine and the D-vine copulas. Section 3 is the estimation procedures. Section 4 presents the data used and provides the empirical results and the final section gives conclusions.

2   Methodology

2.1   Copula

According to Sklar's theorem, an n-dimensional copula $C(u_1, ..., u_d)$ is a multivariate distribution function in $[0, 1]^d$ whose marginal distribution (u_i) is uniform in the $[0, 1]$ interval. In addition, Skalar [16] showed a link between multivariate distribution functions and their marginal distribution functions, that any joint distribution $H(x_1, ..., x_d)$ can be related to the marginal distributions $F_1(x_1), ..., F_d(x_d)$ by an appropriate copula C:

$$H(x_1, ..., x_d) = C(F_1(x_1), ..., F_d(x_d)). \tag{1}$$

The copula density c is obtained by differentiating Eq. (1); thus, we get

$$c(F_1, (x_1), ..., F_d(x_d)) = c(u_1, ..., u_d) = \frac{\partial^d C(u_1, ..., u_d)}{\partial u_1, ..., \partial u_d}. \tag{2}$$

2.2   ARMA-GARCH Models for Univariate Distributions

In this study, we use a univariate ARMA(p,q)-GARCH(m,n) specification which is often chosen to model the marginal distribution of data. It can be described as

$$x_t = \phi_0 + \sum_{i=1}^{p} \phi_i x_{t-1} + \sum_{j=1}^{q} \theta_j \varepsilon_{t-j} + \varepsilon_t, \tag{3}$$

$$\varepsilon_t = h_t z_t, \tag{4}$$

$$h_t^2 = \alpha_0 + \sum_{i=1}^{m} \alpha_i \varepsilon_{t-i}^2 + \sum_{j=1}^{n} \beta_j h_{t-j}^2, \tag{5}$$

where Eqs. (3) and (4) are, respectively, the conditional mean and variance equations, given past information. ϕ, θ, α, and β are the estimated parameters of the model. ε_t is the residual term, h_t is conditional variance and, z_t is the standardized residual which is assumed to have a Student-t distribution. Then, the best-fit marginal distribution provides a standardized ARMA(p,q)-GARCH(m,n) residual which is transformed into a uniform distribution in (0, 1). This step is the first step of the estimation procedure; thus, it is necessary to choose the best-fit ARMA(p,q)-GARCH(m,n) to obtain standardized residuals which are plugged into the vine copulas further.

2.3   C-vine and D-vine Copulas

Before we describe the general form of C-vine and D-vine copulas, lets start with pair copula construction (PPCs) which is introduced by Aas et al. [1]. Lets consider the d random variables with marginal functions $F_1, ..., F_d$ and corresponding copula density (c). By recursive conditioning d dimension joint density can be written as

$$f(x_1, ..., x_d) = f(x_1) \cdot f(x_2 \,|x_1) \cdot f(x_3 \,|x_1, x_2) \cdot, ..., \cdot f(x_d \,|x_1, x_2, ..., x_{d-1}) \qquad (6)$$

According to Sklars theorem, for the example of a three-dimensional random variables case, we know that

$$f(x_2 \,|x_1) = c_{12}(F_1(x_1), F_2(x_2))f_2(x_2) , \qquad (7)$$

$$f(x_3 \,|x_1, x_2) = c_{23|1}\left(F(x_2 \,|x_1), F(x_3 \,|x_1)\right) \cdot c_{13}(F_1(x_1), F_3(x_3))f_3(x_3) , \qquad (8)$$

thus, we get

$$f(x_1, x_2, x_3) = c_{23|1}\left(F(x_2, |x_1), F(x_3 \,|x_1)\right) c_{12}(F_1(x_1), F_2(x_2))$$
$$c_{13}(F_1(x_1), F_3(x_3))f_1(x_1)f_2(x_2)f_3(x_3) \qquad (9)$$

and $c_{23|1}$ are the pair copulas which can be chosen separately and do not depend on the other pair- copulas; $F_i(\cdot)$ is the cumulative distribution function (cdf) of x_i; and $c_{ij}(\cdot)$ is the copula density of (x_i, x_j). In the vine copula approach, Bedford and Cooke [4] are concerned that in Eq. (6) exist many such iterative PCCs, and thus they introduced a graphical model called C-vine tree and D-vine tree. In the C-vine tree, this structure shows the dependency between each variable and the first root nod which are modeled by bivariate copula. Thus, the joint probability density function of d-dimension for C-vine can be formed as

$$f(x_1, ..., x_d) = \prod_{k=1}^{d} f_k(x_k) \prod_{i=1}^{d-1} \prod_{j=1}^{d-j}$$
$$c_{i,i+j|1:(i-1)}\left(F(x_i \,|x_1, ..., x_{i-1}), F(x_{i+j} \,|x_1, ..., x_{i-1}) \,\middle|\, \theta_{i,i+j|1:(i-1)}\right) . \qquad (10)$$

On the other hand, in the D-vine tree, it is constructed by choosing a specific order of the variables. It shows a connection of first and second variable, second and third variable, third and fourth, and so on. Thus, the joint probability density function of d-dimension for D-vine can be formed as

$$f(x_1, ..., x_d) = \prod_{k=1}^{d} f_k(x_k) \prod_{j=1}^{d-1} \prod_{i=1}^{d-j}$$
$$c_{j,j+i|1,...,j-1}\left(F(x_j \,|x_j \,|x_1, ..., x_{j-1}), F(x_{j+i} \,|x_1, ..., x_{j-1}) \,\middle|\, \theta_{i,i+j|(i+1):(j+i-1)}\right) , \qquad (11)$$

where $f_k(x_k)$ denotes the marginal density of x_k, $k = 1, ..., d$; $c_{i,i+j|i+1,...,i+j-1}$ and $c_{j,j+i|1,...,j-1}$ are the bivariate copula densities of each pair copula in C-vine and D-vine, respectively. Finally, the conditional distribution function for the d-dimensional vector v can be written as

$$F(x \,|v, \theta) = F(x \,|v) = \frac{\partial C_{xv_j|v_{-j}}\left(F(x \,|v_{-j}), F(v_j \,|v_{-j}) \,|\theta\right)}{\partial F(v_j \,|v_{-j})}, \qquad (12)$$

where the vector v_j is an arbitrary component of vector v and the vector v_{-j} is vector v excluding v_j [10]. Furthermore, $C_{xv_i|v_{-j}}$ is a bivariate conditional distribution of x and v_j is conditioned on v_{-j} with parameters (θ) specified in the tree.

2.4   Value at Risk, Expected Shortfall and Optimization of Portfolio

Value at risk (VaR) and conditional Value at Risk or Expected Shortfall (ES) have been widely used to measure risk since 1990s. The VaR of portfolio can be written as

$$VaR_\alpha = \inf\{l \in R : P(L > l) \le 1 - \alpha\} \tag{13}$$

where α is a confidence level with value $[0, 1]$ which presents the probability of Loss L to exceed 1 but not larger than $(1 - \alpha)$. While an alternative method, ES, is to extend the VaR concept in order to remedy two conceptual problems of VaR [7]. Firstly, VaR measures only percentiles of profit-loss distribution and it is difficult to control for non-normal distribution. Secondly, VaR is not sub-additive. ES can be written as

$$ES_\alpha = E(L\,|\,L > VaR_\alpha) \tag{14}$$

To find the optimal portfolios, Rockafellar and Uryasev [15] introduced the port-folio optimization by calculating VaR and extend VaR to optimize ES. The app-roach focused on the minimizing of ES to obtain the optimal weight of the large number of instruments. In other words, we can write the problem as in the following. The objective function is

$$MinES_\alpha = E(L\,|\,L > VaR_\alpha) \tag{15}$$

Subject to

$$R_p = \sum_{i=1}^{d} w_i \cdot r_i, \qquad \sum_{i=1}^{d} w_i = 1, \qquad 0 \le w_i \le 1, \tag{16}$$

where R_p is the expected return of the portfolios, w_i is a vector of portfolio weights, and r_i is the return of each instrument.

2.5   Estimation

In the first step of the estimation procedure, the data series are transformed to be in log difference form. The Augmented-Dickey Fuller (ADF) test is used to confirm the stationarity of the data series. The ARMA-GARCH model is employed to obtain the standardized residuals of each series where student-t marginal distribution is assumed. All series are, then, transformed using empir-ical cumulative distribution (ecdf). For the diagnostic tests, Berkowitz uniform test and the LjungBox test, are employed to test uniform distribution and auto-correlation.

The crucial question is how to order the sequence of variables in the C-vine and the D-vine models. Czado et al. [5] suggested that ordering may be given by choosing the strongest correlation in terms of absolute empirical values of pairwise Kendall's τ as the first node. For the D-vine copula, the ordering may be given by the order that has the largest value of the sum of empirical Kendall's τ.

In this section, we provide general form of C-vine and D-vine copulas. For three-dimensional C-vine and D-vine, we can rewrite the joint density function of Eqs. (10) and (11) as

$$f(x_1, x_2, x_3) = f(x_1)f(x_2 \,|x_1)f(x_3 \,|x_1, \ x_2). \tag{17}$$

Then the joint probability density function of three-dimension for C-vine can be formed as

$$\begin{aligned} f(x_1, x_2, x_3) &= f(x_1)f(x_2)f(x_3) \cdot c_{1,2}(F_1(x_1), \\ F_2(x_2)) \cdot c_{1,3}(F_1(x_1), F_3(x_3)) \ \cdot c_{2,3|1}\left(F_{2|1}\left(x_2 \,|x_1\right), F_{3|1}\left(x_3 \,|x_1\right)\right) \end{aligned} \tag{18}$$

where $c_{1,2}$, $c_{1,3}$, $c_{2,3|1}$ represent the densities of the bivariate copulas $C_{1,2}$, $C_{1,3}$ and $C_{2,3|1}$, respectively, and the joint probability density function of three-dimension for D-vine can be formed as

$$\begin{aligned} f(x_1, x_2, x_3) &= f(x_1)f(x_2)f(x_3) \cdot c_{1,2}(F_1(x_1), \\ F_2(x_2)) \cdot c_{2,3}(F_2(x_2), F_3(x_3)) \ \cdot c_{1,3|2}\left(F_{1|2}\left(x_1 \,|x_2\right), F_{3|2}\left(x_3 \,|x_2\right)\right) \end{aligned} \tag{19}$$

where $c_{1,2}$, $c_{2,3}$ and $c_{2,3|1}$ represent the densities of the bivariate copulas $C_{1,2}$, $C_{2,3}$ and $C_{2,3|1}$, respectively. The two-stage estimation method called Inference Function Margins (IFM) is used, following Czedo et al. [5]. In the first step, the sequential maximum likelihood estimation is conducted in order to obtain the initial parameter value for the C-vine copula and the D-vine copula. In this study, the Gaussian copula, T copula, Clayton copula, Frank copula, Gumbel copula, Joe copula, BB1 copula, BB6 copula, BB7 copula, BB8 copula, and rotate copulas are bivariate copula families which are proposed for each conditional pair of variables. To choose the best fit family for each pair variable, the lowest Akaike information criterion (AIC) and Bayesian information criterion (BIC) are preferred. In the second step, the maximum likelihood estimation is applied to update the initial parameters which are obtained from the first step. In the model evaluation, the C-vine copula and the D-vine copula are compared in order to choose an appropriate vine copula structure, using AIC, BIC, and the Vuong test.

Finally, the study has constructed the VaR and ES consisting of 4 steps as follows:

1. Simulate jointly-dependent uniform variates from the estimated C-vine or D-vine copulas.
2. Transform the uniform variates derived from the first step to daily returns $(r_{d,i})$ via the inverse CDF of student-t or normal distribution.
3. Generate portfolio returns from $\sum_{i=1}^{3} w_i r_{d,i}$.
4. The VaR and ES can be calculated at 1%, 5% and 10% levels.

We, then, find an optimal portfolio weight that gives us a minimum risk for a certain level of return.

3   Empirical Result

The data consist Stock Exchange of Thailand index (SET), Thai Baht gold (GOLD), and Treasury 10-year bond yield (BOND). The data cover a time period from January 5, 2009 to July 6, 2015, covering 1,696 observations. The price returns are calculated by using the differences between the logarithms of the close price.

Table 1. Summary statistics

	SET	BOND	GOLD
Mean	0.0005	0.00001	0.0004
Median	0.0575	0.1046	0.0481
Maximum	−0.0581	−0.0788	−0.0881
Minimum	0.0114	0.014	0.0106
Std. Dev.	−0.326	1.1086	−0.5751
skewness	6.2049	12.9408	7.8868
Kurtosis	8.1534	12.7549	10.9659
JarqueBera	750.9698	7283.082	1769.502
Prob.	0.0000	0.0000	0.0000
Sum	1.1449	0.0936	0.2838
ADF-test. (Prob.)	0.0000	0.0000	0.0000

Source: Calculation

Table 1 provides the descriptive statistics of the returns of stock, bond and gold. The table shows that bond has a lower mean return than stock and gold. However, the standard deviation of bond is higher than stock and gold. We can see that all the variables have a kurtosis above 8, and that their skewness is positive. This means that the marginal distribution of our data has a heavy tail to the right rather than to the left. In addition, normality is rejected by the Jarque Bera test, prob. = 0. Thus, normal distribution might not be appropriate for our data. Consequently, we assume the marginal to have student-t distribution. The ADF test confirmed that all return series are stationary at level.

Table 2 shows the results of the marginal distribution of the student-t distribution ARMA(p,q)-GARCH(1,1) model for all the variables. We choose the best specifications for the marginal based on the AIC. The result shows that ARMA(1,1)-GARCH(1,1) is the best specification for the SET, ARMA(10,6)-GARCH(1,1) is the best specification for Bond and ARMA(3,2)-GARCH(1,1) is the best specification for Gold.

In addition, the Berkowitz (BERK) uniform test is used for the marginals. The result shows that none of the BERK tests accepts the null hypothesis. Therefore, it is evident that all the marginal distributions are uniform. Moreover, the LjungBox test, which is used as the autocorrelation test on standard-

Table 2. Results of marginal distribution: ARMA(p,q)-GARCH(1,1)

	SET	BOND	GOLD
Mean equation			
AR(1)	0.0011***	0.0009	0.0002**
AR(2)	0.7937***	0.4736***	0.4853***
AR(3)		0.0505	−0.9221***
AR(4)		−0.3016***	−0.0957***
AR(5)		0.4353***	
AR(6)		0.2981***	
AR(7)		0.0268	
AR(8)		−0.0279	
AR(9)		−0.0235	
AR(10)		0.0069	
MA(1)		0.0541***	
MA(2)	−0.8001***	−0.3881***	−0.5843***
MA(3)		−0.0293	0.9733***
MA(4)		0.2879***	
MA(5)		−0.4026***	
MA(6)		−0.3895***	
MA(7)		−0.0742***	
Variance equation			
α_0	0.0000001	0.000003***	0.00001***
α_1	0.0394***	0.1340***	0.0426***
β	0.9589***	0.8604***	0.9515***
Berk test (prob.)	0.9716	0.9891	0.9527
Q (prob.)	0.5106	0.7587	0.7089
(T-DIST.DOF)	5.701***	−3.5014***	3.789***
AIC	−7.1199***	6.2785	−7.9521***

Source: Calculation

Note: *, **, and *** denote rejections of the null hypothesis at the 10%, 5%, and 1% significance levels, respectively.

Table 3. Sum of Empirical Kendall's τ

	SET	BOND	GOLD
SET	1	0.0072	−0.0229
BOND	0.0072	1	−0.1456
GOLD	−0.0229	−0.1456	1
Sum	1.669877	1.727713	1.344198

Source: Calculation

ized residuals, confirms that there is no rejection of the null hypothesis, i.e. no autocorrelation in any of the series.

Table 3 provides the result of the sum of Kendall's τ and shows that Bond has the strongest dependency in terms of the empirical value of pairwise Kendall's τ; thus, we determine Gold as the first root node, and the order should be as follows: Bond (order 1), SET (order 2), Gold (order 3) for C-vine. For D-vine, we find that the following order of SET, Gold, and Bond presents the highest value of the sum of Kendall's τ. The result shows that C-vine and D-vine have different structure of pair copulas; thus, we calculate both C-vine and D-vine in order to find the best structure for our analysis.

Table 4. Results of C-vine copulas

Copula family	Parameter	Lower and upper tail dependence	AIC	BIC
$C_{bond,set}$ Student-t	0.0109	0.0779, 0.0779	24.852	13.994
$C_{bond,gold}$ Student-t	−0.0243***	0.0705, 0.0705	8.107	2.75
$C_{set,gold|bond}$ Student-t	−0.0311***	0.06917, 0.06917	−8.805	2.052
Log-likelihood	−10.266	SUM	−41.765	−9.192

Source: Calculation

Note: *, **, and *** denote rejections of the null hypothesis at the 10%, 5%, and 1% significance levels, respectively.

Table 5. Results of D-vine copulas

Copula family	Parameter	Lower and upper tail dependence	AIC	BIC
$C_{set,gold}$ Student-t	−0.0231***	0.06915, 0.06915	18.497	29.355
$C_{gold,bond}$ Student-t	−0.0312***	0.0707, 0.0707	4.901	15.758
$C_{set,bond|gold}$ Student-t	0.00986	0.0776, 0.0776	3.085	13.943
Log-likelihood	−7.242	SUM	26.4843	59.057

Source: Calculation

Note: *, **, and *** denote rejections of the null hypothesis at the 10%, 5%, and 1% significance levels, respectively.

The estimated results of the C-vine copula and the D-vine copula are presented in Tables 4 and 5, respectively. In C-vine, we can be observe that student-t presents the best copula families in terms of the lowest AIC and BIC values for $C_{bond,set}$, $C_{bond,gold}$, and $C_{set,gold|bond}$. For D-vine, student-t copula family also shows the lowest AIC and BIC for $C_{set,gold}$, $C_{gold,bond}$, and $C_{set,bond|gold}$. To identity the most appropriate structure, the sum of AIC and BIC is used as a model selection method. According to the sum of AIC and BIC, we observe that AIC and BIC values of C-vine are lower than AIC and BIC values of D-vine thus, it indicates that C-vine is more appropriate than D-vine. Apart from these two criteria, the Vuong test [19] is also implemented in this study to determine the best fit vine copula structure and the results are presented in Table 6, the results show that p-values of Vuong Test are larger than 0.05 thus, we cannot reject the null hypothesis that C vine and D-vine are statistically equivalent. Therefore, both the C-vine copula and the D-vine copula are considered to be interpreted in this study. Considering the dependency of the pair variables, it is evident that there is significant co-movement and tail dependence in SET-Gold and Gold Bond pairs in both C vine and D-vine. Among the significant variable pairs, it is seen that the pair Bond and Gold accounts for the greatest upper tail and lower tail dependency, which includes tail dependence 0.0707 and 0.0705 in D-vine and C-vine, respectively. Moreover, we notice that the tail dependence of $C_{set,gold|bond}$ and $C_{set,gold}$ have some economic interpretation. If bond returns is given as condition, the tail dependence of SET and Gold increases by 0.029%. From the above results, we can conclude that SET, bond, and gold have a small dependence between each other, thus it implies that these assets move independently of each other and investors can make a diversification benefits from this portfolio.

Table 6. Results of comparison between two non-nested parametric models

Test	Statistic	Statistic Akaike	Statistic Schwarz	p-value	p-value Akaike	p-value Schwarz
Vuong test	0.7528	0.7528	0.7528	0.4515	0.4515	0.4515

Source: Calculation

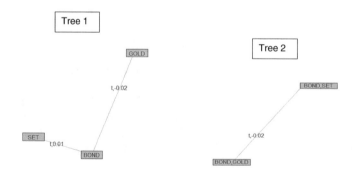

Fig. 1. The C-vine tree plot

Upon observing Figs. 1 and 2, it can be seen that the tree plot shows us the two layers of the C-vine and D-vine copula model, comprising tree 1 and tree 2. Each tree presents a combination of the dependency of the variable pairs by Kendall's τ, as follows:

For C-vine tree plot, the first and second tree show that the Kendall's τ of $C_{bond,set}$, $C_{bond,gold}$, and $C_{set,gold|bond}$ are 0.01, −0.02, −0.02. For D-vine,

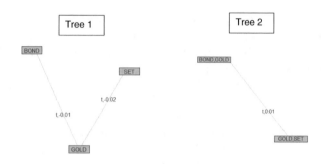

Fig. 2. The D-vine tree plot

the first and second tree demonstrate three pairs dependence, which include the Kendall's τ of $C_{set,gold}$, $C_{gold,bond}$, and $C_{set,bond|gold}$. We found that the Kendall's τ of SET Gold and Bond-Gold are -0.02 and -0.01. Thus, we can say that if the Bond is given as the condition of SET and Gold dependence, the Kendall's τ remain the same. This means that the movement of Bond seems to have no effect on the SET and Gold.

Table 7. Equally weighted portfolios value at risk based on C-vine

Equally weighted portfolio	Risk value		
	1%	5%	10%
VaR	-1.87%	-0.99%	0.68%
ES	-2.58%	-1.54%	1.18%

Source: Calculation

We, then, simulated 10,000 jointly-dependent uniform variates from the estimated C-vine copulas for being more appropriate than D-vine (consider the lowest AIC and BIC). Table 7 reports the VaR and ES at different confidence levels for equally weighted portfolio. We can see that the estimated VaR converges to -1.87%, -0.99%, and -0.68% at 1%, 5%, and 10% level in period, respectively. While ES converges to 2.58%, -1.54%, and -1.18% at 1%, 5%,

Table 8. Optimal weighted portfolios based on C-vine

	Port 1	Port 2	Port 3	Port 4	Port 5	Port 6	Port 7	Port 8	Port 9	Port 10
SET	0.372	0.439	0.498	0.558	0.618	0.674	0.747	0.82	0.907	1
Bond	0.262	0.23	0.188	0.146	0.105	0.058	0.033	0.009	0	0
Gold	0.367	0.331	0.314	0.296	0.277	0.268	0.22	0.171	0.093	0
Return	0.0004	0.0004	0.0005	0.0005	0.0006	0.0006	0.0006	0.0007	0.0008	0.0008
Risk	0.026	0.026	0.027	0.028	0.029	0.031	0.033	0.035	0.038	0.042

Source: Calculation

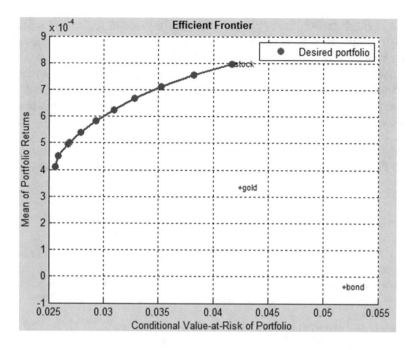

Fig. 3. Efficient frontier (1%) based on C-vine

and 10% level in period, respectively. Then, we tried to constructed the optimal weighted portfolio or efficient frontier based on C-vine copulas and the result is illustrated in Fig. 3. The result shows that a desired portfolio allocation is 49.8% of SET, 18.8% of Bond, and 31.4% of Gold where risk and return of the portfolio are 2.7% and 0.05%, respectively. (See Port 3 in Table 8).

4   Concluding Remarks

This paper proposes the C-vine and D-vine copulas approach to estimate the co-movement and dependence between Stock Exchange of Thailand index, Thai Baht gold, and Treasury 10-year bond yield. According to the AIC and BIC results, the findings confirm that the C-vine structure is more appropriate than the D-vine structure. However, the likelihood ratio Vuong test suggests us that C-vine and D-vine copulas are statistically equivalent. As for the estimated C-vine and D-vine, we find that the dependence of each pair tends to have both a symmetric structure and a hierarchical structure. Additionally, it is found that there is a weak significant dependency between Stock Exchange of Thailand index and Thai Baht gold and dependency between Treasury 10-year bond yield and Thai Baht gold. Then, we generate 10,000 dependent uniform variables based on C-vine copulas since it show the lowest AIC and BIC. We use the Monte Carlo simulation to measure the VaR and ES in this portfolio. Finally, the study optimized the portfolio using minimization of Expected shortfall method in order to obtain the efficient frontier or optimal weight for asset allocation. We find that if these 3 assets are hold in the portfolios, those risk avert investors should hold 49.8% of stock, 18.8% of Bond, and 31.4% of Gold. As far as further studies are concerned, the paper can expand the scope of analysis of the co-movement in other markets such as futures market and exchange rate market, as well as try to extend the vine copula structure in order to find the appropriate dependence structure of the high dimension variables.

Acknowledgements. The authors are grateful to Puay Ungphakorn Centre of Excellence in Econometrics, Faculty of Economics, Chiang Mai University for the financial support.

References

1. Aas, K., Czado, C., Frigessi, A., Bakken, H.: Pair-copula constructions of multiple dependence. Insur. Math. Econ. **44**(2), 182–198 (2009)
2. Autchariyapanitkul, K., Chanaim, S., Sriboonchitta, S.: Portfolio optimization of stock returns in high-dimensions: a copula-based approach. Thai J. Math. **12**, 11–23 (2014)
3. Baur, D.G., Lucey, B.M.: Is gold a hedge or a safe haven? An analysis of stocks, bonds and gold. Finan. Rev. **45**(2), 217–229 (2010)
4. Bedford, T., Cooke, R.M.: Vines: a new graphical model for dependent random variables. Ann. Stat. **30**, 1031–1068 (2002)
5. Czado, C., Schepsmeier, U., Min, A.: Maximum likelihood estimation of mixed C-vines with application to exchange rates. Stat. Model. **12**(3), 229–255 (2012)
6. Gencer, G., Musoglu, Z.: Volatility transmission and spillovers among gold, bonds and stocks: an empirical evidence from Turkey. Int. J. Econ. Finan. Issues **4**(4), 705–713 (2014)
7. Halulu, S.: Quantifying the risk of portfolios containing stocks and commodities. Doctoral dissertation, Bogazii University (2012)
8. Kang, L.: Modeling the Dependence Structure Between Bonds and Stocks: A Multidimensional Copula Approach. Indiana University, Bloomington (2007)
9. Kaura, V.: Portfolio optimisation using value at risk. Doctoral dissertation, Masters thesis, Imperial College, London (2005)
10. Kumar, D.: Return and volatility transmission between gold and stock sectors: application of portfolio management and hedging effectiveness. IIMB Manag. Rev. **26**(1), 5–16 (2014)
11. Markowitz, H.: Portfolio selection. J. Finan. **7**(1), 77–91 (1952)
12. Ortobelli, S., Biglova, A., Rachev, S.T., Stoyanov, S.: Portfolio selection based on a simulated copula. J. Appl. Funct. Anal. **5**(2), 177–193 (2010)
13. Pastpipatkul, P., Yamaka, W., Sriboonchitta, S.: Co-movement and dependency between New York stock exchange, London stock exchange, Tokyo stock exchange, oil price, and gold price. In: Proceedings of International Symposium on Integrated Uncertainty in Knowledge Modelling and Decision Making, pp. 362–373. Springer, Cham (2015)
14. Rachev, S.T., Racheva-Iotova, B., Stoyanov, S.V., Fabozzi, F.J.: Risk management and portfolio optimization for volatile markets. In: Handbook of Portfolo Construction, pp. 493–508. Springer, New York (2010)

15. Rockafellar, R.T., Uryasev, S.: Conditional value-at-risk for general loss distributions. J. Bank. Finan. **26**(7), 1443–1471 (2002)
16. Sklar, M.: Fonctions de repartition an dimensions et leursmarges. Publ. Inst. Statist. Univ. Paris **8**, 229–231 (1959)
17. Sriboonchitta, S., Liu, J., Kreinovich, V., Nguyen, H.T.: A vine copula approach for analyzing financial risk and co-movement of the Indonesian, Philippine and Thailand stock markets. In: Modeling Dependence in Econometrics, pp. 245–257. Springer, Cham (2014)
18. Venter, G., Barnett, J., Kreps, R., Major, J.: Multivariate copulas for financial modeling. Variance **1**(1), 103–119 (2007)
19. Vuong, Q.H.: Likelihood ratio tests for model selection and non-nested hypotheses. Econometrica **57**(2), 307–333 (1989)

Determinants and Stability of Demand for Money in Vietnam

Pham Dinh Long[1](✉) and Bui Quang Hien[2]

[1] Faculty of Economics and Public Management, Ho Chi Minh Open University,
97 Vo Van Tan, Ho Chi Minh City 700000, Vietnam
long.pham@ou.edu.vn
[2] Graduate School, Ho Chi Minh City Open University, 97 Vo Van Tan,
Ho Chi Minh City 700000, Vietnam
hienbq.16ae@ou.edu.vn

Abstract. This paper explores the determinants and stability of money demand in Vietnam. By applying cointegration techniques, we indicate the role of gold price and real effective exchange rate on the money demand. Real deposit interest rate is not an interesting investment channel compare with holding other assets. CUSUM and CUSUM squared tests do not confirm the stability of money demand function in the investigating period.

1 Introduction

The determinants and stability of the money demand function are important to establish and implement an effective monetary policy. According to Kumar [11], Riyandi [15], and Sriram [17], researchers have been investigating money demand for many years. Following the meta-analysis results and literature surveys, the previous studies of this topic have been considering with many aspects as well as arguments. In contrast, there are just a few empirical researches in Viet Nam [6,13,14]. For developing countries, the demand for money function is more complicated than developed countries. One of the reason is the opportunity cost of holding money, which changes rely on economic properties of each country. It is not only for transaction purpose but also for precautionary and speculative motivation. Therefore, this studying is extremely necessary, especially with a developing country like Viet Nam.

Many conclusions had been given by various estimation method. The researchers have often concentrated problem of modeling selection to find the best appropriate model. To forecast money demand quantity accurately, the model estimation should be evaluated and updated corresponding to the actual economic situation for each country. The remaining concerns should be continuing in further work. This mention suggests further studying money demand to reflect more and more accuracy. In Vietnam, a number of quantitative studies are still limited. One of the main obstacles is estimation method. The method only specified one model to analyze, and propose implication policies [6,13,14].

© Springer International Publishing AG 2018
L. H. Anh et al. (eds.), *Econometrics for Financial Applications*, Studies in Computational Intelligence 760, https://doi.org/10.1007/978-3-319-73150-6_56

It implies that policy makers need much more information to make a decision regarding monetary policy. Money demand function was concluded stable over the period 1996–2006, 1999–2010. Do these factors influence to the money demand function and whether its stability exist or not compared to the previous period studied? The purpose of this study is to investigate factor of money demand and stability over 2003Dec to 2014Dec. What is not yet clear about the impact of real gold price and real exchange rate on especially money demand? The cointegration technique persuades to exam the existence of new relationships. This is the main reason why the paper continues to inspect in detail the determinants of money demand in Viet Nam. In addition, this paper will confirm the stability question over 2003Dec to 2014Dec.

The article is outlined as follow. Section 1 introduces some highlights of money demand. Section 2 presents literature review, and previous researches. Section 3 describes methodology including model, technique, and data. In Sect. 4, the results are presented and discussed. Finally, a conclusion is summarized.

2   Literature Review

The previous studying assumed money market equilibrium criteria to estimate money demand [3]. Money market equilibrium occurs when money quantity that people hold equal to money quantity that people want to hold. Some theories presented such as Irving Fisher, the Cambridge school, neo-classical, Keynes, Minton Friedman,... In fact, the literature review identified that almost researchers used Keynes's theory or Friedman's theory to explain and develop money demand [1,5,7,10,11,15,17,18].

Babic [2] and Kumar et al. [12] summarized three main factors of holding money including transaction, precautionary and speculative motives that Keynes implied. Regarding to transaction, people want to keep the money to exchange in daily and its motive is part is proportional to income. Besides, people want to hold money for unexpected needs. The precautionary motivation is approached following this way. The amount of money that people want to hold is determined by their transaction level in the future. This part also is proportional to income. For speculative motive, the interest rate is an important role that determines money demand for speculative activity. The interest rate has an inverse relationship with holding money [12]. The Keynesian theory also notices a difference between real and nominal money quantity. The main result is emphasized is that there is a negative relationship between money demand and interest rate. By combining these components; the model of demand for money (1) following Keynesian approach presents includes income (Y_+) and interest rate (i_-). Where $(+/-)$ signs are positive affect and negative affect.

$$M^d = f(Y_+, i_-) \tag{1}$$

Friedman approached money demand function based on modern money quantity theory. Economists used widely this theory when performing research of money demand function. The money is integrated as an asset classification and money demand function is influenced by some factors related to ownership needs of other assets. From this, money demand function was developed base on asset demand theory. The demand of money is a function which of the availability of resources for personal and expected returns of other asset classes compared with the expected return of the currency. Friedman considered that the velocity of money is high. Therefore, the demand for money function is highly stable and less sensitive to the interest rate [12]. This argument implies that the quantity of money demand can be predicted accurately by money demand. By this approach, he proposed money demand function by formula (2).

$$\frac{M^d}{P} = f(Y_+, r_b - r_{m-}, r_e - r_{m-}, \pi_e - r_{m-}) \qquad (2)$$

where is $\frac{M^D}{P}$: demand for real money balances, Y_p: a measure of wealth, π_m: expected return on money, π_b: expected return on bonds, r_e: expected return on equity, π_e: expected inflation rate.

There are various arguments about the impact of interest rate on the stability of the money demand function. Fisher said that money demand function is stable because velocity money is stable. Friedman also accepted to Fisher viewpoint. Meanwhile, Keynes considered that money demand function is not stable due to unstable velocity money [13]. In general, previous researches have established that the demand for money has a positive relationship with scale variables (S) and a negative relationship with the opportunity cost variables (OC). Sriram [18] summarized the Eq. (3) in a survey of theory literature for money demand. Recently, Kumar [11] mentioned some contrasting results between advanced and developing countries that are also figured out cost of holding money. In financial reforms processing, the definitions of monetary aggregates have a quite different result that produces income elasticity estimates [11]. The selection of these variables depends on characteristics of the economic situation for each country [1,7,10,11,15–17].

$$\frac{M^d}{P} = f(S_+, OC_-) \qquad (3)$$

Previous Studies of Money Demand in Some Countries and in Vietnam
The major findings of empirical studying summarize in Table 1. In the past, there were studies of money demand function in Vietnam by Hoa [6], Nguyen and Pfau [13], and Lan [14]. Most studies selected industrial production value or the gross domestic product to represent the scale variable. The opportunity cost variables of holding money are variously selected. These variables were chosen including deposit interest rate, treasury bills interest rate, consumer price index, exchange rate, and stock price. Gold price and real effective exchange rate have not been reflected in the money demand function in Vietnam. The studying on over the world often focuses on the selection method estimates to make a consistent model [7,9,10,21]. Currently, there has been little quantitative analysis like Vietnam.

Regarding to Vietnam money demand modeling, the empirical studies applied vector error correction model or autoregressive distributed lag to estimate. There is no comparison among these result estimates. It is hard to convince policy makers about implications in future work. On the other hand, some previous studies outside Vietnam applied complicated estimation such as cointegration technique (including FMOL, DOLS, CCR) [7,9], time-varying cointegration [21], neural network [1,10]. It helps analyze and explore the influence factors of money demand clearly. To link actual Vietnam economic situation, the research needs to update such as data source, new explanation variables, improving points of model econometrics to achieve one reliable result. Besides, the most insight of money demand is stable or not. The CUSUM and CUSUM-square result of Hoa [6] and Nguyen and Pfau [13] concluded the money demand stability over the investigating period. This paper will consider cointegration technique including FMOLS, DOLS, CCR in order to expect the new determinants of money demand. Above-mentioned issues will be focused during construct research model construction.

Table 1. Summary of studies on money demand

Author	Money quantity/Period	Country	Variables	Method	Main findings
Alsahafi [1]	M1, M2 93.1-06.3 (q)	Saudi Arabia	GDP, FIDR, EX, FITB, FIGB	VECM, ANN. Johansen cointegration test	GDP (+), FIDR (−), FITB (+). ANN result shows better than VECM
Dogan [5]	M1 02.1-14.12 (m)	Turkey	IP, EX, CPI	ADRL, CUSUM, CUSUMSQ	CPI (−), IP (+), EX (+). M1 is stable
Hoa [6]	M1, M2 94.12-06.12 (m)	Vietnam	IP, CPI, FITB, EX	VECM, CUSUM, CUSUMSQ	IP (+), FIDR (−), FITB (−), EX (−). M1, M2 is stable
Hamdi et al. [7]	M2 80.1-11.4 (q)	Bahrain, Kuwait, Oman, Qatar, Arabia & UAE	GDP, FIDR, REER, FITB	FMOLS, DOLS, CCR, PMGE. Johansen cointegration test	GDP (+), FIDR (−). M2 is stable
Inoue and Hamori [9]	M1, M2, M3 80.1-07.12 (m) 76.1-07.4 (q)	India	(m): IP, FIDR (q): GDP, FIDR	DOLS	IP (+), GDP (+), FIDR (−). M1, M2 is stable
Joseph et al. [10]	M2 97.5-13.2 (m)	United States	GDP, FIDR, CPI, FITB	Linear regression, ANN	GDP (+), FIDR (−), FITB (−). ANN result shows better than the linear model
Kumar et al. [12]	M1 60-08 (y)	Nigeria	GDP, FIDR, EX, CPI	Gregory & Hansen cointegration technique	GDP (+), FIDR (−), EX (−), CPI (−). M2 is stable
Nguyen and Pfau [13]	M2 9.1-09.4 (q)	Vietnam	GDP, CPI, EX, FIDR, St	VECM	GDP (+), CPI (−), St (+), FIDR (−). M2 is stable
Lan [14]	M2 98.1-10.4 (q)	Vietnam	GDP, FIDR, EX	VECM	GDP (+), FIDR (−), EX (−). M2 is stable

Notes: (+/−): positive impact/negative impact.
Abbreviate words: GDP (gross domestic product), CPI (consume price index), FIDR (nominal interest deposit rate), EX (exchange rate), REER (real effective exchange rate), FITB (interest treasure bill), FIGB (interest government bond), St (stock index), IP (industrial production). The period abbreviation including (y), (q), and (m) is defined as yearly, quarterly, and monthly period.

3   Methodology

3.1   Model

In order to analyze the determinants of money demand, this paper considers formula (3) with assuming $M^d = M^s$. Real money supply $M1$ ($M1_r^d$) is selected for the dependent variable. For empirical analysis, scale variable is selected by domestic industrial value (Y_p). The opportunity cost variables included real domestic gold price (G_s), VN-Index (St_s), real effective exchange rate ($REER_s$), real deposit rate (r_s), treasury bills interest rate ($Tbill_s$) and consumer price index (π_s). The research proposes the model by Eq. (4).

$$\ln M1_r^d = \ln Y_p + \ln G_s + St_{vni} + REER_s + r_s + Tbill_s + \pi_s. \qquad (4)$$

3.2   Method

This study uses quantitative analysis method including four steps. The first step is unit root test and unit root test with breakpoint structure to check the stationary characteristic of all variables. The second step is Johansens cointegration test to confirm the long-run relationship between these variables. In the third step, model estimation uses different techniques: vector error correction model (VECM), cointegration regression model. Inside cointegration regression model, there are three models including fully modified ordinary least square (FMOLS), dynamic ordinary least square (DOLS), canonical cointegration regression (CCR). The difference between these models is the calculated way of the long-run covariance matrix to remedy model diseases. Finally, the stability of demand function will be confirmed by CUSUM and CUSUM-squared test.

3.2.1   Unit Root Tests

The equation is described by formulas (5), (6), and (7). The difference between three formulas is none variables, intercept variables (α) and "intercept and trend" (γT) variables. The null hypothesis is $H_0(\delta = 0)$ and the alternative hypothesis is $H_1(\delta < 0)$. In case, statistical value is smaller than the critical value. We reject the null hypothesis of a unit root and conclude that Y_t is a stationary process. In case, the null hypothesis cannot reject, we will conclude that Y_t is a non-stationary process. We perform the null hypothesis of a unit root against the alternative hypothesis of stationarity using the Augmented Dickey-Fuller (ADF) and Augmented Dickey-Fuller with breakpoint structure. The null hypothesis is set that the variable has unit root (not stationary). For ADF with breakpoint structure test, we use Zivot and Andrews [20] (see [8]) method to test the unit root hypothesis under structural breaks. All tests have a markup and no tendency in the series.

$$\Delta Y_t = \delta Y_{t-1} + \sum_{i=1}^{p} \beta_i \Delta Y_{t-i} + u_t \qquad (5)$$

$$\Delta Y_t = \alpha + \delta Y_{t-1} + \sum_{i=1}^{p} \beta_i \Delta Y_{t-i} + u_t \tag{6}$$

$$\Delta Y_t = \alpha + \gamma T + \delta Y_{t-1} + \sum_{i=1}^{p} \beta_i \Delta Y_{t-i} + u_t \tag{7}$$

According to IHS Global Inc. [8], Perron figured out structural change and unit root is a close relationship. The researchers considered that unit root test can be biased and does not reject unit root hypothesis when the data are trend stationary with a structural break. This method was added variable structure breaks to conclude wrong stationary level. Perron-Vogelsang and Clemente-Montanes-Reyes suggested testing model units with structural change [8]. Some variables are defined before processing. First, an intercept break variable $DU_t(T_b) = 1, (t \geq T_b)$ that takes the value 0 for all dates prior to the break, and one thereafter. Second, a trend break variable $DU_t(T_b) = 1, (t \geq T_b)$, which takes the value 0 for all dates prior to the break, and is a break date re-based trend for all subsequent dates. Third, a one-time break dummy variable $DU_t(T_b) = 1, (t = T_b)$ that takes the value of one only on the break date and zeroes otherwise. The unit root test with structural changes described by the formula (8), (9), (10), and (11). Similarly, the null hypothesis is $H_0 (\delta = 0)$ and the alternative hypothesis is $H_1 (\delta < 0)$. If the statistical value is smaller than the critical value, we reject the null hypothesis of a unit root and conclude that Y_t is a stationary process. Otherwise, if we cannot reject the null hypothesis, we conclude that Y_t is a non-stationary process.

Model 1: Non-trending data with intercept break.

$$y_t = \mu + \theta DU_t(T_b) + \omega D_t(T_b) + \alpha y_{t-1} + \sum_{i=1}^{k} c_i \Delta y_{t-i} + u_t \tag{8}$$

Model 2: Trending data with intercept break.

$$y_t = \mu + \beta_t + \theta DU_t(T_b) + \omega D_t(T_b) + \alpha y_{t-1} + \sum_{i=1}^{k} c_i \Delta y_{t-i} + u_t \tag{9}$$

Model 3: Trending data with intercept and trend break.

$$y_t = \mu + \beta_t + \theta DU_t(T_b) + \gamma DT_t(T_b) + \omega D_t(T_b) + \alpha y_{t-1} + \sum_{i=1}^{k} c_i \Delta y_{t-i} + u_t \tag{10}$$

Model 4: Trending data with trend break.

$$y_t = \mu + \beta_t + DT_t(T_b) + \alpha y_{t-1} + \sum_{i=1}^{k} c_i \Delta y_{t-i} + u_t \tag{11}$$

where are y_t: data series over test period; k: lag lengths; u_t: white noise; β_t: trending variable and get value from $[1, n]$.

3.2.2   Cointegration Technique

The existence of cointegration is considered when $(y_t, \Delta x_t)$ is stationary at first difference and (u_t, w_t) is stationary at zero difference. In the linear model, the endogenous phenomenon will be affected to bias coefficient estimates of the explanatory variables. The correlation between u_t and w_t causes a problem. The serial correlation phenomenon will be impacted inaccurate the significance level. When using OLS regression, some diseases caused bias estimation relates to β_1, β_2 coefficients (12). ECM model corrects β_1, β_2 coefficients using error correction coefficient (14). The accuracy estimation will be improved.

$$Y_t = \beta_1 + \beta_2 X_t + u_t \tag{12}$$

$$\Longrightarrow u_t = Y_t - \beta_1 - \beta_2 X_t$$

$$\Delta Y_t = \beta_1 + \beta_2 \Delta X_t + u_t \tag{13}$$

$$\Delta Y_t = \alpha_0 + \alpha_1 \Delta X_t - \pi \left[u_{t-1} \right] + \varepsilon_t$$

$$\Delta Y_t = \alpha_0 + \alpha_1 \Delta X_t - \pi \left[Y_{t-1} - \beta_1 - \beta_2 X_{t-1} \right] + \varepsilon_t \tag{14}$$

Another technique is FMOLS and CCR. The main idea of this method is modified each part of OLS regression using the long-run covariance matrix (15), (16), and (17) to prevent OLS diseases. Meanwhile, the issue of endogenous and serial correlation is resolved. There are three steps to apply this method following [19].

Step 1: Calculate residual u_{1t} from OLS equation and residual u_{2t} from difference equation of Δx_t.

$$y_t = \beta_0 + \beta_1 x_t + u_{1t}$$

$$\Delta x_t = \mu + u_{2t}$$

Step 2: Calculate long-run covariance matrix (LRCOV) of the residuals from combine between residual u_{1t} and residual u_{2t}. Denote $\xi(\Omega) = (u_{1t}, u_{2t})'$. Assume the innovations $ut = (u_{1t}, u_{2t})'$ are strictly stationary and ergodic with zero means (see [19]), contemporaneous covariance matrix Σ, one-sided LRCOV matrix Λ, and nonsingular LRCOV matrix Ω. The long-run covariance matrix is divided into three parts $\Omega = \Sigma + \Lambda + \Lambda'$.

$$\Sigma = E(u_t u_t') = \begin{pmatrix} \sigma_{11} & \sigma_{12} \\ \sigma_{21} & \Sigma_{22} \end{pmatrix} \tag{15}$$

$$\Lambda = \sum_{j=0}^{\infty} E(u_t u_{t-j}') = \begin{pmatrix} \lambda_{11} & \lambda_{12} \\ \lambda_{21} & \Lambda_{22} \end{pmatrix} \tag{16}$$

$$\Omega = \sum_{j=-\infty}^{\infty} E(u_t u_{t-j}') = \begin{pmatrix} \omega_{11} & \omega_{12} \\ \omega_{21} & \Omega_{22} \end{pmatrix} \tag{17}$$

Step 3: While FMOLS method adjusts base on the matrix Ω and Λ, CCR method require another part. It is a matrix Σ. The purpose aims to improve accuracy after modified model. The FMOLS and CCR model describe by (18) and (19).

$$\text{FMOLS:} \quad y_t^+ = \beta_0 + \beta_{fmols} x_t + u_{1t}^+ \tag{18}$$

where coefficient β_{fmols} is calculated y_t^+, λ_{12}^+ and $\omega_{1,2}$.

$$y_t^+ = y_t - \omega_{12} \Omega_{22}^{-1} u_{2t}$$

$$\lambda_{12}^+ = \lambda_{12} - \omega_{12} \Omega_{22}^{-1} \Lambda_{22} \quad \text{(Adjustment coefficient)}$$

$$\omega_{1,2} = \omega_{11} - \omega_{12} \Omega_{22}^{-1} \omega_{21} \quad \text{(Estimates from covariance } u_{1t} \text{ under } u_{2t} \text{ condition).}$$

$$\text{CCR:} y_t^+ = \beta_0 + \beta_{ccr} x_t + u_{1t}^+ \tag{19}$$

where coefficient β_{ccr} is calculated by y_t^+, x_t^+.

$$y_t^+ = y_t - \Sigma^{-1} \Lambda_2 \beta + \left(0, \omega_{12} \Omega_{22}^{-1}\right) u_t$$

$$x_t^+ = x_t - \left(\Sigma^{-1} \Lambda_2\right)' u_t$$

In contrast to FMOLS and CCR, DOLS uses both lags and leads to adjusting estimated coefficients (20). $M = [c, \alpha, \beta, \gamma]$ contains a set of coefficients for each explanatory variable. $X = [1, P_t, Y_t, A_t]$ contains a set of explanatory variables. m, n, k denote the length between lags and leads.

$$Q_t = X_t M' + \sum_{i=-m}^{i=m} \Phi \Delta P_{1\_t} + \sum_{i=-n}^{i=n} \Psi \Delta Y_{1-t} + \sum_{i=-k}^{i=k} \theta_i A_{t-i} + \varepsilon_t \tag{20}$$

3.3 Data

This study uses monthly times series data from 2003 to 2014. The data source includes narrow money supply, deposits interest rate, treasury bills interest rate, consumer price index are collected directly from International Monetary Fund (IMF). Due to more opportunity cost of demand for money in Vietnam, difference data are obtained from General Statistics Office (GSO), vietstocks and [4]. They are gold price index, industrial production value (from GSO), nominal gold price, Vietnam stock index (from Vietstocks), real effective exchange rate from Darvas [4]. To standard and adjust data, this paper computes and presents in natural logarithm form for $M1_r^d$, $\ln G_s$, Y_p. Then, these variables also are deseasonalized using seasonal dummies. Others (St_s, $REER_s$, r_s, $Tbill_s$, π_s are presented by percentage. The real interest rate (r_s and $Tbill_s$) are measured by the formula $i_r = (i_n - \pi)/(1+\pi)$, where, i_r, i_n, π are the real interest rate, expected inflation, and nominal interest rate. From 2003Dec to 2009Mar, SJC nominal gold price is not available. The research infers the data following $G_{t-1} = (G_t * 100)/gold_{index}$. This index ($gold_{index}$) means the current month gold price index compared to the previous month. When using quarterly data, previous research recommended use gross domestic product or gross national product to get a better evaluation. Variable Y_p is selected to present scale variable.

4   Empirical Findings and Discussion

Unit Root Test with Breakpoint Structure. Table 2 summarizes the outcomes of ADF and ADF with breakpoint structure unit root tests on the level and first differences of the variables. The null hypothesis cannot reject that all the variables are unit root at zero difference. However, all of them are stationary with 1% significance level when conducted the first difference. The result suggests that all variables are I(1), which supports the use of the cointegration approach. At first difference, the break date gap shows smaller around 2007–2010 period. It is consistent with the world economy crisis in 2008. Some break dates did not match in 2008. The research considers that there is a latency when this crisis affects to Vietnam economy.

Table 2. Result of unit root test

Variables	Augmented Dicky Fuller (ADF) test statistics		Augmented Dicky Fuller (ADF) with breakpoint structure test statistics			
	Level	1st difference	Level	Break date	1st difference	Break date
$\ln M1_r^d$	−2,582	−5,920*[1]	−3,812	2011m2	−8,845*[1]	2008m2
$\ln G_s$	−0,522	−7,859*[1]	−3,569	2011m2	−10,484*[1]	2006m6
$\ln Y_p$	−1,959	−13,315*[1]	−4,081	2010m11	−13,576*[1]	2011m8
St_s	−2,532	−6,587*[1]	−5,458	2007m11	−7,895*[1]	2007m3
$REER_s$	2,334	−8,891*[1]	−3,379	2010m7	−9,926*[1]	2008m12
r_s	−0,518	−10,342*[1]	−3,728	2007m12	−14,997*[1]	2007m1
$Tbill_s$	−0,923	−7,532*[1]	−3,261	2008m2	−10,214*[1]	2008m7
π_s	−2,220	−4,758*[1]	−3,205	2010m11	−6,351*[1]	2010m9

Note: Zt statistic tests is the null hypothesis that the variables contain a unit root. (*[1]), (*[5]), and (*[10]) reject the null at 1, 5, and 10 %, respectively. ADF with breakpoint structure test is selected using the Andrews and Zivot method.
ADF: 1% = −4,022 (*[1]), 5% = −3,443 (*[5]), 10% = −3,143 (*[10]).
ADF with breakpoint structure: 1% = −5,57 (*[1]), 5% = −5,08 (*[5]), 10% = −4,82 (*[10]).

Table 3. Result of lag selection

Lags	LR	FPE	AIC	HQIC	SBIC
0	-	655,313	24,583	24,655	247,603
1	2781,6	7,60E−09	401,216	466,072	5,60833*5
2	223,63	$3,7 \times 10^{9}$*5	3,27081*5	4,49586*5	62,858
3	93,37	$4,90 \times 10^{9}$	353,926	534,081	797,307
4	100,86*5	$6,40 \times 10^{9}$	374,963	612,767	960,227

Note: (*[1]) and (*[5]) denote the significance at the 1%, 5%, and 10% level, at respectively.

Table 4. The result of Johansen tests for cointegration.

Max rank	Eigen value	Trace statistic	5% critical value	1% critical value
0	-	2,120,584	170,8	182,51
1	0,43647	136,9243*1	136,61	146,99
2	0,27784	94,2828*5	104,94	114,36
3	0,233	595,332	77,74	85,78
4	0,16278	362,584	54,64	61,21
5	0,1177	198,535	34,55	40,49

Note: (*[1]) and (*[5]) denote the significance at the 1%, 5%, and 10% level, at respectively.

Cointegration Test. After investigating the time series properties of all the variables, the research identifies the existence of long-run equilibrium. First, Table 3 gives the optimal lag though FPE, AIC, HQIC, SBIC criteria. The lag selection result supports to select two lags for model estimation. The result of Johansen test is tabulated in Table 4. The critical value (146.99) exceeds the trace test (136.92) at 1% significance level. It implies to reject the null hypothesis. At 5% significance level, the null hypothesis cannot reject (94.28 < 104.94). For more than one cointegration vector, the maximum rank indicates that there are one cointegration equations at 1% significance level and two cointegration equations at 5% significance level. Because all variables are stationary at 1% significance level respectively, the research chooses one cointegration equation to analyze next step.

Short-Run Estimates. For short-run, the estimation result reports in Table 5, which indicates that four parameter estimates (St_s, $REER_s$, r_s, $Tbill_s$) affect to demand for $M1$ money and had significance at 1% level, except at 5% level. The research cannot prove the impact of $\ln G_s$, $\ln Y_p$, and r_s to $M1$ money demand. There is a short-run relationship between dependent variables and independent variable because the error correction coefficient shows -0.034. When real $M1$ money demand deviated from equilibrium status, it will be adjusted up about 3.4% in the next term to recover equilibrium status. The model in short-run is

Table 5. Estimation result of M1 money demand in the short-run.

Variables	Coefficient	Standard error	p-value
$ECT(-1)$	$-0{,}0340634$	$0{,}0167926$	$0{,}043^*$
$\Delta \ln M1_r^d(-1)$	$0{,}0617752$	$0{,}0852173$	$0{,}469$
$\Delta \ln G_s(-1)$	$0{,}0471475$	$0{,}0829781$	$0{,}570$
$\Delta \ln Y_p(-1)$	$0{,}0270579$	$0{,}0206414$	$0{,}190$
$\Delta St_s(-1)$	$0{,}0001469$	$0{,}0000563$	$0{,}009^{***}$
$\Delta REER_s(-1)$	$0{,}4521859$	$0{,}2108798$	$0{,}032^{**}$
$\Delta r_s(-1)$	$-0{,}0007669$	$0{,}0019962$	$0{,}701$
$\Delta Tbill_s(-1)$	$-0{,}0058979$	$0{,}0024104$	$0{,}014^{***}$
$\Delta \pi_s(-1)$	$-0{,}0211578$	$0{,}0042211$	$0{,}000^{***}$
Constant	$0{,}0157713$	$0{,}0042211$	$0{,}000^{***}$

Note: (*), (**), and (***) denote the significance at the 1%, 5%, and 10% level, at respectively. Lag character is denoted in parentheses (-1).

written by the formula (21). Where lag character is denoted in parentheses (-1). $ECT(-1)$ means error correction term with one lag.

$$
\begin{aligned}
\Delta \ln M_r^d(-1) = {} & -0{,}0340634 ECT(-1) + 0{,}0617752 \Delta \ln M_r^d(-1) \\
& + 0{,}0471475 \Delta \ln G_s(-1) + 0{,}0270579 \Delta \ln Y_p(-1) \\
& + 0{,}0001469 \Delta St_{vni}(-1) + 0{,}4521859 \Delta REER_s(-1) \\
& - 0{,}0007669 \Delta r_s(-1) - 0{,}0058979 \Delta Tbill_s(-1) \\
& - 0{,}0211578 \Delta \pi_s(-1) + 0{,}0157713.
\end{aligned}
\tag{21}
$$

First, contrary to expectation, the negative relationship between gold price and $M1$ money demand has no meaning in short-term. In some periods, gold price increased dramatically. Vietnamese people rushed to buy gold but their expectation could not reply as much as possible. A possible explanation for this might be that there is the lack of domestic gold supply, the limited of selling gold quantity by speculators, and the commitment of the state government bank about stabilized local currency and managed the gold market. Second, there is no evidence to figure out the effect of industrial production value determinant to $M1$ money demand. Third, variable provided statistically significant and positive effect. This relationship is given against Friedman's theory that was mentioned the negative relationship between money demand and expected return. However, the empirical research of Nguyen and Pfau [13] supports for this determinant. It is difficult to explain this result, but it might be related to the weak confidence of the people in Vietnam stock market. Due to the instability of the stock market, limited of the transparency, people tend to prefer trading style rather than dividend payment yearly. This rate of return comes from the arbitrage between the purchase price and the sale price. Hence, holding money and holding stock have a positive relationship. The stock elasticity of demand for $M1$ money is very slight. In case, other factors are constant, real $M1$ money demand will increase 0.015% when the stock changes 1%. This minor impact reveals that a promising alternative asset does not attract domestic investors. Fourth, the real effective exchange rate is positively at 5% significance level and strongly influenced by $M1$ money demand (0.45). Regarding holding money, the exchanges rate are one of the opportunity costs variable and is expected to reverse with $M1$ money demand. However, according to the alternative asset approach, when local currency depreciated, people will hold more assets [6]. It implies that domestic goods are cheaper than foreign goods. People tend to hold more local currency in order to buy more assets that are domestic. Fifth, the deposit interest rate has no statistically significant. It was same as Friedman's conclusion that money demand is less sensitive than the interest rate. For developing country, the inflation is often high and fluctuates. It leads to the negative real interest rate in some periods. People balance between savings channel and other investment channels. In addition, the state bank of Vietnam decided deposit interest rate ceiling to control lending interest rate of commercial banks. The attraction of saving money can be decreased. The research considered that this result is appropriate in short-term for Vietnam. Sixth, the treasury bills interest rate are negatively at 1% significance level. This result is similar Friedmans conclusion that money demand is negative with the bond return. By assuming, others factors are constant, $M1$ money demand will fall 0.59% when treasury bills rise 1%. Although this investment channel is considered very low risk, holding treasury bill is small. The lower risk causes lower profit. Therefore, it can explain why this effect has a small influence. Seventh, consumer price index is negative with real $M1$ money demand. The coefficient is small. In case other factors constant, the growth rate of real $M1$ money demand decreased 2.12% when the consumer price index increases 1%. Vietnam is still developing country. To reduce the risk, people balance between holding other alternative assets and holding money. Hence, this thing may explain the relatively correlation between the consumer price index and demand for money.

Long-Run Estimates. In long-run, the results present in Table 6 that four cointegration models given different results relate to sign expectation, model explanation (R-square). While VECM uses error correct coefficient to adjust, FMOLS and CCR use long-run covariance matrix. Conversely, DOLS uses lead and lag to adjust. These methods corrected standard OLS regression relate to serial correlation and endogeneity of regressors. The outcomes of VECM, FMOLS, CCR and DOLS reveals that all the dependent variables are statistically significant. Following the comparable results, CCR model is well determined because its outputs display higher meaning explanation and significance level than other techniques. In long-term, $M1$ demand function is written by the formula (22).

$$\Delta \ln M^{d}_{r_{CCR}} = 105,9 - 0,245\Delta \ln G_s + 0,282\Delta \ln Y_p + 0,00031\Delta St_{vni}$$
$$- 0,753\Delta REER_s - 0,0245\Delta r_s + 0,0117\Delta Tbill_s - 0,00757\Delta \pi_s$$
$$(22)$$

Table 6. Estimation result of M1 money demand in the long-run.

Variables	VECM $\ln M1_r^d$	FMOLS $\ln M1_r^d$	CCR $\ln M1_r^d$	DOLS $\ln M1_r^d$
$\ln G_s$	−0,4595683**	−0,242***	−0,245***	−0,147*
	(0,2336489)	(0,0510)	(0,0520)	(0,0840)
$\ln Y_p$	0,5403968***	0,282***	0,282***	0,324***
	(0,2014496)	(0,0360)	(0,0478)	(0,0860)
St_s	−0,0013442***	0,000308***	0,000310***	0,000412***
	(0,0001489)	(2,93e−05)	(3,45e−05)	(6,40e−05)
$REER_s$	−4,324238***	−0,772***	−0,753***	−0,441
	(0,914772)	(0,163)	(0,199)	(0,380)
r_s	0,0843073***	−0,0245***	−0,0245***	−0,0383***
	(0,009864)	(0,00197)	(0,00235)	(0,00397)
$Tbill_s$	−0,1001702***	0,0117***	0,0117***	0,0278***
	(0,0107419)	(0,00209)	(0,00250)	(0,00460)
π_s	−0,0089349**	−0,00745***	−0,00757***	−0,00738***
	(0,0049226)	(0,000995)	(0,00111)	(0,00192)
Constant	395,3594	107,8	105,9	71,31
Observe	131	132	132	129
R^2	0,3782	0,841	0,869	0,971

Note: (*), (**), and (***) denote the significance at the 1%, 5%, and 10% level, at respectively.
The standard errors are in parentheses ().

As our expected, the estimated coefficient of $\ln G_s$, $REER_s$, r_s, and π_s that is negative and significant at 1% level of significance. In contrast, $\ln Y_p$, St_s, and $Tbill_s$ are positive at the same significance level. First, industrial production value has the same direction with M1 money demand. This result is consistent with the theory of Friedman, previous studies of [6]. Second, the domestic interest rate represents the opportunity cost of holding money. When the money supply increases, the interest rate will fall. It encourages people willing to hold more money in cash. However, when interest rate rises, the people prefer holding more financial assets such as treasury bills, bonds, etc. Third, the positive sign of between treasury bills interest and the demand for M1 money. This result does not match Friedman's conclusion. Assuming money market balance, if the public holds less than the amount of money, they will hold more bonds. It causes a bond price increase. When the price of bonds rises, the interest rate will fall. Fourth, regarding gold price in the long-run relationship, the negative effect strongly influences to money demand. The coefficient is large. It reveals that holding gold needs is still high in addition to holding money. Fifth, the real effective exchange rate does not consistent the result in short-term. This proves that people tend to less holding money when the local currency decreases. The coefficient of $\ln G_s$

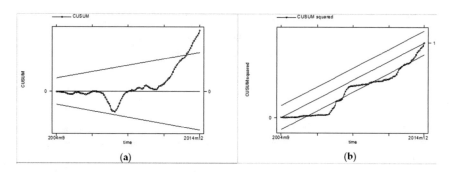

Fig. 1. The result of CUSUM and CUSUM-square for M1 money demand. (a) Description of CUSUM result; (b) Description of CUSUM-square result.

(-0.245) and $REER_s$ (-0.753) show the considerable impact on real M1 money demand function. In fact, the holding gold and foreign currency are two main channels in parallel holding money. According to the above results, there is two main factors impact to money demand in Vietnam. They are $\ln G_s$ and $REER_s$. These facts are important for state government to control money demand more effective. Seventh, π_s variable also affect negatively to money demand same as the short-run relationship.

Stability Tests. We use CUSUM and CUSUMSQ test to confirm whether stable or not for 2003Dec to 2014Dec periods. The CUSUM test provides the recursive estimates of residuals. It calculated CUSUM statistic. Under the null hypothesis, the statistic is drawn from a distribution. If the calculated CUSUM statistics was put inside the drawn line of CUSUM distribution, we reject the null hypothesis (meaning model stability). The null hypothesis is that parameters are stable (constant). Previous studies by Hoa [6], Nguyen and Pfau [13] concluded the stable of M1 money demand function. The stability test of money demand in Vietnam was described in Fig. 1 though CUSUM and CUSUMQ test. In contrast to earlier findings, the result illustrates that M1 demand function does not conclude stable over the 2003Dec–2014Dec period.

5 Conclusions

This paper has investigated the factors, which affect to narrow money demand (M1) using cointegration technique from 2003Dec to 2014Dec. By empirical analysis, we figure out the role of gold price and real effective exchange rate with strongly significant in money demand. Obviously, cointegration technique using FMOLS, CCR, and DOLS demonstrate the outcomes better than VECM in long-run equilibrium. The model regression is modified using the long-run covariance matrix. The data is estimated with caution because this technique prevents the issue of endogenous and serial correlation. Therefore, the new findings of demand for money in Vietnam have identified. The real interest rate is not an interesting investment channel compare with holding other assets. In developing financial market such as Vietnam, the asset price is one of the extremely important points in money demand. The next studying direction should be investigating the effect of these factors in Vietnam monetary policy. In addition, future work can be study money demand function modeling by combining linear and nonlinear method.

References

1. Alsahafi, M.: Linear and non-linear techniques for estimating the money demand function: the case of Saudi Arabia. Doctoral dissertation, University of Kansas (2009)
2. Babic, A.: The Monthly Transaction Money Demand in Croatia. Working Papers, W-5, Zagreb, Croatian National Bank, September 2000
3. Blanchard, O.: Macroeconomics, 7th edn. Pearson, New York (2016)
4. Darvas, Z.: Real Effective Exchange Rates for 178 Countries: A New Database. Bruegel Working Paper, Bruegel, 15 March 2012
5. Dogan, B.: The demand for money during transition from high to low inflation in Turkey in the period 2002–2014. Int. Res. J. Appl. Finan. 4(3), 141–151 (2015)
6. Hoa, H.H.: The demand for money and the consequence of monetary policy in Vietnam. Doctoral dissertation, National Economics University (2008)
7. Hamdi, H., Said, A., Sbia, R.: Empirical evidence on the long-run money demand function in the GCC countries. Int. J. Econ. Finan. 5(2), 603–612 (2015)
8. IHS Global Inc.: EVIEWS 9 Users Guide II (2015)
9. Inoue, T., Hamori, S.: An empirical analysis of the money demand function in India. Econ. Bull. 29(2), 1224–1245 (2009)
10. Joseph, A., Larrain, M., Ottoo, R.: Comparing the forecasts of money demand. Procedia Comput. Sci. 20(1), 478–483 (2013)
11. Kumar, S.: Money demand income elasticity in advanced and developing countries: new evidence from meta-analysis. Appl. Econ. 46(16), 1873–1882 (2014)
12. Kumar, S., Webber, D.J., Fargher, S.: Money demand stability: a case study of Nigeria. J. Policy Model. 35(6), 978–991 (2013)
13. Nguyen, H., Pfau, W.: The Determinants and Stability of Real Money Demand in Vietnam 1999–2009, GRIPS Discussion Papers, GRIPS Policy Research Center, Japan, 1 September 2010
14. Lan, N.P.: Money demand under the relationship between inflation and monetary policy in Vietnam. J. Bank. 1(19), 1–5 (2011)
15. Riyandi, G.: Meta-analysis of money demand in Indonesia. Bull. Monetary Econ. Bank. 15(1), 41–62 (2012)
16. Siregar, R., Nguyen, T.K.C.: Inflationary Implication of Gold Price in Vietnam. Working Papers, Centre for Applied Macroeconomic Analysis, Australian National University, 4 December 2013
17. Sriram, S.: A survey of recent empirical money demand studies. IMF Staff Pap. 47(3), 334–365 (2001)
18. Sriram, S.: Survey of Literature on Demand for Money: Theoretical and Empirical Work with Special Reference to Error-Correction Models. IMF Working Paper, No. 99/64, pp. 1–77 (1999)
19. Wang, Q., Wu, N.: Long-run covariance and its applications in cointegration regression. Stata J. 12(3), 515–542 (2012)
20. Zivot, E., Andrews, D.: Further evidence on the great crash, the oil-price shock, and the unit-root hypothesis. J. Bus. Econ. Stat. 10(3), 251–270 (1992)
21. Zuo, H., Park, S.Y.: Money demand in China and time-varying cointegration. China Econ. Rev. 22(3), 330–343 (2011)

The Effects of Foreign Bank Entry, Deregulation on Bank Efficiency in Vietnam: Stochastic Frontier Analysis Approach

Pham Dinh Long[1](✉) and Luong Cong Hoang[2]

[1] Faculty of Economics and Public Management, Ho Chi Minh Open University,
97 Vo Van Tan, Ho Chi Minh City 700000, Vietnam
`long.pham@ou.edu.vn`
[2] VNP Programme, University of Economics HCMC,
Ho Chi Minh City 700000, Vietnam
`hoang.lc@vnp.edu.vn`

Abstract. This study examines the effects of foreign banks entry on the efficiency of domestic banks in Vietnam following the program of financial deregulation initiated by the government. We also review the bank efficiency in term of bank size and ownership structure. There are two main approaches to assess the efficiency of a bank: data envelopment analysis (DEA) and stochastic frontier analysis (SFA). In this paper, we apply the SFA method which is suggested by Berger et al. (2009), Ahn et al. (2001) to estimate the cost and profit efficiency of three groups: 100% foreign-owned, big four state-owned and other domestic banks. Based on the initial sample which is collected from the BankScope database of Bureau van Dijk and Fitch Ratings and annual reports of 37 banks in Vietnam for the period 2009–2015. Results indicate that big four state-owned banks are seemly efficiency on both cost and profit approach while the 100% foreign-owned banks are not the most efficiency overall. However, the 100% foreign-owned banks are able to gain economy of scale in revenue while the big four state-owned and other domestic banks hardly take an advantage of economic of scale. All group of bank obtain the economy of scope but the level is different. Other domestic banks may exploit economy of scope greater than big four and foreign banks.

Keywords: Foreign banks entry · Financial deregulation
Bank efficiency · SFA

JEL Classification: F63 · F65 · G21

1 Introduction

Banks are notable institutions in any society since they importantly contribute to the development of the economics system. Via operations, banks connect

L. H. Anh et al. (eds.), *Econometrics for Financial Applications*, Studies in Computational Intelligence 760, https://doi.org/10.1007/978-3-319-73150-6_57

the surplus and deficit capital economy economic agents. Due to the economic boom, globalization, deregulation and technological innovation, there are many empirical studies trying to explore the ways to enhance the bank efficiency and stability.

Rosengard and Du (2009) describes financial deregulation as the transition from a closed to a competitive financial system. To more specific, with banking system, it means that the transfer of bank from monopoly or oligopoly, where have a restriction on competition, bank entry, expansion, and diversify, to an open and more competitive in banking system. The government makes an effort to create "a level playing field" not only between privately owned banks and public sector banks, but also between foreign banks and domestic banks.

Related to WTO commitments, Vietnam government issued the decree No 22/2006/ND-CP to allow a typical foreign institutional investor owning a 15% shares in a domestic bank. However, a maximum of 30% shares in a particular domestic bank to be owned by all foreign holdings. Further, the foreign institutional investor can be categorized as "strategic" investors. Thus, they can increase their shares in a particular domestic bank from 15% to 20%. Moreover, of each foreign strategic investor may be increased to 20%. Besides, it was the first time 100% foreign-invested banks have been allowed in Vietnam from April 1, 2007. To be precise, HSBC Vietnam is the first foreign bank which set up a wholly-owned foreign bank in early 2009. From 2009 to 2017, Vietnam consists of seven foreign banks with 100% foreign-invested, namely HSBC (Hong Kong), ANZ (Australia), Shinhan (South Korea), Standard Chartered (UK), Hong Leong (Malaysia), Public Bank (Malaysia) and Woori Bank (South Korea).

The next deregulation is bank ownership diversification. It means that the privatization has been highly supported by the government. The central bank encourages transforming state ownership to private ownership. Following the Government for Notice No.03/TB-VPCP declared that the state ownership has been decreased by 49% by 2010.

The research objective of this paper is that examining the impact of bank deregulation on Vietnamese bank efficiency. We consider which bank efficiency has improved after financial deregulation. To more specific, we investigate the economic efficiency of the Vietnamese banking system under the impact of Viet nam's financial deregulation since 2007. Furthermore, we also discuss bank efficiency in term of different ownership structures and bank size.

Following the previous research, there are two main approaches to assessing the efficiency of banks, data envelope analysis (DEA) and stochastic frontier analysis (SFA). The first, founded by Charnes et al. (1978), DEA is a non – parametric technique applying the linear programming technique to set the best practices. The second, in meanwhile, developed by Aigner et al., Meeusen and Van den Broeck (1977), SFA is a parametric technique which distinguishes the residuals from an estimated function into a stochastic error term and an efficiency. Instead of using the DEA approach, we apply the SFA estimation procedures suggested by Berger et al. (2009), Ahn et al. (2001), Good and Sickles (1995), Schmidt and Sickles (1984). We estimate the cost and profit efficiency depend on a given inputs and outputs factors of foreign, big four and other domestic banks.

Based on the dataset from Bankscope (Bureau van Dijk) and some missing value has been obtained by the annual report of 37 banks for the post-WTO period of 2009–2015, we find that foreign banks are not the most efficiency overall but big four is seemly efficiency on both cost and profit approach.

The rest of the paper is organized as follows: The second section provides a brief review of previous studies. Section 3 describes methodology and data. The empirical results will be discussed in Sect. 4. The last section presents the conclusions.

2   Literature Review

In this section, the related theories and the empirical evidence have been introduced. On the theoretical literature part, we focus on three main problems. The first, related theories of efficiency have been clarified. The second, we identify the stochastic frontier analysis (SFA) to estimate the efficiency of the bank. The third, theory of economic of scope and economic of scale have been introduced to explain the difference among banks. On the empirical research, the valuable empirical evidences of Vietnam and other countries have been presented to support for the related theory.

2.1   The Theoretical Literature

2.1.1   Theory of Efficiency

Efficiency refers to the production frontier which is defined as the ideal relationship between input and output. Farrell (1957), who is the first person introducing the theory of efficiency. He clarified 3 kinds of efficiency: technical efficiency (TE), allocative efficiency (AE) and economics efficiency (EE). To more specific, To more specific, technical efficiency (TE) is explained as the ratio of the input that has been used by a totally efficient firm producing the same output vector to the input usage of the firm under consideration (Chen et al. 2005). To succeed in reaching the desired output, we make an effort to minimize the waste of resources, namely materials, labor, energy and time. Therefore, technological innovations in the industry will be reflected by production frontier. So to speak, if the firms are considered as technically efficient, they may operate on this frontier. On the other hand, the firms are technically inefficient if they operate under the frontier. The further distance, the more inefficiency the firms are.

The efficiency may be measured by the proportion of total outputs to total inputs. If the information on the price of inputs and outputs is available, and an assumption of behavior such as profit maximization and cost minimization is suitable, the mix of this information may be used to measure the performance of a firm. In that case, the allocative efficiency (AE), in addition to technical efficiency (TE) may be recommended. The allocative efficiency (AE) is defined as an optimal utilization for the cost minimizing combination of inputs. Thus, in order to totally efficient, the firm must be obtained both technically and allocatively efficient. The multiplication of allocative and technical efficiency reflects the overall economic efficiency (EE).

$$EE = TE * AE.$$

2.1.2   Stochastic Frontier Analysis (SFA)

Time-Invariant Models. Meeusen and van den Broeck (1977) and Aigner et al. (1977) who are the pioneer of this method. From then, this model has become popular to estimate the efficiency of the firms. A lot of research has produced and developed many formula and extension from original models. In the development of methodology, it is started with a general formula of the stochastic frontier cross-sectional model and then improvement and expansion to the panel-sectional model.

Because of the richer set of information in panel data, it is considered highly accurate inefficiencies. The first generalization of panel model was presented by Pitt and Lee (1981) that applying the maximum likelihood estimation of the half-normal distribution with time invariant when u is fixed by time and differed among banks. It means that u_i may be defined as follows:

$$y_{it} = x'_{it}\beta + \varepsilon_{it} = x'_{it}\beta - u_{it} + v_{it}, \quad i = 1, \ldots, N; \quad t = 1, \ldots, T.$$

where $v_{it} \sim N\left(0, \sigma_v^2\right); u_{it} \sim N^+\left(0, \sigma_u^2\right)$.

Time-Varying Models. Because of the limitation of time-invariant model, researcher such as Cornwell (1990), Kumbhakar (1990), Battese and Coelli (1992) and Greene (2005a) suggested the models which allow its time-varying. In order to overcome the limitation, Greene (2005a) recommended applying a time-varying stochastic frontier half normal model with a specifying intercept of each unit. The formula is specified as follow

$$y_{it} = \alpha + x'_{it}\beta + \varepsilon_{it}.$$

More suitable than the previous model, this specification allows separating time-varying efficiency from unit detailed time invariant unobserved heterogeneity. Following the assumptions on the unobserved unit specific heterogeneity, Greene (2005a) used these model as "true" fixed effects (TFE) and "true" random effects (TRE). The maximum likelihood estimation of the "true" fixed effects time-variant depends on two main issues related to the measurement of nonlinear panel data models. Firstly, purely computational because of the large dimension of the parameters space. Therefore, Greene represented a Maximum likelihood dummy variable (MLDV) specification which is computationally feasible also existed a large number of parameters α_i ($N > 1000$). Secondly, the problem with incidental parameters appearance when a number of units is relatively large in term of the length of the panel. In these situations, the $alpha_i$ is not consistently measured as $N \to \infty$ with fixed T. Meanwhile, the measurement of the "true" random effects specification may be probably conducted by applying simulated maximum likelihood techniques.

2.1.3   Economy of Scope and Economy of Scale

Economy of Scope. Economic of scope is an economic theory which states the lower costs or higher profits of production as a result of increasing a variety of goods produced.

Coelli (2005) stated that the estimation of economic of scope may be conducted by comparing the costs from producing all outputs of a diversified bank and producing each output separately of a focused bank. If this level is greater than zero, cost of the multiproduct firm will smaller than cost of firm with a single output. It means that the firm exploits the economies of scope. In contrast, the firm exists diseconomies of scope when this measure less than zero.

$$\text{Economy of scope}_{cost} = \frac{\sum_{i}^{n} \text{costs}\,(y_i, 0) - \text{costs}\,(y_1, y_2, \dots, y_n)}{\text{costs}\,(y_1, y_2, \dots, y_n)}$$

Economy of Scale. Economic of scale is the reduction of a company's average costs because of increasing the size or scale of a company's operations. We can conduct the revenue scale economy. According to the Cobb – Douglas type production, we have the form as follows:

$$\ln\,(outputs)_{it} = \beta_0 + \beta_i \sum_{i=1}^{n} \ln\,(inputs)_{it} + \alpha_i + \gamma_t + \mu_{it}$$

Where α_i indicates firm-specific effects and γ_t indicates the time - specific effects. The null hypothesis which denotes constant return to scale: $H_0 = \sum \beta_i = 1$. We apply the F-statistic to test when the sum of β_i is significantly different from the H_0. To more specific, the firm will exhibit decreasing, constant or increasing return to scale when sum of β_i is less, equal or greater than one.

2.2   Empirical Studies in Vietnam

In term of analyzing the Vietnamese banking system, very few researchers have been conducted because of data available limitation. There are few publications which were published in regarding to the Vietnamese banking system. Most researchers are interested in evaluating the deregulation process of the Vietnamese financial system and considering the factors which determine the performance of banks. There were no researches before that have showed the negative or positive effect of foreign bank entry to domestic banks efficiency.

Nguyen (2007) evaluated the cost efficiency of 13 commercial banks in Vietnam in the period from 2001 to 2003. The Malmquist DEA approach was employed to assess the productivity, efficiency and technical changes. As a consequence, both technical and allocative efficiency were not obtained in this period. However, there are some limitation in this study because of constraints of time and number of bank.

Another research by Nguyen (2011) focused on estimating the efficiency 20 commercial banks in Vietnam in the 3 year-period (2007–2010) with more efficiency determinants to be evaluated. To more specific, inputs include labor, fixed assets, and deposits while outputs include interest and non-interest income. The conclusion proved that the state-owned banks had lower efficiency scores as compared to their join-stock commercial bank's competitors. Later, a study of Giang & Nguyen (2012) expanded their research to 32 commercial banks (in the period of 2001–2005) through the slacks-based model DEA, argued that there are remaining a room to improve the efficiency of those banks.

Nguyen (2011) and Lieu and Vo (2012) evaluated the efficiency of commercial banks in Vietnam. Labor, total deposits and various kinds of expense were used as inputs to measure the total loans as outputs. The authors showed that the larger asset size of banks has exhibited 11 times higher efficiency than the small ones.

Vu and Turnell (2010) used a Bayesian stochastic frontier approach and recognized that the cost efficiency of Vietnamese banking system slightly reduced in period of 2000–2006 and it is indifferent cost efficiency of stated-owned and joint stock banks.

Ngo (2010) applied DEA approach to estimate the efficiency of 22 Vietnamese commercial banks in 2008 (banks which were ranked top 500 largest enterprises by revenue ranking board (VNR500) in 2009). He concluded that banks still had an opportunity to enhance efficiency though the average score was completely high.

Recently, Nguyen et al. (2014) focused on evaluating the cost and profit efficiency of Vietnamese banks. As the results, they stated that banks were good at managing their cost (with average cost efficiency of 0.9) while it was difficult to have an advantage in profit efficiency (with average profit efficiency score of 0.75). However, both scores increased during the period of 1995–2011. In addition, they found that, at 5% level of significance, SOCBs conducted more efficiency than JSCBs.

In summary, researching on the efficiency of banking system in Vietnam still has several limitation and has just been attracted the researcher's attention over the last decade. In general, previous study indicated that the unclear trend of bank's efficiency before 2010. However, the upward trend may be seen slightly after 2010. Thanks to the previous literature, this study tries to examine the efficiency score and the efficiency gap among three groups of bank (Big four, foreign and other domestic banks).

3   Data and Methodology

3.1   Research Methodology

3.1.1   Stochastic Frontier Analysis (SFA)

Stochastic frontier analysis (SFA), a parametric technique, which was found by Aigner et al., Meeusen and Van den Broeck in 1977. This technique distinguishes the residuals into an inefficiency (u) and a stochastic error term (v). A stochastic error term or statistical noise reflects the effects of an accident, weather, luck, and so on. Let y_{it} be some observed performance measure for bank i in period t.

$$y_{it} = x'_{it}\beta + \varepsilon_{it} = x'_{it}\beta - u_{it} + v_{it} \tag{1}$$

Where x_{it} is the factors that may affect the performance of banks. The "stochastic frontier" is clarified as: $y_{it}^* = x_{it}'\beta + v_{it}$. The one-sided error $u_{it} \geq 1$ represents for the inefficiency. If there are no inefficiency, meaning $u_{it} = 0$, the frontier of outputs should be:

$$y_{it}^* = x_{it}'\beta + \varepsilon_{it} = x_{it}'\beta + v_{it}$$

There are several functions are employed in SFA model, but Cobb-Douglas and translog function are widely used to estimate efficiency. If we assume model take a log-linear Cobb-Douglas functional form, the equation may be written as:

$$\ln(y_i) = \beta_0 + \sum_n \beta_n \ln(x_{ni}) - u_{it} + v_{it} \tag{2}$$

In line with Berger et al. (2009), we apply the translog function (Christensen et al. 1973) to measure cost and profit efficiency of 37 banks in Vietnam. The cost frontier approach is able to measure how to the banks minimize the cost with the same outputs. In contrast, profit frontier approach attempts to estimate how to the banks maximize the profit with the same inputs. If the dependent variable y_{it} is in logarithmic form, the cost efficiency score of bank i at time t are measured as $\exp(-u_{it})$. Therefore, it will range from 0 to 1. It means that the closer it is to 1, the greater bank efficiency is.

For easy comparison, we divide into three groups of banks, including foreign banks, big four state-owned commercial banks, and other domestic banks.

Following Matthews and Thompson (2008), there are two main approaches to specifying inputs and outputs of financial institutions, namely intermediation approach and production approach. Sealey and Lindley (1977) is known as a pioneer in intermediation approach, which considers a financial institution as an intermediary between lenders and borrowers. It means that deposits and purchased fund have been transformed into loans and other earning assets. Compared to intermediate approach, production approach treats that banks develop accounts of various sizes by processing deposits and loans. The choice depends on the purpose of research and availability of data. In this study, we follow the intermediation approach classified by Berger et al. (2009).

Model 1: A detailed description of the cost translog function for bank i at time t is as follows:

$$\ln(\frac{TC}{A}) = \alpha_0 + \sum_{s=1}^{3} \alpha_s \ln(w_s) + \sum_{m=1}^{3} \beta_m \ln(\frac{y_m}{A}) + \frac{1}{2}\sum_{s=1}^{3}\sum_{k=1}^{3} \gamma_{sk} \ln(w_s)\ln(w_k)$$

$$+ \frac{1}{2}\sum_{m=1}^{3}\sum_{l=1}^{3} \delta_{ml} \ln(\frac{y_m}{A})\ln(\frac{y_l}{A}) + \sum_{s=1}^{3}\sum_{m=1}^{3} \varphi_{sm} \ln(w_s)\ln(\frac{y_m}{A}) - u + v \tag{3}$$

Where the dependent variable, TC/A, is measured by total cost (TC) to total assets (A). Three input price includes the price of funds which is measured by total interest expenses divided by total deposits and borrow funds (w_1), the price of fixed assets is measured by the ratio of other operating expenses to fix assets (w_2), and lastly, price of labor is measured by total personal expenses divided by number of employees of banks (w_3). The three outputs are net loans (y_1), other earning assets (y_2), and total non-interest income (y_3).

Model 2: The efficiency of banks is measured by model 3 in turning the cost into profit. The specification for the profit translog function is similar but applying the logarithm of profit before tax Π divided by total assets as the dependent variable. Moreover, the price of outputs is used instead of quantity of outputs. To more specific, the price of loans (z1) is estimated by the ratio of interest income on loans to total loans, the price of other earning assets (z1) is estimated by the ratio of the interest income from other earning assets to other earning assets, and lastly, the price of non-traditional services (z3) is estimated by the ratio of total non-interest income to total assets. The price of inputs has not changed as the cost translog function.

$$\ln\left(\frac{\pi}{A}\right) = \alpha_0 + \sum_{s=1}^{3} \alpha_s \ln(w_s) + \sum_{m=1}^{3} \beta_m \ln\left(\frac{z_m}{A}\right) + \frac{1}{2}\sum_{s=1}^{3}\sum_{k=1}^{3} \gamma_{sk} \ln(w_s)\ln(w_k)$$

$$+ \frac{1}{2}\sum_{m=1}^{3}\sum_{l=1}^{3} \delta_{ml} \ln\left(\frac{z_m}{A}\right)\ln\left(\frac{z_l}{A}\right) + \sum_{s=1}^{3}\sum_{m=1}^{3} \varphi_{sm} \ln(w_s)\ln\left(\frac{z_m}{A}\right) - u + v$$

$$(4)$$

The inefficiency of a bank (u_{it}) is estimated by two specifications including:

The first generation of panel data was founded by Pitt and Lee (1981) and developed by Schmidt and Sickles (1984); Battese and Coelli (1988), when we assumed the efficiency score is time-invariant. It means that efficiency is fixed through time and only varying among banks ($u_{it} = u_i$). Therefore, Eq. (1) may be rewritten as follows:

$$y_{it} = x'_{it}\beta + \varepsilon_{it} = x'_{it}\beta - u_{it} + v_{it}$$

where $i = 1, \ldots, N$; $t - 1, \ldots, T$.

The second case, we assumed that the efficiency of a bank do not stay constant over time. Following Greene (2005), we apply "true" fix effect and "true" random effect to measure efficiency score. The objective of these model is to separate bank effects from efficiency or bank heterogeneity.

Related to "true" fix effect, the formula is described as:

$$y_{it} = \alpha_{it} + x'_{it}\beta - u_{it} + v_{it}$$

Where α_1 is the bank specific constant. Instead of least squares estimation, we obtain maximum likelihood estimation to measure β. Meanwhile, "true" random effect may be specified as:

$$y_{it} = (\alpha_i + \omega_i) + x'_{it}\beta - u_{it} + v_{it}$$

$$u_{it} \sim N^+ \left(0, \sigma_{it}^2\right) = N^+ \left(0, \exp\left(z'_{u,it}\omega_u\right)\right)$$

$$v_{it} \sim N\left(0, \sigma_v^2\right)$$

$$\omega_i \sim N\left(0, \sigma_\omega^2\right)$$

where ω_i is a random constant term varying between banks.

Mester (2008) has recognized the limitation of Collie's method if the level of output is less than zero. Because of non-existent logarithm of number which is less than zero. Therefore, he innovated a form for economics of scope as follow:

$$\text{Scope economies} = \frac{\sum_{i=1}^{N=3} \text{costs}\left(y_i\left(N-1\right)y_i^{\min}, y_{-i}^{\min}\right) - \cos ts(y_1, y_2, y_3)}{\cos ts(y_1, y_2, y_3)}$$

Where costs (y) is the relevant cost function for each of the three specification. Economies of scope may be evaluated by comparing costs of the multiproduct bank and costs of the banks which focus on a single line of business. If this measure is positive, the cost of the specialized banks may be greater than costs of the banks, which are able to offer more products.

3.1.2   Economy of Scale

To test whether the size of bank indeed matters in efficiency, we apply the economies of scale in revenue. A Cobb-Douglas function can be employed as follows:

$$\ln\left(revenue\right)_{it} = \beta_0 + \beta_1 \ln\left(deposit\right)_{it} + \beta_2 \ln\left(fix\,assests\right)_{it} + \beta_2 \ln\left(number\,employee\right)_{it}$$
$$+ \alpha_{it} + \gamma_t + u_{it}$$

Where α_i represents the bank - specific effects and γ_t indicates the time - specific effects. The logarithm of the output may be seen as the logarithm of revenue from different lines of bank products while the logarithm of the input may be estimated approximately by the logarithm of the total deposits, fixed assets and total number employees of banks.

3.2   Data Descriptions

The analysis is based on a sample which exported from the BankScope database from Bureau van Dijk and FitchRatings and annual reports of 37 banks over the period 2009 to 2015. As a result, our dataset is a set of unbalanced panel data. They are six foreign banks, four state-own banks and twenty-seven domestic banks. The reasons why period this time was selected are: (i) after joining the WTO 2007, Vietnam had to fulfill WTO regulations. Before 2007, foreign banks were operated as the branch office, subsidiary or representative office. There-fore, there are restrictions on the operations and services of foreign banks. After WTO, Vietnam has fully opened up the banking system. The first reform is that the 100% foreign banks are allowed to establish in Vietnam from the 01 April

2007. It leads to the international competitive environment. The second reform according to the Government Notice No. 03/TB-VPCP declares that the state aimed to decrease its ownership in banks with the level at less than 49% by 2010. (ii) Most foreign banks were established in 2009 and they were beginning to issue the annual report at that time. It may be useful to fill some missing values from BankScope.

The cost and profit efficiency may be estimated for each bank for each year. To make a summary report, we group domestic banks into two categories as the big four state-owned (Big four) and other domestic banks. The reason is the state bank is still a majority shareholder in the Big Four. Specifically, the level of state ownership of Agribank, Bidv, Vietinbank, Vietcombank in 2015 is 100%, 75.76%, 64.46%, and 77.1%, respectively. In the Vietnam banking industry, the Big Four banks play a vital role with the level of total assets is nearly 54% in 2009 and 50% in 2015. The other domestic banks have mainly focused on the retail market. There are 27 domestic banks for concerned with exception of banks which are disappeared by mergers and acquisitions by 2015 such as MDBank, WesternBank, DaiaBank, Ficombank, TinNghia Bank, SouthernBank, Habubank, MHB. We also excluded GPbank, NCB and Oceanbank because of the change of ownership, 100% stated owned bank in 2015. Although the number of other domestic banks is larger, the market share of total assets is smaller than the Big Four.

The foreign banks are defined as banks with greater than 50% foreign ownership. As usual, foreign banks operate as a wholly-owned subsidiaries. There are 6 foreign banks that we concern in this study namely Hong Leong, Shinhan, HSBC, ANZ, Standard Chartered and VID public. All of the banks which were transformed from a branch office to a 100% foreign ownership commercial banks after relaxed restriction by the decree No 22/2006/ND-CP. In spite of ongoing development and expansion of the presence, the market share of foreign banks is still low, from 3% in 2009 to 3.29% in 2015.

We also excluded 03 joint-ventured (Indovina, Vinasiam, Vietnam-Russia), cooperative and 2 policy banks out of the dataset.

4   Results and Discussion

4.1   Empirical Results

In this section, based on the data and methodology in this Sect. 3, the empirical results have been presented. We define and analyze the cost and profit efficiency of 3 group of banks namely foreign, big four and other domestic banks in the period of 2009–2015. Moreover, thanks to examining the economy of scope and economy of scale, we will determine which group of bank should diversify product or further expand their size to gain more efficient in the future.

Although the market share of total assets of Big Four banks has decreased and Other Domestic banks has increased in the period of 2009–2015, the Big Four banks are still the largest. In the meanwhile, the average market share of foreign banks has just approximately at the level of 3.7%. Compared with other banks, the level of market share in foreign banks is the smallest in the period of 2009–2015 (Table 1).

Table 1. Market share of total assets, 2009–2015 (%)

	2009	2010	2011	2012	2013	2014	2015
Big four	53.86	48.65	45.81	48.99	49.67	49.17	49.97
Domestic banks	42.41	47.93	50.59	47.05	46.59	46.84	46.74
Foreign banks	3.73	3.42	3.60	3.96	3.74	3.99	3.29

The mean value of efficiency scores has been evaluated based on two specification including time-invariant (denoted as pl81) and time-varying (denoted as tre).

Table 2. Mean value of cost and profit efficiency of bank industry in Vietnam, 2009–2015.

	Time-invariant			Time-variant		
	Obs	Mean value	Std. Dev.	Obs	Mean value	Std. Dev.
Cost efficiency	259	0.907348	0.067379	194	0.955751	0.035523
Profit efficiency	259	0.961689	0.025139	182	0.958911	0.036072

Table 2 provides a summary of the efficiency scores of bank industry in Vietnam from 2009–2015 following both time-invariant and time-varying specification. In line of Hausman and Taylor (1981) recommended that the random effect models are dominant specification analysis, so we focus on cost and profit efficiency in time-varying approach. It is practical significance in case of Vietnam. As the results, the average scale efficiency of cost and profit is 95.6% and 95.9% in the period of 2009–2015, respectively. It means that the Vietnamese banking system is able to cut inputs down by approximately 4.4% without changing output levels. In contrast, we may increase the output at the level of 4.1% to maximize efficiency without changing input levels.

Based on varying time approach, we permit a time trend to impact the banks efficiency to reflect the influence of a change of technology and time. Table 3 shows the mean value of the cost efficiency of each group of banks based on time-invariant. As the results, average cost efficiency of foreign banks has been lower than Big Four and Other Domestic banks.

Table 4 presents the mean value of the profit efficiency. The value of Other Domestic banks is the lowest of overall banks over a period of 2009–2015. So to speak, the efficiency estimates of state - owned banks (Big Four) are significantly higher than Other Domestic and Foreign banks.

Table 3. Mean value of the measured overall cost efficiency, 2009–2015.

Variable	Obs	Mean	Std. Dev.	Min.	Max.
Foreign	17	0.935893	0.076556	0.660328	0.984705
Big four	26	0.960193	0.019351	0.888576	0.987526
Other domestic	151	0.957222	0.029763	0.764315	0.989795

Table 4. Mean value of the measured overall profit efficiency, 2009–2015.

Variable	Obs	Mean	Std. Dev.	Min.	Max.
Foreign	17	0.96877	0.021744	0.909944	0.985041
Big four	25	0.975893	0.009272	0.950728	0.989664
Other domestic	140	0.954682	0.039285	0.781397	0.99369

Figures 1 and 2 show the changes of cost and profit efficiency over time of each type of bank. As might be seen, the cost and profit efficiency fluctuated with an unclear trend in the period of 2009–2015. Efficiency in the bank industry ambiguously increases or decreases every year.

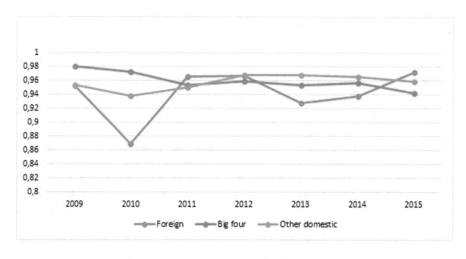

Fig. 1. The changes of cost efficiency over time.

To more precise, the cost efficiency of big four almost steadily decreases over time. In meanwhile, the value of other domestic banks goes up with a small rate. In contrast, the cost efficiency of foreign banks fluctuates with wide variation from 0.87 in 2010 to 0.97 in 2015.

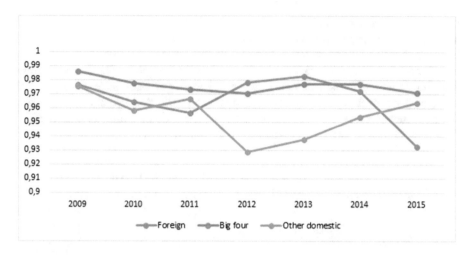

Fig. 2. The changes of profit efficiency over time.

Profit efficiency of Foreign and Other Domestic banks has been lower than that Big Four, except foreign banks in 2012–2013. In 2009, the gap between the average scores of profit efficiency for foreign banks and the highest profit efficiency (Big Four) is 0.01, but in 2015, it increases to negligible 0.038. In meanwhile, it is surprising that the gap between the average indexes of cost efficiency for the foreign bank is lower than the highest cost efficiency (Big four) is approximately 0.03 in 2009 but foreign banks make a progress with the highest efficiency in 2015. Hence, the cost efficiency of the foreign bank is smaller than big four banks in 2009, but it is greater in 2015.

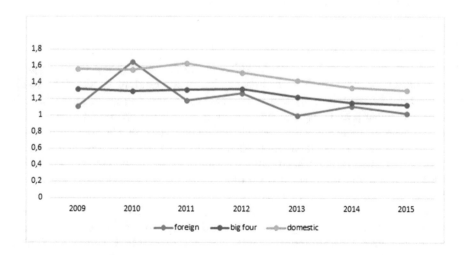

Fig. 3. The economy of scope of banks.

Economy of Scope. Following the Mester (2008), the economies of scope is measured according to the cost function. As can be seen from Fig. 3, all of bank have attained economy of scope but the degree of each group has a difference. The domestic banks are able to take the higher advantage of economies of scope than big four and foreign banks. However, the level of economies of scope of all bank may be decreased in the period of 2009–2015. Besides, the difference of the economy of scope for cost between the domestic banks and foreign banks is 0.45 in 2009, but it narrowed 0.28 in 2015 (Table 5).

Table 5. Regression analyses of economics of scale

	Foreign banks	Big four	Other domestic banks	Big four & other domestic	Total
ln_deposit	0.187	0.426**	0.271***	0.304***	0.268***
	−1.71	−2.3	−2.74	−3.42	−3.81
ln_fixasset	0.188*	0.276*	−0.233	−0.198	0.0698
	−1.76	−2.06	(−1.61)	(−1.46)	−1.38
ln_emp	0.836***	0.0831	0.880***	0.858***	0.722***
	−5.62	−0.5	−4.9	−5.32	−5.57
_cons	−0.936	1.213	−0.865	−1.225***	−1.282**
	(−1.59)	−0.55	(−1.47)	(−2.71)	(−2.03)
N	23	25	163	188	211

*, **, and *** significant at 10%, 5%, and 1%, respectively

Economy of Scale. It is surprising that foreign banks are able to take the advantage of economy of scale. The null hypothesis of constant return to scale is H_0 which is defined: $\beta_1 + \beta_2 + \beta_3 = 1$. In fact, the sum of $\beta_1, \beta_2, \beta_3 = 1$ is 1.21 which is tested by the F-statistic. Moreover, the p-value equals 0.08, signifying that foreign banks are more opportunity to gain an advantage of economy of scale in revenue at the 10% significance level.

4.2   Discussion

Overall, the average cost and profit efficiency of Big four group is the highest in the period of 2009–2015 according to the "true" random effects. The main reasons may be explained by the large size banks and the government have a strong relationship at various levels. Chang et al. (2014) proved that under the financial system regulations has been strongly supervised by the government like Vietnam, the more banks develop a relationship with the government, the more supports and monetary authority may be obtained from the government. This advantage helps Big four bank to produce more outputs, save more inputs, or both.

Following the Government Circular No.13/2010/TT-NHNN stated the stipulating prudential ratios in operation of credit institutions from 2010. All banks must be increased the minimum capital adequacy ratio (CAR, the ratio of total capital to risk-weighted assets) from 8% in 2005 to 9% in 2010 and met the requirements of the loan to deposit ratio (LDR) at the level of 80%. Obviously, this policy impact on mainly other domestic banks since the LDR of this group is high as usual, sometime more than 100%. In meanwhile, the LDR of Big four and foreign banks is lower degree. While the main profit of other domestic and big four banks primary come from loans and deposits, the revenue of foreign banks from interest income and non-interest income may be almost the same. Mainly non-interest income comes from products and services such as: card, wire transferring, overseas remittance, foreign currency trading, etc. This is the reason why the changing of this policy leads to profit efficiency of other domestic banks decreased quickly from 2011.

However, the other domestic bank's group have achieved economy of scope at the highest level among three group, followed big four and foreign banks. It means that other domestic banks are able to gain efficiency in the future by increasing the number of different lines of bank products. Big four and foreign banks also take an advantage of diversify but their level is lower than other domestic banks.

After WTO, the number of foreign banks increases leading competitive pressure of whole banking system. Therefore, domestic banks are seriously concerned about investing heavily in advanced technology and management. As a consequence of which cost goes up and profit efficiency of other domestic banks goes down.

Claessens et al. (2001) believed that in developing countries, foreign banks are more efficient than the domestic banks. In this research, we found that foreign banks do not necessarily mean best with the least average cost efficiency. Moreover, the profit efficiency of foreign banks tends to decrease from 2013. The cost efficiency of the foreign bank in 2010 is the lowest overall. It may be explained by the big total cost in 2010 of group foreign banks. Most of the foreign banks were registered and run business in 2009. At that time, they were expending large cost in opening more branches. Moreover, these foreign banks are located primarily in metropolitan where operating cost is more than another place.

While the regulation on domestic and foreign banks do not seem very different, the difference in cost and profit between domestic and foreign bank may come from the competitive advantages of domestic banks compared to foreign banks. Several advantages including:

- Domestic banks have a wide trading network and a large number of employees. Ho (2014) stated that the banks with more employee and branches will be preferred by customers. Compared to domestic banks, the market share in term of total assets, total loans, total costs of foreign banks are small. To more specific, the market share of total assets of foreign banks has accounted for a low proportion of 3.29% in 2015.

- The main profit of foreign banks come from fee-based services while domestic banks also collect more interest on deposits and mortgages. In fact, the outcome of foreign banks from interest income and non-interest income may be almost the same.
- It is difficult for a new entrance, foreign banks, to attract customers who transact with domestic banks. In Asia culture, they appreciate the long-term relationship. Moreover, switching banks may take the small cost (Ho 2014). In term of economy of scale, foreign banks have more opportunity to improve the efficiency of production in the future by increasing the size, output and activity level.

  To gain the advantage of profit and reduce the cost, it is recommended that the large banks like Big four banks should diversify while small banks, like foreign banks, may give attention to a single line of business. The foreign bank entry has positive effects on the efficiency of bank industry. However, the market share of foreign banks is small. It may take more time for the spillover effects. If foreign banks want to take the greatest advantage of economy of scope and economy of scale, they have to expand in size substantially.

5   Conclusions

Applying the SFA method, we estimated cost and profit efficiency of 37 banks in Vietnam for the period from 2009–2015. It is convenient to categorize according to type and ownership of banks. We divide banks into 3 groups: Big four, other domestic and foreign banks. The results revealed that foreign bank group is not the most efficiency but Big four is the most efficiency in term of both cost and profit overall. Although all group of bank may obtain the economics of scope, the level of each group has the difference. It is surprising that other domestic bank's group has the most capability to exploit economy of scope, followed by big four and foreign banks. We find that other domestic banks have more opportunity to diversify products than big four and foreign banks while foreign bank's group has able to exploit economy of scale in revenue.

References

Ahn, S.C., Lee, Y.H., Schmidt, P.: GMM estimation of linear panel data models with time-varying individual effects. J. Econ. **101**(2), 219–255 (2001)

Berger, A.N., Hasan, I., Zhou, M.: Bank ownership and efficiency in China: what will happen in the world's largest nation? J. Banking Finan. **33**(1), 113–130 (2009)

Battese, G.E.: Frontier production functions and technical efficiency: a survey of empirical applications in agricultural economics. Agric. Econ. **7**(3), 185–208 (1992)

Coelli, T.J., Rao, D.S.P., O'Donnell, C.J.: An Introduction to Efficiency and Productivity Analysis. Springer Science & Business Media, New York (2005)

Lensink, R., Meesters, A., Naaborg, I.: Bank efficiency and foreign ownership: do good institutions matter? J. Banking Finan. **32**(5), 834–844 (2008)

Rosengard, J.K., Du, H.T.: Funding Economic Development: A Comparative Study of Financial Sector Reform in Vietnam and China. Harvard John F. Kennedy School of Government (2009)

Vu, H.T., Turnell, S.: Cost efficiency of the banking sector in Vietnam: a Bayesian stochastic frontier approach with regularity constraints. Asian Econ. J. **24**(2), 115–139 (2010)

The Impact of Ownership on Net Interest Margin of Commercial Bank in Vietnam

An H. Pham[1(✉)], Loan K. T. Vo[2], and Cuong K. Q. Tran[1]

[1] Faculty of Economics, Van Hien University,
665 - 667 - 669 Dien Bien Phu Street, District 3, Ho Chi Minh City, Vietnam
hoangan.tcnh@gmail.com
[2] HCM City Open University, 97 Vo Van Tan Street, District 3,
Ho Chi Minh City, Vietnam

Abstract. This study analyses the impact of the Ownership on Net Interest Margin of commercial banks in Vietnam. The panel data regression method is applied to analyse the second data of 26 banks together with 234 observations during the period of 2008 to 2016. The empirical results firstly indicates that lending scale, operating cost, liquidity risk, equity ratio have a positive and significant effect on net interest margin of the commercial banks in Vietnam. Secondly, Credit Risk has no significant relationship with net interest margin. Finally, there is no significant difference in the net interest margins of state-owned commercial banks and those of joint-stock commercial banks in Vietnam.

1 Introduction

Net Interest Margin (NIM) is a measure for the effectiveness as well as profitability, and a very important performance indicator because it often accounts for about 70–85% the total income of a bank. The higher this ratio is, the higher income of the banks will be. It indicates the ability of the Board of Directors and employees in maintaining the growth of incomes (mainly from loans, investments and service fees) compared with the increase in cost (mainly from interest cost for deposits, monetary markets debts) (Rose 1999).

The effect of bank operation is the issue that the bank managers have concerned seriously because effective management will create the sustainable profit that makes banks sharply develop and compete in the international environment.

A competitive banking system will create a higher efficiency and a lower NIM (Sensarma and Ghosh 2004). High profit return ratio causes significant obstacles to intermediaries, such as more savings encouraged by lower borrowing interest rate and reduced investment opportunities of the banks as a result of higher lending rate (Fungáčová and Poghosyan 2011). Therefore, banks are expected to perform their intermediate functionality with the possible lowest cost to promote economic growth.

The research of Sensarma and Ghosh (2004) in India, Ugur and Erkus (2010) in Turkey, Fungáčová and Poghosyan (2011) in Russia; Hamadi and Awdeh (2012)

© Springer International Publishing AG 2018
L. H. Anh et al. (eds.), *Econometrics for Financial Applications*, Studies in Computational Intelligence 760, https://doi.org/10.1007/978-3-319-73150-6_58

in Lebanon show that net interest margin varies among ownerships. In Vietnam, there are few studies that have evaluated the impact of the ownership on banks NIM. Thus, it is necessary to learn how the ownership has influence on net interest margin of commercial bank. The result of this study can serve as a scientific basis for bank managers to make suitable decisions, bring good efficiency and increase the attractiveness of their stocks.

2   Literature Review

2.1   Net Interest Margin

To calculate the operating effect of any bank, we often analyse Return on Equity (ROE), Return on Asset (ROA), Net interest margin (NIM) and interest spread (Rose 1999). Hempel et al. (1986) stated that NIM is helpful in measuring changes in interest spread and comparing profit between banks.

Net interest margin ratio is one of the most important measurements to quantify financial effectiveness in an intermediary institution (Golin 2001). Net interest margin (NIM) is defined by net interest income over total earning asset.

$$Net\ Interest\ Margin\ (NIM) = \frac{Interst\ Income - Interest\ Expense}{Total\ earning\ asset} \quad (1)$$

2.2   Factors Influencing Net Interest Margin

Based on previous research in Russia, Turkey, China, Lebanon and Fiji, the authors define similarities between Vietnam and these nations, and thereby suggest some factors which have impacts on net interest margin, including:

Ownership. According to the researches of Sensarma and Ghosh (2004) in India, Fungáčová and Poghosyan (2011) in Russia, there is a difference in NIM between State ownership bank and other banks. We suppose Ownership is an X that is given to check the discrepancy in NIM among many ownerships. Consequently, Ownership is 1 if its State Bank and 0 if it is Join-stock bank.

Hypotheses: Researches of Sensarma and Ghosh (2004), Fungáčová and Poghosyan (2011) has shown the ownership is correlation with NIM, thus a hypotheses is being posted:

H_0: there is no correlation between ownership and banks NIM;
H_1: ownership is negative correlation with NIM.

Size. Studies of Maudos and Guevara (2004), Ugur and Erkus (2010) find positive relation between lending scale and banks net interest margin, where large average operating scale leads to higher market risk and credit risk, increasing the possibility of losses. Meanwhile, researches of Fungáčová and Poghosyan (2011), Hamadi and Awdeh (2012) show the negative effect of bank size on NIM, where large banks with high credit ratings earn their profit from economy of scale and have low NIMs. In Vietnam, large banks size have advantages they can utilize to raise capital at low cost, such as: large network of operation with many branches, wide variety of products and service, etc. to make higher profit.

Operating Cost (OC). Brock and Suarez (2000) argue that in a stable banking structure, raising any operating cost will higher the spread interest rate, not lower the dividend. Maudos and Guevara (2004) state that even in case there is no impact of market or any kind of risk, the banks still have to cover their operating cost and achieve higher interest income.

Credit Risk (CR). Credit risk is the risk that customers are not able to repay the debt at its maturity. Angbazo (1997) states that credit risk impacts banks interest income in a positive relation. Banks which are lending out more money face higher credit risk and thus have to maintain more reserve; this situation forces them to charge more interest on their loans in order to make up for the expected losses, causing a positive relation (Garza-García 2010). More studies have found the positive relation between credit risk and net interest margin (Maudos and Guevara 2004; Dolient 2005; Maudos and Solís 2009; Kasman et al. 2010; Gounder and Sharma 2012; Tarus et al. 2012).

Liquidity Risk (LIQ). Liquidity risk is a risk that bank is lack of money to satisfy the need of withdrawing deposits, allocating loans and other cash demands because they cant change immediately assets into money or borrow from other financial institutions. To face Liquidity risk, a bank may need to borrow money with very high interest rate to suffer urgent needs, therefore it reduces the banks profit (Rose 1999). According to Angbazo (1997), when cash and cash equivalents proportion increases, the liquidity also rises that makes NIM decrease.

Equity Capital (CAP). According to the IMF (2006), the ratio of equity over total asset is used as one of the recommended indicators to assess the financial health of a commercial bank. Most studies have found positive correlation between CAP and NIM (Brock and Suarez 2000; Saunders and Schumacher 2000; Maudos and Guevara 2004; Doliente 2005; Hawtrey and Liang 2008; Maudos and Sollís 2009; Garza-Garclía 2010; Ugur and Erkus 2010; Kasman et al. 2010; Fungáčová and Poghosyan 2011). Raising thc capital will increase the mediate cost of keeping equity more than loans due to taxes and diluting shareholders rights. The increase in mediate cost is often recovered through an increase in the interest rate spread. Whenever capital is too high, the manager is pressured to increase profit margin.

3   Methodology and Data

3.1   The Model

Based on research models of Sensarma and Ghosh (2004), Ugur and Erkus (2010), Fungáčová and Poghosyan (2011), Hamadi and Awdeh (2012), this study applies the following model:

$$NIM = \beta_0 + \beta_1 \times OWNERSHIP + \beta_2 \times SIZE + \beta_3 \times OC$$
$$+\beta_4 \times CR + \beta_5 \times LIQ + \beta_6 \times CAP + u_{it} \tag{2}$$

where NIM: Net interest margin; Ownership; SIZE: Lending Size; OC: Operating Cost; CR: Credit risk; LIQ: Liquidity Risk; CAP: Equity Capital.

3.2 Variable Measurements

The description of how to calculate variables and the expected signs are detailed in Table 1.

Table 1. Describing table for variables and expected signs

Variable	Description	Measurement	Expected sign
Dependent			
NIM	Net interest margin	(Interest income – Interest expense)/Total asset	
Independent			
OWNERSHIP	Ownership	= 1 if it is State Own Bank; 0 if it is Join-stock Bank	−
SIZE	Lending size	Logarit of total loans	+
OC	Operating cost	Operating cost/Total asset	+
CR	Credit risk	Loss Provision for credit risk/Total loans	+
LIQ	Liquidity risk	Liquidity asset/Total asset	−
CAP	Capital ratio	Equity/Total asset	+

3.3 The Data

Data in this study was taken from the audited financial statements of Vietnamese Banks in the period 2008–2016. Until 31/12/2016, Vietnam has a total of 35 Commercial Banks. Data is collected after eliminating banks with lacking or unclear information. The result is a random balance panel data involving 26 banks and 234 observations, which accounts for about 70% of the Vietnamese banking system. Hence, one can say that those selected banks represent commercial banks in Vietnam. Table 2 describes mean, standard deviation, min and max of variables.

Table 2. Describing observed variables

Variable	Mean	Standard deviation	Min.	Max.
NIM	0.0263	0.0113	−0.0063	0.0663
OWNERSHIP	0.1153	0.3201	0	1
SIZE	17.3067	1.2915	14.0723	20.3999
OC	0.0159	0.0050	0.0032	0.0306
CR	0.0130	0.0057	0.0021	0.037
CAP	0.1111	0.0624	0.0426	0.4624
LIQ	0.2174	0.1070	0.0462	0.611

4   Empirical Result and Discussion

4.1   Empirical Result

The study examines the possibility of multicollinearity between the variables by setting up a correlation matrix of the variables and calculating VIF indicators, as presented in Table 3.

Results shows that none of the correlation coefficient between pairs of variables exceeds 0,8. The largest VIF index of the independent variables in this study is 3.71, less than 5 (Gujarati 2004). Therefore, the multicollinearity phenomenon in this research models is negligible.

Table 3. Matrix of correlation between the variables

	SIZE	CR	CAP	OC	LIQ	OWNER	VIF
SIZE	1						**3.71**
CR	0.3509	1					1.20
CAP	−0.6900	−0.2723	1				2.41
OC	−0.0923	−0.0224	0.3016	1			1.36
LIQ	−0.3241	−0.0165	0.0861	−0.3760	1		1.43
OWNER	0.6095	0.3443	−0.2639	−0.0836	0.0914	1	1.87

The test results of the variance of the constant error (White test), Prob.Chi - Square $= 0.000$, less than 0.05 and the result of error autocorrelation (Breusch-Godfrey test), Prob.F $(1,25) = 0.000$, also less than 0.05. These results show that the model has both the phenomenon of changing variance and autocorrelation of errors. According to Wooldridge (2002), the solution to changing variance errors & error autocorrelation is applying the regression model with the generalized least squares method (Generalized Least Squares - GLS). Table 4 presents the regression results of using GLS method to estimate the regression coefficients.

Table 4. Regression result (GLS Method)

Variable	NIM		
	Coefficient	t-value	p-value
Constant	−0.0352	−2.56	0.010
OWNERSHIP	0.0019	1.05	0.296
SIZE	0.0018***	2.58	0.010
OC	1.3476***	13.81	0.000
CR	−0.1145	−1.55	0.121
LIQ	0.0165***	3.54	0.000
CAP	0.0479***	4.37	0.000

Note: *** have Statistical significance at $p < 0.01$.

4.2 Discussion

In this section, the research focuses on the results of the regression model using GLS method.

The findings reveal no significant difference between NIM of State Own Banks and Join Stock in Vietnam. This result is explainable due to the priority of lending activities, which increase profit for most of the banks, even in State Own Banks or Join-stock Banks.

Lending scale (SIZE), shares a positive correlation with NIM. The more Vietnam commercial banks enlarge their lending scale, the higher NIM is. These results are consistent with previous findings of Maudos and Guevara (2004) in Europe, Maudos and Solís (2009) in Mexico and Hamadi and Awdeh (2012) in Lebanon. In Vietnam, lending makes up the most traditional and major activities of banks (about 70–80% of bank operations). Therefore, most banks tend to focus on lending activities, their main channel of profits.

Operating Cost (OC) has a positive correlation with NIM of Vietnam commercial banks. Increasing operating cost will lead to the higher spread interest rate, And higher NIM is. These results are consistent with previous research findings: Hawtrey and Liang (2008), Zhou and Wong (2008), Maudos and Solís (2009), Garza-García (2010), Kasman et al. (2010), Ugur and Erkus (2010), Gounder and Sharma (2012), Tarus et al. (2012).

Liquidity Risk (LIQ) has positive correlation with NIM. In Vietnam, when liquidity asset to total asset increase, which helps banks not to lend cash to pay for current needs. As a result, the NIM gets higher. This result is opposite with the one found by Fungáčová and Poghosyan (2011), Hamadi and Awdeh (2012).

Equity Capital (CAP) has a positive correlation with NIM of Vietnam commercial banks, demonstrating the importance of scale of equity in improving the banks' NIM. This study shows that better - capitalized banks face lower risk of default. Moreover, a strong capital structure is essential for banks operating in developing economies, as it provides more power for banks to survive during times of financial crisis and increase the level of security provided to depositors when faced with the conditions of macroeconomic instability. These results are consistent with previous research findings: Brock and Suarez (2000), Saunders and Schumacher (2000), Maudos and Guevara (2004), Doliente (2005), Hawtrey and Liang (2008), Maudos and Solís (2009), Garza-Garcia (2010), Kasman et al. (2010), Ugur and Erkus (2010), Fungáčová and Poghosyan (2011).

5   Conclusion and Suggestion

To quantify the influence of Ownership on net interest margin of bank through experiment, this study uses the panel data of 26 commercial banks from 2008 to 2016. In this study, the difference between net interest margin of State Own Banks and Join-stock Banks has not been found. In Vietnam, lending activities bring up higher profit for banks, therefore, the bank CEOs always consider the balance between cost and expense. Furthermore, they care for the equity capital activities so that they can spend money in the most effective way.

The empirical results indicate that Lending scale, Operating Cost, Liquidity Risk and Equity Capital has a positive correlation with net interest margin of Vietnam commercial banks. However, Credit Risk is statistically insignificant to net interest margin.

Due to the limitations of data, this study didnt consider whether the ownership of foreign banks has an impact on net interest margin and level of impact is higher or lower than domestic banks.

To sum up, through the collection of the researches from different countries, the study has chosen the suitable model for Vietnam by adjusting and testing the effect of ownership on Net interest margin of bank via the specific database.

References

Angbazo, L.: Commercial bank net interest margins, default risk, interest rate risk and off-balance sheet banking. J. Bank. Finan. **21**(1), 55–87 (1997)

Brock, P.L., Suarez, L.R.: Understanding the behavior of bank spreads in Latin America. J. Dev. Econ. **63**(1), 113–134 (2000)

Doliente, J.S.: Determinants of bank net interest margins in Southeast Asia. Appl. Finan. Econ. Lett. **1**(1), 53–57 (2005)

Fungáčová, Z., Poghosyan, T.: Determinants of bank interest margins in Russia: does bank ownership matter? Econ. Syst. **35**(4), 481–495 (2011)

Garza-García, J.G.: What influences net interest rate margins? Developed versus developing countries. Banks Bank Syst. **5**(4), 32–41 (2010)

Golin, J.: The Bank Credit Analysis Handbook: A Guide for Analysts, Bankers and Investors. John Wiley & Sons (Asia) Pre Ltd., Singapore (2001)

Gounder, N., Sharma, P.: Determinants of bank net interest margins in Fiji, a small Island developing state. Appl. Financ. Econ. **22**(19), 1647–1654 (2012)

Gujarati, D.: Basic Econometrics, 4th edn. Tata McGraw Hill, India (2004)

Hamadi, H., Awdeh, A.: The determinants of bank net interest margin: evidence from the Lebanese banking sector. J. Money Investment Bank. **23**(3), 85–98 (2012)

Hawtrey, K., Liang, H.: Bank interest margins in OECD countries. North Am. J. Econ. Financ. **19**(3), 249–260 (2008)

Hempel, G., Coleman, A., Simonson, D.: Bank Management: Text and Cases, 2nd edn. Wiley, New York (1986)

IMF: Financial Soundness Indicators Compilation Guide (2006). http://www.imf.org/external/pubs/ft/fsi/guide/2006/index.htm. Accessed 15 June 2013

Kasman, A., Tunc, G., Vardar, G., Okan, B.: Consolidation and commercial bank net interest margins: evidence from the old and new European Union members and candidate countries. Econ. Model. **27**(3), 648–655 (2010)

Maudos, J., Guevara, J.F.D.: Factors explaining the interest margin in the banking sectors of the European Union. J. Bank. Financ. **28**(9), 2259–2281 (2004)

Maudos, J., Solís, L.: The determinants of net interest income in the Mexican banking system: an integrated model. J. Bank. Financ. **33**(10), 1920–1931 (2009)

Rose, P.S.: Commercial Bank Management. Irwin/McGraw-Hil, Boston (1999)

Sensarma, R., Ghosh, S.: Net interest margins: does ownership matter? Vikalpa: J. Decis. Makers **29**(1), 41–47 (2014)

Saunders, A., Schumacher, L.: The determinants of bank interest margins: an international study. J. Int. Money Financ. **19**(6), 813–832 (2000)

Tarus, D.K., Chekol, Y.B., Mutwol, M.: Determinants of net interest margins of commercial banks in Kenya: a panel study. Procedia Econ. Financ. **2**, 199–208 (2012)

Ugur, A., Erkus, H.: Determinants of the net interest margins of banks in Turkey. J. Econ. Soc. Res. **12**(2), 101–118 (2010)

Wooldridge, J.: Econometric Analysis of Cross Section and Panel Data. MIT Press, Cambridge (2002)

Zhou, K., Wong, M.C.S.: The determinants of net interest margins of commercial banks in Mainland China. Emerg. Markets Financ. Trade **44**(5), 41–53 (2008)

An Alternate Internal Credit Rating System for Construction and Timber Industries Using Artificial Neural Network
Case Study of Bank for Development and Investment

Pham Quoc Hai[1]([⊠]), Truong Thuy Lan Ngoc[2], and Bui Do Thanh Phuong[2]

[1] University of Economics and Finance, Ho Chi Minh City, Vietnam
phamquochai1990@gmail.com
[2] International University Ho Chi Minh City, Ho Chi Minh City, Vietnam

Abstract. Internal credit scoring and rating play an essential role for bank in credit risk management and in pricing loans as well as assigning appropriate lending policy for each class of customers; and also for determining the level of regulatory capital reserve. This implies the importance of deeply understanding about the internal rating models and the respective approaches in execution for bank risk managers.

In 2005, the State Bank of Vietnam promulgated Decision 493/2005/QĐ-NHNN and later Decision 18/2007/QĐ-NHNN in 2007 about debt classification, determining provision and reserve. In addition, the Decisions required commercial banks to establish an internal credit rating system that aligned with Basel Accord II of Basel Committee and Banking Supervision. In 2006, Bank for Development and Investment Vietnam (BIDV) officially adopted the new Internal Credit Rating System (ICRS), which is approved by the State Bank of Vietnam, and was consulted by Ernst & Young Audit firm. The IRCS includes 54 criteria (10 financial and 40 non-financial criteria) to asses a firm creditability.

The ICRS is widely applied for the whole BIDV system including the head quarter and all level of BIDV branches. Credit analysts from different BIDV branches may have different point of view in assessing non-financial (qualitative) criteria and thus lead to different credit rating for the same company in the system. To disregard the difference between branches, this thesis is conducted using data that BIDV consider best practiced credit score rated by BIDV's credit analyst, whose competence are equivalent and whose point of view mostly aligned with BIDV strategy. The research aims to investigate the most important ICRS criteria by using artificial learning machine method, specifically, the Artificial Neural Network. The researches primarily focused on analyzing the ICRS applied for economic organizations with the historical data collected from 33 companies from 14 industries in the year 2015. After a deliberate research process, the research has evaluated several constrains of the outstanding model aiming to constructively contribute for improvement in the future rating activities at BIDV.

© Springer International Publishing AG 2018
L. H. Anh et al. (eds.), *Econometrics for Financial Applications*, Studies in Computational Intelligence 760, https://doi.org/10.1007/978-3-319-73150-6_59

1   Introduction

1.1   Significance of the Research

The construction of internal credit rating model for internal control in accordance with international practices according and borrowers' specific business activities in Vietnam is necessary. However, it is not yet entirely built because the criteria in the ranking system include quantitative and quantitative variables. Therefore, actual credit scoring practice is still significantly affected by subjective factors and professional qualifications of the examiners.

For commercial banks, in addition to building a high quality and stable internal credit rating model, finding approached and mechanisms to control the authenticity of result is necessary as well because this will help accurately assess the quality of the debt and the possibility of losses in credit operations. Those losses are carried out by the customer and their inability to perform obligations under the commitment. Controlling the authenticity of internal credit rating result will help early detection of problematic debts, evaluate degree of debt risk achieve better debt classification. This also helps to determine required risk provisions made for losses that may occur due to customer's failure or not fully implemented their obligations under the commitment.

With urgent requirement of finding a monitoring mechanism to renovate and upgrade internal credit system ratings, this research will include a number of studies with references to credit scoring system ratings which are currently applied in prestigious international agencies. In combination with some models of enterprise and BIDV current rating model, the research will provide some critical points of credit rating system to help management and supervision mechanism in banking system get suitable credit scoring for customers. This research will also propose a number of recommendations for internal credit rating system of BIDV.

1.2   Research Questions

- How does Artificial Neural Network (ANN) predict credit rating?
- Which type of ANN is best replicate the BIDV internal credit rating system for Construction and Timber industries?
- Which credit rating criteria are more important than others?

1.3   Research Limitation

The research containing 37 companies collected from both HoSE and HNX with can be extending to bigger portfolio as well as other stock exchange such as UPCOMP and OTC. In addition, the research only uses artificial intelligence to predict credit rating, which can be also conducted by using statistical model for comparative research.

2   Literature Review

2.1   Artificial Intelligence (AI)

Artificial intelligence (AI) is a field of computer science, in which, the machine is programmed to learn and react like humans. AI is able to deal with core problems such as:

- Knowledge
- Reasoning
- Problem solving
- Perception
- Machine Learning
- Planning
- Ability to manipulate and move objects

For economic application, AI is expected to focus on machine learning, in which the machine is able to learn from the inputs and outputs of a particular question and thrive to predict a particular output for the same question with new input.

Internal credit rating is a process that assessing both quantitative and qualitative factors of a specific firm. "Rating agencies and some researchers have emphasized the importance of subjective judgment in the bond-rating process and criticized the use of simple statistical models and other models derived from AI techniques to predict credit ratings" – Huang et al. 2004. Credit rating learning process can be done two ways, learning with and without supervision (output). With given output (credit score/credit class), the machine thrives to achieve the prediction by classification and numerical regression. On another hand, without output the machine is required an ability to draw the patterns for the streams of inputs.

Historically, many researchers utilize AI in predicting bond rating. There were many types of artificial intelligence applied to predicting bond rating. Specifically, the researchers have used artificial neural networks, rule-based systems, inductive learning/decision Trees and case-based reasoning system. Table 1 describes difference type of AIs used to analyze bond rating.

From Table 1, it has been suggested that artificial neural network had the highest accurate prediction on bond rating in comparison with rule-based systems, inductive learning/decision Trees and case-based reasoning system. Furthermore, some other research papers also wrote comparative research on neural network and other type of artificial intelligence as well as with statistical model, the result enhanced the ability of neural network on predicting credit rating, with the flexibility and high level of accuracy of neural network. Neural network and other types of artificial intelligence: "Neural networks have powerful pattern classification and prediction capabilities and have been successfully used for a variety of tasks in many fields of business, industry, and science" – Widrow (1994). Neural network and statistical method: "Forecasts from neural networks

Table 1. Empirical studies using artificial intelligence

Study	Bond rating categories	Method	Accuracy	Data	Sample size	Benchmark statistical methods
Dutta and Shekhar (1988)	Two (AA vs. non-AA)	BP	83.30%	US	30/17	LinR (64.7%)
Singleton and Surkan (1990)	Two (Aaa vs. A1, A2 or A3)	BP	88%	US - Bell firms	126	MDA (39%)
Garwaglia (1991)	Three	BP	84.90%	US SP	797	N/A
Kim (1993)	Six	BP, RBS	55.17% (BP)	US S&P	110/58/60	LinR (36.21%), MDA (36.20%), LogR (43.10%)
Moody and Utans (1995)	Sixteen	BP	36.2% , 63.8% (5 classes), 85.2% (3 classes)	US S&P	N/A	N/A
Maher and Sen (1997)	Six	BP	70% (7), 66.67% (5)	US Moody's	299	LogR (61.66%), MDA (58–61%)
Kwon et al. (1997)	Five	BP	71–73% (with OPP), 66–67% (without OPP)	Korean	126	MDA (58–62%)
Kwon and Lim (1998)	Five	ACLS, BP	59.9% (ACLS), 72.5% (BP)	Korean	126	MDA (61.6%)
Chaveesuk et al. (1999)	Six	BP, RBF, LVQ	56.7% (BP), 38.3% (RBF), 36.7% (LVQ)	US S&P	60/60 (10 for each category)	LogR (53.3%)
Shin and Han (2001)	Five	CBR, GA	75.5% (CBR, GA combined)	Korean	3886	MDA (58.4–61.6%)

Source: Huang et al. (2003)

BP: Back-propagation Neural Networks, RBS: Rule-based System, ACLS: Analog Concept Learning System, RBF: Radial Basis Function, LVQ: Learning Vector Quantization, CBR: Case-based Reasoning, GA: Genetic Algorithm, MDA: Multiple Discriminant Analysis, LinR: Linear Regression, LogR: Logistic Regression, OPP: Ordinary Pairwise Partitioning. Sample size: Training/tuning/testing.

outperform implied volatility forecasts and are not found to be significantly different from realized volatility" - Hamid (2003).

For the evidence of flexibility and highly accuracy, this research aim to use Artificial Neural Network to replicate the internal credit rating system at BIDV.

2.2   Artificial Neural Network

Artificial neural network are algorithms try to mimic the brain. Neural network is a type of artificial intelligence and used to learn non-linear hypothesis (the pattern of data set). Neural network thrives to classify different types of labels (categorical dependent variable) or to predict the numerical dependent variable.

Neural network itself is a network or neurons usually called "nodes". It performs the brain function to process the data and, in turn, it sends out a response. To illustrate, the paper will briefly describe the processing data progress in human brain. Figure 1 displays an image of a neuron in the network.

A single neuron has two parts that allows it to receive the inputs (data) from other locations and to send the message to other neurons. The "dendrites" are known as "input wires" and "axons" are known as "output wires". At simplicity level, a single neuron is a computational unit that gets a number of inputs

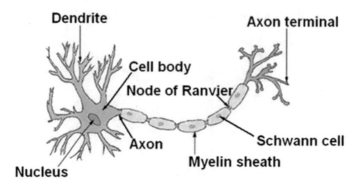

Fig. 1. A neuron in brain

through its input wires, does some computation, and then it sends the output via its "Axon" to other neurons (nodes) in the brain.

In a neural network, or rather in an artificial neural network that is implemented in a computer, it uses a very simple model of what a neuron does. When the neural network receives some inputs (x1, x2, x3), the node processes the data and send out an output h0(x). Figure 2 represents a single node in a neural network.

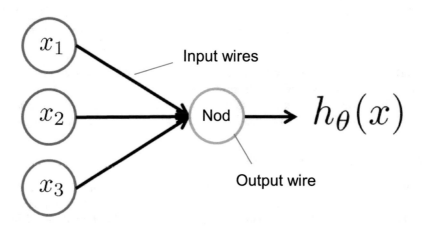

Fig. 2. A single node in neural network

In a neural network, a group of nodes communicate with each other like the neurons in human brain. Figure 3 shows a network of neurons.

The first neuron receives the inputs through its dendrites, does some computation, and send out the output through it axon. The axon of the first neuron connects with the dendrite of the second neuron, then, the second neuron accept the incoming message, does some computation and then, in turn, the second

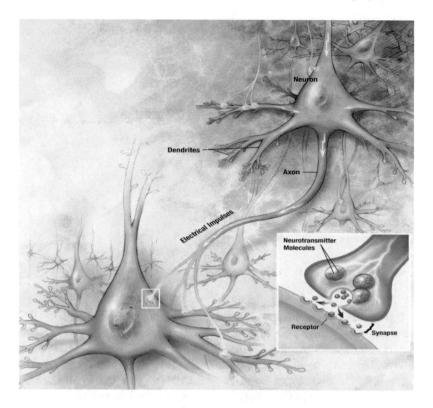

Fig. 3. A network of neurons in brain

neuron may decide to communicate the message with other neurons. This process is exactly what happen to a person when he or she thinking. Neurons in the network do computation on the inputs and passing message to other neurons. Similarly, when a person wants to move his or her muscles, the neurons send the pulses of electricity to his or her muscles and that cause the muscles to react in a particular way. In turn, if the muscles "feel" hurt, it will send the message back to his or her brain by the same path (the muscles send pulse of electricity to the neurons).

Similarly, the neural network owns a network of nodes, in which, each node does some computation from the input and communicate with the others. In a neural network, there are 3 types of layer. The first layer is the input layer, where the neural network receives the parameters (independent variables). The final layer is the output layer consists of a neuron that computes the final value (dependent variable). The middle layers are hidden layers, it is because the nodes secretly do computation itself and it is not seeable. The hidden layers either consist of one or more layers. Figure 4 shows a network of nodes.

Input layer   **Hidden layer**   **Output layer**

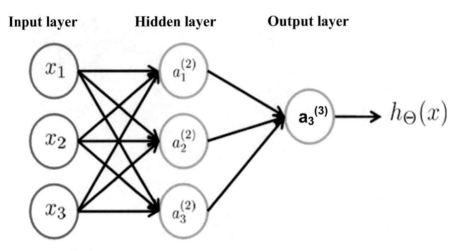

Fig. 4. A simple model of neural network

The progress that a neural network learns the data set is also called the "training progress". In the training progress, the input layer receives the incoming stimuli (independent variables). The input nodes, according to a specific function called transfer function, selected and distribute the result to the next level of nodes. Then the input nodes forward the information to all the nodes in hidden layer. The information sent, have been weighed from the input layer. It means that the result obtained from each node is sized due to the weight of the connection between the two nodes.

Specifically, as shown in Fig. 5 the weight of connection is represented by Wji. Each neuron is characterized by a transition function and a threshold value. The threshold is the minimum value that input must have to activate the neuron. In the hidden layer, each node sums the inputs that are presented to its incoming connections.

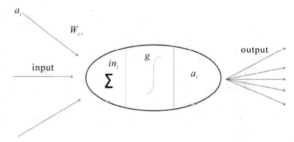

Fig. 5. Processing element

In mathematical terms, "each neuron performs the summation of inputs, which are the product of output neurons of the first layer and weight of the connection. The result of this sum is again drawn on the basis of the transfer function of each neuron" – Pacelli (2010). The results obtain from this layer will then send to the node in the last layer, which are multiplied by the way between nodes.

"Before the neural network can be applied to solve problem, a specific tuning of its weights has to be done. This task is accomplished by the learning algorithm which trains the network and iteratively modifies the weights until a specific condition is verified. In most applications, the learning algorithm stops when the discrepancy (error) between desired output and the output produced by the network falls below a predefined threshold" – Pacelli (2010).

For neural network, there are 3 types of learning mechanism: supervised learning, unsupervised learning and reinforced learning.

- Supervised learning is featured by a training set of correct examples used to train the network. In which, the training set includes the pairs of inputs and desired outputs. The error produced by the network then is used to change the weights. A typical application is classification. A given input has to be inserted in one of the defined categories.
- Unsupervised learning is the case in which the network is only provided with a set of inputs and no desired output is given. The algorithm guides the network to self-organize and adapt its weights. This kind of learning is used for tasks such as data mining and clustering, where some regularity in a large amount of data has to be found.
- Reinforced learning trains the network by introducing prizes and penalties as a function of the network response. Prizes and penalties are then used to modify the weights. Reinforced learning algorithms are applied, for instance, to train adaptive systems which perform a task composed of a sequence of actions.

In this research, for the purpose of replicating the internal credit rating system of BIDV, the author intends to use two neural networks include the multilayer feed-forward, and probabilistic neural network with Supervised learning as the learning mechanism.

2.3   Multi-layer Feed-Forward Neural Network (MLF)

As a type of artificial intelligence, a MLF neural network also consists of neurons, which are ordered into layers (Fig. 6). The first layer is called the input layer; the last layer is called the output layer, and the layers between are hidden layers.

The behavior of the net is determined by:

- Its topology (the number of hidden layers and the numbers of nodes in those layers)
- The "weight" of each connection (a parameter assigned to each connection) and bias terms (a parameter assigned to each node)
- An activation/transfer function, used to convert the inputs of each node into its output

The MLF neural network behavior is determined by quantity of hidden layers as well as of neuron. The "weight: of each connection (assigned), bias unit

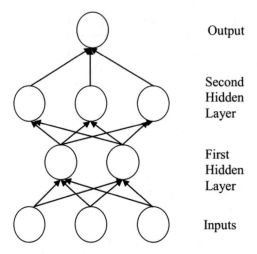

Fig. 6. Multi-layer feed-forward neural network architecture

(assigned to each node) and a transfer function, together, allow the neural network to process the inputs into the output (for each node). Specifically, a hidden neuron with n inputs first computes a weighted sum of its inputs:

$$\text{sum} = \text{in}_1 \cdot w_0 + \text{in}_1 \cdot w_1 + \ldots + \text{in}_n \cdot w_n + \text{bias}$$

In the formula, in0 to inn are outputs of nodes in the previous layer, while w0 to wn are connection weights and each node has its own bias value. Then the activation function is applied to Sum to generate the output of the node. A sigmoid (also called S-shaped) function is used as the activation function in hidden layer nodes. Neural nets are sometimes constructed with sigmoid activation functions in output nodes. Sigmoid functions have restricted an output range from (-1) to 1 and there will typically be dependent values outside the range. Thus, using a sigmoid function in the output node would force an additional transformation of output values before passing training data to the net. When MLF nets are used for classification, they have many output nodes and each of them corresponding to each possible dependent category. A net classifies a case by computing its numeric outputs; the selected category is the one corresponding to the node that outputs the largest value.

2.4 Probabilistic Neural Network (PNN)

Consider the following training data set with two independent numeric variables, and a dependent variable with two categories:

The circles represent training cases in one category, while the squares are those that belong to other categories. It is desirable to predict the category of the case shown with the question mark. As a human, it is quickly to answer that question (Fig. 7). For a PNN it takes time to decide that the case is more likely

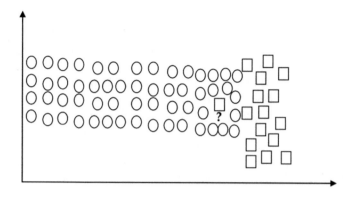

Fig. 7. A sample for classification of PNN

in the circle category than the square category. However, many classification methods will not be able to reach the same conclusions. Especially, methods that require linear reparability of categories will less likely to have right prediction. For PNN, a method called "Nearest neighbor" will assign the unknown case into the square category, in which, PNN focus on the central tendencies, where the square with the question mark is much closely to the center of the whole group of square and more farther from the center of circle group. By measuring the distance, PNN is able to recognize the square question mark is belonging to square group.

A Probabilistic Neural Net is structured as shown in Fig. 8, assuming that there are two independent numeric variables, two dependent categories, and five training cases.

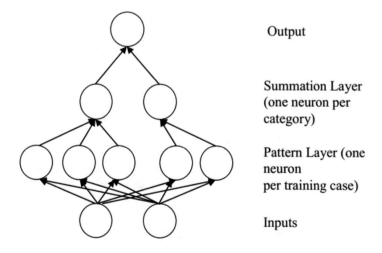

Fig. 8. Probabilistic neural network architecture

When a case (observation/dependent variable) is put into the net, each node in the pattern layer computes the distance between the training case represented by the node, and the input case. The value passed to summation layer nodes is a function of the distance and smoothing factors, in which each input has its own smoothing factor and these factors determine how rapidly the significance of training cases decreases with distance. For summation layer, there is one node per dependent category, in which each node sums the output values for the nodes corresponding to the training cases in that category. In addition, the output values of the summation layer nodes can be interpreted as probability density function estimates for each class. Finally, the output node selects the category with the highest probability density function value as the predicted category.

2.5   Empirical Study of Artificial Intelligence in Credit Rating

2.5.1   Global Study

The application of neural network to solve daily problem exploded in the 90s. "Neural networks have powerful pattern classification and prediction capabilities and have been successfully used for a variety of tasks in many fields of business, industry, and science" – Widrow (1994). Back to 1988, Dutta and Shekkar started a research to investigate the applicability of neural network on bond rating. The research used back-propagation neural network to predict the bond rating, in which the back-propagation network had supervised learning (the inputs and desired outputs are given). The task was to classify between two labels (dependent variables): "AA" and "non-AA". The prediction reached 83.3% correct with 47 data point (30-AA and, 17-non AA). The research's benchmark is Linear Regression (statistical method) reaches only 67.4% of accuracy. In 1990, Singleton and Surkan also used back-propagation neural network to classify bond-rating, their research was to classify two labels: "AAA" or "A1", "A2", "A3". The best neural network - two hidden layers achieve the accuracy level of 88% with them 126 observations. A comparative statistical method, the multiple discriminant analysis, for the same set of data, it achieved 39% correct.

In 1995, Moody and Utans used neural network to predict 16 different labels and Standard and Poors. And the result got 36.2% accurate. In addition, they tested the system with 5 labels and 3 labels prediction and got 63.8% and 85.2% respectively. The further researches clarifies that the less labels (bond rating classes) the more level of accuracy achieved.

Some comparative researches were conducted for popular statistical method with artificial neural network. The result of neural network was outperforming the statistical model. Kim (1993), Maher and Shen (1997) both use back-propagation to represent for the neural network approach. For statistical methods, popular models were used including logistic regression and multiple discriminant analysis. In either Kim or Maher and Shen research, the conclusion was the outperforming of neural network compared to its statistical method benchmark. The highest level accuracy predicted by artificial neural network was 70% and the lowest reached

31.03%. In term of statistical method, the highest accuracy level was 61.6% (Logistic Regression) and the lowest was 36.2% (Multi discriminant analysis).

More recently, Bahrammirzaee (2010) had reviewed 281 papers from 1989–2009 on Artificial Intelligence (AI) applications in finance: Artificial Neural Networks, Expert System and Hybrid Intelligence Systems. In his research, he compared many different methods to investigate which method performs better, and the result was summarized in table below. In short, this research showed that the accuracy of these AI methods is superior to the traditional methods, especially for the nonlinear patterns of the financial problems (Table 2).

Table 2. Comparison of model's performance

Research	Method compared with	Result
1. Baesens et al. (2003)	BPNN with LR, MDA, LS-SVMs	LS-SVMs and BPNN performs better
2. Abdou et al. (2008)	PNN and MLP with DA, Profit and LR	PPN and MLP performs better
3. Desai et al. (1997)	MLP and Modular NN with LDA and LR	ANN performs better than LDA and almost the same as LR
4. West (2000)	5 NN models: MLP, MOE, RBF, LVQ, and fuzzy adaptive resonance with LDA, LR, k-nn, kernel density estimation and CART	LR performs better than ANN
5. Ong et al. (2005)	GP with ANN, C4.5, CART, rough sets and LR	GP performs better
6. Lee (2007)	BPNN with SVM, MDA, and CBR	SVM performs better

Source: Bahrammirzaee (2010)

Abbreviation: *BPNN Backpropagation neural networks; C4.5 Extension of CART and ID3; CART Classification and regression trees; CBR Case-based reasoning; DA Discriminant analysis; FA Factor analysis; K-nn K nearest neighbor; LDA Linear discriminant analysis; LR Logistic regression; LS-SVMs Least squares support vector machines; LVQ Learning vector quantization; MDA Multiple discriminant analyses; MLP Multilayer perceptron; MOE Mixture-of-experts; PNN Probabilistic neural networks; RBF Radial basis functions; SVM Support vector machine.*

2.5.2   Domestic Study

In Vietnam, the study of internal credit rating by using artificial intelligence is limited. There is no specific research paper describes the use of artificial intelligence in Vietnam market. The limitation may because of the available information in Vietnam data market. In addition, the technique for analyze the credit risk or to predict the credit rating was reached statistical method.

Ha (2011), made a research on internal credit rating system of Habubank. The research conducted in a sample of 50 companies in Construction and Trading. It applied the logistic regression to determine the independent variable that has high coefficient to the dependent variable. The research found out a new regression that includes most affected independent variables on the credit score.

The research was not intended to focus on testing the level of accuracy for the model.

In 2012, Chi conducted a research for the case of BIDV Binh Dinh. The research focused on improving the operational risk, human resource management and managing of information system.

Although there were not many researchers investigated on analyzing specific internal credit rating system, there were a few papers that intended to predict the credit rating. For example, Giang (2014) investigated the capital adequacy level for Vietnam banking system. The research emphasized the use of Collateralized Debt Obligation, Merton model and Monte Carlo Simulation. The result allowed the paper to predict credit rating of well-known public firms such as Vinamilk, Masan Corporation, Hoang Anh Gia Lai Corporation, etc. Similarly, Tram (2015) did a research to quantify the credit loss of Vietnam banking system in 5 years by computing Value at risk (VaR) and was also successful in predicting the credit rating of others public firms.

3   Research Methodology

3.1   Research Method

The research primarily employs quantitative method to answer the research question. The type of learning for artificial neural network is supervised style (output desired are known).

3.2   Data Collection

Training and Validating Data
The training data collected belongs to 33 firms in 14 different industries and is collected from BIDV Binh Duong for the year of 2014. Each firm data includes 54 independent variables (14 financial variables, 40 non-financial variables) and a dependent variable (credit score). For non-financial variables, the author eliminates the variables that are not belonging to the common criteria of the data set. Thus the numbers of independent variables reduce to 37 variables (14 financial variables and 23 non-financial variables).

Validating data made up with 10% of the training data and are intended to be split up from the training data for validating function i.e. training data and validating data are independent.

Reclassifying Credit Score
For small date set, it is recommended to re-classify the credit rating categories. In addition, less category increases the prediction ability. Dutta and Shekhar (1998) reclassified 8 credit rating categories to "AA" and "non-AA", number of data point was 47. Similarly, Simon (2009) reclassified 8 credit rating categories into 4 type of categories "T", "L", "H", "M", the number of data point was 205 companies. As mentioned in Research Methodology, 9 levels of credit score is

reclassified into **Group I** (Very good credit score): AAA, AA; **Group II** (non-I, not desirable credit score): A, BBB, BB, B, CCC, CC, C.

For the purpose of the research internal credit rating score, financial parameter, and non-financial have the following abbreviations (Tables 3 and 4):

Table 3. Abbreviation for financial variables

Liquidity ratios	
FL1	Current ratio
FL2	Quick ratio
FL3	Cash ratio
Turnover ratios	
FT1	Account receivable turnover
FT2	Inventory turnover
FT3	Account payable turnover
FT4	Fixed asset turnover
Leverage	
FLE1	Debt to asset ratio
FLE2	Long-term debt to equity ratio
Profitability	
FP1	Gross profit margin
FP2	Operating profit margin
FP3	Return on equity
FP4	Return on asset
FP5	Interest coverage

For this research, the group of non-financial factor indicates the Credit Relationship with BIDV is eliminated for training data. This group of non-financial factors will be ignored.

Testing Data
Testing data is collected from vietstock.vn. The data set are consisted of 37 companies from Construction (19 companies) and Timber (18 companies) industries. Due to the complicated in assessing non-financial parameters of different industries, the author aims to investigate on two industries that made up roughly 60% of the loan outstanding at BIDV Binh Duong.

3.3 Data Preprocessing

Before creating the representative model, we have to prepare the data that could show the traits of the model. The preparation of the data follow 3 step: removing outliers; replacing missing value; replacing correlated variables.

Table 4. Abbreviation for non-financial variables

Ability to repay debt by cash flows	
NFCF1	Ability to repay the medium and long-term principal debt
NFCF2	Source of repayment of the client according to the credit officers
Qualified management and internal environment	
NFIC1	Judicial record of the head of the enterprise/chief accountant
NFIC2	Professional experience of the director of enterprises
NFIC3	Educational attainment of the director of enterprises
NFIC4	Operating competence of the director of enterprises, according to the LO
NFIC5	Relations of the Board with the relevant authorities
NFIC6	The dynamics and the responsiveness of business leaders with the change of the market (according to the LO)
NFIC7	Internal control environment of enterprises, according to the LO
NFIC8	HR internal environment of the enterprise
NFIC9	Vision, business strategy of the company during the period 2–5 years
The external factors	
NFEF1	Future prospects of sector
NFEF2	The ability for market entry of new firms, according to the LO
NFEF5	The protectionist policies/incentives of state
NFEF7	The dependence of the business activities of companies in the natural conditions
The other operational characteristics	
NFOF1	Dependence on a few suppliers (input source)
NFOF2	Dependence on a small number of consumers (outputs)
NFOF3	The average growth rate of sales of the company in the last 3 years
NFOF4	The average growth rate in the profit (after tax) of businesses in the last 3 years
NFOF7	The prestige of the enterprise to the consumer
NFOF9	The influence of the fluctuation of personnel to operate their businesses in the last 2 years
NFOF10	Access to capital
NFOF11	Development prospects of the company, according to the LO

Cleaning Outliers

Outliers can be caused by measurement errors or are genuine observations that differ significantly from the rest of the observations for unknown reasons. Outliers can severely distort the results if they are included in a model. To remove outlying values from the data set, we set up upper limit and lower limit for each category. The upper limit is three standard deviations plus the mean while the lower limit is the mean subject three standard deviations. Each value will be compare with their category upper and lower limit. Because of the limitation on number of data point, the outlying value will be replaced with the nearest upper or lower limit. This cleaning technique had proved itself to be useful by Samarsinghe (2006). See Appendix B. After cleaning the outliner, the number of data point still remains at 33.

Replacing Missing Values

Financial data usually have incomplete data which lead to missing value. According to Frank (2009) and Angelini et al. (2008) and Fethi and Pasiouras (2010), companies with more than 10% of their financial data missing will be removed from the data set. If we choose to keep these data, we have to replace it with the mean data of that category. See Appendix C.

Missing value can arise due to unavailable measurements for specific observations.

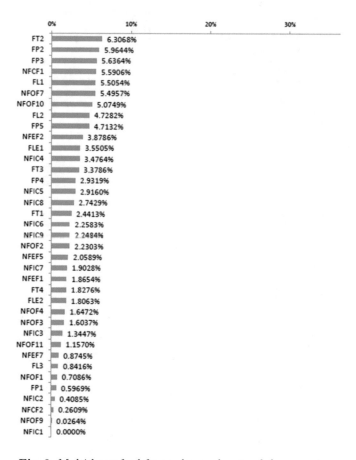

Fig. 9. Multi-layer feed-forward neural network long version

Remove Highly Correlated Data
Highly correlated variable $(0.7 < |\text{correlation}| < 1)$ is eliminated. The eliminated variables are ones that has weaker impact compares to its correlating variables. See Appendix D.

After determining the correlation between independent variables, the correlated variables will be removed by using variables impact. After training the artificial neural networks, each neural network will generate a variable impact table, which measures the "strength" of each variable on the artificial neural network. The variable impact of neural network is similar to the coefficient of a regression. The results of the analysis can help in the selection of a new set of independent variables, one that will allow more accurate predictions. For example, a variable with a low impact value can be eliminated in favor of some new variable (Figs. 9 and 10).

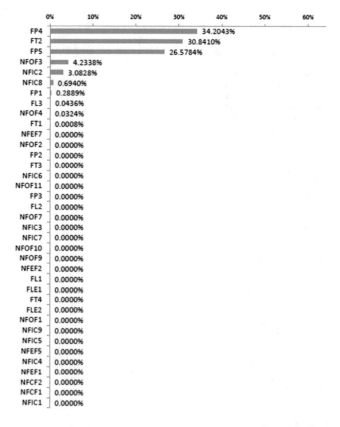

Fig. 10. Probabilistic neural network long version

- *Variable impact is an analysis created to measure the sensitivity of net predictions to changes in independent variables.*
- *Long version of Artificial Neural Network implies the artificial neural network before the omitting of highly correlated variable (long version or more independent variables). The new artificial neural network is assigned as short version.*

According to correlation value, a set of highly correlated variable (correlation value > 0.7) is collected. Table 5 represents a brief correlation matrix of highly correlated variables. Full correlation matrix is shows in Appendix D.

Table 5. Brief summary of highly correlated variables

	FL2	NFCF2	FP4	NFIC8	NFOF10
FL1	0.981				
FL3	0.763				
FP3		0.718	0.791		
FP4		0.73			
NFIC4				0.822	
NFOF11					0.755

Table 6 displays a set of omitted variables for each artificial neural network.

Table 6. Elimination of highly correlated variable of MLF and PNN

Variable Impact on MLF		Variable Impact on PNN	
FP3	5.6364%	FP4	34.2043%
FL1	5.5054%	NFIC8	0.6940%
NFOF10	5.0749%	FL3	0.0436%
FL2	4.7282%	NFOF11	0.0000%
NFIC4	3.4764%	FP3	0.0000%
FP4	2.9319%	FL2	0.0000%
NFIC8	2.7429%	NFOF10	0.0000%
NFOF11	1.1570%	FL1	0.0000%
FL3	0.8416%	NFIC4	0.0000%
NFCF2	0.2609%	NFCF2	0.0000%

Crossed Variables are ones to be eliminated. The rule is starting from the variable that has smallest impact and compare with its highly correlated variable the smaller will be omitted. A removed variable will be considered to have no highly correlated relationship with other variables. After omitting highly correlated variables, the number of independent variables reduces to 31 for each neural network. The new artificial neural network is known as short version.

3.4   Research Model

The research model consists of 3 stages:

1. *Training:* Train the neural network with the given data set from BIDV.
2. *Validating:* test the accurate of the neural network by validating 10% of the given data set.
3. *Testing:* collect the inputs on a number of firms that operate in Construction and Timber. Then, using the trained artificial neural network to predict/ classify the internal credit rating score.

The process of training, validating and testing data is operated in Neural-Tools 7. In testing stage, the author got support from a Risk Management expert at BIDV in assessing non-financial data of testing firm.

3.5   Stage 1 – Training Artificial Neural Network

After collecting data and preprocessing it. The new data set is trained by Neu-ralTool with 2 type of artificial neural network (Multi-layer feed-forward and Probabilistic Neural Network). The trained neural network, then, will validate on the rest 10% of untrained data set (Table 7).

Table 7. Setting for training process

Setting	Multi-layer feed-forward	Probabilistic neural network
Number of nodes	MLF long version: 7 nodes MLF short version: 6 nodes	No setting required
Number of layers	Automatically detected	Automatically detected
Create data set	The input data have to be classified as dependent or independent, numerical or categorical type	
Selected sample case	The sample case account for 10% of given data set	
Choose type of neural network	Multi-layer feed-forward and probabilistic neural network is training separately with both long and short variable data set *Long variable data set:* data set without removing highly correlated variables *Short variable data set:* data set with removing highly correlated variables	
Setting up run times or number of trials	Number of trials set up is 1,000,000	Automatically detected
Calculate variable impact	Require NeuralTools to including variable impact computation	

3.6   Stage 2 – Testing

Preparing Testing Data
The trained neural network will be automatically saved by NeuralTools.

Testing data set includes 14 financial variables and 37 non-financial variables. 14 financial variables and one non-financial variable are calculated from the Financial Statement of 37 firms (19 Construction companies, 18 Timber companies). For financial variables, the author computes the financial ratios for each variable respect to the regulation of BIDV. For non-financial variables, the author assessing this aspect with the supports of a credit risk management expert from BIDV (Table 8).

Table 8. Settings for testing process

Settings	Multi-layer feed-forward	Probabilistic neural network
Create data set	Similar to training stage, the testing stage also required the variables to be classified as dependent or independent, numerical or categorical type	
Choose trained neural network	The trained neural network are saved by NeuralTools under specific name will be chose to test on the testing data set. There are 4 types of trained neural network, Multi-layer Feed-forward long and short version, Probabilistic Neural Network long and short version	

4   Data Analysis and Discussion

4.1   Data Set Overview

Training Data
See (Table 9).

Table 9. Training data summary

	Quantity	Loan outstanding (%)
Construction	7	30
Infrastructures	3	5
Food and Beverage	1	6
Heavy Industry	3	5
Stationaries	2	1
Building Materials Producing	1	5
Light Industry	2	5
Timber	5	27
Transportation	1	6
Real Estate	3	5
BOT	1	2
Fertilizer	1	0.50
Leather Shoe Producing	1	1
Electric Devices	2	2
Total	33	100

Testing Data
See Appendix A.

4.2 Descriptive Finding

Figure 11 shows level of accuracy for each artificial neural network on a specific industry. Base on the testing result, it is considered that Multi-layer Feed-forward short version predicts the bond rating at highest level of accuracy at 86% in overall.

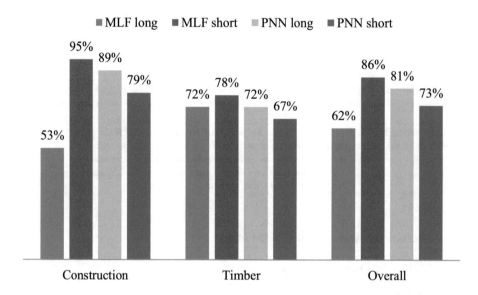

Fig. 11. Descriptive finding

4.3 Result Analysis

The result is analyzed on both the long version and short version for each neural network. After eliminating several variables by variable impact analysis, the author intentionally to discover whether the artificial neural network does better or does worse.

4.4 Accuracy Level

Level of accuracy is measure base on the number of right prediction on the number on predictions. The number of right predictions is determined by matching the predicting result with Credit Information Center (CIC) result. CIC is a state-owned external credit rating agency that operates under the management of the State Bank of Vietnam. Every year, CIC rates credit score for around 1000 firms either listed or unlisted firm from different industries. CIC assesses the companies by using the information of the whole banking system. Every time, a company decides to have a credit relationship with a commercial bank (both private and state-owned), all the information that the company applies to a specific bank, including the value of loan-outstanding and the amount of bad debt exists, will be sent through a system that supervised by the State Bank of Vietnam.

In Vietnam, the internal credit rating system of "big" commercial bank such as Vietinbank, Vietcombank, Agribank and BIDV have to be accepted by the State Bank of Vietnam. A qualify internal credit rating system has to follow the regulation and qualified the standard set up by the State Bank of Vietnam. Because of this characteristic, this external credit rating agency CIC is different from famous external credit rating agency such as Moody, S&P or Fitch, in which it make the internal credit rating score of BIDV to be comparable with ones of CIC. Thus, in the research the accuracy level is determined by comparing the number of matching result with CIC.

4.5   Summary of Training Data

Figures 12 and 13 graphically displays the training data set before and after reclassifying.

Rating category	Quantity
AAA	1
AA	14
A	15
BBB	2
B	1
Total	*33*

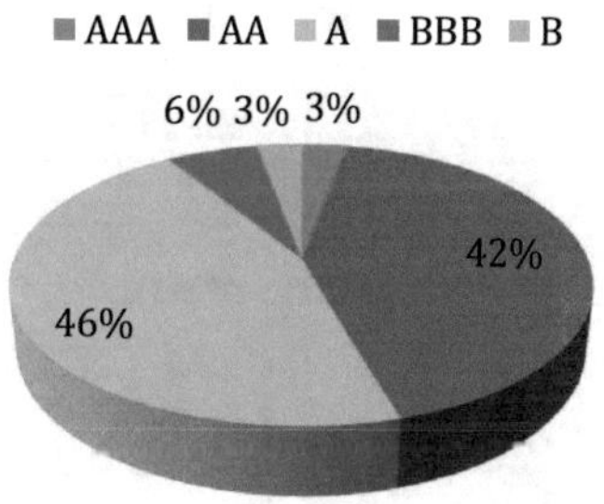

Fig. 12. Summary of training data

Rating category	Quantity
I	15
II	18
Total	*33*

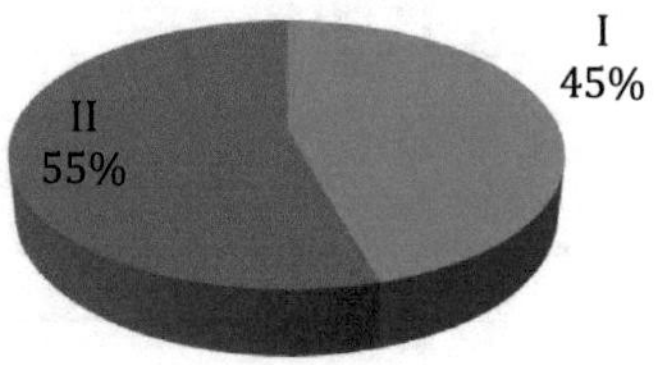

Fig. 13. Summary of training data after reclassifying

4.6   Prediction Result

The training process includes two small stages: training and validating. The data set of 33 data point will be split into 2 group, 30 (90%) of them are belong to training group and the remaining 3 (10%) of them are belong to validating group. In training process, while the neural network learns the pattern of the data set, it also tests on the training data set itself to figure out the "best" case of algorithms to make the prediction. After that internal testing, the "best" of training is applied to classify the validating group. Finally, the appropriate is defined and saved for the testing process.

Construction

Table 10. Testing result for construction sector

	MLF long	MLF short	PNN long	PNN short
% Bad Predictions	47.37%	5.26%	10.53%	21.05%
Mean Incorrect Probability			12.81%	24.25%
Std. Deviation of Incorrect Probability			31.42%	40.41%

Table 11. Detail testing result for construction sector.

Company	CIC	Reclassified CIC	MLF long	MLF short	PNN long	PNN short
CTD	AAA	I			Bad	
HBC	BB	II				
TDC	BB	II	Bad			
CTX	B	II	Bad			
THG	A	II		Bad		Bad
SC5	BB	II				Bad
PXI	BB	II				
CDC	CCC	II	Bad			
CSC	B	II	Bad			
UDC	B	II				Bad
HU1	CCC	II	Bad			
D11	BB	II	Bad		Bad	Bad
DC4	BB	II				
DIH	BB	II	Bad			
CIG	B	II				
DLR	C	II	Bad			
BHT	BBB	II				
CX8	BB	II				
DC2	B	II	Bad			

Timber

Table 12. Testing result for Timber sector

	MLF long	MLF short	PNN long	PNN short
% Bad Predictions	27.78%	22.22%	27.78%	33.33%
Mean Incorrect Probability			23.09%	31.74%
Std. Deviation of Incorrect Probability			38.76%	45.17%

Table 13. Detail testing result for Timber sector

Company	CIC's	Reclassified CIC	MLF long	MLF short	PNN long	PNN short
GDT	AA	I				Bad
AAA	AA	I			Bad	
BMP	AAA	I	Bad		Bad	
BRC	AAA	I	Bad			
SRC	AA	I			Bad	
TTF	CC	II	Bad			
DLG	BB	II				Bad
DCS	BBB	II				
SAV	B	II				
ALT	BBB	II	Bad	Bad		
DNP	BBB	II		Bad		Bad
DPC	A	II			Bad	Bad
CSM	A	II				
DAG	BBB	II				
DRC	A	II				Bad
DTT	A	II	Bad	Bad		Bad
RDP	BBB	II				
TPC	BBB	II		Bad		

Tables 10, 11, 12, and 13 display the predicting result for both Construction and Timber sector. For MLF, the short version of it, reduce the % bad prediction compare to the original MLF. However, after removing highly correlated variable for PNN, mean and standard deviation of Incorrect Probability increases approximately 10% for each sector. In PNN, the mean and standard deviation indicate the level of distance that is wrongly predicted there for affecting the prediction result.

Comparison of Multi-Layer Feed-Forward and Probabilistic Neural Network

See (Table 14).

Table 14. Comparison of MLF and PNN

Multi-layer feed-forward neural network	Probabilistic neural network
• Faster prediction • More reliable in predicting variable that out of range • They are capable of generalizing from very small training sets	• Faster training • Not require setting layers • Enable to calculate the probability

4.7   Bond Rating Prediction of Construction Sector

For Construction, Multi-layer Feed-forward neural network has the highest accuracy level in predicting the bond rating for this industry. See Fig. 14.

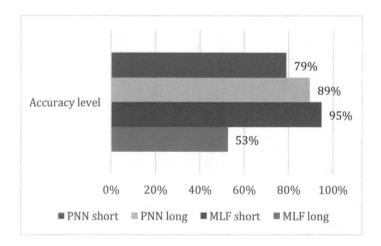

Fig. 14. Accuracy level for predicting bond rating of Construction sector

4.8   Bond Rating Prediction of Timber Sector

It is surprising that Multi-layer feed forward short version has high level of accuracy in prediction for both construction and timber. See Fig. 15.

Table 15 shows the research result in compare with empirical studies that using the similar technique and number of classifying labels. The research result accuracy level of prediction falls between the lowest and highest value, which can be reliable and reasonable Atiya (2011).

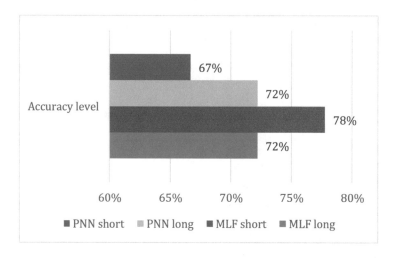

Fig. 15. Accuracy level for predicting bond rating for Timber sector

Table 15. Comparison of result

Empirical studies	Label	Method	Sample size	Data	Accuracy
Dutta and Shekhar (1988)	2	BP	47	US	83.30%
Singleton and Surkan (1990)	2	BP	126	US	88.00%
Result	2	BP	33	VN	86.49%

In overall, the prediction ability of Multi-layer Feed-Forward neural network is the most powerful to accurately compare the prediction power between statistical methods and Artificial Neural Networks on current research data, additional statistical methods can be added such as Linear Regression or Multiple Discriminant Analysis (Atiya 2011; Boyacioglu et al. 2009; Cherubini et al. 2014). However, the similarity between the research data and the data set from Dutta and Shekkar (1988) and Singleton and Surken (1990) studies suggested lower rates of accuracy prediction which were 64.7% (Linear Regression) and 39% (Multiple Discriminant Analysis) compares to Probabilistic Neural Network. Especially, MLF short version is the most outperforming network with the accuracy level on prediction of Construction, Timber and overall reaches the highest at 95%, 78% and 86%, respectively. Therefore, Multi-layer Feed-forward short version is the most suitable artificial neural network to replicate the internal credit rating system of BIDV.

4.9   Variables Impact Analysis of Multi-layer Feed-Forward Neural Network Short Version

Figure 16 graphically displays the variable impact on the Multi-layer Feed-forward short version.

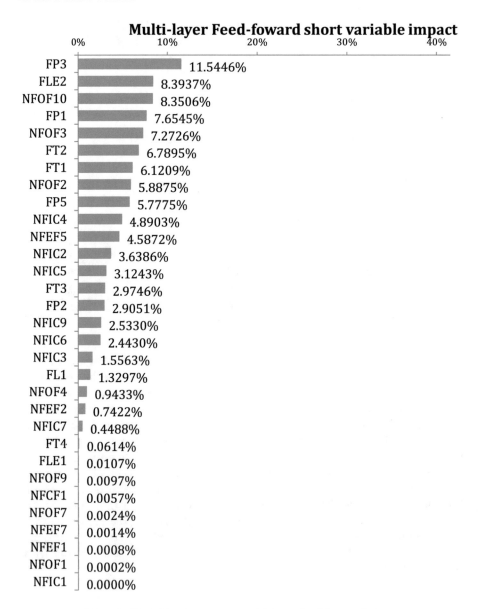

Fig. 16. Variables impact on multi-layer feed-forward short

The variable impact on the MLF short is distributed as follow:

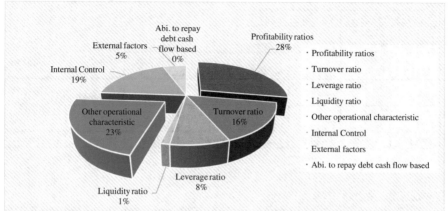

Hot colored areas represent for financial criteria and green ones are non-financial criteria. The "total impact" of financial and non-financial criteria is summed up as follow:

Table 16. Weight of financial and non-financial criteria of MLF short and BIDV regulation

Criteria	MLF short	BIDV
Financial criteria	53.56%	35%
Non-financial criteria	46.44%	65%
Total	100.00%	100%

Table 16 displays the weight of financial and non-financial criteria of MLF short and BIDV regulation. According to BIDV regulation, the weight of non-financial criteria has to be heavy than financial criteria. However, the variable impact analysis show the different result, in which financial criteria and non-financial one are comparable and even the weight of financial criteria is heavier. From the result, it is inferred that the information of non-financial criteria required for assessing is not widely available in Vietnam data market, which is reasonable. BIDV has to conduct a research of the customer itself for all available information, which is costly and they have to devote their time for each customer. The internal credit rating process required the latest update of information of either old or new customer. Because of that features, BIDV may not fully digging the non-financial information of a specific company. Therefore, the variable impact result is understandable for Vietnam data market.

Financial criteria

Base on the group variable impact in Table 17, BIDV raises their concern of priority in the order: ability of generating profit, the effectiveness in using current asset, debt structure and the ability to convert current asset into cash.

Table 17. Group variable impact of financial criteria

Profitability ratios		Turnover ratio		Leverage ratio		Liquidity ratio	
FP3	11.5446 %	FT2	6.7895%				
FP1	7.6545 %	FT1	6.1209%				
FP5	5.7775 %	FT3	2.9746%	FLE2	8.3937%		
FP2	2.9051 %	FT4	0.0614%	FLE1	0.0107%	FL1	1.3297%
Total	**27.88%**	**Total**	**15.95%**	**Total**	**8.40%**	**Total**	**1.33%**

Profitability Ratios

Profitability ratio is most concerned by bank disregard of the industry. For BIDV, the regulation for profitability ratios for construction and timber is showed in Table 18.

Table 18. Weight of each variable in profitability ratios from Construction and Timber sector

		Construction	Timber
FP3	Return on equity	4%	3%
FP1	Gross profit margin	6%	5%
FP5	Interest coverage	5%	4%
FP2	Operating profit margin	6%	5%

According to BIDV regulation, FP1 and FP2 have the highest impact on the group of profitability ratio. FP5 comes second and final is FP2. However, the variable impact analysis – analyze from the actual sample of BIDV, the order of priority is different, specifically, FP3 is the most important, then FP1- FP5-FP2. This can be inferred that the assessing of FP3 (return on equity) is important in assessing the ability of a firm position in competitive market. Normally, through ROE, BIDV can assess the "investment opportunity" of a specific firm, especially, comparing with its peers or the average of that industry. A good ROE should be greater than the cost of equity capital. The second aspect that BIDV give their attention is FP1 (gross profit margin). Gross profit margin derives its meaning by comparison the performance of the company over the past year and industry average as well as to consider the trend of the profit over a period. The purpose of this ratio is to determine whether a firm has sustainable growth or has a declining the ability to making profits. By assessing this ratios, BIDV can detect "potential customer" itself. After assessing "investment opportunity" and "potential customer", BIDV wants to know whether the customer's financial ability is stable enough to cover the interest payment. That is the reason why the FP5 (interest coverage ratio) comes after. Finally, BIDV takes care of the operating profit of the firm as well as a means to determine the "amount of investment" on that firm.

Turnover Ratios

Turn over ratios measure the efficiency of the firm in its daily operation. Table 17 shows the order of Turnover ratios to be assessed by BIDV.

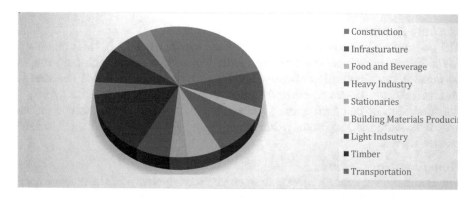

Fig. 17. Quantity of company in percentage of training data set

The order for each variable e on BIDV internal credit rating score is found as follow: (1) FT2: Inventory turnover, (2) Account Receivable turnover, (3) Account payable turnover, (4) Fixed asset turn over. The impact of inventory turnover and account receivable are equivalent and reach the highest impact compares to others in the Turnover ratio group.

BIDV pay attention to the pace of "clean up" or converting the inventory and account receivable into cash. At their first assessing to a firm (all the firms for training and testing was taken out the Relationship with BIDV criteria), it is understandable.

BIDV pay more intention in Account Payable turnover than the fixed asset turnover. The account payable turnover primarily presents a firm ability to negotiate with their supplier or to assess any "bargaining power" of the client, or in another aspect, it can be considered to the ability of the firm to repay their supplier. There are more accounts involved to detect what the situation is. One common method is to assess the Cash Conversion Cycle (CCC) of the firm. The impact of fixed asset turn over can be considered as no impact. There may be is a explanation for this number.

Figure 17 displays the portion of quantity in the training data. Industry that has more company involve with BIDV are not very concern with the assessing of fixed asset turnover. Therefore, the impact of that variable on the internal credit score is not significant.

Leverage Ratios

Following Table 17 the level of significant is from Long-term debt to equity ratio (FLE2) down to Debt to asset ratio (FLE1). Debt to asset ratio can be inferred as not-relevant to the internal credit score. For Long-term debt to equity ratio, BIDV want to know further in the ability of a firm to survive in long-term period. Small FLE2 indicates that the firm has the ability to raise the debt capital if it needed, which is made it has a strong financial ability and well survive under bad condition of economy.

Liquidity Ratios

FL1 stands for the current ratio (current asset/current liability), this ratio itself, has 1.3% of impact on the internal credit score. The draw back here maybe because of the elimination stage, other liquidity ratios was picked out due to high correlation. However, current ratio is a primary ratio or in a funny way is the "first step of being accepted", for which BIDV will raise stereotype on the firm if its value is less than 1. For most firms, the current ratios are always greater than 1.

Non-financial Criteria

The non-financial criteria mostly base on the incompetence of the credit expert and seem to be difficult to explain. Therefore, for non-financial criteria, the author intends to analyze the variable that has the impact level greater than 1%.

In overall, the group with the highest impact is "Other operational characteristic" and the client "Internal Control" and last is the "External factors". Two mostly non-relevant variables such as "External Factor" and "the ability of the firm to repay the debt base on it cash flow" will not be mentioned in the evaluation.

Table 19. Group of non-financial criteria impact on MLF short

Other operational characteristic		Internal Control		External factors		Abi. to repay debt cash flow based	
NFOF10	8.3506%	NFIC4	4.8903%	NFEF5	4.5872%	NFCF1	0.0057%
NFOF3	7.2726%	NFIC2	3.6386%	NFEF2	0.7422%	**Total**	0.01%
NFOF2	5.8875%	NFIC5	3.1243%	NFEF7	0.0014%		
NFOF4	0.9433%	NFIC9	2.5330%	NFEF1	0.0008%		
NFOF9	0.0097%	NFIC6	2.4430%	**Total**	5%		
NFOF7	0.0024%	NFIC3	1.5563%				
NFOF1	0.0002%	NFIC7	0.4488%				
Total	22.47%	NFIC1	0.0000%				
		Total	19%				

Other Operational Characteristic

The operational characteristic is accounted for 22.47% of the variable impact, which is understandable. While this group of criteria assessing the outside business environment of the firm, the stabilization and the growth of specific economic segment are important and sensitivity of the firm with the change of the economy.

As displaying on Table 19, other operational characteristic includes accessibility to capital (NFOF10), the average growth rate of sales of the company in the last 3 years (NFOF3), dependence on a small number of customers (NFOF2).

The accessibility to capital and the average growth rate of sales of the company is approximately affects the internal credit scoring of the same level. BIDV desires it customer to have the ability to access more sources of capital, which is a way to reduce the credit risk for BIDV. The company that has more ability to access capital can be considered as an "active" firm, which can process the debt in time, or in other aspect, the firm has its potential for growing fast in future or being supported by some Government policy and regulation. The average growth rate of sales (NFOF3) is important for assessing the firm ability to have a growth in sale. This type of criteria is mostly for the assessing of stable growth in sales as well as the financial ability. It can also be a evidence for comparing with the actual growth.

Internal Control

Internal control represents a group of criteria that evaluating the control inside a potential client. Table 19 shows the variable impact in order of affection. Beside the sensitivity of the firm with outside environment, BIDV desired to know how strong the firm can stand by itself when the outside environment changes. The ability of the board of director in managing heavily assessed. The variable impact for the competency of the director (NFIC4), professional experience of the director (NFIC2), the relationship of the director with local Government (NFIC5) is account for 11.65%. In business, the experience is top priority for the management, specifically; the education level of director (NIFC3) has the lowest variable impact in this group (regardless of one that has less than 1% affected). BIDV assess the ability of the director primarily through their experience and their relationship in their field of business as well as with other Government Agencies.

The vision and the business strategy in long-term 2 to 5 years (NFIC9) and the dynamic of the director under the change of the market (NFIC6) comes second as a pairs of determining how strong the strategy of the firm for long-term period. The strategy should be specific enough to let the firm have a position in the market and able to adapt different type of working culture. Furthermore, BIDV observes the response of the director under specific economic condition or hard time of the director. This assessment is also an important criterion to evaluate the stabilization of the enterprise.

5   Conclusion

5.1   Discussion of the Result

The research carried out to replacing the internal credit rating system of Bank for Development and Investment Vietnam for Construction and Timber sector. The research result indicated that the short version of Multi-layers Feed-forward neural network was successfully classified two group of companies. Group I includes companies that have high credit rating class (AAA, AA), group II consists of medium and low credit rating class (A, BBB, BB, CCC, CC, C, D) – also referred as non-I. Multi-layer neural network short version was successfully predicting the bond rating at 86.49% accuracy which can be considered as a highly accurately prediction (Lopez and Saidenberg 2000; Nazari et al. 2000; Ghatge et al. 2013; Khemakhem et al. 2015).

Specifically, the research utilizes the artificial neural network in predict internal credit score of 37 companies that specialize in Construction and Timber. The use of artificial neural network on predicting bond includes complicating process. Especially, the pre-processing stage of data required a careful step of removing outliner and replace missing data. The ability of a net to predict accurately is primarily base on pre-processing stage. When the data is ready, it will be put into the artificial neural network. Afterward, the artificial neural network learns the pattern of the input data and generates a neural network – also called trained net to be ready for testing the result. In testing stage, the trained neural network is used to classify the new data set with the same variables. The research not only gives its prediction on bond rating but also reflect the impact of specific rating criteria on the credit score. In which, it is believed that the financial criteria plays an important role than it is supposed to be and the non-financial criteria are overstated by the BIDV regulation. The difference between the regulation and the actual result can be inferred that the non-financial criteria actually have a strong impact on the credit score than financial ones do. However, Vietnam data market is not well organized and has a lots of asymmetric information there for the use of non-financial criteria is not effective but it still has strong impact on the credit score. The method of using Artificial Neural Network (ANN) does have some limitations. (1) available data, the accuracy of an (ANN) depends on the data set where in Vietnam such type of data is not easily accessible (2) long time of training, the accuracy of the net proportionately increases with the number of observations, normally, number of observation reaches 1,000,000 or more, which requires weeks of training the neural network.

5.2   Recommendation

BIDV should find a firm that specialized in searching non-financial information of firm to reduce the cost as well as the time.

The Internal credit rating system of BIDV does not have high level of classification for small sector between within an industry as well as updating new business sector for the internal credit rating system. The rules for assessing financial criteria should be more involve to non-financial criteria instead of evaluating separately. For example, new business sector, with the product can be changeable and has short life cycle such as manufacturing of electricity devices or laptop, the weight that measure the ability of management of the board of director has to be higher than that of traditional sector.

BIDV currently fixes the "proportion" of the specific criteria to the credit score, which means that specific criteria will be account that amount of important under any conditions. To make the credit scoring system more objective and appropriate in prediction, the proportion of specific criteria should increase of decrease its significance under specific condition.

5.3   Further Research

Artificial Neural Network is applied to assess the internal credit rating of Bank for Development and Investment Vietnam (BIDV). The artificial neural network applied includes the Multi-Layer Feed-Forward neural network and the Probabilistic Neural Network. The research is conducted with the internal data set that provided by BIDV Binh Duong and the result is compared with the credit score of Credit Information Center (CIC) of Vietnam. Therefore, this is the first time to apply the artificial neural network on predicting bond rating. The further study can utilize the technique of using artificial neural network on prediction or making a comparative research, investigating either the prediction ability of artificial neural network (or other type of artificial intelligence) with the tradition statistical method.

Appendices

Appendix A

Construction Companies	
CTD	Cotec Construction Joint Stock Company
HBC	Hoa Binh Construction & Real Estate Corporation
TDC	Binh Duong Trade And Development JSC
CTX	Vietnam Investment Construction and Trading JSC
THG	Tien Giang Investment And Construction JSC
SC5	Construction Joint Stock Copany No 5
PXI	Petroleum Industrial & Civil Construction JSC
CDC	Chuong Duong Joint Stock Company
CSC	Thanh Nam Contruction and Investment JSC
UDC	Urban Development & Construction Corporation
CTD	Cotec Construction Joint Stock Company
HBC	Hoa Binh Construction & Real Estate Corporation
TDC	Binh Duong Trade And Development JSC
CTX	Vietnam Investment Construction and Trading JSC
THG	Tien Giang Investment And Construction JSC
SC5	Construction Joint Stock Copany No 5
PXI	Petroleum Industrial & Civil Construction JSC
CDC	Chuong Duong Joint Stock Company
CSC	Thanh Nam Contruction and Investment JSC
UDC	Urban Development & Construction Corporation

Timber company	
GDT	Duc Thanh Wood Processing JSC
TTF	Truong Thanh Furniture Corporation
DLG	Duc Long Gia Lai Group Joitn Stock Company
DCS	Dai Chau Group Joint Stock Company
SAV	Savimex Corporation
AAA	An Phat Plastic & Green Environment JSC
ALT	Alta Company
DNP	Dongnai Plastic Joint – Stock Company
DPC	Da Nang Plastic Joint Stock Company
BMP	Binh Minh Plastic Joint-Stock Company
BRC	Ben Thanh Rubber Joint Stock Company
CSM	The Southern Rubber Industry JSC
DAG	Dong A Plastic Joint Stock Company
DRC	Danang Rubber Joint Stock Company
DTT	Do Thanh Technology Corporation
RDP	Rang Dong Plastic JSC
SRC	Sao Vang Rubber Joint Stock Company
TPC	Tan Dai Hung Plastic Joint Stock Company

Appendix B

APPENDIX B

Credit class	Statistic	FL1	FL2	FL3	FT1	FT2	FT3	FT4	FLE 1	FLE2
I	Mean	2.87	1.86	0.33	1.79	16.17	5.84	10.39	57.94	40.56
	Std. Dev.	5.68	4.42	0.50	1.67	51.67	5.29	14.00	22.02	65.09
	Upper Limit	19.90	15.12	1.83	6.80	171.19	21.71	52.40	124.01	235.82
	Lower Limit	-14.17	-11.40	-1.16	-3.22	-138.85	-10.04	-31.61	-8.13	-154.71
II	Mean	2.87	1.86	0.33	1.79	16.17	5.84	10.39	57.94	40.56
	Std. Dev.	5.68	4.42	0.50	1.67	51.67	5.29	14.00	22.02	65.09
	Upper limit	19.90	15.12	1.83	6.80	171.19	21.71	52.40	124.01	235.82
	Lower limit	-14.17	-11.40	-1.16	-3.22	-138.85	-10.04	-31.61	-8.13	-154.71

Credit class	Statistic	FP1	FP2	FP3	FP4	FP5	NFCF 1	NFCF 2	NFI C1	NFIC2
I	Mean	17.82	3.22	6.86	2.21	7.27	4.97	4.88	5.00	4.88
	Std. Dev.	19.90	21.24	21.25	5.38	12.60	0.17	0.70	0.00	0.33
	Upper Limit	77.52	66.94	70.62	18.34	45.08	5.49	6.97	5.00	5.87
	Lower Limit	-41.88	-60.50	-56.91	-13.93	-30.54	4.45	2.79	5.00	3.88
II	Mean	17.82	3.22	6.86	2.21	7.27	4.97	4.88	5.00	4.88
	Std. Dev.	19.90	21.24	21.25	5.38	12.60	0.17	0.70	0.00	0.33
	Upper limit	77.52	66.94	70.62	18.34	45.08	5.49	6.97	5.00	5.87
	Lower limit	-41.88	-60.50	-56.91	-13.93	-30.54	4.45	2.79	5.00	3.88

Credit class	Statistic	NFIC 3	NFIC 4	NFI C5	NFIC 6	NFIC7	NFIC 8	NFIC 9	NFE F1	NFEF 2
I	Mean	3.55	4.61	4.45	4.39	4.36	4.52	3.85	4.06	4.12
	Std. Dev.	0.90	0.61	0.90	0.79	0.65	0.67	1.00	0.66	0.74
	Upper Limit	6.26	6.43	7.17	6.76	6.32	6.52	6.86	6.04	6.34
	Lower Limit	0.83	2.78	1.74	2.03	2.41	2.51	0.84	2.08	1.90
II	Mean	3.55	4.61	4.45	4.39	4.36	4.52	3.85	4.06	4.12
	Std. Dev.	0.90	0.61	0.90	0.79	0.65	0.67	1.00	0.66	0.74
	Upper limit	6.26	6.43	7.17	6.76	6.32	6.52	6.86	6.04	6.34
	Lower limit	0.83	2.78	1.74	2.03	2.41	2.51	0.84	2.08	1.90

Credit class	Statistic	NFE F5	NFE F7	NFO F1	NFO F2	NFO F3	NFO F4	NFO F7	NFO F9	NFOF 10	NFOF 11
I	Mean	3.67	4.55	4.45	4.21	3.67	4.00	3.76	4.15	4.36	4.30
	Std. Dev.	0.69	0.51	1.03	1.11	1.55	1.58	0.71	1.00	0.55	0.64
	Upper Limit	5.74	6.06	7.56	7.55	8.33	8.74	5.88	7.16	6.01	6.21
	Lower Limit	1.59	3.03	1.35	0.88	-1.00	-0.74	1.63	1.14	2.72	2.39
II	Mean	3.67	4.55	4.45	4.21	3.67	4.00	3.76	4.15	4.36	4.30
	Std. Dev.	0.69	0.51	1.03	1.11	1.55	1.58	0.71	1.00	0.55	0.64
	Upper limit	5.74	6.06	7.56	7.55	8.33	8.74	5.88	7.16	6.01	6.21
	Lower limit	1.59	3.03	1.35	0.88	-1.00	-0.74	1.63	1.14	2.72	2.39

Appendix C

APPENDIX C

Class	FL1	FL2	FL3	FT1	FT2	FT3	FT4	FLE1	FLE2
I	1.03	0.13	0.07	0.30	0.22	5.52	6.61	84.51	82.12
I	1.09	0.85	0.16	**2.27**	**16.32**	2.62	**14.29**	64.08	135.34
I	1.40	0.74	0.14	1.78	2.90	4.97	10.40	61.64	4.81
I	1.72	1.15	0.28	2.45	6.47	6.57	9.15	45.29	0.00
I	1.43	0.74	0.15	2.55	4.71	6.04	5.60	45.59	0.00
I	1.39	0.70	0.03	0.80	1.47	1.73	31.74	63.06	0.00
I	1.21	0.33	0.02	0.45	0.55	1.30	16.17	71.56	5.38
I	1.47	0.59	0.16	**2.27**	**16.32**	0.63	**14.29**	76.36	111.55
I	2.83	2.83	1.83	**2.27**	**16.32**	**6.64**	**14.29**	14.17	5.54
I	3.05	1.49	0.72	1.82	2.90	18.41	48.10	31.68	0.00
I	2.47	1.74	0.06	2.66	6.55	4.72	13.30	33.32	0.00
I	1.46	0.90	0.26	3.88	9.94	11.11	8.83	48.74	0.00
I	2.51	0.35	0.04	**2.27**	**16.32**	**6.64**	**14.29**	37.42	19.74
I	1.47	0.55	0.06	**2.27**	**16.32**	**6.64**	**14.29**	69.84	128.39
I	1.09	0.63	0.09	2.28	4.98	5.25	26.00	81.63	0.12
II	1.04	0.77	0.20	4.33	11.34	13.93	43.71	78.52	53.67
II	0.70	0.02	0.00	0.48	0.47	17.53	46.55	61.77	0.00
II	19.90	15.12	1.83	0.35	2.10	0.63	0.73	73.70	235.82
II	1.05	0.78	0.34	1.83	6.43	3.21	2.87	63.60	7.29
II	4.52	1.89	0.15	1.45	2.60	3.38	**14.29**	19.74	0.00
II	0.48	0.29	0.03	4.41	10.11	9.67	2.23	62.03	0.00
II	0.86	0.10	0.02	0.88	0.79	14.36	1.46	77.78	3.53
II	19.86	11.63	1.27	**2.27**	**16.32**	0.58	**14.29**	3.46	0.00
II	0.69	0.68	0.47	4.54	118.54	11.47	1.15	70.81	124.07
II	1.07	0.08	0.01	0.53	0.76	4.05	4.31	85.95	25.97
II	2.66	1.47	0.08	4.61	12.06	7.86	**14.29**	34.11	0.13
II	0.61	0.61	0.17	**2.27**	**16.32**	**6.64**	**14.29**	52.42	79.10
II	1.56	0.90	0.44	1.71	3.56	5.02	17.50	57.39	0.00
II	2.15	1.95	1.04	1.06	10.24	2.16	8.76	36.74	0.00
II	0.75	0.62	0.25	1.72	7.02	3.24	7.92	74.50	88.19
II	1.04	0.86	0.07	4.49	20.62	6.08	21.14	76.37	2.44
II	0.95	0.94	0.42	5.22	171.19	14.67	6.97	60.18	22.18
II	0.44	0.27	0.02	2.56	5.73	5.87	1.75	94.10	166.60

Appendix D

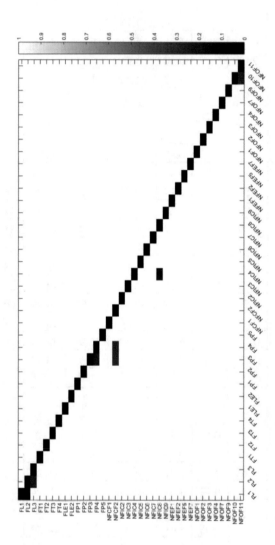

References

Abdou, H., Pointon, J., El-Masry, A.: Neural nets versus conventional techniques in credit scoring in Egyptian banking. Expert Syst. Appl. **35**(3), 1275–1292 (2008)

Angelini, E., Tollo, G.D., Roli, A.: A neural network approach for credit risk evaluation. Q. Rev. Econ. Finan. **48**(4), 733–755 (2008)

Atiya, A.F.: Bankruptcy prediction for credit risk using neural networks: a survey and new results. IEEE Trans. Neural Netw. **12**(6), 929–935 (2011)

Baesens, B., Van Gestel, T., Viaene, S., Stepanova, M., Suykens, J., Vanthienen, J.: Benchmarking state-of-the-art classification algorithms for credit scoring. J. Oper. Res. Soc. **54**(6), 627–635 (2003)

Boyacioglu, A.M., Kara, Y., Baykan, K.O.: Predicting bank financial failures using neural networks, support vector machines and multivariate statistical methods: a comparative analysis in the sample of Savings Deposit Insurance Fund (SDIF) transferred banks in Turkey. Expert Syst. Appl. **36**(2), 3355–3366 (2009)

Cherubini, U., Lunga, G.D.: Liquidity and credit risk. J. Finan. **61**(5), 79–95 (2014)

Desai, V., Crook, J., Overstreet, G.: Credit scoring models in the credit union environment using neural networks and genetic algorithms. IMA J. Math. Appl. Bus. Ind. **8**(4), 232–256 (1997)

Dutta, S., Shekhar, S.: Bond rating: a non-conservative application of neural networks. In: IEEE International Conference on Neural Networks, San Diego, 24–27 July 1988, vol. 2, pp. 443–450. IEEE (1988)

Fethi, D.M., Pasiouras, F.: Assessing bank efficiency and performance with operational research and artificial intelligence techniques: a survey. Eur. J. Oper. Res. **204**(2), 189–198 (2010)

Ghatge, A., Halkarnikar, P.: Ensemble neural network strategy for predicting credit default evaluation. Int. J. Eng. Innovative Technol. (IJEIT) **202**(207), 223–225 (2013)

Huang, Z., Chen, H., Hsu, C.-J., Chen, W.-H., Wu, S.: Credit rating analysis with support vector machines and neural networks: a market comparative study. Decis. Support Syst. **37**(4), 543–558 (2004)

Khemakhem, S., Boujelbene, Y.: Credit risk prediction: a comparative study between discriminant analysis and the neural network approach. Account. Manag. Inf. Syst. **14**(1), 60–78 (2015)

Lee, Y.C.: Application of support vector machines to corporate credit rating prediction. Expert Syst. Appl. **33**(1), 67–74 (2007)

Lopez, J.A., Saidenberg, M.R.: Evaluating credit risk models. J. Bank. Finan. **24**(1–2), 151–165 (2000)

Ong, C.-S., Huang, J.-J., Tzeng, G.-H.: Building credit scoring models using genetic programming. Expert Syst. Appl. **29**, 41–47 (2005)

Singleton, J.C., Surkan, A.J.: Modeling the judgment of bond rating agencies. In: Midwest Finance Association Annual Meeting, 28–31 March, Chicago, IL (1990)

West, D.: Neural network credit scoring models. Comput. Oper. Res. **27**(11–12), 1131–1152 (2000)

Zero Interest Rate for the US Dollar Deposit and Dollarization: The Case of Vietnam

Pham Thi Hoang Anh[1,2(✉)]

[1] Banking Research Institute, Banking Academy of Vietnam, 12, Chua Boc Street,
Dong Da District, Hanoi, Vietnam
anhpth@hvnh.edu.vn
[2] The State Bank of Vietnam, Hanoi, Vietnam

Abstract. The paper's objective is to assess the impact of the ceiling and zero interest rate for US dollar deposits on dollarization index and, more generally, to find factors affecting dollarization in Vietnam. By employing the VAR model with six variables, the paper found empirical evidence that the ceiling on the interest rate for US dollar deposits and the interest rate differentials are important factors affecting deposit dollarization in Vietnam. These results are in agreement with the fact that after the ceiling rate for US dollar deposits was decreased from three percent to zero while maintaining a positive interest rate differential, the deposit dollarization ratio decreased significantly. Inflation, parallel market premium, and international reserves are also found to be important determinants of dollarization status in Vietnam.

1 Introduction

The paper's objective is to assess the impact of the ceiling and zero interest rate for US dollar deposits on dollarization ratio in Vietnam, and more generally, to determine the factors affecting dollarization. The general term "dollarization" indicates that a foreign currency, usually the US dollar, is held by domestic resident of another country (Berg and Borensztein 2000; Goujon 2006). Dollarization in the narrow sense of this word, namely, the circulation and use of US dollars, has a long history, but was observed in a downward trend recently in Vietnam. According to the conventional monetary theory, dollarization has both positive and negative impacts on the economy. Vietnam could not benefit from dollarization because dollarization might be a productive solution for rather small, open economies having close trade[1] and financial ties with the country providing the anchor (substituting) currency (Hauskrecht and Hai 2004).

Most studies on dollarization tried to find out determinants of dollarization or reasons behind uses of foreign currencies instead of domestic currency in the

This research is funded by Vietnam National Foundation for Science and Technology Development (NAFOSTED) under grant number 502.99 - 2016.01.

[1] Vietnam is an open economy with openness ratio of more than 100% since 2000s. In 2014, its openness ratio was 161%. In addition, Vietnam has trade and investment relationship with more than 80 countries in the world.

L. H. Anh et al. (eds.), *Econometrics for Financial Applications*, Studies in Computational Intelligence 760, https://doi.org/10.1007/978-3-319-73150-6_60

economy. Balino et al. (1999) and Civcir (2003) argued that the consumer's portfolio selection model is a basic analytical framework for this topic. Specifically, the model suggests that dollarization is determined by the relative rate of return of domestic and foreign currency denominated assets. By applying VAR model, Civcir (2003) have found that interest rate differential and the expected exchange rates are the dominant variables in determining dollarization in Turkey during 1986–1999. In addition, Neanidis and Savva (2009) used a monthly dataset for 11 transition economies in Central and East Europe (Armenia, Bulgaria, Czech, Estonia, Georgia, Kyrgyz, Lativia, Poland, Romania, Russia, and Ukraine) to assess short-term determinants of both deposit and loan dollarization. They found that both dollarization indexes were affected by interest rate differentials and deviations from desired dollarization. These findings are supported by Catão and Terrones (2000) and Basso et al. (2007), Vega (2013).

Dollarization in developing countries is usually one of the ultimate consequences of high and variable inflation (Calvo and Vegh 1992; Civcir 2003; Hauskrecht and Nguyen 2004; SBV 2003, 2007; Pham et al. 2014). During the period of high inflation, individuals, households and other economic entities preferred to store valuable assets such as foreign currencies (e.g. the US dollar, the Euro[2], etc.) and gold rather than domestic currency. In addition to conventional factors, there were other important determinants of dollarization status such as exchange rate risk and credibility (Civcir 2003). The author found real exchange rate misalignment based on PPP as a proxy for exchange rate risk as a positive determinant of dollarization in Turkey.

Dollarization and de-dollarization in Vietnam has attracted interest from a number of researchers. They found that dollarization in Vietnam resulted from some main causes such as inflation (Hauskrecht and Nguyen 2004; SBV 2003, 2007; Pham et al. 2014); a poor coordination between the exchange rate and interest rate policies (Nguyen 2009); and low effectiveness of monetary policy implementation (SBV 2003; Nguyen 2009, 2011; Pham et al. 2014). This paper differs from other papers on dollarization in Vietnam in the following ways. First, this is the first paper that employs an econometric model, especially VAR, in assessing determinants of dollarization in Vietnam. Second, besides conventional variables (i.e. interest differential, inflation), this is also the first paper that uses ceiling rate for US dollar deposit (which was imposed since April 2011 at 3% and then at zero since September 2015) and parallel market premium as variables in the VAR model.

Our main findings are as follows. By employing the VAR model with six variables, the paper found empirical evidence that ceiling interest rate for US dollar deposit rate and interest rate differentials are important causes of deposit dollarization in Vietnam. These results are in agreement with the fact that after the ceiling rate for US dollar deposits was decreased from three percent to zero while maintaining a positive interest rate differential, the deposit dollarization ratio decreased significantly. Second, unlike most previous studies, the paper suggests that parallel market premium is the most important determinant of

---

[2] Similar to dollarization, a country could be suffered from euroization status.

deposit dollarization in Vietnam because of Vietnamese residents' psychology in favor of foreign currencies as well as gold, especially in period of turbulences in foreign exchange market. Third, inflation and international reserves are also found to be important determinants of dollarization status in Vietnam.

The remainder of the paper is organized in four sections. Section 2 will review dollarization development in Vietnam. In Sect. 3, the paper analyses chronology of interest rate policy in Vietnam, especially on the period 2011–2016 in which the SBV imposed ceiling on the US dollar deposit rate; the ceiling was set to zero in September 2015. Section 4 employs VAR model to evaluate impact of the zero interest rate for US dollar deposit on dollarization index as well as determinants of dollarization in Vietnam. The last part will suggest policy recommendations for the SBV to de-dollarize the economy.

2   Overview of Dollarization Development in Vietnam

In Vietnam, dollarization, the use and circulation of US dollars, has a long record. In fact, the use of the US dollars has been in parallel with the local currency in Vietnam since 1960s. During the Vietnam War, US dollars were widely used and stored by residents in the South where the US army bases were located. In contrast, foreign currencies was totally forbidden in the North (according to Decree 102/CP approved by the Government, dated 06 July 1963) (Kubo 2017; Pham 2017). Although the term of dollarization was not popular in the economy in general, and in the banking and financial field at that time, the government tried their best to improve public confidence in dong's value, in particular, to reduce the amount of money in circulation, so that it could curb the country's hyperinflation.

Economic difficulties in Vietnam before 1988 showed that most Vietnamese government efforts in reforming the local currency resulted in failure. Inconvertibility of the dong and the loss of public confidence in dong's value led to the fact that residents preferred to use gold and foreign currencies (mostly in the US dollars) as valuable assets and means of payments. Even though foreign currencies were not legally allowed in circulation, prices for almost all durable goods such as motorbike, radios, televisions, refrigerators, and real estate, were quoted in gold and/or in US dollars. Because of strict regulations on foreign exchange, the parallel market developed and played an important role of supplying foreign currencies not only to individuals but also to institutions. Like any dollarized economy, dollarization in Vietnam could be presented in some forms including (i) asset substitution – this term refers to a situation in which foreign currency plays a function as a store of value; (ii) currency substitution – the process of substituting a foreign currency for local currency to fulfill the essential functions of money as a medium of exchange (Feige 2003). In this part, the paper will explain one-by-one forms of dollarization in Vietnam.

In terms of asset substitution, dollarization in Vietnam was represented in the form of foreign currency denominated liabilities and assets in the banking system. As seen in the Fig. 1, in terms of the ratio between the Foreign Currency Deposits

(FCD) and the overall money supply (M2), we could see the downward trend in the development of deposit dollarization in Vietnam, but not sustainable. Specifically, the FCD to M2 ratio increased sharply and reached the top of 41.2% in 1991 meaning that the country suffered from high dollarization status. After that, it decreased significantly from the peak to 30.25% in 1992, 23% in 1993, and then the bottom of 20.3% in 1996. The reverse trend of dollarization was seen during the Asian financial crisis when the ratio increased significantly to 23.6% in 1997, 24.6% in 1998 and 31.7% in 2001.

Once again, the trend of dollarization was reversed when deposit dollarization decreased significantly during 2002–2014 in spite of unexpected fluctuations on Vietnam's macroeconomic background, especially in the global financial crisis. Specifically, the FCD to M2 ratio dropped remarkably from 28.4% in 2002 to the bottom of 13.75% in 2010. After a slight increase to 14.6% in 2011, FCD to M2 ratio dropped significantly and reached the bottom of 8.84% in October 2016.

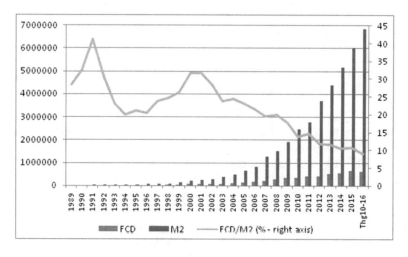

Fig. 1. FCD, M2, and FCD/M2 ratio in Vietnam, 1989-Oct, 2016. Sources: IMF's International Financial Statistics and SBV's website.

In addition to the FCD to M2 ratio, we also show the ratio of FC loans to total loans (local currency and FC loans), an index of loan dollarization. As shown in Fig. 2, the loan dollarization index exhibits a fluctuating but downward trend. From a peak of 38.6% in 1994, loan dollarization decreased to 19.3% in 2001. After a slight increase from 2002–2005, it declined gradually to 9.71% in 2015 and to 8.03% in 2016.

In summary, dollarization has decreased in Vietnam from 1989–2016, dropping to a low of 8.84% (for deposit dollarization) and 8.03% (for loan dollarization). According to IMF's dollarization criteria (Balino et al. 1999), the country can be classified as a moderately dollarized one. A reduction in both deposit and loan dollarization might be resulted from a package of de-dollarization measures

(Kubo 2017) such as imposing limits on FC loans, ceiling on FC deposits, FC reserves requirements, etc.

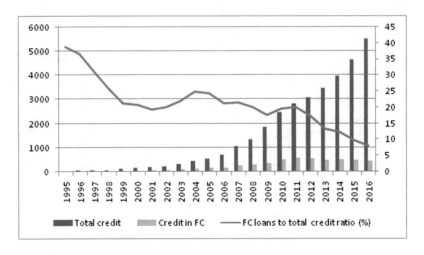

Fig. 2. FC loans, total credit, and FCL/total credit ratio in Vietnam, 1994–2016. Sources: IMF's International Financial Statistics and SBV's website.

However, in a partly dollarized economy with a dual monetary system like Vietnam (Hauskrecht and Nguyen 2004; Bellocq and Silve 2008), FCD to M2 ratio, and FCL to total credit ratio may not reflect all aspects of dollarization status because of the exclusion of foreign currency cash outside the banking system and foreign currency borrowings from foreign entities. Moreover, we realize that absolute levels in terms of US dollar, FCD and FCL have continued to increase and remained high from 2009 to 2015, and slight decrease in 2016.

While asset substitution in the form of FCD as percentage of M2 became notable in 1991, the USD had been used as a medium of exchange and unit of account in the country since as early as the Vietnam War in the 1960s[3]. One of the functions of money is to serve as a unit of account, and in Vietnam foreign currencies (the US dollar in most cases) were used in quotation for goods and services. For example, before 2010, prices for most durable goods (such as automobiles[4], motorbikes, radios, televisions, refrigerators, laptops, and real

---

[3] During the Vietnam War, due to a large amount of the US dollar aide from the US and to the existence of US army bases, residents in the South preferred to use the US dollar for buying goods and services instead of local currency. According to Dacy (1986), during 20 years of the Vietnam War, the South received about 8.5 billion USD as non-refundable aid (Table 10.2, p. 200), not including borrowing from the US government for infrastructure and economic development and an annual allowance for the US army located in the South.

[4] http://www.tinmoi.vn/Cac-hang-oto-niem-yet-gia-bang-USD-vi-VND-bat-tien-0112186.html accessed on 4 July, 2015.

estate[5], etc.) and services (such as hotel rooms, spa services, etc.) were quoted in US dollars. Even tuition fees at some universities[6] and English training centers were quoted in the US dollars. Due to the expected depreciation of the dong, even when residents took out a loan in dong, they still converted their money into an equivalent amount of the US dollar or gold and asked borrowers to repay with an equivalent amount in the US dollar and gold at the maturity date. The exchange rate applied was the parallel exchange rate. These types of transactions were applied to lending-borrowing transaction between individuals, and in urban areas.

3   Overview of Zero Interest Rate Policy for Dollar Deposit in Vietnam

One of significant historical characteristics of Vietnamese economy in the history is that its inflation rate was usually higher than those of other Asian economies (Jongwanich and Park 2009, and Fig. 3), and the domestic currency tended to be depreciated overtime. Therefore, during period 1990–2016, the priority and primary objective of Vietnam's monetary policy was price stability, focusing on curbing inflation rate (including exchange rate stability). In addition, the State Bank of Vietnam (SBV) also focused on stabilizing macroeconomic environment and ensuring social welfare subject to specified economic period, especially during the time of global or regional financial crises. In order to achieve these objectives, the SBV has used a variety of instruments including credit limit (no longer imposed since 2005), reserve requirements, open market operations (which took effect on July 2000), interest rate instruments (the base interest rate, discount rate, financing rate), the official exchange rate (functioned as an indirect instrument and sometimes as a target for monetary policy).

Among them, interest rate was found to be a flexible and effective tool of monetary policy in Vietnam (Pham 2013; Le and Nghia 2013), especially in reducing negative impacts of the global financial crisis and domestic economic turbulences. For example, the SBV had to immediate increase the base interest rate from 8.75 to 12% on 30 May and further to 14% in another 10 days (Takagi and Pham 2011). In addition, for a partly dollarized economy like Vietnam (Pham et al. 2014), the SBV is not only concerned about the interest rate on dong, but also in the interest rate on foreign currencies, especially on the US dollar.

After period of interest rate liberalization since August 2000, the SBV had to re-impose ceiling on interest rate on foreign currency at 3% and dong denominated deposit rate at 14% (April 2011, and October 2011, respectively) (Table 1). After nearly 2 months, the US dollar deposit rate decreased gradually to 2%

---

[5] http://m.tin247.com/van_qua_nhieu_doanh_nghiep_mat_hang_niem_yet_gia_bang_usd-3-21429458.html accessed on 4 July, 2015.

[6] FPT University was found to quote tuition fees in the US dollar during 2010–2011http://laodong.com.vn/kinh-doanh/dai-hoc-fpt-bi-phat-500-trieu-dong-vi-niem-yet-hoc-phi-bang-usd-7863.bld accessed on 4 July, 2015.

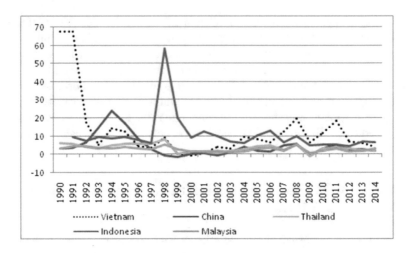

Fig. 3. General statistics of Vietnam, key indicators for Asia and the Pacific.

(June 2011), 1.25% (June 2013), 1% (March 2014), and then to zero rate (September 2015). A reduction in the US dollar deposit ceiling rate was explained as follows (Fig. 4):

(i) Maintaining higher returns on Vietnam dong denominated deposit rate in compared with foreign currency ones. By imposing a ceiling rate on foreign currency deposits (US dollar in this case) at a lower rate than market rate and announcing a commitment in maximum rate of dong's devaluation, the SBV made domestic assets more attractive than foreign assets. Moreover, whenever the SBV reduces the ceiling rate for dong deposit, it also made US dollar deposit rate smaller, so that the SBV could maintain attractiveness of dong deposits. In other words, the SBV gave an incentive to residents to move from foreign assets to domestic ones, so that it could reduce dollarization status in Vietnam.

(ii) Preventing interest rate races (namely interest rate war) among commercial banks in Vietnam. According to the conventional theory on banking, beside safety, interest rate is considered a very important determinant of deposit inflows. During period of illiquidity, Vietnam's small commercial banks had to increase deposit rate to attract inflow of funds and to meet liquidity ratios imposed by the SBV. In this case, residents tend to withdraw their money from bank offering a lower rate and put all in the bank with a higher rate. As a result, banks increase their rates, which leads to an interest rate race and to an instability in the money market.

Table 1. Chronology on ceiling interest rate, 1990–2016 (in percent).

Effective date	US Dollar ceiling deposit rate	VND ceiling deposit rate	Ceiling lending rate
11 July 1991		24	72
6 Dec 1993			7.5 (USD)
25 Mar 1994			8.5 (USD)
28 April 1994			25.2 (VND short-term) 20.4 (VND long-term)
19 Oct 1994			9.0 (USD)
1 Jan 1996			21 (VND short-term) 20.4 (VND long-term)
16 July 1996			19.2 (VND short-term) 20.4 (VND long-term)
1 Sep 1996			18 (VND short-term) 18.6 (VND long-term)
1 Oct 1996			15 (VND short-term) 18 (VND long-term)
28 June 1997			12 (VND short-term) 13.2 (VND long-term) 8.5 (USD)
2000–2011 Interest rate liberalization			
9 April 2011	3		
2 June 2011	2		
6 Oct 2011		14	
12 Mar 2012		13	
11 April 2012		12	
8 May 2012			15
25 May 2012		11	
8 June 2012		9	13
21 Dec 2012		8	12
25 Mar 2013		7.5	11
10 May 2013			10
28 June 2013	1.25	7	9
18 Mar 2014	1	6	8
29 Oct 2014	0.75 (ind.) 0.25 (inst.)	5.5	7
29 Sep 2015	0.25 (ind.) 0 (inst.)	5.5	7
17 Dec 2015	0		

Source: Author compiles from the SBV's regulations.

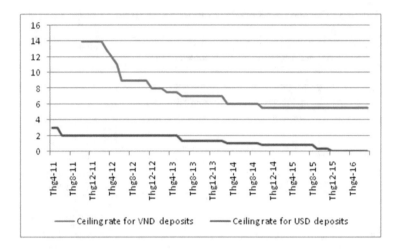

Fig. 4. Ceiling rates for dong deposits and ceiling rate for US dollar deposit, 2011-July 2016 (in percent). Source: the State Bank of Vietnam.

4 Zero Interest Rate for Dollar Deposits, and Currency Depreciation and Dollarization in Vietnam

4.1 Model Specification

The impact of ceiling rate policy on dollar deposits, especially zero interest rate[7], on dollarization in the country can be use of macroeconometric models for policy analysis suggested by Lucas (1976), Catão and Terrones (2000), Basso et al. (2007), and Neanidis and Savva (2009). Before the 1980s, simultaneous equation system was used widely in forecasting and analyzing macroeconomic variables. The macroeconometric models had been criticized by Lucas (1976), since the assumptions of invariant behavioral equations were shown to be inconsistent with dynamic maximizing behavior. Sims (1980) changes the focus of the society of econometricians. He argued that all macroeconomic variables are endogenous, in essence, they are all interrelated. Therefore, he proposed a symmetric model in which all variables play an equal role, and all are endogenous – the Vector Auto-Regressive model (VAR). The VAR models may not satisfy Lucas's criteria for policy intervention, but they are useful for finding impacts – in our case, to find the impact of the ceiling rate policy on dollar deposits (especially of zero interest rate) on dollarization.

Suppose that Y_t is an $n \times 1$ vector of macroeconomic variable whose dynamic behavior is governed by the following model:

$$B_0 Y_t = \gamma + B_1 Y_{t-1} + \ldots + B_p Y_{t-p} + u_t, \tag{1}$$

---

[7] April 9, 2011, was the first time the SBV imposed a ceiling of 3% on the US dollar deposit interest rates. At that time, market rates for US dollar deposits were 5–6%.

where γ is a constant, B_i is a $n \times n$ matrix of coefficient, and u_t is a $n \times 1$ vector of white noise structural disturbances, with covariance matrix Σ.

A reduced form of Y_t can be written as:

$$Y_t = \delta + \alpha_1 Y_{t-1} + \ldots + \alpha_p Y_{t-p} + e_t, \tag{2}$$

where $\delta = B_0^{-1}\gamma$, $\alpha_i = B_0^{-1}B_i$, and $e_t = B_0^{-1}u_t$ is a white noise process, with nonsingular covariance matrix Ω. This equation can be rewritten as

$$\alpha(L)Y_t = \delta + e_t, \tag{3}$$

where $LYy \overset{\text{def}}{=} Y_{t-1}$ and $\alpha(L) = I - \alpha_1 L - \alpha_2 L^2 - \ldots - \alpha_p L^p$.

It is usually assumed that the covariance matrix Σ for u_t is diagonal, and that the matrix B_0 has 1s on its main diagonal (but elsewhere is unrestricted). This implies that each component of Y_t is assigned its own structural equation, which ensures that the shocks can be given an economic interpretation.

The α_i's and Ω can be estimated by applying least squares (OLS) to the reduced form (2). However, if the B_i's are unrestricted, we cannot estimate B_0, since the α_i's contain pn^2 known elements and there are $(p+1)n^2$ unknown elements in the B_i's. Instead, one finds B_0 by using the formula:

$$\Omega = \text{cov}(e_t) = \text{cov}(B_o^{-1}u_t) = B_o^{-1}\Sigma(B_o^{-1})' \tag{4}$$

There are $n(n+1)/2$ distinct covariances in Ω. Our assumption that Σ is diagonal and thus, contains only n unknown elements. It means that $n(n-1)/2$ restrictions are needed to solve this system of equations.

Assuming that Y_t is a covariance stationary vector, (3) can be written as:

$$Y_t = \phi(L)e_t \tag{5}$$

where $\phi(L) = \phi_0 + \sum_{i=1}^{\infty} \phi_i L^i$, with $\phi_0 = 1$. To identify the system, we choose any nonsingular matrix P, such that the positive definite symmetric matrix Ω has the form $\Omega = PP'$. Rewriting (5) gives:

$$Y_t = \sum_{i=0}^{\infty} \phi_i PP^{-1}e_{t-i} = \sum_{i=0}^{\infty} C_i \varepsilon_{t-i} \tag{6}$$

where $C_i = \phi_i P$ and $\varepsilon_t = P^{-1}e_t$. One of applications of the VAR model is to establish the impulse response function. In terms of above notations, the matrix C describes the effect of a unit increase in each of the variables at time t on the values of all the variable from Y at time t_s: $C_s = \dfrac{\partial Y_{t+s}}{\partial \varepsilon_t'}$. To go from the reduced form to the structural model, a set of identifying restrictions must be imposed. As all variables defined in Y_t are stationary, Y_t is a covariance-stationary vector process.

The VAR model specified here, focuses on six variables that collected on monthly basis as follows:

- Deposit dollarization (measured by ratio of foreign currency deposit to money supply – FCD_M2): This ratio is calculated by taking data on foreign currency deposit (from International Financial Statistics, IMF) and M2-money supply (from the State Bank of Vietnam's website);
- Inflation (CPI): Month-by-month rate that collected from General Statistics Office of Vietnam;
- International reserves (RES): this variable is expressed in the form of growth based on primary data from the International Financial Statistics (provided by the IMF);
- Parallel market premium (PMP): this variable is calculated by taking percentage difference between selling rate at Hanoi parallel market and the official rate (since January 4, 2016, this rate is called the central rate);

$$PMP = \frac{\text{Selling rate at Hanoi-Official rate}}{\text{Official rate}} \times 100.$$

Here, the official rate (or central rate) is collected from the State Bank of Vietnam's website, while selling parallel rate obtained from www.vangsaigon. com.vn.

- Interest rate differential (IRD): following the Covered Interest Parity (CIP) as follows

$$\text{Interest rate differentials} = R - (R^* + E(\Delta ER)),$$

where R and R^* are interest rates for Vietnam dong deposit and US dollar deposit, respectively, and $E(\Delta ER)$ is the appreciation level of US dollar against Vietnam dong, which is calculated as follows:

$$E(\Delta ER) = \frac{ER_t - ER_{t-1}}{ER_{t-1}}.$$

There are five data sets including official rate[8], buying and selling rate at commercial banks, buying and selling rate at Hanoi parallel market. The paper employs all five data set for checking robustness of empirical results.

- Dollar deposit rate (USRATE): the ceiling interest rate for dollar deposits set by the State Bank of Vietnam

All variables are collected for period from April 2011 to October 2016 on monthly basis, giving a total of 67 observations. We did not extend backward, because the ceiling rate on FCD was re-imposed in 2011.

By taking the unit root test, we found that FCD_M2, CPI, PMP, RES are stationary at level I(0). The other variables (IRD and US RATE) are stationary at first difference. By employing diagnosis test to check autocorrelation and heteroskedasticity, we found that the VAR model is appropriate for assessing determinants of dollarization index in Vietnam.

---

[8] Official rate is announced by the SBV on daily and was replaced by central rate since 4th January 2016.

4.2   Estimating the Model

4.2.1   Impulse Response Function

The above residual tests showed that VAR model for six variables (FCD-M2, CPI, RES, PMP, DUS, DIRD) are adequate for estimating impacts of foreign currency deposit ceiling rate and the interest rate differential on deposit dollarization. By considering the taking impulse response function (Fig. 5), we found the following.

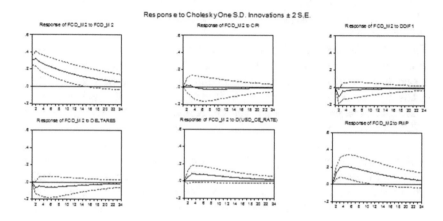

Fig. 5. Impulse response functions.

First, the VAR impulse response function reveals a very interesting evidence that a shock to US dollar deposit ceiling rate has a significantly positive impact on deposit dollarization in Vietnam. Specifically, one percent increase in UD dollar deposit rate might lead to a 0.05% in FCD to M2 ratio after one month. This finding could be explained by the conventional investment theory that investor will not be attracted by an asset/investment with lower return. Therefore, when the SBV decreases ceiling interest rate for US dollar deposit (while domestic interest rate remains constant), individual/institution might shift from foreign assets into domestic asset leading to decrease in deposit dollarization. Figure 2 illustrates that when the SBV imposed zero percent for US dollar deposit, the FCD-M2 ratio went to the bottom of 8.03%, absolute values of foreign currency deposit decrease by 5.18% from 640,000 billion of dong at the end of 2015 to 606,842 billion of dong in October 2016.

Second, the model provided an empirical evidence that parallel market premium is the most important determinant of deposit dollarization in Vietnam in recent period. In detail, one percent shock to parallel market premium leads to an increase of 0.14% in dollarization index. This result might be caused by the Vietnamese residents' psychology in favor of foreign currencies and gold. In practice, if there is a rumor about domestic currency values (i.e. devaluation, replacement of new currency for old currency), residents will want to buy foreign currencies, focusing on the US dollar, to keep as valuable assets. However, according to the Ordinance on Foreign Exchange and other regulations, Vietnamese residents could officially buy foreign currencies only to meet needs of current transactions (Article 4). Therefore, for other purposes, they have to go to the parallel market (e.g., exchange bureaus or gold shops). This movement led to high demand of foreign currencies in parallel market, and resulted in a large parallel market premium. In turn, when spread between official and parallel rates became wider, Vietnamese residents (both individuals and institutions) might be afraid of a further depreciation in future. They would prefer to buy and keep foreign currencies and gold. And the vicious circle was repeated, increasing the dollarization index.

Third, impulse response function illustrates that a shock of one percent in interest rate differential reduces the Vietnam's dollarization degree by 0.02% after one month. In other words, the higher the interest rate differential, the lower the deposit dollarization ration in Vietnam. This finding is consistent with the findings of Catão and Terrones (2000), Nguyen (2009), Civcir (2003), Basso et al. (2007), Neanidis and Savva (2009), Vega (2013). In practice, by maintaining significant positive interest rate differential, the SBV made domestic assets more attractive than foreign assets (Fig. 6). Moreover, whenever the SBV reduces the ceiling rate for dong deposit, it also makes US dollar deposit rate smaller, so that the SBV maintains attractiveness of dong deposits. In other words, the SBV provides incentives to residents to move from foreign assets to domestic ones, so as to reduce Vietnam's dollarization index.

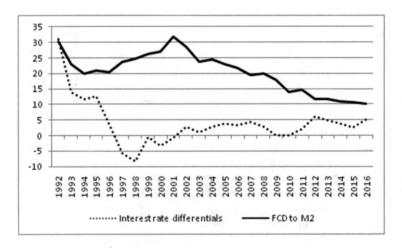

Fig. 6. Interest rate differential and deposit dollarization in Vietnam, 1992–2016 (in percent). Source: SBV, author's calculation.

Fourth, inflation rate is found to be another important determinant of deposit dollarization in Vietnam. Our model suggested that a one percent increase in inflation rate results in a nearly 0.02% increase in deposit dollarization ratio after one month. Our finding is supported by previous studies such as Hauskrecht and Nguyen (2004), SBV (2003, 2007), Nguyen (2009, 2011), and Pham et al. (2014). The finding could be explained by a decrease in public confidence in periods of high inflation. During that time, residents prefer to keep hard foreign currencies or gold as valuable assets than domestic ones. If this is the case, dollarization index should increase. Figure 7 shows similar trends in inflation, FCD to M2 (deposit dollarization) as well as FCL to credit loans (loan dollarization) in Vietnam during 1989–2016.

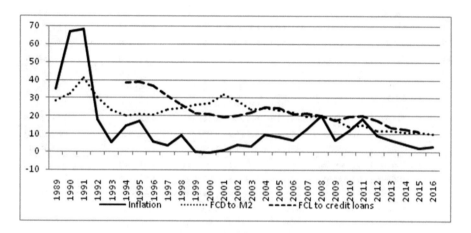

Fig. 7. Inflation and dollarization, 1989–2016 (in percent). Source: SBV, author's calculation.

Last but not least, international reserves is considered as one of determinants of dollarization in Vietnam when we found that a one percent shock to international reserves results in a decrease in deposit dollarization of 0.066%. This finding could be explained as follows. The higher the international reserves, the higher the public confidence in macroeconomic stability and dong's value. Therefore, residents might shift from foreign assets to domestic ones, leading to a decrease in dollarization index.

4.2.2 Variance Decomposition

While impulse response functions trace the effects of a shock to one endogenous variable on other variables, variance decomposition decomposes the variation in an endogenous variable into component shocks attributed to different variables. Specifically, the variance decomposition based on the SVAR estimates provides a quantitative measure of the relative importance of each shock in forecasting the error variance in dollarization index over a 24-month horizon. As we can see in the Table 2, dollarization's own shock and shock to parallel market premium play an important role in explaining the variance of the forecast error of dollarization index in Vietnam. For instance, expectation factor explains the largest portion

Table 2. Variance decomposition of dollarization index.

Period	S.E.	FCD_M2	CPI	DDIF1	DELTARES	D(USD_CE_RATE)	PMP
1	0.299688	100.0000	0.000000	0.000000	0.000000	0.000000	0.000000
2	0.478848	84.27755	0.169234	4.839726	1.897646	0.965105	7.850744
3	0.599229	75.93847	0.178084	3.577422	1.849533	2.787773	15.66872
4	0.692242	70.82416	0.139976	2.973852	2.044645	3.397335	20.62003
5	0.768146	67.32130	0.153923	2.574556	2.259637	3.846631	23.84395
6	0.830165	64.83826	0.199681	2.308616	2.457414	4.133393	26.06263
7	0.881210	63.02060	0.254116	2.121368	2.623838	4.336092	27.64398
8	0.923367	61.65902	0.306999	1.985136	2.759376	4.482733	28.80673
9	0.958303	60.62007	0.354049	1.883288	2.868381	4.592145	29.68207
10	0.987341	59.81504	0.394227	1.805505	2.955782	4.675548	30.35390
11	1.011546	59.18308	0.427836	1.745062	3.025974	4.740280	30.87776
12	1.031772	58.68139	0.455672	1.697418	3.082560	4.791259	31.29170
13	1.048712	58.27924	0.478633	1.659417	3.128395	4.831897	31.62242
14	1.062929	57.95417	0.497557	1.628805	3.165711	4.864620	31.88914
15	1.074880	57.68950	0.513171	1.603940	3.196243	4.891190	32.10596
16	1.084944	57.47265	0.526081	1.583602	3.221342	4.912919	32.28341
17	1.093428	57.29403	0.536781	1.566869	3.242063	4.930795	32.42946
18	1.100590	57.14623	0.545673	1.553033	3.259237	4.945573	32.55025
19	1.106641	57.02344	0.553082	1.541545	3.273520	4.957843	32.65057
20	1.111758	56.92109	0.559270	1.531973	3.285435	4.968067	32.73417
21	1.116088	56.83552	0.564450	1.523973	3.295400	4.976611	32.80404
22	1.119755	56.76382	0.568796	1.517270	3.303754	4.983769	32.86259
23	1.122862	56.70361	0.572447	1.511642	3.310771	4.989780	32.91175
24	1.125495	56.65295	0.575520	1.506907	3.316675	4.994836	32.95311

(nearly 100%) of the variance of the forecast error in dollarization status in the first month, though its relative importance declines over time to 56.6% at the end of horizon. Parallel market premium is the second most important factor for the whole period of 24-month horizon because it accounts for 7.8% to 33%.

5   Concluding Remarks

The paper aims at assessing impact of the zero interest rate for US dollar deposit on dollarization index as well as determinants of dollarization in Vietnam. By employing the VAR model with six variables including deposit dollarization ratio, ceiling interest rate for US dollar, interest rate differential, parallel market premium, inflation and international reserves, the paper shed light on in several ways as follows.

Ours is the first paper that assesses impact of zero interest rate for foreign currency deposit on the financial dollarization status in an emerging country like Vietnam. The paper found empirical evidence that the lower the ceiling rate, the lower the deposit dollarization ratio in Vietnam. This conclusion is in a good accordance with the empirical fact that, by decreasing ceiling interest rate for US dollar deposits from 3% to zero percent, the State Bank of Vietnam succeeded in drastically decreasing Vietnam's dollarization index. Our findings imply that the State Bank of Vietnam should pay more attention on maintaining attractive returns on domestic assets in comparison with foreign assets by lowering return on US dollar deposit as one of effective de-dollarization measures. The model provided an empirical evidence that parallel market premium is the most important determinant of deposit dollarization in Vietnam because of Vietnamese residents' psychology in favor of foreign currencies as well as gold, especially in period of turbulences in foreign exchange market. In practice, if parallel market premium becomes larger, residents tend to move their assets to foreign related ones, leading to high dollarization ratio. In addition, inflation and international reserves are also found to be important determinants of dollarization status in Vietnam. These findings suggest that Vietnam's monetary authority should stabilize the macroeconomic environment including curbing inflation, and increasing international reserves. By doing that, the economy should be de-dollarized.

Appendix

Selecting Lag Length of Model. Lag length of the model is 1 period. This was selected is based on five criteria including LR, FPE (Final prediction error), AIC (Akaike information criterion), SC (Schwarz information criterion), HQ (Hannan-Quinn information criterion), as seen in the following table.

Lag	LogL	LR	FPE	AIC	SC	HQ
0	−711.6017	NA	665.5948	23.52793	23.73555	23.60930
1	−562.2012	264.5124	16.25998*	19.80988	21.26327*	20.37947*
2	−535.2061	42.48411	22.59353	20.10512	22.80427	21.16294
3	−497.9090	51.35995*	23.65951	20.06259	24.00750	21.60864
4	−459.8026	44.97808	26.39573	19.99353	25.18420	22.02780
5	−417.0835	42.01875	29.11211	19.77323*	26.20967	22.29573

Diagnosis Tests. In order to check the appropriateness of the estimated VAR model, we estimate AR roots (aka as the inverse roots) of the characteristic AR polynomial. Below table and figure show that the estimated VAR is stable because all roots have modulus less than one, and thus lie inside the unit circle (Fig. 8).

We also performed the Portmanteau Tests for Autocorrelations. Portmanteau Test computes the multivariate Box-Pierce/Ljung-Box Q-statistics for residual serial correlation up to the specified order. We found that with different lags, p-values of Q-statistics are greater than 5%. We thus accept the null hypothesis of no serial correlation up to lag 2.

In order to test for a range of specifications of heteroskedasticity in the residuals of VAR equation, we employed the White's test. Results of White's Heteroskedasticity Test show that we can accept the null hypothesis of homoskedasticity at one percent level of significance.

Root	Modulus
0.921797	0.921797
0.626391	0.626391
0.231398	0.231398
0.208391	0.208391
−0.175961	0.175961
−0.157069	0.157069

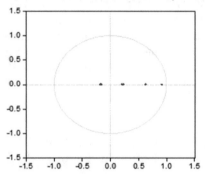

Fig. 8. Checking stationarity.

Lags	Q-Stat	Prob.	Adj Q-Stat	Prob.	df
1	18.01580	NA*	18.29729	NA*	NA*
2	42.23825	0.2194	43.28872	0.1882	36

* The test is valid only for lags larger than the VAR lag order.
df is degrees of freedom for (approximate) chi-square distribution.

VAR Residual Heteroskedasticity Tests: No Cross Terms (only levels and squares)		
Date: 06/30/17 Time: 08:35		
Sample: 2011M04 2016M10		
Included observations: 65		
Joint test:		

Chi-sq	df	Prob.
284.1719	252	0.0799

References

Basso, H.S., Calvo-Gonzalez, O., Jurgilas, M.: Financial dollarization: the role of banks and interest rates. European Central Bank Working Series, Paper 748 (2007)

Berg, A., Borensztein, E.: Full dollarization: the pros and cons. Economic Issues No. 24, IMF (2000)

Balino, T.J.T., Bennett, A., Borensztein, E.: Monetary policy in dollarized economies. Occasional Paper 171. International Monetary Fund, Washington, D.C. (1999)

Bellcocq, F.X., Silve, A.: The banking system of Vietnam after the accession to WWTO: transition and its challenges. Working paper 77, December 2008. Agence Française de Developpement (2008)

Calvo, G., Vegh, C.: Currency Substitution in Developing Countries: An Introduction. International Monetary Fund 1992 Paper 92/40 (1992)

Catão, L., Terrones, M.: Determinants of dollarization: the banking side. IMF working paper, WP/00/146 (2000)

Civcir, I.: Dollarization and its long-run determinants in Turkey. Middle East Economics Series (2003)

Dacy, D.C.: Foreign Aid, War, and Economic Development: South Vietnam, 1955–1975. Cambridge University Press, New York (1986)

Feige, E.L.: The dynamics of currency substitution, asset substitution and de facto dollarization and euroization in transition countries. Working paper presented at the Eight Dubrovnik Economic Conference (2003)

Goujon, M.: Fighting inflation in a dollarized economy: the case of Vietnam. J. Comp. Econ. 34, 564–581 (2006)

Hauskrecht, A., Nguyen, T.: Dollarization in Vietnam. Paper prepared for the 12th Annual Conference on Pacific Basin Finance, Economics, Accounting, and Business, Bangkok, 10–11 August 2004

Jongwanich, J., Park, D.: Inflation in developing Asia. J. Asian Econ. 20(5), 507–518 (2009)

Kubo, K.: Dollarization and De-dollarization in Transitional Economies of Southeast Asia. Springer Verlag (2017)

Le, T.T.N., et al.: Project on "Exchange rate and interest rate: the case of Vietnam". Ministerial level project, research funded by the State Bank of Vietnam (2013)

Lucas Jr., R.: Econometric policy evaluation: a critique. Carnegie-Rochester Conf. Ser. Public Policy 1(1), 19-46 (1976)

Neanidis, K.C., Savva, C.S.: Financial dollarization: short-run determinants in transition economies. Centre for growth and business cycle research, Discussion paper series (2009)

Nguyen, T.H.: Dollarization of financial assets and liabilities of the household sector, the enterprises sector and the banking sector in Vietnam. This paper was presented at the JICA-State Bank of Vietnam Joint Research Project Workshop on Dollarization and its Effect on Monetary and Foreign Exchange Rate Policies and the Development of Financial System, Vietnam, Lao PDR and Cambodia, Hanoi, 5 July 2002 (2002)

Nguyen, T.H · Dollarization and monetary policy: the case of Vietnam. Banking Review, No. 5/2011 (2011). (in Vietnamese)

Pham, T.H.A.: Effectiveness of monetary policy through asset prices channel: The case of Vietnam. Research project funded by the Banking Academy of Vietnam (2013)

Pham, T.H.A.: Dollarization and De-dollarization Policies: The Case of Vietnam. In: Kubo, K. (ed.) Chapter 3, pp. 131–166 (2017)

Pham, M.D., et al.: Foreign exchange regulation on de-dolarization in Vietnam. Research grant at the Vietnam's Economic Committee of Congress (2014)

State Bank of Vietnam: Project on Improving convertible degree of Vietnamese Dong and reducing dollarization degree - together with Decision 98/2007/Q-TTg approved by the Prime Minister (2007)

State Bank of Vietnam: Project on De-dollarization in Vietnam, funded by the State Bank of Vietnam (2013)

Takagi, S., Pham, T.H.A.: Responding to the global financial crisis: Vietnamese exchange rate policy, 2008–2009. J. Asian Econ. 22(6), 507–517 (2011)

Vega, H.: Financial Frictions and the Interest rate differential in a dollarized economy. Working paper series, Banco Central de Reserva del Peru (2012)

Foreign Reserve Accumulation and Sterilization Effectiveness in Vietnam

Pham Thi Tuyet Trinh[✉] and Le Phan Ai Nhan[✉]

Banking University of Ho Chi Minh City, 36 Ton That Dam Street, District 1,
Ho Chi Minh City, Vietnam
{trinhptt,nhanlpa}@buh.edu.vn

Abstract. After Asia financial crisis, Vietnam was a persistent foreign reserve accumulator with two episodes of accumulation interrupted by the global financial crisis. The study explores the effectiveness of sterilizing monetary impact of reserve accumulation in both episodes. Employing two-stage least squares (2SLS) method to estimate simultaneous equations of central bank reaction and balance of payments functions, the study finds overall partial effectiveness of sterilization at 78%. The explanations of incomplete effectiveness sterilization are relied on high capital mobility and inappropriate scale and timing of interventions. Besides, the result of rolling estimation with 30 quarter-sample indicates the degree of sterilization tends to increase and higher in the second episode of reserve accumulation, which can be explained by the improvement of sterilization operation.

Keywords: Foreign reserve accumulation · Sterilization effectiveness
Vietnam

JEL: E52 · E58 · F31 · F41

1 Introduction

Foreign reserves have proved to have important role in open economies. The Asia financial crisis 1997 (AFC) and global financial crisis 2008 (GFC) showed evidence of shock-absorber role of foreign reserves as countries holding higher foreign reserves showed less volatile economic growth rate (Stephane and Cyril 2011). Besides, if central banks try to maintain some degree of both exchange rate stability and monetary independence while continue to integrate wider and deeper into global financial market, foreign reserves holding is a necessary component to keep this pattern stable (Aizenman et al. 2010). However, reserve accumulation causes monetary supply to increase, leading to inflation pressure if central bank does not conduct sterilization interventions. Sterilization is defined as actions taken by central bank to neutralize the impact of its interventions on the foreign exchange market to the money supply. Unsurprisingly, recent empirical studies found monetary sterilization was used in various high reserve

© Springer International Publishing AG 2018
L. H. Anh et al. (eds.), *Econometrics for Financial Applications*, Studies in Computational Intelligence 760, https://doi.org/10.1007/978-3-319-73150-6_61

countries though the effectiveness of sterilization is not complete (for example Brissimis et al. 2002; He et al. 2005; Aizenman and Glick 2009; Hashmi 2011; Ouyang and Rajan 2011; Chung et al. 2014).

Like most Asian economies, Vietnam consecutively increased reserve accumulation since AFC. After interrupted in 2009–2010 due to GFC, foreign reserves of Vietnam have been massively accumulated again since 2012 thanks to large capital inflows and current account surplus. At the beginning of 2016, Vietnam had USD40 billion foreign reserves, equivalent to twelve weeks of imports and still seemed to be a persistent reserve accumulator. Monetary authorities in Vietnam also relied on sterilization interventions to prevent monetary impact of foreign exchange purchase, nonetheless, Vietnam's inflation rose sharply during the periods of massive accumulation. The high correlation of foreign reserves and inflation poses a question about the effectiveness of sterilization. So far empirical studies did not take the case of Vietnam into account, except Anh and Phu (2013) whose results have not been justified as the employed methodology is criticized to have endogenous problem. Therefore, this paper aims to investigate sterilization effectiveness in Vietnam during the process of massive reserve accumulation by employing 2SLS method to estimate simultaneous equations of central bank reaction and balance of payments functions constructed based on loss function of central bank developed by Brissimis et al. (2002).

2   Literature Review

In a central bank's balance sheet, the asset side consists of net foreign assets (NFA) and net domestic assets (NDA) while the liability side is composed of monetary base (MB), thus MB is equal to the sum of NFA and NDA (MB = NFA + NDA) (Table 1). When central bank conducts purchase transactions in foreign exchange to increase reserves, NFA in the asset side will increase. If there are no sterilization interventions, MB in the liability side will increase equivalently, leading to increase in aggregate money supply (MS) which equals money multiplier (MM) times MB, and resulting in inflation pressure.

Table 1. Central bank's balance sheet

Assets	Liabilities
Net Foreign Assets (NFA = Foreign Assets – Foreign Liabilities)	Monetary base (MB)
Net Domestic Assets (NDA)	- Currency in circulation
- Net domestic claims	- Reserves of commercial banks
+ Net claims on government	
+ Net claims on commercial banks	
+ Net claims on private sector	
- Others	

Sterilization can be examined in narrow and broad senses depending on the transactions designed to diminish the monetary impact of foreign exchange interventions (Takagi and Esaka 2001). Sterilization is narrowly defined as actions taken by central bank in order to neutralize the impact of its interventions on the foreign exchange market to the monetary base. Open market operation (OMO) (i.e. selling treasury bonds or central bank papers to decrease NDA) is the most popular measure for narrow sterilization used by central banks. In addition, central banks in practice also use other measures to decrease NDA and control monetary base. For example, shifting government deposits at commercial banks to central bank was used effectively in Malaysia in 1992; controlling credit to commercial bank through discount window was used in Korea (in 1986–1988), Thailand (in 1989–1990), Malaysia (in 1995–1996) (Takagi and Esaka 2001). The effectiveness of sterilization relies on whether an increase in NFA is accompanied by equivalent decrease in NDA (Aizenman and Glick 2009). Sterilization is completely effective when NDA decrease is equal NFA increase, thus leaving MB unchanged ($\Delta MB = \Delta NDA + \Delta NFA = 0$). If NDA decrease is lower than NFA increase, sterilization is partially effective, implying monetary influence of foreign reserve accumulation on the economy to some extent.

Sterilization is more broadly defined as actions to keep the money supply unaffected following foreign exchange interventions (Takagi and Esaka 2001). Facing up to the increase of monetary base due to foreign exchange purchases, central bank can raises the required reserve ratio to make money multiplier go down and leave aggregate money supply unchanged. The central bank of China frequently increased required reserve ratio to control the impact of foreign reserve accumulation during 2008–2010 (Zhang 2011). Many south Asia central banks such as Malaysia (in 1989–1992, 1994, 1996), Korea and Philippines (in 1990), Thailand (1995–1996) and Indonesia (in 1996) also used this measure when facing with massive capital inflows (Takagi and Esaka 2001).

One of the main reasons of partial effectiveness of sterilization is the degree of financial openness. Sterilization interventions in both senses cause higher relative interest rate. If the economy has high financial openness, higher domestic interest rate will attract more capital inflows which can diminish effectiveness of previous sterilization interventions as central bank continues to accumulate foreign reserves and keep stable exchange rate (Frankel 2007). Thus, there is a contemporaneous relationship between reserve accumulation and sterilization reflected by change in the structure of money supply, in particularly, NDA and NFA. A rise in NFA as reserves are purchased reduces NDA when central bank conducts sterilization; conversely, sterilization transactions lead to subsequent increase in NFA, eroding the effectiveness of sterilization as reserves are accumulated continuously.

Argy and Porter (1974) first examined this framework to explore the effectiveness of sterilization by simultaneously estimating the monetary policy reaction function and the balance of payment function. The monetary policy reaction function in Eq. (1) describes the operation of sterilization to offset foreign exchange purchases in the structure of money supply.

$$\Delta NDA_t = \alpha_0 + \alpha_1 \Delta NFA_t + \alpha_2 X_t + u_t \tag{1}$$

where ΔNDA_t is the change in the central bank's net domestic assets; ΔNFA_t is the change in the central bank's net foreign assets and X_t represents other variables that could affect monetary policy operations.

In Eq. (1), α_1 is the sterilization coefficient which is expected to have negative sign, presenting an increase in NFA causes NDA to decrease. When α_1 is equal to -1, the impact of increase in NFA to decrease in NDA is one to one, implying sterilization of central bank has complete effectiveness. When α_1 is higher than -1 and lower than 0 $(-1 < \alpha_1 < 0)$, increase in NFA can not be eliminated by decrease in NDA, implying central bank partially sterilizes the impact of the change in NFA on money supply. When α_1 is equal to zero, central bank does not sterilize foreign exchange purchase or has ineffective sterilization. In addition, Aizenman and Click (2009) argue sterilization coefficient can be lower than -1 as central bank has stronger reaction against inflation pressure.

The balance of payment function in Eq. (2) describes reverse impact of sterilization on capital inflows depending on the degree of capital openness reflected by the subsequent reserve accumulation of central bank.

$$\Delta NFA_t = \beta_0 + \beta_1 \Delta NDA_t + \beta_2 Z_t + v_t \tag{2}$$

In Eq. (2), Z_t represents other variables that could affect ΔNFA_t and β_1 is the offset coefficient reflecting degree of capital mobility. β_1 is also expected to have negative sign, presenting a decrease in NDA causes NFA to increase. β_1 is equal to -1 in case of perfect capital mobility and 0 in case of no capital mobility. The name of this coefficient simply implies its magnitude offsets that of sterilization coefficient. In general, the higher the capital mobility is (higher β_1), the less effective the sterilization policy is (lower α_1).

Besides, the magnitudes of sterilization and offset coefficients also indicate the degree of monetary independence according to the trilemma developed by famous Mundell (1960) and Flemming (1962). A central bank not only accumulates foreign reserves but also obtain stable exchange rate when intervening into foreign exchange market. If the economy also has high degree of capital openness, it has to sacrifice a degree of monetary independence. Contrarily, a small offset coefficient and a large sterilization coefficient in absolute value imply a high degree of monetary independence (Ouyang et al. 2010).

The above simultaneous equations take into account the relationship between NFA and NDA, but ignore the operation of sterilization in broad definition, i.e. required reserve ratio. As most central banks use OMO as the main sterilization tools, this framework is used popularly to investigate the effectiveness of sterilization in previous empirical studies which can be divided into 3 groups.

The first group estimates the monetary policy reaction function separately by Ordinary Least Squares (OLS) such as Aizenman and Glick (2009), Cavoli and Rajan (2006), Anh and Phu (2013), Takagi and Esaka (2001). Takagi and Esaka (2001) estimated the sterilization coefficients of five countries including Philippines, Indonesia, Malaysia, Thailand and Korea and founded that apart from Philippines, coefficients of four remaining countries are not significant. Meanwhile, the results of Cavoli and Rajan (2005) indicated the significant degree of effectiveness of sterilization interventions in these 5 countries was very high and close to complete. Aizenman and Glick (2009) examined the sterilization effectiveness of 12 countries in Asia and Latin America by using 40 quarter rolling samples for each country from Q2/1984 to Q4/2007 and GDP growth as a control variable. The results of sterilization coefficient ranged from -0.6 to -1.4, reflecting different levels of effectiveness of sterilization among countries. The study also found an increase in the extent of sterilization through time, which implies a potential pressure on inflation and a great burden of cost. A limitation of OLS approach is assuming ΔNFA_t are exogenously determined, while the endogeneity problem exists. Consequently, the results taken from OLS estimation may be biased and inconsistent.

The second group employs vector autoregression (VAR) to estimate monetary policy reaction and balance of payment function simultaneously. Studies in this group usually include other variables (besides NFA and NDA) in the VAR system such as domestic interest rate (Christensen 2004), foreign interest rate (He et al. 2005), price level and exchange rate (Moreno 1996). Analyzing the relationship between capital inflows and effectiveness of sterilization interventions in the Czech Republic from 1993 to 1996, Christensen (2004) suggested that sterilization had achieved initial success. However, over time, higher domestic interest rates attracted more capital inflows, the cost of sterilization increased and ultimately this policy became unsustainable. He et al. (2005) indicated that for the case of China, the sterilization coefficient was equal to -1. Moreno (1996) showed that in the period 1981–1994, sterilization interventions in Korea and Taiwan were absolutely effective. In particular, the independence of monetary policy in Korea is higher than in Taiwan because Korea had more restrictive capital controls and sterilized more fully than Taiwan did. Although VAR is enabling to circumvent the endogeneity problem, it is criticized to only estimate coefficients of lagged variables and fail to explore contemporary effect.

The third group estimates a set of simultaneous equations by 2SLS, 3SLS or GMM, therefore can overcome both endogeneity problem and contemporary effect. Among studies in this group, Brissimiss et al. (2002) is the pioneer in constructing a set of simultaneous equations based on the framework of minimizing a loss function of the central bank. The study employs 2SLS and 3SLS on data of Germany in period M2/1980-M2/1992 and find very high sterilization effectiveness (0.74% by 2SLS and 0.96% by 3SLS) and rather low offset coefficient (0.22% by 2SLS and 0.4% by 3SLS). Recent studies usually follow the approach of Brissimiss et al. (2002) to examine sterilization policy in various countries such as China (Ouyang et al. 2010; Zhang 2011; Chung et al. 2014), Singapore and Taiwan (Ouyang and Rajan 2011) and find very high effectiveness of sterilization at 0.79% and above.

3 Foreign Reserve Accumulation and Sterilization Interventions in Vietnam

3.1 Foreign Reserve Accumulation in Vietnam

In the process of integration into regional and global market, Vietnam has two episodes of accumulating foreign reserves, interrupted by the GFC (Table 2). The first episode saw consecutively accumulated in period 2000–2008 thanks to massive capital inflows. As a result of economic integration attempts recorded by bilateral and multilateral agreements such as Bilateral Trade Agreement with US in 2001 and official members of WTO in 2007, capital inflows surged impressively, not only financing large current account deficit but also leading to overall balance surplus, especially since 2006. Foreign exchange interventions allowed the State Bank of Vietnam (SBV) to obtain both exchange rate stability and foreign reserve accumulation. At the end of 2008, foreign reserves reached peak at USD 23.8 billion, equivalent to 25.1% of GDP.

Table 2. Balance of payment and foreign reserves (million USD) in 2000–2014

	2000	2001	2002	2003	2004	2005	2006	2007
Current account	1,106	682	-604	-1,931	-957	-560	-164	-6.953
Financial account	-316	371	2,090	3,279	2,807	3,087	3,088	17.730
Error and omission	-680	-847	-1038	798	-915	-397	1,400	-324
Overall balance	110	206	448	2,146	935	2,130	4,325	10.263
Foreign reserves	3416	3660	4121	6222	7041	9051	13,384	23,479
Reserves /GDP (%)	11.2	11.5	11.8	15.9	15.5	15.8	20.2	30.3
	2008	2009	2010	2011	2012	2013	2014	
Current account	-10,823	-6,608	-4,276	236	9,062	9,471	8,896	
Financial account	12,341	7,172	6,201	6,390	8,275	151	5,826	
Error and omission	-1,044	-9,022	-3,690	-5,477	-5,470	-8,763	-6,342	
Overall balance	474	-8,458	-1,765	1,149	1,1867	859	8,380	
Foreign reserves	23,890	16,447	11,2467	13,539	25,573	25,893	34,273	
Reserves /GDP (%)	25.1	15.9	10.6	9.8	16.1	15.0	17.5	

Source: IMF (2015)

Foreign reserves dropped down dramatically in period 2009–2010 due to sudden stop and reversal of capital flows. Because of GFC 2008, foreign direct investment (FDI) fell significantly and portfolio investment reversed shortly, reducing financial balance surplus to a lower level than current account deficit. SBV's interventions to keep stable exchange rate led to diminishing foreign reserves at USD12 billion, equivalent to 11.3% of GDP and 1.6 months of next-year imports.

The second episode of reserve accumulation has started since 2011 explained by the return of capital inflows and current account surplus. Foreign reserves increased rapidly and reached peak at USD 34.273 billion by 2014, much higher than the peak of the first sub-period, its ratio to GDP was much lower at 17.5%.

3.2 Sterilization Interventions in Vietnam

Sterilization was conducted mainly by OMO, other measures including required reserve ratio, shifting government deposit to central bank were used occasionally. OMO was first used as an instrument of monetary policy in July 2000. During 2000–2005, the role of OMO as a sterilization measure was unapparent because reserve accumulation was still limited and OMO was not fluently used for managing liquidity of banking system until 2003 (Table 3). Since 2006, OMO has been used for sterilizing more often as reserve accumulation was accelerated and interbank market was more developed. For example, in 2006 SBV sold VND 87.4 trillion government securities, making net withdrawal of VND 50.6 trillion in 2006; in 2007, the selling volume was VND 356.844 trillion and the net withdrawal was made up to VND 356.8 trillion. According to World Bank (2007), SBV withdrew out of circulation about VND 11–14 trillion each week since May 2007; this figure increased to VND 15–16.5 trillion by the end of the year. In March 2008, SBV also issued VND 20.3 trillion compulsory bills to strongly reduce liquidity of banking systems for the purpose of controlling inflation pressure which occurred since the second half of 2007.

Table 3. OMO operation in 2000–2013

Year	Total sessions	Volume (trillion VND)	
		Purchase	Sell
2000	17	1.354	550
2001	48	3.314	620
2002	85	7.246	1.900
2003	107	9.844	11.340
2004	123	60.986	950
2005	158	100.679	1.800
2006	162	36.833	87.402
2007	355	61.133	356.844
2008	393	947.206	77.005
2009	329	961.773	102
2010	490	2.101.421	–
2011	431	2.801.253	–
2012	378	449.922	174.000
2013	418	179.386	254.863

Source: 2000–2006: Huyen (2011); 2007–2013: SBV (2008-2013)

In the period of losing foreign reserves because of GFC's impact, SBV largely purchased government securities and SBV's compulsory bills through OMO to inject money to banking system for the purpose of preventing the economy from recession with net injections at 961.7 trillion VND, 2,101.4 trillion VND and 2,801.2 trillion VND in 2009, 2010, 2011 respectively. When foreign reserves were accumulated in the second episode, OMO has been organized to have both purchase and sell sessions in each transaction day since the beginning of 2012 (previously, SBV only purchased or sold in each transaction date). Since March 2013, SBV has started to issue its own bills for all selling sessions through OMO which allowed SBV to no longer depend on government securities for sterilization. SBV's bills had the same short maturities with other government's debt securities such as treasury bills, including 28, 56 and 91 days. Issued SBV's bills increased rapidly with the growth of foreign reserve accumulation (Fig. 1). Outstanding SBV's bills were VND 11.5 trillion, VND 59.4 trillion and VND 140.8 trillion VND by the end of 2012, 2013 and 2014 respectively.

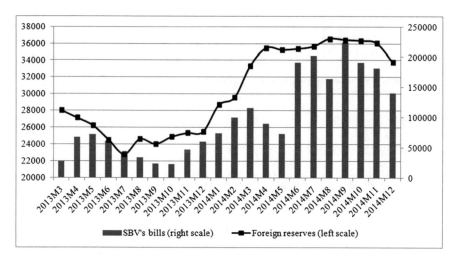

Fig. 1. Outstanding SBV's bills (billion VND) and foreign reserves (million USD)
Source: IMF (2015)

SBV also used reserves requirement in conducting monetary policy but not frequently, therefore, the role of this instrument as a sterilization tool is negligible. In the first episode of foreign reserve accumulation, required reserve ratios were adjusted upward two times to reduce liquidity of banking system: In June 2007, required reserve ratios applied to all terms and depository institutions were raised by twice; In February 2008, required reserve ratios of all terms were raised by 1% and applied to all types of deposits (previously, required reserves were applied to demand deposit and deposit terms less than 24 months). Regardless such substantial adjustments, sterilization by this instrument did not take effect. Large foreign currency purchases since 2005 provided commercial banks with very high liquidity which allowed them to keep much higher reserve-to-deposits ratio than required reserve ratio. Therefore, they did not increase reserve-to-deposits ratios but changed the composition of excess and required reserves in total reserves in response to the upward change of required reserve ratio. As a result, money multiplier was not reduced, reflecting inefficiency of sterilization by this measure (Fig. 2). In the period of diminishing foreign reserves, required

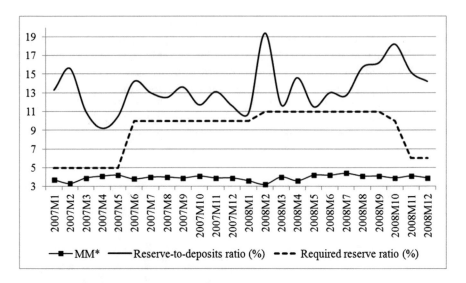

Fig. 2. Money multiplier, reserve-to-deposits ratio and required reserve ratio in 2007–2008 *MM = M2/reserves*
Source: Authors' calculation from IMF (2015)

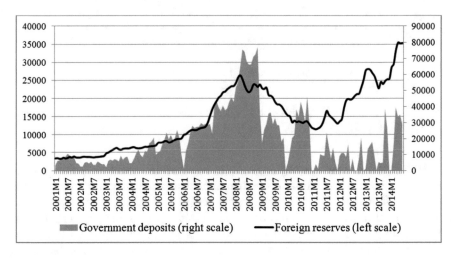

Fig. 3. Government deposits (billion VND) at SBV and foreign reserves (million USD)
Source: IMF (2015)

reserve ratio was cut down significantly to provide baking systems with liquidity for the purpose of preventing recession. After four downward adjustments, required reserve ratio applied to deposits less than 12 months was reduced to 3% in March 2009 and kept unchanged till the end of 2014. Although reserve accumulation was accelerated since 2012, this instrument was not used for sterilization again, reflecting this was not a main but supplement measure for sterilization.

Besides, in early 2008, SBV had another measure for sterilization as the Prime Minister issued instruction to transfer government deposits in commercial banks to SBV (Dispatch 319/TTg-KTTH). Actually, the move of government deposits to SBV was a supplemental resolution to curb inflation arising in the second half of 2007. This move also had effect of sterilization as contributing to reduce SBV's NDA in the second half of 2008. However, the usage of this tool was unclear after 2008 as government deposits did not increase when foreign reserves increase in the third sub-period (Fig. 3).

4 Methodology

4.1 Model Specification

We use two simultaneous equations of monetary policy reaction and balance of payment functions developed by Ouyang and Rajan (2011) to estimate the effectiveness of the sterilization in Vietnam. There are two reasons for this application. First, as OMO is the main sterilization measure in Vietnam, the model evaluating the extent of sterilization in narrow sense is useful. Second, following the approach of Brissimiss et al. (2002), Ouyang and Rajan (2011) also assume central bank has a loss function reflecting its failure to obtain monetary policy objectives. In our model, the loss function includes four components: the squared deviation of price level from its target, the squared deviation of real income from its trend, the volatility of interest rates and exchange rates. By defining each components of the loss function, we specify other macroeconomic variables affecting NDA and NFA in Eqs. (1) and (2), in other words Xt and Zt. These assumptions are very appropriate in the case of Vietnam as the country pursues monetary policy with multiple objectives including inflation control, economic growth, interest rate and exchange rate stability, indicated in Law on SBV 1997 and verified in many studies such as Carmen (2006), Economic Committee of the National Assembly and UNDP in Vietnam (2012). The set of simultaneous equations are presented in Eqs. (3) and (4) as follows:

$$\Delta NDA_t^* = \alpha_0 + \alpha_1 \Delta NFA_t^* + \alpha_2 \Delta MM_t + \alpha_3 \Delta CPI_{t-1} + \alpha_4 GAP_{t-1} \qquad (3)$$
$$+ \alpha_5 \Delta REER_{t-1} + \alpha_6 \Delta (r_*^t + E_t e_{t+1}) + \alpha_7 (d_1 - 1) SDr_{t-1} + u_t$$

$$\Delta NFA_t^* = \beta_0 + \beta_1 \Delta NDA_t^* + \beta_2 \Delta MM_t + \beta_3 \Delta CPI_{t-1} + \beta_4 GDP_{t-1} \qquad (4)$$
$$+ \beta_5 \Delta REER_{t-1} + \beta_6 \Delta (r_t^* + E_t e_{t+1}) + \beta_7 (d_2 - 1) SDe_{t-1} + v_t$$

Where:

ΔNDA_t^* represents the change in the adjusted net domestic assets scaled by GDP and ΔNFA_t^* represents the change in adjusted net foreign assets scaled by GDP. Our main interests are the sterilization coefficient (α_1) and the offset coefficient (β_1) which are expected to be negative.

ΔMM_t represents the change in money multiplier. When money multiplier rises, domestic money supply increases, interest rate decreases, thus the extents of capital inflows and foreign exchange reserves reduce. Therefore, β_2 is expected to be negative. Responding to the increase of money multiplier, central bank implements contractionary monetary policy, hence, the expected sign of α_2 is negative.

ΔCPI_t represents the change in consumer price index and its coefficients in both equations (α_3 and β_3) may be negative. Higher inflation rate leads to capital flight because of concerns about the instability of the economy, hence causing NFA to decrease. To control inflation, central bank should tighten monetary operations, therefore NDA decreases.

GAP_t represents the output gap. The expected sign of its coefficient in Eq. (3) (α_4) is ambiguous because the increase in GAP encourages capital inflows (NFA increases) in one hand and in the other hands, worsens the current account due to the income effect (NFA decreases). If its impact on capital inflows dominates that of current account, NFA will increase and vice versa. Meanwhile, GAP is expected to have negative impact on NFA in Eq. (4) (β_4) as central bank usually conducts counter-cyclical monetary policy.

$\Delta REER_t$ is the change in real effective exchange rate. Similarly to GAP_t, the impact of REER on NFA may be positive or negative. The decrease in REER, implying the appreciation of domestic currency against foreign currency basket, can cause detrimental impact on trade balance, thus, the central bank need to expand monetary conditions in response. Therefore, its impact on NDA may be negative.

$\Delta(r_t^* + E_t e_{t+1})$ represents the change in foreign interest rate plus expected nominal exchange rate. Either the increase in foreign interest rate or the decrease in expected exchange rate leads to capital withdrawals, hence β_6 is expected to be smaller than 0. If $(r_t^* + E_t e_{t+1})$ rises, monetary policy stance should be tighter in order to maintain the stability of exchange rate, thus α_6 is expected to be smaller than 0.

SDr_t represents the deviation in the change in domestic interest rate. d_1 is a dummy variable which takes a value of 2 in case of surplus in domestic money market ($\Delta NDA_t^* < 0$) and a value of 0 in case of deficit ($\Delta NDA_t^* > 0$). The central bank should withdraw (or inject) money supply to reduce interest rate volatility. The more volatile interest rate is, the larger the scale of central bank's intervention is. For example, the rise in SDr and the value 0 of d_1 result in the rise in ΔNDA. Therefore, the coefficient of the volatility of interest rate in Eq. (3) is expected to be negative.

SDe_t represents the deviation in the change in exchange rate VND/USD. d_2 is a dummy variable which takes a value of 2 in case of excess demand for foreign currency $(\Delta NFA_t^* < 0)$ and a value of 0 in case of excess supply $(\Delta NFA_t^* > 0)$. In case of excess supply (or excess demand) for foreign exchange, the central bank has to buy (or sell) foreign currency to stabilize exchange rate. The more volatile exchange rate is, the larger the scale of central bank's intervention is. For example, the rise in SDe and the value 0 of d_2 result in the rise in ΔNFA. Therefore, the coefficient of the volatility of exchange rate in Eq. (4) is expected to be negative.

Additionally, to estimate the variation of offset and sterilization coefficients, we conduct rolling estimation. Initially, we estimate coefficients by using 31 quarter- sample from Q3/2000 to Q1/2008. Then we add the next observation and eliminate the first observation so that the size of sample remains unchanged (e.g., from Q4/2000–Q2/2008, Q1/2001–Q3/2008 ...) and in each time of changing observations, we re-estimate coefficients.

4.2   Data Description

We use quarterly data from Q3/2000 to Q4/2014 for estimation. The choice of this period relies on the fact that SBV officially introduced the usage of OMO in July 2000. Table 4 summarizes the measurements of the variables and sources of the various time series data used for estimation. All data are collected mainly from International Financial Statistics of International Monetary Fund (IMF 2015). In case of data unavailability from this source, we use other reliable substitute sources. In detail, data on GDP is taken from Datastream. For REER calculation, we collect data on exports, imports of Vietnam with 17 trading partners from General Statistics Office of Vietnam (GSO 2015) while data on exchange rate of VND against USD and on other currencies against USD and consumer price index (CPI) of 17 trading partners of Vietnam from IFS.

Following Ouyang and Rajan (2011), we adjust NFA by excluding the revaluation effect. NFA in balance sheets of SBV are denominated in domestic currency and SBV often revalues its assets and liabilities at the end of the accounting period. Consequently, the values of the NFA can be changed due to exchange rate fluctuations even if foreign reserves neither increase nor decrease. Therefore, we need to exclude the revaluation effect before estimations. Ideally, the adjustment is based on the currency composition of reserves. Because of unavailability data, we assume that 100% foreign exchange reserves are held in USD. Besides, we ignore other factors influencing on NFA value, such as interest earnings earned from reserve accumulation. Therefore, change in the adjusted NFA is measured by formula (5). Noticed that the values of the MB are not impacted by the revaluation, the change in the adjusted NDA is measured by formula subtracting MB from adjusted NFA (6).

We employ the Augmented Dickey-Fuller (ADF) and Phillips-Perron (PP) to test stationarity of series. The results reported in Table 5 indicate that all series in first difference are significantly stationary at 10% and above. This allows us to use 2SLS method to estimate equation of NFA and NDA simultaneously.

Table 4. Variable measurements and data sources

Variables	Measurements	Data sources
ΔNFA_t^*	$$\frac{NFA_t - NFA_{t-1}\left(\frac{E_t}{E_{t-1}}\right)}{GDP_t^N} \quad (5)$$ Where: NFA_t = Foreign Assets (FA_t) – Foreign Liabilities (FL_t); GDP_t^N is Nominal GDP; E_t is end-period nominal exchange rate of VND/USD	IMF (2015), Datastream
ΔNDA_t^*	$$\frac{\Delta MB_t}{GDP_t^N} \Delta NFA_t^* \quad (6)$$ Where: MB_t is monetary base	IMF (2015), Datastream
ΔMM_t	$Ln(\frac{M2_t}{MB_t}) - Ln(\frac{M2_{t-1}}{MB_{t-1}})$ Where: $M2_t$ is money supply	IMF (2015)
ΔCPI_t	$Ln(CPI_t) - Ln(CPI_{t-1})$ Where: CPI_t is consumer price index	IMF (2015)
GAP_t	$Ln(GDP_t^R) - Ln(GDP_t^P)$ Where: GDP_t^R is Real GDP; GDP_t^P is Potential GDP calculated by using Hodrick-Prescott (HP) method with smoothing parameter of 1600	IMF (2015), Datastream (2015)
$\Delta REER_t$	$Ln(REER_t) - Ln(REER_{t-1})$ Where: $REER_t$ is real effective exchange rate, calculated based on geometric mean formula: $REER_t = \prod_k^{t=1} RER_{it}^{w_{it}}$	IMF (2015), GSO (2015)
$\Delta(r_t^* + Ee_{t+1})$	$(r_t^* + ln(e_{t+1})) - (r_{t-1}^* + ln(e_t))$ Where: r_t^* is interest rate on US Treasury bills; $e_{t+1} = average - period\,nominal\,exchange$ rate of VND/USD at time t+1	IMF (2015)
d_1 and d_2	$d_1 = 2$ if $\Delta NDA_t^* < 0$ and $d_1 = 0$ if $\Delta NDA_t^* > 0$ $d_2 = 2$ if $\Delta NFA_t^* < 0$ and $d_2 = 0$ if $\Delta NFA_t^* > 0$	
SDr_t	$(1/5) * \sqrt{\sum_{i=-2}^2 (\Delta r_{t+i} - \Delta r)^2}$ Where: rt is domestic interest rate; $\Delta r = (\frac{1}{5}) * \sum_{i=-2}^2 \Delta r_{t+i}$	IMF (2015)
SDe_t	$(1/5) * \sqrt{\sum_{i=-2}^2 (\Delta e_{t+i} - \Delta e)^2}$ Where: $e_t = $ average – period nominal exchange rate of VND/USD $\Delta e = (\frac{1}{5}) * \sum_{i=-2}^2 \Delta e_{t+i}$	IMF (2015)

Source: Authors

Table 5. ADF and PP test results

Variables	ADF test statistic	PP test statistic
ΔNFA_t^*	−5.005***	−4.999***
ΔNDA_t^*	−3.051***	−7.912***
ΔMM_t	−8.374***	−8.416***
ΔCPI_{t-1}	−2.659***	−2.874***
GAP_{t-1}	−2.366**	−3.703***
$\Delta REER_{t-1}$	−9.465***	−9.554***
$\Delta(r_t^* + E_t e_{t+1})$	−4.710***	−4.680***
$(d_1 - 1)SDr_{t-1}$	−3.514***	−7.161***
$(d_2 - 1)SDe_{t-1}$	−1.802*	−4.208***

Note: ***. ** and * denote respectively the 1%. 5% and
10% level of significance
Source: Authors' calculation

5   Estimation Results

Estimation result of Eqs. (3) and (4) are reported in Table 6. Results of diagnostic tests on residuals including White's test for heteroscedasticity, Breusch
Godfrey LM test for autocorrelation and Jaque-Bera test for normality distribution allow us to reject null hypothesis of heteroscedasticity but fail to reject that
of autocorrelation and normal distribution. We correct the autocorrelation by
adding AR (1) in Eq. (4) and obtain the post-corrected model which is suggested
to be well-behaved by diagnostic tests.

Table 6 indicates that coefficients of sterilization and offset are both statistically significant at 1% and have expected negative sign. The estimated sterilization coefficient is −0.775, implying an increase in NFA by 1% causes NDA to
decrease simultaneously by 77.5%. The sterilization effectiveness in Vietnam is
also incomplete, similar to sterilization effectiveness of other countries founded
in previous studies. However, in term of specific degree, the sterilization effectiveness of Vietnam is lower than that of the others which is usually founded to
range from 0.79 to 0.96. Meanwhile, the estimated offset coefficient is −0.903,
implying 1% decrease in NDA is offset by 90.3% increase in NFA. This result
indicates the country still maintains restrictions on capital flows though financial
market is opened to some certain extent.

The partial effectiveness of sterilization in Vietnam can be explained by two
reasons. First, monetary authorities did not have appropriate concerns to the
impacts of reserve accumulation on the economy. Figure 4 shows obvious negative
relationship between NFA and NDA, however, large changes of NFA were offset
by smaller changes of NDA, implying inappropriate scale of sterilization with the
volume of foreign exchange purchases. In addition, the opposite changes of NDA
to offset the changes of NFA were not always simultaneous, indicating inappropriate time of sterilization (Fig. 5). In this circumstance, reserve accumulation could

Table 6. Results of model estimation and diagnostic tests

	Equation (3)	Equation (4)
C	0.037*** (0.008)	0.049*** (0.006)
ΔNDA_t	–	−0.903*** (0.065)
ΔNFA_t	−0.775*** (0.132)	–
ΔMM_t	−0.690*** (0.057)	−0.697*** (0.062)
ΔCPI_{t-1}	−0.181 (0.140)	−0.319* (0.196)
GAP_{t-1}	0.361 (0.345)	0.250 (0.500)
$\Delta REER_{t-1}$	−0.164* (0.091)	−0.116 (0.085)
$\Delta(r_t^* + E_t e_{t+1})$	−0.215 (0.329)	−0.455 (0.293)
$(d_1 - 1) * SDr_{t-1}$	−2.381*** (0.917)	–
$(d_2 - 1) * SDe_{t-1}$	–	−1.563* (0.870)
$AR(1)$	–	0.359*** (0.141)
R2 adjusted	0.935	0.911
White test (Probability Chi–Square)	0.794	0.331
Breusch Godfrey LM test (Probability Chi–Square)	0.686	0.650
Jarque–Bera test (Probability)	0.596	0.211

Source: Authors' calculation

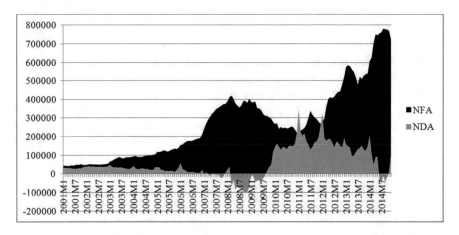

Fig. 4. Scale of NFA and NDA *Source: Authors' calculation from IMF (2015)*

have impact on money supply and inflation. Figure 6 shows very high correlation in fluctuation between NFA growth and inflation. Pham Thi Tuyet Trinh (2015) also indicated inflation increases from the third quarter and settles new equilibrium at 1.1% after an NFA shock.

Second, the high capital mobility, reflected by offset coefficient, negatively influences the effectiveness of sterilization. The KAOPEN index of Chinn and Ito (2014) also indicates Vietnam has higher degree of capital openness than

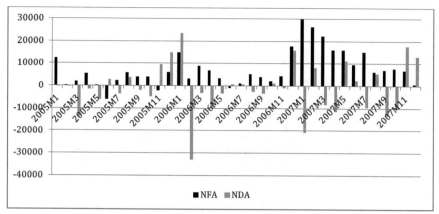

a. Period 2005 - 2007

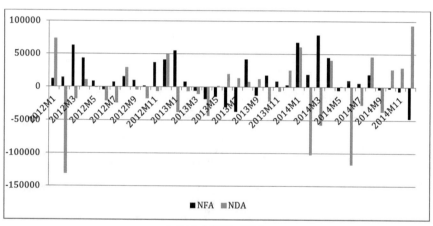

b. Period 2012 - 2014

Fig. 5. Monthly changes of NFA and NDA *Source: Authors' calculation from IMF* (2015)

many other high reserve countries which was founded to have higher sterilization coefficient (Fig. 7). Ha Thi Thieu Dao (2010) also argued that compared with China, while Vietnam has opened to short-term capital flows into and out of the stock market quite quickly, authorities do not have appropriate control measures. This explains why the sterilization effectiveness of Vietnam is lower than that of China which founded to be above 83% (Ouyang et al. 2010; Zhang 2011; Chung et al. 2014).

Besides, estimated results in Table 5 also indicate coefficients of money multiplier (ΔMM_t) are statistically significant in both equations with expected negative sign. This suggests that rise in money multiplier will reduce capital inflow; countering to this problem, SBV tends to implement tight monetary policy. The coefficient of price level (ΔCPI_{t-1}) is significantly negative in Eq. (3) implying higher

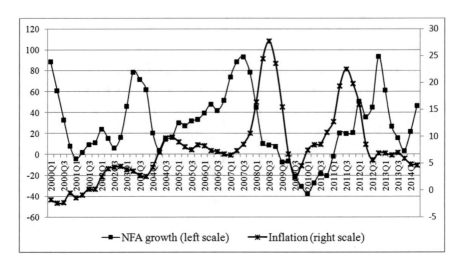

Fig. 6. NFA growth and inflation *Source: Authors' calculation from IMF (2015)*

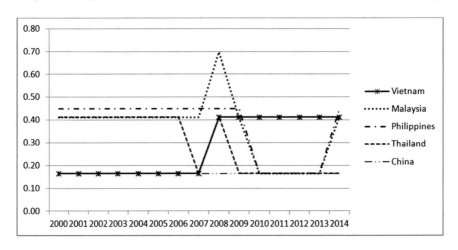

Fig. 7. Capital openness of Vietnam and high reserve countries *Source: Aizenman et al. (2014)*

inflation reduces capital inflows as investors worry about the stability of macroeconomic condition. However, the coefficient of price level is not significant in Eq. (4) indicating SBV does not response to the fluctuation of inflation. The reason is that SBV is not independent enough in conducting monetary policy. In some circumstances, to attain the objectives of economic growth set up by government, SBV is forced to keep expansionary monetary policy regardless of high inflation pressure. Indeed, among four types of central bank autonomy – goal autonomy, target autonomy, instrument autonomy and limited autonomy, SBV is classified into "limited autonomy" (Giang 2010). The significant coefficient of interest rate volatility $((d_1 - 1) * SDr_{t-1})$ and exchange rate volatility $((d_2 - 1) * SDe_{t-1})$ demonstrates that SBV really concerns about stability of interest rate and exchange rate.

Rolling estimation results in Fig. 8 shows that sterilization coefficient tends to increase slightly. In detail, the sterilization coefficient fluctuates around -0.5 before keeping stable at -0.6% from 2012. Meanwhile, offset coefficient seems to be unchanged during the period 2008–2014, reflecting that the degree of capital mobility is very stable. This result implies that the improvement of sterilization effectiveness is not explained by more restrictive capital control. As Vietnam

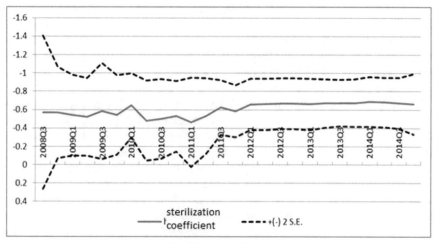

a. Sterilization coefficient

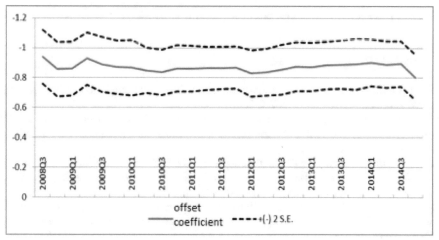

b. Offset coefficient

Fig. 8. Rolling estimation results *Source: Authors' calculation from IMF (2015)*

has two episodes of reserve accumulation, separated by the GFC, the fact that the increase in sterilization coefficient occurred in the second episodes reflects the improvement in sterilization interventions of monetary authorities which can mainly be explained by a more developed and active OMO with large size, both buy and sell sessions in trading date and issued SBV's bill.

6   Conclusions and Recommendations

On examining the effectiveness of sterilizing the monetary impact of foreign reserve accumulation in Vietnam by using quarterly data in period 2000–2014, we find that sterilization interventions were not complete but partial effective at 77.5%, lower than the effectiveness degrees of high reserve countries founded in previous studies. The reasons for the incomplete effectiveness of sterilization are: (i) the high capital mobility; and (ii) inappropriate scale and timing of intervention. This result implies reserves accumulation as one of the sources of high inflation in Vietnam.

Besides, the effectiveness degree of sterilization tends to slightly increase and higher in the second episode of reserve accumulation. While coefficient of capital mobility is stable, higher degree of sterilization effectiveness reflects effort of monetary authorities to improve of sterilization operation.

It seems that Vietnam will continue to accumulate foreign reserves in the future because of self-protection demand in the process of integration. Monetary authorities should pay more concerns to the effectiveness of sterilization interventions to prevent unexpected impacts of reserve accumulation on monetary conditions and price levels. Particularly, scale and timing of sterilization should be more appropriate with those of foreign reserve purchases. In addition, measures of sterilization should be diversified by using other tools fluently such as required reserves ratio, swaps facilities, discount window.

References

Aizenman, J., Glick, R.: Sterilisation, monetary policy, and global financial integration. Rev. Int. Econ. **17**(4), 777–801 (2009)

Aizenman, J., Chinn, M., Ito, H.: The emerging global financial architecture: tracing and evaluating the new patterns of the trilemma's configurations. J. Int. Money Financ. **29**(4), 615–641 (2010)

Argy, V., Kouri, P.J.K.: Sterilization policies and the volatility of international reserves. In: Aliber, R.Z. (ed.) National Monetary Policies and the International Financial System. University of Chicago Press, Chicago (1974)

Brissimis, S.N., Gibson, H.D., Tsakalotos, E.: A unifying framework for analysing offsetting capital flows and sterilization: Germany and the ERM. Int. J. Financ. Econ. **7**(1), 63–78 (2002)

Carmen, U.: Monetary policy in Vietnam: the case of a transition country, in BIS. In: Monetary Policy in Asia: Approaches and Implementation. BIS Papers, vol. 31, pp. 232–252 (2006)

Cavoli, T., Rajan, R.: Have Exchange Rate Regimes in Asia Become More Flexible Post Crisis? Re-visiting the Evidence, Centre for International Economic Studies Working Papers 2005–06, University of Adelaide, Centre for International Economic Studies (2005)

Cavoli, T., Rajan, R.S.: The capital inflows problem in selected Asian economies in the 1990s revisited: the role of monetary sterilization. Asian Econ. J. **20**, 409–423 (2006)

Christensen, J.: Capital inflows, sterilization, and commercial bank speculation: the case of Czech Republic in the mid 1990s. IMF Working paper, 04/218 (2004)

Chung, C., Hwang, J., Wang, C.: The effectiveness and sustainability of the sterilization policy in China. Econ. Res. Int. **2014**, 14 (2014). Article ID 509643

Economic Committee of the National Assembly and UNDP in Vietnam, 2012, Targeting and The Implications for Monetary Policy Framework in Vietnam. http://www.vn.undp.org/content/dam/vietnam/docs/Publications/Inflation_targeting_and_the_implications_for_monetary_policy_framework_in_Vietnam-Eng.pdf. Accessed Sep 2015

Fleming, J.M.: Domestic financial policies under fixed and under floating exchange rates. IMF Staff Pap. **9**(3), 369–379 (1962)

Frenkel, R.: The Sustainability of Sterilization Policy. Center for Economic and Policy Research, Washington D.C. (2007). http://www.itf.org.ar/pdf/documentos/58-2007.pdf. Accessed Sep 2015

GSO (General Statistics Office of Vietnam), Foreign trade data in Vietnam, GSO, Hanoi (2015). http://gso.gov.vn. Accessed Sept 2015

Dao, H.T.T.: Than trong trong xay dung lo trinh tu do hoa giao dich von cua Viet Nam (Prudent route to liberalize financial accounts of Vietnam). J. Econ. Dev. 232 (2010)

Hashmi, M.S.: Monetary policy reaction function and sterilization of captial inflows: an analysis of Asia countries. Asian Soc. Sci. **7**(9), 19–32 (2011)

He, D., Chu, C., Shu, C., Wong, A.: Monetary management in Mainland China in the face of large capital inflows. Research Memorandum no. 7, Hong Kong Monetary Authority, Hongkong (2005). http://www.info.gov.hk/hkma/eng/research/RM07-2005.pdf. Accessed Sep 2015

International Monetary Fund, International Financial Statistics, IMF (2015). http://data.imf.org/?sk=5DABAFF2-C5AD-4D27-A175-1253419C02D1. Accessed Sep 2015

Moreno, R.: Intervention, sterilization, and monetary control in Korea and Taiwan. Econ. Rev. **3**, 23–33 (1996). Federal Reserve Bank of San Francisco

Mundell, R.A.: The monetary dynamics of international adjustment under fixed and flexible rates. Quart. J. Econ. **74**, 227–257 (1960)

Giang, N.H.: Su doc lap cua NHTW va mot so goi y chinh sach cho Viet Nam (Independence of central bank and policy recommendation for Vietnam). Bank. Rev. **23**, 15–23 (2010)

Ouyang, A.Y., Rajan, R.: Reserve accumulation and monetary sterilization in Singapore and Taiwan. Appl. Econ. **43**(16), 2015–2031 (2011)

Ouyang, A.Y., Rajan, R.S., Willett, T.D.: China as a reserve sink: the evidence from offset and sterilization coefficients. J. Int. Money Financ. **29**, 951–972 (2010)

Anh, P.T.H., Phu, B.D.: Danh gia muc do can thiep trung hoa cua NHNN tren thi truong ngoai hoi bang mo hinh tuyen tinh va phi tuyen tinh (Evaluating degree of sterilization of SBV on foreign exchange market by linear and non-linear model). Bank. Rev. **16**, 8–14 (2013)

Huyen, P.T.T.: Nghiep vu thi truong mo sau hon 10 nam thuc hien (Open market operations after 10 years operating). Bank. Rev. **20**, 30–35 (2011)

Ito, H., Chinn, M.: The rise of the "redback" and china's capital account liberalization: an empirical analysis on the determinants of invoicing currencies. In: Proceedings of the ADBI ConferenceCurrency Internationalization: Lessons and Prospects for the RMB, 8 August 2013 (2014)

Trinh, P.T.T.: Tac dong cua tich luy du tru ngoai hoi den lam phat: Tiep can bang mo hinh VAR (The impact of reserves accumulation on inflation: A VAR approach). J. Econ. Dev. **26**(4), 46–48 (2015)

Stephane, C., Cyril, R.: Emerging countries' foreign exchange reserves and accumulation strategies. Treso-Economics, 87, No 87 (2011). www.tresor.economie.gouv.fr/File/327938. Accessed Sep 2015

Takagi, S., Esaka, T.: Sterilization and the capital inflow problem in East Asia. In: Ito, T., Krueger, A.O., (eds.) Regional and Global Capital Flows: Macroeconomics Causes and Consequences, NBER-EASE, vol. 10, pp. 197–232 (2001)

The State Bank of Vietnam, Annual report. Information and Publication Informing House, SBV, Hanoi (2008–2013)

The State Bank of Vietnam, Database (2015). http://sbv.gov.vn. Accessed Sep 2015

Zhang, C.: Sterilization in China: Effectiveness and cost. Working papers, University of Pennsylvania, Wharton School, Weiss Center (2011)

A Study on Optimal Outsourcing Service Nation in the East and Southeast Asian Region: A Comparison Between AHP and Fuzzy AHP Approach

Pham Van Kien[1]([✉]) and Nguyen Ngoc Thach[2]

[1] International Economic Faculty, Banking University of Ho Chi Minh City,
36 Ton That Dam Street, District 1, Ho Chi Minh City, Vietnam
kienpv@buh.edu.vn
[2] Head of Research Institute, Banking University of Ho Chi Minh City,
36 Ton That Dam Street, District 1, Ho Chi Minh City, Vietnam

Abstract. To sustain competitive advantage, nowadays more than ever, outsourcing is crucially important for international and domestic firms. However, making the right outsourcing decision is not a just a simple multiple-factors decision making problem.

The aim of this paper is to construct an outsourcing hierarchy model based on the concept of the analytic hierarchy process with four levels of the most concerned attributes: competitiveness, human resources, business environment, and government policies. With respect to these factors, six typical alternatives in the Southeast Asian region were selected to compare with China. In this study, the authors attempt to investigate the best outsourcing service destination among these seven typical outsourcing service countries.

This study carries out the comparison between the AHP and Fuzzy-AHP approaches to see which one is more acceptable. Theoretically, the higher the priorities weight, the more important the factor or the alternative would be. The findings show that China is the optimal outsourcing service destination. Besides China, other Southeast Asian nations like Vietnam, Thailand, or the Philippines are emerging to be attractive locations. Regarding the main factors to select such the best country, cost competitiveness is the most important attribute. Furthermore, the comparison between the AHP and Fuzzy AHP show some significant differences which allow the study to conclude that the FAHP technique is more suitable to use for the research problem.

This study bridge a gap to enrich the existing literature to provide managers and policy makers a comprehensive view on how to construct an outsourcing hierarchy model based on the analytic hierarchy process and a Fuzzy-AHP approach with competitiveness, human resources, business environment, and government policies.

L. H. Anh et al. (eds.), *Econometrics for Financial Applications*, Studies in Computational Intelligence 760, https://doi.org/10.1007/978-3-319-73150-6_62

1   Introduction

The trend for globalization, tense competition among companies, and the quick change of technology are now putting companies into a dilemma of how to sustain their competitive advantages with a tight budget. In doing so, a firm should focus on its core competencies and outsource the rest to other companies (Mohr et al. 2011). Outsourcing is a very important strategic decision for management of information systems, management of human resource, accounting, management of supply chain, and other organizational functions. In addition, the previous studies indicate that outsourcing enables a firm to not only reduce costs, but focus on maintaining its internal functions and lend impetus to its process of innovation and knowledge transferring (Florin et al. 2005; Graf and Mudambi 2005). It is particularly valid for business organizations in Southeast Asian countries regarding their attempt to restructure or rationalize their processes.

Globalization means that firms, especially in developed countries do not hesitate anymore to step beyond their borders when deciding to outsource (Chen et al. 2016). But among Asian countries, where is the right place to go? And what are the key factors that companies should use to evaluate and select an outsourcing service country? These two questions become the major issues in the study. Moreover, making a decision where the company should outsource is not easy, but requires very complicated methods as evaluating the outsourcing destination is a process including many different factors, decision makers have to make the right choice between these tangible and intangible based on their evaluations (Ghodsypour and O'Brien 1998; Chen et al. 2016). In other words, this is the problem of multiple factors decision making (MCDM). Existing literatures on the outsourcing field have attempted to use various methods based on MCDM-concept to help decision makers to evaluate and select an outsourcing vendor such as an interactive group decision-making approach under multiple factors: Fuzzy TOPSIS method (Kahraman et al. 2009), Fuzzy AHP, Preference Ranking Organization Method, etc. Unfortunately, most of those only focus on ranking and valuating outsourcing vendors based on specific fields at company level like their information system and information technology (Yang and Peng 2012). The aim of this paper is to conduct a comprehensive evaluation at the country level to evaluate the optimal outsourcing service country based on the AHP theory. We select the AHP due to its simplicity and usefulness. However, the limitation of AHP is that it cannot measure well for fuzzy data, while human judgments over attributes given in the research model cannot be reflected by exact numbers due to the nature of subjective and qualitative evaluations (Chan and Kumar 2007). In this paper we also apply fuzzy approach because it takes care of fuzzy data with variables which are either not well defined or ambiguous (Kahraman et al. 2004).

For these reasons, the authors attempt to look at both techniques (AHP and FAHP) to figure out the differences between the two. In order to do so, we divide the study into three main stages. The first stage is to apply the single AHP introduced by Saaty in 1971. The second stage combines the AHP with the fuzzy theory for the purpose of reducing uncertain information proposed by

Chang (1996). The final stage is the comparison between the AHP and the FAHP method. To complete the mentioned tasks, the authors carry out the empirical study which involves six typical countries in Southeast Asia including Vietnam, Indonesia, Malaysia, Thailand, the Philippines, and Singapore to compare with China. As we all know, China has already successfully built its reputation in providing outsourcing service to foreigner partners. The position of China is in no need to discuss, but in our study, could this position be maintained while other Southeast Asian countries are on the rise to become more and more attractive? Hopefully, the findings of this study will not only help to address the said research gaps, but also provide interesting discussions for readers. For scholars, the study would be an important reference to check the significance of the AHP and Fuzzy AHP method. Eventually, decision makers may look at this paper as the useful document before performing an outsourcing decision.

2   Literature Review

2.1   Outsourcing Theory

Technological innovation improves productivity that expands the US economy. However, US economic prosperity has resulted in labor a shortage that motivates organizations to adopt outsourcing strategy (Konrad and Deckop 2001). Outsourcing strategy is an effective cost-saving strategy used by a company to cut down their production costs by transferring part of work to vendors that rents its technology, manpower, skills, knowledge, and service (Adler 2003; Bartholemy 2003; Kakabadse and Kakabadse 2005; Sharma and Loh 2009). Narayanan and Narasimhan (2009) suggest that there are four main reasons to outsource: reduce number of staff, improve business performance, get better control of payment, and improve cash flow. It is difficult for any company to maintain its competitive advantage if it does not adopt the outsourcing strategy because there is strong competition to meet what customers' demand: lower price, high quality, and better technology of a product. In addition, Gupta et al. (2009) explore different strategic concepts to find that outsourcing activities could reduce innovation process obstacles.

Outsourcing became popular because of a recent change in business philosophy. In order to diversify their business and reduce risk, before 1990, many companies liked to merge and acquire other related and unrelated companies. However, many companies found out that it is not wise to run a large number of not-so-related businesses, and changed their business philosophy to focus on one or a few closely related businesses. Companies with core competence would have competitive advantage in their industry. Thus, the main activities of the company's operations are around the core competence, and any activity that is not core competence will then be outsourced.

Companies could decide to adopt different degrees of outsourcing. For example, they could adopt total outsourcing, by dismantling entire departments and outsourcing their functions to an outside vendor, so that they do not take any responsibility in the production. They could also choose selective outsourcing to outsource only one or a few time-consuming task to outside specialists. Vendors providing outsourcing services include Application Service Provider (ASP) and Business Process Outsourcing (BPO), see, for example, Currie and Seltsikas (2001). The companies using the BPO model will transfer major resources to the vendors. Thus, total outsourcing will use BPO. On the other hand, if they choose the ASP model, vendors will provide selected services to several clients and thus, selective outsourcing will use ASP.

Outsourcing is increasing its important role in many industries as well as organizations nowadays. There are many reasons to make outsourcing decision and many investigations have been conducted to study this interesting issue. The studies conclude that the advantage of outsourcing is widely recognized by both scholars and decision makers. However, is outsourcing always good? If so, why do decision makers need to have a careful evaluation on this strategy? These two questions are addressed in this section by looking into the advantages and disadvantage of outsourcing.

Advantages of Outsourcing. The advantage of outsourcing is to realize gains in business profitability and efficiency. Companies embrace outsourcing so that the companies can cut cost and have better control over the outsourced function (Gupta and Gupta 1992). Companies could reduce cost through outsourcing could reinvest to obtain better profit and gain competitive advantage (Jiang and Qureshi 2006). The vendor the companies outsourced specializes in some areas could performs better and more efficiently. The disadvantage of keeping all activities in a company will result in using outdated technology and not complying with government regulations in some of their activities. Thus, another advantage of outsourcing is that the company could focus on its core competence and outsourcing other business to other companies that can use updated technology and comply with government regulations. Outsourcing could lead companies to cut cost, gain knowledge, resulting in leading the company to get new products and technologies. Companies outsourcing could reduce headcount and reduce fluctuations in staffing that may happen because of changes in demand for a product or service. They could also reduce the workload on their existing employees and provide better development opportunities for their staff. They could reduce distractions and focus on their core competencies. Companies could lease some employees so that the company does not have to take care of recruiting, hiring, and training employees. They do not have to take care of employees' health care coverage and other benefits. This is another advantage of outsourcing. Well-developed core competencies adopting outsourcing strategy can concentrate on their investment to maximize returns on internal resources and form barriers against competitors to expand their business into the core interest of the company (Quinn and Hilmar 1994). Quinn (1999) concludes that companies increasing profits through adopting outsourcing strategy is benefited from to today's knowledge and service-based economy.

Disadvantages of Outsourcing. Drawbacks of outsourcing includes poor quality control or even loss of control over the product or service, loss of strategic alignment, decreased company loyalty, and a lengthy bid process. Kremic et al.

(2006) comment that the potential risks of outsourcing include cultural problems and security issues while Lonsdale and Cox (1997) warn managers that there are many outsourcing-related risks. Quinn and Hilmar (1994) comment that markets could be imperfect, face a lot of risks, and operation could be ineffective with a lot of friction, and obtain unnecessary transaction costs when companies carry forward outsourcing practice. In addition, some employees could be laid off as a result of outsourcing, leading to a fall in employee morale (Belcourt 2006).

2.2 Critical Factors for Outsourcing Service Country

As outsourcing is one of the most strategic and complicated business decisions including a variety of critical dimensions, many studies have focused on determining factors affecting the selection of an outsourcing country. According to Ozcan and Suzan (2011), in order to select a good outsourcing service supplier, both qualitative and quantitative factors should be considered. In addition, Ghodsypour and O'Brien (1998) suggested that the best outsourcing vendors are selected by making a tradeoff between tangible and intangible factors. So which are those factors? Since outsourcing as a whole is related to every aspect of culture language, economic and business environment, infrastructure, politics and government policies and so on (King 2005; Vestring et al. 2005; Beaumont and Sohal 2004; Venkatraman 2004; Carmel 2003; Apte and Mason 1995), it would include numerous attributes. Further, Ramingwong and Sanjeev (2007) note that cross-cultural factor is another serious risks to outsourcing when outsourcing to different countries. According to Nahar and Kuivanen (2010), three categories of factors have become critical elements in the success of outsourcing contracts in the developing countries. One of these factors is global and country level factor. At the country level, the researcher proposed some main factors in the conceptual framework to evaluate an outsourcing service provider including different government policies, different cost level, different cultural and language, infrastructure, and human resource, among other factors.

Based on the above discussions and the proposed methodology of this study, all possible factors are drawn from the existing literature and divided into four main groups based on the four most common reasons for an outsourcing performance, namely government policies, business environment, human resources, and cost competitiveness. Each group represents the most concerned factors or sub-factors that have been widely discussed in the relevant literature for the selection and evaluation of the best outsourcing services country (Table 1).

2.3 Outsourcing Service in Southeast Asian

As mentioned, this study selected seven leading countries in the Southeast Asian region including China, Singapore, Malaysia, Vietnam, The Philippines, Indonesia and Thailand. This section discusses shortly the general outsourcing perspective in each country to demonstrate our selection.

China: China has been long well-known as an attractive destination to outsource (Asakawa and Som 2008). Nowadays, everything seems to be "made in China".

Table 1. Factors and sub-factors for evaluating and selecting an outsourcing service country

Factors and sub-factors		Explanations
Cost competitiveness		
F11	Freight costs	Transportation cost to deliver cargo from one point to another, inventory cost, and insurance costs
F12	Labor costs	Amount of money paid to employees
F13	Taxes/Tariffs	Corporate, profit, turnover, property, revenue, labor, and other taxes, and social contributions
F14	Production costs	Costs incurred by a business when manufacturing or producing a product or service
Human resources		
F21	Workforce size	Size of labor force, rigidity of employment, and cost of laying off employee
F22	Education level	Total annual university graduates
F23	Technology readiness	Technology capabilities of labor forces
F24	English ability	Adult literacy and familiarity with English
F25	Culture	Differences of culture, habits and customs
Business environment		
F31	Stability	Politic stability, economic stability, safe business environment, and rare natural disasters
F32	Infrastructure	The state of electricity, road, water, and communication infrastructure as well as transportation system
F33	Corruption	Amount of "brribes" necessary to enforce an outsourcing contract
F34	Full outsourcing service	Ability to provide a full-service outsourcing from raw materials to finished products
Government policies		
F41	Regulations	Overall ease of doing business covers construction permits, registering property
F42	Fair-trade protection	Fair trade is to seek fair equity in international trade among all trading partnerships obtained by respect, transparency, and dialogue, that It contributes to sustainable development by offering better trading conditions to, and securing the rights of, marginalized producers and workers - EFTA
F43	IP protection	Copyright protection, software piracy
F44	Tax/tariff incentives	Including tax cuts, tax exclusion or exemption

Source: Apte and Mason (1995), Bahli and Rivard (2005), Nahar and Kuivanen (2010), Ghodsypour and O'Brien (1998), Carmel (2003), King (2005), Venkatraman (2004), Vestring et al. (2005), Beaumont and Sohal (2004), Brown and Wilson (2005), Collier (1985), Bahli and Rivard (2005), Jain and Song (2002), Jennex (2003), Kumar and Palvia (2002), Rajkumar and Mani (2001), Gattiker et al. (2000), Raval (1999), Tan and Leewongcharoen (2005), Dedrick and Kraemer (2001), WFTO (2009), Cloete et al. (2002).

Therefore it is not surprising when China is also popular with the name "the world's factory" (Lee 1998).

Singapore: The success of Singapore has been widely recognized all over the world and it is no surprise when Singapore is the outsourcing home to a number of "giants" such as IBM, Microsoft, Seagate, Citibank, and so on. However, recently, the world has witnessed the rise of other emerging nations like China or India. These countries are proving to be strong competitors in the outsourcing battle (Kuruvilla et al. 2002; Kuruvilla and Rodney 2000).

Indonesia: Indonesia is the largest country in Southeast Asia. It is one of the G-20 major economies and a member of OPEC. Recently Indonesia works in the direction of not-to-depend on exports to move to modern economy providing outsourcing services (Sameer et al. 2009).

Malaysia: Malaysia is ranked the highest "Business environment that allows". In order to attract more direct foreign investment, Government carries out several reforms to the services sector. Since Government supports industry expertise in BFSI domain knowledge, oil, gas, and logistics, outsourcing industry in Malaysia is growing (Suhaimi et al. 2007; Sohail et al. 2006).

Philippines: the country is leading in both outsourcing services and IT services including software development, web design, maintenance programs, animation, etc. (Yap and Balboa 2008; Mark and Dianne 2006).

Thailand: Thailand is famous in tourism, automotive industry, and information technology industry. It is the sixth top outsourcing countries in 2013 though it only get very limited support from government and lacks outsourcing infrastructure. Thailand government plans to build a four-billion-baht outsourcing industry, including the Thai IT Outsourcing Association, that is a collaboration of ten leading international outsourcing companies today (Thipchutar et al. 2004; Thanapol et al. 2013).

Vietnam: In recent years, Vietnam has become one of the potential outsourcing destinations in the Southeast Asia region due to its great reform policies and benefits gained from the rising cost and growing problems in China. Many big companies from The United States and European countries have moved their factories in China to Vietnam overtime (Jenkins 2004; Gallaugher and Stoller 2004).

3 Data and Methodology

3.1 Questionnaire Design and Data Collection

In order to make all of the possible pair-wise comparisons among the factors and sub-factors, the study designed the questionnaire based on a typical nine-point Likert Scale as shown in Table 3. All of the selected factors in the structure model (refer to Table 2, including four major factors and seventeen sub-factors) were applied to the pair-wise comparisons. The questionnaire is set by experts who have good knowledge in data collection process and in the outsourcing domain. A total of 102 copies of the questionnaire were delivered to 102 experts for their own opinions on the pair-wise comparisons. These experts come from both industry and academia hold MBA's or Ph.D.'s degrees. For those come from the industry, they are CEO, mangers, or other professional staffs from different companies who outsource parts of their business to Asian countries including Vietnam, China and others. For those from academia, they have been studying in various institutes with topics relating to the outsourcing issue. Therefore, participants joined in the questionnaire have expertise in the outsourcing sector as well as deeply understanding about the countries to which they are considering for an outsourcing decision. The data collection process was done by both email and direct interviews.

Table 2. Expert's information.

Working/studying field	Number of experts	Percent	Country	Number of experts	Percent
IT	21	20.6	Vietnam	15	14.7
Semiconductor	15	14.7	Taiwan	15	14.7
Service	12	11.8	India	6	5.9
Accounting	36	35.3	French	3	2.9
Other	18	17.6	USA	5	4.9
Education			Japan	4	3.9
Bachelor	0	0	The Philippines	3	2.9
MBA	63	62	Indonesia	12	11.8
Ph.D	39	38	Malaysia	4	3.9
Job Position			The UK	12	11.8
CEO	2	1.96	Korea	5	4.9
Production Manager	12	11.8	Thailand	5	4.9
Operations Manager	15	14.7	Singapore	5	4.9
Marketing Manager	14	13.7	China	5	4.9
Accountant	14	13.7	Canada	3	2.9
Professional Staff	25	24.5			
Professors	20	19.6			

In consequence, as summarized in Table 2, among 102 experts there are 15 representatives from Taiwan and 15 from Vietnam accounting for 29.4%, following by respondents from Indonesia (11.8%) and the UK (11.8%). With regard to working and studying experiences, this factor is also diversified with 20.6% for IT, 14.7% for semiconductor, 11.8% for service, 35.3% for accounting, and 17.6% for other fields. Regarding to the educational level, most of the experts hold MBA's degrees (62%), the remaining 38% hold Ph.D.'s degree. Eventually, with regard to the job position, while only two CEOs directly joined in the questionnaire, there are up to 41 functional managers (accounting for 40.2%), followed by the professional staffs with 24.5% in total. Professors who are teaching in universities take up 19.6%. The remaining 13.7% are accountants. Based on the analysis of experts' information in Table 2, we have a good reason to believe that participants joining in the survey are experts in the given field.

3.2 The Analytic Hierarchy Process (AHP)

The Analytic Hierarchy Process (Saaty 1980) is a multifactors decision making approach in which factors are arranged in a hierarchic structure. In order to apply the AHP method, it is necessary to construct a hierarchy expressing the relative values of a set of attributes. Decision-makers evaluate the relative importance of the attributes in each level based on the AHP scale. This scale, in turn, is used to direct decision-makers to express their preferences in each pairwise comparison. They are required to select whether each element is of equal importance, somewhat more important, much more important, very much more important or absolutely important to another. These important intensities are, respectively, converted to numeral values in the AHP Scale as 1, 3, 5, 7, 9, and 2, 4, 6, 8 are intermediate values (see Table 3). By using this scale, the qualitative judgments of evaluators are converted into the quantitative values, and thus, construct out a pairwise comparison matrix.

The pairwise comparison matrix is made for all elements to be considered in the construct hierarchy and the results from these comparisons are used to calculate a list of relative weights and importance of the factors (eigenvector) based on the rapid application development (RAD) method (Martin 1991) as follows:

1. Calculate the weight of the factor i:

$$\overline{w_i} = \sqrt[n]{\prod_{j=1}^{n} a_{ij}}, \quad i = 1, 2, \ldots,$$

where the $\overline{w_i}$ is a vector of priorities,
2. Normalize $\overline{w_i}$ to be w_i:

$$w_i = \frac{\overline{w_l}}{\sum_{i=1}^{n} \overline{w_l}}, \quad i = 1, 2, \ldots,$$

3. Compute the maximal eigenvalue:

$$\lambda_{\max} = \sum_{i=1}^{n} \frac{\sum_{j=1}^{n} a_{ij} w_j}{\sum_{j=1}^{n} w_j},$$

4. Calculate the consistency index (CI) introduced by Saaty (1971):

$$CI = \frac{\lambda_{\max} - n}{n - 1}$$

5. Compute the consistency ratio (CR):

$$CR = \frac{CI}{RI}$$

where RI is the average random conformance rate index of judgment matrix, and

6. Obtain the global priority P_i for the factor i:

$$P_i = \sum_{j=1}^{n} w_j l_{ij},$$

where l_{ij} is the local priority.

We note that the judgment matrix is created by experts' judgment over factors. Because there will be inevitable mismatch in the judgment, the individual judgment is not used directly. Instead, the consistency index (CI) is used and it becomes a consistency index of judgment matrix to measure the critical thinking consistency of the decision maker. In order to measure the consistency of the satisfaction in different judgment matrices, the average random conformance rate index (RI) of judgment matrix is used. Readers may read Saaty (1977) for more information. The consistency index CI is used to measure whether the pairwise comparison matrix is consistent:

- If the CR is smaller than 0.1, it is considered to be consistent enough.
- If the CR is much bigger than 0.1, the comparison matrix is considered to be unreliable.

Finally, the global priority P_i is used to synthesize and determine the most important factor (global priority) among a set of given attributes by adopting an additive aggregation with normalization of the sum of the local priorities to unity.

Uncertainty in the AHP. As discussion early, the AHP approach shows some uncertainties in nature (Saaty 2008). Additionally, various scholars (Kabir and Hasin 2011; Sarami et al. 2009; Buckley 1985) have challenged this method by raising many questions about certainty in the ratio used in the AHP. Particularly, they are concerned how decision makers may express their uncertain feelings when considering different factors or alternatives. To circumvent this limitation, academics recommend using fuzzy numbers due to fuzzy numbers' ability to take uncertainty into consideration. The Fuzzy AHP method is discussed in next section.

3.3   The Fuzzy-AHP Theory

The theory of fuzzy systems (Zadeh 1965) extends the classical set theory (Cantor 1874) by dealing with vague, ambiguous, and imprecise information. The classical set theory is used when each element either belongs or does not belong to the set. An element that belongs/does not belong to the set is called member/non-member. The classical set theory is only meaningful for a precise and clear boundary of information. However, in reality, information is commonly available in the appearance of vague, ambiguous or imprecise form, and thus, the classical set theory could not be used (Liu and Zhong 2001). To circumvent the limitation of the classical set theory, fuzzy sets can be used effectively to model uncertain system in industry and solve multi-factors decision making problems when information gathered is incomplete and imprecise. Fuzzy sets overcome the limitations of classical set theory, since they can accept partial membership in which an element partially belongs to a fuzzy set described by a membership function valued from $[0, 1]$, with 0 represents complete non-membership and 1 represents complete membership in which values between 0 and 1 are intermediate degrees of membership.

3.4   MCDM-Method

Selection of the best outsourcing destination among a huge range of providers is an important multi-factors problem in the literature. There are several approaches can be used in the selection, including Foring approach (Lucas and Moore 1976) ranking method (Buss 1983), and linear goal-programing model (Buffa and Jackson 1983). Liu et al. (2000) and Ghodsypour and O'brien (2001) proposed a system of decision by combining the linear programing with AHP technique. Though the AHP is useful in handling both quantitative and qualitative factors of multi-factors decision making problems, judgments, fuzziness and vagueness may result in getting imprecise judgments of decision makers in the conventional AHP approaches (Bouyssou et al. 2000). Buckley (1985) and Chang (1996) show that the fuzzy AHP provides a more adequate description for the decision making process compared to the traditional AHP methods. This study utilizes comparison of the AHP and the Fuzzy AHP method by using optimal outsourcing service country evaluation system due to its ability to construct complex and multi-objective problems hierarchically, analyze each level of the hierarchy independently, and thereby synthesize the final results.

Computing the weights of factors, sub-factors, and alternatives. 55 pairwise comparisons among the main factors, sub-factors, and alternatives based on the typical nine-point scale are combined with fuzzy numbers as shown in Table 3. The next step is to deal with calculating the priority weights of factors, sub-factors, and alternatives by adopting FAHP approach that has been introduced previously. The idea of calculating the priority weights of attributes is based on the pair-wise comparisons given in the questionnaire. In doing so, a set of comparison questions were made to ask the experts for their valuations of one over another. The higher the evaluation, the greater the importance of a factor will be.

Corresponding to four level of the hierarchical model, the experts first evaluated the five main factors in the second level with respect to the overall goal. Then, they compared the sub-factors in the third level with respect to the main factors. Eventually, in the fourth level, pair-wise comparisons of alternatives were made with respect to the overall goal. The linguistic variables were used to direct experts giving their rates, and then these rates in turn were translated into triangular fuzzy numbers as shown in Table 3. As mentioned early, a total 102 questionnaires were administered to 102 different experts from 15 different countries in order to gather their opinions about outsourcing service country selection problem. As a consequence, 102 questionnaires returned accounting for 100% of total sample.

Table 3. AHP and FAHP scale.

Intensity of the AHP scale	Linguistic variable	Positive TFN	Positively reciprocal TFN
1	The same important	(1, 1, 1)	(1, 1, 1)
3	Weakly more important	(2, 3, 4)	(1/4, 1/3, 1/2)
5	Fairly more important	(4, 5, 6)	(1/6, 1/5, 1/4)
7	Strongly more important	(6, 7, 8)	(1/8, 1/7, 1/6)
9	Absolutely more important	(8, 9, 10)	(1/10, 1/9, 1/8)
2, 4, 6, 8	Intermediate values		

Aggregation of Decision Makers' Evaluations. After all pair-wise comparisons were rated by the evaluators, it is crucial to aggregate the decision makers' evaluations. To complete this task, several methods have been proposed. For example, Büyüközkan et al. (2008) and Chang et al. (2009) suggested using the following algorithm to combine fuzzy pair-wise comparison:

$$l_j = \min_k \{w_j^k | k = 1, 2, ..., K\}; \; m_j = \left[\prod_{k=1}^{K} w_j^k\right]^{1/K} ; \; u_j = \max_k \{w_j^k | k = 1, 2, ..., K\}.$$

This approach seems to be not efficient when using min-max values if the sample has a wide range of lower and upper bandwidths. For this reason, this study applied geometric mean for both l_i and u_j.

A geometric mean approach was suggested by Saaty (1990), Dyer and Forman (1992), and Davies (1994) to integrate the individual judgments. The geometric mean is able to deliver satisfying fuzzy group weighting. The geometric average is applied to combine the fuzzy weight of decision makers.

$$\overline{w_l} = \left(\prod_{k=1}^{K} \overline{w_1^k} \right)^{1/K},$$

where:

- $\overline{w_l}$ is the combined fuzzy weight of decision element ℓ over all K decision makers,
- $\overline{w_1^k}$ is the fuzzy weight of decision element l for the k-th decision maker.

Approximation of Fuzzy Priorities. After the evaluations of 102 decision makers were combined and had already passed the consistency test in the previous section, the study then estimates the fuzzy priorities adopting the most common method proposed by Chang (1996). This method is known as the extend analysis method which is defined as follows:

In an outsourcing destination selection and evaluation problem, let $X = \{x_1, \ldots, x_n\}$ indicates an objective goal set, and $G = \{g_1, \ldots, g_n\}$ be a goal set. According to the method of Chang's extent analysis, each object is taken and extent analysis for each goal performed respectively. Therefore, m extent analysis values for each object can be obtained, with the following signs:

$$M_{g_1}^j, M_{g_2}^j, \ldots, M_{g_i}^j,$$

where $j = 1, 2, \ldots, m$.

3.5   Structuring the Hierarchy Model

The hierarchy model for the AHP and FAHP approach (Fig. 1) is designed based on the extensive review of literature from various sources. It is divided into four levels and arranged in descending order. The first level presents the overall goal which is the selection of the optimal outsourcing service destination for outsourcing companies. It therefore is located at the top of the hierarchy. In the second level, four major factors drawn from the previous works are inserted into the model, namely cost competitiveness, human resources, business environment, and government policies. Each factor itself includes several sub-factors in the third level of the hierarchy. For instance, cost competitiveness factor is explained by four sub-factors as freight costs, labor costs, taxes/tariffs, and production costs. Human resources in turn include workforce efficiency, education level, technology readiness, English ability, and culture differences. Similarly, business environment consists of four sub-factors: stability, infrastructure, corruption, and full outsourcing service. The last factor includes regulations, fair-trade protection, IP protection, and tax/tariffs incentives belonging to government policies. In proportion to the overall goal in the top of the hierarchy, seven alternative countries selected from the Southeast Asian regions are presented at the bottom. These alternatives include China, Indonesia, Malaysia, the Philippines, Singapore, Thailand, and Vietnam.

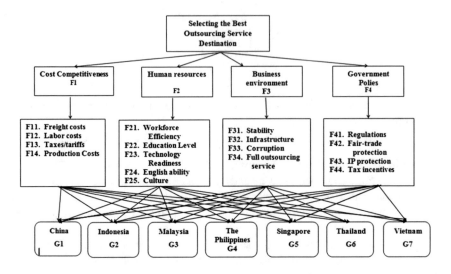

Fig. 1. The hierarchy model for selecting the best outsourcing service country.

4 Results and Discussion

4.1 The Results of the AHP

Synthesizing the Factors and Sub-factors. The study applied the Expert Choice evaluation system to convert the pairwise comparison judgment into the corresponding largest eigenvalue problem and determine the normalized priority weights as illustrated in Table 4. The consistency ratio of each pairwise comparison judgment matrix is 0.00 (<0.01). It means that the consistency of each matrix is very good and the weights calculated are usable. When combining 102 participants together, with respect to the overall goal, the results show that C1 or cost competitiveness has the highest priority weight (0.339). So, it can be concluded that among 4 factors, Cost competitiveness is the most important factor for decision-makers when selecting an outsourcing vendor, followed by C4 or government policies (0.202) and human resources (0.201). The least important factor is C3 or business environment (0.197), but differs little from C2 and C4. Although the priority measurements of sub-factors have been calculated, it is impossible to evaluate which is the most important sub-factor among a total of seventeen sub-factors given in the study. Therefore, this step synthesizes the local priorities across all factors to explore the global priorities. Local priorities in here can be understood as the measurements obtained from each group of sub-factors and these weights in total will be 1.000. Similarly, global priorities are the amounts computed from the local priorities and the global amounts in total will be equal as the weight of their factor. Therefore, the local priority weights calculated for each pairwise judgment matrix are used to synthesize all factors and sub-factors together in order to obtain the global priority weight.

After coming up with the global priority measurements of sub-factors, each sub-factor in turn has been ranked in order of intensity importance (Table 4). The findings show that cost competitiveness is the most important factor and it results in the dominance of its four sub-factors. Specifically, labor costs (F12) rank as the highest priority; followed by taxes/tariffs (F13), production costs (F14), and freight costs (F11), respectively. Then, educational level (F22) and workforce efficiency (F21) rank number five and number six, respectively in the list of priority. These two sub-factors belong to human resources factor, while at the same time English ability (F24), cultural differences (F25) and technology readiness (F26) under human resources are three least important sub-factors. A government's tax/tariff incentives element (F44) ranks seventh in importance, followed by infrastructure (S32) which belong to business environment factor.

Table 4. Composite priority weights for factors and sub-factors

Factors ranking	Factors	Original weight		Sub-factors ranking		Local weight	Global weight
1	F1	0.399	F11	Freight costs	4	0.188	0.075
			F12	Labor costs	1	0.317	0.127
			F13	Taxes/tariffs	2	0.290	0.116
			F14	Production costs	3	0.205	0.082
2	F2	0.201	F21	Workforce efficiency	6	0.297	0.060
			F22	Education level	5	0.301	0.061
			F23	Technology readiness	15	0.160	0.032
			F24	English ability	17	0.109	0.022
			F25	Culture differences	16	0.133	0.027
4	F3	0.197	F31	Stability	9	0.278	0.055
			F32	Infrastructure	8	0.290	0.057
			F33	Corruption	12	0.239	0.047
			F34	Full outsourcing service	14	0.193	0.038
3	F4	0.202	F41	Government policies	11	0.238	0.048
			F42	Fair-trade protection	10	0.254	0.051
			F43	IP protection	13	0.218	0.044
			F44	Taxes/tariffs incentives	7	0.290	0.059

CR = 0.00

Synthesizing the Alternatives. With respect to the alternatives, experts' responses were based on 4 major factors and 17 sub-factors given in the hierarchy and they were asked to rank their priority over the alternatives. In the study, seven countries are selected to be the alternatives and arranged in the matrix to make pairwise comparison. As summarized in Table 5, after normalizing, the priority weights of the alternative were found as

$$(0.21, 0.095, 0.112, 0.158, 0.150, 0.133, 0.142).$$

The final results are presented in percentage, in which China is considered the most important country with the highest percentage of priority (21.9%) and it is ranked much more important than the others six Southeast Asian destinations. Vietnam (16.4%) is the second most attractive country, followed by the Philippines (14.8%), Singapore (14.6%), and Thailand (13.4%), respectively. The differences among these four nations are not big if compared to China. Finally, Indonesia and Malaysia with 9.5% are the countries that have the lowest ranks and it is about more than two times less important than China. Therefore, we can conclude that with the exception of China in the top and Indonesia as well as Malaysia in the bottom, four other typical countries in the Southeast Asian region have almost the same competitive advantage in providing outsourcing services to foreign partners and this advantage may differ from country to country. For example, while the Philippines have its advantage of English ability, Vietnam recently has gained much attention when opening up its economy. Then, if Singapore has an enabling business environment with a strong nation's infrastructure, Thailand has an advantage because of its human resources where loyalty is valued and the favor of concentrated work. These findings strengthen the notion that China is still the world's leading country in terms of outsourcing service provision, while other Southeast Asian nations are new attractive destination for decision makers.

Table 5. Relative importance with respect to alternatives

Alternatives	G1	G2	G3	G4	G5	G6	G7	Priority	Ranking
China (G1)	1	2.17	1.98	1.39	2.05	1.55	1.34	0.219	1
Indonesia (G2)		1	1.02	0.71	0.54	0.74	0.57	0.095	6
Malaysia (G3)			1	0.69	0.54	0.78	0.50	0.095	6
The Philippine (G4)				1	1.07	1.00	1.07	0.148	3
Singapore (G5)					1	1.05	0.87	0.146	4
Thailand (G6)						1	0.80	0.134	5
Vietnam (G7)							1	0.164	2
CR = 0.001									

4.2 The Results of the Fuzzy-AHP

The Fuzzy Comparison Matrix with Respect to the Overall Goal. By using the geometric mean, we obtained the fuzzy comparison matrix as indicated in Table 6. Table 6 also shows that the consistency ratio is 0.03 which is less than the suggested value of 0.1. This means that the matrix can be considered as having an acceptable consistency in the first level of the hierarchy. Accordingly, in order to find the priority weights of main factors, Eq. (2) was used to calculate the fuzzy synthesis values. The different values of four different factors were labeled as F_1, F_2, F_3, and F_4.

Table 6. The fuzzy comparison matrix of factors with respect to the overall goal

F	F1	F2	F3	F4	Sum
F1	(1, 1, 1)	(0.7, 1.1, 1.5)	(0.5, 0.83, 1.17)	(0.74, 1.07, 1.42)	(2.94, 4.0, 5.09)
F2	(0.39, 0.48, 0.59)	(1, 1, 1)	(1.09, 1.32, 1.55)	(0.79, 0.92, 1.05)	(3.27, 3.72, 4.19)
F3	(0.35, 0.39, 0.45)	(0.64, 0.76, 0.92)	(1, 1, 1)	(1.15, 1.34, 1.53)	(3.14, 3.49, 3.9)
F4	(0.33, 0.39, 0.47)	(0.95, 1.09, 1.27)	(0.65, 0.74, 0.87)	(1, 1, 1)	(2.93, 3.22, 3.61)
Sum	(2.07, 2.26, 2.51)	(3.29, 3.95, 4.69)	(3.24, 3.89, 4.59)	(3.68, 4.33, 5)	(12.28, 14.43, 16.79)

Consistency ratio: 0.03

$$F_1 = (2.94, 4.0, 5.09) \times (1/16.79, 1/14.43, 1/12.28) = (0.18, 0.28, 0.41);$$
$$F_2 = (3.27, 3.72, 4.19) \times (1/16.79, 1/14.43, 1/12.28) = (0.19, 0.26, 0.34);$$
$$F_3 = (3.14, 3.49, 3.9) \times (1/16.79, 1/14.43, 1/12.28) = (0.19, 0.24, 0.32);$$
$$F_4 = (2.93, 3.22, 3.61) \times (1/16.79, 1/14.43, 1/12.28) = (0.17, 0.22, 0.29).$$

Then, the Eqs. (6) and (7) were adopted to estimate the possibility degree that F_i is preferable to F_j $(i, j = 1, 2, 3, 4, 5; i \neq j)$ as below:

$$V(F_1 \geq F_2) = 1, \quad V(F_1 \geq F_3) = 1, \quad V(F_1 \geq F_4) = 1;$$
$$V(F_2 \geq F_2) = 0.89, \quad V(F_2 \geq F_3) = 1, \quad V(F_2 \geq F_4) = 1;$$
$$V(F_3 \geq F_1) = 0.58; \quad V(F_3 \geq F_2) = 0.87; \quad V(F_3 \geq F_4) = 1;$$
$$V(F_4 \geq F_1) = 0.65; \quad V(F_4 \geq F_2) = 0.71; \quad V(F_4 \geq F_3) = 0.83.$$

After comparing those above fuzzy numbers, the minimum degree of possibility or the priority weight was given by using the Eq. (9) as follows:

$$d'(F_1) = \min(1, 1, 1) = 1;$$
$$d'(F_2) = \min(0.89, 1, 1) = 0.89,$$
$$d'(F_3) = \min(0.58, 0.87, 1) = 0.58,$$
$$d'(F_4) = \min(0.65, 0.71.0.83) = 0.65.$$

Finally, the weight vector is determined as $W' = (1, 0.89, 0.58, 0.65)^T$. This weight vector was normalized with the aim to determine the priority weights (i.e., eigenvalues) of the main factors with respect to the overall goal. As a result, the weight vector of the main factors which include cost competitiveness, human resources, business environment, and government policies was calculated as

$$(0.32, 0.29, 0.19, 0.21).$$

The final results shows that among five main factors, decision makers rank cost competitiveness are the most important factor accounting for 32% of their decision, followed by human resources (29%), government policies (21%), and business environment (19%), respectively as shown in Fig. 7. Additionally, the same process was carried out in the next sections for the other pair-wise comparison matrices and the priority weights of each sub-factor and alternative with respect to each main factor.

The above process is applied to calculate the priority weights for the following sections including costs, human resources, business environment, and government policy, and alternatives. The final results of the calculations are combined in Table 7.

Table 7. The fuzzy comparison matrix of sub-factors with respect to C1, C2, C3, and C4

F1	Sum	F2	Sum	F3	Sum	F4	Sum
F11	(3.18, 3.57, 4.01)	F21	(3.99, 5.19, 6.51)	F31	(4.52, 5.18, 5.91)	F41	(3.83, 5.18, 5.96)
F12	(4.93, 5.6, 6.22)	F22	(3.61, 4.94, 6.32)	F32	(4.53, 5.31, 6.03)	F42	(3.73, 5.31, 7.04)
F13	(4.17, 4.75, 5.53)	F23	(4.55, 5.27, 6.01)	F33	(3.84, 4.37, 5.23)	F43	(2.75, 3.73, 4.4)
F14	(3.97,4.45,5.12)	F24	(3.5,4.29, 5.03)	F34	(3.53,3.29,5.02)	F44	(2.6,2.89,4.73)
		F25	(3.73, 5.8, 7.84)				
Sum	(16.25, 18.37, 20.88)	Sum	(19.38, 25.49, 31.71)	Sum	(16.42, 18.15, 22.19)	Sum	(12.91, 17.11, 22.13)
Consistency ratio is 0.00							

The Priority Weights with Respect to Costs (F1). Table 7 shows that the consistency ratio is 0.00, smaller than the standardized value of 0.1. This means that the matrix can be considered as having a good consistency in the second level of the hierarchy. The different values of four different sub-factors were denoted as $S_{F11}, S_{F12}, S_{F13}$, and S_{F14}. After calculating, the weight vector was given as $W' = (0.08, 1, 0.71, 0.57)^T$. After Normalizing, the weight vector of sub-factors (set-up costs, freight costs, administrative costs, labor costs, taxes/tariffs, production costs) was calculated as $(0.03, 0.43, 0.30, 0.24)$. Based on the findings, we can see that freight costs account for only 3% in decision maker's evaluations, while labor costs become the most important factor which account for 43%. The second most important factor is taxes/tariffs accounting for 30%, followed by production costs. Therefore, it can be concluded that when considering costs in an outsourcing service location, the most influential dimension is labor costs. Besides that, taxes/tariffs and production costs also play significant roles, while freight costs seem to be a small problem in making an outsourcing performance.

The Priority Weights with Respect to Human Resources (F2). As indicated in Table 7, the consistency ratio is 0.00 less than 0.1, which means that the matrix can be considered as having an acceptable consistency in the second level of the hierarchy. The different values of five different sub-factors (workforce size efficiency, English ability, culture differences, education level, skilled workers, and technology readiness) with respect to human resources were represented as $S_{F21}, S_{F22}, S_{F23}, S_{F24}$ and S_{F25}. From the calculations, the weight vector was given as $W' = (0.88, 0.85, 0.91, 0.71, 1)^T$. This weight vector after normalization becomes $(0.2, 0.2, 0.21, 0.16, 0.23)$. Thus, it can be concluded that among five sub-factors under human resources, culture differences play the most remarkable role in an outsourcing decision, which accounts for 23%. The second most important sub-factor belongs to technology readiness, followed by workforce efficiency and educational level with 20%. Finally, English ability shows the less important role with 16%. In general, there are small differences among five sub-factors, which mean that decision makers attach much importance to every aspect of human resources, especially culture differences.

The Priority Weights with Respect to Business Environment (F3). The fuzzy comparison matrix is presented in Table 7. In here, the consistency ratio is 0.00, which is less than the suggested value of 0.1. This means that the matrix can be considered as having an acceptable consistency in the second level of the hierarchy. The four different sub-factors of business environment attribute were labeled as $S_{F31}, S_{F32}, S_{F33}$, and S_{F34}. From the calculations, the weight vector of the business environment attribute, which includes stability, infrastructure, corruption, and full outsourcing service, was determined as $W' = (1, 1, 0.71, 0.5)^T$. Then, this weight vector was converted to the normalized weight vector

$$(0.31, 0.31, 0.22, 0.16).$$

As a consequence, when considering the business environment of an outsourcing service country, the stability and state of infrastructure are the most important elements accounting for 31%, followed by the corruption situation (22%) and full outsourcing service (16%) respectively.

The Priority Weights with Respect to Government Policies (F4). Table 7 illustrates that the matrix consistency ratio of the fuzzy comparison is 0.0 lower than 0.1. This means that the matrix has a good consistency. The different values of four different sub-factors were denoted as $S_{F41}, S_{F42}, S_{F43}$, and S_{F44}. After calculating, the weight vector was given as $W' = (1, 1, 0.68, 0.61)^T$. This weight vector then is normalized to draw the priority weights of the main factors with respect to government policies. Eventually, the weight vector of the main factors, which includes quality of roads, quality of electric supply, transport system, the communication system, and quality of water supply, was calculated as

$$(0.3, 0.3, 0.21, 0.19).$$

Therefore, we can conclude that the most important element with respect to government policies in outsourcing selection and evaluation process is government regulations and the protection of fair trade which accounts for 30%. The second important element is IP protection accounting for 21%, followed by taxes/tariffs incentives (19%).

The Priority Weights of Alternatives. Finally, the fuzzy comparison matrix and the priority weights of alternatives with respect to the main factors were obtained in this section. Table 8 displays the consistency ratio is 0.03, which is smaller than 0.1. Therefore, the matrix can be considered as having a good consistency in the fourth level of the hierarchy. Alternatives in this study consist of eight representatives in the Asian region, which are China, India, Indonesia, Malaysia, Singapore, Thailand, The Philippines, and Vietnam. The different values of different alternatives were designated as $S_{G1}, S_{G2}, S_{G3}, S_{G4}, S_{G5}, S_{G6}$ and S_{G7}, respectively. As a consequence, the weight vector was determined as $W' = (1, 0.15, 0.35, 0.93, 0.85, 0.67, 0.92)^T$. After normalizing, the priority weights of the alternative was found as

$$(0.21, 0.03, 0.07, 0.19, 0.17, 0.14, 0.19).$$

After converting the priority weights into percentage, we can see that China is considered the most important country with highest the priority weight (21%), followed by The Philippines and Vietnam with the same percentage share of 19%. Singapore comes third with 17%, followed by Thailand with 14%. The last two countries, Malaysia and Indonesia only account for 7% and 4% respectively. We therefore can conclude that China is the optimal country in providing outsourcing services to outsourcers. Besides that, Vietnam and The Philippines are emerging to be attractive destinations for outsourcing services nowadays.

Table 8. The fuzzy comparison matrix of alternatives with respect to the main factors

G	G1	G2	G3	G4	G5	G6	G7
Sum	(6.16, 8.25, 10.22)	(4.91, 5.75, 6.89)	(4.61, 5.5, 6.75)	(6.25, 7.56, 9.53)	(6.06, 7.35, 8.85)	(5.15, 6.17, 7.4)	(6.36, 7.63, 8.83)

4.3 Discussion on the AHP and FAHP Approach

This study uses the quantity method in order to solve the multi-factors decision-making problem based on the concept of the analytic hierarchy process and the fuzzy analytic hierarchy process. This comparison aims at figuring out a numerical illustration on how some factors affect the difference given by the two approaches. Although many previous papers state that the FAHP method may reflect better results (Kabir and Hasin 2011; Bozbura et al. 2007; Kahraman et al. 2004; Chang 1996), to the best of the author's knowledge, there is very few evidence in the existing literature that can confirm this statement. Recently, several papers have been published that compare AHP and FAHP (Debmallya and Bani 2013; Remica and Singh 2013; Saeed et al. 2012; Kabir and Hasin 2011). For example, Debmallya and Bani (2013) worked on the comparison of AHP and Fuzzy AHP in order to evaluate private technical institutions in India. They concluded that between non-fuzzy (AHP) and fuzzy (FAHP) process, there are some differences in terms of priority weights corresponding to some individual sub factors, but in case of the weights corresponding to the factors and sub factors in aggregate there is hardly any difference. In the same year, Remica and Singh (2013) used the AHP and extent Fuzzy AHP approach for prioritization of performance measurement attributes. The authors state that by using the fuzziness, scholars can effectively deal with the linguistic values. Earlier, Kabir and Hasin (2011) carry out a comparative analysis of AHP and FAHP model for multi-factors inventory classification. After the empirical comparison, they come up with a conclusion that conventional AHP and fuzzy AHP are different at different conditions. They also suggest the AHP would be used when information is certain, whereas the fuzzy approach should be preferred if information is not certain. This finding is also supported by Saeed et al. (2012). So, through the review of four recent studies, we can see that there does not have any conclusion or statement about the better method between the AHP and FAHP. Generally, previous studies propose that there exist some differences between two methods. Thus, when information is vague, there is a need to combine fuzziness with the AHP.

In this paper, the research problem is to evaluate the optimal outsourcing service nation among various alternatives by considering many different factors and sub-factors. As the author already mentioned, in order to solve the research problem, we need to perform a complex process based on human judgments over various factors which include cost competitiveness, human resources, business environment, and government policies. These factors seem to be all very important for evaluate an outsourcing service country. So, how to consider this factor is more important than another and with respect to the more important attribute which country should be selected are not easy jobs. To apply the AHP approach in the first study, the authors had to assume that information given by experts is certain and thus, that there are n relationships among factors and sub-factors. However, in reality, these relationships do exist. For instance, a country's cost competitiveness (C1) is very much depended on government policies (C2), and business environment also has a strong connection with government policies. Hence, it is essential to apply fuzzy method. As a consequence, the empirical comparison in this paper shows some differences between two methods in terms of the important level of each factor, sub-factor, and alternative. Particularly, in the AHP approach, the distance between China and other emerging countries like Vietnam, The Philippines, and Thailand is very big (21.9% compared to 16.4%, 14.8%, and 13.4%, respectively), while the FAHP indicates smaller distance (21% compared to 19%, 19%, and 17%). This trend is also happened for main factors when cost competitiveness reduced its share from 39.9% in the AHP to 32% FAHP. Obviously, these evidences allow us to conclude that the fuzzy AHP is more suitable to deal with the research problem due to its ability to reduce uncertainties in human's judgment. However, if we look at the whole picture, we can see that there is no major difference between two approaches in terms of the paper's main goal: namely, both studies show China is an ideal outsourcing service destination. Furthermore, cost competitiveness is considered the most importance attribute in both studies. These findings support the notion that the AHP and FAHP are not different in nature, but FAHP is only an extension of AHP used in cases of uncertain information (Kabir and Hasin 2011; Bozbura et al. 2007; Kahraman et al. 2004; Chang 1996). In conclusion, the use of fuzziness in the second study helps to strengthen the results found in the Study 1. Therefore, this project brings a meaningful approach to deal with the research problem.

5   Conclusion

The study successfully converted the outsourcing country selection problem to the MCDM problem and solved it based on the comprehensive model by using the advanced methodology to calculate the priority of factors and alternatives. Thus, it has brought the more comprehensive view of the given field to readers and practitioners whereby they could easily and quickly understand the factors that support success when making an outsourcing decision. The key findings indicate that cost competitiveness and human resources are among the most important factors to attractive foreign outsourcers. Moreover, among various players in the Southeast Asian region, China is known as a brightest candidate. Therefore, the findings in this paper can also be utilized as a good reference for policy-makers in Southeast Asian countries to know the distance between their countries and China in terms of the ability of outsourcing provision. Finally, managers in both outsourcing and outsourced companies may also apply the paper's model and findings to their decisions. For these reasons, the paper's contributions are presented under three respects namely existing literature, managers, and policy makers. The following will discuss each aspect in more detailed.

To the best of the author's knowledge, existing literature only empirically ranks and valuates outsourcing providers in various specific fields at the company level, ignoring the country level. Furthermore, selecting the best outsourcing destination is a complex process including a lot of conflicting factors, so it requires a comprehensive framework with the aim of aiding decision makers in performing outsourcing contracts. Unfortunately, such a comprehensive framework in the given field has never been done before. Finally, although many previous works mention that the FAHP method may lead to better results than the AHP, there is no empirical evidence in the existing literature that can prove those discussions. For these reasons, these three research gaps motivate the authors to construct a decision-making model in which macro factors at country level are selected and make the comparison between the AHP and FAHP with the aim of figuring out a numerical illustration on how some factors affect the difference given by the two approaches. As a result, the paper successfully built the decision-making model based on the AHP theory and empirically proved that FAHP is more suitable to use in case of information ambiguity and uncertainty. Overall, the paper did provide extended insight in the given field.

As mentioned earlier, the study focuses on solving the research problem at the country level instead of company level like previous works. Therefore, governments in Southeast Asia might apply the findings of this project to their decisions. Particularly, the decision making models in this project mentioned a total of four factors at the country level which include a country's cost competitiveness, human resources, business environment, and government policies. Thus, in order to attract outsourcing contracts from foreign companies, governments need to put more actions to improve their competitive ability. The final findings in this study shows that China is always known as an optimal outsourcing destination in Asia and the gap between China and other Southeast Asian countries is quite big. Therefore, it requires the government in each country like Vietnam, Thailand or Indonesia to form suitable policies and actions to make this gap shorter. Besides that, the rise of Southeast Asian countries meanwhile brings big challenges to China because the costs in China are in the trend of rising while cost competitiveness is the most important factor in general.

Therefore, it creates a good chance to shorten the gap with China for other countries. However, the competition is not only between China and other countries but also the competition among those other countries. Overall, the game for attracting outsourcing contract is the game of governments and the results in this paper should be a good document for government's reference.

By using the hierarchy model in this study, managers in outsourcing seeking companies now have possibility to better structure as well as gain the best outcome from outsourcing performances. The theories applied in this paper bring overall insights for managers in the given field. They would evaluate and select the ideal outsourcing service vendor for their decision based on the main factors, sub-factors, and alternatives given in the study. The findings in this paper might also direct managers to the right place. China or other Southeast Asian countries like Vietnam or Thailand, among those, where is the best destination? The answer is hard to answer as it depends on each industry's feature. For instance, managers in an IT company may prefer to select a country that has a high standard human resources, while in a textile company, managers would care more about a cheap labor force with low skill. However, this project still has its value when providing an overall insight over the outsourcing sector. Besides, managers in outsourcing supply companies may also use the paper's result to enhance outsourcing services at the company level.

Limitations and Future Research Direction. Although the research models was constructed based on the extensive review of literature, it is not yet a perfect solution for the problem of outsourcing vendor selection. In addition, although the advantages and usefulness of the AHP and FAHP approach have been extensively recognized by both researchers and practitioners over the last decades, it still shows some unexpected limitations.

First, in order to apply the AHP and FAHP approach, this study assumed that there were independent relationships among factors in the hierarchy. In fact, it is not that simple, but existing interdependent relationships within the network instead of the hierarchy. For instance, cost competitiveness is closely related to business environment since a better infrastructure system helps to reduce costs in general etc. Thus, it is necessary to apply a method that enables to construct all relationships in the network. Secondly, the study's approach is to reply on the experts' judgments from some different fields to draw results, so it may cause some unexpected biases because they might not represent for all the industries. In reality, the result is probably different depending on which industry (ICs, ICT, pharmacy, etc.) practitioners are outsourcing and on whether practitioners are manufacturing, designing, distributing, etc. Finally, the paper selects a total of four main factors and seventeen sub-factors to construct the research hierarchy. Furthermore, six typical countries in the Southeast Asian region coupled with China are chosen to be seven alternatives. However, there existing many other elements worth considering like the natural disaster or type of economy of a country. Additionally, other countries in the Southeast Asian region including Laos, Cambodia, and Myanmar are also attempting to improve the national conditions in order to welcome foreign outsourcing contracts. These nations have achieved given positions in the race with others in the same region. Therefore, there is a wide gap if ignoring them.

Based on the above discussions, it can be concluded that it is still valuable for future research to improve the MCDM model for evaluating and selecting the optimal outsourcing destination in the Southeast Asian region, meanwhile adopt more advanced methods to reduce the uncertainty and consider the inter-relationships. Specifically, the author suggests future research should adopt the fuzzy analytic network process (FANP) to construct all relationships of factors and sub-factors in the network as well as reduce the uncertainty in calculating the relative weights. Eventually, future research would focus on a specific industry like semiconductors or information and communication technology (ICT) in order to improve the application of the research model and draw a better outcome for the purpose of determine the best outsourcing service country.

References

Adler, P.S.: Making the HR outsourcing decision. MIT Sloan Manag. Rev. **45**(1), 53–60 (2003)

Apte, U.M., Mason, R.O.: Global disaggregation of information intensive services. Manag. Sci. **4**(7), 1250–1262 (1995)

Asakawa, K., Som, A.: Internationalization of R&D in China and India: conventional wisdom versus reality. Asia Pac. J. Manag. **25**(3), 375–394 (2008)

Bahli, B., Rivard, S.: Validating measures of information technology outsourcing risk factors. Omega **33**(2), 175–187 (2005)

Bartholemy, J.: The hard and soft sides of IT outsourcing management. Eur. Manag. J. **21**(5), 539–548 (2003)

Beaumont, N., Sohal, A.: Outsourcing in Australia. Int. J. Oper. Prod. Manag. **24**(7), 688–700 (2004)

Belcourt, M.: Outsourcing - the benefits and the risks. Hum. Resour. Manag. Rev. **16**(2), 269–279 (2006)

Bouyssou, D., Marchant, T., Pirlot, M., Perny, P., Tsoukias, A., Vincke, P.: Evaluation and Decision Models: A Critical Perspective. Kluwer Academic Publishers, Dordrecht (2000)

Brown, D., Wilson, S.: The Black Book Outsourcing, How to Manage the Changes, Challenges, and Opportunities. Wiley, Hoboken (2005)

Bozbura, F.T., Beskese, A., Kahraman, C.: Prioritization of human capital measurement indicators using fuzzy AHP. Expert Syst. Appl. **32**(4), 1100–1112 (2007)

Buckley, J.J.: Fuzzy hierarchical analysis. Fuzzy Sets Syst. **17**(3), 233–247 (1985)

Buffa, F.P., Jackson, W.M.: A goal programming model for purchase planning. J. Purch. Mater. Manag. **19**(13), 27–34 (1983)

Buss, M.D.J.: How to rank computer projects. Harvard Bus. Rev. **61**(1), 118–125 (1983)

Büyüközkan, G., Feyzioglu, O., Nebol, E.: Selection of the strategic alliance partner in logistics value chain. Int. J. Prod. Econ. **11**(31), 148–158 (2008)

Cantor, G.: Ueber eine Eigenschaft des Inbegriffes aller reellen algebraischen Zahlen. Journal für die reine und angewandte Mathematik **77**, 258–262 (1874)

Carmel, E.: The new software exporting nations: success factors. Electron. J. Inf. Syst. Dev. Countries **13**(4), 1–12 (2003)

Carmel, E.: Taxonomy of new software exporting nations. Electron. J. Inf. Syst. Dev. Countries **13**(2), 1–6 (2003)

Chan, F.T.S., Kumar, N.: Global supplier development considering risk factors using fuzzy extended AHP based approach. Omega **35**(4), 417–431 (2007)

Chang, C.W., Wu, C.R., Lin, H.L.: Applying fuzzy hierarchy multiple attributes to construct an expert decision making process. Expert Syst. Appl. **36**(4), 7363–7368 (2009)

Chang, D.Y.: Applications of the extent analysis method on FAHP. Eur. J. Oper. Res. **95**(3), 649–655 (1996)

Chen, S., Pham, V.K., Chen, J.K.: Evaluating and selecting the best outsourcing service country in East and Southeast Asia: an AHP approach. J. Test. Eval. **44**(1), 89–101 (2016)

Cloete, E., Courtney, S., Fintz, J.: Small businesses' acceptance and adoption of e-commerce in the Western-Cape province of South-Africa. Electron. J. Inf. Syst. Dev. Countries **10**(4), 1–13 (2002)

Collier, D.A.: Service Management: The Automation of Services. Reston Publishing, Reston (1985)

Currie, W.L., Seltsikas, P.: Exploring the supply-side of IT outsourcing: evaluating the emerging role of application service providers. Eur. J. Inf. Syst. **10**(3), 123–134 (2001)

Debmallya, C., Bani, M.: Potential hospital location selection using AHP: a study in rural India. Int. J. Comput. Appl. **71**(17), 1–7 (2013)

Davies, M.A.P.: A multi-criteria decision model application for managing group decisions. J. Oper. Res. Soc. **45**(1), 47–58 (1994)

Dedrick, J., Kraemer, K.L.: China IT report. Electron. J. Inf. Syst. Dev. Countries **6**(2), 1–10 (2001)

Dyer, R.F., Forman, E.H.: Group decision support with the analytic hierarchy process. Decis. Support Syst. **8**(2), 99–124 (1992)

Florin, J., Bradford, M., Pagach, D.: Information technology outsourcing and organizational restructuring: an explanation of their effects on RM value. J. High Technol. Manag. Res. **16**, 241–253 (2005)

Gallaugher, J., Stoller, G.: Software outsourcing in Vietnam: a case study of a locally operating pioneer. Electron. J. Inf. Syst. Dev. Countries **17**(1), 1–18 (2004)

Gattiker, U.E., Perlusz, S., Bohmann, K.: Using the internet for B2B activities: a review and future directions for research. Internet Res. **10**(2), 126–140 (2000)

Ghodsypour, S.H., O'Brien, C.: The total cost of logistics in supplier selection, under conditions of multiple sourcing, multiple criteria and capacity constraint. Int. J. Prod. Econ. **73**(1), 15–27 (2001)

Ghodsypour, S.H., O'Brien, C.: A decision support system for supplier selection using an integrated analytic hierarchy process and linear programming. Int. J. Prod. Econ. **56**(57), 199–212 (1998)

Graf, M., Mudambi, S.M.: The outsourcing of IT-enabled business processes: a conceptual model of the location decision. J. Int. Manag. **11**(2), 253–268 (2005)

Gupta, S., Woodside, A., Chris, D., Bradmore, D.: Diffusing knowledge-based core competencies for leveraging innovation strategies: modelling outsourcing to knowledge process organizations (KPOs) in pharmaceutical networks. Ind. Mark. Manag. **38**, 219–227 (2009)

Gupta, U.G., Gupta, A.: Outsourcing the IS function: is it necessary for your organization? Inf. Syst. Manag. **9**(3), 44–47 (1992)

Jain, H.K., Song, J.: Location economics and global software development centers. In: Palvia, P.C., Palvia, S.C.J., Roche, E.M. (eds.) Global Information Technology and Electronic Commerce: Issues for the New Millennium, pp. 447–462. Ivy League Publishing Limited, Marietta (2002)

Jennex, M.E.: IT in the energy sectors of Ukraine, Armenia, and Georgia. Commun. AIS **11**, 413–437 (2003)

Jenkins, R.: Vietnam in the global economy: trade, employment and poverty. J. Int. Dev. **16**(1), 13–28 (2004)

Jiang, B., Qureshi, A.: Research on outsourcing results: current literature and future opportunities. Manag. Decis. **44**(1), 44–55 (2006)

Kabir, G., Hasin, M.A.A.: Evaluation of customer oriented success factors in mobile commerce using fuzzy AHP. J. Ind. Eng. Manag. **4**(2), 361–386 (2011)

Kabir, G., Hasin, M.A.A.: Comparative analysis of AHP and fuzzy AHP models for multicriteria inventory classification. Int. J. Fuzzy Logic Syst. **1**(1), 1–16 (2011)

Kahraman, C., Cebeu, U., Ruan, D.: Multi-attribute comparison of catering service companies using fuzzy AHP: the case of Turkey. Int. J. Prod. Econ. **87**(16), 171–184 (2004)

Kahraman, C., Engin, O., Kabak, Ö., Kaya, I.: Information systems outsourcing decisions using a group decision-making approach. Eng. Appl. Artif. Intell. **22**, 832–841 (2009)

Kakabadse, A., Kakabadse, N.: Outsourcing: current and future trends. Thunderbird Int. Bus. Rev. **47**(2), 183–204 (2005)

King, W.R.: Outsourcing becomes more complex. Inf. Syst. Manag. **22**(2), 89–90 (2005)

Konrad, A.M., Deckop, J.: Human resource management trends in the USA-challenges in the midst of prosperity. Int. J. Manpower **22**(3), 269–278 (2001)

Kremic, T., Tukel, O.I., Rom, W.O.: Outsourcing decision support: a survey of benefits, risks, and decision factors. Supply Chain Manag. Int. J. **11**(6), 467–482 (2006)

Kumar, N., Palvia, P.: A framework for global IT outsourcing management: key influence factors and strategies. J. Inf. Technol. Cases Appl. **4**(1), 56–75 (2002)

Kuruvilla, S., Rodney, C.: How do nations increase workforce skills? Factors influencing the success of the Singapore skills development system. Glob. Bus. Rev. **1**(1), 11–49 (2000)

Kuruvilla, S., Erickson, C., Hwang, A.: An assessment of the Singapore skills development system: lessons for developing countries. World Dev. **30**(8), 1461–1476 (2002)

Lee, C.K.: Gender and the South China Miracle: Two Worlds of Factory Women. University of California Press, Berkeley (1998)

Liu, C., Zhong, N.: Rough problem settings for ILP dealing with imperfect data. Comput. Intell. **17**(3), 446–459 (2001)

Liu, J., Ding, F.Y., Lall, V.: Using data envelopment analysis to compare suppliers for supplier selection and performance improvement. Supply Chain Manag. Int. J. **5**(3), 143–150 (2000)

Lonsdale, C., Cox, A.: Outsourcing: risk and rewards. Supply Manag. **3**, 32–34 (1997)

Lonsdale, C., Cox, A.: The historical development of outsourcing: the latest fad? Ind. Manag. Data Syst. **100**(9), 44–50 (2000)

Lucas, H.C., Moore, J.R.: A multiple-criterion scoring approach to information system project selection. Inf. Syst. Oper. Res. **14**(1), 1–12 (1976)

Mark, M.J., Dianne, H.B.W.: Outsourcing in the IT industry: the case of the Philippines. Int. Entrepreneurship Manag. J. **2**(1), 111–123 (2006)

Martin, J.: Rapid Application Development. Macmillan Publishing, New York (1991)

Mohr, J.J., Sengupta, S., Slater, S.S.: Mapping the outsourcing landscape. J. Bus. Strategy **32**(1), 42–50 (2011)

Nahar, N., Kuivanen, L.: An integrative conceptual model of Vietnam as an emerging destination. J. Int. Technol. Inf. Manag. **19**(3), 39–74 (2010)

Narayanan, S., Narasimhan, R.: In search of outsourcing excellence. J. Transp. Res. Board **13**(2), 36–42 (2009)

Ozcan, K., Suzan, A.O.: Fuzzy AHP approach for supplier selection in a washing machine company. Expert Syst. Appl. **38**(8), 9656–9664 (2011)

Quinn, J.B.: Strategic outsourcing: leveraging knowledge capabilities. MIT Sloan Manag. Rev. **40**(4), 9–21 (1999)

Quinn, J.B., Hilmar, F.G.: Strategic outsourcing. MIT Sloan Manag. Rev. **35**(4), 43–55 (1994)

Rajkumar, T.M., Mani, R.V.S.: Offshore software development, the view from Indian suppliers. Inf. Syst. Manag. **18**(2), 63–73 (2001)

Ramingwong, S., Sanjeev, A.S.M.: Offshore outsourcing: the risk of keeping mum. Commun. ACM **50**(8), 101–103 (2007). Please check and confirm the inserted article title for "Ramingwong and Sanjeev (2007)"

Raval, V.: Seven secrets of successful offshore software development, information strategy. Executive's J. **15**(4), 34–39 (1999)

Remica, A., Singh, A.: AHP and extent fuzzy AHP approach for prioritization of performance measurement attributes. Int. J. Soc. Hum. Sci. Eng. **7**(1), 43–48 (2013)

Saaty, T.L.: Decision making with the analytical hierarchy process. Int. J. Serv. **1**(1), 83–98 (2008)

Saaty, T.L.: How to make a decision: the analytic hierarchy process. Eur. J. Oper. Res. **48**(1), 9–26 (1990)

Saaty, T.L.: The Analytical Hierarchy Process. McGraw-Hill, New York (1980)

Saaty, T.L.: A scaling method for priorities in a hierarchical structure. J. Math. Psychol. **15**, 234–281 (1977)

Saaty, T.L.: On polynomials and crossing numbers of complete graphs. J. Comb. Theory **10**(2), 183–184 (1971)

Saeed, N.A., Mansoor, K.M., Jahromi, A.R.M., Sayareh, J.: Comparison of AHP and FAHP for selecting yard gantry cranes in marine container terminals. J. Persian Gulf **3**(7), 59–70 (2012)

Sameer, K., Jennifer, M., Mark, T.N.: Is the offshore outsourcing landscape for US manufacturers migrating away from China? Supply Chain Manag. Int. J. **14**(5), 342–348 (2009). https://doi.org/10.1108/13598540910980251

Sarami, M., Mousavi, S.F., Sanayei, A.: TQM consultant selection in smes with TOPSIS under fuzzy environment. Expert Syst. Appl. **36**(2), 2742–2749 (2009)

Sharma, A., Loh, P.: Emerging trends in sourcing of business service. Bus. Process Manag. J. **15**(2), 149–165 (2009)

Sohail, M.S., Bhatnagar, R., Sohal, A.S.: A comparative study on the use of third party logistics services by Singaporean and Malaysian firms. Int. J. Phys. Distrib. Logistics Manag. **36**(9), 690–701 (2006)

Suhaimi, M.A., Hussin, H., Mustaffa, M.: Information systems outsourcing - motivations and implementation strategy in a Malaysian bank. Bus. Process Manag. J. **13**(5), 644–661 (2007)

Tan, B.F., Leewongcharoen, K.: Factors contributing to IT industry success in developing countries: the case of Thailand. Inf. Technol. Dev. **11**(2), 161–194 (2005)

Thanapol, O., Settapong, M., Navneet, M.: Risk analysis of it outsourcing case study on public companies in Thailand. J. Econ. Bus. Manag. **1**(4), 365–370 (2013)

Venkatraman, N.V.: Offshoring without guilt. MIT Sloan Manag. Rev. **45**(3), 14–16 (2004)

Vestring, T., Rouse, T., Reinert, U.: Hedge your offshoring bets. MIT Sloan Manag. Rev. **46**(3), 27–29 (2005)

World Fair Trade Organization (WFTO): Charter of Fair Trade Principles. World Wide Web (2009). http://www.wfto.com/index.php?option=com_content&task=view&id=1082&Itemid=12. Accessed 12 Dec 2012

Yang, L.J., Peng, J.L.: Comprehensive evaluation for selecting IS/IT outsourcing vendors based on AHP. J. Inf. Comput. Sci. **9**(9), 2515–2525 (2012)

Yap, J., Balboa, J.: Why has the Philippines lagged. The East Asian Bureau Econ. Res. (EABER) Newsletter (2008)

Zadeh, L.A.: Fuzzy sets. Inf. Control **8**(3), 338–353 (1965)

Expectile Kink Regression: An Application to Service Sector Output

Varith Pipitpojanakarn[1], Paravee Maneejuk[1,2], Worapon Yamaka[1,2(✉)], and Songsak Sriboonchitta[1,2]

[1] Faculty of Economics, Chiang Mai University, Chiang Mai 50200, Thailand
oakvarith@gmail.com, mparavee@gmail.com, woraphon.econ@gmail.com,
songsakecon@gmail.com
[2] Center of Excellence in Econometrics,
Chiang Mai University, Chiang Mai 50200, Thailand

Abstract. In this study, we propose a non-linear model for explaining the relationship between the dependent and the independent variables beyond the conditional mean. We extend the kink approach to expectile regression thus the model provides a more flexible means to explain the non-linear relationship in the model across different expectile indices. We also introduce the sup-F statistic test for the existence of kink effect in each expectile. The simulation and application studies are also proposed to examine the performance of our model. We apply our methodology to study the input factor affecting service sector growth in Asian economy. The use of this model allows us to identify and explore the non-linear labour effect on the service output. We can find both labour effect and kink effect present over a range of expectiles in the service output in this application.

1 Introduction

It is challenging for the researcher to find the relationship between the dependent and the independent variables beyond the conditional mean obtained from linear regression. The expectile regression of Aigner et al. [1] and Newey and Powell [8] was proposed and became a nice tool for estimating the conditional expectiles of independent variable given a set of independent variables. As the expectile regression is analogous to quantile regression, the studies of Koenker [5] and Yao and Tong [15] tried to make comparison between these two models and found that both have their advantages over each other. While quantile regression is more robust against outliers in the response measurements than expectile regression, the computation of expectile regression is much easier and the calculation of the asymptotic covariance matrix of the multiple linear expectile regression estimator does not involve calculating the values of the density function. In addition, Sobotka and Kneib [7] also claimed that expectile regression is more efficient when applied to the real data since it relies on the distance of observations from predictor variables while quantile regression only relies on

© Springer International Publishing AG 2018
L. H. Anh et al. (eds.), *Econometrics for Financial Applications*, Studies in Computational Intelligence 760, https://doi.org/10.1007/978-3-319-73150-6_63

the information on whether an observation is above or below the predictor variables. Thus, in this study, we adopt the expectile regression. In the context of this model, it can be written as

$$Y = X\beta^\tau + \varepsilon, \tag{1}$$

where Y is the dependent variable, X is vector of independent variables, and β^τ is the estimated coefficient at given τ-expectile. Newey and Powell [8] showed that one could estimate the conditional τ-expectile by minimizing the empirical check loss where the check loss function is defined as

$$e_\tau(\varepsilon|\tau) = |I(\varepsilon \le 0) - \tau| \varepsilon^2 \tag{2}$$

The expectile regression seems to be flexible and workable to study the relationship between dependent and independent variables beyond conditional mean. However, some economic phenomena involve a structural change which sometime is explained by two line segments with different slopes [10,11]. Therefore, this study aims to improve the linear expectile regression model by allowing for a nonlinear relationship between dependent and independent variables at any expectile. In the literature, some findings confirmed the nonlinear relationship of the economic data and the linear models fail to explain an extreme market conditions that mostly entail large changes. Recently, a number of nonlinear models have been developed in the expectile fashion, to capture the nonlinear relationship between dependent and independent variables, which represents spline smoothing approach to non parametric regressions. (See, [4,6]). Although these approaches worked well to predict the response variable at any expectile levels, they cannot provide the threshold value of the relationship between dependent and independent variables. Therefore, the threshold model of Tong [12,13] has been applied to expectile regressions to find the threshold value of the models. For a brief review of the threshold expectile regression, refer to Zhang and Li [16].

From the review of the above literature, although threshold model has previously been built into expectile regression models, virtually no works were found investigating the expectile kink regressions except for Zhang and Li [16] who developed a continuous threshold expectile regression. In this study, we employ a kink regression approach of Hansen [3] instead of using threshold model since it has continuous function but the slope has a discontinuity at a threshold point, called "kink". In addition, this approach is more flexible since it allows us to have a kink point for either some or any independent variables. Thus, in this study, we develop a expectile kink regression model to explain the non-linear structure and also make an assessment of the complete conditional distribution of responses even in the presence of heteroscedastic errors.

In the estimation procedures, a least asymmetrically weighted squares (LAWS) method is employed for expectile kink regression model. Hence, the contribution of this study is three-fold. First, we develop expectile regression models by applying a kink regression hence the models have an ability to estimate non-linear relationship between dependent and independent variables at given τ-expectile. Second, we apply a recently developed inference method by Hansen [3] in expectile kink regression for testing the existence of structural change at a given expectile level, for this test only requires fitting the model under the null hypothesis in the absence of a threshold and thus it is computationally efficient.

For the empirical application, we apply our model and inference procedure to study the input factor affecting service sector growth in Asian economy. The aim is to understand the nature and direction of relationship between the service sector growth of Asia and its determinants since the change in service sector can give a huge impact on nearly everyone within 45 Asian countries.

The remainder of the study is organized as follows: in the second section we describe the basic ideas of expectile kink regressions. We then introduce the estimation techniques; and the statistical test for kink effect in the model. The third section contains simulation study in order to confirm the accuracy of the model. In Sect. 4, we will describe the data set on service sector in Asian countries and investigate the impact of the different covariates at different expectiles as well as different economic states (regimes). In the last section, we provide the conclusions with possible future extensions.

2 Methodology

In this section, let us explain the basic ideas of expectile kink regression model. Then, we will explain about the estimation technique and the testing for kink effect.

2.1 Geoadditve Expectile Kink Regression

2.1.1 Model Structure

Newey and Powell [8] showed that one can estimate the conditional expectile of dependent and independent variables. Here, we are applying a kink regression approach to expectile regression, thus we rewrite Eq. (1) as

$$
\begin{aligned}
y_j = {} & \beta_{\tau,1}^-(x'_{1,j} - \gamma_{\tau,1})_- + \beta_{\tau,1}^+(x'_{1,j} - \gamma_{\tau,1})_+ + \\
& + \cdots + \beta_{\tau,k}^-(x'_{k,j} - \gamma_{\tau,k})_+ + Z_{i,j}\phi_{\tau,i} + \varepsilon_j
\end{aligned}
\tag{3}
$$

where $j = 1, ..., N$ Suppose we have N countries in the model, the term $y_{i,j}$ is a response variable and $x_{i,j}$ is $(N \times k)$ matrix of k predictor variables. $(\beta_{\tau,1}^-, \cdots, \beta_{\tau,k}^-)$ and $(\beta_{\tau,1}^+, \cdots, \beta_{\tau,k}^+)$ and are the coefficients of lower and upper regimes, respectively. Following Hansen [3], we use $(x'_{k,j})_- = \min\{x'_{k,j}, 0\}$ and $(x'_{k,j})_+ = \max\{x'_{k,j}, 0\}$ to separate $x'_{k,j}$ into two regimes. The parameter $\gamma_{\tau,k}$ is called "kink" or "threshold" parameter which is in the interior of the support of $x'_{k,j}$. In Eq. (3), the slope with respect to variable $x'_{k,j}$ equals $\beta_{\tau,k}^-$ for value of $x'_{k,j} < \gamma_{\tau,k}$ while β_k^+ is the coefficient of the variable $x'_{k,j} > \gamma_{\tau,k}$. In addition, the model also includes vector of covariates Z_i whose relationship with y_j is linear and ϕ which are regime independent coefficients of $Z_{i,j}$. The error term of the model is ε_j where the distribution depends on the $\tau^{th}(0 < \tau < 1)$ expectile level.

2.2   Estimation of Expectiles

Prior to explaining the estimation that we employ in this study, let us briefly review expectiles. Let $\Theta_\tau = \left\{ \beta^-_{\tau,k}, \beta^+_{\tau,k}, \phi_{\tau,k}, \gamma_{\tau,k} \right\}$, the estimated parameter of the all conditional expectile level is based on least asymmetrically weighted squares (LAWS) on the following key observation.

$$\Theta_\tau = \arg\min_{\Theta_\tau} \frac{1}{N} \sum_{j=1}^{N} e_\tau \left(y_j - \zeta(x_j, \Theta_\tau) \right), \tag{4}$$

where $\zeta(x_j, \Theta_\tau)$ is formulated as a non-linear function of parameters in Eq. (3); the function $e_\tau(\cdot)$ is the check function of expectile, respectively where

$$e_\tau(\varepsilon) = \begin{cases} (1 - \tau)\varepsilon_j^2 & , \ y \le \varepsilon_j, \\ \tau \varepsilon_j^2 & , \ y > \varepsilon_j, \end{cases} \tag{5}$$

respectively. As mentioned above, these two models are different but closely related. Newey and Powell [8] suggested that expectiles are determined by tail expectations.

For the minimization problem in Eq. (5), there is no closed form solution since the objective function is not convex in kink parameter, $\gamma_{\tau,k}$. Hence, it is very difficult to reach a global minimization. Thus, in this study, we can consider the profile expectile kink regression. Here, the selected estimation approach is a combination of concentration and grid search strategy. To proceed, we rewrite the objective function Eq. (4) with respect to $\theta = (\beta^-_{\tau,k}, \beta^+_{\tau,k}, \phi_{\tau,i})$ as

$$\theta_\tau(\gamma_\tau) = \arg\min_{\theta_\tau} S_{N\tau}(\theta_\tau, \gamma_\tau) = \arg\min_{\theta_\tau} \frac{1}{N} \sum_{j=1}^{N} e_\tau \left(y_j - \zeta(x_j, \theta_\tau) \right), \tag{6}$$

The criteria function $S_{N\tau}(\theta_\tau, \gamma_\tau)$ is the concentrated sum-of-squared errors function in θ_τ. To find γ_τ, we can obtain by

$$\gamma_\tau = \arg\min_{\gamma_\tau \in \Gamma} \min_{\theta_\tau} S_{N\tau}(\theta_\tau, \gamma_\tau). \tag{7}$$

Where Γ is the range of all possible $\gamma_{i\tau}$ in the grid search. At each grid point for $\gamma_{i\tau}$ we estimate the least squares coefficients and compute $S_{N\tau}(\theta_\tau, \gamma_\tau)$. Then we can find the γ_τ at the minimum $S_{N\tau}(\theta_\tau, \gamma_\tau)$. After γ_τ is found, the estimated parameter θ_τ are obtained by least asymmetrically weighted squares of y_j on $x_j(\gamma_\tau)$ where

$$x_j(\gamma_\tau) = \begin{pmatrix} (x_j - \gamma_\tau)_- \\ (x_j - \gamma_\tau)_+ \\ Z \end{pmatrix}. \tag{8}$$

2.3   Testing for a Kink Effect at a Given Expectile

It is important to check whether there exists a kink at τ-expectile before fitting our proposed model. If a kink does not exist, a linear expectile regression is preferred. To test the existence of a kink or threshold effect, let us consider a simple non-linear equation

$$y_j = \beta_{\tau,1}^-(x_{1,j}' - \gamma_{\tau,1})_- + \beta_{\tau,1}^+(x_{1,j}' - \gamma_{\tau,1})_+ + \varepsilon_j, \tag{9}$$

against a linear equation

$$y_j = x_{1,j}'\beta_1^\tau + \varepsilon_j \tag{10}$$

We test null (H_0) and alternative (H_1) hypotheses where

$$H_0 : \beta_{\tau,1} = \beta_{\tau,1}^+ \text{ for any } \gamma_{\tau,1} \in \Gamma$$

$$H_1 : \beta_{\tau,1} \neq \beta_{\tau,1}^+ \text{ for some } \gamma_{\tau,1} \in \Gamma$$

These hypotheses are composite hypotheses, which are designed to test the existence of a kink effect at each expectile point τ. For different $\gamma_{\tau,1}$, our test allows different kink values to exist under our hypothesis.

To construct our test statistic, we take an approach in spirit similar to the test for kink effect in regression in Hansen [3]. Because the kink or threshold value is unknown, we need to search through all the possible values in $x_{1,j}'$. Therefore, we employ the F-statistic to be

$$T_n = \sup_{y \in \Gamma} \frac{1}{\tau(1-\tau)} G(\gamma, \tau)\Omega(\gamma, \tau)^{-1}G(\gamma, \tau) \tag{11}$$

where $G(\gamma)$ is a zero-mean normal with covariance kernel

$$\mathbb{E}\left(G(\tau_1, \gamma_1)G(\tau_2, \gamma_2)'\right) = \mathbb{E}\left(x_t(\gamma_1)x_t(\gamma_2)'\varepsilon_j^2\right) \tag{12}$$

and, $\Omega(\gamma, \tau) \equiv \mathbb{E}\left(x_j x_j' 1\{x_i \leq \gamma\}f(\Theta_\tau^0 x_j | x_j)\right)$, where $f(\Theta_\tau^0 x_j | x_j)$ is the conditional probability density function y_j of x_j given x_j. This statistic test shows that the asymptotic null distribution of the kink F-statistic can be written as the supremum of a stochastic process. Following Hansen [3], we test expectile kink with unknown threshold and simulate the critical values for the sup T_n statistic with the following algorithm:

1. Generate iid $\{u_1, \cdots, u_N\}$ from $N(0, 1)$.
2. Set $\varepsilon_j^* = \varepsilon_j^0 u_j$ where ε_j^0 are the estimated residuals obtained from fitted linear expectile regression under the null hypothesis: linear model.
3. Calculate the F-statistic for every γ

$$T_n^\tau = N(\tilde{\sigma}^\tau - \sigma^\tau)/\sigma^\tau \tag{13}$$

where $\tilde{\sigma}^\tau = \varepsilon_j^2 = (\varepsilon_j^* - x_j'\beta_\tau^*)^2$ and β_τ^* are obtained from least asymmetrically weighted squares of ε_j^* on x_j of linear expectile regression. $\sigma^\tau = \varepsilon_j^2 = (\varepsilon_j^* - \beta_\tau^-(x_{1,j}' - \gamma_\tau))^2$, where β_τ^+ and β_τ^- are obtained from least asymmetrically weighted squares of ε_j^* on x_j of kink expectile regression.

4. Repeat this B times, so as to obtain a sample of simulated F-statistics $T_n^{\tau(1)}, \cdots, T_n^{\tau(B)}$.

5. Compute the p-value as the percentage of simulated F-statistics which exceed the actual value:

$$p_n = \frac{1}{B} \sum_{b=1}^{B} (T_n^{\tau(b)} \geq T_n^{\tau}) \tag{14}$$

where $T_n^{\tau} = N(\tilde{\sigma}^{*\tau} - \sigma^{*\tau})/\sigma^{*\tau}$; and $\tilde{\sigma}^{*\tau}$ and $\sigma^{*\tau}$ are sum of square error of fitted linear and kink equations, respectively, obtaining from expectile.

6. Reject H_0 at the significant α if $p_n < \alpha$.

7. For other expectile levels, we can use this same algorithm.

3   Simulation Study

In this section, we conduct a Monte Carlo simulation study to assess and compare the finite sample performance of our two proposed models.

3.1   Design

The model used for simulations is defined as

$$y_j = 1 + 2(x'_{1,j} - 3)_- - 1(x'_{1,j} - 3)_+ + \varepsilon_j, \tag{15}$$

where are $x'_{1,j}$ independent and identically distributed (IID) and generated from a normal distribution $x'_{1,j} \sim N(0, 10)$. In addition, as mentioned in the introduction, our models have an ability to explain a complete conditional distribution of responses even in the presence of heteroscedastic errors. To make a fair comparison, we consider two scenarios: (1) IID ε_j is assumed to be asymmetric Laplace $\varepsilon_j \sim ALD(0, 1, \tau)$ and (2) Heteroscedasticity $\varepsilon_j \sim ALD(0, (0.5 + x'_{1,j})^2, \tau)$. For this simulation study, we consider three expectile levels $\tau = 0.1$, $\tau = 0.5$, and $\tau = 0.9$. We consider both sample sizes of $n = 200$ and $n = 300$. The number of repetitions is set at 500.

3.2   Simulation Results

The quality of the results will be measured in terms of a Bias for our estimated parameters which is defined as:

$$\text{Bias} = \frac{1}{M} \sum_{r=1}^{M} (\tilde{\Theta}_t^{\tau} - \Theta_t^{\tau}),$$

where M is the number of bootstrapping; $\tilde{\Theta}_t^{\tau}$ and Θ_t^{τ} and are the estimated value and true value, respectively. According to the results of our two scenarios in Tables 1 and 2, respectively, we can observe that the estimated biases are

Table 1. Bias and coverage probability of expectile kink regressions, for $\varepsilon_j \sim ALD(0, 1, \tau)$

τ	Parameter	N = 200			N = 300		
		TRUE	Expectiles kink		TRUE	Expectiles kink	
			Bias	CP		Bias	CP
0.1	$\beta_{\tau,0}$	1	0.9193	0.945	1	0.9314	0.924
	$\beta_{\tau,1}^-$	2	−0.0017	0.89	2	−0.0022	0.894
	$\beta_{\tau,1}^+$	−1	0.0006	0.864	−1	−0.0055	0.902
	γ	3	−0.0163	0.878	3	0.0165	0.936
0.5	$\beta_{\tau,0}$	1	0.0077	0.888	1	−0.0053	0.866
	$\beta_{\tau,1}^-$	2	−0.0009	0.901	2	0.002	0.874
	$\beta_{\tau,1}^+$	−1	−0.0011	0.879	−1	0.0005	0.894
	γ	3	0.0154	0.883	3	−0.0044	0.874
0.9	$\beta_{\tau,0}$	1	−0.9037	0.955	1	0.946	0.251
	$\beta_{\tau,1}^-$	2	0.0044	0.88	2	−0.0011	0.876
	$\beta_{\tau,1}^+$	−1	−0.0079	0.882	−1	−0.0034	0.866
	γ	3	0.0105	0.884	3	0.0319	0.874

Source: Calculation.

Table 2. Bias and coverage probability of expectile kink regressions, for $\varepsilon_j \sim ALD(0, (0.5 + x'_{1,j})^2, \tau)$

τ	Parameter	N = 200			N = 300		
		TRUE	Expectiles kink		TRUE	Expectiles kink	
			Bias	CP		Bias	CP
0.1	$\beta_{\tau,0}$	1	0.7826	0.650	1	0.8181	0.692
	$\beta_{\tau,1}^-$	2	−0.0729	0.846	2	−0.1757	0.794
	$\beta_{\tau,1}^!$	−1	−0.0049	0.838	−1	0.1525	0.836
	γ	3	0.0354	0.864	3	−0.0547	0.834
0.5	$\beta_{\tau,0}$	1	0.0062	0.888	1	0.1704	0.888
	$\beta_{\tau,1}^-$	2	0.0007	0.888	2	0.0311	0.886
	$\beta_{\tau,1}^+$	−1	0.0001	0.912	−1	0.0038	0.854
	γ	3	−0.9461	0.900	3	0.0115	0.824
0.9	$\beta_{\tau,0}$	1	−0.9461	0.900	1	−0.7679	0.664
	$\beta_{\tau,1}^-$	2	−0.0011	0.879	2	0.1819	0.904
	$\beta_{\tau,1}^+$	−1	−0.0034	0.866	−1	−0.0303	0.914
	γ	3	0.0319	0.874	3	0.0465	0.835

Source: Calculation.

mostly small in both scenarios. This indicates that the proposed expectile kink regression estimator is asymptotically consistent. In addition, we also observe two interesting results from this simulation study. First, the Biases are generally larger when τ is close to 0 or 1 than when τ is 0.5. Second, as the sample size increases from 200 to 300, we find that the Biases decrease, which is consistent with the $\sqrt{n}$-consistency of the coefficient estimators. Moreover, the coverage probabilities (CP) of the each estimated parameter are close to the nominal level 90%. Therefore, we can conclude that both proposed models have a good finite sample performance.

4 Application

4.1 Service Output and Employment Data

In this section we show an analysis of data from the 2013 Asian service sector. The data set contains information on 45 countries of Asia. As dependent variable, we consider the service sector output of these countries. Our data is a cross section data which was collected from the World Bank database in 2013. The dataset consist Service sector output (Y_j) and labour (L_j) of country j. Our study expect that there might be a nonlinear relationship between output (Y_j) and labour (L_j). We consider the following production model. The model primarily takes the form of Cobb-Douglas production function where labour (L_j) is input.

$$Y_j = \alpha L_j^\beta \tag{16}$$

Then, we transform Eq. (16) into a translog production frontier which takes the form as

$$\ln Y_j = \alpha_\tau + \beta_\tau \ln L_j + \varepsilon_j. \tag{17}$$

In this study, we consider expectile kink regressions with $\tau = 0.1, 0.2, 0.5, 0.8, 0.9$.

4.2 Test for Presence of Expectile Kink Effect

We first consider the F-statistic test, as mentioned in Sect. 2.3, for the presence of expectile kink effect at the three significance levels, namely, 1%, 5%, and 10%. To test whether there exist kink effect at given expectile levels, we proceed the test as explained in Sect. 2.3. We consider testing the null hypothesis of no kink effect for all expectile levels (i.e. $\tau = 0.1$, 0.2, 0.5, 0.8, 0.9). Here the number of bootstrap replication B is set to be $B = 1000$. The resulting estimates are shown in Table 3. The results report that the simulated p-value for sup-F test based $= 0.1$, 0.2, 0.5, 0.8, 0.9 of expectile kink regressions are less than 0.100, which indicate an existence of kink or jump behavior in the labour of service sector. The results also suggest that the estimated kink points occur around 2.8884–2.887 for expectile kink regression. Generally speaking, this indicates that expectile kink regression model can be used for the consideration of unobserved heterogeneities in kink point across expectiles. However, the kink values seem not to change across expectiles.

Table 3. Test for kink effect.

τ	γ	p-value
Expectile kink		
0.1	2.8887	0.011
0.2	2.8884	0
0.5	2.8885	0
0.8	2.8886	0
0.9	2.8887	0

Source: Calculation.

Table 4. Estimated results

Parameter	$\tau = 0.1$	$\tau = 0.2$	$\tau = 0.5$	$\tau = 0.8$	$\tau = 0.9$
Expectile kink regression					
α_τ	25.7021	25.9059	26.7141	27.6009	28.0487
	(-0.9131)	(-0.6802)	(-0.6318)	(-0.7073)	(-0.7252)
β_τ^-	10.3521	9.0301	8.2311	7.9421	7.5719
	(-3.7893)	(-2.7245)	(-1.9634)	(-2.5883)	(-2.6895)
β_τ^+	-1.7229	-1.4035	-1.5976	-2.1137	-2.4791
	(-1.7782)	(-1.4904)	(-1.1598)	(-1.0882)	(-1.0146)
SSE	256.1137	182.3995	136.8625	174.0531	228.0985

Source: Calculation.

4.3 Estimation Results

According to Table 4, the estimated coefficients are presented. We can see that the signs of the estimated coefficients of the different regressors considered in Eq. 17 do not change across the τ - expectile. However, the results also show that the magnitude effect of the coefficients differ across the τ - expectile. The estimated coefficient of the labour is always positive in the first regime where the labour is lower than kink point around 2.8884–2.887. On the other hand, The estimated coefficient of the labour is always negative in the second regime where the labour is greater than kink point around 2.8888. These results indicate that the output of service sector increases with more labour when the number of log(Labour) is less than their kink points. However, if log(Labour) is larger than their kink points, labour has a negative effect on service sector output. Surprisingly, we find a negative effect of labour on service sector. According to Gordon [2], he suggested that wage compression introduced by unions in many countries can bring a negative effect to service sector. We expect that the higher labour power in service sector leads the labour unions to ask for a wage higher than the equilibrium level. As a result, the service firms will face with higher cost which leads to the lower service sector output.

Then, we display a scatter plot of $\ln Y_j$ and $\ln L_j$ along with the fitted expectile curves in Fig. 1. The curves are plotted with the expectile kink regressions for

Fitted expectile curves

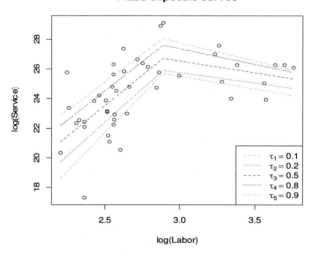

Fig. 1. Expectile curves.

$\tau = 0.1$, 0.2, 0.5, 0.8, 0.9 respectively. We can see that the fitted curves provide a similar pattern in every expectile.

5   Conclusions

We use expectile regression models to check whether or not there exists a kink or threshold effect for all expectiles. In this paper, we proposed expectile kink regression models to explain non-linear structure and also made assessment of the complete conditional distribution of responses even in the presence of heteroscedastic errors. We applied as well a grid search method and F-test statistic of Hansen [3] to test the structural change and to estimate the unknown kink or threshold at a given expectile. The simulated data sets were generated to evaluate the accuracy of the model. The results show a good finite sample performance.

Finally, the proposed models have been illustrated using real data to study the effect of labour force on service sector in Asia. We found a kink effect on our data in every expectile. Our application result showed labour has a positive effect on service sector output and there is a switch to a negative effect when labour exceeds kink point at all expectile levels. This indicates that in Asian countries, with both high and low service output, there exists a nonlinear effect of labour on service output. We expect that the higher demand for service labour will lead to a greater strength of labour unions that in turn will hurt the economy by demanding service business to pay higher wage. Therefore, these firms will tend to hire fewer workers or decrease the scale of operations, resulting in the lower service output.

References

1. Aigner, D.J., Amemiya, T., Poirier, D.J.: On the estimation of production frontiers: maximum likelihood estimation of the parameters of a discontinuous density function. Int. Econ. Rev. **17**, 377–396 (1976)
2. Gordon, R.: Is there a tradeoff between unemployment and productivity growth in unemployment policy? In: Snowereds, D., de la Dehese, G. (eds.) Unemployment Policy. Cambridge University Press, Cambridge (1997)
3. Hansen, B.E.: Regression kink with an unknown threshold. J. Bus. Econ. Stat. **35**(2), 228–240 (2017)
4. Kim, M., Lee, S.: Nonlinear expectile regression with application to value at-risk and expected shortfall estimation. Comput. Stat. Data Anal. **94**, 1–19 (2016)
5. Koenker, R.: When are expectiles percentiles? (solution). Econ. Theor. **9**(03), 526–527 (1993)
6. Schnabel, S., Eilers, P.: Optimal expectile smoothing. Comput. Stat. Data Anal. **53**, 4168–4177 (2009)
7. Sobotka, F., Kneib, T.: Geoadditive expectile regression. Comput. Stat. Data Anal. **56**, 755–767 (2012)
8. Newey, W., Powell, J.: Asymmetric least squares estimation and testing. Econometrica **55**, 819–847 (1987)
9. Schnabel, S.K., Eilers, P.H.: Optimal expectile smoothing. Comput. Stat. Data Anal. **53**, 4168–4177 (2009)
10. Pastpipatkul, P., Maneejuk, P., Sriboonchitta, S.: Testing the validity of economic growth theories using copula-based seemingly unrelated quantile kink regression. In: Robustness in Econometrics, pp. 523–541. Springer International Publishing (2017)
11. Sriboochitta, S., Yamaka, W., Maneejuk, P., Pastpipatkul, P.: A generalized information theoretical approach to non-linear time series model. In: Robustness in Econometrics, pp. 333–348. Springer International Publishing (2017)
12. Tong, H.: On a threshold model. In: Chen, C.H. (ed.) Pattern Recognition and Signal Processing, pp. 575–586. Sijthoff and Noordhoff, Amsterdam (1978)
13. Tong, H.: Threshold Models in Nonlinear Time Series Analysis. Lecture Notes in Statistics. Springer, New York (1983)
14. Xing, J.J., Qian, X.Y.: Bayesian expectile regression with asymmetric normal distribution. Commun. Stat. Theor. Methods **46**(9), 4545–4555 (2017)
15. Yao, Q., Tong, H.: Asymmetric least squares regression estimation: a nonparametric approach. J. Nonparametric Stat. **6**(2–3), 273–292 (1996)
16. Zhang, F., Li, Q.: A Continuous Threshold Expectile Model. arXiv preprint arXiv:1611.02609 (2016)

Adjusting Beliefs via Transformed Fuzzy Prices

Tanarat Rattanadamrongaksorn[1]($^{(\boxtimes)}$) (iD), Duangthip Sirikanchanarak[1,2],
Jirakom Sirisrisakulchai[1,3], and Songsak Sriboonchitta[1,3]

[1] Faculty of Economics, Chiang Mai University, Chiang Mai 50200, Thailand
tanarat_ra@cmu.ac.th, doungtis@gmail.com, sirisrisakulchai@hotmail.com,
songsakecon@gmail.com
[2] Bank of Thailand, Bangkok 10200, Thailand
[3] Center of Excellence in Econometrics,
Chiang Mai University, Chiang Mai 50200, Thailand

Abstract. In the situation that the *gut feeling* tells otherwise, incorporating information from expert opinions can significantly improve the accuracy of standard estimation and prediction methods, which rely only on observed data. To cope with this problem, we propose the fusion of data under the Bayesian framework by transforming price estimates into initial beliefs of assets. The proposed methodology focuses on modeling the price expectation by linguistic terms and mathematically extending them to other parameters like the standard deviation. On five sample assets from different markets, our method was experimented and compared with the method of the ARMA-GARCH beyond the points of structural change. The problems are multi-dimensional but conveniently solved by the Metropolis-Hastings algorithm. The results show the significant impacts of expert opinions on the posterior.

Keywords: Bayesian inference · Possibility-probability transformation
Fuzzy price · Fuzzy deviation · Data fusion · MCMC

1 Introduction

Financial econometrics depends much on the applications of statistical data. However, this is insufficient in some situations like when the structural change occurs. The changes in the important factors such as political policies, administrative styles, technological constraints, market conditions, etc. might affect the problem at the structural level and shift the properties of economic entities from the current positions. These could move the prices of the financial assets away from their previous means, and the moves cannot be captured if we only analyze the previous prices of these assets. Hence, the analysis only by statistical data may be misleading totally in this case.

Instead, the experts in the field could very possibly have their *gut feelings* that, by some clues, enable them to predict the changes. For instance, a rice exporter may have a good perception of the new regulations of his competitors'

© Springer International Publishing AG 2018
L. H. Anh et al. (eds.), *Econometrics for Financial Applications*, Studies in Computational
Intelligence 760, https://doi.org/10.1007/978-3-319-73150-6_64

countries so that he is aware of the significant increases on the next transactions before the price starts changing. Furthermore, because his company is one of the top merchandises, the government may ask his opinion on the proper price level of the rice after the regulations were enforced. His response might be either one of the choices between (1) the returns for rice has its mean at 12.5% with variance at 5.5% or (2) the estimated returns is the most pessimistic at 2.5%, the most possible at 15% and the most optimistic at 25%. The second statement seems to be more practical and realistic in representing belief of estimation. The representation of knowledge by the expert should be in the form of linguistic terms, not the parameters of probability distribution. We feel that the subjective knowledge could be acquired more easily and naturally by fuzzy numbers than by random variable. The expert opinions are inferred by the experiences and relevant circumstances without systematic evidence. The numerical data can be extracted verbally from the expert opinions but, however, the exact number is rarely given due to the uncertain nature of the conclusive thought. This uncertainty can be handled well by fuzzy set theory.

While the expert opinions are useful in this consideration, the data available in the past should not be neglected. It is too relaxed to utilize only the expert opinions and too restricted to rely solely on the historical data. To be more specific, the subjective data are usually high in variations since the human estimations tend to stay on the safe side. On the other hand, the objective data do not reflect the expected foreseeable results in the future caused by the abrupt changes. In this research, we take into account both because throwing away either of them is a loss of opportunity. We, therefore, propose the fusion of data using the Bayes' updating mechanism with possibility-to-probability transformation.

In the following sections, we shall discuss the proposed methodology including the necessary concepts of fuzzy set theory and Bayesian statistics and how to apply these methods to our problems (Sect. 2); the fuzzy and statistical data, the related tests and results of the example (Sect. 3); and the conclusions of the research's contributions, advantages, disadvantages, and future research topics (Sect. 4).

2 Proposed Methodology

In this study, the price of asset, x, is considered following the data generating mechanism of the normal model which is characterized by price expectation and variance, $x \sim \mathcal{N}orm(\mu, \sigma^2)$ where mu, σ, and σ^2 represents mean, standard deviation, and variance respectively.

The proposed methodology contains the following steps:

Step 1: Formulate the fuzzy prices from expert opinions which are expressed by linguistic terms (Sect. 2.1);

Step 2: Determine the distributions of standard deviations by the mathematical operations (Sect. 2.2);

Step 3: Transform the expert opinions into probability distributions (Sect. 2.3);

Step 4: Analyze the results as per the conventional Bayesian statistics (Sect. 2.5) by computational-statistics method (Sect. 2.6).

2.1 Fuzzy Price

Fuzzy price is an estimated price that reflects the expert's imprecise intuition. This type of uncertainty about the future has its source from the incomplete knowledge about the expected value of the asset and the price is measured by the possibility. It is estimated verbally by asking the qualified person in the field with the question like "What is your opinion on the expectations for the price of the particular asset to change?". The common answers may be "at least 2.5%", "between 5 and 7.5%", "the most possible at 12%" and so on. These linguistic terms contain fuzziness because the exact solution might not be given. These answers are called expert opinions and can be represented efficiently by fuzzy numbers which are the basic elements of fuzzy set theory [15].

Fuzzy set theory is the generalization of set theory for representing the non-statistical uncertainties. The formal definition of fuzzy set, $\mathcal{A}$, is the mapping from the set of objects to the unit interval, $\mathcal{A} = \{\theta \in \Omega | \pi(\theta) \in [0, 1]\}$, The fuzzy set is characterized by a set of pairs - an element in the universe of discourse and its possibility, $[\theta, \pi(\theta)]$. In this study, the element, is a set of parameters i.e. price expectation and standard deviation, $\theta = (\mu, \sigma)$. The value $\pi(\theta)$ is the degree to which θ is possible: $\pi(\theta) = 1$ means that θ is the most possible, $\pi(\theta) = 0$ means that θ is not possible.

Previously, a series of fuzzy numbers was proposed for translating the Return On Asset (ROA) into fuzzy return in the style of, but not limited to [7], random variable [12]. In the same sense, the fuzzy price can be represented by the same set of fuzzy numbers. There are two fuzzy numbers depicted in Fig. 1 that are commonly used in practice and this research.

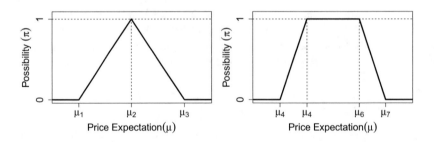

Fig. 1. Triangular fuzzy price (left) and Trapezoidal fuzzy price (right)

As a matter of fact, there are many forms of fuzzy numbers but we consider only (a) triangular and (b) trapezoidal fuzzy numbers. When the prices of assets are represented by these forms, they are called the triangular and trapezoidal fuzzy prices accordingly. The triangular fuzzy price could be constructed by three-point estimation - the most pessimistic, (μ_1); the most possible, (μ_2); and the most optimistic, (μ_3) estimations. These fuzzy numbers could match with probability distributions like triangle distribution, which is used intensively in other fields, e.g. project management [8,13]. The trapezoidal fuzzy price is the interval estimation by approximating the price ranges of the largest possible, $(\mu_4 - \mu_7)$, and the most possible, $(\mu_5 - \mu_6)$. The representation of both fuzzy prices can be written by the simplified expression:

$$\mathcal{A} = (a, b, c, d),\tag{1}$$

whose prices are expected to be a, b, c, and d and their corresponding possibilities are $(0, 1, 1, 0)$ respectively but omitted.

The linguistic mapping of fuzzy number is not restrictive to the provided definitions but, rather, flexible to various arrangements. For instance, in the latter section (Sect. 3.1), we could estimate the least possible prices by point estimations i.e. the most pessimistic and the most optimistic as in the triangular fuzzy price and the most possible prices by the interval estimation as in the trapezoidal fuzzy price. The mixture of linguistic terms forms a better presentation as a result (see this example in Table 1).

2.2 Fuzzy Deviation

Data distribution in the normal model is characterized by mean, and standard deviation, $\theta = (\mu, \sigma)$. Although it is possible to repeat the expert questioning for the standard deviations, the process seems exhaustive and impractical. Alternatively, extracting the secondary information from the existing fuzzy prices sounds more practical. This leads to fuzzy deviation:

$$\widetilde{\sigma}(\mu) = (a - \widetilde{\mu}, b - \widetilde{\mu}, c - \widetilde{\mu}, d - \widetilde{\mu}),\tag{2}$$

where $\widetilde{\mu}$ corresponds to the centroid or the expectation of fuzzy price obtained by taking moment against the possibility axis:

$$\widetilde{\mu} = \frac{\int \theta \pi(\theta)\, d\theta}{\int \pi(\theta)\, d\theta},\tag{3}$$

indicates the possible deviation of the price from its expectation which is calculated by subtracting the centroid from the fuzzy price. It is easily seen that the fuzzy deviation can be either positive or negative that is interpreted as possible gain or loss from the expectation respectively but only the absolute values are considered in our implementation.

2.3 Possibility-to-Probability Transformation

When expert opinions and historical observations are both available, they could be used separately because they are based on the different theories. In order to be utilized as homogeneous information, they need to be transformed into either possibility or probability. In this research, we choose to transform the expert opinions into probability and combine them under the Bayesian framework due to the loss of information from the inverse operation.

Transformation of possibility to probability is guided by three principles. The process should (1) add or, at least, retain information, (Probability-possibility consistency), $P(\theta) \leq \pi(\theta)$; (2) maintain the order of possibility and probability (Preference preservation), $\pi(\theta) > \pi(\theta') \Leftrightarrow P(\theta) > P(\theta')$; and (3) contain as much uncertainty as possible (Least commitment). These criteria can be met by the procedure in the following steps:

Step 1: Select the possibility which is called alpha-cut, α, randomly from the uniform distribution, $\alpha \sim Unif(0,1)$, that gives the alpha-cut fuzzy number, $\mathcal{A}_\alpha$, which is a subset of the original fuzzy number, $\mathcal{A}_0$, whose all possibilities are greater than or equal to this value, $\mathcal{A}_\alpha = \{\theta \subseteq \mathcal{A}_0 | \pi(\theta) \geq \alpha\}$;

Step 2: Choose the alpha-cut parameter, θ_α, randomly from the uniform distribution on the support of the alpha-cut fuzzy number, $\theta_\alpha \sim Unif(\min(\mathcal{A}_\alpha), \max(\mathcal{A}_\alpha))$;

Step 3: Obtain the possibility, $\pi_a(\theta)$, corresponding to the parameter from the previous step; $\pi_a(\theta) = \{\pi(\theta) | \theta = \theta_\alpha\}$;

Step 4: Repeat the first three steps for T possibilities;

Step 5: Sort the possibility from the previous steps, π_a, into the ordered possibilities, $\pi_j(\theta)$, with the transforming order, j, descending in the values of possibilities, $max(\pi_a(\theta)) = \pi_1(\theta) \geq \pi_2(\theta) \geq \ldots \pi_j(\theta) \ldots \geq \pi_{T-1}(\theta) \geq \pi_T(\theta) = min(\pi_a(\theta))$.

Step 6: Transform the ordered possibility, $\pi_j(\theta)$, into the transformed probability, $P_i(\theta)$, by the accumulation of the differences between adjacent possibilities (the current- and the next-order possibility to be exact), $\pi_j(\theta) - \pi_{j+1}(\theta)$, divided by their transforming orders, j, ranging from the possibility in the considered order, $\pi_{j=i}(\theta)$, to the one before the last ordered possibility, $\pi_{T-1}(\theta)$, expressed mathematically as [4]:

$$P_i(\theta) = \sum_{j=i}^{T-1} \left(\frac{\pi_j(\theta) - \pi_{j+1}(\theta)}{j} \right),$$
(4)

where $i = 1, \ldots, T-1$.

This equation was modified slightly on the domain of the parameter from the original version. The results from these steps are the discrete probabilities that increase with accelerating rate and leverage on both sides of distributions.

The joint transformed probability of expectation and deviation can be calculated by the rule of conditional probability:

$$P(\mu, \sigma) = P(\sigma|\mu)P(\mu),$$
(5)

as we interpret the probabilities of transformed deviations exclusively from the price expectations as conditional probabilities, $P(\sigma|\mu)$. The results are treated as external knowledge and, therefore, processed as priors into the Bayes' theorem.

2.4   Bayes' Theorem

Bayesian framework offers another insight of statistics in addition to the Frequentist. Rather than finding the fixed and frequency-based properties, the outcome of Bayesian is the degree of belief on random parameter which is called posterior probability or, just, posterior. The posterior, $P(\theta|X)$ is proportioned to the product of prior, $P(\theta)$, and likelihood, $P(X|\theta)$. The result can be normalized by marginal likelihood, $P(X)$, but usually neglected in the computation. The prior is the initial belief of the parameter (price expectation and standard deviation in our study) before seeing data and the likelihood is the joint probability of data. Bayes' theorem [1] is:

$$P(\theta|X) = \frac{P(\theta)P(X|\theta)}{P(X)}, \tag{6}$$

where X is the observed data and θ is the parameter of interest. All relating concepts necessary for making Bayesian inference center around this formula (Eq. 6)—with the exception of non-parametric Bayesian.

2.5 Predictive Posterior

In our experiment, the proposed method was also evaluated by time-series analysis comparing with ARMA-GARCH model. The prediction is produced by the predictive posterior sampled from the predictive posterior distribution that is computed by marginalizing out both parameters:

$$P(X_{t+1}|X_t) = \int_{\mu,\sigma} P(X_{t+1}|\mu,\sigma)P(\mu,\sigma|D_t)\,d\mu\,d\sigma, \tag{7}$$

where X_{t+1} and X_t are the next-period and the current-period data respectively. The predictive posterior distribution, $P(X_{t+1}|X_t)$, is the distribution of the unobserved data, X_{t+1}, conditional on the observed data, X_t.

2.6 Metropolis-Hastings

The distributions can be approximated by the computational-statistics. We describe one of the methods in the Markov-chain Monte-Carlo families we employed in this section but other methods are also viable by users' choices. The Metropolis-Hastings (MH) algorithm [6] is a class of Markov-Chain Monte-Carlo ($MCMC$) that relies on the Ergodic theorem [2,9] stating briefly that the process will run through every possibility given long-enough period of time. The technique is widely used in high-dimension analysis, on the condition that the samples can be drawn from the distribution, because in that case the analytic solution is extremely difficult.

The MCMC iterates through the simulations by generating and testing a sample based on the previous result i.e. Markov chain. The idea is to gradually explore the probability regions and move toward, or at least stay on, the high probability path. The MH improves the MCMC by the asymmetric distribution preventing the algorithm from being trapped in only one side of the proposal distribution.

To avoid the complexity in high dimension, we choose the component-wise type of MCMC so that the problem could be solved in the univariate scheme; in other words, one parameter is examined at a time, as described by the following procedure:

Step 1: Initialize a value for each parameter, $\theta^{(k)}$, from the pdf of the expected posterior or the target distribution, $\Gamma^{(k)}(\theta)$, where (k) indicates the order of the parameters;

Step 2: Generate a sample called proposal, θ', from the desired sampling or the proposal distribution, $\tau^{(k)}(\theta'|\theta_{i-1})$, and compute the acceptance probability:

$$\Lambda(\theta', \theta_{i-1}^{(k)}) = \min\left(1, \frac{\Gamma(\theta')}{\Gamma(\theta_{i-1}^{(k)})} \frac{\tau(\theta_{i-1}^{(k)}|\theta')}{\tau(\theta'|\theta_{i-1}^{(k)})}\right), \tag{8}$$

which is one of the most popular formulas from many variations. The acceptance probability, $\Lambda(\theta', \theta_{i-1}^{(k)})$ or Λ, is based on two decision criteria: the ratio of target probabilities, $\Gamma(\theta')/\Gamma(\theta_{i-1}^{(k)})$, keeping samples generated in the high probabilities and that of proposal probabilities, $\tau(\theta_{i-1}^{(k)}|\theta')/\tau(\theta'|\theta_{i-1}^{(k)})$, allowing the MH to return to the previous instantiation;

Step 3: Compare the acceptance probability with the random probability, u, which is sampled from the uniform distribution, $u \sim \mathcal{U}nif(0,1)$;

(3.1) Accept the proposal as a new observation, $\theta_i = \theta'$, with the probability Λ if the probability of acceptance is higher than that of the randomness, $\Lambda > u$;

(3.2) Reject the proposal with the probability $1 - \Lambda$ and assign the new observation with the previous one, $\theta_i = \theta_{i-1}$, otherwise;

Step 4: Repeat the Steps 2 and 3 for each parameter;

Step 5: Repeat the Step 4 for M simulations.

The MH can also be attached with the common techniques such as the random walk and thinning (not included in the above procedure). The random walk perturbs the previous observations and the thinning mitigates the effect of auto-correlation by filtering for every fifth observation. However, these techniques appeared performing well in some cases but inefficient in the others.

3   Examples

This section contains the descriptions of data used in the examples (Sect. 3.1), data pretest (Sect. 3.2), the traditional time-series predicting model as a baseline method (Sect. 3.3), and the results (Sect. 3.4).

3.1   Data

To demonstrate the proposed methodology, we selected five sample assets i.e. forex, gold, rice, domestic stock and foreign stock from different markets during the years 2001–2016. Instead of the ROA as in the common financial analyses, the data were demonstrated by price because the structural change can be observed more clearly. The basic properties of each asset are shown in Tables 1 and 2.

Table 1. Fuzzy data from expert opinions

No.	Asset name	Most negative	Most possible	Most positive	μ-Centroid	π-Centroid
1	Forex	41.2	[41.8, 43.7]	44.5	42.8	0.22
2	Gold	1602	[1639, 1675]	1680	1648	0.27
3	Rice	16.6	[17.2, 17.8]	18.0	17.4	0.27
4	Foreign stock	36.3	37.5	38.5	37.4	0.50
5	Domestic stock	8.1	8.6	9.1	8.6	0.50

Remark: The π-Centroid has no application in this research.

Table 2. Basic statistics of data observations

No.	Asset name	Mean	Variance	Min.	Max.	Median	Mode	Confidence interval	Size
1	Forex	41.1	1.3	36.9	43.8	40.8	40.4	[39.3, 43.4]	1930
2	Gold	792	226099	256	1896	650	362	[266, 1743]	3158
3	Rice	7.0	4.9	3.4	11.3	7.1	7.3	[3.6, 10.5]	1737
4	Foreign stock	19.3	113.2	3.6	41.0	16.3	12.2	[4.7, 39.3]	2032
5	Domestic stock	8.3	1.2	6.3	10.2	8.6	9.3	[6.6, 9.9]	344

3.2   Unit Root Tests and Breakpoint

When one works on time-series data, the first step recommended is to check the stationarity and break date of data. We choose the Dickey-Fuller t-statistic [11] and the modified augmented Dickey-Fuller tests [10] for the tests of these properties respectively. Moreover, the results were also verified by the specialist.

3.3   ARMA-GARCH

We applied the functions for estimating and forecasting various univariate GARCH-type time series models in the conditional variance and ARMA specification in the conditional mean [14]. The model specification was selected based on Akaike Information Criterion (AIC). In order to take the heteroskedastic effects into account, the AutoRegressive Conditional Heteroskedastic model (ARCH) [5] and, later improved to, Generalized AutoRegressive Conditional Heteroskedastic model (GARCH) [3], were selected as a baseline method in our experiments.

The ARMA(p, q) process of the autoregressive order, p, and the moving average order, q, with the variance equation of GARCH(m, n) can be described as:

$$X_t = \vartheta + \sum_{i=1}^{p} \alpha_i X_{t-i} + \sum_{j=1}^{q} \beta_j \xi_{t-j} + \xi_t,$$

$$\xi_t = z_t \varrho_t,$$

$$z_t \sim D_v(0, 1),$$

$$\varrho_t^2 = \gamma_0 + \sum_{i=1}^{m} \delta_i \xi_{t-i}^2 + \sum_{j=1}^{n} \gamma_j \varrho_{t-j}^2,$$

(9)

where X_t, ϑ, α_i, and β_j represent the asset prices or observed data at time t, mean, autoregressive coefficient, and moving average coefficient respectively. The ξ_t term in the ARMA mean is the innovations of the time series process. ϱ_t^2 is variance at time t and $D_v(0, 1)$ is the probability density function of the residuals with zero mean and unit variance. Optionally, v is additional distribution parameters to describe the skewness and the shape of the distribution. In this study, we used the GARCH(1, 1) model to estimate the parameters. Given the model for the conditional mean, variance, and observed univariate price series, the Maximum Log-Likelihood Estimation was employed to fit the model.

3.4   Results

The fuzzy prices of sample assets were constructed by the expert opinions shown in Table 1 and depicted in Fig. 2 (dashed line). The examples consists of three trapezoidal fuzzy prices of the forex, gold, and rice and two triangular fuzzy prices of the foreign and domestic stocks. The transformed prices (solid lines) are also illustrated in the same plot in order to compare the before- and after-transformation. The number of transformations was set equal to the size of the observations in each asset. Generally, the shapes of the transformed prices are similar to but smaller than those of fuzzy prices. The transformed trapezoids are much steeper than the transformed triangles because more possibilities could be identified and, as a result, the higher probabilities were induced through the calculation by Eq. 4. The fuzzy deviations as seen in Fig. 2 were obtained by Eq. 2. Both fuzzy and transformed deviations are in similar shapes as their corresponding prices but simply shifted by their expectations to zero. It is understandable that the estimates of fuzzy prices are inclined more to the positive side such as in the case of the gold and rice because people always stay on the safe side by giving the conservative opinions, even the experts.

The results in bivariate distributions of the joint distributions of priors, likelihoods, and posteriors may be too difficult to explain so that we shall make the discussions based on the marginal distributions instead. Tables 3, 4 and 5 are used in conjunction with Figs. 3, 4 and 5 respectively. Each table contains the basic parameters for general inference i.e. mean, variance, minimum, maximum, median, mode, and credible interval. Because our priors are not in the parametric form and do not depend on the analytical method, the results of each

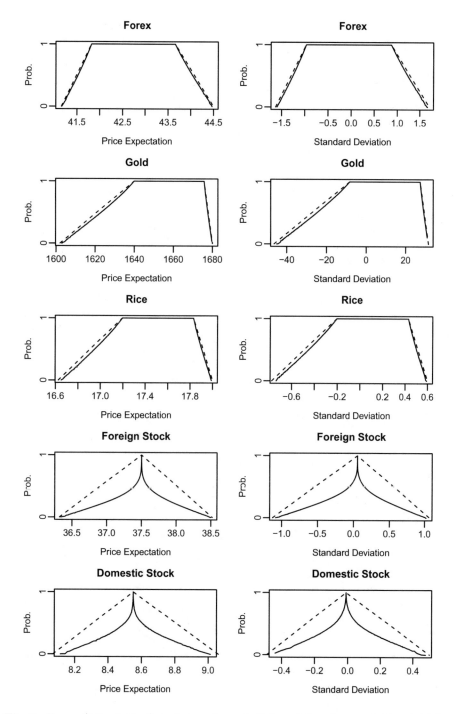

Fig. 2. Fuzzy (left, dashed) and transformed (left, solid) prices and Fuzzy (right, dashed) and transformed (right, solid) deviations

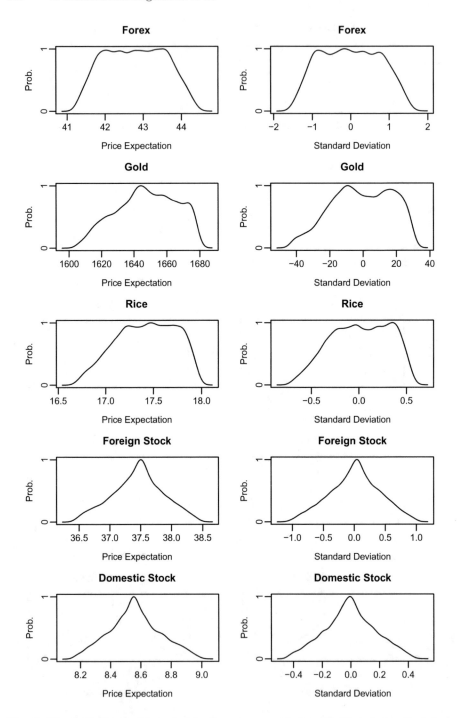

Fig. 3. Marginal distributions of prior for price expectations (left) and standard deviations (right)

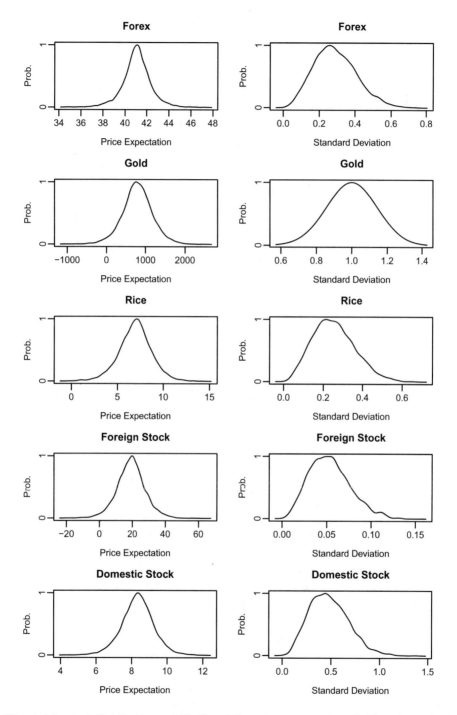

Fig. 4. Marginal distributions of likelihood for price expectations (left) and standard deviations (right)

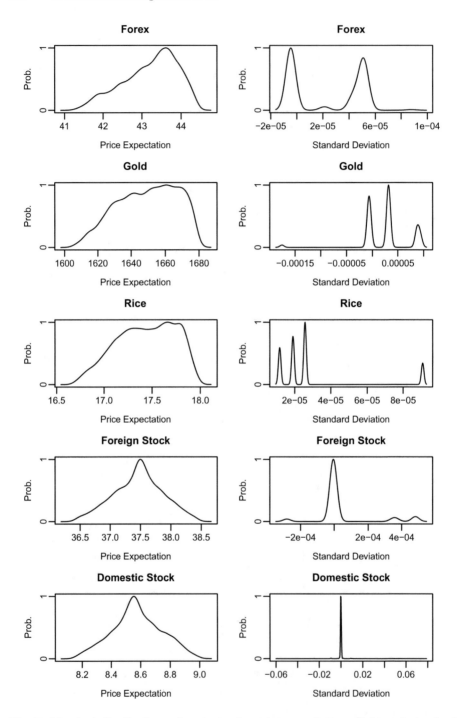

Fig. 5. Marginal distributions of posterior for price expectations (left) and standard deviations (right)

distribution are not well-smoothly distributed. They inclined more to the shapes of fuzzy priors.

Although, the distributions of likelihoods could be built by the sequences of the parameters, we made them portrayed by the MCMC to give their looks and feels not too contrast with the others. While the distributions of likelihoods are as expected, those of priors and posteriors are more interesting. For the prior distributions, their shapes of distributions still retain the configurations of fuzzy priors. However, some of them tend to turn into multiple modes. For example, in the case of gold, its priors (Table 3), have two modes on the marginal distribution of standard deviation so that making inference by the mean (−0.78) might not be the best choice since both modes, by visual inspection, have higher probabilities. The highest one (−9.30) might be more appropriate for inference. The same results also occur but less obvious in the case of forex (the mean and mode are −0.01 and −0.16 respectively). This scenario seems to happen more easily on the trapezoid because the probabilities are more constant or do not change much in the transformed probabilities (Fig. 3).

The posteriors result in the compromise between priors and likelihoods. Most of them also develop multiple modes as well. The shapes of posteriors are mixed by both distributions. It is probable that the biases of posteriors are influenced more by the priors but the variations are affected more by the likelihoods. In the case of forex, for instance, the posterior (Fig. 5) results in the same shape as the likelihood (Fig. 4). The higher probabilities on the positive side is as biased as its prior (Fig. 2) and distribution of the asset is as wide as the range of its likelihoods (Fig. 4). By considering the numerical results, the means for prior, likelihood, and posterior are 42.8, 41.1, and 43.2 respectively and the variance are 0.6, 1.6, and 0.5 respectively; therefore these confirm our earlier conclusions. Also the foreign stock also gives the similar results in that the mean is closer to the prior (the means are 37.5, 19.2, and 37.5 in the same order as previously) and the variance is more similar to its likelihood (0.2, 73.8, and 0.2 in the same order as previously). The standard deviations, on the other hand, move in very small ranges with high fluctuations in all assets. This may give some difficulties if one wants to make the inference on them.

In Table 6, the results of estimating predictive distribution are compared to those of the Maximum Likelihood Estimation (MLE) for the reason that MLE could represent both the Frequentist and the uninformative-prior Bayesian. All predictive distributions shifted from the positions estimated by the MLE due to the influences mainly by their priors. This corresponds to the results of the joint distributions discussed earlier. For example, the posterior means of foreign stock are 18.2 by the MLE and 37.5 by the proposed method (Table 6) that correspond to the values of 19.3 of the mean and 37.4 of the centroid (Table 1). In another example, the domestic stock whose distributions from both methods with the smallest gap has the mean and centroid of 8.3 and 8.6 comparing to the modes of 8.3 and 8.6 that are not too hard to guess the outcome.

Table 3. Statistics of joint priors' marginal distributions for price expectations and fuzzy deviations

No.	Name	Mean	Variance	Min.	Max.	Median	Mode	Credible interval
Price expectation								
1	Forex	42.8	0.6	41.2	44.5	42.8	43.5	[41.5, 44.1]
2	Gold	1647	324	1603	1679	1647	1644	[1612, 1676]
3	Rice	17.4	0.1	16.7	18.0	17.4	17.5	[16.8, 17.9]
4	Foreign stock	37.5	0.2	36.4	38.5	37.5	37.5	[36.6, 38.2]
5	Domestic stock	8.6	0.0	8.1	9.0	8.6	8.6	[8.2, 8.9]
Standard deviation								
1	Forex	−0.01	0.57	−1.59	1.67	−0.02	−0.16	[−1.30, 1.33]
2	Gold	−0.78	316	−43.97	30.92	−0.65	−9.30	[−35.98, 27.70]
3	Rice	0.01	0.09	−0.73	0.59	0.02	0.35	[−0.57, 0.50]
4	Foreign stock	0.01	0.16	−1.07	1.01	0.03	0.04	[−0.81, 0.79]
5	Domestic stock	0.00	0.03	−0.44	0.46	−0.01	−0.01	[−0.35, 0.34]

Table 4. Statistics of likelihoods' marginal distributions for price expectations and standard deviations

No.	Name	Mean	Variance	Min.	Max.	Median	Mode	Credible interval
Price expectation								
1	Forex	41.1	1.6	34.6	47.4	41.1	41.1	[38.4, 43.7]
2	Gold	791	145954	−1020	2525	788	757	[12, 1571]
3	Rice	6.9	3.2	−0.5	14.5	7.0	7.1	[3.3, 10.5]
4	Foreign stock	19.2	73.8	−20.6	64.1	19.2	19.7	[1.9, 36.7]
5	Domestic stock	8.3	0.8	4.3	12.1	8.4	8.4	[6.5, 10.1]
Standard deviation								
1	Forex	0.29	0.01	0.01	0.75	0.28	0.26	[0.08, 0.55]
2	Gold	1.00	0.00	1.00	1.00	1.00	1.00	[1.00, 1.00]
3	Rice	0.25	0.01	0.00	0.68	0.25	0.21	[0.07, 0.48]
4	Foreign stock	0.05	0.00	0.00	0.15	0.05	0.05	[0.02, 0.10]
5	Domestic stock	0.48	0.04	0.01	1.39	0.46	0.44	[0.15, 0.92]

All time-series data were tested for the unit root and structural break (Sect. 3.2). Some asset like the rice had a few breaks but only the latest one were considered (Fig. 6).

The results of the prediction are shown after the breaks separated from the training data by vertical marks (Fig. 7). Each plot consists of three series: the actuals (gray line), the predicts by ARMA-GARCH (dashed line), and the predicts by the proposed method (solid line). While the movements of data from

Table 5. Statistics of posteriors' marginal distributions for price expectations and standard deviations

No.	Name	Mean	Variance	Min.	Max.	Median	Mode	Credible interval
Price expectation								
1	Forex	43.2	0.5	41.2	44.5	43.3	43.6	[41.7, 44.3]
2	Gold	1649	304	1605	1679	1650	1660	[1614, 1676]
3	Rice	17.4	0.1	16.7	18.0	17.5	17.7	[16.8, 17.9]
4	Foreign stock	37.5	0.2	36.4	38.5	37.5	37.5	[36.6, 38.3]
5	Domestic stock	8.6	0.0	8.1	9.0	8.6	8.6	[8.2, 8.9]
Standard deviation								
1	Forex	2.3e−05	0.0e+00	5.0e−06	8.8e−05	2.1e−05	5.0e−06	[5.0e−06, 5.1e−05]
2	Gold	2.6e−05	0.0e+00	1.76e−4	9.4e−05	3.2e−04	3.2e−05	[0.0e+00, 9.4−04]
3	Rice	2.9e−05	0.0e+00	1.2e−05	9.1e−05	1.9e−05	2.6e−05	[1.2e−05, 9.1e−05]
4	Foreign stock	3.9e−05	0.0e+00	2.8e−04	4.8e−04	5.0e−05	5.0e−06	[0.0e+00, 4.8e−04]
5	Domestic stock	5.9e−05	5.0e−06	5.9e−02	7.8e−02	2.4e−05	3.9e−05	[0.0e+00, 2.3e−04]

Table 6. Statistics of predictive posteriors' marginal distributions by Maximum Likelihood Estimation and the proposed method

No.	Name	Method	Mean	Variance	Min.	Max.	Median	Mode	Credible interval
1	Forex	MLE	41.1	1.3	36.9	43.8	41.1	41.1	[41.0, 41.1]
2	Forex	Proposed	43.2	0.5	41.2	44.5	43.3	43.6	[41.7, 44.3]
3	Gold	MLE	792	226099	256	1896	792	792	[775, 808]
4	Gold	Proposed	1649	304	1605	1679	1650	1660	[1614, 1676]
5	Rice	MLE	7.0	4.9	3.4	11.3	7.0	7.0	[6.9, 7.1]
6	Rice	Proposed	17.4	0.1	16.7	18.0	17.5	17.7	[16.8, 17.9]
7	Foreign	MLE	19.3	113.2	3.6	41.0	19.3	19.3	[18.8, 19.8]
8	Foreign	Proposed	37.5	0.2	36.4	38.5	37.5	37.5	[36.6, 38.3]
9	Domestic	MLE	8.3	1.2	6.3	10.2	8.3	8.3	[8.2, 8.4]
10	Domestic	Proposed	8.6	0.0	8.1	9.0	8.6	8.6	[8.2, 8.9]

the baseline method smoothly continue by data trends, those from our method jump at the breaks and stay close to their expected values. In our experiment, the fuzzy prices were given only once. If the expert opinions could be provided on daily basis, the predicted lines could be smoother and more realistic. Most of the times, except for the foreign stock, it seemed that the proposed method could adjust the movement of the data according to the breaks. This could be explained by the quality of the opinions provided. In the case of foreign stock, the subjective estimates were possibly less accurate because the person might be influenced by other factors.

Our test is numerically confirmed by the Mean Squared Errors (MSE) which are included on the top of each individual plot. The MSE from the proposed

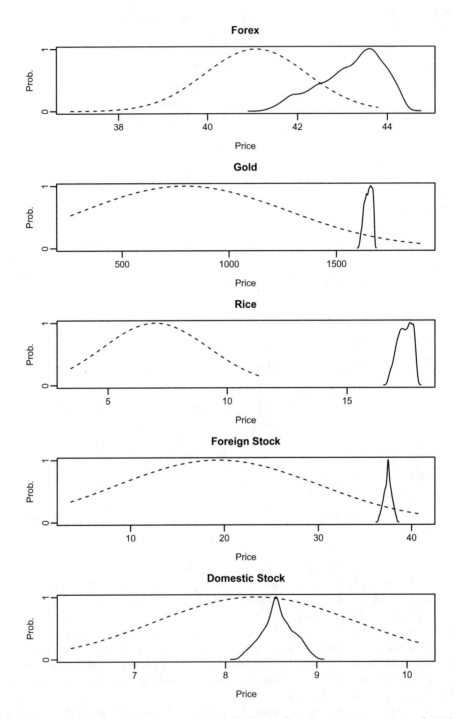

Fig. 6. Price distributions computed by MLE (dashed) and predictive posterior (solid)

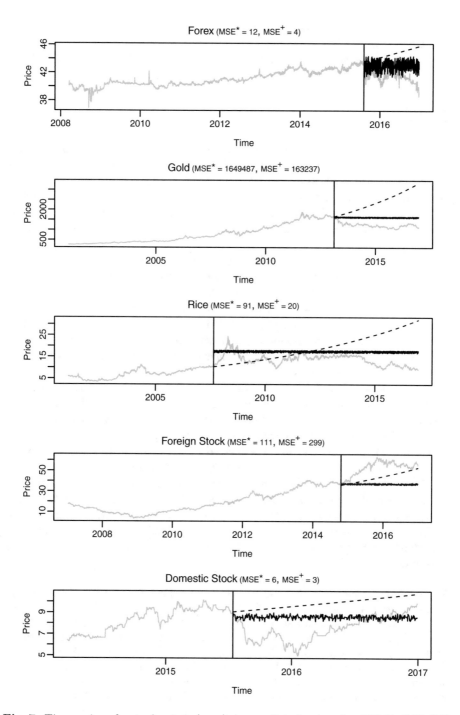

Fig. 7. Time-series of actual prices (gray) vs. predicted prices by ARMA-GARCH* (dashed), predicted prices by the proposed methodology+ (solid)

methodology is smaller than that from the baseline method for four assets indicating our method had fewer errors of the prediction from the actuals in this particular circumstance. Although one asset (foreign stock) is poorer in terms of MSE, the overall performances are quite satisfactory.

4   Conclusions

In this applied research, we propose the alternative approach in combining different types of financial assets by including those acquired from expert opinions. The methodology is based on the fuzzy prices that could be elicited from linguistic terms and, further, be evaluated for other parameters such as the standard deviation. After transforming the expert opinions, the transformed knowledge and the corresponding past data could, then, be merged simply by the Bayes' updating mechanism.

The experiments show the effects of structural changes to occur in every financial asset. No doubt that the impacts of these changes are high and unpredictable by the sole statistical methods. Fortunately, the experts in the field can use their intuitions by inferring other relevant factors proactively. By the same sense, we introduced the method that can adjust the beliefs about the problem by extracting expert opinions systematically. The proposed methodology demonstrated and is able to avoid the misleading outcomes that the traditional approaches may produce.

Our approach has specific usages in the situation such as the structural change where existing statistical data are inefficiently sufficient. However, it must also be careful in that (1) the method is based on fuzzy data and Bayesian statistics thus the results must be carefully interpreted especially in the case of fuzzy deviation, (2) the fuzzy deviation is derived by the concept of either minitive or maxitive and, as a result, quite conservative for data inference.

Demonstrated by this research, there still exist rooms for the possibility theory in the social sciences. The future topics of this branch are numerous in both theoretical and applied researches. Currently, the authors are working on (1) testing a proper ratio between possibility and probability data in making financial inference and (2) applying the methodology to the optimization of economic problems.

Acknowledgements. The first author is grateful to Prof. Dr. Hung T. Nguyen for his invaluable helps and supports; and Prof. Dr. Vladik Kreinovich for his lots of efforts spending with the comments and suggestions that so much improve the quality of this work. He is also pleased with the kind discussions on some ambiguities arising during the problem formulation with Dr. Supanika Leurcharusmee and Asst. Prof. Dr. Napat Harnpornchai. The second author is thankful to the Bank of Thailand for the academic opportunities and all other supports. This research is financed by the Center of Excellence in Econometrics and the Faculty of Economics, Chiang Mai University.

References

1. Bayes, T.: An essay towards solving a problem in the doctrine of chances. Phil. Trans. Roy. Soc. Lond. **53**, 370–418 (1763)
2. Birkhoff, G.D.: Proof of the Ergodic theorem. Proc. Nat. Acad. Sci. **17**(12), 656–660 (1931)
3. Bollerslev, T.: Generalized autoregressive conditional heteroskedasticity. J. Econ. **31**(3), 307–327 (1986)
4. Dubois, D., Prade, H., Sandri, S.: On possibility/probability transformations. In: Lowen, R., Roubens, M. (eds.) Fuzzy Logic: State of the Art, pp. 103–112. Springer, Netherlands, Dordrecht (1993)
5. Engle, R.F.: Autoregressive conditional heteroscedasticity with estimates of the variance of United Kingdom inflation. Econ. J. Econ. Soc. **50**(4), 987–1007 (1982)
6. Hastings, W.K.: Monte Carlo sampling methods using Markov chains and their applications. Biometrika **57**(1), 97–109 (1970)
7. Lorterapong, P., Moselhi, O.: Project-network analysis using fuzzy sets theory. J. Constr. Eng. Manag. **122**(4), 308–318 (1996)
8. Miller, R.W.: Schedule, Cost, and Profit Control with PERT. McGraw-Hill, New York (1963)
9. Neumann, J.V.: Proof of the Quasi-Ergodic hypothesis. Proc. Nat. Acad. Sci. **18**(1), 70–82 (1932)
10. Perron, P.: The great crash, the oil price shock, and the unit root hypothesis. Econ. J. Econ. Soc. **57**, 1361–1401 (1989)
11. Perron, P., Vogelsang, T.J.: Nonstationarity and level shifts with an application to purchasing power parity. J. Bus. Econ. Stat. **10**(3), 301–320 (1992)
12. Rattanadamrongaksorn, T., Sirisrisakulchai, J., Sriboonjitta, S.: Usages of fuzzy returns on Markowitz's portfolio selection. In: Proceedings of the Integrated Uncertainty in Knowledge Modelling and Decision Making: 5th International Symposium, IUKM 2016, Da Nang, Vietnam, 30 November–2 December 2016, pp. 124–135. Springer, Heidelberg (2016)
13. Sapolsky, H.: The Polaris System Development: Bureaucratic and Programmatic Success in Government. Harvard University Press, Cambridge (1972)
14. Wurtz, D., Chalabi, Y., Luksan, L.: Parameter estimation of ARMA models with GARCH/APARCH errors an R and SPlus software implementation. J. Stat. Softw. **55**, 28–33 (2006)
15. Zadeh, L.: Fuzzy sets. Inf. Control **8**(3), 338–353 (1965)

On Characterizations of Bivariate Schur-constant Models and Applications

Bao Q. Ta[1(✉)], Dong S. Le[2], Minh B. Ha[2], and Xuan D. Tran[3]

[1] Department of Mathematical Economics, Banking University of Ho Chi Minh City,
39 Ham Nghi St., District 1, Ho Chi Minh City, Vietnam
baotq@buh.edu.vn
[2] Faculty of Management Information System,
Banking University of Ho Chi Minh City, 39 Ham Nghi St., District 1,
Ho Chi Minh City, Vietnam
{dongls,minhhb}@buh.edu.vn
[3] University of Science of Ho Chi Minh City, 227 Nguyen Van Cu Street,
Ho Chi Minh City, Vietnam
trandongxuan@gmail.com

Abstract. We study some properties of the family of copulas which are generated from the Laplace transform of bivariate Schur-constant models. The applications of these models in life insurance and in telecommunication are also discussed.

1 Introduction

Let (X, Y) be a positive random vector. It is said to be *Schur-constant* if the joint survival function has the following property

$$\mathbb{P}(X > x, Y > y) = S(x + y) \tag{1}$$

for all $x, y > 0$, where S is some appropriate function. It is seen that X, Y are identical in law with the survival functions $S(x), S(y)$, i.e., $S(x) = \mathbb{P}(X > x)$, $S(y) = \mathbb{P}(Y > y)$, respectively. The Schur-constant models play an important role in studying the analysis of lifetimes (see, Barlow and Mendel [1,2], Caramellino and Spizzichino [4,5]). It is shown in Nelsen [14] that the class of survival copulas generated from bivariate Schur-constant models has a variety of useful properties, e.g., one of the most important properties is that survival copulas of Schur-constant vector random variable (X, Y) are Archimedean. We refer to [14, p. 32] for a definition of survival copula.

Recently, some applications of Schur-constant models in Insurance Mathematics are scrutinized by Chi et al. [7], and Castañer et al. [6]. Furthermore, Schur-constant models are used to investigate the relationship between the starting time and ending time of the busy period of diffusion local time storage (see Kozlova and Salminen [10, p. 213–215] for a concept of diffusion local time storage, Salminen and Vallois [17] and Salminen et al. [18]). In particular, Ta and Pham Van [20] study the Laplace transform of two-dimensional Schur-constant

© Springer International Publishing AG 2018
L. H. Anh et al. (eds.), *Econometrics for Financial Applications*, Studies in Computational Intelligence 760, https://doi.org/10.1007/978-3-319-73150-6_65

random vector and prove that the transform induces a new class of copulas. It is also shown that there is a connection with reflecting Brownian motion. Furthermore, they investigate a new class of Achimedean copulas generated from bivariate Schur-constant models associated the starting time and ending time of the on going-busy period (or excursion straddling) of the Ornstein-Uhlenbeck process.

In the next section we provide structure and some basic properties of Schur-constant random variables. In Sect. 3 we consider copulas generated from the Laplace transform of Schur-constant vector random variable and study properties of these copulas. In Sect. 4 we discuss some applications of Schur-constant models in Insurance and Telecommunication.

2   General Properties

In this section we give some important properties of Schur-constant distributions. In particular, we introduce some results related to the Laplace transform of Schur-constant vector (X, Y) (see, [20] for more details and applications). First of all we recall the definitions of copula, Archimedean copula. Copula theory has a long history in the theory of probability, the theory has been initiated in fifties by the works of Frechet [9] and Sklar [19]. In the stochastic world, modelling the dependence of two (or more) random variables plays a crucial role not only in probability theory, but also in applications, e.g., finance, insurance and telecommunication. The classical and natural dependence measure is known as linear correlation. However, dependence between random variables is very complicated. Copulas provide a powerful tool to model complicated dependence structure of random variables. We refer to Nelsen [15] for an excellent introduction to copula theory, Durante and Sempi [8] for a completed review of important properties of copulas.

Definition 1. *A function $C : [0, 1]^2 \to [0, 1]$ is a copula if it has following properties*

(i) $C(u_1, 0) = C(0, u_2) = 0$.
(ii) $C(u_1, 1) = u_1$ and $C(1, u_2) = u_2$.
(iii) C is 2-increasing in $[0, 1]^2$, i.e.,

$$\Delta_C(u_1, u_2; v_1, v_2) := C(u_1, v_1) - C(u_1, v_2) - C(u_2, v_1) + C(u_2, v_2) \geq 0$$

for all $0 \leq u_1 < u_2 \leq 1$ and $0 \leq v_1 < v_2 \leq 1$.

Let X, Y be random variables with marginals F_1, F_2, respectively. It is proved by Sklar [19] that there is a copula C such that the joint distribution F of X, Y connects with the margins F_1, F_2 via C as follows

$$F(x_1, x_2) = C(F_1(x_1), F_2(x_2)). \tag{2}$$

Furthermore, if F_1 and F_2 are continuous then copula C is determined uniquely, for more details see Nelsen [15, Theorem 2.10.9]. It is seen from (2) that copula C

contains the information about dependence structure between random variables X, Y. Next we consider a special class of copulas, the so-called Archimedean copulas. This class of copulas is very useful in empirical modelling and has a wide range of applications. From practical point of view this class allows us to generate multivariate joint distributions from specified marginals.

Definition 2. *A copula C is called Archimedean if it admits the form*

$$C(u_1, u_2) = \psi(\psi^{-1}(u_1) + \psi^{-1}(u_2)), \quad 0 \leq u_1, u_2 \leq 1,$$

where the function $\psi : [0, \infty) \to [0, 1]$ satisfies the following conditions

(i) ψ is a continuous, non-increasing and strictly decreasing on $[0, \inf\{x : \psi(x) = 0\}]$,
(ii) $\psi(0) = 1$ and $\lim_{x \to \infty}(\psi(x)) = 0$.

The inverse ψ^{-1} is defined as $\psi^{-1}(t) := \inf\{x > 0 : \psi(x) = t\}$. The function ψ is called generator of Achimedean copula. From the structure of Archimedean copulas we see that they are easily derived and used for capturing wide ranges of dependence.

In many cases we are interest of the *concordance* between extreme values of random variables. More precisely, let (x_i, y_i) and (x_j, y_j) be two observations from (X, Y). These observations are called concordant if $(x_i - x_j)(y_i - y_j) > 0$, otherwise, they are called *discordant*. It is proved that the probability $\mathbb{P}((X_i - X_j)(Y_i - Y_j) > 0)$ and the difference $Q = \mathbb{P}((X_i - X_j)(Y_i - Y_j) > 0) - \mathbb{P}((X_i - X_j)(Y_i - Y_j) < 0)$ can be represented via copula of X, Y. For more details and further reading, we refer to [15, p. 157–162]. The *tail dependences* are very important in measuring dependence of two random variables in the upper-right and lower-left quadrant of $[0, 1]^2$. Roughly speaking, these dependence measures are the conditional probability that one random variable exceeds some level given that another exceeds some value (see Nelsen [15, Sect. 5.4] for further reading).

Definition 3. *The upper and lower tail dependence λ_U and λ_L of random variables X_1, X_2 with copula C are defined as*

$$\lambda_U := \lim_{t \to 1^-} \mathbb{P}(X_1 > F_1^{-1}(t) | X_2 > F_2^{-1}(t)),$$

$$\lambda_L := \lim_{t \to 0^+} \mathbb{P}(X_1 \leq F_1^{-1}(t) | X_2 \leq F_2^{-1}(t)),$$

if the respective limit exits.

The tail dependence can be simply calculated via copula as follows (see, e.g., [15, Theorem 5.4.2]).

Proposition 1. *Let C be the copula of X_1 and X_2. If the upper and lower tail dependence λ_U and λ_L exist, then*

$$\lambda_U = \lim_{x \to 1^-} \frac{1 - 2x + C(x,x)}{1 - x},$$

and

$$\lambda_L = \lim_{x \to 0^+} \frac{C(x,x)}{x}.$$

Now consider Schur-constant random variables X, Y with the joint survival function given by (1). Then X and Y can be represented as $X = UV$ and $Y = (1 - U)V$, where U is uniformly distributed random variable and independent of an arbitrary positive random variable V (see, e.g., [7,17]). We have the proportion $X/(X + Y)$ is independent of the sum (or the total lifetime) $X + Y$ and $X/(X + Y) \sim U(0, 1)$. Similarly for $Y/(X + Y)$. In particular, the ration Y/X has Pareto distribution with the density $f(x) = (1 + x)^{-2}$, $x > 0$ and Y/X is independent of $V \overset{(d)}{=} X + Y$. Denote $\bar{F}_V$ the survival function of V. Then it is easy seen that the function S is determined by $\bar{F}_V$ via the following differential equation (see [14, Theorem 4])

$$uS'(u) - S(u) + \bar{F}_V(u) = 0.$$

The following result provides a necessary and sufficient conditions such that a random vector (X, Y) is Schur-constant. It also shows that distribution of Schur-constant random variables X, Y is determined by the Laplace transform of V. We refer to [10, Proposition 5.7] and [20, Proposition 2.5] for the completed proof.

Proposition 2. *Let (X, Y) be a positive random vector. Then (X, Y) is Schur-constant if and only if there exists a positive random variable V such that for all $\alpha, \beta > 0$ and $\alpha \neq \beta$ it holds*

$$\mathbb{E}(e^{-\alpha X - \beta Y}) = \frac{1}{\alpha - \beta} \int_{\beta}^{\alpha} \mathbb{E}(e^{-tV}) dt. \tag{3}$$

In particular, V has the same distribution as $X + Y$.

One of the most important properties related to copulas of Schur-constant random variables is provided by the following result (see, Nelsen [14, Theorem 2]).

Theorem 1. *Let (X, Y) be a Schur-constant vector random variable with the joint survival function $\mathbb{P}(X > x, Y > y) = S(x + y)$ for all $x, y > 0$. Then copula of X and Y is an Archimedean copula with its generator S, i.e.,*

$$C(u, v) = S(S^{-1}(u) + S^{-1}(v)). \tag{4}$$

Example 1. Let $V \sim \Gamma(2, \gamma)$. We have the Laplace transform

$$\mathbb{E}(e^{-tV}) = \frac{\gamma^2}{(\gamma + t)^2},$$

and, hence,

$$\mathbb{E}(e^{-\alpha X - \beta Y}) = \frac{\gamma^2}{(\gamma + \alpha)(\gamma + \beta)},$$

$$\mathbb{E}(e^{-\alpha X}) = \frac{\gamma}{(\gamma + \alpha)}.$$

Implying that

$$\mathbb{E}(e^{-\alpha X - \beta Y}) = \mathbb{E}(e^{-\alpha X})\mathbb{E}(e^{-\beta Y}).$$

So X and Y are independent and copula $C(u, v) = uv$.

3   The Laplace Transform

Now let (X_1, X_2) have Schur-constant distribution. Recall that X_1 and X_2 can be represented as $X_1 = UV$ and $X_2 = (1-U)V$, where U is uniformly distributed random variable and independent of an arbitrary positive random variable V.

Consider the functions $H(\alpha, \beta) := \mathbb{E}(e^{-\alpha X_1 - \beta X_2})$ and $H_1(\alpha) := \mathbb{E}(e^{-\alpha X_1})$, where $\alpha, \beta > 0$. It is seen that $H_1 : [0, \infty) \to [0, 1]$ is strictly decreasing and

$$\lim_{\alpha \to \beta} H(\alpha, \beta) = \mathbb{E}(e^{-\beta V}) =: \phi_V(\beta).$$

So there exists the inverse function G of H_1. From (3) we obtain

$$H_1(\alpha) = \frac{1}{\alpha} \int_0^\alpha \mathbb{E}(e^{-tV}) dt. \tag{5}$$

Now let us define a function $\bar{C} : [0, 1] \times [0, 1] \to [0, 1]$ as follows

$$\bar{C}(u, v) := \mathbb{E}(e^{-G(u)X_1 - G(v)X_2}). \tag{6}$$

The following result shows that the function $\bar{C}$ is a copula. We refer to [20] for a detailed proof.

Proposition 3. *The function H is 2-increasing, i.e., for $(\alpha_1, \beta_1), (\alpha_2, \beta_2) \in [0, \infty) \times [0, \infty)$ with $\alpha_1 \leq \alpha_2, \beta_1 \leq \beta_2$*

$$H(\alpha_2, \beta_2) - H(\alpha_2, \beta_1) - H(\alpha_1, \beta_2) + H(\alpha_1, \beta_1) \geq 0.$$

Furthermore, the function $\bar{C}$ defined by (6) is a copula.

The copula $\bar{C}$ is now named *Laplace transform copula*. We have

$$H(\alpha, \beta) = \frac{\alpha H_1(\alpha) - \beta H_1(\beta)}{\alpha - \beta}, \tag{7}$$

and

$$\bar{C}(u, v) = \frac{uG(u) - vG(v)}{G(u) - G(v)}. \tag{8}$$

Example 2. Let V be a random variable with the Laplace transform

$$\mathbb{E}(e^{-xV}) = r\frac{a^2}{(1+x^r)^2} + (1-r)\frac{a}{1+x^r}, \quad 0 < r \leq 1, \quad 0 < a.$$

We have

$$H_1(\alpha) = \frac{a}{a + x^r},$$

and, hence,

$$G(u) = H_1^{-1}(u) = a^{1/r}(\frac{1-u}{u})^{1/r}.$$

So we obtain the family of Laplace copulas

$$\bar{C}_r(u,v) = \frac{uv^{1/r}(1-u)^{1/r} - vu^{1/r}(1-v)^{1/r}}{v^{1/r}(1-u)^{1/r} - u^{1/r}(1-v)^{1/r}}.$$

It is seen that with $r = 1$ we get $\bar{C}(u,v) = uv$. Now let $r = 1/2$ we have

$$\bar{C}_{1/2}(u,v) = \frac{uv(1-uv)}{u+v-uv} = \frac{uv}{u+v-uv} - \frac{(uv)^2}{u+v-uv}. \tag{9}$$

Notice that the first term in the right-hand side of (9) is Mikhail-Haq copula which is Archimedean.

Next we consider some concepts and basis results of *Schur-convex, Schur-concave* functions and then we show that the Laplace copula has Schur-concave property.

Definition 4. *A function $\phi : A \subset \mathbb{R}^2 \rightarrow \mathbb{R}$ is called Schur-convex function if for all $x = (x_1, x_2), y = (y_1, y_2) \in A$ such that x is majorized by y, denoted by $x \prec y$, i.e., $\max(x_1, x_2) \leq \max(y_1, y_2)$ and $x_1 + x_2 = y_1 + y_2$ then*

$$\phi(x) \leq \phi(y).$$

If $\phi(x) \geq \phi(y)$ then function ϕ is called Schur-concave.

The following result provides conditions such that a function is Schur-convex. This result is due to Schur-Ostrowski (see, e.g., [12]).

Theorem 2. *Let $\phi : A \subset \mathbb{R}^2 \rightarrow \mathbb{R}$ be a continuously differentiable function. Then ϕ is Shur-convex on A if*

(i) S is symmetric
(ii) For all $x \in A$

$$(x_1 - x_2)\left(\frac{\partial\phi}{\partial x_1} - \frac{\partial\phi}{\partial x_2}\right) \geq 0.$$

Using Theorem 2 we have the following result.

Proposition 4. *Let* $\psi : (0, \infty) \to [0, 1]$ *be a completely monotonic function. Define*

$$\tilde{F}(\alpha, \beta) := \frac{1}{\alpha - \beta} \int_\beta^\alpha \psi(t) dt.$$

Then $\tilde{F}$ *is joint distribution function of some Schur-constant vector random variable* (X, Y). *Furthermore,* $\tilde{F}$ *is Schur-convex.*

Proof. Since ψ is complete monotonic function then there exists a positive random variable V such that for all $t > 0$

$$\mathbb{E}(e^{-tV}) = \psi(t),$$

From Proposition 2 there exists a Schur-constant vector random variable (X, Y) such that $\tilde{F}$ is its joint distribution function. We have

$$\frac{\partial \tilde{F}}{\partial \alpha} = -\frac{1}{(\alpha - \beta)^2} \int_\beta^\alpha \psi(t) dt + \frac{\psi(\alpha)}{\alpha - \beta},$$

$$\frac{\partial \tilde{F}}{\partial \beta} = \frac{1}{(\alpha - \beta)^2} \int_\beta^\alpha \psi(t) dt - \frac{\psi(\beta)}{\alpha - \beta},$$

and, hence,

$$(\alpha - \beta) \left(\frac{\partial \tilde{F}}{\partial \alpha} - \frac{\partial \tilde{F}}{\partial \beta} \right) = -\frac{2}{(\alpha - \beta)} \int_\beta^\alpha \psi(t) dt + \psi(\alpha) + \psi(\beta)$$

$$= -2\tilde{F}(\alpha, \beta) + \psi(\alpha) + \psi(\beta).$$

Notice that $\lim_{u \to v} \tilde{F}(u, v) = \psi(v)$. Since $\tilde{F}$ is 2-increasing then for all (α, β) we have

$$\psi(\alpha) + \psi(\beta) \geq 2\tilde{F}(\alpha, \beta).$$

So we get

$$(\alpha - \beta) \left(\frac{\partial \tilde{F}}{\partial \alpha} - \frac{\partial \tilde{F}}{\partial \beta} \right) \geq 0.$$

The proof is complete.

Proposition 5. *Let* $\hat{F}$ *be a function defined as follows*

$$\hat{F}(\alpha, \beta) = \frac{1}{\varphi(\alpha) - \varphi(\beta)} \int_{\varphi(\beta)}^{\varphi(\alpha)} \psi(t) dt,$$

and $\lim_{\alpha \to \beta} \hat{F}(\alpha, \beta) = \psi(\varphi(\beta))$, *where* $\varphi : [0, 1] \to (0, \infty)$ *is convex, decreasing and* $\psi : (0, \infty) \to [0, 1]$ *is completely monotonic. Then* $\hat{F}$ *is Schur-concave, and, hence, Laplace copula* $\bar{C}$ *is Schur-concave.*

Proof. It is shown that the function

$$\tilde{F}(\alpha, \beta) = \frac{1}{\alpha - \beta} \int_{\beta}^{\alpha} \psi(t)dt,$$

is Schur-convex and $\tilde{F}(\alpha, \alpha) = \psi(\alpha)$. We have $\hat{F}(\alpha, \beta) = \tilde{F}(\varphi(\alpha), \varphi(\beta))$. Since φ is convex then we have

$$\varphi(\lambda\alpha + (1 - \alpha)\beta) \leq \lambda\varphi(\alpha) + (1 - \lambda)\varphi(\beta),$$

and

$$\varphi((1 - \lambda)\alpha + \alpha\beta) \leq (1 - \lambda)\varphi(\alpha) + \lambda\varphi(\beta),$$

for all $\lambda \in [0, 1]$. On the other hand, $\tilde{F}$ is decreasing in each argument, we have

$$\tilde{F}(\varphi(\lambda\alpha + (1 - \alpha)\beta), \varphi((1 - \lambda)\alpha + \alpha\beta))$$
$$\geq \tilde{F}(\lambda\varphi(\alpha) + (1 - \lambda)\varphi(\beta), (1 - \lambda)\varphi(\alpha) + \lambda\varphi(\beta)).$$

Using Schur-convex property of $\tilde{F}$ we get

$$\hat{F}(\lambda\alpha + (1 - \alpha)\beta, (1 - \lambda)\alpha + \alpha\beta) \geq \hat{F}(\alpha, \beta),$$

and, hence, $\tilde{F}$ is Schur-concave. Consequently copula $\bar{C}$ is Schur-concave.

4   Applications

In recent years the theory of copula and its applications have been widely used in a variety of fields, especially, in finance, econometric, actuarial science, risk management (see, [13]) and telecommunication. In this section we review some applications of copula theory in life insurance and in telecommunication.

4.1   In Life Insurance

Life insurance involves estimating the survival probability of each individual, and more complicated, the joint survival probability of a group of individuals. In classical viewpoint the joint survival probability of group of individuals can be estimated simply since it based on the assumption that each individuals in life insurance portfolio are independent. However, the situation is more complicated in the case that each individuals in portfolio are NOT independent. The theory of copula will help in this case, for example, see the following arguments we review from [7].

In a life insurance portfolio there are n individuals at ages $x_i, i = 1, \ldots, n$. Let $X_i, i = 1, \ldots, n$ and $\tau_i = X_i - x_i, i = 1, \ldots, n$, be their lifetimes and their residual lifetimes, respectively. For any $(t_1, \ldots, t_n) \in \mathbb{R}^n_+$, we should estimate the following joint survival probability:

$$\mathbb{P}(\tau_1 > t_1, \ldots, \tau_n > t_n)$$

Another application comes from the viewpoint of insurer: to estimate how much money would be pay for their customers. Assume that each customers in portfolio have unit payment and the interest rate is δ, then the present value of payment of insurer would be

$$S_n = \sum_{i=1}^n e^{-\delta \tau_i},$$

and hence, the average payment of life insurance portfolio is $\frac{S_n}{n}$. Let $d > 0$ be retention level of insurer. Then the following quantity

$$\mathbb{E}\left(\frac{S_n}{n} - d\right)_+$$

is called the stop-loss premium of insurer. Calculating this quantity is very important task for insurer. However, this task is not easy to compute accurately in general case and it would be nice if one can find an upper bound for that.

In the case that the lifetimes $(X_1, \ldots, X_n)$ is Schur-constant, we will see that the first question can be answered easily (see, e.g., [7])

$$\mathbb{P}(\tau_1 > t_1, \ldots, \tau_n > t_n)$$
$$= \mathbb{P}(X_1 > t_1 + x_1, \ldots, X_n > t_n + x_n \mid X_1 > x_1, \ldots, X_n > x_n)$$
$$= \frac{S\left(\sum_{i=1}^n t_i + \sum_{i=1}^n x_i\right)}{S\left(\sum_{i=1}^n x_i\right)},$$

or equivalently, $(\tau_1, \ldots, \tau_n)$ is Schur-constant with survival distribution given as above.

The second question also can be answered in the case that the lifetimes $(X_1, \ldots, X_n)$ is Schur-constant. By [7, Proposition 6.1] the upper bound is given as follows

$$\mathbb{E}\left(\frac{S_n}{n} - d\right)_+ \leq \mathbb{E}\left(\frac{1}{n}\sum_{i=1}^n U_i^{\frac{\delta T_n}{n}} - d\right)_+,$$

where $T_n = \sum_{i=1}^n \tau_i$ and $U_i, i = 1, \ldots, n$ are uniformly i.i.d. random variables. Simulation results shown in [7, p. 406] point out that the upper bound is tight, especially well approximated in the case that n is large. In what follows we run some simulations to see that things. We first assume that $\frac{T_n}{n} \sim \text{Gamma}(a, 1)$ and $\tau_i \sim \text{Exp}(1/a)$, and then we generate 10000 random samples to compute the stop-loss premium and the upper bound. The results are given in the following table (Table 1).

4.2   In Telecommunication

In this subsection the applications of copula theory in telecommunication are presented. We revise some applications of copula theory in *fluid queues (or storage processes)*. More precisely, we study structure dependence of the starting

Table 1. Stop-loss premium and upper bound when $\frac{T_n}{n} \sim$ Gamma$(a, 1)$, $\tau_i \sim$ Exp$(1/a)$, and $n = 1000$.

δ	d	a	$\mathbb{E}\left(\frac{S_n}{n} - d\right)_+$	$\mathbb{E}\left(\frac{1}{n}\sum_{i=1}^{n} U_i^{\frac{\delta T_n}{n}} - d\right)_+$
0.05	0.1	20	0.4001272	0.4063313
		30	0.2999662	0.3049161
		40	0.2332146	0.2369343
		50	0.1856149	0.1886476
		60	0.1499636	0.1524426
	0.2	20	0.299962	0.3061238
		40	0.1332609	0.1370832
		60	0.05002032	0.05248253
0.02	0.1	20	0.6142805	0.6172535
		40	0.4554959	0.4581796
		60	0.3544156	0.3568509
	0.2	20	0.5143545	0.5172148
		40	0.355656	0.3583195
		60	0.2544879	0.2568361

time and ending time of storage process via copula. We will find more explicit expression for distributions and other characteristics for this fluid queue.

Let B_t be Brownian motion on the time axis $\mathbb{R}$ and $\mu = \sqrt{2c} > 0$. Then the *storage or fluid queue process* with Brownian input, constant service rate μ is defines as

$$Q_t := \sup_{-\infty < s \leq t} \{W_t - W_s - \mu(t - s)\}, \quad t \in \mathbb{R}.$$

It is proved in [16, Proposition 2.8] that the process Q_t is identical in law to a reflecting Brownian motion with drift $-\mu$, say, BM$(-\mu)$, starting at time 0 with the initial distribution $\hat{m}(dx) = 2\mu e^{-2\mu x} dx$. Now let us define definition of starting time and ending time.

Definition 5. *Let* $g_t := \sup\{s < t : Q_s = 0\}$ *and* $d_t := \inf\{s > t : Q_s = 0\}$ *for all* $t \in \mathbb{R}$. *Then* g_t *and* d_t *are called starting time and ending time, respectively, of the on going-busy period at time* t *or excursion straddling* t.

In Examples 1 and 2 with $r = 1$, equivalently $V \sim \Gamma(2, a)$ we see that Laplace copula $\bar{C}$ is the product copula, and, hence, it is Archimedean, e.g., X_1, X_2 are independent. So, a natural question now arises of does there exist a random variable V such that Laplace transform copula is Archimedean in which X_1 and X_2 are not independent? It is shown in [20] that there exists a random variable V such that the corresponding Laplace copula is Archimedean. Furthermore, there is a connection with starting time and ending time of the *on going-period* (or *excursion straddling* t) $Q^{(t)} := \{Q_s : g_t \leq s \leq d_t\}$.

Proposition 6. *If $V \sim \Gamma(1/2, c)$, i.e., V has the density function*

$$f_V(x) = \sqrt{\frac{c}{\pi}} x^{-1/2} e^{-cx} \mathbb{1}_{\{x>0\}},$$

then the corresponding Laplace transform copula $\bar{C}$ is Ali-Mikhail-Haq copula which is Archimedean, i.e.,

$$\bar{C}(u, v) = \frac{uv}{u + v - uv},$$

and X_1, X_2 are identical in law to the remaining lengths $t - g_t$ and $d_t - t$ of the busy period at time t of reflecting Brownian motion with drift $-\sqrt{2c}$.

We see that the length $t - g_t$ and $d_t - t$ are Schur-constant random variables, and, hence, their copula is, in fact, Archimedean. So we consider copula of $t - g_t$ and $d_t - t$ in case the input is *Ornstein-Uhlenbeck process*. We refer to [11, p. 225] for a more detailed discussion of the process. Now let $X = (X_t)_{t \geq 0}$ be a linear regular diffusion taking values in $\mathbb{R}_+$. We refer to [3] for the general theory of diffusion processes. Assuming that the process X starts at 0, then for all $t > 0$ the starting time and ending time of the on going-busy period at time t

$$g_t = \sup\{s < t : X_s = 0\} \quad \text{and} \quad d_t = \inf\{s > t : X_s = 0\}.$$

Proposition 7. *Let X be Ornstein-Uhlenbeck process with parameter $r > 0$, i.e., X is the solution of the stochastic differential equation*

$$dX_t = -rX_t dt + dB_t \quad \text{with} \quad X_0 = 0,$$

where B_t is Brownian motion. Then the copula of the respective random variables $t - g_t$ and $d_t - t$ is

$$C(u, v) = \frac{2}{\pi} \arcsin\left(\sin\left(\frac{\pi u}{2}\right) \sin\left(\frac{\pi v}{2}\right)\right).$$

This is Archimedean with the generator

$$\psi(x) = \frac{2}{\pi} \arcsin(e^{-rx}).$$

Furthermore, the dependence of upper and lower quadrant tails of starting time and ending time are

$$\lambda_U = \lim_{x \to 1^-} \frac{1 - 2x + C(x, x)}{1 - x} = 2 - 2 \lim_{y \to 0^+} \frac{S'(2y)}{S'(y)} = 2 - \sqrt{2},$$

$$\lambda_L = \lim_{x \to 0^+} \frac{C(x, x)}{x} = 2 \lim_{y \to \infty} \frac{S'(2y)}{S'(y)} = 0.$$

The complete proof can be found in [20]. It is interesting to see that the copula C does not depend on r of the Ornstein-Uhlenbeck process X.

References

1. Barlow, R.E., Mendel, M.B.: de Finetti-type representations for life distributions. J. Am. Stat. Assoc. **87**(420), 1116–1122 (1992)
2. Barlow, R.E., Mendel, M.B.: Similarity as a probabilistic characteristic of aging. In: Reliability and Decision Making, Siena 1990, pp. 233–245. Chapman & Hall, London (1993)
3. Borodin, A.N., Salminen, P.: Handbook of Brownian Motion—Facts and Formulae. Probability and its Applications, 2nd edn. Birkhäuser Verlag, Basel (2002)
4. Caramellino, L., Spizzichino, F.: Dependence and aging properties of lifetimes with Schur-constant survival functions. Probab. Eng. Inf. Sci. **8**, 103–111 (1994)
5. Caramellino, L., Spizzichino, F.: WBF property and stochastical monotonicity of the Markov process associated to Schur-constant survival functions. J. Multivar. Anal. **56**(1), 153–163 (1996)
6. Castañer, A., Claramunt, M.M., Lefèvre, C., Loisel, S.: Discrete Schur-constant models. J. Multivar. Anal. **140**, 343–362 (2015)
7. Chi, Y., Yang, J., Qi, Y.: Decomposition of a Schur-constant model and its applications. Insur. Math. Econ. **44**(3), 398–408 (2009)
8. Durante, F., Sempi, C.: Copula theory: an introduction. In: Copula Theory and its Applications, vol. 198, Lecture Notes in Statistics Proceedings, pp. 3–31. Springer, Heidelberg (2010)
9. Fréchet, M.: Sur les tableaux de corrélation dont les marges sont données. Ann. Univ. Lyon. Sect. A **3**(14), 53–77 (1951)
10. Kozlova, M., Salminen, P.: Diffusion local time storage. Stoch. Process. Appl. **114**(2), 211–229 (2004)
11. Le Gall, J.F.: Brownian Motion, Martingales, and Stochastic Calculus. Springer, Heidelberg (2016)
12. Marshall, A.W., Olkin, I., Arnold, B.C.: Inequalities: Theory of Majorization and its Applications. Springer Series in Statistics, 2nd edn. Springer, New York (2011)
13. McNeil, A.J., Frey, R., Embrechts, P.: Quantitative Risk Management: Concepts, Techniques and Tools. Princeton Series in Finance. Princeton University Press, Princeton (2005)
14. Nelsen, R.B.: Some properties of Schur-constant survival models and their copulas. Braz. J. Probab. Stat. **19**(2), 179–190 (2005)
15. Nelsen, R.B.: An Introduction to Copulas. Springer Series in Statistics, 2nd edn. Springer, New York (2006)
16. Salminen, P., Norros, I.: On busy periods of the unbounded Brownian storage. Queueing Syst. **39**(4), 317–333 (2001)
17. Salminen, P., Vallois, P.: On first range times of linear diffusions. J. Theor. Probab. **18**(3), 567–593 (2005)
18. Salminen, P., Vallois, P., Yor, M.: On the excursion theory for linear diffusions. Jpn. J. Math. **2**(1), 97–127 (2007)
19. Sklar, M.: Fonctions de répartition à n dimensions et leurs marges. Publ. Inst. Stat. Univ. Paris **8**, 229–231 (1959)
20. Ta, B.Q., Van, C.P.: Some properties of bivariate Schur-constant distributions. Stat. Probab. Lett. **124**, 69–76 (2017)

Time-Varying Beta Estimation in CAPM Under the Regime-Switching Model

Roengchai Tansuchat[(✉)], Sukrit Thongkairat, Woraphon Yamaka,
and Songsak Sriboonchitta

Faculty of Economics, Centre of Excellence in Econometrics,
Chiang Mai University, Chiang Mai, Thailand
roengchaitan@gmail.com

Abstract. The objectives of this study are to analyze the risk of investment and to examine the structural change in the CAPM. To there ends,
the Markov Switching dynamic regression is employed to construct the
time varying beta risk when the market exhibits structural change. The
model is applied to the Thai stock return data. The empirical results
show a strong evidence of structural change in CAPM for four out of five
Thai stocks of large market capitalization. We observe that the movement of Thai stocks fluctuated widely during the market turbulence,
especially at the time of Thai financial crisis.

Keywords: Time-varying beta · Markov switching · Kalman filter
Thai stock

1 Introduction

Investors are currently investing in securities because of capital gains from security price change and dividend. Investment in securities is a risky venture compared to other investments. To invest in securities, one must take into account
many factors such as economic conditions, interest rates, inflation, etc. that
affect the expected return and risk. In addition, investing in high-return securities always has a higher risk. In financial model, there are many models indicating
the correlation between expected return and risks. The most widely used model
for pricing of risky securities is the Capital Asset Pricing Model (CAPM) [20,24]
which describes the relationship between systematic risk and expected return for
assets, particularly stocks.

In CAPM model, the expected return of a security equals the risk-free rate
of the security plus a risk premium. The beta coefficient (β) of CAPM model
is viewed as the volatility, or systematic risk, of a security in comparison to the
market as a whole. If the security's price moves with the market, beta equals
to one. If beta is greater than one, it demonstrates that the stock moves larger
than the market does in the same direction so the security's price is theoretically
more volatile or riskier than the market, but potentially more profitable. On the

© Springer International Publishing AG 2018
L. H. Anh et al. (eds.), *Econometrics for Financial Applications*, Studies in Computational
Intelligence 760, https://doi.org/10.1007/978-3-319-73150-6_66

other hand, the beta is less than one, this indicates that the stock moves less than the market does in the same direction so the security's price is theoretically less volatile or less risky than the market.

Traditionally, the beta of risky asset calculated from CAPM by using simple regression analysis, is constant over time [3]. However, due to the microeconomic factors, such as business environment and business operation and/or macroeconomic factors such as inflation, interest rate, and macroeconomic policies, the stock's beta coefficients tend to vary significantly over time [2,4,6,7,9–14]. Therefore, when betas vary over time the standard linear regression leads to inconsistency problem and misspecified inference. Thus, linear regression is not appropriate for estimating the CAPM [1].

In the literature, there are many nonlinear models used to estimate beta from CAPM model in the last decade. There are two groups of beta estimation technique. First, the beta risk is estimated by bivariate GARCH models such as BEKK, DCC (see, [5,25]). Second, the beta risk is directly estimated from CAPM equation with nonlinear estimation technique like Markov switching regression model, or state-space model such as McKenzie et al. [22], Korkmaz et al. [18,19], Prukumpai [23].

In the case of Thai stock market, Korkmaz et al. [18] tested Markov switching model of International CAPM (MS-ICAPM) of 23 emerging markets which include Thailand, and found regime switching between high and low volatility regime. In addition, by likelihood ratio test, the results confirmed that MS-ICAPM is superior to linear ICAPM. Prukumpai [23] investigated the time-varying beta behavior of industrial portfolio returns in the Stock Exchange of Thailand by applying the two-regime Markov-switching model. The result showed that conditional betas are unstable over time, and there dynamics can be classified into two regimes. Thus, the systematic risk of industrial portfolios is time-varying and regime-dependent.

The purpose of this paper is to estimate time varying-beta of CAPM model from the Stock Exchange of Thailand (SET) data by using Markov Switching dynamic regression model. The model allows us to capture the full dynamics of beta risk in the CAPM and to understand the movement among the beta estimates of various stocks in two different regimes, namely market upturn and downturn [21]. Our contributions to financial economics are as follows: First, a manifest advantage of the Markov Switching dynamic regression model is firstly applied to capture the structure change in the CAPM. Second, we are the first make use of this model to quantify the beta risks of stocks in SET.

The structure of the remainder of the paper is as follows. Section 2 discusses the econometrics models to be estimated. Section 3 describes the data, descriptive statistics, and unit root test. Section 4 analyzes the empirical estimates from empirical modeling. Some concluding remarks are given in Sect. 5.

2   Methodology

In this section, we introduce a dynamic linear models also known as linear state space model of West and Harrison [26]. This model allows for a time-varying nature of the linear relation between the dependent and independent variables over time. In addition, since this study also considers the phenomena of structural changes over time, we also introduce a regime switching dynamic linear models or a Markov-switching state space model of Kim [17] and investigate the performance of these two models using Akaiki information criterion (AIC).

2.1   Dynamic Linear Regression Models

With the notations following Harvey [16], the dynamic linear regression is specified by the following equations:

$$y_t = \beta_t x'_t + \varepsilon_t, \tag{1}$$

$$\beta_t = F_t \beta_{t-1} + \eta_t, \tag{2}$$

where y_t is an observed dependent variable and x'_t is an observed matrix of independent variable, while β_t is a generally unobservable k dimensional vector of time varying coefficient at time t. The disturbance ε_t is observation errors and η_t is unobserved evolution errors. F_t is known $k \times k$ matrix of integers. These two errors are assumed to be independent and have normal distribution, $\varepsilon_t \sim N(0, R_t)$, and $\eta_t \sim N(0, Q_t)$, respectively. R_t and Q_t are matrices of the dimension 1 by k, and k by k, respectively. In this study, we are aware that time varying equation in Eq. (2) implies that the coefficients follow the AR(1) process.

2.2   Markov Switching Dynamic Regression Model

In this study, the two regimes Markov-switching model is considered since we focus on two different regimes or states namely, stock market upturn (Bull market) and stock market downturn (Bear market). Thus, Eqs. (1) and (2) can be rewritten in the Markov Switching form,

$$y_t = \beta_{t,(S_t)} x'_t + \varepsilon_t, \tag{3}$$

$$\beta_{t,(S_t)} = F_{t,(S_t)} \beta_{t-1,(S_t)} + \eta_t, \tag{4}$$

where $\beta_{t,(S_t)}$ is regime dependent time varying coefficient which is allowed to vary over time according to Eq. (4). S_t represents a state variable which is governed by first order Markov chain; thus, transition probability (P) can be defined by

$$P_{ij} = (S_{t+1} = j \,|\, S_t = i) \text{ and } \sum_{j=1}^{k} p_{ij} = 1; \ i, j = 1, ..., k \tag{5}$$

where p_{ij} is the probability of regime i followed by regime j. The two errors are assumed to have normal distribution, $\varepsilon_t \sim N(0, E_{t,(S_t)})$, and $\eta_t \sim N(0, Q_{t,(S_t)})$, respectively. The parameter matrices of the model $\beta_{t,(S_t)}$, $F_{t,(S_t)}$, $Q_{t,(S_t)}$, and $R_{t,(S_t)}$ may be known under different regimes or states, but in some circumstances a particular element of a parameter matrix may take on different values which are unknown.

2.3   The Kalman Filter and Estimation of the Model

The Kalman filter is employed to filter the time varying coefficient using the algorithm which explained in West and Harrison [26] and Kim [17].

(1) Estimating the starting value of time varying coefficient at time for both two regimes using the conventional ordinary least squares (OLS) estimator.

$$\beta_{t=1} = (x'_t x_t)^{-1} x'_t y_t \tag{6}$$

(2) Prediction step
  We estimate the state dependent vector $\beta_{t,(S_t)}$ and their covariance matrix $\Gamma_{t|t-1,(S_t)}$ at time t with all information available at time $t-1$, w_{t-1}, and the prediction algorithm is as follows:

$$\beta_{t|t-1,(S_t)} = F_{(S_t)}\beta_{t-1|t-2,(S_t)}, \tag{7}$$

$$\Gamma_{t|t-1,(S_t)} = F_{(S_t)}\Gamma_{t-1|t-2,(S_t)}F'_{(S_t)} + Q_{(S_t)}, \tag{8}$$

$$v_{t|t-1,(S_t)} = y_t - x_t\beta_{t|t-1,(S_t)}, \tag{9}$$

$$f_{t|t-1,(S_t)} = x_t\Gamma_{t|t-1,(S_t)}x'_t + R_{(S_t)}, \tag{10}$$

where $v_{t|t-1,(S_t)}$ and $f_{t|t-1,(S_t)}$ are state dependent prediction error and conditional variance of the prediction error, respectively. In Eq. (7), an inference on $\beta_{t|t-1,(S_t)}$ given information up to time $t-1$ is a function of an inference on $\beta_{t-1|t-2,(S_t)}$ given information up to time $t-2$.

(3) Updating step

$$\beta_{t|t,(S_t)} = \beta_{t|t-1,(S_t)} + K_{t,(S_t)}v_{t|t-1,(S_t)}, \tag{11}$$

$$\Gamma_{t|t} = \Gamma_{t|t-1,(S_t)} - P_{t|t-1,(S_t)}K_{t,(S_t)}x_t\Gamma_{t|t-1,(S_t)}, \tag{12}$$

where $K_{t,(S_t)} = \Gamma'_{t|t-1,(S_t)}x'_t f^{-1}_{t|t-1,(S_t)}$ is Kalman Gain, which determines the weight assigned to new information about $\beta_{t|t,(S_t)}$ contained in the prediction error.

Repeating the steps 2–3 for $t = 2, ..., T$, we obtain the time varying coefficients for two regimes. Then, the log likelihood for the Markov-switching dynamic regression model can be formed as

$$L(\theta|y,x) = \sum_{S_t=1}^{2}\left(\sum_{t=1}^{T}\left(\log\left(2\pi\left|f_{t,(S_t)}\right|\right) + \left(\frac{1}{2}\varepsilon'_{t,(S_t)}f^{-1}_{t,(S_t)}\varepsilon_{t,(S_t)}\right)\right)\left(\left(S_t\middle|w_t\right)\right)\right) \tag{13}$$

where $e_{t,(S_t)} = y_t - x'_t\beta_{t,(S_t)}$, θ is all unknown parameters of the model, and $(\Pr(S_t = k|w_t))$ is the filter probabilities obtained from Hamilton filtering [15].

2.4 Hamilton's Filter

According to Eq. (13), the filter probabilities $\Pr(S_t = k\,|w_t)$ is an important process for filtering the estimated coefficient and variance parameters into two different regimes. One of the most famous filtering approach is Hamilton's filter, and it is determined using the following algorithm.

1. Given an initial guess of transition probabilities P which are the probabilities of switching between regimes; thus the transition probabilities of two regimes is,

$$P = \begin{bmatrix} p_{11} & p_{12} \\ p_{21} & p_{22} \end{bmatrix} \tag{14}$$

2. Updating the transition probabilities of each state with all past information including the parameters in the system equation and observations in order to compute the log likelihood function in each state at time t. Let w_t be all information in the past, the probability of each state is to be updated by the following formula.

$$Pr(S_t = 1, 2\,|w_t) =$$

$$\left[\frac{f(y_t, x_t\,|S_t = 1, w_{t-1})(S_t = 1\,|w_{t-1})}{\sum_{k=1}^{2} f(y_t, x_t\,|S_t = k, w_{t-1})(S_t = k\,|w_{t-1})}, \quad 1 - Pr(S_t = 1\,|w_t) \right] \cdot P \tag{15}$$

where $f(y_t, x_t\,|S_t = k, w_{t-1})$ is the normal density function of the model in regime k.

3. Iterating step 1 and 2 for $t = 1, ..., T$. Finally, the maximum likelihood estimation can therefore be implemented easily by maximizing the log likelihood Eq. (13) respect to θ.

3 Data

In this study, we employ weekly stock prices of five companies consisting KBANK of banking sector; PTT and SCG of energy & utilities sector; CPALL of commerce sector and ADVANC of information & communication sector, which are the top 5 stock market capitalization of the Stock Exchange of Thailand (SET). These weekly stock prices and SET50 index is collected between August 13, 1997 and August 31, 2017. The total 1,044 weekly observations are obtained from Thomson Reuter's database. The weekly stock price and market index are converted into rate of return, $r_{it} = \ln(p_{it}/p_{it-1})$, where p_{it} and p_{it-1} are closing price of stock price i for day t and $t - 1$, respectively.

Table 1 presents the descriptive statistics for the returns series. The average returns of stocks are similar, excepting the return of AOT which show the highest average return. The kurtosis statistic of returns are exceed than 3, indicating a high kurtosis of the returns. For normality test, Jarque–Bera statistics of all returns are statistically significant, thereby meaning that the distributions of these returns are non-normal, which may be due to the presence of extreme observations. In addition, we perform the stationary test of the data using Augmented-Dickey Fuller test (ADF) and the result shows that all returns are stationary at the level at 1% significant level.

Table 1. Data descriptive statistics

	SET	PTT	ADVANC	SCG	CPALL	KBANK
Mean	0.00060	0.0008	0.00066	0.00059	0.00183	0.00092
Median	0.00168	0	0	0	0	0
Maximum	0.07575	0.10334	0.05874	0.08867	0.10473	0.10537
Minimum	−0.08317	−0.12366	−0.09382	−0.071948	−0.094976	−0.103804
Std. dev.	0.01265	0.01886	0.01736	0.01558	0.01843	0.01867
Skewness	−0.67475	−0.30240	−0.43783	0.290425	0.38167	0.065001
Kurtosis	9.59089	8.29047	5.39984	6.36687	7.87556	6.27806
Jarque-Bera	1359.718***	851.8295***	196.0537***	350.6836***	731.6289***	323.3275***
Sum	0.43581	0.60549	0.48078	0.43199	1.32081	0.66863
Sum sq. dev.	0.11536	0.25631	0.21708	0.1749	0.24462	0.25113
ADF-test						
	−15.7640**	−30.0768***	−35.4665***	−15.2339***	−31.2448***	−34.4393***

Source: Calculation
*, **, and *** significant at 10%, 5%, and 1%, respectively

Table 2. Model selection

	AIC(2 regime)	AIC(1 regime)
PTT	6647.927	5927.035
ADVANC	5202.7748	6951.915
SCG	5567.4346	7091.872
CPALL	5035.4198	5075.953
KBANK	5217.702	6776.078

Source: Calculation

4   Empirical Results

4.1   Number of Regime Selection

The study employs the CAPM to quantify the beta risk of the stock returns in SET50. However, before we further investigate the risk of stock returns, we need to check whether a structural change exist in The CAPM or not. To achieve our purpose, the study consider the Akaiki information criterion (AIC) to check the number of regime. Table 2 reports the AIC values of both linear and Markov Switching time varying dynamic regression models for our CAPM. We observe that among the trial runs for CAPM, the results provide evidence that the Markov Switching dynamic CAPM(2 regimes) for KBANK, SCG, CPALL and ADVANC present the lowest AIC. However, in the case of PTT, we find that the Markov Switching dynamic CAPM does not show the higher performance compared to linear dynamic CAPM(1 regime) since the AIC of linear CAPM is lower than Markov Switching dynamic CAPM.

Table 3. Transition probability matrix

ADVANC		Regime 1	Regime 2
	Regime 1	0.9999	0.0001
	Regime 2	0.0229	0.9771
SCG		Regime 1	Regime 2
	Regime 1	0.9989	0.0011
	Regime 2	0.0024	0.9976
CPALL		Regime 1	Regime 2
	Regime 1	0.9532	0.0468
	Regime 2	0.0001	0.9999
KBANK		Regime 1	Regime 2
	Regime 1	0.9687	0.0313
	Regime 2	0.0082	0.9918

Source: Calculation

Note: For the CAPM of PTT return, the model is not exist a structural change (see, Table 2)

4.2 Transition Probabilities

The transition matrices of the four CAPM which obtained from the Markov Switching dynamic CAPM are presented in Table 3. The result shows that regime 1 and regime 2, are persistent because the probability of staying in each of these regimes is more than 95%, while the probability of moving between these regimes is only nearly 4%. These results suggest that it is only an extreme event that can switch the stock returns switching between the regimes.

4.3 Time-Varying Beta for Five Stocks

In this section, we illustrate the time-varying betas and the filtered probabilities for the Markov Switching dynamic CAPM of KBANK, SCG, CPALL and ADVANC while a linear dynamic CAPM is illustrated for PTT return.

(1) Two regimes time-varying Beta of ADVANC

The empirical results show the time-varying beta of ADVANC in two regimes. Here, we interpret state or regime 1 as the market upturn regime, or there is an expansion of ADVANC stock, while state or regime 2 is interpreted as market downturn regime, or there is a recession of ADVANC stock. As we can observe form Fig. 2, the estimated betas vary significantly over time, particularly in early 1997 during the Thai financial crisis. The betas for the ADVANC are likely to show a high fluctuation during this period while those for the rest of time show less volatility. That is, the stocks of ADVANC became more risky during the crisis. Furthermore, considering the difference between the movements of the

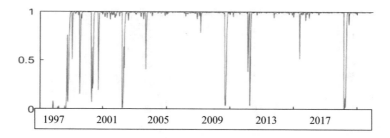

Fig. 1. Filtered probabilities of market upturn

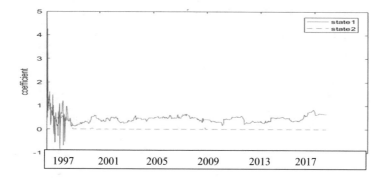

Fig. 2. Two regimes time-varying beta of ADVANCE

betas in two regimes, we observe that the time varying beta of market downturn is lower and less volatile when compared to market upturn. This indicates that ADVANC has a higher risk when it is in market upturn. To classify the betas whether they belong to the period of market upturn or downturn, we can consider the filtered probabilities as shown in Table 1. The results of filtered probabilities of CAPM for ADVANC show that the probability values are vary between 0 and 1, and then the beta is classified in the state or regime 1 if $\Pr(S_t = 1 | \beta_t) > 0.5$ and in the state or regime 2 of $\Pr(S_t = 1 | \beta_t) \leq 0.5$. Figure 1 shows that the filtered probabilities of the model is mostly stay in the market upturn. However, we observe that the probability of the market upturn is low in 1997 which corresponding to the Thai financial crisis. This result confirms that our model can detect the Thai financial crisis in 1997 and the risk of ADVANC is high in that period.

(2) Two regimes time-varying Beta of SCG

The empirical results show the time-varying beta of SCG in two regimes. Here, we also interpret the meaning of each state the same as before. Considering the time varying betas in Fig. 4, we can observe that the estimated betas vary significantly over time. The betas for the ADVANC is show a high fluctuation during 1997–2001, corresponding to the recession and recovery periods of the Thai financial crisis. We expect that SCG stock, which is classified in the construction index,

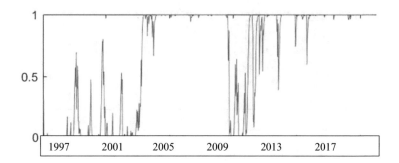

Fig. 3. Filtered probabilities of market upturn

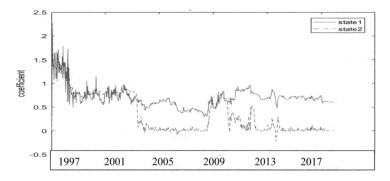

Fig. 4. Two regimes time-varying beta of SCG

might need time for their business to recover and the fragile Thai economic recovery might weaken the confidence of investors who understood that the economic conditions were worse than expected. However, after 2002, SCG seems to show less volatility. This indicates that SCG was more risky during the crisis. In addition, from the difference between the movements of the betas in two regimes, we can say that the time varying beta of market downturn is lower and less volatile compared to market upturn. This indicates that SCG has a higher risk when it is in market upturn. From filtered probabilities in Fig. 3, we observe some interesting movement pattern between filtered probability and time varying beta. We can see that the betas of SCG in market upturn and downturns are move together during the period of financial crisis (see, Fig. 3 during 1997–2001 and 2009), while the rest show the high difference between beta of market upturn and downturn.

(3) Two regimes time-varying Beta of CPALL

The empirical results show the time-varying betas which describe the behavior of the CPALL returns during 1997–2017. We can observe that the movements of beta in regime 1 and 2 are similar. The results of CPALL case seem to differ form those of ADVANC and SCG. The difference between the betas in these two regimes are close to zero, meaning that the betas in both regimes are move together. When we consider the filtered probabilities, the probability of staying

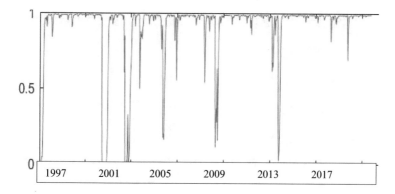

Fig. 5. Filtered probabilities of market upturn

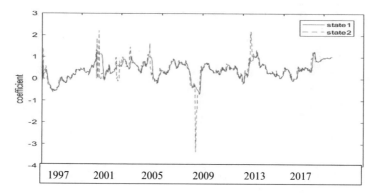

Fig. 6. Two regimes time-varying beta of CPALL

in market upturn is longer than in market downturn. This indicates that the CPALL returns mostly stay in the market upturn and that the financial crisis had little impact on this stock. However, if we look at the movement of the beta, we find that the betas vary between −1 to 1, indicating that Thai stock market can render both positive and negative effects to CPALL. Therefore, investors should be careful when considering invest in this stock (Figs. 5 and 6).

(4) Two regimes time-varying beta of KBANK

The empirical results show the time-varying betas of KBANK in two regimes. Regime 1 and 2 are also interpreted as the market upturn and downturn regime, respectively. As we can observe form Fig. 7, the estimated betas vary with high volatility over time, particularly in early 1997 during the Thai financial crisis. The betas for the ADVANC show a high fluctuation. We can observe that the beta values are mostly positive, excepting in the time of Thai financial crisis in 1997. The values of betas are negative during this period while those for the rest of time are positive. That is, Thai stock market gave a negative effect on KBANK during the financial crisis. The stock of KBANK becomes more risky

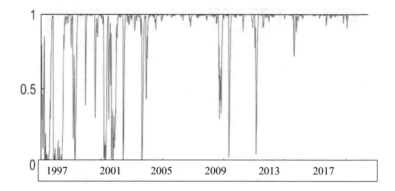

Fig. 7. Filtered probabilities of market upturn

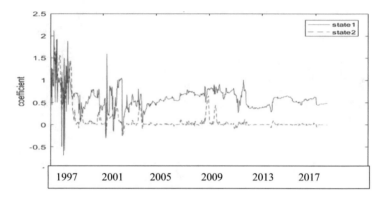

Fig. 8. Two regimes time-varying beta of KBANK

during the crisis. Furthermore, considering the difference between the movements of the betas in two regimes, we observe that the time varying beta of market downturn is mostly lower and less volatile compared to market upturn, except for the 1997–2002 period. We expect that Bank sector might face with the severe negative effect form the crisis during that period of time, and the confidence of people become lower. The movement of KBANK returns probably reflects the sensitivity to news or the shock of Thai economy. As we see can see in Fig. 7 which shows the filtered probabilities of the KBANK, there is a high fluctuation along 1997–2002 (Fig. 8).

(5) One regime time-varying beta of PTT

Finally, the only one regime time-varying betas estimated in this study is for PTT stock. We cannot find the structural change in CAPM in the case of PTT, therefore, we cannot distinguish the beta of PTT into two regimes. As we can observe from Fig. 9, the estimated betas vary positively significantly over time. This indicates that SET market always has a positive effect on PTT stock. We can observe two high peaks of the beta movement, corresponding to 1999 and

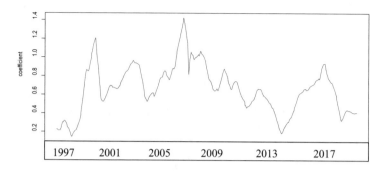

Fig. 9. One regime time-varying beta of PTT

2006. We find that the world oil price is relatively high and greater than 1 in that two periods. This indicates that PTT becomes an aggressive stock when the oil price is high. Conversely, the beta of PTT reached the lowest in 1997 and 2013, corresponding to the reduction in world oil price.

5   Conclusions

In this paper, we consider the Markov Switching CAPM with time-varying to quantify the investment risk when there exists a structural change in the CAPM. Compared to the earlier works, our model allows us to capture the full dynamics of beta in the CAPM and to understand the movement among the beta estimates of various stocks in two different regimes, namely market upturn and downturn. The empirical results confirm such structural change among the beta estimates of stocks in Thailand's market and show that there is a huge increase in values of beta during market upturn. Moreover, the beta estimates show high volatility and the values of beta are not much different during Thai financial crisis.

Regarding further research, we suggest several interesting topics for consideration. First, the proposed methodology can be readily applied to other prominent asset pricing models, such as the Fama and French [8] three-factor model, to gain further insight on the dynamics of another two factors. Second, the model is based on the normal distribution assumption which might not be appropriate to quantify the risk based on the financial data. Finally, the study can expand to the scope of international CAPM to quantify the risk of the overall capital market.

Acknowledgements. We are grateful for financial support from Puay Ungpakorn Centre of Excellence in Econometrics, Faculty of Economics, Chiang Mai University.

References

1. Ang, A., Chen, J.: CAPM over the long run: 1926–2001. J. Empir. Finan. **14**, 1–40 (2007)
2. Blume, M.E.: On the assessment of risk. J. Finan. **26**(1), 1–10 (1971)
3. Bos, T., Newbold, P.: An empirical investigation of the possibility of stochastic systematic risk in the market model. J. Bus. **57**, 35–41 (1984)
4. Chen, S.N.: Beta nonstationarity, portfolio residual risk and diversification. J. Financ. Quant. Anal. **16**(1), 95–111 (1981)
5. Choudhry, T.: Time-varying beta and the Asian financial crisis: evidence from Malaysian and Taiwanese firms. Pac. Basin Finan. J. **13**, 93–118 (2005)
6. Fabozzi, F., Francis, J.: Beta as a random coefficient. J. Financ. Quant. Anal. **13**, 101–116 (1978)
7. Fama, E.F., French, K.R.: The cross-section of expected stock returns. J. Finan. **47**(2), 427–465 (1992)
8. Fama, E.F., French, K.R.: Common risk factors in the returns of stocks and bonds. J. Financ. Econ. **33**, 3–56 (1993)
9. Ferson, W.E., Harvey, C.R.: The variation of economic risk premiums. J. Polit. Econ. **99**(2), 385–415 (1991)
10. Ferson, W.E., Harvey, C.R.: The risk and predictability of international equity returns. Rev. Financ. Stud. **6**(3), 527–566 (1993)
11. Ferson, W.E., Korajczyk, R.A.: Do arbitrage pricing models explain the predictability of stock returns. J. Bus. **68**(3), 309–349 (1995)
12. Garcia, R., Ghysels, E.: Structural change and asset pricing in emerging markets. J. Int. Money Finan. **17**(3), 455–473 (1998)
13. Ghysels, E.: On stable factor structures in the pricing of risk: do time-varying betas help or hurt? J. Finan. **53**(2), 549–573 (1998)
14. Groenewold, N., Fraser, P.: Time-varying estimates of CAPM betas. Math. Comput. Simul. **48**, 531–539 (1999)
15. Hamilton, J.D.: A new approach to the economic analysis of nonstationary time series and the business cycle. Econometrica **57**, 357–384 (1989)
16. Harvey, A.C.: Forecasting Structural Time Series Models and the Kalman Filter. Cambridge University Press, Cambridge (1989)
17. Kim, C.J.: Dynamic linear models with Markov-switching. J. Econom. **60**, 144–165 (1994)
18. Korkmaz, T., Evik, E.I., Birkan, E., Zata, N.: Testing CAPM using Markov switching model: the case of coal firms. Ekonomska Istraivanja **23**(2), 44–59 (2010)
19. Korkmaz, T., Evik, E.I., Grkan, S.: Testing of the international capital asset pricing model with Markov switching model in emerging markets. Invest. Manag. Financ. Innov. **7**(1), 37–49 (2010)
20. Lintner, J.: The valuation of risk assets and the selection of risky investments in stock portfolios and capital budgets. Rev. Econ. Stat. **47**(1), 13–37 (1965)
21. Maneejuk, P., Pastpipatkul, P., Sriboonchitta, P.: Analyzing the effect of time-varying factors for Thai rice export. Thai J. Math. **14**, 201–213 (2016)
22. McKenzie, M.D., Brooks, R.D., Faff, R.W.: The use of domestic and world market indexes in the estimation of time-varying betas. J. Multinatl. Financ. Manag. **10**(2000), 91–106 (2000)
23. Prukumpai, S.: Time-varying industrial portfolio betas under the regime-switching model: evidence from the stock exchange of Thailand. Appl. Econ. J. **22**(2), 54–76 (2015)

24. Sharpe, W.F.: Capital asset prices: a theory of market equilibrium under conditions of risk. J. Finan. **19**(3), 425–442 (1964)
25. Tsai, H.-J., Chen, M.-C., Yang, C.-Y.: A time-varying perspective on the CAPM and downside betas. Int. Rev. Econ. Finan. **29**, 440–454 (2014)
26. West, M., Harrison, J.: Bayesian Forecasting and Dynamic Models, 2nd edn., p. 682. Springer, New York (1997)

Interval-Valued Estimation for the Five Largest Market Capitalization Stocks in the Stock Exchange of Thailand by Markov-Switching CAPM

Karn Thamprasert[1(✉)], Pathairat Pastpipatkul[1,2], and Woraphon Yamaka[1,2]

[1] Faculty of Economics, Chiang Mai University, Chiang Mai 50200, Thailand
iw.karn@gmail.com, ppthairat@hotmail.com, woraphon.econ@gmail.com
[2] Center of Excellence in Econometrics, Chiang Mai University,
Chiang Mai 50200, Thailand

Abstract. The paper aims to quantify risk using Capital Asset Pricing Model (CAPM) of the 5 top-traded stocks in Thailand's Stock Exchange market as well as to investigate existence of structural change in CAPM. Thus, in this paper, we compare non-linear Markov-switching CAPM with linear CAPM. In addition, we use interval-valued data instead of conventional single-valued data because of its ability to capture the whole period rather than a point of time. This paper, therefore, introduces an approach to fit both Markov-switching and linear CAPM to interval-valued data. Interval value of each stock return is retrieved from its midpoint to fit in Markov-switching and linear regression estimations which apply the midpoint of interval value of market return. From empirical analysis, the results are satisfactory as AIC judged that our Markov-switching CAPM outperforms the linear benchmark four out of five stocks.

1 Introduction

Investment plays a significant role in present day human life. However, risk is always involved. Investment risks may lead to financial downfall of individuals. This research paper aims to quantify relationship between expected asset return and its risk by using Capital Asset Pricing Model (CAPM). As widely known, the CAPM especially puts emphasis on two components which are the risk-free rate and beta. To compare asset's risk and return with risk-free investment, this paper uses the Thai government 10-year bond's yield and Thai five top-traded stocks as test subjects. Key assumptions of our research are that the use of interval-valued return is more appropriate and conventional linear CAPM is likely to be inadequate for capturing stocks behavior.

This paper uses the highest and the lowest weekly price of five top Thai securities, return of 10-year bond, and index of the Stock Exchange of Thailand (SET). All of which are later transformed into weekly interval-valued return data. There are reasons behind our preference of choosing weekly interval-valued return data.

© Springer International Publishing AG 2018
L. H. Anh et al. (eds.), *Econometrics for Financial Applications*, Studies in Computational Intelligence 760, https://doi.org/10.1007/978-3-319-73150-6_67

First, the weekly data should represent not so fast or so slow timeframe while daily data may seem too fluctuating and monthly data may seem to outdate investment decisions. Second, interval-valued or symbolic data should capture stocks behavior better than some points of time especially the closing price; for example, a price last week between high and low sounds more relevant than the closing price on Friday evening. After Diday [3], Diday et al. [4], Emilion [7] and Diday and Emilion [5,6] who were pioneers to use symbolic interval-valued data in several papers, there has been other recent extension of the CAPM to the interval data, as seen in Piamsuwannakit et al. [10] and Phochanachan et al. [11]. Their work is about using the interval data to predict the return and quantify the risk of the stock by applying CAPM. They found that using the interval data in CAPM can provide a better result than using the closing price.

While data is all set, we expect that return and risk of these stocks may not behave in single linear trend. At least, there are uptrend and downtrend known as "Boom" and "Bust". The model should not assume that the stocks would walk in the straight path of everlasting growth. Markov-switching modeling or regime-switching method of Hamilton [8] which we choose is one of the most popular non-linear time series model. Specifically, it enables researchers to construct multiple equations for time series in different regimes, capture complex dynamic patterns, categorize data into created conditioned structure, and allow switching of data between one equation in some range of time then jump to another equation in some other periods. Great feature of Markov-switching is that the mechanism of switching works by kind of unobservable state that matches the first-order Markov chain. In this case we use two regimes of uptrend and downtrend.

Finally, we compare the results of conventional linear regression and non-linear Markov switching model. The strong normal assumption of the CAPM is what we are concerned since the financial data are rather symmetric about their means, but the tails are fatter than what would be expected with normal distributions. To overcome this assumption, we consider student's t-distribution, and skewed student's t-distribution as alternative assumptions for our models. To choose the best model, the lowest Akaike Information Criterion (AIC) is considered in this study.

The contents of this paper are as follows. In Sect. 2, we describe components of the model which are CAPM model, Markov-Switching Model by using center method, and the Hamilton's filter. Section 3 posts the process to estimate the model. In Sect. 4, we select the best model for each stock, derive parameters, and interpret results. Conclusions are in Sect. 5.

2   Interval-Valued Markov-Switching CAPM

2.1   Review of Capital Asset Pricing Model (CAPM)

Capital Asset Pricing Model or as known as CAPM is a model for examining relationship between expected asset return and its risk. The concept is the trade-off compensation of risk and return while comparing to the market return, or how the asset return deviates from group. By setting expected asset return equal to risk-free asset return plus a multiplication of market risk premium and systematic risk, the model can be easily expressed in an equation of

$$E(R_i) = R_f + [E(R_m) - R_f] \cdot \beta_i, \tag{1}$$

where

$E(R_i)$ represents expected asset return

R_f represents risk-free asset return (usually referred to government bond yield or fixed bank interest rate)

R_m represents expected market return

β_i represents systematic risk of the asset as known as "Beta Coefficient"

$E(R_i) - R_f$ represents market risk premium

The beta Coefficient represents relationship between selected asset's risk and the risk of the whole market or in another meaning is an indicator for systematic risk of an asset as follows.

(1) Whenever β_i is larger than 1, the risk of the asset i is more fluctuating than the market and we can expect the return of asset i to be more than market's return, this asset will be classified as aggressive asset.

(2) Whenever β_i is less than 1 means the risk of the asset i is less fluctuate than the market and we can expect the return of asset i to be less than market's return, this asset will be classified as defensive asset.

(3) Whenever β_i is equal to 1 means the risk of the asset i is equal to the market and we can expect the return of asset i to be equal to market's return, and this asset will be classified as neutral asset.

2.2   Markov Switching Model with the Center Method

Merging core basis of Billard and Diday's [1,2] interval-valued regression using center method (see [9]) with Hamilton's [8] Markov switching model, Markov chains concept can be explained easily through many games of chance by rolling between points. The children's game "Snakes and Ladders" is a good example of how Markov chains work. At each turn, the player starts in each state (on a given square) and from there has fixed odds of moving to certain other states (squares). Normally, the model is separated into 2 different regimes, mostly distinguished as upturn and downturn. The Markov Switching interval-valued regression model is expressed by

$$\begin{aligned} y_t^l &= \beta_{0,s(t)}^l + \beta_{1,s(t)}^l X_t^l + \varepsilon_{t,s(t)}^l, \\ y_t^h &= \beta_{0,s(t)}^h + \beta_{1,s(t)}^h X_t^h + \varepsilon_{t,s(t)}^h, \end{aligned} \tag{2}$$

where $\beta_{0,s(t)}^l$ and $\beta_{1,s(t)}^l$ are the regime dependent intercept term and beta risk coefficient of the lowest value of the interval-valued data while $\beta_{0,s(t)}^h$ and $\beta_{1,s(t)}^h$

are the regime dependent intercept term and beta risk coefficient of the highest value of the interval-valued data. $\varepsilon^l_{t,s(t)}, \varepsilon^h_{t,s(t)} \sim i.i.d.\ N(0,\sigma^2)$, $y_t = [y^l_t, y^h_t]$ are the lowest and highest interval-valued of dependent variables and $X_t = [X^l_t, X^h_t]$ is the lowest and highest interval-valued of independent variable. $s(t) = i$, $i = 1, \cdots, k$ is state or regime latent variable. Consequently, it is possible to put vector estimations of parameters into matrix notation intervals for the center method, simplified as

$$y^c_t = \beta^c_{0,s(t)} + \beta^c_{s(t)} X^c_t + \varepsilon^c_{s(t)}, \tag{3}$$

where $y^c_t = (y^l_t + y^h_t)/2$ and $X^c_t = (X^l_t + X^h_t)/2$. In the center method, $\beta^c_{0,s(t)}$ and $\beta^c_{1,s(t)}$ are the regime dependent intercept term and beta risk coefficient which are based on the midpoint of the intervals. The assumption is that Markov matrix(Q) which is a matrix used to describe the transitions of a Markov chain is governed by the first order Markov chain.

$$p(s(t) = j | s(t-1) = i) = p_{ij}, \quad \sum_{j=1}^{k} p_{ij} = 1, \forall i = 1, \cdots, k. \tag{4}$$

Consider that this simplifying process can only be applicable for two regimes cases; then, the first order Markov process could be rewritten as

$$\begin{aligned}
p(s(t) = 1 | s(t-1) = 1) &= p_{11} \\
p(s(t) = 1 | s(t-1) = 2) &= p_{12} \\
p(s(t) = 2 | s(t-1) = 1) &= p_{21} \\
p(s(t) = 2 | s(t-1) = 2) &= p_{22}
\end{aligned} \tag{5}$$

To avoid the strong normal assumption, we, thus, consider student's t-distribution, and skewed student's t-distribution as alternative assumptions for our model. Let $\theta = (\beta_{s(t)}, \sigma_{s(t)}, Q)$ be denoted as vector of model parameters. Our likelihood function of Markov Switching model can be written as the following:

(1) Normal likelihood

$$L_c(\theta_{s(t)} | y_t, X_t) = \sum_{j=1}^{2} \left(\prod_{i=1}^{n} \left(\frac{1}{\sqrt{2\pi\sigma^2_{s(t)=j}}} \exp \left[\frac{(\varepsilon^l_{t,s(t)=j} + \varepsilon^h_{t,s(t)=j})^2}{2(\sigma_{s(t)=j})} \right] \right) Pr(s_t = j | \theta_{s(t-1)=j}) \right) \tag{6}$$

where $\varepsilon^l_{t,s(t)} = y^l_t - \beta^l_{0,s(t)} - \beta^l_{1,s(t)} X^l_t$ and $\varepsilon^h_{t,s(t)} = y^h_t - \beta^h_{0,s(t)} - \beta^h_{1,s(t)} X^h_t$.

(2) Student's t likelihood

$$L_c(\theta_{s(t)} | y_t, X_t) = \sum_{j=1}^{2} \prod_{i=1}^{n} \left(\frac{\Gamma(\frac{d+1}{2})}{\sqrt{(d-2)\pi}\Gamma(\frac{d}{2})} \left(1 + \frac{\varepsilon^l_{t,s(t)=j} + \varepsilon^h_{t,s(t)=j}}{(d-2)\sigma_{s(t)=j}} \right)^{-\frac{d+1}{2}} (\frac{1}{\sigma_{s(t-1)=j}}) \right)$$

$$\times Pr(s_t = j | \theta_{s(t-1)=j}) \tag{7}$$

where d is degree of freedom. $\Gamma(\cdot)$ is a gamma distribution.

(1) Skewed student's t likelihood

$$
L_c(\theta_{s(t)}|y_t, X_t) =
\begin{cases}
\displaystyle\sum_{j=1}^{2}\prod_{i=1}^{n_1}\left(\cfrac{2}{\xi_{s(t)=j} + \cfrac{1}{\xi_{s(t)=j}}}\, f(\xi_{s(t)=j}x_{s(t-1)=j})\right) \times Pr(s_t = j|\theta_{s(t-1)=j}), \\
\text{for } x_{s(t)=j} < 0 \\[2ex]
\displaystyle\sum_{j=1}^{2}\prod_{i=1}^{n_2}\left(\cfrac{2}{\xi_{s(t)=j} + \cfrac{1}{\xi_{s(t)=j}}}\, f(\cfrac{\xi_{s(t)=j}}{x_{s(t-1)=j}})\right) \times Pr(s_t = j|\theta_{s(t-1)=j}), \\
\text{for } x_{s(t)=j} > 0
\end{cases}
$$

$$(8)$$

where ξ is skew parameter, x is the regime dependent expected mean of the model and $f(\cdot)$ is density of student's t-distribution, while n_1 and n_2 are the number of observations in each case. The model will be estimated to get beta parameter value $(\beta^c_{1,s(t)})$ which is the objective of this process.

2.3 Hamilton's Filter

The popular filter approach is introduced by Hamilton [8] in 1989 which is called Hamilton's filter. The filter operates by these 3 steps.

1. Initiate a first guess of transition probabilities Q which is a matrix set of chance to jump between regimes. Thus, the transition probabilities matrix for two regimes is,

$$
Q = \begin{bmatrix} p_{11} & p_{12} \\ p_{21} & p_{22} \end{bmatrix}
\tag{9}
$$

2. Update the transition probabilities Q of each next state with the previous in alternation including parameters in system equation $\theta_{s(t-1)}$ and Q. Then immediately calculate likelihood function for each state at time t. After that, the probability of each state is to be derived by this equation

$$
Pr(s(t) = j|\theta_{s(t)}) = \frac{f(y_t, X_t|s(t) = j, \theta_{s(t-1)})Pr(s(t) = j|\theta_{s(t-1)})}{\sum_{j=1}^{k} f(y_t, X_t|s(t) = j, \theta_{s(t-1)})Pr(s(t) = j|\theta_{s(t-1)})}
\tag{10}
$$

where $f(y_t, X_t|s(t) = j, \theta_{s(t-1)})$ is likelihood function and $Pr(s(t) = j, \theta_{s(t-1)})$ is filtered probability value.

3. Repeat steps 1 and 2 from $t = 1$ until the end of the series $t = T$.

3 Data and Model Specification

Collected are weekly highest price and weekly lowest price time series data of Petroleum Authority of Thailand (PTT), Siam Cement Group (SCC), Airports of Thailand (AOT), Advanced Info Service (ADVANC), Siam Commercial Bank

(SCB), Stock Exchange of Thailand Index (SET), and 10-year Thai Government Bond Yield. The study will use secondary data since the first week of January 2006 through the last week of April 2016 with the total of 522 observations in 10 yr and 4 months. Data collected are assigned into these vectors

S_{ptt}^h : weekly highest security price of PTT
S_{ptt}^l : weekly lowest security price of PTT
S_{scc}^h : weekly highest security price of SCC
S_{scc}^l : weekly lowest security price of SCC
S_{aot}^h : weekly highest security price of AOT
S_{aot}^l : weekly lowest security price of AOT
S_{adv}^h : weekly highest security price of ADVANC
S_{adv}^l : weekly lowest security price of ADVANC
S_{scb}^h : weekly highest security price of SCB
S_{scb}^l : weekly lowest security price of SCB
S_m^h : weekly highest SET index
S_m^l : weekly lowest SET index
$\hat{r}_f^h$: weekly highest annual return of Thai-10Y bond
$\hat{r}_f^l$: weekly lowest annual return of Thai-10Y bond.

Data collected for Thai bond are already in return intervals but they are annual return, which need to be transformed into weekly risk free return, r_f^h and r_f^l, by $r_f^h = \sqrt[52]{1 + \hat{r}_f^h} - 1$ and $r_f^h = \sqrt[52]{1 + \hat{r}_f^l} - 1$ respectively. Then, the data also need to be transformed into highest and lowest weekly return, r_i^h and r_i^l, respectively. Then we transform the interval prices of PTT, SCC, AOT, ADVANC, SCB by

$$r_{i,t}^h = \frac{S_{i,t}^h - S_{i,t-1}^{avg}}{S_{i,t-1}^{avg}} \tag{11}$$

$$r_{i,t}^l = \frac{S_{i,t}^l - S_{i,t-1}^{avg}}{S_{i,t-1}^{avg}} \tag{12}$$

where $S_{i,t-1}^{avg} = (S_i^h + S_i^l)/2$ and this study will use high and low data in weekly return intervals of PTT, SCC, AOT, ADVANC, SCB, SET Index, and Thai-10Y Bond from January 2006 until April 2016. After that, we integrate CAPM component by replacing variables in CAPM equation as in the following:

$$r_i^h - r_f^h = \alpha_{1,s(t)}^h + \beta_{1,s(t)}^h (r_m^h - r_f^h) + \varepsilon^h \tag{13}$$

$$r_i^l - r_f^l = \alpha_{1,s(t)}^l + \beta_{1,s(t)}^l (r_m^l - r_f^l) + \varepsilon^l \tag{14}$$

Then, the study will test for stationarity by using Augmented Dickey Fuller (ADF) Unit Root Test. The descriptive statistics and ADF-test of our return intervals are shown in Table 1. The results show that our data series are stationary since the p-value of ADF-test shows a statistical significance at 0.01.

Table 1. Descriptive statistics

	SET^h	SET^l	PTT^h	PTT^l	SCC^h	SCC^l
Mean	0.018792	−0.01722	0.029499	−0.02866	0.02688	−0.02489
Median	0.017591	−0.01027	0.024299	−0.02073	0.020654	−0.02007
Maximum	0.183314	0.067793	0.301536	0.108845	0.267142	0.082901
Minimum	−0.06959	−0.24375	−0.06748	−0.24192	−0.06519	−0.17621
Std. Dev.	0.022856	0.030912	0.038769	0.041863	0.035721	0.033174
Skewness	1.072749	−2.13879	1.285496	−1.42388	1.623885	−1.12276
Kurtosis	8.786604	11.95891	8.020913	6.682916	8.700773	5.703896
Jarque-Bera	853.805	2209.384	713.2885	485.8499	964.9663	276.922
ADF(prob)	0	0	0	0	0	0
	AOT^h	AOT^l	$ADVANC^h$	$ADVANC^l$	SCB^h	SCB^l
Mean	0.036856	−0.02899	0.029843	−0.02892	0.032387	−0.02935
Median	0.02585	−0.021	0.026834	−0.0225	0.026455	−0.02371
Maximum	0.364184	0.085916	0.251062	0.125136	0.24478	0.113134
Minimum	−0.10619	−0.40647	−0.12845	−0.25019	−0.05102	−0.28771
Std. Dev.	0.047257	0.044818	0.035971	0.037356	0.0389	0.040307
Skewness	1.724091	−2.07973	0.632561	−1.46131	1.234108	−1.24244
Kurtosis	9.007777	13.8563	6.372744	9.056419	6.118837	8.274909
Jarque-Bera	1075.627	3029.846	290.8773	1013.725	354.6145	762.15
ADF(prob)	0	0	0	0	0	0

Source: Calculation

4   Estimation Results

4.1   Model Selection

In this section, we compared various model specifications. We employed the
Markov-Switching Model by using center method to apply with CAPM equations
in five stocks in SET market. The comparison of models is based on Akaike information criterion (AIC) which measures how fit the model is. We also consider
three different likelihood distributions to estimate each stock which are normal
distribution, student's t distribution, and skewed student's t distribution. After
the runs of several alternative distribution functions, we obtained the results as
presented in Table 2 indicating that skewed student's t distribution provides the
lowest AIC for CAPM equations of PTT, SCC, and AOT while student's t distribution provides the lowest AIC for CAPM equations of ADVANC and SCB.
In addition, we also examined structural break in CAPM by comparing Markov
switching CAPM models with linear CAPM models. According to AIC, CAPM
equations of PTT, SCC, AOT, and ADVANC are likely to confirm a structural
break and the Markov-switching model is preferred in these CAPM equations.
However, we found that there is no structural change in SCB stock since AIC of
linear regression with student's t distribution remains the lowest. Hence, these
results are consistent with the expectation that the "beta" relationship exists
non-linear. The stock risk would be affected by the structural change in Thai
stock exchange market.

Table 2. Model comparison

CAPM Equation	PTT	SCC	AOT	ADVANC	SCB
Markov switching-normal distribution	−2621.329	−2665.172	−2222.501	−2348.974	−2598.044
Markov switching-student's t distribution	−2633.767	−2672.442	−2244.337	−2358.125	−2598.831
Markov switching-skewed student's distribution	**−2653.517**	**−2688.356**	**−2277.811**	−2326.863	−2614.276
Linear model - normal distribution	−2637.331	−2679.817	−2236.991	−2306.365	−2598.814
Linear model - student's t distribution	−2650.055	−2686.893	−2258.975	−2338.973	**−2615.057**
Linear model - student's t student's distribution	−2635.818	−2680.316	−2258.57	−2291.052	−2606.81

Source: Calculation

4.2 Estimated Parameter Results

From Table 3, we can observe the range of estimated beta risks for individual stocks are between −0.0114 and 1.6337. The beta risks for 4 individual stocks are fitted by Markov-switching CAPM and 1 individual stock is fitted by linear

Table 3. Estimates of parameters

	PTT	SCC	AOT	ADV	SCB
$\alpha^c_{1,s(t)=1}$	−0.0114	0.0007	0.0112	-0.0039	0
Standard error	0.0046	0.001	0.0049	0.0027	0.0009
$\beta^c_{1,s(t)=1}$	**1.2341**	**0.8428**	**1.6337**	**0.3194**	**1.1768**
Standard error	0.0744	0.0441	0.1229	0.0766	0.0397
$\sigma^c_{1,s(t)=1}$	0.0226	0.0193	0.0445	0.0226	0.0026
Standard error	0.0035	0.0011	0.0058	0.0019	0.0009
$\alpha^c_{1,s(t)=2}$	0.0057	0.0017	0.003	0.0064	
Standard error	0.0013	0.0031	0.0014	0.0045	
$\beta^c_{1,s(t)=2}$	**1.2128**	**1.3069**	**0.9408**	**1.1523**	
Standard error	0.0468	0.0993	0.0465	0.1029	
$\sigma^c_{1,s(t)=2}$	0.0183	0.0306	0.0286	0.0299	
Standard error	0.0013	0.0041	0.0015	0.0023	
p_{11}	0.6097	0.9114	0.9869	0.8405	
Standard error	0.1221	0.0631	0.009	0.0717	
p_{22}	0.7968	0.9732	0.9484	0.8226	
Standard error	0.0567	0.0223	0.0309	0.0901	

Source: Calculation

Note : Degree of freedom and skew parameters are not shown in this table

CAPM. Among the first 4 individual stocks, the beta risks are separated into two regimes, $\beta^c_{1,s(t)=1}$ (downtrend) and $\beta^c_{1,s(t)=2}$ (uptrend). We can observe the heterogeneous results from these 4 individual stocks. The beta risks in downtrend regime of PPT and AOT are higher than uptrend regime whereas the beta risks of SCC and ADV are higher in uptrend regime. However, these 4 individual stocks show the same positive significance at 1% level. Consider uptrend regime, among these 4 individual stocks, the highest beta risk attainable stock is AOT ($\beta^c_{1,s(t)=1} = 1.6337$) and the lowest beta risk attainable stock is ADV ($\beta^c_{1,s(t)=1} = 0.3194$). On the contrary, consider downtrend regime, among these 4 individual stocks, the highest beta risk attainable stock is SCC ($\beta^c_{1,s(t)=1} = 1.3069$) and the lowest beta risk attainable stock is AOT ($\beta^c_{1,s(t)=1} = 0.9408$). For the linear CAPM, the beta risk of SCB is $\beta^c_1 = 1.1768$.

From the results, we notice that some stocks are acting as aggressive in uptrend situation but acting defensive in a downtrend and vice versa. If we did not post a model to separate into these 2 regimes, we should have missed something. Also, these data are all in interval-valued data and we did not compare with single-valued data. By the way, we compare our model with linear model and our model is slightly better by the judgment of AIC.

5   Conclusions

This paper notices advantages of interval-valued data, especially for financial application. We also apply to CAPM which is commonly used to measure risk and return relationship between single securities with corresponded market. Based on most research literatures, CAPM uses closing price or spot price which is single-valued data. We argue that spot price at some point of time should not represent behavior of the securities in the period well enough. To describe a period, interval-valued data should match the job as the paper suggests considering all the prices which should capture all fluctuations. With the use of highest and lowest prices, then nothing will be left behind. Moreover, we are not fully convinced that the behavior of stocks is all behaving in linear trend. We consider "Boom" and "Bust" situations which are rising and downfall trends. Hence, we come up with the Markov switching interval-valued regression model and apply it to CAPM. The accuracy and performance of our model can beat linear model in 4 out of 5 securities (see the results of model selection). An issue that has not been considered in this paper is its extension to the Markov switching interval VAR and viewing the system as a matrix. Alternatively, researchers should consider other time series models such as ARMA or GARCH models which should be pursued further.

References

1. Billard, L., Diday, E.: Regression analysis for interval-valued data. In: Data Analysis, Classification, and Related Methods, pp. 369–374. Springer, Heidelberg (2000)
2. Billard, L., Diday, E.: Symbolic regression analysis. In: Classification, Clustering, and Data Analysis, pp. 281–288. Springer, Heidelberg (2002)

3. Diday, E.: Probabilist, possibilist and belief objects for knowledge analysis. Ann. Oper. Res. **55**, 227–276 (1995)
4. Diday, E., Emilion, R.: Lattices and capacities in analysis of probabilist objects. In: Diday, E., Lechevallier, Y., Opilz, O. (eds.) Studies in Classification, pp. 13–30 (1996)
5. Diday, E., Emilion, R.: Capacities and credibilities in analysis of probabilistic objects by histograms and lattices. In: Hayashi, C., Ohsumi, N., Yajima, K., Tanaka, Y., Bock, H.-H., Baba, Y. (eds.) Data Science, Classification, and Related Methods, pp. 353–357 (1998)
6. Diday, E., Emilion, R., Hillali, Y.: Symbolic Data Analysis of Probabilistic Objects by Capacities and Credibilities, pp. 5–22. Societeà Italiana di Statistica (1996)
7. Emilion, R.: Differentiation des Capacities. Comptes Rendus de l'Academie des Sciences - Series I - Mathematics **324**, 389–392 (1997)
8. Hamilton, J.D.: A new approach to the economic analysis of nonstationary time series and the business cycle. Econometrica J. Econom. Soc. 357–384 (1989)
9. Neto, E.D.A.L., de Carvalho, F.D.A.: Constrained linear regression models for symbolic interval-valued variables. Comput. Stat. Data Anal. **54**(2), 333–347 (2010)
10. Piamsuwannakit, S., Autchariyapanitkul, K., Sriboonchitta, S., Ouncharoen, R.: Capital asset pricing model with interval data. In: Integrated Uncertainty in Knowledge Modelling and Decision Making, pp. 163–170. Springer, Cham (2015)
11. Phochanachan, P., Pastpipatkul, P., Yamaka, W., Sriboonchitta, S.: Threshold regression for modeling symbolic interval data. Int. J. Appl. Bus. Econ. Res. **15**(7), 195–207 (2017)

The Roles of Perceived Risk and Trust on E–Payment Adoption

Thanh D. Nguyen[1,2(✉)] ⓘ and Phuc A. Huynh[1]

[1] Banking University of Ho Chi Minh City, Ho Chi Minh City, Vietnam
thanhnd@buh.edu.vn, huynhanhphuc@gmail.com
[2] Bach Khoa University, Ho Chi Minh City, Vietnam

Abstract. E–payment is one of the major constituents of e–commerce, which assists to enhance user efficiency and smarten intention to use of e–commerce in the digital era. This study investigates the roles of perceived risk and trust on e–payment adoption. Data is collected from respondents who have used or intend to use e–payments for e–commerce in Ho Chi Minh City. The structural equation modelling (SEM) is analyzed on a total convenient sampling of 200 respondents. Interestingly, research results externalize that perceived risk and trust have the principal roles of the structural model of e–payment adoption. The research model accounts for 38% of e–payment adoption.

Keywords: E–commerce · E–payment · Perceived risk · IT adoption
Trust

1 Introduction

With the presto development of the Internet, e–payment systems in banking services have met the customer expectations when participating in e–commerce [32]. In the e–commerce context, e–payment is closely related to electronic transactions, e–payment is understood as the payment process made without the paper instrument usage [24, 37]. The e–payment systems cover the variety of e–channels (e.g., debit/credit card, online banking, m–banking, e–wallet, e–cash, e–check, online storage value) [21, 24, 44]. According to e–commerce and IT department in 2015, the proportion of cash flow on total payment forms is only about 12%, and 97% of enterprises accepted payment via funds transfer, 16% payment cards, 4% e–wallet [11]. According to Vietnam e–commerce association in 2016, the obstacles in user e–commerce usage are the quality of products and services in e–commerce business [21]. Currently, the habit of paying in cash is the biggest obstacle for the development of e–payments, although according to e–commerce and IT department, up to 45% of the population use the Internet, revenue from e–payment only reached roughly 5% [11]. Which provided that customers are still frightened to use e–payment when participating in e–commerce. There are many theoretical models of the IT adoption such as TAM [9, 10], TAM2 [39], TAM2' [38], TAM3 [40], and UTAUT [41] – these are the typical theories for measuring the information systems' behavioral intention and actual use by users. Furthermore, there are several studies related to e–commerce around the world (e.g., e–CAM of Park et al. [28], intention to use e–payment of Bankole and Bankole [3],

© Springer International Publishing AG 2018
L. H. Anh et al. (eds.), *Econometrics for Financial Applications*, Studies in Computational Intelligence 760, https://doi.org/10.1007/978-3-319-73150-6_68

Cabanillas et al. [7], Gao and Waechter [17], Gefen et al. [18]), and in Vietnam (e.g., e–banking adoption of Nguyen and Cao [26]). Nevertheless, there are few studies on the e–payment adoption in a potential market as Vietnam (except, e.g., Nguyen and Nguyen [27]). Consequently, in the e–commerce context, research on e–payment adoption is exceedingly essential and meaningful study.

The objective of this study investigates the role of perceived risk and trust on e–payment adoption, measures the impact of influence factors on e–payment adoption of e–commerce users. Data is collected from individual customers who have used or intend to use e–payment in Ho Chi Minh City. The relationships in the research model were analyzed by structural equation modelling (SEM). Research results not only provide information to the business in developing e–commerce products and services, and e–payment system, but also add knowledge to the theory of IT adoption.

2 Background

2.1 E–Payment

The rapid development of IT facilitates the e–payments' development [36]. With the propensity growth of the e–payment systems, demonstrating its potential, it will change from a cash–based economy to a cashless or non–cash economy, but it's hard to transform a non–cash economy, so current trading activities are still firmly compacted in cash–based [44]. According to Telle [37, p. 17], Junadi [21, p. 215], e–payment is defined as "*the payment process made without the paper instruments usage*". The e–payment systems cover the variety of e–channels, including debit/credit card, funds transfer, e–banking, online banking, m–banking, e–wallet, e–cash, e–check, online storage value, digital accumulating balance, wireless payment [21, 24, 37, 44]. The stakeholders in e–payment systems include customers, merchants, consumer banks, merchant banks, payment instrument issuers, and payment service providers [24].

2.2 E–Payment Benefits and Disadvantages

E–payment Benefits
For Customers. The e–payment systems allow clients to make their payments from anytime and anywhere [44]. It is very convenient – the greatest e–payments benefit [35] to handle their transactions (e.g., users only have bank accounts for logging–on the browsers with online banking or smartphone applications with m–banking) [44]. Besides, it may increase the benefits of cashless payments without using retail payments [18] (e.g., customers have credit/debit cards for access to card systems [44]). *For Organizations.* The e–payment cost is another benefit [35], for most of the vendors, businesses and merchants – it is free of charge [44]. On the flip side, e–payment systems may commit the e–commerce quality services due to serve better the customer's needs. Accordingly, every transaction is encrypted, making monitoring more

convenient and effective[1]. Thus, for credit institutions providing e–payment services have to invest enormous amounts of money for IT infrastructures. *For the Economy.* E–payment has many positive impacts on the economy [18]. It is committed to price and quality guarantee. E–payment eliminates gradually using cash–based, which limits illegal activities (e.g., counterfeit money, tax evasion, corruption...), so increasing the transparency [18] and the law enforcement capacity. E–payment reduces cash in monetary circulation and driving economic benefits [18].

E–payment Disadvantages

Online Security. The e–payment systems have the lack of authentications – the biggest e–payments problem [35]. Meanwhile, there are no solutions to verify or authenticate who entering the information e–payment systems [35] that are not a criminal [33]. Thus, it comes critical to dispute the fraudulent actions have been made by using bank accounts or debit/credit cards. *Missed Errors.* The merchandise arrives with the mistakenly pay and ordered in e–commerce, so it cannot be used, and customers have been out their money [33]. Hence, it is must be to return the unessential merchandise and wait for a long time to replace another. *Fees.* Some vendors of e–payment systems require the customers or merchants pay a service fee [33] (e.g., Paypal, Visa, Master...). Furthermore, One of e–payments downside is repudiation of charges, because no transaction information of the e–payments, customers may have an exceedingly hard time debating this cost [35]. Thus, customers should know how to protect in place concerning fees applied to the incorrect processing of e–payment systems.

3   Research Model

3.1   Literature Review

IT Adoption

Fishbein and Ajzen's theory of reasoned action (TRA) [14] is a fundamental theory of human behavior. Ajzen's theory of planned behavior (TPB) [1] is inherited TRA and integrated a construct of perceived behavioral control to improve intention and behavior. Davis [9] and Davis et al. [10] developed technology acceptance model (TAM), which based on TRA and TPB with characteristic features of perceived usefulness and perceived ease of use. TAM provides a shrewdness view to predicting systems that effect on attitude and behavior use of information technology. The technology acceptance model extension, TAM2 of Venkatesh and Davis [39], TAM2' of Venkatesh [38], TAM3 of Venkatesh and Bala [40] emphasize the perceived usefulness and perceived ease of use in the intention to use and usage behavior, respectively. Venkatesh et al. [41] proposed the unified theory of acceptance and use of technology (UTAUT) for explaining the behavioral intention and use behavior of information systems. Especially, TAM [10] and UTAUT [41] are two models of information

---

[1] E–payments in Vietnam: Need to exploit all potential (2016) at http://tapchitaichinh.vn.

systems have the most cited in the world (TAM: 35,825 and UTAUT: 18,647 – updated on August 10, 2017, from Google scholar).

Perceived Risk

The perceived risk theory explains user behavior on the risks [5]. The negative causalities may emphasize from the user actions lead to a significantly established concept in consumer behavior, namely perceived risk [5]. Meanwhile, perceived risk is a natural conception of as sense unreliability regarding possible use negatively related a service or product [13]. According to Bauer [5, p. 13] perceived risk is *"a combination of uncertainty plus seriousness of outcome involved"*. Besides, perceived risk has been divided into two dimensions, called perceived risk in the online transaction (PRT) and perceived risk with a service or product (PRP) [28]. In which, PRT is a possible the risk of transactions that users can face when revealed to e–commerce [28] and RP is the overall account to anxiety observed or uncertainty by a user in a particular service or product when they use e–commerce [28]. A typical model for perceived risk is e–commerce adoption model (e–CAM) of Park et al. [28].

Trust

Together with perceived risk, trust is one of two critical challenges of e–payment systems [2]. Besides, the trust is the subjective belief that a party will fulfil its obligations according to the expectations of stakeholders as the goodwill [25, 29]. According to Pavlou [29, p. 106] trust is *"a defining feature of the major social and economic interactions in which uncertainty is present"*. Furthermore, the trust is also built from the reputation. Meanwhile, reputation is the faith in the transaction environment [29]. Many authors have studied about two of perceived risk and trust on e–services (e.g., Featherman and Pavlou [13], Gefen et al. [18]), e–commerce (e.g., Park et al. [28], Pavlou [29]), e–banking (e.g., Gao and Waechter [17], Nguyen and Nguyen [27], Pham et al. [30]), and evidenced that the perceived risk and the trust are the most protrusive elements in e–payment. In which, a comparison between traditional payment and e–payment systems provided that the more consumers would rather use e–payment channel, the trust higher level in the e–payments [43].

3.2   Research Model

From the theoretical basis of IT adoption as TAM, TAM2, TAM2', UTAUT, the theories of perceived risk and trust, and the previous studies, it proposes a model of perceived risk and trust on e–payment adoption for measuring the relationships among the constructs (Fig. 1). The research model is integrated from the relevant theoretical models. In which, the concepts of perceived risk and trust are based on Bauer [5], Pavlou [29], e–CAM of Park et al. [28] the concepts of ease of use and perceived usefulness are based on TAM of Davis [9] and Davis et al. [10], TAM2 of Venkatesh and Davis [39], TAM2' of Venkatesh [38], TAM3 of Venkatesh and Bala [40]. The details of all concepts and relationships among these concepts are interpreted as follows:

Research Concepts

E–payment Adoption (EPA) is understood as the intention to use by the current user or intention continue to use in the future [1]. The EPA dimension harmonizes with the foundational theories of behavioral intention is considered in IT adoption [9, 10, 38–41], as the basis for the relationship from antecedent and intermediate to the outcome as behavioral intention. The e–payment adoption has been referenced from the TAM by Davis [9] and Davis et al. [10], TAM2 by Venkatesh and Davis [39], TAM2' by Venkatesh [38], TAM3 by Venkatesh and Bala [40], UTAUT by Venkatesh et al. [41], and previous studies such as Barkhordari et al. [4], Francisco et al. [15], Phonthanukitithaworn et al. [31], Yaokumah et al. [44]. This research tests the structural model with the relationships between antecedent elements (perceived risk, and trust), intermediate elements (perceived usefulness, and ease of use), and outcome element (e–payment adoption).

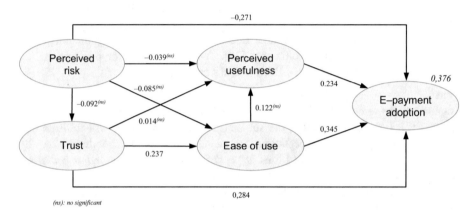

Fig. 1. Research model and testing result

Perceived Risk (PER) is a structure that reflects the emotions of customer uncertainty about the possible negative impact on using new technology [5]. The PER dimension has been referenced from the perceived risk theory by Bauer [5], e–CAM by Park et al. [28]. The perceived risk is also referenced from previous empirical studies about e–services such as Featherman and Pavlou [13], Pavlou [29], e–payment adoption as Cabanillas et al. [7, 8], Francisco et al. [15], Phonthanukitithaworn et al. [31], Yang et al. [43]. In this study, perceived risk is understood as the risks of online transactions and the security issues of the e–payment systems.

Trust (TRU) is the belief in individual interactions that cannot be sure of the outcome, or which the other party will act appropriately responsibly [29]. Trust also increases the perception of certainty associated with expected behavior when using e–payment and reduces the fear of the risks [22]. The TRU element has been referenced from the theory of perceived risk and trust by Pavlou [29]. The trust is also referenced from previous empirical studies on e–services such as Featherman and Pavlou [13], Gefen et al. [18], e–payment as Barkhordari et al. [44], Cabanillas et al. [7, 8], Francisco et al. [15], Yang et al. [43]. In this research, trust is understood for the safety and reliability of e–payments. Despite the risk, but customers still ready use the e–payment systems.

Perceived Usefulness (PEU) is the degree to which a person believes that using a particular system will enhance their performance [9]. The PEU dimension has been referenced from the TAM by Davis et al. [9] and Davis et al. [10], TAM2 by Venkatesh and Davis [39]. The perceived usefulness is also referenced from previous empirical studies about e–payment adoption such as Cabanillas et al. [7, 8], Francisco et al. [15], Gefen et al. [18], Phonthanukitithaworn et al. [31]. In this study, perceived usefulness is understood as the value that users receive when using e–payment.

Ease of Use (EOU) is the level to which a person believes that using a particular system without much effort [9]. The EOU element has been referenced from the TAM of Davis et al. [9] and Davis et al. [10], TAM2' by Venkatesh [38], TAM3 by Venkatesh and Bala [40]. The ease of use is also referenced from previous empirical studies on e–payment adoption as Cabanillas et al. [7, 8], Gefen et al. [18], Phonthanukitithaworn et al. [31], Yang et al. [43]. In this study, ease of use is understood as the ease accomplishing an e–payment in e–commerce transactions.

Hypotheses
Some scholars provided that perceived risk and trust have the parallel relationship (e.g., Featherman and Pavlou [13], Francisco et al. [15]). Others noted that the relationship between the perceived risk and the trust is serial, and trust as a perceived risk's function (e.g., Cabanillas et al. [7], Francisco et al. [15], Pavlou [29]). Moreover, Kim and Benbasat [22] mentioned about risk concerning trust as the less risk would exhort the more trust in e–commerce. Hence, under e–payment systems, we propose a hypothesis:

- *H1: The perceived risk's lower level is related to the trust's higher level*

The theoretical models and relevant studies pointed out the positive impact of the trust on the perceived usefulness such as Francisco et al. [15], Gefen et al. [18], Pavlou [29] and ease of use as Cabanillas et al. [7, 8], Pavlou [29]. The other side, perceived risk has a negative relationship with the perceived usefulness such as Featherman et al. [12], Featherman and Pavlou [13], Yang et al. [43] and ease of use such as Cabanillas et al. [7, 8], Lacan and Desmet [23], Yang et al. [39]. Hence, under e–payment systems, we propose these hypotheses:

- *H2_a: Perceived risk has a negative effect on ease of use.*
- *H2_b: Trust has a positive effect on ease of use.*
- *H2_c: Perceived risk has a negative effect on perceived usefulness.*
- *H2_d: Trust has a positive effect on perceived usefulness.*

The models of TAM [8], TAM2 [39], TAM2' [38] indicated the positive impact of the ease of use on the perceived usefulness. Furthermore, the relevant studies have confirmed this relationship (e.g., Featherman et al. [12], Francisco et al. [15], Gefen et al. [14]; Park et al. [22], Phonthanukitithaworn et al. [31]). Thus, under e–payment systems, we propose a hypothesis:

- *H3: Ease of use has a positive impact on perceived usefulness.*

The positive influence of the perceived usefulness and the ease of use on intention to use or behavioral intention is the main point of the IT adoption models, namely TAM [10], TAM2 [39], TAM2' [38]. Furthermore, Bankole and Bankole [3], Gefen et al.

[18], Park et al. [28], Phonthanukitithaworn et al. [31] have also tested and evidenced about the relationship between the perceived usefulness, the ease of use and the behavioral intention. Hence, under e–payment systems, we propose these hypotheses:

– *H4: Perceived usefulness has a positive impact on e–payment adoption.*
– *H5: Ease of use has a positive impact on e–payment adoption.*

Moreover, perceived risk has a negative impact on the intention to use such as Barkhordari et al. [4], Cabanillas et al. [7, 8], Francisco et al. [15], Park et al. [28], Phonthanukitithaworn et al. [31]. Aside from, trust also has a positive impact on the intention to use as Gefen et al. [16], Gao and Waechter [17], Phonthanukitithaworn et al. [31], Yang et al. [43]. Thus, under e–payment systems, we propose these hypotheses:

– *H6: Perceived risk has a negative impact on e–payment adoption.*
– *H7: Trust has a positive impact on e–payment adoption.*

3.3   Research Methods

Research process
This study is conducted in two phases – the first phase is a preliminary research with the qualitative method, and the second phase is a formal analysis with the quantitative method. Firstly, it is begun from the theoretical basis and the relevant studies, for constituting a draft scale. Next, the discussions with the experts on e–commerce and banking systems are implemented – especially, e–payment systems, for ensuring the accuracy of the scale's contents. Then, the adjusted scale is used as a measurement scale for the formal research. This study uses a 5–point Likert questionnaire with the levels: (1) strongly disagree – (2) disagree – (3) undecided – (4) agree – (5) strongly agree, for measuring the assessment levels of the indicators. The details of the scale are designated in Table 1. Meanwhile, in the measurement scale, there are five items of perceived risk, three items of trust, five items of perceived usefulness, four items of ease of use, and five items of in e–payment adoption dimensions.

Data is collected by a convenient sampling method, based on the accessibility of the surveyees. The questionnaires are sent to respondents who have used or intended to use all kinds of e–payments in Ho Chi Minh City – Vietnam by online via Google docs, and offline via hard copy. Finally, the collected data are cleaned, coded, and analyzed structural equation modelling (SEM) by IBM software with AMOS and SPSS. All of the 200 valid samples out of a total of 215 samples (15 invalid samples) of 20 indicators in the measurement scale are used in this study.

Data description
Age. the age groups of 16–22 and 23–30 are the majority with 40.0% and 34.5%, respectively, followed by the age group of 31–45 to 19.0%, and over 45 years old is the lowest proportion 6.5%. *Education.* the university degree is the highest percentage of 44.0%, there are 30.0% of respondents have postgraduate qualifications, intermediate/college and high school account for 22.0% and 4.0%, respectively. *Income.* the level of below VND 5 million has the most interest in e–payment with 41.5%, followed by the level of VND 5–10 million accounts for 29.5%, similarities

Table 1. Data description

	Frequency	Percentage (%)		Frequency	Percentage (%)
Age			*Gender*		
Ages 16–22	80	40.0	Male	94	47.0
Ages 23–30	69	34.5	Female	106	53.0
Ages 31–45	38	19.0	*Job position*		
Over age 45	13	6.5	Students	89	44.5
Education			Worker/Jobholder	34	17.0
High school	8	4.0	Officer	39	19.5
Intermediate/College	44	22.0	Entrepreneur/Manager	30	15.0
University degree	88	44.0	Others	8	4.0
Postgraduate	60	30.0	*E–payment type*		
Income			Debit card	52	14.6
Below VND 5 mil.	83	41.5	Credit card	84	23.7
VND 5–10 mil.	59	29.5	E–wallet	4	1.2
VND 10.1–15 mil.	29	14.5	Online banking	123	34.6
Over VND 15 mil.	29	14.5	Mobile banking	92	25.9

exist between the levels of VND 10.1–15 million and over VND 15 million are roundly 14.5% respondents. *Gender.* it is no large difference with 47.0% male and 53.0% female. *Job Position.* 44.5% of students, and officer, worker/jobholder and entrepreneur/manager diminutively amount to 19.5%, 17.0% and 15.0%. The detail of data sample description is presented in Table 1.

E–payment Type. the most respondents use online banking with 34.6%, mobile banking and credit card account for 25.9% and 23.7%, other types of e–payment has a lower rate. Furthermore, the respondents have used more than two types of e–payment up to 33.5%. Hence, it can be seen that e–payment is being used widely, but more focusing mainly on e–banking and credit card than other types.

4   Research Results

4.1   Exploratory and Confirmatory Factor Analyses

Exploratory Factor Analysis
The first exploratory factor analysis (EFA) eliminates two indicators, namely PEU_2 and EOU4 of *perceived usefulness (PEU)* and *ease of use (EOU)* dimensions whereas the EFA's factor loading <0.50 [20]. Then, the second EFA has extracted five elements from 18 indicators. The variables coalesce into five groups of factors in the rotated component matrix as in the theoretical model, including perceived risk, trust, perceived usefulness, ease of use, and e–payment adoption factors. The EFA factor loading of all indicators ranges from 0.714 to 0.912, details in Table 2.

Table 2. Measurement scales and analysis results

Talent		Indicator	Factor loading		CR	AVE
			EFA	CFA		
Perceived risk	PER$_1$	There may be leaked information online transactions	0.810	0.688	0.879	0.646
	PER$_2$	There may be caused error in the process of online transactions	0.714	Eliminated		
	PER$_3$	There may be caused fraud or lost money when using e–payments	0.836	0.815		
	PER$_4$	There may be accessed into unauthorized personal data by hackers	0.887	0.798		
	PER$_5$	E–payment transactions may not be secure	0.900	0.901		
Trust	TRU$_1$	Despite perceived risk, but still ready to use e–payments	0.765	0.657	0.753	0.506
	TRU$_2$	Believe that using e–payment systems, whereas safety and reliability	0.859	0.759		
	TRU$_4$	Believe that using e–payment systems will bring many benefits	0.796	0.712		
Perceived usefulness	PEU$_1$	Using e–payment systems unnecessary carry cash	0.927	0.963	0.930	0.895
	PEU$_2$	Using e–payment systems to help control spending	Eliminated	–		
	PEU$_3$	Using e–payment systems to improve payment efficiency	0.912	0.948		
	PEU$_4$	Using e–payment systems to make the transaction faster	0.911	0.929		
	PEU$_5$	Using e–payment systems to make the transaction easier	0.865	0.784		
Ease of use	EOU$_1$	E–payment systems is ease of use	0.822	0.771	0.757	0.516
	EOU$_2$	E–payment systems is evident and easy to understand	0.800	0.740		
	EOU$_3$	Ease of use of e–payment systems proficiently	0.797	0.627		
	EOU$_4$	E–payment transactions may be used every where and every time	Eliminated	–		
E–payment adoption	EPA$_1$	Having a inention to use e–payment systems	0.789	0.828	0.798	0.590
	EPA$_2$	Having a plan to use e–payment systems in the future	0.832	0.720		
	EPA$_3$	Willing use regularly e–payment systems in the future	0.810	0.712		

AVE: Average variance extracted; CR: Composite reliability

Furthermore, the test indexes of Kaiser Meyer Olkin (KMO) and Bartlett result as KMO measure of sampling adequacy = 0.782; Chi–square (χ^2) = 212.207; Bartlett test of sphericity, dF = 136 (p = 0.000). Those provide that the EFA of the all observational variables is appropriate [20], so the measurement scale is valuable. Nevertheless, the total variance extracted (TVA) = 75.72% that should well explain the difference in the data roughly 75.72%. Hence, the measurement scale may use for the further analysis, including confirmatory factor analysis (CFA), structural equation modelling (SEM), and Bootstrap analysis.

Confirmatory Factor Analysis

The first confirmatory factor analysis (CFA) continues to eliminate one observational variable, namely PER_2 of *perceived risk (PER)* dimension, whereas the CFA's factor loading <0.50 [20]. The second CFA with 17 indicators demonstrates that the indexes of the measurement scale as χ^2/dF = 1.215; GFI = 0.943; TLI = 0.986; CFI = 0.989; RMSEA = 0.033, so the measurement model is compatible with the market data [6]. The CFA factor loading of all indicators ranges between 0.688 and 0.948, details in Table 2. Besides, the composite reliability (CR) of these factors as perceived risk, trust, perceived usefulness, ease of use, and e–payment adoption ranges from 0.753 to 0.930 (>0.5), details in Table 2.

Table 3. Data, average variance extracted, and square correlation coefficient

	Mean	S.D	PER	TRU	PEU	EOU	EPA
PER	3.512	0.545	0.646*				
TRU	3.347	0.858	0.008	0.506*			
PEU	3.753	0.823	0.003	0.002	0.895*		
EOU	3.585	0.912	0.032	0.012	0.001	0.516*	
EPA	3.570	0.867	0.038	0.097	0.064	0.159	0.590*

*S.D: Std. deviation; AVE: Average variance extracted

In addition, the values of the average variance extracted (AVE) are from 0.506 to 0.895, details in Table 3, so the measurement scale gains the convergence value [6]. Moreover, the value of the AVE for each factor is also larger than the square correlation coefficient (r^2), respectively – details are indicated in Table 3, so the measurement scale is discriminant value [16].

Structural Equation Modelling

The structural equation modelling (SEM) is proceeded by the maximum likelihood (ML) estimation provides that the theoretical scale indexes as χ^2/dF = 1,185; GFI = 0,936; TLI = 0,987; CFI = 0,988; RMSEA = 0,036. Whereby, the theoretical model is adequate fit with the market data [6]. The SEM in the estimates is presented in Table 4.

Meanwhile, antecedent factors – *perceived risk (PER)* has the negative impact on *trust (TRU), perceived usefulness (PEU), ease of use (EOU),* and *e–payment adoption (EPA)* factors. Notwithstanding, there is only a negative relationship between *PER* and

EPA is statistical significance with γ coefficient = –0.271 (p < 0.001), the relationships between *PER* and the others are not statistical significance (p > 0.05). Thus, a hypothesis *H6* is supported, and the hypotheses *H1, H2$_a$* and *H2$_c$* are rejected. Hence, with perceived risk under e–payment systems, "*perceived risk has the negative effect e–payment adoption*" is a significant statement. Likewise, *TRU* factor has the positive impact on *EOU, PEU*, and *EPA*, but the positive paths from *TRU* to *EOU* and *EPA* are statistical significance with γ = 0.237 (p < 0.01) and 0.284 (p < 0.001), respectively. Thus, the hypotheses *H2$_b$* and *H7* are supported. The relationship between *TRU* and *PEU* are not statistical significance (p > 0.05). Thus, a hypothesis *H2$_d$* is rejected. Hence, with trust under e–payment systems, "*trust has the positive effect on ease of use and e–payment adoption*" is a significant statement.

Furthermore, intermediate factors – the positive relationship between *EOU* and *PEU* is not statistical significance (p > 0.05), so a hypothesis *H3* is also rejected. In addition, the data support the positive paths from *PEU* and *EOU* to *EPA* with γ coefficients = 0.345 (p < 0.001) and 0.234 (p < 0.01), respectively, Thus, the hypotheses *H4* and *H5* are supported. Hence, with e–payment adoption under e–payment systems, "*perceived usefulness and ease of use have the positive effect on e–payment adoption*" is a significant statement. Interestingly, all paths which have the relationships with e–payment adoption are strongly supported. The SEM in the paths is shown in Table 4.

Table 4. Path and hypothesis testing results

H	Path	Estimate	S.E.	C.R.	p–value	Result
H1	TRU ← PER	–0.092	0.131	–0.961	0.337	Rejected
H2$_a$	EOU ← PER	–0.085	0.106	–1.001	0.317	Rejected
H2$_b$	EOU ← TRU	0.237	0.092	2.413	0.008	Supported
H2$_c$	PEU ← PER	–0.039	0.142	–0.519	0.604	Rejected
H2$_d$	PEU ← TRU	0.014	0.103	0.166	0.868	Rejected
H3	PEU ← EOU	0.122	0.136	1.243	0.214	Rejected
H4	EPA ← PEU	0.345	0.173	3.731	***	Supported
H5	EPA ← EOU	0.234	0.096	2.914	0.002	Supported
H6	EPA ← PER	–0.271	0.133	–3.407	***	Supported
H7	EPA ← TRU	0.284	0.163	3.521	***	Supported

*** p < 0,001

According to the SEM results, the indexes of the standard error (S.E.) – standard deviation of the sampling distribution, and the critical ratio (C.R.) – dividing regression weight by the standard error estimate of each path in the research model. Both S.E. and C.R. indexes are detailed in Table 4.

Bootstrap Analysis

The Bootstrap analysis is used to validate the estimates of the research model with the repeat and replace samples [34]. In this study, the quantity of reduplicate sample has been artificial as N = n*5 = 1000. The results of Bootstrap estimates are calculated by the difference of the critical ratio (C.R.), the detail indexes are externalized in Table 5.

The indexes include the estimate, standard error (S.E.), S.E.–S.E., mean, bias, S.E.–bias, and |C.R.|. According to Bootstrap results, all paths of the theoretical model have | C.R.| < 2, so the difference is very assumption and no significant (p > 0.05) [20]. Thus, the estimates of the research model are trustworthy.

4.2   Result Discussions

The research results manifested that the influences of the trust – positive and the perceived risk – negative on the e–payment adoption are relatively large, with the coefficients of $\gamma = 0.284$ and -0.271, respectively. That confirmed the relationships as in e–CAM of Park et al. [28] and validated the substantial roles of the perceived risk and the trust on the e–payment adoption. On the other hand, the trust also has a significant impact on an intermediate – ease of use with a relatively large influence ($\gamma = 0.237$), but not statistical significance with another intermediate – perceived usefulness. This result indicated that, under e–payment systems, the data rejected the relationship between the trust and the perceived usefulness as in the theories of Gefen et al. [18], Pavlou [29]. There are non–significant for a path from the perceived risk to the perceived usefulness and the ease of use. Incidentally, under e–payment systems, the data did not support a negative relationship between perceived risk and trust as in Kim and Benbasat [22], Pavlou [27]. Differently, although the e–payment is ease of use, uncertain the perceived usefulness because the data did not support the path from the ease of use to the perceived usefulness as in the TAM. Furthermore, the impacts of two intermediates on the e–payment adoption are significant. In particular, the perceived usefulness and the ease of use had statistical significance with the e–payment adoption, the coefficients of $\gamma = 0.234$ and 0.345, respectively. Which demonstrated that the research model is propriety with the IT adoption models as the TAM series [9, 10, 38–40]. Meanwhile, the path from the ease of use to the e–payment adoption is the largest coefficient in the structural model. Therefore, besides the roles of the perceived risk and the trust, together with the perceived usefulness, the ease of use are the critical factors in the e–payment adoption.

Table 5. Bootstrap analysis results

Path	Estimate	Bootstrap				
		S.E.	Mean	Bias	|C.R.|	p–value
TRU ← PER	−0.092	0.137	−0.089	0.003	0.750	ns
EOU ← PER	−0.085	0.110	−0.091	−0.006	1.500	ns
EOU ← TRU	0.237	0.164	0.268	−0.005	1.667	ns
PEU ← PER	−0.039	0.152	−0.040	−0.001	0.200	ns
PEU ← TRU	0.014	0.115	0.008	−0.006	1.500	ns
PEU ← EOU	0.122	0.156	0.125	0.003	0.600	ns
EPA ← PEU	0.345	0.198	0.340	−0.005	1.667	ns
EPA ← EOU	0.234	0.168	0.237	0.003	0.750	ns
EPA ← PER	−0.271	0.177	−0.265	0.006	1.200	ns
EPA ← TRU	0.284	0.178	0.252	−0.002	0.500	ns

ns: non–significant

These results exposed that, under e–payment systems, the trust has the direct and indirect (through ease of use) impacts on e–payment adoption. The path of the trust – ease of use – e–payment adoption has the maximum value and statistical significance. Which demonstrated that the perceived trust of e–payment systems (e.g., safety and reliability, e–payment benefits) and the perceived ease of use of e–payment systems (e.g., clear and easy to understand, use systems proficiently…) have a large influence on e–payment adoption. Although the perceived risk supported the direct relation with the e–payment adoption, not supported with the other factors as trust, perceived usefulness, and ease of use. Hence, with the perceived risk, under e–payment systems, which evidenced that the customers are not interested in the perceived usefulness (e.g., don't improve payment efficiency, don't make transactions faster and easier); the ease of use (e.g., don't think e–payments is easy to understand and use); and the trust (e.g., don't convince when using e–payments, not yet ready to use e–payments, because perceived risk…). Interestingly, along with the IT adoption based – the perceived usefulness and the ease of use, the perceived risk and the trust have been the principal roles of the structural model of e–payment adoption.

Summary, there are five out of ten hypotheses have been supported. The structural equation modelling (SEM) also denoted that the antecedent elements – perceived risk and trust, and intermediate elements – perceived usefulness and ease of use may be explained roughly 38% ($R^2 = 0.376$) the outcome factor – intention to use or e–payment adoption. The findings are comparable with TAM of Davis [9] and Davis et al. [10], and UTAUT of Venkatesh et al. [41] amounted to 40% and 56% in behavioral intention, respectively. The explanations in this study are modest when compared with the IT adoption models. Besides, this study also identified the integrated dimensions as perceived risk and trust in the structural model (Fig. 1). Which are the contributions to the theoretical side of the perceived risk, the trust, and the IT adoption.

5   Conclusion and Future Work

The research results demonstrated that the measurement scale of the antecedent elements, the intermediate elements, and e–payment adoption are ensured reliability. The exploratory factor analysis and the confirmatory factor analysis of all indicators externalized that the factor loading of all items loaded high value relatively. Hence, the measurement scale is performed discriminant and convergence values. The structural equation modelling with the maximum likelihood provided a structural relationship among the dimensions, including perceived risk, trust, perceived usefulness, ease of use and e–payment adoption. Whereby, the Bootstrap analysis indicated that the research model's estimates are trusted. In addition, the model with a structural relationship among these dimensions accounted for roughly 36.7% of the intention to use or e–payment adoption. Interestingly, along with the IT adoption based – the perceived usefulness and the ease of use, the perceived risk and the trust have been the principal roles of the structural model of e–payment adoption. Which are the contributions to the theoretical side of the perceived risk, the trust, and the IT adoption.

In future work, it may be possible to add more elements that impact on the intention to use or e–payment adoption (e.g., habit, organizational culture, social influence…), literature the theory on behavioral intention and actual use of information systems. Furthermore, it's also possible to consider demographic as a moderator variable for the e–payment systems' acceptance and use.

References

1. Ajzen, I.: The theory of planned behavior. Organ. Behav. Hum. Decis. Process. **50**(2), 179–211 (1991)
2. Aladwani, A.M.: Online banking: a field study of drivers, development challenges, and expectations. Int. J. Inf. Manag. **21**(3), 213–225 (2001)
3. Bankole, F., Bankole, O.: The effects of cultural dimension on ICT innovation: empirical analysis of mobile phone services. Telemat. Inform. **34**(2), 490–505 (2017)
4. Barkhordari, M., Nourollah, Z., Mashayekhi, H., Mashayekhi, Y., Ahangar, M.: Factors influencing adoption of e–payment systems: an empirical study on Iranian customers. Inf. Syst. e–Bus. Manag. **14**(3), 89–116 (2016)
5. Bauer R.A.: Consumer behavior as risk taking. In: AMA Proceedings, Chicago (1960)
6. Byrne, B.: Structural Equation Modeling with AMOS. Routledge, New York (2016)
7. Cabanillas, F., Fernandez, J., Leiva, F.: The moderating effect of experience in the adoption of mobile payment tools in virtual social networks: the m-payment acceptance model in virtual social networks (MPAM-VSN). Int. J. Inf. Manage. **34**(2), 151–166 (2014)
8. Cabanillas, F., Leiva, F., Fernandez, J.: A global approach to the analysis of user behavior in mobile payment systems in the new electronic environment. Serv. Bus. **11**(1), 1–40 (2017)
9. Davis, F.: Perceived usefulness, perceived ease of use, and user acceptance of information technology. MIS Q. **13**(3), 319–340 (1989)
10. Davis, F., Bagozzi, R., Warshaw, P.: User acceptance of computer technology: a comparison of two theoretical models. Manag. Sci. **35**(8), 982–1003 (1989)
11. E–commerce and IT department: 2015 e–commerce report. Hanoi (2016)
12. Featherman, M., Miyazaki, A., Sprott, D.: Reducing online privacy risk to facilitate e–service adoption: the influence of perceived ease of use and corporate credibility. J. Serv. Mark. **24**(3), 219–229 (2010)
13. Featherman, M., Pavlou, P.: Predicting e–services adoption: a perceived risk facets perspective. Int. J. Hum.-Comput. Stud. **59**(4), 451–474 (2003)
14. Fishbein, M., Ajzen, I.: Belief, Attitude, Intention and Behavior: An Introduction to Theory and Research. Addison–Wesley, Reading (1975)
15. Francisco, L., Francisco, M., Juan S.: Payment systems in new electronic environments: consumer behavior in payment systems via SMS. Int. J. Inf. Technol. Decis. Mak. **14**(2), 421–449 (2015)
16. Fornell, C., Larcker, D.: Evaluating structural equation models with unobservable variables and measurement error. J. Mark. Res. **18**(1), 39–50 (1981)
17. Gao, L., Waechter, K.: Examining the role of initial trust in user adoption of mobile payment services: an empirical investigation. Inf. Syst. Front. **19**(3), 525–548 (2017)
18. Gefen, D., Karahanna, E., Straub, D.: Trust and TAM in online shopping: an integrated model. MIS Q. **27**(1), 51–90 (2003)
19. Goczek, L., Witkowski, B.: Determinants of card payments. Appl. Econ. **48**(16), 1530–1543 (2016)
20. Hair, J., Black, W., Babin, B., Anderson, R., Tatham, R.: Multivariate Data Analysis. Pearson (2014)

21. Junadi, S.: A model of factors influencing consumer's intention to use e–payment system in Indonesia. In: ICCSCI 2015 Proceedings, pp. 214–220. Indonesia (2015)
22. Kim, D., Benbasat, I.: The effects of trust–assuring arguments on consumer trust in internet stores: application of Toulmin's model of argumentation. Inf. Syst. Res. **17**(3), 286–300 (2006)
23. Lacan, C., Desmet, P.: Does the crowdfunding platform matter? Risks of negative attitudes in two–sided markets. J. Consum. Mark. **34**, 472–479 (2017)
24. Laudon, K., Traver, C.: E–commerce: Business Technology Society. Pearson (2016)
25. Lu, Y., Yang, S., Chau, P., Cao, Y.: Dynamics between the trust transfer process and intention to use mobile payment services: a cross–environment perspective. Inf. Manag. **48** (8), 393–403 (2011)
26. Nguyen, T.D., Cao, T.H.: Structural model for adoption and usage of e–banking in Vietnam. J. Econ. Dev. **220**, 116–135 (2014)
27. Nguyen, V.T.T., Nguyen, T.D.: Perceived risk in the e–payment adoption via social network. J. Econ. Dev. **27**(12), 66–81 (2016)
28. Park, J., Lee, D., Ahn, J.: Risk–focused e–commerce adoption model: a cross–country study. J. Glob. Inf. Technol. Manag. **7**, 6–30 (2004)
29. Pavlou, P.A.: Consumer acceptance of electronic commerce: integrating trust and risk with the technology acceptance model. Int. J. Electron. Commer. **7**(3), 101–134 (2003)
30. Pham, L., Cao, N.Y., Nguyen, T.D., Tran, P.T.: Structural models for e–banking adoption in Vietnam. Int. J. Enterp. Inf. Syst. **9**(1), 31–48 (2013)
31. Phonthanukitithaworn, C., Sellitto, C., Fong, M.: An investigation of mobile payment (m–payment) services in Thailand. Asia-Pac. J. Bus. **8**(1), 37–54 (2016)
32. Poon, W.C.: Users' adoption of e–banking services: the Malaysian perspective. J. Bus. Ind. Mark. **23**(1), 59–69 (2007)
33. Rahman, S.: Introduction to E–Commerce Technology in Business. GRIN (2014)
34. Schumacker, R., Lomax, R.: A Beginner's Guide to Structural Equation Modeling. Routledge, New York (2016)
35. Swick, N.K.: Benefits & risks of electronic payment systems (2010). https://thatcreditunionblog.wordpress.com
36. Tee, H., Ong, H.: Cashless payment and economic growth. Financ. Innov. **2**(1), 2–4 (2016)
37. Tella, A.: Determinants of e–payment systems success: a user's satisfaction perspective. Int. J. E-Adopt. **4**(3), 15–38 (2012)
38. Venkatesh, V.: Determinants of perceived ease of use: integrating perceived behavioral control, computer anxiety and enjoyment into the technology acceptance model. Inf. Syst. Res. **11**(4), 342–365 (2000)
39. Venkatesh, V., Davis, F.: A theoretical extension of the technology acceptance model: four longitudinal field studies. Manag. Sci. **46**(2), 186–204 (2000)
40. Venkatesh, V., Bala, H.: Technology acceptance model 3 and a research agenda on interventions. Decis. Sci. **39**(2), 273–315 (2008)
41. Venkatesh, V., Morris, M., Davis, F.: User acceptance of information technology: toward a unified view. MIS Q. **27**, 425–478 (2003)
42. Vietnam e–commerce association: Vietnam e–commerce index report. HCMC (2016)
43. Yang, Q., Pang, C., Liu, L., Yen, D., Tarn, J.: Exploring consumer perceived risk and trust for online payments: an empirical study in China's younger generation. Comput. Hum. Behav. **50**, 9–24 (2015)
44. Yaokumah, W., Kumah, P., Okai, E.: Demographic influences on e–payment services. Int. J. E-Bus. Res. **13**(1), 44–65 (2017)

Modelling Exchange Rate Volatility Using GARCH Model: An Empirical Analysis for Vietnam

Thi Kim Dung Nguyen[✉]

Faculty of Accounting and Finance, Nha Trang University,
Nha Trang City 650000, Vietnam
dungntk@ntu.edu.vn

Abstract. This paper empirically investigates the nature of exchange rate volatility in the context of Vietnam FX market. The study uses monthly data on exchange rates of Vietnamese Dong in term of major currencies such as US Dollar, British Pound, Japanese Yen and Canadian Dollar. The empirical analysis has been carried out for the period from Jan 1990 to Jun 2017, for a total of 330 observations. The exchange rate volatility of Vietnamese Dong against foreign currencies are estimated using GARCH models. Results show that ARMA(1,0)-GARCH(1,2) models with Student-t error distribution are well adequate models to capture the mean and volatility process of USD-VND and GBP-VND exchange rate returns, while ARMA(1,0)-GARCH(1,1) models with Student-t distribution are reasonably adequate models to capture the mean and volatility process of JPY-VND and CAD-VND exchange rate returns. The results also show that the exchange rate returns are rejected to follow Gaussian distribution at 1% significant level, and all four exchange rate return series maintain high persistence and volatility clustering.

1 Introduction

An international exchange rate, also known as foreign exchange (FX) rate, is the rate at which one currency can be exchanged for another currency. Foreign exchange rate, in fact, are one of the most important determinants of a country's relative level of economic health. It has strong impact on economic developments, foreign direct investment flows, international trade and capital mobility (see e.g., [1,8,15,18]). Therefore, measuring volatility, which is the dispersion of exchange rate returns, has useful and practical applications for risk management, and policy evaluation, academics, policymakers, regulators, and market practitioners.

More practically, understanding and estimating exchange rate volatility is important for exchange rate pricing, portfolio allocation, and risk management. Traders and regulators must consider not only the expected return from their trading activity but also risk exposure during volatile periods since traders' performance is highly affected by the accuracy of volatility forecast.

© Springer International Publishing AG 2018
L. H. Anh et al. (eds.), *Econometrics for Financial Applications*, Studies in Computational Intelligence 760, https://doi.org/10.1007/978-3-319-73150-6_69

Though volatility estimation and forecast have attracted attention of numerous researchers and academics all over the world, there is only modest number of studies conducted on this problem in Vietnam, especially for foreign exchange market. Hence, results in this study will contribute to empirical analysis of exchange rate volatility in Vietnam FX market as well as build a foundation for volatility forecast, portfolio allocation and risk management, which is extremely beneficial for traders, regulators, policy makers, and portfolio managers in Vietnamese FX market.

The structure of this paper is as follows: Sect. 2 provides literature review, Sect. 3 introduces methodology and data, Sect. 4 presents empirical analysis, and Sect. 5 provides conclusion.

2   Literature Review

A considerable amount of research have been done to model volatility in stock exchange and foreign exchange market. Noticeably, Engle [4] developed an Autoregressive Conditional Heteroscedasticity (ARCH) model to estimate financial volatilities. Bollerslev [2] first extended ARCH model to a widely known and used GARCH model. Under this class, the conditional variance is allowed to depend not only on previous squared shocks but also on its previous own lags [3].

Since then, applications of ARCH/GARCH models as well as their extensions in modelling and forecasting the asset return volatility have been very common in finance and economics. For example, Pilbeam and Langeland [14] forecasted volatility in foreign exchange market using different specification of GARCH models namely GARCH(1,1), EGARCH(1,1) and GJR-GARCH(1,1). Similarly, Kamal et al. [6] examined the performance of GARCH family models including symmetric GARCH-M, asymmetric EGARCH and TGARCH models in forecasting volatility behavior of Pakistan Forex market.

Mondal [9] used bivariate GARCH model to capture the volatility spillover between Reserve Bank of India's intervention and exchange rate to examine the effectiveness of RBI's intervention policy in foreign exchange market. Kamble and Honrao [7] used GARCH(1,1) model on monthly data of Indian Rupee-US dollar bilateral exchange rate to explain Indian foreign exchange rate volatility. Morekwa and Misati [10] used GARCH, TGARCH and EGARCH to investigate volatility of equity returns in Nairobi Stock Exchange, Kenya. Nor and Gharleghi [12] used Neural Network Autoregressive with Exogenous Input (NNARX) as a dynamic non-linear neural network, Artificial Neural Network (ANN) as a static neural network, Generalized Autoregressive Conditional Heteroscedasticity (GARCH) as a non-linear econometric model and Autoregressive Integrated Moving Average (ARIMA) as a linear econometric model to forecast the Singaporean dollar over US dollar exchange rate in three times horizon.

Another study done by Jabeen and Ismail [5] investigated the volatility dynamics of Pak Rupee exchange rates in term of major currencies in Pakistan. They found that GARCH models, i.e. GARCH-M and GJR-GARCH-M models are adequate in capturing the volatility dynamics in Pak Rupee exchange rates returns evidenced by the diagnostics test. They also revealed that there are no evidence of asymmetry and risk premium in Pak Rupee exchange rates except PKR-USD.

In Vietnam, there is limited number of studies using financial time series to model volatility. Pham [13] used ARMA(4,1)-GARCH(1,1) model to estimate volatility of VN Index in Vietnamese stock exchange. Tran and Ho [16] empirically analyzed the spillover effects between stock market and foreign exchange market in Vietnam using Vector Autoregressive (VAR) model and GARCH model. Nguyen and Bui [11] used the combination of ARIMA and GARCH model to forecast volatility of VN-Index and compared results with the two single models. They concluded that using hybrid ARIMA(1,1,1) – GARCH(1,1) model provide better prediction than the single models of ARIMA(1,1,1) and GARCH(1,1).

However, all of them have been done for stock exchange only, and there are no studies conducted to investigate the volatility dynamics of exchange rate in Vietnam. Nguyen [11] used to apply ARIMA(11,1,25) model to analyze and forecast the USD/VND exchange rate in the short term. However, using ARIMA model separately, the author only focused on the linear relationship between the current exchange rate and its past value and cannot captured the conditional standard deviation of the exchange rate returns. Therefore, this study is the first study estimating volatility of foreign exchange rate in Vietnamese FX market, which combined ARMA model for mean equation and GARCH model for volatility equation with assumption of Student-t error distribution.

3   Methodology and Data

The process of model estimation is composed of 4 steps. First, a mean equation is specified based on the test of ARIMA effects. Next, use the results of the mean equation to test for ARCH effect. Third, specify a volatility model and perform a joint estimation of the mean and volatility equations. Finally, the estimated models are checked carefully and refined where necessary.

3.1   ARMA Models

Autoregressive Moving Average (ARMA) models are used in this study for mean equation.

A general ARMA(p,q) model is in the form:

$$r_t = \emptyset_0 + \sum_{i=1}^{p} \emptyset_i r_{t-i} + a_t - \sum_{i=1}^{q} \emptyset_i a_{t-i}$$

where $\{a_t\}$ is a white noise series and p and q are nonnegative integers.

3.2 GARCH Models

Clustering is one of the main features of volatility, in which the large shocks or volatility have a tendency to be clustered around the great shocks and the small shocks tend to follow by a small changes in volatility [2]. In this paper, Generalized Autoregressive Conditional Heterocedasticity (GARCH) models are employed for modeling exchange rate volatility and for testing market efficiency. Under GARCH models, the conditional variance is allowed to be dependent not only on the square of previous shocks but also on its previous own lags [3]. The GARCH model captures the clustering in volatility and also leptokurtic behaviour in the distribution tails as the tail of the distribution of a GARCH process is fatter and heavier than a Gaussian distribution [17]. In particular, the conditional variance equation of a GARCH (p, q) model is as:

$$a_t = \sigma_t \epsilon_t$$

$$\sigma_t^2 = \alpha_0 + \sum_{i=1}^{p} \alpha_i a_{t-i}^2 + \sum_{j=1}^{q} \beta_j \sigma_{t-j}^2$$

σ_t^2 is calculated from the most recent p observations on a^2 and the most recent q estimates of the variance rate. GARCH(1,1) is by far the most popular of the GARCH models,

Assumptions:

- The error terms $\{\epsilon_t\}$ are independently and identically distributed (iid) with zero mean and variance of 1.
- $\alpha_0 > 0$, $\alpha_i \geq 0$, $\beta_i \geq 0$ (non-negativity constraint).
- $\sum_{i=1}^{\max(p,q)} (\alpha_i + \beta_i) < 1$ where $\alpha_i = 0$ for $i > p$ and $\beta_j = 0$ for $j > q$. This assumption is made in order to ensure that the variance process is stationary and finite.

3.3 Data

In this empirical analysis, the monthly data extracted from Global Economic Monitor (WorldBank) from January 1990 to June 2017 for a total of 330 observations are used. Bilateral Vietnamese Dong official exchange rates in the term of major currencies including US dollar, British Pound, Japanese Yen, and Canadian Dollar are examined. Additionally, direct quote is applied in this paper, meaning all the foreign exchange rates are quoted as Vietnamese Dong per unit of the foreign currency.

4 Empirical Analysis and Results:

4.1 Summary Statistics and Diagnostic Check

As can be seen in Fig. 1, there is a general upward trend in all four exchange rates over the sample period. Furthermore, the plot of log return series show no

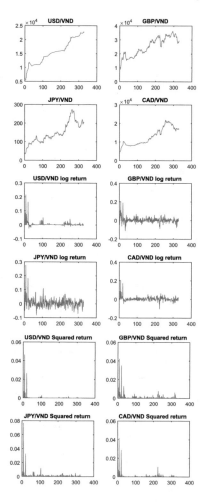

Fig. 1. Monthly VN Dong exchange rates, log returns and squared returns

definite pattern in the log return series. The plot also reveals that the variances change over time and volatility tends to be cluster. The exchange rate returns are complying with the mean reverting and volatility clustering stylized facts. The squared returns are taken for the proxy of volatility. Figure 1 also indicates the variation in volatility over the sample period.

In this study, the time series data is employed. One key assumption in empirical work based on time series data is that the underlying time series is weakly stationary, which is both the mean of r_t and the covariance between r_t and r_{t-l} are time invariant, where l is an arbitrary integer [17]. However, many studies have found that majority of time series variables are non-stationary and using non-stationary time series in a regression analysis may lead to spurious regression [3,17]. Hence before doing any empirical analysis, we need to check the stationarity assumption first. Among a number of unit root tests available for stationarity analysis, Augmented Dickey-Fuller (ADF test) and Philip-Perron (PP) test are most popular used by researchers. In this study, Dickey-Fuller test (ADFtest) are employed to test the stationarity of the exchange rate series and the log return series. Results in Table 1 show that stationary existence in the first difference series (i.e. log return series) at 1% significant level with all p-values of 0.0010. That is all the exchange rate return series are stationary at 1% level.

Table 1 reports the summary statistics for the monthly exchange rate return series. All the exchange rate return series are positively skewed, and have extremely fat tails (kurtorsis > 3). This indicates a departure from normality, which is also confirmed by the quantile-quantile plot in Fig. 2 and by Jarque-Bera test (JB test). Result of JB test is shown in Table 1, indicating that the null hypothesis of normal distribution for monthly exchange rate returns is rejected at 1% significant level for all underlying exchange rate returns.

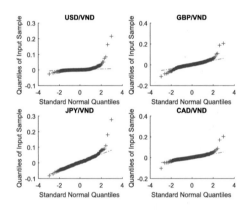

Fig. 2. QQ plot of return series

For model selection, Ljung-Box (LB) test with 15 df is conducted to check whether AR and ARCH effects exist. For AR effects, the null hypothesis of the test is $H_o : p_1 = ... = p_m = 0$ where p is the linear correlation between serial monthly log returns. For ARCH effects, the null hypothesis is the same, apart from that p is the linear correlation between serial squared log returns. The significance level of 5% is used for the two tests. P-values and test statistics of the two tests are summarized in Table 1. All p-value of LB test for log return series and squared log returns are less than 0.01 indicating the null is rejected at 1% significant level. Consequently, AR and ARCH effects do exist in these series. This is also confirmed by the autocorrelation and partial autocorrelation plot of returns and squared returns in Fig. 3. As a consequence, taking the AR effects and ARCH effects into consideration, ARMA-GARCH models are used in this study.

Table 1. Summary statistics and diagnostic check of return series

	USD/VND	GBP/VND	JPY/VND	CAD/VND
Mean	0.0044	0.0036	0.0052	0.0040
Median	0.0009	0.0021	0.0038	0.0020
Max	0.2151	0.2049	0.2764	0.2036
Min	-0.0258	-0.1127	-0.0820	-0.1029
Std. dev	0.0177	0.0298	0.0329	0.0241
Skewness	7.9628	1.4564	2.3011	2.6825
Kurtosis	83.2046	14.2333	18.4636	24.0223
JB test statistic	91660 (0.0000)**	1846.1 (0.0000)**	3568.3 (0.0000)**	6452.8 (0.0000)**
Dickey Fuller test (ADFtest) statistic	-15.0361 (0.0010)**	-12.5452 (0.0010)**	-12.7550 (0.0010)**	-13.6869 (0.0010)**
LB-Q(15) test	120.7735 (0.0000)**	72.2128 (0.0000)**	91.8268 (0.0000)**	58.4513 (0.0003)**
LB$-Q(15)^2$ test	69.7724 (0.0000)**	100.7955 (0.0000)**	41.7551 (0.0002)**	75.0438 (0.0000)**

Note: p-values are in parentheses, ** indicates significant at 1%, * indicates significant at 5%

4.2   Model Estimation

As shown in Fig. 3, significant serial correlations do exist in the first 12 lags of the series. In this study, ARMA(1,0) specification is chosen to reflect the autocorrelation in the return series as supported by ACF and PACF plots. Therefore, an ARMA (1,0) model is selected for the mean equation of the GARCH model for all four exchange rate returns. Additionally, as proven above, normality is rejected

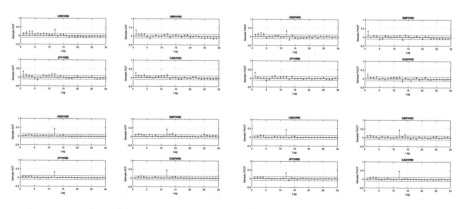

Fig. 3. Autocorrelation and partial autocorrelation of returns and squared returns

in all four exchange rate return series, therefore Student-t error distribution is employed in this study when specifying GARCH models.

The order of GARCH models are determined based on information criteria AIC and SIC. The rule is selecting a particular order k that has the minimum AIC and SIC value. For USD-VND, AIC suggested ARMA(1,0)-GARCH(1,4)-t while SIC suggested ARMA(1,0)-GARCH(1,2)-t. ARMA(1,0)-GARCH(1,2)-t is suggested by SIC is chosen in this study because of relatively less parameters. For GBP-VND, AIC suggested ARMA(1,0)-GARCH(1,2)-t while SIC suggested ARMA(1,0)-GARCH(1,1)-t. For JPY-VND and CAD-VND, both AIC and SIC suggested ARMA(1,0)-GARCH(1,1)-t.

Using Maximum Likelihood (ML) method, the models are estimated. Outputs of ML method are summarized in Table 2.

All AR parameter estimates are significantly different from 0 at 5% significant level with t-statistics greater than 2. The ARCH and GARCH parameters are significant except GARCH(1) for USD-VND and GBP-VND, and ARCH(1) for JPY-VND. In addition, the non-negativity constraints condition does satisfied. The constant term in the variance equation and estimated values of parameters are not smaller than zero (as demonstrated by t-statistics). Moreover, the estimated persistence coefficients are 0.9999, 0.8987, 0.9187 and 0.9161 for USD-VND, GBP-VND, JPY-VND and CAD-VND respectively (see Table 3), which are very close to 1 indicating an extremely strong volatility persistence and a very slow mean reversion to its estimated long run unconditional variance.

4.3   Assess the Fitness of the Models

As Student-t distribution is assumed in this study, before conducting a diagnostic checking, the standardised residuals are transformed back from a Student-t to a Gaussian residuals. The desired properties of the transformed standardized residuals series are that it should be independently and identically distributed (i.e. iid) and it follows a normal distribution.

Table 3 summarizes test statistics for checking adequacy of mean equation, volatility equation and normality distribution of transformed standardized residuals.

Mean equation adequacy. Ljung-Box Q test for transformed standardised residuals are conducted to check the adequacy of mean equation. Using m of 15 with degree of freedom adjusted for different models, all p-values are greater than 0.05 except for JPY-VND having p-value of 0.0042 (see Table 3). This means there are no remaining significant AR effects in the residual series for USD-VND, GBP-VND and CAD-VND. In other words, the mean equations are adequate for USD-VND, GBP-VND and CAD-VND.

Volatility equation adequacy. Ljung-Box Q test for squared transformed standardised residuals are conducted to check the adequacy of volatility equation. Using m of 5 and 15, all p-values are greater than 0.05 except for CAD-VND having p-value of 0.01 for $LB - Q(15)^2$ test and 0.1037 for $LB - Q(5)^2$

Table 2. ML estimates from ARMA–GARCH–Student-t models for exchange rate returns

USD-VND

Mean: ARMA(1,0); Variance: GARCH(1,2) Conditional Probability Distribution: t

Parameter	Value	Standard Error	T Statistic
C	0.000600968	0.000104978	5.72471
AR(1)	0.288415	0.0433595	6.65171
K	1.43519e−06	9.23516e−07	1.55405
GARCH(1)	0.113779	0.0822864	1.38272
GARCH(2)	0.281669	0.0661128	4.26044
ARCH(1)	0.604551	0.174725	3.46001
DoF	2.60724	0.241046	10.8164

GBP-VND

Mean: ARMA(1,0); Variance: GARCH(1,2) Conditional Probability Distribution: t

Parameter	Value	Standard Error	T Statistic
C	0.00192482	0.00111621	1.72442
AR(1)	0.216143	0.0598281	3.61273
K	7.79702e−05	3.91278e−05	1.99271
GARCH(1)	0.115152	0.123074	0.935631
GARCH(2)	0.516617	0.157969	3.27036
ARCH(1)	0.266916	0.102872	2.59464
DoF	4.77979	1.40819	3.39428

JPY-VND

Mean: ARMA(1,0); Variance: GARCH(1,1) Conditional Probability Distribution: t

Parameter	Value	Standard Error	T Statistic
C	0.00170927	0.00138718	1.23219
AR(1)	0.267531	0.0478407	5.59212
K	5.92588e−05	7.14384e−05	0.829509
GARCH(1)	0.901916	0.110509	8.16151
ARCH(1)	0.0168169	0.0204307	0.823118
DoF	5.9883	1.21638	4.92306

CAD-VND

Mean: ARMA(1,0); Variance: GARCH(1,1) Conditional Probability Distribution: t

Parameter	Value	Standard Error	T Statistic
C	0.00111966	0.000837154	1.33746
AR(1)	0.291251	0.0512852	5.67903
K	3.12028e−05	1.56235e−05	1.99718
GARCH(1)	0.797154	0.0648712	12.2883
ARCH(1)	0.118942	0.0529281	2.24723
DoF	4.67379	1.01779	4.59211

Table 3. Models fitness checking

	USD-VND	GBP-VND	JPY-VND	CAD-VND
Model	ARMA(1,0)-GARCH(1,2)-t	ARMA(1,0)-GARCH(1,2)-t	ARMA(1,0)-GARCH(1,1)-t	ARMA(1,0)-GARCH(1,1)-t
$\alpha + \beta$	0.999999	0.898685	0.9187329	0.916096
Skewness	0.5114	0.0360	0.3713	0.1937
Kurtosis	3.1707	2.9947	3.2342	3.1413
Jarque-Bera test	14.7415 (0.0047)**	0.0716 (0.5000)	8.3099 (0.0215)*	2.3302 (0.2769)
LB-Q(15) test	16.8652 (0.3270)	20.3190 (0.1600)	33.3659 (0.0042)**	23.8081 (0.0684)
LB$-Q(15)^2$ test	16.7498 (0.3340)	15.7173 (0.4011)	20.0688 (0.1693)	30.4536 (0.0104)*
LB$-Q(5)^2$ test	6.2120 (0.2861)	4.8085 (0.4397)	4.4617 (0.4850)	0.1375 (0.1037)

Note: P-values are in parentheses, ** indicates significant at 1%; * indicates significant at 5%

test (see Table 3). Therefore, there is insufficient evidence to reject the independence null hypothesis of these series. There is no remaining significant ARCH effect in the series. In other words, the volatility fits the data really well.

Normality checking. As shown in the histogram of the transformed standardized residuals (Fig. 4), apart from a positive outlier of 4%, the distribution is spread out between −3% and 3%. Moreover, there is a small deviation of the distribution from the expected quantile line in the tails, especially for USD-VND. Furethermore, p-values of the JB test for normality are greater than 0.05 for GBP-VND and CAD-VND, indicating the null hypothesis of normal distribution for transformed standardized residuals could not be rejected at 5% significant level. This suggests that the Student-t is a good choice of error

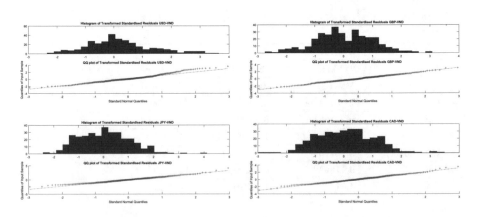

Fig. 4. Histogram and QQ plot of transformed standardized residuals

distribution for GBP-VND and CAD-VND data. For USD-VND and JPY-VND, though JB tests reject the null at 1% and 5% significant level respectively, the skewness of around 0 and kurtosis of just above 3 seem nice for transformed standardised residuals. Indeed, these results under Student-t errors are much better than under Gaussian error assumption.

5 Conclusion

This paper empirically investigates the nature of volatility of Vietnamese Dong exchange rate in term of US dollar, British Pound, Japanese Yen and Canadian Dollar through using the combination of ARMA(1,0) and GARCH(1,2)/GARCH(1,1) models with assumption of Student-t error. Results show that the combination of linear and non-linear model in this study well capture volatility dynamics of all foreign exchange rate series, whereas the mean equations are reasonably adequate to capture the mean return series, except for JPY-VND. Moreover, results found that model specification with Student-t error is a better choice than Gaussian error assumption as it well captures non-normality of the series. Additionally, high estimated persistence parameters in these models show that the exchange rate series have an extremely strong volatility persistence and a very slow mean reversion to its estimated long run unconditional variance.

References

1. Appuhamilage, K.S.A., Alhayky, A.A.A.: Exchange rate movements' effect on Sri Lanka-China trade. J. Chin. Econ. Foreign Trade Stud. **3**(3), 254–267 (2010)
2. Bollerslev, T.: Generalized autoregressive conditional heteroscedasticity. J. Econom. **31**(3), 307–327 (1986)
3. Brooks, C.: Introductory Econometrics for Finance. Cambridge University Press, New York (2014)
4. Engle, R.: Autoregressive conditional heteroscedasticity with estimates of the variance of United Kingdom inflation. Econometrica **50**(4), 987–1007 (1982)
5. Jabeen, M., Ismail, A.: Exchange rate volatility and market efficiency: evidence from Pakistan. J. Econ. Coop. Dev. **36**(3), 67–92 (2015)
6. Kamal, Y., ul Haq, H., Ghani, U., Khan, M.M.: Modeling the exchange rate volatility, using generalized autoregressive conditionally heteroscedastic (GARCH) type models evidence: from Pakistan. Afr. J. Bus. Manag. **6**(8), 2830–2838 (2012)
7. Kamble, G., Honrao, P.: Time-series analysis of exchange rate volatility of Indian Rupee/US dollar - an empirical investigation. J. Int. Econ. **5**(2), 17–29 (2014)
8. Korkmaz, M., Alacahan, N.D., Aytaç, A., Aksoy, M., Germİr, H.N., Karta, N.: The relation between real exchange rate in Turkey and foreign trade: an applied analysis. Int. Ref. Acad. Soc. Sci. J. **18**(6), 84–104 (2015)
9. Mondal, L.: Volatility spillover between the RBI's intervention and exchange rate. Int. Econ. Econ. Policy **11**(4), 549–560 (2014)
10. Morekwa, E.N., Misati, R.: Modelling the time-varying volatility of equities returns in Kenya. Afr. J. Econ. Manag. Stud. **1**(2), 183–196 (2010)

11. Nguyen, T., Bui, Q.: Application of combined ARIMA-GARCH model to forecast the VN-Index. J. Sci. Technol. Danang Univ. **04**(89) (2015)
12. Nor, A.H.S.M., Gharleghi, B.: Application of dynamic models for exchange rate prediction. Int. J. Innov. Manag. Technol. **2**(6), 459–463 (2011)
13. Pham, C.: Volatility forecast of Vietnamese stock exchange using ARCH/GARCH models. Review of Finance (2017). http://tapchitaichinh.vn/nghien-cuu-trao-doi/du-bao-bien-dong-gia-chung-khoan-qua-mo-hinh-archgarch-117763.html
14. Pilbeam, K., Langeland, K.N.: Forecasting exchange rate volatility: GARCH models versus implied volatility forecasts. Int. Econ. Econ. Policy **12**(1), 127–142 (2015)
15. Ramasamy, R., Munuswamy, S., Helmi, M.H.M.: Comparative predictive accuracy of GARCH, GJR and EGARCH models applied on select exchange rates. Int. J. Arts Sci. **5**(1), 105–119 (2012)
16. Tran, N., Ho, T.: The spillover effect between the stock market and foreign exchange market in Vietnam. Dev. Integr. **21**(31), 34–39 (2015)
17. Tsay, R.S.: Analysis of Financial Time Series. Wiley, Hoboken (2010)
18. Ullah, S., Haider, S.Z., Azim, P.: Impact of exchange rate volatility on foreign direct investment: a case study of Pakistan. Pak. Econ. Soc. Rev. **50**(2), 121–138 (2012)

Pricing Assets with Higher Co-moments and Value-at-Risk by Quantile Regression Approach: Evidence from Vietnam Stock Market

Toan Luu Duc Huynh[✉], Sang Phu Nguyen, and Duy Duong

Faculty of Finance, Banking University of Ho Chi Minh City,
Ho Chi Minh City, Vietnam
toanhld@buh.edu.vn

Abstract. This paper examines the role of higher co-moments of the shape of return distribution in capturing secondary data for 274 non-financial firms listed in the Vietnam Index, considered as one of the emerging stock markets, during the period from July 2006 to June 2016. We employ Fama-French model combined with higher co-moments, particularly co-skewness and co-kurtosis, and value-at-risk (VaR) to explain the return-generating process. Quantile regression is also used in descending order with the two methods of equally weighted and value-weighted portfolios. The findings show that investors could maximize their portfolio return by holding more stocks with the positive co-skewness and restricting the large co-kurtosis ones. It implies that in addition to co-momentum effects other determinants such as size, value and maximal value of losses also have a strong influence on stock return.

JEL Classification: G11 · G12 · G14

Keywords: Co-skewness · Co-kurtosis · Fama and French factors
Value-at-Risk · Vietnam

1 Introduction

In theory, portfolio management has been constructed in fundamental concepts of mean-variance as the first and second moment framework such as Markowitz [28], Sharpe [36] and Lintner [27]. The invention of the Capital Asset Pricing Model (CAPM) has motivated much research into nonlinear and abnormal distribution models due to highly restrictive conditions applied in studies as Kraus and Litzenberger [26] or Friend and Westerfield [15]. Until now, there have been many criticizes regarding CAPM which may not be applicable with single factor. For many years, the two determinants, market risk premium and expected return, have always been mutually influenced. Fama and French [11] shows that there are also more idiosyncratic factors on stock size and book value to market value of individual equity without the existence of any co-moments.

© Springer International Publishing AG 2018
L. H. Anh et al. (eds.), *Econometrics for Financial Applications*, Studies in Computational Intelligence 760, https://doi.org/10.1007/978-3-319-73150-6_70

In the same year, Bali and Cakici [1] asserts that value at risk element mainly contributes into expected return with the assumption for normal return distribution. However, these studies are excluded with the higher co-moments in pricing models at that time.

The returns are plotted into non-normal distribution and there exists a non-quadratic utility function by investors. Therefore, the investors might care about the moments as the implications by Rubinstein [34]. From the market perspective, Harvey and Siddique [18] indicates that only co-skewness factor conveyed in market portfolio has an explanation to US return by an average premium up to 3.6% annually. In addition, this study emphasizes that the larger the negative co-skewness is, the more profitable the portfolio is. To support this result, the co-skewness factor conducted by Harvey and Siddique [17] demonstrates that the market tends to outperform with 5% per year when the correlation is positively skewed whereas it easily underperform 2.81% annually with negative skewness. Furthermore, investors have a tendency to consciously behave in aversive ways with negative skewness and excess kurtosis (Scott and Horvath [35]). Moreover, when 'extreme events' happen unexpectedly, the asset return might have a deviation from normal distribution and get far away from the value of mean, median or mode (Jondeau and Rockinger [22]).

Most current research tests how prominent co-skewness and co-kurtosis appear in risk-adjusted performance for the US equity funds (Moreno and Rodríguez [32]), which proves the similar outcome as well as explanation by Ding and Shawky [8]. In emerging stock markets, the role of higher co-moments is underestimated because of omitting the marginal change in taking stocks in portfolio. By adding the value-at-risk factor, many researches show that a strong lower tail, which represents losses, generates the large negative co-skewness as the same result by Chabi-Yo et al. [5]. The test of tail dependence simulates not only equity assets but also debt securities in many aspects. Garcia and Tsafack [16] highlights that the less return because of strong dependence on tail, the larger co-skewness it creates in international bond analysis.

From the aforementioned studies, the authors focus on testing quantitative inference that higher co-moments would be key factors to drive the gain or loss in the stock markets. Thus the first research question is whether the higher co-moments including co-skewness and co-kurtosis and value-at-risk impact statistically Vietnamese stock return or not. We approach this question by sorting each stock into portfolio based on two method equally weighted and value-weighted portfolios, for example, from the lowest indicator (co-skewness and co-kurtosis) to the highest one. In the following step, we test the correlation by using quantile regression in order to avoid the phenomenon of heteroscedasticity. Especially, the Vietnamese stock exchange is an emerging market, which is quite sensitive with any changes as well as risk contagion from the bigger one during the financial crisis. The second research question examines the direction and degree of these components' effect on stock return. This could be done by using coefficient extracting from quantile regression to determine the low ($\tau = 0.25$), medium ($\tau = 0.5$) and high ($\tau = 0.25$) quantile. We also use characteristic-sorted portfolio to evaluate more accurate. The last research question proposed by authors is to determine the difference in each portfolio constructed by level of co-skewness and co-kurtosis estimation. These research questions still lack of empirical evidence from newly emerging stock market, Vietnam.

2   Literature Review

There are some researches that estimate the nonlinear shape in risk measurement as Rubinstein [34] and Kraus and Litzenberger [26] regarding modern portfolio theory. They also emphasize that the relationship between the expected return of an asset and unconditional variance and co-skewness in terms of risk aversion and weighted classification under systematic risks. In the following studies, the existing accounts fail to resolve the contradiction between co-moments and expected return because it represents the misspecification of ignoring the co-skewness impact on the volatility of these assets' return in researches of Harvey and Siddique [18] and Barone Adesi et al. [3]. The studies done by Fang and Lai [12], Hwang and Satchell [20], Conrad et al. [6] prove that the significance of return is more explanatory with the second and third co-moments compared to the first one. Most empirical findings show that the premium stock return must compensate only if a decrease in market portfolio skewness as Moreno and Rodríguez [31] and an increase in its kurtosis as Fang and Lai [12].

In this field, there are some development regarding return estimation based on moment-built investible factors. They divide the dataset into two categories, which have different traits. In particular, one has highest co-moments occupied 30% and the rest is the opposite side accounted for 30%. Interestingly, some researches such as Harvey and Siddique [18], Jarjir [21], Kole and Verbeek [24], Moreno and Rodríguez [31] and Kostakis et al. [25] construct the portfolios with the same characteristics such as co-skewness and co-kurtosis to estimate the level of impact on stock returns.

In order to overcome the previous drawbacks, many researches including Kat and Miffre [23] extend the portfolios formalization into six units by intersection of co-variance and co-kurtosis as well as co-variance and co-skewness. The findings are different between average high and low scoring portfolio arrangements and inversely. It demonstrates the relationship between risk factor premium and stock return. However, its limitation is unbalanced portfolios and some dataset omitted due to lack of observation (Tables 1 and 2).

Interestingly, the previous researches have an attention to risk factors to measure the impact on stock returns. However, most journals do not focus on quantile regression to avoid effect of outliers and to observe the entire distribution of return as well as to restrict heteroscedasticity in regression model. This presents a the scientific gap. Therefore, the authors decide examine this phenomenon in the Vietnam stock exchange, one of the emerging stock markets in the world. The authors come into conclusion that based on theoretical and empirical framework, higher co-moments with basic elements has considerable explanatory ability over the cross-sectionof co-skewness and co-kurtosis

Table 1. Key researches related to the relationship between stock return and risk factors

Significant	Source	Data span	Empirical approach	Remarks
+	Fang and Lai [12]	United State, NYSE (January 1969–December 1988)	The ordinary least square (OLS) and the instrumental variable estimation (IVE)	The expected excess rate of return is related not only to the systematic variance but also to the systematic skewness and systematic kurtosis
+	Harvey and Siddique [18]	United State, CRSP NYSE0AMEX and NASDAQ (July 1963–December 1993)	ARMA (2,0) joint full-information maximum likelihood, month-by-month cross-sectional regressions	Conditional skewness helps explain the cross-sectional variation of expected returns across assets and is significant even when factors based on size and book-to-market are included
+	Kostakis et al. [25]	United Kingdom, London Stock Exchange (1986–2008)	GMM with Newey-West standard errors corrected for heteroscedasticity and serial correlation	Higher co-moment asset pricing model can have significant explanatory power in the cross-section of stock returns
+	Barone-Adesi et al. [2]	French (July 1963 to December 2000)	GMM, Monte Carlo simulation	Portfolios of small (large) firms have negative (positive) Co-skewness with the market

(Source: The authors)

portfolios' return. Besides, many researches regarding positive and negative impact, there is also an equilibrium position for pricing with these risk factors. To be more specific, Mitton and Vorkink [30] indicates that the price will be in equilibrium when investors expect heterogeneity in their preferences and there exists idiosyncratic skewness. In addition, Boyer et al. [4] concludes that this conjecture is completely held in UK market. The aforementioned idiosyncratic skewness can be priced in this market and it focuses on both co-skewness as well as co-kurtosis for pricing models.

3   Asset Pricing and Higher Co-moments

Formally, application with modern studies regarding non-linear models in stock pricing such as Kraus and Litzenberger [26], Harvey and Siddique [18],

Table 2. Researches related to the relationship between stock return and risk factors, particularly co-kurtosis

Significant	Source	Data span	Empirical approach	Remarks
+	Kostakis et al. [25]	United Kingdom (1986–2008)	GMM with Newey-West standard errors corrected for heteroscedasticity and serial correlation	Higher co-moment asset pricing model can have significant explanatory power in the cross-section of stock returns
+	Fang and Lai [12]	United State, NYSE (January 1969–December 1988)	The ordinary least square (OLS) and the instrumental variable estimation (IVE)	The expected excess rate of return is related not only to the systematic variance but also to the systematic skewness and systematic kurtosis

(Source: The authors)

Dittmar [9], Barone Adesi et al. [3] is very necessary to upgrade the traditional CAPM by Sharpe [36]

$$R_{i,t} - r_{f,t} = \beta_{(0,i)} + \beta_{1,i}[R_{M,t} - r_{f,t}] + \beta_{2,i}[R_{M,t} - \overline{R}_{M,t}]^2 + \beta_{3,i}[R_{M,t} - \overline{R}_{M,t}]^3 + \varepsilon_{i,t} \tag{1}$$

where it denotes that $R_{i,t}; r_{f,t}$ and $R_{M,t}$ are return calculated by monthly approach of security ith, the interest rate extracted from Treasury Bill in a month and total securities on market portfolio, respectively. Moreover, $\overline{R}_{M,t}$ is the average market return from 36 months prior to the determined date. The coefficient indicators, extracted from the regression (1), are determined as proxies for beta, co-skewness and co-kurtosis on the market premium $(\beta_{1,i})$, the square of market exceeding return $(\beta_{2,i})$, the cube of ones $(\beta_{3,i})$. However, it is mainly based on historical data to estimate these figures and the mismatched results in the choice between long-term data series and short-term series. In the recent study, Kole and Verbeek [24] reveals that there is considerable difference in using a set of 120-month data in comparison with 60-month one.

Furthermore, Kostakis et al. [25] continues their researches with figuring out the major concern of valid Stochastic Discount Factor (SDF) called by letter 'M', which conveys quadratic and cubic market returns in terms of co-skewness and co-kurtosis. Then, the marginal rate of substitution in model is concluded here

$$M_{t+1} = 1 + \tilde{b}R_{m,t+1} + \tilde{c}R_{m,t+1}^2 + \tilde{d}R_{m,t+1}^3. \tag{2}$$

Afterward, this paper also gives an account of non-linear functional form for central asset pricing equation from (2) as follows with determination of the required premium

$$E_t(R_{t+1}) - R_t^f = -(1 + R_t^f)\tilde{b}Cov(R_{m,t+1}, R_{t+1}) - (1 + R_t^f)\tilde{c}Cov(R_{m,t+1}^2, R_{t+1})$$
$$- (1 + R_t^f)\tilde{d}Cov(R_{m,t+1}^3, R_{t+1}). \tag{3}$$

This research incorporates co-skewness and co-kurtosis risk by measuring the standardization with the relevant aspect of market return over the period from $t - 60$ to t via two formula

$$CSK_i = \frac{E[\varepsilon_{i,t}\varepsilon_{m,t}^2]}{\sqrt{E[\varepsilon_{i,t}^2]}E[\varepsilon_{m,t}^2]} \tag{4}$$

$$CKT_i = \frac{E[\varepsilon_{i,t}\varepsilon_{m,t}^3]}{\sqrt{E[\varepsilon_{i,t}^2]}E[\varepsilon_{m,t}^3]}. \tag{5}$$

It denotes that $\varepsilon_{i,t}$ is the residual part, which is extracted from the CAPM regression and $\varepsilon_{m,t}$ is the deviation between market return in month t from the benchmark of average value over the period from $t-1$ to t under specific research time.

4   Data and Methodology

4.1   Data

We collect all the secondary data from 274 non-financial firms on Hochiminh Stock Exchange (HoSE) from Vietnam Securities Depository over the course of 10 years for which the following information is available. Firstly, the official daily closing price adjusted for the Friday transaction date with the 102,305 observations for 274 shares is collected. The Vietnamese stock market index (VN-Index) is used in applying logarithm return calculation by Miller [29]. Secondly, the number of common shares, book value and other indicators are collected from each audited financial statement firm. Thirdly, we also take into account the securities event schedules. Finally, the Vietnamese interbank interest rate is collected as the risk-free rate as the method by Kostakis et al. [25].

Noticeably, our dataset consists of all Vietnam listed and de-listed common shares available in Vietnam Securities Depository and we reconcile with the dataset from Thomson Reuters in 10 years. Because our data included both listed and dead firms, we establish the requirements of removing biased figures. We excluded firms with financially oriented traits such as unit trusts or investment trusts, non-60-consecutive monthly transactions to ensure the reasonable calculation for co-skewness and co-kurtosis. In Vietnam, there is no firm, which issues Global Depository Receipts (GDRs) as well as American Depository Receipts (ADRs). This is followed by conventional method of collecting data by Florackis et al. [14] and Fletcher and Kihanda [13]. Therefore, our sample covers 86.16 % of number of stocks trading in the Hochiminh stock exchange with the 62.28 % of market value amount. Interestingly, in emerging market particularly Vietnam, the banking system is occupied by the large market cap and has specific traits, which is different from manufacturing sectors. Hence, the authors excluded it from our samples.

4.2   Methodology

We employ the approach of Kraus and Litzenberger [26] to calculate the degree of
co-skewness and co-kurtosis for each common share, which is listed in Hochiminh
Stock Exchange at given month t as follows:

$$\text{Coskewness}_i = \frac{E[\{R_i - E(R_i)\}\{R_m - E(R_m)\}^2]}{\{R_m - E(R_m)\}^3} \qquad (6)$$

$$\text{Cokurtosis}_i = \frac{E[\{R_i - E(R_i)\}\{R_m - E(R_m)\}^3]}{\{R_m - E(R_m)\}^4}. \qquad (7)$$

After having necessary data to estimate the two aforementioned figures for each
share i during the period of t, we construct decile portfolios by ascending order
sorted by the value of two statistical factors. Hence, Portfolio 1 (P1) includes
the shares with the lowest standardized co-skewness and co-kurtosis value by the
above calculation methods, respectively, whereas the Portfolio 10 (P10) contains
the shares with the highest values, respectively. One observer has already drawn
attention to the paradox in portfolio classification. Hung et al. [19] and Pettengill
et al. [33] researches show that the method of sorting portfolios by size and value
as Fama and French [10] does not capture all the risks. Cremers et al. [7] arranges
these shares into 10×10 size and book-to-market value portfolios in two-step as
the same of Fama and French [10] cross-sectional data based on regression basis.
For robustness, we calculate both equally weighted and value-weighted portfolios
returns by quantile regression.

In this research, we estimate the full sets of variables employed from Fama
and French [10] with three factors such as the value-at-risk premium, co-skewness
premium, co-kurtosis premium

$$R_{i,t} - R_f^t = \alpha_i + \beta_{i,MKT}(R_{m,t} - R_f^t) + \beta_{i,SMB}SMB_t + \beta_{i,HML}HML_t$$
$$+ \beta_{i,HVARL}HVARL_t + \beta_{i,CSK}CSK_t + \beta_{i,CKT}CKT_t + c_{i,t} \qquad (8)$$

where $R_{i,t}$ is the return of portfolio in week t, R_f^t is the risk-free rate for week t
and the figure $(R_{m,t} - R_f^t)$ stands for the excess market portfolio return in week t.
We also include SMB_t and HML_t variables from Fama and French [10], which
means the size and value risk factors respectively. However, we focus on the
three risk factors including value-at-risk premium $(HVARL)$, co-skewness premium (CSK) and co-kurtosis premium (CKT), which influence stock returns. In
order to mitigate the errors-in-variable problems, we also use system-based estimation as well as the other factors such as size and value risk. Additionally, we
mainly employ quantile regression to correct heteroscedasticity, and its regression method will help fully consider the effect of outliers to observe the entire
distribution of return, not merely its conditional mean as OLS. The specific

quantile regression estimator for the chosen quantile q minimizes the objective function here

$$Q(\beta_q) = \sum_{i:return_i \geq risk\ factors_i'(\beta)}^{N} q|return_i - risk\ factors_i'(\beta)|$$

$$+ \sum_{i:return_i < risk\ factors_i'(\beta)}^{N} (1-q)|return_i - risk\ factors_i'(\beta)|.$$

We choose three quantile levels with 0.25; 0.5 and 0.75 for regression in accordance with the classification of stock returns into low, medium and high.

4.3   Test Hypotheses

Tests are conducted on the values of the estimated coefficient for each factors affecting stock returns. The following hypotheses are tested.

Hypothesis 1. *The risk factors including value-at-risk premium, co-skewness premium and co-kurtosis premium will earn higher profit when it comes to positive function of these associated elements; for example, $E(\beta_{i,HVARL})$, $E(\beta_{i,CSK})$ and $E(\beta_{i,CKT})$ are expected to be positive value and are significant at 1%, 5% or 10% level in statistical test.*

Hypothesis 2. *For each quantile, the effect of risk factors is different and it might capture non-linearity risk.*

Hypothesis 3. *The difference in each portfolio sorting by the level of co-skewness and co-kurtosis estimation.*

5   Data Description

The table below shows the statistical summary for the variables inputted into the models by quantile regression (Table 3).

6   Empirical Results

6.1   The Basis Analysis for Quantile Regression by Two Methods with Ten Portfolios

Based on our regression results illustrated in Table 4, market risk premium factor has significantly affected the required rate of returns on 10 co-skewness portfolios with statistical significance at 1% level, grouping three quantiles 0.25; 0.5; 0.75. In particular, the magnitude of market coefficients has ranged from 0.7794 to 0.9505 in ascending tri-quantile and match with CAPM theory, reflecting the investors' requirement for market risk. Besides, these above results appropriate with the features of quantile regression method. The higher quantiles

Table 3. Summary statistics

Variable	Rm-Rf	SMB	HML	CSK	CKT	HVARL
Observation	519	519	519	519	519	519
Mean	−0.00098	0.006707	0.010565	−0.07721	−0.53997	−0.01404
Std. Dev	0.041244	0.029154	0.040042	0.231968	1.805802	0.032624
Min	−0.17896	−0.09306	−0.1182	−1.21467	−11.6894	−0.19094
Max	0.16566	0.14859	0.19159	0.13774	0.2103	0.05249
Median	−0.00085	0.00086	0.00203	−0.00503	−0.00208	−0.00661
Skewness	−0.16867	1.322506	1.737616	−2.70894	−3.60968	−1.8593
Kurtosis	5.901233	6.589818	7.016634	8.898775	15.80937	7.224471

This table above shows the statistical description for independent variables including market risk premium, size risk premium, value risk premium, co-skewness premium, co-kurtosis premium and the maximal loss risk premium. This summarizes for the total 519 observations calculated by the weekly collected in the period from July 2006 to June 2016. By results presented, the authors concluded that the volatility of co-kurtosis premium is quite large, which implies that the contribution to stock returns when regressing in the proposed models. In addition, the median is concentrated on the approximate 'zero' figure, which demonstrates that the distribution is around this value. Most of variables are disclosed with the shape of lepto kurtosis whereas up to four variables over six are left-tail distribution with negative value received.

(Source: The authors)

regress; the bigger impact. Regarding the HVARL factor, it has also affected these co-skewness portfolios with statistical significance at 1% level, grouping tri-quantiles. The effect of HVARL factor is scattered and ranged from 0.2422 to 0.9751. These coefficients are all positive and reflect the impact in the same direction. In addition, the coefficients are irregular magnitude due to portfolios are sorted by co-skewness level. Regarding the HCSKL factor, there are 8 over the 30 beta coefficients which have statistical significance at 1% or more, grouping three quantiles. As mentioned, 10 co-skewness portfolios having the value of the CSK coefficient ranges from negative to positive. Therefore, P1 co-skewness portfolio has a significant influence, below −0.91 in tri-quantile, on declining required rate of returns on stocks with statistical significance at 1% level. For P10 co-skewness portfolio, this influence is positive, above 0.07 in two quantiles 0.5 and 0.75. This indicates that the co-skewness portfolio having negative CSK coefficient reflects the risk proposition. By contrast, the co-skewness portfolio having positive CSK coefficient reflects the amplitude to get abnormal return. As regards the HCKTL factor, because co-skewness portfolios are sorted by CSK levels, regression results of HCKTL factor are arranged randomly and reflects inaccurately. So, the effect and statistical significance of HVARL factor will be presented in the following part. However, according to Table 5, regression results indicate that market risk premium factor has affected on 10 co-kurtosis portfolios with statistical significance at 1% level.

Table 4. Performance of the decile co-skewness portfolios (by equally weighted) returns by high co-moments impact

$\tau = 0.25$					
CSK portfolios	P1	P2	P3	P4	P5
MRP	0.8176***	0.8310***	0.7794***	0.8838***	0.8389***
	[0.0472]	[0.0327]	[0.0237]	[0.0365]	[0.0274]
HCKTL	−0.005	−0.002*	−0.002 ∗ ∗	0.0009	−0.003**
	[0.0100]	[0.0014]	[0.0010]	[0.0016]	[0.0015]
HCSKL	−0.928***	−0.040**	−0.009	−0.049**	−0.000
	[0.0918]	[0.0158]	[0.0115]	[0.0237]	[0.0164]
HVARL	0.5481***	0.2748***	0.3717***	0.4050***	0.6603***
	[0.0828]	[0.0591]	[0.0669]	[0.1115]	[0.0719]
R-Square	0.7667	0.5947	0.5748	0.5265	0.4860
CSK portfolios	P6	P7	P8	P9	P10
MRP	0.8365***	0.8183***	0.8574***	0.8131***	0.8176***
	[0.0360]	[0.0468]	[0.0564]	[0.0421]	[0.0439]
HCKTL	−0.003	−0.004**	−0.004*	−0.004*	−0.005**
	[0.0022]	[0.0017]	[0.0021]	[0.0022]	[0.0025]
HCSKL	0.0138	0.0321**	−0.036	−0.030	0.0719**
	[0.0198]	[0.0159]	[0.0338]	[0.0282]	[0.0300]
HVARL	0.4539***	0.5224***	0.7968***	0.9751***	0.5481***
	[0.0903]	[0.0694]	[0.2014]	[0.1126]	[0.0995]
R-Square	0.5331	0.5158	0.5082	0.5109	0.5134
$\tau = 0.5$					
CSK portfolios	P1	P2	P3	P4	P5
MRP	0.8740***	0.8507***	0.8013***	0.8631***	0.8707***
	[0.0452]	[0.0260]	[0.0117]	[0.0349]	[0.0479]
HCKTL	−0.006***	−0.001	−0.002	−0.002*	−0.005***
	[0.0019]	[0.0010]	[0.0018]	[0.0014]	[0.0016]
HCSKL	−0.910***	−0.060***	−0.012	−0.057***	−0.013
	[0.0221]	[0.0121]	[0.0162]	[0.0199]	[0.0142]
HVARL	0.5263***	0.2830***	0.3993***	0.5202***	0.8105***
	[0.0930]	[0.0784]	[0.0640]	[0.0894]	[0.0824]
R-Square	0.8336	0.5733	0.5380	0.4816	0.4409
CSK portfolios	P6	P7	P8	P9	P10
MRP	0.8508***	0.8226***	0.8786***	0.8739***	0.8740***
	[0.0415]	[0.0263]	[0.0273]	[0.0402]	[0.0439]
HCKTL	−0.005***	−0.004**	−0.006***	−0.002	−0.006***
	[0.0014]	[0.0018]	[0.0016]	[0.0019]	[0.0015]
HCSKL	0.0036	0.0168	−0.013	−0.027	0.0892***
	[0.0155]	[0.0155]	[0.0190]	[0.0191]	[0.0212]
HVARL	0.4149***	0.5595***	0.7851***	0.7815***	0.5262***
	[0.0634]	[0.0849]	[0.0846]	[0.1203]	[0.1236]
R-Square	0.4781	0.4735	0.4795	0.4889	0.4796

(continued)

Table 4. (*continued*)

$\tau = 0.75$					
CSK portfolios	P1	P2	P3	P4	P5
MRP	0.9296***	0.9043***	0.8589***	0.8475***	0.9158***
	[0.0496]	[0.0244]	[0.0278]	[0.0610]	[0.0622]
HCKTL	−0.005**	−0.001**	−0.004**	−0.003*	−0.006**
	[0.0021]	[0.0004]	[0.0018]	[0.0020]	[0.0025]
HCSKL	−0.913***	−0.061***	0.0106	−0.044**	0.0007
	[0.0154]	[0.0233]	[0.0236]	[0.0184]	[0.0199]
HVARL	0.4363***	0.2422***	0.3596***	0.4437***	0.6670***
	[0.1383]	[0.0840]	[0.0853]	[0.1085]	[0.1346]
R-Square	0.8932	0.5883	0.5274	0.4695	0.4231
CSK portfolios	P6	P7	P8	P9	P10
MRP	0.8356***	0.7957***	0.9505***	0.8825***	0.9295***
	[0.0592]	[0.0554]	[0.0391]	[0.0395]	[0.0387]
HCKTL	−0.005***	−0.005**	−0.006***	−0.003	−0.005***
	[0.0015]	[0.0023]	[0.0020]	[0.0024]	[0.0016]
HCSKL	−0.014	−0.008	−0.008	−0.022	0.0862***
	[0.0216]	[0.0207]	[0.0276]	[0.0175]	[0.0233]
HVARL	0.4264***	0.7060***	0.6577***	0.8010***	0.4362***
	[0.0938]	[0.1047]	[0.1326]	[0.0855]	[0.1279]
R-Square	0.4729	0.4586	0.4843	0.5040	0.4561

This table reports the three-quantile regression results including $\tau = 0.25, 0.5, 0.75$ for the Model (8) on EW CSK portfolios (equally weighted co-skewness returns portfolios). The above table is conducted 102,305 observations (compatible with weekly collected data) of 274 non-financial corporates listed on HoSE in the period from July 2006 to June 2016. These portfolios are sorted in ascending order according to their co-skewness (CSK) value estimated via the period from July 2006 to June 2016 and they are assigned to 10 portfolios. In which, P1 is the lowest CSK portfolio and P10 is the highest CSK portfolio. Portfolios is maintained in the research process to measure effects of co-skewness on stock returns. The risk factors are estimated by quantile regression on excess portfolio returns and risk free rate on the corresponding market, size, and value, value at risk and co-skewness as well as co-kurtosis risk factors. It is noted that (*), (**), (***) reflected statistically significant of the corresponding coefficients at 1%, 5% and 10% level. Standard errors of the corresponding coefficients are reflected in square brackets. The last row reports R2 of the model including 4 control variables (MRP-market risk premium, HVARL, HCSK, HCSKT) and 2 non-control variables (SMB, HML).

In particular, the magnitude of market coefficients has larger amplitude than 10 co-skewness portfolios, and ranged from 0.4863 to 1.1144. The coefficients of HVARL factor have statistical significance at 1% level or more in 10 co-kurtosis

Table 5. Performance of the decile co-kurtosis portfolios (by equally weighted) returns by high co-moments impact

$\tau = 0.25$					
CSK portfolios	P1	P2	P3	P4	P5
MRP	1.1000***	0.4863***	0.6295***	0.6427***	0.7508***
	[0.0636]	[0.0684]	[0.0376]	[0.0197]	[0.0331]
HCKTL	−1.005***	−0.004	−0.000	−0.002	−0.003
	[0.0019]	[0.0029]	[0.0016]	[0.0023]	[0.0028]
HCSKL	−0.129***	−0.012	0.0301*	0.0166	0.0301
	[0.0339]	[0.0249]	[0.0171]	[0.0284]	[0.0261]
HVARL	0.9170***	0.4984***	0.3088***	0.4487***	0.3747***
	[0.1489]	[0.1161]	[0.0882]	[0.0713]	[0.1268]
R-Square	0.9500	0.2680	0.4066	0.4614	0.5163
CSK portfolios	P6	P7	P8	P9	P10
MRP	0.8524***	0.8831***	0.8681***	0.9732***	1.0918***
	[0.0457]	[0.0276]	[0.0283]	[0.0340]	[0.0627]
HCKTL	−0.004***	−0.003	−0.004*	−0.002***	−0.005*
	[0.0014]	[0.0025]	[0.0026]	[0.0009]	[0.0026]
HCSKL	0.0245	0.0192	−0.043*	−0.065***	−0.125***
	[0.0167]	[0.0199]	[0.0219]	[0.0099]	[0.0320]
HVARL	0.5233***	0.5029***	0.8661***	0.6333***	0.9158***
	[0.0964]	[0.0952]	[0.0814]	[0.0818]	[0.1520]
R-Square	0.5623	0.5650	0.5896	0.6243	0.5796
$\tau = 0.5$					
CSK portfolios	P1	P2	P3	P4	P5
MRP	1.1144***	0.5148***	0.5818***	0.6544***	0.7850***
	[0.0408]	[0.0354]	[0.0461]	[0.0343]	[0.0456]
HCKTL	−1.006***	−0.002	−0.002*	−0.002*	−0.002
	[0.0019]	[0.0020]	[0.0015]	[0.0017]	[0.0021]
HCSKL	−0.096***	−0.003	0.0390***	−0.001	0.0337*
	[0.0222]	[0.0129]	[0.0127]	[0.0176]	[0.0201]
HVARL	0.7775***	0.4006***	0.2201***	0.5064***	0.3145***
	[0.0915]	[0.0747]	[0.0726]	[0.0842]	[0.1014]
R-Square	0.9675	0.2524	0.3358	0.4124	0.4650
CSK portfolios	P6	P7	P8	P9	P10
MRP	0.8544***	0.8750***	0.9457***	0.9967***	1.1133***
	[0.0402]	[0.0203]	[0.0369]	[0.0267]	[0.0345]
HCKTL	−0.006***	−0.004**	−0.005***	−0.002**	−0.006***
	[0.0014]	[0.0019]	[0.0019]	[0.0013]	[0.0016]
HCSKL	0.0240	0.0200	−0.033**	−0.063***	−0.098***
	[0.0223]	[0.0166]	[0.0165]	[0.0113]	[0.0147]
HVARL	0.5776***	0.5065***	0.7025***	0.5515***	0.7891***
	[0.0716]	[0.0788]	[0.1064]	[0.0827]	[0.0500]
R-Square	0.5174	0.5174	0.5453	0.6023	0.5536

(*continued*)

Table 5. (*continued*)

$\tau = 0.75$					
CSK portfolios	P1	P2	P3	P4	P5
MRP	1.0919***	0.5112***	0.6345***	0.6974***	0.7929***
	[0.0569]	[0.0441]	[0.0423]	[0.0325]	[0.0431]
HCKTL	−1.003***	−0.002	−0.004***	−0.004***	−0.003
	[0.0017]	[0.0026]	[0.0017]	[0.0015]	[0.0022]
HCSKL	−0.107***	−0.001	0.0634**	0.0195	0.0385*
	[0.0179]	[0.0210]	[0.0260]	[0.0133]	[0.0225]
HVARL	0.7354***	0.3621***	0.0863	0.4589***	0.3621***
	[0.0886]	[0.1169]	[0.1024]	[0.1045]	[0.0715]
R-Square	0.9811	0.2262	0.3348	0.4167	0.4584
CSK portfolios	P6	P7	P8	P9	P10
MRP	0.8273***	0.9146***	0.9526***	1.0617***	1.0884***
	[0.0603]	[0.0400]	[0.0297]	[0.0230]	[0.0530]
HCKTL	−0.004*	−0.003**	−0.009***	−0.001	−0.003***
	[0.0025]	[0.0017]	[0.0023]	[0.0015]	[0.0012]
HCSKL	0.0137	0.0349	−0.029	−0.061***	−0.104***
	[0.0276]	[0.0234]	[0.0192]	[0.0174]	[0.0210]
HVARL	0.5937***	0.3891***	0.7915***	0.4909***	0.7196***
	[0.0955]	[0.1341]	[0.0841]	[0.0804]	[0.1200]
R-Square	0.5118	0.4924	0.5359	0.6039	0.5480

This table reports the three-quantile regression results including $\tau = 0.25, 0.5, 0.75$ for the Model (8) on EW CKT portfolios (equally weighted returns portfolios). The above table is conducted 102,305 observations (compatible with weekly collected data) of 274 non-financial corporates listed on HoSE in the period from July 2006 to June 2016. These portfolios are sorted in ascending order according to their co-kurtosis (CKT) value estimated via the period from July 2006 to June 2016 and they are assigned to 10 portfolios. In which, P1 is the lowest CKT portfolio and P10 is the highest CKT portfolio. Portfolios are maintained in the research process to measure effects of co-kurtosis on stock returns. The risk factors are estimated by quantile regression on excess portfolio returns and risk free rate on the corresponding market, size, and value, value at risk and co-skewness as well as co-kurtosis risk factors. (*), (**), (***) reflected statistically significant of the corresponding coefficients at 1%, 5% and 10% level. Standard errors of the corresponding coefficients are reflected in square brackets. The last row reports R2 of the model including 4 control variables (MRP-market risk premium, HVARL, HCSK, HCSKT) and 2 non-control variables (SMB, HML).

portfolios excluding Portfolio 3 (has non-statistical significance at quantile 0.75). The magnitude of these coefficients are ranged from 0.2201 to 0.917. The mean of both coefficient factors are above indicated. In terms of the HCKTL factor, there are 12 over the 30 coefficients which have statistical significance at 1% level in tri-quantiles. Especially, the portfolios are sorted by co-kurtosis level form low to high. Therefore, the portfolio P1 has significantly affected on investors' rate of return, approximately lower than -1 with statistical significance at 1% level in tri-quantiles. Regarding HCSKL factor, because the portfolios are sorted by co-kurtosis level, regression results of HCSKL factor are irregularly distributed and reflect inaccurately. Therefore, the magnitude as well as the statistical significance of co-skewness factor is accurately expressed in the aforementioned content. Interestingly, in terms of value weighted approach, market risk premium factor of 10 CSK portfolio has lower coefficients than equally weighted approach (between 0.1184 and 0.7149) and has statistical significance at 1% in tri-quantiles. This indicates that portfolios' return are modified and suit to value weighted approach. Therefore, there are no impacts on outlier variables, and this decreases the effect of market risk premium factor on portfolios' return. There are 16 over 30 portfolios which have statistical significance at 1% level. To be more detailed, HVARL factor affects in the same as well as reveres direct on portfolios' return. Regarding lower Co-skewness portfolios (from P1 to P3), the risk premium of maximal loss in investment capital has opposite side effect on portfolios' return due to negative Co-skewness. It results to higher Value-at-Risk than positive Co-skewness portfolios. Therefore, HVARL factor by value weighted approach accurately affects than equally weighted approach. In terms of HCSKL factor, just only portfolio 10 has statistical significance at 1% level in quantile 0.25. This indicates that HCSKL factor has no significance level in the context of reflecting on the risk of maximal loss in investment capital. In comparison with equally weighted approach, this methodology demonstrates the impact of HCSKL factor has lower significance level based on result Table 6.

Lastly, it is worth mentioning that value weighted co-kurtosis portfolios, market risk premium and HVARL factors have same impact on require rate of return of 10 CKT portfolios which is mentioned in the pervious part. In terms of HCKTL, the factor has statistical significance at 1% level in Portfolio 1 and 10 in quantile 0.75. This reflects that using value weighted approach has inaccurately effect on investors' return to be compared with equally weighted approach. *In conclusion*, as the first hypothesis, the risk elements such as value-at-risk premium, co-skewness premium and co-kurtosis have statistical significance at positive value to stock return at 1%, 5% and 10%. This means that these risk factors can clearly explain the fluctuation in the return of different stocks in the Vietnam securities market with different quantiles.

6.2   Risk Factors are Captured as Non-linearity Shape

Under the risk factors' effect, four tables below indicate that the shape of risk factors which are captured by quantile regression is non-linearity. Interestingly, most market risk premiums are considered as linear line in relative relationship to stock return in accordance with many previous researches such as Fang and Lai [12], Barone Adesi et al. [3], Harvey and Siddique [18] and Kostakis et al. [25]. However, when it comes to heteroscedasticity fixing and the error in OLS and the other regression, the results show that most risk factors exists one broken

Table 6. Performance of the decile co-skewness portfolios (by valued-weighted) returns by high co-moments impact

$\tau = 0.25$					
CSK portfolios	P1	P2	P3	P4	P5
MRP	0.6481***	0.6001***	0.1978***	0.6759***	0.2942***
	[0.0599]	[0.0379]	[0.0119]	[0.0690]	[0.0380]
HCKTL	0.0026	−0.001	−0.000	−0.000	−0.002
	[0.0023]	[0.0025]	[0.0005]	[0.0023]	[0.0024]
HCSKL	0.0043	0.0172	−0.002	−0.045*	−0.013
	[0.0221]	[0.0218]	[0.0080]	[0.0267]	[0.0197]
HVARL	−0.243**	−0.039	−0.015	0.2989**	0.2734***
	[0.1018]	[0.1300]	[0.0267]	[0.1373]	[0.0982]
R-Square	0.3611	0.3487	0.1835	0.3845	0.2583
CSK portfolios	P6	P7	P8	P9	P10
MRP	0.6004***	0.5157***	0.4235***	0.2424***	0.1184***
	[0.0293]	[0.0321]	[0.0343]	[0.0259]	[0.0121]
HCKTL	−0.000	−0.004**	−0.003**	−0.002*	−0.001***
	[0.0020]	[0.0020]	[0.0018]	[0.0012]	[0.0005]
HCSKL	0.0026	−0.026	−0.026	−0.024**	0.0102**
	[0.0271]	[0.0251]	[0.0340]	[0.0123]	[0.0047]
HVARL	0.1426	0.4767***	0.5228***	0.3093***	0.0875***
	[0.0884]	[0.0896]	[0.1702]	[0.0881]	[0.0281]
R-Square	0.4468	0.3466	0.3499	0.2226	0.1514
$\tau = 0.5$					
CSK portfolios	P1	P2	P3	P4	P5
MRP	0.6601***	0.6131***	0.2070***	0.6905***	0.2774***
	[0.0525]	[0.0389]	[0.0119]	[0.0332]	[0.0285]
HCKTL	0.0040**	−0.000	−0.000	−0.000	−0.001
	[0.0019]	[0.0020]	[0.0004]	[0.0014]	[0.0011]
HCSKL	0.0035	0.0188	−0.003	−0.031**	−0.001
	[0.0172]	[0.0202]	[0.0048]	[0.0124]	[0.0102]
HVARL	−0.355***	−0.105	0.0177	0.1325***	0.1508***
	[0.0769]	[0.1356]	[0.0428]	[0.0459]	[0.0469]
R-Square	0.3355	0.3306	0.1866	0.3727	0.2210
CSK portfolios	P6	P7	P8	P9	P10
MRP	0.6269***	0.5006***	0.4339***	0.2486***	0.1195***
	[0.0271]	[0.0442]	[0.0314]	[0.0190]	[0.0083]
HCKTL	−0.001	−0.003***	−0.004**	−0.002***	−0.000
	[0.0009]	[0.0013]	[0.0016]	[0.0005]	[0.0004]
HCSKL	−0.002	−0.028	−0.024	−0.018**	0.0120***
	[0.0151]	[0.0217]	[0.0148]	[0.0087]	[0.0029]
HVARL	0.1597**	0.4623***	0.4501***	0.3341***	0.0356**
	[0.0773]	[0.0629]	[0.0736]	[0.0567]	[0.0154]
R-Square	0.4078	0.2897	0.3225	0.2166	0.1560

(continued)

Table 6. (*continued*)

τ = 0.75					
CSK portfolios	P1	P2	P3	P4	P5
MRP	0.7149***	0.6214***	0.2408***	0.6972***	0.3121***
	[0.0724]	[0.0286]	[0.0185]	[0.0312]	[0.0326]
HCKTL	0.0042	−0.001	−0.001***	−0.001	−0.000
	[0.0027]	[0.0011]	[0.0002]	[0.0013]	[0.0018]
HCSKL	0.0206	0.0210	0.0120	−0.026**	−0.027*
	[0.0251]	[0.0181]	[0.0086]	[0.0123]	[0.0139]
HVARL	−0.517***	−0.148	−0.048	0.1391	0.1487**
	[0.0848]	[0.1308]	[0.0423]	[0.0963]	[0.0719]
R-Square	0.3544	0.3657	0.1776	0.3837	0.2269
CSK portfolios	P6	P7	P8	P9	P10
MRP	0.6420***	0.5452***	0.4433***	0.2397***	0.1284***
	[0.0442]	[0.0379]	[0.0275]	[0.0415]	[0.0056]
HCKTL	−0.000	−0.002	−0.002**	−0.001	−0.000***
	[0.0012]	[0.0020]	[0.0010]	[0.0015]	[0.0003]
HCSKL	−0.008	−0.033*	−0.059***	−0.038**	0.0080**
	[0.0153]	[0.0173]	[0.0152]	[0.0184]	[0.0038]
HVARL	0.1335	0.5199***	0.6162***	0.3805***	0.1177***
	[0.0838]	[0.0882]	[0.0626]	[0.1015]	[0.0330]
R-Square	0.4129	0.3151	0.3430	0.2127	0.1377

This table reports the three-quantile regression results including $\tau = 0.25, 0.5, 0.75$ for the Model (8) on VW CSK portfolios (value-weighted returns portfolios). The above table is conducted 102,305 observations (compatible with weekly collected data) of 274 non-financial corporates listed on HoSE in the period from July 2006 to June 2016. These portfolios are sorted in ascending order according to their co-skewness (CSK) value estimated via the period from July 2006 to June 2016 and they are assigned to 10 portfolios. In which, P1 is the lowest CSK portfolio and P10 is the highest CSK portfolio. Portfolios is maintained in the research process to measure effects of co-skewness on stock returns. The risk factors are estimated by quantile regression on excess portfolio returns and risk free rate on the corresponding market, size, and value, value at risk and co-skewness as well as co-kurtosis risk factors. (*), (**), (***) reflected statistically significant of the corresponding coefficients at 1%, 5% and 10% level. Standard errors of the corresponding coefficients are reflected in square brackets. The last row reports R2 of the model including 4 control variables (MRP-market risk premium, HVARL, HCSK, HCSKT) and 2 non-control variables (SMB, HML).

point. From the illustration perspectives, the majority of these factor risk line is asymptotic with each specific threshold. Hence, the investors could consider how much these elements contribute to the stock return and the maximal value in each quantile.

Table 7. Performance of the decile co-kurtosis portfolios (by valued-weighted) returns by high co-moments impact

$\tau = 0.25$					
CSK portfolios	P1	P2	P3	P4	P5
MRP	0.0117	0.0585***	0.1338***	0.4022***	0.7349***
	[0.0085]	[0.0183]	[0.0204]	[0.0421]	[0.0316]
HCKTL	−0.000	−0.001*	−0.000	−0.002*	−0.000
	[0.0010]	[0.0009]	[0.0008]	[0.0012]	[0.0022]
HCSKL	−0.064***	0.0050	−0.001	−0.025*	0.0333**
	[0.0103]	[0.0087]	[0.0103]	[0.0146]	[0.0145]
HVARL	0.0517***	0.0699	0.1108*	0.3573***	−0.270***
	[0.0172]	[0.0577]	[0.0588]	[0.0811]	[0.0831]
R-Square	0.2480	0.0817	0.0957	0.2934	0.4180
CSK portfolios	P6	P7	P8	P9	P10
MRP	0.3816***	0.2933***	0.6352***	0.4258***	0.7932***
	[0.0155]	[0.0383]	[0.0300]	[0.0169]	[0.0344]
HCKTL	−0.001	−0.003	−0.002*	−0.000	−0.003
	[0.0013]	[0.0023]	[0.0016]	[0.0006]	[0.0025]
HCSKL	0.0167	0.0418**	−0.050***	−0.002	−0.067**
	[0.0145]	[0.0200]	[0.0188]	[0.0106]	[0.0282]
HVARL	0.1042*	−0.054	0.5022***	0.0030	0.6110***
	[0.0554]	[0.0955]	[0.0974]	[0.0625]	[0.1055]
R-Square	0.3892	0.1285	0.4911	0.3792	0.4862
$\tau = 0.5$					
CSK portfolios	P1	P2	P3	P4	P5
MRP	0.0182***	0.0442***	0.1146***	0.4217***	0.7473***
	[0.0041]	[0.0083]	[0.0222]	[0.0362]	[0.0356]
HCKTL	0.0001	0.0002	0.0001	−0.003**	−0.000
	[0.0003]	[0.0006]	[0.0007]	[0.0016]	[0.0019]
HCSKL	−0.068***	−0.007	0.0012	−0.007	0.0337***
	[0.0033]	[0.0055]	[0.0052]	[0.0109]	[0.0124]
HVARL	0.0374**	0.0483**	0.0157	0.2202***	−0.288***
	[0.0166]	[0.0204]	[0.0343]	[0.0256]	[0.0800]
R-Square	0.3643	0.0658	0.0854	0.2547	0.3991
CSK portfolios	P6	P7	P8	P9	P10
MRP	0.3795***	0.2015***	0.6381***	0.4326***	0.7847***
	[0.0246]	[0.0311]	[0.0234]	[0.0144]	[0.0409]
HCKTL	−0.000	0.0000	−0.003***	0.0000	−0.004**
	[0.0010]	[0.0016]	[0.0010]	[0.0004]	[0.0020]
HCSKL	−0.004	0.0117	−0.038**	0.0015	−0.078***
	[0.0127]	[0.0146]	[0.0193]	[0.0062]	[0.0153]
HVARL	0.1097***	−0.023	0.3698***	−0.045	0.4988***
	[0.0415]	[0.0400]	[0.0851]	[0.0445]	[0.0937]
R-Square	0.3433	0.1078	0.4656	0.3730	0.4640

(*continued*)

Table 7. (*continued*)

τ = 0.75					
CSK portfolios	P1	P2	P3	P4	P5
MRP	0.0516***	0.0659***	0.1877***	0.4345***	0.7179***
	[0.0154]	[0.0152]	[0.0277]	[0.0443]	[0.0427]
HCKTL	0.0014***	0.0012**	−0.000	−0.002	−0.001
	[0.0004]	[0.0005]	[0.0017]	[0.0022]	[0.0022]
HCSKL	−0.080***	−0.016	0.0173	−0.046**	0.0274
	[0.0090]	[0.0112]	[0.0198]	[0.0207]	[0.0173]
HVARL	0.0839	0.0865*	0.0111	0.3531***	−0.277***
	[0.0661]	[0.0501]	[0.0992]	[0.0872]	[0.0919]
RSquare	0.4762	0.0896	0.0996	0.2532	0.4135
CSK portfolios	P6	P7	P8	P9	P10
MRP	0.3831***	0.3242***	0.6728***	0.4849***	0.7807***
	[0.0225]	[0.0389]	[0.0311]	[0.0253]	[0.0511]
HCKTL	−0.000	0.0082***	−0.003	−0.000	−0.005***
	[0.0014]	[0.0021]	[0.0020]	[0.0005]	[0.0017]
HCSKL	−0.002	−0.046	−0.058***	0.0167	−0.109***
	[0.0128]	[0.0308]	[0.0138]	[0.0107]	[0.0262]
HVARL	0.0830***	−0.098	0.3701***	−0.055	0.6327***
	[0.0305]	[0.1150]	[0.0842]	[0.0620]	[0.1211]
RSquare	0.3298	0.1329	0.4769	0.3591	0.4747

This table reports the three-quantile regression results including $τ =$ 0.25, 0.5, 0.75 for the Model (8) on VW CKT portfolios (value-weighted returns portfolios). The above table is conducted 102,305 observations (compatible with weekly collected data) of 274 non-financial corporates listed on HOSE in the period from July 2006 to June 2016. These portfolios are sorted in ascending order according to their co-kurtosis (CKT) value estimated via the period from July 2006 to June 2016 and they are assigned to 10 portfolios. In which, P1 is the lowest CKT portfolio and P10 is the highest CKT portfolio. Portfolios is maintained in the research process to measure effects of co-kurtosis on stock returns. The risk factors are estimated by quantile regression on excess portfolio returns and risk free rate on the corresponding market, size, and value, value at risk and co-skewness as well as co-kurtosis risk factors. (*), (**), (***) reflected statistically significant of the corresponding coefficients at 1%, 5% and 10% level. Standard errors of the corresponding coefficients are reflected in square brackets. The last row reports R2 of the model including 4 control variables (MRP-market risk premium, HVARL, HCSK, HCSKT) and 2 non-control variables (SMB, HML).

Hence, our study emphasizes that risk factors are captured as non-linearity shape under the different quantiles. As regards the second hypothesis, this paper examines both figures which are extracted from the estimated equation and the simulation graphs to conclude. The magnitude by risk factors' effect is different and it is definitely non-linearity even in terms of the confidence intervals around each coefficient. Therefore, the quantile regression are approached by bootstrap standard error produced for reliable graphs herein.

6.3   The Distinguished Features in Different Portfolios from Influence from Risk Factors

From Tables 4, 5, 6 and 7 above, apparently, when sorting out by the co-skewness method, the authors conclude that portfolios being under-zero value (or left-tail distribution) are effected by negatively co-skewness risk premium with significance level at 1%. In contrast, the portfolios with upper-zero receive the positive influence by the same risk factors with significance level at 1%. This is completely appropriate for the fundamental theory in statistics of risk management.

Furthermore, it emphasizes that the portfolios' return constructed by co-kurtosis benchmark are randomly reflected by risk factors. In terms of equally weighted, the lowest co-kurtosis portfolios have the negative impact with the large magnitude, which is similar to the risk factors (HCKTL). Meanwhile, the larger portfolios sorted by co-kurtosis reach at the zero impact by risk elements. When it comes to value weighted, the ones which range in low co-kurtosis, have the same-side effect on stock return by HCKTL. Noticeably, the distribution is recorded by random spread. It is quite suitable with the theorem by statistics by risk management. It could be explained that lepto kurtosis (fat tails) generates more risk in comparison with platy kurtosis. However, these concepts do not conclude for two-tail to exactly know the losses in which tail. *To conclude the third hypothesis*, the authors see that different portfolios with providing the different impact on stock return as proposed.

7   Conclusion, Implications and Limitation

This research contributes quantile regression, a new methodology and empirical evidence in stock return explanation in Vietnam emerging stock market, to a variety of risk models. In fact, there is a lack of previous studies in using the higher co-moments (with the movement by markets) to explain the effect on stock returns. In addition, the authors also sort out by the clear criteria (co-skewness and co-kurtosis benchmark) as well as various method (equally weighted and value weighted) with definite numbers (10) to consider the level of impact. Taking into consideration methodology approach, the authors decide to choose quantile regression to avoid the heteroscedasticity and the other outliers' impact. This is also a new contribution to this field research.

As regards implications, the authors suggest investors to construct the portfolios with higher co-skewness level (positive) and lower co-kurtosis one. This might increase stock returns and limit the risk, which could be incurred by investors. The authors also test the other determinants such as the size, value and maximal value of losses, strongly impacting on stock return beside co-momentum effects.

However, this research face some limitations, which could be improved in further studies. First, it is only focused on one market, particularly country research. Hence, the following paper could expand the quantity of research nations. Second, Vietnam is early stock market with only up to 10 years; therefore, the observation is quite limited.The authors recommend to choose the appropriate

Table 8. Non-linearity shape of the decile co-skewness portfolios (by equally weighted) returns by high co-moments impact

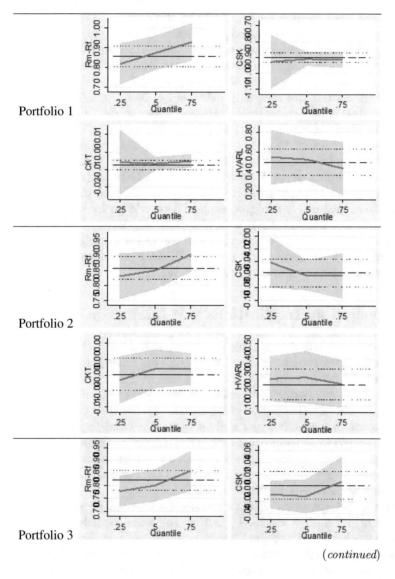

(*continued*)

Table 8. (*continued*)

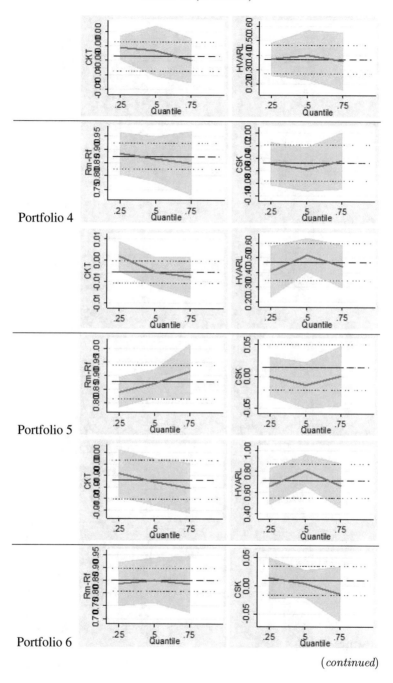

Portfolio 4

Portfolio 5

Portfolio 6

(*continued*)

Table 8. (*continued*)

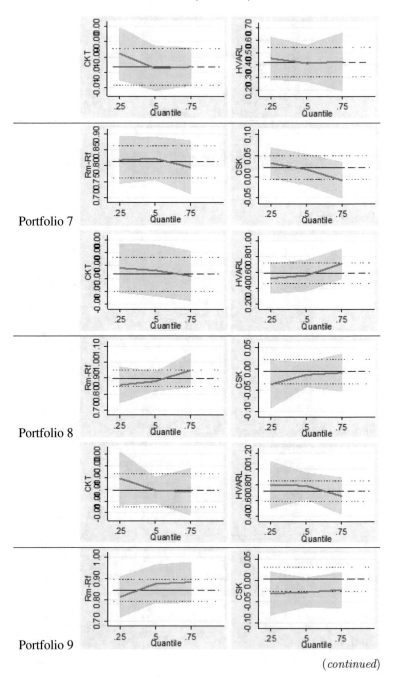

Portfolio 7

Portfolio 8

Portfolio 9

(*continued*)

Table 8. (*continued*)

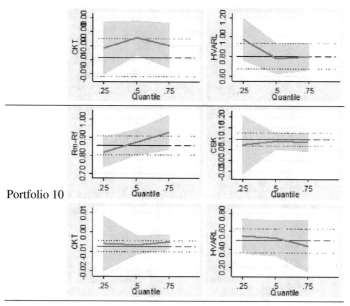

Portfolio 10

(*Source: The authors*)

Table 9. Non-linearity shape of the decile co-kurtosis portfolios (by equally weighted) returns by high co-moments impact

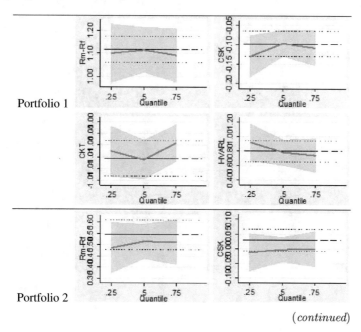

Portfolio 1

Portfolio 2

(*continued*)

Table 9. (*continued*)

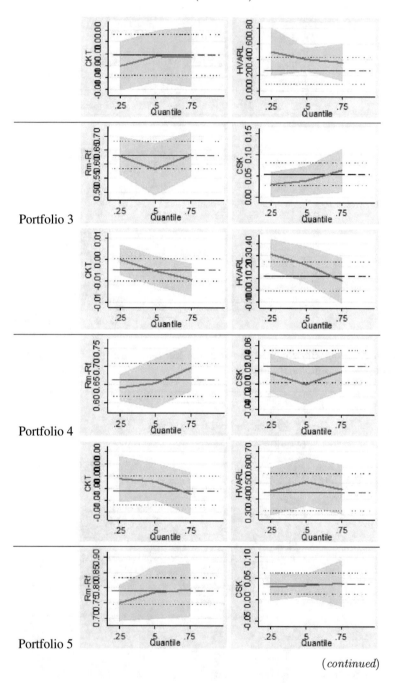

Portfolio 3

Portfolio 4

Portfolio 5

(*continued*)

Table 9. (*continued*)

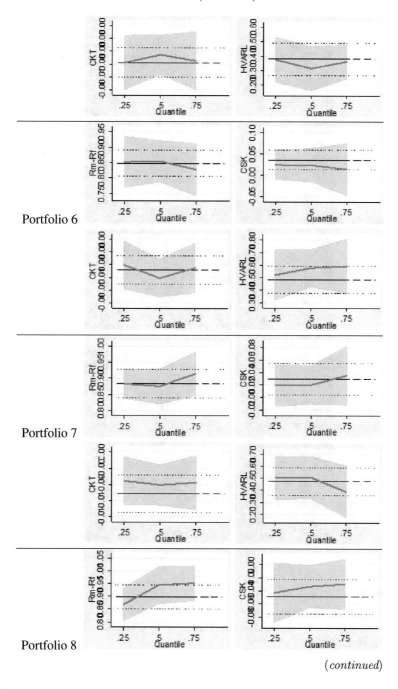

Portfolio 6

Portfolio 7

Portfolio 8

(*continued*)

Table 9. (*continued*)

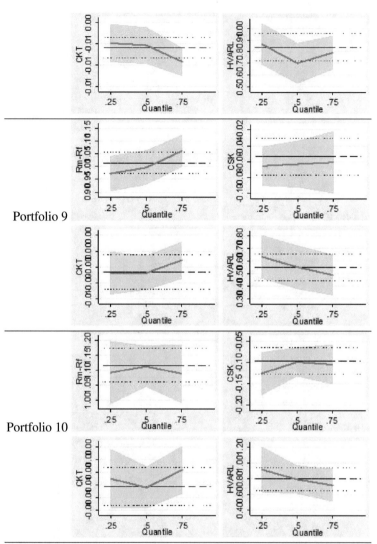

Portfolio 9

Portfolio 10

(*Source: The authors*)

markets with efficiency time-span. *Last*, this paper could be better when comparing with the other same feature markets to estimate the different impact (Tables 8, 9, 10 and 11).

Table 10. Non-linearity shape of the decile co-skewness portfolios (by valued-weighted) returns by high co-moments impact

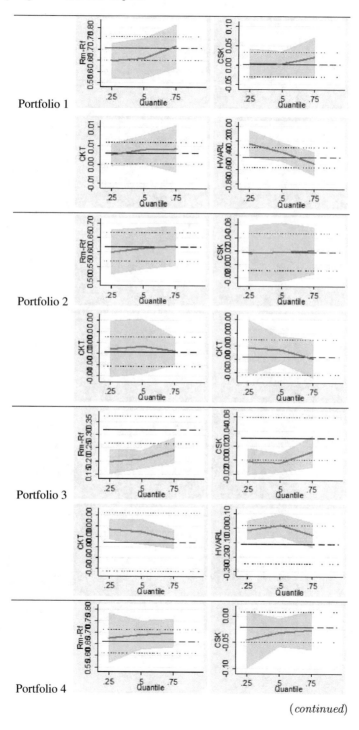

(*continued*)

Table 10. (*continued*)

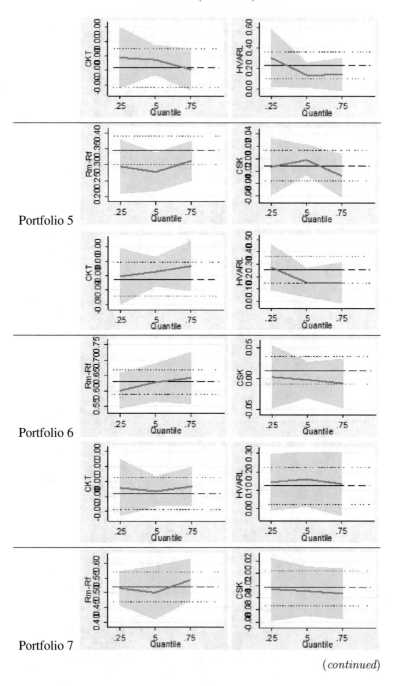

(*continued*)

Table 10. (*continued*)

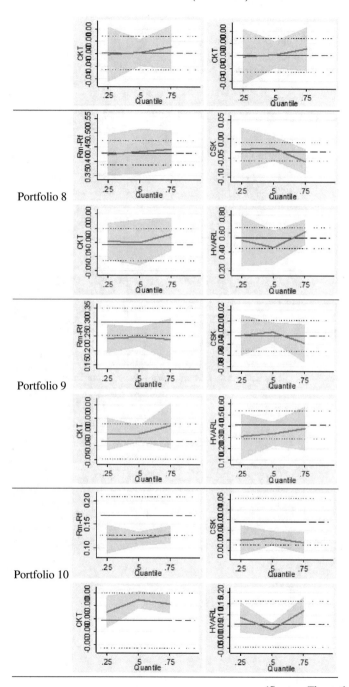

(*Source: The authors*)

Table 11. Non-linearity shape of the decile co-kurtosis portfolios (by valued-weighted) returns by high co-moments impact

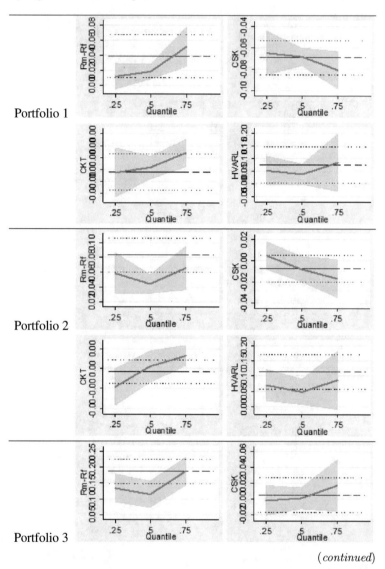

Portfolio 1

Portfolio 2

Portfolio 3

(*continued*)

Table 11. (*continued*)

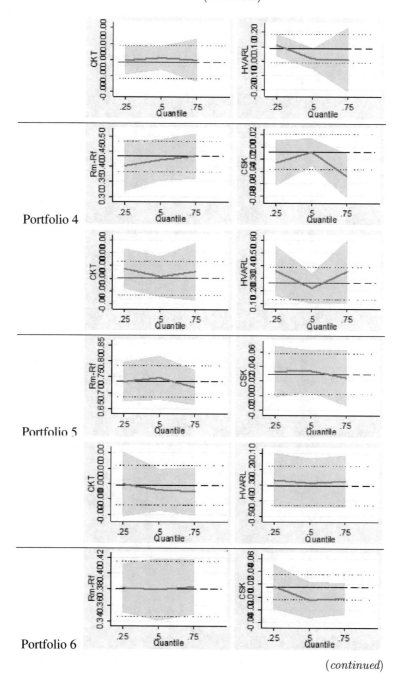

(*continued*)

Table 11. (*continued*)

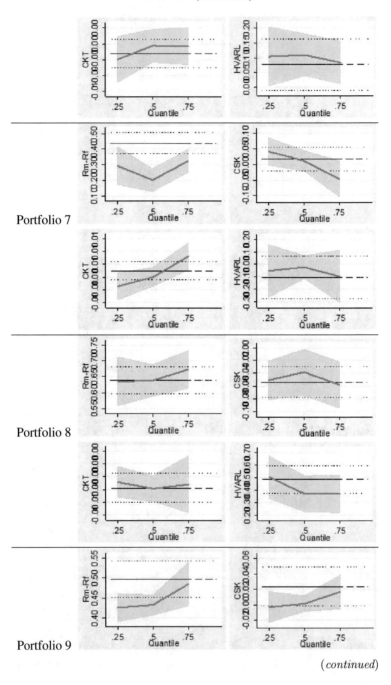

(*continued*)

Table 11. (*continued*)

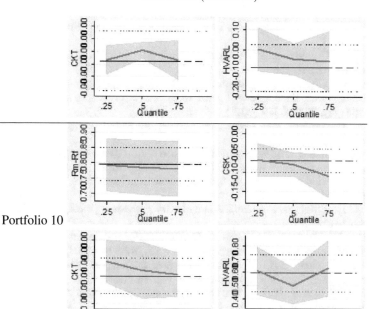

Portfolio 10

(*Source: The authors*)

References

1. Bali, T.G., Cakici, N.: Value at risk and expected stock returns. Financ. Anal. J. **60**(2), 57–73 (2004)
2. Barone Adesi, G., Gagliardini, P., Urga, G.: Homogeneity hypothesis in the context of asset pricing models: the quadratic market model (2000)
3. Barone Adesi, G., Gagliardini, P., Urga, G.: Testing asset pricing models with coskewness. J. Bus. Econ. Stat. **22**(4), 474–485 (2004)
4. Boyer, B., Mitton, T., Vorkink, K.: Expected idiosyncratic skewness. Rev. Financ. Stud. **23**(1), 169–202 (2009)
5. Chabi-Yo, F., Ruenzi, S., Weigert, F.: Crash sensitivity and the cross-section of expected stock returns (2015)
6. Conrad, J., Dittmar, R.F., Ghysels, E.: Ex ante skewness and expected stock returns. J. Financ. **68**(1), 85–124 (2013)
7. Cremers, M., Petajisto, A., Zitzewitz, E.: Should benchmark indices have alpha? Revisiting performance evaluation (2010)
8. Ding, B., Shawky, H.A.: The performance of hedge fund strategies and the asymmetry of return distributions. Eur. Financ. Manag. **13**(2), 309–331 (2007)
9. Dittmar, R.F.: Nonlinear pricing kernels, kurtosis preference, and evidence from the cross section of equity returns. J. Financ. **57**(1), 369–403 (2002)
10. Fama, E.F., French, K.R.: Common risk factors in the returns on stocks and bonds. J. Financ. Econ. **33**(1), 3–56 (1993)

11. Fama, E.F., French, K.R.: The capital asset pricing model: theory and evidence. J. Econ. Perspect. **18**(3), 25–46 (2004)
12. Fang, H., Lai, T.Y.: Co-kurtosis and capital asset pricing. Financ. Rev. **32**(2), 293–307 (1997)
13. Fletcher, J., Kihanda, J.: An examination of alternative CAPM-based models in UK stock returns. J. Bank. Financ. **29**(12), 2995–3014 (2005)
14. Florackis, C., Gregoriou, A., Kostakis, A.: Trading frequency and asset pricing on the London Stock Exchange: evidence from a new price impact ratio. J. Bank. Financ. **35**(12), 3335–3350 (2011)
15. Friend, I., Westerfield, R.: Co-skewness and capital asset pricing. J. Financ. **35**(4), 897–913 (1980)
16. Garcia, R., Tsafack, G.: Dependence structure and extreme comovements in international equity and bond markets. J. Bank. Financ. **35**(8), 1954–1970 (2011)
17. Harvey, C.R., Siddique, A.: Autoregressive conditional skewness. J. Financ. Quant. Anal. **34**(4), 465–487 (1999)
18. Harvey, C.R., Siddique, A.: Conditional skewness in asset pricing tests. J. Financ. **55**(3), 1263–1295 (2000)
19. Hung, D.C.H., Shackleton, M., Xu, X.: CAPM, higher comoment and factor models of UK stock returns. J. Bus. Financ. Acc. **31**(12), 87–112 (2004)
20. Hwang, S., Satchell, S.E.: Modelling emerging market risk premia using higher moments. Return Distributions in Finance, vol. 75 (1999)
21. Jarjir, S.L.: Size and Book to Market Effects vs. Co-skewness and Co-kurtosis in Explaining Stock Returns (2004)
22. Jondeau, E., Rockinger, M.: Testing for differences in the tails of stock-market returns. J. Empir. Financ. **10**(5), 559–581 (2003)
23. Kat, H.M., Miffre, J.: The impact of non-normality risks and tactical trading on hedge fund alphas. J. Altern. Invest. **10**(4), 8–21 (2008)
24. Kole, E., Verbeek, M.: Crash risk in the cross section of stock returns. Netherlands Working Paper, Erasmus University Rotterdam (2006)
25. Kostakis, A., Muhammad, K., Siganos, A.: Higher co-moments and asset pricing on London Stock Exchange. J. Bank. Financ. **36**(3), 913–922 (2012)
26. Kraus, A., Litzenberger, R.H.: Skewness preference and the valuation of risk assets. J. Financ. **31**(4), 1085–1100 (1976)
27. Lintner, J.: The valuation of risk assets and the selection of risky investments in stock portfolios and capital budgets. Rev. Econ. Stat. **47**(1), 13–37 (1965)
28. Markowitz, H.: Portfolio selection. J. Financ. **7**(1), 77–91 (1952)
29. Miller, M.B.: Mathematics and Statistics for Financial Risk Management. Wiley, Hoboken (2013)
30. Mitton, T., Vorkink, K.: Equilibrium underdiversification and the preference for skewness. Rev. Financ. Stud. **20**(4), 1255–1288 (2007)
31. Moreno, D., Rodríguez, R.: The coskewness factor: implications for performance evaluation. J. Bank. Financ. **33**(9), 1664–1676 (2009a)
32. Moreno, D., Rodríguez, R.: The value of coskewness in mutual fund performance evaluation. J. Bank. Financ. **33**(9), 1664–1676 (2009b)
33. Pettengill, G.N., Sundaram, S., Mathur, I.: The conditional relation between beta and returns. J. Financ. Quant. Anal. **30**(1), 101–116 (1995)
34. Rubinstein, M.E.: The fundamental theorem of parameter-preference security valuation. J. Financ. Quant. Anal. **8**(1), 61–69 (1973)
35. Scott, R.C., Horvath, P.A.: On the direction of preference for moments of higher order than the variance. J. Financ. **35**(4), 915–919 (1980)
36. Sharpe, W.F.: Capital asset prices: a theory of market equilibrium under conditions of risk. J. Financ. **19**(3), 425–442 (1964)

Contagion Risk Measured by Return Among Cryptocurrencies

Toan Luu Duc Huynh$^{(\boxtimes)}$, Sang Phu Nguyen, and Duy Duong

Faculty of Finance, Banking University of Ho Chi Minh City,
36 Ton That Dam Street, District 1, Ho Chi Minh City, Vietnam
toanhld@buh.edu.vn

Abstract. This paper examines the movement of cryptocurrencies' return based on price. This volatility can spread to others of the same kind. Currently, the more cryptocurrencies are traded in market, the more chances are available for investors. The author wonders whether contagion risk among these cryptocurrencies happens or not in the event of crashing. We also introduce one empirical evidence of the mutual influence on these cryptocurrencies using Copulas approach. The findings show that all pairs have the structure dependence with Kendall-plots, particularly strong left tail dependence with Chi-plots. It also means the existence of contagion risk among these cryptocurrencies. The three methodologies namely Kendall-plots, Chi-plots and Copulas estimation produce consistent results. Therefore, the investors should carefully perform portfolio diversification to avoid contagious phenomenon.

1 Introduction and Literature Review

Need for estimation. The analysis of cryptocurrencies has been recently drawn much attention from many scholars. These features are evaluated as innovation, transparency, anonymity and popularity (Urquhart [19]). Meanwhile, not only policy makers but also economists are afraid of controlling them under their own authorization as well as warn users about a tool for money laundering (Dyhrberg [6]).

Literature Review. Some studies conclude that Bitcoin is mainly used as a financial asset for investing rather than a currency for trading products (Dyhrberg [5]). Baek and Elbeck [1] indicates that Bitcoin and the other asset should be seen as speculative commodity instead of currency. Hence, all of them are speculatively traded, which possibly causes the bubble phenomenon with virtual value (Grinberg [13]). Thus, many investors should be careful in choosing when to invest as well as which kind of cryptocurrencies to buy in order to avoid the collapse of exchange market and possible risks.

In context of considering these cryptocurrencies as a tool investment, they could be placed into portfolio management and financial markets theorem (Dyhrberg [6]). There are over 1,000 types of virtual currencies' family for investors to put in portfolios (coinmarketcap.com accessed on 9th August 2017).

© Springer International Publishing AG 2018
L. H. Anh et al. (eds.), *Econometrics for Financial Applications*, Studies in Computational Intelligence 760, https://doi.org/10.1007/978-3-319-73150-6_71

In addition, Urquhart [19] asserts that the Bitcoin's price is reflected by weak efficiency over the full sample period by random walk in pricing.

However, the authors concern how transmission happens among these cryptocurrencies under a weak efficient market. The literature on cryptocurrencies are quite limited due to a concentration on safety, legacy, and code of ethics such as corruption, money laundering, etc. Cheah and Fry [2] assert that Bitcoin (a type of cryptocurrencies) is a means of storing value and is comprised in true unit as well as account.

This study suggests that the expression for volatility has no display in bubbles and crash. Dwyer [4] figures out the comparison between Bitcoin and gold volatility for average and relative value. Cheung et al. [3] evaluate that there is a bubble phenomenon in cryptocurrencies over the research period. In addition, this journal also illustrates three giant bubbles under the unlimited short existing bubbles. Once again, Dyhrberg [5] and Dyhrberg [6] prove that Bitcoin has the same characteristics to become a hedging instrument, which is better than gold and dollar. Thus, the investors could employ them as a tool for risk management.

The terminology of "contagion risk" is comprised of interdependence, co-movement and dependence structure. Forbes and Rigobon [11] show that the phenomenon, which involves high market correlation level and less co-movement after any shocks, is considered as a strong dependence and interdependence. Hence, dependence structure means the same movement among financial markets or investment assets but there does not exist linear correlation or conditional interdependence which is based on non-linearity (El Hedi Arouri et al. [7]). The previous publication regarding negative bubbles between Bitcoin and Ripple is based on economic-physics model (Fry and Cheah [12]). The number of studies regarding co-movement in price by these cryptocurrencies is limited in the content of the development of these instruments.

2   Data and Methodology

Data. There are many cryptocurrency-trading exchanges, which are available for users to collect data each day. Thus, all input data, collected from coinmarketcap.com, represent most cryptocurrencies in terms of price, volume as well as market capitalization. In particular, we employed Bitcoin (BTC), Ripple (XRP) and Lite-coin (LTC) with two main reasons: Firstly, these cryptocurrencies have the approximate 1.500 observations during the period from August 2013 to August 2017, which is considered as the long time series and sufficient data for any econometric model. Our data consists of daily closing for each currency calculated in US. Dollar. Secondly, until 5th August 2017, the total market capitalization for all kinds is around 113 billion whereas our sample is covered 50% of value. Therefore, this sample can generalize the cryptocurrencies market.

Methodology. These currencies show non-parametric data, which is considered as limitation for many model in price dataset. Thus, the determination of dependence structures is done by covariance as well as correlation, which can mislead the result. We decide to apply some new methodologies by following the steps hereinafter. The first phase is to approach chi-plots by Fisher and Switzer [9] as well as Fisher and Switzer [10] and Kendall-plots to find the dependence structure among BTC, XRP and LTC. In order to ensure the result, this study also use the results from running Copulas of chi-square and Kendall-square as p-value to test the dependence. Next, we use the good-of-fitness to estimate how fit Copulas are applied. The last stage is to apply the three copulas of Gumbel, Clayton and Normal (Gaussian) to investigate the structure, capturing the right tail, left tail and no tail structure, respectively.

Chi-plot is a method to examine the presence and pattern of dependence structure between a set of two variables. The function including joint continuous distribution formed by a set of two random vectors (X, Y) is as follows, and let $I(A)$ be the indicator function of one sudden event A and let $(x_1; y_1), \ldots, (x_n; y_n)$ be random samples of the remaining observation from H below. For one data point picked from sample as $(x_i; y_i)$, then H_i can be illustrated as

$$H_i = \sum_{j \neq i} I(X_j \leq X_j, Y_j \leq Y_i)/(n-1)$$

$$F_i = \sum_{j \neq i} I(X_j \leq X_i)/(n-1), G_i = \sum_{j \neq i} I(Y_j \leq Y_i)/(n-1).$$

wherein three formulas above, in the event that A occurs, the indicator function $I(\cdot)$ will receive the value of 1 and 0 otherwise.

$$S_i = \text{sign}\left\{ \left(F_i - \frac{1}{2}\right)\left(G_i - \frac{1}{2}\right) \right\}.$$

Then, this results in the function

$$X_i = \frac{H_i - F_i G_i}{\sqrt{F_i(1 - F_i)G_i(1 - G_i)}}$$

$$\lambda_i = 4S_i \max\left\{ \left(F_i - \frac{1}{2}\right)^2, \left(G_i - \frac{1}{2}\right)^2 \right\}.$$

The Chi-plot, solely estimated by the shape in joint distribution of copula, is a scatter plot of the pairs (λ_i, χ_i) within the range of rectangle $[-1; 1] \times [-1; 1]$. In addition, this method also employs pseudo-confidence which is mostly based on $\chi = \pm c_p/n^{1/2}$ by estimating Monte Carlo simulations with $p \times 100\%$ of the pairs (λ_i, χ_i) lie between the lines. For the positive marginal dependence, the couple of (λ_i, χ_i) has a trend of spreading out the line above.

Kendall-plots. Based on the publication by Genest and Boies [14], Kendall-plot will propose dependence between variables, known as alternative ranked-based procedure, which uses the theorem of a probability plot (Q-Q plot). The form of the bivariate distribution of K_0 is illustrated below

$$K(\omega) = K_0(\omega) = Probability\{UV \leq \omega\} = \omega - \log(\omega), 0 \leq \omega \leq 1.$$

The couple of data (X_i, Y_i) will transform into $(W_i : n, H(i))$ with $i = 1, 2, \ldots, n$. Furthermore, the value of $H(i)$ is defined as follows

$$W_i : n = \omega k_0(\omega)\{K_0(\omega)\}^{i-1}\{1 - K_0(\omega)\}^{n-1}d\omega.$$

Assume U and V have traits of independent uniform random variables within the interval ranging from 0 to 1. Thus, $W_{(i:n)}$ as defined by Genest and Boies [14], is formed as follows:

$$W_{(i:n)} = n\binom{n-1}{i-1}\int_0^1 \omega\{K_0(\omega)\}^{i-1}\{1 - K_0(\omega)\}^{n-1}dK_0(\omega).$$

The amount of curvature of the graph is an element for estimating the degree of association in data. The findings show a level of inefficiency in linearity in the graph. To be more detail, the closer the Kendall plot is to the 45-degree line, the less dependent the too variables are.

Copulas method. In scope of this paper, we only examine the dependence structure of variables using three families of Copulas such as Gumbel, Clayton and Normal. Nelsen [16] demonstrates that if $F(x, y)$ is a joint density function with margin $F(X)$ and $F(Y)$. Hence, there exists one Copulas for all $x, y \in [-\infty, +\infty]$, which is $F(x, y) = C(FX(x), FY(y))$.

Note that C exists only if X and Y share similarities in continuous random variables. The copulas C of these variables is strictly increasing transformations of marginal distribution of $F(X)$ and $F(Y)$. Hence, this is an important characteristic for Copulas to estimate dependence structure.

Gumbel Copulas. This formula will indicate this kind of Copulas

$$C_\theta(u, v) = e^{-[(-\ln u)^\theta + (-\ln v)^\theta]}$$

where $\phi(t) = (-\ln t)^\theta, \theta \geq 1$ with left (λ_L) and right tail (λ_U) dependence: $(\lambda_U) : (\lambda_L) = 0, (\lambda_U) = 2 - 2^{1/\theta}$.

It will capture the upper tail dependence, with the tail index of λ_U demonstrating the right tail, which means that the two events might incur simultaneously in the positive return.

Clayton Copulas

$$C_\theta(u, v) = (u^{-\theta} + v^{-\theta} - 1)^{\frac{-1}{\theta}} \text{ and } C_0(u, v) = \pi = uv,$$

where $(t) = \dfrac{t^{-\theta} - 1}{\theta}, \theta \geq 0$, with left-tail and the right-tail $(\lambda_U) : (\lambda_L) = 2^{\frac{-1}{\theta}}, (\lambda_U) = 0$.

It will show the lower tail dependence, with the tail index of λ_L measuring the lower tail dependence, which means that two events might happen in the worst-case scenarios, resulting in negative return.

Normal Copulas. It does not display lower or upper tail dependence, and its parameter will lie into a range of $0 \leq \theta \leq 1$

$$C_\theta(u, v) = \int_{-\infty}^{\phi^{-1}(u)} dx \int_{-\infty}^{\phi^{-1}(v)} dy \frac{1}{2\pi\sqrt{1-\theta^2}} \exp\left\{ -\frac{x^2 - 2\theta xy + y^2}{2(1-\theta)^2} \right\}.$$

This maximum pseudo likelihood method is to estimate Copulas parameters. Firstly, the data will be filtered out using GARCH $(1,1)$; therefore, the marginal distributions are quite practical illustration. Secondly, the maximum likelihood estimation method is applied to calculate the Copulas parameter.

$$L(\theta) = \sum_{i=1}^{n} \ln\left(C_\theta\left(\frac{R_i}{n+1}; \frac{S_i}{n+1}\right) \right)$$

where n is the sample size and θ is vector of parameters in the model. The two factor $\dfrac{R_i}{n+1}, \dfrac{S_i}{n+1}$ will determine corresponding values of empirical marginal distributional functions of random variables X and Y.

In this paper, we employ log-likelihood to rank and choose the fittest function of estimated model. The largest figure of log-likelihood will prove the most appropriate Copulas for estimation. In order to test the good-of-fitness for these Copulas models, we employ the methodologies by Prokhorov [17] based on p-value of each model. If the result is greater than the alpha threshold (normally level of significance of 0.01), then the estimated Copulas matches the empirical Copulas. Thus, we can use the results for determining the dependence structure.

3   Empirical Results

Descriptive Statistics

From Table 1, it can clearly be seen that the average return could be earned in the highest level from Ripple (0.0024) even though the price is quite low in relative comparison with Bitcoin. The discrepancy among them is not quite large when it comes to mean value. Both kurtosis values are quite high. In addition, two out of three items have positive skewness whereas the remaining one is negative. Thus, they only share similarities in terms of an increase in probability to have losses in statistical summary. However, Bitcoin might suffer a loss in the magnitude rather than the other two cryptocurrencies. The other criteria in this tables are demonstrated that the data is quite appropriate with the methodology in Copulas approach, which is proposed above. It is noticeable that the heavy tail appears (by excess kurtosis). Therefore, Copulas estimation is supposed to produce reliable findings (Fig. 1).

This figure demonstrates that the return generated by each currency is quite mutually dependent during the observed time. However, the listed time for each cryptocurrency is different; therefore, we choose the well-matched time to have balanced data in order to avoid the missing observation phenomenon.

Table 1. Descriptive statistics for variables

Variable	Obs.	Mean	Std. Dev.	Min	Max	Median	Kurtosis	Skewness
Bitcoin (BTC)	1462	0.0023	0.0422	−0.2662	0.3575	0.0019	13.9644	−0.1606
Ripple (XRP)	1462	0.0024	0.0772	−0.6163	1.0274	−0.0025	35.4133	2.0949
Lite-coin (LTC)	1462	0.0019	0.0681	−0.5139	0.8290	0	34.3230	2.1442

(*Source: The authors*)

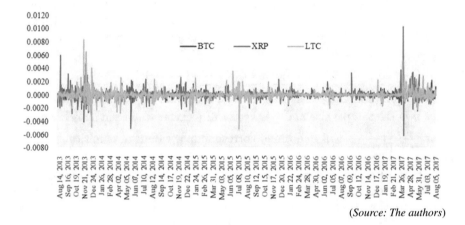

(*Source: The authors*)

Fig. 1. Time-series graph the price of the daily price for three kinds of cryptocurrencies

We employed Miller [15] calculation natural logarithm return: $R_t = \ln\left(\dfrac{P_t}{P_{t-1}}\right)$ in which P_t is the current price whereas P_{t-1} is the previous price of one cryptocurrency. In fact, there are some crashes and some similarities in some previous studies such as Fry and Cheah [12]. This paper asserts that in the first stage of this currency has considerably higher return in the full sample and in accordance with expectation with investors.

The dependence of Bitcoin, Lite-coin and Ripple. This graph hereinafter demonstrates the dependence structure by each couple of cryptocurrencies using non-parametric approach.

We employ Kendall-plots theorem in order to measure the dependence structure of Bitcoin, Lite-coin and Ripple by R-Studio program. In case of Kendall-plots, the stronger the deviation from the center of the main diagonal, the more positive association between variable pairs. Based on this theorem, the results from Fig. 2 show that there are associations between these variables when witnessing the phenomenon in divergence of dependence structure from the diagonal line. In particular, the association between Bitcoin and Lite-coin is more accurately reflected and associated than other couples due to the highest deviation from the diagonal line for correlation of Bitcoin and Lite-coin in comparison with the other two. When indicating the correlation of other two pairs,

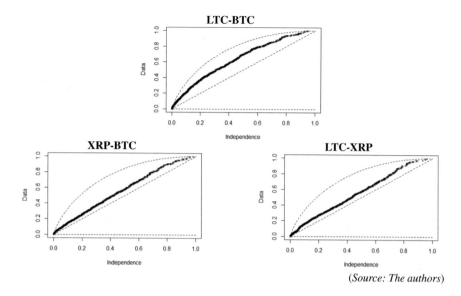

(*Source: The authors*)

Fig. 2. Kendall-plots illustrated by each couple of cryptocurrencies

we noticeably conclude that there is higher associated tail dependence than the other side. From that result, the authors have a part of condition to establish tails-dependence model.

From another perspective, the dependence of two cryptocurrencies is also analyzed by Chi-plots. The result shown in Figs. 3, 4 and 5 clearly indicates that the variables are tail-dependent when the vast majority of distribution points plotted beyond the control lines (+/- 0.05). When considered in pair, Lite-coin and Bitcoin have a strong left-tail dependence with a number of scatterplots on far away from the control lines (between +0.1 to +0.5). In addition, it shares the similarity in distributed shape between cryptocurrency pairs namely XRP-BTC, LTC-XRP when left tail-dependence simultaneously appears. From the graphical diagonal approach, the authors notice that there is a strong dependence on left tail between the aforementioned objects. This is a fundamental conclusion to continue this research and employ copula model to measure the effect level of cryptocurrencies (Table 2).

Table 2. Kendall-τ for determining dependence structure

No	Pairs	τ	p-value
1	XRP-BTC	0.209***	2.22E-16
2	LTC-BTC	0.503***	2.22E-16
3	LTC-XRP	0.205***	2.22E-16

(*Source: The authors*)

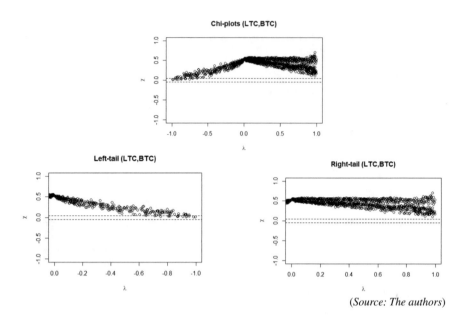

(*Source: The authors*)

Fig. 3. Chi-Plots illustrated by each couple of cryptocurrencies

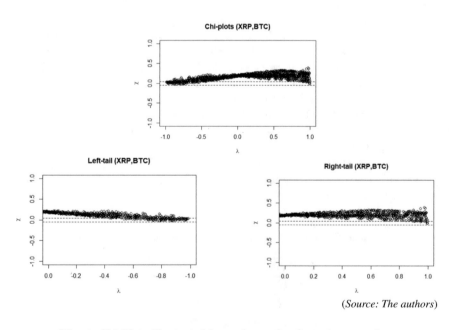

(*Source: The authors*)

Fig. 4. Chi-Plots illustrated by each couple of cryptocurrencies

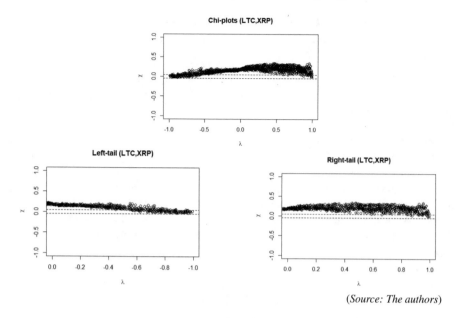

(*Source: The authors*)

Fig. 5. Chi-Plots illustrated by each couple of cryptocurrencies

The null and alternative hypothesis for above testing is $\begin{cases} H_0 : \tau_B = 0 \\ H_A : \tau_B \neq 0 \end{cases}$.

To be more precise, the alternative hypothesis is that the variables are correlated, and τ_B is non-zero. (***), (**), (*) reflected statistically significant of the corresponding coefficients at 1%, 5% and 10% level.

Based on the result table above, we strongly conclude that all pairs of cryptocurrencies face the dependence structure. This test is an alternative test to confirm the findings which are shown and analyzed from the graphical methodology.

Good-of-fitness test for Copulas. This table below illustrates the results extracted from R-programing to confirm the good-of-fitness of the previous run Copulas. Interestingly, most of statistical figures above does not meet the requirements in specific Copulas due to the low indicated gof-p-value. However, Embrechts [8] indicates that up to 99.9% of Copulas approach will pass through the good-of-fitness to capture the parameter for each Copulas and choose the appropriate family of Copulas by pseudo - maximum likelihood method (Table 3).

The null and alternative hypothesis for above testing is $\begin{cases} H_0 : C \in C_0 \\ H_A : C \neq C_0 \end{cases}$

with C_0 is a specific Copulas. (***), (**), (*) reflected statistically significant of the corresponding coefficients at 1%, 5% and 10% level.

Table 3. Good-of-fitness for the three Copulas such as Clayton, Gumbel and Normal

Pairs	Data	Clayton	Gumbel	Normal
XRP-BTC	Gof-Parameter	0.88195(***)[0.77176]	1.2577(***)[0.14966]	0.32972(***)[0.091162]
LTC-BTC	Gof-Parameter	1.6792(***)[0.26809]	1.891(***)[0.26809]	0.6721(***)[0.26809]
XRP-LTC	Gof-Parameter	0.86936(***)[0.73138]	1.2524(***)[0.73138]	0.31048(***)[0.73138]

(*Source: The authors*)

Tail-dependence determination. We employ Copulas approach based on Forbes and Rigobon [11], Reboredo [18], Genest and Favre [14] and Grégoire et al. [13] to choose which Copulas matches the result. The table below shows the estimated parameters of three Copula families, namely Clayton, Gumbel and Normal. The results indicate that there exists contagion risk between these pairs of cryptocurrencies. Where copula function is chosen based on highest log-likelihood value. In general, it is noticeable that Clayton is chosen as the best representative for fittest copulas for left-tail dependence between the pairs of cryptocurrencies due to highest log-likelihood.

When considering dependence structure between Lite-coin and Bitcoin, we notice that log-likelihood value of Clayton Copula is highest (488.6) similarly with Ripple-Bitcoin and Litecoin-Ripple, which have Log-likelihood value is 122.3 and 114.7. Based on Copula dependence theory, we found that there is existent contagion risk between cryptocurrency pairs due to existent left-tail dependence (Clayton Copula) (Table 4).

Table 4. The estimated parameter for cryptocurrencies

Pairs	Parameters	Clayton	Gumbel	Normal
LTC-BTC	θ	**1.679**	1.891	0.3105
	Loglikelihood	488.6	436.8	72.98
XRP-BTC	θ	**0.8819**	1.258	0.3297
	Loglikelihood	122.3	85.53	82.89
LTC-XRP	θ	**0.8694**	1.252	0.3105
	Loglikelihood	114.7	81.89	72.98

(*Source: The authors*)

Particularly, when a Bitcoin decreases, it leads to investors' withdrawing their money from the Cryptocurrency market and the prices of other currencies significantly decrease, reducing investors' return. The result can be used to generalize the trend of other cryptocurrencies. In summary, based on measuring dependence structure between cryptocurrencies, we notice that there is existent contagion risk between two cryptocurrencies (the left-tail dependence). The results from copula approach, are completely consistent with the non-parametric ones.

4   Conclusion

This paper uses daily data from 2013 to 2017 and employs Chi-plots, Kendall (K)-plots and three different copula functions to examine the tail dependence between Bitcoin and Ripple; Bitcoin and Lite-coin; Ripple and Lite-coin in order to test the contagion effect with balanced dataset. The above analysis shows that all cryptocurrencies depend on left-tail, which means that the spread of contagion risk among them.

It is suggested that investors should carefully consider to invest in these cryptocurrencies due to the spreading co-movement in price. It could be inferred that a decrease in price of Bitcoin might lead to a loss for both Ripple and Lite-coin, which results in a decline in these cryptocurrencies' prices. We conclude that the prices are mutually correlated in non-linearity form and their probability in dropping will be witnessed. It seems to be more contagious rather than prone to crashing among these cryptocurrencies. Therefore, there is a potential risk when diversification is misleading.

This also implies the efficiency of cryptocurrencies regarding price movement. In other words, a cryptocurrency's fluctuation will result in a decrease in prices of others. These conclusions are new and have not been documented in the existing literature about cryptocurrencies. However, this research also faces limitations such as short dataset (only 4 year), with only three main currencies (Bitcoin, Ripple and Lite-coin) not covering all the market, etc. Thus, we also suggest that further research aim at testing contagion risk in this alarming markets.

References

1. Baek, C., Elbeck, M.: Bitcoins as an investment or speculative vehicle? A first look. Appl. Econ. Lett. **22**(1), 30–34 (2015)
2. Cheah, E.-T., Fry, J.: Speculative bubbles in Bitcoin markets? An empirical investigation into the fundamental value of Bitcoin. Econ. Lett. **130**, 32–36 (2015)
3. Cheung, A., Roca, E., Su, J.-J.: Crypto-currency bubbles: an application of the Phillips-Shi-Yu (2013) methodology on Mt. Gox bitcoin prices. Appl. Econ. **47**(23), 2348–2358 (2015)
4. Dwyer, G.P.: The economics of Bitcoin and similar private digital currencies. J. Financ. Stab. **17**, 81–91 (2015)
5. Dyhrberg, A.H.: Bitcoin, gold and the dollarA GARCH volatility analysis. Finance. Res. Lett. **16**, 85–92 (2016a)
6. Dyhrberg, A.H.: Hedging capabilities of bitcoin. Is it the virtual gold? Finance. Res. Lett. **16**, 139–144 (2016b)
7. El Hedi Arouri, M., Bellalah, M., Nguyen, D.K.: The comovements in international stock markets: new evidence from Latin American emerging countries. Appl. Econ. Lett. **17**(13), 1323–1328 (2010)
8. Embrechts, P.: Copulas: a personal view. J. Risk Insur. **76**(3), 639–650 (2009)
9. Fisher, N., Switzer, P.: Chi-plots for assessing dependence. Biometrika **72**(2), 253–265 (1985)
10. Fisher, N., Switzer, P.: Graphical assessment of dependence: Is a picture worth 100 tests? Am. Stat. **55**(3), 233–239 (2001)

11. Forbes, K.J., Rigobon, R.: No contagion, only interdependence: measuring stock market comovements. J. Finance **57**(5), 2223–2261 (2002)
12. Fry, J., Cheah, E.-T.: Negative bubbles and shocks in cryptocurrency markets. Int. Rev. Financ. Anal. **47**, 343–352 (2016)
13. Grégoire, V., Genest, C., Gendron, M.: Using copulas to model price dependence in energy markets. Energy Risk **5**(5), 58–64 (2008)
14. Genest, C., Boies, J.C.: Detecting dependence with Kendall plots. Am. Stat. **57**(4), 275–284 (2003)
15. Miller, M.B.: Mathematics and Statistics for Financial Risk Management. Wiley, Hoboken (2013)
16. Nelsen, R.B.: An Introduction to Copulas. Springer, New York (2006)
17. Prokhorov, A.: A goodness-of-fit test for copulas, MPRA Paper No. 9998 (2008)
18. Reboredo, J.C.: How do crude oil prices co-move?: A copula approach. Energy Econ. **33**(5), 948–955 (2011)
19. Urquhart, A.: The inefficiency of Bitcoin. Econ. Lett. **148**, 80–82 (2016)

The Effect of Macroeconomic Factors on Investor Purchase Decision: The Case Study of HOSE

Tran Anh Tung[1], Pham Van Kien[2(✉)], and Ho Thanh Phong[3]

[1] Department of Business, Saigon Institute of Technology,
Ho Chi Minh City, Vietnam
tungta@saigontech.edu.vn
[2] International Economic Faculty, Banking University HCM City,
Ho Chi Minh City, Vietnam
kienpv@buh.edu.vn
[3] Industrial and System Engineering Department,
International University - VNUHCM, Ho Chi Minh City, Vietnam
htphong@hcmiu.edu.vn

Abstract. **Purpose:** The aim of this paper is to examine how macroeconomic factors affect investors purchase decision based on the volatile trading value of VN-index in the Ho Chi Minh Stock exchange.

Methodology: This study applies the quantitative approach as the major method. The quantitative analysis of this research will conduct by using R-software and SPSS with statistical technique including simple time series analysis, stationary testing parameters, model specification, and binary logistic regression analysis. The study uses the secondary data collected online from the period of 2008 to 2015. Specifically, this paper uses a total of 84 data points which are plenty for the effective regression analysis. This study attempts to provide the better understanding of investors' behavior under effect of macroeconomic factors.

Findings: All measurement parameters in this study demonstrate as stationary factors and macroeconomic factors show the significant effects on investor purchase decision with regard to Inflation rate, Crude oil price, and Exchange rate. As a result, this research indicates the result of binary regression model related to the odd ratio equation and the level of accurate prediction of the model.

Value: There are various studies carried out to examine the macroeconomic effects on investor purchase decision, but very few of them observed the said effects with the empirical approach. For this reason, this study applies the new method of binary logistic regression in order to calculate the proportion of purchase decision.

Keywords: Investor purchase decision · Macroeconomic effect
Stock exchange market · HOSE · Vietnam

© Springer International Publishing AG 2018
L. H. Anh et al. (eds.), *Econometrics for Financial Applications*, Studies in Computational Intelligence 760, https://doi.org/10.1007/978-3-319-73150-6_72

1 Introduction

In 2014, based on the report of state securities committee of Viet Nam, the Viet Nam stock exchange market was demonstrated a very positive performance in terms of the growth aspect index, degree of stability, volume and clarity which reflected the positive changes in the economy. The development of stock exchange market is considered as a significant growth, though it also has the down trend adjusted among them. The VN-Index was set a new record as 640.75 after six years existing on the stock market, increased 13.3% comparing to the same period of the year 2013. In the third quarter in 2016, according to the report of BSC (BIDV joint stock Company), Viet Nam stock market still remained the long-term uptrend since the beginning of 2016, therefore, that leads to the conclusion as the Viet Nam stock exchange market shows not only the potential growth but also the attractive place for foreign investors.

Nevertheless, In Vietnam, behavioral finance needs to study extensively as it is a viable model to explain on how investors make decisions (Kim and Nofsinger 2008). Thuy and Duong (2015) state that despite Vietnam has a table macroeconomic environment, the Viet Nam stock market is still dominant by the concerns of macroeconomic factors changes in investors' sentiment. Consequently, the potential growth of the stock market is constrained. In general, the Viet Nam stock exchange market still shows the shortenings of modern instruments.

Therefore, this research's aim is to provide the specific observation and econometric approach to examine the significant effects of macroeconomic factors on investor purchase decision in Ho Chi Minh stock exchange market (HOSE). More specifically, the purposes of this research to address the following questions:

RQ1: Does the data of HOSE collected from online sources display as stationary or reliable data for the research?
RQ2: What do macroeconomic factors possess the significant effect on investor purchase decision in the Ho Chi Minh Stock Exchange Market (HOSE)?
RQ3: How much proportion is considered as reliable for investors to make a purchase decision?

2 Literature Review

First of all, VN-index is the average index value and an overview of people interested in since it reflects an average value of all the shares listing on the Viet Nam market. Phan and Zhou (2014) studies the weak-form efficiency of VN-index with weekly observations and daily observations of 5 oldest stocks listed from 2000 to 2013. The result shows that the volatile of VN-index is not random and under pressure of instability sentiment of investors. Therefore, understanding investor purchase decision plays an important role to figure out the investment trend in the Viet Nam stock market.

2.1 Investor Purchase Decision

Quan and Cong (2016) conducts a study on the Viet Nam stock exchange market. The findings imply that Vietnamese investors do not always make rational decisions, but they are also influenced by psychological factors in their decision-making process. In addition, Thuy and Duong (2015) states that the change of macroeconomic policies usually occurs suddenly, which may effect on the psychology of investors. Therefore, analyzing the impact of macroeconomic factors in the economy in general and the stock market in particular becomes an essential topic.

Furthermore, Daniel et al. (1998) state that investors' sentiment plays as the core function of investment outcomes. In line with this, the findings of Malcolm and Jeffrey (2007) also point out that the investor purchase decision is usually established from the investors' sentiment variables which illustrate the valuation of stock and hard to arbitrage. More specifically, investors could be categorized in three aspects of sentiment namely high, moderate, and low sentiment. Another study also prove the canonical correlation between investor purchase decision and sentiment, it reveals not only investment variable aspects could influence to their sentiments behavior, but also the certain demographic could do it, respectively (Charles and Kasilingam 2014). As a result, the relationship between investors' sentiment and purchase decision are observed and other connections between investors' sentiment and the value of stock are illustrated in the previous works.

2.2   Macroeconomic Factors that Effect on Investor Purchase Decision

Hardouvelis (1987) implies that the activities of stock price are a reflection of market perception and the sensitiveness of financial companies stocks is also manipulated by monetary news. Specifically, he points out that the fluctuation of stock price occurred in two major factors including expected real interest rate and expected inflation rate. Inheriting the research of Hardouvelis, many other macroeconomic factors are brought to observe their influences on stock market. The impulse response analysis of the Greek stock market return shows that all macroeconomic factors regarding industrial production, interest rate, real oil price, nominal exchange rate, and consumer price index are important to explain the movement of stock price (Hondroyiannis and Papapetrou 2001).

In Viet Nam, the change in macro policies together with the number of macroeconomic factors often have a strong impact (both positive and negative) on the psychology of investors and the stock market finally. Therefore, the study on the relationship between the macroeconomic factors and the volatility of the Vietnam stock exchange market is very important (Kieu et al. 2013). As a result, five factors of macroeconomic including inflation rate, exchange rate, interest rate, crude oil price and FDI are widely discussed in the previous researches. Thus, in this study we also select the mentioned five factors to understand investor purchase behavior.

Inflation rate or consumer price index
Inflation is defined as the changing of percentage in value of wholesale and measured based on the Consumer Price Index (CPI) (Elliott 2007). Smith (2009) indicates that inflation plays an important role in the market trend forecasting and orientation. As a result, high inflation rates increase the costs of living and transfer the resources from the stock market tools into consumables. This trend leads to the reducing trading volume on the stock market. Consequently, the value of stocks are decreased. Therefore, inflation is expected to have a negative impact on the stock market. For this reason, we come up with the first hypothesis (H1).

H1. There is a negative effect of inflation on investor purchase decision.

Exchange rate
El-Masry (2009) indicates that exchange rate demonstrates a high significant relationship with industrial stock. In addition, Suthar (2010) states that there is no economy can be stable or immune with the impact of external factors like exchange rate. Moreover, risks due to the impact of exchange rate on investments make investors forecast the currency may be discounted in future. Thus, investors would decide not to invest in stocks or seeking to replace stocks by foreign currency assets since the stock value reduced. Therefore, exchange rate can be considered one of macroeconomic factors which possesses the significant impact on the stock market as well as investor purchase decision. Based on the above discussion, we propose the second hypothesis as follows:

H2. There is a positive effect of exchange rate on investor's purchase decision.

Interest rate
Joseph (2002) finds that the changes of two economic factors like exchange rate and interest rate have created the adverse on the stock return. On the other hand, an increase of the market interest rates in general and banks in particular has close ties to the volatility of the stock price. This is the most evident in the increase or decrease of the stock and bond value. Thuy and Duong (2015) also indicate that when interest rate falls, it will have a positive impact on stock price indices because the cheaper in capital costs makes it easier for companies to mobilize capital and carry out investment projects. Additionally, this trend helps to reduce costs for large financial leverage, improve company profits and increase company stock prices. Whereas, when interest rates rise, it will have a negative impact on the overall performance of the economy. Finally, investors recognize the positive signal from interest rate – as it fall – they obviously make a buying decision and in contrast, they make a selling decision. Hence, interest rate can be observed as an important factor that directly affect investor purchase decision as stated in the third hypothesis (H3).

H3. There is a positive effect of interest rate on investor's purchase decision.

Crude oil price
Crude oil price is one of the most powerful instruments to develop the national economy. By adjusting different levels of crude oil price, it will lead to the result that the economy could generate output and revenue in the large capacity when the world crude oil price increased (Saari et al. 2007). The empirical research of Bhunia (2012) find that the fluctuation of crude oil price consequently causes the great volatile swings on the stock market in both developed and developing countries. In 2015, according to the report of Bao Viet security, oil price movements have both positive and negative effects on the profit prospects of companies listed on the Vietnam stock market. Companies that have input materials are petroleum products (such as PLC - Petrolimex Petrochemicals, BMP - Binh Minh Plastics …) and companies with direct input from petrol and oil (PVT - Transportation Petroleum) will benefit from this trend. On the contrary, companies operating in oil and gas services (such as PVD - Petroleum Drilling and Well Services, PVS - Petroleum Technical Services, PVC - Petroleum

Construction …) have been negatively impacted by the drop in oil prices. Based on these analyses, the fourth hypothesis is given as follows:

H4. There is a positive effect of crude oil price on investor's purchase decision.

Foreign direct investment (FDI)
Fifeková and Nemcová (2015) indicates that FDI is played as an accelerator vehicle in order to narrow the gap between nations on the world. Therefore, we can observe that the important of FDI in the macroeconomic environment is not only a resource of development for nation, but also an image to prove the attractive economic environment for worldwide investors. In line with this, Ray (2012) find that there is a positive impact of foreign direct investment on stock prices, however, this impact is limited. Therefore, we have a good reason to propose the fifth hypothesis as follows:

H5. There is a positive effect of FDI on investor purchase decision.

3   Model Construction

In order to examine determine the macroeconomic factors that affect investor purchase decision in the Ho Chi Minh stock market, this research model is conducted with five independent variables namely inflation rate, exchange rate, interest rate, crude oil rice, and FDI as shown in Fig. 1. These factors are based on the comprehensive review of the given literature.

The research model is tested based on the daily fluctuation of VN-index price, if the percentage in price fluctuation more than 0, then the value of that month would be denoted as 1 (buying decision more than selling decision), in the other hand, if the percentage was negative the value of that month would be denoted as 0, respectively.

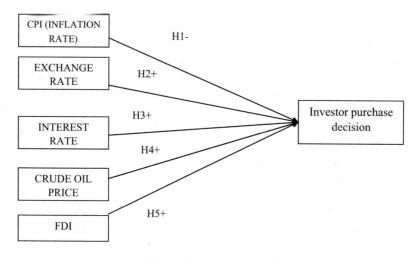

Fig. 1. The research model

4 Research Method

4.1 Binary Regression Approach

The quantitative analysis of this research is conducted by using R-software and SPSS with statistical techniques including stationary tests, model specification, and binary logistic regression analysis.

Based on the foundation of the logistic regression model developed by Cox (1972), in general, there are many phenomena we need to predict the (modeling) capability of the incident which we are interested (probability events) in economic activities. This incident is the result of a choice between distinct modes, such as decisions to buy or not to buy certain products, decisions to lend or not to lend, borrowers have to repay the debt or not, poor or not poor, moving or not moving. In the research model, the dependent variables are a non-continuous variables (separately) shows the result of an event, as expressed by the probability, would get value = 1, if such things happen and get value = 0. If things do not happen (this variable called variable alternately known as binary variables). This model is often applied to study factors affecting the decision whether or not to have the form of questions yes or no.

- Kleinbaum and Klein (2002) indicate that binary regression is a special kind of regression as the dependent variable is a binary variable (just get a value of 1 or 0). This kind of regression model is used to anticipate the probability of the event occurrence (the dependent variable) based on the information of the independent variables in the model.
- This regression method is applied extensively in the study of economics, business, society, and medicine to predict the likelihood of an event that somehow we are interested. Also, the logistic regression analysis also enables us to assess the relative importance and impact of independent variables on the probability that an event occurred.

Data collection and analysis
The secondary data in this study is extracted from the statistical report of VN-index of HOSE on website http://data.vietstock.vn/, under the period of 2008–2015. This research makes use of 84 data points which are plenty for the effective regression analysis. The data are obtained consistently with the unity and reliable online source. The observation takes as different scenarios at close of stock price in order to extract the variety of influence of macroeconomic on a specific stock value. The data was filtered by using the mathematic functions in excel including average, what-if function and count if based on the following steps:

Step 1: calculating the average monthly value of VN-index, CPI, exchange rate, interest rate, crude oil price, and FDI. Specially, the value of VN-index from 2008 to 2015 was greater than 1000 numbers with daily data which are converted into monthly data with the 84 numbers only.

Step 2: based on the daily fluctuation of VN-index price, if the percentage in price fluctuation more than 0, then the value of that month would be denoted as 1 (buying decision more than selling decision), in the other hand, if the percentage was negative the value of that month would be denoted as 0, respectively.

Step 3: Using Count-if function in order to count the number of 1 or 0, if the number of 1 more than 0, it would be denoted as 1 for total value of that month.

Measurement model

To exam the relationship between predetermined macroeconomic factors and the VN-index stock price performance, this study employs the macro-econometric model, according to modified model of Hondroyiannis and Papapetrou (2001). The original model is an estimate of the investment equation that does not restrict the types of accelerators generally including cash flow, leverage - debt-capital and the ratio of debt to equity as multiple regression model. Therefore, we determined the following empirical model to be estimated (1):

$$VNI = \alpha + \alpha_1 CPI_t + \alpha_2 Exr_t + \alpha_3 Ir_t + \alpha_4 Ov_t + \alpha_5 FDI + \varepsilon_t \tag{1}$$

where, VNI – the stock price of VN-index; CPI – the consumer price index stand for inflation rate; Exr – the exchange rate; Ir – the deposit interest rate of commercial bank; Ov – the crude oil price, and FDI – foreign direct investment; α's represented for coefficient of the independent variables and ε is the error item (white noise). This study used to test the ADF to perform unit root tests should only focus on theoretical models. Specifically, according to Dickey and Fuller (1981) model unit root tests ADF.

Dickey and Fuller (1981) have given accreditation Dickey and Fuller (DF) test the expanded Dickey and Fuller (ADF), in order to exam the stationary or non-stationary of time series data. This study used to test the ADF to perform unit root tests should only focus on theoretical models. Specifically, according to Dickey and Fuller (1981) model unit root tests ADF extended form:

$$\Delta y_t = \alpha_0 + \beta y_{t-1} + \sum_{(j=1)}^{k} \delta_j \cdot \Delta y_{t-j} + \varepsilon_t \tag{2}$$

$$\Delta y_t = \alpha_0 + \delta t + \beta y_{t-1} + \sum_{(j=1)}^{k} \delta_j \cdot \Delta y_{t-j} + \varepsilon_t \tag{3}$$

The results of ADF test are very sensitive to the selection of length of lag time k, therefore the AIC standard (Akaike's Information Criterion) of Akaike (1973) was used to choose the optimal k in ADF model test. Specifically, the value of k is chosen such that the smallest AIC. The hypothesis testing was raised as null hypothesis stood for the time series data was non-stationary, and the hypothesis was stationary.

The statistical Z_t of Phillips and Perron (1988) is a variation of the Dickey - Fuller t which enables statistical autocorrelation and random conditional variables error in term of the Dicky – Fuller regression. This is based on the equation:

$$\Delta x_t = \alpha_0 + \alpha_1 \cdot t + \alpha_2 \cdot x_{t-1} + \varepsilon_t \tag{4}$$

Next, Unit root tests were performed on the variable of data preparation for testing and verification at for both co-integrated and causal effect. When the unit roots tests are clarified and data was set up following the order of integration of variables, therefore, the co-integrated was confirmed by using the Johansen co-integrated test:

$$\Delta X_t = a_0 + a_1 X_{t-1} + a_2 t + \sum_{i=2}^{p} b_i \Delta X_{t-i+1} + u_t \tag{5}$$

When regression models with time-series variables as they require that arises has to be stationary, the model estimates the dependence of the rate of change of one variable to another variable was stated that vector error correction model, which can be described as a kind of general VAR model, used in case of data was non stationary and contained co-integrated relationship. The following equation of VECM:

$$\Delta x_t = \pi \cdot x_t \gamma_1 \Delta x_{t-1} + \ldots + \gamma_{p-1} \Delta x_{t-p+1} + u_t \tag{6}$$

The unit of analysis is the monthly level of each factor. The target population is the value of VN-index from 2008 to 2014, and coding if the signal of the index is green (increased) or red (decreased), it will define as yes (1) or no (0). In addition, the coding was based on the fluctuation of VN-index price on HOSE, the increased or decreased were also calculated in the monthly average price value. As a result, the logistic regression model equation in this research could present as following:

$$VNI(0, 1) = B + B_1 CPI_t + B_2 Exr_t + B_3 Ir_t + B_4 Ov_t + B_5 FDI\varepsilon_t \tag{7}$$

In conclusion, the ultimate outcome of this research is justified the proportion of selling and buying point, therefore, the formation after regression model will be formed as following:

$$Pi = \frac{e^{\beta_0 + \beta_1 X_1 + \beta_2 X_2 + \ldots + \beta_k Xk}}{1 + e^{\beta_0 + \beta_1 X_1 + \beta_2 X_2 + \ldots + \beta_k Xk}} \tag{8}$$

Pi: the proportion of VN-index at that time performed the power of buyer of seller

$P = 1$:   the proportion of selling point

$P = 0$:   the proportion of buying point

X_k: the independent variables (macroeconomic factors).

5   Results and Findings

Unit root test using ADF technique
    Firstly, to obtain the regression analysis mean, this study performs a unit root test to determine their stationary or the variables from the use of time-series data. According to Gujarati (2004), a time series was seen as stationary when the mean, variance, covariance (at different lags) remains constant; whether the series of data is determined at any time of lag.

Using ADF (Augmented Dickey-Fuller) with AIC (Akaike information criterion) to test the stationary of each variable from dependent variable VN-index to independent variables as inflation rate, exchange rate, interest rate, crude oil price, FDI, in addition, the Philips-Perron test also conducted with the type "Z-tau" and model "constant" in order to give the variation of unit root test. Therefore, the result is displayed in the following Table:

Table 1. Summary of Unit-root test

Variables	ADF test	p-value/difference	Supported
Index	−4.2691	0.0004074	***
Cpi	−2.5159	0.03535	*
Ex	−1.6238	0.03378	*
Ir	−2.6228	0.002323	**
Ov	−3.4357	0.002919	**
Fdi	−5.9143	0.0000	***
Variables	Philips-Perron test (Z-tau)	p-value/difference	Supported
Index	−4.1781	0.000	***
Cpi	−2.2196	0.000	***
Ex	1.4912	0.000	***
Ir	−2.4802	0.000	***
Ov	−3.4677	0.000	***
Fdi	−5.4843	0.0000295	***

Note: significant codes: 0 '***' 0.001 '**' 0.01 '*' 0.05 '.' 0.1

According to the Table 1, the summary of the result of Unit-root test, results obtained indicate that all variables are stationary at the value of significant level lower than 0.05. Especially in terms of lag order, all factors are stationary at difference 0. Nevertheless, the result also implies that there is presence of unit root in some of data series possessed the existence of possible long run relationship among the variables.

- *Co-integration test using Johansen test.*

Next, the semi-strong form test is conducted. This includes the Johansen and Juselius (1990) co-integration test. Unit root tests are performed on the variable of data preparation for testing and verification at for both co-integrated and causal effect. When the unit roots tests are clarified and data was set up following the order of integration of variables, therefore, the co-integrated was confirmed by using the Johansen co-integrated test

In order to confirm that existence, the next stage of estimation is conducted through co-integrated test (or coitegration test), Tables 2 and 3 present the results of those tests.

The first step is to use the linear regression technique to estimate the spread of dependent variable (index) and independent variables (Cpi, Ex, Ir, Ov, Fdi) then ADF to test if the spread is stationary, which in the method also tests a co-integration among

Table 2. Coefficients summary

Variables	Estimate	Std. Error	t-value	Sig
(Intercept)	9.390e+02	1.520e+02	6.179	2.74e−08 ***
Cpi	−2.457e+01	1.992e+02	−0.123	0.902177
Ex	−2.689e−02	8.317e−03	−3.233	0.001796 **
Ir	−1.424e+03	4.890e+02	−2.912	0.004675 **
Ov	2.426e+00	7.011e−01	3.461	0.000877 ***
Fdi	−7.417e+00	4.196e+00	−1.767	0.081071

Note: significant codes: 0 '***' 0.001 '**' 0.01 '*' 0.05 '.' 0.1

Table 3. Cointegartion test summary

Parameter	Statistic	P value
Lag Order: 1	Dickey-Fuller: −4.3353	0.01***

In adftest (reg$residuals, type = "nc"):
P-value smaller than printed p-value

variables. Therefore, the result of Table 3 indicates that there are existing a co-integration among those variables (p-value = 0.01, significant proved).

According to Engel and Granger (1987), based on the theory of co-integrated of variables, since it is founded, the result is concluded that there is a long run relationship among variables. In order to capture that behavior of long run, the Johansen test is conducted with type trace and spec long run. The following coding in R-software and summary result of long run co-integrated Johansen test (Table 4).

$coRes = ca.jo(data, type = "trace", K = 2, ecdet = "none", spec = "longrun")$
Test type: trace statistic, with linear trend.

Table 4. Values of test statistic and critical values of test

	Test	10pct	5pct	1pct
r ≤ 5	3.06	6.50	8.18	11.65
r ≤ 4	8.86	15.66	17.95	23.52
r ≤ 3	19.81	28.71	31.52	37.22
r ≤ 2	43.27	45.23	48.28	55.43
r ≤ 1	71.81	66.49	**70.60**	78.87
r = 0	135.13	85.18	90.39	104.20

In terms of r was described as the rank of given matrix of time series data with available significance level. Therefore, Johansen test function was estimating that rank, from the r = 0 (there is no co-integration at all) till r ≤ 5 (n − 1, where n = 6 in the research's sample). As a result, at r ≤ 1, there was a presence of long run co-integration because of the value of test (71.81) was greater than a confidence level's value (say 5%, 70.60).

Table 5. Variance decomposition of VN-index

Period	S.E.	VNI	CPI	EX	IR	OV	FDI
1	75.09999	100.0000	0.000000	0.000000	0.000000	0.000000	0.000000
2	100.7882	97.87056	0.000950	0.009387	1.152470	0.259676	0.706956
3	113.0043	96.27898	0.320448	0.369586	0.927384	0.476421	1.627179
4	124.7765	92.95418	0.408282	0.503111	0.910363	0.484198	4.739862
5	137.4596	90.47186	0.539408	0.429314	0.915086	1.082611	6.561720
6	148.0825	89.80134	0.718335	0.502460	0.940968	1.662902	6.373990
7	157.2311	89.65902	0.942710	0.714098	0.917100	1.895718	5.871352
8	166.2868	89.78968	1.114825	0.754710	0.828035	1.984937	5.527810
9	175.2976	89.81574	1.192333	0.717396	0.745719	2.045061	5.483753
10	183.7035	89.67960	1.234794	0.698383	0.691201	2.096804	5.599214

- *Variance decomposition test*

Table 5 below illustrates the Variance Decomposition in order to measure the effect of macroeconomic variables on the changing value of VN-index within ten periods:

VN-index greatly influents shocks from volatility by itself. The level of self-explanatory VN-index to 10th period was 89.67%, which regarding to CPI contribution 1.23%, exchange rate 0.698%, interest rate 0.691%, crude oil price 2.09%, and FDI 5.59% of the shocks on VN-index. As the observation from Table 5, over time the role of the CPI and oil prices showing a gradual increase significantly in explaining the variation of the VN-Index, while the remaining variables are the abnormalities in each period. Additionally, we also observe a very small impact of interest rate and exchange rate on VN-index, explain to the exchange rate and interest rate have little impact on the VN-Index may be due to exchange rate management policy in Vietnam with huge intervention from state banks and not really interesting guide for foreign investors.

The results of variance decomposition show that the Vietnam stock market is not yet as good as a barometer of the economy like other countries, investors do not really make good use on macroeconomic data, should be really careful when using macro variables to predict the change in stock prices.

- *Binary regression model.*

According to the Table 6, there are 3 out of 5 variables regarding to Cpi (inflation rate), Ov (crude oil price), and Ex (exchange rate) possessed a statistical significant at 90% confidence level and satisfied the expectation trend (positive or negative effect).

The Table 6 also includes significant test for each coefficient in the logistic regression model. In fact, the small sample of t-value is invalid and Wald statistics should be used instead, Wald is basically that t^2 Chi-Square distribution with df = 1.

From the beginning of the research, the expectation is set as interest rate and foreign direct investment might change the purchase decision of investors. Especially, the increasing of interest rate – deposit rate would draw the intention of investors to put more money in their bank account instead of investing in stock exchange. In addition, the foreign direct investment might affect the purchase decision as the increasing of value of FDI would draw the intention of foreign investors and would cause the

Table 6. Summary regression result

Variables	B (Coeffients)	Wald	Sig.
Cpi	−8.573	2.975	0.085
Ov	0.030	2.843	0.092
Ir	−16.369	1.693	0.193
Ex	0.00376	3.016	0.082
Fdi	0.034	0.103	0.748
Constant	7.714	3.756	0.053

Table 7. Classification table

Observed		Predicted		
		Index		Percentage
		Sell	Buy	correct
Index	Sell	11	19	36.7
	Buy	5	49	90.7
Overall percentage				71.4

significant increasing of VN-index. Nevertheless, the results indicate that there are no correlation between the binary number of VN-index and those 2 variables. As a result, we come up with the function of regression model as follows:

$$E(Y/X) = \frac{e^{7.714-8.573*Cpi+0.03*Ov+0.00376*Ex}}{1+e^{7.714-8.573*Cpi+0.03*Ov+0.00376*Ex}}$$

- *The level of accurate prediction of the model.*

Assuming if boundary classification is taken with random probability is 0.5 percent if the stock price is greater than 0, the classification is decided by investors purchase with probability 0.5 (Y = 1).

Table 7 shows that: In 30 (11 + 19) selling decision, the proportion of correct prediction of model is 36.7%. In 54 (5 + 49) buying decision, the proportion of correct prediction of model is 90.7%. Therefore, the proportion of correct prediction of the entire pattern is 71.4%

6 Conclusion

The purpose of the study is to empirically examine the relationships existing between macroeconomic factors including inflation rate, exchange rate, interest rates, crude oil price, FDI and investor purchase decision in HOSE under the period of 2008 to 2015. The results of the study indicate that all data is stationary and are shown as follows:

Firstly, there has been existing the long-term relationships between macroeconomic variables on the value changes of VN-Index with the exception of two variables including (1) interest rates and (2) FDI. Specifically, most clearly expressed about the degree of influence is inflation rate which affects negatively the value of VN-index, followed by exchange rate and crude oil prices, respectively. In addition, the inspection in the data analysis section shows that causation between consumer price index and VN-index possesses the Granger Causality. Therefore, the adjustment of inflation rate could affect investors' sentiment negatively. In fact, from the early of 2008 to 2015, the actual inflation rate at the end of the year always has a huge difference compared to the goal that the Congress stated at the beginning of the year, even if this goal has been amended several times. Obviously, if there is no improvement on the level of market confidence in terms of policies, we will not have any magical tools to make the inflation forecast to be the actual approach. This, in turn, involves the rational caution when building the target to ensure a compliance with the capacity to achieve the economies of calculating interests and costs. A prudent route and not hasty to be considered when managing actors and aiming for long-term benefits rather than short-term. Viet Nam can be described as an economy with more than 90% of small and medium enterprises being low-tech and dependent on external sources such as raw materials and capital (both for short, medium and long term). As a result, it will inevitably lead to sacrifice very high rate when pursuing the inflation target. Indeed Vietnam market has demonstrated this effect.

Secondly, results from variance decomposition show that the VN-index is less sensitive to shocks from the macroeconomic information and tends to react the imbalance when shocks occurred. The leading reason may be the government has spent too much administrative intervention which limits the amplitude in the market and restricts short selling. That makes little macro variables reflecting the substance of economic activity.

Finally, in forecasting perspective, from the results of the logistic regression, only three variables namely CPI, exchange rates and crude oil prices are capable for forecasting price movements of shares on the stock market. The remaining two variables are FDI and interest rate which mostly do not have any influences on investor purchase decision. Therefore, under this perspective, the stock market is showing a quite separate operation with financial markets. For this reason, there is still a lot of potential risks when the stock market volatility has a slow response to the information of macro-economic factors. The control on information publishing and transparency in securities transactions are still very limited. The irregularities in the management of the stock market is only discovered at the most basic level such as violations of reporting, information disclosure of listed companies, violation reports in Trading insiders, and improper content of business operation. Furthermore, the price manipulation of shares shall not be clarified. This is true as the competent authority and supervision of the State Securities Commission are limited.

In conclusion, CPI, exchange rates and crude oil prices are capable to forecast the value of shares on the HOSE, the other two variables (FDI and interest rate) mostly do not prove any influences. Hence, the psychology of Viet Nam investors may describes the trend of short-term investment or surfing investment. In more detailed, FDI information only affects investors who possess the related shares of the field being received the investment, in the other hand, the effect of FDI is partial and clustered but impact on the whole market. In addition, although the effect of interest rate proves a significant effect with the share value, under general aspect of wholesale investors, especially financial firms – who mostly possess a huge volume of shares in the market, that effect is probably considered as a small impact. The changing value of interest rate is so small ranging from 1–2%, so it will not really make investors withdraw their investments in the stock market.

From the analysis of effected and non-effected factors, this study suggest two method to minimize the risk and maximize the utility of macro-economic information. (1) the market regulation mechanism of equilibrium is not always possible. Therefore, the automatic cut loss/take profit order should be applied to aid investors feel not too passive under the "unreasonable" situation of the market. (2) Investors should limit the use of high margin if possible. The use of financial leverage can easily cause investors to fall into excessive optimism or excessive pessimism when market fluctuates.

Eventually, although the study has successfully examined the effects of macro factors on investor purchase decision, there have some main unobserved issues. Firstly, this study only conducts the case study of VN-index in the HOSE, so it does not really express the whole picture of the Viet Nam stock market. In addition, this study also does not pay much attention in share value manipulating by big companies. As a result, those obstacles will be addressed in future studies.

References

Akaike, H.: Information theory and an extension of the maximum likelihood principle. In: Petrov, B.N., Csaki, F. (eds.) Proceedings of the 2nd International Symposium on Information Theory, Budapest Akademiai Kiado, pp. 267–281 (1973)

Bhunia, A.: The shock of the shock of crude price, stock price and selected macro-economic variables on the growth of Indian economy. Sumedha J. Manag. 1(2), 82–89 (2012)

Charles, A., Kasilingam, R.: Does market sentiments influence investor's investment decisions? Anvesha 7(3), 24–33 (2014)

Cox, D.R.: Regression models and life tables. J. R. Stat. Soc. Ser. B 34(2), 187–220 (1972)

Daniel, K., Hirshleifer, D., Subrahmanyam, A.: Investor psychology and security market under- and overreactions. J. Financ. 53, 1839–1885 (1998)

Dickey, D.A., Fuller, W.A.: Likelihood ratio statistics for autoregressive time series with a unit root. Econometrica 49(4), 1057–1072 (1981)

Fifeková, E., Nemcová, E.: Impact of FDI on economic growth: evidence from V4 countries. Periodica Polytechnica 23(1), 7–14 (2015)

Elliott, K.: An empirical identification of an appropriate inflation definition and an inflation-targeting monetary policy (Order No. 3251363). Available from ProQuest Central (304767934) (2007). https://search.proquest.com/docview/304767934?accountid=63189

El-Masry, A.: The Exchange Rate Exposure of UK Nonfinancial Companies: Industry-Level Analysis. Social Science Research Network, Rochester (2009)

Engel, R.F., Granger, C.W.J.: Cointegration and error correction: representation, estimating and testing. Econometric 55, 251–276 (1987)

Gujarati, D.N.: Basic Econometrics, 4th Economy edn. Tata McGraw-Hill Publishing Ltd., New Delhi (2004)

Hardouvelis, G.A.: Macroeconomic information and stock prices. J. Econ. Bus. **39**(2), 131 (1987)

Hondroyiannis, G., Papapetrou, E.: Macroeconomic influences on the stock market. J. Econ. Financ. **25**(1), 33–49 (2001)

Johansen, S., Juselius, K.: Maximum likelihood estimation and inference on cointegration—with applications to the demand for money. Oxford Bull. Econ. Stat. **52**(2), 169–210 (1990)

Joseph, N.L.: Modelling the impacts of interest rate and exchange rate changes on UK stock returns. Deriv. Use Trading Regul. **7**(4), 306–323 (2002)

Kieu, N.M.D., Nguyen, V., Tam, L.H.N.: (2013). http://tapchitaichinh.vn/nghien-cuu–trao-doi/trao-doi-binh-luan/cac-yeu-to-kinh-te-vi-mo-va-bien-dong-cua-thi-truong-chung-khoan-viet-nam-39026.html. Accessed Sept 2015

Kim, K., Nofsinger, J.: Behavioral finance in Asia. Pac.-Basin Financ. J. **16**(1–2), 1–7 (2008)

Kleinbaum, D.G., Klein, M.: Logistic Regression-A Self-learning Text, 2nd edn. Springer Science and Business Media, Heidelberg (2002)

Malcolm, B., Jeffrey, W.: Investor sentiment in the stock market. J. Econ. Perspect. **21**, 129–151 (2007)

Phan, K.C., Zhou, J.: Market efficiency in emerging stock markets: a case study of the Vietnamese stock market. IOSR J. Bus. Manag. **16**(4), 61–73 (2014)

Phillips, P.C.B., Perron, P.: Testing for a unit root in time series regression. Biometrika **75**(2), 335–346 (1988). Oxford University Press

Quan, V.D.H., Cong, B.C.: (2016). http://tapchitaichinh.vn/nghien-cuu-trao-doi/nghien-cuu-dieu-tra/yeu-to-and-huong-den-quyet-dinh-cua-nha-dau-tu-chung-khoan-86193.html. Accessed Sept 2016

Ray, S.: Testing granger causal relationship between macroeconomic variables and stock price behavior: evidence from India. Adv. Appl. Econ. Financ. **3**(1), 470–481 (2012). http://worldsciencepublisher.org/journals/index.php/AAEF/issue/view/79

Saari, M.Y., Puasa, A.F., Hassan, K.H.: The impact of world crude oil price changes on the Malaysian economy: an input-output analysis. Malays. J. Econ. Stud. **44**(1), 1–12 (2007)

Smith, G.W.: The Missing Links: Better Measures of Inflation and Inflation Expectations in Canada. Commentary - C.D. Howe Institute (287), 0_1, 0_2, pp. 1–19. C.D. Howe Institute, Toronto (2009)

Suthar, M.H.: Understanding exchange rate expectations in India IUP J. Appl. Econ. **9**(1), 106–126 (2010)

Thuy, T.T.T., Duong, V.T.T.: Sự tác động của các nhân tố kinh tế vĩ mô đến các chỉ số giá cổ phiếu tại HOSE. Tạp chí nghiên cứu và trao\ đổi, Đại Học Kinh Tế, số 24, trang 34 (2015)

The Inflation-Economic Growth Relationship: Estimating the Inflation Threshold in Vietnam

Tuyen H. Tran[✉]

Banking Academy of Vietnam, 12 Chua Boc, Dong Da, Hanoi, Vietnam
tuyenth@hvnh.edu.vn

Abstract. Utilizing annual data over the period 1990–2015 and employing the threshold model developed by Sarel (1996) with some modifications, this paper estimates the inflation threshold in Vietnam and simultaneously examines the linkage between inflation and economic growth. The findings show that the estimated inflation threshold stays at 3%–4%, above which the positive effect of inflation on economic growth vanishes, and this effect starts fading at 5.5%–7.5%. The findings of this paper regarding the inflation threshold and the inflation-economic relationship are expected to help the State Bank of Vietnam (SBV) in conducting its monetary policy, especially the inflation policy. Moreover, the findings also help the government evaluate the role of the gross domestic product's determinants in promoting the economic growth.

1 Introduction

Economic growth and inflation are both the central topics of monetary policy and two important economic objectives of all nations. Most central banks in the world pursue a monetary policy that tries to simultaneously achieve low inflation and the high economic growth targets. Thus, understanding the linkage between inflation and output growth plays a vital role in a country's monetary policy framework, and Vietnam is not an exception. This linkage, however, is not straightforward and difficult to grasp because, in fact, inflation is a monetary phenomenon whereas economic growth is a real one. Additionally, characteristics of this relationship might be different across different countries as well as different periods. Therefore, it is necessary for policymakers to have the right conception of the inflation and economic growth relationship in their own country.

After the Asian financial crisis in 1997, inflation has been regarded as one of the four major factors that resulted in the macroeconomic instability in Vietnam, together with the exchange rate, budget deficit and current account deficit (Nguyen et al. 2010). During this period, inflation has been one of the hottest macroeconomic issues, and has been widely discussed in Vietnam among economists as well as policy makers because of its high volatility, unpredictability, persistence as well as its effect on economic growth. Apart from the period 2000–2003 when the inflation rate was quite low and stable, the inflation rate in Vietnam is often high and unpredictable in comparison with other countries in ASEAN. This circumstance has had adverse effects on the economy. Thus, Vietnam's government and SBV changed their primary objectives from maintaining high economic growth to controlling and stabilizing inflation since 2011 when

L. H. Anh et al. (eds.), *Econometrics for Financial Applications*, Studies in Computational Intelligence 760, https://doi.org/10.1007/978-3-319-73150-6_73

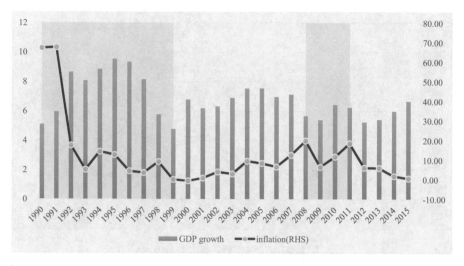

Fig. 1. The inflation and economic growth (1990–2015). *Source: IFS & GSO*

Resolution 11 was adopted. This was considered as a suitable change in light of the macroeconomic instability experienced during the period leading to 2011. As a result, the economy has since experienced a period of relatively low and stable inflation, together with moderate and increasing economic growth (Fig. 1).

To date, only a limited number of research about the inflation-economic relationship and the estimation of inflation threshold have been conducted in Vietnam, and most of them are cross-country studies. This paper is one of the first studies for investing the relationship between inflation and economic growth focusing on Vietnam, and simultaneously estimating the inflation threshold over the period 1990–2015.

The structure of paper consists of six sections that are described as follows. The first section begins the paper by introducing the rationale of the study, the statement of the problem, and the organization of the paper. Section 2 gives a brief overview of the evolution of inflation. Section 3 provides a review of previous studies relating to the issues that are analyzed in the paper. Section 4 explains the framework of empirical analysis which are used in this paper. Section 5 presents and analyzes the main findings. Finally, the final section offers the conclusion and policy implications derived from the findings of this study.

2 The Evolution of Inflation

Before 1986, Vietnam applied Soviet Union's distribution system through food stamp and rice book which used to be a major way that people could buy their goods and services. In this planned economy where goods and services were distributed rather than traded, the

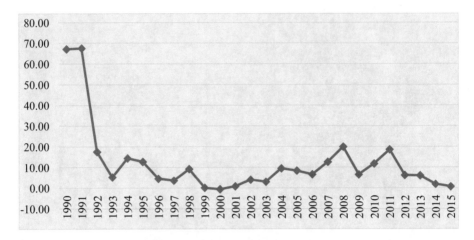

Fig. 2. Inflation rate (1990–2015). *Source: IFS*

authorities did not pay much attention to inflation. Since Doi Moi[1] in 1986, Vietnam officially dropped such a distribution system. Prices of goods and services were allowed to be determined by their supply and demand in the market. This major shift together with serious food shortages and production stagnation resulted in a skyrocketing consumer price index for the period 1986–1992. Consequently, Vietnam experienced hyperinflation with approximately 800% per year in 1986. High inflation in this period had several adverse impacts on the economy. For example, Vietnamese citizens' living standards substantially declined as their income's purchasing power weakened considerably (Vuong 2004). Furthermore, other reasons of high inflation in this period came from bad weather, the weak financial system, and poor governance of the authority.

With the aim of restraining inflation, SBV tightened monetary policy by increasing interest rates, reducing money supply growth rates, and pegging rigidly exchange rates against the USD in the beginning of 1990s. These policies resulted in a strong decline in inflation from around 67% in 1991 and 1992 to only 12.7% in 1995. After a moderate inflation period from 1995–2006 with the rate of inflation always below 10%, even the first modest deflation in Vietnam with −0.6% in 2000, high and volatile inflation came back to the Vietnam economy in the period from 2007–2011 with double-digit levels (except for 2009) (Fig. 2).

In comparison with its major trading partners, Vietnam suffered higher and more volatile inflation rates in this period, especially in 2008 and 2011 when inflation in these years reached almost 20% (Fig. 3). Similar to previous high inflation periods, high inflation led to high interest rates which had many negative effects on the Viet-namese economy. In terms of sources, the sharp surge of inflation in this period stemmed from both internal and external factors, including (i) the government increased the minimum wage, (ii) the central bank implemented monetary policy easing

---

[1] Doi Moi is the name given to the economic reforms initiated in Vietnam in 1986 with the goal of creating a "socialist-oriented market economy".

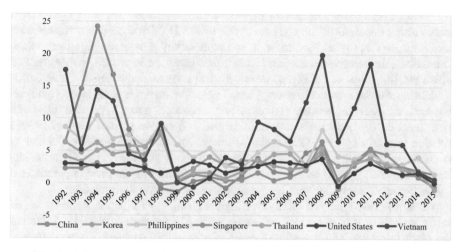

Fig. 3. Inflation rates of Vietnam and selected countries (1992–2015). *Source: IFS*

and fixed exchange rate regime, (iii) Vietnam had more open economy after becoming an official WTO member in 2006, (iv) prices of international commodities (mostly crude oil and materials) strongly surged. Especially, in 2008 and 2011, Vietnam experienced severe floods and storms which substantially affected the production of crops. These negative supply shocks dramatically affected foods prices, and indirectly inflation. The susceptibility to weather conditions is accentuated in Vietnam since food accounts for around 40% of the consumption basket which is used to calculate CPI.

To prevent high inflation, SBV changed its monetary policy stance to tightening by increasing the reserve requirement ratio for foreign currency deposits, selling securities in the open market, increasing policy rates and applying ceiling deposit rates. After experiencing a high level of inflation in 2007–2008 and 2011, SBV has been focusing much more on controlling inflation and keeping economic growth at a reasonable level. Consequently, in the period from 2012–2015, the inflation rate in Vietnam returned to a more moderate level with a relatively low and stable rate. Remarkably, the inflation in 2015 was only 0.63%: the lowest inflation rate recorded over the last 14 years. More-over, low inflation in this period was induced by low consumption and investment levels which raise concerns for authorities about slowing economic growth in the future.

3   Literature Review

a. The Inflation and Economic Growth Relationship

In recent decades, the relationship between inflation and economic growth has long been a controversial debate among researchers as well as policymakers. While some studies suggest that there is a linear relationship (positive or negative) between the two variables (Fischer 1993, Barro, 1995, Gregorio 1993), others argue that this relationship is non-linear (Ghosh and Phillips 1998, Eggoh and Khan 2014, Bruno and Easterly 1996). And a few argue that the relationship is unclear and insignificant (Paul et al. 1997).

Fischer (1993) in his influential paper implicitly indicated that there is a negative relationship between inflation and economic growth. He also argued that in addition to economic growth, capital accumulation and productivity growth are negatively related to inflation. He presented empirical results that suggest an important channel through which inflation adversely affects economic growth is by reducing capital accumulation.

Having similar empirical results, Barro (1995) confirmed a statistically significant negative relationship between inflation and economic growth on one hand and investment on the other for a sample of around 100 countries from 1960 to 1990. If inflation increases by 10%, GDP and investment would decrease by 0.2 to 0.3 and 0.4 to 0.6%, respectively. Other studies (Kormendi and Mequire 1985, Gylfason and Herbertsson 2001), based on a cross-country database, also reconfirmed the presence of a negative relationship between long-term economic growth and inflation.

Gregorio (1993) focused on discovering the channels of inflation's effects on economic growth in twelve Latin American countries in the period 1950–1985. He was especially interested in investigating the impact of inflation on economic growth through the productivity of capital and the rate of capital accumulation. His conclusion suggested that inflation has a negative effect on growth primarily through the productivity of capital rather than the rate of capital accumulation. In addition, he concluded that not only inflation but also its variance has adverse effects on economic growth.

Using a database of 70 countries in the period from 1960 to 1989, Paul et al. (1997) applied the Granger methodology to test both the direction and pattern of causality between inflation and economic growth. The empirical results did not find any evidence to support a significant relationship between inflation and economic growth. Moreover, Paul et al. (1997) pointed out that this relationship is non-uniform across countries. Specially, 40% of countries in the sample show no causality; around 30% present unidirectional causality (which causes which differs from country to country); and about 20% of the sample exhibit bidirectional causality.

In another study that tested the relationship between inflation and economic growth, Bruno and Easterly (1998) in a sample of 127 countries between 1961 and 1994 chose a 40% inflation rate to split observations of inflation into two groups: high and low inflation rate. Above this level, inflation negatively affects growth, but inflation has no impact on economic growth when inflation is smaller than 40%. Therefore, a threshold exists at 40%, at which the impact of inflation on economic growth changes from unclear to a negative relationship. However, since inflation of 40% is too high in most cases, according to this study, this relationship is not meaningful.

Agreeing with the conclusion of non-linear relationship between inflation and GDP growth, the study of Ghosh and Phillips (1998), which covers 145 International Monetary Fund (IMF) member countries over 1960–1996, also found that at very low levels of inflation (around two to three percent), the relationship is positive. At all other levels, a negative correlation was found, and the marginal effect of inflation on growth declines as inflation increase. For example, a decrease in economic growth as inflation increases from twenty percent to thirty percent is larger than when it increases from thirty to forty percent.

In summary, based on a series of cross-country studies, although the relationship between inflation and economic growth is not straightforward, the majority of studies conclude that there is a non-linear relationship between inflation and economic growth. They have indicated that there exists at least one inflation threshold at which impact of inflation on economic growth alters. A review of further studies on the possible presence of an inflation threshold is provided in the next section.

b. Estimation of the Inflation Threshold

Although, to date, most economists concur that a significant non-linear relationship exists between inflation and growth, what remains unclear is the threshold level of inflation, above which the impact of inflation on growth changes its magnitudes or even signs. Also unclear is whether there is any difference in the inflation threshold level between developed and developing countries.

Empirical studies have used different econometric methodologies with the objective of estimating inflation threshold level for a group of countries or an individual country. While the former used cross-country data to determine a common threshold level for a group of countries, the latter focused on a country's data to detect inflation threshold. One of the big advantages of the first approach is a large number of observations. This approach may not be so reasonable, however, because it assumes that different countries with different characteristics have the same level of inflation threshold. On the other hand, even though the second approach gives more meaningful implications for the country in question, it must deal with the problem of limited data as one of the main obstacles. For example, in developing countries, the number of observations of macroeconomic variables typically is not larger than 40 years.

Ideally, to capture specific characteristics of each country, inflation threshold level should be estimated for each individual country separately. But because of limited data, as mentioned above, the majority of studies have used panel techniques (Espinoza et al. 2010). This explains the reason why most of the research about inflation threshold are cross-country studies. The following sections cover both individual and cross-country studies.

i. Cross-Country

Fischer (1993) (considered to be one of the first economists to pay attention to the possibility of a nonlinear relationship between economic growth and inflation) used cross-sectional as well as panel regression for a large sample of 101 countries in the period from 1960 to 1989. In order to test for the possibility of nonlinear effects of inflation on economic growth, the author broke inflation rates into three categories, namely less than 15%; between 15% and 40%: and larger than 40%. The findings show that whatever the category is, economic growth always negatively relates to inflation. However, this relationship is non-linear, and the effect of inflation on growth is stronger at low and moderate levels than at high inflation[2]. Put differently, this relationship becomes weaker when inflation increases. Moreover, he also showed that high inflation has adverse effects on growth through channels of reducing capital accumulation and productivity growth rate.

The existence of such a non-linearity of the linkage between inflation and economic growth has been confirmed by other studies, such as Ghosh and Phillips (1998), Bruno and Easterly (1998), Christoffersen and Doyle (1998), Burdekin et al. (2004), Gillman

---

[2] See Fisher (1993. pp. 503).

and Kejak (2005). The characteristics of this association in terms of the estimated inflation-growth threshold and the marginal effect of inflation on growth, nevertheless, are very different for different studies.

Sarel (1996) used annual data of 87 countries in the period from 1970 to 1990 to examine the nonlinear effect in the relationship between inflation and economic growth. His paper suggests a significant structural beak in the economic growth and inflation relationship which occurs at an inflation rate of 8%. When inflation rate is smaller than 8%, inflation does not have any impact on economic growth, and the effect even is a little positive. In contrast, when inflation is above that rate, the estimated result is negative and highly significant. The methodology that he used to measure the point of the structural break is to estimate the regression with different values of inflation threshold in the given interval. The inflation threshold is a point at which the sum of squared residuals from the regression is minimized. In other words, the value of inflation that maximizes R-squared is the inflation threshold. Many other studies (Ghosh and Phillips 1998; Judson and Orphanides 1999) also re-confirmed similar structural breaks in the growth and inflation linkage.

Additionally, using the dataset of 140 countries which includes both developed and developing countries through the period 1960 to 1998, Khan and Senhadji (2001) re-estimated the existence of threshold effects. Their methodology basically is an extension of the study of Sarel (1996), and employed threshold estimate technique of Hansen (2000), together with some new economic techniques that aim to provide the more suitable procedure. The empirical results indicated that inflation threshold had significantly existed in both groups (developed and developing countries), and estimated at 1–3% for industrial countries and 11–12% for developing countries.

Another study by Espinoza et al. (2010) also used cross-country data from 165 countries over the period 1960 to 2007 to re-examine the classic relationship between inflation and economic growth. The Logarithm Smooth Transition model (LSTR) was firstly applied with several control variables as determinants of growth to measure the speed of transition at which inflation becomes harmful to growth after it is greater than the threshold. In addition, by applying this model, it is possible to capture similar effects on the economic growth of multiplicative shocks[3] to inflation for all levels of initial inflation. Their results identified that a threshold level for the entire sample of data is about 9%. Nevertheless, when the entire sample is divided into several country groups, (including advanced countries, emerging countries, a group of oil producers, and non-oil producers), the estimated inflation threshold level for each group is relatively different. While that for advanced countries is much lower, the adverse effect of high inflation on economic growth in oil producers is much stronger than others.

More recently, using both the Panel Smooth Transition model (PSTR) specification and Generalized Method of Moments (GMM) procedures for panel data on a wide sample of 44 countries, consisting of high income countries, upper middle-income countries, lower middle-income countries and emerging countries, López-Villavicencio and Mignon (2011) detected strong evidence that inflation non-linearity effects on

---

[3] In the log model, whatever level of initial inflation is, a doubling of the inflation rate will have the same effect on economic growth (Espinoza et al. 2010).

growth. Besides, the threshold level is considerably different between industrialized and emerging countries. Specifically, the inflation threshold is 2.7% for industrialized countries and 17.5% for emerging ones[4]. The authors explained that this difference might stem from Balassa-Samuelson effect[5], the indexation systems, the exchange rate policies and the high levels of inflation encountered by some emerging countries.

With the aim of observing a nonlinear relationship between inflation and economic growth for a group of 32 Asian countries in the period from 1980 to 2009, Vinaya-gathasan (2013) employed a dynamic panel regression that allows for fixed effects and endogeneity. The empirical results identified that there is an inflation threshold of 5.43% at the 1% significance level. Inflation impedes economic growth when it's above the threshold, but doesn't affect growth below this threshold level.

Also, for Asian countries, Su (2015) examined the inflation threshold for ASEAN-5 countries (Indonesia, Malaysia, Philippines, Thailand and Vietnam) over the period 1980–2011. The author employed the Panel Smooth Transition Regression model to estimate the threshold, and GMM-IV specification was applied to check robustness. The results indicated that the inflation threshold for ASEAN-5 is 7.84% above which there is a statistically significant negative relationship between inflation and growth, and inflation becomes harmful to economic growth.

ii. Individual Country

Few studies have tried to test the non-linear relationship between inflation and economic growth in a specific country, as well as measure its value of inflation threshold, such as Singh and Kaliappa (2003); Lee and Wong (2005); Chowdhury and Kalirajan (2009).

The study of Chowdhury and Kalirajan (2009) tried to answer the question of what level the Bangladesh Bank should set its inflation target through estimating an inflation threshold value above which inflation negatively affects economic growth. Although the study did not identify any inflation threshold for the Bangladesh economy, it shed new light on the causes of inflation in Bangladesh. Furthermore, using the vector autoregressive (VAR) model with Bangladesh annual data for the period 1976 to 2005 helped discover some important channels of inflation. In line with the empirical result, the paper suggested that because an increase in any level of inflation is always harmful to economic growth, the central bank should focus monetary policy towards maintaining price stability.

Interestingly, Singh and Kaliappa (2003) obtained similar results for the case of India. In their study, they basically applied the threshold model developed by Sarel (1996) for India for the period of 1971–1998. Many of variables that are partially correlated with economic growth have been included in the model as explanatory variables. Their findings have clearly shown that there is no threshold level of inflation

---

[4] This group is composed of upper-middle income and lower-middle to low-income countries.

[5] The Balassa-Samuelson effect suggests that in developing economies where labor productivity catch-up is taking place in manufacturing, the equilibrium inflation rate tends to be higher than for developed countries.

in India. In other words, an increase in inflation from any level always has an adverse effect on economic growth.

Lee and Wong (2005) employed the "Threshold Autoregressive" (TAR) model proposed by Tong (1978) and Hansen (1996) to investigate inflationary threshold in the financial development and economic growth relationship in Japan and Taiwan. The empirical results identified that there are two inflation thresholds in Japan (2.53% and 9.66%) and one in Taiwan (7.25%). Some other studies for individual countries also have obtained significant results, such as Mubarik (2005) for Pakistan; Salami and Kelikume (2010) for Nigeria.

This section has reviewed the important findings of the studies relating to the estimation of the inflation threshold in the sample of both cross-country and individual-country data. Most of the studies (except for the studies of Chowdhury and Kalirajan 2009; and Singh and Kaliappa 2003) successfully estimated an optimal inflation threshold at which the effect of inflation on growth substantially changes. The estimated values of inflation threshold, nevertheless, are non-uniform among the studies. Moreover, because cross-country studies are based on the assumption of different countries having a common inflation threshold, the results may not be applicable to a specific country case.

In addition, according to the literature presented above, most of the cross-country studies don't include Vietnam in the sample[6], which is likely because of limited data. And with the individual - country studies, there is very few empirical analysis of this topic in Vietnam compared to other developing countries. Thus, this study can be seen as one of the first efforts for estimating the inflation threshold for the case of Vietnam.

4   Empirical Framework

The methodology which is used for the purpose of estimating the level of inflation threshold is mainly based on the simple technique developed by Sarel (1996) with some modifications relating to forms of variables. At first, a series of cross-sectional regressions will be run to generate sums of squared residuals corresponding with different dummy variables. Then, the maximization of the sum of squared residuals methodology will be applied to determine the best value for the inflation threshold. The remainder of this section will discuss this methodology in greater detail.

The main regression takes the form as shown in Eq. (1). The dependent variable is the growth rate of real gross domestic product in year t(GDP_t). On the right-hand side of the regression, independent variables consist of inflation rate in year t(INF_t), extra inflation which is measured by the difference between actual inflation rate and a threshold $(INF_t - INF_i^*)$, together with other explanatory variables which are presented by a vector of control variables (X_t).

Obviously, the threshold must lie between the maximum and minimum value of the series. In the sample, except for two outliers of inflation in 1990 and 1991, the inflation

---

[6] Only the studies of Espinoza (2010), Vinayagathasan (2012), and Su (2015) include Vietnam in their cross-country sample.

rate fluctuates from −0.6% to 19.89% over the period 1990–2015 (Fig. 2). Hence, the range from 1% to 20% will be given for the values of inflation threshold (INF_i^*) in the regression procedure. With the aim of finding the breaking point in the relationship between inflation and economic growth, a dummy variable would be defined as $DUM_i = 1$ if $INF_t > INF_i^*$, 0 otherwise (Eq. 2).

$$GDP_t = \beta_0 + \beta_1 INF_t + \beta_2 DUM_i * \left(INF_t - INF_i^*\right) + \Theta'X_t + \varepsilon_t, \qquad i = 1, \ldots, 20 \quad (1)$$

$$DUM_i = \begin{cases} 1 & \text{if } INF_t > INF_i^* \\ 0 & \text{if } INF_t \leq INF_i^* \end{cases} \quad (2)$$

Based on Eq. (1), the effect of inflation on economic growth is measured by the coefficient of INF_t or (β_1) when inflation is lower than a threshold value, and measured by the sum of coefficients of INF_t and $DUM_i * \left(INF_t - INF_i^*\right)$ or $(\beta_1 + \beta_2)$ when inflation is higher than a threshold value. In other words, the coefficient β_2 reflects the difference in the impact of inflation on growth between two sides of the structural break, and its t-statistic value shows whether a threshold is significant or not. To determine whether the sum of $(\beta_1 + \beta_2)$ is statistically different from zero, we also will conduct the Wald Test (Coefficient Restrictions).

With different DUM_i variables corresponding to different values of INF_i^*, Eq. (1) is iterated with each dummy variable to generate a sequence of sums of squared residuals (RSS) by using the Ordinary Least Squares (OLS) method.

Nevertheless, the OLS estimator of the regression coefficients might be biased due to the endogeneity issue which happens when one or more explanatory variables are correlated with the error term. Such a correlation, in turn, can stem from omitted variables or simultaneous causality. Equation (1) possibly involves this issue because omitted variables are an inevitable problem of most studies based on the regression analysis. Moreover, not only independent variables cause the dependent variable, but the causality also runs in the opposite direction (the dependent variable causes independent variables). For example, many studies confirmed that FDI and economic growth are important determinants of each other. In addition to omitted variables, Eq. (1) therefore easily faces the simultaneous causality threat. One possible measure of handling the endogeneity is to apply the Two-Stage Least Squares (2-SLS) method with instrumental variables. Specifically, lagged values of variables are used as instruments.

Technically, the inflation threshold would be an inflation rate that minimizes the sum of squared residuals (RSS) value or maximizes R^2. The threshold estimate level INF^* is chosen to minimize $S_1(INF)$, that is:

$$INF^{threshold} = argmin\{S_1(INF^*), INF^* = 1, \ldots, 20\} \quad (3)$$

In order to check the robustness of the empirical findings, different model specifications with different samples will be used. As described in Sect. 2, the full sample of 1990–2015 includes observations of two years (1990, 1991) with very high inflation levels (67.5% and 67.1% respectively). They are regarded as outliers which partly affect the slope coefficients of the regression. To test the influence of these outliers as well as ensure the robustness of the empirical result, we repeat the same process of estimating the inflation threshold for a sub-sample that excluding observations of 1990 and 1991. In addition, for comparative purposes, the procedure of estimating the inflation threshold also is applied for the 1995–2015 period.

Furthermore, another traditional way to discover the nonlinear relationship between inflation and economic growth without using the threshold model is to employ quadratic specification. This method is particularly useful in testing whether or not the effect on economic growth of a change in inflation depends on the level of the inflation rate. Additionally, this technique allows us to identify the peak value of inflation, at which the impact of inflation on economic growth changes. GDP will be regressed on inflation (INF) and squared inflation (INF^2), plus with other explanatory variables. The regression takes the following form:

$$GDP_t = \beta_0 + \beta_1 INF_t + \beta_2 INF_t^2 + \Theta' X_t + \varepsilon_t \tag{4}$$

The value of peak inflation can be calculated by the following equation:

$$INF^{peak} = \frac{\beta_1}{2 * \beta_2} \tag{5}$$

Besides, the estimated coefficients of explanatory variables will help understand more about the effects of determinants on the economic growth. Based on previous studies together with Vietnam's economic situation and the availability of data, some variables have been chosen as important determinants of economic growth.

5   Empirical Results

a. Unit Root Test

Prior to employing the threshold model, it is important to test whether the variables that will be used in the regression are stationary or not. This step makes sure that the variables are not subject to spurious correlation. The Augmented Dickey-Fuller (ADF) and the Phillips-Perron (PP) are employed to test the hypothesis of a unit root of the variables for the full sample period of 1990–2015[7]. The number of lags in ADF test was selected according to Akaike Information Criterion (AIC) and Schwarz Information Criterion (SIC).

The entire results of unit root test are given in Table 1. The ADF test with an intercept but without time trend suggests that the null hypothesis of unit root is rejected at 10% or higher significance level for most variables used in the model (except for INV and AGR). If both intercept and time trend are included, these variables are also stationary at 5% level. In the Phillips-Perron test, most variables (except for GDP, POP,

---

[7] The results of unit root test for the sub-sample periods (1992–2015 and 1995–2015) suggest that all variables are stationary.

Table 1. Results of unit root tests

Variables	Augmented Dickey-Fuller (ADF)		Phillips-Perron (PP)	
	Intercept without time trend	Intercept with time trend	Intercept without time trend	Intercept with time trend
GDP	−2.6922*	−4.6030 * **	−2.4898	−3.1300
INF	−4.0311 * **	−3.5768*	−8.6056 * **	−7.1108 * **
ROT	−5.0740 * **	−4.7722 * **	−4.6710 * **	−4.6195 * **
REX	−4.8013 * **	−5.0487 * **	−4.8157 * **	−5.0608 * **
POP	−2.8453*	−0.1460	−2.5742	−0.3412
FDI	−3.2976 * *	−3.1741	−3.2801 * *	−3.0802
INV	−1.7894	−3.8536 * *	−1.7475	−1.8843
AGR	−0.8992	−5.1592 * **	−3.7006 * *	−7.0797 * **

Source: Author's calculation using EViews 8
Note: (i) The ADF and PP tests are based on the null hypothesis of unit roots. ***, **, *
indicate significant at 1%, 5%, and 10% levels respectively, based on the critical t statistics as
computed by MacKinnon (1996)
(ii) The variables are all in growth rates

and INV) are found stationary either with or without time trend at 10% or higher significance level. Although ADF and PP tests are not entirely conclusive, we proceed, assuming that all variables in the study are stationary.

b. **Main results**

i. **Ordinary Least Squares (OLS) Method**

This section will show and analyze the empirical results of estimating the inflation threshold for the sample of 1990–2015 and 1992–2015[8] by using the OLS method.

The period 1990–2015

First of all, the threshold model is estimated for the full sample period of 1990–2015. As illustrated in Fig. 4, it can be easily seen that the sum of squared residuals attains the lowest value with $INF^* = 4$. As explained in Sect. 4, the estimated inflation threshold, in this case, is 4%. According to Table 2, the coefficient of inflation (β_1) and the coefficient of extra inflation (β_2) are statistically significant at 1% and 5% level respectively. β_1 is 0.3372 whereas (β_2) is −0.3896, so their sum is negative.

And when using the Wald Test for testing coefficient restrictions, we can reject the null hypothesis that $\beta_1 + \beta_2 = 0$ at 1% level (Table 3). This indicates that inflation has a positive impact on output growth when it is below 4%. In contrast, when inflation is higher than 4%, this linkage becomes slightly negative $(\beta_1 + \beta_2 = 0.3372 - 0.3896 = -0.0524)$. Hence, the estimated results reinforce a non-linear relationship between inflation and economic growth in Vietnam in the period 1990–2015, with the breaking point occurring at 4%.

---

[8] The estimated result of 1995–2015 is not statistically significant.

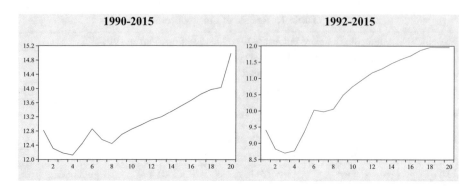

Fig. 4. OLS-value of sum of squared residuals. *Source: Authors' own calculations by using Eviews*

Table 2. OLS estimation

Variable	Coef.	
	1990–2015	1992–2015
INF	0.3372**	0.4528**
	(0.1301)	(0.1720)
DUM*(INF-INF*)	0.3896***	−0.4668**
	(0.1351)	(0.1904)
ROT	−0.0399**	−0.0388**
	(0.0188)	(0.0171)
REX	0.0632***	0.0764***
	(0.0194)	(0.0198)
POP	2.3115**	3.2839***
	(0.9956)	(1.0244)
INV	0.0465**	0.0434**
	(0.0213)	(0.0194)
AGR	−0.1104	−0.2551*
	(0.1277)	(0.1383)
Intercept	2.4205**	1.6779
	(1.0905)	(1.0318)
Inflation threshold	4%	3%
Number of observations	26	24
Adjusted R-squared	0.62	0.69

Source: Authors' own calculations by using Eviews 8
Note: Dependent variable is GDP growth rate; ***,
**, * indicate significant at 1%, 5% and 10% levels
respectively

Let us now consider the effect of other variables on the economic growth. All coefficients of explanatory variables are statistically significant at 5% level (except for AGR). In particular, estimated coefficients of REX, POP, and INV are positive, which

Table 3. Wald test-null hypothesis $\beta_1 + \beta_2 = 0$ (The inflation-growth linkage)

Method	Period	Value	Std. Err	t-statistic	F-statistic	Chi-square
OLS	1990–2015	−0.0524	0.0176	−2.9768	8.8615	8.8615
	1992–2015	−0.0139	0.0365	−0.3807	0.1449	0.1449
2-SLS	1992–2015	−0.0119	0.0370	−0.3237	0.1048	0.1048
	1995–2015	0.0029	0.0396	0.0744	0.0055	0.0055

Source: Author's calculation using Eviews 8

implies that growth rate of export ratio, population and investment positively affect the growth rate of the real gross domestic product. These results are in line with theories as well as the situation in Vietnam where both export and investment play a very important role in the economy. Nonetheless, estimated result of ROT is counter-intuitive and statistically significant at 5% level.

The period 1992–2015
Similar to the previous regression with the full sample 1990–2015, the output of this period also strongly suggests that there is a non-linear relationship between inflation and economic growth. The estimated inflation threshold is 3% which is slightly lower than the threshold suggested in the previous regression. This difference is likely due to high levels of inflation rate in 1990 and 1991 which partly influence estimated outcomes. Corresponding to this value of inflation threshold, the regression results are also presented in Table 2, which shows that below the inflation threshold of 3%, inflation and output growth have a positive linkage with the estimated coefficient of the inflation rate (β_1) is 0.4528. This linkage turns to slightly negative when inflation exceeds 3% level $(\beta_1 + \beta_2 = 0.4528 - 0.4668 = -0.0139)$. However, in this case, the Wald Test suggests that the null hypothesis $\beta_1 + \beta_2 = 0$ cannot be rejected at 10% level (Table 3). In other words, the total of $\beta_1 + \beta_2$ is statistically insignificantly different from zero. Thus, the findings don't provide enough evidence to confirm whether an increase in inflation rate above 3% will decrease or increase economic growth rate. In summary, although the estimated results provide strong evidence that there exists a nonlinear relationship between inflation and economic growth, they fail in identifying the effect of inflation on economic growth when it is above the estimated threshold.

The other slope coefficients all continue to be highly statistically significant at 5% significance level or higher (except for AGR at 10% level). An increase of 1% in REX, POP, and INV will, on average, boosts economic growth by 0.07%, 3.28%, and 0.04% respectively. However, the estimated coefficients of both ROT and AGR are still negative.

ii. Two-Stage Least Squares (2-SLS) Method

The period 1992–2015
Turning now to the 2-SLS method, Table 4 shows the results obtained from the 2-SLS method for the sample of 1992–2015. The findings are qualitatively the same as the case of OLS, and clearly indicate that the level of inflation threshold is 3%, which means that the sum of the squared residuals from the regression is minimized at

Table 4. 2-SLS estimation

Variable	Coef.	
	1992–2015	1995–2015
INF	0.4523**	0.3580*
	(0.1750)	(0.1806)
DUM*(INF–INF*)	−0.4643**	−0.3551*
	(0.1938)	(0.1970)
ROT	−0.0394**	−0.0247
	(0.0173)	(0.0227)
REX	0.0774***	0.0459*
	(0.0210)	(0.0256)
POP	3.3009***	3.7660***
	(1.0258)	(1.2297)
INV	0.0436**	0.1087**
	(0.0196)	(0.0377)
AGR	−0.2620*	−0.2603
	(0.1405)	(0.1574)
Intercept	1.6734	0.4247
	(1.0346)	(1.1962)
Inflation threshold	3%	3%
Number of observations	24	21
Adjusted R-squared	0.690	0.716

Source: Authors' own calculations by using Eviews 8
Note: Dependent variable is GDP growth rate; ***,
** and * stand for significance at the 1, 5 and 10%
levels, respectively

$INF^* = 3$ (Fig. 5). Corresponding to this value of inflation threshold, the entire regressed results are presented in Table 4. In particularly, the estimated coefficients of INF and DUM3*(INF–INF*) are statistically significant at the 5% level, with the value of 0.4523 and −0.4643 respectively. Moreover, the result of the Wald test indicates that total of $\beta_1 + \beta_2$ is statistically insignificantly different from zero (Table 3).

Regarding the other explanatory variables, the estimated coefficients of REX, POP and INV variables continue to be statistically significant with expected signs at 5% significance level. In addition, the coefficients of ROT and AGR indicate somewhat counterintuitive relationships with growth, with their p-values at 5% and 10% significance level respectively.

The period 1995–2015

In comparison with the period 1992–2015, we reached similar results for the period 1995–2015 with the inflation threshold obtained from this regression being 3% (Fig. 5), and most estimated coefficients are statistically significant at 10% level (except for constant term, ROT and AGR). Corresponding to this value of inflation threshold, the regression results are provided in Table 4. Particularly, the absolute value of $\beta_2 (-0.3551)$ is somewhat smaller than the value of $\beta_1 (0.3580)$. This indicates

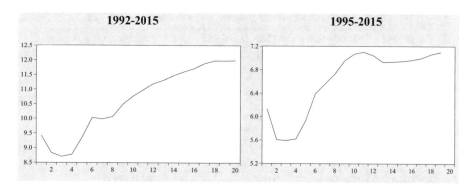

Fig. 5. 2-SLS - value of sum of squared residuals. *Source: Authors' own calculations by using Eviews 8*

that the overall effect of high inflation (i.e. inflation rate higher than 3%) on output growth is slightly positive ($\beta_1 + \beta_2 = 0.3580 - 0.3551 = 0.0029$) instead of negative as previous regressions. Nevertheless, based on the result of the Wald test for coefficient restrictions (Table 3), there is not enough evidence to affirm that inflation becomes harmful or beneficial to economic growth when it is higher than 3%. Instead, we can only conclude that inflation has an ambiguous effect on economic growth above the inflation rate of 3%.

In sum, the empirical findings in the OLS and 2-SLS sections successfully confirm that there exists a nonlinear relationship between inflation and economic growth in Vietnam over the period 1990–2015. The findings also suggest the existence of the threshold at 3–4% above which the positive effect of inflation on growth disappears.

iii. Quadratic Specification

Thus far, the empirical result strongly suggests a non-linear relationship between inflation and economic growth together with the estimated inflation threshold around 3–4%. Let us see whether the result of quadratic specifications can reconfirm these empirical findings.

The result of quadratic specifications for the period 1992–2015 and 1995–2015[9] shown in Tables 5 and 6 strongly suggests that the relationship between inflation and GDP growth is positive for low values of the inflation rate, but gradually diminishing as inflation rate increases. The rest of this section describes the estimated results more in detail.

The period 1992–2015
In the case of 1992–2015 period, the estimated coefficient of both INF(β_1) and INF^2 (β_2) are highly statistically significant at 1% level with the value of $\beta_1 = 0.2318$ and $\beta_2 = -0.0156$. In addition, the results of other explanatory variables are significant at the 5% level (except for INV at 10% level). These findings offer clear evidence of a non-linear linkage between inflation and economic growth.

---

[9] The quadratic specification of the 1990–2015 period does not provide significant results.

Table 5. Quadratic and linear inflation effect estimation (1992–2015)

Parameter	No inflation effect	Linear inflation effect	Quadratic inflation effect
INF		−0.0427	0.2318***
		(0.0385)	(0.0755)
INF²			−0.0156***
			(0.0039)
REX	0.0694***	0.0724***	0.0912***
	(0.0209)	(0.0209)	(0.0161)
INV	0.0432	0.0387	0.0355*
	(0.0260)	(0.0262)	(0.0193)
REX (−1)	0.0410**	0.0386*	0.0421***
	(0.0186)	(0.0186)	(0.0137)
INV (−1)	0.0499*	0.0540*	0.0545**
	(0.0265)	(0.0266)	(0.0196)
FDI (−1)	0.0075	0.0122	0.0188**
	(0.0085)	(0.0094)	(0.0071)
Constant	4.4475***	4.6823***	3.8677***
	(0.4942)	(0.5348)	(0.4452)
Observations	24	24	24
Adjusted R²	0.587	0.593	0.779

Source: Authors' own calculations by using Eviews 8
*Note: Dependent variable is GDP growth rate. ***, ** and * stand for significance at the 1, 5 and 10% levels, respectively*

Furthermore, according to the nature of the quadratic function, we can calculate the value of the peak inflation impact on growth. Using β_1 and β_2 values obtained from Table 5, the peak inflation is 7.42%.

$$INF^{peak} = \frac{\beta_1}{2 * \beta_2} = \frac{0.2318}{2 * 0.0156} = 7.42\%$$

In addition, in order to assess the fitness of quadratic specification, GDP also is regressed with and without inflation as illustrated in Table 5. The estimated result shows that INF is not statistically significant even at 10% level in the linear inflation effect specification. Also, adjusted R-squared of quadratic specification (77.9%) is greater than that of the linear inflation effect specification (59.3%) and the no inflation effect specification (58.7%). That is, the quadratic regression explains the inflation-economic growth relationship better than linear specification. Once again, empirical findings reconfirm the non-linear relationship between inflation and economic growth.

The period 1995–2015

Regarding the period 1995–2015, the null hypothesis that the relationship between inflation and economic growth is linear against the alternative that such relationship is non-linear can be tested by testing the null hypothesis that $\beta_2 = 0$ against the alternative that $\beta_2 \neq 0$. Since the value of $\beta_2 = -0.0091$ and its $p-value = 0.0714$, we can reject

Table 6. Quadratic and linear inflation effect estimation (1995–2015)

Parameter	No inflation effect	Linear inflation effect	Quadratic inflation effect
INF		−0.0645*	0.1022
		(0.0309)	(0.0895)
INF^2			−0.0091*
			(0.0047)
REX	0.0346	0.0353	0.0614**
	(0.0231)	(0.0209)	(0.0232)
INV	0.1295***	0.1209***	0.0872**
	(0.0355)	(0.0323)	(0.0341)
REX (−1)	0.0284	0.0180	0.0231
	(0.0213)	(0.0199)	(0.0183)
INV (−1)	0.0810**	0.1111***	0.0969**
	(0.0377)	(0.0370)	(0.0345)
FDI (−1)	−0.0059	−0.0019	0.0074
	(0.0095)	(0.0088)	(0.0093)
Constant	3.0779	3.1868***	3.2141***
	(0.6072)	(0.5515)	(0.5028)
Observations	21	21	21
Adjusted R^2	0.689	0.745	0.789

Source: Authors' own calculations by using Eviews 8
*Note: Dependent variable is GDP growth rate. ***, ** and * stand for significance at the 1, 5 and 10% levels, respectively*

the null hypothesis at the 10% significance level. Thus, similar to the previous specification of 1992–2015 period, the estimated results again support a non-linear relationship between inflation and economic growth with the breaking point occurring at 5.61%.

$$INF^{peak} = \frac{\beta_1}{2*\beta_2} = \frac{0.1022}{2*0.0091} = 5.61\%$$

Additionally, by comparing adjusted R-squared, we can see that the quadratic regression fits the data better than the linear models, with 78.9% against 74.5% and 68.9% (Table 6).

All in all, regardless of the methods or periods, all methods highly confirm the non-linear relationship between inflation and economic growth in Vietnam over the period 1990–2015. On the other hand, the results relating to inflation threshold are mixed and insufficient to answer the second research question. While the values of estimated threshold obtained from the threshold model are 3% or 4%, the inflation peak values calculated from the quadratic specification are about 5.5%–7.5%. Hence, with the different methods, the implication of the results varies. Whereas the threshold model implies that above 3-4%, the inflation' effect becomes neutral to economic growth, the quadratic specification suggests that the positive impact of inflation on economic growth starts diminishing at 5.5%–7.5%. Taken together, the possible conclusion is that the effect of inflation on economic growth in Vietnam is positive when inflation is below 3%–4%, but this effect disappears when inflation is above these levels. And when inflation reaches 5.5%–7.5%, inflation begins to negatively affect economic growth.

6 Conclusions and Policy Implications

This paper was undertaken primarily to estimate the inflation threshold in Vietnam over the period 1990–2015. Simultaneously, this study examined characteristics of the relationship between inflation and economic growth. The paper has the following main findings:

Firstly, the study strongly confirmed that there exists a nonlinear relationship between inflation and economic growth in Vietnam over the period 1990–2015. At low levels of inflation, inflation positively relates to economic growth. In contrast, when inflation is higher than a certain level, this positive effect virtually disappears. And, if inflation continues increasing to higher levels, the effect of inflation on economic growth will diminish.

Secondly, the results of the threshold estimation showed that the estimated inflation threshold level consistently stays at 3%–4%, at which the positive effect of inflation on economic growth vanishes. The study from the quadratic model provided evidence that this effect starts fading at inflation levels somewhere between 5.5%–7.5%.

Thirdly, the present study confirmed previous findings and contributed additional evidence that suggests exports, investment, and FDI have played an essential role in promoting the Vietnamese economy.

These findings have significant implications for implementing monetary policy in Vietnam as follows:

Firstly, SBV should keep inflation at around 4% in order to take advantage of the inflation's positive effect on economic growth. However, in the case of sudden supply shocks, strong fluctuations in international prices, or other adverse circumstances, SBV could consider the higher inflation target of 5%. Yet, based on the empirical findings, in any case, inflation should not be greater than 5.5% because above this level, the inflation's positive effect on output growth begins to diminish.

Secondly, the presence of a nonlinear relationship between inflation and output growth implies that monetary policy might affect economic growth differently depending on the inflation level. Thus, when the economy encounters high inflation, monetary policy should be formed to prioritize restraining inflation as soon as possible in order to avoid its adverse effects on output growth.

Thirdly, Vietnam should enhance the domestic investment as well as the international trade activities to promote the economy. Moreover, Vietnam should continue attracting more FDI which significantly contributes to the growth of the Vietnamese economy.

References

Barro, R.J.: Determinants of Economic Growth: A Cross-Country Empirical Study. Working Paper (National Bureau of Economic Research) No. 5698 (1996)

Barro, R.J.: Inflation and economic growth. Fed. Reserve Bank St. Louis Rev. **78**, 153–169 (1995)

Bruno, M., Easterly, W.: Inflation crises and long-run growth. J. Monetary Econ. **41**(1), 3–26 (1998)

Burdekin, R., Denzau, A.T., Keil, M.W., Sitthiyot, T., Willett, T.D.: When Does Inflation Hurt Economic Growth? Different Nonlinearities for Different Economies. Claremont Colleges Working Papers (2004)

Christoffersen, P., Doyle, P.: From inflation to growth: eight years of transition. IMF Working Pap. **100**(98), 1–36 (1998)

Chowdhury Hayat, Z.U., Kalirajan, K.P.: Is there a threshold level of inflation for Bangladesh? J. Appl. Econ. Res. **3**(1), 1–20 (2009)

Eggoh, J., Khan, M.: On the nonlinear relationship between inflation and economic growth. Res. Econ. **68**(2), 133–143 (2014)

Espinoza, R.A., Hyginus L., Prasad, A.: Estimating the inflation-growth nexus: a smooth transition model. IMF Working Papers WP/10/76 (2010)

Fischer, S.: The role of macroeconomic factors in growth. J. Monetary Econ. **32**(3), 485–512 (1993)

Ghosh, A., Phillips, S.: Warning: inflation may be harmful to your growth. IMF Staff Pap. **45**(4), 672–710 (1998)

Gillman, M., Kejak, M.: Contrasting models of the effect of inflation on growth. J. Econ. Surv. **19**(1), 113–136 (2005)

De Gregorio, J.: Inflation, taxation, and long-run growth. J. Monetary Econ. **31**(3), 271–298 (1993)

Gylfason, T., Herbertsson, T.: Does inflation matter for growth? Jpn. World Econ. **13**(4), 405–428 (2001)

Judson, R., Orphanides, A.: Inflation, Volatility and Growth. Int. Finan. **2**(1), 117–138 (1999)

Khan, M.S., Senhadji, A.S.: Threshold effects in the relationship between inflation and growth. IMF Staff Pap. **48**(1) (2001)

Kormendi, R., Meguire, P.: Macroeconomic determinants of growth: cross-country evidence. J. Monetary Econ. **16**(2), 141–163 (1985)

Lee, C.-C., Wong, S.Y.: Inflationary threshold effects in the relationship between financial development and economic growth: evidence from Taiwan and Japan. J. Econ. Dev. **30**, 49–69 (2005)

López-Villavicencio, A., Mignon, V.: On the impact of inflation on output growth: does the level of inflation matter? J. Macroecon. **33**(3), 455–464 (2011)

MacKinnon, J.G.: Numerical distribution functions for unit root and cointegration tests. J. Appl. Econometrics **11**(6), 601–608 (1996)

Mubarik, A.Y.: Inflation and growth: an estimate of the threshold level of inflation in Pakistan. SBP-Res. Bull. **1**(1), 35–44 (2005)

Hang N.T.T., Thanh D.N.: Macroeconomic Determinants of Vietnam's Inflation 2000–2010: Evidence and Analysis. Vietnam Centre for Economic and Policy Research, Hanoi (2010)

Hansen, B.: Inference when a nuisance parameter is not identified under the null hypothesis. Econometrica **64**(2), 413–430 (1996)

Hansen, B.E.: Sample splitting and threshold estimation. Econometrica **68**, 575–603 (2000)

Paul, S., Kearney, C., Chowdhury, K.: Inflation and economic growth: a multi-country empirical analysis. Appl. Econ. **29**(10), 1387–1401 (1997)

Salami, D., Kelikume, I.: An estimation of inflation threshold for Nigeria 1970–2008. Int. Rev. Bus. Res. Pap. **6**(5), 375–385 (2010)

Sarel, M.: Nonlinear effects of inflation on economic growth. IMF Staff Papers **43**(1), 199–215 (1996)

Singh, K., Kalirajan, K.: The inflation-growth nexus in India: an empirical analysis. J. Policy Model. **25**, 377–396 (2003)

Su, T.D.: Threshold effects of inflation on growth in the ASEAN-5 countries: a panel smooth transition regression approach. J. Econ. Finan. Adm. Sci. **20**, 41–48 (2015)

Tong, H.: On a threshold model. In: Chen, C.H. (ed.) Pattern Recognition and Signal Processing, Sijhoff and Nordhoff, Amsterdam, pp. 575–586 (1978)

Vinayagathasan, T.: Inflation and economic growth: a dynamic panel threshold analysis for Asian economies. J. Asian Econ. **26**, 31–41 (2013)

Vuong, Q.H.: The Vietnam's transition economy and its fledgling financial markets: 1986–2003. CEB Working Paper 4 (32) (2004)

Factors Affecting the Level of Financial Information Transparency – Evidence from Top 30 Listed Companies in Singapore, Philippines, and Vietnam

Tran Quoc Thinh[(✉)]

Banking University of Ho Chi Minh City, Ho Chi Minh City, Vietnam
thinhtq@buh.edu.vn

Abstract. The level of financial information transparency (FIT) is important for listed companies on the stock markets of countries, since it ensures that the users get quality information. Because of this importance, the authors use the panel data methodology to examining the factors affecting to the FIT of the top 30 listed companies in Singapore, Philippines and Vietnam for the period 2012–2016. As a result, we find all the factors – related both to corporate governance and to financial indicators – that affect the FIT in these countries. It turns out that the factors are largely the same for all three countries, but the numerical effect of each factor on the FIT is somewhat different in different countries.

1 Introduction

FIT is increasingly receiving the attention of users to develop the capital markets of the country, especially developing countries. Recently, the ASEAN Economic Community (AEC) has been established, opening up many opportunities for ASEAN countries to attract capital through the stock market. As a result, FIT has become even more important for developing countries that form ASEAN. According to OECD (2004), to protect the interests of investors in making business decisions, it is important to improve the FIT level. FIT is influenced by a variety of factors including corporate governance and financial indicators (Bushman et al. 2004; Cheung et al. 2005; Desoky and Mousa 2012).

FIT studies have looked at many countries, both developed and developing (Bushman et al. 2004). Most studies evaluating and analyzing FIT focus on a single country, but a comparison of different countries is also important (Desoky and Mousa 2012). Cheung et al. (2005) compared two ASEAN countries with similar economic characteristics: Hong Kong and Thailand, to see which of the two countries is, in this aspect, more favorable for investors. In this study, the authors examine three other ASEAN countries with similar economic characteristics: Singapore, Philippines and Vietnam. These three countries are easier to study that other ASEAN members since they mostly use English in their financial reports.

© Springer International Publishing AG 2018
L. H. Anh et al. (eds.), *Econometrics for Financial Applications*, Studies in Computational Intelligence 760, https://doi.org/10.1007/978-3-319-73150-6_74

To get a more adequate description of FIT, instead of selecting a few listed companies – as in most previous studies – the authors selected the top 30 of listed companies in each country. To get an accurate idea of each company's FIT, we considered its FIT scores over a 5-year time period. The resulting need to process tim series prompted us to use panel data methodology, a methodology that is known to be well-suited for processing economic time series.

This paper consists of five sections. Section 2 presents theoretical background and an overview of prior studies. Section 3 describes the methodology of our study. Section 4 describes and analyzes the results of our research. Finally Sect. 5 contains conclusions.

2   Theoretical Background and an Overview of Prior Studies

2.1   The Concept of FIT

Bushman et al. (2004) view FIT as the availability of specific company information for investors and outside shareholders. Kulzick (2004) define FIT as public availability of information which is accurate, consistent, relevant, complete, clear, timely and convenient. Zarb (2006) argues that for FIT to be useful and timely, published information should be reliable, comparable, universal, and transparent. To achieve transparency in financial reporting, information must be of high quality, in particular, it must be in accordance with generally accepted accounting principles.

2.2   Background Theory

The topic of information transparency encompasses two fundamental theories: useful information theory and asymmetric information theory. The useful information theory by Ijiri and Jaedicke (1996) states that for the information to be useful, it must be timely, reliable, relevant, and include all the needed accounting data. The asymmetric information theory was first introduced by Akerlof et al. (1970). This theory studies the asymmetry (dis-balance) between the information available to information holders (e.g., companies) and information users (investors, shareholders, etc.). Such a dis-balance occurs if the information provided to the users is unclear and/or incomplete, and thus not sufficient to make informed business decisions.

2.3   Previous Studies

Bushman et al. (2004) studied information transparency of 1,000 listed companies in 41 countries in 2001. The authors used the cross-data method. Research results show that economic policies and enterprise size have a positive impact on FIT, and that transparency in governance information can be improved by using appropriate regulatory mechanisms.

Cheung et al. (2005) examined the FITs of 168 listed companies in Hong Kong and 337 listed companies in Thailand in 2002. The authors used regression analysis to analyze the effect of financial indicator and corporate governance on FIT. The results shown that if we jointly consider companies from both countries, the FIT improves with the increase in the size of the business and the size of the board, but decreases when the board's structure gets more complex. If we consider only Hong Kong companies, then FIT increases with an increase in the size of the business, with efficiency of asset use, and with the amount of fixed assets, and decreases when the board structure becomes more complex. For Thailand, the only two factors that statistically significantly impact FIT the size of the board (the larger the board, the better FIT) and the complexity of the board (the more complex the board, the worse FIT).

Desoky and Mousa (2012) investigated the influence of factors on FIT at 100 listed companies in India in 2010. The study used the panel data method. This research has shown that FIT is higher for locally-owned companies than for the foreign-owed ones. Somewhat surprisingly, in India, on average, FIT slightly decreased with the size of the company.

3   Research Design

3.1   Describe the Research Sample

In our research, in each of the three selected countries, we the top 30 listed companies – the 30 largest in terms of combined capitalization and liquidity. For companies listed in Singapore and Philippines we used listings provided by ASEAN UP, while got Vietnam, we used information available in relation to the listing of the company on the Vietnam stock market. To ensure the usefulness of the study, the authors selected the most recent five-year period 2012–2016.

3.2   Research Models

To gauge financial information transparency, the authors used the Governance and Transparency Index (GTI) developed in Singapore. The GTI methodology also measures government information transparency and several other important characteristics. In our study, we only used the part of this methodology that deals with FIT. This part gauges FIT by looking at several issues such as timeliness of releasing financial information, timeliness of the company'e reply to requests, shareholder participation level, etc.; see Appendix for details. To each of the issues is assigned the number of points. These points are assigned if the corresponding information is available. The maximum possible number of points is 25. The overall transparency index $TRAN_i$ is then obtained by dividing the overall number of points by 25.

Based on the the prior research on factors affecting FIT, the authors use the following model:

$$TRAN_i = \beta_0 + \beta_1 \cdot SIZE_i + \beta_2 \cdot PROFIT_i + \beta_3 \cdot TURNOVER_i + \beta_4 \cdot FIX_i$$
$$+ \beta_5 \cdot DEBT_i + \beta_6 \cdot CONC_i + \beta_7 \cdot BSIZE_i + \beta_8 \cdot BEXC_i$$
$$+ \beta_9 \cdot OUTSIDE_i + \beta_{10} \cdot FLIQUI_i + \beta_{11} \cdot AUDITF_i + \epsilon_i,$$

where:

- $TRAN$ is the level of financial information transparency (FIT) is measured as above
- $SIZE$ is the natural logarithm of total amount of assets
- $PROFIT$ is the ration of after-tax profit to the total amount of assets
- $TURNOVER$ is the net sales volume divided by the total amount of assets
- FIX is the total amount of fixed assets (i.e., assets prone to depreciation) divided by total amount of assets
- $DEBT$ is the total liabilities divided by total assets
- $CONC$ is the proportion of major shareholders, i.e., shareholders holding at least 5% of total shares
- $BSIZE$ is the size of the board, i.e., the number of directors forming the board
- $BEXC$ is the percentage of executive directors on the board
- $OUTSIDE$ is 0 if the same person combines the positions of the CEO and the Chairman of the Board, 1 otherwise
- $FLIQUI$ the amount of short-term assets divided by the amount of short-term debt
- $AUDIT$ is 1 if the company audited by by one of the Big Four auditing firms, else 0.

4   Results

4.1   Do We Need to Apply This Model?

In the study, the authors use the adjusted R^2 to check whether there is a need for a multivariate regression model Table 1 shows that the each factor explains at most 11.9% of the variance in the dependent variable $TRAN_i$ (the adjusted R^2 is 0.119), so the multivariate regression is indeed needed.

Table 1. Model summary

Model summary				
Model	R	R Square	Adjusted R square	Std. error of the estimate
1	.381[a]	.145	.119	1.553

[a]Predictors: (Constant), AUDIT, CONC, PROFIT, FIX, BEXC, BSIZE, DEBT, TURNOVER, FLIQUI, SIZE, OUTSIDE

4.2 Suitability of the Model

The paper examines the fit of the model to see if there is indeed a statistically significant linear relationship between the dependent variable and the independent variables. In this analysis, we use two hypotheses:

H_0: $\beta_i = 0$ The variables introduced into the model do not affect FIT.

H_1: $\beta_i \neq 0$ The variables introduced into the model affect FIT. The results, as presented in Table 2, show the significance level 0.000, smaller than 0.005, so the hypothesis H_0 is rejected. Thus, it makes sense to use the linear regression model.

Table 2. ANOVA.

ANOVA[b]

Model		Sum of squares	df	Mean square	F	Sig
1	Regression	148.859	11	13.533	5.609	.000[a]
	Residual	875.851	363	2.413		
	Total	1024.709	374			

[a]Predictors: (Constant), AUDIT, CONC, PROFIT, FIX, BEXC, BSIZE, DEBT, TURNOVER, FLIQUI, SIZE, OUTSIDE
[b]Dependent Variable: TRANSP

4.3 Regression Results

The results of analyzing the effects of different factors on FIT are presented in the following table:

Table 3. Regression result

Coefficients[a]

Model		Unstandarized coefficients		Standardized coefficients	t	Sig.	Collinearity statistics	
		B	Std. error	Beta			Tolerance	VIF
1	(Constant)	12.992	1.527		8.507	.000		
	SIZE	.302	.152	.135	1.990	.047	.515	1.942
	PROFIT	.505	1.392	.022	.363	.717	.626	1.598
	TURNOVER	.308	.166	.112	1.860	.064	.653	1.531
	FIX	.336	.391	.044	.860	.390	.893	1.120
	DEBT	−.567	.534	−.065	−1.062	.289	.629	1.590
	CONC	−1.392	.451	−.196	−3.089	.002	.583	1.716
	BSIZE	.086	.036	.159	2.913	.004	.792	1.262
	BEXC	−2.745	.786	−.248	−3.494	.001	.467	2.144
	OUTSIDE	−.651	.262	−.172	−2.483	.013	.489	2.047
	FLIQUI	−.086	.078	−.065	−1.105	.270	.684	1.462
	AUDIT	.882	.267	.165	3.303	.001	.946	1.057

[a]Dependent Variable: TRANSP

According to the results listed Table 3, there are no hyperbolic factors (with VIF less than 2), so all the results are consistent in time.

If we combine the companies of all three countries, then in the resulting regression, there are 6 factors affecting the FIT at significance level <0.05. The resulting regression model is as follows:

$$TRAN = 12.992 + 0.302 \cdot SIZE - 1.392 \cdot CONC + 0.086 \cdot BSIZE$$
$$- 2.745 \cdot BEXC - 0.651 \cdot OUTSIDE + 0.882 \cdot AUDIT.$$

This result is similar to that of Cheung et al. (2005) in describing the effect of corporate governance factors such as SIZE, BSIZE, BEXC and Bushman et al. (2004), Desoky and Mousa (2012) in describing the effects of financial indicator factors such as SIZE. In all these studies, an increase SIZE, BSIZE, and AUDIT increases FIT, while an increase in CONC, BEXC, and OUTSIDE decreases FIT.

This behavior makes perfect economic sense. For example, listed companies that have large assets tend to have increased transparency to attract investors. The larger the size of the board, the larger the difference of opinion between board members, and the more likely that more information will be provided to resolve these differences. The proportion of large shareholders (who own more than 5% of shares) in the total number of shareholders tends to reduce FIT because if the company's strategy is decided by a small group of large share-holders, there is no incentive to provide detailed information to others. Similarly, if a significant proportion of the board consists of full-time executive directors, they make all the decisions, and there is no incentive for them to share information with others. For audits, the Big Four's audits generally require higher transparency disclosure requirements for listed companies.

The authors also performed a similar an analysis for each of the three countries separately.

Singapore

Table 4 shows that, with significance level less than 0.05, five variables affect FIT. The resulting regression model is as follows:

$$TRAN = 26.165 - 1.010 \cdot SIZE - 2.458 \cdot TURNOVER + 1.930 \cdot DEBT$$
$$- 2.997 \cdot CONC + 0.225 \cdot BSIZE.$$

This result shows that FIT is affected by two corporate governance factors and three financial indicator factors impact on FIT. An increase in DEBT or BSIZE increases FIT, while an increase in SIZE, TURNOVER, and CONC decreases FIT. We can also see that the FIT of Singapore companies is much more affected by the financial indicators than by the corporate governance.

For financial indicator factors, our analysis shows that for Singapore companies, high levels of conventional debt force them to be more transparent – to meet the needs of creditors. On the other hand, companies with large assets and high asset utilization efficiency tend to reduce information transparency.

Table 4. Regression result for Singapore companies.

Coefficients[a]

Model	Unstandarized Coefficients		Standardized Coefficients	t	Sig.	Collinearity Statistics	
	B	Std. Error	Beta			Tolerance	VIF
1 (Constant)	12.992	1.527		8.507	.000		
SIZE	.302	.152	.135	1.990	.047	.515	1.942
PROFIT	.505	1.392	.022	.363	.717	.626	1.598
TURNOVER	.308	.166	.112	1.860	.064	.653	1.531
FIX	.336	.391	.044	.860	.390	.893	1.120
DEBT	−.567	.534	−.065	−1.062	.289	.629	1.590
CONC	−1.392	.451	−.196	−3.089	.002	.583	1.716
BSIZE	.086	.030	.159	2.913	.004	.792	1.262
BEXC	−2.745	.786	−.248	−3.494	.001	.467	2.144
OUTSIDE	−.651	.262	−.172	−2.483	.013	.489	2.047
FLIQUI	−.086	.078	−.065	−1.105	.270	.684	1.462
AUDIT	.882	.267	.165	3.303	.001	.946	1.057

[a]Dependent Variable: TRANSP

This may be explained by competitive pressures: the companies do not want yo disclose any information that can be used by their competitors.

Philippines

Table 5 shows that only 5 variables affect FIT with significance level less than 0.05. The resulting model is as follows:

$$TRAN = 8.696 + 0.450 \cdot SIZE + 1.816 \cdot FIX - 2.755 \cdot DEBT$$
$$+ 0.153 \cdot BSIZE + 0.825 \cdot AUDIT.$$

Table 5. Regression result for Philippines companies.

Coefficients[a]

Model	Unstandarized coefficients		Standardized coefficients	t	Sig.	Collinearity statistics	
	B	Std. error	Beta			Tolerance	VIF
1 (Constant)	8.696	2.553		3.406	.001		
SIZE	.450	207	.202	2.172	.032	.711	1.407
PROFIT	−705	2.774	−.026	−.254	.800	.611	1.636
TURNOVER	−.025	.327	−.007	−.075	.940	.698	1.433
FIX	1.816	.590	.306	3.077	.003	.621	1.610
DEBT	−2.755	1.164	−.270	−2.367	.020	.472	2.117
CONC	.936	1.187	.069	.788	.432	.807	1.239
BSIZE	.153	.048	.325	3.199	.002	.595	1.680
BEXC	2.049	1.576	.184	1.301	.196	.307	3.261
OUTSIDE	.052	.410	.017	.127	.899	.335	2.981
FLIQUI	.046	.159	.032	.287	.774	.510	1.962
AUDIT	.825	.382	.212	2.161	.033	.639	1.566

[a]Dependent Variable: TRANSP

Similar to Singapore, this result shows that FIT is affected by two corporate governance factors and three financial indicator factors. When SIZE, FIX, BSIZE, and AUDIT increase, FIT increases, while when DEBT increases, FIT decreases.

These results can be explained in the same way as the results of analyzing all three countries together. For example, the effect of the FIX variable is interpreted similarly to effect of the SIZE variable of the overall results: companies with large assets tend to be more transparent.

Vietnam

Results from Table 4 show that, with a significance level of 5%, there were 4 factors affecting FIT. The resulting model is as follows:

$$TRAN = 20.419 + 0.590 \cdot TURNOVER - 4.950 \cdot CONC - 4.341 \cdot BEXC$$
$$+ 2.510 \cdot AUDIT.$$

These results can be explained similarly to the results of the overall analysis.

It is worth mentioning that, in contrast to Singapore and the Philippines, for Vietnam corporate governance factors affect FIT more than the financial indicator factor. This shows that in Vietnam, the effects of corporate governance dominate FIT (Table 6).

Table 6. Regression result for Vietnam companies.

Coefficients[a]

Model		Unstandarized coefficients		Standardized coefficients	t	Sig.	Collinearity statistics	
		B	Std. error	Beta			Tolerance	VIF
1	(Constant)	20.419	5.015		4.072	.000		
	SIZE	−.476	.522	−.100	−.913	.363	.499	2.005
	PROFIT	2.739	2.122	.138	1.291	.200	.526	1.902
	TURNOVER	.590	.243	.249	2.433	.017	.569	1.759
	FIX	1.956	1.122	.174	1.743	.084	.600	1.668
	DEBT	1.734	1.511	.172	1.147	.254	.266	3.759
	CONC	−4.950	1.155	−.440	−4.285	.000	.566	1.767
	BSIZE	−.139	.102	−.129	−1.369	.174	.670	1.492
	BEXC	−4.341	1.592	−.278	−2.727	.008	.576	1.735
	OUTSIDE	.236	1.080	.023	.218	.828	.517	1.936
	FLIQUI	.000	.216	.000	.002	.998	.279	3.589
	AUDIT	2.510	.614	.412	4.091	.000	.587	1.702

[a]Dependent Variable: TRANSP

5   Conclusions

FIT is an important issue. By examining the FIT of the top 30 listed companies in Singapore, Philippines and Vietnam, the results show that both the corporate governance factors and the financial indicators affect FIT. Specifically, an increase in SIZE, BSIZE, and/or AUDIT increase FIT, while an increase in CONC, OUTSIDE, and/or BEXC decreases FIT.

For the separate studies conducted in each market, the results are similar, but there are also differences due to the economic and culture characteristics of each country. The results of the study in the Philippines and Vietnam are quite similar to the overall effect of variables on FIT compared to Singapore. The results also show in Singapore and Philippines, financial indicator factors are more important, while in Vietnam, the corporate governance factors are more important. This difference between different markets is in line with the view of Cheung et al. (2005). As a result, while it is important to analyze the overall economic market, specific policies should take into account specific conditions of each country.

How can this research be further expanded? The article focuses on three countries of Southeast Asia. It is desirable to extend this analysis to all ASEAN countries. It may also be beneficial to consider how other factors such as institutional ownership, foreign ownership, growth rate, sales, etc. affect FIT.

Appendix

See Table 7.

Table 7. List of items for Governance and Transparency Index (GTI).

1	**Timeliness of release of results**	
	Were financial year results released on time	5
2	**Corporate website**	
	Is the link provided on the SGX website and/or annual report?	1
	Does the website have a clearly dedicated IR link instead of providing the financial information under links such as"News" or "Announcements"?	1
	Are the latest financial results available on the website?	1
	Is the latest annual report available on the website?	1
	Is the IR contact given on the website/annual report?	1
3	**Does the company respond to calls/emails requesting information quickly (i.e. within a week)**	2
4	**Results briefings**	
	Did the Board stated in the company's annual report the steps it has taken to solicit and understand the views of the shareholders?	2
	Are the powerpoint slides/webcast from the briefing available on SGX or the corporate website?	1
	Did the company have an effective investor relations policy to regularly convey pertinent information to shareholders?	1
5	**Shareholder participation**	
	Does the company allow shareholders who hold shares through nominees to attend AGMs as observers without being constrained by the two-proxy rule?	2
	Is there enough time from the notice of AGM sent to shareholders to the date of the AGM	1
	Is detailed information on each item in the agenda for the AGM disclosed in the Notice?	1
	Does the company conduct voting by show of hands or by ballot/poll?	1
	Does the company publish detailed information of the vote results?	1
6	**Dividend payment**	
	If the company have paid dividend, did the company disclose a policy on payment of dividends? If dividends are not paid, did the company disclose the reasons?	2
		25

References

Akerlof, G.A., Spence, A.M., Stiglitz, J.E.: Asymmetric information. Ann. Rev. Sociol. **2**, 263–284 (1970)

Bushman, R.M., Piotroski, J.D., Smith, A.J.: What determines corporate transparency? J. Acc. Res. **42**(2), 207–252 (2004)

Cheung, S.Y.-L., Connelly, J.T., Limpaphayom, P., Zhou, L.: Determinants of corporate disclosure and transparency: evidence from Hong Kong and Thailand. Int. Corp. Responsib. J. **3**, 313–342 (2005)

Desoky, A.M., Mousa, G.A.: Corporate governance practices: transparency and disclosure-evidence from the Egyptian exchange. J. Acc. Finan. Econ. **2**(1), 49–72 (2012)

Ijiri, Y., Jaedicke, R.K.: Reliability and objectivity of accounting measurements. Acc. Rev. **41**, 474–483 (1996)

Kulzick, R.S.: Sarbanes-Oxley: effects on financial transparency. Adv. Manag. J. **69**(1), 43–49 (2004)

Organization for Economic Cooperation and Development: OECD Principles of Corporate Governance. OECD Publications, Paris (2004)

Zarb, J.N.: The quest for transparency in financial reporting: certified public accountant. CPA J. **76**(9), 30–33 (2006)

Evaluating the Impact of Factors on the Shift of Economic Structure in Vietnam

Huong Thi Thanh Tran and Huyen Thanh Hoang[(✉)]

Banking Academy, 12 Chua Boc Street, Dong Da District, Hanoi, Vietnam
{huongttt76,huyenht}@hvnh.edu.vn

Abstract. For the economic development goals of Vietnam by 2020, the economic structure is moving towards increasing the proportion of construction industry group and service group, and gradually reducing the share of agriculture group in GDP. Accordingly, the shift of the economic structure leading to the contribution of sectors to the output of GDP has also changed. In this paper, the authors examine the impact of factors on the shift of economic structure in Vietnam. Quantifying the impact of factors that affect the shift of economic structure is also measuring the positive impacts on the increase of VA proportion of construction industry group and service group in GDP. Our findings indicated that labor proportion and labor productivity of construction industry and service group have strongly positive impact on the VA share of non-agricultural sector and an increase in labor proportion of service group will have a greater impact on the shift of economic structure.

1 Introduction

Economic structure formation, development and shifting is a process of objective and comply with the laws and rules. However, the results of shifting, time shifting, as well as efficient shifting of economic structure in each country also differ. Because the economic restructuring affected by various factors and the level of influence as well as the exploitation of these factors are different.

Economic structure can be classified in many different ways. According to the criteria, it can be arranged by economic sector, type of ownership, or region. As a basis for calculating, the economic structure can be categorized as the input indicators (labor structure, capital structure) and output indicators (GDP). To assess the impact of these factors on the shift of economic structure, we selected the dependent variable following the category of output (GDP).

In this paper, we focus on five issues: (1) Identify the factors that affect the shift of economic structure; (2) Identify indicators (variables) representing the influencing factors; (3) Building a model of the impact assessment factor to the shift of economic structure; (4) Evaluate the impact of these factors on the VA proportion in GDP of non-agricutural group, as well as the shift of economic structure of Vietnam. The results are estimated by data panel regression with data array of 63 provinces and cities under central authority in the period 2004–2012; (5) From the empirical study, given the conclusions and recommendations.

© Springer International Publishing AG 2018
L. H. Anh et al. (eds.), *Econometrics for Financial Applications*, Studies in Computational Intelligence 760, https://doi.org/10.1007/978-3-319-73150-6_75

In this study, we conducted a regression model with data sets from 2004 to 2012. We use the updated data set for 2012 for the following reasons: (1) The data used to analyze the effects of structural change on economic growth and development are determined by the relative prices. Compare (to eliminate the effect of price fluctuations). However, the data series from 1986 to 2012, GSO published at 1994 constant prices, from 2013 until now GSO only released data at 2010 comparable prices. (2) GDP target by current prices from 2012 back to GSO announced at production prices; Starting from 2013 until now announced at basic prices. To obtain data from 63 provinces and centrally-run cities to 2016 (standardized by 2010 prices and basic prices), much depends on the standardization and publication of GSO data, as well as 63 Provincial Statistics Offices. Therefore, the authors will update the data up to 2016 in our subsequent studies.

2 Literature Review

Market size. Market size is a factor that directly influences the formation and economic structure. The magnitude of market size will decide on the scale of allocation and movement of input resources for different production areas. In general, the market size (quantity demanded) is determined by the size of the population and the level of income. Population size and people's intellectual level have a great influence on the size and structure of market demand. The larger size and better quality of the population are the better the conditions for forming and developing the qualified economic structure with sectors, areas that can breakthrough, bring high socio-economic efficiency (Tat 2005). When the income of the population is low, most of the income is spent on essential commodities such as rations, food, etc. However, when the income of the population increases, the structure of consumption begins to change in the direction that the rate of spending on essential commodities decreases, the rate of spending on high-end products begins to increase. Signs of demand structure transformation with the ability to pay have the effect of guiding the business investment direction of employers and thereby have a significant impact on the formation of economic structure (Tat 2005).

Investment capital resources. The important factor that has a direct impact on the formation and shift of a country's economic structure is the investment capital. Investment capital resources used properly will promote economic restructure in the direction of implementing properly the strategy, planning and socio-economic development plan. The invested economic sector, investment scale, effectiveness of using capital investment, etc. affect the speed of development and the ability to grow facilities of each economic sector, thereby promote economic restructuring in a positive direction. Through investment activities, many new economic sectors and fields will appear. Some sectors are strongly stimulated and developed by the investment, but many sectors are not paid attention, increasingly eroded, resulting wiped out.

Human resources. When considered at the input aspect of the production process, human resources (working capacity) have long been regarded as a decisive factor in the production process. The impact of this factor on the formation and transformation of economic structure is considered in some major aspects as follows:

The size of human resources is one of the important factors contributing to the formation of economic structure. In order to achieve economic efficiency by scale for production and business activities, in certain science and technology conditions, it must have an appropriate workforce. For some countries, the small labor size has been the cause of difficulty in some areas, even they have to "import labor". In contrast, many developing countries have the phenomenon of labor "superabundance". Therefore, the formation of economic structure with ability to "fully use laborers" is one of their top priorities (Tat 2005). The quality of human resources (showed in the level of culture, technical qualification, etc.) is the most important factor for the formation of economic structure, especially with the sectors, fields requiring trained and highly skilled laborers, as some areas of services associated with modern technology, the field of precision engineering manufacturing and electronic industry, ... Economic sectors with effective operation, high investment, high labor productivity will attract the labor shift. If the laborers transform into an economic sector, the share of economic sectors in GDP also changes.

Science and technology. Science and technology is the main motivation which promotes the division of social laborers. The advancement of science and technology not only accelerates the pace of development of business and production facilities but also creates new production capabilities as well as new products that are suitable for the market, thereby increases their share in the overall economy. The advancement of science and technology also creates new needs, requires the emergence of some new economic sectors, thereby also makes the changes of economic structure. The development of science and technology is the decisive factor to transform from economic structure developed in width into economic structure developed in depth (the structure of taking the equipment in place of laborers, replacing the traditional industries consuming many muscular laborers and materials by the high-tech industries consuming less laborers and high intellectual laborer). The advancement of science and technology and the speed of technological improvements have the direct impact on the formation and development of economic structure; make the change of scale and quality of development of the industries leading to change the economic structure (Ngoc 2002).

Group factor of institution and policy. The economic policies of the State have a strong impact on the movement tendency of restructuring the economy. Encouragement or discouragement will affect the increase in growth or restraint of the development of some economic sectors. In the process of economic structure transformation, the State plays a decisive role in planning guidelines and policies to promote the economic structure transformation. The state develops and decides socio-economic development strategies and plans to achieve the country's socio-economic objectives. Actually, it is the orientation of development, resource allocation and investment by sector. With this orientation, the State has influenced the formation and trend of restructuring in economy. (Ngoc 2002).

3   Empirical Study

Choice of variables. In order to represent the factor groups above, the authors propose to use the following indicators:

- **Market size.** We use the population index to illustrate factor group market size. Population refers to all people living within a given territory (a country, a territory, an administrative unit, etc.) at a certain time or time period. Data source: Statistical yearbook of General Statistics Office of Viet Nam and provinces and cities under the central government.
- **Investment capital.** We propose to use two indexes: (1) The investment capital for developing the whole society is spententire capital (expenditure) to increase or maintain production capacity and resources for improving the material and spiritual living standards of the whole society in a certain period of time (month, quarter, year). (2) Capital productivity of the whole economy: is an indicator reflecting the relationship between the production results (GDP) and total investment capital of the whole society. This indicator pointed out the amount of copper results obtained (GDP) when spending a copper. Data source: Statistical yearbook of General Statistics Office of Viet Nam and provinces and cities under the central government.
- **Human resource.** For the purpose of representing the factor group of human resource, we use 6 indexes: (1) Number of laborers working in the economy: includes people at the age of 15 and above in reference period (one week), belongs to one of the following categories: (1) Working with payment of salary/wage; (2) Self-employed or holding the ownes ship. (2) Proportion of laborers of the construction industry group: is the comparison rate between the number of laborers at the age of 15 and above working in the construction industry group and total number of laborer of the whole economy. (3) Proportion of laborers of service group: is the comparison rate between the number of laborers at the age of 15 and above working in the service group and total number of laborer of the whole economy. (4) Proportion of laborers of agriculture, forestry and fishery group: is the comparison rate between the number of laborers at the age of 15 and above working in the agriculture, forestry and fishery group and total number of laborer of the whole economy. (5) Labor productivity of the construction industry group: is the indicator showing comparative relations between production results (Value added -VA) of construction industry group and the number of laborers of construction industry group. (6) Labor productivity of the service group: is the indicator showing comparative relations between production results (Value added -VA) of service group and the number of laborers of service group. Data source: Statistical yearbook of provinces and cities under the central government.
- **Science and technology.** In order to illustrate this group, we put forward to use two indexes: (1) Number of inventions granted a protection certificate:

is the inventions granted a protection certificate by competent state agencies in order to establish domestic and foreign industrial property rights. Data source: General statistical reporting regimes applied to the Ministry of Science and Technology. (2) Expenses for technological innovation in enterprises: is the investment amount of the enterprise for product innovation and production process innovation, such as the purchase of inventions, the purchase of equipment, the production of management equipment, the new production line, the purchase of trademark rights, etc. Data source: Basic statistical reporting regimes applied to State enterprises, enterprises with foreign investment capital; Non-State Enterprise Survey of GSO; Survey of individual business and production facilities of GSO.

- **Group factor of policy.** In order to represent the factor group of policy, we use two indexes: (1) PCI (Provincial Competitiveness Index) This index assesses and ranks the governments of the provinces and cities of Vietnam in the quality of economic governance and building business environment that facilitate the development of private enterprises. PCI of Vietnameses a scale from 10 (highest) to 1 (lowest) for the following 10 indexes: Market access; land access; Transparency; Cost of time; Unofficial expense; Incentives to foreign enterprises; Dynamism; Private economic development policies; Labor training; Legal institution. Data resource:Vietnam Chamber of Commerce and Industry (VCCI). (2) PAPI (Provincial Governance and Public Administration Performance Index). PAPI towards improving the efficiency of serving the people of local government in order to better meet the increasing requirements of the people through two complementary mechanisms: (i) Creating the healthy competition habit and culture of learning experience among local governments; and, (ii) Creating opportunities for people to improve their capacity to evaluate the performance of their government and encourage the government to improve their serving method for people. Data resource: http://papi. org.vn/.

Selected Model. Within the research scope of the article and actual data conditions of Vietnam, we conducted the Regression Model for Panel Data Analysis to quantify the impact of factors on the VA proportion in GDP of construction industry and service group (VAP_NA) as follows:

$$VAP\_NA_{ij} = \beta_0 + \beta_1 \ln PS_{ij} + \beta_2 \ln IC_{ij} + \beta_3 \ln P_i c_{ij} + \beta_4 \ln LB_{ij}$$
$$+ \beta_5 LBA_{ij} + \beta_6 LBI_{ij} + \beta_7 LBS_{ij} + \beta_8 LPI + \beta_9 LPS + c_i + u_{ij}. \quad (1)$$

In which:

VAP_NA_{ij}: The VA proportion of construction industry and service group in GDP of province i in year j. VAP_NA_{ij} degined as the proportion between VA of construction industry and service group and GDP (calculated at current prices) of province i in year j. Unit: percent.
PS: Population of province i in year j. Unit: thousand people.

IC: Investment capital of province i in year j. Unit: billion dong (compared with 1994).

P_ic: invesment capital productivity. Unit: times.

LB: Number of laborers of province i in year j. Unit: thousand people.

LBA_{ij}: The labor proportion of agriculture, forestry and fishery group of province i in year j. Unit: percent.

LBI_{ij}:The labor proportion of construction industry group of province i in year j. Unit: percent.

LBS_{ij}: The labor proportion of service group of province i in year j. Unit: percent.

LPI: Labour productivity of construction industry group.

LPS: Labour productivity of service group.

Data sources for calculating these indexes are taken from Statistical Yearbook of General Statistics Office of Viet Nam and 63 provinces and cities under central authority.

Due to existing data conditions, in this model, the authors have not taken the variables representing the science and technology group and policy group. With the science and technology group, the indexes reflect the number of inventions with protection certificate and costs for scientific and technological innovation in the enterprise have no adequate data from 63 provinces and cities from 2004 to present. With policy and institution group, PCI has been calculated since 2005 for 42 provinces/cities and 63 provinces/cities since 2006. PAPI has been calculated since 2009 for only some provinces and since 2011 for 63 provinces/cities.

Based on the existence of perfect multicollinearity between variables: LBA, LBI, LBS, the model (1) can be rewritten as follows:

$$VAP\_NA_{ij} = \beta_0 + \beta_1 \ln PS_{ij} + \beta_2 \ln IC_{ij} + \beta_3 \ln P_i c_{ij}$$
$$+ \beta_4 \ln LB_{ij} + \beta_5 LBI_{ij} + \beta_6 LBS_{ij} + \beta_7 LPI + \beta_8 LPS + c_i + u_{ij}. \quad (2)$$

It is known that, when the population size, capital investment, capital productivity, laborer, labor proportion of construction industry and service group, productivity of construction industry and service group increase, then the VA proportion of non-agricultural group also grows up.

As a result of the geographical inequality, the socio-economic factor between localities such as: infrastructures, socio-economic policies, geographic location, natural resources, ..., many factors are not observed or have no compatibility data. In such conditions, the application of the Regression Model for Panel Data Analysisis is the most appropriate choice to handle this inconsistency (Wooldridge 2002).

In order to choose a model in the form of combined OLS or random effect, we use Breusch-Pagan test. The hypothetical pair for this test is:

H_0: **No existence of random effect** ($\delta_u^2 = 0$)
H_1: **The existence of random effect**

If P value of the test $\chi^2 < 0,0000$ rejects the hypothesis H_0, it means that the model exists random effect, then the model in the form of combined OLS should not be used and the model in the form of random effect shall be used.

The inconsistency among provinces is showed by c_i. In most cases, when c_i is not correlated with independent variables in the model, $tv_{ij} = c_i + u_{ij}$ can be considered as a synthetic random error of the model, then random effect model (REM). However, in other cases, if c_i correlates with independent variables, it cannot combine this element with random error factor, then the model is called a fixed effect model (FEM). In general, if the panel number is taken out or almost complete whole, the FEM is more suitable. And when the panels are selected from the large whole, the REM may be appropriate. The choice of FEM or REM depends on the Hausman test result. The hypothetical pair for the test is:

H_0: **There is no significant difference between the estimated values from the two models of FEM and REM**
H_1: **There are significant differences between the estimated values from the two models of FEM and REM**

This test is guided as follows: If there is no significant difference between the estimated values from the two models of FEM and REM, it is a sign that c_i does not correlate with explanatory variables, then the REM is the appropriate choice; on the contrary, if there are significant differences between the estimated values from the two models of FEM and REM, it is a sign that there is a correlation between c_i and explanatory variables, then the FEM is the appropriate choice.

One of the assumptions of FEM are random errors u_{ij} with constant variance under j and variables are not correlated. When this hypothesis is violated, the estimates obtained are not effective estimates. Therefore, the statistical inferences from the obtained estimates will not be reliable. Modified Wald test used to show that the variance of panel error is the same or not. The hypothetical pair of Wald test is:

H_0: **The variance of the error is homogeneous among the panels $(\delta_i^2 = \delta^2)$ for all i**
H_1: **The variance of the error is not homogeneous among the panels.**

If $Pvalue$ of the test $\chi^2 > 0.05$, accept the hypothesis H_0, it means that the variance of the error is homogeneous among the panels.

For FEM, the hypothesis $cor(u_{ij}, u_{i,j-s}) = 0$ with all $s \neq 0$; it means that there is no serial correlation between panels. If the model has this defect, the standard error of the coefficients is smaller than the actual one and make R^2 greater. To test the serial correlation between the panels, the Wooldridge test is used, hypothetical pair as follow:

H_0: **The error of FEM with none auto correlation**
H_1: **The error of FEM with auto correlation**

If P Value of the test $\chi^2 > 0,05$, accept the hypothesis H_0, it means that there is no auto correlation between the random errors of FEM.

Result of the Empirical Study. Let us now to evaluate the impact factors on the shift of economic structure wih the model (2). We use the data set from 63 provinces/cities under the central government from 2004 to 2012.

According to test results in Fig. 1, the appendix, it shows that the variables of IC, P_ic have coefficient $Pvalue > 0.05$, there is no evidence to show that investment capital and capital productivity of the whole economy affect the Vietnamese economic structure shift in the phase 2004–2012. After eliminating the variables of IC and P_ic out of the model, the results are shown in Fig. 2. Accordingly, the remaining variables have $Pvalue < 0.05$. It means that these variables have statistical significance.

To find an appropriate model in the form of combined or random effect, we consider Breusch-Pagan test. Results of Breusch-Pagan test (Fig. 3, Appendix) show that P value of statistics χ^2 is very small ($P = 0,0000$), then the hypothesis H_0 is rejected, the model has a random effect, so do not use the model of combined OLS.

To choose a random effect model (REM) or fixed effect model (FEM), we use Hausman test. Figure 4 (Appendix) shows the results of Hausman test, P value of statistics χ^2 is very small ($P = 0,0000$), then the hypothesis H_0 is rejected. Thus, there is a correlation between c_i and the explanatory variables in the model. It is therefore reasonable to select the fixed effect model (FEM) and the estimates obtained are consistent estimates.

In most cases, the fixed effect model (FEM) used to determine the estimated results obtained from the model are consistent estimates. It is also required to be composited with test serial correlations between the panels and changed error variance. Thus, Wald test and Wooldridge test are selected for this purpose. The results of Wald test (Fig. 6, Appendix) and Wooldridge test (Fig. 5, Appendix) show that the model exists the serial correlation between the panels and changed error variance. To overcome the above phenomenal of the model, the authors choose the fixed impact regression model with standard error of Robust.

Applying Robust regression model with data set of 63 provinces and cities from $2004 - 2012$, estimated results as follows:

$$VAP\_NA_{ij} = 49.2944 - 0.00836PS + 0.00002LB + 0.2862LBI$$
$$+ 0.2723LBS + 0.1245LPI + 0.1539LPS.$$

The results show that with the variable LPS, regression coefficient <0 is not consistent with initial expectation. During this period, the population has reverse effect on the VA proportion of non-agricultural group. With the remaining variables, the sign of the estimated coefficients are consistent with initial expectation and variables have statistical significance.

The finding indicates the positive correlations between five independent variables (LB, LBI, LBS, LPI, LPS) and the dependent variable (VAP_NA_{ij}) (coefficient $\beta > 0$). It implies that in order to increase the VA proportion of non-agricultural group, it is necessary to increase the total number of laborers and labor proportion of construction industry group and service group.

The test and comparison of estimated coefficients also show that the coefficient of LBS variable is greater than coefficient of LBI variable with significance level of 5 percent. This implies that an increase in labor proportion of service group will have a greater impact on economic restructuring. Estimated results also indicate that labor productivity of construction industry and service group has a positive impact on the VA proportion in GDP of non-agricultural group (Fig. 7).

4   Conclusion

The purpose of this study is to examine the impact of factors on the shift of economic structure in Vietnam in the period 2004–2012. In order to achieve a positive economic structure shift in 2020, Vietnam needs to increase the share of VA of non-agricultural sector and gradually reduce the proportion of the agriculture sector in GDP. Evaluating the impact factors on the increase of VA proportion of non-agricultural sector in GDP means measuring the impacts on the shift of economic structure by sector in a positive direction.

It is shown that the VA proportion of non-agricultural sector is influenced by a variety of factors, such as population size, labor force, labor productivity, investment capital,.... And most of these factors have a positive effect on the VA share in GDP of non-agricultural sector.

The outcomes of this paper also indicate that the labor proportion and labor productivity of construction industry and service group have a strongly positive impact on VA proportion of non-agricultural group in GDP. Accordingly, Vietnam needs to promote labor structure transformation by economic sector group. Specifically, to continue to prioritize the development of construction industry and service group in the direction of exporting, using more laborers, especially rural laborers; having proper investment in training, improving the quality of human resources; making the socialization of vocational training, connecting with enterprises and employers; enhancing the application of modern science and technology to improve labor productivity; liberating labor capacity, creating the motivation for labor structure transformation between the economic sector groups and in each internal sector as well as each economic sector.

The empirical study also shows that the population size has reverse effect on VA proportion of non- agricultural sector. Therefore, to promote the economic structure in the positive direction, for increase VA proportion of construction industry and service group, Vietnam needs to focus on solutions to control population size as well as improve the quality of population.

Appendix

```
Random-effects GLS regression            Number of obs        =       567
Group variable: matinh                   Number of groups     =        63

R-sq:  within  = 0.4854                   Obs per group: min =         9
       between = 0.4755                                   avg =       9.0
       overall = 0.4713                                   max =         9

                                          Wald chi2(8)        =     510.41
corr(u_i, X)   = 0 (assumed)              Prob > chi2         =     0.0000
```

VAP_NA	Coef.	Std. Err.	z	P>|z|	[95% Conf.	Interval]
PS	-.0057137	.0020145	-2.84	0.005	-.009662	-.0017655
IC	.0001818	.0000977	1.86	0.063	-9.58e-06	.0003732
PC	.0709344	.0489791	1.45	0.148	-.0250628	.1669317
LB	.0000114	3.43e-06	3.32	0.001	4.68e-06	.0000181
LBI	.3528628	.0488573	7.22	0.000	.2571043	.4486213
LBS	.283975	.0444863	6.38	0.000	.1967834	.3711666
LPI	.1231814	.0131419	9.37	0.000	.0974236	.1489391
LPS	.1523189	.0412285	3.69	0.000	.0715125	.2331253
_cons	46.3779	1.793897	25.85	0.000	42.86193	49.89387

sigma_u	8.2363704	
sigma_e	3.1227977	
rho	.87431489	(fraction of variance due to u_i)

Fig. 1. Estimation result

```
Random-effects GLS regression          Number of obs     =       567
Group variable: matinh                 Number of groups  =        63

R-sq:  within  = 0.4811                Obs per group: min =         9
       between = 0.4848                              avg =       9.0
       overall = 0.4770                              max =         9

                                       Wald chi2(6)      =    506.71
corr(u_i, X)   = 0 (assumed)           Prob > chi2       =    0.0000
```

VAP_NA	Coef.	Std. Err.	z	P>|z|	[95% Conf. Interval]	
PS	-.0061388	.0020084	-3.06	0.002	-.0100752	-.0022024
LB	.0000124	3.40e-06	3.64	0.000	5.72e-06	.000019
LBI	.3481189	.047951	7.26	0.000	.2541366	.4421011
LBS	.2914667	.0436979	6.67	0.000	.2058204	.3771129
LPI	.1244681	.0130166	9.56	0.000	.0989561	.1499802
LPS	.1616215	.039529	4.09	0.000	.084146	.239097
_cons	46.83394	1.909364	24.53	0.000	43.09166	50.57622
sigma_u	9.724205					
sigma_e	3.1382627					
rho	.90567186	(fraction of variance due to u_i)				

Fig. 2. Eliminated variable estimation result

Breusch and Pagan Lagrangian multiplier test for random effects

VAP_NA[matinh,t] = Xb + u[matinh] + e[matinh,t]

Estimated results:

	Var	sd = sqrt(Var)
VAP_NA	233.1296	15.26858
e	9.848693	3.138263
u	94.56016	9.724205

Test: Var(u) = 0

chibar2(01) = 1625.13

Prob > chibar2 =   0.0000

Fig. 3. Breusch-Pagan test result

	—— Coefficients ——			
	(b)	(B)	(b-B)	sqrt(diag(V_b-V_B))
	mohinhfe	mohinhre	Difference	S.E.
PS	-.0083607	-.0061388	-.0022219	.0008112
LB	.0000153	.0000124	2.97e-06	9.99e-07
LBI	.286216	.3481189	-.0619029	.0104266
LBS	.2723006	.2914667	-.0191661	.0121592
LPI	.1242648	.1244681	-.0002033	.002728
LPS	.1539219	.1616215	-.0076996	.0033979

b = consistent under Ho and Ha; obtained from xtreg

B = inconsistent under Ha, efficient under Ho; obtained from xtreg

Test: Ho: difference in coefficients not systematic

chi2(5) = (b-B)'[(V_b-V_B)^(-1)](b-B)

= 31.74

Prob>chi2 =   0.0000

(V_b-V_B is not positive definite)

Fig. 4. Hausman test result

Wooldridge test for autocorrelation in panel data

H0: no first-order autocorrelation

F(1,   62) =   26.591

Prob > F =   0.0000

Fig. 5. Wooldridge test result

Modified Wald test for groupwise heteroskedasticity

in fixed effect regression model

H0: sigma(i)^2 = sigma^2 for all i

chi2 (63) =   8242.33

Prob>chi2 =   0.0000

Fig. 6. Wald test result

| Fixed-effects (within) regression | | | | Number of obs | = | 567 |
| Group variable: matinh | | | | Number of groups | = | 63 |

R-sq: within = 0.4842   Obs per group: min =   9
       between = 0.4448   avg =   9.0
       overall = 0.4371   max =   9

                                       F(6,62)   =   18.37
corr(u_i, Xb) = 0.2905   Prob > F   =   0.0000

(Std. Err. adjusted for 63 clusters in matinh)

VAP_NA	Coef.	Robust Std. Err.	t	P>|t|	[95% Conf. Interval]	
PS	-.0083607	.0037998	-2.20	0.032	-.0159563	-.000765
LB	.0000153	7.14e-06	2.15	0.035	1.08e-06	.0000296
LBI	.286216	.0854268	3.35	0.001	.1154502	.4569817
LBS	.2723006	.0688835	3.95	0.000	.1346044	.4099968
LPI	.1242648	.0481996	2.58	0.012	.0279151	.2206145
LPS	.1539219	.07875	1.95	0.055	-.0034971	.311341
_cons	49.29444	2.501273	19.71	0.000	44.29447	54.29442
sigma_u	11.65288					
sigma_e	3.1382627					
rho	.93237568	(fraction of variance due to u_i)				

Fig. 7. Robust estimation result

References

Wooldrige, J.M.: Econometric Analysis of Cross Section and Panel Data. The MIT press, Cambridge (2002)

Tat, T.B.: The Transformation of Economic Structure. Social Science Publishing House, Hanoi (2005)

Ngoc, D.P.: The economic restructuring of the agro-industrial sector in the Red River Delta: current situation and solutions. Ph.D. thesis, National Political Academy (2002)

Testing the Evidence of Purchasing Power Parity for Southeast Asia Countries

M. A. Truong Thiet Ha[⊠]

Ho Chi Minh city Institute for Development Studies, Ho Chi Minh City, Vietnam
thietha_1703@yahoo.com

Abstract. This paper tests the validity of purchasing power parity (PPP) hypothesis using panel methods for nine countries in Southeast Asia in US Dollar and Japanese Yen. The results show that the absolute PPP is rejected by the panel unit root test for Southeast Asia countries over the January 1995 to February 2017. However, when we use developed panel unit root that accounts for structural breaks in the data, and test the PPP hypothesis over the July 1997 to August 2008, the PPP proposition seems to hold for after the Asian financial crisis period 1997 and before the global financial crisis 2008. In addition, this paper has used recent developed panel cointegration tests and found the long-run relationship between the nominal exchange rate and the relative prices – the relative PPP – and the results offer more evidence in Japanese Yen based in favor of cointegration in long-run compared with US Dollar is the base currency.

Keywords: Purchasing power parity (PPP) · Panel data · Unit root Cointegration · Southeast Asia countries

1 Introductuon

Purchasing power parity theory – was developed by Gustav Cassel in 1918 – analyzes the relationship between inflation and the exchange rate. There are two kinds of purchasing power parity: the absolute PPP – also known as the Law of One price – and the relative PPP. The Southeast Asia countries has many similarities on the economic conditions. This also supports the validity of purchasing power parity hypothesis within the region. However, because of the presence of exogenous shocks affect each particular country, PPP theory does not hold.

Inflation and its effect on the exchange rate have always been interested by many researchers over the world. Besides, after many years establishment of the Association of Southeast Asian Nations (ASEAN), it is important to investigate whether goods markets in these countries had been more integrated, towards the establishment of a monetary union in the future. Therefore, this paper tests the validity of purchasing power parity in Southeast Asian countries, namely Vietnam, Laos, Cambodia, Thailand, Malaysia, Singapore, Myanmar,

© Springer International Publishing AG 2018
L. H. Anh et al. (eds.), *Econometrics for Financial Applications*, Studies in Computational Intelligence 760, https://doi.org/10.1007/978-3-319-73150-6_76

Indonesia, and the Philippines from January 1995 to February 2017. This is done by comparing the PPP proposition between two numeraire currencies – US Dollar and Japanese Yen – as based currencies by using panel unit root test and panel cointegration test.

2   Literature Review

The theory of purchasing power parity has been tested in many countries around the world; of which, the PPP holds or not is still debated fiercely. In particular, several studies find that the relative PPP holds in long-term (Zhou (2013)). However, many other researchers as Caporale and Gil-Alana (2010) have strongly rejected the PPP hypothesis, and they also offer explanations for that matter.

Besides, a number of researchers have discovered two PPP Puzzles. Specifically, the first PPP Puzzle statement that although the absolute purchasing power parity exists, we also uncertain that the relative purchasing power parity holds. Besides, the second PPP Puzzle statement that PPP holds in the long run also suggested that the speed at which real exchange rates adjust to the PPP exchange rate was extremely slow (Huizinga (1987)); in addition, some researchers also proposed some solutions of this second PPP puzzle (Becmann (2013)).

On the other hand, a number of studies have been undertaken to test the validity of PPP in the Southeast Asia countries, they show that many base currencies are used in the data. Since then, according to some studies as Ridzuan and Ahmed (2011), have concluded that we will have different results when using different based currencies. However, some researches show that despite any base currencies, the testing results remain unchanged (Kim et al. (2009)). In addition, the testing with the presence of the structural breaks in real exchange rate is also made, such as the Asian financial crisis in 1997. And they conclude that the existence of purchasing power parity is different in different times, before and after the structural breaks (Choudhry (2005)). Purchasing power parity is also tested by unit root tests, and most of them could not find evidence in favour of PPP. Besides, cointegration tests are also applied to examine the PPP hypothesis, and they show that results will vary depending on the study. Over the last decade, the empirical unit root and/or cointegration tests of the long run purchasing power parity relationship have shifted from a linear towards a nonlinear setup (Bec and Zeng (2012)).

However, there are few studies use this method with data of Southeast Asia countries. Yet, as stressed by Kim et al. (2009), the PPP assumption has a special meaning to Southeast Asian countries. Therefore, this paper tests the validity of purchasing power parity in Southeast Asian countries.

3   Methodology and Data

3.1   Empirical Metholodogy

In this paper, we employ the panel data methods. There are two approach to study purchasing power parity, the monetary approach (panel cointegration tests) and real exchange rate approach (panel unit root tests).

Both tests are conducted by using Eviews 8.0 software.

3.2   Data

The empirical results of this study produced by using monthly data, including the nominal exchange rate and consumer price index for nine Southeast Asia countries, namely Vietnam, Laos, Cambodia, Thailand, Myanmar, Malaysia, Singapore, Indonesia, and the Philippines over the period January 1995 until February 2017. We do not test the PPP hypothesis in Brunei Darussalam and Timor-Leste because of the limitations of data. Besides, the monthly consumer price index of Japan and United State are also used.

The nominal exchange rate used in this study are pegged into two major currencies; one is US Dollar and the other one is Japanese Yen, to check whether research results are inconsistent.

These data can be obtained for website Fxtop, and the International Financial Statistic published by International Monetary Fund. Each of the consumer price index and nominal exchange rate series was transformed into natural logarithms before the econometric analysis.

As mentioned in the content above, we will test the PPP hypothesis with the monetary approach and real exchange rate approach. So, we use the nominal exchange rate and consumer price index to calculate the real exchange rate.

The real exchange rate is defined as the nominal exchange rate adjusted for changes in the home and foreign price levels, is given by the following formula:

$$R_{it} = (E_{it} p_t^*)/P_{it}$$

Where R_{it} is the real exchange rate for country i at time t, E_{it} is the nominal exchange rate for country i at time t, P_{it} is the domestic price index for country i at time t, P^*_t is the foreign price index (USA or Japan) at time t, and i is an index for Vietnam, Laos, Cambodia, Thailand, Myanmar, Malaysia, Singapore, Indonesia, and the Philippines.

Using lowercase to denote variables in their natural logarithm form yields:

$$r_{it} = e_{it} - p_{it} + p_t^*$$

Where r_{it} is the natural logarithm of the real exchange rate for country i at time t, e_{it} is the natural logarithm of the nominal exchange rate for country i at time t, p_{it} is the natural logarithm of the domestic price index for country i at time t, p^*_t is the natural logarithm of the foreign price index (USA or Japan) at time t, and i is an index for Vietnam, Laos, Cambodia, Thailand, Myanmar, Malaysia, Singapore, Indonesia, and the Philippines.

3.3   The Sequence of Testing

– <u>Step 1</u>: This study employs the panel unit root tests with the real exchange rate over the period January 1995 to February 2017 in order to test the absolute PPP.
– <u>Step 2</u>: With two major structural changes occur at the Asian financial crisis in 1997 and the global financial crisis in 2008, the same panel unit root tests were re-run with the real exchange rate by using the data set from July 1997 to August 2008 (respectively after the Asian financial crisis in 1997 and before the global financial crisis in 2008), and the data set from September 2008 to February 2017 (respectively after the global financial crisis of 2008 onwards), to examine whether differences in the existence of the absolute PPP before and after these structural breaks. There are two reasons for choosing these structural breaks, including economic theories and literature review. The Asian financial crisis in 1997 and the global financial crisis in 2008 are two crises that affect negatively many Southeast Asia countries. In addition, July 1997, the Asian financial crisis started in Thailand.
– <u>Step 3</u>: We apply traditional panel unit root tests with a data set of nominal exchange rate and relative prices over the period January 1995 to February 2017 in order to prepare for panel cointegration tests.
– <u>Step 4</u>: We test for a long run relationship between nominal exchange rate and relative prices, which known as the relative PPP, over the period January 1995 to February 2017.

4   Results

4.1   Panel Unit Root Tests

Results for panel unit root tests of real exchange rates for two difference base numeraire currencies from January 1995 to February 2017, are reported in Table 1.

The panel unit root tests fail to reject the null of a unit root in level of data set from January 1995 to February 2017 (except for the test which advocated by Levin et al. (2002) for US Dollar base cannot be rejected at 1% significance level). Therefore, the results strongly indicate the presence of unit root in real exchange rates for Southeast Asia countries over the period estimation. There are many reasons why the absolute PPP does not hold: the difference in interest rates, income levels, government strategies or substitutes for imported goods and services. The difference in calculation the price index is also a reason to explain this matter, namely the difference of the selected items of goods and services in CPI "basket".

To examine the purchasing power parity hypothesis aftermath financial crises. Results for panel unit root tests of real exchange rates with the presence structural breaks for two difference base numeraire currencies from July 1995 to August 2008, are reported in Table 2.

Table 1. Panel unit root tests of real exchange rates

Common root	Individual root	Individual root
Levin, Lin and Chu t-stat	Im, Pesaran and Shin W-stat	ADF – Fisher Chi-square
US Dollar = base currency		
−3.976*** (0.000)	−1.267 (0.103)	25.795 (0.105)
Japanese Yen = base currency		
−0.709 (0.225)	−1.709 (0.126)	26.353 (0.105)

Note: ***, ** and * indicates significant at 1%, 5% and 10% significance levels respectively; "Common root" indicates that the tests are estimated assuming a common AR structure for all of the series; "Individual root" is used for tests which allow for different AR coefficients in each series. Exogenous variables: Individual effects, individual linear trends. Newey-West bandwidth selection using Bartlett kernel. Probabilities for Fisher tests are computed using an asymptotic Chi-square distribution. All other tests assume asymptotic normality. () indicates p-value, respectively.

Table 2. Panel unit root tests of real exchange rates with the presence of structural breaks

	Common root	Individual root	Individual root
Period	Levin, Lin and Chu t-stat	Im, Pesaran and Shin W-stat	ADF – Fisher Chi-square
US Dollar = base currency			
7/1997 – 8/2008	−3.289*** (0.000)	−2.003** (0.022)	55.214*** (0.000)
9/2008 – 6/2013	1.886 (0.970)	3.192 (0.999)	4.535 (0.999)
Japanese Yen = base currency			
7/1997 – 8/2008	−1.929** (0.027)	1.991** (0.023)	34.800*** (0.010)
9/2008 – 6/2013	−3.118 (0.491)	−3.847 (0.127)	49.142*** (0.000)

Note: ***, ** and * indicates significant at 1%, 5% and 10% significance levels respectively; "Common root" indicates that the tests are estimated assuming a common AR structure for all of the series; "Individual root" is used for tests which allow for different AR coefficients in each series. Exogenous variables: Individual effects, individual linear trends. Newey-West bandwidth selection using Bartlett kernel. Probabilities for Fisher tests are computed using an asymptotic Chi-square distribution. All other tests assume asymptotic normality. () indicates p-value, respectively.

During the period from July 1997 to August 2008, empirical results show that even though sample span is short (compared with the data set from January 1995 to February 2017), purchasing power parity hypothesis seems to hold for nine

Southeast Asia countries in post Asian financial crisis and pre global financial crisis period. This reinforced the earlier findings, that is, the behaviour of real exchange rate after Asian financial crisis as a group is noticeably different from pre-crises period as discussed by Ridzuan and Ahmed (2011).

This matter can be explained as follows: After the Asian financial crisis occurred, the Southeast Asian countries have not maintained the anchor currency as in earlier periods anymore, example the national governments change policies, improve the competitiveness of goods and services, reduce monopolies and trade barriers.

During the period September 2008 to February 2017, the null hypothesis of unit root for real exchange rate cannot be rejected for nine Southeast Asia countries (except for the test which advocated by Maddala and Wu (ADF – Fisher) for Japanese Yen base can be rejected at 1% significance level). Therefore, the real exchange rate seem failed to find evidence supporting validity of PPP for post global financial crisis 2008 period.

4.2   Panel Cointegration Tests

Results for panel unit root tests with a data set of nominal exchange rate and relative prices over the period January 1995 to February 2017 in order to prepare for panel cointegration tests, are reported in Table 3.

Table 3 indicates that the unit root null could not be rejected (except for the test which advocated by Levin, Lin and Chu for US Dollar base cannot be rejected at 1% significance level, and the test which advocated by Levin, Lin

Table 3. Panel unit root tests for nominal exchange rate and relative prices

	US Dollar based		Japanese Yen based	
	Nominal exchange rate	Relative price	Nominal exchange rate	Relative price
Methods	Statistic	Statistic	Statistic	Statisitc
Levin, Lin and Chu t-stat	−5.052*** (0.000)	−3.295*** (0.000)	−1.843** (0.033)	−2.482 (0.139)
Im, Pesaran and Shin W-stat	−2.262 (0.206)	0.027 (0.511)	−0.627 (0.265)	1.244 (0.893)
ADF-Fisher Chi-square	34.076 (0.328)	14.133 (0.720)	17.166 (0.512)	12.175 (0.838)

Note: ***, ** and * indicates significant at 1%, 5% and 10% significance levels respectively; "Common root" indicates that the tests are estimated assuming a common AR structure for all of the series; "Individual root" is used for tests which allow for different AR coefficients in each series. Exogenous variables: Individual effects, individual linear trends. Newey-West bandwidth selection using Bartlett kernel. Probabilities for Fisher tests are computed using an asymptotic Chi-square distribution. All other tests assume asymptotic normality. () indicates p-value, respectively.

and Chu for Japanese Yen base cannot be rejected at 5% significance level), and hence these two series are generated by a I(1) process despite US or Japan being base country. Therefore, the panel cointegration test can be applied.

Results for the Pedroni (1999, 2004) panel cointegration regression are presented in Table 4.

Table 4. Panel cointegration tests for nominal exchange rate and relative prices

	US Dollar based real exchange rates		Japanese Yen based real exchange rates	
	Constant	Constant + Trend	Constant	Constant + Trend
Alternative hypothesis: common AR coefs. (within-dimension)				
Panel v-statistics	−2.088 (0.982)	6.296*** (0.000)	−2.296 (0.989)	8.636*** (0.000)
Panel Rho-statistics	1.005 (0.843)	−2.609** (0.005)	0.881 (0.811)	−4.594*** (0.000)
Panel PP-statistics	−0.259 (0.398)	−2.077** (0.019)	0.033 (0.513)	−3.381*** (0.000)
Panel ADF-statistics	0.256 (0.601)	1.134 (0.871)	0.123 (0.549)	−2.979** (0.001)
Alternative hypothesis: individual AR coefs. (between-dimension)				
Group Rho-statistics	2.292 (0.989)	0.617 (0.731)	2.249 (0.988)	−0.989 (0.161)
Group PP-statistics	1.437 (0.925)	1.045 (0.852)	2.207 (0.986)	−0.424 (0.336)
Group ADF-statistics	1.346 (0.911)	1.364 (0.914)	1.821 (0.966)	−1.588* (0.056)

Note: ***, ** and * indicates significant at 1%, 5% and 10% significant levels respectively. Trend assumption based on no deterministic trend and deterministic intercept and trend. Automatic lag selection based on AIC with 16 maximum lag. Newey-West bandwidth selection using Bartlett kernel. () indicates p-value, respectively.

Table 4 shows that only three statistics (i.e., Panel v-statistics, Panel Rho-statistics, Panel PP-statistics) out of seven are able to reject the null of non-cointegration in US Dollar based real exchange rate of nine Southeast Asia countries. In particular, most of statistics favour the relative purchasing power parity hypothesis in Japanese Yen based real exchange rate, because the null hypothesis is rejected most at 1% significant level; while US Dollar is base currency, most of the null hypothesis is rejected at 5% significant level. There is vary between different numeraire currencies, similar to previous studies. Besides, results seem to support the existence of a long-run relationship between nominal exchange rate, domestic and foreign prices for full panel of Southeast Asia countries – known as the relative PPP – although the absolute PPP does not hold over the period January 1995 to February 2017.

5   Conclusion

The results show that the absolute PPP is rejected by the panel unit root test for Southeast Asia countries over the January 1995 to February 2017. However, when we use developed panel unit root that accounts for structural breaks in the data, and test the PPP hypothesis over the July 1997 to August 2008, the PPP proposition seems to hold for after the Asian financial crisis period 1997 and before the global financial crisis 2008. In addition, this paper has used recent developed panel cointegration tests and found the long-run relationship between the nominal exchange rate and the relative prices – the relative purchasing power parity – and the results offer more evidence in Japanese Yen based in favor of cointegration in long-run compared with US Dollar is the numeraire currency.

Indeed, some researchers argue that a long-run PPP is a valid equilibrium relationship if Japanese Yen is used as the numeraire currency which mainly due to close trade and financial linkages among the Southeast Asia countries. The PPP hypothesis is important to economists not only because it is the centrepiece of many exchange rate models including the monetary model of exchange rate determination, but also because of its policy implications. If the purchasing power parity proposition hold in long run then national monetary authorities will be able successful to conduct independent monetary policy and simultaneously control the movement of exchange rates. Otherwise, invalid PPP will create high possibility unbounded gains from arbitrage in traded goods (Kapetanios et al. (2003)), disqualifies monetary approach to exchange rate determination and so on.

In addition, we can test purchasing power parity hypothesis by allowing for nonlinear dynamics in real exchange rate adjustment, because of transactions costs in international arbitrage, in order to explain the failure of linear models, thus solving the PPP puzzles. These challenges remain on the agenda for future research.

Appendix

US Dollar Based Real Exchange Rates

Jan 1995 to Feb 2017

> Group unit root test: Summary
> Series: R_CAM_USD, R_IND_USD, R_LAO_USD, R_MAL_USD,
>   R_MYA_USD, R_PHI_USD, R_SIN_USD, R_THA_USD, R_VIE_USD
> Date: 06/20/17   Time: 15:40
> Sample: 1995M01 2017M02
> Exogenous variables: Individual effects, individual linear trends
> Automatic selection of maximum lags
> Automatic lag length selection based on SIC: 0 to 12
> Newey-West automatic bandwidth selection and Bartlett kernel

Method	Statistic	Prob.**	Cross-sections	Obs
Null: Unit root (assumes common unit root process)				
Levin, Lin and Chu t*	−3.97563	0.0000	9	2359
Breitung t-stat	0.28439	0.6119	9	2350
Null: Unit root (assumes individual unit root process)				
Im, Pesaran and Shin W-stat	−1.26698	0.1026	9	2359
ADF - Fisher Chi-square	25.7954	0.1045	9	2359
PP - Fisher Chi-square	15.8372	0.6039	9	2385

** Probabilities for Fisher tests are computed using an asymptotic Chi-square distribution. All other tests assume asymptotic normality.

Jul 1997 to Aug 2008

Group unit root test: Summary
Series: R_CAM_USD, R_IND_USD, R_LAO_USD, R_MAL_USD,
    R_MYA_USD, R_PHI_USD, R_SIN_USD, R_THA_USD, R_VIE_USD
Date: 06/19/17   Time: 11:26
Sample: 1997M07 2008M08
Exogenous variables: Individual effects, individual linear trends
Automatic selection of maximum lags
Automatic lag length selection based on AIC: 0 to 12
Newey-West automatic bandwidth selection and Bartlett kernel
Balanced observations for each test

Method	Statistic	Prob.**	Cross-sections	Obs
Null: Unit root (assumes common unit root process)				
Levin, Lin and Chu t*	−3.28666	0.0005	9	1206
Breitung t-stat	4.52276	1.0000	9	1197
Null: Unit root (assumes individual unit root process)				
Im, Pesaran and Shin W-stat	−2.00361	0.0226	9	1206
ADF - Fisher Chi-square	55.2138	0.0000	9	1206
PP - Fisher Chi-square	66.8510	0.0000	9	1206

** Probabilities for Fisher tests are computed using an asymptotic Chi-square distribution. All other tests assume asymptotic normality.

Sep 2008 to Feb 2017

Group unit root test: Summary
Series: R_CAM_USD, R_IND_USD, R_LAO_USD, R_MAL_USD,
    R_MYA_USD, R_PHI_RSD, R_SIN_USD, R_THA_USD, R_VIE_USD
Date: 06/19/17 Time: 11:32
Sample: 2008M09 2017M02
Exogenous variables: Individual effects, individual linear Trends

Automatic selection of maximum lags
Automatic lag length selection based on AIC: 0 to 11
Newey-West automatic bandwidth selection and Bartlett kernel
Balanced observations for each test

Method	Statistic	Prob.**	Cross-sections	Obs
Null: Unit root (assumes common unit root process)				
Levin, Lin and Chu t*	1.88556	0.9703	9	918
Breitung t-stat	1.87541	0.9696	9	909
Null: Unit root (assumes individual unit root process)				
Im, Pesaran and Shin W-stat	3.19208	0.9993	9	918
ADF - Fisher Chi-square	4.53506	0.9994	9	918
PP - Fisher Chi-square	6.80443	0.9917	9	918

** Probabilities for Fisher tests are computed using an asymptotic Chi-square
distribution. All other tests assume asymptotic normality.

Japanses Yen Based Real Exchange Rates

Jan 1995 to Feb 2017

Group unit root test: Summary
Series: R_CAM_JPY, R_IND_JPY, R_LAO_JPY, R_MAL_JPY, R_MYA_JPY,
R_PHI_JPY, R_SIN_JPY, R_THA_JPY, R_VIE_JPY
Date: 06/19/17   Time: 11:39
Sample: 1995M01 2017M02
Exogenous variables: Individual effects, individual linear trends
Automatic selection of maximum lags
Automatic lag length selection based on AIC: 0 to 12
Newey-West automatic bandwidth selection and Bartlett kernel

Method	Statistic	Prob.**	Cross-sections	Obs
Null: Unit root (assumes common unit root process)				
Levin, Lin and Chu t*	−0.75693	0.2245	9	2367
Breitung t-stat	−3.20534	0.0007	9	2358
Null: Unit root (assumes individual unit root process)				
Im, Pesaran and Shin W-stat	−1.70880	0.1264	9	2367
ADF - Fisher Chi-square	26.3532	0.1045	9	2367
PP - Fisher Chi-square	27.8788	0.0639	9	2385

** Probabilities for Fihser tests are computed using an asymptotic Chi-square
distribution. All other tests assume asymptotic normality.

Jul 1997 to Aug 2008

Group unit root test: Summary
Jeries: R_CAM_JPY, R_IND_JPY, R_LAO_JPY, R_MAL_JPY, R_MYA_JPY,
    R_PHI_JPY, R_SIN_JPY, R_THA_JPY, R_VIE_JPY
Date: 06/19/17   Time: 11:42
Sample: 1997M07 2008M08
Exogenous variables: Individual effects, individual linear trends
Automatic selection of maximum lags
Automatic lag length selection based on AIC: 0 to 12
Newey-West automatic bandwidth selection and Bartlett kernel
Balanced observations for each test

Method	Statistis	Prob.**	Cross-sections	Obs
Null: Unit root (assumes common unit root process)				
Levin, Lin and Chu t*	−1.92886	0.0269	9	1206
Breitung t-stat	−0.03302	0.4868	9	1197
Null: Unit root (assumes individual unit root process)				
Im, Pesaran and Shin W-stat	−1.99167	0.0232	9	1206
ADF - Fisher Chi-square	34.8009	0.0100	9	1206
PP - Fisher Chi-square	41.6382	0.0012	9	1206

** Probabilities for Fisher tests are computed using an asymptotic Chi-square
    distribution. All other tests assume asymptotic normality.

Sep 2008 to Feb 2017

Group unit root test: Summary
Series: R_CAM_JPY, R_IND_JPY, R_LAO_JPY, R_MAL_JPY, R_MYA_JPY,
    R_PHI_JPY, R_SIN_JPY, R_TIIA_JPY, R_VIE_JPY
Date: 06/19/17   Time: 11:44
Sample: 2008M09 2017M02
Exogenous variables: Individual effects, individual linear trends
Automatic selection of maximum lags
Automatic lag length selection based on AIC: 0 to 8
Newey-West automatic bandwidth selection and Bartlett kernel
Balanced observations for each test

Method	Statistic	Prob.**	Cross-sections	Obs
Null: Unit root (assumes common unit root process)				
Levin, Lin and Chu t*	−3.11784	0.4910	9	918
Breitung t-stat	−1.19804	0.1155	9	909
Null: Unit root (assumes individual unit root process)				
Im, Pesaran and Shin W-stat	−3.84732	0.1270	9	918
ADF - Fisher Chi-square	49.1419	0.0001	9	918
PP - Fisher Chi-square	36.3889	0.0063	9	918

** Probabilities for Fisher tests are computed using an asymptotic Chi-square
distribution. All other tests assume asymptotic normality.

US Dollar Based Nominal Exchange Rates

Group unit root test: Summary
Series: CAM_USD, IND_USD, LAO_USD, MAL_USD, MYA_USD, PHI_USD,
    SIN_USD, THA_USD, VIE_USD
Date: 06/19/17   Time: 11:47
Sample: 1995M01 2017M02
Exogenous variables: Individual effects, individual linear trends
Automatic selection of maximum lags
Automatic lag length selection based on AIC: 0 to 13
Newey-West automatic bandwidth selection and Bartlett kernel

Method	Statistic	Prob.**	Cross-sections	Obs
Null: Unit root (assumes common unit root process)				
Levin, Lin and Chu t*	−5.05152	0.0000	9	2332
Breitung t-stat	−0.22853	0.4096	9	2323
Null: Unit root (assumes individual unit root process)				
Im, Pesaran and Shin W-stat	−2.26243	0.2060	9	2332
ADF - Fisher Chi-square	34.0760	0.3280	9	2332
PP - Fisher Chi-square	12.1710	0.8383	9	2385

** Probabilities for Fisher tests are computed using an asymptotic Chi-square
distribution. All other tests assume asymptotic normality.

Japanese Yen Based Nominal Exchange Rates

Group unit root test: Summary
Series: CAM_JPY, IND_JPY, LAO_JPY, MAL_JPY, MYA_JPY, PHI_JPY,
    SIN_JPY, THA_JPY, VIE_JPY
Date: 06/19/17   Time: 11:48
Sample: 1995M01 2017M02
Exogenous variables: Individual effects, individual linear trends
Automatic selection of maximum lags
Automatic lag length selection based on AIC: 0 to 12
Newey-West automatic bandwidth selection and Bartlett kernel

Method	Statistic	Prob.**	Cross-sections	Obs
Null: Unit root (assumes common unit root process)				
Levin, Lin and Chu t*	−1.84337	0.0326	9	2367
Breitung t-stat	−1.11867	0.1316	9	2358
Null: Unit root (assumes individual unit root process)				
Im, Pesaran and Shin W-stat	−0.62710	0.2653	9	2367
ADF - Fisher Chi-square	17.1660	0.5117	9	2367
PP - Fisher Chi-square	11.5363	0.8702	9	2385

** Probabilities for Fisher tests are computed using an asymptotic Chi-square distribution. All other tests assume asymptotic normality.

Relative Prices (Between Southeast Asia Countries and USA)

Group unit root test: Summary
Series: USA_CAM, USA_IND, USA_LAO, USA_MAL, USA_MYA, USA_PHI, USA_SIN, USA_THA, USA_VIE
Date: 06/19/17   Time: 11:50
Sample: 1995M01 2017M02
Exogenous variables: Individual effects, individual linear trends
Automatic selection of maximum lags
Automatic lag length selection based on AIC: 5 to 15
Newey-West automatic bandwidth selection and Bartlett kernel

Method	Statistic	Prob.**	Cross-sections	Obs
Null: Unit root (assumes common unit root process)				
Levin, Lin and Chu t*	−3.29551	0.0005	9	2283
Breitung t-stat	1.74157	0.9592	9	2274
Null: Unit root (assumes individual unit root process)				
Im, Pesaran and Shin W-stat	0.02687	0.5107	9	2283
ADF - Fisher Chi-square	14.1325	0.7204	9	2283
PP - Fisher Chi-square	8.15953	0.9762	9	2385

** Probabilities for Fisher tests are computed using an asymptotic Chi-square distribution. All other tests assume asymptotic normality.

Relative Prices (Between Southeast Asia Countries and Japan)

Group unit root test: Summary
Series: JP_CAM, JP_IND, JP_LAO, JP_MAL, JP_MYA, JP_PHI, JP_SIN, JP_THA, JP_VIE
Date: 06/19/17   Time: 11:52

Sample: 1995M01 2017M02
Exogenous variables: Individual effects, individual linear trends
Automatic selection of maximum lags
Automatic lag length selection based on AIC: 5 to 15
Newey-West automatic bandwidth selection and Bartlett kernel

Method	Statistic	Prob.**	Cross-sections	Obs
Null: Unit root (assumes common unit root process)				
Levin, Lin and Chu t*	−2.48195	0.1390	9	2275
Breitung t-stat	2.65437	0.9960	9	2266
Null: Unit root (assumes individual unit root process)				
Im, Pesaran and Shin W-stat	1.24354	0.8932	9	2275
ADF - Fisher Chi-square	12.1754	0.8381	9	2275
PP - Fisher Chi-square	7.08272	0.9894	9	2385

** Probabilities for Fisher tests are computed using an asymptotic Chi-square
distribution. All other tests assume asymptotic normality.

Panel Cointegration Tests for Nominal Exchange Rate and Relative Prices (USD = base currency)

Trend Assumption Based on no Deterministic Trend

Pedroni Residual Cointegration Test
Series: CPI_USA NER_USD
Date: 06/19/17   Time: 15:02
Sample: 1995M01 2017M02
Included observations: 2394
Cross-sections included: 9
Null Hypothesis: No cointegration
Trend assumption: No deterministic trend
Automatic lag length selection based on SIC with a max lag of 15
Newey-West automatic bandwidth selection and Bartlett kernel

Alternative hypothesis: common AR coefs. (within-dimension)

	Statistic	Prob.	Weighted Statistic	Prob.
Panel v-Statistic	-2.087924	0.9816	-2.216234	0.9867
Panel rho-Statistic	1.005046	0.8426	1.655940	0.9511
Panel PP-Statistic	-0.259800	0.3975	1.118510	0.8683
Panel ADF-Statistic	0.256004	0.6010	1.116704	0.8679

Alternative hypothesis: individual AR coefs. (between-dimension)

	Statistic	Prob.
Group rho-Statistic	2.291816	0.9890
Group PP-Statistic	1.436580	0.9246
Group ADF-Statistic	1.346219	0.9109

Trend assumption based on deterministic intercept and trend

Pedroni Residual Cointegration Test
Series: CPI_USA NER_USD
Date: 06/19/17   Time: 15:03
Sample: 1995M01 2017M02
Included observations: 2394
Cross-sections included: 9
Null hypothesis: No cointegration
Trend assumption: Deterministic intercept and trend
Automatic lag length selection based on SIC with a max lag of 15
Newey-West automatic bandwidth selection and Bartlett kernel

Alternative hypothesis: common AR coefs. (within-dimension)

	Statistic	Prob.	Weighted Statistic	Prob.
Panel v-Statistic	6.295889	0.0000	-0.744054	0.7716
Panel rho-Statistic	-2.608702	0.0045	1.863090	0.9688
Panel PP-Statistic	-2.077368	0.0189	1.764214	0.9612
Panel ADF-Statistic	1.134335	0.8717	1.864502	0.9689

Alternative hypothesis: individual AR coefs. (between-dimension)

	Statistic	Prob.
Group rho-Statistic	0.616626	0.7313
Group PP-Statistic	1.044948	0.8520
Group ADF-Statistic	1.363748	0.9137

Panel Cointegration Tests for Nominal Exchange Rate and Relative Prices (JPY = Base Currency)

Trend Assumption Based on no Deterministic Trend

Pedroni Residual Cointegration Test
Series: CPI_JAP NER_JPY
Date: 06/19/17   Time: 15:05
Sample: 1995M01 2017M02
Included observations: 2394
Cross-sections included: 9
Null hypothesis: No cointegration
Trend assumption: No deterministic trend
Automatic lag length selection based on SIC with a max lag of 15
Newey-West automatic bandwidth selection and Barltett kernel

Alternative hypothesis: common AR coefs. (within-dimension)

	Statistic	Prob.	Weighted Statistic	Prob.
Panel v-Statistic	-2.296345	0.9892	-2.787317	0.9973
Panel rho-Statistic	0.881330	0.8109	1.505624	0.9339
Panel PP-Statistic	0.032539	0.5130	1.492832	0.9323
Panel ADF-Statistic	0.122709	0.5488	1.113075	0.8672

Alternative hypothesis: individual AR coefs. (between-dimension)

	Statistic	Prob.
Group rho-Statistic	2.249583	0.9878
Group PP-Statistic	2.206913	0.9863
Group ADF-Statistic	1.820547	0.9657

Trend assumption based on deterministic intercept and trend

Pedroni Residual Cointegration Test
Series: CPI_JAP NER_JPY
Date: 06/19/17   Time: 15:08
Sample: 1995M01 2017M02
Included observations: 2394
Cross-sections included: 9
Null hypothesis: No cointegration
Trend assumption: Deterministic intercept and trend
Automatic lag length selection based on SIC with a max lag of 15
Newey-West automatic bandwidth selection and Bartlett kernel

Alternative hypothesis: common AR coefs. (within-dimension)

	Statistic	Prob.	Weighted Statistic	Prob.
Panel v-Statistic	8.635564	0.0000	1.945228	0.0259
Panel rho-Statistic	-4.593599	0.0000	0.764611	0.7777
Panel PP-Statistic	-3.381250	0.0004	0.543331	0.7065
Panel ADF-Statistic	-2.979405	0.0014	-0.936730	0.1744

Alternative hypothesis: individual AR coefs. (between-dimension)

	Statistic	Prob.
Group rho-Statistic	-0.989463	0.1612
Group PP-Statistic	-0.424091	0.3357
Group ADF-Statistic	-1.588199	0.0561

References

Bec, F., Zeng, S.: Are Southeast Asian real exchange rates mean reverting? Int. Fin. Mark. Inst. Money **23**, 265–282 (2012)

Becmann, J.: Nonlinear adjustment, purchasing power parity and the role of nominal exchange rates and prices. North Am. J. Econ. Financ. **24**, 176–190 (2013)

Caporale, G.M., Gil-Alana, L.A.: Testing PPP for South America Rand/US Dollar Exchange Rate at Different Frequencies. Brunel University, West London, Working Paper No. 10–11 (2010)

Choudhry, T.: Asian currency crisis and the generalized PPP: evidence from the Far East. Asian Econ. J. **19**(2), 137–157 (2005)

Huizinga, J.: An Empirical Investigation of the Long-Run Behavior of Real Exchange Rates. Carnegie-Rochester Conference Series on Public Policy **27**, 149–215 (1987)

Kapetanios, G., Shin, Y., Snell, A.: Testing for a unit root in the nonlinear STAR framework. J. Econometrics **112**(2), 359–379 (2003)

Kim, B.H, Kim, II.K., Oh, K.Y.: The purchasing power parity of Southeast Asian currencies: a timevarying coefficient approach. Econ. Model. **26**, 96–106 (2009)

Levin, A., Lin, C.-F., Chu, C.-S.J.: Unit root tests in panel data: asymptotic and finite-sample properties. J. Econometrics **108**(1), 1–24 (2002)

Pedroni, P.: Critical values for cointegration tests in heterogeneous panels with multiple regressors. Oxford Bull. Econ. Stat. **61**(S1), 653–670 (1999)

Pedroni, P.: Panel cointegration: asymptotic and finite sample properties of pooled time series tests with an application to the PPP hypothesis. Econometric theor. **20**(3), 597–625 (2004)

Ridzuan, R., Ahmed, E.M.: Testing the evidence of purchasing power parity for ASEAN-5countries using panel estimation. Int. J. Econ. Bus. Model. **2**(1), 42–56 (2011)

Zhou, S.: Purchasing power parity and real effective exchange rates. University of Texas at San Antonio, Working Paper No. 0005ECO-106-2013 (2013)

Author Index

© Springer International Publishing AG 2018
L. H. Anh et al. (eds.), *Econometrics for Financial Applications*, Studies in Computational Intelligence 760, https://doi.org/10.1007/978-3-319-73150-6

Printed in the United States
By Bookmasters